# PERIODIC TABLE OF THE ELEMENTS

**Legend:**
- Metals (main-group)
- Metals (transition)
- Metals (inner-transition)
- Metalloids
- Nonmetals

MAIN-GROUP ELEMENTS

MAIN-GROUP ELEMENTS

TRANSITION ELEMENTS

Period

| Group | IA (1) | IIA (2) | IIIB (3) | IVB (4) | VB (5) | VIB (6) | VIIB (7) | VIIIB (8) | VIIIB (9) | VIIIB (10) | IB (11) | IIB (12) | IIIA (13) | IVA (14) | VA (15) | VIA (16) | VIIA (17) | VIIIA (18) |
|---|---|---|---|---|---|---|---|---|---|---|---|---|---|---|---|---|---|---|
| 1 | 1 **H** 1.008 | | | | | | | | | | | | | | | | | 2 **He** 4.003 |
| 2 | 3 **Li** 6.94 | 4 **Be** 9.012 | | | | | | | | | | | 5 **B** 10.81 | 6 **C** 12.01 | 7 **N** 14.01 | 8 **O** 16.00 | 9 **F** 19.00 | 10 **Ne** 20.18 |
| 3 | 11 **Na** 22.99 | 12 **Mg** 24.31 | | | | | | | | | | | 13 **Al** 26.98 | 14 **Si** 28.09 | 15 **P** 30.97 | 16 **S** 32.06 | 17 **Cl** 35.45 | 18 **Ar** 39.95 |
| 4 | 19 **K** 39.10 | 20 **Ca** 40.08 | 21 **Sc** 44.96 | 22 **Ti** 47.87 | 23 **V** 50.94 | 24 **Cr** 52.00 | 25 **Mn** 54.94 | 26 **Fe** 55.85 | 27 **Co** 58.93 | 28 **Ni** 58.69 | 29 **Cu** 63.55 | 30 **Zn** 65.38 | 31 **Ga** 69.72 | 32 **Ge** 72.63 | 33 **As** 74.92 | 34 **Se** 78.97 | 35 **Br** 79.90 | 36 **Kr** 83.80 |
| 5 | 37 **Rb** 85.47 | 38 **Sr** 87.62 | 39 **Y** 88.91 | 40 **Zr** 91.22 | 41 **Nb** 92.91 | 42 **Mo** 95.95 | 43 **Tc** (98) | 44 **Ru** 101.1 | 45 **Rh** 102.9 | 46 **Pd** 106.4 | 47 **Ag** 107.9 | 48 **Cd** 112.4 | 49 **In** 114.8 | 50 **Sn** 118.7 | 51 **Sb** 121.8 | 52 **Te** 127.6 | 53 **I** 126.9 | 54 **Xe** 131.3 |
| 6 | 55 **Cs** 132.9 | 56 **Ba** 137.3 | 57 **La** 138.9 | 72 **Hf** 178.5 | 73 **Ta** 180.9 | 74 **W** 183.8 | 75 **Re** 186.2 | 76 **Os** 190.2 | 77 **Ir** 192.2 | 78 **Pt** 195.1 | 79 **Au** 197.0 | 80 **Hg** 200.6 | 81 **Tl** 204.4 | 82 **Pb** 207.2 | 83 **Bi** 209.0 | 84 **Po** (209) | 85 **At** (210) | 86 **Rn** (222) |
| 7 | 87 **Fr** (223) | 88 **Ra** (226) | 89 **Ac** (227) | 104 **Rf** (265) | 105 **Db** (268) | 106 **Sg** (271) | 107 **Bh** (270) | 108 **Hs** (277) | 109 **Mt** (276) | 110 **Ds** (281) | 111 **Rg** (280) | 112 **Cn** (285) | 113 **Nh** (284) | 114 **Fl** (289) | 115 **Mc** (288) | 116 **Lv** (293) | 117 **Ts** (294) | 118 **Og** (294) |

INNER-TRANSITION ELEMENTS

| | | | | | | | | | | | | | | | |
|---|---|---|---|---|---|---|---|---|---|---|---|---|---|---|---|
| 6 Lanthanides | 58 **Ce** 140.1 | 59 **Pr** 140.9 | 60 **Nd** 144.2 | 61 **Pm** (145) | 62 **Sm** 150.4 | 63 **Eu** 152.0 | 64 **Gd** 157.3 | 65 **Tb** 158.9 | 66 **Dy** 162.5 | 67 **Ho** 164.9 | 68 **Er** 167.3 | 69 **Tm** 168.9 | 70 **Yb** 173.0 | 71 **Lu** 175.0 |
| 7 Actinides | 90 **Th** 232.0 | 91 **Pa** 231.0 | 92 **U** 238.0 | 93 **Np** (237) | 94 **Pu** (244) | 95 **Am** (243) | 96 **Cm** (247) | 97 **Bk** (247) | 98 **Cf** (251) | 99 **Es** (252) | 100 **Fm** (257) | 101 **Md** (258) | 102 **No** (259) | 103 **Lr** (262) |

# THE ELEMENTS

| Element | Symbol | Atomic Number | Relative Atomic Mass* | Element | Symbol | Atomic Number | Relative Atomic Mass* |
|---------|--------|---------------|----------------------|---------|--------|---------------|----------------------|
| Actinium | Ac | 89 | (227) | Mendelevium | Md | 101 | (258) |
| Aluminum | Al | 13 | 26.98 | Mercury | Hg | 80 | 200.6 |
| Americium | Am | 95 | (243) | Molybdenum | Mo | 42 | 95.95 |
| Antimony | Sb | 51 | 121.8 | Moscovium | Mc | 115 | (288) |
| Argon | Ar | 18 | 39.95 | Neodymium | Nd | 60 | 144.2 |
| Arsenic | As | 33 | 74.92 | Neon | Ne | 10 | 20.18 |
| Astatine | At | 85 | (210) | Neptunium | Np | 93 | (237) |
| Barium | Ba | 56 | 137.3 | Nickel | Ni | 28 | 58.69 |
| Berkelium | Bk | 97 | (247) | Nihonium | Nh | 113 | (284) |
| Beryllium | Be | 4 | 9.012 | Niobium | Nb | 41 | 92.91 |
| Bismuth | Bi | 83 | 209.0 | Nitrogen | N | 7 | 14.01 |
| Bohrium | Bh | 107 | (270) | Nobelium | No | 102 | (259) |
| Boron | B | 5 | 10.81 | Oganesson | Og | 118 | (294) |
| Bromine | Br | 35 | 79.90 | Osmium | Os | 76 | 190.2 |
| Cadmium | Cd | 48 | 112.4 | Oxygen | O | 8 | 16.00 |
| Calcium | Ca | 20 | 40.08 | Palladium | Pd | 46 | 106.4 |
| Californium | Cf | 98 | (251) | Phosphorus | P | 15 | 30.97 |
| Carbon | C | 6 | 12.01 | Platinum | Pt | 78 | 195.1 |
| Cerium | Ce | 58 | 140.1 | Plutonium | Pu | 94 | (244) |
| Cesium | Cs | 55 | 132.9 | Polonium | Po | 84 | (209) |
| Chlorine | Cl | 17 | 35.45 | Potassium | K | 19 | 39.10 |
| Chromium | Cr | 24 | 52.00 | Praseodymium | Pr | 59 | 140.9 |
| Cobalt | Co | 27 | 58.93 | Promethium | Pm | 61 | (145) |
| Copernicium | Cn | 112 | (285) | Protactinium | Pa | 91 | (231.0) |
| Copper | Cu | 29 | 63.55 | Radium | Ra | 88 | (226) |
| Curium | Cm | 96 | (247) | Radon | Rn | 86 | (222) |
| Darmstadtium | Ds | 110 | (281) | Rhenium | Re | 75 | 186.2 |
| Dubnium | Db | 105 | (268) | Rhodium | Rh | 45 | 102.9 |
| Dysprosium | Dy | 66 | 162.5 | Roentgenium | Rg | 111 | (280) |
| Einsteinium | Es | 99 | (252) | Rubidium | Rb | 37 | 85.47 |
| Erbium | Er | 68 | 167.3 | Ruthenium | Ru | 44 | 101.1 |
| Europium | Eu | 63 | 152.0 | Rutherfordium | Rf | 104 | (265) |
| Fermium | Fm | 100 | (257) | Samarium | Sm | 62 | 150.4 |
| Flerovium | Fl | 114 | (289) | Scandium | Sc | 21 | 44.96 |
| Fluorine | F | 9 | 19.00 | Seaborgium | Sg | 106 | (271) |
| Francium | Fr | 87 | (223) | Selenium | Se | 34 | 78.97 |
| Gadolinium | Gd | 64 | 157.3 | Silicon | Si | 14 | 28.09 |
| Gallium | Ga | 31 | 69.72 | Silver | Ag | 47 | 107.9 |
| Germanium | Ge | 32 | 72.63 | Sodium | Na | 11 | 22.99 |
| Gold | Au | 79 | 197.0 | Strontium | Sr | 38 | 87.62 |
| Hafnium | Hf | 72 | 178.5 | Sulfur | S | 16 | 32.06 |
| Hassium | Hs | 108 | (277) | Tantalum | Ta | 73 | 180.9 |
| Helium | He | 2 | 4.003 | Technetium | Tc | 43 | (98) |
| Holmium | Ho | 67 | 164.9 | Tellurium | Te | 52 | 127.6 |
| Hydrogen | H | 1 | 1.008 | Tennessine | Ts | 117 | (294) |
| Indium | In | 49 | 114.8 | Terbium | Tb | 65 | 158.9 |
| Iodine | I | 53 | 126.9 | Thallium | Tl | 81 | 204.4 |
| Iridium | Ir | 77 | 192.2 | Thorium | Th | 90 | 232.0 |
| Iron | Fe | 26 | 55.85 | Thulium | Tm | 69 | 168.9 |
| Krypton | Kr | 36 | 83.80 | Tin | Sn | 50 | 118.7 |
| Lanthanum | La | 57 | 138.9 | Titanium | Ti | 22 | 47.87 |
| Lawrencium | Lr | 103 | (262) | Tungsten | W | 74 | 183.8 |
| Lead | Pb | 82 | 207.2 | Uranium | U | 92 | 238.0 |
| Lithium | Li | 3 | 6.94 | Vanadium | V | 23 | 50.94 |
| Livermorium | Lv | 116 | (293) | Xenon | Xe | 54 | 131.3 |
| Lutetium | Lu | 71 | 175.0 | Ytterbium | Yb | 70 | 173.0 |
| Magnesium | Mg | 12 | 24.31 | Yttrium | Y | 39 | 88.91 |
| Manganese | Mn | 25 | 54.94 | Zinc | Zn | 30 | 65.38 |
| Meitnerium | Mt | 109 | (276) | Zirconium | Zr | 40 | 91.22 |

*All relative atomic masses are given to four significant figures. Values in parentheses represent the mass number of the most stable isotope

# Introduction to Chemistry

## Fifth Edition

**Richard C. Bauer**
*Arizona State University*

**James P. Birk**
*Arizona State University*

**Pamela S. Marks**
*Arizona State University*

McGraw Hill Education

INTRODUCTION TO CHEMISTRY, FIFTH EDITION

Published by McGraw-Hill Education, 2 Penn Plaza, New York, NY 10121. Copyright © 2019 by McGraw-Hill Education. All rights reserved. Printed in the United States of America. Previous editions © 2016, 2013, and 2010. No part of this publication may be reproduced or distributed in any form or by any means, or stored in a database or retrieval system, without the prior written consent of McGraw-Hill Education, including, but not limited to, in any network or other electronic storage or transmission, or broadcast for distance learning.

Some ancillaries, including electronic and print components, may not be available to customers outside the United States.

This book is printed on acid-free paper.

1 2 3 4 5 6 7 8 9 LWI 21 20 19 18

ISBN 978-1-259-91114-9
MHID 1-259-91114-4

Portfolio Manager: *David Spurgeon, Ph.D.*
Product Developers: *Lead: Robin Reed; Digital: Marisa Moreno*
Marketing Manager: *Matthew Garcia*
Content Project Managers: *Laura Bies, Rachael Hillebrand, Sandy Schnee*
Buyer: *Sandy Ludovissy*
Design: *David W. Hash*
Content Licensing Specialists: *Melissa Homer*
Cover Image: *Frozen methane bubbles at Abraham lake at sunrise with beautiful sky.*
        *©Naphat Photography/Moment/Getty Images*
Compositor: *Aptara®, Inc.*

All credits appearing on page or at the end of the book are considered to be an extension of the copyright page.

**Library of Congress Cataloging-in-Publication Data**

Names: Bauer, Richard C., 1963 November 24- author. | Birk, James P., author. |
   Marks, Pamela, author.
Title: Introduction to chemistry / Richard C. Bauer, Arizona State University, James P. Birk, Arizona State
   University, Pamela S. Marks, Arizona State University.
Description: Fifth edition. | New York, NY : McGraw-Hill Education, 2019. |
   Includes index.
Identifiers: LCCN 2017036029 | ISBN 9781259911149 (alk. paper)
Subjects: LCSH: Chemistry—Textbooks.
Classification: LCC QD33.2 .B38 2019 | DDC 540—dc23 LC record available at https://lccn.loc.gov/2017036029

The Internet addresses listed in the text were accurate at the time of publication. The inclusion of a website does not indicate an endorsement by the authors or McGraw-Hill Education, and McGraw-Hill Education does not guarantee the accuracy of the information presented at these sites.

mheducation.com/highered

*In memory of my parents. Their love of reading inspired my academic interests.*

　　—Rich Bauer

*To my wife, Kay Gunter, who encouraged me through battles with blank pages and shared the joys of completed chapters; and in memory of my parents, Albert and Christine Birk, who taught me to love books enough to see blank pages as a worthwhile challenge.*

　　—Jim Birk

*To my husband, Steve, for his love and support, and to my children, Lauren, Kelsey, and Michael, for helping me see aspects of the world differently; also to my mother, Jewel Nicholls, who inspired my love of chemistry at a young age.*

　　—Pam Marks

# About the Authors

**Richard Bauer** was born and raised in Saginaw, Michigan, and completed his B.S. degree in chemistry at Saginaw Valley State University. While pursuing his undergraduate degree he worked at Dow Chemical as a student technologist. He pursued master's and PhD degrees in chemistry education at Purdue University under the direction of Dr. George Bodner. After Purdue, he spent 2 years at Clemson University as a visiting assistant professor. He is currently the Faculty Head for Science, Mathematics, and Social Science at the Downtown Phoenix Campus of Arizona State University. He was the General Chemistry Coordinator on the Tempe Campus where he implemented an inquiry-based laboratory program. He has taught introductory and general chemistry courses for over 25 years, and has taught a methods of chemistry teaching course. He is especially fond of teaching introductory chemistry because of the diversity of students enrolled. In addition to general chemistry lab development, he has interests in student visualization of abstract, molecular-level concepts; TA training; and methods of secondary school chemistry teaching. In addition to his scholarly interests, he plays the piano, sings, and directs choirs.

**James Birk** is Professor Emeritus of Chemistry and Biochemistry at Arizona State University. Born in Cold Spring, Minnesota, he received a B.A. degree in chemistry from St. John's University (Minnesota) and a PhD in physical chemistry from Iowa State University. After a postdoctorate at the University of Chicago, he started his academic career at the University of Pennsylvania, where he was appointed to the Rhodes-Thompson Chair of Chemistry. Initially doing research on mechanisms of inorganic reactions, he switched to research on various areas of chemical education after moving to Arizona State University as Coordinator of General Chemistry. Dr. Birk's teaching responsibilities have been in general chemistry, introductory chemistry, chemistry for engineers, inorganic chemistry, methods of teaching chemistry, and graduate courses on inorganic reaction mechanisms, chemical education, and science education. He has received several teaching awards, including awards for Distinction in Undergraduate Teaching, Teaching Innovation awards, the National Catalyst Award, and the President's Medal for Team Excellence. He has been a feature editor for the *Journal of Chemical Education*, editing the columns: Filtrates and Residues, The Computer Series, and Teaching with Technology. Recent research has focused on visualization (such as dynamic visualization in chemistry and the hidden earth), on inquiry-based instruction, and on misconceptions (Chemistry Concept Inventory).

**Pamela Marks** is a Principal Lecturer in the School of Molecular Sciences at Arizona State University where her main focus has been teaching introductory chemistry, general chemistry, and chemistry for engineers the past 22 years. She has been involved in improving inquiry-based learning in the general chemistry program, and has recently modified her introductory chemistry course to a flipped classroom format. She has also taught in the general chemistry program at the College of St. Benedict and St. John's University in Minnesota. Previous education publications include a multimedia-based general chemistry education curriculum. She received her B.A. in chemistry from St. Olaf College in 1984 and her M.A. in inorganic chemistry at the University of Arizona in 1988. She spends her free time with her family and hiking with her Rhodesian Ridgeback.

# Brief Contents

**1** Matter and Energy 1

**2** Atoms, Ions, and the Periodic Table 56

**3** Chemical Compounds 89

**4** Chemical Composition 127

**5** Chemical Reactions and Equations 171

**6** Quantities in Chemical Reactions 214

**7** Electron Structure of the Atom 259

**8** Chemical Bonding 302

**9** The Gaseous State 344

**10** The Liquid and Solid States 395

**11** Solutions 441

**12** Reaction Rates and Chemical Equilibrium 484

**13** Acids and Bases 528

**14** Oxidation-Reduction Reactions 569

**15** Nuclear Chemistry 615

**16** Organic Chemistry 651

**17** Biochemistry 696

## Appendices

**A** Useful Reference Tables and Figures A-1

**B** Math Toolboxes A-2

**C** Answers to Consider This Questions and Practice Problems A-3

**D** Answers to Selected Questions and Problems A-14

Glossary G-1

Index I-1

# Contents

Preface    xiii

# 1  Matter and Energy    1

### 1.1  Matter and Its Classification    3
Composition of Matter    3
Representations of Matter    9
States of Matter    11

### 1.2  Physical and Chemical Changes and Properties of Matter    13
Physical Properties    13
Physical Changes    23
Chemical Changes    24
Chemical Properties    24

### 1.3  Energy and Energy Changes    27

### 1.4  Scientific Inquiry    30
Observations    31
Hypotheses    31
Laws    33
Theories    33
Scientific Inquiry in Practice    33

Math Toolbox 1.1  Scientific Notation    35
Math Toolbox 1.2  Significant Figures    37
Math Toolbox 1.3  Units and Conversions    41
Chapter Review    45
Questions and Problems    47

# 2  Atoms, Ions, and the Periodic Table    56

### 2.1  Dalton's Atomic Theory    58

### 2.2  Structure of the Atom    60
Subatomic Particles    60
The Nuclear Atom    62
Isotopes, Atomic Number, and Mass Number    64

### 2.3  Ions    69

### 2.4  Atomic Mass    72

### 2.5  The Periodic Table    75
Classification of Elements    75
Ions and the Periodic Table    78

Chapter Review    80
Questions and Problems    81

# 3  Chemical Compounds    89

### 3.1  Ionic and Molecular Compounds    91

### 3.2  Monatomic and Polyatomic Ions    96
Monatomic Ions    96
Polyatomic Ions    98

### 3.3  Formulas for Ionic Compounds    101

### 3.4  Naming Ionic Compounds    104

### 3.5  Naming and Writing Formulas for Molecular Compounds    110

### 3.6  Acids and Bases    112

### 3.7  Predicting Properties and Naming Compounds    116

Chapter Review    118
Questions and Problems    119

# 4  Chemical Composition    127

### 4.1  Percent Composition    130

### 4.2  Mole Quantities    132
Moles and Particles    132
Molar Mass    135

### 4.3  Determining Empirical and Molecular Formulas    140
Empirical and Molecular Formulas    140
Determining Empirical Formulas    143
Empirical Formulas from Percent Composition    143
Empirical Formulas for Compounds Containing More Than Two Elements    145
Empirical Formulas with Fractional Mole Ratios    146
Molecular Formulas from Empirical Formulas    147
Determining Percent Composition    149

**4.4   Chemical Composition of Solutions   150**
Concentration   151
Percent by Mass   151
Molarity   152
Dilution   156

**Math Toolbox 4.1** Mole Conversions   158
Chapter Review   163
Questions and Problems   164

# 5   Chemical Reactions and Equations   171

**5.1   What Is a Chemical Reaction?   173**

**5.2   How Do We Know a Chemical Reaction Occurs?   174**

**5.3   Writing Chemical Equations   176**

**5.4   Predicting Chemical Reactions   182**
Decomposition Reactions   186
Combination Reactions   188
Single-Displacement Reactions   190
Double-Displacement Reactions   193
Combustion Reactions   199

**5.5   Representing Reactions in Aqueous Solution   200**
Chapter Review   202
Questions and Problems   203

# 6   Quantities in Chemical Reactions   214

**6.1   The Meaning of a Balanced Equation   217**

**6.2   Mole-Mole Conversions   218**

**6.3   Mass-Mass Conversions   219**

**6.4   Limiting Reactants   223**

**6.5   Percent Yield   233**

**6.6   Energy Changes   235**
Law of Conservation of Energy   235
Energy Changes That Accompany Chemical Reactions   236
Quantities of Heat   238

**6.7   Heat Changes in Chemical Reactions   244**
Chapter Review   247
Questions and Problems   248

# 7   Electron Structure of the Atom   259

**7.1   Electromagnetic Radiation and Energy   261**
Properties of Electromagnetic Radiation   262
Atomic Spectra   266

**7.2   The Bohr Model of the Hydrogen Atom   267**

**7.3   The Modern Model of the Atom   270**
Orbital Diagrams for Multielectron Atoms   272
Electron Configurations   276

**7.4   Periodicity of Electron Configurations   277**

**7.5   Valence Electrons for the Main-Group Elements   282**

**7.6   Electron Configurations for Ions   284**

**7.7   Periodic Properties of Atoms   286**
Chemical Reactivity and Electron Configurations   286
Ionization Energy   288
Atomic Size   292
Sizes of Ions   293
Chapter Review   295
Questions and Problems   296

# 8   Chemical Bonding   302

**8.1   Types of Bonds   304**
Ionic and Covalent Bonding   305
Polar and Nonpolar Covalent Bonds   307
Electronegativity   307

**8.2   Ionic Bonding   309**
Lewis Symbols   310
Structures of Ionic Crystals   312

**8.3   Covalent Bonding   313**
The Octet Rule   313
Lewis Structures for the Diatomic Elements   314
Valence Electrons and Number of Bonds   315
Structures of Covalent Molecules   317
Exceptions to the Octet Rule   322

**8.4   Bonding in Carbon Compounds   323**
Hydrocarbons   323
Functional Groups   324

**8.5   Shapes of Molecules   326**
The Valence-Shell Electron-Pair Repulsion Theory   327
Polarity of Molecules   333
Chapter Review   335
Questions and Problems   337

# 9  The Gaseous State    344

**9.1  The Behavior of Gases    347**
Temperature and Density    347
Pressure    348

**9.2  Factors That Affect the Properties of Gases    351**
Volume and Pressure    351
Volume and Temperature    355
Volume, Pressure, and Temperature    358
Gay-Lussac's Law of Combining Volumes    360
Avogadro's Hypothesis    360

**9.3  The Ideal Gas Law    364**
Calculations with the Ideal Gas Law    365
Dalton's Law of Partial Pressures    368

**9.4  Kinetic-Molecular Theory of Gases    370**
Postulates of Kinetic-Molecular Theory    370
Diffusion and Effusion    372

**9.5  Gases and Chemical Reactions    373**
Product Volume from Reactant Volume    373
Moles and Mass from Volume    375

**Math Toolbox 9.1**  Graphing    376
**Math Toolbox 9.2**  Solving Simple Algebraic
    Equations    379
Chapter Review    380
Questions and Problems    382

# 10  The Liquid and Solid States    395

**10.1  Changes of State    398**
Liquid-Gas Phase Changes    399
Liquid-Solid Phase Changes    403
Solid-Gas Phase Changes    403
Cooling and Heating Curves    405
Energy Changes    407

**10.2  Intermolecular Forces    410**
London Dispersion Forces    411
Dipole-Dipole Forces    413
Hydrogen Bonding    414
Trends in Intermolecular Forces    417

**10.3  Properties of Liquids    420**
Density    420
Viscosity    421
Surface Tension    422

**10.4  Properties of Solids    424**
Crystals and Crystal Lattices    424
Types of Crystalline Solids    426

Chapter Review    432
Questions and Problems    433

# 11  Solutions    441

**11.1  The Composition of Solutions    443**
**11.2  The Solution Process    447**
**11.3  Factors That Affect Solubility    451**
Structure    451
Temperature    452
Pressure    454

**11.4  Measuring Concentrations of Solutions    455**
Percent by Mass    457
Percent by Volume    458
Mass/Volume Percent    459
Parts per Million and Parts per Billion    460
Molarity    461
Molality    462

**11.5  Quantities for Reactions That Occur in Aqueous Solution    463**
Precipitation Reactions    463
Acid-Base Titrations    467

**11.6  Colligative Properties    469**
Osmotic Pressure    469
Vapor Pressure Lowering    471
Boiling Point Elevation    472
Freezing Point Depression    473
Colligative Properties and Strong Electrolytes    474

Chapter Review    475
Questions and Problems    476

# 12  Reaction Rates and Chemical Equilibrium    484

**12.1  Reaction Rates    487**
**12.2  Collision Theory    488**
**12.3  Conditions That Affect Reaction Rates    491**
Concentration and Surface Area    491
Temperature    492
Catalysts    494

**12.4  Chemical Equilibrium    498**

**12.5 The Equilibrium Constant 500**
The Equilibrium Constant Expression 500
Predicting the Direction of a Reaction 504
Heterogeneous Equilibrium 506

**12.6 Le Chatelier's Principle 509**
Reactant or Product Concentration 510
Volume of the Reaction Container 512
Temperature 515
Catalysts 516
Increasing Product Yield 517

Chapter Review 518
Questions and Problems 519

# 13 Acids and Bases 528

**13.1 What Are Acids and Bases? 530**
Acid and Base Definitions 530
Conjugate Acid-Base Pairs 532
Acidic Hydrogen Atoms 534

**13.2 Strong and Weak Acids and Bases 534**
Strong Acids 535
Strong Bases 535
Weak Acids 536
Weak Bases 537

**13.3 Relative Strengths of Weak Acids 541**
Acid Ionization Constants 541
Polyprotic Acids 542

**13.4 Acidic, Basic, and Neutral Solutions 544**
The Ion-Product Constant of Water 544
Calculating $H_3O^+$ and $OH^-$ Ion Concentrations 545

**13.5 The pH Scale 547**
Calculating pH 547
Calculating pOH 551
Calculating Concentrations from pH or pOH 551
Measuring pH 553

**13.6 Buffered Solutions 555**

**Math Toolbox 13.1** Log and Inverse
Log Functions 558
Chapter Review 561
Questions and Problems 562

# 14 Oxidation-Reduction Reactions 569

**14.1 What Is an Oxidation-Reduction Reaction? 572**
**14.2 Oxidation Numbers 576**
**14.3 Batteries 581**

**14.4 Balancing Simple Oxidation-Reduction Equations 588**
**14.5 Balancing Complex Oxidation-Reduction Equations 592**
**14.6 Electrochemistry 597**
Voltaic Cells 598
Electrolytic Cells 599
**14.7 Corrosion Prevention 603**
Chapter Review 605
Questions and Problems 606

# 15 Nuclear Chemistry 615

**15.1 Radioactivity 617**
Nuclear Decay 617
Radiation 618

**15.2 Nuclear Reactions 619**
Equations for Nuclear Reactions 619
Particle Accelerators 626
Predicting Spontaneous Nuclear Decay
Reactions 626

**15.3 Rates of Radioactive Decay 630**
Detection of Radiation 630
Half-Lives 631
Archeological Dating 633
Geological Dating 634

**15.4 Medical Applications of Isotopes 635**
Power Generators 635
Medical Diagnoses 635
Positron Emission Tomography 636
Cancer Therapy 637

**15.5 Biological Effects of Radiation 637**
Radiation Exposure 637
Radon 639

**15.6 Nuclear Energy 640**
Uranium-235 Fission 640
Chain Reactions 642
Fission Reactors 642
Fusion Reactors 644

Chapter Review 646
Questions and Problems 647

# 16 Organic Chemistry 651

**16.1 Representations of Organic Molecules 655**
**16.2 Hydrocarbons 659**
Classes of Hydrocarbons 659
Petroleum 661

**16.3  Acyclic Hydrocarbons   662**
Alkanes   662
Alkenes and Alkynes   669

**16.4  Cyclic Hydrocarbons   673**
Cycloalkanes and Cycloalkenes   673
Aromatic Hydrocarbons   674

**16.5  Alcohols and Ethers   676**
Alcohols   676
Ethers   679

**16.6  Aldehydes and Ketones   679**
Aldehydes   679
Ketones   680

**16.7  Carboxylic Acids and Esters   680**
Carboxylic Acids   680
Esters   681

**16.8  Amines   684**

**16.9  Organic Nomenclature   684**
Alkanes   685
Alkenes and Alkynes   685
Aromatic Hydrocarbons   685
Other Naming Conventions   686

Chapter Review   687
Questions and Problems   688

**17   Biochemistry   696**

**17.1  Proteins   698**
Composition of Proteins   699
Hydrolysis of Proteins   705
Structure of Proteins   708
Denaturation of Proteins   713

**17.2  Nucleic Acids   713**
Structure of Nucleic Acids   714
Deoxyribonucleic Acid and Replication   717
Ribonucleic Acid, Transcription, and Translation   718

**17.3  Carbohydrates   722**
Simple Carbohydrates   723
Complex Carbohydrates   725

**17.4  Lipids   729**
Chapter Review   734
Questions and Problems   736

**Appendices   A-1**

A   **Useful Reference Tables and Figures   A-1**
B   **Math Toolboxes   A-2**
C   **Answers to Consider This Questions and Practice Problems   A-3**
D   **Answers to Selected Questions and Problems   A-14**

**Glossary   G-1**

**Index   I-1**

# Preface

The fifth edition of *Introduction to Chemistry* continues to build on our belief that students learn best when the text and our classroom presentations focus on a conceptual approach to chemistry. Our class meetings are significantly different from traditional lecture presentations in many ways. Beginning with the first week of classes and continuing through the rest of the semester, we follow a sequence of topics that allows us to explain macroscopic phenomena from a molecular perspective. This approach places emphasis on conceptual understanding over algorithmic problem solving. To help students develop conceptual understanding, we use numerous images, animations, video clips, and live demonstrations. Roughly a third of each class period is devoted to explaining chemical phenomena from a conceptual perspective. During the remaining time, students work in groups to discuss and answer conceptual and numerical questions.

Our desire to create a conceptually based text stems from our own classroom experience, as well as from educational research about how students learn. This book is grounded in educational research findings that address topic sequence, context, conceptual emphasis, and concept-embedded numerical problem solving. Throughout the text, we have made an effort to relate the content to students' daily lives and show them how chemistry allows us to understand the phenomena—both simple and complex—that we encounter on a regular basis. Students' initial exposure to chemical concepts should be in the realm of their personal experience, to give context to the abstract concepts we want them to understand later. This text presents macroscopic chemical phenomena early and uses familiar contexts to develop microscopic explanations.

This textbook is designed for the freshman-level Introductory Chemistry course that does not have a chemistry prerequisite and is suitable for either a one-semester course or a two-semester sequence. The book targets introductory courses taken by non-physical science majors who may be in allied health, agriculture, or other disciplines that do not require the rigor of a science major's General Chemistry course, or for students fulfilling university liberal arts requirements for science credits. In addition, students who lack a strong high school science background often take the course as a preparation for the regular general chemistry sequence.

## NEW! Student Hot Spot and Student-Centered Refinements Using Heat Maps

Using heat maps from the adaptive reading tool SmartBook®, and the detailed analysis of student performance it provides, we were able to target specific learning objectives for minor re-wording, further explanation, or better illustration. Because SmartBook is a dynamic learning tool, we have a multitude of live data that show us exactly where students have been struggling with content; and we have direct insight into student learning that may not always be evident through other assessment methods. The data, such as average time spent answering each question and the percentage of students who correctly answered the question on the first attempt, revealed the learning objectives that students found particularly difficult. For example, the heat map below indicates that students struggled with questions related to the concept of density.

---

**Density** The density of an object is the ratio of its mass to its volume. While mass and volume both depend on the size of the object or sample, density does not. Density is an unvarying property of a substance no matter how much of it is present, as long as temperature and pressure are constant. The densities of a few substances are listed in Table 1.6.

---

Analysis of results like this allowed our revisions to be truly student-centered. For example, given specific known topics where students are struggling, we are able to clarify our description of density as shown below.

> **Density** The **density** of an object is the ratio of its mass to its volume. While mass and volume both depend on the size of the object or sample, density does not. Density is an unvarying property of a substance no matter how much of it is present, as long as temperature and pressure are constant. For example, the density of water at 4°C is 1.00 g/mL. It doesn't matter if we have 10 mL or 10 L; the ratio of mass to volume would be the same, 1.00 g/mL. However, if the temperature increases, the water would expand to a larger volume, while the mass stays the same. The density of liquid water would decrease as the temperature increases. The densities of water and a few other substances are listed in Table 1.6.

Further, armed with this powerful insight into the places many students struggle with content, we are able to provide strategically timed access to additional learning resources. In the text, we have identified areas of particularly difficult content as "Student Hot Spots"—and use them to direct students to a variety of learning resources specific to that content. Students will be able to access over 800 digital learning resources throughout this text's SmartBook. These learning resources present summaries of concepts and worked examples, including over 200 videos of chemistry faculty solving problems or modeling concepts that students can view over and over again.

**Student Hot Spot**

Student data indicate you may struggle with writing balanced equations for combination reactions. Access the SmartBook to view additional Learning Resources on this topic.

In the electronic version of the text, learning resources for these Student Hot Spots are embedded with the content for immediate access.

Guided by student results of conceptual understanding, we edited the content in all the chapters. While some changes are subtle, others are more extensive. Our ability to employ live student-assessment data for revisions to address areas of common misunderstanding is unprecedented and has afforded us the opportunity to change how we provide the best possible learning materials.

# Features of this Text

Learning theory indicates that we should start with the concrete, macroscopic world of experience as the basis for developing student understanding of abstract, microscopic concepts. This textbook follows a topic sequence typically found in many general chemistry texts. That is, macroscopic ideas about chemical behavior are discussed before descriptions of abstract, molecular-level concepts associated with electron structure. The macroscopic ideas that begin chapters or sections are grounded in real-life experiences. Where appropriate, the macroscopic to molecular-level progression of ideas is carried over to topic sequence within individual chapters or sections in addition to the general sequence of chapters.

Each chapter begins with a chapter-opening outline and an **opening vignette** that personalizes the content by telling a story about chemical phenomena encountered by students. These applications help students see how chemistry relates to their daily lives.

## Questions for Consideration

**9.1** What are some general properties of gases?
**9.2** How does the behavior of gases vary with changes in pressure, temperature, volume, and number of molecules (atoms)?
**9.3** What are the mathematical relationships between volume, pressure, temperature, and amount of gas?
**9.4** What is the theory that explains the behavior of gases in terms of atomic or molecular motion?
**9.5** How can quantities of gases in chemical reactions be calcula

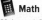 **Math Tools Used in This Chapter**

Units and Conversions (Math Toolbox 1.3)
Mole Quantities (Math Toolbox 4.1)
Graphing (Math Toolbox 9.1)
Solving Simple Algebraic Equations (Math Toolbox 9.2)

A can of diet cola floats in water, but a can of regular cola sinks. Suggest a reason why. How can you use this information to quickly select your preferred type of soft drink from a cooler filled with ice water at a party?

©Brian Moeskau/Moeskau Photography

The chapter then offers some guiding questions typical of inquiry instruction. These **Questions for Consideration** serve as a guide in topic development through the chapter. **Margin notes** contain further explanations and chemical applications, combined with visuals, to help students conceptualize lessons.

### 5.1 What Is a Chemical Reaction?

Each of the students described in the introduction was carrying out some kind of chemical reaction. For example, when Antonio put iron metal into a solution of copper(II) chloride, a chemical reaction occurred. What is a chemical reaction?

A chemical reaction is the conversion of one substance or set of substances into another. Any substance that we start with is a **reactant.** A new substance that forms during the reaction is a **product.** Products are different from reactants in the *arrangement* of their component atoms. Chemical reactions neither destroy atoms nor create new atoms. They occur because the bonds that hold atoms together can be broken and rearranged. (You'll read more about chemical bonding in Chapter 8.)

Consider the reaction of hydrogen gas with oxygen gas, as shown in Figure 5.5. Hydrogen and oxygen both occur naturally as diatomic molecules. If they mix, they react slowly; but if ignited, the reaction is vigorous, even explosive. In either case, the reactants form the same product: gaseous water molecules, each containing two hydrogen atoms and one oxygen atom. Figure 5.6 shows the rearrangement of the

Chemical reactions are not the same as nuclear reactions, in which new elements may form. You'll learn about those in Chapter 15.

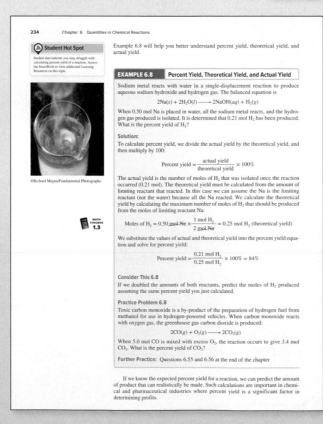

**FIGURE 5.5** When ignited, hydrogen molecules and oxygen molecules in the air react explosively to form gaseous water.
©Charles D. Winters

Reactants          Products

**FIGURE 5.6** The hydrogen atoms from hydrogen molecules combine with the oxygen atoms from oxygen molecules to form gaseous water molecules.

---

We believe that an introductory chemistry textbook should maintain a focus on *chemistry,* rather than on math. Students' interest must be captured early if they're going to persevere in the class. Early in this text, we introduce chemical reactions from macroscopic perspectives. A general fundamental knowledge of chemical behavior on a macroscopic level facilitates further development of molecular-level ideas, such as atomic structure.

We believe that the best approach to incorporating math involves development of associated math on an as-needed basis with an emphasis on concepts that problems are trying to illustrate. This text integrates need-to-know mathematical ideas that are important to chemists into conceptual discussions. **Math toolboxes** include a thorough explanation of the math, examples, worked-out solutions, and practice problems.

---

### Math Toolbox 4.1    Mole Conversions

The mole (mol) unit describes a quantity of particles and is usually used to describe numbers of atoms or formula units such as molecules.

$$1 \text{ mole} = 6.022 \times 10^{23} \text{ particles}$$

The number of particles in 1 mol is often referred to as Avogadro's number.

It is useful to describe quantities in mole units, and it is also sometimes important to know the mass or number of molecules or atoms in a given sample. The following diagram shows the importance of the mole in converting between the macroscopic level (grams and moles) and the molecular level (formula units, molecules, atoms, and ions), and shows the individual steps involved in converting between these quantities.

Mass in grams → molar mass → Moles → Avogadro's number → Number of formula units → chemical formula → Number of atoms or ions

In the sections that follow, we will look at examples of individual calculations and multistep calculations involving these quantities.

**Converting Between Moles and Number of Particles**

When we are given the moles of a substance, the mole quantity describes the number of particles indicated by the substance's formula:

$$1 \text{ mol Cu} = 6.022 \times 10^{23} \text{ Cu atoms}$$
$$1 \text{ mol O}_2 = 6.022 \times 10^{23} \text{ O}_2 \text{ molecules (or formula units)}$$
$$1 \text{ mol CO}_2 = 6.022 \times 10^{23} \text{ CO}_2 \text{ molecules (or formula units)}$$
$$1 \text{ mol NaCl} = 6.022 \times 10^{23} \text{ NaCl formula units}$$

For molecular compounds or elements, we can refer to the particles as either molecules or formula units. However, for ionic compounds, we must refer to the particles only as formula units since ionic compounds do not exist as molecules.

The relationship $1 \text{ mole} = 6.022 \times 10^{23}$ formula units can be used to convert between moles and the number of formula units for any substance.

Moles ← $1 \text{ mole} = 6.022 \times 10^{23}$ formula units → Formula units

The conversion ratio for $CO_2$, for example, will have one of the following two forms, depending on whether the conversion is to formula units or to moles:

$$\frac{6.022 \times 10^{23} \text{ CO}_2 \text{ formula units}}{1 \text{ mol CO}_2} \quad or \quad \frac{1 \text{ mol CO}_2}{6.022 \times 10^{23} \text{ CO}_2 \text{ formula units}}$$

Conversion *ratios* are similar to the conversion *factors* we used in Chapter 1 to convert measured values to different units.

**EXAMPLE 4.20   Converting Between Moles and Formula Units**

Use Avogadro's number for the following conversions:
(a) Convert 0.042 mol $CO_2$ to formula units (molecules) of $CO_2$.
(b) Convert $4.4 \times 10^{24}$ NaCl formula units to moles.

---

### 234    Chapter 6   Quantities in Chemical Reactions

**Student Hot Spot**

Student data indicate you may struggle with calculating percent yield of a reaction. Access the SmartBook to view additional Learning Resources on this topic.

©Richard Megna/Fundamental Photographs

MATH TOOLBOX 1.3

Example 6.8 will help you better understand percent yield, theoretical yield, and actual yield.

**EXAMPLE 6.8   Percent Yield, Theoretical Yield, and Actual Yield**

Sodium metal reacts with water in a single-displacement reaction to produce aqueous sodium hydroxide and hydrogen gas. The balanced equation is

$$2Na(s) + 2H_2O(l) \longrightarrow 2NaOH(aq) + H_2(g)$$

When 0.50 mol Na is placed in water, all the sodium metal reacts, and the hydrogen gas produced is isolated. It is determined that 0.21 mol $H_2$ has been produced. What is the percent yield of $H_2$?

**Solution:**

To calculate percent yield, we divide the actual yield by the theoretical yield, and then multiply by 100:

$$\text{Percent yield} = \frac{\text{actual yield}}{\text{theoretical yield}} \times 100\%$$

The actual yield is the number of moles of $H_2$ that was isolated once the reaction occurred (0.21 mol). The theoretical yield must be calculated from the amount of limiting reactant that reacted. In this case we can assume the Na is the limiting reactant (not the water) because all the Na reacted. We calculate the theoretical yield by calculating the maximum number of moles of $H_2$ that should be produced from the moles of limiting reactant Na:

$$\text{Moles of H}_2 = 0.50 \text{ mol Na} \times \frac{1 \text{ mol H}_2}{2 \text{ mol Na}} = 0.25 \text{ mol H}_2 \text{ (theoretical yield)}$$

We substitute the values of actual and theoretical yield into the percent yield equation and solve for percent yield:

$$\text{Percent yield} = \frac{0.21 \text{ mol H}_2}{0.25 \text{ mol H}_2} \times 100\% = 84\%$$

**Consider This 6.8**

If we doubled the amounts of both reactants, predict the moles of $H_2$ produced assuming the same percent yield you just calculated.

**Practice Problem 6.8**

Toxic carbon monoxide is a by-product of the preparation of hydrogen fuel from methanol for use in hydrogen-powered vehicles. When carbon monoxide reacts with oxygen gas, the greenhouse gas carbon dioxide is produced:

$$2CO(g) + O_2(g) \longrightarrow 2CO_2(g)$$

When 5.0 mol CO is mixed with excess $O_2$, the reaction occurs to give 3.4 mol $CO_2$. What is the percent yield of $CO_2$?

**Further Practice:** Questions 6.55 and 6.56 at the end of the chapter

If we know the expected percent yield for a reaction, we can predict the amount of product that can realistically be made. Such calculations are important in chemical and pharmaceutical industries where percent yield is a significant factor in determining profits.

---

Toolboxes are referenced with toolbox icons, where appropriate. As problem solving is developed within the text, emphasis is placed on the underlying concepts, letting the numerical solutions emerge from conceptual understanding. Numerical-type problems often ask students to estimate answers and to consider the physical meaning of calculated quantities.

The problem-solving approach used in this text is supported by worked example boxes that contain the following steps:

▶ question(s)
▶ solution
▶ consider this question
▶ practice problems
▶ further practice

Problem solving in chemistry is much more than algorithmic number crunching. It involves applying principles to solve conceptual as well as numerical problems. Conceptual problems are those that require students to apply their understanding of concepts instead of applying an algorithm. This text emphasizes the underlying concepts when discussing numerical problems within in-chapter worked examples. Many end-of-chapter problems also emphasize conceptual problem solving.

There are several other features of this textbook that support student learning. End-of-chapter materials include **math toolboxes** (when appropriate), **key concepts summary, key terms list,** and **key relationships.** Each chapter has extensive **end-of-chapter questions and problems** that range in difficulty and conceptual/quantitative emphasis. Most of the questions and problems are sorted by section and are paired, with selected answers appearing in Appendix D. There are also **vocabulary identification questions** at the beginning of the end-of-chapter problems, as well as many questions involving interpretation of molecular-level images. The **concept review questions** provide students with practice at reasoning through multiple-choice questions. To support the text's problem-solving approach, video tutorials in the online homework system (Connect) provide assistance to students struggling with a particular question. These tutorials guide students through an approach to a similar type of question and emphasize the conceptual foundation and the process that leads to a reasonable answer.

---

### Screenshot 1 (Example page)

Chapter 10 The Liquid and Solid States

**EXAMPLE 10.3  Cooling and Heating Curves**

The following diagrams represent two different changes of state.

(a)   (b)

For each change of state, although heat is being added, the temperature does not change. Identify where each change of state would be found on the following heating curve.

**Solution:**

(a) This molecular diagram shows the liquid and gaseous states coexisting. Since the temperature does not change during the process, this corresponds to any point on the *DE* plateau in the heating curve.

(b) This molecular diagram shows solid and liquid together, which corresponds to any point on the *BC* plateau in the heating curve.

**Consider This 10.3**

If heat were being removed from a substance, how would the graph and diagram from the example differ?

**Practice Problem 10.3**

The following diagrams represent the physical states of a substance.

(a)   (b)   (c)

Identify where each state would predominate on the heating curve given in the example.

**Further Practice:** Questions 10.31 and 10.32 at the end of the chapter

---

### Screenshot 2 (Chapter Review)

Key Relationships  **561**

## CHAPTER REVIEW

### KEY CONCEPTS

- According to Brønsted-Lowry theory, an acid is a substance that donates an $H^+$ ion to another substance, and a base is an $H^+$ acceptor.
  - Brønsted-Lowry acids react with water in a process called ionization to produce hydronium ions, $H_3O^+$.
  - Brønsted-Lowry bases ionize or dissociate in water to produce hydroxide ions, $OH^-$.
- Acids and bases have varying strengths.
  - Strong acids and bases ionize or dissociate completely in water.
  - When weak acids and bases dissolve in water, only a small percentage of the molecules ionizes; the equilibrium lies to the left toward the reactant molecules.
  - Equilibrium constants for the reactions of acids with water are called acid ionization constants, $K_a$. These values can be used to compare the strengths of different weak acids. The stronger the acid, the greater the $K_a$ value.
- The relative concentrations of $H_3O^+$ and $OH^-$ ions in an aqueous solution are determined by the ion-product constant for water, $K_w$, which is the equilibrium constant for the reaction $2H_2O(l) \rightleftharpoons H_3O^+(aq) + OH^-(aq)$.
  - $K_w$ is equal to the product of the hydronium and hydroxide ion concentrations: $K_w = [H_3O^+] \times [OH^-]$.
  - The value of $K_w$ is $1.0 \times 10^{-14}$ at 25°C.
  - Neutral solutions have equal concentrations of $H_3O^+$ and $OH^-$, each equal to $1.0 \times 10^{-7}$ M at 25°C. Acidic solutions contain a greater concentration of $H_3O^+$ than $OH^-$. Basic solutions contain a greater concentration of $OH^-$ than $H_3O^+$. Because $[H^+] \times [OH^-] = 1.0 \times 10^{-14}$, as one concentration increases the other decreases.
- The acidity of a solution is commonly expressed in terms of pH, the negative logarithm of the hydronium ion concentration: $pH = -\log[H_3O^+]$.
  - At 25°C, acidic solutions have [ ... ] greater than 7, and neutral solut[ ... ]
  - Indicators and pH meters are [ ... ] solution.
- Buffered solutions contain a weak [ ... ] and its conjugate acid) in similar [ ... ] changes in pH by reacting with sm[ ... ]

### KEY RELATIONSHIPS

| Relationship |
| --- |
| The self-ionization constant for water equal[ ... ] concentration and the hydroxide ion concen[ ... ] $1.0 \times 10^{-14}$. |
| The pH of a solution equals the negative log[ ... ] The pOH of a solution equals the negative [ ... ] |
| The hydronium ion concentration equals th[ ... ] The hydroxide ion concentration equals the[ ... ] |
| The pH and pOH of a solution total 14.00. |

---

### Screenshot 3 (Key Terms and Questions)

**562**   Chapter 13  Acids and Bases

### KEY TERMS

| | | | |
| --- | --- | --- | --- |
| acid ionization constant, $K_a$ (13.3) | basic solution (13.4) | ion-product constant of water, $K_w$ (13.4) | self-ionization (13.4) |
| acidic solution (13.4) | Brønsted-Lowry theory (13.1) | neutral solution (13.4) | strong acid (13.2) |
| amphoteric substance (13.1) | buffer (13.6) | pH (13.5) | strong base (13.2) |
| Arrhenius model of acids and bases (13.1) | conjugate acid (13.1) | polyprotic acid (13.3) | weak acid (13.2) |
| | conjugate base (13.1) | | weak base (13.2) |
| | hydronium ion (13.1) | | |

### QUESTIONS AND PROBLEMS

The following questions and problems, except those in *Additional Questions* and *Concept Review Questions*, are paired. Questions in a pair focus on the same concept. Answers to selected questions and problems are in Appendix D.

**Matching Definitions with Key Terms**

13.1 Match the key terms with the descriptions provided.
  (a) a base that ionizes or dissociates completely when dissolved in water
  (b) a theory that defines an acid as a substance that donates an $H^+$ to another substance in solution and a base as a substance that accepts an $H^+$ in solution
  (c) a substance that forms after a base gains an $H^+$ ion
  (d) an acid containing more than one acidic hydrogen
  (e) a solution in which the $H_3O^+$ concentration is greater than the $OH^-$ concentration; a solution with a pH less than 7
  (f) a base that does not completely ionize when dissolved in water
  (g) a substance that can act as either an acid or a base
  (h) the equilibrium constant for the self-ionization of water; the product of the $H_3O^+$ ion concentration and the $OH^-$ ion concentration in any aqueous solution
  (i) a process in which one molecule transfers an $H^+$ to another molecule of the same substance; water does this to a very small extent

13.2 Match the key terms with the descriptions provided.
  (a) an acid that ionizes completely when dissolved in water
  (b) a model that describes acids as substances that generate $H^+$ ions in solution and bases as substances that generate $OH^-$ ions in solution
  (c) an aqueous hydrogen ion; $H_3O^+(aq)$
  (d) a solution in which the $OH^-$ concentration is greater than the $H_3O^+$ concentration; a solution with a pH greater than 7
  (e) an acid that does not completely ionize when dissolved in water
  (f) a substance that forms after an acid loses an $H^+$ ion
  (g) an equilibrium constant for the ionization of an acid in water; a value that expresses the strength of a weak acid
  (h) a measure of the acidity of aqueous solutions; the negative logarithm of the $H_3O^+$ concentration
  (i) a combination of a weak acid and its conjugate base (or a weak base and its conjugate acid) in similar amounts; when in solution, something that helps prevent large changes in pH when small amounts of $H_3O^+(aq)$ or $OH^-(aq)$ are added

**Math Toolbox Questions**

13.3 Use your calculator to find the log of the following numbers.
  (a) $10^{-5}$
  (b) $1 \times 10^{-11}$
  (c) $7.4 \times 10^3$
  (d) $10^3$
  (e) 1

13.4 Use your calculator to find the log of the following numbers.
  (a) $10^4$
  (b) $1 \times 10^{-6}$
  (c) $1.7 \times 10^4$
  (d) $10^{-8}$
  (e) 10

13.5 Use your calculator to find the inverse log of the following numbers.
  (a) 1.20
  (b) −6.20
  (c) 0

13.6 Use your calculator to find the inverse log of the following numbers.
  (a) 12.7
  (b) −9.4
  (c) 1

**What Are Acids and Bases?**

13.7 What are some properties of acids?

13.8 What are some properties of bases?

13.9 List some common foods or household products that are bases.

13.10 List some common foods or household products that are acids.

13.11 In terms of the Arrhenius concept of acids and bases, how does an acid behave when dissolved in water? How does an Arrhenius base behave?

13.12 How does the Arrhenius concept of acids and bases emphasize that they are electrolytes?

13.13 How is the Brønsted-Lowry theory of acids and bases different from the Arrhenius model?

13.14 Why is $H_3O^+(aq)$ a more accurate representation of an aqueous $H^+$ ion than $H^+(aq)$?

## The Art Program

A conceptual understanding of chemistry requires students to visualize molecular-level representations of macroscopic phenomena, as well as to connect macroscopic and molecular-level understandings to symbolic representations. To help students connect verbal descriptions to molecular-level representations, this book has an extensive **art program.** You'll notice many examples of zoomed art, where pictures or other macroscopic images have close-ups that show the particular phenomena at a molecular level.

**FIGURE 3.6** For each of the solutions, inspect the molecular-level images in which each type of particle is labeled. If any of these substances dissociate into ions, separate particles of single atoms (or atom groups) are visible. Which of the compounds dissociate completely in solution? Which only partially dissociate? Which do not dissociate at all? The acids do not simply dissociate, but form $H_3O^+$ ions in solution.
©McGraw-Hill Education/Stephen Frisch

**FIGURE 1.8** A copper pipe consists of a regular array of copper atoms.
©Thinkstock/Getty Images

**FIGURE 1.9** Helium atoms are present inside the balloon.
©Jules Frazier/Getty Images

**FIGURE 4.18** The color intensity of a colored solution decreases as the concentration of the solute decreases. Compare the concentrations of ions in the solution for these copper(II) sulfate solutions. Which has more copper(II) ions? More sulfate ions? Which solution has the greatest concentration?
© Jim Birk

Students who enroll in an introductory chemistry course often take an associated lab. Most of the experiments these students conduct involve working with solutions. To enhance this lab experience, a brief **introduction to solution behavior** appears early in the textbook (Chapter 4). This early introduction will allow students to better understand what they experience in the lab, as well as understand the multitude of solutions we encounter on a daily basis.

# Detailed List of Changes

## All Chapters

Since the fourth edition of this book, elements 113, 115, 117, and 118 have been named: nihonium (Nh), moscovium (Mc), tennessine (Ts), and oganesson (Og), respectively. Where applicable, periodic tables have been revised to include these elements names or symbols.

## Chapter 1  Matter and Energy

Based on Smartbook Heat Maps, the following changes were made for clarification of important concepts:
- Description of metal properties
- Description of compound
- General descriptions of physical and chemical properties
- Converting units, including Example 1.5
- Description of volume units
- Description of density and relationship to temperature
- Description of electric energy to include discussion of a battery
- Description of metric conversions and negative exponents

A new margin note was added making a distinction between heating, cooking, and burning.

An in-text calculation was added for determining the density of a cube of an unknown metal.

A Student Hot Spot was added for unit conversions.

## Chapter 2  Atoms, Ions, and the Periodic Table

Based on Smartbook Heat Maps, the following changes were made for clarification of important concepts:
- Description of isotopes
- Transition to Example 2.4 for interpreting isotope symbols
- Description of ions

A new margin note was added, using grade point average as an example of a weighted average.

Concept Review Question 2.147 was modified to remove premature reference to ionic radius.

Student Hot Spots were added for relative atomic mass, classifying elements in the periodic table, and predicting ionic charges.

## Chapter 3  Chemical Compounds

Based on Smartbook Heat Maps, the following changes were made for clarification of important concepts:
- Figure 3.1 caption
- General descriptions of electrolyte and nonelectrolyte
- General descriptions of ionic and molecular compounds and their similarities and differences in properties and molecular-level structure
- Difference between atoms and monatomic ions
- Naming ionic compounds, including Example 3.9
- Naming binary and oxoacids using bulleted points

Practice Problem 3.2 was modified to ask about melting point instead of boiling point. The compounds were also changed to emphasize the difference between ionic and molecular compounds.

A transition to Example 3.5 was added to emphasize that ion ratios remain the same when an ionic compounds dissolves in water.

The difference in naming hydrogen halides such as HCl as molecular compounds or as acids when dissolved in water was noted with a new margin note.

Student Hot Spots were added for writing formulas for ionic compounds, writing formulas for ionic compounds that contain polyatomic ions, and naming an ionic compound from its formula.

## Chapter 4  Chemical Composition

Based on Smartbook Heat Maps, the following changes were made for clarification of important concepts:
- In-text problem calculating number of ions
- General description of molar mass and description of molar mass determination for a compound
- General description of empirical and molecular formulas
- Description of empirical formulas from percent composition
- Description of molecular formulas from empirical formulas
- Description of ionic compound in solution, including the molecular-level image shown in Example 4.14
- Description of percent by mass and molarity
- Approach for molarity calculations, including dilution problems
- Rewording in Math Toolbox 4.1

An in-text calculation was added for converting moles to mass for a compound as in Example 4.6.

The ionic compound in Example 4.22 was changed from NaCl to $MgCl_2$.

Student Hot Spots were added for converting grams to formula units and molarity.

## Chapter 5  Chemical Reactions and Equations

Based on Smartbook Heat Maps, the following changes were made for clarification of important concepts:
- Predicting products of combination reactions between a metal and a nonmetal
- Description of single-displacement reactions
- Predicting the precipitate in precipitation reactions
- General description of gas-forming reactions
- Description of an ionic equation

Some photos in Figure 5.8 were updated.

Figure 5.12 was updated to include rubidium in the series of alkali metals reacting with water.

The photo in end-of-chapter Question 5.15 was updated.

Context was added to Practice Problem 5.14.

Student Hot Spots were added for writing a balanced equation for a combination reaction between a metal and a nonmetal, predicting and writing equations for single-displacement reactions, and writing net ionic equations and determining spectator ions.

## Chapter 6  Quantities in Chemical Reactions

Based on Smartbook Heat Maps, the following changes were made for clarification of important concepts:

- Discussion of a balanced equation in Example 6.1
- Magnesium was changed to potassium in the box on exact amounts of reactants
- Example 6.9 was modified to indicate that the energy produced is electricity

A box was added to reference the concept of moles from Chapter 4.

Figure 5.12 was updated to include rubidium in the series of alkali metals reacting with water.

Consider This 6.11 was replaced with a new question.

Student Hot Spots were added for interpreting chemical equations, limiting reactants at a molecular level, determining the mass of product when there is a limiting reactant, calculating percent yield of a reaction, and calorimetry calculations.

## Chapter 7  Electron Structure of the Atom

Based on Smartbook Heat Maps, the following changes were made for clarification of important concepts:

- Description of electromagnetic radiation, frequency, and wavelength
- Description of Bohr model
- Description of valence level

The caption for Figure 7.6 was revised for clarity.

Students Hot Spots were added for calculations involving photon energy, wavelength, and frequency; and the Bohr model of the atom.

## Chapter 8  Chemical Bonding

Based on Smartbook Heat Maps, the following changes were made for clarification of important concepts:

- General descriptions of ionic and covalent bonding, including the caption to Figure 8.2
- Descriptions of covalent bonds and unshared electron pairs
- Drawing Lewis structures for polyatomic ions and oxoacids
- Molecules and ions with similar Lewis structures
- Determining the shape of molecule or polyatomic ion from its Lewis structure
- Determining the structures of molecules with more than one central atom

The key term **Lewis formula** was changed to **Lewis structure** throughout chapter and later chapters.

The Consider This question that follows Example 8.7 was modified to emphasize when resonance is not possible.

Table 8.6 was reorganized to emphasize the importance of electron domains in determining molecular shapes.

Modified text, including captions in Figures 8.32 and 8.24 and explanation in Example 8.11.

Student Hot Spots were added for describing the difference between polar and nonpolar covalent bonds, drawing Lewis structures,

determining shape of a molecule from its Lewis structure, and determining if a molecule is polar from its molecular shape.

## Chapter 9  The Gaseous State

Based on Smartbook Heat Maps, the following changes were made for clarification of important concepts:

- Specified that the given density of water is for the liquid
- Margin note added on air density and temperature
- A reference to Math Toolbox 9.2 was added to the discussion of combined gas law calculations.
- The Avogadro's Hypothesis discussion and the caption to Figure 9.20 were modified to indicate that the simple direct relationship works only at constant temperature and pressure.

A margin note was added to emphasize that temperature must be in units of kelvins in gas law calculations.

Student Hot Spots were added for volume-temperature relationships, pressure-volume-temperature relationships, application of Dalton's law, and relative rates of diffusion.

## Chapter 10  The Liquid and Solid States

Based on Smartbook Heat Maps, the following changes were made for clarification of important concepts:

- Description of liquid-gas phase changes
- Description of solid-gas phase changes
- Description of surface tension

Table 10.2 was revised for clarity.

Student Hot Spots were added for energy for phase changes and the relationship between strength of intermolecular forces and boiling point.

## Chapter 11  Solutions

Based on Smartbook Heat Maps, the following changes were made for clarification of important concepts:

- Description of solution composition, including a transition for Example 11.1
- General description of factors that affect solubility
- Margin note added to provide context to the definition of solubility
- Descriptions of unsaturated and supersaturated solutions
- Description of precipitation reactions

The margin note about the ppm unit was revised to include the ppb unit with examples.

The brief description of normality and its use in some commercial labs was modified and moved into the text.

In Example 11.8, the balanced equation for the reaction was added and the solution was modified to improve clarity.

The definition of the term **saturated solution** was modified to be more precise and applicable to usage in this textbook.

A margin note was added about the solubility of sucrose in water.

Student Hot Spots were added for representing a dissolving process with a balanced equation, understanding the changes in attractive forces when an ionic compound dissolves, and predicting solubility in polar and nonpolar solvents.

## Chapter 12  Reaction Rates and Chemical Equilibrium

Based on Smartbook Heat Maps, the following changes were made for clarification of important concepts:

- In Example 12.3, "hydronium ions" was changed to "aqueous hydrogen ions."
- Description of effect of temperature on the position of equilibrium and the direction of a shift

New photos were used in the chapter introduction and Figures 12.2 and 12.20.

Student Hot Spots were added for collision theory, identifying intermediates and catalysts, equilibrium constant expressions, reactions quotients and predicting the direction of a reaction, and applying Le Chatelier's principle.

## Chapter 13  Acids and Bases

Based on Smartbook Heat Maps, the following changes were made for clarification of important concepts:

- Description of weak acid and its conjugate base
- Description of weak base and its conjugate acid
- Clarified solution to Example 13.5
- Clarified caption to Figure 13.21

The description of an amphoteric substance was modified to include $HSO_3^-$.

Student Hot Spots were added for determining hydronium and hydroxide concentration for solutions of strong acids or strong bases and determining hydronium and hydroxide concentration from pH or pOH.

## Chapter 14  Oxidation-Reduction Reactions

Based on Smartbook Heat Maps, the following changes were made for clarification of important concepts:

- The wordings to the caption of Figure 14.6 and to the solution of Example 14.1 were modified.
- The solution to Example 14.4 was modified.
- The solution to Example 14.7 was modified.
- The cost of corrosion repair and prevention was updated.

New photos were used in the chapter introduction and Figures 14.2, 14.9, and 14.26.

Student Hot Spots were added for assigning oxidation numbers, identifying oxidation-reduction reactions and changes in oxidation number, balancing simple oxidation-reduction reaction equations, balancing complex oxidation-reduction reaction equations, and predicting spontaneous reactions in voltaic cells.

## Chapter 15  Nuclear Chemistry

Based on Smartbook Heat Maps, the following changes were made for clarification of important concepts:

- A transition was added before Example 15.2 to clarify the usage of beta radiation as electron and positron particles.
- The solution to Example 15.2 was modified.

The caption to Figure 15.23 was modified to include a brief description of the purpose of the control rods and moderator in the reactor core in a fission reactor.

Student Hot Spots were added for writing balanced equations for radioactive decay processes, predicting the method of radioactive decay by which an unstable nuclide will undergo, and using half-lives to determine the amount of nuclide remaining after a given amount of time.

## Chapter 16  Organic Chemistry

A margin note was added to help students recall that carbon generally forms four covalent bonds because carbon atoms have four valence electrons and needs eight to obtain an octet.

Student Hot Spots were added for representing organic molecules in different ways, determining if a compound is saturated or unsaturated from its formula, writing names for alkanes and drawing structural formula, and determining names and structures for alkenes and alkynes.

## Chapter 17  Biochemistry

Based on Smartbook Heat Maps, the following changes were made for clarification of important concepts:

- Caption for Figure 17.5
- Caption for Figure 17.9
- Caption for Figure 17.16
- Description of glycosidic linkage formation

The solution for Example 17.4 was modified for clarity.

Student Hot Spots were added for hydrolysis of peptide bonds and converting the open-chain form of a monosaccharide to the ring form.

McGraw-Hill Connect® is a highly reliable, easy-to-use homework and learning management solution that utilizes learning science and award-winning adaptive tools to improve student results.

# Homework and Adaptive Learning

- Connect's assignments help students contextualize what they've learned through application, so they can better understand the material and think critically.

- Connect will create a personalized study path customized to individual student needs through SmartBook®.

- SmartBook helps students study more efficiently by delivering an interactive reading experience through adaptive highlighting and review.

Over **7 billion questions** have been answered, making McGraw-Hill Education products more intelligent, reliable, and precise.

## Connect's Impact on Retention Rates, Pass Rates, and Average Exam Scores

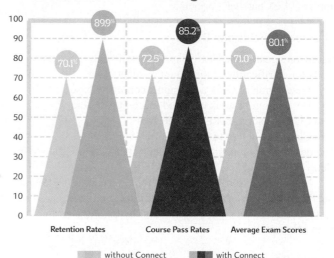

| Retention Rates | Course Pass Rates | Average Exam Scores |

70.1% 89.9% 72.5 85.2% 71.0% 80.1%

without Connect    with Connect

Using **Connect** improves retention rates by **19.8%**, passing rates by **12.7%**, and exam scores by **9.1%**.

73% of instructors who use **Connect** require it; instructor satisfaction **increases** by 28% when **Connect** is required.

# Quality Content and Learning Resources

- Connect content is authored by the world's best subject matter experts, and is available to your class through a simple and intuitive interface.

- The Connect eBook makes it easy for students to access their reading material on smartphones and tablets. They can study on the go and don't need internet access to use the eBook as a reference, with full functionality.

- Multimedia content such as videos, simulations, and games drive student engagement and critical thinking skills.

# Robust Analytics and Reporting

©Hero Images/Getty Images

- Connect Insight® generates easy-to-read reports on individual students, the class as a whole, and on specific assignments.
- The Connect Insight dashboard delivers data on performance, study behavior, and effort. Instructors can quickly identify students who struggle and focus on material that the class has yet to master.
- Connect automatically grades assignments and quizzes, providing easy-to-read reports on individual and class performance.

## Impact on Final Course Grade Distribution

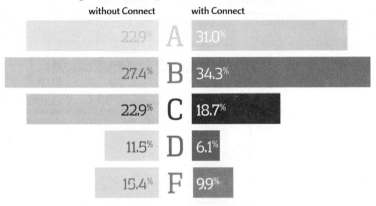

| | without Connect | with Connect |
|---|---|---|
| A | 22.9% | 31.0% |
| B | 27.4% | 34.3% |
| C | 22.9% | 18.7% |
| D | 11.5% | 6.1% |
| F | 15.4% | 9.9% |

More students earn **As** and **Bs** when they use **Connect**.

# Trusted Service and Support

- Connect integrates with your LMS to provide single sign-on and automatic syncing of grades. Integration with Blackboard®, D2L®, and Canvas also provides automatic syncing of the course calendar and assignment-level linking.
- Connect offers comprehensive service, support, and training throughout every phase of your implementation.
- If you're looking for some guidance on how to use Connect, or want to learn tips and tricks from super users, you can find tutorials as you work. Our Digital Faculty Consultants and Student Ambassadors offer insight into how to achieve the results you want with Connect.

# Instructor and Student Resources

A robust set of questions, problems, and interactive figures are presented and aligned with the textbook's learning goals. The integration of **ChemDraw by PerkinElmer,** the industry standard in chemical drawing software, allows students to create accurate chemical structures in their online homework assignments. As an instructor, you can edit existing questions and write entirely new problems. Track individual student performance—by question, assignment, or in relation to the class overall—with detailed grade reports. Integrate grade reports easily with Learning Management Systems (LMS), such as WebCT and Blackboard—and much more. Also available within Connect, our adaptive SmartBook has been supplemented with additional learning resources tied to each learning objective to provide point-in-time help to students who need it. To learn more, visit **www.mheducation.com.**

Instructors have access to the following instructor resources through Connect.

- **Art** Full-color digital files of all illustrations, photos, and tables in the book can be readily incorporated into lecture presentations, exams, or custom-made classroom materials. In addition, all files have been inserted into PowerPoint slides for ease of lecture preparation.
- **Animations** Numerous full-color animations illustrating important processes are also provided. Harness the visual impact of concepts in motion by importing these files into classroom presentations or online course materials.
- **PowerPoint Lecture Outlines** Ready-made presentations that combine art and lecture notes are provided for each chapter of the text.
- **Computerized Test Bank** Over 2,000 test questions that accompany *Introduction to Chemistry* are available utilizing the industry-leading test generation software TestGen. These same questions are also available and assignable through Connect for online tests.
- **Instructor's Solutions Manual** This supplement contains complete, worked-out solutions for *all* the end-of-chapter problems in the text.

Fueled by LearnSmart—the most widely used and intelligent adaptive learning resource—**LearnSmart Prep** is designed to get students ready for a forthcoming course by quickly and effectively addressing prerequisite knowledge gaps that may cause problems down the road. By distinguishing what students know from what they don't, and honing in on concepts they are most likely to forget, LearnSmart Prep maintains a continuously adapting learning path individualized for each student, and tailors content to focus on what the student needs to master in order to have a successful start in the new class.

Based on the same world-class, superbly adaptive technology as LearnSmart, **McGraw-Hill LearnSmart Labs** is a must-see, outcomes-based lab simulation. It assesses a student's knowledge and adaptively corrects deficiencies, allowing the student to learn faster and retain more knowledge with greater success. First, a student's knowledge is adaptively leveled on core learning outcomes: Questioning reveals knowledge deficiencies that are corrected by the delivery of content that is conditional on a student's response. Then, a simulated lab experience requires the student to think and act like a scientist: Recording, interpreting, and analyzing data using simulated equipment found in labs and clinics. The student is allowed to make mistakes—a powerful part of the learning experience! A virtual coach provides subtle hints when needed, asks questions about the student's choices, and allows the student to reflect on and correct those mistakes. Whether your need is to overcome the logistical challenges of a traditional lab, provide better lab prep, improve student performance, or make your online experience one that rivals the real world, LearnSmart Labs accomplishes it all.

## McGraw-Hill Create™

With **McGraw-Hill Create,** you can easily rearrange chapters, combine material from other content sources, and quickly upload content you have written, like your course syllabus or teaching notes. Find the content you need in Create by searching through thousands of leading McGraw-Hill textbooks. Arrange your book to fit your teaching style. Create even allows you to personalize your book's appearance by selecting the cover and adding your name, school, and course information. Order a Create book and you'll receive a complimentary print review copy in 3–5 business days or a complimentary electronic review copy (eComp) via email in minutes. Go to ww.mcgrawhillcreate.com today and register to experience how McGraw-Hill Create empowers you to teach *your* students *your* way. **www.mcgrawhillcreate.com**

## My Lectures—Tegrity®

**McGraw-Hill Tegrity** records and distributes your class lecture with just a click of a button. Students can view anytime/anywhere via computer or mobile device. It indexes as it records your PowerPoint® presentations and anything shown on your computer so students can use keywords to find exactly what they want to study. Tegrity is available as an integrated feature of McGraw-Hill Connect Chemistry and as a standalone.

## Student Solutions Manual

This separate printed manual contains detailed solutions and explanations for all odd-numbered problems and all concept review questions in the text.

# Acknowledgments

We want to thank all those who helped in this team effort. We extend a special thank you to all the copy editors, proofreaders, reviewers, and accuracy checkers. To the amazing staff at McGraw-Hill, we extend our deepest appreciation. Robin Reed, Lead Product Developer; Laura Bies, Senior Core Content Project Manager; Shirley Hino, Senior Director of Digital Content; Matthew Garcia, Senior Marketing Manager; and David Spurgeon, Director of Chemistry, got us started and kept us going on the fifth edition. Thanks for guidance on the new authoring platform used for this edition. We appreciate everyone's commitment to our vision and their help in getting this edition into production. We also wish to acknowledge our families. They assumed there would be a nice break between completion of the fourth edition and work on the fifth. Unfortunately, the break was shorter than we anticipated. We appreciate their guidance, support, and patience as we forged through the fifth edition.

Thanks also to David G. Jones, who helped write and review learning goal-oriented content for fifth edition SmartBook.

We also wish to thank the many reviewers over the course of four editions who have influenced this text with their thoughtful comments and suggestions. In particular, we want to thank those who reviewed in advance of the fifth edition:

Shadi Assaf, *Moraine Valley Community College*
Todd Clements, *Mt. San Antonio College*
Louis Fadel, *Ivy Tech Community College*
Kamatchi Lakshminarayanan, *California State University—Northridge*
Shawn Llopis, *Delgado Community College*

Kathryn Plath, *Colorado University*
Erin Rennells, *Hudson Valley Community College*
Brandon Tenn, *Merced College*
Susan Thomas, *University of Texas—San Antonio*
Raymund Torralba, *Harry S. Truman College*
Kamruz Zaman, *Community College of Baltimore County*

# Matter and Energy

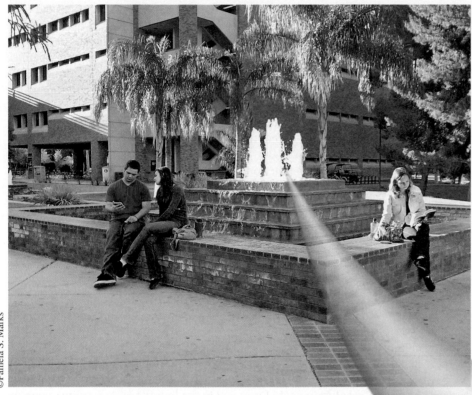

©Pamela S. Marks

1.1 Matter and Its
    Classification

1.2 Physical and Chemical
    Changes and Properties
    of Matter

1.3 Energy and Energy
    Changes

1.4 Scientific Inquiry

    Math Toolbox 1.1
    Scientific Notation

    Math Toolbox 1.2
    Significant Figures

    Math Toolbox 1.3
    Units and Conversions

    Chapter Review

    Questions and Problems

Anna and Bill are college students enrolled in an introductory chemistry course. For their first assignment, the professor has asked them to walk around campus, locate objects that have something to do with chemistry, and classify the things they find according to characteristics of structure and form.

Anna and Bill begin their trek at the bookstore. They spot a fountain, a large metallic sculpture, a building construction site, and festive balloons decorating the front of the store. They notice water splashing in the fountain and coins that have collected at the bottom. The metallic sculpture has a unique color and texture. At the building construction site they notice murals painted on the wooden safety barricade. Through a hole in the fence, they see a construction worker doing some welding.

Bill and Anna make a list of the things that attracted their attention and start trying to classify them. Inspecting the fountain, they notice that it appears to be composed of pebbles embedded in cement. As water circulates in the fountain, it travels in waves on the water's surface. The coins in the fountain, mostly pennies, vary in their shininess. Some look new, with their copper color gleaming in the bright sunshine. Others look dingy, brown, and old. The metal sculpture has a unique, modern design, but it's showing signs of age. A layer of rust covers its entire surface. Anna and Bill decide to classify the sculpture as a metal, like the coins in the fountain. They also conclude that the water, pebbles, and concrete in the fountain are not metals.

As they approach the construction site, Anna and Bill examine the painted mural on a safety fence that surrounds the site. Through the peephole in the mural, they see gravel, cinder blocks, metallic tubes for ductwork, steel beams, and copper pipe. They add more nonmetals and metals to their list. A welder is joining two pieces of metal. Sparks are flying everywhere. Anna and Bill wonder what is in the sparks. Since the sparks are so small and vanish so rapidly, they don't know how to classify them.

Electric carts are an ideal way to get across a large campus. They run using an electric battery instead of fuel.

©Steven M. Marks

As they continue their walk, they pass the intramural fields and the gym where they see students using tennis rackets, baseball bats, bicycles, and weight belts. They wonder how they will classify these items. For lunch, Bill and Anna buy pizza. They sip soft drinks from aluminum cans. They settle on a bench to enjoy their lunch in the sunshine and watch students playing volleyball in a sandpit. As they put on their sunscreen, they wonder how they might classify sunlight. After lunch, they hurry off to an afternoon class. On the way, they notice a variety of vehicles on campus. Some are gasoline-powered cars and buses, but others have signs on them saying they operate on alternative fuels. Trucks lumber by, exhaust fumes spewing from their tailpipes. Bill and Anna feel the hoods of parked cars. Some are still warm from their engine's heat.

How are Bill's and Anna's observations related to chemistry? What characteristics have they identified that they can use for classification purposes? They have started their classification with metals and nonmetals. What other categories should they devise?

Now it's your turn. Make a list of things relevant to chemistry in the location where you are reading this. How will you classify the things on your list? What characteristics will you use to organize the items into categories? Most important, why bother to classify things at all?

In this chapter we will explore some answers to these questions. As you learn what chemistry is, you'll begin to develop explanations for how substances look, change, and behave.

## Questions for Consideration

**1.1**  What characteristics distinguish different types of matter?

**1.2**  What are some properties of matter?

**1.3**  What is energy and how does it differ from matter?

**1.4**  What approaches do scientists use to answer these and other questions?

**Math Tools Used in This Chapter**

Scientific Notation (Math Toolbox 1.1)
Significant Figures (Math Toolbox 1.2)
Units and Conversions (Math Toolbox 1.3)

> This icon refers to a Math Toolbox that provides more detail and practice.

## 1.1 Matter and Its Classification

All the *things* that Anna and Bill observed on campus are examples of matter. The fountain, the metal sculpture, the construction site, the balloons outside the bookstore, the exhaust fumes from buses, the pizza they had for lunch, even Bill and Anna themselves—all are matter. **Matter** is anything that occupies space and has mass. **Mass** is a measure of the quantity of matter. The interaction of mass with gravity creates weight, which can be measured on a scale or balance.

Some of Bill's and Anna's observations, however, were not of matter. Sunlight, the light from welding, and the heat of automobile engines are not matter. They do not occupy space, and they have no mass. They are forms of energy. *Energy* is the capacity to move an object or to transfer heat. We'll discuss energy in Section 1.3, but for now, let's focus on matter.

All of Anna's and Bill's observations are relevant to chemistry, because chemistry is the study of matter and energy. Since the entire physical world is matter and energy, chemistry would be an overwhelming subject of study if we did not classify phenomena in manageable ways. Anna and Bill used characteristics like shininess and hardness when they decided some materials were metals and others were not. Let's explore some other characteristics that can be used to classify matter.

### Composition of Matter

One way to classify matter is by its chemical composition. Some types of matter always have the same chemical composition, no matter what their origin. Such matter is called a **pure substance** or more briefly, a *substance*. A pure substance has the same composition throughout and from sample to sample. It cannot be separated into components by physical means.

Some pure substances can be observed. For example, the aluminum in Anna's soda can is pure. It is not combined with any other substances, although it is coated with plastic and paint. Consider also the sandpit where Bill and Anna watched the volleyball game. The sand is not a pure substance, but if we removed all the dirt, minerals, and other contaminants, it would be the pure substance, silica, which is one kind of sand (Figure 1.1). Grains of silica differ in size, but they all have the same chemical composition, which can be determined in the laboratory.

In contrast to pure substances, other materials are mixtures. A **mixture** consists of two or more pure substances and may vary in composition. The fountain, for example, is made from a mixture of gravel, concrete, and pebbles. Even the water in the fountain is not a pure substance since small amounts of gases and minerals are dissolved in it. Like sand, however, it could be made pure if all the other substances were removed.

There are two types of pure substances: elements and compounds. We will discuss these first, and then we'll describe types of mixtures.

**Elements**   All matter consists of pure substances or mixtures of substances. Pure substances, in turn, are of two types: elements and compounds. An **element** is a substance that cannot be broken down into simpler substances *even by a chemical reaction*. For example, suppose we first purified the water in a fountain to remove contaminants. Then we used a chemical process called *electrolysis* to separate it into its component elements. Water can be broken down by chemical means into hydrogen and oxygen, as shown in Figure 1.2, so water is not an element. The hydrogen and oxygen, however, are elements. We cannot break them down into any simpler

> Are there any things where you are now that might be pure substances? Actually, pure substances are rare in our world. Most things are mixtures of some kind. Pure substances are found most often in laboratories where they are used to determine the properties and behavior of matter under controlled conditions.

**FIGURE 1.1**  Sand is composed of a mineral, silica. It contains the elements silicon and oxygen in specific proportions.
©Comstock/Getty Images

**FIGURE 1.2** When electric current is passed through water, the water decomposes into the elements hydrogen and oxygen. The hydrogen (left) and oxygen (right) can be seen bubbling to the tops of the tubes.
©McGraw-Hill Education/Stephen Frisch

substances using heat, light, electricity, or any chemical process. We can convert them into more complex substances, but not into simpler ones.

Elements are the building blocks of all matter. Of the 118 elements that have been given names, 83 can be found in natural substances and in sufficient quantity to isolate. The many examples of matter that we use, see, and read about are all built up of different elements in different combinations. The elements that are not isolated from natural sources on Earth have been synthesized by scientists. Some are so unstable that they have only a fleeting existence. To classify elements, chemists use a *periodic table,* like that shown in Figure 1.3. The elements in each column, called *groups* or *families* of elements in the periodic table, share similar characteristics, or *properties.*

Elements are generally classified into two main categories: metals and nonmetals. Generally, a **metal** can be distinguished from a **nonmetal** by its luster (shininess), ability to conduct heat (thermal conductivity), and ability to conduct electricity (electrical conductivity). Copper, aluminum, iron, and other metals are good conductors of electricity. Nonmetal elements, such as carbon (in the form of diamond), chlorine, and sulfur, normally are not. Note the difference in appearance of the metals and nonmetals shown in Figure 1.4. Not all elements fit neatly into such categories. In Chapter 2 we'll discuss elements that have properties somewhere between metals and nonmetals.

**FIGURE 1.3** The periodic table organizes the known elements according to their properties. The letters are symbols for the names of the elements.

Phosphorus    Copper    Bromine    Nickel    Lead

Gold    Carbon    Aluminum    Sulfur    Tin

**FIGURE 1.4** Some elements. Which of these are metals?

(Phosphorus, Bromine, Nickel, Lead, Aluminum, Sulfer & Tin): ©McGraw-Hill Education/Stephen Frisch; (Copper): ©Jim Birk; (Gold): ©Digital Vision/Getty Images; (Carbon): ©Photodisc/Getty Images.

## EXAMPLE 1.1    Metals and Nonmetals

Which of the elements pictured are metals? Why do you think so?

Iron

Carbon

Magnesium

Sulfur

Aluminum

(Iron): ©Sinclair Stammers/Science Source; (Carbon): ©Photodisc/Getty Images; (Magnesium, Sulfur & Aluminum): ©McGraw-Hill Education/Stephen Frisch.

**Solution:**

Notice that three of the elements—iron, aluminum, and magnesium—have a luster; that is, they shine. They are metals. If you could handle and test the substances, you could use other properties, such as electrical conductivity, to distinguish between metals and nonmetals.

**Consider This 1.1**

What if you were given element properties and had to determine the identity of the element from a list of possibilities? Suppose an element is rather dull in appearance, a poor conductor of electricity, and a gas at room temperature. Would this element be zinc, platinum, or chlorine?

**Practice Problem 1.1**

Identify the nonmetals in Figure 1.4. Explain the characteristics you considered in making your decision.

**Further Practice:** Questions 1.31 and 1.32 at the end of the chapter

**TABLE 1.1** Symbols of Selected Elements

| English Name | Original Name | Symbol | English Name | Original Name | Symbol |
|---|---|---|---|---|---|
| copper | cuprum | Cu | potassium | kalium | K |
| gold | aurum | Au | silver | argentum | Ag |
| iron | ferrum | Fe | sodium | natrium | Na |
| lead | plumbum | Pb | tin | stannum | Sn |
| mercury | hydrargyrum | Hg | tungsten | wolfram | W |

To become familiar with the periodic table, you should learn the names and symbols for the first 36 elements, as well as the symbols for silver, tin, gold, mercury, and lead. Your instructor may ask you to learn others.

To avoid having to write out the name of an element every time we refer to it, we use a system of symbols. An **element symbol** is a shorthand version of an element's longer name. Often, the symbol is one or two letters of the element's name (C for carbon, He for helium, Li for lithium). The first letter is uppercase, and the second letter, if present, is lowercase. When the names of two elements start with the same two first letters (magnesium and manganese, for example), the symbol uses the first letter and a later letter to distinguish them (Mg for magnesium, Mn for manganese).

For a few elements, the symbols are based on their Latin names or on names from other languages. These are listed in Table 1.1. Some recently synthesized elements have been named for famous scientists or locations. See the periodic table and list of elements on the first page of this book for a complete list of the modern names and symbols.

---

**EXAMPLE 1.2**    **Element Symbols**

Potassium is a soft, silver-colored metal that reacts vigorously with water. Write the symbol for the element potassium.

**Solution:**

The symbol for potassium is K. In the periodic table, potassium is element 19 in group (column) IA (1) of the periodic table.

**Consider This 1.2**

What if you instinctively identified the element symbol as P or Po? Why are these symbols incorrect for potassium?

**Practice Problem 1.2**

(a) Lead is a soft, dull, silver-colored metal. Write the symbol for the element lead.

(b) The symbol for a common element used to make jewelry is Ag. What is the name of this element?

**Further Practice:** Questions 1.39 and 1.40 at the end of the chapter

---

Iron pyrite

Sulfur            Iron

**FIGURE 1.5** Iron pyrite is composed of the elements iron and sulfur. Iron is magnetic and can be separated from sulfur when the two exist as elements mixed together. Iron pyrite, a compound of iron and sulfur, is not magnetic.

(left): ©McGraw-Hill Education/Doug Sherman; (right): ©McGraw-Hill Education/ Stephen Frisch

**Compounds**    A **compound,** sometimes called a *chemical compound,* is a pure substance composed of two or more elements combined in definite proportions. A compound has properties different from those of its component elements. For example, iron pyrite can be broken down into its component elements, iron and sulfur, but its characteristics are different from both (Figure 1.5). Anna and Bill saw many compounds during their trek around campus. These, and all compounds, can be chemically separated into their component elements. Sand is a compound of silicon and oxygen. Water, as discussed earlier, is composed of hydrogen and oxygen. The cheese on their pizza contains many complex compounds, but each of the compounds contains carbon, hydrogen, oxygen, nitrogen, and a few other elements.

Chemists represent compounds with formulas based on the symbols for the elements that are combined in the compound. (Chemical formulas are not the same as the mathematical formulas that may be familiar to you, such as $A = \pi r^2$ for the area of a circle.) A **chemical formula** describes the composition of a compound, using the symbols for the elements that make up the compound. Subscript numbers show the relative proportions of the elements in the compound. If no subscript number is given for an element in a formula, then you may assume that the element has a relative proportion of one. For example, water is known to consist of one unit of oxygen and two units of hydrogen. This compound is represented by the formula $H_2O$. Sodium chloride, the chemical compound commonly called table salt, contains equal portions of the elements sodium and chlorine. Its formula is therefore $NaCl$. We will discuss formulas in detail in Chapter 3.

**Mixtures**   Some forms of matter, such as pencil lead, do not have the same composition in every sample. (Pencil lead isn't the element lead. It is a *mixture* of graphite and clay.) A mixture consists of two or more elements or compounds. It is possible to separate mixtures into their component pure substances. The separation can be done physically, using procedures such as grinding, dissolving, or filtering. Chemical processes are not needed to separate mixtures.

We can illustrate the difference between pure substances and mixtures by looking at salt water. Water that has been purified is a pure substance that is composed of hydrogen and oxygen, always in the same proportions. Salt water, on the other hand, is water mixed with salt and many other substances in varying proportions. For example, the Great Salt Lake in Utah is approximately 10% salt, while the Dead Sea is about 30% salt. In either case, we can readily separate salt from water by evaporating the water (Figure 1.6).

Mixtures differ in uniformity of composition. A **homogeneous mixture** has a uniform composition throughout and is often called a **solution.** Most solutions that we commonly encounter are composed of compounds dissolved in water. They are often clear. For example, a well-mixed sample of salt water prepared in a kitchen is uniform in appearance. The salt dissolved in it is invisible. Furthermore, any microscopically small portion of the sample would have the same composition as any other. The particles in the mixture might not be arranged in exactly the same pattern, but each sample, regardless of size, would have the same components in the same proportions.

A mixture that is not uniform throughout—a mixture of salt and pepper, for instance—is a **heterogeneous mixture.** Different samples have their components present in different proportions. Which of the things that Bill and Anna had for lunch is a homogeneous mixture? Which is heterogeneous? How about your own lunch? How can you tell?

We have considered a number of classes and subclasses of matter: mixtures, homogeneous mixtures, heterogeneous mixtures, pure substances, compounds, elements, metals, and nonmetals. A method for classifying matter into these categories is outlined in Figure 1.7. Note in the figure that yes or no answers to several questions distinguish one type of matter from another. First, we ask if the material can be separated physically. If so, then it is a mixture. If not, it must be a pure substance. If this substance can be decomposed (broken down into simpler substances) by chemical reactions, it is a compound. If it cannot, it is an element.

Graphite leaves a mark similar to that made by dragging a rod of lead along a surface, so it was called lead. A hardness number indicates the relative amounts of graphite and clay in a pencil lead. A number 2 pencil is fairly soft, while a number 6 pencil is quite hard. Which has more graphite?

**FIGURE 1.6**   To collect salt, water is diverted into large ponds. The water evaporates, leaving solid salt behind.
©Science Photo Library/Alamy Stock Photo

Not all solutions are liquids. For example, consider air that has been filtered to remove suspended solid particles. Filtered air is a gaseous solution containing a mixture of primarily oxygen and nitrogen. Solid solutions also exist and are called alloys. For example, brass is a solution of zinc and copper.

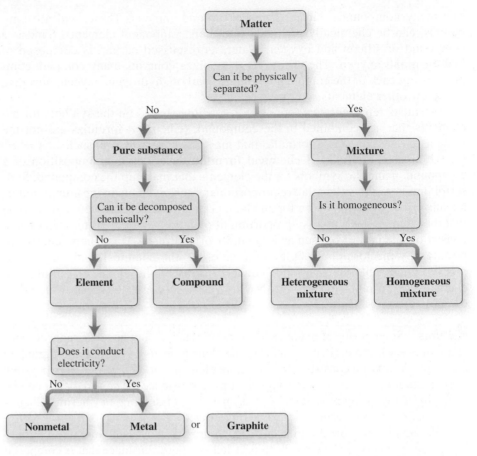

**FIGURE 1.7** We can classify matter by answering the short series of questions in this flowchart.

---

**EXAMPLE 1.3**     **Elements, Compounds, and Mixtures**

Which of the following pictures represent pure substances?

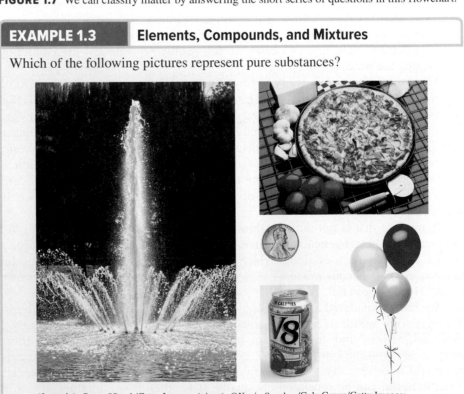

(fountain): ©vora/iStock/Getty Images; (pizza): ©Kevin Sanchez/Cole Group/Getty Images;
(coin): ©Randy Allbritton/Getty Images; (balloons): ©Jules Frazier/Getty Images
(drink): ©Brian Moeskau/Moeskau Photography

**Solution:**

The copper on the outside of the coin and the helium inside the balloons are pure substances. (However, the helium and balloons considered together provide an example of a mixture.)

**Consider This 1.3**

Why isn't the water in the fountain considered a pure substance?

**Practice Problem 1.3**

Which of the pictures represent mixtures? Which are heterogeneous? Which are homogeneous?

**Further Practice:** Questions 1.45 and 1.46 at the end of the chapter

## Representations of Matter

Chemists and other scientists view the world on several different levels. So far we have considered matter on a macroscopic scale. That is, we've discussed matter and phenomena we can see with our eyes. But simple observation is limited. Sometimes we cannot classify things merely by looking at them as Anna and Bill did. What do we do then? Chemists try to make sense of the structure of matter and its behavior on a scale that is much, much smaller than what we can see with our eyes.

Consider the copper pipe at the construction site, for example. If we could enlarge the tiniest unit that makes up the pipe, what would we see? Experimental evidence tells us copper is made up of discrete, spherical entities that all appear to be identical (Figure 1.8). Chemists identify these entities as atoms. An **atom** is the smallest unit of an element that has the chemical properties of that element. For example, we can imagine the helium inside a balloon as many, many atoms of helium, which we represent symbolically as He. In Figure 1.9, each sphere represents a single helium atom. Similarly, if we could magnify the structure of water, we would find two small hydrogen atoms bound separately to a single larger oxygen atom (Figure 1.10). Such a combination of elemental units is a **molecule.** Molecules are made up of two or more atoms bound together in a discrete arrangement. Several molecules of water, $H_2O$, are shown in Figure 1.10, where the central red sphere represents an oxygen atom and the two smaller, white spheres stand for hydrogen atoms. (Some compounds do not exist as molecules. We will discuss them in Chapter 3.)

Although chemists generally use color coding to distinguish between atoms of different elements in representations, the atoms themselves do not have colors. Macroscopic samples of matter may have color, but these colors do not usually match those used to represent atoms. In accurate representations, the sizes of the spheres change to reflect the relative differences in the sizes of atoms of different elements.

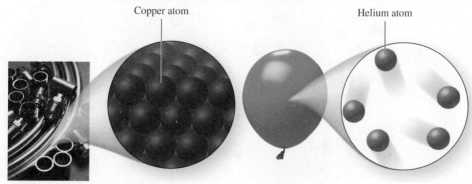

Copper atom

Helium atom

**FIGURE 1.8**  A copper pipe consists of a regular array of copper atoms.
©Thinkstock/Getty Images

**FIGURE 1.9**  Helium atoms are present inside the balloon.
©Jules Frazier/Getty Images

Oxygen atom

Hydrogen atom

**FIGURE 1.10**  Molecules containing hydrogen atoms and oxygen atoms make up the water in the fountain. Note: The molecular-level image does not include dissolved matter that is present in the fountain water.
©Glowimages/Getty Images

**FIGURE 1.11**  Oxygen molecules are made up of two interconnected oxygen atoms and are represented symbolically as $O_2$.

In addition to molecules of compounds, molecules can also be formed by the combination of atoms of only one element. For example, as shown in Figure 1.11, the oxygen we breathe consists of molecules of two oxygen atoms joined together. We represent oxygen molecules symbolically as $O_2$.

Chemists use many different ways to represent matter. Some for water are shown in Figure 1.12. Element symbols with subscripts represent a ratio of elements in a compound. One example is Figure 1.12B. To describe how the atoms are attached to one another, chemists often use lines and element symbols as shown in Figure 1.12C. In Figure 1.12D, spheres represent the atoms, and sticks show how they are connected. Figure 1.12E represents how the atoms fit together and their relative sizes. Macroscopic, molecular-level, and symbolic representations like these all have their advantages, and sometimes one is more convenient than another. You'll use them all as you progress through this course.

**FIGURE 1.12**  Different ways of representing water: (A) macroscopic, (B and C) symbolic, and (D and E) molecular.
©Royalty-Free/Corbis

| EXAMPLE 1.4 | **Representations of Matter** |

(a)  Which of these images best represents a mixture of elements?

(b)  If image A represents nitrogen, write its formula.

**Solution:**

(a) There are two mixtures represented in the images. Since the spheres (representing atoms) in image C have different colors and sizes, we can conclude that image C is a mixture of two elements. Image D is also a mixture, but it is a mixture of an element and a compound.

(b) The formula of the substance represented in image A is $N_2$. Note that two atoms are connected in the molecule.

**Consider This 1.4**

Which combination of entities shown in the images would represent a mixture of compounds?

**Practice Problem 1.4**

(a) Which of the images represents an element that exists as a molecule?

(b) If image E represents a compound of oxygen (red) and sulfur (yellow), what is its formula? (Write the symbol for sulfur first.)

**Further Practice:** Questions 1.51 and 1.52 at the end of the chapter

**FIGURE 1.13** Dry ice is the solid state of carbon dioxide. It converts from a gas to a solid at a very low temperature.
©Charles D. Winters/Science Source

## States of Matter

Earlier we considered the classification of matter based on composition. Let's look at a different way to classify matter: by its physical state. A **physical state** is a form that matter can take. The three most familiar to us are *solid, liquid,* and *gas.* Some substances, including some of those Anna and Bill observed, can be found in all three states under more or less ordinary conditions. Water, for example, can be a solid (ice), a liquid (flowing water), or a gas (water vapor) at environmental temperatures.

Other substances require extreme conditions to change from one state to another. For example, while carbon dioxide is a gas under normal conditions, it becomes a solid, called dry ice, at very low temperatures (Figure 1.13).

How do we know if a substance is in the solid, liquid, or gaseous state? Each state has characteristics that we can observe with our eyes and characteristics that are detectable or measurable at the molecular level. These characteristics are summarized in Table 1.2.

 **ANIMATION:** Three States of Matter

**TABLE 1.2** Characteristics of the Physical States of Matter

| Solid | Liquid | Gas |
|---|---|---|
| fixed shape | shape of container (may or may not fill it) | shape of container (fills it) |
| its own volume | its own volume | volume of container |
| no volume change under pressure | slight volume change under pressure | large volume change under pressure |
| particles are fixed in place and tend to be in a regular (crystalline) array | particles are randomly arranged and free to move about until they bump into one another | particles are widely separated and move independently of one another |

**FIGURE 1.14** The liquid and solid states of iron. Notice that the liquid atoms are randomly arranged and are free to move around each other. In the solid, the atoms are fixed in a regular array.

Liquid iron

Solid iron

A **solid** has a fixed shape that is not related to the shape of the container holding it. When you place an iron pipe in a box, the pipe does not change shape. Some solids can be made to change shape if enough force is applied. However, if you try to squeeze a solid to make it smaller, you'll fail. A solid cannot be compressed because its particles are arranged in a tightly packed, highly ordered structure that does not include much free space into which they might be squeezed. Note the closely packed particles in the solid state of iron shown in Figure 1.14.

A **liquid** is different from a solid in that it has no fixed shape. It takes the shape of the filled portion of its container, and it can be poured. Although they touch, the particles in a liquid are not arranged in ordered structures like those in a solid; they are free to move past one another. A liquid can be compressed slightly because its particles have a little free space between them. Note the differences between the liquid and solid states of iron shown in Figure 1.14.

A **gas** has no fixed shape; it adopts the shape of its container, expanding to fill the available space completely. A gas is easily compressed. When squeezed, gases can undergo large changes in volume. The particles of a gas are widely separated with much empty space between them. When a gas is compressed, the amount of space between the particles is reduced. This happens when pressure is applied, such as when a bicycle tire is filled with air, as shown in Figure 1.15. Another characteristic of gases is that they move through space quickly. When Bill and Anna smelled the pizza they had for lunch, they were detecting particles that migrated as gases from the source of the food to their noses. When gases cool sufficiently, they become liquids or even solids. This occurs, for example, when water vapor in the air liquefies on the surface of a cold glass. Note the differences between the liquid and gaseous states of water shown in Figure 1.16.

It is often convenient to show the physical state of a substance when representing it symbolically. For example, solid, liquid, and gaseous water can be represented as $H_2O(s)$, $H_2O(l)$, and $H_2O(g)$, respectively. The symbol $(aq)$ represents an **aqueous solution,** a solution in which a substance is dissolved in water. A salt and water solution, for instance, can be written as $NaCl(aq)$. These symbols for physical state are listed in Table 1.3.

Some solids, called amorphous solids, do not have the high order that most crystalline solids have.

Does $NaCl(aq)$ represent a pure substance or a mixture?

FIGURE 1.15   At the same temperature, a gas under high pressure has particles closer together than at low pressure. Notice that the composition (1 $O_2$:4 $N_2$) does not change with an increase in pressure.

FIGURE 1.16   Water condenses from a gas to a liquid on a cold surface. Air molecules (e.g., oxygen and nitrogen) are not shown.

©Brian Moeskau/Moeskau Photography

**TABLE 1.3** Symbols for Physical State

| Physical State | Symbol | Example (bromine) |
|---|---|---|
| solid | ($s$) | $Br_2(s)$ |
| liquid | ($l$) | $Br_2(l)$ |
| gas | ($g$) | $Br_2(g)$ |
| aqueous (dissolved in water) | ($aq$) | $Br_2(aq)$ |

# 1.2 Physical and Chemical Changes and Properties of Matter

Bill and Anna observed some of the *properties* of matter, including *changes* in matter. Observations can be either *qualitative,* based on some quality of the matter; or *quantitative,* based on a numerical value. When making qualitative observations, Bill and Anna described color, shape, texture, shininess, and physical state. Quantitative observations are different. They are numbers or measurements, and they must be carefully made and carefully reported.

Since quantitative data used to describe matter can involve both very large and very small numbers, it is often useful to express such numbers in *scientific* or *exponential notation.* Math Toolbox 1.1 (located at the end of this chapter) provides a review of this notation. In addition, it is necessary to express numbers in such a way as to indicate how accurately the value is known and how precisely it has been measured. The use of *significant figures* to properly express numerical values is presented in Math Toolbox 1.2.

MATH
TOOLBOX
**1.1**

MATH
TOOLBOX
**1.2**

## Physical Properties

When reporting qualitative data, we can classify properties as either physical or chemical. When Bill and Anna observed the color, shape, texture, shininess, and physical state of things around them, they were noting their physical properties.

A **physical property** is a characteristic that we can observe or measure without changing the composition of a substance. Other examples of physical properties are odor, taste, hardness, mass, volume, density, magnetism, conductivity, and the temperatures at which a substance changes from one physical state to another. (Later in this section we'll discuss *chemical properties,* which include reactivity and flammability.) Let's take a close look at mass, volume, density, and temperature. These four properties are *quantitative;* they involve numerical values.

**Mass**    Recall that mass is a measure of the quantity of matter. We usually measure the mass of an object by weighing it on a balance. In chemistry, masses are often reported in units of grams (g). Large masses, like people or elephants, may be reported in units of kilograms (kg); and small masses, such as salt crystals or impurities in water, may be reported in units of milligrams (mg) or micrograms (μg), as shown in Figure 1.17. (Math Toolbox 1.3 summarizes the relationships among units such as these.) Sometimes the mass of something is reported in grams, but we might want to know the mass in another mass unit such as milligrams or kilograms. We can easily convert a measurement from one unit to another if we know the relationship between the units. Tables 1.4 and 1.5 summarize common relationships between metric and English units. Example 1.5 shows how to convert between mass units. (See Math Toolbox 1.3 for more information on unit conversions.)

**MATH TOOLBOX 1.3**

**TABLE 1.4** Metric Conversions

| Prefix | Factor | Symbol |
|--------|--------|--------|
| giga | $10^9$ | G |
| mega | $10^6$ | M |
| kilo | $10^3$ | k |
| deci | $10^{-1}$ | d |
| centi | $10^{-2}$ | c |
| milli | $10^{-3}$ | m |
| micro | $10^{-6}$ | μ |
| nano | $10^{-9}$ | n |
| pico | $10^{-12}$ | p |

**TABLE 1.5** Some English-Metric Conversions

| English Unit | Metric Unit |
|--------------|-------------|
| 1 lb = 16 oz | 453.6 g |
| 1 in | 2.54 cm (exactly) |
| 1 yd | 0.9144 m |
| 1 mi | 1.609 km |
| 1 fluid oz | 29.57 mL |
| 1 qt | 0.9464 L |
| 1 gal | 3.785 L |
| 1 ft³ | 28.32 L |

Mass: 50 mg, 0.05 g, or $5 \times 10^{-5}$ kg

Mass: $7 \times 10^7$ mg, $7 \times 10^4$ g, or 70 kg

**FIGURE 1.17** A salt crystal has a mass of about 50 mg, while a person has a mass of about 70 kg.
(left): ©Jim Birk; (right): ©Doug Menuez/Getty Images

---

**EXAMPLE 1.5**    **Units of Mass**

Anna and Bill notice that there are 50.0 mg of sodium in the soda they bought to go with their lunch. How many grams of sodium are present in the can of soda? How many pounds?

**Solution:**

One way to solve this problem uses the dimensional-analysis approach. Consult Math Toolbox 1.3 for details. The general approach to solving the first part of the problem can be summarized by the following diagram:

Mass in milligrams  →  ?  →  Mass in grams

**MATH TOOLBOX 1.3**

The mass in milligrams has to be converted to the mass in grams. We need to find a relationship between these two quantities: $10^{-3}$ g = 1 mg (obtained from Table 1.4). Another way of expressing this relationship is 1 g = 1000 mg. Using this relationship, we get the following conversion:

We use the equivalence to set up possible conversion ratios:

$$\frac{1 \text{ g}}{1000 \text{ mg}} \quad \text{and} \quad \frac{1000 \text{ mg}}{1 \text{ g}}$$

To convert milligrams to grams, we can multiply 50.0 mg by the ratio (conversion factor) that will allow like units to cancel:

$$\text{Mass in g} = 50.0 \text{ mg} \times \frac{1 \text{ g}}{1000 \text{ mg}} = 0.0500 \text{ g}$$

Note that the milligram units cancel to leave the appropriate unit of grams. Also notice that the answer is reported to three significant figures because the measured quantity (50.0 mg) is reported to three significant figures and the other numbers in the calculation (1 g and 1000 mg) are exact quantities. (Consult Math Toolbox 1.2 for details about significant figures.) For convenience, we could report the answer in scientific notation: $5.00 \times 10^{-2}$. (See Math Toolbox 1.1 for a discussion of scientific notation.)

MATH
TOOLBOX
**1.2**

MATH
TOOLBOX
**1.1**

The second part of the question asks you to convert milligrams to pounds:

There isn't a direct relationship between milligrams and pounds listed in Tables 1.4 and 1.5. However, Table 1.5 lists a relationship between pounds and grams: 1 lb = 453.6 g. We can convert the grams we found in the first part of this example to pounds using the relationship summarized in the following diagram:

The ratios for converting between grams and pounds are

$$\frac{1 \text{ lb}}{453.6 \text{ g}} \quad \text{and} \quad \frac{453.6 \text{ g}}{1 \text{ lb}}$$

To convert grams to pounds, we multiply 0.0500 g by the ratio (conversion factor) that will allow like units to cancel:

$$\text{Weight in pounds} = 0.0500 \text{ g} \times \frac{1 \text{ lb}}{453.6 \text{ g}} = 1.10 \times 10^{-4} \text{ lb}$$

Does this answer make sense? Yes, it does. There are a lot of grams (453.6) in a pound, so we would expect the answer to be very small.

Without a single conversion from milligrams to pounds, the problem we just solved involves multiple steps:

The sequence of steps can be summarized as:

$$\text{Weight in pounds} = 50.0 \cancel{\text{mg}} \times \frac{1 \cancel{g}}{1000 \cancel{\text{mg}}} \times \frac{1 \text{ lb}}{453.6 \cancel{g}} = 1.10 \times 10^{-4} \text{ lb}$$

### Consider This 1.5

Why wouldn't an answer of 22.7 lb for the second part of the question make sense?

### Practice Problem 1.5

Anna and Bill see an aluminum recycling truck pass by on their way to class. If there are 765 lb of aluminum in the truck, how many grams are there? How many kilograms?

**Further Practice:** Questions 1.67 and 1.68 at the end of the chapter

If you need to determine the volume of a sphere, the relationship between volume and radius is $V = \frac{4}{3}\pi r^3$.

**Volume**  **Volume** is the amount of space a substance occupies. We can determine the volume of a rectangular solid such as a cube by measuring its length, width, and height and then multiplying them. For example, the volume of a cube that is 2.0 centimeters (cm) on each side is 8.0 cubic centimeters ($cm^3$):

$$\text{Volume of a cube} = \text{length} \times \text{width} \times \text{height}$$
$$\text{Volume} = 2.0 \text{ cm} \times 2.0 \text{ cm} \times 2.0 \text{ cm} = 8.0 \text{ cm}^3$$

Notice that the units are $cm^3$, consistent with a three-dimensional quantity. One cubic centimeter is equal to 1 mL (1 $cm^3$ = 1 mL), so the volume of 8.0 $cm^3$ could also be reported as 8.0 mL.

The volumes of liquids are usually measured in units of liters (L) or milliliters (mL), as shown in Figure 1.18. Larger volumes, such as big bottles of soda, are usually reported in liters. A 1-L bottle of soda contains 1000 mL. Example 1.6 shows how to convert between volume units.

**FIGURE 1.18** Some 500-mL, 1-L, and 250-mL containers.
©Brian Moeskau/Moeskau Photography

| EXAMPLE 1.6 | **Units of Volume** |
|---|---|

For lunch, Anna and Bill had 12-ounce (oz) cans of soda. What is the volume of a 12.0-oz can of soda in units of milliliters? What is its volume in units of liters?

**Solution:**

To solve this problem using the dimensional-analysis approach (see Math Toolbox 1.3), we determine if there is a relationship between fluid ounces and milliliters:

To convert fluid ounces to milliliters, we use the following relationship from Table 1.5: 1 oz = 29.57 mL.

We use the equivalence to set up possible conversion ratios:

$$\frac{29.57 \text{ mL}}{1 \text{ oz}} \quad \text{and} \quad \frac{1 \text{ oz}}{29.57 \text{ mL}}$$

To convert ounces to milliliters, we can multiply 12.0 oz by the ratio (conversion factor) that will allow like units to cancel:

$$\text{Volume in milliliters} = 12.0 \text{ \cancel{oz}} \times \frac{29.57 \text{ mL}}{1 \text{ \cancel{oz}}} = 355 \text{ mL}$$

The answer is reported to three significant figures, because the quantity we're given (12.0 oz) has three significant figures. Consult Math Toolbox 1.2 for details.

The second part of this problem asks you to convert milliliters to liters:

To convert volume in milliliters to volume in liters, we use the following relationship from Table 1.4: 1 mL = $10^{-3}$ L or 1000 mL = 1 L.

This is a unit conversion similar to the conversion we just did for mass. The ratios for converting between milliliters and liters are

$$\frac{1 \text{ L}}{1000 \text{ mL}} \quad \text{and} \quad \frac{1000 \text{ mL}}{1 \text{ L}}$$

To convert from milliliters to liters, we can multiply 355 mL by the conversion factor that allows like units to cancel:

$$\text{Volume in L} = 355 \text{ \cancel{mL}} \times \frac{1 \text{ L}}{1000 \text{ \cancel{mL}}} = 0.355 \text{ L}$$

Without a single conversion from ounces to liters, the problem we just solved involves multiple steps:

The sequence of steps can be summarized as:

$$\text{Volume in liters} = 12.0 \, \cancel{oz} \times \frac{29.57 \, \cancel{mL}}{1 \, \cancel{oz}} \times \frac{1 \, L}{1000 \, \cancel{mL}} = 0.355 \, L$$

### Consider This 1.6

What if Anna and Bill shared a 1-L bottle of water over lunch? How many ounces is this?

### Practice Problem 1.6

Anna and Bill saw some balloons outside the bookstore. The volume of gas inside one of the helium balloons was 4.60 L. What is the volume of gas in units of milliliters? In units of cubic centimeters? In units of gallons (4 qt = 1 gal)?

**Further Practice:**  Questions 1.71 and 1.72 at the end of the chapter

**FIGURE 1.19** The densities of antifreeze, corn oil, dish detergent, maple syrup, shampoo, and water in g/mL are 1.13, 0.93, 1.03, 1.32, 1.01, and 1.00, respectively. Which layer is which substance?

©Richard Megna/Fundamental Photographs

**FIGURE 1.20** Gold (Au) has a greater density than aluminum (Al) because gold has a greater mass per unit volume.

**Density**     The **density** of an object is the ratio of its mass to its volume. While mass and volume both depend on the size of the object or sample, density does not. Density is an unvarying property of a substance no matter how much of it is present, as long as temperature and pressure are constant. For example, the density of water at 4°C is 1.00 g/mL. It doesn't matter if we have 10 mL or 10 L; the ratio of mass to volume would be the same, 1.00 g/mL. However, if the temperature increases, the water would expand to a larger volume, while the mass stays the same. The density of liquid water would decrease as the temperature increases. The densities of water and a few other substances are listed in Table 1.6.

As Anna and Bill noted when they observed the fountain, a copper coin sinks in water. It sinks because copper (and the other metals in a penny) have a greater density than water. Conversely, air bubbles, just like other gases, rise to the top of water because gases are less dense than liquids. Oil floats on water for this same reason.

The density column in Figure 1.19 shows a variety of liquids with different densities. Which liquid has the greatest density? Which is the least dense?

If we compare *equal volumes* of two different substances, such as aluminum and gold, as shown in Figure 1.20, the substance with the greater mass has the greater

**TABLE 1.6**  Densities of Some Common Substances

| Substance | Physical State | Density (g/mL)* |
|---|---|---|
| helium | gas | 0.000178 |
| oxygen | gas | 0.00143 |
| cooking oil | liquid | 0.92 |
| water | liquid | 1.00 |
| mercury | liquid | 13.6 |
| gold | solid | 19.3 |
| copper | solid | 8.92 |
| zinc | solid | 7.14 |
| ice | solid | 0.92 |

*At room temperature and at normal atmospheric pressure, except gases at 0 degrees Celsius (°C) and water at 4°C.

density. How, though, can we compare densities if we do not have equal volumes? The mathematical relationship of mass, volume, and density reveals the answer:

$$\text{Density} = \frac{\text{mass}}{\text{volume}}$$

For example, a 1.0-cm$^3$ sample of copper has a mass of 8.9 g. An 8.0-cm$^3$ sample of copper has a mass of 71 g. A 27-cm$^3$ sample of copper has a mass of 240 g. In all these samples (Figure 1.21), the mass of copper divided by its volume is 8.9 g/cm$^3$. This is the density of copper.

If we know the mass and volume of an object, we can determine its density by substituting directly into the density equation. For example, suppose we have a cube of an unknown metal with a mass of 178 g and an edge length of 2.92 cm. The volume of the cube is 24.9 cm$^3$:

$$\text{Volume} = 2.92 \text{ cm} \times 2.92 \text{ cm} \times 2.92 \text{ cm} = 24.9 \text{ cm}^3$$

The density can then be calculated by taking the ratio of the mass to volume:

$$\text{Density} = \frac{\text{mass}}{\text{volume}}$$

$$\text{Density} = \frac{178 \text{ g}}{24.9 \text{ cm}^3} = 7.15 \text{ g/cm}^3$$

Consulting Table 1.6, we see that the unknown metal could be zinc.

Additionally, if we know the density of a substance and its mass in our sample, we can determine its volume. For example, suppose we want to know the volume occupied by 100 g of copper. Should the volume be greater than or less than 100 cm$^3$? There are many approaches to this problem. One way is to rearrange the density equation to solve for volume. Another way is to solve for the unknown volume in a set of equivalent ratios because density is a ratio of mass and volume that is constant for a given substance at a particular temperature. Both of these methods are shown in Example 1.7.

**8.9 g** **71 g** **240 g**
**1.0 cm$^3$** **8.0 cm$^3$** **27.0 cm$^3$**

**FIGURE 1.21** The density of copper is 8.9 g/cm$^3$. All three samples have the same ratio of mass to volume.

**ANIMATION:** Density of Liquids and Solids

These samples of metals have the same mass. Which has the greater density?

©Jim Birk

---

| **EXAMPLE 1.7** | **Density, Volume, and Mass** |
|---|---|

What is the volume of 100.0 g of copper? The density of copper is 8.9 g/cm$^3$.

**Solution:**

We need to carry out the following conversion:

The relationship between mass and volume is given by density:

First, we rearrange the density equation to get volume on one side by itself. This manipulation involves cross multiplication, which is described in Math Toolbox 1.3 (*Ratio Approach*). In the expression for density there is an implied 1:

$$\text{Density} = \frac{\text{mass}}{\text{volume}}$$

$$\frac{\text{Density}}{1} = \frac{\text{mass}}{\text{volume}}$$

**MATH TOOLBOX 1.3**

A can of diet cola floats in water, but a can of regular cola sinks. Suggest a reason why. How can you use this information to quickly select your preferred type of soft drink from a cooler filled with ice water at a party?

©Brian Moeskau/Moeskau Photography

Cross multiplying this density expression, we get:

$$\text{Density} \times \text{volume} = \text{mass} \times 1$$

Since we are trying to find the volume, we want to isolate it on one side of the equation. We can do this by dividing both sides by the density. (We'll also drop the "× 1" because any quantity times 1 is that quantity.)

$$\frac{\cancel{\text{Density}} \times \text{volume}}{\cancel{\text{Density}}} = \frac{\text{mass}}{\text{density}}$$

Now we have an expression that solves for the volume:

$$\text{Volume} = \frac{\text{mass}}{\text{density}}$$

Then, we substitute the known values of mass and density into the equation and solve for the value of volume:

$$\text{Volume} = \frac{100.0 \text{ g}}{8.9 \text{ g/cm}^3} = 11 \text{ cm}^3$$

In a second approach to this problem, consider that since the density of copper is always the same, the ratio of mass to volume is the same for both what we know and what we don't:

$$\frac{8.9 \text{ g}}{1 \text{ cm}^3} = \frac{100.0 \text{ g}}{x \text{ cm}^3}$$

Cross multiply to solve for *x:*

$$x \text{ cm}^3 = \frac{(1 \text{ cm}^3) \times (100.0 \cancel{\text{g}})}{8.9 \cancel{\text{g}}} = 11 \text{ cm}^3$$

In both approaches, the gram units cancel to give the expected volume unit of $\text{cm}^3$.

There is yet another approach to solving this problem that involves using density as a conversion factor:

$$\text{Volume} = 100.0 \cancel{\text{g}} \times \frac{1 \text{ cm}^3}{8.9 \cancel{\text{g}}} = 11 \text{ cm}^3$$

Does the answer make sense? Yes. The density tells us that 8.9 g of copper occupy a volume of 1 $\text{cm}^3$. The mass given, 100.0 g, is over 10 times greater than 8.9 so we would expect it to occupy a volume that is over 10 times greater than 1 $\text{cm}^3$.

**Consider This 1.7**

What if the 100.0-g piece of metal were gold? Would you expect its volume to be greater or less than 11 $\text{cm}^3$? How much greater or less?

**Practice Problem 1.7**

Solve the following problems.

(a)  The density of pure gold is 19.3 $\text{g/cm}^3$. What is the volume of 1.00 g of pure gold?
(b)  14-Carat gold is a homogeneous mixture of metals containing 58% gold by mass. The other 42% is a mixture of silver and copper. Silver and copper are both less dense than gold. Which of the following could be the mass of 1.00 $\text{cm}^3$ of 14-carat gold: 16.0 g, 19.3 g, or 23.0 g?

**Further Practice:** Questions 1.77 and 1.78 at the end of the chapter

> The density 8.9 $\text{g/cm}^3$ can be expressed in fractional form as:
> $$\frac{8.9 \text{ g}}{1 \text{ cm}^3}$$

> Water is unique among liquids because its solid form (ice) floats on its liquid form. This results from the relatively open structure adopted by water molecules in the solid state. What would happen to fish during the winter if water were like other substances that sink in their solid form?

Why do substances have different densities? Gases, in general, have very low densities because gas particles spread out and occupy large volumes. Metals tend to have high densities because their atoms pack together efficiently. Because ice floats on water, we can infer that water in its solid form must have a lesser density than water in its liquid form. Example 1.8 shows how to use molecular pictures to predict relative densities.

ANIMATION: Unique Properties of Water

## EXAMPLE 1.8   Explanations for Density

How do the molecular diagrams of ice and water help explain why ice is less dense than water?

Ice                Liquid water

**Solution:**

In ice, the $H_2O$ molecules have more space between them than in liquid water. The total volume occupied by a given number of molecules is greater in ice. Because density is a ratio of mass to volume, the larger volume accounts for the lower density.

**Consider This 1.8**

What if hexane, with a density of 0.659 g/cm$^3$, were carefully added to a glass of ice water? Where would the hexane be relative to the ice and liquid water?

**Practice Problem 1.8**

Helium balloons rise in air, which is a mixture of oxygen and nitrogen molecules, so we know helium is less dense than air. Look at the molecular-level diagrams of helium and carbon dioxide. Predict whether a helium balloon rises or falls in an atmosphere of carbon dioxide.

Helium              Carbon dioxide

**Further Practice:** Questions 1.81 and 1.82 at the end of the chapter

**Temperature**   Bill and Anna weren't happy with their lunches. The pizza was cold and their sodas were warm. When we make such comparisons, we are observing relative temperatures. **Temperature** is a measure of how hot or cold something is relative to some standard. We measure temperature with a thermometer.

In the United States, we often use the Fahrenheit scale to measure body temperature and air temperature. Fahrenheit is rarely used in science. Two other temperature scales are standard: the Celsius scale and the Kelvin scale. The relationships between the three temperature scales, Fahrenheit (°F), Celsius (°C), and Kelvin (K), are shown in Figure 1.22.

Another property of matter that is independent of sample size is the temperature at which the substance changes from one physical state to another. The *boiling point* is the temperature at which the liquid form of a substance changes to the gaseous form.

Temperatures are written differently for the different scales. While Celsius and Fahrenheit use the superscript ° to indicate degrees, the Kelvin scale does not. The unit is written as K (the capital letter), but temperatures are measured in kelvins (lowercase).

**FIGURE 1.22** The Fahrenheit, Celsius, and Kelvin temperature scales.

Lowest possible temperature:   −460°F   −273.15°C   0 K

At the *melting point,* the substance changes from a solid to a liquid. Between these two temperatures, the substance is normally in its liquid state. For example, on the Celsius scale, the boiling point of water is 100°C. Water melts (or freezes, depending on its original state) at 0°C. On the Kelvin scale, these values are 373.15 K and 273.15 K, respectively. On the Fahrenheit scale, they are 212°F and 32°F, respectively.

There are no negative values on the Kelvin scale. It is an *absolute temperature scale* because its zero point is the lowest possible temperature observable in the universe. This value is absolute zero, which is equivalent to −273.15°C. The temperature increments on the Kelvin scale are the same as those on the Celsius scale. The *difference* in temperature between the boiling point of water and the freezing point of water is 100 in both the Celsius (100°C − 0°C) and Kelvin (373.15 K − 273.15 K) scales, while the difference is 180 on the Fahrenheit scale (212°F − 32°F). Because the temperature in kelvins is always 273.15 greater than the temperature in degrees Celsius, we can easily convert between them:

$$T_K = T_{°C} + 273.15$$

When converting between the Fahrenheit and Celsius scales, the calculation is more complicated, because the degree increments are not equal:

$$T_{°F} = 1.8(T_{°C}) + 32$$

The equation can be rearranged, solving for degrees Celsius:

$$T_{°C} = \frac{T_{°F} - 32}{1.8}$$

| **EXAMPLE 1.9** | **Units of Temperature** |
| --- | --- |

The melting point of copper is 1083°C. Above what temperature, in kelvins and degrees Fahrenheit, is copper a liquid?

**Solution:**

Copper becomes a liquid above its melting point. In units of kelvin this temperature is

$$T_K = T_{°C} + 273.15$$

We substitute the value of the Celsius temperature into the expression and solve for the temperature in kelvins:

$$T_K = 1083 + 273.15 = 1356 \text{ K}$$

To convert degrees Celsius to degrees Fahrenheit, we use the equation:

$$T_{°F} = 1.8(T_{°C}) + 32$$

Substituting the temperature in degrees Celsius, we get:

$$T_{°F} = 1.8(1083) + 32 = 1981°F$$

**Consider This 1.9**

What is the physical state of copper at 1000°C?

**Practice Problem 1.9**

(a) The boiling point of acetylene is −28.1°C. Below what temperature, in kelvins and degrees Fahrenheit, is acetylene a liquid?

(b) The boiling point of helium is 4 K. Below what temperature, in degrees Celsius, is helium a liquid?

(c) Human body temperature is normally 98.6°F. What is this temperature on the Celsius and Kelvin scales?

**Further Practice:** Questions 1.85 and 1.86 at the end of the chapter

## Physical Changes

A process that changes the physical properties of a substance *without changing its chemical composition* is a **physical change.** For example, we can change liquid water to water vapor by heating it. This change from a liquid to a gas, called *boiling* or *vaporization,* is a physical change, since both forms involve the same chemical substance, water ($H_2O$).

To represent such changes, we can refine the symbolic representations we developed for elements and for compounds. We write the chemical formula for the initial condition and composition of the matter we are considering, then an arrow, and finally the chemical formula for the final condition and composition. The arrow is used to show that a change has occurred and in which direction. Using this symbolism, the change of water from a liquid to a gas would be represented as

$$H_2O(l) \longrightarrow H_2O(g)$$

The molecular and symbolic representations in Figure 1.23 show that the water molecules do not themselves change, but their physical state does. All the processes that change water from one physical state into another are summarized in Figure 1.24.

Another example of a physical change is the separation of different substances in a mixture. For example, a magnet divides magnetic materials from nonmagnetic materials without changing their identities, as shown earlier in Figure 1.5. A filter separates solid materials from liquid substances without changing either one chemically.

The most common type of heating we encounter involves things we do in the kitchen. When we heat something up, we're typically not changing the chemical composition. However, the process of cooking or baking often involves a combination of physical and chemical changes. Of course, if you were to overcook something and burn it, that would be a chemical change.

$$H_2O(l) \longrightarrow H_2O(g)$$

**FIGURE 1.23** Molecular-level and symbolic representations of the evaporation of water.

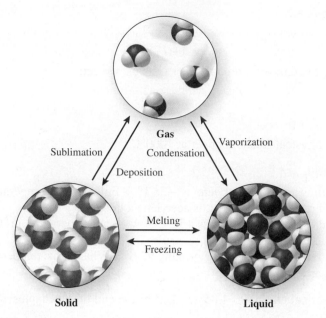

**FIGURE 1.24** The physical states of solid, liquid, and gas can all change into one another either directly or by going through two changes of state. The names of these processes are shown here next to arrows that designate the direction of the change.

## Chemical Changes

Remember the pennies in the fountain that Anna and Bill observed? Some were shiny and others looked dingy and brown. They might describe these less-shiny pennies as "tarnished." The pennies have undergone a **chemical change,** a process in which one or more substances are converted into one or more new substances. When pennies tarnish, some of the copper and zinc metal atoms in them combine with oxygen, forming compounds called metal oxides. The compounds are chemically different from either of the elements that formed them.

Suppose we clean a tarnished penny. Is the process a physical or a chemical change? It can be either. If you simply rub off the metal oxide coating with an eraser, the change is physical. Most penny collectors, however, prefer a chemical change that removes less metal. Rubbing ketchup on a penny is a great way to make it shiny. The vinegar in the ketchup reacts chemically with the metal oxides, freeing them from the surface of the penny. When the penny is rinsed, the result of the chemical change is easy to see.

Anna and Bill observed other examples of chemical change during their campus walk. When gasoline-powered cars burn fuel, a chemical change occurs. The gasoline reacts with oxygen to form carbon dioxide and water vapor. This chemical change releases the energy that runs the car. Chemical changes that involve burning are often accompanied by the release of energy. Anna and Bill also observed vehicles that run on alternative fuels. In hydrogen-powered vehicles, the hydrogen fuel combines with oxygen to form water vapor—and to release a lot of energy. The molecular-level and symbolic representations for this chemical change are shown in Figure 1.25. A chemical change is often called a **chemical reaction.** What are some examples of chemical reactions that you can observe around you?

## Chemical Properties

The copper and zinc in a penny, the gasoline in a car, and the hydrogen in an alternative-fuel vehicle all share a common chemical property: They react with oxygen. However, they differ in how they react and what products they form. Only the latter two release sufficient energy rapidly enough to make their use as fuels possible.

Many metals combine with oxygen to form a metal oxide compound at the surface of the metal. When this occurs with iron, we call it *rust*.

$$2H_2(g) + O_2(g) \longrightarrow 2H_2O(g)$$

**FIGURE 1.25** A chemical change occurs when the atoms in $H_2$ and $O_2$ rearrange to form $H_2O$.

A                                       B

**FIGURE 1.26** (A) The Hindenburg was a giant, rigid balloon filled with hydrogen gas. In 1937, it was destroyed when its hydrogen caught fire. (B) Today, blimps are filled with helium, an inert gas that will not explode.

(a): ©Bettmann/Getty Images; (b): ©David R. Frazier/Alamy Stock Photo

A **chemical property** of a substance is defined by what it is composed of and what chemical changes it can undergo. For example, let's compare hydrogen and helium. Although they have similar physical properties (colorless gases, similar densities), their chemical properties are very different. While hydrogen reacts with many other elements and compounds, helium is considered *inert* (Figure 1.26). It has not yet been shown to react with any other element or compound. Other terms used to describe chemical properties include reactivity and flammability.

---

**EXAMPLE 1.10**     **Physical and Chemical Changes**

Which of the following are physical changes and which are chemical changes?

(a) evaporation
(b) burning methane gas to form carbon dioxide and water
(c) using a magnet to separate metal and plastic paperclips
(d) rusting (the conversion of iron to iron oxide)

**Solution:**

(a) Evaporation is a physical change because it involves only a change of state.
(b) Burning methane gas is a chemical change because new substances form.
(c) Separating components of a mixture is a physical change.
(d) Rusting is a chemical change because a new substance forms.

**Consider This 1.10**

Is digesting the pizza Anna and Bill ate for a lunch a chemical or a physical change?

Sometimes simple observation cannot tell us whether a change is chemical or physical. For example, bubbles appear when baking soda and vinegar mix. Bubbles also appear when water boils, but the changes that produce the bubbles are different in these two cases. Baking soda and vinegar release bubbles because a chemical change takes place. They react to form carbon dioxide gas. However, when we warm water in a pan on the stove, small bubbles rise due to the release of dissolved air (mostly oxygen and nitrogen gas) from the water (before the water starts to boil). This process is only a physical change. If we could look at the nitrogen and oxygen molecules, as shown in Figure 1.27A, we would see that they are the same whether they are dissolved in water or not. When these molecules are dissolved in water, they form a homogeneous mixture with it. Heating the water merely separates the oxygen and nitrogen molecules from the water molecules. If we continue to heat the water to boiling, larger bubbles form and then rise from the bottom, as shown in Figure 1.27B. These bubbles are gaseous water, or water vapor. The result of the physical change can be represented symbolically as

$$H_2O(l) \longrightarrow H_2O(g)$$

Water molecules surround solute particles in aqueous solutions. For clarity in seeing the particles, this book shows water molecules faded in the background as in Figure 1.27A.

**FIGURE 1.27** Water contains small amounts of dissolved nitrogen and oxygen gases. (A) When heated, these molecules go to the gaseous state in bubbles that rise to the surface. (B) When the water begins to boil, it no longer contains dissolved gases, and the bubbles contain gaseous water.

©Brian Moeskau/Moeskau Photography

| **EXAMPLE 1.11** | **Physical and Chemical Changes** |
|---|---|

Do the following molecular-level images represent a chemical change or a physical change?

Before          After

**Solution:**

The substances after the change have a different composition than the substances before the change. Therefore, this is a chemical change.

**Consider This 1.11**

How do atom attachments differ for each of the elements in the change shown in the example?

**Practice Problem 1.11**

Do the following molecular-level images represent a chemical change or a physical change?

Before                    After

**Further Practice:** Questions 1.95 and 1.96 at the end of the chapter

## 1.3  Energy and Energy Changes

Physical and chemical changes involve energy. Energy is hard to define, but we see and feel evidence of it when something moves or changes temperature. At the construction site, Anna and Bill saw a worker pushing a wheelbarrow up a ramp. If released at the top of the ramp, the wheelbarrow would roll back down, converting energy from one form to another in the process. This release of energy is related to the spontaneous process of rolling down the ramp. (A spontaneous process is one that doesn't have to be forced to occur after it gets started.) But returning the wheelbarrow to the top of the ramp is not spontaneous. It requires a continuous energy input. Similarly, chemical and physical changes are usually accompanied by energy changes. Some chemical reactions are spontaneous. They happen on their own. Others need a continuous energy input from an external source. Consider the reaction (Figure 1.28) of hydrogen and oxygen gases to form water vapor, represented by this equation:

$$2H_2(g) + O_2(g) \longrightarrow 2H_2O(g)$$

**FIGURE 1.28**  In the reaction shown here, a balloon was filled with appropriate amounts of hydrogen and oxygen gas. When a lit candle touched the balloon, the hydrogen and oxygen reacted explosively to form water vapor.
©Charles D. Winters

This reaction is spontaneous and explosive. It releases a tremendous amount of energy. But the opposite reaction—the breakdown of water into hydrogen and oxygen gases—is not spontaneous. It occurs only if sufficient energy is continuously added, such as by passing electricity through liquid water. This process, called electrolysis, was shown in Figure 1.2, and can be described symbolically as follows:

$$2H_2O(l) \xrightarrow{\text{electrolysis}} 2H_2(g) + O_2(g)$$

But what is energy? **Energy** is the capacity to do work or to transfer heat. **Work,** usually taken to mean mechanical work, occurs when a force acts over a distance. For example, work is done when the construction worker pushes the wheelbarrow up a ramp. Work is done when compressed gases, resulting from the combustion of a fuel, push the piston in the cylinder of an automobile engine. Not all reactions can be made to do work directly, but heat energy can be harnessed to do work. For example, boiling water produces steam, which turns the turbines in power plants. The turbines spin copper coils inside a magnetic field in a generator to produce an electric current.

Energy takes many different forms, and it can be converted from one form to another. Scientists describe two types of energy: kinetic energy and potential energy. **Kinetic energy** is the energy of motion. The wheelbarrow rolling down a ramp possesses kinetic energy. **Potential energy** is energy possessed by an object because of its position. The wheelbarrow resting at the top of the ramp has potential energy. If it did not contain this stored energy, it could not release energy when it rolled down the ramp. Any object in a position to be rolled, dropped, or otherwise allowed to move spontaneously has potential energy that will be converted to kinetic energy once the motion starts.

To distinguish kinetic from potential energy, consider the volleyball game Anna and Bill watched while they ate lunch. After they ate, Anna and Bill joined the game. When Anna served the ball, kinetic energy was transferred from her hand to the volleyball (Figure 1.29). As it ascended, the ball transferred some of its kinetic

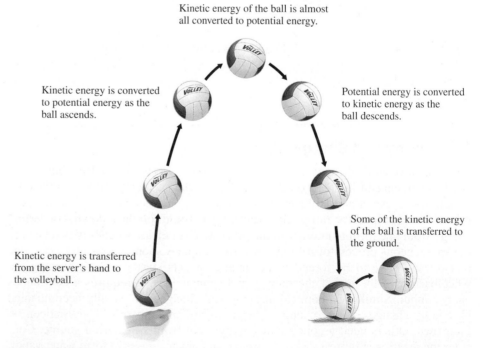

Kinetic energy of the ball is almost all converted to potential energy.

Kinetic energy is converted to potential energy as the ball ascends.

Potential energy is converted to kinetic energy as the ball descends.

Kinetic energy is transferred from the server's hand to the volleyball.

Some of the kinetic energy of the ball is transferred to the ground.

**FIGURE 1.29** Kinetic energy is transferred from the server to the volleyball when it is served. As the volleyball rises in the air, its kinetic energy is converted to potential energy. When it descends, potential energy is converted to kinetic energy.

energy to the surrounding air molecules, but most of its kinetic energy was converted to potential energy. Nearly all of the kinetic energy was converted to potential energy when the ball reached the top of its ascent. As the volleyball descended, its potential energy was converted to kinetic energy. Most of its kinetic energy was transferred to the ground when Bill missed the ball. The ball's remaining kinetic energy allowed it to bounce back up, where again its kinetic energy was converted to potential energy during its ascent.

Other forms of energy—chemical, mechanical, electric, and heat energy, for example—are really just forms of kinetic or potential energy. For example, chemical compounds can release chemical energy, the energy associated with a chemical reaction. Chemical energy is potential energy arising from the positions of the atoms and molecules in the compounds. A compound releases its potential energy when it undergoes a spontaneous chemical reaction that forms substances with less potential energy. For example, the explosive material TNT (trinitrotoluene) contains considerable potential energy that is released as kinetic energy when it reacts. Chemical compounds can also have kinetic energy. Molecules move faster as the temperature rises. The motion of molecules or atoms is associated with heat energy, or the kinetic energy that increases with increasing temperature. The fast-moving gases produced by the explosion of TNT have high kinetic energy.

---

### EXAMPLE 1.12    Molecular Motion and Kinetic Energy

Which of these two samples of argon gas has more kinetic energy?

A                    B

We will use the lengths of trails behind atoms or molecules to depict their relative speeds. Longer trails correspond to higher speeds.

**Solution:**

The atoms that are moving faster have the greater kinetic energy. Thus, the atoms in A have the greater kinetic energy.

**Consider This 1.12**

How would the trail lengths have to be changed if the kinetic energy of A decreased and B increased?

**Practice Problem 1.12**

Which of the two samples of argon gas is at a lower temperature?

**Further Practice:** Questions 1.103 and 1.104 at the end of the chapter

---

Electric energy is associated with the passage of electricity, generally through metals. The electric energy from a battery arises from a chemical reaction. The chemical energy stored in the compounds that make up the battery is converted to electric energy. Electric current passed through the filament of a lightbulb causes the metal to glow red and increases the motion of the atoms. This is a conversion of electric energy to kinetic energy. A lightbulb also gives off light energy. Nuclear energy involves both light and heat. Energy is released when one element is converted to another, as in a nuclear reactor or in the Sun, where hydrogen atoms fuse to form helium.

### EXAMPLE 1.13 | Forms of Energy

Identify examples of potential and kinetic energy in this picture.

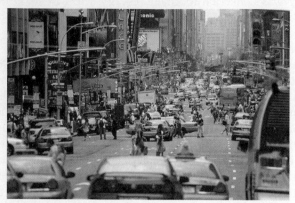

©Grant V. Faint/Getty Images

**Solution:**

Anything that might move in the picture has potential energy. Kinetic energy is evident in the moving people and vehicles.

**Consider This 1.13**

What are some examples of electric energy in the picture?

**Practice Problem 1.13**

Identify three additional forms of energy in the photograph.

**Further Practice:** Questions 1.107 and 1.108 at the end of the chapter

---

Common units of energy are calories, Calories (with a capital C), and joules. The energy input you need per day is about 2000 Calories, 2 million calories, or 8 million joules. We will study ways of measuring energy changes in Chapter 6.

All these forms of energy can be converted into one another. For example, the welder at the construction site starts a gasoline engine that runs a generator that makes the electricity the welder uses to join two pieces of metal together. The chemical energy in the gasoline is converted to mechanical energy that turns the generator. The mechanical energy is converted to electric energy by the generator. The electric energy is converted to heat in an arc that is formed between the welding rod and the metal to be welded. This heat melts the metal and creates the weld. Some of the electric energy is also converted to light. What energy conversions can you observe going on around you right now?

## 1.4 Scientific Inquiry

We began this chapter by describing some of the things that Anna and Bill saw around their campus. To classify the items, they observed similarities and differences in properties. They were making observations to help them understand nature.

Observation is one of the tools of scientific inquiry, but it is not the only one. The **scientific method** is an approach to asking questions and seeking answers that employs a variety of tools, techniques, and strategies. Although the scientific method is often explained as a series of steps and procedures, it is more accurately described as a way of looking at the world that differs from nonscience forms of inquiry. Scientists, like all humans, use intuition. They generalize about the world, sometimes with insufficient data. Chemists, especially, make inferences about atoms and molecules from data obtained from instruments that aren't quite capable of showing these tiny particles.

Scientists differ from professionals in nonscience disciplines in at least three important ways: (1) They test ideas by experimentation, (2) they organize their findings in particular (often mathematical) ways, (3) and they try to explain *why* things happen. Scientists use what is already known or believed about particular phenomena to gain insight into new observations from their experiments. Careful reasoning and insightful analogy are often employed, but sometimes intuition and luck play a part. Good scientists have an ability to couple objective scientific thinking with creative problem solving. In addition, a scientist must be curious enough to pursue the study of a seemingly trivial observation that can sometimes—albeit rarely—lead to a major advance in understanding.

Practicing scientists employ a variety of approaches to generate new knowledge or solve problems. Scientific inquiry generally includes observations, hypotheses, laws, and theories.

## Observations

Scientific inquiry begins with ideas, knowledge, and curiosity in the minds of scientists. To look for answers to their questions, scientists collect data. Data may derive from the observation of a naturally occurring event or from deliberate experimentation. When experimenting, scientists set conditions, allow events to occur, and observe the result. This procedure permits scientists to examine events under controlled conditions not found in nature. In addition, scientists repeat one another's experiments and compare observations, thus checking the accuracy of their findings. A common experimental design involves the isolation of one factor at a time to determine which of many variables influences the outcome. The results of experiments may be qualitative (descriptive) or quantitative (numeric).

Consider this example. Suppose Anna and Bill want to experiment to find a way to clean the dull, brown coating from the pennies they saw in the fountain. After reading the results of previous studies, they think an acid might be a good agent to use. They try several, always being sure to work under the same conditions of temperature, light, ventilation, degree of tarnish, and so on. The pennies look a little cleaner after being placed in acetic acid, but the dull coating remains unaffected. With nitric acid, a reaction occurs. A blue-green liquid and a red gas form. The tarnish disappears, but so do the pennies (Figure 1.30). Hydrochloric acid appears to work best because the surface tarnish vanishes quickly, leaving the penny intact. However, Anna and Bill notice that one of the pennies appears to react with the hydrochloric acid to form bubbles. After a time, this penny begins to float (Figure 1.31). Upon closer examination of this penny, they notice that bubbles are coming from scratches on its surface. Their descriptions of the behavior of pennies in acids are observations made from deliberate and controlled experimentation.

## Hypotheses

To organize and correlate multiple observations and sets of collected data, scientists propose hypotheses. A **hypothesis** is a tentative explanation for the properties or behavior of matter that accounts for a set of observations and can be tested. Because hypotheses are usually starting points in the explanation of natural phenomena, they normally lead to further experimentation. In practice, hypotheses are intuitive guesses that may be based on small amounts of data. Often a hypothesis is modified repeatedly in light of the results of additional experimentation. The building of scientific knowledge involves a cyclic interplay between observations and the making and testing of hypotheses (Figure 1.32).

Consider the experiments with pennies and acids. Earlier, Bill and Anna observed that one of the pennies reacted with hydrochloric acid to form a colorless gas. That gas formed along scratches on the penny's surface. Their observation might lead them to form a hypothesis: Pennies react with hydrochloric acid to form bubbles of a gas because they have scratches that make the reaction faster. How might they

**FIGURE 1.30** Copper reacts vigorously with nitric acid to form copper nitrate in solution and gaseous nitrogen dioxide.
©Charles D. Winters/Science Source

**FIGURE 1.31** When some pennies were cleaned with hydrochloric acid, one began to bubble and eventually floated.
©Richard Megna/Fundamental Photographs

| Obtain data from observations or experiments | → | Recognize patterns or trends | → | Propose and test hypotheses | → | Propose theory |

**FIGURE 1.32** Observations and hypotheses are linked in a cyclic fashion during the development of scientific knowledge.

**FIGURE 1.33** Some of the pennies react with hydrochloric acid at the location of the scratches, while others do not.
©Richard Megna/Fundamental Photographs

**FIGURE 1.34** Pennies are marked with the year and city where they were minted.
©Brian Moeskau/Moeskau Photography

test this hypothesis? They could scratch several pennies, dropping them into hydrochloric acid and checking for gas bubble formation. Suppose they performed this experiment with 10 pennies (Figure 1.33) and found that 7 of them formed bubbles at the scratches, but 3 did not. They would have to conclude that their hypothesis was only partially correct. Why? Because some of the pennies did not form bubbles as predicted by their hypothesis. Therefore, they might think further, asking themselves what other characteristics of pennies could influence the outcome.

While all pennies appear the same, close observation reveals detectable differences. For example, mints located in different cities produce pennies. The first letter of the city name appears just below the year in which the penny was minted (Figure 1.34). A *D* is stamped on pennies minted in Denver; *S* is used for San Francisco; and pennies minted in Philadelphia have no identifying mark. Given this new information, Anna and Bill might state a new hypothesis: Pennies react with hydrochloric acid to form bubbles if they are scratched *and* if they come from a specific mint—perhaps because that mint formulates its pennies differently. However, examination of the pennies from the previous experiment indicates that the mint location doesn't matter. Some scratched pennies from all the mints react and some don't. The amended hypothesis is incorrect.

As Bill and Anna set out to amend their hypothesis again, they realize they must isolate and assess one variable at a time. If they try to test too many ideas at once, they'd end up in a muddle of effects with several possible causes. So, while it's clear that scratches are necessary for bubbling to occur, some other variable or variables must be involved. As a next step, Anna and Bill reexamine the pennies from their previous experiment. They find that all the reactive pennies were minted in 1984 or more recently, while all the nonreactive pennies were minted in 1982 or before. This observation leads them to another amended hypothesis: Scratched pennies minted in 1982 or before do not react with hydrochloric acid to form bubbles, while scratched pennies minted since 1984 do—perhaps because the metal used to make the pennies was changed between 1982 and 1984.

How could they test their modified hypothesis? First, they must make sure that all old and new pennies respond in the same way as their original 10. They must then rule out any effects of chance by testing dozens, maybe even hundreds, of pennies. Assuming they get the results their hypothesis predicted, they may go on to cut some pennies open, perhaps verifying with direct observation that the insides of new and old pennies contain different materials.

Such findings would allow them to predict the behavior of pennies in hydrochloric acid, but they would not have a complete explanation for the chemical behavior. They would need to carry the process of scientific inquiry even further.

## Laws

When the behavior of matter is so consistent that it appears to have universal validity, we call this behavior a *law*. A scientific law describes the way nature operates under a specified set of conditions. For example, in the late eighteenth century, observations of the amounts of materials consumed and produced in chemical reactions led to the formulation of the *law of conservation of mass*. This law states that the mass of products obtained from a chemical reaction equals the mass of the substances that react. Every known chemical reaction that has been studied follows this law. For example, we could measure the mass of a penny and the mass of the nitric acid before allowing the reaction in Figure 1.30 to proceed. Upon completion of the reaction, we would find that the mass of the blue-green liquid plus the mass of the red gas is equal to the masses of our starting materials.

When nuclear reactions were discovered—such as the fusion reactions that power our Sun—scientists realized that processes that release very large amounts of energy do not conserve mass. To accommodate this new information, the *law of conservation of mass* had to be modified to become the *law of conservation of mass and energy.*

## Theories

Earlier we developed a hypothesis about the behavior of pennies in hydrochloric acid. In the Laws section we described an example of a scientific law. Hypotheses and laws only describe *how* nature works, not *why*. *Theories* explain why observations, hypotheses, or laws apply under many different circumstances. For example, the *atomic theory,* which we will discuss in Chapter 2, explains many aspects of the behavior of matter, including the law of conservation of mass. Theories often employ mathematical or physical models that, if correct, explain the behavior of matter. Like hypotheses, a theory fits known observations. If new facts become known, theories may have to be modified or amended.

We will discuss the reactions of metals with acid in Chapter 5.

Let's return to the behavior of pennies in hydrochloric acid. Anna and Bill need to propose a theory that explains the behavior of all the pennies. Their Internet and library research reveals that before 1982 all pennies were minted from a copper alloy containing 5% zinc. After 1984 all pennies were copper-coated disks of zinc. The newer pennies contain approximately 97.5% zinc. If Anna and Bill can find out how pure zinc and pure copper react with hydrochloric acid, they might develop an explanation for the chemical behavior of pennies in the presence of hydrochloric acid. A test with pieces of copper and zinc (Figure 1.35) indicates that zinc does indeed react with hydrochloric acid to release bubbles, while copper does not. They now have an explanation for the behavior of pennies in hydrochloric acid that is consistent with all the relevant observations. They can even explain why some of the pennies float. If the zinc is completely removed by reaction with hydrochloric acid, the copper shell that remains may fill with gas and rise to the surface of the liquid.

## Scientific Inquiry in Practice

There is a perception that scientists have wild hair, dress funny, wear pocket protectors, lack social graces, and work in isolation in a laboratory. This stereotype is often perpetuated in movies, but is it real? Do you know any scientists who fit this description? Perhaps you do, but the stereotype is best left to the movies. Scientists are people, and they vary as much in their appearance, personalities, and preferences as

**FIGURE 1.35** Zinc, but not copper, reacts with hydrochloric acid, producing hydrogen gas.
©Jim Birk

**FIGURE 1.36** Chemists are a diverse group of people who work in a variety of environments.
(left): ©PeopleImages/Getty Images; (right): ©YinYang/Getty Images

people in any other vocation (Figure 1.36). Although many scientists do indeed work independently in the lab, research groups who share common interests do most scientific work collaboratively. Sometimes they make discoveries that follow long periods of painstaking work. At other times, new insights arise rapidly through serendipity. These are accidental, fortunate breakthroughs, such as Alexander Fleming's discovery of penicillin or Henri Becquerel's discovery of radioactivity. Such serendipitous events happen only occasionally, and they happen to individuals who can recognize their importance. Both Fleming and Becquerel noticed anomalies that less skilled observers might have overlooked. As Louis Pasteur stated, "In the fields of observation, chance favors only the prepared mind."

While most of scientific inquiry proceeds through hypothesis testing, this is not the only form the scientific method takes. A new approach, especially useful in the search for new drugs, is *combinatorial chemistry*. A series of related chemical compounds is systematically prepared and tested for effectiveness in disease treatment. Many different combinations are tried, using techniques involving miniaturization, robotics, and computer control. The compounds are screened as possible candidates for drug action. For example, if a drug is needed that binds to a particular enzyme to produce a biological effect, then the various compounds would be added to that enzyme to see if they do bind. If so, additional testing would be carried out.

An alternate approach involves the mixing of many chemicals, producing many different products in one container. These are then tested, either in the mixture or after separation. With current techniques, it is possible for one laboratory to synthesize and test as many as 100,000 new compounds in a month.

To avoid generating a lot of chemical waste from the synthesis of compounds tested by combinatorial techniques or waste generated at the end of testing, scientists run these experiments on a microlevel scale. The machine like the one shown in Figure 1.37

**FIGURE 1.37** Many chemicals can be tested at once using equipment such as that shown here. Each tip dispenses a very small quantity of chemical into a corresponding well where the chemical's action as a potential pharmaceutical agent can be tested.
©Anthony Bradshaw/Getty Images

can dispense as little as 10 to 50 µL in each sample well. An important initiative within the chemistry community is to develop chemical processes that prevent pollution and reduce the amount of natural resources used to manufacture various consumer, research, and industrial products. *Green chemistry* and *sustainability* are terms often given to describe these efforts.

One of the primary principles of green chemistry is to prevent waste. Industrial and research facilities are working diligently to develop processes that do not generate waste that must be treated or disposed of. One way of eliminating waste is to design products that break down to harmless materials after they are used. Another principle of green chemistry is to design processes that do not deplete natural resources. Instead, green-chemistry principles recommend using raw materials that are renewable—that is, raw materials that come from agricultural sources or from the waste products of other processes. Fossil fuels are the primary source of depleting raw materials.

---

#  Math Toolbox 1.1   Scientific Notation

Many numbers used in chemistry are either very large, such as 602,200,000,000,000,000,000,000 atoms in 12 g of carbon, or very small, such as 0.0000000001 centimeter per picometer (cm/pm). It is very easy to make mistakes with such numbers, so we express them in a shorthand notation called exponential or scientific notation. The numbers used in the examples are written as $(6.022 \times 10^{23}$ atoms$)/(12$ g C$)$ and $1 \times 10^{-10}$ cm/pm in scientific notation. In this form, the numbers are easier to write, and it is easier to keep track of the position of the decimal point when carrying out calculations.

A number in scientific notation is expressed as $C \times 10^{n}$, where $C$ is the *coefficient* and $n$ is the *exponent*. The coefficient $C$ is a number equal to or greater than 1 and less than 10 that is obtained by moving the decimal point the appropriate number of places. The exponent $n$ is a positive or negative integer (whole number) equal to the number of places the decimal point must be moved to give $C$. For numbers greater than 1, the decimal point is moved to the left and the exponent is positive:

$$523 = 5.23 \times 10^{2}$$

For numbers smaller than 1, the exponent is negative, since the decimal point must be moved to the right:

$$0.000523 = 5.23 \times 10^{-4}$$

---

### EXAMPLE 1.14   Converting Between Decimal Form and Scientific Notation

(a) Convert 453,000 to scientific notation.
(b) Convert 0.00052 to scientific notation.
(c) Write the value $4.0 \times 10^{-3}$ in decimal form.
(d) Write the value $8.657 \times 10^{4}$ in decimal form.
(e) Place the following numbers in order from smallest to largest: $10^{3}, 1, 0, 10^{-3}, 10^{-9}$.

### Solution:

(a) The number 453,000 is greater than 1, so our exponent $n$ in $10^{n}$ will be a positive value. To convert to scientific notation, we move the decimal point between the 4 and 5. The decimal point

is moved five decimal places to the left, so the exponent is 5. The value 453,000 in scientific notation is therefore $4.53 \times 10^{5}$.
(b) The number 0.00052 is less than 1, so our exponent $n$ in $10^{n}$ will be a negative value. To convert to scientific notation, we move the decimal point between the 5 and 2. The decimal point is moved four decimal places to the right, so the exponent is $-4$. The value 0.00052 in scientific notation is therefore $5.2 \times 10^{-4}$.
(c) The number $4.0 \times 10^{-3}$ has a negative exponent, so it is less than 1. To convert to decimal form, we move the decimal point to the left three places: 0.0040.
(d) The number $8.657 \times 10^{4}$ has a positive exponent, so it is greater than 1. To convert to decimal form, we move the decimal point to the right four places: 86,570.
(e) Zero is the lowest value since none of the numbers are negative in value. The values with negative exponents are between 0 and 1. The value with a positive exponent is the only value greater than 1. The order is $0 < 10^{-9} < 10^{-3} < 1 < 10^{3}$.

### Practice Problem 1.14

(a) Convert 0.0123 to scientific notation.
(b) Convert 123 to scientific notation.
(c) Write the value $2.55 \times 10^{4}$ in decimal form.
(d) Write the value $1 \times 10^{-3}$ in decimal form.
(e) Place the following numbers in order from smallest to largest: $10^{-6}, 0, 1, 10^{4}, -2$.

**Further Practice:** Questions 1.3 and 1.5 at the end of the chapter

### Using Your Calculator

Such operations can be carried out more easily with a scientific calculator. Methods of use vary with different brands and models, so users should always consult their manual for specific instructions. However, the general approach is as follows. First enter the coefficient, including the decimal point. Then press the appropriate button for entering the exponent followed by the value of the exponent. The button is generally labeled EE, EXP, or 10ˣ. If the exponent has a negative value, press the change-sign button before entering the

**Math Toolbox 1.1 (continued)**

value of the exponent. Don't confuse this button with the subtraction button. The change-sign button is usually labeled $+/-$ or $(-)$.

Generally, it takes fewer key strokes to input a number in exponential form using the $\boxed{EE}$ or $\boxed{EXP}$ buttons. For a number in the format $C \times 10^n$, the sequence of calculator buttons is $C$ $\boxed{EE}$ $n$ or $C$ $\boxed{EXP}$ $n$. For example, the number $3.4 \times 10^{-5}$ is input as 3.4 $\boxed{EE}$ $\boxed{+/-}$ 5. To test your calculator, try the mathematical operation $(8 \times 10^6) \div 2$. We input this as 8 $\boxed{EE}$ 6 $\div$ 2. The answer is $4 \times 10^6$, so 4 $\boxed{EE}$ 6 should appear on your calculator.

## Rules of Exponents

When you carry out mathematical operations with numbers written in scientific notation, you need to know how to work on the exponential part of the number. You can usually do this with your calculator, but you should always check whether an answer from a calculator makes sense. To do this, verify the calculation by hand or make an estimate in your mind. The following instructions show how to work without a calculator. You should also perform these manipulations with your calculator to assure yourself that you can do them all.

In several of the examples, the coefficient is understood to be 1. That is, $1.0 \times 10^4$ is the same as $10^4$. Since 1 multiplied or divided by 1 still equals 1, the coefficient does not need to be written. However, if you want to enter $10^4$ into your calculator using the $\boxed{EE}$ button, you would enter it as 1 $\boxed{EE}$ 4.

To multiply exponential notation numbers, multiply the coefficients and add the exponents:

$$10^4 \times 10^7 = 10^{(4+7)} = 10^{11}$$
$$10^3 \times 10^{-5} = 10^{(3+-5)} = 10^{-2}$$
$$10^{-4} \times 10^4 = 10^{(-4+4)} = 10^0 = 1$$
$$(2 \times 10^{-4}) \times (4 \times 10^4) = (2 \times 4) \times 10^{(-4+4)} = 8 \times 10^0 = 8$$

To divide numbers written in exponential notation, divide the coefficients and subtract the exponents:

$$\frac{10^6}{10^4} = 10^{(6-4)} = 10^2$$

$$\frac{10^{-3}}{10^{-8}} = 10^{(-3--8)} = 10^5$$

$$\frac{8 \times 10^6}{2 \times 10^4} = \frac{8}{2} \times 10^{(6-4)} = 4 \times 10^2$$

To raise an exponential number to a power, raise both the coefficient and the exponent to the power by multiplying:

$$(10^4)^3 = 10^{(4 \times 3)} = 10^{12}$$
$$(10^{-2})^{-3} = 10^{(-2 \times -3)} = 10^6$$
$$(2 \times 10^4)^3 = (2)^3 \times 10^{(4 \times 3)} = 8 \times 10^{12}$$

To extract the root of an exponential number, take the root of both the coefficient and the exponent:

$$\sqrt{10^8} = (10^8)^{1/2} = (10)^{8/2} = 10^4$$
$$\sqrt[9]{10^{27}} = (10^{27})^{1/9} = (10)^{27/9} = 10^3$$
$$\sqrt{4 \times 10^8} = (4 \times 10^8)^{1/2} = (4)^{1/2} \times (10)^{8/2} = 2 \times 10^4$$

If the exponent is not evenly divisible by the root, we rewrite the number so the exponent is evenly divisible:

$$(10^7)^{1/2} = (10 \times 10^6)^{1/2} = 10^{1/2} \times (10^6)^{1/2} = 3.16 \times 10^3$$

To add or subtract numbers in exponential notation, first express both numbers with identical exponents. To do so, shift the decimal point in one of the numbers so its exponential part is the same as that of the other number. Then simply add or subtract the coefficients and carry the exponential part through unchanged. For example, suppose we want to add the following quantities:

$$5.2 \times 10^4 + 7.0 \times 10^3$$

We first change one number so they both have the same exponents:

$$52. \times 10^3 + 7.0 \times 10^3$$

Then we can add the two coefficients:

$$(52. + 7.0) \times 10^3$$

If necessary, we shift the decimal point to return the number to proper scientific notation:

$$59. \times 10^3 = 5.9 \times 10^4$$

## EXAMPLE 1.15   Mathematical Operations with Numbers Written in Scientific Notation

Perform the following operations without using your calculator and report your answers in scientific notation.

(a) $(2.0 \times 10^{-2}) \times (2.5 \times 10^4)$   (c) $(3.0 \times 10^4)^2$
(b) $(4.0 \times 10^{-2}) \div (2.0 \times 10^3)$   (d) $(4.3 \times 10^2) + (6.90 \times 10^3)$

**Solution:**

(a) To multiply numbers in scientific notation, multiply the coefficients and add the exponents:

$$(2.0 \times 2.5) \times (10^{-2+4}) = 5.0 \times 10^2$$

(b) To divide numbers in scientific notation, divide the coefficients and subtract the exponents:

$$(4.0 \div 2.0) \times (10^{-2-3}) = 2.0 \times 10^{-5}$$

(c) To raise a number in scientific notation to a power, raise both the coefficient and the exponent to the power:

$$(3.0 \times 10^4)^2 = (3.0)^2 \times (10^4)^2$$

When we raise an exponent to a power, we multiply the exponent by the power:

$$(3.0)^2 \times (10^4)^2 = 9.0 \times 10^8$$

(d) To add two numbers in scientific notation, we must first express both numbers with identical exponents:

$$(4.3 \times 10^2) + (6.90 \times 10^3) = (0.43 \times 10^3) + (6.90 \times 10^3)$$

Then we add the coefficients:

$$(0.43 + 6.90) \times 10^3 = 7.33 \times 10^3$$

Now, confirm your answers with a calculator. If you don't get the answers shown, you're not inputting numbers in scientific notation correctly.

**Practice Problem 1.15**

Perform the following operations without using your calculator and report your answers in scientific notation.

(a) $(4.0 \times 10^6) \times (1.5 \times 10^{-3})$
(b) $(3.0 \times 10^8) \div (1.0 \times 10^{-7})$
(c) $(2.0 \times 10^{-2})^3$
(d) $(4.10 \times 10^{10}) - (2.0 \times 10^9)$

**Further Practice:** Questions 1.7 and 1.8 at the end of the chapter

# Math Toolbox 1.2    Significant Figures

Numbers used in chemistry can be placed into two categories. Some numbers are *exact*. They are established by definition or by counting. Defined numbers are exact because they are assigned specific values: 12 in = 1 ft, 2.54 cm = 1 in, and 10 mm = 1 cm, for example. Numbers established by counting are known exactly because they can be counted with no errors. Since defined and counted numbers are known precisely, there is no uncertainty in their values.

Other numbers are not exact. They are numbers obtained by measurement or from observation. They may also include numbers resulting from a count if the number is very large. There is always some uncertainty in the value of such numbers because they depend on how closely the measuring instrument and the experimenter can measure the values.

## Precision and Accuracy

Uncertainty in numbers can be described in terms of either precision or accuracy. The *precision* of a measured number is the extent of agreement between repeated measurements of its value. *Accuracy* is the difference between the value of a measured number and its expected or correct value. If repetitive measurements give values close to one another, the number is precise, whether or not it is accurate. The number is accurate only if it is close to the true value.

We usually report the precision of a number by writing an appropriate number of significant figures. *Significant figures* in a number are all the digits of which we are absolutely certain, plus one additional digit, which is somewhat uncertain. For example, if we measure the height of a line on a graph calibrated with a line every 10 cm, we might obtain a value between 10 and 20 cm, which we estimate to be 18 cm. We are certain of the 1 in 18 cm, but not of the 8. Both of these digits are considered significant; that is, they have meaning. Unless we have other information, we assume that there is an uncertainty of at least one unit in the last digit.

With a graph calibrated more finely, in centimeters, we see that the line height is indeed somewhat more than 18 cm. The line appears to reach about three-tenths of the distance between the 18-cm and 19-cm marks, so we can estimate the height at 18.3 cm. The number of significant figures is three, two of which are certain (1 and 8) and one of which is somewhat uncertain (3).

## Determining the Number of Significant Figures

Every number represents a specific quantity with a particular degree of precision that depends on the manner in which it was determined. When we work with numbers, we must be able to

Good accuracy, good precision

Poor accuracy, good precision

Poor accuracy, poor precision

**Math Toolbox 1.2** *(continued)*

recognize how many significant figures they contain. We do this by first remembering that nonzero digits are always significant, no matter where they occur. The only problem in counting significant figures, then, is deciding whether a zero is significant. To do so, use the following rules:

- A zero alone in front of a decimal point is not significant; it is used simply to make sure we do not overlook the decimal point (**0**.2806, **0**.002806).
- A zero to the right of the decimal point but before the first nonzero digit is simply a place marker and is not significant (0.**00**2806).
- A zero between nonzero numbers is significant (28**0**6, 0.002806).
- A zero at the end of a number and to the right of the decimal point is significant (0.002806**0**, 2806.**0**).
- A zero at the end of a number and to the left of the decimal point (2806**0**) may or may not be significant. We cannot tell by looking at the number. It may be precisely known, and thus significant, or it may simply be a placeholder. If we encounter such a number, the best we can assume is that any trailing zeros are not significant.

The following table summarizes the significant figures in the numbers we just considered. The digits that are significant are highlighted.

| Number | Count of Significant Figures |
|---|---|
| 0.2806 | 4 |
| 0.002806 | 4 |
| 2806 | 4 |
| 0.0028060 | 5 |
| 2806.0 | 5 |
| 28060 | Assume 4 |

To avoid creating an ambiguous number with zeros at its end like 28060, we write the number in scientific notation (see Math Toolbox 1.1), so that the troublesome zero occurs to the right of the decimal point. In this case, it is simple to show whether the zero is significant ($2.8060 \times 10^4$) or not ($2.806 \times 10^4$). The power of ten, $10^4$, is not included in the count of significant figures, since it simply tells us the position of the decimal point.

### EXAMPLE 1.16  Number of Significant Figures

Determine the number of significant figures in the following:

(a) 0.060520        (b) 5020.01

**Solution:**

(a) Zeros before the first nonzero digit are not significant. Zeros at the end of a number and to the right of the decimal point are significant. Zeros between nonzero digits are significant. Thus, all but the first two zeros are significant, giving five significant figures.

(b) Zeros between nonzero digits are significant, so all digits in this number are significant. There are six significant figures in this number.

### Practice Problem 1.16

Determine the number of significant figures in the following:

(a) 14800.0        (b) 0.00076

**Further Practice:** Questions 1.9 and 1.10 at the end of the chapter

### Significant Figures in Calculations

We must also be concerned about the proper expression of numbers calculated from measured numbers. Calculators will not do this for us, so we have to modify their output. The rules for determining the proper number of significant figures in mathematical manipulations of measured numbers are simple. To determine the proper number of significant figures in the answer to a calculation, we consider only measured numbers. We need not consider numbers that are known exactly, such as those in conversions such as 1 foot (ft) = 12 inches (in). In the following examples, we will not use units, so that we can focus on the manipulation of the numbers. Remember, however, that a unit must accompany every measured number.

### Multiplication and Division

In a multiplication or division problem, the product or quotient must have the same number of significant figures as the least precise number in the problem. Consider the following example:

$$2.4 \times 1.12 = ?$$

The first number has two significant figures, and the second has three. Since the answer must match the number with the fewest significant figures, it should have two:

$$2.4 \times 1.12 = 2.7 \ (\text{not } 2.69)$$

### EXAMPLE 1.17  Significant Figures in Multiplication and Division

Express the answers to the following operations with the proper number of significant figures:

(a) $5.27 \times 3.20$
(b) $(1.5 \times 10^6) \times 317.832$
(c) $6.0/2.9783$
(d) $(2.01)^3$

**Solution:**

(a) Each number has three significant figures, so the answer should have three:

$$5.27 \times 3.20 = 16.9$$

(b) The first number has two significant figures, while the second has six. The answer should have two significant figures, the same as the number with fewer significant figures:

$$(1.5 \times 10^6) \times 317.832 = 4.8 \times 10^8$$

(c) The numerator has two significant figures; the denominator has five. The answer should have two significant figures:

$$\frac{6.0}{2.9783} = 2.0$$

**Math Toolbox 1.2 (*continued*)**

(d) A number raised to a power is the same as multiplying that number by itself as many times as indicated by the power, so the answer should have the same number of significant figures as the number:

$$(2.01)^3 = 2.01 \times 2.01 \times 2.01 = 8.12$$

**Practice Problem 1.17**

Express the answers to the following operations with the proper number of significant figures:

(a) $5.01 \times 1.5$
(b) $0.009/3.00$
(c) $3.60 \times 4.2 \times 10^{10}$
(d) $(1.85)^2$

**Further Practice:** Questions 1.11 and 1.12 at the end of the chapter

**Addition and Subtraction**

A sum or difference can only be as precise as the least precise number used in the calculation. Thus, we round off the sum or difference to the first uncertain digit. If we add 10.1 to 1.91314, for instance, we get 12.0, not 12.01314, because there is uncertainty in the tenths' position of 10.1, and this uncertainty carries over to the answer. In this case, it is not the number of significant figures that is important but rather the place of the last significant digit. Consider another example:

$$0.0005032 + 1.0102 = ?$$

The first number has four significant figures and the second number has five. However, this information does not determine the number of significant figures in the answer. If we line these numbers up, we can see which digits can be added to give an answer that has significance:

$$
\begin{array}{r}
0.0005032 \\
+\ 1.0102\phantom{000} \\
\hline
1.0107032
\end{array}
$$

The digits after the 5 in the first number have no corresponding digits in the second number, so it is not possible to add them and obtain digits that are significant in the sum. The answer is 1.0107, which has five significant figures, dictated by the position of the last significant digit in the number we added with the greatest uncertainty (1.0102).

**EXAMPLE 1.18  Significant Figures in Addition and Subtraction**

Express the answers to the following operations with the proper number of significant figures:

(a) $24 + 1.001$
(b) $24 + 1.001 + 0.0003$
(c) $428 - 0.01$
(d) $14.03 - 13.312$

**Solution:**

(a) The first number has no digits after the decimal point, while the second has three. The digits after the decimal point will have no significance:

$$24 + 1.001 = 25$$

(b) Only the numbers to the left of the decimal point will have significance in the answer:

$$24 + 1.001 + 0.0003 = 25$$

(c) The second number has no digits to the left of the decimal point, while the first number has all its significant digits to the left of the decimal point, so the subtracted number does not affect the answer:

$$428 - 0.01 = 428$$

(d) The first number has two digits to the right of the decimal point; the second has three. The answer should have only two digits past the decimal:

$$14.03 - 13.312 = 0.72$$

**Practice Problem 1.18**

Express the answers to the following operations with the proper number of significant figures:

(a) $1.003 + 1.02 + 1.0$
(b) $1.0003 - 0.10$
(c) $2.5020 - 2.50$
(d) $7.10 + 13.473$

**Further Practice:** Questions 1.13 and 1.14 at the end of the chapter

**Calculations Involving Multiple Steps**

When calculations involve multiple steps, the number of significant figures in subsequent steps requires us to know the number of significant figures in the answers from the previous steps. Therefore, we must keep track of the last significant figure in the answer to each step. Let's consider a two-step calculation involving both subtraction and division:

$$\frac{(12.26 - 10.1)}{2.45} = ?$$

We always begin with steps set off in parentheses. In this case the first step is the subtraction step:

$$
\begin{array}{r}
12.26 \\
-\ 10.1\phantom{0} \\
\hline
2.16
\end{array}
$$

The significant figure rules for subtraction tell us that this answer should be rounded to one decimal place (2.2 with two significant figures). However, to prevent rounding errors, we do not yet want to round. To keep note that the answer to our first step has only two significant figures, we will underline the significant digit farthest to the right:

$$12.26 - 10.1 = 2.1\underline{6}$$

**Math Toolbox 1.2 *(continued)***

Now we can use this value for the division step. With only two significant figures, the result from the previous step limits the number of significant figures in the second step to two significant figures:

$$\frac{2.1\underline{6}}{2.45} = 0.88$$

---

### EXAMPLE 1.19  Significant Figures in Calculations Involving Multiple Steps

Express the answers to the following calculations with the proper number of significant figures:

(a) $\dfrac{1202}{(14.7 + 76)}$

(b) $\dfrac{(14)^2}{201} - 0.888$

(c) $(8.9 \times 10.2) + 4.25$

(d) $\dfrac{(1.0 \times 10^{-2} \text{ g} - 1.1 \times 10^{-3} \text{ g})}{(1.496 \times 10^{-1} \text{ cm})^3}$

**Solution:**

(a) We begin with the step in parentheses, the one in the denominator. This gives a value to the ones place:

$$\frac{1202}{(14.7 + 76)} = \frac{1202}{(9\underline{0}.7)}$$

When we divide a number with four significant figures by a number with two significant figures, we get an answer with two significant figures:

$$\frac{1202}{(9\underline{0}.7)} = 13$$

(b) Squaring the value in parentheses gives a value with two significant figures with the last significant figure in the tens' place:

$$\frac{(14)^2}{201} - 0.888 = \frac{19\underline{6}}{201} - 0.888$$

The result of the division step gives a value with two significant figures with the last significant figure in the one-hundredths' place:

$$\frac{19\underline{6}}{201} - 0.888 = 0.975\underline{1} - 0.888$$

The sum of these values will have two decimal places:

$$0.97\underline{5}1 - 0.888 = 0.0\underline{8}712$$

$$= 0.09$$

(c) The result of multiplication gives a value with two significant figures:

$$(8.9 \times 10.2) + 4.25 = 9\underline{0}.78 + 4.24$$

The sum gives a value with the last significant digit in the ones' place:

$$90.78 + 4.24 = 9\underline{5}.03$$
$$= 95$$

(d) In this problem, we need to consider the number of decimal places in each value in the numerator. The numbers in this problem also have units we'll want to consider. Rewriting the numbers in the numerator in decimal form shows us that the answer must have the last significant figure in the 0.001 place. Also, the unit on the subtraction is grams.

$$\frac{(1.0 \times 10^{-2} \text{ g} - 1.1 \times 10^{-3} \text{ g})}{(1.496 \times 10^{-1} \text{ cm})^3} = \frac{(0.010 \text{ g} - 0.0011 \text{ g})}{(1.496 \times 10^{-1} \text{ cm})^3}$$

$$= \frac{0.008\underline{9} \text{ g}}{(1.496 \times 10^{-1} \text{ cm})^3}$$

For the denominator, we have to cube both the number and the cm units:

$$\frac{0.0089 \text{ g}}{(1.496 \times 10^{-1})^3 \times (\text{cm})^3} = \frac{0.0089 \text{ g}}{3.348 \times 10^{-3} \text{ cm}^3}$$

Dividing a value with one significant figure by a value with three significant figures gives a result with just one significant figure:

$$\frac{0.008\underline{9} \text{ g}}{3.348 \times 10^{-3} \text{ cm}^3} = 2.658 \text{ g/cm}^3$$

$$= 3 \text{ g/cm}^3$$

**Practice Problem 1.19**

Express the answers to the following calculations with the proper number of significant figures:

(a) $\dfrac{8.9 \times 10^4}{1.25 \times 10^5} + 0.135$

(b) $\dfrac{(52.50 - 2.0)}{3.00}$

(c) $\dfrac{9.25}{(1.9)^3} - \dfrac{19.64}{(2.50)^3}$

(d) $(6.2 \text{ g/mL} \times 1.95 \text{ mL}) + 0.6 \text{ g}$

**Further Practice:** Questions 1.15 and 1.16 at the end of the chapter

### Rounding Off Numbers

When a number contains more digits than are allowed by the rules of significant figures, we drop the digits after the last significant figure, using the following procedure:

- If the first digit being dropped is less than 5, leave the last significant figure unchanged. (For example, rounding to three significant figures, 6.073 becomes 6.07, and 6073 becomes $6.07 \times 10^3$.)

**Math Toolbox 9.2 (*continued*)**

- If the first digit being dropped is greater than 5 or is 5 followed by digits other than zero, increase the last significant figure by one unit (rounding to three significant figures, 6.077 becomes 6.08, and 60,751 becomes $6.08 \times 10^4$).
- If the first digit being dropped is 5 followed only by zeros or by no other digits, then increase the last significant figure by one unit if it is odd (6.075 becomes 6.08) but leave it unchanged if it is even (6.085 becomes 6.08).

**EXAMPLE 1.20   Rounding Numbers**

Round each of the following numbers to three significant figures:

(a) 3245
(b) 12.263
(c) 0.001035
(d) 312,486
(e) 312,586

**Solution:**

(a) With four significant figures, the last digit must be dropped to convert to three significant figures. Since the second to the last digit is even, it remains unchanged: 3240 or $3.24 \times 10^3$.

(b) The last two digits must be dropped. Since the first one dropped is greater than 5, the previous digit is increased by 1: 12.3.

(c) The last digit must be dropped. Since this digit is a 5 and the previous digit is odd, the previous digit is increased by 1: 0.00104.

(d) The last three digits must be dropped. Since the first one dropped is less than 5, the previous digit is unchanged: 312,000 or $3.12 \times 10^5$.

(e) The last three digits must be dropped. Since the first one dropped is 5 followed by nonzero digits, the previous digit must be increased by 1: 313,000 or $3.13 \times 10^5$.

**Practice Problem 1.20**

Round each of the following numbers to two significant figures:

(a) 1.502
(b) 13.50
(c) 0.001056
(d) 1200.0
(e) 0.2035

**Further Practice:** Questions 1.17 and 1.18 at the end of the chapter

 # Math Toolbox 1.3   Units and Conversions

All experimental sciences are based on observation. To be useful to the experimenter and to others, observations often must involve making measurements. Measurement is the determination of the size of a particular quantity—the number of nails, the mass of a brick, the length of a wall. Measurements are always defined by both a *quantity* and a *unit,* which tells what it is we are measuring.

But what system of units do we use? Most countries use the metric system, while the United States still uses primarily the English system. The English system is based on various units that are not related to one another by a consistent factor. Common units of length, for example, are inches, feet (12 in), yards (3 ft), and miles [1760 yards (yd)]. The metric system, on the other hand, uses units that are always related by a factor of 10, or by some power of 10. Some units of length are centimeters ($10^{-2}$ m), decimeters (10 cm or $10^{-1}$ m), meters, and kilometers (1000 m). Scientists in all countries use the metric system in their work.

The metric system is convenient to use because its units are related by powers of 10. Converting between related units is simply a matter of shifting a decimal point. The metric system adds a further convenience. It does not use arbitrary names like the English system, but rather, it defines *base units* of measure, and any multiple or fraction of these base units is defined by a special prefix. Some of the most common prefixes are listed here.

| Prefix | Factor | Symbol | Example |
|--------|--------|--------|---------|
| giga | $10^9$ | G | 1 Gg = 1,000,000,000 g |
| mega | $10^6$ | M | 1 Mg = 1,000,000 g |
| kilo | $10^3$ | k | 1 kg = 1000 g |
| deci | $10^{-1}$ | d | 1 dg = 0.1 g |
| centi | $10^{-2}$ | c | 1 cg = 0.01 g |
| milli | $10^{-3}$ | m | 1 mg = 0.001 g |
| micro* | $10^{-6}$ | μ | 1 μg = 0.000001 g |
| nano | $10^{-9}$ | n | 1 ng = 0.000000001 g |
| pico | $10^{-12}$ | p | 1 pg = 0.000000000001 g |

*The microgram (μg) is sometimes abbreviated mcg in the health professions.

For factors that have a negative exponent, alternative relationships can be derived. For example, 100 cg = 1 g, 1000 mg = 1 g, 1,000,000 μg = 1 g, and so forth.

Many science and health-related disciplines require conversions between different systems. Some of the relationships listed here may be useful in converting between English and metric units.

| English Unit | Metric Unit |
|--------------|-------------|
| 1 lb = 16 oz | 453.6 g |
| 1 in | 2.54 cm (exactly) |
| 1 yd | 0.9144 m |
| 1 mi | 1.609 km |
| 1 fluid oz | 29.57 mL |
| 1 qt | 0.9464 L |
| 1 gal | 3.785 L |
| 1 ft$^3$ | 28.32 L |

**Math Toolbox 1.3 *(continued)***

## Metric Base Units and Derived Units

In 1960 an organization of scientists met in France to determine standards for scientific measurements. This group established the SI units (from the French, *Système Internationale*) listed here.

| Unit | Symbol | Quantity |
|------|--------|----------|
| meter | m | length |
| kilogram | kg | mass |
| second | s | time |
| ampere | A | electric current |
| kelvin | K | temperature |
| mole | mol | amount of substance |
| candela | cd | luminous intensity |

All other units, called *derived units*, are based on these by multiplication or division and then applying the appropriate prefix or combination of units. For example, if the mass of an object is measured in grams, it could be converted to the SI unit of mass, the kilogram, by dividing the measured mass in grams by 1000 and adding the prefix *kilo* to the quantity. (We'll discuss conversions of units in the next section.) An example of a derived unit that involves a combination of units is volume. The SI-derived unit for volume is cubic meters ($m^3$), which for a box would mean measuring length, width, and height in meters, and then multiplying them together. Another derived unit, discussed in Section 1.2, is density. Density is a ratio of mass to volume for a given substance. What is the SI-derived unit for density?

## Conversion of Units

Measurements must be interpreted, often by mathematical manipulation of the data. This manipulation often involves converting one set of units into another using relationships between them. Two approaches to such conversions are *ratios* and *dimensional analysis*. In both cases, an analysis of the units provides clues to the correct solution of the problem.

Consider a problem that can be solved almost automatically: "Convert 30 min into hours." You probably gave the answer without thinking: 30 min is one-half hour (0.50 h). But how did you know this answer? Let's systematize the problem-solving process. First—even if we don't always consciously think about it—we must decide what the problem is asking for. This problem requires converting a number of minutes into the corresponding number of hours. Next, we must know the number of minutes in 1 h. From this information, write a mathematical expression that shows the equivalence of two quantities having different units:

$$1 \text{ h} = 60 \text{ min}$$

We can use this expression in two ways to solve this problem. First, we will describe the ratio approach.

## Ratio Approach

In this approach, we use a known relationship to compare with our unknown relationship. We know how many minutes make up 1 h, but we want to know how many hours there are in 30 min. From the known relationship, we develop a ratio of equivalent quantities that is equal to 1 and that has different units in the numerator and denominator. To develop a ratio to convert minutes into hours, we use the equivalent quantities and divide 1 h by 60 min.

$$\frac{1 \text{ h}}{60 \text{ min}}$$

The ratio for our unknown relationship should equal the known ratio as long as we use the same units, so we can set the unknown ratio equal to the known ratio:

$$\frac{x \text{ h}}{30 \text{ min}} = \frac{1 \text{ h}}{60 \text{ min}}$$

We can cross multiply to solve for $x$:

$$x \text{ h} \times 60 \text{ min} = 1 \text{ h} \times 30 \text{ min}$$

We can then solve for $x$ by dividing both sides by 60 min:

$$\frac{x \text{ h} \times 60 \text{ min}}{60 \text{ min}} = \frac{1 \text{ h} \times 30 \text{ min}}{60 \text{ min}}$$

We then carry out the mathematical operations:

$$x \text{ h} = \frac{1 \text{ h} \times 30 \text{ min}}{60 \text{ min}} = 0.5 \text{ h}$$

Notice that the units of minutes cancel out, leaving the units of hours, which is the unit called for in the problem.

## Dimensional Analysis

We also use the ratio of equivalent quantities in the dimensional-analysis approach. In this approach, however, we get to the form of the manipulated ratio more quickly. We multiply the known quantity by the ratio, so that units cancel and we get the unknown quantity with the desired units:

$$\text{Time in h} = 30 \text{ min} \times \frac{1 \text{ h}}{60 \text{ min}} = \frac{30 \times 1 \times \text{min} \times \text{h}}{60 \text{ min}} = 0.50 \text{ h}$$

Notice how the starting units (minutes) cancel out, leaving the desired units (hours) to accompany the calculated number. In the dimensional-analysis approach, units in the numerator and denominator of a fraction are treated exactly the same as numbers—canceled out, multiplied, divided, squared, or whatever the mathematical operations demand. We set up the conversions so that desired units are introduced and beginning units cancel out.

Now reexamine the conversion just developed. From the same equivalence expression, 1 h = 60 min, we can derive another ratio by dividing 60 min by 1 h:

$$\frac{60 \text{ min}}{1 \text{ h}}$$

What happens when the original quantity is multiplied by this conversion factor?

$$\text{Time in h} = 30 \text{ min} \times \frac{60 \text{ min}}{1 \text{ h}} = \frac{30 \times 60 \times \text{min} \times \text{min}}{1 \times \text{h}}$$
$$= 1800 \frac{\text{min}^2}{\text{h}}$$

This answer is *clearly wrong* because the units make no sense. There are always two possible ratios of the known equivalent quantities, but only one gives the proper units in the answer.

**Math Toolbox 9.2 (*continued*)**

Which ratio is appropriate to use depends on how the problem is stated. But in every case, using the proper ratio cancels out the old units and leaves intact the desired units. Using the improper ratio results in nonsensical units (like min²/h). This is a clear warning that the ratio was upside down or otherwise incorrect.

### EXAMPLE 1.21  Converting Units

Use the dimensional-analysis approach to convert 75.0 km to meters.

**Solution:**

We begin by determining if there is a relationship between kilometers and meters:

From the table of metric units we see that 1 km = 1000 m, so there is a relationship between these quantities:

From the equivalence expression, 1 km = 1000 m, we can derive two ratios (conversion factors), one of which we'll use in the calculation:

$$\frac{1\ km}{1000\ m} \quad or \quad \frac{1000\ m}{1\ km}$$

Only one of these expressions, when multiplied by the given quantity in kilometers, will correctly cancel the units. What are the units if we multiply 75.0 km by the first ratio?

$$75.0\ km \times \frac{1\ km}{1000\ m} = 0.0750\ \frac{km^2}{m}$$

The units tell us that our first attempt leads to an answer that is nonsense. Applying the other conversion factor, units cancel and lead to an answer that makes sense:

$$75.0\ \cancel{km} \times \frac{1000\ m}{1\ \cancel{km}} = 75,000\ m$$

Notice that our given quantity has three significant figures. To show that we know the answer in the conversion to three significant figures, we should express it in scientific notation: $7.50 \times 10^4$ m. (See Math Toolboxes 1.1 and 1.2 for details about scientific notation and significant figures.)

**Practice Problem 1.21**

Convert 0.0276 kg to grams.

**Further Practice:** Questions 1.19 and 1.20 at the end of the chapter

### An Approach to Problem Solving

The processes just discussed can be generalized by the following diagram:

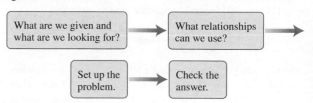

Consider the following problem: A liquid fertilizer tank has a volume of 6255 ft³. How many gallons of liquid fertilizer can fit into this tank?

1. Decide what the problem is asking for.

First, read the problem carefully. If you are not sure what a term means, look it up. If it's necessary to use an equation to solve the problem, be sure you understand the meaning of each symbol. Look for clues in the problem itself—words or phrases such as *determine, calculate, what mass, what volume,* or *how much.* After deciding what quantity the problem is asking for, write the units in which this quantity must be stated. In the fertilizer tank example, we want to find the volume of the tank in units of gallons from units of cubic feet.

2. Decide what relationships exist between the information given in the problem and the desired quantity.

If necessary, recall or look up equivalence relationships (such as 1 h = 60 min) inside the back cover. This information may not be given in the problem itself. In the fertilizer tank example, we need to convert volume in cubic feet to volume in gallons, so we need an equivalence relationship between those two quantities. The expression is 1 ft³ = 7.481 gal.

The additional information that describes the relationship between two quantities won't always be a direct equivalence. A series of equivalence expressions and their derived conversion factors may be needed. Sometimes a mathematical equation is required to express the equivalence between two quantities [e.g., $T_{°C} = (T_{°F} - 32)/1.8$]. At other times, some chemical principle may have to be applied.

A word of caution: Just as a problem may contain less information than you need to solve it, it may also contain more information than you need. Never assume that you must use all of the information. Examine all information critically and reject any that is not pertinent.

3. Set up the problem logically, using the relationships decided upon in step 2.

Starting with the relationships you obtained in step 2, develop the ratios needed to arrive at the final answer. Be sure to set up the ratios so that the old units cancel out and the desired units are introduced in the appropriate positions.

In the fertilizer tank example, the ratio is (7.481 gal)/(1 ft³). The conversion is as follows:

$$Volume\ in\ gal = 6255\ \cancel{ft^3} \times \frac{7.481\ gal}{1\ \cancel{ft^3}} = 46,790\ gal$$

**Math Toolbox 1.3 *(continued)***

If you need a series of ratios to find the final answer, it is often helpful to "map out" the route you will follow to get there. Suppose you need to know how many centimeters are in 1.00 mile (mi). This conversion can be summarized by a diagram:

$$\boxed{\text{mi}} \xrightarrow{\ ?\ } \boxed{\text{cm}}$$

Unfortunately you can only find a table that gives the number of feet in a mile, the number of inches in a foot, and the number of centimeters in an inch. The diagram to get from miles to centimeters might look like this:

$$\boxed{\text{mi}} \xrightarrow{\ 1\ \text{mi} = 5280\ \text{ft}\ } \boxed{\text{ft}} \xrightarrow{\ 1\ \text{ft} = 12\ \text{in}\ }$$

$$\boxed{\text{in}} \xrightarrow{\ 1\ \text{in} = 2.54\ \text{cm}\ } \boxed{\text{cm}}$$

The problem setup would follow this progression:

$$\text{Distance in cm} = 1.00\ \cancel{\text{mi}} \times \frac{5280\ \cancel{\text{ft}}}{1\ \cancel{\text{mi}}} \times \frac{12\ \cancel{\text{in}}}{1\ \cancel{\text{ft}}} \times \frac{2.54\ \text{cm}}{1\ \cancel{\text{in}}}$$

$$= 1.61 \times 10^5\ \text{cm}$$

4. Check the answer to make sure it makes sense, both in magnitude and in units.

This step is just as important as the others. You must develop an intuitive feeling for the correct magnitude of physical quantities. Suppose, for example, you were calculating the volume of liquid antacid an ulcer patient needed to neutralize excess stomach acid, and the result of your calculation was 32 L. Now, units of liters look reasonable, since the liter is a valid unit of volume; but it should be obvious that something is wrong with the numerical answer. Since a liter is about the same volume as a quart, this patient would need to drink 8 gal of antacid! The source of the problem is probably an arithmetic error, or possibly omission of the metric prefix *milli-* ($10^{-3}$) somewhere, since 32 mL would be an appropriate volume to swallow. In any event, a result such as this one should be recognized as unacceptable and the setup and calculations checked for errors.

In the fertilizer tank example, the units cancel properly, so the answer should be correct, if there are no arithmetic errors. Since there are about 7.5 gal in 1 ft³, the answer should be greater than the volume in cubic feet by something less than a factor of 10. The answer is indeed within this expected range.

### EXAMPLE 1.22  Problem Solving

If a laser beam fired from the moon takes 1.30 s to reach Earth, what is the distance in meters between the moon and Earth? Light travels in a vacuum at a speed of $3.00 \times 10^{10}$ cm in 1 s or $3.00 \times 10^{10}$ cm/s.

**Solution:**

We begin by determining if there is a relationship between seconds and meters:

There isn't, but the speed given, $3.00 \times 10^{10}$ cm/s, is a relationship between time and distance in centimeters. We could use this relationship to calculate the distance in centimeters and then convert to meters.

The ratios for converting seconds to centimeters can be expressed in two ways:

$$\frac{3.00 \times 10^{10}\ \text{cm}}{1\ \text{s}} \quad \text{or} \quad \frac{1\ \text{s}}{3.00 \times 10^{10}\ \text{cm}}$$

When multiplied by the time, 1.30 s, the first one cancels units of seconds and gives an answer in centimeters. To convert centimeters to meters, we use the equivalence expression

$$1\ \text{cm} = 10^{-2}\ \text{m}$$

Rather than working with negative exponents, it is sometimes useful to work in whole numbers. Dividing both sides of the expression by the quantity of the negative exponent provides a whole number ratio that may be more convenient:

$$\frac{1\ \text{cm}}{10^{-2}} = \frac{\cancel{10^{-2}}\ \text{m}}{\cancel{10^{-2}}}$$

$$100\ \text{cm} = 1\ \text{m}$$

Two ratios can be written to describe this relationship:

$$\frac{100\ \text{cm}}{1\ \text{m}} \quad \text{or} \quad \frac{1\ \text{m}}{100\ \text{cm}}$$

When multiplied by the distance in centimeters, the second expression will correctly cancel units and give an answer in meters.

Multiplying the given time by the two conversion factors provides a sequence of steps that allows us to cancel the time units (seconds) and distance in centimeters, giving the distance from the moon to Earth in meters:

$$\text{Distance in meters} = 1.30\ \cancel{\text{s}} \times \frac{3.00 \times 10^{10}\ \cancel{\text{cm}}}{1\ \cancel{\text{s}}} \times \frac{1\ \text{m}}{100\ \cancel{\text{cm}}}$$

$$= 3.90 \times 10^8\ \text{m}$$

Does this answer make sense? Yes, because we would expect there to be a great distance between the moon and Earth. If we had inversed the conversion factors, we would have gotten $4.33 \times 10^{-9}$ and units that do not make sense. The answer, $3.90 \times 10^8$ m, has three significant figures, which is the same as the number in the given quantity.

### Practice Problem 1.22

A tablet of a typical pain reliever contains 200 mg of active ingredient. How many pounds of active ingredient are in a tablet? (Assume there are three significant figures in the 200-mg quantity.)

**Further Practice:** Questions 1.21 and 1.22 at the end of the chapter

**Math Toolbox 1.3** *(continued)*

---

### EXAMPLE 1.23  Problem Solving with More Complex Units

Amoxicillin is an antibiotic commonly used to treat bacterial infections. When given to infants, the dosage must be carefully determined. The usual daily dosage for infants under 3 months old is 30.0 milligrams (mg) per kilogram (kg) of body weight. What mass of amoxicillin in milligrams should be given to an 8.00 lb infant in one day?

**Solution:**

We begin by determining a plan. If we knew the body weight of the infant in kg, then we could use the daily dosage per 1 kg body weight given to determine the mass of the daily dose:

However, we were given the body weight in pounds (lb), not kg, so we must first convert 8.00 lb to units of kg:

Now we can set up the two-step calculation, starting with the given body weight in units of lbs, and then multiplying by the conversion ratios written such that units cancel:

$$\text{Daily dosage in mg} = 8.00 \text{ lb body wt} \times \frac{1 \text{ kg body wt}}{2.205 \text{ lb body wt}}$$

$$\times \frac{30.0 \text{ mg amoxicillin}}{1 \text{ kg body wt}}$$

$$= 109 \text{ mg amoxicillin}$$

**Practice Problem 1.23**

The TGV POS high-speed train in France has a clocked speed of 574.8 km/h. Convert this speed to the following units:

(a) miles per hour (mi/h)
(b) meters per second (m/s)

**Further Practice:** Questions 1.23 and 1.24 at the end of the chapter

---

# CHAPTER REVIEW

## KEY CONCEPTS

- Matter can be classified in different ways based on its characteristics. Matter can be classified as an element, a compound, or a mixture of one or both.
  - An element is a pure substance that cannot be broken down into substances by a chemical reaction.
  - Elements generally fall into two main categories: metals and nonmetals.
  - A compound is a pure substance composed of two or more elements combined in definite proportions.
  - A mixture consists of two or more pure substances and may vary in composition. A homogeneous mixture (solution) has a uniform composition, while a heterogeneous mixture does not.
  - An atom is the smallest unit of an element that has the chemical properties of that element.
  - Molecules are made up of two or more atoms bound together in a discrete arrangement. Molecules can be either elements or compounds.
  - Elements and compounds are commonly represented using element symbols and chemical formulas.
  - Matter can also be classified by its physical state: solid, liquid, or gas.

- A change in a substance can be classified as either physical or chemical.
  - In a physical change, the identity of the substance(s) remains unchanged.
  - A physical property is a characteristic that we can observe or measure without changing the composition of a substance. Examples include color, odor, physical state, mass, volume, density, and temperature.
  - In a chemical change (chemical reaction), atoms rearrange to form new substances.
  - A chemical property is determined by what it is composed of and what chemical changes it can undergo.
- Physical and chemical changes involve energy, the capacity to do work or to transfer heat.
  - The energy of an object is a combination of its kinetic and potential energy.
  - Energy can change from one form to another, and it can be transferred as heat.
- The scientific method includes a variety of methods of inquiry employed by scientists.
  - Observation (or data collection) frequently is a necessary part of the scientific method, and frequently involves the design of carefully controlled experiments.
  - Hypotheses are tentative explanations for the results of experiments.
  - Laws result when observations appear to have universal validity, and theories are explanations for laws.

## KEY RELATIONSHIPS

| Relationship | Equation |
|---|---|
| The density of an object is the ratio of its mass and volume. | $Density = \dfrac{mass}{volume}$ |
| The absolute temperature in kelvins is offset from the Celsius temperature by 273.15. | $T_K = T_{°C} + 273.15$ |
| The Fahrenheit temperature has degrees that are 1.8 times as large as Celsius degrees. The freezing point of water is set at 32°F and 0°C. | $T_{°F} = 1.8(T_{°C}) + 32$ |

## KEY TERMS

aqueous solution (1.1)
atom (1.1)
chemical change (1.2)
chemical formula (1.1)
chemical property (1.2)
chemical reaction (1.2)
compound (1.1)
density (1.2)
element (1.1)

element symbol (1.1)
energy (1.3)
gas (1.1)
heterogeneous mixture (1.1)
homogeneous mixture (1.1)
hypothesis (1.4)
kinetic energy (1.3)
liquid (1.1)
mass (1.1)

matter (1.1)
metal (1.1)
mixture (1.1)
molecule (1.1)
nonmetal (1.1)
physical change (1.2)
physical property (1.2)
physical state (1.1)
potential energy (1.3)

pure substance (1.1)
scientific method (1.4)
solid (1.1)
solution (1.1)
temperature (1.2)
volume (1.2)
work (1.3)

## QUESTIONS AND PROBLEMS

The following questions and problems, except those in *Additional Questions* and *Concept Review Questions,* are paired. Questions in a pair focus on the same concept. Answers to selected questions and problems are in Appendix D.

### Matching Definitions with Key Terms

**1.1** Match the key terms with the following descriptions.
(a) a measure of the quantity of matter
(b) a characteristic of a substance involving the possible transformations that the substance can undergo to produce a new substance
(c) a combination of two or more substances that can be separated by physical means
(d) a pure substance that cannot be broken down into simpler stable substances in a chemical reaction
(e) the capacity to do work or to transfer heat
(f) a characteristic of a substance that can be observed without changing its composition
(g) the physical state in which matter has no characteristic shape but takes the shape of the filled portion of its container
(h) the ratio of the mass of a substance to its volume
(i) a mixture with uniform composition
(j) the physical state of matter characterized by a fixed shape and low compressibility

**1.2** Match the key terms with the following descriptions.
(a) the smallest particle of an element that retains the characteristic chemical properties of that element
(b) a change in which substances are converted into new substances that have compositions and properties different from those of the original substances
(c) anything that occupies space and is perceptible to the senses
(d) a substance composed of two or more elements combined in definite proportions
(e) a combination of atoms of one or more elements
(f) a process characterized by changes only in the physical properties of a substance, not in its composition
(g) the physical state in which matter has no fixed shape or volume but expands to fill its container completely
(h) the energy possessed by an object because of its position
(i) a tentative explanation for the properties or behavior of matter that accounts for a set of observations and can be tested
(j) the energy possessed by an object because of its motion

### Math Toolbox Questions

**1.3** Convert each of the following values to scientific notation.
(a) 29,500
(b) 0.000082
(c) 650,000,000
(d) 0.0100

**1.4** Convert each of the following values to scientific notation.
(a) 0.00010
(b) 4500
(c) 90,100,000
(d) 0.0000079

**1.5** Convert each of the following values from scientific notation to decimal form.
(a) $1.86 \times 10^{-5}$
(b) $1 \times 10^{7}$
(c) $4.53 \times 10^{5}$
(d) $6.1 \times 10^{-3}$

**1.6** Convert each of the following values from scientific notation to decimal form.
(a) $8.2 \times 10^{3}$
(b) $2.025 \times 10^{-6}$
(c) $7 \times 10^{-2}$
(d) $3.0 \times 10^{8}$

**1.7** For each of the following, carry out the mathematical operation and report answers in scientific notation.
(a) $(3.1 \times 10^{5}) \times (2.0 \times 10^{-2})$
(b) $(7.0 \times 10^{9}) \div (2.0 \times 10^{2})$
(c) $(2.8 \times 10^{-4}) \div (9.6 \times 10^{-2})$
(d) $(5.0 \times 10^{-4})^{2}$
(e) $(8.50 \times 10^{5}) - (3.0 \times 10^{4})$
(f) $(6.4 \times 10^{-3}) \div (4.0 \times 10^{3})$

**1.8** For each of the following, carry out the mathematical operation and report answers in scientific notation.
(a) $(8.2 \times 10^{6}) \div (4.1 \times 10^{-2})$
(b) $(3.5 \times 10^{3})^{4}$
(c) $(6.6 \times 10^{-5}) \div (2.2 \times 10^{5})$
(d) $(4.23 \times 10^{2}) \times (2.0 \times 10^{-8})$
(e) $(8.44 \times 10^{5}) + (1.2 \times 10^{4})$
(f) $(4.58 \times 10^{5}) \div (3.64 \times 10^{-3})$

**1.9** Determine the number of significant figures in the following.
(a) 0.0950
(b) 760
(c) 1.005
(d) 0.0052
(e) $3.46 \times 10^{15}$

**1.10** Determine the number of significant figures in the following.
(a) $1.48 \times 10^{-2}$
(b) 156.0
(c) 0.04350
(d) 70.25
(e) 140

**1.11** Express the answers to the following operations with the proper number of significant figures.
(a) $1.2 \times 1.216$
(b) $3.000/2.0$
(c) $(1.201 \times 10^{3}) \times (1.2 \times 10^{-2})$
(d) $(1.44)^{1/2}$

**1.12** Express the answers to the following operations with the proper number of significant figures.
(a) $(1.600 \times 10^{-7})(2.1 \times 10^{3})$
(b) $(1.33)^{3}$
(c) $1.93 \times 2.651$
(d) $4.4/2.200$

**1.13** Express the answers to the following operations with the proper number of significant figures.
(a) $1.6 + 1.15$
(b) $41.314 - 41.03$
(c) $1.8 + 1.022 + 0.001$
(d) $0.050 - 0.0012$

**1.14** Express the answers to the following operations with the proper number of significant figures.
(a) $87.5 + 1.3218$
(b) $422 - 410.2$
(c) $1.2316 - 1.01$
(d) $1.5678 + 0.23 + 0.0006$

1.15 Express the answers to the following calculations with the proper number of significant figures and with the correct units.

(a) $\dfrac{(20.90\ \text{kg} - 12.90\ \text{kg})}{10.00\ \text{L}}$

(b) $\left[\dfrac{45.82\ \text{g}}{(3.0\ \text{cm})^3} - \dfrac{0.64\ \text{g}}{(0.859\ \text{cm})^3}\right] \div 2$

(Assume that the 2 with no units is an exact number.)

(c) $(0.35\ \text{m} \times 0.55\ \text{m}) + 25.2\ \text{m}^2$

1.16 Express the answers to the following calculations with the proper number of significant figures and with the correct units.

(a) $(0.25\ \text{m/s}) \times (45.77\ \text{s}) + 5.0\ \text{m}$

(b) $\dfrac{2.523\ \text{lb}}{(62.9\ \text{gal} - 58.9\ \text{gal})}$

(c) $(9.0\ \text{cm} \times 15.1\ \text{cm} \times 10.5\ \text{cm}) + 75.7\ \text{cm}^3$

1.17 Round each of the following numbers to three significant figures.

(a) 1.2126  (d) 42.186
(b) 0.2045  (e) 0.007101
(c) 1.8351

1.18 Round each of the following numbers to three significant figures.

(a) 0.020450  (d) 16.225
(b) $1.3602 \times 10^4$  (e) 1.0001
(c) 13.475

1.19 Carry out the following conversions. Report your answers to the correct number of significant figures.

(a) 36 mm to m  (g) 168 lb to g
(b) 357 kg to g  (h) 914 qt to L
(c) 76.50 mL to L  (i) 44.5 cm to in
(d) 0.0084670 m to cm  (j) 236.504 g to lb
(e) 597 nm to m  (k) 2.0 L to qt
(f) 36.5 in to cm

1.20 Carry out the following conversions. Report your answers to the correct number of significant figures.

(a) 75.5 km to m  (g) 25.6 lb to g
(b) 25.7 g to mg  (h) 4.005 qt to L
(c) 0.516 L to dL  (i) 934 cm to in
(d) 5.2 cm to m  (j) 155 g to lb
(e) 0.000000450 m to nm  (k) 22.4 L to qt
(f) 12 in to cm

1.21 Carry out the following conversions. Report your answers to the correct number of significant figures.

(a) 947 m to mi  (g) 689 lb to kg
(b) 6.74 kg to lb  (h) 0.5 qt to mL
(c) 250.4 mL to gal  (i) 125 cm to ft
(d) 2.30 dL to mL  (j) 542 mg to lb
(e) 0.000450 cm to nm  (k) 25 nL to gal
(f) 37.5 in to m

1.22 Carry out the following conversions. Report your answers to the correct number of significant figures.

(a) 32 cm to ft  (g) 2.7 lb to kg
(b) 0.579 kg to lb  (h) 8.320 qt to mL
(c) 22.70 μL to qt  (i) 375 km to ft
(d) 9212 mm to km  (j) 62 g to oz
(e) 465 nm to mm  (k) 752 mL to gal
(f) 4 ft to cm

1.23 Carry out the following conversions. Report your answers to the correct number of significant figures.

(a) 375 m/s to ft/min
(b) 24.5 $\text{cm}^3$ to $\text{in}^3$
(c) 19.3 g/mL to $\text{lb/in}^3$

1.24 Carry out the following conversions. Report your answers to the correct number of significant figures.

(a) 27 ft/s to cm/min
(b) 2764 $\text{ft}^3$ to $\text{m}^3$
(c) 0.927 g/mL to lb/gal

## Matter and Its Classification

1.25 How would you classify the following items observed by Bill and Anna?

(a) water in a fountain containing dissolved dye
(b) a copper pipe
(c) the contents of a balloon after blowing it up by mouth
(d) a slice of pizza

1.26 How would you classify the following items observed by Bill and Anna?

(a) sand in a volleyball court
(b) a baseball bat made entirely of aluminum
(c) the contents of a balloon filled from a helium tank
(d) a glass filled with a soft drink

1.27 Which of the following are examples of matter?

(a) sunlight
(b) gasoline
(c) automobile exhaust
(d) oxygen gas
(e) iron pipe

1.28 Which of the following are not examples of matter?

(a) light from a fluorescent bulb
(b) sand
(c) wheelbarrow rolling down a ramp
(d) helium balloons
(e) heat from a welding torch

1.29 How are elements distinguished from compounds?

1.30 How are homogeneous mixtures distinguished from heterogeneous mixtures?

1.31 List characteristics of metals.

1.32 List characteristics of nonmetals.

1.33 Name the following elements.

(a) Ti  (b) Ta  (c) Th  (d) Tc  (e) Tl

1.34 Name the following elements.

(a) C  (e) Cu
(b) Ca  (f) Cl
(c) Cr  (g) Cs
(d) Co

1.35 Name the following elements.

(a) B  (b) Ba  (c) Be  (d) Br  (e) Bi

1.36 Name the following elements.

(a) S  (b) Si  (c) Se  (d) Sr  (e) Sn

1.37 Name the following elements.

(a) N  (c) Mn  (e) Al
(b) Fe  (d) Mg  (f) Cl

1.38 Name the following elements.

(a) Be  (c) Ni  (e) Ti
(b) Rb  (d) Sc  (f) Ne

1.39 What are the symbols for the following elements?
   (a) iron          (d) gold
   (b) lead          (e) antimony
   (c) silver

1.40 What are the symbols for the following elements?
   (a) copper        (d) sodium
   (b) mercury       (e) tungsten
   (c) tin

1.41 A chemical novice used the symbol Ir to represent iron. Is this an acceptable symbol for the element? If not, what is the correct symbol for iron?

1.42 A chemical novice used the symbol SI to represent silicon. Is this an acceptable symbol for the element? If not, what is the correct symbol?

1.43 The symbol NO was used by a student to represent nobelium, an unstable, synthetic element. Is this an acceptable symbol for the element? If not, what is the correct symbol?

1.44 A student used the symbol CO to represent cobalt, an element found in vitamin $B_{12}$. Is this an acceptable symbol for the element? If not, what is the correct symbol?

1.45 Classify each of the following as a pure substance, a homogeneous mixture (solution), or a heterogeneous mixture: hamburger, salt, soft drink, and ketchup.

©Patricia Brabant/Cole Group/Getty Images

1.46 Classify each of the following as a pure substance, a homogeneous mixture (solution), or a heterogeneous mixture: sand, boardwalk, ocean, and roller coaster.

©trekandshoot/iStock/Getty Images

1.47 Elemental hydrogen normally exists as two hydrogen atoms bound together. Write the formula for this molecule, and draw a picture to represent how it might look on a molecular level.

1.48 Elemental chlorine normally exists as two chlorine atoms bound together. Write the formula for this molecule, and draw a picture to represent how it might look on a molecular level.

1.49 This image is a representation for a compound containing nitrogen and oxygen. Write the formula for this compound.

1.50 This image represents a compound containing phosphorus and chlorine. Write the formula for this compound.

1.51 Which of the images represents a mixture of an element and a compound?

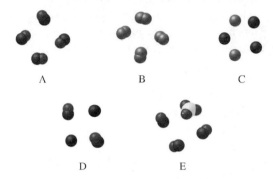

1.52 Which of the images in Question 1.51 represents a pure substance that is a compound?

1.53 Classify each of the following as an element or a compound.
   (a) $O_2$        (c) $P_4$       (e) NaCl
   (b) $Fe_2O_3$    (d) He          (f) $H_2O$

1.54 Classify each of the following as an element or a compound.
   (a) hydrogen gas      (d) nitrogen dioxide
   (b) water             (e) aluminum chloride
   (c) salt              (f) neon

1.55 Under normal conditions, mercury is a liquid. Draw molecular-level pictures of what mercury atoms might look like in the liquid and solid states.

1.56 Under normal conditions, bromine is a liquid. Draw molecular-level pictures of what bromine, $Br_2$, might look like in the liquid and gaseous states.

1.57 What type of matter expands to fill its container and can be compressed to a smaller volume?

1.58 What type of matter is composed of particles that do not move past one another?

1.59 Identify the physical state of each of the following elements from their symbols.
   (a) $Cl_2(g)$
   (b) $Hg(l)$
   (c) $C(s)$

1.60    Identify the physical state of each of the following compounds from their symbols.
   (a) NaCl($s$)
   (b) CH$_3$OH($l$)
   (c) CO$_2$($g$)

1.61    What physical state is represented in this diagram?

1.62    Draw a picture of the gaseous state of the substance shown in Question 1.61.

1.63    How might you symbolically represent a homogeneous mixture of oxygen, O$_2$, and water?

1.64    Why does the symbol H$_2$O($aq$) make no sense?

## Physical and Chemical Changes and Properties of Matter

1.65    At the beginning of the chapter, Anna and Bill made many observations about what they saw. Some of their observations included color, texture, and shininess. Are these physical or chemical properties?

1.66    At the beginning of the chapter, you were asked to classify processes like a truck running on gas or welding pieces of metal. After reading the chapter, would you classify these processes as physical or chemical changes?

1.67    A slice of Swiss cheese contains 45 mg of sodium.
   (a) What is this mass in units of grams?
   (b) What is this mass in units of ounces (oz)?
        (16 oz = 453.6 g)
   (c) What is this mass in pounds (lb)?
        (1 lb = 453.6 g)

1.68    A package of Swiss cheese has a mass of 0.340 kg.
   (a) What is this mass in grams?
   (b) What is this mass in ounces (oz)?
        (16 oz = 453.6 g)
   (c) What is this mass in pounds (lb)?
        (1 lb = 453.6 g)

1.69    A grain of salt has a mass of about $1.0 \times 10^{-4}$ g. What is its mass in the following units?
   (a) milligrams
   (b) micrograms
   (c) kilograms

1.70    If a dog has a mass of 15.2 kg, what is its mass in the following units?
   (a) grams
   (b) milligrams
   (c) micrograms

1.71    If you drank 1.2 L of a sports drink, what volume did you consume in the following units?
   (a) milliliters
   (b) cubic centimeters
   (c) cubic meters

1.72    If the volume of helium in a balloon is 145 cm$^3$, what is its volume in the following units?
   (a) milliliters
   (b) liters
   (c) cubic meters

1.73    If the length, width, and height of a box are 8.0 cm, 5.0 cm, and 4.0 cm, respectively, what is the volume of the box in units of milliliters and liters?

1.74    If a cubic box (all sides the same length) has a volume of 1.0 L, what is the length of each side of the box?

1.75    A slice of cheese has a mass of 28 g and a volume of 21 cm$^3$. What is the density of the cheese in units of g/cm$^3$ and g/mL?

1.76    Two stones resembling diamonds are suspected of being fakes. To determine if the stones might be real, the mass and volume of each were measured. Both stones have the same volume, 0.15 cm$^3$. However, stone A has a mass of 0.52 g and stone B has a mass of 0.42 g. If diamond has a density of 3.5 g/cm$^3$, could the stones be real diamonds? Explain.

1.77    If the density of a sugar solution is 1.30 g/mL, what volume of this solution has a mass of 50.0 g?

1.78    The density of a certain type of plastic is 0.75 g/cm$^3$. If a sheet of this plastic is 10.0 m long, 1.0 m wide, and 1 cm thick, what is its mass?

1.79    Why do liquids have greater densities than gases?

1.80    When a balloon filled with air is heated, the balloon increases in volume. Does the density of the air in the balloon increase, decrease, or remain the same?

1.81    A piece of plastic sinks in oil but floats in water. Place these three substances in order from lowest density to greatest density.

1.82    What special molecular-level feature of ice explains why ice floats in water?

Ice                              Liquid water

1.83    Acetone, a component of some types of fingernail polish remover, has a boiling point of 56°C. What is its boiling point in units of kelvin?

1.84    The boiling point of liquid nitrogen is 77 K. What is its boiling point in units of degrees Celsius?

1.85    What is the difference in temperature between the boiling point of water and the freezing point of water in each of the following temperature scales?
   (a) Celsius scale
   (b) Kelvin scale
   (c) Fahrenheit scale

1.86   If the temperature of a cup of coffee decreases from 60.0°C to 25.0°C, what is the decrease in temperature in units of degrees Celsius and kelvin?

1.87   Does the boiling point of a substance depend on how much of this substance you have?

1.88   Does the melting point of a substance depend on how much of this substance you have?

1.89   Identify each of the following as a physical property or a chemical property.
    (a) mass
    (b) density
    (c) flammability
    (d) resistance to corrosion
    (e) melting point
    (f) reactivity with water

1.90   Identify each of the following as a physical property or a chemical property.
    (a) boiling point
    (b) reactivity with oxygen
    (c) resistance to forming compounds with other elements
    (d) volume

1.91   Identify each of the following as a physical change or a chemical change.
    (a) boiling acetone
    (b) dissolving oxygen gas in water
    (c) combining hydrogen and oxygen gas to make water
    (d) burning gasoline
    (e) screening rocks from sand
    (f) the conversion of ozone to oxygen, $2O_3(g) \longrightarrow 3O_2(g)$

1.92   Identify each of the following as a physical change or a chemical change.
    (a) condensation of ethanol
    (b) combining zinc and oxygen to make the compound zinc oxide
    (c) dissolving sugar in water
    (d) burning a piece of paper
    (e) combining sodium metal with water, producing sodium hydroxide and hydrogen gas
    (f) filtering algae from water

1.93   Write a symbolic representation and a molecular-level representation for the change that occurs during the condensation of chlorine, $Cl_2$.

1.94   Write a symbolic representation and a molecular-level representation for the process of freezing oxygen, $O_2$.

1.95   Do the changes shown in this diagram represent a physical or chemical change?

Before           After

1.96   Do the changes shown in this diagram represent a physical or chemical change?

Before           After

1.97   Draw a picture that shows $CH_4$ (shown in the "After" image in Question 1.95) condensing from a gas to a liquid. Does this picture represent a physical or a chemical change?

1.98   Draw a picture that shows water boiling. Does this picture represent a physical or chemical change?

1.99   The image shows what happens when iodine, $I_2$, is placed in the bottom of a beaker and heated. Is this a physical or chemical change? (Note that the photo shows an inner flask containing ice.)

©Charles D. Winters/Science Source

1.100   The picture shows natural gas ($CH_4$) burning. In the process, carbon dioxide and water form. Is this a physical change or a chemical change?

©Sami Sarkis/Getty Images

## Energy and Energy Changes

1.101   Anna and Bill saw a construction worker welding pipe. Classify the forms of energy they were observing.

1.102   Bill and Anna watched students playing volleyball in the sunshine. Classify the forms of energy they were observing.

1.103 Which of these two samples of carbon dioxide gas has more kinetic energy? Explain your answer.

**A**          **B**

1.104 Which of these two samples of methane gas is at a higher temperature? Explain your answer.

**A**          **B**

1.105 Give examples of potential energy that you can find in your room.

1.106 Give examples of kinetic energy that you can find in your room.

1.107 Distinguish between the different types of energy shown in the following picture.

©John Foxx/Getty Images

1.108 Distinguish between the different types of energy shown in the following picture.

©Digital Vision/Getty Images

1.109 Describe how some common types of energy can change from one type to another.

1.110 If energy cannot be created, what is the source of the energy that is released when gasoline is burned?

1.111 In terms of kinetic and potential energy, describe what happens when a home run is hit on a softball field.

1.112 In terms of kinetic and potential energy, describe what happens when a basketball is dribbled down a basketball court and shot to score three points.

1.113 Body mass index (BMI) is a number calculated from a person's weight and height. BMI provides a reliable indicator of body fatness for most people and is used to screen for weight categories that may lead to health problems. The formula for a person's BMI is

$$BMI = \frac{weight\ (kg)}{[height\ (m)]^2}$$

For an adult, a BMI value between 18.5 and 24.9 is healthy and considered normal.
   (a) Calculate the BMI for a 6-ft, 2-in man who weighs 169 lb.
   (b) Is this person underweight, at a healthy weight, or overweight?

1.114 Consider propane fuel being burned to power a vehicle. How would you classify the transformation of energy in this process in terms of kinetic and potential energy?

1.115 Consider water falling into a fountain. How would you classify the transformation of energy in this process in terms of kinetic and potential energy?

## Scientific Inquiry

1.116 Explain the difference between a hypothesis and a theory.

1.117 Explain how a hypothesis is used in scientific research.

1.118 Classify each of the following as an observation, hypothesis, law, or theory.
   (a) When wood burns, oxygen is consumed.
   (b) Heavier-than-air objects always fall toward the center of Earth.
   (c) Matter is composed of atoms.
   (d) Crime rates increase when the moon is full.

1.119 Classify each of the following as an observation, an hypothesis, law, or theory.
   (a) Bad luck results from walking under a ladder.
   (b) Oil floats on water.
   (c) Oil floats on water because it is less dense.
   (d) Wood burns.

1.120 You observe coins in a fountain and propose the hypothesis that fountains are built with coins in them. Suggest experiments to test this hypothesis.

1.121 You observe a piece of balsa wood floating on water and propose the hypothesis that all wood floats on water because wood is less dense than water. Suggest experiments to test this hypothesis.

## Additional Questions

**1.122** Rank the following measurements in order from smallest to largest: $1.0 \times 10^{-4}$ m, $2.0 \times 10^{-5}$ m, $3.0 \times 10^{-6}$ km, $4.0 \times 10^{2}$ mm, 0.0 m, 1.0 m.

**1.123** The density of air in a balloon is less at high altitudes than at low altitudes. Explain this difference.

**1.124** If the temperature in a room increases from 20.0°C to 30.0°C, what is the temperature change in units of kelvin?

**1.125** If you have a sample of zinc and a sample of copper, and both have the same mass, which has the greatest volume?

**1.126** Give the symbols for potassium and phosphorus.

**1.127** Give the symbols for the following noble gas elements:
(a) helium
(b) neon
(c) argon
(d) krypton
(e) xenon
(f) radon

**1.128** The red blood cell (RBC) count for a normal female is about 5 million RBCs per μL (1 μL = 1 mm³). How many red blood cells are in a gallon of normal female blood?

**1.129** Recycling facilities around the world use a variety of techniques to separate different types of recyclable waste. Wet separation is a method that involves processing certain kinds of waste (for example, glass, sand, and metal) based on their densities relative to water. Is this type of separation based on the physical or chemical properties of the recyclable waste?

**1.130** These samples of metals have the same mass. Which has the greater density?

A          B
©Jim Birk

**1.131** The typical dose of epinephrine at a particular concentration administered to a patient under cardiac arrest is 0.1 mg per kilogram of body weight. If a patient weighs 180 lb, how much epinephrine should be administered?

**1.132** About 70 million tons of paper are used per year in the United States. If the U.S. population is about 301.6 million people, how many kilograms of paper are used by the average U.S. citizen?

**1.133** During a typical physical exam, blood tests to measure cholesterol levels are usually run. A total cholesterol level of 240 mg/dL or above is considered high. What is this threshold level in units of pounds per fluid ounce? If a typical adult has 4 to 6 L of blood, how many pounds of cholesterol are present in the blood of a patient with a total cholesterol level of 260 mg/dL?

**1.134** The densities of antifreeze, corn oil, dish detergent, maple syrup, shampoo, and water in g/mL are 1.13, 0.93, 1.03, 1.32, 1.01, and 1.00, respectively. Which layer is which substance in the figure?

©Richard Megna/Fundamental Photographs

**1.135** The lowest possible temperature is the temperature at which a substance would have no kinetic energy (no motion). What is the lowest possible temperature in each of the following temperature units?
(a) Kelvin scale
(b) Celsius scale
(c) Fahrenheit scale

**1.136** Classify the substance in the molecular-level image as:
(a) an element, compound, or mixture
(b) atoms or molecules

©Jules Frazier/Getty Images

**1.137** Classify each of the following as a symbolic representation for a pure element, pure compound, or mixture.
(a) $NaNO_3(s)$
(b) $N_2(g)$
(c) $NaCl(aq)$

**1.138** Titanium is a strong metal with a low density that is shiny with a white-metallic color. It is relatively resistant to corrosion by acids and chlorine and resists tarnishing by air by forming a protective oxide layer. It is described as a nontoxic, inert biomaterial, which means that it does not interact unfavorably with human tissues and fluids. Identify all the chemical and physical properties of titanium in this description.

1.139   Blood is a water-based liquid in which solids (blood cells and platelets) are suspended. Classify blood as an element, compound, homogeneous mixture, or heterogeneous mixture.

1.140   Explain how a fuel such as gasoline has both kinetic and potential energy.

1.141   Convert 10.0 m$^3$ to units of cm$^3$ using the conversion factor 1 m = 100 cm.

1.142   The average blood volume in the human body is about 4.8 L. Convert this volume to the following units:
   (a) milliliters (mL)
   (b) cubic centimeters (cm$^3$)
   (c) cubic meters (m$^3$)

1.143   The average density of human blood is 1060 kg/m$^3$. What is the mass of blood in a person who has a blood volume of 0.00500 m$^3$? Report your answer in units of kilograms and in units of pounds.

1.144   What is the name for the change in physical state described in each of the following processes?
   (a) $Fe(s) \longrightarrow Fe(l)$
   (b) $CH_3OH(l) \longrightarrow CH_3OH(g)$
   (c) $H_2O(g) \longrightarrow H_2O(l)$
   (d) $H_2O(l) \longrightarrow H_2O(s)$

1.145   A car traveling at 29.1 m/s drives for 2.5 hours then goes another 75 km. How far did the driver travel in miles?

1.146   A large chunk of metal weighing 2.00 lb was found in a mine and appears to be gold. The metal was placed in 1.0 qt of water and the final volume was 38.4 fl oz. What is the density of this metal in g/cm$^3$? Could it be gold?

## Concept Review Questions

1.147   Which of the following is an example of a mixture?
   A. aluminum
   B. lake water
   C. sodium chloride
   D. iron
   E. carbon dioxide

   For those that are not mixtures, classify them as elements or compounds.

1.148   Which of the following is a pure substance that is composed of molecules?

A          B          C

D          E

   For the correct answers, are they elements or compounds? How would you classify the incorrect answers?

1.149   Which of the following statements regarding liquids is correct?
   A. They consist of particles that are relatively far apart.
   B. They can be easily compressed.
   C. The symbol for a liquid is (*aq*).
   D. Their particles can move around each other.
   E. They have a fixed shape.
   Modify each incorrect statement to make it a true statement.

1.150   A rectangular block of an unknown metal with a mass of 456 g is measured to have a length, width, and height of 10.2 cm, 5.08 cm, and 3.26 cm, respectively. What is the likely identity of the metal?
   A. zinc, density = 7.14 g/cm$^3$
   B. lead, density = 11.3 g/cm$^3$
   C. aluminum, density = 2.70 g/cm$^3$
   D. nickel, density = 8.91 g/cm$^3$
   E. titanium, density = 4.51 g/cm$^3$
   What would be the masses of the same size blocks of the other metals?

1.151   Which of the following statements is correct?
   A. Natural gas burning to produce water and carbon dioxide is an example of a chemical change.
   B. In a physical change, atoms rearrange to form new substances.
   C. When a substance undergoes a chemical change, its chemical composition does not change.
   D. Formation of rust on an old car is an example of a physical change.
   E. The boiling point of nitrogen is −196°C. At −210°C, nitrogen is expected to be a gas.
   Modify each incorrect statement to make it a true statement.

1.152   Which of the following statements is correct?
   A. Heat coming off a car engine is an example of matter.
   B. Potential energy is the energy of motion.
   C. When a compound undergoes a spontaneous chemical reaction, it releases its potential energy.
   D. Exhaust released from the tailpipe of an old truck is an example of energy.
   E. A skateboarder rolling down a hill is converting kinetic energy to potential energy.
   Modify each incorrect statement to make it a true statement.

1.153   Which of the following is an example of a hypothesis?
   A. The mass of products obtained in a chemical reaction always equals the mass of the substances that react.
   B. Zinc reacts with hydrochloric acid but copper doesn't because zinc is more reactive than copper.
   C. Silver tarnishes.
   D. Oil and water don't mix.
   E. All matter is composed of atoms.
   How are the incorrect statements classified?

1.154 The number 0.00063780 correctly expressed in scientific notation is:

A. $6.38 \times 10^{-4}$

B. $6.378 \times 10^{-4}$

C. $6.3780 \times 10^{-4}$

D. $6.3780 \times 10^{4}$

E. $6.3780 \times 10^{-5}$

Write each incorrect answer in decimal form and compare to the value given in the problem.

1.155 Which of the following mathematical operations would have two significant figures in the answer?

A. $(7.10 \times 10^{-3}) + (8.3 \times 10^{-4})$

B. $\dfrac{(6.32 \times 10^{4}) \times (4.8 \times 10^{-3})}{0.05}$

C. $\dfrac{(473 + 27)}{0.134}$

D. $\dfrac{7.460 \times 10^{-3}}{(1.85 \times 10^{2}) \times (0.050)}$

E. $2 \times \dfrac{(4.70 \times 10^{-7})}{(3.00 \times 10^{3})}$

Carry out each mathematical operation and express your answers to the correct number of significant figures.

1.156 Which of the following has the largest mass?

A. $4.1 \times 10^{-2}$ kg

B. $1.8 \times 10^{4}$ mg

C. 25 g

D. $6.5 \times 10^{6}$ μg

E. $7.6 \times 10^{-1}$ g

Arrange these quantities in order of increasing mass.

1.157 A bicyclist is traveling at 6.7 meters per second. Which of the following mathematical operations would correctly determine the speed in miles per hour?

A. $\dfrac{6.7 \text{ m}}{\text{s}} \times \dfrac{1000 \text{ km}}{1 \text{ m}} \times \dfrac{0.621 \text{ mi}}{1 \text{ km}} \times \dfrac{60 \text{ s}}{1 \text{ min}} \times \dfrac{60 \text{ min}}{1 \text{ hr}}$

B. $\dfrac{6.7 \text{ m}}{\text{s}} \times \dfrac{1 \text{ km}}{1000 \text{ m}} \times \dfrac{1 \text{ km}}{0.621 \text{ mi}} \times \dfrac{60 \text{ s}}{1 \text{ min}} \times \dfrac{60 \text{ min}}{1 \text{ hr}}$

C. $\dfrac{6.7 \text{ m}}{\text{s}} \times \dfrac{1 \text{ km}}{1000 \text{ m}} \times \dfrac{0.621 \text{ mi}}{1 \text{ km}} \times \dfrac{60 \text{ s}}{1 \text{ min}} \times \dfrac{60 \text{ min}}{1 \text{ hr}}$

D. $\dfrac{6.7 \text{ m}}{\text{s}} \times \dfrac{1 \text{ km}}{1000 \text{ m}} \times \dfrac{0.621 \text{ mi}}{1 \text{ km}} \times \dfrac{1 \text{ min}}{60 \text{ s}} \times \dfrac{1 \text{ hr}}{60 \text{ min}}$

E. $\dfrac{6.7 \text{ m}}{\text{s}} \times \dfrac{1000 \text{ m}}{1 \text{ km}} \times \dfrac{0.621 \text{ mi}}{1 \text{ km}} \times \dfrac{1 \text{ min}}{60 \text{ s}} \times \dfrac{1 \text{ hr}}{60 \text{ min}}$

Explain what is wrong with the incorrect responses.

# 2

# Atoms, Ions, and the Periodic Table

2.1 Dalton's Atomic Theory

2.2 Structure of the Atom

2.3 Ions

2.4 Atomic Mass

2.5 The Periodic Table

Chapter Review

Questions and Problems

© Andrew Dunn/Alamy Stock Photo

When Anna and Bill went looking for things related to chemistry (in Chapter 1), they classified nearly everything they saw as matter. Now Anna, who is writing a history paper, wants to know how our modern ideas about the atom arose. She learned in a philosophy class that the ancient Greeks thought the world was made up of four kinds of matter: earth, air, fire, and water. They called them elements. How did we come to understand matter as being composed of atoms of the different elements in the periodic table?

As Anna sets out to learn more, she discovers that the concept of the atom dates back to 450 BC. The Greek philosopher Democritus believed that there was a limit to how far matter could be broken down into smaller pieces. A piece of sand, for example, could be split into smaller pieces, and those pieces into even smaller pieces. But it made no sense to Democritus that a piece of sand could be broken down into smaller pieces forever. Democritus gave the smallest components of matter the name *atomos* (*atom* in English), which is a Greek word meaning "unbreakable," to describe the smallest components of matter.

The concept of the atom did not become popular until centuries later. In the 1700s, chemists realized that they must look for elements by breaking down matter into simpler substances. Once they had something that could not be broken down any further, they called it an *element*. About 30 different elements had been discovered by the end of the 1700s. Throughout the 1700s, few chemists believed that elements were made of atoms. The experiments of many chemists, however, provided evidence for the atomic nature of matter.

Anna learns that the science and math teacher John Dalton offered the first convincing argument for the existence of atoms, based on well-established experimental results, which he published in a book in 1808. So persuasive were his arguments that the idea came to be called *Dalton's atomic theory*. As Anna continues her research, she learns that the acceptance of Dalton's atomic theory by the scientific community paved the way for our current understanding of the atom—the *modern model of the atom*. Many additional questions pop into Anna's head now. How are atoms of different elements, such as carbon and hydrogen, different from one another? Do they have different masses? Are atoms composed of even smaller particles?

Anna isn't the only one who's finding that chemistry relates to other subjects. Anna's roommate Megan, a nutrition major, is quick to realize that the human body is composed of matter, and that everything we are made of is just different combinations of elements. Although we are about 93% carbon, hydrogen, and oxygen, small amounts of many other elements are essential to proper functioning. Many of these are called *essential minerals*. Minerals are crucial to the growth and production of bones, teeth, hair, blood, nerves, and skin, not to mention the enzymes and hormones that living cells and tissues need to function. The essential minerals that we need in the greatest quantity are calcium, phosphorus, potassium, sodium, chlorine, magnesium, and iron. We get them from the foods we eat and drink. For example, calcium, which makes bones strong, is found in significant amounts in milk products. Potassium, an important regulator of cell fluids, is found in many fruits and vegetables. Iron, a building block for the oxygen-carrying hemoglobin in our blood, comes from fruits, vegetables, and red meat.

Megan notices that many of the minerals are classified as metals on the periodic table. She wonders if we ingest *metals* in the foods we eat. Generally, no. We take them in as components of compounds in which a metal such as iron exists as an ion. An *ion* is an atom with an electrical charge. The properties of metal ions are different from the properties of pure metal elements. Ions are important to the chemical processes that occur in our bodies. After learning about ions as nutrients, Megan begins to wonder how ions differ from atoms and why they have different properties.

In this chapter we will learn about the atoms that compose the elements and about the similarities and differences among them. We will answer Anna's and Megan's questions and many of your own.

Democritus, a philosopher predating Socrates, wrote about mathematics, cosmology, nature, and ethics in addition to atoms.

©SuperStock/Getty Images

Some essential minerals are involved in many processes. Magnesium, for example, plays important roles in more than 300 chemical processes that occur in the human body.

The common iron supplement iron(II) sulfate contains iron in an ion form.

©McGraw-Hill Education/John Flournoy

## Questions for Consideration

**2.1** What evidence suggests that matter is composed of atoms?

**2.2** How can the composition of atoms differ?

**2.3** How do ions differ from the atoms of elements?

**2.4** How can we describe the mass of the atoms of an element?

**2.5** How does the periodic table relate to the structure and behavior of atoms?

 **Math Tools Used in This Chapter**

Scientific Notation (Math Toolbox 1.1)

Significant Figures (Math Toolbox 1.2)

## 2.1 Dalton's Atomic Theory

In Anna's research, she learned that before the nineteenth century, many believed that matter was continuous—that is, that it could be divided infinitely (Figure 2.1). That notion began to change in 1808, when John Dalton published his atomic theory. Dalton based his argument on experimental evidence that had begun accumulating about three decades earlier. For example, Antoine Lavoisier's experimental results led to the **law of conservation of mass,** which he published in 1787. His experiments showed that no measurable change in mass occurs during a chemical reaction. The mass of the products of a reaction always equals the mass of the reacting substances.

When we observe chemical changes, it may seem in some cases that the law of conservation of mass does not apply. After wood is burned in a bonfire, for example, there is less mass left in the fire pit. When an iron nail rusts, the rusted nail has a greater mass than the original nail. However, if we carry out each of these reactions in a closed container and find the mass of the closed container and its contents before and after the change, the mass does not change. With an open container, the gases produced when wood burns escape into the atmosphere, so they are not included in the final measurement. The gases consumed when a nail rusts are not included in the original mass. The example in Figure 2.2 shows the conservation of mass when a reaction occurs in a closed system. Lavoisier did experiments like these in which he was careful not to let any matter enter or escape.

Joseph Proust did similar experiments, mostly reacting metals with oxygen. He found that the oxygen content of the products was always fixed at one or two values, rather than showing a range of all possible values. His findings, published between 1797 and 1804, led to the **law of definite proportions.** This law states that all

As you will learn in Chapter 15, mass is not conserved in reactions that involve the nucleus of the atom. In nuclear fusion reactions that occur in the Sun, for example, matter is converted into energy. Nuclear reactions were not known in Lavoisier's and Dalton's time.

**FIGURE 2.1** Even when viewed under an optical microscope (center), the particle composition of the copper wire is not obvious. It seems continuous rather than particulate. As shown by the model (right), it would be necessary to use an electron microscope to observe that copper metal, and all matter, is composed of particles.
©imagebroker/Alamy Stock Photo; (inset): ©Jim Birk

Cu atom

**FIGURE 2.2** When sodium carbonate contained in the balloon is placed in a solution of hydrochloric acid, a reaction forms carbon dioxide gas. Because the carbon dioxide is not allowed to dissipate into the atmosphere, its mass is included in the final mass in the experiment shown here. How would the result be different if the container were open?
©Jim Birk

samples of the same compound always contain the same proportions by mass of the component elements. For example, pure water is always composed of oxygen and hydrogen with a mass ratio of 8 g of oxygen for every 1 g of hydrogen. Conversely, when water is broken down into its elements using electricity, oxygen and hydrogen are formed in this same mass ratio of 8:1.

Dalton reasoned that the law of conservation of mass and the law of definite proportions could be explained only if matter was composed of atoms. The following postulates summarize **Dalton's atomic theory:**

1. All matter is composed of exceedingly small, indivisible particles, called atoms.
2. All atoms of a given element are identical both in mass and in chemical properties. However, atoms of different elements have different masses and different chemical properties.
3. Atoms are not created or destroyed in chemical reactions.
4. Atoms combine in simple, fixed, whole-number ratios to form compounds.

A chemical reaction, according to Dalton's atomic theory, is a rearrangement of atoms into a new combination, resulting in the formation of one or more new chemical substances (Figure 2.3). Although Dalton's theory has been modified over the

In addition to developing atomic theory, John Dalton also studied meteorology and color blindness.

©GeorgiosArt/Getty Images

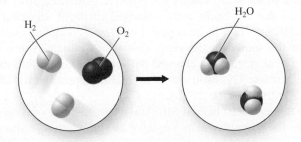

**FIGURE 2.3** Two $H_2$ molecules combine with one $O_2$ molecule and rearrange to form two $H_2O$ molecules. A chemical reaction is simply the rearrangement of atoms into new combinations.

**FIGURE 2.4** This scanning tunneling microscope (STM) image shows atoms on the surface of copper. Because atoms are too small to be seen with an optical microscope, images could not be produced until the invention of the STM in the 1980s.
©Drs. Ali Yazdani & Daniel J. Hornbaker/ Science Source

past 200 years, it still provides the basis for understanding how atoms are the building blocks of matter. As with most theories, atomic theory has been amended as new evidence has been found. We now know that some of Dalton's postulates are not quite correct. The first statement, for example, is not strictly true. Atoms are composed of even smaller particles, called *subatomic particles*. The second statement is not quite accurate either. As you will learn in Section 2.2, atoms of a given element actually can vary in mass.

Still, Dalton's central idea has stood the test of time and experimentation—although for nearly two centuries no one had a way to see the atoms that Dalton proposed. Because atoms are so small, they cannot be seen using an optical microscope. However, with the invention of the scanning tunneling microscope (STM) in 1981, scientists can now observe what we interpret to be single atoms on the surface of a material (Figure 2.4).

An optical microscope cannot resolve images smaller than the size of the waves making up the light used to examine the object.

## 2.2  Structure of the Atom

When reading nutritional labels, Megan noticed that a food supplement contains both iron and selenium. She wondered how the atoms of these elements differ. Scientists have also asked this question about atoms in general and found ways to probe their composition. They have studied the smaller particles that actually make up atoms.

Selenium is an essential mineral involved in the function of the thyroid gland.

### Subatomic Particles

The existence of the **electron,** a negatively charged subatomic particle, was demonstrated by J. J. Thomson in 1897. He conducted a series of experiments with cathode-ray tubes (Figure 2.5). In a partially evacuated cathode-ray tube, a voltage is applied by connecting each end of the tube to a battery. Electricity then flows from one end of the tube to the other in the form of a ray. The invisible rays can be observed when they cause certain materials coated on the glass to glow. Thomson found that, in a magnetic or electric field, the rays bent toward a positively charged plate and were deflected away from a negatively charged plate outside the tube. He knew that like electrical charges repel each other and opposite charges attract each other. The bending of the beam toward the positive plate (and away from the negative plate) showed that the beam was composed of negatively charged particles. Thomson showed that the rays had a negative electrical charge no matter what material was used for the source of the rays. This result indicated that the rays were composed of identical, negatively charged particles common to all matter. We call these particles electrons. Thomson was also able to determine the charge-to-mass ratio of the electron from such experiments.

Cathode-ray tubes (CRTs) are the fundamental components of older television picture tubes and computer monitors. The screen contains chemical compounds that glow when struck by fast-moving electrons. Different chemicals that glow different colors provide a color picture.

Building on these results, Robert Millikan experimented with oil droplets in an effort to measure the strength of the negative charge on an electron (Figure 2.6).

How did the discovery of electrons make it necessary to change one of Dalton's postulates?

**FIGURE 2.5** J. J. Thomson's experiments with cathode-ray tubes led to the discovery of the electron. (A) In a normal cathode-ray tube, the cathode ray travels in a straight path when there is no external field. (B) When Thomson applied an external electric field to the cathode-ray tube, the cathode ray bent toward the positive plate. It was known that like charges repel each other and opposite charges attract each other. The bending of the beam toward the positive plate (and away from the negative plate) indicated that the beam was composed of negatively charged particles.

**FIGURE 2.6** In Millikan's oil-drop experiment, the electric field strength required to suspend an oil droplet depended upon the number of extra electrons on it. These experiments allowed Millikan to determine the charge of a single electron and calculate its mass.

When he exposed the droplets to radiation, they took on an electrical charge. By measuring the magnitude of the electric field necessary to cause the droplets to hang suspended in air, Millikan determined that the charge on an electron is $-1.6022 \times 10^{-19}$ coulombs (C) (a *coulomb* is a unit of electrical charge). He then calculated the mass of an electron to be $9.1094 \times 10^{-28}$ g from its charge and the charge-to-mass ratio determined by Thomson. Since even the lightest atoms have a mass greater than $10^{-24}$ g, an electron contributes only a small part to the mass of an atom. In fact, the mass of an electron is 1836 times less than the mass of one hydrogen atom, the lightest of all the elements. Because of the electron's small mass, it was originally thought that thousands of them must lie inside a single hydrogen atom. We now know there is only one.

The discovery of the electron stimulated many more experiments in search of other subatomic particles. Since atoms are electrically neutral, scientists reasoned that atoms must contain positively charged particles to counter the negatively charged electrons. The positively charged particle, called a **proton,** has a charge equal in magnitude to the electron but opposite in sign, $+1.6022 \times 10^{-19}$ C. To be electrically neutral, an atom must have equal numbers of protons and electrons. To make it easier to deal with electrical charges in matter, we usually express the charges as a multiple of the charge of an electron or of a proton, instead of in units of coulombs. Expressed in this way, the charge of an electron is $1-$, and the charge of a proton is $1+$.

Proton therapy is a radiation technique used to treat various forms of cancer. During treatment a beam of radiation (protons) is targeted directly at tumor cells. The radiation ionizes molecules in the cells of tumors, causing permanent damage to the DNA of the cells.

## The Nuclear Atom

How might protons and electrons be arranged in an atom? Thomson's model of atomic structure, called the "plum pudding" model, assumed that protons and electrons were evenly distributed throughout the atom (Figure 2.7). Ernest Rutherford designed an experiment to test this model, and his associate, Hans Geiger, carried it out. The experiment involved bombarding thin gold foil with alpha particles. *Alpha particles* ($\alpha$) were known at the time as positively charged particles thousands of times greater in mass than electrons. (Today we know them as helium atoms that have lost their electrons.) According to the plum pudding model, none of the alpha particles should have been affected by the dispersed bits of positive and negative charge in the gold atoms. They should have zipped right through the gold foil, and most did. However, some were deflected slightly, and a few actually bounced backwards, as shown in Figure 2.8. The result was quite unexpected. It was as if you fired a bullet at a sheet of tissue paper and it came back and hit you! The deflection of these relatively massive alpha particles suggested that most of the mass of the atom had to be concentrated in a positively charged core, which Rutherford called the **nucleus.** The electrons, he reasoned, had to be dispersed in the large volume outside of the nucleus. The large electron space was the area penetrated by most of the alpha particles. Only if an alpha particle came close enough to the incredibly dense nucleus would it be deflected from its original path. The alpha particles that hit the nucleus of a gold atom head-on were deflected backwards.

Rutherford's experiment was the basis for the *nuclear model of the atom* (Figure 2.9), developed in 1907. The model suggested that the nucleus contains the protons and most of the mass of the atom. The electrons exist outside the nucleus in what is often called an "electron cloud."

The diameter of the nucleus is about $10^{-14}$ m, and the diameter of the atom is about $10^{-10}$ m. These relative sizes are comparable to a flea in the center of a domed stadium. While the volume of the nucleus is about 10,000 times smaller than the entire atom, the mass of the nucleus accounts for most of the mass of the atom. This is because the nucleus contains heavy particles, such as protons, that are packed closely together.

**ANIMATION:** α-Particle Scattering

**ANIMATION:** Rutherford's Gold Foil Experiment

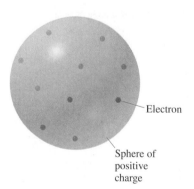

**FIGURE 2.7** Thomson's model suggested that electrons in the atom might be embedded in a sphere of positive charge, like raisins in plum pudding.

Although a penny contains about $1 \times 10^{22}$ atoms, most of the penny is empty space. Why?

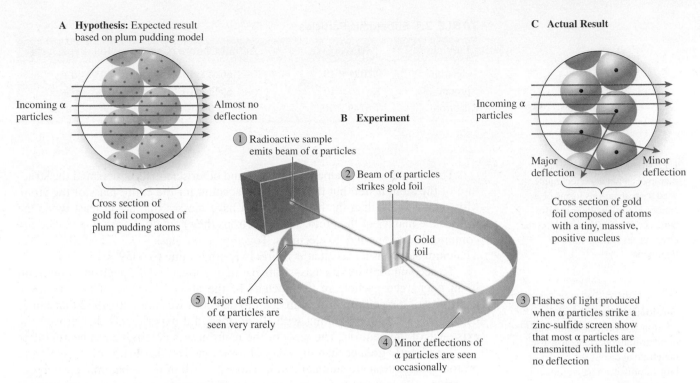

**FIGURE 2.8** Rutherford's gold foil experiment led to the nuclear model of the atom. A beam of positively charged alpha (α) particles was directed through a thin layer of gold. A zinc-sulfide screen detected the alpha particles by producing a flash of light upon contact. (A) According to the plum pudding model, all alpha particles should have penetrated straight through the gold atoms. (B) In the experiment, many of the alpha particles penetrated straight through the atom, but some were deflected. (C) The nuclear model of the atom explains the experimental results. The positively charged protons are packed tightly together in the very center of the nucleus. When the nucleus is in the path of an alpha particle, the alpha particle is deflected from its original path.

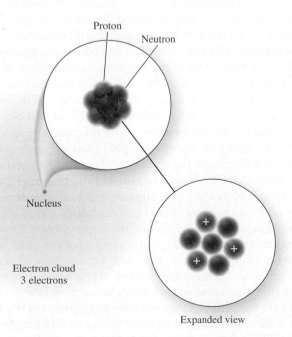

**FIGURE 2.9** In the nuclear model of the atom, protons (blue spheres) and neutrons (red spheres) are located in a tiny nucleus at the center of the atom. The space outside the nucleus is occupied by the electrons.

**TABLE 2.1** Subatomic Particles

| Particle | Mass (g) | Actual Charge (C) | Relative Charge |
|---|---|---|---|
| electron | $9.1094 \times 10^{-28}$ | $-1.6022 \times 10^{-19}$ | $1-$ |
| proton | $1.6726 \times 10^{-24}$ | $+1.6022 \times 10^{-19}$ | $1+$ |
| neutron | $1.6749 \times 10^{-24}$ | $0$ | $0$ |

Like proton therapy, neutron therapy is a radiation technique used to treat various forms of cancer. However, the scientific evidence is inconclusive about the effectiveness of this form of treatment.

Due to its neutral charge, the neutron is an effective nuclear bombardment particle now used to penetrate and split the nuclei of large atoms in energy-producing fission reactions.

79
**Au**

**FIGURE 2.10** The atomic number of each element is indicated on the periodic table in this book just above the element symbol.

How does the existence of isotopes make it necessary to change one of Dalton's postulates?

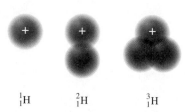

$^{1}_{1}H$ $\qquad$ $^{2}_{1}H$ $\qquad$ $^{3}_{1}H$

**FIGURE 2.11** These images show the subatomic particles in the nuclei of hydrogen isotopes. Isotopes of hydrogen vary in the number of neutrons. In these representations, the blue spheres with a plus in them represent protons. The red spheres represent neutrons. How many neutrons are in each isotope? What do the letters and numbers below each diagram represent?

These experiments helped Rutherford and other scientists understand the structure of the atom better, but they could not account for the entire mass of the atom. Most atoms other than the hydrogen atom have masses that are at least twice the sum of the masses of the protons and electrons they contain. For example, calcium contains 20 protons and 20 electrons. Together, their mass is $3.3471 \times 10^{-23}$ g. Yet a calcium atom has a mass that is nearly twice this value ($6.6359 \times 10^{-23}$ g).

To account for the extra mass, Rutherford hypothesized the neutron. A **neutron** is an uncharged particle in the nucleus of the atom. Because of the electrical neutrality of the neutron, it was difficult to detect. It was not until 1932 that James Chadwick, a scientist working with Rutherford, did experiments that proved the existence of the neutron. The mass of the neutron was determined to be $1.6749 \times 10^{-24}$ g, slightly greater than the mass of a proton. The properties of the electron, proton, and neutron are summarized in Table 2.1. All of the subatomic particles in the table are important to the nuclear model of the atom as we understand it today (Figure 2.9).

In Chapters 7 and 8 you will learn how electrons relate to chemical reactivity. In Chapter 15, you will learn how neutrons relate to nuclear radioactivity. In this chapter, we will focus on how protons determine the identity of an atom, neutrons help determine its mass, and electrons determine its charge.

## Isotopes, Atomic Number, and Mass Number

Recall that Anna was wondering how atoms of carbon and hydrogen differ. What is it about an atom of an element that distinguishes it from an atom of some other element? The same scientists who identified the subatomic particles found an answer to this question. They learned, for example, that all hydrogen atoms contain just one proton, and that any atom containing just one proton is a hydrogen atom. In a similar fashion, any atom that contains two protons is a helium atom. An atom with three protons is a lithium atom, and so on. *The number of protons in an atom's nucleus determines the identity of that element.* The number of protons in the nucleus of each atom of an element is the **atomic number** ($Z$) of that element. On the periodic table in this book, elements are shown with the atomic number just above the element symbol, as shown for the element gold (Au) in Figure 2.10.

How can we determine the number of electrons and neutrons in an atom? Atoms are electrically neutral. This means that the number of electrons in an atom equals the number of protons—the atomic number. For example, the atomic number of gold (Au) is 79, so an atom of gold has 79 protons and 79 electrons. Not all atoms of an element contain the same number of neutrons, however. An **isotope** of an element is an atom that contains a specific number of neutrons. Most of the elements consist of more than one isotope with different numbers of neutrons. For example, most hydrogen atoms contain no neutrons. Their nucleus consists only of a single proton. Not all hydrogen atoms are alike, however. Some contain one neutron and some even have two. All three are isotopes of hydrogen. They are still hydrogen atoms because they have one proton, but their nuclei differ in the number of neutrons (Figure 2.11). The isotopes of an element have essentially identical chemical properties, but their physical properties, such as melting point and boiling point, may differ slightly.

One way to distinguish between isotopes is by their *mass number*. The **mass number** (*A*) of an isotope is the sum of the number of protons (*Z*) and the number of neutrons (*N*) in its nucleus:

Mass number = number of protons + number of neutrons

$$A = Z + N$$

The three isotopes of hydrogen have different mass numbers (*A* = 1, 2, and 3) because they have different numbers of neutrons, also called the **neutron number** (*N* = 0, 1, and 2). The mass number is not an actual mass; it is a count of the number of particles in the nucleus.

Example 2.1 shows how you can determine the atomic number and mass number for an isotope of an element by looking at an atomic-level representation.

---

**EXAMPLE 2.1**     **Determining Atomic Number and Mass Number**

For the atom represented in the following diagram,

(a)  Determine the number of protons and neutrons.
(b)  Identify the atomic number and the element.
(c)  Determine the mass number for this isotope.

5 electrons

Expanded view

**Solution:**

(a)  The protons and neutrons make up the nucleus. The protons are positively charged and neutrons have no charge. There are five protons and six neutrons.

(b)  The atomic number, the number of protons, is 5. From the periodic table we see that the element with atomic number 5 is boron.

(c)  The mass number, the sum of the protons and neutrons, is 11.

**Consider This 2.1**

If the number of protons were one less than shown in the diagram, and the number of neutrons were one greater, how would the atomic number and mass number change and would this represent the same element or a different element?

**Practice Problem 2.1**

For the atom represented in the following diagram,

(a)  Determine the number of protons and neutrons.
(b)  Identify the atomic number and the element.
(c)  Determine the mass number for this isotope.

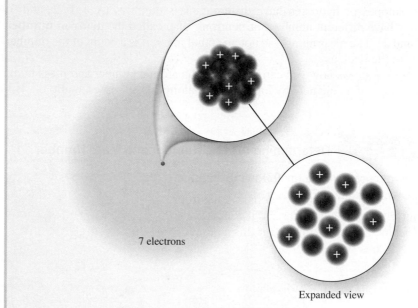

7 electrons

Expanded view

**Further Practice:**  Questions 2.29 and 2.30 at the end of the chapter

As shown by Example 2.2, you can determine the number of each type of subatomic particle when given both the atomic number and the mass number.

| EXAMPLE 2.2 | **Numbers of Subatomic Particles** |
| --- | --- |

The only stable isotope of naturally occurring fluorine has a mass number of 19. How many protons, electrons, and neutrons are in a fluorine atom?

**Solution:**

The periodic table shows the atomic number of fluorine is 9. Because the atomic number is equal to the number of protons, the number of protons in a fluorine atom is 9. The number of electrons in an atom equals the number of protons, so fluorine has 9 electrons. The number of neutrons can be determined from the known mass number and known number of protons. The mass number, the total number of protons and neutrons, is 19. There are 9 protons, so there must be 10 neutrons to give a mass number of 19.

**Consider This 2.2**

If, on another planet, there were another stable isotope of fluorine with a mass number of 17, how would the numbers of protons, electrons, and neutrons differ?

**Practice Problem 2.2**

A rare isotope of carbon has a mass number of 14. How many protons, electrons, and neutrons are in this isotope of carbon?

**Further Practice:** Questions 2.33 and 2.34 at the end of the chapter

**FIGURE 2.12** An isotope symbol consists of the element symbol with a subscript to denote the number of protons and a superscript to denote the sum of the number of protons and neutrons, the mass number.

The mass number and atomic number of an isotope are often represented by the notation called an **isotope symbol,** as shown in Figure 2.12. In an isotope symbol, $X$ is the element symbol shown on the periodic table, $A$ is the mass number, and $Z$ is the atomic number (number of protons). For example, the isotope symbol for the isotope that contains one proton and no neutrons in its nucleus is $^{1}_{1}\text{H}$, while hydrogen with one proton and one neutron is represented as $^{2}_{1}\text{H}$ (see Figure 2.11). The symbol for carbon with six protons and eight neutrons is $^{14}_{6}\text{C}$. Because each element can only have one atomic number $Z$, sometimes the atomic number is left off the isotope symbol. For example, the isotope of carbon with eight neutrons can also be represented as $^{14}\text{C}$. Other common representations use the name of the element followed by the mass number (carbon-14), or the element symbol followed by the mass number (C-14).

---

**EXAMPLE 2.3** | **Writing Isotope Symbols**

Write two representations for the following isotope.

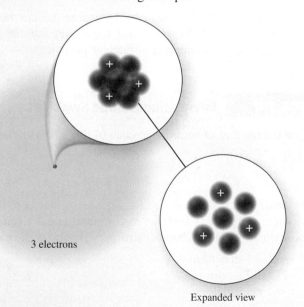

3 electrons

Expanded view

**Solution:**

There are three protons, so the atomic number is 3, corresponding to the element lithium (Li). There are four neutrons, so the mass number is 7, the sum of the protons and neutrons. This isotope can be represented as $^{7}_{3}\text{Li}$, $^{7}\text{Li}$, lithium-7, or Li-7.

**Consider This 2.3**

If there were one less neutron in the image of the nucleus, how would the isotope representation differ?

**Practice Problem 2.3**

Write two representations for the following isotope.

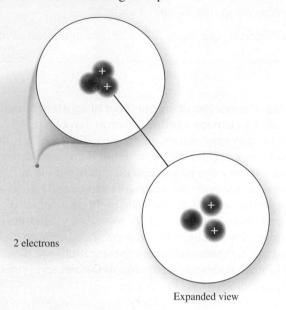

2 electrons

Expanded view

**Further Practice:**  Questions 2.35 and 2.36 at the end of the chapter

As shown by Example 2.4, you can determine the number of neutrons in an atom from its isotope symbol by subtracting the atomic number (number of protons) from the mass number (total number of nuclear particles).

## EXAMPLE 2.4    Interpreting Isotope Symbols

Determine the number of neutrons in each of the following isotopes.

(a)  $^{238}_{92}U$          (b)  $^{23}Na$          (c)  hydrogen-3

**Solution:**

The number of neutrons is equal to the difference between the mass number ($A$) and the atomic number ($Z$). If the atomic number is not given in the symbol, look on the periodic table for the number that appears above the element symbol.

(a)  $238 - 92 = 146$ neutrons
(b)  $23 - 11 = 12$ neutrons
(c)  $3 - 1 = 2$ neutrons

**Consider This 2.4**

If the mass numbers were omitted from the isotope symbols provided, could you still determine the number of neutrons in each isotope?

**Practice Problem 2.4**

Determine the number of neutrons in each of the following isotopes.

(a)  $^{131}_{53}I$          (b)  $^{37}Cl$          (c)  carbon-13

**Further Practice:**  Questions 2.37 and 2.38 at the end of the chapter

The isotope of hydrogen that has a mass number of 2 ($_1^2H$) has a special name, *deuterium.* Deuterium (sometimes represented by the symbol D) accounts for less than 1% of naturally occurring hydrogen. It is used frequently by chemists to observe specific hydrogen atoms in experiments. When deuterium ($D_2$) combines with oxygen, deuterated water ($D_2O$) forms. It behaves chemically just like ordinary water, but it has different physical properties. The photo in Figure 2.13 shows two ice cubes in water. One ice cube is made of $H_2O$ and the other of $D_2O$. What property is different for these ice cubes?

Since one ice cube sinks and the other does not, we can conclude that they have different densities. Deuterated water ($D_2O$) is denser than $H_2O$ because deuterium has two particles in its nucleus and hydrogen-1 only has one. In its solid state, the density of $D_2O$ is greater than 1, so it sinks in water. For this reason $D_2O$ is sometimes called *heavy water.*

Isotopes have been extensively applied in various fields, from archeology to medicine to art. Radioactive isotopes (those that emit radiation) can be used to date objects of archeological or geological significance. Radioactive isotopes can also be used to generate electricity, such as in lightweight portable power packs used on space vehicles or in heart pacemakers. Isotopes find extensive use in medical studies. For example, thyroid problems can be diagnosed by ingestion of radioactive iodine-131. Isotopes can be used to treat cancer as well as to detect it. A higher dose of iodine-131 destroys thyroid tumors. Isotopes don't always have to be radioactive to be useful. The techniques known as nuclear magnetic resonance or magnetic resonance imaging locate hydrogen-1 atoms for the purpose of determining molecular structures, or for imaging body parts. In a technique known as neutron activation analysis, isotopes can be detected in a nondestructive analysis of objects such as old paintings and other valuable works of art to determine whether the distribution of elements is like those in common use by a supposed artist, thereby authenticating the painting.

**FIGURE 2.13** Which ice cube contains deuterium (hydrogen-2)? How can you tell?
©Tom Pantages

## 2.3 Ions

In the introduction, Megan was interested in the essential minerals that our bodies need. She noted that the minerals we consume in our food are mostly metal elements in their *ion* (electrically charged) form. She was especially interested in sodium because her doctor told her to reduce it in her diet. Dietary sodium, found in large quantities in snack foods and sports drinks, is also in an ion form. How are ions different from neutral atoms?

In a neutral (uncharged) atom, the number of electrons equals the number of protons, so the overall charge on the atom is zero. This is because the negative charge on each electron is equal in magnitude and opposite in sign to the positive charge on each proton. What happens when the protons and electrons are *not* equal? Then the overall charge is not zero. The overall charge is positive if the atom contains fewer electrons than protons. It is negative if there are more electrons than protons. When an atom contains more or less electrons than protons, it has a charge and is called an **ion.**

Many elements exist in nature as ions. The elements sodium and calcium, for example, are generally *not* found in nature as neutral atoms. Instead, they exist as ions—either combined with other ions in compounds or dissolved in water. Seawater, for example, contains dissolved sodium and calcium ions along with chloride ions. Supplies of elemental sodium and calcium that contain neutral atoms are made from the naturally occurring ionic forms by chemical reactions. Some elements, such as gold, do exist naturally as neutral atoms, but can be made into ions by chemical reactions.

Ions can be classified as *cations* or *anions.* A **cation** is a positively charged ion that contains fewer electrons than the number of protons in the nucleus. The overall charge is represented as a superscript to the right of the element symbol. An example is a calcium ion, $Ca^{2+}$, which contains 20 protons and only 18 electrons. An **anion** is a negatively charged ion that contains more electrons than the number of protons in the nucleus. An example is $Cl^-$, which contains 17 protons and 18 electrons.

12 protons
12 neutrons

12 electrons

Mg atom

12 protons
12 neutrons

10 electrons

$Mg^{2+}$ cation

**A**

7 protons
7 neutrons

7 electrons

N atom

7 protons
7 neutrons

10 electrons

$N^{3-}$ anion

**B**

**FIGURE 2.14** In an ion, the numbers of protons and electrons are not equal. (A) The cation $Mg^{2+}$ forms when a magnesium atom loses two electrons. (B) The anion $N^{3-}$ forms when a nitrogen atom gains three electrons.

Ions do *not* form by gaining or losing protons or neutrons. Changing the number of protons would change the identity of the element! This does not occur under conditions other than nuclear reactions.

Although many ions exist naturally, they can also be formed from their neutral atoms. When a magnesium atom loses two electrons, it forms an $Mg^{2+}$ cation. When a nitrogen atom gains three electrons, it forms an $N^{3-}$ anion. Figure 2.14 shows the formation of cations and anions from neutral atoms. The charge on the ion is the net charge that results from unequal numbers of protons and electrons. As shown in Example 2.5, an ion symbol lets us determine the number of protons and electrons in the ion.

---

**EXAMPLE 2.5**    **Ions**

How many protons and electrons compose the following ions? Identify each as a cation or anion.

(a)  $Na^+$          (b)  $O^{2-}$          (c)  $Cr^{3+}$

**Solution:**

(a)  The number of protons is determined from the atomic number. The atomic number of sodium is 11, so there are 11 protons in $Na^+$. It has a positive charge because it does not have enough electrons to balance the positive charges from the protons. A 1+ charge shows that there is one less electron than the number of protons, so its total number of electrons must be 10. The $Na^+$ ion is a cation because it is positively charged.

(b) The atomic number for oxygen is 8, so there are 8 protons. A negative charge results when there are more electrons than protons. Since the charge is 2−, there are two more electrons than protons, so there are 10 electrons. The $O^{2-}$ ion is an anion because it is negatively charged.

(c) The atomic number of chromium is 24, so there are 24 protons in $Cr^{3+}$. A 3+ charge indicates that there are three fewer electrons than the number of protons, so there are 21 electrons. The $Cr^{3+}$ ion is a cation because it is positively charged.

**Consider This 2.5**

Which of the three ions has the same number of electrons as a noble gas atom?

**Practice Problem 2.5**

How many protons and electrons compose the following ions? Identify each as a cation or anion.

(a) $F^-$          (b) $Mg^{2+}$          (c) $N^{3-}$

**Further Practice:** Questions 2.49 and 2.50 at the end of the chapter

If we know the number of protons and electrons in an ion, we can write an ion symbol for it. If we also know the number of neutrons, we can write a combination isotope-ion symbol by writing an isotope symbol and adding the overall charge as the right-hand superscript. This is shown in Example 2.6.

**EXAMPLE 2.6    Writing Isotope Symbols for Ions**

Write the isotope symbols for ions that have the following numbers of protons, neutrons, and electrons.

(a) 35 protons, 44 neutrons, and 36 electrons
(b) 13 protons, 14 neutrons, and 10 electrons
(c) 47 protons, 62 neutrons, and 46 electrons

**Solution:**

(a) The atomic number (the number of protons) is 35, indicating the element symbol is Br. The mass number is the sum of the protons and neutrons, which is 79. The charge is 1− because there is one more electron than protons. The symbol is $^{79}_{35}Br^-$.

(b) The atomic number is 13, so the element symbol is Al. The mass number is 27 (13 + 14). The charge is 3+ because there are three fewer electrons than protons. The symbol is $^{27}_{13}Al^{3+}$.

(c) The atomic number is 47, so the element is silver (Ag). The mass number is 109 (47 + 62). The charge is 1+ because the number of electrons is one less than the number of protons. The symbol is $^{109}_{47}Ag^+$.

**Consider This 2.6**

If you omitted the left-hand subscripts on your isotope symbol representations, would they still be correct?

**Practice Problem 2.6**

Write the isotope symbols for ions that have the following numbers of protons, neutrons, and electrons.

(a)  16 protons, 18 neutrons, and 18 electrons
(b)  11 protons, 12 neutrons, and 10 electrons
(c)  20 protons, 20 neutrons, and 18 electrons

**Further Practice:**  Questions 2.51 and 2.52 at the end of the chapter

## 2.4 Atomic Mass

The term *relative atomic mass* is used by the International Union of Pure and Applied Chemistry (IUPAC), the worldwide governing body that determines official names in chemistry. Relative atomic mass is used to describe the weighted average of the atomic masses of the isotopes that compose an element as it is found in nature. This term is synonymous with the older term *atomic weight,* which is still commonly used.

How can we describe the mass of an atom of an element? Although single atoms cannot be weighed on a balance, modern techniques such as *mass spectrometry* (Figure 2.15) can be used to determine individual atomic masses accurately. For example, the mass of a single hydrogen-1 atom is $1.67380 \times 10^{-24}$ g. The mass of a carbon-12 atom is $1.99272 \times 10^{-23}$ g, about 12 times the mass of a hydrogen-1 atom. Numbers as small as these are difficult to remember and use. It is more convenient to think of a carbon-12 atom as being about 12 times the mass of a hydrogen-1 atom. For this reason, scientists have devised a method for expressing masses of atoms in a more convenient way. They use the atomic mass unit (amu). This mass scale uses carbon-12 ($^{12}C$), the most abundant isotope of carbon, as the standard to which all other atoms are compared. Carbon-12 is assigned an atomic mass of exactly 12 atomic mass units, or 12 amu. One **atomic mass unit (amu)** is equal to one-twelfth the mass of a carbon-12 atom:

$$1 \text{ amu} = \tfrac{1}{12} \times \text{mass of 1 C-12 atom} = 1.6606 \times 10^{-24}\text{g}$$

Atomic mass units make it easy to compare masses of atoms. For example, if a carbon atom has a mass 12 times that of a hydrogen atom, then the mass of the hydrogen atom is about 1 amu. A hydrogen-2 atom is one-sixth the mass of the carbon-12 atom, so it has a mass of approximately 2 amu. A hydrogen-3 atom is one-fourth the mass of a carbon atom, so its mass is approximately 3 amu.

Note that there are three different, naturally occurring isotopes of hydrogen, each with a different mass. How do we describe the mass of the atoms in a sample of an element if it is composed of isotopes of different masses? From mass spectrometry (Figure 2.15), we can find the mass of each isotope and measure how much of each

**FIGURE 2.15**  A mass spectrometer measures the masses of individual atoms. Electrons are removed from atoms (or molecules) to produce ions that are accelerated through a magnetic field. The degree of bending in the path of the ions is related to the mass and the charge of the ions. Mass spectrometry is also used to determine the relative amounts of different isotopes in a sample of an element.

is present in a sample. This allows us to take a *weighted average.* The average of the mass of the individual isotopes, taking into account the naturally occurring relative abundance of each, is the **relative atomic mass** of the element. The relative atomic mass of hydrogen, to four significant figures, is 1.008 amu. The relative atomic mass of hydrogen is closer to that of hydrogen-1 than the other isotopes because the hydrogen-1 isotope is present in the largest abundance (99.99%) in Earth's crust and atmosphere. The relative atomic mass of carbon is 12.01 amu because carbon-12 is most abundant (98.93%), with smaller amounts of carbon-13 and carbon-14.

To find the relative amounts of each isotope that compose an element using a mass spectrometer, samples of that element from many parts of the world are analyzed. For example, suppose a sample of neon is injected into a mass spectrometer (Figure 2.15). The detector indicates that there are three isotopes of neon: neon-20, neon-21, and neon-22. The detector also shows that the isotopes are present in different amounts. The percent abundance of each isotope is determined and an average is taken. The percent abundances from many different locations are similar.

On the periodic table on the inside front cover of this book, the relative atomic mass of each element is listed below its element symbol. Elements such as hydrogen and carbon have relative atomic mass values very close to the mass of a particular isotope. Others, like silver (Ag), have relative atomic mass values that are averages based on significant contributions from more than one isotope. For example, natural silver has a relative atomic mass of 107.9 amu (Figure 2.16). At first it may seem that natural silver is composed mostly of the isotope $^{108}$Ag. However, silver has been found to consist of two naturally occurring isotopes with different percent abundances: 51.82% $^{107}$Ag and 48.18% $^{109}$Ag. The exact mass of $^{107}$Ag is measured with a mass spectrometer to be 106.9051 amu, and the exact mass of $^{109}$Ag is 108.9048 amu. To obtain the relative atomic mass of silver, we calculate the weighted average using the masses of Ag-107 and Ag-109. To do this, we multiply the mass of each isotope by its relative abundance, expressed in decimal form (percent divided by 100). This gives the mass contribution from each isotope. Summing the mass contributions gives the weighted average, which is the relative atomic mass listed on the periodic table:

$$\text{Isotope mass} \times \text{abundance} = \text{mass contribution from isotope}$$
$$^{107}\text{Ag } 106.9051 \text{ amu} \times 0.5182 = 55.40 \text{ amu}$$
$$^{109}\text{Ag } 108.9048 \text{ amu} \times 0.4818 = 52.47 \text{ amu}$$

$$\overline{\hphantom{xxxxxxxxx}107.87 \text{ amu (relative atomic mass of Ag)}}$$

The relative atomic mass of silver listed on the periodic table is 107.9 amu, rounded to four significant figures.

If only two major isotopes comprise an element, we can usually determine which isotope is most abundant by referring to the relative atomic mass of the element on the periodic table. This is shown in Example 2.7.

An example of a weighted average is your grade point average (GPA). Commonly an A grade is worth 4 points, a B grade is worth 3 points, etc. The GPA is calculated by multiplying 4 by the fraction of course credits that were earned with an A, multiplying 3 by the fraction of course credits that were earned with a B, etc. Then these numbers are added to obtain the GPA, which is the average of the grade points weighted by the fraction of credits that received each grade.

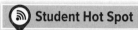

**Student Hot Spot**

Student data indicate you may struggle with relative atomic mass. Access the SmartBook to view additional Learning Resources on this topic.

**MATH TOOLBOX 1.2**

| 47 |
| **Ag** |
| 107.9 |

**FIGURE 2.16** The relative atomic mass of each element is indicated on the periodic table in this book just below the element symbol.

---

**EXAMPLE 2.7** | **Relative Atomic Mass**

Naturally occurring chlorine consists of $^{35}$Cl (34.9689 amu) and $^{37}$Cl (36.9659 amu). Which isotope is most abundant?

**Solution:**

The relative atomic mass of chlorine is 35.45 amu, as shown on the periodic table. Since the relative atomic mass is closer to the mass of $^{35}$Cl (34.9689 amu) than the mass of $^{37}$Cl (36.9659 amu), the $^{35}$Cl isotope must be most abundant. (The actual percentages are 75.77% $^{35}$Cl and 24.23% $^{37}$Cl.)

**Consider This 2.7**

If, on a different planet, the relative atomic mass of chlorine were 36.31 amu with the same two isotopes, which isotope would be more abundant?

**Practice Problem 2.7**

The copper mined from Earth's crust consists of $^{63}$Cu (62.93 amu) and $^{65}$Cu (64.93 amu). Which isotope is most abundant?

**Further Practice:** Questions 2.75 and 2.76 at the end of the chapter

Example 2.8 gives you practice calculating the relative atomic mass for an element given the masses and percent abundances of its isotopes.

| EXAMPLE 2.8 | Calculating Relative Atomic Mass |

The element magnesium is composed of three isotopes: magnesium-24, magnesium-25, and magnesium-26. The mass and percent abundance of each are listed in the following table. Calculate the relative atomic mass of magnesium, and compare your calculated value to the value in the periodic table.

| Isotope | Mass (amu) | Natural Abundance (%) |
|---------|-----------|----------------------|
| $^{24}$Mg | 23.985 | 78.99 |
| $^{25}$Mg | 24.986 | 10.00 |
| $^{26}$Mg | 25.983 | 11.01 |

**Solution:**

MATH TOOLBOX 1.2

To determine the mass contribution of each isotope, multiply its mass by its percent abundance in decimal form. To find the relative atomic mass, total the mass contributions of the isotopes:

$^{24}$Mg 23.985 amu × 0.7899 = 18.95 amu
$^{25}$Mg 24.986 amu × 0.1000 =  2.499 amu
$^{26}$Mg 25.983 amu × 0.1101 =  2.861 amu

24.31 amu (calculated relative atomic mass)

The calculated relative atomic mass, 24.31 amu, is the same as that on the periodic table to four significant figures.

**Consider This 2.8**

What could account for a relative atomic mass of magnesium that was less than 24.31 amu on a different planet?

**Practice Problem 2.8**

The element lithium is composed of two isotopes, lithium-6 and lithium-7. The mass and percent abundance of each is listed in the following table. Calculate the relative atomic mass of lithium, and then compare your answer to the value listed on the periodic table.

| Isotope | Mass (amu) | Natural Abundance (%) |
|---------|-----------|----------------------|
| $^{6}$Li | 6.01512 | 7.793 |
| $^{7}$Li | 7.01600 | 92.21 |

**Further Practice:** Questions 2.77 and 2.78 at the end of the chapter

The mass of an isotope, as measured by mass spectrometry, is similar in value to its mass number. A magnesium-24 atom, for example, has a mass of 23.985 amu. Its mass number, 24, is a whole number that is close to its atomic mass. Remember that the mass number is not an actual mass; it is a count of the number of particles in the nucleus. The mass of an atom must be measured. Because the mass number and the atomic mass of an isotope are always similar in value, for convenience we often approximate the mass of an isotope as a whole number that is also its mass number. The approximate mass of a magnesium-24 atom, therefore, is about 24 amu. Its more exact mass (23.985 amu) is measured experimentally and is provided in published data tables.

We use relative atomic mass when we refer to an "average atom" of an element, or a group of atoms. For example, the mass of 100 magnesium atoms can be determined by multiplying 100 atoms by the relative atomic mass of magnesium, 24.31 amu per atom, to get a total mass of 2431 amu. What is the mass of 1000 lithium atoms?

## 2.5  The Periodic Table

As Anna continued her historical research, she read about the Russian scientist Dmitri Mendeleev. He developed and published the basic arrangement of the periodic table between 1869 and 1871 (Figure 2.17), decades before the discovery of protons. Mendeleev did not arrange the elements in order of atomic number. He arranged the 63 known elements in order of increasing relative atomic mass, and he grouped elements with similar properties into columns and rows so that the properties of the elements varied in a regular pattern (periodically). Once he arranged all of the elements that were known, he predicted the existence and properties of three elements that were unknown at the time: gallium (Ga), scandium (Sc), and germanium (Ge). He also placed two pairs of elements (Co/Ni and Te/I) in an order that did not match their relative atomic masses. Using what he thought was the correct order based on the properties of these elements, Mendeleev thought that their calculated relative atomic mass values must be in error. We now know that the periodicity of the properties of the elements occurs in the order of atomic number (number of protons), not relative atomic mass.

### Classification of Elements

The modern version of the **periodic table,** in which all the known elements are arranged in columns and rows to emphasize periodic properties, is shown in Figure 2.18 and on the Periodic Table. In the periodic table, the elements are in the order of their atomic number (number of protons) from smallest to largest. There are many ways to use the classification system the periodic table provides. One way is to look at elements in the same vertical column. Because they have similar properties, we call them a **family** or a **group.** Each group has a Roman numeral (I through VIII) and a letter (A or B) associated with it. A newer system (shown in parentheses after the Roman numeral/ letter combination) uses only Arabic numbers (1 through 18) to designate each group.

|  |  |  | Ti = 50 | Zr = 90 | ? = 180 |
|---|---|---|---|---|---|
|  |  |  | V = 51 | Nb = 94 | Ta = 182 |
|  |  |  | Cr = 52 | Mo = 96 | W = 186 |
|  |  |  | Mn = 55 | Rh = 104,4 | Pt = 197,4 |
|  |  |  | Fe = 56 | Ru = 104,4 | Ir = 198 |
|  |  | Ni = Co = 59 |  | Os = 199 |
| H = 1 |  |  | Cu = 63,4 | Ag = 108 | Hg = 200 |
|  | Be = 9,4 | Mg = 24 | Zn = 65,2 | Cd = 112 |  |
|  | B = 11 | Al = 27,4 | ? = 68 | Ur = 116 | Au = 197? |
|  | C = 12 | Si = 28 | ? = 70 | Sn = 118 |  |
|  | N = 14 | P = 31 | As = 75 | Sb = 122 | Bi = 210? |
|  | O = 16 | S = 32 | Se = 79,4 | Te = 128? |  |
|  | F = 19 | Cl = 35,5 | Br = 80 | J = 127 |  |
| Li = 7  Na = 23 |  | K = 39 | Rb = 85,4 | Cs = 133 | Ti = 204 |
|  | Ca = 40 | Sr = 87,6 | Ba = 137 | Pb = 207 |
|  | ? = 45 | Ce - 92 |  |  |
|  | ?Er = 56 | La = 94 |  |  |
|  | ?Yt = 60 | Di = 95 |  |  |
|  | ?In = 75,6 | Th = 118? |  |  |

**FIGURE 2.17** Mendeleev based his original periodic table on the elements that were known at the time. He used it to predict the existence of other, then unknown, elements. The numbers accompanying the element symbols are relative atomic mass values that were known at the time.

**FIGURE 2.18** The periodic table helps us to classify elements in a variety of ways.

Metalloids such as silicon and germanium are commonly used in semiconductor devices.

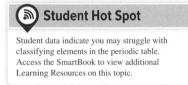
**Student Hot Spot**

Student data indicate you may struggle with classifying elements in the periodic table. Access the SmartBook to view additional Learning Resources on this topic.

A horizontal row of elements in the periodic table is a **period.** Elements in the same period have properties that tend to vary in a regular fashion. Periods are labeled with numbers (1 through 7). Figure 2.19 shows a picture of elements in period 3 and in group VA (15). Note the gradual variation in the physical appearance of these elements.

Look for the stair-step line that begins at boron and moves down the periodic table to astatine. This line separates the *metals* (to its left) from the *nonmetals* (to its right). (The properties of metals and nonmetals were discussed in Chapter 1.) A **metalloid,** or *semimetal,* is an element that has physical properties resembling a metal but chemical reactivity more like that of a nonmetal. The metalloids, identified by green boxes, lie along the stair-step line. Some metals, nonmetals, and metalloids are shown in Figure 2.20.

Any element in one of the eight groups labeled with the letter A is a **main-group element,** or a *representative element.* An element in any of the 10 groups labeled with the letter B is a **transition metal.** The 14 members of the *lanthanide* and *actinide* series of elements are usually placed on separate lines at the bottom of the table to conserve space, but they belong immediately after elements La (Z = 57) and Ac (Z = 89), respectively. Any lanthanide or actinide element is an **inner-transition metal.**

Several of the groups have descriptive names that are frequently used instead of their group numbers. Any member of group IA (1), except hydrogen, is an **alkali**

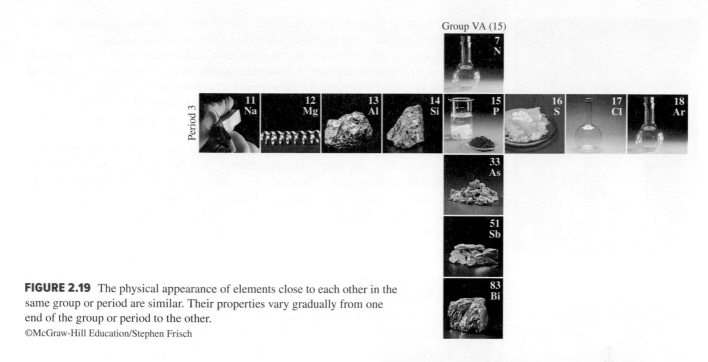

**FIGURE 2.19** The physical appearance of elements close to each other in the same group or period are similar. Their properties vary gradually from one end of the group or period to the other.
©McGraw-Hill Education/Stephen Frisch

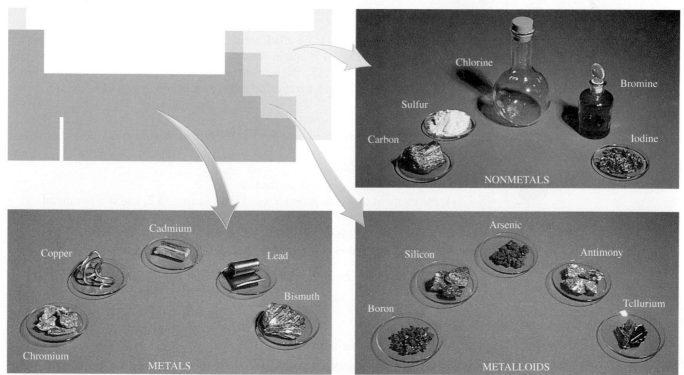

**FIGURE 2.20** The metals, on the left side of the periodic table, have a shiny appearance. They are malleable and ductile (can be pulled into thin wires). They conduct heat and electricity. The nonmetals, on the right side of the periodic table, are gases, liquids, or powder or crystalline solids. They are insulators (nonconductors of heat and electricity). The metalloids have properties between the metals and the nonmetals.
©McGraw-Hill Education/Stephen Frisch

**metal.** Any element in group IIA (2) is an **alkaline earth metal.** An element in group VIIA (17) is a **halogen.** On the far right are the group VIIIA (18) elements. An element in this group is called a **noble gas.** These names remind us that elements in a group display similar properties. The alkali metals, for example, are said to be reactive. This means they react readily with many other elements and compounds,

**ANIMATION:** Properties of Alkali and Alkaline Earth Materials

**FIGURE 2.21** Potassium, an alkali metal, reacts violently with water, producing flammable hydrogen gas. ©McGraw-Hill Education/Stephen Frisch

| | | 1<br>H |
|---|---|---|
| 7<br>N | 8<br>O | 9<br>F |
| | | 17<br>Cl |
| | | 35<br>Br |
| | | 53<br>I |

**FIGURE 2.22** Seven elements occur as diatomic molecules.

The alkali metals (group IA) got their name because they react with water to form an alkaline (basic) solution. The word *alkali* comes from Arabic and means "ashes," the material that was originally used to make alkaline solutions. The name for the alkaline earth metals (group IIA) is derived from their presence in metal oxides, which were once called *earths*.

Astatine (At), a halogen, is found in very small amounts in Earth's crust. It is radioactive and very short lived. Since astatine has properties most similar to iodine, scientists believe astatine also exists as diatomic molecules.

including water (Figure 2.21). The alkaline earth metals are less reactive than the alkali metals, but are more reactive than most of the transition metals.

The halogens are similar in several ways. One is that they occur as elements in the form of diatomic molecules. A **diatomic molecule** is a molecule consisting of two atoms. When we represent the halogens as elements, we use a subscript to show that they are diatomic molecules: $F_2$, $Cl_2$, $Br_2$, $I_2$. Other elements that exist naturally as diatomic molecules are hydrogen ($H_2$), oxygen ($O_2$), and nitrogen ($N_2$). Elements that occur as diatomic molecules are shown in Figure 2.22.

The noble gases are unique in that they exist naturally in their elemental form as single atoms. They are said to be inert; that is, they do not react chemically with other elements or compounds. The noble gases were not known until the late 1800s, although many of them rank as highly abundant elements in the universe and in Earth's atmosphere. As you will learn later, the unique stability of the noble gases helps us understand the chemical reactivity of other elements.

---

**EXAMPLE 2.9**  **Classification of Elements**

Classify each of the following elements by group number, group name (if applicable), and period, and as a metal, nonmetal, or metalloid.

(a) sodium    (b) silicon    (c) bromine    (d) copper

**Solution:**
(a) Na is in group IA (1), the alkali metal group, and in period 3, and is a metal.
(b) Si is in group IVA (14) and in period 3, and is a metalloid.
(c) Br is in group VIIA (17), the halogen group, and in period 4, and is a nonmetal.
(d) Cu is in group IB (11), a transition metal group, and in period 4, and is a metal.

**Consider This 2.9**

What element is the only nonmetal located on the left side of the period table?

**Practice Problem 2.9**

Identify the element that is described.
(a) the element that is in group VA (15) and in period 2, and is a nonmetal
(b) the element that is a noble gas and in period 3, and is a nonmetal
(c) the element that is an alkaline earth metal and in period 4
(d) the element that is in group IB (11), is a transition metal, and is in period 5

**Further Practice:** Questions 2.85 and 2.86 at the end of the chapter

---

## Ions and the Periodic Table

In Section 2.3, we saw that ions have charges because they have a different number of electrons than there are protons in the nucleus. Megan noticed that some of the essential nutritional elements are ions that have the same charge. For example, sodium and potassium both exist as ions with a 1+ charge. Similarly, calcium and magnesium exist as 2+ ions. She found out that the pattern is no coincidence. The position of an element in the periodic table helps us to predict the charge on its ion. To understand this trend, let's start by looking at the elements that do not form ions—the noble gases, group VIIIA (18).

The noble gases are the most stable—the least reactive—of all the elements. Their stability is associated with the number of electrons they contain. To achieve a similar stability, *many atoms of the main-group elements gain or lose electrons to form ions with the same electron count as the nearest noble gas.* The nonmetals

**FIGURE 2.23** Many of the main-group elements form ions that have charges that can be predicted by their positions in the periodic table. The ions shown here are sometimes called *common ions* because these elements rarely form ions of other charges.

usually *gain* electrons to form anions that have a noble gas electron count. Most main-group metals *lose* electrons to form cations that have a noble gas electron count. For example, a sodium atom, with 11 electrons, forms an $Na^+$ ion by losing one electron. When it does, it has the same number of electrons as neon, 10. How many electrons does a magnesium atom lose when it forms an ion? When oxygen forms $O^{2-}$ ions, each atom, with eight electrons, gains two electrons for a total of 10, like neon. How many electrons will a nitrogen atom gain to form an ion?

Because elements in the same group have an electron count that differs from that of the nearest noble gas by the same number, elements in the same group often form ions of the same charge. Like sodium, all group IA (1) elements lose one electron to form a cation with a charge of 1+. Group IIA (2) elements lose two electrons to form cations with a charge of 2+. Aluminum atoms of group IIIA (3) lose three electrons to form 3+ ions. Group VIIA (17) elements gain one electron to form anions with a charge of 1−. Group VIA (16) elements gain two electrons to form anions with a charge of 2−. Nitrogen and phosphorus of group VA (15) gain three electrons to form anions with a charge of 3−. The charges for ions that can be predicted from the periodic table are labeled on the abbreviated periodic table in Figure 2.23. Most of the other elements form two or more different ions of different charges. The charges of these ions cannot be predicted from the periodic table.

Although no compounds of noble gases exist naturally, a few stable compounds of krypton and of xenon have been prepared. A few compounds of argon are known, but they are stable only at low temperatures. This picture shows one synthetic compound of a noble gas, $XeF_2$, adhering to the interior of the reaction vessel.

©Gary J. Schrobilgen/McMaster University

---

**EXAMPLE 2.10** | **Predicting Charges on Ions**

Write the symbol for the ion that each of the following elements is predicted to form.

(a) magnesium          (b) bromine          (c) nitrogen

**Solution:**

These ions can be predicted by their positions in the periodic table.

(a) Magnesium is in group IIA (2), so it will lose two electrons to form $Mg^{2+}$, giving it the same number of electrons as neon.

(b) Bromine is in group VIIA (17), so it will gain one electron to form Br⁻, giving it the same number of electrons as krypton.

(c) Nitrogen is in group VA (15), so it will gain three electrons to form $N^{3-}$, giving it the same number of electrons as neon.

**Consider This 2.10**

Based on its position on the periodic table, what charge would you predict for the hydrogen ion?

**Practice Problem 2.10**

Write the symbol for the ion that each of the following elements is predicted to form.

(a) lithium      (b) sulfur      (c) aluminum

**Further Practice:**   Questions 2.105 and 2.106 at the end of the chapter

In our bodies, ions usually exist dissolved in body fluids. Ions can also exist as components of compounds. In Chapter 3 you will learn about these *ionic compounds*. You will learn about their properties and how to name them. You will also learn how to determine charges on ions that cannot be predicted from the periodic table.

## CHAPTER REVIEW

## KEY CONCEPTS

- Our current understanding of the atom began with Dalton's atomic theory two centuries ago. It was not until the early 1900s that there was experimental evidence that showed atoms consist of protons, neutrons, and electrons.
  - Protons are the positively charged particles in the nucleus. The number of protons in an atom is unique to each element and is called the atomic number ($Z$). The atomic number is listed above the element symbol on the periodic table in this book.
  - Neutrons, also in the nucleus, have no charge. Atoms of an element can differ in the number of neutrons ($N$), which results in isotopes. The symbol for an isotope includes the mass number ($A$), which is the sum of the numbers of protons and neutrons. The mass number is written as a left-hand superscript on the element symbol to distinguish the isotopes of an element.
  - Electrons are negatively charged and exist in the relatively large space outside the nucleus. In a neutral atom, the number of electrons is equal to the number of protons.
- Ions differ from atoms in that they have a positive or negative charge.
  - The charge on an ion is written as a right-hand superscript on the element symbol.
  - A positively charged ion, called a cation, contains fewer electrons than protons.
  - A negatively charged ion, called an anion, contains more electrons than protons.
  - When an ion is formed from a neutral atom, electrons are lost or gained, resulting in the net charge on the ion.
  - The properties of ions differ from elements containing neutral atoms.
- The mass of an atom is usually reported in atomic mass units (amu).
  - The mass of an atom, usually known to many decimal places, is not the same as the mass number, although they are similar in value.
  - The relative atomic mass is the weighted average of the masses of all the isotopes of an element, taking into account their relative abundances in a naturally occurring sample. This value is listed under the element symbol on the periodic table in this book.

- The periodic table is organized by elements with similar properties, which allows us to determine or predict certain properties of elements.
  ○ Elements in the same vertical column are in the same group and display similar chemical properties. For example, elements in group IA, the alkali metals, all form ions with a 1+ charge.
  ○ Elements can also be classified as metals, nonmetals, or metalloids, or as main-group elements, transition metals, lanthanides, or actinides.

## KEY TERMS

alkali metal (2.5)

alkaline earth metal (2.5)

anion (2.3)

atomic mass unit (amu) (2.4)

atomic number (2.2)

cation (2.3)

Dalton's atomic theory (2.1)

diatomic molecule (2.5)

electron (2.2)

family (2.5)

group (2.5)

halogen (2.5)

inner-transition metal (2.5)

ion (2.3)

isotope (2.2)

isotope symbol (2.2)

law of conservation of mass (2.1)

law of definite proportions (2.1)

main-group element (2.5)

mass number (2.2)

metalloid (2.5)

neutron (2.2)

neutron number (2.2)

noble gas (2.5)

nucleus (2.2)

period (2.5)

periodic table (2.5)

proton (2.2)

relative atomic mass (2.4)

subatomic particle (2.2)

transition metal (2.5)

## QUESTIONS AND PROBLEMS

The following questions and problems, except those in *Additional Questions* and *Concept Review Questions,* are paired. Questions in a pair focus on the same concept. Answers to selected questions and problems are in Appendix D.

### Matching Definitions with Key Terms

2.1    Match the key terms with the descriptions provided.
  (a) an electrically neutral subatomic particle
  (b) a law stating that the mass of the substances produced in a reaction equals the mass of the substances that reacted
  (c) a positively charged subatomic particle
  (d) a member of one of the A groups of elements in the periodic table
  (e) the average mass of an atom of an element, taking into account the masses and abundances of all the naturally occurring isotopes
  (f) the sum of the protons and neutrons in an atom
  (g) an atom of an element with a specific number of neutrons
  (h) an ion with a positive charge
  (i) any particle found within an atom
  (j) any element in group IA (1)
  (k) a chart of all the known elements ordered by increasing atomic number and arranged in columns and rows to emphasize similar properties

2.2    Match the key terms with the descriptions provided.
  (a) a member of one of the B groups of elements in the periodic table
  (b) a law stating that all samples of a pure compound always contain the same proportions of the component elements
  (c) a negatively charged subatomic particle
  (d) an ion with a negative charge

  (e) a neutral particle consisting of two bound atoms
  (f) any element in group VIIIA (18)
  (g) a set of elements in the same horizontal row in the periodic table
  (h) the number of protons in an atom
  (i) the unit of mass used to describe single atoms or small numbers of atomic particles, which is equal to one-twelfth the mass of a carbon-12 atom
  (j) a set of elements in the same vertical column in the periodic table
  (k) any element in group IIA (2)
  (l) the central core of the atom, which contains the protons and neutrons, and most of the mass of the atom

### Dalton's Atomic Theory

2.3    Which laws did Dalton use to argue that matter consists of atoms?

2.4    What modern technique allows us to "see" the surface of atoms?

2.5    How does Dalton's atomic theory describe atoms of different elements?

2.6    How does Dalton's atomic theory explain the law of conservation of mass?

2.7    How does Dalton's atomic theory explain the law of definite proportions?

2.8    How does Dalton's atomic theory explain the formation of compounds from elements?

2.9     Does the following diagram represent a chemical reaction that obeys the law of conservation of mass? If it does not, how should the diagram be modified to represent the reaction correctly?

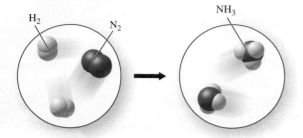

2.10    Does the following diagram represent a chemical reaction that obeys the law of conservation of mass? If it does not, how should the diagram be modified to represent the reaction correctly?

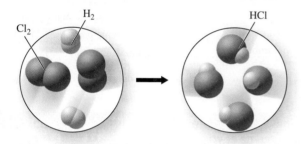

## Structure of the Atom

2.11    Which experiment showed that all atoms contain negatively charged particles with masses far less than the mass of a hydrogen atom?

2.12    What information about the structure of the atom did Rutherford's gold foil experiment provide?

2.13    Which subatomic particles are negatively charged?

2.14    Which subatomic particles are positively charged?

2.15    Helium is used in balloons and blimps because of its lifting power. Draw a nuclear model of a helium atom with a mass number of 4 (helium-4). Show the locations of the nucleus, protons, neutrons, and electrons. Indicate the number of each type of subatomic particle.

2.16    Hydrogen-3 (tritium) is used as a tracer in chemical, biological, and medical research. Draw a nuclear model of a hydrogen atom with a mass number of 3 (hydrogen-3). Show the locations of the nucleus, protons, neutrons, and electrons. Indicate the number of each type of subatomic particle.

2.17    Which subatomic particle has approximately the same mass as a proton?

2.18    Which of the following best describes about how many times greater in mass a proton is than an electron: 200, 2000, or 20,000?

2.19    Explain why the mass of a carbon atom is about twice the mass of the protons in a carbon atom. What accounts for the difference in mass?

2.20    Which subatomic particle was discovered last? Why was it difficult to detect?

2.21    What is the atomic number for each of the following elements?
   (a) hydrogen       (b) oxygen       (c) silver

2.22    How many protons are in each atom of each of the following elements?
   (a) helium         (b) argon        (c) lead

2.23    The number of which subatomic particle determines the identity of an element?

2.24    How do isotopes of an element differ?

2.25    What information do you need to determine the atomic number of an atom?

2.26    What information do you need to determine the mass number of an atom?

2.27    Which of the following are always the *same* for atoms of the same element?
   (a) mass number
   (b) atomic number
   (c) neutron number
   (d) mass of an atom

2.28    Which of the following are *different* for isotopes of an element?
   (a) mass number
   (b) atomic number
   (c) neutron number
   (d) mass of an atom

2.29    Given the composition of the nuclei, what are the atomic number, neutron number, and mass number of each isotope of hydrogen shown?

$^1_1H$          $^2_1H$          $^3_1H$

2.30    Identify the element from the atom shown. What are the atomic number, neutron number, and mass number of this isotope?

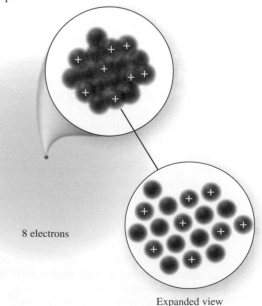

8 electrons

Expanded view

2.31    What are the atomic number, neutron number, and mass number of the following isotopes?

(a) $^{36}_{18}Ar$          (b) $^{38}_{18}Ar$          (c) $^{40}_{18}Ar$

2.32    Identify the element from the atom shown. What are the atomic number, neutron number, and mass number of this isotope?

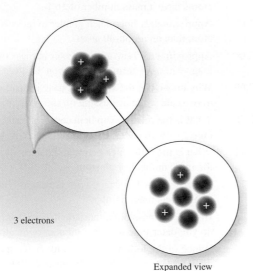

3 electrons

Expanded view

2.33    How many protons, neutrons, and electrons are in each of the following?

(a) an oxygen atom with a mass number of 15

(b) a silver atom with a mass number of 109

(c) a chlorine atom with a mass number of 35

2.34    How many protons, neutrons, and electrons are in each of the following?

(a) a hydrogen atom with a mass number of 1

(b) a magnesium atom with a mass number of 26

(c) a lithium atom with a mass number of 6

2.35    What is the isotope symbol for atoms that contain the following numbers of subatomic particles?

(a) 1 proton, 1 electron, and 2 neutrons

(b) 4 protons, 4 electrons, and 5 neutrons

(c) 15 protons, 15 electrons, and 16 neutrons

2.36    What is the isotope symbol for atoms that contain the following numbers of subatomic particles?

(a) 2 protons and 1 neutron

(b) 47 protons and 62 neutrons

(c) 82 protons and 125 neutrons

2.37    How many protons and neutrons are in an atom represented by the following?

(a) $^{56}_{26}Fe$

(b) $^{39}K$

(c) copper-65

2.38    How many protons and neutrons are in an atom represented by the following?

(a) $^{11}_{5}B$

(b) $^{68}Zn$

(c) iodine-127

2.39    A study of blood flow in heart muscle uses nitrogen-13 incorporated into a chemical agent. How many protons and neutrons are found in the N-13 nucleus?

2.40    Phosphorus-32 can be used to treat eye tumors. How many protons and neutrons are found in the P-32 nucleus?

2.41    Complete the following table for the designated atoms.

| Isotope Symbol | Number of Protons | Number of Neutrons | Number of Electrons |
|---|---|---|---|
| $^{23}_{11}Na$ | | | |
| | 25 | 31 | |
| | | 10 | 8 |
| $^{19}_{9}F$ | | | |

2.42    Complete the following table for the designated atoms.

| Isotope Symbol | Number of Protons | Number of Neutrons | Number of Electrons |
|---|---|---|---|
| $^{14}_{6}C$ | | | |
| | 12 | 11 | |
| | | 14 | 16 |
| $^{30}_{14}Si$ | | | |

## Ions

2.43    How do an atom and an ion of the same element differ?

2.44    What changes when an ion forms from an atom, or when an atom forms from an ion?

2.45    What forms when a neutral atom undergoes each of the following changes?

(a) It gains one electron.

(b) It loses two electrons.

2.46    What forms when an ion with a 1+ charge undergoes each of the following changes?

(a) It gains one electron.

(b) It loses two electrons.

2.47    Write the symbol for the ion that is formed after each of the following changes. Identify each as a cation or an anion.

(a) A zinc atom loses two electrons.

(b) A phosphorus atom gains three electrons.

2.48    Write the symbol for the ion that is formed after each of the following changes. Identify each ion as a cation or an anion.

(a) A selenium atom gains two electrons.

(b) A mercury atom loses two electrons.

2.49    How many protons and electrons are found in each of the following?

(a) $Zn^{2+}$

(b) $F^-$

(c) $H^+$

2.50    How many protons and electrons are found in each of the following?

(a) $P^{3-}$

(b) $Al^{3+}$

(c) $O^{2-}$

2.51     Complete the following table for the designated ions.

| Isotope Symbol | Number of Protons | Number of Neutrons | Number of Electrons |
|---|---|---|---|
| $^{37}_{17}Cl^-$ | | | |
| | 12 | 13 | 10 |
| | 7 | 6 | 10 |
| $^{40}Ca^{2+}$ | | | |

2.52     Complete the following table for the designated ions.

| Isotope Symbol | Number of Protons | Number of Neutrons | Number of Electrons |
|---|---|---|---|
| $^{81}_{35}Br^-$ | | | |
| | 38 | 50 | 36 |
| | 1 | 1 | 2 |
| $^1H^+$ | | | |

2.53     What element has 18 electrons when it forms a cation with a 1+ charge?

2.54     What element has 18 electrons when it forms an anion with a 2− charge?

2.55     What element has 27 electrons when it forms a cation with a 2+ charge?

2.56     What element has 46 electrons when it forms a cation with a 1+ charge?

2.57     How do $^7Li^+$ and $^6Li$ each differ from a neutral lithium-7 atom? Which differs from lithium-7 by the greatest mass?

2.58     How do $^{79}_{35}Br^-$ and $^{81}_{35}Br$ differ in their numbers of subatomic particles? Which should have the greater mass?

2.59     Potassium citrate is a compound found in some sparkling beverages. The potassium in this compound is a cation with a 1+ charge. How many protons are found in the nucleus of a potassium ion? How many electrons surround this nucleus?

2.60     Calcium citrate is a compound found in some calcium supplement medications. The calcium in this compound consists of ions containing 18 electrons. What is the charge of the calcium ions? How many protons are found in the nucleus of a calcium ion?

## Atomic Mass

2.61     What is the basis for the atomic mass unit (amu) scale?

2.62     What is the mass of a carbon-12 atom on the amu scale? Is this value exact or approximate?

2.63     What is the *approximate* mass, in atomic mass units, of the following isotopes?
         (a) $^2_1H$
         (b) $^{238}_{92}U$

2.64     Estimate the combined mass (in atomic mass units) of 10 cobalt-59 atoms.

2.65     Approximately how much greater in mass is a $D_2$ molecule than an $H_2$ molecule?

2.66     Approximately how much greater in mass is a $D_2O$ molecule than an $H_2O$ molecule? (Assume the oxygen atoms have a mass number of 16.)

2.67     Approximately how much greater in mass is a krypton-80 atom than an argon-40 atom?

2.68     Approximately how much greater in mass is a magnesium-24 atom than a carbon-12 atom?

2.69     Why do we use the amu mass scale instead of the gram mass scale when discussing masses of atoms?

2.70     What is the relationship between grams and atomic mass units? How many atomic mass units are in 1 g?

2.71     What is the difference between the mass of an atom and the mass number of an atom?

2.72     Which isotope has exactly the same mass (in atomic mass units) as its mass number?

2.73     How is the mass of an individual atom determined? How do you determine the mass number of an atom?

2.74     Which of the following is most likely the mass of a nickel-62 atom: 62.0000000 amu, 62.495654 amu, 61.5871338 amu, or 61.9283461 amu. Why?

2.75     Naturally occurring calcium is composed of two isotopes: calcium-40 and calcium-44. Which isotope is most abundant?

2.76     Naturally occurring silicon is composed mostly of one isotope. Write its symbol.

2.77     An unknown element (X) discovered on a planet in another galaxy was found to exist as two isotopes. Their atomic masses and percent abundances are listed in the following table. What is the relative atomic mass of the element?

| Isotope | Mass (amu) | Natural Abundance (%) |
|---|---|---|
| $^{22}X$ | 21.995 | 75.00 |
| $^{20}X$ | 19.996 | 25.00 |

2.78     Suppose that the isotope abundance of magnesium on another planet has been determined to be different from that on Earth, as shown in the following table. Calculate the relative atomic mass of magnesium on this planet.

| Isotope | Mass (amu) | Natural Abundance (%) |
|---|---|---|
| $^{24}Mg$ | 23.985 | 20.00 |
| $^{25}Mg$ | 24.985 | 20.00 |
| $^{26}Mg$ | 25.983 | 60.00 |

2.79    The mass spectrum of nickel is shown.

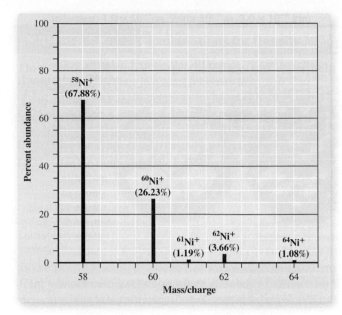

(a) What is the most abundant isotope of nickel?
(b) What is the least abundant isotope of nickel?
(c) Will the relative atomic mass of nickel be closer to a value of 58 or of 60?
(d) List the number of protons, electrons, and neutrons in each of the ions shown in the mass spectrum.

2.80    The mass spectrum of magnesium is shown.

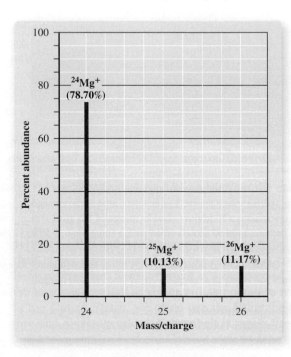

(a) What is the most abundant isotope of magnesium?
(b) What is the least abundant isotope of magnesium?

(c) Will the relative atomic mass of magnesium be closer to a value of 24 or 25 or 26?
(d) List the number of protons, electrons, and neutrons in each of the ions shown in the mass spectrum.

2.81    What is the mass in amu of 1000 boron atoms?

2.82    What is the mass in amu of 1000 mercury atoms?

2.83    Which contains more atoms, 2500 amu of boron atoms or 2500 amu of mercury atoms?

2.84    A sample of pure silver and a sample of pure gold have the same mass. Which contains the greatest number of atoms?

## The Periodic Table

2.85    Identify which of the elements Br, K, Mg, Al, Mn, and Ar can be classified in each of the following ways.
(a) alkali metal
(b) halogen
(c) transition metal
(d) alkaline earth metal
(e) noble gas
(f) main-group element

2.86    Identify which of the elements He, Zn, Pb, I, Ca, and Na can be classified in each of the following ways.
(a) alkali metal
(b) halogen
(c) transition metal
(d) alkaline earth metal
(e) noble gas
(f) main-group element

2.87    Name the element that is a halogen in period 3.

2.88    Name the element that is an alkaline earth metal in period 5.

2.89    What element is in group IVB (4) and in period 4?

2.90    What element is in group IVA (14) and in period 2?

2.91    Identify each of the following elements as a metal, nonmetal, or metalloid.
(a) calcium
(b) carbon
(c) potassium
(d) silicon

2.92    Identify each of the following elements as a metal, nonmetal, or metalloid.
(a) phosphorus
(b) chromium
(c) arsenic
(d) sodium

2.93    Identify the following elements as a main-group element, transition metal, lanthanide, or actinide.
(a) oxygen
(b) magnesium
(c) tin
(d) uranium
(e) chromium

2.94    Identify the following elements as a main-group element, transition metal, lanthanide, or actinide.
(a) sulfur
(b) iron
(c) plutonium
(d) calcium
(e) xenon

2.95 In which group of the periodic table do all the elements occur as diatomic molecules?

2.96 Seven elements occur as diatomic molecules. Write molecular formulas for each.

2.97 Which of the following elements does *not* occur as a diatomic molecule: nitrogen, fluorine, or neon?

2.98 Which of the following elements does *not* occur as a diatomic molecule: sulfur, hydrogen, or oxygen?

2.99 In which group of the periodic table do all the elements exist naturally as gases of uncombined atoms?

2.100 The noble gases are sometimes called "inert gases." Why? Why do they not form ions?

2.101 The ions of many of the main-group elements have the same number of _____ as the noble gas nearest to them in the periodic table.

2.102 In which group of the periodic table do all the elements not form ions?

2.103 Identify the groups of the periodic table in which all the elements form ions of the charge indicated.
(a) 1+ cations
(b) 2+ cations
(c) 1– anions
(d) 2– anions

2.104 For each group listed, identify the charge on the ions that they form.
(a) group VIA (16)
(b) halogens
(c) alkali metals
(d) group IIA (2)

2.105 For each element listed, write the symbol for the ion it forms.
(a) sodium
(b) oxygen
(c) sulfur
(d) chlorine
(e) bromine

2.106 For each element listed, write the symbol for the ion it forms.
(a) nitrogen
(b) phosphorus
(c) magnesium
(d) potassium
(e) aluminum

2.107 Sodium reacts vigorously with water to form hydrogen gas and a compound containing sodium ions. Which other elements are expected to react with water in a similar way?

2.108 Chlorine gas reacts with potassium to form a compound with properties similar to table salt. Which other elements are expected to react with potassium in a similar way?

## Additional Questions

2.109 When iron rusts, a compound forms with the formula $Fe_2O_3$. When iron nails rust, should their mass increase, decrease, or remain the same? Explain your answer.

2.110 Is the law of conservation of mass obeyed in the reaction shown? Explain.

2.111 A 100-g sample of zinc sulfide contains 67.1 g zinc and 32.9 g sulfur. If a 1.34-g sample of zinc is heated with excess sulfur, 2.00 g of zinc sulfide form. Show how these data are in agreement with the law of definite proportions.

2.112 A 100-g sample of HgO contains 92.6 g mercury and 7.4 g oxygen. What mass of mercury atoms and what mass of oxygen atoms are contained in any 100-g sample of HgO?

2.113 What properties of electrons contributed to why they were the first subatomic particles to be discovered?

2.114 What experimental evidence led to each of the following advances in our understanding of atomic structure?
(a) discovery of the electron and its charge-to-mass ratio
(b) determination of the mass and charge of the electron
(c) the nuclear model of the atom

2.115 Which isotope has a mass number of 60 and an atomic number of 28?

2.116 Why is it impossible for an element to have a mass number less than its atomic number?

2.117 How many protons and neutrons are in a potassium-39 atom?

2.118 What number of each type of subatomic particle is found in most $H^+$ ions?

2.119 Explain why the relative atomic mass of cobalt is greater than that of nickel, although the atomic number of nickel is one greater than that of cobalt.

2.120 Which element has an isotope that has a mass approximately three times greater than its most abundant isotope?

2.121 Some tables list relative atomic mass to as many significant figures as can be reported accurately. The relative atomic mass of argon, 39.948 amu, is known to five significant figures, while that of fluorine, 18.9984032 amu, is known to nine significant figures. Natural samples of argon contain three isotopes, but natural samples of fluorine contain only one. Explain why their relative atomic masses differ in the number of significant figures reported.

2.122 How many atoms are in each of the following samples of carbon-12?
(a) 120 amu     (b) 12,000 amu     (c) $7.22 \times 10^{24}$ amu

2.123 Naturally occurring iodine is composed of only one isotope. What is its mass number?

2.124 How many atoms are in each of the following samples of naturally occurring iodine? (Iodine has only one naturally occurring isotope.)
(a) 127 amu     (b) 12,700 amu     (c) $7.22 \times 10^{24}$ amu

2.125 A sample of pure carbon and a sample of pure iodine have the same mass. Which has the greatest number of atoms?

2.126 Two equal-volume balloons contain the same number of atoms. One contains helium and one contains argon. Comment on the relative densities of the gases in these balloons.

2.127 Naturally occurring boron comprises two isotopes, boron-10 and boron-11. The atomic mass of boron-10 is 10.013 amu. The atomic mass of boron-11 is 11.009 amu. Which of the following is the best estimate of the percent abundance of each isotope of boron? Why?
50.0% boron-10 and 50.0% boron-11
20.0% boron-10 and 80.0% boron-11
80.0% boron-10 and 20.0% boron-11
95.0% boron-10 and 5.0% boron-11
5.0% boron-10 and 95.0% boron-11

2.128 Sodium metal reacts with water to form the compound sodium hydroxide and hydrogen gas. What is the formula for the hydrogen gas? Would you expect a similar reaction when potassium is added to water? How about copper or silver? Explain why.

2.129 Bromine is a reddish-brown liquid at room temperature. Write the formula that represents liquid bromine.

2.130 Nitrogen and oxygen are the main components of the air we breathe. Write formulas to represent these gases.

2.131 Which element is in group IA (1) but is not an alkali metal?

2.132 Which element is the only nonmetal in group IVA (14)?

2.133 Why couldn't we form a cation of an element by adding a proton to its atom?

2.134 Compare and contrast Mendeleev's periodic table with the modern periodic table.

2.135 Complete the following table for the designated subatomic particles.

|  | Particle | Mass (g) | Relative Charge |
|---|---|---|---|
| (a) |  | $1.6749 \times 10^{-24}$ | 0 |
| (b) |  | $9.1094 \times 10^{-28}$ |  |
| (c) |  | $1.6726 \times 10^{-24}$ |  |

2.136 Atoms of the element shown in this photo contain 28 protons. What is the element?

©McGraw-Hill Education/Stephen Frisch

2.137 Positron emission tomography (PET) is used in medicine to produce an image of functions inside the body and can identify tumors. To make a tumor visible in the scan, a compound containing an isotope of a certain element is injected into the patient. The most common isotope used for PET scans is fluorine-18. How many protons, neutrons, and electrons are present in an atom of this isotope?

2.138 When silver tarnishes, a chemical reaction occurs in which a compound of silver and sulfur forms on the surface of the silver metal. The silver in the compound with sulfur is in the form of an ion with a 1+ charge. In the process of going from a neutral Ag atom to an $Ag^+$ ion, what happens to the number of protons and electrons?

2.139 The iron found in the human body is in the form of $Fe^{2+}$ or $Fe^{3+}$ ions. How do neutral atoms of iron and the $Fe^{2+}$ and $Fe^{3+}$ ions differ?

2.140 Which has the greater density, an ice cube containing the hydrogen-1 isotope or an ice cube containing the hydrogen-2 isotope (see Figure 2.13)? Why?

2.141 Iodine is an essential trace element required by the body and is associated with the function of the thyroid gland. To which group and period on the periodic table does iodine belong?

2.142 After he conducted cathode ray tube experiments proving the existence of negatively charged particles we now call electrons, Thomson proposed a model of the atom called the plum pudding model. Describe how Rutherford's gold foil experiment disproved Thomson's model.

2.143 Analysis of carbon-14 content is used to date fossils up to about 60,000 years old. There are two other naturally occurring isotopes of carbon, $^{12}C$ and $^{13}C$. Given the relative atomic mass of carbon, 12.01 amu, which is the most likely abundance of carbon-14 in nature: approximately 99%, approximately 33%, less than 0.1%?

2.144 A naturally occurring isotope of potassium has a mass number of 40. One of argon's naturally occurring isotopes also has a mass number of 40. How does an atom of $^{40}K$ differ from an atom of $^{40}Ar$?

## Concept Review Questions

2.145 Which of the following has a positive charge?
A. a sodium atom
B. a sodium ion
C. a proton
D. an electron
E. both a sodium ion and a proton
For those that do not have a positive charge, identify whether they have a negative charge or no charge.

2.146 A monatomic ion with a 4+ charge has the same number of electrons as an argon atom. What is the formula for this ion?
A. $C^{4+}$
B. $Ti^{4+}$
C. $Si^{4+}$
D. $Ar^{4+}$
E. $Sc^{4+}$
For those ions that do not have the same number of electrons as argon, determine the number of electrons that they do have.

2.147   Which of the following has the greatest number of electrons?

A. $Ca^{2+}$

B. $K^+$

C. $Cl^-$

D. $S^{2-}$

E. These all have the same number of electrons.

Which has the greater number of protons?

2.148   Chlorine gas is a yellow-green gas that is highly toxic. What is the formula for chlorine gas?

A. $Cl_2$

B. $Cl_2^-$

C. $Cl_2^{2-}$

D. $Cl^-$

E. $Cl$

Which of these is the correct formula for chloride ion?

2.149   Atoms that are isotopes of the same element differ in

A. number of neutrons

B. number of protons

C. number of electrons

D. charge

E. atomic number

For each incorrect answer, identify a pair of substances that would differ in that way.

2.150   Which of the following statements about atoms and subatomic particles is true?

A. The nucleus of an atom has no mass.

B. Neutrons are negatively charged.

C. Electrons exist in the nucleus.

D. The nucleus makes up most of the mass of an atom.

E. A cation has more electrons than protons.

Modify each incorrect statement to make it correct.

2.151   Elements in which group always form monatomic ions with a negative 1 (1−) charge?

A. alkali metals

B. alkaline earth metals

C. halogens

D. noble gases

E. transition metals

For each incorrect group, identify the charge those elements would have when they form ions. Note if any groups do not form ions that are not predictable or if they do not form ions at all.

2.152   Identify the number of protons and neutrons in an atom of the isotope $^{64}Zn$.

A. 64 protons, 34 neutrons

B. 64 protons, 30 neutrons

C. 30 protons, 30 neutrons

D. 30 protons, 34 neutrons

E. More information is needed.

For zinc-64, to which of the following do the values 30, 34, and 64 correspond: mass number, atomic number, neutron number?

2.153   Naturally occurring gallium (Ga) is made of two isotopes: gallium-69 and gallium-71. Which of the following statements is true?

A. Gallium's relative atomic mass is 70.00 amu.

B. Both isotopes have the same mass: 69.72 amu.

C. The isotopes are present in the same percentages (50% and 50%).

D. Gallium-71 is present in the largest percent abundance.

E. Gallium-69 is present in the largest percent abundance.

Explain why the incorrect answers are false.

2.154   Which of the following elements is classified as a nonmetal in period 3?

A. potassium

B. phosphorus

C. bromine

D. silicon

E. krypton

For each incorrect answer, identify the period the element is in and whether the element is a metal, nonmetal, or metalloid.

# Chemical Compounds

Source: Peggy Greb/USDA

3.1 Ionic and Molecular Compounds

3.2 Monatomic and Polyatomic Ions

3.3 Formulas for Ionic Compounds

3.4 Naming Ionic Compounds

3.5 Naming and Writing Formulas for Molecular Compounds

3.6 Acids and Bases

3.7 Predicting Properties and Naming Compounds

Chapter Review

Questions and Problems

Some students taking a chemistry course went on a canoe trip on a local river. They collected water samples from several places along the river for analysis back in their lab. As they paddled downstream, two of the students, Jeff and Megan, began talking about all the different kinds of water they encounter in their daily lives: bottled, mineral, mountain spring, seltzer, tonic, carbonated, lake, ocean, pond, hard, soft, and more. In their precollege science courses, Jeff and Megan recalled learning that water is one of the most important and abundant substances on Earth's surface and that it supports the existence of all living things on this planet. As a liquid, it occurs in the rivers, lakes, and oceans that cover over 70% of Earth's surface. It is also found underground. Groundwater provides about three-fourths of the water used by cities in the United States. Water also exists on Earth in solid form as snow and ice at the polar ice caps and in other cold regions. It is found in the atmosphere as gaseous water vapor and clouds.

The orange juice Jeff and Megan drank as they collected samples along the river is mostly water. The salads they ate for lunch are mostly water, too; and because the day was warm and humid, water vapor condensed on the outside of their iced tea glasses. Before they left for the trip, they showered and brushed their teeth using water. When they rode the bus to the canoe launch site, the vehicle released water into the air from its exhaust pipe. The water formed as a product of combustion in the engine. As they paddled the canoe, Megan and Jeff perspired. "Even our sweat is mostly water," Megan joked.

To a chemist, pure water exists as $H_2O$ molecules in which two hydrogen atoms are connected to one oxygen atom (Figure 3.1). In this form, it is colorless, odorless, and tasteless. In nature, water is rarely pure because it has the unique ability to dissolve so many other substances. One of Megan and Jeff's assignments was to analyze their water samples for various materials—suspended and dissolved—that had been carried down the river.

To understand where those materials come from, let's consider how water makes its way from a cloud over Minnesota to the surface of Earth, across the land to the Mississippi River, down the river to New Orleans and the Gulf of Mexico, and ultimately out into the Atlantic Ocean. The water picks up many different substances as it travels, such as minerals and organic matter that will later be absorbed or ingested by plants and animals. Falling through the air as rain, it dissolves gases from the atmosphere. As it percolates through the ground and runs down streambeds, it dissolves a variety of minerals and gases. As it flows in smaller rivers that feed the Mississippi, it gathers more minerals. Finally it becomes "salty" as it mixes with seawater in the Gulf of Mexico. Water from the surface enters the atmosphere as puddles and ponds evaporate. Evaporation from rivers, lakes, and oceans also contributes to atmospheric water. Transpiration (water loss) from plants adds water to the atmosphere, too. In the air, gaseous water condenses to form clouds, beginning the process all over again (Figure 3.2).

The human body is about 70% water.

$H_2O$

**FIGURE 3.1** Pure water contains only water molecules. Each water molecule consists of two hydrogen atoms attached to an oxygen atom.

**FIGURE 3.2** Water is transported from place to place on Earth in a series of processes called the water cycle. As water flows across the land or falls through the air, many substances end up suspended or dissolved in it.

From rain cloud to ocean, water is always in the form of a solution. Numerous substances dissolved in the water provide nutrients to plants and animals. For example, many of the minerals needed by living organisms are present in water as ions. They include ions of iron, calcium, potassium, and magnesium. The ability of water to dissolve other substances is very important. Many of the topics we discuss throughout this book involve the chemical behavior of substances dissolved in water.

How does *dissolving* differ from *melting*?

Many of the substances that dissolve in water can be classified into categories of chemical compounds. The categories vary greatly in their physical and chemical properties. In this chapter we'll examine the characteristics and behavior of these compounds and look at different ways of naming them.

### Questions for Consideration

**3.1** How do ionic compounds differ from molecular compounds?

**3.2** What kinds of ions are in ionic compounds?

**3.3** What do formulas for ionic compounds represent?

**3.4** How are ionic compounds named?

**3.5** What do formulas for molecular compounds represent and how are they named?

**3.6** What are some common acids and bases and how are they named?

**3.7** How do names of compounds communicate their classification and properties?

## 3.1 Ionic and Molecular Compounds

When they returned to the lab, Jeff and Megan performed several tests on their samples, including one to determine the water's electrical conductivity. They used a simple device like that shown in Figure 3.3 to see if substances dissolved in the river water allow a complete electric circuit to operate. Notice in Figure 3.3B the snipped wire that leads down from the lightbulb. If the snipped wires were to come into contact, the circuit would be complete and the bulb would glow. Another way of completing the circuit is to immerse the wires in a solution that conducts electricity. For an electric current to pass through a solution, ions must be present.

The purity of distilled water or of deionized water can be determined by measuring the electrical conductivity of the water.

Megan and Jeff determined that the water conducted an electric current. To investigate further, they tested several different materials in their lab. Consider the four images shown in Figure 3.4. When the electrical conductivity of pure water is

| **A** Circuit complete | **B** Circuit not complete | **C** Circuit complete |

**FIGURE 3.3** (A) When electricity flows around a complete circuit, the bulb lights. (B) Snip a wire and the circuit is broken. (C) When the ends of the snipped wire are placed in a solution of a substance that contains ions, the bulb glows again. This is a simple procedure for testing the conductivity of a substance dissolved in water.

(A–B): ©Jim Birk; (C): ©McGraw-Hill Education/Stephen Frisch

**A** Pure water

**B** Aqueous sugar solution

**C** Aqueous NaCl solution

**D** Aqueous $CH_3CO_2H$ solution

**FIGURE 3.4** When electrical conductivity is tested with a lightbulb apparatus as shown in Figure 3.3, (A) pure water and (B) a sugar solution do not appear to complete the circuit, but (C) a sodium chloride solution does—the bulb glows. (D) With a solution that contains relatively few ions, such as that of acetic acid, the bulb glows only faintly.
©McGraw-Hill Education/Stephen Frisch

**FIGURE 3.5** Sodium chloride, NaCl, is an electrolyte because it separates into its ions ($Na^+$ and $Cl^-$) when dissolved in water.

🔄 **ANIMATION:** Strong Electrolyte, Weak Electrolyte, and Nonelectrolyte

tested, the lightbulb does not glow. Pure water is not a good enough conductor to complete the circuit, and neither is a solution of sugar in water. However, when electricity is passed through a sodium chloride, NaCl, solution, the bulb glows brightly. For electric current to flow through an aqueous solution, ions must be present. NaCl provides them, while pure water and sugar water do not—at least not in sufficient quantities to carry the electric current. A substance like NaCl that separates into ions when dissolved in water is an **electrolyte** (Figure 3.5). One that stays intact and does not separate into ions, such as sugar, is a **nonelectrolyte.** Electrolyte solutions conduct electricity, while nonelectrolyte solutions do not.

We can be more specific by distinguishing among strong electrolytes, weak electrolytes, and nonelectrolytes. Consider the solutions shown in Figure 3.6. Which compounds appear to break apart as ions in solution? Do some of the compounds produce more ions than others? The process by which a compound dissolves in water to produce ions is called *dissociation* or *ionization,* depending on the type of compound dissolving. Because a **strong electrolyte** dissociates extensively in water, it conducts electricity well. Notice in Figure 3.6 that sodium chloride, NaCl, and hydrochloric acid, HCl, dissociate extensively. They are good conductors of electricity (Figure 3.4C). Other substances, such as acetic acid, $CH_3CO_2H$, dissociate only partially in water. Such a substance does not conduct electricity well (Figure 3.4D) and is called a **weak electrolyte.** The solution containing methanol, $CH_3OH$, does not conduct electricity. Methanol does not dissociate into ions at all. It is a nonelectrolyte.

Many experiments, including those conducted by Jeff and Megan, have shown that, in general, compounds can be classified into two categories. An **ionic compound** (sometimes generically referred to as a *salt*) consists of oppositely charged cations and anions in proportions that give a net zero charge. Ionic compounds are easily identified because they usually consist of ions from a *metal* with ions from a *nonmetal.* Sodium chloride, which is table salt, and is composed of $Na^+$ and $Cl^-$ ions, is an example of an ionic compound (Figure 3.7A). A **molecular compound,** on the other hand, is not composed of ions. It is composed of atoms from two or more different *nonmetals.* It exists as a discrete unit of atoms held together in a molecule. Carbon dioxide, $CO_2$, is a molecular compound (Figure 3.7B). Molecular elements consist of atoms of the same element. Elemental oxygen, $O_2$, is a molecular element (Figure 3.7C).

**FIGURE 3.6**  For each of the solutions, inspect the molecular-level images in which each type of particle is labeled. If any of these substances dissociate into ions, separate particles of single atoms (or atom groups) are visible. Which of the compounds dissociate completely in solution? Which only partially dissociate? Which do not dissociate at all? The acids do not simply dissociate, but form $H_3O^+$ ions in solution.
©McGraw-Hill Education/Stephen Frisch

---

**EXAMPLE 3.1**          **Ionic and Molecular Compounds**

Based on their formulas, which of the following compounds are ionic?

(a) KCl          (b) $CO_2$          (c) CaO          (d) $CCl_4$

**Solution:**

We can determine if a compound is ionic by looking at the elements that compose it. An ionic compound is usually composed of ions from a metal and a nonmetal. Two of the compounds, KCl and CaO, meet this criterion.

**Consider This 3.1**

Would a compound containing phosphorus and fluorine be ionic or molecular?

**Practice Problem 3.1**

Which of the compounds listed in the example are molecular?

**Further Practice:** Questions 3.7 and 3.8 at the end of the chapter

**A** NaCl

**B** $CO_2$

**C** $O_2$

**FIGURE 3.7**  (A) Sodium chloride, NaCl, is an ionic compound. (B) Carbon dioxide, $CO_2$, is a molecular compound. (C) Elemental oxygen, $O_2$, is a molecular element.

**FIGURE 3.8** In solid NaCl, the ions are arranged in a crystal lattice. Once sodium chloride dissolves in water, the ions of sodium and chlorine separate and become surrounded by water molecules. When all the NaCl is completely dissolved, the solution becomes clear.
©Brian Moeskau/Moeskau Photography

**ANIMATION:** Sodium Chloride Dissolving in Water

Like charges repel. Opposite charges attract.

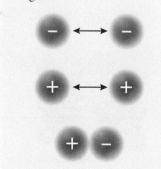

Ozone is actually a triatomic oxygen molecule, $O_3$. The ozone layer within the stratosphere (the atmospheric layer that extends from 10 to 50 km above Earth) helps protect living things on the surface by absorbing much of the harmful ultraviolet radiation in sunlight.

To see how ionic and molecular compounds differ, let's first consider what happens to them in aqueous solution. Ionic compounds that dissolve in water separate into the independent ions that make them up. When a compound composed of ions separates into its component ions in water, the process is described as *dissociation*. In contrast, molecular elements and compounds generally retain their molecular structure. The atoms in a molecule usually stay together when the substance dissolves in water.

Ionic compounds are solids at room temperature, melt at very high temperatures, and are brittle. What causes these special properties? Notice in Figure 3.8 the ordered arrangement of cations and anions in the ionic solid sodium chloride, NaCl. This ordered arrangement is one type of solid structure called a *crystal structure* or *crystal lattice*. The strong attraction between the oppositely charged ions in an ionic compound—along with the pattern in which they arrange themselves—gives ionic solids their unique properties.

The crystal lattice of ionic solids breaks down when they dissolve in water. The solid dissociates into ions. Compare the molecular-level image of solid NaCl in Figure 3.8 to the representation of it dissolved in water. How do the ions of sodium and chlorine differ in the pictures?

We represent the process of sodium chloride dissolving with the equation

$$\text{NaCl}(s) \xrightarrow{\text{H}_2\text{O}} \text{Na}^+(aq) + \text{Cl}^-(aq)$$

where the $H_2O$ written above the arrow indicates that water participates in the process but is not permanently changed. The ($aq$) after $Na^+$ and $Cl^-$ indicates that these ions are in aqueous solution.

In contrast to ionic compounds, most molecular compounds remain as molecules when dissolved in water. When not in solution, molecular compounds can be gas, liquid, or solid at room temperature. They have much lower melting and boiling points than ionic compounds. Carbon dioxide, $CO_2$, is a molecular compound. Because carbon dioxide is composed of molecules of uncharged atoms, not ions, its appearance and behavior are different from those of an ionic compound (Figure 3.9). Molecular elements have properties that are similar to molecular compounds. For example, elemental oxygen is diatomic; that is, it exists as two oxygen atoms bound

**FIGURE 3.9** While molecular compounds like carbon dioxide are composed of molecules, not ions, some can also form solids. At room temperature, $CO_2$ readily goes from the solid phase directly to a gaseous state. (Other gas particles are not shown in the molecular-level image of the gaseous state.)
©Jim Birk

together in a molecule, represented as $O_2$. When it dissolves in water, the oxygen atoms remain together as a molecule (Figure 3.10).

Other differences between ionic and molecular compounds are summarized in Table 3.1. To explain these differences, we can compare the molecular-level structures of ionic and molecular compounds. For example, compare solid NaCl to solid $NH_3$ in Figure 3.11 and imagine what is involved when these two compounds melt. The process of melting for any substance requires the dispersion of the components of the solid into a more random arrangement. In the case of NaCl, the components are cations and anions, and its melting point is 801°C. In the case of $NH_3$, the components are individual $NH_3$ molecules, and its melting point is only −77.7°C. The greater heat energy required to melt the ionic compound indicates that the force of attraction between oppositely charged ions is much stronger than the attractive forces between molecules when they are in the solid state.

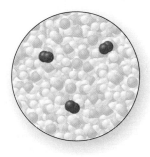

**FIGURE 3.10** Many molecular substances such as oxygen, $O_2$, remain as intact molecules when dissolved in water.

**TABLE 3.1** Properties of Ionic and Molecular Compounds Compared

| Ionic Compounds | Molecular Compounds |
|---|---|
| crystalline solid | gas, liquid, or solid |
| hard, brittle solid | soft solid |
| very high melting point | low melting point |
| very high boiling point | low boiling point |
| high density | low density |
| strong electrolyte in aqueous solution | weak electrolyte or nonelectrolyte in aqueous solution |
| electrical conductivity good when compound is molten | electrical conductivity poor in pure form |

A

B

**FIGURE 3.11** (A) Strong attractive forces between oppositely charged cations and anions must be overcome for solid NaCl to melt. (B) Molecules of $NH_3$ in the solid state are not held together by oppositely charged particles. Instead, weak forces between the $NH_3$ molecules hold them together in the crystal.

---

**EXAMPLE 3.2** | **Properties of Ionic and Molecular Compounds**

Which would you expect to have the lower melting point, KCl or $CCl_4$?

**Solution:**

Ionic compounds consist of oppositely charged ions that have a strong force of attraction between them. These compounds require significantly higher temperatures for melting. It would take a lot more energy to break apart the ions of KCl than to separate $CCl_4$ molecules from each other, so the molecular compound $CCl_4$ should have the lower melting point.

**Consider This 3.2**

Which of the compounds should be a strong electrolyte?

**Practice Problem 3.2**

Which would you expect to have the higher melting point, $Al_2O_3$ or $N_2O_3$?

**Further Practice:** Questions 3.9 and 3.10 at the end of the chapter

---

Binary compounds are composed of only two elements, but often are composed of more than two atoms (e.g., $N_2O_5$).

In Example 3.2, the compounds listed are composed of only two elements. They are binary compounds. A **binary compound** is a compound containing atoms or ions of only two elements. Many chemical compounds are binary. Ionic compounds that contain more than two elements usually contain ions that consist of multiple atoms. You'll learn more about these special ions in Section 3.2.

## 3.2 Monatomic and Polyatomic Ions

When Jeff and Megan tested their samples from the river, they discovered a lot of sodium, magnesium, and iron in the water. These elements were not present in the form of solid metals. They were dissolved in the water as cations. In addition, Megan and Jeff found several anions. When they completely boiled off all the water from a sample, a variety of compounds remained as a solid residue in the flask. Most of them were ionic compounds.

Hard water contains ions of $Ca^{2+}$, $Mg^{2+}$, and $Fe^{3+}$, along with anions such as $Cl^-$ and $SO_4^{2-}$. Soft water contains only low concentrations of these ions.

### Monatomic Ions

To better understand the composition of ionic compounds, let's look more closely at the kinds of ions that make them up. In Chapter 2 we described how the charges on the ions of some main-group elements can be predicted from their positions in the periodic table. For example, find magnesium and sulfur in the periodic table. Based on their positions, what charge would you expect for their ions? An Mg atom has two more electrons than an atom of its nearest noble gas, Ne. A magnesium ion has the same number of electrons as a neon atom, which is two less than the magnesium atom. We represent that ion as $Mg^{2+}$. In contrast, an S atom contains two fewer electrons than the nearest noble gas, Ar. A sulfur ion has the same number of electrons as an argon atom, which is two more than a sulfur atom. With two extra electrons, the charge on the sulfur ion is 2−, and the sulfur ion is represented as $S^{2-}$. The ions of magnesium and sulfur are monatomic. A **monatomic ion** is an ion of a single atom. The compound MgS consists of monatomic ions.

Several common monatomic ions are shown in Figure 3.12. Many ions of the main-group elements, those on the far left or far right of the periodic table, can be predicted from their positions. However, the charges of ions for elements in the middle, especially the transition metals, cannot be predicted from their group numbers. These elements often form more than one kind of ion.

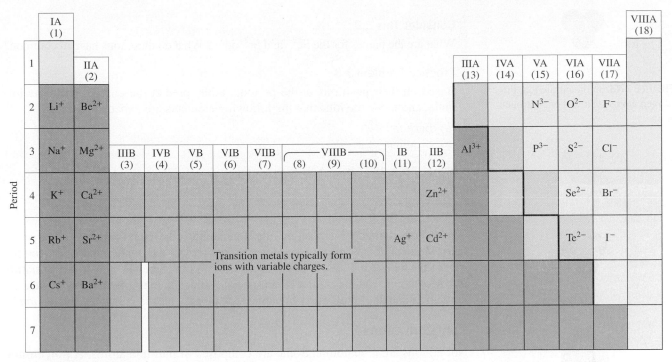

**FIGURE 3.12** The charges on the common monatomic ions of many main-group elements correlate with the elements' positions in the periodic table. Charges on the other elements are not easily predicted from their positions. A few of the transition metals exhibit only one charge.

**TABLE 3.2** Common Monatomic Ions

| Formula of Ion | Name of Ion | Formula of Ion | Name of Ion | Formula of Ion | Name of Ion |
|---|---|---|---|---|---|
| $O^{2-}$ | oxide ion | $Al^{3+}$ | aluminum ion | $K^+$ | potassium ion |
| $Na^+$ | sodium ion | $Mg^{2+}$ | magnesium ion | $F^-$ | fluoride ion |
| $N^{3-}$ | nitride ion | $Cl^-$ | chloride ion | | |

Not all monatomic cations are named this way. If an element can form monatomic cations with different charges, a Roman numeral is used in the naming system to indicate charge. This system is discussed in Section 3.4.

*Nomenclature* is a system of naming. To distinguish one chemical entity from another, unique names are assigned to all elements, ions, and compounds. Just as elements have unique names and symbols, ionic compounds and ions have their own names—and a special set of rules for correct naming. Several common monatomic ions are listed in Table 3.2. Try to determine some of the rules that were used to name them.

Did you figure out how to name ions? Did you determine when the *–ide* ending is used and how to use it? Now try to apply your rules in Example 3.3.

---

**EXAMPLE 3.3**     **Formulas and Names of Ions**

Based on their positions in the periodic table, predict the charge, write the formula, and name the ions that the following elements are expected to form.

(a) lithium
(b) sulfur

**Solution:**

(a) An atom of lithium is expected to lose one electron to have the same number as helium. Its ion is represented with the formula $Li^+$ and is named *lithium ion.*

(b) Sulfur is expected to gain two electrons to have the same number as argon. The representation for the ion of sulfur is $S^{2-}$ and is named *sulfide ion.*

$NO_3^-$

**FIGURE 3.13** In the nitrate ion, three oxygen atoms surround one nitrogen atom.

$S^{2-}$
Sul*fide*

$SO_3^{2-}$
Sul*fite*

$SO_4^{2-}$
Sul*fate*

**FIGURE 3.14** Anions containing sulfur differ in the number of oxygen atoms attached to the sulfur. All have a 2− charge.

In a formula for a binary molecular compound or polyatomic ion, the element present in the least amount is usually positioned central to the atoms of the other element. The central atom is usually listed first in the formula.

Many commercial and domestic water-softening products and devices are available for removing the $Ca^{2+}$, $Mg^{2+}$, and $Fe^{3+}$ ions in hard water.

**FIGURE 3.15** Over time, solid buildup of carbonates can close off pipes.
©Jim Birk

**Consider This 3.3**

What are the names for the $Se^{2-}$ and $Sr^{2+}$ ions? What do these ions have in common?

**Practice Problem 3.3**

Based on their positions in the periodic table, predict the charge, write the formula, and name the ions that the following elements are expected to form.

(a) barium
(b) bromine

**Further Practice:** Questions 3.11 and 3.12 at the end of the chapter

If you had trouble establishing rules for naming ions, the following summary may help. Monatomic anions are named as the first part of the element name (called the root of the name) and –*ide* is added as a suffix. For example, $S^{2-}$ is known as *sulfide ion*. Monatomic cations are not given special endings to their root names, so we refer to them as an ion of their element name; for example, $Na^+$ is known as *sodium ion.*

## Polyatomic Ions

Not all ions are monatomic. Some are composed of multiple atoms. A **polyatomic ion** is an ion containing two or more atoms, usually of more than one element. An example is the nitrate ion, $NO_3^-$. Nitrate ions are discrete units of three oxygen atoms and one nitrogen atom, as represented in Figure 3.13. Although the formula for nitrate ion might suggest that the oxygen atoms are attached to each other, they are actually attached to a central nitrogen atom. This polyatomic ion bears a charge of 1− because it has one more electron than the total number provided by the nitrogen and oxygen atoms.

The most common polyatomic ions are anions that contain oxygen attached to some other element. Such an anion is called an **oxoanion.** Oxoanions are typically combinations of oxygen with a nonmetal, although some contain metals. For a given nonmetal, there are often two or more common oxoanions. When this is the case, the two oxoanions differ in the number of oxygen atoms surrounding the nonmetal atom. For example, there are two known oxoanions of sulfur, both with a 2− charge. One contains a central sulfur atom with three oxygen atoms attached to it. The other has a central sulfur atom with four oxygen atoms. These oxoanions are represented in Figure 3.14 with the formulas $SO_3^{2-}$ (sul*fite* ion) and $SO_4^{2-}$ (sul*fate* ion).

Some water supplies contain a variety of aqueous ions, including sulfate, carbonate, phosphate, and chloride ions, as well as ions of calcium, magnesium, iron, and aluminum. Water containing the cations $Ca^{2+}$, $Mg^{2+}$, and sometimes $Fe^{3+}$, along with anions such as $Cl^-$, $SO_4^{2-}$, and $HCO_3^-$, is called *hard water*. When Megan and Jeff boiled away their water samples, they noticed a residue left on the glassware. They tested it and found that carbonate ions, $CO_3^{2-}$, were present. The hard water deposits were mostly metal carbonates, $CaCO_3$ and $MgCO_3$. They had noticed similar residues at home on cooking pans and drinking glasses from which water had boiled or evaporated. Carbonate deposits also build up in pipes (Figure 3.15). Although their formulas are more complex than those of binary compounds, $CaCO_3$ and $MgCO_3$ behave like binary ionic compounds. When $CaCO_3$ dissolves in water, $Ca^{2+}$ and $CO_3^{2-}$ ions form (Figure 3.16). When the water is removed by evaporation or boiling, metal carbonates in solid form stay behind.

Nearly all polyatomic ions are negatively charged. An exception is the polyatomic cation $NH_4^+$ (called ammonium ion). Like monatomic ions, polyatomic ions have their own rules for naming. Several common ones are shown in Table 3.3. Can you figure out the rules used to name them?

Now apply your rules to naming the polyatomic ions in Example 3.4.

**FIGURE 3.16**  A solution of $CaCO_3$ contains a small amount of $Ca^{2+}$ and $CO_3^{2-}$ ions. Boiling some of the water from the solution leaves behind solid deposits of calcium carbonate.
©David A. Tietz/Editorial Image, LLC

**TABLE 3.3**  Common Oxoanions

| Formula of Ion | Name of Ion |
| --- | --- |
| $NO_3^-$ | nitrate ion |
| $NO_2^-$ | nitrite ion |
| $SO_4^{2-}$ | sulfate ion |
| $SO_3^{2-}$ | sulfite ion |
| $ClO_3^-$ | chlorate ion |
| $ClO_2^-$ | chlorite ion |

## EXAMPLE 3.4          Formulas and Names of Polyatomic Ions

There are two oxoanions of arsenic. One has three oxygen atoms attached to an arsenic atom and has a 3− charge. The other has the same charge and four oxygen atoms attached to an arsenic atom. Write the formula for the arsenite ion.

**Solution:**

From Table 3.3 we can tell that for a particular nonmetal oxoanion, the one that contains fewer oxygen atoms gets an *–ite* ending. We represent the arsenite ion as $AsO_3^{3-}$.

**Consider This 3.4**

Two oxoanions of bromine are $BrO_3^-$ and $BrO_2^-$. Which one is the formula for the bromate ion?

**Practice Problem 3.4**

Write the formula for the arsenate ion.

**Further Practice:** Questions 3.23 and 3.24 at the end of the chapter

Too much nitrogen and phosphorus, in the form of nitrates and phosphates, in water—from fertilizers, sewage, or livestock runoff—stimulates the growth of algae. These algal growth spurts, or blooms, soon die off. Bacteria decompose the dead plants, consuming oxygen in the process. The reduction of dissolved oxygen in the water kills fish and other aquatic organisms.

If you had trouble establishing rules for naming oxoanions, the following summary may help. If there are only two oxoanions of an element, the one with the greater number of oxygen atoms is named by combining the root of the nonmetal element name and the ending –*ate*. The oxoanion with the lesser number of oxygen atoms uses the ending –*ite*. Thus, sulf*ate* is the sulfur oxoanion with the formula $SO_4^{2-}$ and sulf*ite* is the name of $SO_3^{2-}$. Formulas and charges for several of the most common oxoanions are shown in Figure 3.17. Do you see any patterns in the ions shown?

Some nonmetals form more than two oxoanions. For example, chlorine forms four known oxoanions. They are named as before with the –*ate* or –*ite* suffixes. However, the prefix *per*– is placed before the root of the element name and –*ate* is added to the end for the oxoanion that contains the most oxygens. The prefix *hypo*– and the ending –*ite* are used for the anion with the fewest oxygens. These, plus other polyatomic ions whose names do not fit regular patterns, are listed in Table 3.4.

**FIGURE 3.17** Formulas and charges of some oxoanions and monatomic anions, arranged by the position of their central atom in the periodic table. Do you see any patterns?

| **B** $BO_3^{3-}$ borate | **C** $CO_3^{2-}$ carbonate | **N** $NO_3^-$ nitrate $NO_2^-$ nitrite $N^{3-}$ nitride | **O** $O_2^{2-}$ peroxide $O^{2-}$ oxide | **F** No oxoanions $F^-$ fluoride |
|---|---|---|---|---|
| | **Si** $SiO_4^{4-}$ silicate | **P** $PO_4^{3-}$ phosphate $P^{3-}$ phosphide | **S** $SO_4^{2-}$ sulfate $SO_3^{2-}$ sulfite $S^{2-}$ sulfide | **Cl** $ClO_4^-$ perchlorate $ClO_3^-$ chlorate $ClO_2^-$ chlorite $ClO^-$ hypochlorite $Cl^-$ chloride |
| | | **As** $AsO_4^{3-}$ arsenate $AsO_3^{3-}$ arsenite $As^{3-}$ arsenide | **Se** $SeO_4^{2-}$ selenate $SeO_3^{2-}$ selenite $Se^{2-}$ selenide | **Br** $BrO_4^-$ perbromate $BrO_3^-$ bromate $BrO_2^-$ bromite $BrO^-$ hypobromite $Br^-$ bromide |
| | | | **Te** $TeO_4^{2-}$ tellurate $TeO_3^{2-}$ tellurite $Te^{2-}$ telluride | **I** $IO_4^-$ periodate $IO_3^-$ iodate $IO_2^-$ iodite $IO^-$ hypoiodite $I^-$ iodide |

The phosphite ion, which would have the formula $PO_3^{3-}$, does not exist. The simplest phosphorus ion with three oxygen atoms is the hydrogen phosphite ion, $HPO_3^{2-}$

**TABLE 3.4**  Important Polyatomic Ions

| **1– Ions** | | **2– Ions** | |
|---|---|---|---|
| nitrate | $NO_3^-$ | chromate | $CrO_4^{2-}$ |
| nitrite | $NO_2^-$ | dichromate | $Cr_2O_7^{2-}$ |
| bicarbonate (hydrogen carbonate) | $HCO_3^-$ | sulfate | $SO_4^{2-}$ |
| perchlorate | $ClO_4^-$ | sulfite | $SO_3^{2-}$ |
| chlorate | $ClO_3^-$ | carbonate | $CO_3^{2-}$ |
| chlorite | $ClO_2^-$ | oxalate | $C_2O_4^{2-}$ |
| hypochlorite | $ClO^-$ | peroxide | $O_2^{2-}$ |
| cyanide | $CN^-$ | hydrogen phosphite | $HPO_3^{2-}$ |
| hydroxide | $OH^-$ | **3– Ions** | |
| acetate | $CH_3CO_2^-$ | phosphate | $PO_4^{3-}$ |
| permanganate | $MnO_4^-$ | borate | $BO_3^{3-}$ |
| hydrogen sulfate | $HSO_4^-$ | **1+ Ion** | |
| dihydrogen phosphate | $H_2PO_4^-$ | ammonium | $NH_4^+$ |

**FIGURE 3.18** Ionic compounds containing polyatomic ions such as calcium carbonate, $CaCO_3$, exist in a regular array of cations and anions. Notice the similarities between sodium chloride and calcium carbonate.
©RF Company/Alamy Stock Photo

In their solid form, ionic compounds exist in an organized array of alternating positive and negative ions. Calcium carbonate, $CaCO_3$, the hard water residue that is also a major component of limestone, is another example of an ionic compound. Although its formula might suggest that it is nothing like sodium chloride, it does have similar properties. It is composed of positive and negative ions: the calcium ion, $Ca^{2+}$, and the carbonate ion, $CO_3^{2-}$. Notice in Figure 3.18 that calcium carbonate forms an ordered array of alternating cations and anions similar to sodium chloride's crystal lattice structure. Instead of a monatomic anion, the anion in this case is $CO_3^{2-}$. As indicated by the molecular-level representation, a polyatomic ion behaves as a unit.

> Calcium carbonate is used in antacid tablets, which reduce the amount of $HCl(aq)$ in gastric acid.

## 3.3 Formulas for Ionic Compounds

Some of the ionic compounds that Megan and Jeff isolated from their river water sample were $MgCl_2$, $NaNO_3$, $CaSO_4$, and $FeCl_3$. What do these formulas represent? How are they determined? Unlike molecular compounds such as $CH_4$, ionic compounds such as NaCl do not exist as discrete molecules. As shown in Figure 3.19, the lattice structure of NaCl composed of $Na^+$ and $Cl^-$ ions is different from the molecular structure of $CH_4$. Methane exists as discrete molecules, each composed of one carbon atom and four hydrogen atoms.

If ionic compounds don't exist as molecules, then how can we write formulas for them? The charged cations and anions that make up an ionic compound are present in quantities that result in a net charge of zero. That is, the number of positive charges equals the number of negative charges. Formulas for ionic compounds are written to reflect this fact, using the lowest possible whole numbers. The sum of the positive charges must equal the sum of the negative charges:

$$\begin{array}{ccc} \text{Total positive charge} & + & \text{total negative charge} & = & \text{zero net charge} \\ \text{from cations} & & \text{from anions} \end{array}$$

Why is NaCl the formula for sodium chloride? Based on their positions in the periodic table, we can predict that the charge on sodium ion is 1+ and the charge on chloride ion is 1−. For the compound to be electrically neutral, the two ions have to

NaCl

$CH_4$

**FIGURE 3.19** Ionic compounds such as NaCl do not exist as discrete molecules. The highlighted region of the NaCl structure is the *formula unit* for this compound. Molecular compounds such as $CH_4$ do exist as discrete molecules.

**Student Hot Spot**

Student data indicate you may struggle with writing formulas for ionic compounds. Access the SmartBook to view additional Learning Resources on this topic.

$CO_3^{2-}$

$Ca^{2+}$

**FIGURE 3.20** The formula unit of calcium carbonate is $CaCO_3$.

**Student Hot Spot**

Student data indicate you may struggle with writing formulas for ionic compounds that contain polyatomic ions. Access the SmartBook to view additional Learning Resources on this topic.

be present in a one-to-one ratio, or $Na_1Cl_1$. When there is only one of a given atom or ion in a formula, the subscript 1 is understood, so the formula for sodium chloride can be written simply as NaCl.

Compare this formula to the highlighted region of the sodium chloride structure in Figure 3.19. It has repeating units of one sodium ion and one chloride ion. Because an ionic compound does not exist as a molecule, its formula represents its smallest repeating unit. We call this repeating unit a **formula unit.** The formula unit of sodium chloride is NaCl. How do you know to put the sodium first? Generally, chemical names start with the name of the element that is farther to the left or farther down on the periodic table. Most often, the name or symbol of a metal is written first.

For ionic compounds, the formula represents the *ratio* of ions of each kind, not an exact count of ions. Consider another example, sodium sulfide. Based on their positions in the periodic table, we can predict that the charge on sodium ion is 1+ and the charge sulfide ion is 2−. Two $Na^+$ ions are required to provide the 2+ charge needed to balance the 2− charge of the $S^{2-}$, so these ions will combine to form a compound with the formula $Na_2S$.

Formulas for ionic compounds containing polyatomic ions must also be written so that the sum of the positive charges from the cations equals the sum of the negative charges from the anions. For example, consider calcium carbonate (Figure 3.20). From the periodic table we know that the calcium ion has a charge of 2+ ($Ca^{2+}$). From Table 3.4 we know carbonate ion has a charge of 2− ($CO_3^{2-}$). Therefore, the formula unit is written to show a one-to-one ratio of ions, $CaCO_3$. The calcium appears first in both the name and the formula because it is the cation. We cannot deduce the charge on a polyatomic ion from the periodic table. We must remember it or look it up.

As another example, consider a compound containing calcium ion and nitrate ion. Based on its position in the periodic table we expect the calcium ion to carry a charge of 2+ ($Ca^{2+}$). The nitrate ion is known to have a charge of 1− ($NO_3^-$). Two nitrate ions are required for every calcium ion to write a formula for the compound that is neutral: $Ca(NO_3)_2$. The parentheses around the formula for the polyatomic ion show that the subscript applies to the whole unit. If only one polyatomic ion is represented in a formula, the parentheses are not needed. For examples, see Figure 3.21.

When an ionic compound is dissolved in water, its ion ratio is the same as in the pure compound. Any polyatomic ions present remain intact.

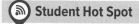

| NaCl | $Ca(NO_3)_2$ | $Na_2SO_4$ | $BaCl_2$ | $Ba_3(PO_4)_2$ |
|---|---|---|---|---|
| $Na^+$    $Cl^-$ | $Ca^{2+}$    $NO_3^-$ | $Na^+$    $SO_4^{2-}$ | $Ba^{2+}$    $Cl^-$ | $Ba^{2+}$    $PO_4^{3-}$ |
| 1+    1− | 2+    2− | 2+    2− | 2+    2− | 6+    6− |

**FIGURE 3.21** Ionic compounds are formed by combinations of cations and anions in proportions that lead to electrical neutrality. The images represent a formula unit of each compound.

| EXAMPLE 3.5 | Ionic Compounds |

The given images represent aqueous solutions of ionic compounds. Match the images to the following formulas: $BaCl_2(aq)$, $Na_2S(aq)$, and $KBr(aq)$.

**Solution:**

Based on their positions in the periodic table, barium is expected to have a charge of 2+ and the chloride ion is expected to be 1−. Image B corresponds to $BaCl_2$. Based on its position in the periodic table, we expect the sodium ion to have a charge of 1+. The sulfide ion has a charge of 2−. Image C corresponds to $Na_2S$. Based on their positions in the periodic table, the potassium ion is expected to have a charge of 1+ and the bromide ion is expected to be 1−. Image A corresponds to KBr. Notice that in all of the images ions are present in appropriate ratios of cations and ions.

**Consider This 3.5**

What if 10 barium ions were displayed in image B? How many chloride ions would have to be shown?

**Practice Problem 3.5**

Suppose an ionic compound containing magnesium and chloride ions is dissolved in water. Beginning with a representation for five magnesium ions, draw a picture of a solution containing magnesium and chloride ions that is electrically neutral. You can omit the water molecules.

**Further Practice:** Questions 3.27 and 3.28 at the end of the chapter

In Example 3.5 we matched formulas with molecular-level representations. We can also write formulas from ion names. First, we determine if a polyatomic ion is present. Next, we determine the charge on the ions that make up the compound. If the ion is monatomic, the charge can often be predicted from the periodic table. If a polyatomic ion is present, we must recall its charge. Once we know the charges on the cations and anions in the compound, we can write a correct formula, as Example 3.6 demonstrates.

| EXAMPLE 3.6 | Formulas for Ionic Compounds |

Write the formulas for compounds containing the following ions:

(a) calcium ion and nitride ion
(b) barium ion and nitrate ion
(c) potassium ion and sulfate ion

**Solution:**

(a) Calcium ion and nitride ion are monatomic. Their charges can be determined from their positions in the periodic table. Calcium ion is expected to have a 2+ charge, and nitride ion is expected to have a 3− charge. Combining three calcium ions with two nitride ions yields a formula that has equal amounts of positive and negative charges: $Ca_3N_2$.

(b) Barium ion is a monatomic cation. Nitrate ion is a polyatomic anion. We can use the periodic table to predict the charge of the cation. The barium ion is expected to have a 2+ charge. From Table 3.4 we know that nitrate has a charge of 1−. It takes two nitrate ions for every one barium ion to give equal amounts of positive and negative charges: $Ba(NO_3)_2$.

(c) Potassium ion is a monatomic cation. Sulfate ion is a polyatomic anion. Based on its position in the periodic table, potassium ion is expected to have a 1+ charge. From Table 3.4 we know that the charge on sulfate ion is 2−. It takes two potassium ions for every one sulfate ion to yield equal amounts of positive and negative charges: $K_2SO_4$.

**Consider This 3.6**

Suppose someone incorrectly wrote the formula for a compound containing sodium ion and sulfite ion as $Na_2S$. What is wrong with the formula for this combination of ions?

**Practice Problem 3.6**

What are the formulas for all the compounds that can be formed by $K^+$, $Fe^{3+}$, $Br^-$, and $SO_4^{2-}$?

**Further Practice:** Questions 3.29 and 3.30 at the end of the chapter

## 3.4 Naming Ionic Compounds

On their canoe trip, Megan and Jeff took along a packet of instant juice mix. As Jeff mixed the powder with water from his canteen, he noticed the list of ingredients on the label. He asked Megan if she knew what any of the long chemical names meant. Although she didn't recognize them all, several names of compounds were familiar to her. Some were ionic compounds, such as potassium iodide, copper sulfate, and magnesium oxide. Megan knew that they provide essential minerals the body needs for good health. Although she didn't know the formulas for some of the others, she recognized them as sources of other metal ions the body needs. They include ferrous fumarate, which contains iron, and sodium selenite, a source of selenium.

Jeff and Megan, like most new students to chemistry, wondered why chemical names have to be so complicated. Actually, they are not as complicated as they look at first. Chemical names have been established by applying a consistent set of rules that allow chemists to communicate accurately and efficiently. The systematic rules of nomenclature ensure that each chemical compound has a unique name that distinguishes it from all other compounds. Formulas and names for several ionic compounds are listed in Table 3.5. Can you determine the rules used for naming them?

The systematic rules of nomenclature are established by the International Union of Pure and Applied Chemistry (IUPAC).

Sodium nitrite, $NaNO_2$, is added to processed meats to prevent botulism.

**TABLE 3.5** Names of Some Ionic Compounds

| Formula | Name | Formula | Name |
|---|---|---|---|
| NaCl | sodium chloride | $Mg(NO_3)_2$ | magnesium nitrate |
| $NaNO_2$ | sodium nitrite | BaO | barium oxide |
| $MgCl_2$ | magnesium chloride | $Li_3N$ | lithium nitride |

The names in the table follow from the rules for naming the component cations and anions explained in Sections 3.2 and 3.3. Apply them as you name the ionic compounds in Example 3.7.

---

**EXAMPLE 3.7**    **Naming Ionic Compounds**

Name the compounds given by the formulas (a) $Na_2O$ and (b) $Ca_3(PO_4)_2$.

**Solution:**

(a) The first compound, $Na_2O$, is composed of monatomic cations and anions whose charges can be predicted from the periodic table. We state the name of the metal ion first, followed by the root of the name for the nonmetal with an *–ide* ending: *sodium oxide*.

(b) The charge on a calcium ion can be predicted from the periodic table. The polyatomic ion is $PO_4^{3-}$, named phosphate ion. The name of the compound is *calcium phosphate*.

**Consider This 3.7**

Megan noticed that Jeff incorrectly named $Ba(NO_3)_2$ as barium nitride. What is wrong with this name?

**Practice Problem 3.7**

Name the compounds $K_2O$ and $MgSO_3$.

**Further Practice:** Questions 3.39 and 3.40 at the end of the chapter

---

If you had trouble establishing rules for naming ionic compounds, the following summary may help. These compounds are named by simply naming the cation followed by the anion. For ionic compounds containing common monatomic ions, we give the name of the cation first, simply using its element name. To name the anion we add the suffix *–ide* to the root of the name of the element that corresponds to the anion. Generally for ionic compounds, a prefix indicating the atomic ratio is not used. (As we'll see later in Section 3.5, prefixes are used for molecular compounds.) Thus, NaCl is composed of $Na^+$ and $Cl^-$ ions and is named sodium chloride. The compound with the formula $AlBr_3$ is composed of $Al^{3+}$ ions and $Br^-$ ions and is named aluminum bromide.

Naming ionic compounds containing polyatomic anions is similar to naming compounds with monatomic anions. The cation is named first, followed by the name of the polyatomic ion. For example, the compound that contains the sodium ion, $Na^+$, and the sulfate ion, $SO_4^{2-}$, has the formula $Na_2SO_4$ and is called sodium sulfate. Some ionic compounds contain the ammonium ion, $NH_4^+$. An example is ammonium chloride, $NH_4Cl$. Some rules for naming ionic compounds are summarized in Figure 3.22.

Names for these ionic compounds do not directly communicate the relative number of ions in the formula unit. This is unnecessary because we can use the ionic charges for cations and anions predicted from the periodic table or recalled from a table of polyatomic ions such as Table 3.4 to write their formulas. For example, the compound aluminum oxide contains an ion of an element that is a metal and an ion of an element that is a nonmetal, so we recognize it as ionic. Aluminum, in group IIIA (13), forms a 3+ ion. Oxygen, in group VIA (16), forms a 2− ion. To be electrically neutral, the formula for aluminum oxide must be $Al_2O_3$ (Figure 3.23).

The previous examples considered ionic compounds with cations capable of just one possible charge. Metals that can form ions of different charges can combine with a nonmetal to form more than one ionic compound. For example, copper can combine with chlorine to form $CuCl$ and $CuCl_2$. In these two compounds, the charge

**Student Hot Spot**

Student data indicate you may struggle with naming an ionic compound from its formula. Access the SmartBook to view additional Learning Resources on this topic.

**FIGURE 3.22** This flowchart can be used to determine names for ionic compounds where the cation can have only one charge.

**FIGURE 3.23** Because aluminum has a 3+ charge and oxygen has a 2− charge, the formula for aluminum oxide is $Al_2O_3$.

**FIGURE 3.24** This periodic table shows some of the metals that exhibit multiple charges. Not all possible charges are shown. Notice that zinc, silver, and cadmium are exceptions. They exhibit only one charge.

on the copper ion differs (1+ and 2+). Many of the metals, especially the transition metals, can exhibit multiple charges (Figure 3.24).

The names we assign to the two compounds of copper and chlorine must be unique so that one can be distinguished from the other. There are two ways of naming compounds of this kind. One is systematic and more current. The other uses common names that were developed—not quite so systematically—in past centuries.

Some formulas and systematic names for ionic compounds containing metals that can have more than one charge are shown in Table 3.6. Try to figure out the rules used to name them.

**TABLE 3.6** Ionic Compounds Containing Metals with Multiple Charges

| Formula | Cation | Systematic Name |
|---------|--------|-----------------|
| $FeCl_2$ | $Fe^{2+}$ | iron(II) chloride |
| $FeCl_3$ | $Fe^{3+}$ | iron(III) chloride |
| $Cu_2O$ | $Cu^+$ | copper(I) oxide |
| $CuO$ | $Cu^{2+}$ | copper(II) oxide |
| $CuSO_4$ | $Cu^{2+}$ | copper(II) sulfate |
| $SnO$ | $Sn^{2+}$ | tin(II) oxide |

A metal's charge is important to its properties. The chromium in $CrO_4^{2-}$, which has a charge of 6+, is carcinogenic. But $Cr^{3+}$ is not. In fact, $Cr^{3+}$, as in $Cr_2O_3(s)$, is found in the green ink used to print U.S. paper money. It is also found in weight-loss supplements sold in health food stores.

Did you figure out that the Roman numeral in a compound name represents the charge on the metal cation? Try to use this rule to name the ionic compounds in Example 3.8.

---

**EXAMPLE 3.8**    **Naming Ionic Compounds Containing Metals That Exhibit Multiple Charges**

Name the compounds given by the formulas (a) $SnCl_2$ and (b) $SnO_2$.

**Solution:**

In both cases the charge on tin is not apparent from its position in the periodic table, so we'll have to deduce its charge in each compound from the anions.

(a) Based on its position in the periodic table, we predict that chloride ion has a 1− charge. For the compound to be electrically neutral, the Sn must have a 2+ charge to counter the negative charges from the two $Cl^-$ ions. We can now write the name for the compound by beginning with the metal name followed by the Roman numeral to represent its charge, tin(II). Then we write the name of the anion. The name of the compound is *tin(II) chloride.*

(b) Based on its position in the periodic table, we predict that oxide ion has a 2− charge. For the compound to be electrically neutral, the Sn must have a 4+ charge to counter the four negative charges from the two $O^{2-}$ ions. We can now write the name for the compound by beginning with the metal name followed by the Roman numeral to represent its charge, tin(IV). Then we write the name of the anion. The name of the compound is *tin(IV) oxide.*

**Consider This 3.8**

What is incorrect about the name iron(II) oxide for a compound with the formula $Fe_2O_3$?

**Practice Problem 3.8**

Name the compounds $Cr_2S_3$ and $FeSO_3$.

**Further Practice:** Questions 3.43 and 3.44 at the end of the chapter

---

If you had trouble establishing rules for naming ionic compounds containing metals that exhibit multiple charges, the following summary may help. The systematic method used to name ionic compounds is called the *Stock system.* It uses a Roman numeral in parentheses after the name of the metal to indicate the charge. There is no space between the metal name and the parentheses containing the Roman numeral. For example, CuCl is named copper(I) chloride because the copper in this compound has a charge of 1+. The other compound of copper and chlorine, $CuCl_2$, is named copper(II) chloride because the copper has a 2+ charge. Rules for naming ionic compounds containing a cation that can exhibit multiple charges are shown in Figure 3.25.

Zinc, silver, and cadmium normally form only one ion, so Roman numerals are not used when naming their compounds.

**FIGURE 3.25** This flowchart can be used to determine names for ionic compounds that contain metals that exhibit multiple charges.

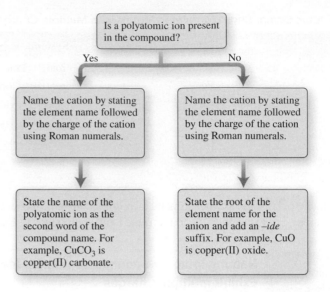

Low chromium(III) levels in the diet can increase blood sugar, triglycerides, and cholesterol, thereby increasing the risk of diabetes and heart disease. Good sources of chromium(III) include whole-grain breads and cereals, cheese, and lean meats.

Constructing formulas from Stock system names is particularly straightforward, since the name gives the charge of the metal. For example, in chromium(III) oxide, the chromium has a charge of 3+. From the position of oxygen in the periodic table, we know that oxide ion has a charge of 2−. For the sum of the positive and negative charges to be equal, there must be three $O^{2-}$ ions for every two $Cr^{3+}$ ions, giving a formula of $Cr_2O_3$. To practice writing formulas from names, try Example 3.9.

---

**EXAMPLE 3.9**    **Formulas for Compounds Containing Metals That Exhibit Multiple Charges**

Write the formulas for the compounds with the following names:

(a)  iron(III) sulfide
(b)  cobalt(II) nitrate

**Solution:**

(a)  The Roman numeral in the name indicates that iron has a 3+ charge, $Fe^{3+}$. Sulfide is the name of the monatomic ion of sulfur that, from its position in the periodic table, we expect to have a charge of 2−, $S^{2-}$. For the sum of the positive and negative charges to be equal, there must be three $S^{2-}$ ions for every two $Fe^{3+}$ ions, giving a formula of $Fe_2S_3$.

(b)  The Roman numeral in the name indicates that cobalt has a 2+ charge, $Co^{2+}$. The nitrate ion a polyatomic ion with the formula $NO_3^-$. For the sum of the positive charges and the sum of the negative charges to be equal, there must be two $NO_3^-$ ions for every one $Co^{2+}$ ion, giving a formula of $Co(NO_3)_2$.

**Consider This 3.9**

What is the difference between cobalt(II) nitrate and cobalt(III) nitrate?

**Practice Problem 3.9**

Write the formulas for the compounds with the following names:

(a)  copper(I) sulfate
(b)  iron(II) oxide

**Further Practice:** Questions 3.49 and 3.50 at the end of the chapter

While the systematic Stock naming sytem directly communicates the charge on the metal in ionic compounds, an older method of naming these compounds is still commonly used in labels of consumer products. In this older method, the name of the metal is changed by adding a suffix, either *–ous* or *–ic,* to the root of the original Latin name for the metal. (See Chapter 1, Table 1.1, for Latin names of some of the elements.) The suffix *–ous* indicates the lower of two charges, and the suffix *–ic* indicates the higher. For example, the two compounds of copper and chlorine described earlier have the formulas CuCl and $CuCl_2$. In the first one, the copper ion has a charge of 1+. This is the lower possible charge, so the compound is called cupr*ous* chloride. The other compound, $CuCl_2$, contains copper ions with the higher possible charge, so it is called cupr*ic* chloride. This older method is not often used in laboratories, because it applies only to metals that exhibit two charges. Some examples of common names are listed in Table 3.7, along with the corresponding Stock system names. Example 3.10 asks you to employ the system of common names.

**TABLE 3.7** Names of Some Ionic Compounds Composed of Metals That Exhibit Multiple Charges

| Formula | Ion | Stock Name | Old Name |
|---|---|---|---|
| $FeCl_2$ | $Fe^{2+}$ | iron(II) chloride | ferrous chloride |
| $FeCl_3$ | $Fe^{3+}$ | iron(III) chloride | ferric chloride |
| $Cu_2O$ | $Cu^+$ | copper(I) oxide | cuprous oxide |
| CuO | $Cu^{2+}$ | copper(II) oxide | cupric oxide |
| SnO | $Sn^{2+}$ | tin(II) oxide | stannous oxide |
| $SnO_2$ | $Sn^{4+}$ | tin(IV) oxide | stannic oxide |

---

**EXAMPLE 3.10**    **Common Names for Ionic Compounds Containing Metals That Exhibit More Than One Charge**

(a) Using common nomenclature, name the two compounds of iron and oxygen with the formulas FeO and $Fe_2O_3$.

(b) Write formulas for the compounds cuprous nitrate and cupric nitrate, given that the possible charges of copper ions are 1+ and 2+.

**Solution:**

(a) From its position in the periodic table we know that the oxygen ions in both compounds have a charge of 2−. In the first compound, FeO, the iron has a charge of 2+ to make the sum of the positive charges and the sum of the negative charges equal. In the second compound, the iron ion has a charge of 3+. The compound with the formula FeO contains the iron ion with the lower of two possible charges, so it is called *ferrous oxide.* The other compound, $Fe_2O_3$, contains the iron ion with the higher charge, so it is called *ferric oxide.*

(b) The *–ous* part of name of the first compound, cuprous nitrate, indicates that it contains copper ions with the lower of two possible charges. 1+. This compound contains $Cu^+$ ions. Nitrate ion has the formula $NO_3^-$. For the sum of the positive charges and the sum of the negative charges to be equal, the formula for cuprous nitrate must be $CuNO_3$. In the second compound, cupric nitrate, the *–ate* part of the name indicates that it contains copper ions with the higher of two possible charges, 2+. This compound contains $Cu^{2+}$ ions. For the sum of the positive charges and the sum of the negative charges to be equal, the complete formula must be $Cu(NO_3)_2$.

Common names are often used on consumer ingredient labels. For example, stannous fluoride, $SnF_2$, is the source of fluoride in some types of toothpaste.

**Consider This 3.10**

To fortify cereals, some manufacturers add a compound containing iron. To avoid chemical names that may frighten consumers, they list the added iron on the ingredient label as "iron" or "reduced iron." The source of this iron in cereal is typically ferrous sulfate. What is the formula of this additive?

**Practice Problem 3.10**

(a) Using common nomenclature, name the compounds with the formulas $Sn(SO_4)_2$ and $SnSO_4$.

(b) Write formulas for the compounds ferric sulfide and ferrous sulfide.

**Further Practice:** Questions 3.51 and 3.52 at the end of the chapter

## 3.5 Naming and Writing Formulas for Molecular Compounds

Among the substances Megan and Jeff found in their samples of river water were carbon dioxide and carbon tetrachloride. These are binary molecular compounds. The naming of such compounds follows rules that differ slightly from those used for naming ionic compounds. Several molecular compounds are listed in Table 3.8. Try to determine the rules used to name them.

**TABLE 3.8** Names of Some Common Molecular Compounds

| Formula | Name | Formula | Name |
|---------|------|---------|------|
| $CO$ | carbon monoxide | $SO_3$ | sulfur trioxide |
| $CO_2$ | carbon dioxide | $N_2O_4$ | dinitrogen tetroxide |
| $CCl_4$ | carbon tetrachloride | $PF_5$ | phosphorus pentafluoride |

Did you figure out how to name binary molecular compounds? Did you determine when and how to use prefixes? Now try to apply the rules to name the molecular compounds in Example 3.11.

---

**EXAMPLE 3.11**    **Formulas and Names for Molecular Compounds**

Name the compounds (a) $P_4O_{10}$ and (b) $NO_2$.

**Solution:**

(a) Since $P_4O_{10}$ is a compound containing two nonmetals, we consider it a molecular compound. We start by naming the first element in the formula. Because there are four phosphorus atoms in the molecule, we add the prefix *tetra–*. Oxygen is named second by taking the root of its name and adding the suffix *–ide*. We must add the prefix *deca–* to the beginning of the element name to indicate that there are 10 oxygen atoms in the molecule. The full name is *tetraphosphorus decoxide*. (When two vowels are adjacent, as in deca-oxide, the vowel at the end of the prefix is often dropped.)

(b) The second compound, $NO_2$, is also molecular, because it contains elements that are nonmetals. We start by naming the first element in the formula. Because there is only one atom of nitrogen in a molecule of the compound, the prefix *mono–* is omitted but implied. The prefix *mono–* is not used for the first element named in a molecular compound. Oxygen is named second by taking the root of its name and adding the suffix *–ide*. We add the prefix *di–* to the beginning of the element name to indicate that there are two oxygen atoms in the molecule. The full name is *nitrogen dioxide*.

**Consider This 3.11**

Laughing gas, $N_2O$, has the older common name nitrous oxide. What is the systematic name for laughing gas?

**Practice Problem 3.11**

Name the compounds $P_4O_6$ and $N_2O_5$.

**Further Practice:** Questions 3.61 and 3.62 at the end of the chapter

If you had trouble establishing rules for naming molecular compounds, the following summary may help. Because some nonmetals can combine to form more than one compound, the name for a molecular compound must communicate the specific number of atoms of each element in the molecule. For example, there are two common compounds containing carbon and oxygen (Figure 3.26). One of them, CO, contains one atom each of carbon and oxygen. Its name is carbon monoxide. (The extra vowel o is dropped from mono-oxide.) The other compound, $CO_2$, contains one atom of carbon and two atoms of oxygen. It is called carbon dioxide.

To name molecular compounds systematically, we name the elements in the order that they appear in the formula and to the root of the second element, we add an *–ide* ending. If we are not provided with a formula, the first element named is generally the one farther to the left and farther down on the periodic table. Unlike ionic compounds, we use prefixes such as *mono–, di–,* and *tri–* to communicate the number of atoms of each element in a molecule. (These and other prefixes are listed in Table 3.9.)

**FIGURE 3.26** Carbon and oxygen form two common molecular compounds.

Molecular compounds are named with prefixes; ionic compounds are not.

**TABLE 3.9** Common Greek Prefixes

| Prefix | Number | Prefix | Number | Prefix | Number |
|--------|--------|--------|--------|--------|--------|
| mono- | 1 | penta- | 5 | octa- | 8 |
| di- | 2 | hexa- | 6 | nona- | 9 |
| tri- | 3 | hepta- | 7 | deca- | 10 |
| tetra- | 4 | | | | |

The prefix *mono–* is generally not used except in carbon monoxide, CO, and nitrogen monoxide, NO.

Let's consider a few compounds to clarify and practice the naming rules. First, consider a compound of sulfur and oxygen in which a molecule contains one sulfur atom and two oxygen atoms. We write the formula as $SO_2$, with sulfur listed first, because sulfur is farther to the left and farther down on the periodic table (Figure 3.27). Similarly, when we name the compound, sulfur will be identified first. Since there is only one sulfur atom in the molecule, a prefix is not necessary. To the root of the name for the second element in the compound, we add an *–ide* ending, as we did for ionic compounds. We call this compound sulfur dioxide to distinguish it from other compounds containing sulfur and oxygen. Another such compound, $SO_3$, is shown in Figure 3.27. What is its name?

As another example, consider a compound with the formula HCl. Using the rules just discussed, its name would be hydrogen monochloride. However, the prefix *mono–* is not used because there are no other compounds with this combination

**FIGURE 3.27** These images represent two molecular compounds containing atoms of sulfur and oxygen.

The hydrogen halides like HCl are named as molecular compounds when in their pure state. When HCl is dissolved in water, it becomes HCl(*aq*) and is named as an acid. Rules for naming acids are discussed in Section 3.6.

Name the first element in the formula by stating the element name. If there is more than one atom present in a molecule, use a Greek prefix.

State the root of the element name for the second element in the formula and add an –*ide* suffix. If there is more than one atom present in a molecule, use a Greek prefix. For example, $SO_2$ is sulfur dioxide.

**FIGURE 3.28** A flowchart can be used to determine the names for binary molecular compounds.

of elements. Its name is simply hydrogen chloride. This pattern of naming is similar for HF, HBr, and HI. Rules for naming binary molecular compounds are summarized in Figure 3.28.

Knowing the process by which compounds are named allows us to write formulas for molecular compounds from chemical names. Unlike the formula for an ionic compound, which reveals the simplest ratio of cations and anions in a formula unit, a **molecular formula** tells us the actual number of atoms present in a single molecule of a compound. The formula $CO_2$, for example, shows that each molecule is made up of one carbon atom and two oxygen atoms. Consider how molecular formulas are written in Example 3.12.

---

**EXAMPLE 3.12**    **Writing Formulas from Names for Molecular Compounds**

What is the formula for a compound with the name chlorine dioxide?

**Solution:**
The first word indicates that chlorine is present in the compound and its symbol should be listed first in the formula. Since there is no prefix preceding its name, one atom is understood. The second element in the formula is oxygen. The prefix *di–* indicates that there are two oxygen atoms in a molecule of this compound. The formula is $ClO_2$.

**Consider This 3.12**
How does $SO_3$ differ from $SO_3^{2-}$?

**Practice Problem 3.12**
What is the formula for a compound with the name dinitrogen trioxide?

**Further Practice:** Questions 3.63 and 3.64 at the end of the chapter

---

In a water molecule, the oxygen atom is the central atom. The formula $H_2O$ is an exception to listing the central atom first in a formula. Other exceptions are $H_2S$, $H_2Se$, and $H_2Te$.

Some binary molecular compounds are known by *trivial,* or nonsystematic, names. Such names often bear no direct relation to the compound's composition. For example, dihydrogen monoxide, $H_2O$, is commonly known by its trivial name, water. Ammonia, $NH_3$, and hydrogen peroxide, $H_2O_2$, are other examples.

## 3.6  Acids and Bases

The "acid" in acid rain comes from oxides of sulfur and nitrogen, released into the air as gases from the burning of coal and other fuels. The oxides react with water in the air to form sulfuric and nitric acids, which can damage grasslands, forests, and buildings, and threaten the health of humans and other animals.

Megan and Jeff tested their river water samples using an instrument called a pH meter. It revealed that some of their samples tended to be acidic, while others were more basic. An **acid,** simply defined, is a substance that when dissolved in water provides hydrogen ions, $H^+$. When a compound that is not ionic provides ions when dissolved in water, the process is called *ionization.*

To understand the behavior of acids, consider hydrogen chloride, HCl. By itself this substance is a gas under normal conditions (room temperature and atmospheric pressure). It is a molecular compound that, like other acids, behaves quite differently from the molecular compounds discussed earlier. When gaseous HCl is bubbled into water, it ionizes, producing hydrogen ions, $H^+$, and chloride ions, $Cl^-$ (Figure 3.29).

Gastric acid, also called stomach acid, is a solution of hydrochloric acid, HCl($aq$).

**FIGURE 3.29** The molecular compound HCl, present in the gas bubbles, ionizes to H$^+$($aq$) and Cl$^-$($aq$) when dissolved in water. H$^+$($aq$) is associated with one or more water molecules.
©Brian Moeskau/Moeskau Photography

In general, acids are compounds that usually contain a hydrogen atom that can be removed as an H$^+$ ion when the compound is dissolved in water. The process can be summarized for hydrogen chloride by the following equation:

$$HCl(g) \xrightarrow{H_2O} H^+(aq) + Cl^-(aq)$$

A hydrogen ion, H$^+$, does not actually exist in solution as an isolated entity. Instead, it is surrounded by several water molecules. An H$^+$ ion in solution is often represented as H$_3$O$^+$ (called *hydronium ion*) to show that the H$^+$ ion is really attached to water molecules.

One important class of acids is the organic (carbon-based) compounds that contain the combination of atoms −CO$_2$H. They are called carboxylic acids. An example is acetic acid, CH$_3$CO$_2$H, the substance present in vinegar that gives it its sour taste. The structure of acetic acid is shown in Figure 3.30. Various organic acids are found in the human body and in many of the foods we eat. For example, citric acid is present in citrus fruits such as lemons, oranges, and grapefruit.

Not all compounds containing hydrogen are acids in water. When hydrogen combines with an element on the far right of the periodic table (excluding the noble gases), the compound is an acid. For example, HF and HCl are both acids, but methane, CH$_4$, and ammonia, NH$_3$, are not.

When writing formulas for acids, we generally place the hydrogen first and write ($aq$) after the formula to indicate that the compound is an acid when dissolved in water. For example, HCl($g$) represents a molecular compound that exists as a gas and we call it hydrogen chloride. The formula HCl($aq$) corresponds to a solution that's been prepared by dissolving gaseous hydrogen chloride in water and obtaining aqueous hydrogen ions and chloride ions. It's called hydrochloric acid.

Hydrogen combined with a polyatomic ion can also behave as an acid. For example, nitric acid, HNO$_3$, ionizes in water according to the equation

$$HNO_3(l) \xrightarrow{H_2O} H^+(aq) + NO_3^-(aq)$$

Many other compounds containing hydrogen and oxoanions are acids called oxoacids. Notice in Figure 3.31 that each hydrogen atom is bound to an oxygen atom in each molecule. When these compounds are dissolved in water, the hydrogen atoms separate from the molecules as H$^+$ ions.

We can represent the ionization process in a few ways:

$$HCl(g) \xrightarrow{H_2O} HCl(aq)$$
$$HCl(g) \xrightarrow{H_2O} H^+(aq) + Cl^-(aq)$$
$$HCl(g) + H_2O(l) \longrightarrow H_3O^+(aq) + Cl^-(aq)$$

These expressions represent the same process, but the last one is the most complete.

Acidic hydrogen

**FIGURE 3.30** A class of organic (carbon-containing) compounds contains a −CO$_2$H group. These compounds are called carboxylic acids. The hydrogen atom attached to an oxygen atom is acidic; the others are not. Shown here are three ways of representing acetic acid.

| $HNO_3$ | $H_2SO_4$ | $H_2CO_3$ |
| Nitric acid | Sulfuric acid | Carbonic acid |

**FIGURE 3.31** In oxoacids, the acidic hydrogen atoms are attached to oxygen atoms.

A **base,** simply defined, is a substance that reacts with an acid in aqueous solution to form water. Most common bases either contain hydroxide ion, $OH^-$, or can provide $OH^-$ ions in solution. For example, sodium hydroxide, NaOH, is an active ingredient in chemical drain cleaners. It is a base that dissolves in water according to the equation

$$NaOH(s) \xrightarrow{H_2O} Na^+(aq) + OH^-(aq)$$

If NaOH were added to a solution containing an acid, the $OH^-$ ion from NaOH would react with the $H^+$ ion from the acid to form water according to the equation

$$H^+(aq) + OH^-(aq) \longrightarrow H_2O(l)$$

An example of a common base that does not contain hydroxide ion is ammonia, $NH_3$, which is used in various household cleaners. Its behavior is not apparent from its formula. When gaseous ammonia is bubbled into water, a reaction occurs in which some of the ammonia molecules produce aqueous ammonium ions, $NH_4^+(aq)$, and hydroxide ions, $OH^-(aq)$ (Figure 3.32).

Many consumer products contain acids or bases. Some examples are shown in Figure 3.33.

Acids and bases are electrolytes that vary in the extent of their ionization or dissociation in water. Strong acids and bases ionize or dissociate completely. They are strong electrolytes. For example, hydrochloric acid is a strong acid used in many industrial and cleaning applications, including some toilet bowl cleaners. It ionizes completely in water to produce ions, so in aqueous solution it is a strong electrolyte. You may encounter several strong acids while working in the laboratory. Some of them are listed in Table 3.10. They must be handled with care, as they can burn skin

Some common bases are NaOH, KOH, $Mg(OH)_2$, and $NH_3$.

**FIGURE 3.32** When ammonia, $NH_3$, is bubbled into water, some of the $NH_3$ molecules react with water to produce ammonium ions, $NH_4^+$, and hydroxide ions, $OH^-$.
©Brian Moeskau/Moeskau Photography

**FIGURE 3.33** Many consumer products exhibit acidic or basic properties. For example, window cleaners often contain ammonia and sodas often contain phosphoric acid or citric acid. The next time you are in the grocery store, look at the labels on some products. How can you determine which contain acids or bases?
©Brian Moeskau/Moeskau Photography

**TABLE 3.10** Some Strong Acids

| Formula | Name | Formula | Name |
| --- | --- | --- | --- |
| HCl(aq) | hydrochloric acid | $H_2SO_4(aq)$ | sulfuric acid |
| $HNO_3(aq)$ | nitric acid | $HClO_4(aq)$ | perchloric acid |

**TABLE 3.11** Names of Some Common Acids

| Formula | Name | Formula | Name |
|---|---|---|---|
| HF($aq$) | hydrofluoric acid | H$_2$SO$_4$($aq$) | sulfuric acid |
| HCl($aq$) | hydrochloric acid | H$_2$SO$_3$($aq$) | sulfurous acid |
| HI($aq$) | hydroiodic acid | HClO$_4$($aq$) | perchloric acid |
| H$_2$S($aq$) | hydrosulfuric acid | HClO$_3$($aq$) | chloric acid |
| H$_2$CO$_3$($aq$) | carbonic acid | HClO$_2$($aq$) | chlorous acid |
| HNO$_3$($aq$) | nitric acid | HClO($aq$) | hypochlorous acid |

Hydrochloric acid

Acetic acid

**FIGURE 3.34** Hydrochloric acid ionizes completely. Most acetic acid molecules, on the other hand, remain intact.

and eyes and eat holes in clothing. Acetic acid, on the other hand, is present in vinegar, which can be consumed safely. In water, relatively few of its molecules ionize. It is both a weak electrolyte and a weak acid. The behaviors of hydrochloric acid and acetic acid are represented in Figure 3.34.

Like the nomenclature of ionic and molecular compounds, the naming of acids follows systematic rules. Table 3.11 shows the formulas and names for many common acids. Can you determine the rules for naming acids?

Did you figure out how to name binary acids? Did you determine how to name acids containing polyatomic ions? Now apply your rules to name the acids in Example 3.13.

---

**EXAMPLE 3.13    Naming Acids**

(a) Name the acid with the formula HBr($aq$).
(b) The compound with the formula HNO$_3$($aq$) is called nitric acid. What is the name for the acid with the formula HNO$_2$($aq$)?

**Solution:**

(a) Since HBr($aq$) contains just hydrogen and a nonmetal, it is a binary acid in aqueous solution. Binary acids have a *hydro–* prefix, the stem of the element name to which the hydrogen is attached, and an *–ic* suffix. The stem of the name derives from the name of the element hydrogen is attached to. In this case, the element bromine provides the stem *brom.* Assembling the prefix, the stem, and the suffix, along with the word *acid,* gives the name *hydrobromic acid.*

(b) In acids containing a nonmetal that can form two oxoanions, the one with the higher number of oxygen atoms is named with the root for the oxoanion with an *–ic* ending, followed by the word *acid.* Thus, HNO$_3$($aq$) is nitric acid, as stated in the problem. The oxoanion with fewer oxygens is named with the root for the oxoanion name and an *–ous* ending. The name ends with the word *acid,* so HNO$_2$($aq$) is *nitrous acid.*

**Consider This 3.13**

How does HBr($g$) differ from HBr($aq$)?

**Practice Problem 3.13**

(a) Name the compound H$_2$Se($aq$).
(b) An acid containing an oxoanion of phosphorus with the formula H$_3$PO$_3$($aq$) is named phosphorous acid. What is the name for the acid with the formula H$_3$PO$_4$($aq$)?

**Further Practice:** Questions 3.67 and 3.68 at the end of the chapter

Replacing the *–ate* ending with *–ic* and the *–ite* ending with *–ous* added to the root of the element name applies to most acids of oxoanions. However, naming oxoacids of sulfur and phosphorus is slightly different. For sulfur oxoacids, the *–ic* and *–ous* endings are added to the full element name as in sulfuric acid. For phosphorus oxoacids, the *–ic* and *–ous* endings are added to the root phosphor– as in phosphorous acid.

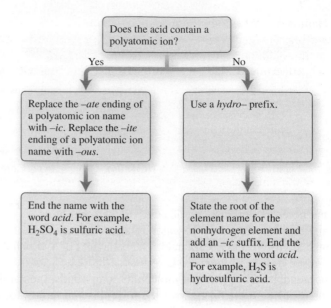

**FIGURE 3.35**  A flowchart can be used to determine the names for acids.

Perhaps the rules you came up with allowed you to complete Example 3.13 successfully. If you had trouble, the following summary for naming acids may help.

- Binary acids in solution are named with the prefix *hydro–* followed by the stem of the name of the nonmetal with the suffix *–ic* and the word *acid* attached. Thus, HCl(*aq*) is named hydrochloric acid.
- Naming acids containing polyatomic ions requires modifying the name of the polyatomic ion. The prefix *hydro–* is not used in naming acids containing polyatomic ions. For example, the compound $H_2CO_3(aq)$, present in many carbonated beverages, contains the carbonate ion. In naming this acid, we remove the *–ate* ending from the name of the polyatomic ion and replace it with *–ic*. Then we add the term *acid* to complete the name. This acid is carbonic acid.
- If there is more than one polyatomic ion for a given nonmetal, the names of the acids must distinguish among them. For example, there are two common polyatomic ions containing sulfur. They are the sulfate ion, $SO_4^{2-}$, and the sulfite ion, $SO_3^{2-}$. The acids of these ions have the formulas $H_2SO_4(aq)$ and $H_2SO_3(aq)$. In naming the acid containing the sulfate ion, the *–ate* ending is replaced with *–ic* added to the element name sulfur and the term *acid* is added at the end. $H_2SO_4(aq)$ is sulfuric acid. Naming the acid containing the sulfite ion requires us to replace the *–ite* ending with *–ous* added to the element name sulfur and add the term *acid* to the end. Therefore, $H_2SO_3(aq)$ is called sulfurous acid.

Rules for naming acids are summarized in Figure 3.35.

Sulfuric acid, $H_2SO_4(aq)$, is commonly used in automobile batteries.

## 3.7 Predicting Properties and Naming Compounds

Jeff and Megan encountered a variety of compounds in their analysis of river water samples. At first they thought that some of the names for compounds were overly complicated, but once they learned the rules, they felt more comfortable. They also realized that by deducing the elements in a compound from its name or formula, they could classify the compound and predict many of its properties. For example, from the compound formula $MgCl_2$ they determine using the periodic table that it contains a metal and a nonmetal. They conclude that $MgCl_2$ is an ionic compound and, therefore, that it probably has some characteristics in common with other ionic compounds. They predict that it might be a brittle solid with a very high melting point. They look up its melting point and find that $MgCl_2$ melts at 712°C. And, because it is an ionic compound, they feel justified in applying the appropriate naming rules and calling it magnesium chloride.

Another ionic compound, $Cr_2(SO_4)_3$, has similar properties. It has the same general physical properties as $MgCl_2$, and its melting point is even higher, 1857°C. Refer back to Table 3.1, where the general properties of ionic and molecular compounds are listed. Now that we know how to name compounds in these classes, we can predict properties based on their formulas or names. In addition to predicting they have high melting points and boiling points, we would correctly predict the ionic compounds $MgCl_2$ and $Cr_2(SO_4)_3$ would be crystalline solids, be strong electrolytes in water, and conduct electricity when molten or when dissolved in water.

Remember that it's important to classify a compound before starting to name it. If you don't know which rule to apply to name a particular compound, the flowchart in Figure 3.36 provides some guidance. The questions are designed to help distinguish among ionic compounds, molecular compounds, and acids because they are named differently. For example, consider $Cr_2(SO_4)_3$. This compound contains a metal, so we follow the "Yes" branch in response to the first question. It is an ionic compound. The metal exhibits more than one charge, so a Roman numeral is necessary in its systematic name. The metal ion is chromium in the 3+ state, so the first part of the name will be chromium(III). The compound consists of the polyatomic ion $SO_4^{2-}$, which is the sulfate ion, so the name ends with sulfate. The name of this compound is chromium(III) sulfate.

The compound $H_2S$ does not contain a metal (or ammonium ion), so it is not an ionic compound. It is composed of nonmetals, and its formula does not show that it is dissolved in water, so we will not name it as an acid. From Figure 3.36 we see that it is a molecular compound, named dihydrogen sulfide. From the properties of molecular compounds listed in Table 3.1, we would predict that $H_2S$ has relatively low

> Although we use prefixes for molecular compounds (e.g., $CCl_4$ is carbon tetrachloride), we never use prefixes for ionic compounds (e.g., $AlCl_3$ is aluminum chloride).

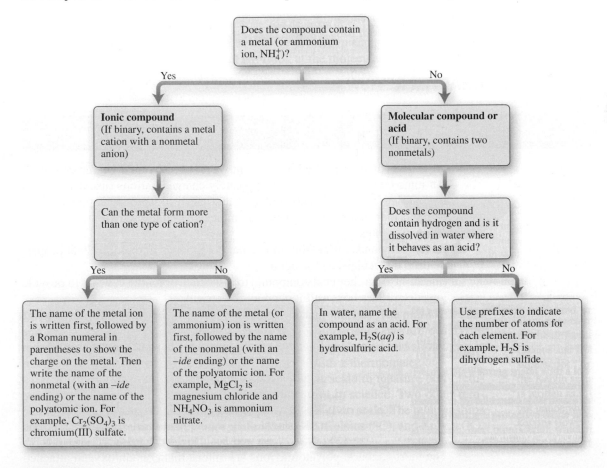

**FIGURE 3.36** A flowchart can be used to determine which rules to apply in naming compounds.

melting and boiling points. Indeed, this is the case, because its melting and boiling points are −85.5°C and −60.7°C, respectively. It is a gas at, and well below, room temperature. When written as $H_2S(aq)$, however, this substance is classified as an acid and is named hydrosulfuric acid. We would correctly predict that it is an electrolyte and therefore its solutions conduct electricity. When naming a compound, it is very important to first classify it. Use the flowchart to name the compounds in Example 3.14.

---

| **EXAMPLE 3.14** | **Using a Flowchart to Name Compounds** |

Classify and name the compounds with the formulas (a) $CCl_4$ and (b) $H_3PO_4(aq)$.

**Solution:**

(a) The compound $CCl_4$ is molecular because it is composed of two nonmetals. We list the carbon first, followed by the prefix *tetra–* to indicate four chlorine atoms. The name of this compound is *carbon tetrachloride*.

(b) The compound $H_3PO_4(aq)$ is an acid containing the phosphate ion. To name this compound we replace the *–ate* ending with *–ic* added to the root *phosphor–* and add the word *acid*. This compound is *phosphoric acid*.

**Consider This 3.14**

What is incorrect about the name dicopper sulfide for a compound with the formula $Cu_2SO_4$?

**Practice Problem 3.14**

Classify and name the compounds with the formulas (a) $PBr_3$, (b) $Na_2S$, (c) $H_2SO_4(aq)$, and (d) $NH_4Cl$.

**Further Practice:** Questions 3.75 and 3.76 at the end of the chapter

---

## CHAPTER REVIEW

## KEY CONCEPTS

- Compounds can be classified in different ways based on their characteristics.
  - An ionic compound consists of oppositely charged cations (usually a metal ion) and anions (usually a nonmetal ion that can be monatomic or polyatomic).
  - A molecular compound consists of molecules composed of two or more nonmetal atoms.
  - Ionic compounds differ from molecular compounds in their physical properties and molecular-level structure.
  - The ability of a chemical compound to dissociate or ionize in water to provide ions determines how well it conducts electricity.
  - Compounds that dissociate or ionize extensively into ions are strong electrolytes.
  - Substances, such as weak acids, that ionize only partially in water are weak electrolytes.
  - Those that do not dissociate or ionize when dissolved are nonelectrolytes.
- Ions can be monatomic or polyatomic.
  - A monatomic ion consists of a single atom.
  - A polyatomic ion consists of two or more atoms, usually of more than one element.
  - The most common polyatomic ions are anions that contain oxygen attached to some other element, usually a nonmetal. These polyatomic ions are called oxoanions.

- ○ Most polyatomic ions are negatively charged, but some have positive charge, like $NH_4^+$.
- Formulas and names distinguish one compound (or ion) from all others.
  - ○ Formulas for ionic compounds can only tell us the ratio of cations to anions in a compound that gives a net charge of zero.
  - ○ Formulas for molecular compounds indicate the number of atoms of each element in a molecule of a compound.
  - ○ Ionic compounds are named by listing the name of the cation first, followed by the name of the anion.
  - ○ For ionic compounds that contain a metal that can exhibit multiple charges, a Roman numeral is inserted after the cation name to indicate the charge.
  - ○ In naming molecular compounds, the first atom in the formula (the element farther down or to the left in the periodic table) is named first with the second element named as if it were an anion. Greek prefixes are used to designate the number of atoms in the molecule.
- Within the broad categories of ionic and molecular compounds, some chemical compounds can be described in terms of their acidic or basic properties.
  - ○ Acids are compounds that provide hydrogen ions in solution.
  - ○ Bases react with acids to form water and often provide hydroxide ions in solution.
  - ○ Binary acids are named as *hydro–* followed by the root of the element name with an *–ic* suffix and the word *acid* placed at the end of the name.
  - ○ Generally, acids containing polyatomic ions are named by taking the root of the polyatomic ion name, replacing *–ate* with *–ic* or replacing *–ite* with *–ous* and adding the word *acid* at the end.

## KEY TERMS

| | | | |
|---|---|---|---|
| acid (3.6) | formula unit (3.3) | monatomic ion (3.2) | strong electrolyte (3.1) |
| base (3.6) | ionic compound (3.1) | nonelectrolyte (3.1) | weak electrolyte (3.1) |
| binary compound (3.1) | molecular compound (3.1) | oxoanion (3.2) | |
| electrolyte (3.1) | molecular formula (3.5) | polyatomic ion (3.2) | |

## QUESTIONS AND PROBLEMS

The following questions and problems, except those in *Additional Questions* and *Concept Review Questions,* are paired. Questions in a pair focus on the same concept. Answers to selected questions and problems are in Appendix D.

### Matching Definitions with Key Terms

3.1    Match the key terms with the descriptions provided.
  (a) the smallest repeating unit in an ionic compound
  (b) a substance that dissociates or ionizes completely into ions in solution and conducts electricity well
  (c) a compound composed of two or more nonmetals that exists in discrete units of atoms held together
  (d) a compound that releases hydrogen ions when dissolved in water
  (e) a substance that retains its molecular identity when it dissolves and does not conduct electricity
  (f) an anion containing oxygen attached to some other element

3.2    Match the key terms with the descriptions provided.
  (a) a formula that indicates the actual number of atoms present in a molecule
  (b) a substance that dissociates partially into ions upon dissolving and conducts electricity to a slight extent

  (c) a compound consisting of cations and anions in proportions that give electrical neutrality
  (d) a substance that reacts with an acid to form water
  (e) a compound containing atoms or ions of two elements
  (f) an ion containing two or more atoms, such as $NO_3^-$

### Ionic and Molecular Compounds

3.3    Identify the following compounds containing sulfur as representations of ionic or molecular compounds.

$K_2S$            $Na_2SO_4$            $SO_2$

3.4    Identify the following compounds containing nitrogen as representations of ionic or molecular compounds.

Na₃N

KNO₃                    NO₂

3.5    Determine whether a compound containing each of the following combinations of elements is an ionic or molecular compound.
   (a) nitrogen and oxygen
   (b) potassium and oxygen
   (c) phosphorus and fluorine
   (d) magnesium and chlorine

3.6    Determine whether a compound containing each of the following combinations of elements is an ionic or molecular compound.
   (a) nitrogen and chlorine
   (b) sulfur and oxygen
   (c) calcium and chlorine
   (d) sodium and nitrogen

3.7    Determine if the following formulas represent ionic or molecular compounds.
   (a) $PCl_5$              (c) $BaCl_2$
   (b) LiF                (d) $N_2O_5$

3.8    Determine if the following formulas represent ionic or molecular compounds.
   (a) $Al_2S_3$            (c) $MgBr_2$
   (b) $P_2O_5$            (d) $K_2O$

3.9    Which of the compounds LiF, $CO_2$, or $N_2O_5$ is expected to have the highest melting point?

3.10   Which of the compounds NaCl, CO, or KF is expected to have the lowest melting point?

## Monatomic and Polyatomic Ions

3.11   Based on their positions in the periodic table, predict the charge, write the formula, and name of the monatomic ion formed by each of the following elements.
   (a) sodium      (b) potassium       (c) rubidium

3.12   Based on their positions in the periodic table, predict the charge, write the formula, and name the monatomic ion formed by each of the following elements.
   (a) fluorine     (b) chlorine        (c) iodine

3.13   Based on their positions in the periodic table, predict the charge, write the formula, and name the monatomic ion formed by each of the following elements.
   (a) calcium      (b) nitrogen        (c) sulfur

3.14   Based on their positions in the periodic table, predict the charge, write the formula, and name the monatomic ion formed by each of the following elements.
   (a) aluminum    (b) barium         (c) phosphorus

3.15   The following image represents an oxoanion of nitrogen with a 1− charge. Write a formula for this ion and name it.

3.16   The following image represents an oxoanion of nitrogen with a 1− charge. Write a formula for this ion and name it.

3.17   Name the following polyatomic ions.
   (a) $SO_4^{2-}$        (b) $OH^-$          (c) $ClO_4^-$

3.18   Name the following polyatomic ions.
   (a) $SO_3^{2-}$        (b) $CN^-$          (c) $ClO_2^-$

3.19   Write formulas for the following ions.
   (a) nitride       (b) nitrate      (c) nitrite

3.20   Write formulas for the following ions.
   (a) chlorite      (b) chloride     (c) chlorate

3.21   Write formulas for the polyatomic ions with the following names.
   (a) carbonate
   (b) ammonium
   (c) hydroxide
   (d) permanganate

3.22   Write formulas for the polyatomic ions with the following names.
   (a) bicarbonate
   (b) acetate
   (c) periodate
   (d) hypochlorite

3.23   This image is a representation for an oxoanion of sulfur with a 2− charge. Write the formula and name this ion.

3.24   This image is a representation for an oxoanion of nitrogen with a 1− charge. Write the formula and name this ion.

3.25   The iodate ion is an oxoanion of iodine bearing a negative one (1−) charge and containing three oxygen atoms. Write the formula for the iodate ion.

3.26   The selenite ion is an oxoanion of selenium bearing a negative two (2−) charge and containing three oxygen atoms. Write the formula for the selenite ion.

## Formulas for Ionic Compounds

**3.27** Suppose an ionic compound containing aluminum and chloride ions dissolves in water. Beginning with a representation for five aluminum ions, draw a picture of a solution that is electrically neutral.

**3.28** Suppose an ionic compound containing magnesium and nitrate ions dissolves in water. Beginning with a representation for five magnesium ions, draw a picture of a solution that is electrically neutral.

**3.29** Write the formula of the compound formed by each of the following sets of ions.
(a) $Ba^{2+}$ and $Cl^-$ 　　(c) $Ca^{2+}$ and $PO_4^{3-}$
(b) $Fe^{3+}$ and $Br^-$ 　　(d) $Cr^{3+}$ and $SO_4^{2-}$

**3.30** Write the formula of the compound formed by each of the following sets of ions.
(a) $Na^+$ and $N^{3-}$ 　　(c) $Ti^{3+}$ and $CO_3^{2-}$
(b) $Cs^+$ and $ClO_3^-$ 　　(d) $NH_4^+$ and $S^{2-}$

**3.31** Identify the ions in the compounds represented in the following formulas.
(a) KBr 　　(c) $Mg_3(PO_4)_2$
(b) $BaCl_2$ 　　(d) $Co(NO_3)_2$

**3.32** Identify the ions in the compounds represented in the following formulas.
(a) NaBr
(b) $AlCl_3$
(c) $Ba_3(PO_4)_2$
(d) $Mn(NO_3)_2$

**3.33** Iron forms two ions, $Fe^{2+}$ and $Fe^{3+}$.
(a) What are the formulas for the compounds the two iron ions can form with oxide ion, $O^{2-}$?
(b) What are the formulas for the compounds the two iron ions can form with chloride ion, $Cl^-$?

**3.34** Two ions of chromium are $Cr^{2+}$ and $Cr^{3+}$.
(a) What are the formulas for the compounds the two chromium ions can form with oxide ion, $O^{2-}$?
(b) What are the formulas for the two compounds the chromium ions can form with chloride ion, $Cl^-$?

**3.35** Sodium ion has a 1+ charge: $Na^+$. In combination with the sulfate ion, it forms the compound $Na_2SO_4$.
(a) What is the charge on the sulfate ion?
(b) What is the charge on a strontium ion if it forms the compound $SrSO_4$?

**3.36** Magnesium has a 2+ charge: $Mg^{2+}$. In combination with the nitrate ion, it forms the compound $Mg(NO_3)_2$.
(a) What is the charge on the nitrate ion?
(b) What is the charge on a potassium ion if it forms the compound $KNO_3$?

**3.37** The listed formulas are *incorrect*. Determine what is wrong with each and correct it.
(a) $NaCl_2$
(b) $KSO_4$
(c) $Al_3NO_3$

**3.38** The listed formulas are *incorrect*. Determine what is wrong with each and correct it.
(a) MgCl
(b) $Na(SO_4)_2$
(c) $K_2(NO_3)_2$

## Naming Ionic Compounds

**3.39** Name the following ionic compounds.
(a) $MgCl_2$ 　　(d) KBr
(b) $Al_2O_3$ 　　(e) $NaNO_3$
(c) $Na_2S$ 　　(f) $NaClO_4$

**3.40** Name the following ionic compounds.
(a) $BaCl_2$ 　　(d) $KNO_2$
(b) $(NH_4)_2S$ 　　(e) $Mg_3(PO_4)_2$
(c) MgO 　　(f) $KClO_2$

**3.41** For which of the following compounds must we specify the charge on the metal when we name the compound: $MgSO_4$, $MnSO_4$, $CaCl_2$, $CoCl_2$, $AgNO_3$?

**3.42** For which of the following compounds must we specify the charge on the metal when we name the compound: $Cr(NO_3)_2$, $Ba(NO_3)_2$, $AlBr_3$, NiS, $ZnCl_2$?

**3.43** Name the following ionic compounds.
(a) $Cu_2O$ 　　(c) $FePO_4$
(b) $CrCl_2$ 　　(d) CuS

**3.44** Name the following ionic compounds.
(a) $TiCl_3$ 　　(c) $Mn_2O_3$
(b) $CoBr_2$ 　　(d) $Fe_3(PO_4)_2$

**3.45** Write formulas for the following ionic compounds.
(a) calcium sulfate
(b) barium oxide
(c) ammonium sulfate
(d) barium carbonate
(e) sodium chlorate

**3.46** Write formulas for the following ionic compounds.
(a) potassium bicarbonate
(b) sodium oxide
(c) ammonium hydroxide
(d) magnesium bromide
(e) sodium hypochlorite

**3.47** Given the following formulas, what is the charge on the metal? What is the name of each compound?
(a) $CoCl_2$
(b) $PbO_2$
(c) $Cr(NO_3)_3$
(d) $Fe_2(SO_4)_3$

**3.48** Given the following formulas, what is the charge on the metal? What is the name of each compound?
(a) $CoBr_3$
(b) $CuCl_2$
(c) $Cr_2O_3$
(d) $FeSO_4$

**3.49** Write formulas for the following ionic compounds.
(a) cobalt(II) chloride
(b) manganese(II) nitrate
(c) chromium(III) oxide
(d) copper(II) phosphate

**3.50** Write formulas for the following ionic compounds.
(a) copper(II) bromide
(b) iron(III) nitrate
(c) chromium(VI) oxide
(d) copper(II) chlorate

**3.51** What are the common names for $Fe(NO_3)_2$ and $Fe(NO_3)_3$?

**3.52** What are the common names for $Cu_2SO_4$ and $CuSO_4$?

3.53  Complete the following table by writing the names and formulas for the compounds formed when the cations listed across the top combine with the anions listed along the side.

|        | $Ca^{2+}$ | $Fe^{2+}$ | $K^+$ |
|--------|-----------|-----------|-------|
| $Cl^-$ |  |  |  |
| $O^{2-}$ |  |  |  |
| $NO_3^-$ |  |  |  |
| $SO_3^{2-}$ |  |  |  |
| $OH^-$ |  |  |  |
| $ClO_3^-$ |  |  |  |

|        | $Mn^{2+}$ | $Al^{3+}$ | $NH_4^+$ |
|--------|-----------|-----------|----------|
| $Cl^-$ |  |  |  |
| $O^{2-}$ |  |  |  |
| $NO_3^-$ |  |  |  |
| $SO_3^{2-}$ |  |  |  |
| $OH^-$ |  |  |  |
| $ClO_3^-$ |  |  |  |

3.54  Complete the following table by writing the names and formulas for the compounds formed when the cations listed across the top combine with the anions listed along the side.

|        | $Ba^{2+}$ | $Fe^{3+}$ | $Na^+$ |
|--------|-----------|-----------|--------|
| $Br^-$ |  |  |  |
| $S^{2-}$ |  |  |  |
| $NO_2^-$ |  |  |  |
| $SO_4^{2-}$ |  |  |  |
| $HCO_3^-$ |  |  |  |
| $ClO_4^-$ |  |  |  |

|        | $Ni^{2+}$ | $Cr^{3+}$ | $Ag^+$ |
|--------|-----------|-----------|--------|
| $Br^-$ |  |  |  |
| $S^{2-}$ |  |  |  |
| $NO_2^-$ |  |  |  |
| $SO_4^{2-}$ |  |  |  |
| $HCO_3^-$ |  |  |  |
| $ClO_4^-$ |  |  |  |

3.55  Complete the following table by writing the formulas for the compounds formed when the cations listed across the top combine with the anions listed along the side.

|             | potassium | iron(III) | strontium |
|-------------|-----------|-----------|-----------|
| iodide      |  |  |  |
| oxide       |  |  |  |
| sulfate     |  |  |  |
| nitrite     |  |  |  |
| acetate     |  |  |  |
| hypochlorite |  |  |  |

|             | aluminum | cobalt(II) | lead(IV) |
|-------------|----------|------------|----------|
| iodide      |  |  |  |
| oxide       |  |  |  |
| sulfate     |  |  |  |
| nitrite     |  |  |  |
| acetate     |  |  |  |
| hypochlorite |  |  |  |

3.56  Complete the following table by writing the formulas for the compounds formed when the cations listed across the top combine with the anions listed along the side.

|           | sodium | chromium(II) | calcium |
|-----------|--------|--------------|---------|
| chloride  |  |  |  |
| sulfide   |  |  |  |
| nitrate   |  |  |  |
| sulfite   |  |  |  |
| hydroxide |  |  |  |
| chlorate  |  |  |  |

|           | ammonium | iron(III) | lead(II) |
|-----------|----------|-----------|----------|
| chloride  |  |  |  |
| sulfide   |  |  |  |
| nitrate   |  |  |  |
| sulfite   |  |  |  |
| hydroxide |  |  |  |
| chlorate  |  |  |  |

3.57  What is the formula for silver chloride?

3.58  What is the formula for zinc chloride?

## Naming and Writing Formulas for Molecular Compounds

3.59 Write formulas for the molecular compounds represented by the following images.

(a)

(b)

(c)

3.60 Write formulas for the molecular compounds represented by the following images.

(a)

(b)

(c)

3.61 Name the following compounds.
   (a) $PF_5$         (c) CO
   (b) $PF_3$         (d) $SO_2$

3.62 Name the following compounds.
   (a) $SO_3$        (c) $NO_2$
   (b) $N_2O_4$      (d) $CS_2$

3.63 Write formulas for the following compounds.
   (a) sulfur tetrafluoride
   (b) tricarbon dioxide
   (c) chlorine dioxide
   (d) sulfur dioxide

3.64 Write formulas for the following compounds.
   (a) carbon disulfide
   (b) dinitrogen pentoxide
   (c) boron nitride
   (d) iodine heptafluoride

## Acids and Bases

3.65 Which of the following images correctly represents phosphoric acid?

A

B

C

3.66 Which of the images shown in Question 3.65 corresponds to phosphorous acid?

3.67 Name the following acids.
   (a) HF(*aq*)
   (b) $HNO_3$(*aq*)
   (c) $H_3PO_3$(*aq*)

3.68 Name the following acids.
   (a) HI(*aq*)
   (b) $HNO_2$(*aq*)
   (c) $H_3PO_4$(*aq*)

3.69 Write formulas for the following acids.
   (a) hydrofluoric acid
   (b) sulfurous acid
   (c) perchloric acid

3.70 Write formulas for the following acids.
   (a) hydroiodic acid
   (b) sulfuric acid
   (c) hypochlorous acid

3.71 When nitric acid ionizes in water, what ions are present in solution?

3.72 When hydrobromic acid ionizes in water, what ions are present in solution?

## Predicting Properties and Naming Compounds

3.73 Which of the following compounds should be crystalline, brittle solids at room temperature and are electrolytes: KI, $Mg(NO_3)_2$, $NO_2$, $NH_3$, $NH_4NO_3$?

3.74 Which of the following compounds should have relatively low melting points and boiling points and be nonconductors of electricity when in the liquid state: $CH_3OH$, $Li_2O$, $CaCO_3$, $H_2O$, $NO_3$?

3.75 Many sunscreens contain tiny particles (nanoparticles) of either titanium(IV) oxide or zinc oxide. These compounds have the ability to filter a broader range of ultraviolet light than other sunscreening agents. Classify these compounds as ionic or molecular, and write their formulas.

3.76 An injection of ammonium chloride can help correct blood pH levels that are too high (alkaline) due to vomiting or the use of diuretics. Classify this compound as ionic or molecular, and write its formula.

3.77 Carbon dioxide and dinitrogen oxide are two principal greenhouse gases that enter the atmosphere because of human activities. Classify these compounds as ionic or molecular, and write their formulas.

3.78 Sulfur trioxide and nitrogen dioxide gases mix with rainwater to form sulfuric acid and nitric acid. Write the formulas for both gases and both acids.

3.79 Name the compounds represented in Question 3.3.

3.80 Name the compounds represented in Question 3.4.

3.81 Name the following compounds or ions. Identify each compound as molecular or ionic.
 (a) $NO_3$      (e) $AlCl_3$
 (b) $NO_3^-$      (f) $PCl_3$
 (c) $KNO_3$      (g) $TiO$
 (d) $Na_3N$      (h) $MgO$

3.82 Name the following compounds or ions. Identify each compound as molecular or ionic.
 (a) $SO_4^{2-}$      (e) $CaCO_3$
 (b) $S^{2-}$      (f) $Fe_2S_3$
 (c) $CO_2$      (g) $CS_2$
 (d) $CO_3^{2-}$      (h) $Na_2S$

3.83 Write the formula for each of the following.
 (a) sodium carbonate      (e) sulfur trioxide
 (b) sodium bicarbonate      (f) copper(II) sulfate
 (c) carbonic acid      (g) sulfuric acid
 (d) hydrofluoric acid      (h) hydrosulfuric acid

3.84 Write the formula for each of the following.
 (a) ammonia      (e) chromium(III) oxide
 (b) ammonium phosphate      (f) magnesium nitrate
 (c) hydrochloric acid      (g) manganese(IV) oxide
 (d) potassium hydroxide      (h) nitrogen dioxide

3.85 How do $HCl(aq)$ and $HCl(g)$ differ? How are they named differently?

3.86 How do $H_2S(aq)$ and $H_2S(g)$ differ? How are they named differently?

3.87 Explain why the name given for each compound is incorrect.
 (a) $Na_2SO_4$, disodium sulfate
 (b) $CaO$, calcium(II) oxide
 (c) $CuO$, copper oxide
 (d) $PCl_5$, phosphorous chloride

3.88 Explain why the name given for each compound is incorrect.
 (a) $NF_3$, nitrogen fluoride
 (b) $NiS$, nickel sulfate
 (c) $Ba(NO_3)_2$, barium(II) nitrate
 (d) $H_2O_2$, dihydrogen dioxide

3.89 Explain why the formula given for each compound is incorrect.
 (a) potassium sulfide, KS
 (b) nickel(II) carbonate, $Ni_2CO_3$
 (c) sodium nitride, $NaNO_3$
 (d) nitrogen triiodide, $N_3I$

3.90 Explain why the formula given for each compound is incorrect.
 (a) calcium hydroxide, $CaOH_2$
 (b) aluminum chloride, $AlCl_2$
 (c) dinitrogen trioxide, $N_3O_2$
 (d) potassium phosphide, $K_3PO_4$

## Additional Questions

3.91 Determine what ions and how many of each are formed for every one formula unit that dissociates in water.
 (a) NaCl      (c) $Na_2SO_4$
 (b) $MgCl_2$      (d) $Ca(NO_3)_2$

3.92 Determine what ions and how many of each are formed for every one formula unit that dissociates in water.
 (a) KBr      (c) $K_3PO_4$
 (b) $NH_4Cl$      (d) $Al(NO_3)_3$

3.93 Indicate whether each of the following substances is an electrolyte or a nonelectrolyte.
 (a) $NaOH(aq)$      (c) $NaCl(aq)$
 (b) $HCl(aq)$      (d) $C_{12}H_{22}O_{11}(aq)$
                       (sucrose solution)

3.94 Indicate whether each of the following substances is an electrolyte or a nonelectrolyte.
 (a) $CH_2Cl_2(aq)$      (c) $KBr(aq)$
 (b) $HNO_3(aq)$      (d) $KOH(aq)$

3.95 Many transition metals form more than one ion and are named using Roman numerals to designate the charge on the ion. Why are Roman numerals unnecessary when naming ionic compounds containing silver, zinc, or cadmium?

3.96 Determine what ions are present when each of the following dissociates or ionizes in water.
 (a) HCl      (d) $KNO_3$
 (b) NaOH      (e) $Na_3PO_4$
 (c) $CaCl_2$

3.97 Write formulas, including charge, for each of the following polyatomic ions.
 (a) nitrate      (e) sulfate
 (b) sulfite      (f) nitrite
 (c) ammonium      (g) perchlorate
 (d) carbonate

3.98 Name the following compounds and mixtures.
 (a) $KBr(s)$      (d) $Na_2SO_4(s)$
 (b) $N_2O_5(g)$      (e) $Fe(NO_3)_3(s)$
 (c) $HBr(aq)$

3.99 Name the following compounds and mixtures.
 (a) $MgBr_2(s)$      (d) $CoCl_3(s)$
 (b) $H_2S(g)$      (e) $KOH(aq)$
 (c) $H_2S(aq)$      (f) $AgBr(s)$

3.100 Name the following compounds.
(a) $NH_3$
(b) $H_2O$
(c) $H_2O_2$

3.101 Write formulas for each of the following compounds.
(a) lead(II) chloride
(b) magnesium phosphate
(c) nitrogen triiodide
(d) iron(III) oxide
(e) calcium nitride
(f) barium hydroxide
(g) dichlorine pentoxide
(h) ammonium chloride

3.102 Suppose you need some iron(III) chloride for an experiment. When you go to the stockroom, there is one bottle labeled $FeCl_2$ and another with the formula $FeCl_3$. Which bottle should you select?

3.103 The active ingredient in baking soda is sodium bicarbonate (also called sodium hydrogen carbonate). What is the formula for this compound?

3.104 An active ingredient in toothpaste is tin(II) fluoride. What is the formula of this compound? What is the common name for this compound?

3.105 Calcium hypochlorite is a compound used in water treatment to kill disease-causing organisms. What is the formula of this compound?

3.106 Calcium hydrogen carbonate, magnesium chloride, and calcium sulfate can cause scale buildup in pipes. What are the formulas for these compounds?

3.107 Aluminum sulfate and calcium oxide are used in water treatment to form a compound that allows solid materials to be removed from water.

$$Al_2(SO_4)_3(aq) + 3CaO(s) + 3H_2O(l) \longrightarrow 2Al(OH)_3(s) + 3CaSO_4(s)$$

Based on their formulas, what substances in this chemical equation are molecular?

3.108 Consider the following formulas and names. Explain what is wrong with each name.
(a) $FeBr_2$, iron dibromide
(b) $CS_2$, copper(IV) sulfide
(c) $Co(NO_3)_3$, cobalt trinitrite
(d) $Mg(OH)_2$, magnesium dihydroxide
(e) $Cu_2O$, dicopper(II) oxide

3.109 When solid copper metal is added to a solution of silver nitrate, solid silver and copper(II) nitrate form. Write formulas for all the substances involved in this process.

3.110 Indicate whether each of the following combinations of elements would form molecular or ionic compounds.
(a) sodium and iodine
(b) selenium and chlorine
(c) oxygen and chlorine
(d) chromium and oxygen

3.111 Write formulas for the following compounds.
(a) ammonia
(b) nitric acid
(c) nitrous acid

3.112 A sports drink contains the ingredients sucrose, whey protein concentrate, fructose, maltodextrin, citric acid, sodium chloride, lecithin, magnesium oxide, vitamin E, and ferrous sulfate. Write formulas for the molecular and ionic compounds present in this product that can be derived from what you learned in this chapter.

3.113 Methane, $CH_4$, is not an acid, but $CH_3CO_2H$ is. Given the following figures, which hydrogen is responsible for the acid properties of the second compound, acetic acid?

Methane                    Acetic acid

3.114 The compounds $CH_4$ and $NH_3$ are not acids, but glycine is. Given the structure shown, which hydrogen is responsible for the acidic properties of this amino acid?

3.115 List similarities and differences between each of the following substances.

3.116 Write formulas for
(a) potassium peroxide
(b) calcium peroxide

3.117 Nasal irrigation (also known as sinus rinse) is used to treat sinus problems. Commercial powders available in drugstores to make a nasal rinse by dissolving in water contain sodium chloride and sodium bicarbonate. Classify these compounds as ionic or molecular, and write their formulas.

3.118 An intravenous injection for patients in shock, called D5W, contains dextrose at 5% in water. Dextrose has the formula $C_6H_{12}O_6$. Classify dextrose as an ionic or a molecular compound and as an electrolyte or a nonelectrolyte.

3.119 A common diaper rash ointment is a creamy mixture containing 10% zinc oxide. Write a formula for zinc oxide, and indicate the charge on each ion in this compound.

3.120 Which of the following compounds is expected to be a crystalline, brittle solid at room temperature: $LiBr$, $FeSO_4$, $H_2CO_3$, $PCl_3$, $CaCO_3$, $CS_2$?

3.121 Which of the following compounds should have a Roman numeral in their names: $ZnCl_2$, $FeCl_2$, $NCl_3$, $CrCl_3$?

3.122 Vanadium forms the compounds $VO$ and $V_2O_3$. What are the charges on the vanadium ions in these compounds?

3.123 Write names for HF in the two physical states, HF(*g*) and HF(*aq*). Which of these two formulas represents an acid? To which class would you assign the substance represented by the other formula?

3.124 Determine the number of ions in one formula unit of each of the following compounds.
(a) $Li_2CO_3$
(b) $NH_4Cl$
(c) $Ca_3(PO_4)_2$
(d) $Na_2O_2$

3.125 Which of the following compounds should have the highest melting point: $PCl_3$, $CrCl_3$, $NCl_3$, $ClO_3$?

3.126 Which of the following formulas represents an acid: $CH_3CO_2H(aq)$, $HBr(aq)$, $NH_3(aq)$, $HCl(g)$?

## Concept Review Questions

3.127 Which of the following statements is correct?
A. Nonelectrolytes are compounds that break apart into ions when dissolved in water and conduct electricity well.
B. Weak electrolytes are compounds that break apart completely into ions when dissolved in water and do not conduct electricity.
C. Compounds that dissociate or ionize completely into ions are considered nonelectrolytes.
D. Strong electrolytes are compounds that break apart into neutral atoms when dissolved in water and do not conduct electricity.
E. Strong electrolytes are compounds that break apart into ions when dissolved in water and conduct electricity well.

Modify each incorrect statement to make it a true statement.

3.128 Which of the following is classified as a molecular compound?
A. $CO_2$           D. $Na_2CO_3$
B. $O_2$            E. $KCl$
C. $CO_3^{2-}$

Classify those that are not molecular compounds.

3.129 Which of the following statements is correct?
A. Molecular compounds are expected to have very high melting and boiling points compared to ionic compounds.
B. Molecular compounds are strong electrolytes in aqueous solution.
C. An ionic compound can be a gas, liquid, or solid at room temperature.
D. Molecular compounds form crystalline solids that are hard and brittle.
E. Ionic compounds are expected to have very high melting and boiling points compared to molecular compounds.

Modify each incorrect statement to make it a true statement.

3.130 Which of the following describes the nitrate ion?
A. one nitrogen atom, two oxygen atoms, and a charge of 1−
B. one nitrogen atom, three oxygen atoms, and a charge of 1−
C. one nitrogen atom and a charge of 3−
D. one nitrogen atom, three oxygen atoms, and a charge of 0
E. one nitrogen atom, three hydrogen atoms, and a charge of 0

Write formulas and names for each response.

3.131 What is the correct formula for aluminum sulfate?
A. $AlSO_3$
B. $AlSO_4$
C. $Al_2S_3$
D. $Al_2(SO_4)_3$
E. $Al_3(SO_4)_2$

What is the correct formula for aluminum sulfide?

3.132 Which of the following compounds contains a metal ion with a charge of 3+?
A. $Cu_3PO_4$
B. $Cr(NO_3)_2$
C. $Mn_2O_3$
D. $PbO_2$
E. $Fe_3(PO_4)_2$

What is the charge on each ion in these compounds?

3.133 What is the correct name for $Cu_2S$?
A. copper sulfide
B. copper(I) sulfide
C. copper(I) sulfite
D. dicopper sulfide
E. copper(II) sulfide

Which of the other names are proper names for other compounds of copper?

3.134 What is the correct formula for dinitrogen trioxide?
A. $NO$            D. $N_2O_3$
B. $NO_3^-$          E. $N_3O_2$
C. $NO_3$

What are the names for the other responses?

3.135 Which of the following behaves as an acid when it dissociates or ionizes in water?
A. $CH_3CO_2H$
B. $NH_3$
C. $CH_4$
D. $NaOH$
E. $KNO_3$

How do the others behave in water?

3.136 Which of the following formula/name pairs is correct?
A. $Cr_2O_3$, chromium(II) trioxide
B. $Fe(NO_3)_3$, iron(III) nitrate
C. $MgOH$, magnesium hydroxide
D. $K_2SO_4$, potassium sulfite
E. $AlSO_3$, aluminum sulfide

What is incorrect about the other responses?

# Chemical Composition

4.1 Percent Composition

4.2 Mole Quantities

4.3 Determining Empirical and Molecular Formulas

4.4 Chemical Composition of Solutions

Math Toolbox 4.1 Mole Conversions

Chapter Review

Questions and Problems

© Royce Bair/Getty Images

**FIGURE 4.1** Many objects we use on a daily basis contain copper. Pipes, coins, and wires are commonly made of copper.

© McGraw-Hill Education/Stephen Frisch

Although the minerals contain large amounts of copper, the ore often contains only small amounts of minerals, so it often contains less than 1% copper by mass. In other words, 100 kg of ore may yield as little as 1 kg of copper.

Nutritional minerals are a subset of geological minerals. The metal elements in nutritional minerals are in ionic forms in foods and vitamins that allow for human consumption.

Copper plays an important role in our daily lives. To see how, let's follow Julio, a college student, through part of his day. Bzzz! The radio alarm goes off and awakens Julio. The circuits in the radio and the wires that connect it to the wall outlet contain copper. Realizing he may be late for class, Julio leaps out of his bed, which has a brass frame. (Brass is a homogeneous mixture of copper and zinc.) He opens the bathroom door and turns on the light. The door handle and hinges are also made of brass. The light switch contains copper contacts, and the base of the light-bulb is made of copper. He turns on the water to wash his face. The sink faucet is brass, coated with nickel and chromium, and copper pipes carry water through the house to his bathroom. Many objects are made of copper (Figure 4.1). In our life-time, most of us will make use of over 1500 lb (680 kg) of copper metal. Do you see any objects made of copper or brass in your environment?

Most of the copper found in nature occurs as part of a mineral. A *mineral* is a naturally occurring element or compound that has a distinct chemical composition and solid form. The most common minerals of copper contain sulfide, oxide, or sili-cate along with other metals (Figure 4.2). Copper is removed from open-pit mines as ore, which is rock containing relatively small amounts of copper minerals.

Pure copper is obtained from the minerals in the ore by a series of physical and chemical processes that depend on the type of mineral: those containing sulfur, such as chalcocite, $Cu_2S$, and those containing oxygen, such as malachite, $Cu_2CO_3(OH)_2$. For example, one step in obtaining copper from copper sulfide minerals involves igniting the mineral in the presence of oxygen to drive off the sulfur as sulfur diox-ide gas. While the percentage of copper in the mined ores starts out very low, the copper obtained after the final stages of processing is nearly 100%.

While Julio is in the bathroom washing his face, he notices that there is a green-blue solid on the faucet opening. This solid (copper oxide and/or carbonate com-pounds) is common and is a result of the interaction of the water with copper pipes as it travels from the water system to his faucet (Figure 4.3). Most households use copper pipes to transport water, and we ingest some copper into our bodies from the water we drink and cook with. Usually the amount of copper we get from water is about 5% of the recommended daily allowance. Copper is an essential mineral to our health and plays an important role in our metabolism. In our bodies, copper is known as a trace mineral because it is present in small amounts: 1.4 to 2.1 mg for each kg of body weight. Copper is found in a variety of enzymes that are required for important functions such as iron utilization, melanin pigment production, main-tenance of healthy bones and connective tissue, and the formation of antioxidants that help prevent disease. The recommended daily allowance (RDA) of copper in

**FIGURE 4.2** Copper is found mostly as various sulfide, oxide, and silicate minerals. The colors visible on the surface of these ore samples are due to the unique chemical compositions of the minerals that make up the ores.

©Cordelia Molloy/Science Source

**FIGURE 4.3** The green-blue solid on this faucet is a copper compound that came from dissolved copper in the water.

©Steven M. Marks

Directions: Adults: One tablet daily, with food.

## Supplement Facts
Serving Size: One tablet

| | Amount Per Serving | % Daily Value | | Amount Per Serving | % Daily Value |
|---|---|---|---|---|---|
| Total Carbohydrate | < 1 g | < 1%* | Vitamin B$_{12}$ | 6 mcg | 100% |
| Vitamin A | 2500 IU | 50% | Biotin | 30 mcg | 10% |
| (20% as beta-carotene) | | | Pantothenic Acid | 5 mg | 50% |
| Vitamin C | 60 mg | 100% | Calcium (elemental) | 450 mg | 45% |
| Vitamin D | 800 IU | 200% | Iron | 18 mg | 100% |
| Vitamin E | 30 IU | 100% | Magnesium | 50 mg | 13% |
| Vitamin K | 25 mcg | 31% | Zinc | 15 mg | 100% |
| Thiamin (B$_1$) | 1.5 mg | 100% | Selenium | 20 mcg | 29% |
| Riboflavin (B$_2$) | 1.7 mg | 100% | Copper | 2 mg | 100% |
| Niacin | 10 mg | 50% | Manganese | 2 mg | 100% |
| Vitamin B$_6$ | 2 mg | 100% | Chromium | 120 mcg | 100% |
| Folic Acid | 400 mcg | 100% | *Percent Daily Values are based on a 2,000 calorie diet. | | |

**INGREDIENTS:** Calcium Carbonate, Cellulose, Magnesium Oxide, Ascorbic Acid, Ferrous Fumarate, Corn Starch, Maltodextrin, dl-Alpha-Tocopheryl Acetate, Acacia, Croscarmellose Sodium, Zinc Oxide, Titanium Dioxide, Dextrin, Hypromellose, Magnesium Stearate, Dicalcium Phosphate, Gelatin, Niacinamide, Silicon Dioxide, D-Calcium Pantothenate, Manganese Sulfate, Polyethylene Glycol, Cupric Sulfate, Dextrose, Pyridoxine Hydrochloride, Glucose, Soy Lecithin, Riboflavin, Thiamine Mononitrate, Vitamin A Acetate, Chromium Chloride, Folic Acid, Beta-Carotene, FD&C Yellow #5 (tartrazine) Lake, FD&C Yellow #6 Lake, Sodium Selenate, Biotin, Phytonadione, FD&C Blue #2 Lake, Cholecalciferol, Tricalcium Phosphate, Cyanocobalamin. **Contains:** Soy.

**WARNING:** Accidental overdose of iron-containing products is a leading cause of fatal poisoning in children under 6. Keep this product out of reach of children. In case of accidental overdose, call a doctor or poison control center immediately.

**FIGURE 4.4** Multivitamins commonly contain copper. Notice that these vitamins provide slightly more than the recommended daily allowance for an adult.
©Steven M. Marks

adults is 0.9 mg/day, and research shows that up to 10 mg daily appears to be safe for consumption in adults. Many vegetables, seeds, and nuts are good sources of copper. Multivitamins contain about 1 to 2 mg of copper, usually in the form of $CuSO_4$ (Figure 4.4).

Copper deficiency is uncommon, but it can be found in people whose bodies do not properly absorb copper, such as those with Crohn's disease or those who take zinc supplements. Copper deficiency can cause anemia, frequent infections, hair loss, fatigue, bone and joint problems, and skin sores.

Copper toxicity can also be a problem. High concentrations of copper interfere with zinc absorption, which is needed to build digestive proteins, resulting in a dislike of meat and other protein-rich foods. While toxic levels commonly are seen in patients with Wilson's disease, a genetic disease that causes copper accumulation in various organs, copper toxicity can also be caused by high concentrations of dissolved copper in drinking water transported through copper pipes. The concentration of copper in drinking water substantially increases in water that is too acidic and has been motionless in the pipes for many hours. If too much copper in drinking water is an issue, flushing the water through the faucet for a little while will produce water with lower copper concentrations. The U.S. Environmental Protection agency (EPA) has determined that copper levels in drinking water (out of the tap) should not exceed 1.3 mg/L. While that value may seem small, it is about 100 times greater than the EPA limit on lead in drinking water, 0.015 mg/L.

In this chapter we will investigate the elemental composition of chemical compounds, such as copper and other elements in copper minerals. We will also learn ways to describe the composition of solutions (homogeneous mixtures).

## Questions for Consideration

**4.1** How can we describe the mass composition of elements in a compound?

**4.2** How can we determine the number of atoms in a given mass of a material? The number of molecules? The number of formula units?

**4.3** How can we use the masses of elements in a compound to determine its chemical formula?

**4.4** How can we express the composition of a solution?

### 📟 Math Tools Used in This Chapter

Scientific Notation (Math Toolbox 1.1)
Units and Conversions (Math Toolbox 1.3)
Mole Conversions (Math Toolbox 4.1)

## 4.1 Percent Composition

The copper in Julio's light switch was processed from the mineral chalcopyrite. Suppose we dissolve a sample of chalcopyrite in acid. We get a blue solution, characteristic of copper(II) ions. The solution gives off the odor of rotten eggs, good evidence that sulfur is also present in the mineral. If we add ammonia to some of the solution, the color gets much more intense, confirming the presence of copper. We also see a brownish-red solid settling out of the solution, which indicates that iron is also present. The tests show that copper, sulfur, and iron are present in chalcopyrite, but how do we determine their amounts? One way is to experiment. Suppose careful separating and measuring revealed that a 3.67-g sample of chalcopyrite contained 1.27 g Cu, 1.12 g Fe, and 1.28 g S. While these numbers are useful, they do not describe all samples, only our original 3.67-g sample. If we examined a larger sample—for example, a sample with a mass of 8.33 g—the composition would be 2.88 g Cu, 2.54 g Fe, and 2.91 g S.

> Even though we cite the mass of each element using the element's symbol, our notation does not necessarily indicate that these elements are present as atoms. Rather, they may be present as monatomic or polyatomic ions.

There is another convenient way to express composition that is valid for samples of any size. We can convert the mass of each component to a percentage of the total mass of the sample. The result is the **percent composition by mass** (or *percent composition*). It is an expression of the portion of the total mass contributed by each element:

$$\% \text{ E (E is any element)} = \frac{\text{mass of E}}{\text{mass of sample}} \times 100\%$$

The percent composition is constant no matter what the size of the sample, consistent with the law of definite proportions (Section 2.1). To calculate the percent copper in our 3.67-g sample of chalcopyrite, divide the mass of copper by the mass of the sample and multiply this ratio by 100% (to convert to a percentage):

$$\% \text{ Cu} = \frac{1.27 \text{ g Cu}}{3.67 \text{ g sample}} \times 100\% = 34.6\% \text{ Cu}$$

Similar calculations for iron and sulfur give 30.5% Fe and 34.9% S. The sum of those percentages is 100% (Figure 4.5).

Let's examine this relationship more closely to see how it arises. Percent means "per hundred." You use percent frequently. For example, if you pay a sales tax, your total expenditure is greater than the purchase price. For example, a total cost of 107 cents may arise from merchandise costing 100 cents with a tax of 7 cents. Although sales tax is usually expressed as a percentage of the purchase price, what is the percentage of the total cost due to the tax? The percentage due to the tax is 6.54% of the total cost. If an item costs $30, you pay $32.10, including $2.10 in taxes, which again is 6.54% of the total cost. You pay $14.00 in taxes for every $200.00 purchased, which is 6.54% of the total cost of $214.00. Although those numbers sound different, they all have the same composition of 7 parts tax and 100 parts of purchase to every 107 parts of total cost. In a similar fashion, if we know the ratio of the mass of copper to the mass of the sample, we can easily find the unknown mass of Cu in a 100-g sample. Why? Because the ratio of the part to the whole is always the same:

$$\frac{1.27 \text{ g Cu}}{3.67 \text{ g sample}} = \frac{x \text{ g Cu}}{100 \text{ g sample}}$$

**FIGURE 4.5** The percent by mass of the component elements must total 100%. How would this graph change if we doubled the amount of chalcopyrite under consideration?

If we rearrange the equation to solve for mass Cu, we get the following:

$$x \text{ g Cu} = \frac{1.27 \text{ g Cu}}{1.67 \text{ g sample}} \times 100 \text{ g sample} = 34.6 \text{ g Cu}$$

MATH
TOOLBOX
**1.3**

Since this is the mass of Cu per 100-g sample, the value corresponds to the % Cu:

$$\% \text{ Cu} = 34.6\%$$

The percent composition by mass for any substance is always independent of the sample size. Use Example 4.1 to convince yourself that the percentages would be the same for an 8.33-g sample of chalcopyrite as for a 3.67-g sample.

---

**EXAMPLE 4.1**     **Percent Composition from Masses**

What is the percent copper in an 8.33-g sample of chalcopyrite that contains 2.88 g Cu?

**Solution:**

Set up the equation for % Cu as the mass Cu divided by the mass of the sample, with the ratio multiplied by 100%:

$$\% \text{ Cu} = \frac{2.88 \text{ g Cu}}{8.33 \text{ g sample}} \times 100\% = 34.6\% \text{ Cu}$$

Note that the answer is the same percentage as for the smaller chalcopyrite sample. This is what we would expect since minerals, like other chemical compounds, have a constant composition throughout.

**Consider This 4.1**

How would the percent copper change if we had a much larger sample?

**Practice Problem 4.1**

What are the percent iron and the percent sulfur in an 8.33-g sample of chalcopyrite that contains 2.54 g Fe and 2.91 g S?

**Further Practice:**  Questions 4.3 and 4.4 at the end of the chapter

---

If we know the percent composition of a particular compound, we can use this information to determine the mass of each element in any size sample of the compound.

---

**EXAMPLE 4.2**     **Mass from Percent Composition**

Magnetite, $Fe_3O_4$, is a mineral containing 72.4% iron. What mass of iron is present in an 837-g sample of magnetite?

**Solution:**

The percent composition of a substance tells us the mass percent of the elements present in any sample regardless of its size. The percent iron in magnetite indicates that for a 100-g sample, 72.4 g of iron is present:

$$\frac{72.4 \text{ g iron}}{100 \text{ g magnetite}}$$

Magnetite

© McGraw-Hill Education/Ken Cavanagh

We can use this relationship as a conversion factor to calculate the amount of iron present in a different-sized sample of the mineral:

$$837 \text{ g magnetite} \times \frac{72.4 \text{ g iron}}{100 \text{ g magnetite}} = 606 \text{ g iron}$$

**Consider This 4.2**

How would the mass of iron change if we had a smaller sample?

**Practice Problem 4.2**

Magnesium, in its ionic form, is a mineral (dietary, not geological) used by the body in over 300 essential reactions. Magnesium carbonate, $MgCO_3$, is a supplement used for treating magnesium deficiency and contains 28.8% magnesium. It is also an active ingredient in some antacids. If an antacid tablet contains 105 mg of magnesium carbonate, how many milligrams of magnesium does the tablet provide?

**Further Practice:**  Questions 4.7 and 4.8 at the end of the chapter

## 4.2  Mole Quantities

Let's consider the copper wires in Julio's computer. They started out as the copper sulfide mineral chalcopyrite. When dissolved in acid, chalcopyrite forms a blue solution that gives off a gas that smells like rotten eggs. This odor arises from the molecular compound hydrogen sulfide, which typically forms as a gas when an acid reacts with an ionic compound containing sulfide ions. Hydrogen sulfide contains 94.1% S by mass and 5.9% H. However, we would like to describe the composition of hydrogen sulfide in terms of its chemical formula, but how can we know how many atoms of each component element are in it? If we could count the atoms of each element in a small sample of hydrogen sulfide, we could determine the relative numbers of each and deduce the chemical formula.

### Moles and Particles

Consider the molecular-level view of hydrogen sulfide gas shown in Figure 4.6. We can identify individual molecules of hydrogen sulfide, with the white spheres representing hydrogen atoms and the yellow spheres representing sulfur atoms. Each molecule is made up of two hydrogen atoms and one sulfur atom. In a collection of

**FIGURE 4.6**  Hydrogen sulfide, $H_2S$, is given off when a sulfide ore reacts with acid.

© Tom Pantages

$H_2S$

molecules, such as the four molecules pictured in Figure 4.6, we can count eight hydrogen atoms and four sulfur atoms, giving a ratio of two hydrogen atoms per sulfur atom, the same ratio as in each molecule.

Unfortunately, we cannot see atoms and molecules in this way because they are too small. Although we can use a scanning tunneling microscope to "see" atoms and molecules on the surface of a solid, as shown in Figure 4.7, this technique reveals only atoms on a surface, not those beneath. Furthermore, because atoms are so small, a dust-sized piece of a solid contains over $10^{16}$ of each type of atom. We certainly wouldn't want to count that many atoms individually, even if we could. Fortunately, we don't have to see atoms to count them. Instead we can find the relative masses of the elements in a compound. From this, we can determine the 2:1 ratio of atoms and deduce the formula $H_2S$.

In Chapter 2, we saw that the average masses of the elements can be expressed *as their relative atomic masses.* For the elements in hydrogen sulfide, the relative atomic masses are 1.008 amu for hydrogen and 32.06 amu for sulfur. While relative atomic masses are very useful for comparing the masses of elements, atomic mass units are not useful to chemists weighing substances to use in chemical reactions like those for processing copper oxides. Common laboratory balances measure masses in grams, not atomic mass units. We cannot weigh one or even a few atoms, molecules, or formula units of a substance, since the masses of most atoms and common molecules are less than $10^{-21}$ g. To work with a mass of about 1 g, we have to measure a number of formula units greater than $10^{21}$. For example, to get 1 g of $H_2S$, we would need almost $2 \times 10^{22}$ molecules. To get an idea of how large this number is, consider a teaspoon of water. There are as many molecules of water in this teaspoon (about $1.7 \times 10^{23}$) as there are teaspoonfuls of water in all the oceans on Earth. Similarly, there are about as many $SiO_2$ formula units in one grain of quartz sand as there are sand grains on all the beaches on Earth. Because numbers this large are too unwieldy to work with easily, we need some way to measure the amount of a substance without counting the number of atoms, formula units, or molecules.

The solution is to use a unit called the mole. A **mole** contains $6.022 \times 10^{23}$ atoms, molecules, ions, or formula units. This number, called **Avogadro's number,** is the same as the number of atoms in exactly 12 g of the isotope carbon-12, or $^{12}C$. Figure 4.8 shows 1 mole (abbreviated mol) of several different substances. Although 1 mol contains different volumes and different masses of different materials, it is always the same number of formula units (atoms, molecules, or ionic compound formula units).

**FIGURE 4.7** The bumps in this image are atoms on the surface of a mica crystal. Mica is a mineral containing potassium and aluminum cations and silicate and hydroxide anions.

© B.L. Ramakrishna/Interactive Nano-visualization for Science and Engineering Education (IN-VSEE) an NSF funded project at ASU

We only have to go to the hardware or grocery store to find many examples of counting by weighing. We buy nails and walnuts by the pound rather than by the piece, for example.

Recall that a formula unit is the smallest repeating unit in an ionic compound. The composition of a formula unit is the same as the composition of the ionic compound.

**FIGURE 4.8** One-mole quantities of various elements and compounds. They have different masses and volumes, but the same number of formula units, $6.022 \times 10^{23}$. For example, 1 mol of aluminum contains $6.022 \times 10^{23}$ aluminum atoms, 1 mol of water contains $6.022 \times 10^{23}$ $H_2O$ molecules, and 1 mol of sodium chloride contains $6.022 \times 10^{23}$ NaCl formula units.

© Jim Birk

We can convert between moles and formula units using the following pathway:

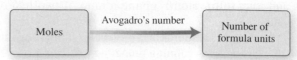

A formula unit can refer to an element, a molecular compound, or an ionic compound. For example, a formula unit of elemental copper is one atom of Cu, a formula unit of elemental oxygen is one $O_2$ molecule, a formula unit of water is one $H_2O$ molecule, and a formula unit of copper(II) chloride is one $Cu^{2+}$ ion and two $Cl^-$ ions together. Therefore, 1 mol of copper contains $6.022 \times 10^{23}$ *atoms* of copper, 1 mol of $O_2$ contains $6.022 \times 10^{23}$ *molecules* of $O_2$, 1 mol of $H_2O$ contains $6.022 \times 10^{23}$ *molecules* of $H_2O$, and 1 mol of $CuCl_2$ contains $6.022 \times 10^{23}$ *formula units* of $CuCl_2$. To get the number of formula units using Avogadro's number, we use the relationship:

$$1 \text{ mol} = 6.022 \times 10^{23} \text{ formula units}$$

We use this equivalence to set up possible conversion ratios:

$$\frac{6.022 \times 10^{23} \text{ formula units}}{1 \text{ mol}} \quad \text{and} \quad \frac{1 \text{ mol}}{6.022 \times 10^{23} \text{ formula units}}$$

We use the abbreviation mol for *mole,* but *molecule* should not be abbreviated.

If we're converting from moles, we use the first relationship. For example, we can calculate the number of elemental copper atoms in exactly 2 mol of copper as follows:

$$2 \text{ mol Cu} \times \frac{6.022 \times 10^{23} \text{ formula units Cu}}{1 \text{ mol Cu}} = 1.204 \times 10^{24} \text{ formula units}$$

 **MATH TOOLBOX 1.1**

A formula unit of Cu is a copper atom, so there are $1.204 \times 10^{24}$ copper atoms in exactly 2 mol of copper. Convince yourself that there would be $3.01 \times 10^{23}$ atoms of Cu in 0.500 mol of copper. (If you need to review mathematical operations using scientific notation, see Math Toolbox 1.1.)

We can also convert between formula units and atoms or ions. To determine the number of atoms or ions in a given number of moles of a substance, we extend the pathway further to include analysis of the chemical formula:

 **MATH TOOLBOX 4.1**

For example, 1 mol of $O_2$ contains $6.022 \times 10^{23}$ molecules of $O_2$, and one molecule (formula unit) of $O_2$ contains two oxygen atoms:

$$1 \text{ mol } O_2 \times \frac{6.022 \times 10^{23} \text{ molecules } O_2}{1 \text{ mol } O_2} \times \frac{2 \text{ O atoms}}{1 \text{ molecules } O_2} = 1.204 \times 10^{24} \text{ O atoms}$$

Consider another example, copper(I) oxide. How many $Cu^+$ and $O^{2-}$ ions does 1 mol of $Cu_2O$ contain? How many total ions are present? Analysis of the chemical formula indicates that a formula unit of $Cu_2O$ contains two $Cu^+$ ions and one $O^{2-}$ ion, as shown in Figure 4.9. This gives us the following:

$$1 \text{ mol } Cu_2O \times \frac{6.022 \times 10^{23} \text{ formula units } Cu_2O}{1 \text{ mol } Cu_2O} \times \frac{2 \text{ Cu}^+ \text{ ions}}{1 \text{ formula unit } Cu_2O}$$

$$= 1.204 \times 10^{24} \text{ Cu}^+ \text{ ions}$$

$$1 \text{ mol } Cu_2O \times \frac{6.022 \times 10^{23} \text{ formula units } Cu_2O}{1 \text{ mol } Cu_2O} \times \frac{1 \text{ O}^{2-} \text{ ions}}{1 \text{ formula unit } Cu_2O}$$

$$= 6.022 \times 10^{23} \text{ O}^{2-} \text{ ions}$$

**FIGURE 4.9** One formula unit of $Cu_2O$, shown highlighted, contains two $Cu^+$ ions and one $O^{2-}$ ion.

$$\text{Total ions} = 1.204 \times 10^{24} \text{ Cu}^+ \text{ ions} + 6.022 \times 10^{23} \text{ O}^{2-} \text{ ions}$$
$$= 1.807 \times 10^{24} \text{ ions}$$

In 1 mol of $Cu_2O$, there is a total of $1.807 \times 10^{24}$ ions. Calculations like these, and other calculations used in this chapter, are summarized in Math Toolbox 4.1.

---

**EXAMPLE 4.3**    **Particles in One Mole**

How many $H_2S$ molecules are in exactly 1 mol of hydrogen sulfide, $H_2S$? How many hydrogen atoms and sulfur atoms are there?

**Solution:**

To convert from moles of $H_2S$ to molecules of $H_2S$, we consider the relationship between moles and formula units for the substance of interest. In this case, the formula unit is an $H_2S$ molecule, so 1 mol of $H_2S$ contains $6.022 \times 10^{23}$ $H_2S$ molecules. To determine the number of hydrogen and sulfur atoms in the sample, we analyze the chemical formula, which tells us there are two atoms of H and one atom of S per molecule of $H_2S$. We can use these quantities as conversion factors to calculate the number of H and S atoms present in exactly 1 mol of $H_2S$:

$$1 \; \text{mol } H_2S \times \frac{6.022 \times 10^{23} \; \text{molecules } H_2S}{1 \; \text{mol } H_2S} \times \frac{2 \; \text{H atoms}}{1 \; \text{molecule } H_2S}$$
$$= 1.204 \times 10^{24} \; \text{H atoms}$$

$$1 \; \text{mol } H_2S \times \frac{6.022 \times 10^{23} \; \text{molecules } H_2S}{1 \; \text{mol } H_2S} \times \frac{1 \; \text{S atom}}{1 \; \text{molecule } H_2S}$$
$$= 6.022 \times 10^{23} \; \text{S atoms}$$

**Consider This 4.3**

How would the number of atoms of hydrogen and of sulfur change if we had less than 1 mole of $H_2S$?

**Practice Problem 4.3**

How many $N_2O_5$ molecules are in exactly 1 mol of dinitrogen pentoxide, $N_2O_5$? How many nitrogen atoms and oxygen atoms?

**Further Practice:** Questions 4.17 and 4.18 at the end of the chapter

---

## Molar Mass

The relative atomic mass is the average mass of atoms of an element in atomic mass units (Section 2.4). This definition describes the average mass of one atom, but we can extend it to larger quantities of matter as well. Recall that the basis of the definition is that one atom of $^{12}C$ has a mass of exactly 12 amu. Avogadro's number has been defined so that the mass of 1 mol of $^{12}C$ (that is, $6.022 \times 10^{23}$ atoms of $^{12}C$) is exactly 12 g. Consequently, the mass of a basic particle of any substance in atomic mass units and the mass of 1 mol of the substance in grams have exactly the same numerical value. The definition of a mole can be restated, then, in terms of the mass of our reference, $^{12}C$: *A mole is the amount of substance that contains as many basic particles (atoms, molecules, or formula units) as there are atoms in exactly 12 g of $^{12}C$*. The term that describes the mass of 1 mol of a substance is **molar mass,** and is expressed in units of grams per mole.

Since the molar mass value for any atom, in units of grams per mole (g/mol), is numerically the same as the relative atomic mass value (in units of amu per atom), we can use the periodic table to determine the molar mass of any element or compound. For example, the average mass of 1 hydrogen atom is 1.008 amu, and the mass of 1 mol of hydrogen atoms is 1.008 g. Similarly, the average mass of 1 sulfur atom is 32.06 amu, and the mass of 1 mol of sulfur atoms is 32.06 g. *The molar mass of a molecular or ionic substance has the same value as the sum of the relative atomic masses of its component elements.* For example, the average mass of a molecule of $H_2S$ is 34.08 amu, and its molar mass is 34.08 g/mol. Molar masses of molecules (or formula units) are obtained by adding the molar masses of the component elements, each weighted by the number of that atom in the molecule (or formula unit). For simplicity, we will abbreviate molar mass as *MM*.

---

### EXAMPLE 4.4    Molar Masses of Substances

When copper sulfide ores are roasted in a furnace, sulfur dioxide gas, $SO_2$, forms. What is the molar mass of $SO_2$?

**Solution:**

The molar mass of a compound is the sum of the molar masses of its component elements—in this case, sulfur and oxygen. A molecule of sulfur dioxide contains one sulfur atom and two oxygen atoms, so a mole of sulfur dioxide contains one mole of sulfur atoms and two moles of oxygen atoms. We must multiply the molar mass of oxygen by a factor of 2. We can find the molar masses of sulfur and oxygen on a periodic table, multiply the molar mass of oxygen by 2, and find the sum:

$$\text{Mass of 1 mol of S} = 1 \text{ mol} \times 32.06 \text{ g/mol} = 32.06 \text{ g}$$
$$\text{Mass of 2 mol of O} = 2 \text{ mol} \times 16.00 \text{ g/mol} = \underline{32.00 \text{ g}}$$
$$\text{Mass of 1 mol of } SO_2 \qquad\qquad\qquad\quad = 64.06 \text{ g}$$

The molar mass of $SO_2$ is 64.06 g/mol.

**Consider This 4.4**

If we formed $SO_3$ instead of $SO_2$, would the molar mass of the product be greater or less?

**Practice Problem 4.4**

Many of the minerals that yielded the copper used to make Julio's brass bed came out of the ground in ores of silicon dioxide, $SiO_2$. What is the molar mass of $SiO_2$?

**Further Practice:**  Questions 4.27 and 4.28 at the end of the chapter

---

Conversion between mass and moles is very common. For example, if we want to figure out how much of a substance is formed by a chemical reaction, we have to convert masses of starting materials to moles and then figure out the moles of the substance formed. Then we can determine the mass of that product. We'll discuss these calculations in Chapter 6.

So far we have seen that composition can be expressed in terms of mass (percent composition), atoms in a molecular compound, or ions in formula units of an ionic compound. We can also express the composition of a substance in terms of the number of moles of each element.

The number of moles of a substance is directly proportional to the mass of the substance. The ratio of mass to number of moles is always the same. It equals the molar mass, which is the mass of 1 mol. Recall the 3.67-g sample of chalcopyrite that contains 1.27 g Cu. Suppose we want to know how many moles of copper are in this amount. We can equate the ratio of 1.27 g of copper to the unknown number of moles of copper and the mass-to-mole ratio for 1 mol of copper (the molar mass):

$$\frac{1.27 \text{ g Cu}}{x \text{ mol Cu}} = \frac{65.55 \text{ g Cu}}{1 \text{ mol Cu}}$$

We rearrange this equation to solve for moles of copper:

$$x \text{ mol Cu} = 1.27 \text{ g Cu} \times \frac{1 \text{ mol Cu}}{63.55 \text{ g Cu}} = 0.0200 \text{ mol Cu}$$

MATH TOOLBOX 1.3

The ratio approach will always work to solve problems such as these, but we can generally solve them more quickly by using dimensional analysis. For a review of the ratio approach and the dimensional-analysis approach, see Math Toolbox 1.3. Conversions specific to mole calculations are shown in Math Toolbox 4.1. The conversion of mass to moles follows this pathway:

$$\boxed{\text{Mass in grams}} \xrightarrow{\text{molar mass}} \boxed{\text{Moles}}$$

In dimensional analysis, we start with an expression that equates two equivalent quantities. Here, we know that 1 mol of Cu has a mass of 63.55 g. The equivalence expression is

$$1 \text{ mol Cu} = 63.55 \text{ g Cu}$$

We use this expression to create a ratio of the two quantities. The ratio can be expressed in two ways:

$$\frac{63.55 \text{ g Cu}}{1 \text{ mol Cu}} \quad \text{and} \quad \frac{1 \text{ mol Cu}}{63.55 \text{ g Cu}}$$

We select the ratio that will cancel the old units (grams in this case) and introduce the new units (moles). We multiply the starting quantity (1.27 g Cu) by the ratio to obtain the moles of copper:

$$\text{mol Cu} = 1.27 \text{ g Cu} \times \frac{1 \text{ mol Cu}}{63.55 \text{ g Cu}} = 0.0200 \text{ mol Cu}$$

This equation is the same as the final equation developed in the ratio approach. We will use dimensional analysis for the remaining examples in this chapter, but remember that the same results can be obtained either way.

Example 4.5 demonstrates the dimensional-analysis approach for calculating the moles of Fe and S in our sample of chalcopyrite.

---

**EXAMPLE 4.5**       **Moles from Mass**

How many moles of iron are in the 1.12 g of Fe found in our sample of chalcopyrite?

**Solution:**

From the periodic table, we obtain the molar mass of iron to use in an equivalence expression:

$$1 \text{ mol Fe} = 55.85 \text{ g Fe}$$

We use this expression to create two possible conversion ratios:

$$\frac{55.85 \text{ g Fe}}{1 \text{ mol Fe}} \quad \text{and} \quad \frac{1 \text{ mol Fe}}{55.85 \text{ g Fe}}$$

Since we want to convert grams to moles, we select the second conversion ratio to carry out the conversion:

$$\text{mol Fe} = 1.12 \text{ g Fe} \times \frac{1 \text{ mol Fe}}{55.85 \text{ g Fe}} = 0.0201 \text{ mol Fe}$$

MATH
TOOLBOX
**4.1**

**Consider This 4.5**

Which conversion ratio would we use if we knew the number of moles of iron and wanted to calculate its mass?

**Practice Problem 4.5**

How many moles of sulfur are in the 1.28 g of S found in our sample of chalcopyrite?

**Further Practice:** Questions 4.37 and 4.38 at the end of the chapter

So far we have calculated 0.0200 mol Cu, 0.0201 mol Fe, and 0.0399 mol S (obtained in Practice Problem 4.5) from grams of each element in the 3.67-g sample of chalcopyrite and the molar mass of the elements. We can calculate the moles of a compound in a similar manner. We use the molar mass of the compound in the same way that we used the molar mass of an element to make the conversion from mass to moles. Suppose we have a 25.2-g sample of $Cu_2O$ and want to calculate the moles of $Cu_2O$ present. First, we consult the periodic table and determine the molar mass of $Cu_2O$:

$$\text{Mass of 2 mol of Cu} = 2 \text{ mol} \times 63.55 \text{ g/mol} = 127.1 \text{ g}$$
$$\text{Mass of 1 mol of O} = 1 \text{ mol} \times 16.00 \text{ g/mol} = \underline{16.00 \text{ g}}$$
$$\text{Mass of 1 mol of Cu}_2\text{O} = 143.1 \text{ g}$$

The molar mass provides two possible conversion ratios:

$$\frac{143.1 \text{ g Cu}_2\text{O}}{1 \text{ mol Cu}_2\text{O}} \quad \text{and} \quad \frac{1 \text{ mol Cu}_2\text{O}}{143.1 \text{ g Cu}_2\text{O}}$$

Converting grams of $Cu_2O$ to moles, we use the second conversion ratio:

$$\text{mol Cu}_2\text{O} = 25.2 \text{ g Cu}_2\text{O} \times \frac{1 \text{ mol Cu}_2\text{O}}{143.1 \text{ g Cu}_2\text{O}} = 0.176 \text{ mol Cu}_2\text{O}$$

The 25.2-g sample of $Cu_2O$ contains 0.176 mol of $Cu_2O$.

If we know the number of moles and need the mass, we can use the molar mass to create a ratio for converting moles to mass. Example 4.6 illustrates this situation.

---

### EXAMPLE 4.6 | Mass from Moles

While Julio puzzles over mole-to-mass conversions, a friend heats water in a copper kettle and makes him a cup of tea. He tells Julio he placed 0.0120 mol of table sugar (sucrose, $C_{12}H_{22}O_{11}$) in the tea. What mass of sugar has he added?

**Solution:**

We use the molar mass of sugar to convert moles to mass. First we obtain the molar mass from the molar masses of the elements, each multiplied by the number of atoms of that element in the formula of the compound $C_{12}H_{22}O_{11}$:

$$\text{Mass of 12 mol of C} = 12 \text{ mol} \times 12.01 \text{ g/mol} = 144.1 \text{ g}$$
$$\text{Mass of 22 mol of H} = 22 \text{ mol} \times 1.008 \text{ g/mol} = 22.2 \text{ g}$$
$$\text{Mass of 11 mol of O} = 11 \text{ mol} \times 16.00 \text{ g/mol} = \underline{176.0 \text{ g}}$$
$$\text{Mass of 1 mol of C}_{12}\text{H}_{22}\text{O}_{11} = 342.3 \text{ g}$$

The molar mass of $C_{12}H_{22}O_{11}$ is 342.3 g/mol. Now we have an equivalence expression:

$$1 \text{ mol C}_{12}\text{H}_{22}\text{O}_{11} = 342.3 \text{ g C}_{12}\text{H}_{22}\text{O}_{11}$$

**MATH TOOLBOX 4.1**

From this expression, we have two possible ratios for conversion between units:

$$\frac{342.3 \text{ g C}_{12}\text{H}_{22}\text{O}_{11}}{1 \text{ mol C}_{12}\text{H}_{22}\text{O}_{11}} \quad \text{and} \quad \frac{1 \text{ mol C}_{12}\text{H}_{22}\text{O}_{11}}{342.3 \text{ g C}_{12}\text{H}_{22}\text{O}_{11}}$$

We know moles and want to calculate mass, so we use the first ratio:

$$\text{g C}_{12}\text{H}_{22}\text{O}_{11} = 0.0120 \text{ mol C}_{12}\text{H}_{22}\text{O}_{11} \times \frac{342.3 \text{ g C}_{12}\text{H}_{22}\text{O}_{11}}{1 \text{ mol C}_{12}\text{H}_{22}\text{O}_{11}} = 4.11 \text{ g C}_{12}\text{H}_{22}\text{O}_{11}$$

This is about the amount of sugar in one teaspoonful.

**Consider This 4.6**

If Julio added only half a teaspoon of sugar to his tea, how would the number of moles of sugar change?

**Practice Problem 4.6**

Julio's friend prefers an artificial sweetener such as aspartame, $C_{14}H_{18}N_2O_5$, in his tea. A packet of aspartame contains 40 mg of sweetener, measured to two significant figures. How many moles of aspartame are in the packet? What mass of aspartame would give 0.0120 mol? How many packets is this?

**Further Practice:**   Questions 4.43 and 4.44 at the end of the chapter

MATH
TOOLBOX
**4.1**

Earlier in this section we showed that if we know how many moles of a substance are in a sample, we can calculate the number of formula units in this sample. For example, because we know the number of moles of each element in our 3.67-g sample of chalcopyrite, we can calculate the number of atoms it contains. Recall that we found 0.0200 mol Cu, 0.0201 mol Fe, and 0.0399 mol S in the sample. To get the number of atoms of Cu, use the relationship between moles and Avogadro's number:

$$1 \text{ mol} = 6.022 \times 10^{23} \text{ atoms}$$

We use this equivalence to set up the possible conversion ratios:

$$\frac{6.022 \times 10^{23} \text{ atoms}}{1 \text{ mol}} \quad \text{and} \quad \frac{1 \text{ mol}}{6.022 \times 10^{23} \text{ atoms}}$$

Since we must convert moles to number of atoms, we use the first ratio:

$$\text{Atoms Cu} = 0.0200 \text{ mol Cu} \times \frac{6.022 \times 10^{23} \text{ atoms Cu}}{1 \text{ mol Cu}} = 1.20 \times 10^{22} \text{ atoms Cu}$$

Similar calculations give us the number of iron and sulfur atoms: $1.21 \times 10^{22}$ Fe atoms and $2.40 \times 10^{22}$ S atoms.

Suppose we do not know the number of moles of the substance, but we know its mass. To find the number of formula units, we first have to convert the mass to moles, using the molar mass. Then we find the number of formula units using Avogadro's number. In the final step, we use the chemical formula to determine the number of atoms or ions. The conversion between mass, moles, number of formula units, and atoms or ions involves molar mass, Avogadro's number, and the chemical formula, and follows this pathway:

Any of these conversions can be carried out in either direction. Each equivalence expression (using molar mass or Avogadro's number) can be used to create two conversion ratios. One of them converts units left to right; the other converts units right to left.

**EXAMPLE 4.7          Number of Molecules from Mass**

A substance named Agorca M5640 (officially 5-nonyl salicylaldoxime) is used for concentrating extracted copper ore. Its molecular formula is $C_{16}H_{25}NO_2$. It has a molar mass of 263.4 g/mol. If you have a 150.0-g sample of Agorca M5640, how many molecules do you have?

MATH
TOOLBOX
**4.1**

**Solution:**

To convert from mass to number of molecules, we will need two conversion steps.
First, we need equivalence expressions for each conversion, using the molar mass
and Avogadro's number:

$$1 \text{ mol } C_{16}H_{25}NO_2 = 263.4 \text{ g } C_{16}H_{25}NO_2$$

$$1 \text{ mol } C_{16}H_{25}NO_2 = 6.022 \times 10^{23} \text{ molecules } C_{16}H_{25}NO_2$$

We use expressions to develop possible conversion ratios:

$$\frac{1 \text{ mol } C_{16}H_{25}NO_2}{263.4 \text{ g } C_{16}H_{25}NO_2} \quad \text{and} \quad \frac{263.4 \text{ g } C_{16}H_{25}NO_2}{1 \text{ mol } C_{16}H_{25}NO_2}$$

$$\frac{6.022 \times 10^{23} \text{ molecules}}{1 \text{ mol}} \quad \text{and} \quad \frac{1 \text{ mol}}{6.022 \times 10^{23} \text{ molecules}}$$

Following the pathway outlined earlier, we start with 150.0 g $C_{16}H_{25}NO_2$ and
convert this quantity to moles. Then we convert moles to molecules. The first of
each pair of conversion ratios are used in the calculation:

$$\text{Molecules } C_{16}H_{25}NO_2 = 150.0 \text{ g } C_{16}H_{25}NO_2 \times \frac{1 \text{ mol } C_{16}H_{25}NO_2}{263.4 \text{ g } C_{16}H_{25}NO_2}$$

$$\times \frac{6.022 \times 10^{23} \text{ molecules } C_{16}H_{25}NO_2}{1 \text{ mol } C_{16}H_{25}NO_2}$$

$$= 3.430 \times 10^{23} \text{ molecules } C_{16}H_{25}NO_2$$

**Consider This 4.7**

If the molecular formula of Agorca M5640 were actually $C_{32}H_{50}N_2O_4$, would the
150.0-g sample contain more or fewer molecules?

**Practice Problem 4.7**

After working through this example, Julio develops a headache and takes an aspirin tablet. One 5-grain aspirin tablet contains 0.324 g of acetylsalicylic acid
($C_9H_8O_4$, molar mass = 180.2 g/mol). How many molecules of acetylsalicylic
acid are in one tablet?

**Further Practice:** Questions 4.51 and 4.52 at the end of the chapter

## 4.3 Determining Empirical and Molecular Formulas

We have seen how to express the composition of a substance in terms of masses,
percent composition, moles, and atoms or molecules of the component elements.
While these are all useful ways to express composition, none of them is very compact. In this section we will see how to convert such information into a chemical
formula, which is an efficient way of representing the composition of a substance.

### Empirical and Molecular Formulas

There are two types of chemical formulas that are related to the composition of a
substance. An **empirical formula** expresses the simplest ratios of atoms in a compound. It is written with the smallest possible whole-number subscripts. It is not
always the same as a *molecular formula* (review Section 3.5), which expresses the
actual number of atoms in a molecule. The molecular formula is either the same as
the empirical formula, or it is some multiple of the empirical formula. Thus, the

molecular and empirical formulas for water, with two hydrogen atoms and one oxygen atom per molecule, are both $H_2O$. The 2:1 ratio of hydrogen atoms to oxygen atom can't be reduced to a simpler ratio. The molecular formula for hydrogen peroxide, which contains two hydrogen atoms and two oxygen atoms, is $H_2O_2$. The 2:2 ratio of hydrogen atoms to oxygen atoms can be reduced to the simpler ratio of 1:1, so the empirical formula of hydrogen peroxide is HO. The difference between molecules of water and hydrogen peroxide is shown in Figure 4.10.

For some substances there is no useful molecular formula, because no specific molecule can be identified in the structure. Quartz sand, for example, is simply networks of silicon and oxygen atoms linked together as shown in Figure 4.11. The atoms occur in a 1:2 ratio, and the substance, silicon dioxide, is commonly represented by the empirical formula $SiO_2$. Similarly, formulas for ionic compounds are usually empirical formulas. As shown in Figure 4.12, a copper(II) oxide crystal contains equal numbers of copper(II) ions and oxide ions, but no molecules. For such ionic compounds, we represent the compound symbolically with its empirical formula.

Whenever possible, chemists like to use molecular formulas. Molecular formulas contain more information—not only the ratios of atoms but also the actual numbers of atoms in a molecule. Consider, for example, the substances benzene and acetylene. They have the molecular formulas $C_6H_6$ and $C_2H_2$, respectively (Figure 4.13). Both have the empirical formula CH. The empirical formula tells us

$H_2O$

$H_2O_2$

**FIGURE 4.10** For the water molecule, the molecular and empirical formulas are the same, $H_2O$. For hydrogen peroxide, the smallest ratio of hydrogen to oxygen atoms is 1:1, but the molecule actually contains two atoms of each, so the molecular and empirical formulas are different— namely, $H_2O_2$ and HO, respectively.

**FIGURE 4.11** In quartz, there are two oxygen atoms for every one silicon atom, but there is no identifiable $SiO_2$ molecule. For such extended structures, there is no molecular formula, only an empirical formula.
©Siede Preis/Getty Images

**FIGURE 4.12** In ionic solids, the empirical formula represents the smallest repeating unit in the crystal, the formula unit. These crystals of tenorite contain equal numbers of copper(II) ions and oxide ions, so the formula is CuO.
©Richard Megna/Fundamental Photographs

$C_6H_6$

$C_2H_2$

**FIGURE 4.13** Benzene, $C_6H_6$, and acetylene, $C_2H_2$, have the same empirical formula, CH.

**TABLE 4.1** Some Empirical and Molecular Formulas

| Substance | Molecular Formula | Empirical Formula |
|---|---|---|
| cyclopentane | $C_5H_{10}$ | $CH_2$ |
| cyclohexane | $C_6H_{12}$ | $CH_2$ |
| ethylene | $C_2H_4$ | $CH_2$ |
| hydrogen sulfide | $H_2S$ | $H_2S$ |
| calcium chloride | There is no molecular formula for an ionic compound. | $CaCl_2$ |

**FIGURE 4.14** For which of these substances is the empirical formula the same as the molecular formula?

only that their molecules contain carbon and hydrogen atoms in equal numbers, or in a 1:1 ratio. It does not distinguish benzene and acetylene. We need molecular formulas to tell them apart. Some other examples of empirical and molecular formulas are given in Table 4.1.

Examine Figure 4.14 and determine the empirical formula for each substance shown. To do this, find the smallest subscript and divide the other subscripts by its value. If the other subscripts are *evenly* divisible by the smallest one, then the division results in the empirical formula. Sometimes, if this fails, all the subscripts may be evenly divisible by some other number, so other choices have to be checked as well. If the subscripts are not evenly divisible by some number other than 1, the molecular formula and the empirical formula are the same.

---

**EXAMPLE 4.8**    **Empirical and Molecular Formulas**

Consider the molecular formula $C_6H_3Cl_3$ (see Figure 4.14). What is the empirical formula of this compound?

**Solution:**

The subscripts in the formula are all evenly divisible by 3, the smallest subscript. Carrying out this division gives us $C_{6/3}H_{3/3}Cl_{3/3}$ or $C_2HCl$ as the empirical formula. The molecular formula is three times the empirical formula.

**Consider This 4.8**

What would be the empirical formula if the molecular formula were $C_6H_2Cl_4$?

**Practice Problem 4.8**

What is the empirical formula of each of the following compounds?

(a) $CH_2Cl_2$          (b) $CH_3CO_2H$          (c) $P_4O_6$

**Further Practice:** Questions 4.67 and 4.68 at the end of the chapter

## Determining Empirical Formulas

We can now determine empirical formulas from data on the chemical composition of a compound. Recall, for example, our 3.67-g sample of chalcopyrite. It contained 1.27 g Cu, 1.12 g Fe, and 1.28 g S. We have converted the masses of the constituent elements to numbers of moles and numbers of atoms. Either of these sets of numbers can give us a chemical formula. Because the numbers are easier to manipulate, let's use the number of moles: 0.0200 mol Cu, 0.0201 mol Fe, and 0.0399 mol S. To write a chemical formula for chalcopyrite, we need only convert these values to whole numbers. Why? Because the relative number of moles of atoms in a sample of a compound must be the same as the relative number of atoms in one formula unit or one molecule of the compound. To get the relative number of atoms, divide the number of moles of each element by the number of moles of the element present in the least amount. In this example, copper is present in the least amount, so we need only divide the moles of Fe and S by the moles of Cu to reduce all the numbers to whole numbers:

$$\frac{0.0201 \text{ mol Fe}}{0.0200 \text{ mol Cu}} = \frac{1.01 \text{ Fe}}{1.00 \text{ Cu}}$$

$$\frac{0.0399 \text{ mol S}}{0.0200 \text{ mol Cu}} = \frac{2.00 \text{ S}}{1.00 \text{ Cu}}$$

Slight errors in the experimental determination of the masses can yield values that are close to whole numbers, but not exactly. In such cases, we can safely round the values to whole numbers.

When rounded, the whole-number values indicate that there are 1 mol Cu, 1 mol Fe, and 2 mol S in 1 mol of chalcopyrite. From these results, we can write the empirical formula of chalcopyrite as $CuFeS_2$. The procedure and the result would be the same if we used the number of atoms of each element instead of the number of moles.

## Empirical Formulas from Percent Composition

The copper in Julio's lights, refrigerator, bed, and computer came from different minerals, but the same procedures we used to determine the empirical formula for chalcopyrite can be used when working with any of the others. That is, if we know the mass of each element in a sample of a compound, we can calculate the moles of each element to determine the empirical formula. If we know the percent composition instead of actual masses, we must convert percent composition to relative masses before we calculate moles. The simplest approach is to use a hypothetical sample mass for the compound of exactly 100 g, so the percent composition is numerically equal to the mass of each component element. We can actually pick any mass. The exact sample mass is not important since we calculate the ratio of moles in the sample, and this ratio is independent of the sample size.

For compounds containing only two elements, A and B, the conversions from percent composition to empirical formula can be represented schematically as follows:

If the compound contains more than two elements, we add a line to the schematic for each additional element.

---

**EXAMPLE 4.9** | **Empirical Formula from Percent Composition**

Determine the empirical formula for the mineral chalcocite, which has the percent composition 79.8% Cu and 20.2% S.

**Solution:**

If we assume the sample of chalcocite has a mass of 100 g, the masses of copper and sulfur will be numerically equal to their percent composition values:

$$79.8\% \text{ of } 100 \text{ g is } 79.8 \text{ g} = \text{mass Cu}$$
$$20.2\% \text{ of } 100 \text{ g is } 20.2 \text{ g} = \text{mass S}$$

Next we convert the mass of each element into its number of moles. The equivalence expressions arise from the molar masses of the elements:

$$1 \text{ mol Cu} = 63.55 \text{ g Cu}$$
$$1 \text{ mol S} = 32.06 \text{ g S}$$

MATH
TOOLBOX
**4.1**

We use these expressions to create possible conversion ratios:

$$\frac{63.55 \text{ g Cu}}{1 \text{ mol Cu}} \quad \text{and} \quad \frac{1 \text{ mol Cu}}{63.55 \text{ g Cu}}$$

$$\frac{32.06 \text{ g S}}{1 \text{ mol S}} \quad \text{and} \quad \frac{1 \text{ mol S}}{32.06 \text{ g S}}$$

Since we need to cancel grams of each element and calculate moles, the second ratio is used in each case:

$$\text{mol Cu} = 79.8 \text{ g Cu} \times \frac{1 \text{ mol Cu}}{63.55 \text{ g Cu}} = 1.26 \text{ mol Cu}$$

$$\text{mol S} = 20.2 \text{ g S} \times \frac{1 \text{ mol S}}{32.06 \text{ g S}} = 0.630 \text{ mol S}$$

Finally, divide by the smaller of the two mole quantities:

$$\frac{\text{mol Cu}}{\text{mol S}} = \frac{1.26 \text{ mol Cu}}{0.630 \text{ mol S}} = \frac{2.00 \text{ Cu}}{1.00 \text{ S}}$$

Since 2 moles of copper are present for each 1 mole of sulfur, the empirical formula is $Cu_2S$.

**Consider This 4.9**

How would the chemical formula change if we had a larger sample of the compound?

**Practice Problem 4.9**

Determine the empirical formula for the mineral covellite, which has the percent composition 66.5% Cu and 33.5% S.

**Further Practice:** Questions 4.73 and 4.74 at the end of the chapter.

## Empirical Formulas for Compounds Containing More Than Two Elements

For a compound that contains more than two elements, we follow the same procedure, but we add a series of conversions for each additional element.

---

| **EXAMPLE 4.10** | **Empirical Formulas for More Than Two Elements** |

The mineral malachite is a beautiful green color, often with swirling bands of different intensity. It is often used in jewelry because of its attractive appearance. An analysis of malachite gives the following composition: 57.48% copper, 5.43% carbon, 0.91% hydrogen, and 36.18% oxygen. Determine the empirical formula of malachite.

Malachite
© RF Company/Alamy Stock Photo

**Solution:**

Consider a 100-g sample of malachite, so the masses of each element are numerically the same as the percent composition: 57.48 g Cu, 5.43 g C, 0.91 g H, and 5.43 g O. Then calculate the moles of each element in 100 g of the compound. We use the molar masses to create equivalence expressions and convert them to ratios for the conversions. As in Example 4.9, the correct ratio has moles divided by grams:

$$1 \text{ mol Cu} = 63.55 \text{ g Cu} \qquad \frac{1 \text{ mol Cu}}{63.55 \text{ g Cu}}$$

$$1 \text{ mol C} = 12.01 \text{ g C} \qquad \frac{1 \text{ mol C}}{12.01 \text{ g C}}$$

$$1 \text{ mol H} = 1.008 \text{ g H} \qquad \frac{1 \text{ mol H}}{1.008 \text{ g H}}$$

$$1 \text{ mol O} = 16.00 \text{ g O} \qquad \frac{1 \text{ mol O}}{16.00 \text{ g O}}$$

We multiply the mass of each element by the corresponding conversion ratio to get moles of this element:

$$\text{mol Cu} = 57.48 \text{ g Cu} \times \frac{1 \text{ mol Cu}}{63.55 \text{ g Cu}} = 0.9045 \text{ mol Cu}$$

$$\text{mol C} = 5.43 \text{ g C} \times \frac{1 \text{ mol C}}{12.01 \text{ g C}} = 0.452 \text{ mol C}$$

$$\text{mol H} = 0.91 \text{ g H} \times \frac{1 \text{ mol H}}{1.008 \text{ g H}} = 0.90 \text{ mol H}$$

$$\text{mol O} = 36.18 \text{ g O} \times \frac{1 \text{ mol O}}{16.00 \text{ g O}} = 2.261 \text{ mol O}$$

To calculate the mole ratios, first identify the element present in the smallest amount in moles. In this example, it's carbon. Then divide the moles of each of the other elements by the moles of carbon:

$$\frac{\text{mol Cu}}{\text{mol C}} = \frac{0.9045 \text{ mol Cu}}{0.452 \text{ mol C}} = \frac{2.00 \text{ Cu}}{1.00 \text{ C}}$$

$$\frac{\text{mol H}}{\text{mol C}} = \frac{0.90 \text{ mol H}}{0.452 \text{ mol C}} = \frac{1.99 \text{ H}}{1.00 \text{ C}}$$

$$\frac{\text{mol O}}{\text{mol C}} = \frac{2.261 \text{ mol O}}{0.452 \text{ mol C}} = \frac{5.00 \text{ O}}{1.00 \text{ C}}$$

When the mole ratios are rounded to the nearest whole number, we see that one formula unit has 1 carbon atom, 2 copper atoms, 2 hydrogen atoms, and 5 oxygen atoms. From this information, we can write the empirical formula $Cu_2CO_5H_2$. [The formula is usually written as $Cu_2CO_3(OH)_2$, to show which polyatomic ions are found in the mineral. Malachite contains two anions, hydroxide, $OH^-$, and carbonate, $CO_3^{2-}$. However, this cannot be deduced from the empirical formula for this compound.]

### Consider This 4.10

How would the procedure change if we had a compound that contained five elements?

### Practice Problem 4.10

Another copper mineral is chrysocolla. An analysis of chrysocolla gives the following composition: 36.18% copper, 15.99% silicon, 2.29% hydrogen, and 45.54% oxygen. Determine the empirical formula of chrysocolla.

**Further Practice:** Questions 4.75 and 4.76 at the end of the chapter

## Empirical Formulas with Fractional Mole Ratios

In Example 4.10, we calculated a mole ratio of 1.99, which we rounded to a value of 2. The decimal values will not always be that close to a whole number, so rounding off is not always appropriate. The ratios may, however, have fractional values that correspond to ratios of small whole numbers, such as 1.25 (5/4), 1.33 (4/3), 1.50 (3/2), and 1.67 (5/3). In such cases, we multiply each ratio by a small whole number to make all the subscripts whole numbers. For example, a compound of nitrogen and oxygen has the following mole ratio:

$$\frac{\text{mol O}}{\text{mol N}} = \frac{2.50 \text{ mol O}}{1.00 \text{ mol N}}$$

Multiplying each number by 2 corresponds to a whole-number ratio of 5/2. Thus, there must be 5 oxygen atoms for every 2 nitrogen atoms in the compound, giving the formula $N_2O_5$.

---

**EXAMPLE 4.11**   **Empirical Formulas with Fractional Mole Ratios**

The copper mineral azurite has the deep-blue color azure. Azurite contains 55.31% copper, 6.97% carbon, 37.14% oxygen, and 0.58% hydrogen. Calculate the empirical formula of azurite.

## Solution:

Calculate the moles of each element in 100 g of the compound, following the same procedure used in Examples 4.9 and 4.10:

$$\text{mol Cu} = 55.31 \text{ g Cu} \times \frac{1 \text{ mol Cu}}{63.55 \text{ g Cu}} = 0.8703 \text{ mol Cu}$$

$$\text{mol C} = 6.97 \text{ g C} \times \frac{1 \text{ mol C}}{12.01 \text{ g C}} = 0.580 \text{ mol C}$$

$$\text{mol H} = 0.58 \text{ g H} \times \frac{1 \text{ mol H}}{1.008 \text{ g H}} = 0.58 \text{ mol H}$$

$$\text{mol O} = 37.14 \text{ g O} \times \frac{1 \text{ mol O}}{16.00 \text{ g O}} = 2.321 \text{ mol O}$$

Azurite
© RF Company/Alamy Stock Photo

Now calculate the ratio of the moles of each element to the moles of the element present in the smallest amount (carbon or hydrogen):

$$\frac{\text{mol Cu}}{\text{mol C}} = \frac{0.8703 \text{ mol Cu}}{0.580 \text{ mol C}} = \frac{1.50 \text{ Cu}}{1.00 \text{ C}}$$

$$\frac{\text{mol H}}{\text{mol C}} = \frac{0.58 \text{ mol H}}{0.580 \text{ mol C}} = \frac{1.0 \text{ H}}{1.00 \text{ C}}$$

$$\frac{\text{mol O}}{\text{mol C}} = \frac{2.321 \text{ mol O}}{0.580 \text{ mol C}} = \frac{4.00 \text{ O}}{1.00 \text{ C}}$$

One of the ratios is not close to a whole number, but it corresponds to a ratio of whole numbers (3/2). We can multiply this ratio by 2 to get a whole number. We must also multiply all the other ratios by 2. Thus, for every 2 mol of C, we have 3 mol of Cu, 2 mol of H, and 8 mol of O. The empirical formula is $Cu_3C_2H_2O_8$.

### Consider This 4.11

If the mole ratio is 1.875, should we round this to 2?

### Practice Problem 4.11

Shattuckite is a fairly rare copper mineral. It has the composition 48.43% copper, 17.12% silicon, 34.14% oxygen, and 0.31% hydrogen. Calculate the empirical formula of shattuckite.

**Further Practice:** Questions 4.77 and 4.78 at the end of the chapter

## Molecular Formulas from Empirical Formulas

The molar mass of a compound can be determined experimentally by a variety of methods. We can also calculate the mass of 1 mol of formula units, using the empirical formula. Comparing the experimentally determined molar mass and the calculated molar mass of the empirical formula allows us to determine and confirm the molecular formula of a compound. In other words, if the experimental molar mass is the same as the calculated one, the molecular formula is the same as the empirical formula. If the experimental molar mass is greater than that calculated from the empirical formula, then the molecular formula is some multiple of the empirical formula. For example, propene has an empirical formula of $CH_2$ and a molar mass of 42.12 g/mol. The molar mass calculated for the empirical formula $CH_2$ is 14.04 g/mol. Taking the ratio of the experimental molar mass to the calculated molar mass for the empirical formula indicates a multiple of 3 for the molecular formula (Figure 4.15). The molecular formula of propene is $C_3H_6$.

All compounds with the same empirical formula have the same percent composition. In such cases, we cannot distinguish between the compounds by comparing either the percent compositions or the empirical formulas.

**FIGURE 4.15** Three formula units of $CH_2$ make up one molecule of $C_3H_6$. The sum of the masses of the formula units (14.04 g/mol) make up 1 molar mass of 42.12 g/mol.

---

### EXAMPLE 4.12     Molecular Formulas

The empirical formula for an acid was determined to be $HCO_2$. If the molar mass of the acid is determined to be about 90.0 g/mol, what is the molecular formula for this acid?

**Solution:**

If the empirical formula is the same as the molecular formula, then the molar mass of $HCO_2$ should be 90.0 g/mol. We calculate the molar mass of the empirical formula $HCO_2$ by summing the weighted molar masses of the constituent atoms:

$$\begin{aligned}
\text{Mass of 1 mol of H} &= 1\ \text{mol} \times 1.008\ \text{g/mol} = 1.008\ \text{g} \\
\text{Mass of 1 mol of C} &= 1\ \text{mol} \times 12.01\ \text{g/mol} = 12.01\ \text{g} \\
\text{Mass of 2 mol of O} &= 2\ \text{mol} \times 16.00\ \text{g/mol} = \underline{32.00\ \text{g}} \\
\text{Mass of 1 mol of } HCO_2 &\qquad\qquad\qquad\qquad\quad\ = 45.02\ \text{g}
\end{aligned}$$

The molar mass calculated for $HCO_2$ is 45.02 g/mol. The ratio of the experimental molar mass for the compound and the calculated molar mass for its empirical formula is

$$\frac{\text{Experimental } MM}{\text{Empirical formula } MM} = \frac{90.0\ \text{g/mol}}{45.02\ \text{g/mol}} = 2.00$$

Since 90.0 g/mol is very close to twice the molar mass calculated from the empirical formula, the subscripts in the empirical formula must all be multiplied by 2, and the molecular formula must be twice the empirical formula: $H_2C_2O_4$. (The structure for this compound is shown in Figure 4.14.)

**Consider This 4.12**

How would the molecular formula change if the experimental molar mass were 180 g/mol?

**Practice Problem 4.12**

The empirical formula for an organic compound was determined to be $CH_2O$. If the molar mass of this compound is determined to be about 90.1 g/mol, what is the molecular formula for this compound?

**Further Practice:** Questions 4.85 and 4.86 at the end of the chapter

## Determining Percent Composition

If we know a chemical formula for a compound, we can use the relationships between moles and mass to determine its percent composition without having to determine masses by experiment. For example, we might compare the copper content of two minerals, such as the cuprite, $Cu_2O$, and chalcocite, $Cu_2S$, shown in Figure 4.16, to determine which would yield more copper.

To simplify converting a chemical formula to a percent composition, we can assume a sample size of 1 mol. Then the mass of an element in 1 mol of the compound equals the number of moles of the element times the molar mass of the element:

The percent composition of a given element then equals the mass of this element divided by the molar mass of the compound (which is the mass of 1 mol) and multiplied by 100% to convert to a percentage:

**FIGURE 4.16** Although they have similar formulas, cuprite, $Cu_2O$ *(top)*, and chalcocite, $Cu_2S$ *(bottom)*, contain different percentages of copper by mass. (top): © Joel Arem/Science Source; (bottom): © Biophoto Associates/Science Source

For example, there are 2 mol of copper in 1 mol of $Cu_2O$, cuprite. Thus, we can determine that 1 mol of the compound contains 2 mol × 63.55 g/mol or 127.1 g of copper. Adding the molar masses of the elements gives 143.1 g/mol for the molar mass of cuprite, so 1 mol has a mass of 143.1 g. The percent copper in the compound is thus

$$\% \ Cu = \frac{127.1 \ g}{143.1 \ g} \times 100\% = 88.82\% \ Cu$$

Since the percent composition of all the component elements must add up to 100%, there must be 11.18% oxygen in cuprite.

---

### EXAMPLE 4.13  Percent Composition from Chemical Formula

Calculate the percent copper in chalcocite, $Cu_2S$.

**Solution:**

For chalcocite, 1 mol of the compound contains 2 mol × 63.55 g/mol or 127.1 g of copper. Adding the molar masses of the elements gives 159.2 g/mol for the molar mass of chalcocite. The percent copper in the compound is thus

$$\% \ Cu = \frac{127.1 \ g}{159.2 \ g} \times 100\% = 79.84\% \ Cu$$

Thus, chalcocite, $Cu_2S$, contains a lower percentage of copper than does cuprite, $Cu_2O$, even though each contains 2 mol of copper per mole of compound. This is because sulfur contributes a greater percentage to the mass of $Cu_2S$ than oxygen contributes to the mass of $Cu_2O$.

**Consider This 4.13**

If Se were substituted for S to form the compound $Cu_2Se$, how would the percent copper change?

**FIGURE 4.17** In this system, dissolved copper(II) sulfate is present in a lesser quantity than water, so copper(II) sulfate is the solute and water is the solvent.

© Brian Moeskau/Moeskau Photography

**Practice Problem 4.13**

Calculate the percent copper in covellite, CuS.

**Further Practice:** Questions 4.87 and 4.88 at the end of the chapter

## 4.4 Chemical Composition of Solutions

Chemical reactions between solids are normally slow, so we usually carry out such reactions by first dissolving the compounds in a liquid to form a solution. For example, one of the processes used to treat oxide ores of copper produces a solution containing copper(II) sulfate. Which of the liquids that you encounter every day are solutions?

Recall from Chapter 1 that a *solution* is any mixture that is homogeneous at the molecular or ionic scale. In a solution, the substance that is dissolved is called the **solute** (usually present in a lesser amount), and the substance doing the dissolving is called the **solvent** (usually present in a greater amount). Figure 4.17 shows a solution of copper(II) sulfate (the solute) dissolved in water (the solvent). This solution is described symbolically as $CuSO_4(aq)$.

| EXAMPLE 4.14 | Solute and Solvent |
| --- | --- |

From this picture of a solution containing $CuSO_4$ and water, identify the solute and the solvent.

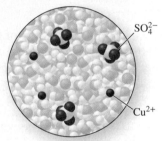

**Solution:**

In this image, the dark red spheres represent $Cu^{2+}$ ions. The yellow and red groups of spheres represent $SO_4^{2-}$ ions. The pale red and white groups in the background represent $H_2O$ molecules. The number of $Cu^{2+}$ and $SO_4^{2-}$ ions is equal and is less than the number of $H_2O$ molecules. Thus, the solute is $CuSO_4$, which yields $Cu^{2+}$ and $SO_4^{2-}$ ions in solution. The solvent is $H_2O$ since it is present in the greater amount.

**Consider This 4.14**

If the image contained six $CH_3OH$ molecules instead of $Cu^{2+}$ and $SO_4^{2-}$ ions, what would be the solute and solvent?

**Practice Problem 4.14**

In the following image of a solution containing hydrogen sulfide and water, identify the solute and the solvent.

**Further Practice:** Questions 4.93 and 4.94 at the end of the chapter

## Concentration

Solutions are homogeneous mixtures, but different solutions can contain varying amounts of solute and solvent. So, how do we express the composition of a solution? One way is to describe its **concentration,** which is the relative amounts of solute and solvent in it. When comparing solutions, we can describe them as either dilute or concentrated. A **dilute solution** contains a relatively small amount of solute, whereas a **concentrated solution** contains a comparatively large amount of solute. These terms are helpful when we compare two solutions of different concentrations, indicating that one contains more or less solute than the other.

We see differences in concentration in everyday solutions. For example, when Julio brews a pot of tea, the color is more intense if the tea is "strong" (concentrated) than if it is "weak" (dilute). What are some other solutions that you see in your environment where color gives you an indication of the concentration? What other ways can you compare concentrations of solutions? The concentrations of sugar solutions could be compared by how slowly they pour or by their density. If you compared a teaspoonful of sugar dissolved in a cup of water to molasses or syrup (also sugar solutions), which would pour more slowly? Which would be denser? Taste is another way we compare concentration in our daily lives (but never in the laboratory), but it is not an accurate measure of concentration. Would you want the nurse in the hospital to taste the saline solution to see if it is the correct concentration before putting it into your IV?

A variety of experimental methods allows us to determine the concentrations of solutions. For example, if the solute is colored and the solvent is colorless, then the intensity of color in the solution is a measure of its concentration. Consider the aqueous solutions of copper(II) sulfate shown in Figure 4.18. The more copper(II) sulfate dissolved in the water, the more intense is the blue color due to the $Cu^{2+}$ ions.

Figure 4.18 also shows, at a molecular level, how the relative amounts of solute and solvent vary with the concentration. If you count the solute particles [copper(II) ions and sulfate ions] and compare them to the number of water molecules, you can see that as the solution becomes more dilute, the number of ions in a measured amount decreases and the number of water molecules in that same measured amount increases. Because the concentration of copper(II) ions decreases relative to the number of water molecules, the color of the blue solution fades.

## Percent by Mass

Just as we could express the composition of a chemical compound as a percent by mass of each component element, we can do the same for the composition of a solution. Since the amount of solute relative to the amount of solution is of most interest,

The terms *strong* and *weak* are used here in their everyday sense, not in their scientific sense. When we discuss strong and weak acids, we are not talking about their concentration, but about their ability to dissociate in solution.

**FIGURE 4.18** The color intensity of a colored solution decreases as the concentration of the solute decreases. Compare the concentrations of ions in the solution for these copper(II) sulfate solutions. Which has more copper(II) ions? More sulfate ions? Which solution has the greatest concentration?
© Jim Birk

$Cu^{2+}$

$SO_4^{2-}$

the percent by mass for a solution equals the mass of the solute divided by the mass of solution, multiplied by 100% to convert the ratio to a percentage:

$$\% \text{ mass} = \frac{\text{mass solute}}{\text{mass solution}} \times 100\%$$

For example, hydrochloric acid is sold in solutions that are 36% by mass HCl, with the remainder of the mass supplied by the solvent, water. The mass of solution is the sum of the masses of solute and solvent. This way of measuring concentration has two advantages. First, it does not depend on temperature. The volume may change with temperature, but the mass does not. Second, although the volume of a solution may be affected by the presence of solute, the percent by mass is unaffected. Thus, we don't need special glassware that measures exact volumes to prepare solutions accurately by mass.

## Molarity

In addition to percent by mass, the concentration of a solution can be expressed in a variety of ways. One of the most common is molarity (*M*). The **molarity** of a solution is the number of moles of *solute* dissolved in 1 L of *solution* (not *solvent*). Recall from Section 4.2 that one mole of any substance contains $6.022 \times 10^{23}$ formula units of that substance. Molarity is a convenient way of describing concentration because it provides us with a direct measure of the count of particles. The molarity of any solution can be calculated by dividing the moles of solute by the liters of solution:

> In Chapter 11 we will see how chemical reactions can be used to determine how much solute is present in a solution.

$$\text{Molarity} = \frac{\text{moles of solute}}{\text{liters of solution}}$$

A solution having a volume of 1.000 L and containing 159.6 g (1.000 mol) of $CuSO_4$ is a 1.000 *M* copper(II) sulfate solution and the solution contains $6.022 \times 10^{23}$ formula units of $CuSO_4$. If the liter of solution had half this much copper(II) sulfate, 79.8 g, its concentration would be 0.500 *M* and $3.01 \times 10^{23}$ formula units of $CuSO_4$ would be present. What would the concentration be if 79.8 g were contained in 0.500 L of solution?

Molarity is calculated from liters of *solution,* not liters of *solvent.* How large the difference is depends on the concentration of the solution. Dilute solutions contain only a small amount of solute. The volume of such solutions is very nearly the same as the volume of solvent. However, the difference can be significant for solutions in which the volume of the solvent is significantly less than the volume of the solution. To ensure accuracy, solutions of the desired molarity are usually prepared in volumetric flasks, as shown in Figure 4.19.

A

B

C

D

**FIGURE 4.19** Solutions of known molarity are usually prepared in volumetric flasks, which are calibrated to contain a specific volume. (A) To prepare 250 mL of a 0.100 *M* solution of $CuSO_4 \cdot 5H_2O$, weigh 6.24 g (which is 0.0250 mol) of the solute. (B) Transfer it to the 250-mL flask and add some solvent, water. Dissolve the solute by swirling. (C) Solvent is added with swirling until the bottom of the liquid surface just matches the marked line on the neck of the flask. (D) Then a stopper is inserted into the flask and the flask is inverted several times to thoroughly mix the contents. Convince yourself that this procedure will indeed result in a solution whose concentration is 0.100 *M*.

If we know the mass of the solute and the volume of the solution in liters, we can calculate the molarity of the solution. To begin, calculate the number of moles of solute contained in the solution from the solute's molar mass. Then divide by the volume, in liters, of the solution.

---

**EXAMPLE 4.15**      **Molarity of Solutions**

A solution is prepared from 17.0 g of NaCl dissolved in sufficient water to give 150.0 mL of solution. What is the molarity of the solution? (The molar mass of NaCl is 58.44 g/mol.)

**Solution:**

We know the volume of solution in mL, but we must find the moles of solute before we can calculate the molarity:

$$\boxed{\text{Grands of solute}} \xrightarrow{MM} \boxed{\text{Moles of solute}} \xrightarrow{\text{volume (L)}} \boxed{\text{Molarity}}$$

The moles of sodium chloride can be calculated from its mass using the molar mass, as we have done previously:

$$\text{mol NaCl} = 17.0 \text{ g NaCl} \times \frac{1 \text{ mol NaCl}}{58.44 \text{ g NaCl}} = 0.291 \text{ mol NaCl}$$

Convert the volume to units of liters:

$$\text{Volume (L)} = 150.0 \text{ mL} \times \frac{1 \text{ L}}{1000 \text{ mL}} = 0.1500 \text{ L}$$

Divide the number of moles by the volume in liters to get molarity:

$$\text{Molarity} = \frac{0.291 \text{ mol NaCl}}{0.1500 \text{ L solution}} = 1.94 \frac{\text{mol NaCl}}{\text{L solution}} = 1.94 \ M$$

A normal saline solution has a molar concentration of 1.54 $M$ NaCl.

**Consider This 4.15**

If the solution volume were increased, would the molarity increase, decrease, or stay the same?

**Practice Problem 4.15**

A solution is prepared from 22.5 g of $H_2S$ dissolved in sufficient water to give 250.0 mL of solution. What is the molarity of the solution?

**Further Practice:**  Questions 4.101 and 4.102 at the end of the chapter

---

We commonly express the concentration of a solution as a molarity of the solute compound. For example, a solution might contain 0.100 $M$ $CuCl_2$. If we are interested in the concentrations of ions in a solution of this ionic compound, we must account for the fact that the solution contains twice as many chloride ions as copper(II) ions (Figure 4.20). For each formula unit of $CuCl_2$ in the solution, we have one $Cu^{2+}$ ion and two $Cl^-$ ions. For each mole of $CuCl_2$ in the solution, we have 1 mol of $Cu^{2+}$ ions and 2 mol of $Cl^-$ ions. Thus, the 0.100 $M$ $CuCl_2$ solution is 0.100 $M$ in $Cu^{2+}$ ions, 0.200 $M$ in $Cl^-$ ions, and 0.300 $M$ in total ions.

If we know the volume of a solution and the molarity, we can calculate the moles of the solute:

**FIGURE 4.20** (A) One formula unit of $CuCl_2$ consists of one $Cu^{2+}$ ion and two $Cl^-$ ions. (B) For every formula unit that dissolves in water, one aqueous $Cu^{2+}$ ion and two aqueous $Cl^-$ ions form.

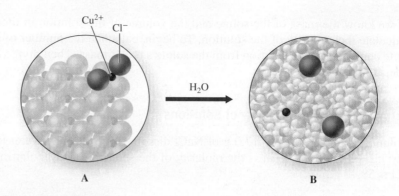

A                                B

---

**EXAMPLE 4.16**   **Moles from Volume and Molarity**

Hypokalemia is a medical condition also known as potassium deficiency. The disease can be treated with potassium chloride supplements administered orally or, in extreme cases, intravenously. However, great caution must be used in intravenous administration because KCl can cause cardiac arrest. How many moles of KCl are present in 1.50 L of a 0.0200 $M$ solution? How many moles of $K^+$ ion are present?

**Solution:**

We know the volume in liters and molarity of the KCl solution and want to calculate the moles of solute present. The molarity of the solution can be expressed in two ways and can be used as a conversion factor:

$$\frac{0.0200 \text{ mol KCl}}{1 \text{ L solution}} \quad \text{and} \quad \frac{1 \text{ L solution}}{0.0200 \text{ mol KCl}}$$

To get from volume to moles, we multiply by the molarity conversion factor that cancels units of liters and gets us to units of moles:

$$\text{mol KCl} = 1.50 \text{ L solution} \times \frac{0.0200 \text{ mol KCl}}{1 \text{ L solution}} = 0.0300 \text{ mol KCl}$$

To determine the moles of $K^+$ ion present, we must inspect the chemical formula. For every mole of KCl, there is 1 mol of $K^+$:

$$\frac{1 \text{ mol KCl}}{1 \text{ mol K}^+} \quad \text{and} \quad \frac{1 \text{ mol K}^+}{1 \text{ mol KCl}}$$

MATH TOOLBOX **1.3**

To get from moles of KCl to moles of $K^+$, we use the conversion factor that cancels the appropriate unit:

$$\text{mol K}^+ = 0.0300 \text{ mol KCl} \times \frac{1 \text{ mol K}^+}{1 \text{ mol KCl}} = 0.0300 \text{ mol K}^+$$

Since 1 mol of $K^+$ ion is present for every mole of KCl, the solution contains 0.0300 mol $K^+$ ion.

**Consider This 4.16**

If the solution contains the same number of moles of $K_2SO_4$ instead of KCl, what would happen to the number of moles of $K^+$?

**Practice Problem 4.16**

Calculate the moles of solute present in 325.0 mL of a 0.200 $M$ NaCl solution.

**Further Practice:** Questions 4.103 and 4.104 at the end of the chapter

If we know the volume and the molarity of a solution, we can also calculate the mass of solute in that volume of solution. This is an extension of the process illustrated in Example 4.16.

---

**EXAMPLE 4.17** **Mass from Volume and Molarity**

A sample of copper ore is soaked in hydrochloric acid, giving a solution that contains some copper(II) chloride. If the solution has a concentration of $3.94 \times 10^{-6}\,M$ $CuCl_2$, what mass of $CuCl_2$ (in grams) is contained in 1500.0 L of the solution? (The molar mass of $CuCl_2$ is 134.5 g/mol.)

**Solution:**

To determine the mass of the solute, first determine the moles of solute by multiplying the molarity by the volume of solution in liters:

$$\boxed{\text{Molarity}} \xrightarrow{\text{volume}} \boxed{\text{Moles of solute}} \xrightarrow{MM} \boxed{\text{Grams of solute}}$$

$$\text{mol } CuCl_2 = 1500.0\,\cancel{L} \times 3.94 \times 10^{-6}\,\frac{\text{mol}}{\cancel{L}} = 5.91 \times 10^{-3}\,\text{mol } CuCl_2$$

We expressed the molarity as mol/L in order to allow the units to cancel. Molar mass is then used to convert moles to mass:

$$\text{g } CuCl_2 = 5.91 \times 10^{-3}\,\cancel{\text{mol } CuCl_2} \times \frac{134.5\,\text{g } CuCl_2}{1\,\cancel{\text{mol } CuCl_2}} = 0.795\,\text{g } CuCl_2$$

**Consider This 4.17**

If the solution had a higher concentration, how would the mass of copper(II) chloride differ?

**Practice Problem 4.17**

Copper compounds such as bluestone are often used to control algae in fish-farming ponds. Bluestone is copper(II) sulfate pentahydrate, $CuSO_4{\cdot}5H_2O$, with a molar mass of 249.7 g/mol. A sample of pond water was found to have a concentration of $6.2 \times 10^{-5}\,M$ copper(II) sulfate. If the pond has a volume of $1.8 \times 10^7$ L, what mass of bluestone did the farmer add to the pond?

**Further Practice:** Questions 4.105 and 4.106 at the end of the chapter

---

Finally, we can calculate the volume of solution required to make a solution of a specific concentration from a given mass of solute. Again, we need to use the gram-to-mole relationship from the solute's molar mass and the conversion factor arising from the definition of molarity.

---

**EXAMPLE 4.18** **Volume and Molarity**

A solution of copper(II) acetate, $Cu(CH_3CO_2)_2$, is used as a green dye for textiles. We want to prepare a 0.150 $M$ solution of copper(II) acetate, starting with 40.0 g of $Cu(CH_3CO_2)_2$. What should be the total volume of the solution? (The molar mass of $Cu(CH_3CO_2)_2$ is 181.6 g/mol.)

Many ionic compounds, especially those of the transition metals, contain not only a metal cation and a nonmetal anion or oxoanion, but also water molecules. Such compounds are called hydrates. The amount of water in a formula unit of the compound is shown by the formula of water preceded by the number of water molecules in a formula unit that is separated from the rest of the formula by a centered dot. An example is $CuSO_4{\cdot}5H_2O$, which, as can be seen in the photo, has different properties than $CuSO_4$. The blue material in the center of the $CuSO_4$ was obtained by adding water.

$CuSO_4$

$CuSO_4{\cdot}5H_2O$

© Brian Moeskau/Moeskau Photography

**MATH TOOLBOX 4.1**

**Student Hot Spot**

Student data indicate you may struggle with calculations involving molarity. Access the SmartBook to view additional Learning Resources on this topic.

**Solution:**

We need a two-step conversion to go from mass of solute to volume of solution:

Convert the mass of solute to moles of solute by using the molar mass (181.6 g/mol):

$$\text{mol Cu(CH}_3\text{CO}_2)_2 = 40.0 \text{ g Cu(CH}_3\text{CO}_2)_2 \times \frac{1 \text{ mol Cu(CH}_3\text{CO}_2)_2}{181.6 \text{ g Cu(CH}_3\text{CO}_2)_2}$$

$$= 0.220 \text{ mol Cu(CH}_3\text{CO}_2)_2$$

Then, using the molarity of the solution, convert moles of solute to volume of solution:

$$\text{Volume of solution} = 0.220 \text{ mol Cu(CH}_3\text{CO}_2)_2 \times \frac{1 \text{ L solution}}{0.150 \text{ mol Cu(CH}_3\text{CO}_2)_2}$$

$$= 1.47 \text{ L solution}$$

**Consider This 4.18**

If we wanted to prepare a more dilute copper(II) acetate solution from the same mass of $Cu(CH_3CO_2)_2$, how would the volume change?

**Practice Problem 4.18**

Copper(II) chlorate, $Cu(ClO_3)_2$, is used to treat fibers before dying them so the dye will adhere. Suppose we wish to prepare a 0.225 *M* solution of $Cu(ClO_3)_2$, starting with 1.89 g of copper(II) chlorate. What should be the total volume of the solution?

**Further Practice:**  Questions 4.107 and 4.108 at the end of the chapter

MATH
TOOLBOX
**4.1**

## Dilution

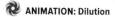

ANIMATION: Dilution

Dilution is a simple way to reduce concentration, but how would you increase the concentration of a solution?

Suppose you taste your lemonade and find that it is too sweet. How might you make it less sweet? The easiest way is to add more water. Adding more water to lessen its concentration dilutes the lemonade. To lower the concentration of a solution, we use the process of **dilution,** or adding more solvent to a solution, as depicted in Figure 4.21. In this process, the relative numbers of solute and solvent particles change. Adding more solvent increases the number of solvent particles and increases the volume. For a given amount of the original solution (often called the *stock* solution), the number of solute particles stays the same, but they are now spread out through a greater volume, so their concentration is less.

   If we dilute a solution of known molarity, we can calculate the molarity of the diluted solution. The number of moles of solute contained in the volume of the more concentrated stock solution ($V_{conc}$) is found by multiplying the molarity ($M_{conc}$) of this solution by its volume in liters:

$$\text{Moles}_{conc} = M_{conc} \times V_{conc}$$

Since we are adding only solvent, the moles of solute do not change upon dilution, so the moles of solute before and after dilution are equal:

$$\text{Moles}_{conc} = \text{moles}_{dil}$$

Concentrated solution

A                    B                    Water                    Dilute solution

                                          C          D          E

**FIGURE 4.21** (A) A pipet is used to accurately measure 25.00 mL of 1.000 $M$ CuSO$_4$ solution and (B) deliver it into a 250.0-mL volumetric flask. (C and D) Water is added with swirling until the diluted solution fills the flask to the marked level. (E) As water is added, the amount of water increases, leading to a lesser concentration of solute. A stopper is placed in the flask, and the flask is inverted several times to completely mix the solution. The number of Cu$^{2+}$ ions and SO$_4^{2-}$ ions in a given volume of solution decreases, but the total number of Cu$^{2+}$ ions and SO$_4^{2-}$ ions in the flask is constant. The new solution has a concentration of 0.1000 $M$.
© Brian Moeskau/Moeskau Photography

The moles of solute in the diluted solution also equal the product of molarity ($M_{dil}$) and volume ($V_{dil}$) in liters, where the volume is the total volume after dilution:

$$\text{Moles}_{dil} = M_{dil} \times V_{dil}$$

Since the number of moles of solute is unchanged, we get an equation that relates the molarity and volume of the diluted solution to the molarity and volume of the original, more concentrated, solution:

$$M_{dil} V_{dil} = M_{conc} V_{conc}$$

We can rearrange this equation to solve for the molarity of the diluted solution. To do this, we divide both sides of the equation by $V_{dil}$:

$$\frac{M_{dil} V_{dil}}{V_{dil}} = \frac{M_{conc} V_{conc}}{V_{dil}}$$

Since the terms cancel on the left side of the equation, we get the following rearranged equation:

$$M_{dil} = \frac{M_{conc} V_{conc}}{V_{dil}}$$

Thus, if 25.00 mL of a 0.1000 $M$ copper(II) sulfate solution is diluted to 250.0 mL, the concentration of the diluted solution is obtained by substituting the molarity of

The dilution equation is sometimes expressed as $M_1 V_1 = M_2 V_2$

the concentrated solution and the two volumes, first converted to liters, into the equation. The molarity of the original concentrated solution ($M_{\text{conc}}$) is 0.1000 $M$, the volume of the concentrated solution ($V_{\text{conc}}$) is 0.02500 L, and the volume of the diluted solution ($V_{\text{dil}}$) is 0.2500 L:

$$M_{\text{dil}} = \frac{0.1000 \, M \times 0.02500 \, \cancel{L}}{0.2500 \, \cancel{L}} = 0.01000 \, M$$

Variations on this calculation can be carried out as long as any three of the four variables in the equation are known.

---

**EXAMPLE 4.19**  **Dilution**

If 85.2 mL of 2.25 $M$ $CuCl_2$ solution is diluted to a final volume of 250.0 mL, what is the molarity of the diluted solution of $CuCl_2$?

**Solution:**

The numbers of moles of $CuCl_2$ are the same before and after dilution, so we can use the equation for dilution:

$$M_{\text{dil}} = \frac{M_{\text{conc}} V_{\text{conc}}}{V_{\text{dil}}}$$

The molarity of the concentrated solution ($M_{\text{conc}}$) is 2.25 $M$, the volume of the concentrated solution ($V_{\text{conc}}$) is 0.0852 L, and the volume of the diluted solution ($V_{\text{dil}}$) is 0.2500 L. Insert the values of the quantities and solve the equation for the molarity of the diluted solution:

$$M_{\text{dil}} = \frac{2.25 \, M \times 0.0852 \, \cancel{L}}{0.2500 \, \cancel{L}} = 0.767 \, M$$

**Consider This 4.19**

If we diluted a smaller volume of solution to the same final volume, would the molarity be higher or lower?

**Practice Problem 4.19**

If 42.8 mL of 3.02 $M$ $H_2SO_4$ solution is diluted to a final volume of 500.0 mL, what is the molarity of the diluted solution of $H_2SO_4$?

**Further Practice:** Questions 4.113 and 4.114 at the end of the chapter

---

 **Math Toolbox 4.1   Mole Conversions**

The mole (mol) unit describes a quantity of particles and is usually used to describe numbers of atoms or formula units such as molecules.

$$1 \text{ mole} = 6.022 \times 10^{23} \text{ particles}$$

The number of particles in 1 mol is often referred to as Avogadro's number.

It is useful to describe quantities in mole units, and it is also sometimes important to know the mass or number of molecules or atoms in a given sample. The following diagram shows the importance of the mole in converting between the macroscopic level (grams and moles) and the molecular level (formula units, molecules, atoms, and ions), and shows the individual steps involved in converting between these quantities.

In the sections that follow, we will look at examples of individual calculations and multistep calculations involving these quantities.

### Converting Between Moles and Number of Particles

When we are given the moles of a substance, the mole quantity describes the number of particles indicated by the substance's formula:

$$1 \text{ mol Cu} = 6.022 \times 10^{23} \text{ Cu atoms}$$
$$1 \text{ mol O}_2 = 6.022 \times 10^{23} \text{ O}_2 \text{ molecules (or formula units)}$$
$$1 \text{ mol CO}_2 = 6.022 \times 10^{23} \text{ CO}_2 \text{ molecules (or formula units)}$$
$$1 \text{ mol NaCl} = 6.022 \times 10^{23} \text{ NaCl formula units}$$

For molecular compounds or elements, we can refer to the particles as either molecules or formula units. However, for ionic compounds, we must refer to the particles only as formula units since ionic compounds do not exist as molecules.

The relationship 1 mole = $6.022 \times 10^{23}$ formula units can be used to convert between moles and the number of formula units for any substance.

The conversion ratio for $CO_2$, for example, will have one of the following two forms, depending on whether the conversion is to formula units or to moles:

$$\frac{6.022 \times 10^{23} \text{ CO}_2 \text{ formula units}}{1 \text{ mol CO}_2} \quad \text{or} \quad \frac{1 \text{ mol CO}_2}{6.022 \times 10^{23} \text{ CO}_2 \text{ formula units}}$$

Conversion *ratios* are similar to the conversion *factors* we used in Chapter 1 to convert measured values to different units.

---

### EXAMPLE 4.20  Converting Between Moles and Formula Units

Use Avogadro's number for the following conversions:

(a) Convert 0.042 mol $CO_2$ to formula units (molecules) of $CO_2$.
(b) Convert $4.4 \times 10^{24}$ NaCl formula units to moles.

**Solution:**

(a) We begin by determining the conversion ratio for converting between moles and formula units:

We set up the conversion with formula units in the numerator and moles in the denominator so that units cancel properly:

$$0.042 \text{ mol CO}_2 \times \frac{6.022 \times 10^{23} \text{ CO}_2 \text{ formula units}}{1 \text{ mol CO}_2} = 2.5 \times 10^{22} \text{ CO}_2 \text{ formula units (or molecules)}$$

This result makes sense because it is less than Avogadro's number, which represents 1 mole.

(b) In this case we are converting from formula units to moles, so we use the conversion ratio with moles in the numerator and formula units in the denominator:

$$4.41 \times 10^{24} \text{ NaCl formula units} \times \frac{1 \text{ mol NaCl}}{6.022 \times 10^{23} \text{ NaCl formula units}} = 7.32 \text{ mol NaCl}$$

This result makes sense because it is greater than 1 mol and we have more than Avogadro's number of particles.

**Practice Problem 4.20**

Use Avogadro's number for the following conversions:

(a) Convert $8.1 \times 10^{20}$ $NH_3$ molecules to moles.
(b) Convert 7.8 mol $CuSO_4$ to formula units of $CuSO_4$.

**Further Practice:** Questions 4.119 and 4.120 at the end of the chapter

## Calculating Number of Atoms or Ions

Sometimes we are interested in the number of atoms or ions in a sample of a compound. For example, one $CO_2$ molecule contains one carbon atom and two oxygen atoms. When we have more than one $CO_2$ molecule, the relationship between the number of atoms and the number of molecules remains the same. The following table illustrates this:

| Number of $CO_2$ Molecules | Number of Carbon Atoms | Number of Oxygen Atoms |
|---|---|---|
| 1 | 1 | 2 |
| 5 | 5 | 10 |
| 200 | 200 | 400 |
| $6.02 \times 10^{23}$ | $6.02 \times 10^{23}$ | $1.20 \times 10^{24}$ |
| 1.000 mol | 1.000 mol | 2.000 mol |

The relationships between the number of atoms of each element and $CO_2$ molecules can be expressed as the following ratios:

$$\frac{1 \text{ C atom}}{1 \text{ CO}_2 \text{ molecule}} \quad \text{and} \quad \frac{2 \text{ O atoms}}{1 \text{ CO}_2 \text{ molecule}}$$

These ratios can be used as conversion ratios for determining the number of atoms in any quantity of $CO_2$. Example 4.21 shows how ratios such as these can be used as conversion ratios for determining the number of atoms or ions in a given number of particles (or moles) of a substance.

---

### EXAMPLE 4.21  Determining the Number of Atoms or Ions

(a) How many hydrogen atoms are in 12 molecules of propane, $C_3H_8$?
(b) How many hydrogen atoms are in 0.250 mol of propane, $C_3H_8$?
(c) How many $Na^+$ ions are in 1.00 mol sodium carbonate, $Na_2CO_3$?
(d) How many total ions ($Na^+$ and $CO_3^{2-}$) are in 1.50 mol sodium carbonate, $Na_2CO_3$?

**Solution:**

(a) Number of hydrogen atoms = $12 \text{ C}_3\text{H}_8 \text{ molecules} \times \dfrac{8 \text{ hydrogen atoms}}{1 \text{ C}_3\text{H}_8 \text{ molecules}}$ = 96 hydrogen atoms

(b) Number of hydrogen atoms = $0.250 \text{ mol C}_3\text{H}_8 \times \dfrac{6.022 \times 10^{23} \text{ C}_3\text{H}_8 \text{ molecules}}{1 \text{ mol C}_3\text{H}_8} \times \dfrac{8 \text{ hydrogen atom}}{1 \text{ C}_3\text{H}_8 \text{ molecules}}$ = $1.20 \times 10^{24}$ hydrogen atoms

(c) 1.00 mole is equal to $6.02 \times 10^{23}$ $Na_2CO_3$ formula units, so we will start the calculation with formula units:

Number of $Na^+$ ions = $6.02 \times 10^{23} \text{ Na}_2\text{CO}_3 \text{ formula units} \times \dfrac{2 \text{ Na}^+ \text{ ions}}{1 \text{ Na}_2\text{CO}_3 \text{ formula unit}}$ = $1.20 \times 10^{24}$ $Na^+$ ions

(d) We start with 1.50 mole of $Na_2CO_3$ to calculate the formula units of $Na_2CO_3$:

$1.50 \text{ mol Na}_2\text{CO}_3 \times \dfrac{6.022 \times 10^{23} \text{ Na}_2\text{CO}_3 \text{ formula units}}{1 \text{ mol Na}_2\text{CO}_3}$ = $9.033 \times 10^{23}$ $Na_2CO_3$ formula units

Each $Na_2CO_3$ formula unit contains two $Na^+$ ions and one $CO_3^{2-}$ ion, so there is a total of three ions in every one $Na_2CO_3$ formula unit. We multiply the number of $Na_2CO_3$ formula units by the ratio of ions to formula units to obtain the total number of ions:

Number of ions = $9.033 \times 10^{23} \text{ Na}_2\text{CO}_3 \text{ formula units} \times \dfrac{3 \text{ ions}}{1 \text{ Na}_2\text{CO}_3 \text{ formula units}}$ = $2.71 \times 10^{24}$ ions

**Practice Problem 4.21**

(a) How many carbon atoms are in 5 molecules of sucrose, $C_{12}H_{22}O_{11}$?
(b) How many total atoms are in 5 molecules of sucrose, $C_{12}H_{22}O_{11}$?
(c) How many nitrate ions are in 1.0 mol aluminum nitrate, $Al(NO_3)_3$?

**Further Practice:** Questions 4.21 and 4.24 at the end of the chapter

## Mole-Mass Relationships

The relationship between the mass of a substance and its mole quantity is defined by the molar mass of the substance. Molar mass is the mass (in grams) of 1 mol of that substance. Molar mass values for the atomic elements can be obtained directly from the periodic table. For example, the molar mass of carbon shown on the periodic table is 12.01 g/mol. This means that the mass of 1.000 mol of carbon atoms is equal to 12.01 grams. Molar masses for compounds or molecular elements are determined by combining the molar masses of the component atomic elements. The molar mass of $CO_2$, 44.01 g/mol, is determined by combining the molar masses of carbon and oxygen and adjusting according to the chemical formula:

$$1 \times \text{molar mass C} = 1 \times 12.01 \text{ g/mol}$$
$$2 \times \text{molar mass O} = \underline{2 \times 16.00 \text{ g/mol}}$$
$$\text{Molar mass } CO_2 = \quad 44.01 \text{ g/mol}$$

Molar mass provides a relationship between mass and moles for a given substance, and can therefore be used as a conversion ratio:

For example, if we are converting between grams and moles for a sample of $CO_2$, we can use the molar mass ratio, or its inverse ratio, depending on whether we are converting to grams or to moles:

$$\frac{44.01 \text{ g } CO_2}{1 \text{ mol } CO_2} \quad \text{or} \quad \frac{1 \text{ mol } CO_2}{44.01 \text{ g } CO_2}$$

Example 4.22 shows conversions from grams to moles and from moles to grams for different substances.

---

### EXAMPLE 4.22  Conversions Between Mass in Grams and Moles

(a) Convert 0.30 mol $CO_2$ to grams.
(b) Convert 10.0 g $MgCl_2$ to moles.

#### Solution

(a) The molar mass of $CO_2$ was calculated prior to this example and is 44.01 g/mol.

$$0.30 \text{ mol } CO_2 \times \frac{44.01 \text{ g } CO_2}{1 \text{ mol } CO_2} = 13 \text{ g } CO_2$$

(b) The molar mass of $MgCl_2$ is: 24.31 g/mol + (2 × 35.45 g/mol) = 95.21 g/mol

$$10.0 \text{ g } MgCl_2 \times \frac{1 \text{ mol } MgCl_2}{95.21 \text{ g } MgCl_2} = 0.105 \text{ mol } MgCl_2$$

#### Practice Problem 4.22

(a) Convert 8.8 g $CH_4$ to moles.
(b) Convert 0.352 mol $N_2$ to grams.

**Further Practice:**  Questions 4.39 and 4.45 at the end of the chapter.

---

## Multistep Calculations Involving Mole Quantities

Because there is no convenient direct relationship between grams and number of particles, these types of calculations involve multiple steps as shown in the diagram at the beginning of this Math Toolbox (reproduced here):

If we need to convert between grams and atoms or ions, then the calculation will involve the three steps shown in the figure. The number of steps would be less if we were given, or asked to determine, moles or number of formula units. We've learned how to do each of these steps in earlier sections of this Math Toolbox, so here we will show how to combine these steps. It is important to determine what quantity you are given and what you are looking for. Think about what quantity you are calculating in each step, and check your units to make sure they have cancelled properly. These steps can be done in a series of individual steps or as a string of combined steps. Example 4.23 illustrates calculations as a string of combined steps.

### EXAMPLE 4.23  Multistep Mole Calculations

(a) Determine the number of $Cu^+$ ions in 6.0 g $Cu_2O$.
(b) Determine the number of hydrogen atoms in 0.22 mol $NH_3$.
(c) Determine the number of $Br_2$ molecules in 1 drop of bromine that has a mass of 0.070 g.

### Solution

(a) In this problem we are given mass in grams, and we must determine the number of $Cu^+$ ions. The conversion from mass in grams to number of ions is a three-step calculation:
Step 1: Convert grams to moles using the molar mass of $Cu_2O$ (143.1 g/mol).
Step 2: Convert moles to formula units (using Avogadro's number).
Step 3: Convert formula units to $Cu^+$ ions (using the ratio of 2 $Cu^+$ ions to 1 $Cu_2O$ formula unit).

$$\text{Number of } Cu^+ \text{ ions} = 6.0 \text{ g } Cu_2O \times \frac{1 \text{ mol } Cu_2O}{143.1 \text{ g } Cu_2O} \times \frac{6.022 \times 10^{23} \text{ } Cu_2O \text{ formula units}}{1 \text{ mol } Cu_2O} \times \frac{2 \text{ } Cu^+ \text{ ions}}{1 \text{ } Cu_2O \text{ formula unit}} = 5.0 \times 10^{22} \text{ } Cu^+ \text{ ions}$$

(b) In this problem we are given moles, and we must determine the number of hydrogen atoms. The conversion from mass in grams to number of ions is a two-step calculation:
Step 1: Convert moles to formula units (molecules) using Avogadro's number.
Step 2: Convert molecules to hydrogen atoms (using the ratio of 3 H atoms to 1 $NH_3$ molecule).

$$\text{Number of H atoms} = 0.22 \text{ mol } NH_3 \times \frac{6.022 \times 10^{23} \text{ } NH_3 \text{ molecules}}{1 \text{ mol } NH_3} \times \frac{3 \text{ H atoms}}{1 \text{ } NH_3 \text{ molecule}} = 4.0 \times 10^{23} \text{ H atoms}$$

(c) In this problem we are given mass in grams, and we must determine the number of $Br_2$ molecules. The conversion from mass in grams to number of molecules is a two-step calculation:
Step 1: Convert grams to moles using the molar mass of $Br_2$ (159.8 g/mol).
Step 2: Convert moles to molecules (using Avogadro's number).

$$\text{Number of } Br_2 \text{ molecules} = 0.070 \text{ g } Br_2 \times \frac{1 \text{ mol } Br_2}{159.8 \text{ g } Br_2} \times \frac{6.022 \times 10^{23} \text{ } Br_2 \text{ molecules}}{1 \text{ mol } Br_2} = 2.6 \times 10^{20} \text{ } Br_2 \text{ molecules}$$

### Practice Problem 4.23

(a) Determine the number of oxygen atoms in 5.21 mol $N_2O_5$.
(b) Determine the number of $O_2$ molecules in an oxygen gas sample that has a mass of 0.12 g.
(c) Determine the number of $Cl^-$ ions in 17.8 g of $MgCl_2$.

**Further Practice:** Questions 4.55 and 4.56 at the end of the chapter

# CHAPTER REVIEW

## KEY CONCEPTS

- Chemical composition often requires us to know the numbers of atoms, molecules, or ions in a sample. Because these values are so large for normal sample sizes, we use units of moles (mol).
  - A mole is defined as Avogadro's number ($6.022 \times 10^{23}$) of particles. This value is used to convert between the number of moles of a chemical substance and the number of formula units (such as molecules) in a substance.
  - The mass of 1 mole of a substance is its molar mass in units of grams per mole (g/mol) and is determined from the periodic table.
  - Molar mass can be used to convert between the mass (in grams) and the number of moles of a pure substance.
  - Because a compound's formula unit often contains more than one atom or ion of an element, the ratio described by the formula can be used to convert the number of formula units (or molecules) to the number of atoms or ions of a given element in a sample.
- The chemical composition of a compound tells us the relative amounts of the elements that are in a compound, which has constant composition.
  - The percent composition by mass is a ratio of the mass of one element to the mass of the entire compound, expressed as a percentage.
  - The empirical formula is an atom ratio (or mole ratio) of the elements in a compound, expressed as the lowest possible whole-number ratio.
  - Empirical formulas can be determined from experimentally measured masses of the component elements or from percent composition by mass.
  - The molecular formula is related to the empirical formula, but shows the exact number of atoms of each element in a formula unit, or molecule.
  - Molecular formulas can be determined from empirical formulas if the molar mass of the compound is known.
- The composition of a solution is described by its concentration.
  - Concentration is commonly expressed in terms of percent by mass or molarity. Both describe the ratio of the amount of solute to the amount of solution.
  - Dilute solutions (of lesser concentration) can be prepared from concentrated solutions (of greater concentration) by adding known amounts of solvent, a process known as dilution.

## KEY RELATIONSHIPS

| Relationship | Equation |
|---|---|
| Percent composition is expressed as the fraction of the sample mass due to a specific element, multiplied by 100 to convert to a percent. | $\% \text{ E (E is any element)} = \dfrac{\text{mass of E}}{\text{mass of sample}} \times 100\%$ |
| Percent by mass composition of a solution is the fraction of the solution mass due to the solute, multiplied by 100 to convert to a percent. | $\% \text{ mass} = \dfrac{\text{mass solute}}{\text{mass solution}} \times 100\%$ |
| The concentration of a solution in units of molarity is the ratio of moles of solute to liters of solution. | $\text{Molarity} = \dfrac{\text{moles of solute}}{\text{liters of solution}}$ |
| When diluting a solution, the moles of solute in the solution remain unchanged. Moles are calculated by multiplying molarity by volume in liters. This product has the same value before and after dilution. | $M_{\text{dil}} V_{\text{dil}} = M_{\text{conc}} V_{\text{conc}}$ |

## KEY TERMS

Avogadro's number (4.2)
concentrated solution (4.4)
concentration (4.4)

dilute solution (4.4)
dilution (4.4)
empirical formula (4.3)

molar mass (4.2)
molarity (4.4)
mole (4.2)

percent composition by mass (4.1)
solute (4.4)
solvent (4.4)

## QUESTIONS AND PROBLEMS

The following questions and problems, except those in Additional Questions and Concept Review Questions, are paired. Questions in a pair focus on the same concept. Answers to selected questions and problems are in Appendix D.

### Matching Definitions with Key Terms

4.1     Match the key terms with the descriptions provided.
(a) the amount of a substance that contains the same number of atoms, molecules, or formula units as there are atoms in exactly 12 g of carbon-12 ($^{12}C$)
(b) the number of fundamental particles (atoms, molecules, or ions) in 1 mol of any substance
(c) a formula written with the simplest ratios of atoms or ions (the smallest whole-number subscripts)
(d) the substance being dissolved; usually, that component of a solution that is present in the lesser amount
(e) the number of moles of solute per liter of solution
(f) a solution that contains a relatively high concentration of solute

4.2     Match the key terms with the descriptions provided.
(a) the mass of 1 mol of any substance, in units of grams per mole
(b) an expression of the percent of the total mass due to each element in a compound
(c) the substance doing the dissolving; usually, that component of a solution that is present in the larger amount
(d) the relative amounts of solute and solvent in a solution
(e) a solution that contains a relatively small concentration of solute
(f) the process of adding more solvent to give a solution of lesser concentration

### Percent Composition

4.3     A 24.7-g sample of limestone, $CaCO_3$, contains 9.88 g of calcium. What is the percent Ca in limestone?

4.4     A 3.75-g sample of limestone, $CaCO_3$, contains 1.80 g of oxygen and 0.450 g of carbon. What is the percent O and the percent C in limestone?

4.5     A common nasal spray used to treat allergies contains a steroid drug. The dose from one spray from the dispenser contains 50.0 μg of active ingredient, in which there is 30.0 μg of carbon. What is the percent carbon in the active ingredient?

4.6     In addition to its more well-known use, a common erectile dysfunction drug has been shown to be effective in treating a rare pulmonary disorder. A 20.0-mg dose of the drug contains 3.54 mg of nitrogen. What is the percent nitrogen in this drug?

4.7     Lithium carbonate, $Li_2CO_3$, contains 18.8% lithium and is used in the treatment of mental illnesses such as bipolar

disorder. What mass of lithium is present in a 1.20-g dose of lithium carbonate?

4.8     Quartz, $SiO_2$, contains 46.7% silicon. What mass of silicon is present in 348 g of quartz?

### Mole Quantities

4.9     Write the formulas for the compounds described by the following compositions.
(a) twice as many hydrogen atoms as sulfur atoms
(b) 1.5 times as many oxygen atoms as nitrogen atoms
(c) one-half as many calcium ions as chloride ions

4.10    Write the formulas for the compounds described by the following compositions.
(a) three times as many hydrogen atoms as phosphorus atoms
(b) 2.5 times as many oxygen atoms as nitrogen atoms
(c) one-third as many aluminum ions as chlorate ions

4.11    What are the formulas of the following molecules?
(a)                     (b)                      (c)

4.12    What are the formulas of the following molecules?
(a)                     (b)                      (c)

4.13    What is the molecular formula of these molecules?

**4.14**    What is the molecular formula of these molecules?

**4.15**    What are the formula units for each of the following?

(a)

(c)

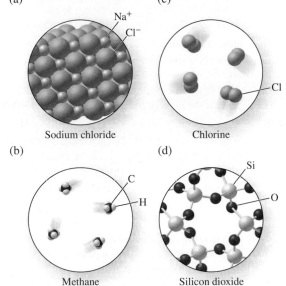

Sodium chloride                Chlorine

(b)

(d)

Methane                Silicon dioxide

**4.16**    What are the formula units for each of the following?

(a)

(c)

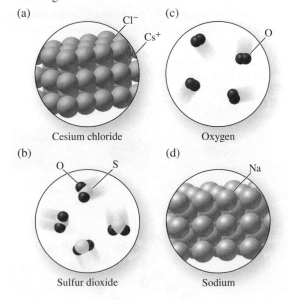

Cesium chloride                Oxygen

(b)

(d)

Sulfur dioxide                Sodium

**4.17**    How many $NH_3$ molecules are in 0.50 mol of $NH_3$? How many nitrogen atoms and hydrogen atoms are there?

**4.18**    How many $SO_2$ molecules are in 1.75 mol of $SO_2$? How many sulfur atoms and oxygen atoms are there?

**4.19**    How many formula units are in 0.5 mol of $Cu_2S$?

**4.20**    How many formula units are in 1.5 mol of $CuSO_4$?

**4.21**    How many sulfur atoms are in 0.2 mol of $SO_2$?

**4.22**    How many oxygen atoms are in 0.2 mol of $SO_2$?

**4.23**    How many calcium ions are in 1 mol of $CaCl_2$?

**4.24**    How many chloride ions are in 2 mol of $CaCl_2$?

**4.25**    Calculate the molar mass of the substances shown in Question 4.15.

**4.26**    Calculate the molar mass of the substances shown in Question 4.16.

**4.27**    Calculate the molar mass of each of the following compounds.

(a) $Hg_2Cl_2$                (c) $Cl_2O_5$
(b) $CaSO_4 \cdot 2H_2O$       (d) $NaHSO_4$

**4.28**    Calculate the molar mass of each of the following compounds.

(a) $K_2SO_4$                (c) $C_2H_4Cl_2$
(b) $NiCl_2 \cdot 6H_2O$       (d) $Mg(NO_3)_2$

**4.29**    Calculate the molar mass of the following substances.

(a) $I_2$          (b) $CrCl_3$          (c) $C_4H_8$

**4.30**    Calculate the molar mass of the following substances.

(a) $P_4$          (b) $CrO_2Cl_2$          (c) $CaF_2$

**4.31**    In general, why must we weigh a sample in order to determine the number of atoms it contains?

**4.32**    What is a mole? Why do chemists need to use the concept?

**4.33**    If 1.00 mol of LiCl has a mass of 42.39 g, what is the average mass of 1 LiCl formula unit in atomic mass units?

**4.34**    If one formula unit of $CuCl_2$ has an average mass of 134.5 amu, what is the mass of 1.00 mol of $CuCl_2$?

**4.35**    If $2.01 \times 10^{23}$ molecules of a substance have a mass of 12.0 g, what is the molar mass of the substance?

**4.36**    If the molar mass of a substance is 98.09 g/mol, what is the mass of $3.01 \times 10^{23}$ molecules of the substance?

**4.37**    Calculate the number of moles in 10.0 g of the following substances.

(a) $KHCO_3$                (c) Se
(b) $H_2S$                (d) $MgSO_4$

**4.38**    Calculate the number of moles in 100.0 g of each of the following compounds.

(a) $SO_2$                (c) $BaSO_4$
(b) $Na_2SO_4$                (d) $KAl(SO_4)_2$

**4.39**    Calculate the moles of the following substances.

(a) 32.5 g NaCl
(b) 250.0 mg aspirin, $C_9H_8O_4$
(c) 73.4 kg $CaCO_3$
(d) 5.47 µg CuS

**4.40**    Calculate the moles of the following substances.

(a) 72.2 g $K_2SO_4$
(b) 160.0 mg oxycodone, $C_{18}H_{21}NO_4$
(c) 2.82 kg magnetite, $Fe_3O_4$
(d) 5.00 µg sucrose, $C_{12}H_{22}O_{11}$

**4.41**    Which element, S, Fe, Au, or C, contains the most moles of atoms in a 1.0-g sample?

**4.42**    Which element, Cu, P, Ag, or B, contains the least moles of atoms in a 5.0-g sample?

4.43   Calculate the mass of 2.50 mol of the following substances.
   (a) $Ba(OH)_2$
   (b) $Cl_2$
   (c) $K_2SO_4$
   (d) $PF_3$

4.44   Calculate the mass of 0.750 mol of the following substances.
   (a) $I_2$
   (b) $Mg(NO_3)_2$
   (c) $SiO_2$
   (d) $Na_3PO_4$

4.45   A chemical reaction requires 2.7 mol of zinc acetate, $Zn(CH_3CO_2)_2$. What mass of zinc acetate is needed?

4.46   A chemical reaction is used to produce 3.4 mol of copper(II) bicarbonate, $Cu(HCO_3)_2$. What mass of copper(II) bicarbonate is produced?

4.47   A sample of ammonia, $NH_3$, weighs 30.0 g. Calculate the following quantities.
   (a) moles of $NH_3$
   (b) number of $NH_3$ molecules
   (c) number of N atoms
   (d) moles of H atoms

4.48   A sample of sodium carbonate, $Na_2CO_3$, weighs 15.0 g. Calculate the following quantities.
   (a) moles of $Na_2CO_3$
   (b) number of $Na_2CO_3$ formula units
   (c) number of $Na^+$ ions
   (d) moles of $CO_3^{2-}$ ions

4.49   Which of these substances has the most atoms per mole? Which has the least?

NaCl

$H_2SO_4$

Na

SO_3

4.50   Which of these substances has the most atoms per mole? Which has the least?

SiO_2

$C_2H_6$

Fe

CO_2

4.51   A raindrop weighs 0.050 g. How many molecules of water are in one raindrop?

4.52   A grain of sand weighs $7.7 \times 10^{-4}$ g. How many formula units of silicon dioxide, $SiO_2$, are in one sand grain?

4.53   How many formula units are in 250.0 g of the following substances?
   (a) $Br_2$
   (b) $MgCl_2$
   (c) $H_2O$
   (d) Fe

4.54   How many formula units are in 375.0 g of the following substances?
   (a) Cu
   (b) NaBr
   (c) $SO_2$
   (d) $NH_4Cl$

4.55   How many atoms (or ions) of each element are in 140.0 g of the following substances?
   (a) $H_2$
   (b) $Ca(NO_3)_2$
   (c) $N_2O_2$
   (d) $K_2SO_4$

4.56   How many atoms (or ions) of each element are in 140.0 g of the following substances?
   (a) $BaSO_4$
   (b) $Mg_3(PO_4)_2$
   (c) $O_2$
   (d) KBr

4.57   What is the mass of $6.4 \times 10^{22}$ molecules of $SO_2$?

4.58   What is the mass of $1.8 \times 10^{21}$ molecules of $H_2SO_4$?

4.59   Which compound, $NH_3$, $NH_4Cl$, $NO_2$, or $N_2O_3$, contains the most nitrogen atoms in a 25.0-g sample?

4.60   Which compound, NaCl, $PCl_3$, $CaCl_2$, or $HClO_2$, contains the most chlorine atoms (or ions) in a 100.0-g sample?

## Determining Empirical and Molecular Formulas

4.61   You have two colorless gases, each made of sulfur and oxygen. If they have different percent compositions, can they be the same substance?

4.62   Describe some uses for the percent composition of a substance.

4.63   What is the difference between an empirical formula and a molecular formula?

4.64   Why do we normally use an empirical formula instead of percent composition to represent the composition of a substance?

4.65   Which of the following molecules have an empirical formula that is different from their molecular formula?

$H_2O_2$

$HCO_2H$

$N_2O_4$

SiO_2

**4.66** Which of the following substances have an empirical formula that is the same as their molecular (or ionic) formula?

$S_2Cl_2$

$N_2O_3$

$C_6H_6$

NaCl

**4.67** What is the empirical formula of each of the following compounds?
(a) $P_4O_{10}$
(b) $Cl_2O_5$
(c) $PbCl_4$
(d) $HO_2CC_4H_8CO_2H$

**4.68** What is the empirical formula of each of the following compounds?
(a) $As_4O_6$
(b) $H_2S_2$
(c) $CaCl_2$
(d) $C_3H_6$

**4.69** Given the following molecular formulas, write the empirical formulas.
(a) $C_6H_4Cl_2$
(b) $C_6H_5Cl$
(c) $N_2O_5$

**4.70** Given the following molecular formulas, write the empirical formulas.
(a) $N_2O_4$
(b) $H_2C_2O_4$
(c) $CH_3CO_2H$

**4.71** Which of the following compounds of nitrogen and oxygen have identical empirical formulas: $N_2O$, $NO$, $NO_2$, $N_2O_3$, $N_2O_4$, $N_2O_5$?

**4.72** Which of the following compounds of carbon and hydrogen have identical empirical formulas: $CH_4$, $C_2H_4$, $C_3H_6$, $C_4H_{12}$, $C_6H_6$?

**4.73** What are the empirical formulas of the compounds with the following compositions?
(a) 72.36% Fe, 27.64% O
(b) 58.53% C, 4.09% H, 11.38% N, 25.99% O

**4.74** What are the empirical formulas of the compounds with the following compositions?
(a) 85.62% C, 14.38% H
(b) 63.15% C, 5.30% H, 31.55% O

**4.75** Eugenol, a chemical substance with the flavor of cloves, consists of 73.19% C, 19.49% O, and 7.37% H. What is the empirical formula of eugenol?

**4.76** One of the compounds in cement has the following composition: 52.66% Ca, 12.30% Si, and 35.04% O. What is its empirical formula?

**4.77** The explosive trinitrotoluene (TNT) has the composition 37.01% C, 2.22% H, 18.50% N, and 42.27% O. What is the empirical formula of TNT?

**4.78** Strychnine (rat poison) has the composition 75.42% C, 6.63% H, 8.38% N, and 9.57% O. What is the empirical formula of strychnine?

**4.79** An unknown organic compound was determined to have the following mass-percent composition: 45.42% carbon, 2.720% hydrogen, and 51.86% oxygen. What is the empirical formula for this compound?

**4.80** A compound consisting of phosphorus and sulfur was determined to have the following mass-percent composition: 35.57% phosphorus, 64.43% sulfur. Use this information to determine the empirical formula for this compound.

**4.81** What information is needed to determine a molecular formula?

**4.82** In what way does a molecular formula represent more information than an empirical formula?

**4.83** A compound with the empirical formula $CH_2O$ has a molar mass of approximately 90 g/mol. What is its molecular formula?

**4.84** A gaseous compound has the empirical formula $NO_2$. If its molar mass is approximately 92 g/mol, what is its molecular formula?

**4.85** A compound has a molar mass of approximately 180 g/mol and a percent composition of 40.00% C, 6.72% H, and 53.29% O. What is the molecular formula?

**4.86** A compound with the molar mass of approximately 142 g/mol has the composition 50.7% C, 9.9% H, and 39.4% N. What is its molecular formula?

**4.87** What is the percent composition of each of the following compounds?
(a) $SO_2$
(b) $CuCl_2$
(c) $Na_3PO_4$
(d) $Mg(NO_3)_2$

**4.88** What is the percent nitrogen in each of the following compounds?
(a) $NaNO_3$
(c) $N_2H_4$
(b) $NH_4Cl$
(d) $N_2O$

**4.89** Consider the following minerals as potential sources of copper metal. Which mineral contains the highest percentage of copper?
(a) chalcocite, $Cu_2S$
(b) malachite, $Cu_2(CO_3)(OH)_2$
(c) cuprite, $CuO$
(d) azurite, $Cu_3(CO_3)_2(OH)_2$

**4.90** Calculate the percent iron in the following iron minerals, which might be found in iron ores. Which contains the highest percentage of iron?
(a) wustite, $FeO$
(b) hematite, $Fe_2O_3$
(c) magnetite, $Fe_3O_4$
(d) siderite, $FeCO_3$

## Chemical Composition of Solutions

**4.91** What is a solution? Give five examples of solutions that can be found in your home or dorm.

**4.92** Distinguish between a solute and a solvent. Identify the solute in five solutions that can be found in a grocery store.

4.93    Identify the solute and solvent in this solution containing calcium chloride and water. Explain your answer.

4.94    Identify the solute and solvent in this solution containing sodium chloride and water. Explain your answer.

4.95    What is the difference between a dilute and a concentrated solution?

4.96    Give an example of two solutions that could be described as dilute and concentrated.

4.97    What relationship is described by concentration?

4.98    Define molarity.

4.99    Which of the following solutions of NaCl is more concentrated?

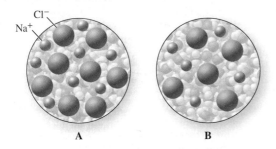

A                          B

4.100   Which of the following solutions of CaCl$_2$ is more dilute?

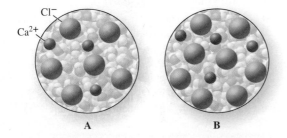

A                          B

4.101   Calculate the molarity of each of the following solutions.
(a) 122 g of acetic acid, CH$_3$CO$_2$H, in 1.00 L of solution
(b) 185 g of sucrose, C$_{12}$H$_{22}$O$_{11}$, in 1.00 L of solution
(c) 70.0 g of hydrogen chloride, HCl, in 0.600 L of solution
(d) 45.0 g of potassium hydroxide, KOH, in 250.0 mL of solution

4.102   Calculate the molarity of each of the following solutions.
(a) 6.30 g HNO$_3$ dissolved in 255 mL of solution
(b) 49.0 g H$_2$SO$_4$ dissolved in 125 mL of solution
(c) 2.80 g KOH dissolved in 525 mL of solution
(d) 7.40 g Ca(OH)$_2$ dissolved in 200.0 mL of solution

4.103   How many moles of Na$_2$SO$_4$ are present in 150.0 mL of a 0.124 $M$ solution? How many moles of Na$^+$ ions are present? How many moles of SO$_4^{2-}$ ions are present?

4.104   How many moles of Mg(NO$_3$)$_2$ are present in 225.0 mL of a 1.20 $M$ solution? How many moles of Mg$^{2+}$ ions are present? How many moles of NO$_3^-$ ions are present?

4.105   Calculate the moles and the mass of solute in each of the following solutions.
(a) 250.0 mL of 1.50 $M$ KCl
(b) 250.0 mL of 2.05 $M$ Na$_2$SO$_4$

4.106   Calculate the moles and the mass of solute in each of the following solutions.
(a) 150.0 mL of 0.245 $M$ CaCl$_2$
(b) 1450 mL of 0.00187 $M$ H$_2$SO$_4$

4.107   Calculate what volume of the following solutions is required to obtain 0.250 mol of each solute.
(a) 0.250 $M$ AlCl$_3$
(b) 3.00 $M$ HCl

4.108   Calculate what volume of the following solutions is required to obtain 0.250 mol of each solute.
(a) 1.50 $M$ H$_2$SO$_4$
(b) 0.750 $M$ NaCl

4.109   How much water had to be added to 10.0 mL of the first solution to obtain the second solution?

4.110   How much water had to be added to 50.0 mL of the first solution to obtain the second solution?

4.111   How much water must be added to 935.0 mL of 0.1074 $M$ HCl to obtain a solution that is exactly 0.1000 $M$?

4.112   If you wish to prepare a 0.055 $M$ solution of NaNO$_3$, to what volume would you have to dilute 25 mL of 3.0 $M$ NaNO$_3$?

4.113   Determine the concentrations of the following diluted solutions.
(a) 24.75 mL of 0.1832 $M$ HCl diluted to 250.0 mL
(b) 125 mL of 1.187 $M$ NaOH diluted to 0.500 L
(c) 10.00 mL of 0.2010 $M$ acetic acid, CH$_3$CO$_2$H, diluted to 50.00 mL

**4.114** Determine the concentrations of the following diluted solutions.
  (a) 35.45 mL of 6.00 $M$ KOH diluted to 150.0 mL
  (b) 250.0 mL of 0.00102 $M$ $NH_3$ diluted to 500.0 mL
  (c) 5.00 mL of 0.8045 $M$ $CaCl_2$ diluted to 250.00 mL

## Additional Questions

**4.115** What is the mass of 0.100 mol of $Cu(OH)_2$?

**4.116** How many (a) moles, (b) grams, and (c) atoms of vanadium are in 52.5 g of $V_2O_5$?

**4.117** What is the average mass of one atom of argon in atomic mass units and in units of grams?

**4.118** How many molecules are present in 15.43 g of butyl alcohol, $C_4H_9OH$?

**4.119** Calculate the number of moles of atoms and the number of atoms in the following quantities.
  (a) 36.1 g of argon
  (b) the 44.5-carat Hope diamond, which consists of carbon (1 carat = 0.200 g)
  (c) 2.50 mL of mercury with a density of 13.6 g/mL

**4.120** What mass of bromine contains the same number of atoms as 50.0 g of iodine?

**4.121** Calcium carbonate is the active ingredient in some antacids. It is also the compound in the mineral known as calcite. How many moles of calcium carbonate are present in a chewable tablet that contains 750.0 mg $CaCO_3$?

**4.122** A gaseous compound has the composition 92.3% C and 7.7% H by mass. What is its empirical formula? If the molar mass of the compound is 78.1 g/mol, what is its molecular formula?

**4.123** Tear gas has the composition 40.25% C, 6.19% H, 8.94% O, and 44.62% Br. What is its empirical formula?

**4.124** A chemical procedure calls for 3.54 mol of dry ice (solid $CO_2$). What mass of $CO_2$ should we add?

**4.125** How many oxygen atoms are contained in 10.00 g of $H_3PO_4$?

**4.126** Calculate the mass percent of each element in $Na_2CO_3$.

**4.127** How many molecules of $CO_2$ are contained in the 960 g of $CO_2$ exhaled daily by an average person?

**4.128** The formula for ibuprofen is $C_{13}H_{18}O_2$. How many moles of ibuprofen are present in a 200.0-mg tablet?

**4.129** Vanillin, the active component of the flavoring vanilla, has the composition 63.15% C, 5.30% H, and 31.55% O.
  (a) What is its empirical formula?
  (b) If the molar mass of vanillin is 152.14 g/mol, what is its molecular formula?

**4.130** Monosodium glutamate, $NaC_5H_8NO_4$, is used extensively as a flavor enhancer. What is its percent composition?

**4.131** If 5.00 g of Al is burned completely in excess oxygen gas, 9.45 g of an aluminum oxide is formed. What is the empirical formula of that oxide?

**4.132** If 4.32 g of Cu combines with 1.09 g of S to form 5.41 g of the bluish-black mineral chalcocite, what is the empirical formula of chalcocite?

**4.133** What mass of $Ca(NO_3)_2$ is required to prepare a solution containing 0.742 mol of $Ca(NO_3)_2$?

**4.134** Determine which of each pair of substances contains the greatest mass percent of carbon.
  (a) $CO_2$ or $CH_4$   (b) $CH_4$ or $C_2H_6$   (c) $C_6H_{12}$ or $C_5H_{10}$

**4.135** Calcium phosphate, $Ca_3(PO_4)_2$, and calcium citrate, $Ca_3(C_6H_5O_7)_2$, are both commonly used as calcium supplements. If you had equal-mass samples of each, which sample would contain the greatest mass of calcium?

**4.136** Calcium citrate is used to treat osteoporosis. Calculate the number of $Ca_3(C_6H_5O_7)_2$ formula units in 5.0 moles.

**4.137** Consider a 1.0-g sample of each of the substances $F_2$, $SF_2$, or $CF_4$. Which sample contains the greatest number of molecules?

**4.138** Consider a 1.0-mol sample of each of the substances $F_2$, $SF_2$, or $CF_4$.
  (a) Which sample contains the greatest number of molecules?
  (b) Which sample contains the greatest number of atoms?

**4.139** Sea salt is formed by the natural evaporation of ocean water and contains 98% sodium chloride. The remaining 2% are natural minerals such as iron and sulfur. Consider a 20.0-g sample of sea salt.
  (a) What is the mass of sodium chloride in the sample?
  (b) What is the mass of sodium (from sodium chloride) in the sample?
  (c) What is the mass of chlorine (from sodium chloride) in the sample?

**4.140** The concentration of citric acid ($H_3C_6H_5O_7$) in citric fruits ranges from 0.005 $M$ to 0.30 $M$. Consider a 10.0-mL sample of pure lime juice with a citric acid concentration of 0.30 $M$.
  (a) How many moles of citric acid are in the sample?
  (b) How many grams of citric acid are in the sample?

**4.141** How many chloride ions are in each of the following?
  (a) 1.0 mol $AlCl_3$
  (b) 0.25 mol $AlCl_3$
  (c) 0.25 mol $MgCl_2$

**4.142** How many moles of nitrate ion are in 1.0 L of a 2.5 $M$ $Mg(NO_3)_2$ solution?

## Concept Review Questions

**4.143** Which hydrocarbon pair has an identical percent composition by mass of carbon?
  A. $C_3H_4$ and $C_3H_6$
  B. $C_2H_4$ and $C_3H_4$
  C. $C_2H_4$ and $C_4H_2$
  D. $C_2H_4$ and $C_3H_6$
  E. $C_4H_8$ and $C_3H_8$
  For those with different percent compositions, indicate which compound in each pair has the lower percent carbon by mass.

**4.144** Which of the following describes 44.0 g of propane, $C_3H_8$?
  A. One mole of propane
  B. The amount of propane that contains 1.0 g of hydrogen
  C. The amount of propane that contains $44.0 \times 6.02 \times 10^{23}$ molecules of propane
  D. The amount of propane that contains $8 \times 6.02 \times 10^{23}$ $H_2$ molecules
  E. The amount of propane that contains 36.0 g of carbon
  For those statements that do not correctly describe 44.0 g of propane, rewrite the statement to make it correct.

4.145    Which of the following samples contains the smallest number of molecules?
A. 1 g of phosphorus, $P_4$
B. 1 g of chlorine, $Cl_2$
C. 1 g of nitrogen, $N_2$
D. 1 g of arsenic, $As_4$
E. 1 g of sulfur, $S_8$
Is it necessary to actually calculate the number of moles or the number of molecules of each substance? What information is needed to compare these substances? Arrange the other four substances in order of increasing number of molecules.

4.146    Which of the following samples contains the largest number of atoms?
A. 1 g of phosphorus, $P_4$
B. 1 g of chlorine, $Cl_2$
C. 1 g of nitrogen, $N_2$
D. 1 g of arsenic, $As_4$
E. 1 g of sulfur, $S_8$
What values can you compare to determine the answer without actually calculating the number of atoms for each substance? Arrange the other four substances in order of increasing number of atoms.

4.147    Which of the following samples contains the largest number of oxygen atoms?
A. 1.0 mol of $H_2O$
B. 0.20 mol of $N_2O_4$
C. 0.60 mol of $O_2$
D. 0.11 mol of $As_4O_{10}$
E. 0.30 mol of $O_3$
Arrange the other samples in order of increasing number of oxygen atoms.

4.148    Of the following, the only empirical formula is
A. $N_2F_2$
B. $N_2F_4$
C. $H_2C_2$
D. $H_2N_2O_2$
E. $HNF_2$
Explain why the other formulas are not empirical formulas and convert each of them to an empirical formula.

4.149    The molecular formula of a compound is always _____ the empirical formula.
A. more complex than
B. different from
C. an integral multiple of
D. the same as
E. a fraction of
For each of the incorrect responses, provide an example that illustrates that it is incorrect.

4.150    Which of the following would give a solution that is 0.20 $M$?
A. Dissolving 0.40 mol NaCl in water to make 0.20 L of solution
B. Dissolving 0.80 mol NaCl in water to make 1.6 L of solution
C. Dissolving 0.10 mol NaCl in water to make 500.0 mL of solution
D. Dissolving 0.20 mol NaCl in water to make 100.0 mL of solution
E. Dissolving 2.0 mol NaCl in water to make 10.0 mL of solution
Modify the number of moles of NaCl in each incorrect answer so the prepared solution will have a concentration of 0.20 $M$.

4.151    Which of the following is true concerning 2.00 L of 0.100 $M$ solution of $Ca_3(PO_4)_2$?
A. This solution contains 0.200 mol of $Ca_3(PO_4)_2$.
B. This solution contains 0.800 mol of oxygen atoms.
C. 1.00 L of this solution contains 0.600 mol of $Ca^{2+}$ ions.
D. There are $6.02 \times 10^{23}$ phosphorus atoms in 500.0 mL of this solution.
E. This solution contains $6.67 \times 10^{-2}$ mol of $Ca^{2+}$.
Modify each incorrect statement to make it a true statement.

4.152    Which of the following solutions contains the largest number of moles of chloride ion?
A. 100.0 mL of 0.200 $M$ NaCl
B. 200.0 mL of 0.100 $M$ $CaCl_2$
C. 150.0 mL of 0.100 $M$ $FeCl_3$
D. 300.0 mL of 0.200 $M$ NaCl
E. 100.0 mL of 0.050 $M$ $CaCl_2$
Rank the other solutions in order of decreasing moles of chloride ion.

# Chemical Reactions and Equations

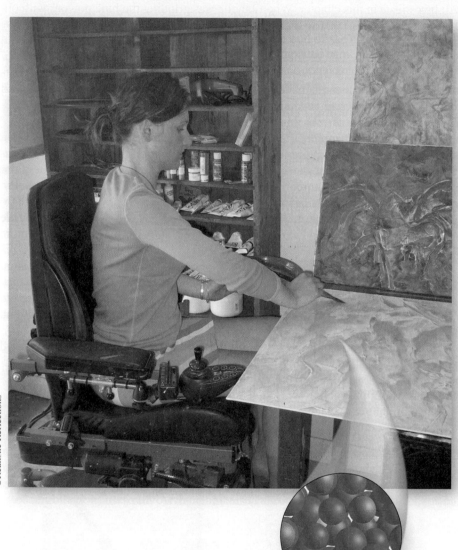

©Realistic Reflections

5.1 What Is a Chemical Reaction?

5.2 How Do We Know a Chemical Reaction Occurs?

5.3 Writing Chemical Equations

5.4 Predicting Chemical Reactions

5.5 Representing Reactions in Aqueous Solution

Chapter Review

Questions and Problems

**FIGURE 5.1** When solutions of iron(II) sulfate and potassium ferricyanide mix, the insoluble blue compound $KFe[Fe(CN)_6]$ forms.
©Tom Pantages

Around 1704, the color manufacturer Diesbach, working in Berlin, accidentally discovered the pigment known as Prussian blue (or Berlin blue). It was prepared by melting dried blood in the presence of potassium salts and adding iron(II) sulfate to an aqueous extract of the mixture. We now have less gory methods of preparing Prussian blue.

Janelle takes the elevator to the second floor of the classroom building and begins her scheduled class. She goes to her workstation and selects two solutions, one of iron(II) sulfate, $FeSO_4$, and the other of potassium ferricyanide, $K_3Fe(CN)_6$. She mixes the two solutions, and a deep blue solid forms (Figure 5.1). She filters the mixture to isolate the solid and places it in an oven to dry. She then grinds the dried solid to a fine powder in oil. What course do you think Janelle is taking?

Janelle just prepared a blue paint for her art class. The deep blue solid that formed is a compound called Prussian blue, which has been used as a pigment since about 1700. Many pigments are ionic compounds made by the kinds of reactions we will investigate in this chapter. When Janelle was grinding the blue powder with oil, she was making paint that will dry to form a film.

Meanwhile Antonio is at his workstation on the first floor of the classroom building. He selects a piece of iron and drops it into a solution of copper(II) chloride, $CuCl_2$. Over a few minutes, the shiny piece of iron takes on a dull brown coating (Figure 5.2). He notices that the blue solution becomes less intense as the brown coating forms. He concludes that copper(II) ions have been removed from the solution. In which course do you think Antonio is enrolled?

Antonio is taking a metal sculpture class and is investigating the reaction between metals under corrosive conditions. His study will help him understand how to construct large metal structures to be displayed outdoors. He has learned that a major problem with metal sculptures is corrosion. An example is the Statue of Liberty, which degraded gradually over a 100-year period due to the corrosive marine environment of New York Harbor. We will examine the kinds of reactions responsible for corrosion in this chapter. Corrosion (and its prevention) will be explored in more detail in Chapter 14.

Next door, Breanna drops pieces of a rock into beakers containing different concentrations of sulfuric acid. The rocks begin to give off bubbles of gas, which rise to the surface (Figure 5.3). She collects some of the gas and bubbles it through a solution of barium chloride, $BaCl_2$. A white solid forms. What course do you think Breanna is taking?

Breanna is taking an art preservation class and is investigating the effect of acid rain on marble sculptures. Many old statues and architectural features, such as gargoyles, have corroded as a result of air pollution. Sulfur and nitrogen oxides released into the atmosphere by the burning of coal and other fuels react with water in clouds and raindrops to form sulfuric acid and nitric acid, causing acid rain. The

**FIGURE 5.3** The calcium carbonate in marble reacts with sulfuric acid to release carbon dioxide gas.
©McGraw-Hill Education/Stephen Frisch

**FIGURE 5.2** When iron metal is placed in a solution of copper(II) chloride, copper metal is deposited as a coating on the iron.
©Jim Birk

acids in the rain react with calcium carbonate in the marble and limestone of statues to release carbon dioxide gas. In this chapter we will examine some reactions of acids with other substances.

Elsewhere, Jennifer, a student enrolled in a chemistry course, takes a ceramic pot and soaks it in a solution of acetic acid, $CH_3CO_2H$. After an hour, she heats the solution to evaporate the acetic acid, leaving a solid residue. She adds some water and potassium iodide, KI, forming a yellow solid (Figure 5.4). She recognizes this solid as lead(II) iodide, $PbI_2$, indicating that the pot contains a lead compound. Before the dangers of lead poisoning were recognized, lead compounds were added to the glazing material on ceramic pots. They lowered the melting temperature of the glaze and made it form a glassy coating. Jennifer's test reveals whether the pot she is studying has lead in its glaze.

Art and chemistry are intimately connected. Just as artists classify works of art as paintings, sculptures of stone or metal, stained glass, or ceramics, chemists classify reactions into different groups. Each of these students' experiments provides an example of a different art form and a different category of reactions.

### Questions for Consideration

**5.1** What happens in a chemical reaction?
**5.2** How do we know whether a chemical reaction takes place?
**5.3** How do we represent a chemical reaction with a chemical equation?
**5.4** How are chemical reactions classified? How can the products of different classes of chemical reactions be predicted?
**5.5** How do we represent chemical reactions in aqueous solution?

## 5.1 What Is a Chemical Reaction?

Each of the students described in the introduction was carrying out some kind of chemical reaction. For example, when Antonio put iron metal into a solution of copper(II) chloride, a chemical reaction occurred. What is a chemical reaction?

A chemical reaction is the conversion of one substance or set of substances into another. Any substance that we start with is a **reactant.** A new substance that forms during the reaction is a **product.** Products are different from reactants in the *arrangement* of their component atoms. Chemical reactions neither destroy atoms nor create new atoms. They occur because the bonds that hold atoms together can be broken and rearranged. (You'll read more about chemical bonding in Chapter 8.)

Consider the reaction of hydrogen gas with oxygen gas, as shown in Figure 5.5. Hydrogen and oxygen both occur naturally as diatomic molecules. If they mix, they react slowly; but if ignited, the reaction is vigorous, even explosive. In either case, the reactants form the same product: gaseous water molecules, each containing two hydrogen atoms and one oxygen atom. Figure 5.6 shows the rearrangement of the

**FIGURE 5.4** When potassium iodide is added to a solution containing lead(II) ions, the yellow solid lead(II) iodide forms.
©McGraw-Hill Education/Stephen Frisch

Chemical reactions are not the same as nuclear reactions, in which new elements may form. You'll learn about those in Chapter 15.

**FIGURE 5.5** When ignited, hydrogen molecules and oxygen molecules in the air react explosively to form gaseous water.
©Charles D. Winters

**FIGURE 5.6** The hydrogen atoms from hydrogen molecules combine with the oxygen atoms from oxygen molecules to form gaseous water molecules.

atoms during the reaction. Note that this representation shows the reactant molecules separated from the product molecules, but they are actually separated by time, not distance. The reactants are in the container where the reaction occurs, and the products are in that same container after the reaction is complete.

Count the number of atoms of hydrogen and oxygen in the reactants and in the product. Has either number changed? No. In every chemical reaction the number of atoms of each element stays constant. This is consistent with the law of conservation of mass (see Chapter 2), which states that matter cannot be created or destroyed by ordinary chemical means.

Now consider the arrangement of atoms in Figure 5.6. What are oxygen atoms attached to in the reactants? In the products? How about hydrogen atoms? In the reactants, each hydrogen atom is attached to another hydrogen atom and each oxygen atom is attached to another oxygen atom. In the products, two hydrogen atoms are attached to one oxygen atom. The number of atoms is the same, but their arrangement is different.

---

### EXAMPLE 5.1    Identify Reactant and Product

Breanna mixed a solution of sulfuric acid and solid calcium carbonate. Carbon dioxide gas bubbled out of the solution as the initial solid disappeared, leaving behind a deposit of calcium sulfate. Identify the reactants and products in this chemical reaction.

**Solution:**

Sulfuric acid and calcium carbonate are the starting substances, the reactants. They are converted into carbon dioxide and calcium sulfate, the products. (Another product is water, which can't be identified from this description.)

**Consider This 5.1**

If two elements combine to form a compound, are the elements reactants or products?

**Practice Problem 5.1**

Antonio placed a piece of iron in a solution of copper(II) sulfate. He obtained a deposit of copper metal and a solution containing iron(II) sulfate. Identify the reactants and products in this chemical reaction.

**Further Practice:** Questions 5.3 and 5.4 at the end of the chapter

---

## 5.2  How Do We Know a Chemical Reaction Occurs?

Just as artists use paint and stone to create works of art, chemists use elements and compounds to create new substances. Both the artist and the chemist are constrained by the materials available to them. Both must work within the rules of nature regarding the properties and behaviors of their materials. Both rely on their materials to produce new creations, but is simply using materials enough? Is putting some paint on canvas art? Is mixing two substances together chemistry?

Consider the two beakers containing clear, colorless liquids shown in Figure 5.7. What happened when the two liquids mixed? Can you tell if a chemical reaction occurred? Just by looking, you can't. You started with two colorless liquids, and their mixture remains colorless. A rearrangement of atoms might or might not have taken place. There's no obvious way to tell.

Without an observable change in properties, it is hard to tell if a reaction has occurred without further tests. In most chemical reactions, however, an observable change in properties does occur. What are some visible clues that would indicate

**FIGURE 5.7** Without an observable change in properties, it is hard to tell if a chemical reaction has occurred.
©Brian Moeskau/Moeskau Photography

**FIGURE 5.8** Do any of these photos of chemical reactions illustrate items on the list you made of nature's clues that a chemical reaction has occurred?

(A) HCl($aq$) + NaOH($aq$) + phenolphthalein($aq$)

(B) Mg($s$) + O$_2$($g$)

(C) Ba(OH)$_2 \cdot$ 8H$_2$O($s$) + NH$_4$Cl($s$)

(D) Cu($s$) + HNO$_3$($aq$)

(E) BaCl$_2$($aq$) + Na$_2$SO$_4$($aq$)

(F) CH$_4$($g$) + O$_2$($g$)

(G) [Ag(NH$_3$)$_2$]$^+$($aq$) + aldehyde

(H) CaCO$_3$($s$) + HCl($aq$)

(a): ©GIPhotoStock/Science Source; (b): ©McGraw-Hill Education/Stephen Frisch; (c): ©SPL/Science Source; (d): ©McGraw-Hill Education/Stephen Frisch; (e): ©Charles D. Winters/Science Source; (f): ©McGraw-Hill Education/Stephen Frisch; (g): ©Martyn F. Chillmaid/SPL/Science Source; (h): ©McGraw-Hill Education/Stephen Frisch

that a chemical reaction may have occurred? Make a list of ideas, and then examine Figure 5.8 to see if any of the photos of chemical reactions match your list of clues.

The clues most often used by chemists are the following:

- Change in color
- Production of light
- Formation of a solid (such as a *precipitate* in solution, smoke in air, or a metal coating)
- Formation of a gas (bubbles in solution or fumes in the gaseous state)
- Absorption or release of heat (sometimes appearing as a flame)

Did you come up with any clues that don't fit one of these categories?

These clues are only indications that a reaction *may* have taken place. Physical changes can also cause such effects. For example, a temperature change might result from dissolving a compound, and a color change might arise from mixing two colored compounds.

**FIGURE 5.9** When a powdered mixture of aluminum metal and iron(III) oxide is heated, it reacts vigorously to form liquid iron metal and aluminum oxide. This reaction is called the thermite reaction.
©Charles D. Winters/Science Source

**FIGURE 5.10** The thermite reaction is used to make molten iron. In the device shown here, the liquid iron runs down from the crucible on top into the mold. Upon solidification, the iron joins two sections of railroad track.
©Paul Mayall/Alamy Stock Photo

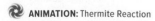 **ANIMATION:** Thermite Reaction

When reading an equation, we replace the arrow with a word such as *yields*, *produces*, or *forms*.

## 5.3 Writing Chemical Equations

Describing chemical reactions in words can be awkward and time-consuming. For this reason, chemists use a chemical equation to show what reactants are involved and what products are formed. A **chemical equation** is a symbolic representation of a chemical reaction.

To see how equations are written, let's look at the reaction shown in Figure 5.9. Two powders—gray aluminum metal, Al, and red iron(III) oxide, $Fe_2O_3$—are mixed and poured into a ceramic container with a small hole in the bottom. A wick in the pile of powder is ignited. The powder smokes and then gives off a shower of sparks. As this spectacular display continues and then subsides, liquid metal runs out of the hole in the container. This liquid is molten iron. Left behind is aluminum oxide, which was also present in the smoke. This reaction is called the *thermite reaction.* It is sometimes used to weld pieces of iron or steel together, as in reinforcing rod (rebar) in concrete buildings or in the rails for high-speed railroad tracks (Figure 5.10).

A verbal description is one way to represent the thermite reaction, but as you can see, it is not very efficient. We might also draw pictures to show the changes that occur at the atomic level, but this is time-consuming and takes up a lot of space, too. The easier, shorter way is to use the symbols and formulas that represent the substances in the reaction.

To write a chemical equation, we first identify the reactants and write their chemical formulas with plus signs (+) between them. We place an arrow after them. The arrow stands for the process of converting one set of substances into another. Then we write the chemical formulas of the products, again with plus signs between them if there is more than one.

The physical state—solid (*s*), liquid (*l*), gas (*g*), or aqueous solution (*aq*)—is shown for each reactant and product. If heat must be supplied for the reaction to take place, or if some other special conditions are needed, we describe them above the arrow. Finally, the equation must be balanced. In a **balanced equation,** *the number of atoms of each element is the same in the products as in the reactants.* We indicate the relative number of molecules or formula units of reactants or products by *coefficients,* which are whole numbers that immediately precede the formula of each substance. If a coefficient has a value of 1, it is omitted but is understood to be there. The process of balancing an equation involves examining and modifying coefficients until the reactants and products contain the same numbers of the component elements. The coefficients are written as the smallest possible whole numbers. The coefficients are whole numbers because parts or fractions of atoms or molecules do not exist.

Let's work through writing and balancing the thermite reaction. In words, it looks like this:

$$\text{aluminum} + \text{iron(III) oxide} \longrightarrow \text{aluminum oxide} + \text{iron}$$

Using the rules developed in Chapter 3, we can substitute formulas for names in the equation. We should also add the physical state of each substance:

$$Al(s) + Fe_2O_3(s) \longrightarrow Al_2O_3(s) + Fe(l)$$

The equation is not yet complete because it is not balanced. The same number of atoms of each element must appear on both sides of the arrow. In its present form, it has the following numbers of atoms:

| Atoms in Reactants $Al(s) + Fe_2O_3(s)$ | | Atoms in Products $Al_2O_3(s) + Fe(l)$ | |
| --- | --- | --- | --- |
| 1 Al | ● | 2 Al | ●● |
| 3 O | ●●● | 3 O | ●●● |
| 2 Fe | ●● | 1 Fe | ● |

The number of oxygen atoms is balanced at 3 in the reactants and 3 in the products, but the other elements are not balanced. The reactants contain only 1 aluminum atom, but the products contain 2. The reactants contain 2 iron atoms, but the products contain only 1. To balance the equation, we must change coefficients until the reactants and products contain the same number of atoms of each element. *It is important to remember that you cannot change the subscript of an element in a compound to balance an equation.* For example, $Al_2O_3$ cannot be changed to $AlO_3$ to balance the number of aluminum atoms. Even if $AlO_3$ existed, it would be an entirely different compound.

When the atoms of the reactants and products have not yet been equalized, the equation is called a *skeletal equation* or an *unbalanced equation*. To balance the skeletal equation for the thermite reaction, we need to do nothing about oxygen. It is already balanced. To balance the aluminum atoms, we can place the coefficient 2 in front of Al on the reactant side of the equation:

$$2Al(s) + Fe_2O_3(s) \longrightarrow Al_2O_3(s) + Fe(l)$$

This balances the aluminum atoms, but the iron atoms remain unbalanced. We can place the coefficient 2 in front of the product Fe to balance them:

$$2Al(s) + Fe_2O_3(s) \longrightarrow Al_2O_3(s) + 2Fe(l)$$

If we count the numbers of each element again, we see that they are now the same on both sides of the equation. The equation is balanced. Mass is conserved.

| Atoms in Reactants $2Al(s) + Fe_2O_3(s)$ | | Atoms in Products $Al_2O_3(s) + 2Fe(l)$ | |
|---|---|---|---|
| 2 Al | ●● | 2 Al | ●● |
| 3 O | ●●● | 3 O | ●●● |
| 2 Fe | ●● | 2 Fe | ●● |

To see how and why equations balance, use a model kit or gumdrops and toothpicks to build models of reactant molecules. Take them apart and rearrange them into product molecules. Use the smallest number of atoms possible to achieve conservation of mass. Reactions don't actually occur this way, but we want to emphasize the conservation of atoms in reactions as a means of balancing equations.

Let's consider this model approach with the reaction of oxygen, $O_2$, and methane, $CH_4$, a gas commonly used as a source of heat in the laboratory and in gas stoves (Figure 5.11). The products of the reaction are carbon dioxide, $CO_2$, and water, $H_2O$. All of the substances are in the gaseous state. Let's start by writing a skeletal equation so we know which models we need:

$$CH_4(g) + O_2(g) \longrightarrow CO_2(g) + H_2O(g)$$

We start with at least one molecule of each reactant:

We can disassemble our models to see what parts we have to work with to form products. We have 1 carbon atom, 4 hydrogen atoms, and 2 oxygen atoms. Note that we have not removed the original models from our diagram so that we can keep track of their numbers:

} Unused model parts

The equation can be written with the reactants in any order on the left side of the arrow and with the products in any order on the right side of the arrow. Any of the elements in the products can be used as a starting point for the balancing process.

If you do not have a model kit, you can make one with some gumdrops and toothpicks. Use different colors of gumdrops to represent different elements. Connect the gumdrops with toothpicks to represent molecules. Build several molecules of each reactant, but use only the minimum necessary to disassemble and build whole numbers of products.

**FIGURE 5.11** In a gas stove, methane reacts with oxygen to form carbon dioxide and water.
©Sami Sarkis/Getty Images

The model for $O_2$ shows two bonds (sticks) holding the two atoms (spheres) together. In Chapter 8, we will explore the nature of bonds between atoms and we will see why two bonds are needed in this molecule.

From the atoms we have available, we can make either 1 carbon dioxide molecule or 2 water molecules. Either approach will work. Let's pick the carbon dioxide to build first:

That leaves us with 4 extra hydrogen atom models. We have not formed any water molecules, although water is a product. According to the law of conservation of mass, all of the reactant atoms must be completely consumed, so we know we need to make some water molecules from the leftover hydrogen atoms. But we need oxygen to do that, and the only way to get it is to add another oxygen molecule to the reactants. That gives us 2 additional oxygen atom models to work with.

To balance an equation, we need equal numbers of *atoms* of each element on each side of the equation but not necessarily equal numbers of *molecules*. Consider the example $N_2(g) + 3H_2(g) \longrightarrow 2NH_3(g)$. In this reaction four molecules of reactants form two molecules of product. It is common for a reaction to yield more or fewer molecules than were initially present, although the total number of atoms always stays the same.

Now we have enough parts to build 2 water molecules from the 2 oxygen atom models and 4 hydrogen atom models. This leaves no unused atoms, so the equation must be balanced.

You can verify what your models show by counting the atoms of each element on each side of the equation. Since the numbers are equal, the equation is indeed balanced. Since we do not write coefficients of 1, the final version of the equation is

$$CH_4(g) + 2O_2(g) \longrightarrow CO_2(g) + 2H_2O(g)$$

Before we start balancing equations using formulas, let's reexamine the equation we balanced using models. What is the skeletal equation?

Are the atoms of each element balanced? The carbon atoms are balanced, but the hydrogen atoms and oxygen atoms are not. Adding another water molecule to the products balances the hydrogen atoms:

To balance the oxygen atoms, we add 1 more oxygen molecule to the reactants:

Now the numbers of each element are equal on both sides of the equation. The equation is balanced. When we balance equations using formulas, we use coefficients in front of the formulas to indicate the number of each element needed.

Balancing this equation was relatively simple, but sometimes changing a coefficient to balance one element unbalances another. In such cases, we must go back and change coefficients. Example 5.2 shows how.

## Steps for Balancing Equations

1. What is the skeletal equation?
   (a) Write correct formulas for reactants on the left side and products on the right side of the arrow.
   (b) Include physical state symbols for all substances.
   (c) The coefficients (the numbers preceding the formulas) are all 1 in a skeletal equation.
2. How are atoms of each element balanced?
   (a) First look for elements that appear only once on each side of the equation. If there are different numbers of atoms of that element on each side of the equation, change the coefficient of one or both substances until there is the same number of atoms on each side of the equation.
   (b) Do not change the subscripts! Changing the subscripts in a formula would change the identity of the substance.
   (c) Continue checking all elements in the equation, changing coefficients when needed.
   (d) The last atoms to balance should be those in elemental substances (such as $O_2$), because changing a coefficient in front of an element does not change the numbers of atoms of any other elements.
3. Is the equation balanced? Make a final check by counting the atoms of each element on both sides of the equation. If they are the same, the equation is balanced.

## EXAMPLE 5.2     Balancing Chemical Equations

Potassium chlorate is a white solid that decomposes when heated, forming solid potassium chloride and oxygen gas. Balance the equation that represents the reaction.

### Solution:

Potassium chlorate is the only reactant. It has the formula $KClO_3$ and is in the solid state. The products are solid potassium chloride, $KCl$, and oxygen gas, $O_2$. The word equation gives the name of the reactant before the arrow and the names of the products, separated by a plus sign, after the arrow:

$$\text{potassium chlorate} \xrightarrow{\text{heat}} \text{potassium chloride} + \text{oxygen}$$

To get the skeletal equation, we substitute the formulas and their physical states for the names:

$$KClO_3(s) \xrightarrow{\text{heat}} KCl(s) + O_2(g)$$

Next we count atoms to see where to begin changing coefficients. Potassium and chlorine occur once on each side of the arrow and thus are balanced already, but there are 3 oxygen atoms on the left and 2 oxygen atoms on the right:

$$KClO_3(s) \xrightarrow{\text{heat}} KCl(s) + O_2(g)$$

| Atoms in Reactants | Atoms in Products |
|---|---|
| 1 K  1 Cl  3 O | 1 K  1 Cl  2 O |

We can balance the oxygen atoms by placing a 2 in front of $KClO_3$ and a 3 in front of $O_2$. Now there are 6 oxygen atoms on each side of the equation:

$$2KClO_3(s) \xrightarrow{\text{heat}} KCl(s) + 3O_2(g)$$

| Atoms in Reactants | Atoms in Products |
|---|---|
| 2 K  2 Cl  6 O | 1 K  1 Cl  6 O |

Now, however, the potassium and chlorine are no longer balanced. The left side of the equation has 2 of each, while the right side has only 1. We can rebalance the K and Cl atoms by placing a 2 in front of KCl on the product side of the equation:

$$2KClO_3(s) \xrightarrow{\text{heat}} 2KCl(s) + 3O_2(g)$$

A final count shows that this is a correct, balanced equation:

| Atoms in Reactants | Atoms in Products |
|---|---|
| 2 K  2 Cl  6 O | 2 K  2 Cl  6 O |

**Consider This 5.2**

If all the coefficients in a chemical equation are divisible by 2, is the equation properly balanced?

**Practice Problem 5.2**

When sodium metal is dropped into liquid water, the mixture releases gaseous hydrogen, leaving behind a colorless solution of sodium hydroxide. Balance the equation that represents this reaction.

**Further Practice:** Questions 5.27 and 5.28 at the end of the chapter

Small amounts of oxygen gas can be produced for laboratory use by heating $KClO_3$. (This reaction has to be run very carefully to prevent the $KClO_3$ from exploding.) If you ran this reaction in the lab, how might you collect the oxygen gas that's produced?

At intermediate stages, coefficients needed to balance an equation may not be whole numbers. Sometimes a temporary balancing with fractional coefficients makes balancing easier. But in the end, the coefficients must be whole numbers. Example 5.3 shows a method for balancing equations this way.

| **EXAMPLE 5.3** | **Balancing Chemical Equations with Fractional Coefficients** |
|---|---|

Ethane, $C_2H_6$, is a gaseous fuel that can be burned with oxygen. Balance the following skeletal equation for this reaction:

$$C_2H_6(g) + O_2(g) \longrightarrow CO_2(g) + H_2O(g)$$

**Solution:**

We can see that this equation is not balanced by counting atoms on each side of the equation:

$$C_2H_6(g) + O_2(g) \longrightarrow CO_2(g) + H_2O(g)$$

| Atoms in Reactants | Atoms in Products |
|---|---|
| 2 C  6 H  2 O | 1 C  2 H  3 O |

Since oxygen occurs in three substances, we should start by balancing the carbon and hydrogen. There are 2 carbon atoms on the left, but only 1 on the right, so we place a 2 in front of $CO_2$:

$$C_2H_6(g) + O_2(g) \longrightarrow 2CO_2(g) + H_2O(g)$$

| Atoms in Reactants | Atoms in Products |
|---|---|
| 2 C  6 H  2 O | 2 C  2 H  5 O |

There are 6 hydrogen atoms on the left and 2 on the right, so we place a 3 in front of $H_2O$:

$$C_2H_6(g) + O_2(g) \longrightarrow 2CO_2(g) + 3H_2O(g)$$

| Atoms in Reactants | Atoms in Products |
|---|---|
| 2 C  6 H  2 O | 2 C  6 H  7 O |

Now all we have left to balance is the oxygen. There are 4 oxygen atoms in 2 $CO_2$ molecules and 3 oxygen atoms in 3 $H_2O$ molecules, making a total of 7 oxygen atoms in the products. Since we cannot balance the oxygen atoms in the reactants with a whole number unless we unbalance everything else, we can use a fractional coefficient, $\frac{7}{2}$ (or 3.5) temporarily:

$$C_2H_6(g) + \tfrac{7}{2} O_2(g) \longrightarrow 2CO_2(g) + 3H_2O(g)$$

| Atoms in Reactants | Atoms in Products |
|---|---|
| 2 C  6 H  7 O | 2 C  6 H  7 O |

The atoms of each element are equal on both sides of the equation, but the coefficients are not all whole numbers. To get whole-number coefficients, we must multiply all coefficients in the equation by 2:

$$2C_2H_6(g) + 7O_2(g) \longrightarrow 4CO_2(g) + 6H_2O(g)$$

This is a correct, balanced equation, as shown by the final atom counts.

| Atoms in Reactants | Atoms in Products |
|---|---|
| 4 C  12 H  14 O | 4 C  12 H  14 O |

### Consider This 5.3

If we left fractional coefficients in an equation for which the atom counts were balanced, why would the equation not be considered to be properly balanced?

### Practice Problem 5.3

Liquid butane is often used as a fuel in cigarette lighters. When the pressure is released, butane evaporates and can burn with oxygen. Balance the following skeletal equation for the reaction of butane with oxygen gas.

$$C_4H_{10}(g) + O_2(g) \longrightarrow CO_2(g) + H_2O(g)$$

**Further Practice:** Questions 5.37 and 5.38 at the end of the chapter

If you have a gas stove or furnace, the fuel that burns is mostly methane, $CH_4$. Small amounts of ethane, $C_2H_6$, are also present in natural gas. Propane, $C_3H_8$, is the fuel sold in tanks for gas grills. Butane, $C_4H_{10}$, is burned when a lighter is struck. What do these *hydrocarbon* compounds have in common?

Let's consider one final example that illustrates how to proceed if part of a reactant is unchanged in the products. This situation usually arises with ionic compounds in which the anion is an oxoanion or some other polyatomic anion. If the anion is unchanged in the products, it can be treated as a unit instead of as a collection of atoms that must be balanced individually.

| EXAMPLE 5.4 | Balancing an Equation with an Unchanged Oxoanion |
|---|---|

When a piece of iron is added to a solution of copper(II) sulfate, the following reaction occurs:

$$Fe(s) + CuSO_4(aq) \longrightarrow Fe_2(SO_4)_3(aq) + Cu(s)$$

Balance this skeletal equation.

**Solution:**

Notice that the sulfate ion, $SO_4^{2-}$, occurs unchanged in the reactants and products. Its component elements, S and O, do not occur anywhere else in the reaction, so the sulfate ion can be treated as a unit. We start by counting atoms and sulfate ions.

$$Fe(s) + CuSO_4(aq) \longrightarrow Fe_2(SO_4)_3(aq) + Cu(s)$$

| Atoms/Ions in Rectants | Atoms/Ions in Products |
|---|---|
| 1 Fe  1 Cu  1 $SO_4^{2-}$ | 2 Fe  1 Cu  3 $SO_4^{2-}$ |

The copper atoms are balanced. Two iron atoms occur in the products, but only 1 in the reactants, so we can change the coefficient of Fe to 2:

$$2Fe(s) + CuSO_4(aq) \longrightarrow Fe_2(SO_4)_3(aq) + Cu(s)$$

| Atoms/Ions in Rectants | Atoms/Ions in Products |
|---|---|
| 2 Fe  1 Cu  1 $SO_4^{2-}$ | 2 Fe  1 Cu  3 $SO_4^{2-}$ |

The sulfate ions are unbalanced. We can balance them by changing the coefficient of $CuSO_4$ to 3:

$$2Fe(s) + 3CuSO_4(aq) \longrightarrow Fe_2(SO_4)_3(aq) + Cu(s)$$

| Atoms/Ions in Rectants | Atoms/Ions in Products |
|---|---|
| 2 Fe  3 Cu  3 $SO_4^{2-}$ | 2 Fe  1 Cu  3 $SO_4^{2-}$ |

This change unbalanced the copper atoms, so we have to change the coefficient of Cu to 3:

$$2Fe(s) + 3CuSO_4(aq) \longrightarrow Fe_2(SO_4)_3(aq) + 3Cu(s)$$

| Atoms/Ions in Rectants | Atoms/Ions in Products |
|---|---|
| 2 Fe  3 Cu  3 $SO_4^{2-}$ | 2 Fe  3 Cu  3 $SO_4^{2-}$ |

The equation is now balanced, as indicated by the final counts.

**Consider This 5.4**

If the charge on iron in the iron sulfate product were 2+ instead of 3+, what would be the formula of the iron compound, and how would the balanced equation change?

**Practice Problem 5.4**

Balance the following skeletal equation for the reaction of zinc metal with nitric acid:

$$Zn(s) + HNO_3(aq) \longrightarrow Zn(NO_3)_2(aq) + H_2(g)$$

**Further Practice:** Questions 5.39 and 5.40 at the end of the chapter

## 5.4  Predicting Chemical Reactions

In all the chemical reactions we have considered so far, the identities of the reactants and products are known. The identity of the reactants is easy enough, since we usually know what we have mixed. But how do we know what products will result? The answer could come from an experiment. We could carry out the reaction and then isolate and identify the products, or if the reaction has been investigated before, we could use or extend the experimental work of others.

We saw in Chapter 2 that the chemical properties of the elements are periodic. That is, elements in the same group in the periodic table undergo similar reactions.

| Lithium | Sodium | Potassium | Rubidium | Cesium |

**FIGURE 5.12** The alkali metals all react with water to form the corresponding alkali metal hydroxide and hydrogen gas. The pattern of rearrangement is the same for all of these reactions. They differ only in their vigor.

(lithium): ©Chemistry at Work/Videodisk/Videodiscovery, Inc.; (sodium) ©Richard Megna/Fundamental Photographs; (potassium): ©Richard Megna/Fundamental Photographs; (cesium): ©Chemistry at Work/Videodisk/Videodiscovery, Inc.

We can use this periodicity to predict the products of chemical reactions. For example, if we place a piece of lithium metal in liquid water, aqueous lithium hydroxide and hydrogen gas form (Figure 5.12):

$$2Li(s) + 2H_2O(l) \longrightarrow 2LiOH(aq) + H_2(g)$$

The principle of periodicity suggests that the other alkali metals will also react with water to form a metal hydroxide and hydrogen gas. As we can see from the sequence of photographs in Figure 5.12, all of these metals react with water. What differs is the vigor of the reaction. With the potassium reaction, the hydrogen always catches fire. For the alkali metals below potassium in the periodic table, the reaction with water is explosive.

In addition to periodic trends, atoms have only a limited number of ways to rearrange themselves. If we can classify reactions into types, we can determine what kinds of atomic rearrangements are theoretically possible for different combinations of chemical substances. For example, suppose we depict elements and compounds with letters and spheres:

Spheres A and B are elements, either monatomic or diatomic. Spheres C, D, E, and F might be single or multiple atoms, monatomic ions, or polyatomic ions arranged to form compounds as shown. How many different types of chemical reactions can you identify by rearranging these atoms or groups of atoms? To keep this exercise from getting too complicated, use only one substance by itself, or any combination of two substances, but no more than two as your reactants and your products. When you have completed this exercise, read on.

To start, let's consider the case in which we have just one reactant. We probably have to heat the substance for something to happen. If we have an element, what could result? A monatomic element can't do anything because there is no way to rearrange a single atom. A diatomic element might break down into its component atoms. We won't encounter examples of such reactions, so we won't consider them further.

If we heat the compound CD, it can break into smaller pieces, such as the elements C and D:

Example: $2HgO(s) \longrightarrow 2Hg(l) + O_2(g)$

Alkaline earth metals react in a similar way, producing a metal hydroxide and oxygen gas. Can you write a balanced equation for the reaction of barium with water?

A compound can also break up into smaller molecules:

Example: $NH_4Cl(s) \longrightarrow NH_3(g) + HCl(g)$

The compound EF could similarly break up into E and F.

Next let's consider combinations of two substances. Let's start with the elements. We can react element A with element B to form a compound:

Example: $Zn(s) + Cl_2(g) \longrightarrow ZnCl_2(s)$

We can combine an element with a compound to make a new compound:

Example: $H_2(g) + CO(g) \longrightarrow H_2CO(g)$

This type of combination could also happen between A and EF, between B and CD, and between B and EF. All four would be the same reaction type.

An element could also react with a compound to form a new compound and release an element from the original compound:

Example: $Zn(s) + CuCl_2(aq) \longrightarrow Cu(s) + ZnCl_2(aq)$

Depending on the identity of the component elements, the products might instead be the new compound CA and the released element D. This type of reaction could also happen with the compound EF, and element B could also react in this way with either compound CD or compound EF.

Finally let's consider the reaction between two compounds. They could combine to make a new compound:

Example: $CaO(s) + SO_2(g) \longrightarrow CaSO_3(s)$

Or two compounds could rearrange to form two new compounds:

Example: $CuCl_2(aq) + Na_2S(aq) \longrightarrow 2NaCl(aq) + CuS(s)$

Similar reactions can occur when an ionic compound reactant is composed of a polyatomic ion:

Example: $AgNO_3(aq) + KCl(aq) \longrightarrow AgCl(s) + KNO_3(aq)$

Did your list of reaction types match these? Did you notice any patterns in the reactants and products? In each of the examples, we had either one or two reactants and either one or two products. They could be either elements or compounds. Let's make a list of the possible combinations based on this classification scheme:

| Reactants | Products |
| --- | --- |
| 1 compound | 2 elements (or smaller compounds) |
| 2 elements or compounds | 1 compound |
| 1 element and 1 compound | 1 element and 1 compound |
| 2 compounds | 2 compounds |

This gives us just four different patterns of reactivity, or reaction classes. Chemists give them the following names: decomposition reactions, combination (or synthesis) reactions, single-displacement reactions, and double-displacement reactions. They are summarized in Table 5.1.

🌀 **ANIMATION:** Types of Chemical Reactions

**TABLE 5.1** Classes of Chemical Reactions

| Class | Reactants | Products | Example |
|---|---|---|---|
| Decomposition | 1 compound | 2 elements (or smaller compounds) | $CD \longrightarrow C + D$ |
| Combination | 2 elements or compounds | 1 compound | $A + B \longrightarrow AB$ |
| Single-displacement | 1 element and 1 compound | 1 element and 1 compound | $A + CD \longrightarrow C + AD$ |
| Double-displacement | 2 compounds | 2 compounds | $CD + EF \longrightarrow CF + ED$ |

We will investigate examples of each class in the remainder of this section. At this point, we can place known chemical reactions into classes by considering the type and number of reactants and products.

---

**EXAMPLE 5.5**    **Classes of Reactions**

Classify each of the following as a decomposition, combination, single-displacement, or double-displacement reaction.

(a) $Na_2S(aq) + 2HCl(aq) \longrightarrow H_2S(g) + 2NaCl(aq)$
(b) $H_2(g) + CuO(s) \longrightarrow Cu(s) + H_2O(g)$
(c) $2Fe(s) + 3Cl_2(g) \longrightarrow 2FeCl_3(s)$
(d) $2Na_2O_2(s) \longrightarrow 2Na_2O(s) + O_2(g)$

**Solution:**

To classify the reactions, look for the patterns outlined in Table 5.1.

(a) This reaction involves two compounds as reactants and two compounds as products. This pattern is called double-displacement.
(b) This is a reaction of an element with a compound, producing a different element and compound. This is a single-displacement reaction.
(c) In this reaction, two elements combine to form a compound. This pattern is found in combination reactions.
(d) Here, one compound is converted into a simpler compound and an element. This is a decomposition reaction.

**Consider This 5.5**

If three elements or compounds react together to form one compound, how would you classify this reaction?

**Practice Problem 5.5**

Classify each of the following as a decomposition, combination, single-displacement, or double-displacement reaction.

(a) $NH_3(g) + HCl(g) \longrightarrow NH_4Cl(s)$
(b) $CuCl_2(aq) + Na_2S(aq) \longrightarrow CuS(s) + 2NaCl(aq)$
(c) $NiSO_3(s) \longrightarrow NiO(s) + SO_2(g)$
(d) $Ca(s) + PbCl_2(aq) \longrightarrow CaCl_2(aq) + Pb(s)$

**Further Practice:** Questions 5.51 and 5.52 at the end of the chapter

## Decomposition Reactions

When a compound undergoes a **decomposition reaction,** it breaks down into the elements of which it is composed or into simpler compounds. The compounds that form are predictable—usually they are common, stable, small molecules, especially gases. If a gas can form as a product, it nearly always does. Water is often produced when the reactant compound contains hydrogen and oxygen. Patterns for predicting the products of decomposition reactions can be formulated for a variety of compounds. Some of these patterns are listed with examples in Table 5.2.

A few binary compounds decompose to their constituent elements upon heating. For example, when red solid mercury(II) oxide is heated, it slowly decomposes. Gaseous oxygen fills the container, and droplets of liquid mercury settle on its cooler parts (Figure 5.13).

The oxide ions lose their negative charge as they are converted to oxygen atoms. The mercury(II) ions lose their positive charge to form mercury atoms. These changes are shown in Figure 5.14.

Decomposition reactions may form more than two products in some cases. Similarly, combination reactions may involve more than two reactants.

**Before reaction**

**During reaction**

**After reaction**

**FIGURE 5.13** When solid mercury(II) oxide, HgO, is heated, it decomposes into liquid mercury metal and oxygen gas.

©McGraw-Hill Education/Stephen Frisch

**TABLE 5.2** Decomposition Reactions That Occur When Compounds Are Heated

- *Oxides* and *halides* of the metals Au, Pt, and Hg decompose to the elements.

$$2HgO(s) \xrightarrow{\text{heat}} 2Hg(l) + O_2(g)$$

$$PtCl_4(s) \xrightarrow{\text{heat}} Pt(s) + 2Cl_2(g)$$

- *Peroxides* decompose to oxides and oxygen gas.

$$2H_2O_2(aq) \xrightarrow{\text{heat}} 2H_2O(l) + O_2(g)$$

- *Metal carbonates,* except those of the group IA (1) metals, decompose to metal oxides and carbon dioxide gas.

$$NiCO_3(s) \xrightarrow{\text{heat}} NiO(s) + CO_2(g)$$

- *Oxoacids* decompose in a similar way to form nonmetal oxides and water. Other compounds that contain the components of water frequently decompose with the elimination of water. The other product is what is left after water is removed.

$$H_2CO_3(aq) \longrightarrow CO_2(g) + H_2O(l)$$

$$Ca(OH)_2(s) \xrightarrow{\text{heat}} CaO(s) + H_2O(g)$$

$$CuSO_4 \cdot 5H_2O(s) \xrightarrow{\text{heat}} CuSO_4(s) + 5H_2O(g)$$

- *Ammonium compounds* lose ammonia.

$$(NH_4)_2SO_4(s) \xrightarrow{\text{heat}} 2NH_3(g) + H_2SO_4(l)$$

Reactions in which the ionic charges change are called *oxidation-reduction reactions*. In the decomposition of HgO, the charge of mercury changes from 2+ to 0 when Hg(l) is formed. The charge of oxygen changes from 2− to 0 when O₂(g) is formed. We will explore oxidation-reduction reactions in more detail in Chapter 14.

$$2HgO(s) \xrightarrow{\text{heat}} 2Hg(l) + O_2(g)$$

**FIGURE 5.14** The regular array of mercury(II) ions and oxide ions breaks down when mercury(II) oxide is heated. The oxide ions form oxygen atoms, which combine to form gaseous oxygen molecules. The mercury(II) ions form mercury metal in the liquid state.

**FIGURE 5.15**  Limestone chunks, when heated by burning fuel in the large rotary kiln in the center of the photograph, decompose to form lime and carbon dioxide.
©David R. Frazier/Corbis

Many metal compounds containing oxoanions, such as carbonates, decompose upon heating. A common example is calcium carbonate.

$$CaCO_3(s) \xrightarrow{\text{heat}} CaO(s) + CO_2(g)$$

Portland cement has calcium oxide (lime) as one of its ingredients. The calcium oxide is made from limestone (calcium carbonate) by decomposition in large kilns at about 1900°C (Figure 5.15).

A number of compounds that contain water or the components of water—hydrates, hydroxides, and oxoacids—lose water when heated. Hydrates—compounds that contain water molecules—lose water to form **anhydrous** compounds, compounds free of molecular water. One example is hydrated calcium sulfate. It can be dehydrated by heating to form a powder:

$$CaSO_4 \cdot 2H_2O(s) \xrightarrow{\text{heat}} CaSO_4(s) + 2H_2O(g)$$

The anhydrous calcium sulfate can be mixed with water to use in making plaster for walls or casts, or as an artist's surface for painting.

Portland cement gets its name from its resemblance to Portland stone found on the Isle of Portland, Dorset, England. Portland stone is found in many famous London buildings, including the Tower of London and Buckingham Palace.

| **EXAMPLE 5.6** | **Decomposition Reactions** |
|---|---|

Michelle, another art student, needs some cobalt(II) oxide to make cobalt glass. She has the following cobalt compounds available to her: $CoCl_2 \cdot 6H_2O$, $CoCO_3$, $CoS$, and $Co(OH)_2$. Which of these compounds can make the cobalt(II) oxide she needs? What products would form if the other compounds were heated?

**Solution:**

In Table 5.2 we can see that Michelle could use $CoCO_3$ or $Co(OH)_2$. Metal carbonates lose carbon dioxide gas when heated, leaving a metal oxide:

$$CoCO_3(s) \xrightarrow{\text{heat}} CoO(s) + CO_2(g)$$

Metal hydroxides lose water to form metal oxides:

$$Co(OH)_2(s) \xrightarrow{\text{heat}} CoO(s) + H_2O(g)$$

©PjrStudio/Alamy Stock Photo

Heating a hydrate results in the loss of gaseous water:

$$CoCl_2 \cdot 6H_2O(s) \xrightarrow{\text{heat}} CoCl_2(s) + 6H_2O(g)$$

We have seen no patterns that suggest that anything would happen to CoS upon heating.

### Consider This 5.6

What product is always formed when a metal carbonate compound undergoes a decomposition reaction?

### Practice Problem 5.6

Janelle wants to make paper pulp to be used in a printing project. She needs magnesium sulfite trihydrate, $MgSO_3 \cdot 3H_2O$, to bleach the paper pulp. This compound decomposes upon heating. Predict the products and write complete, balanced equations for any reactions.

**Further Practice:**  Questions 5.55 and 5.56 at the end of the chapter

Combination reactions in which at least one reactant is an element are also classified as oxidation-reduction reactions because a change in charge occurs. For example, in the formation of $AlBr_3$ from Al and $Br_2$, the charge on aluminum changes from 0 to 3+, and the charge on bromine changes from 0 to 1−.

**FIGURE 5.16** Aluminum reacts with bromine to form aluminum bromide.
©Tom Pantages

**ANIMATION: Sodium Chloride**

## Combination Reactions

In a **combination reaction** (also called *synthesis*), two substances react to produce a single compound. The reactants can be two elements, an element and a compound, or two compounds. Some patterns of combination reactions are listed with examples in Table 5.3.

An important type of combination reaction involves two elements. *When a metal element reacts with a nonmental element, the product is an ionic compound.* The formula for the ionic compound product can be predicted from the charges expected for the cation of the metal and the anion of the nonmetal, as you learned in Chapter 2. For example, the metal aluminum reacts with the nonmetal bromine (Figure 5.16) to form solid aluminum bromide. Because these elements form ions that we can predict from the periodic table ($Al^{3+}$ and $Br^-$), the correct formula for the product can be written ($AlBr_3$). Then we use coefficients to balance the equation:

$$2Al(s) + 3Br_2(l) \longrightarrow 2AlBr_3(s)$$

The reaction of the elements sodium metal and chlorine gas to form the ionic compound sodium chloride is another example where the formula of the product (NaCl) can be produced from the charges expected for the ions ($Na^+$ and $Cl^-$). Once we write the correct formulas for the reactants (Na and $Cl_2$) and product (NaCl), we can then balance the equation with coefficients:

$$2Na(s) + Cl_2(g) \longrightarrow 2NaCl(s)$$

**TABLE 5.3** Summary of Combination Reactions

- Metals can react with nonmetals to form an ionic compound.

$$2Na(s) + Cl_2(g) \longrightarrow 2NaCl(s)$$

- Nonmetals can react with other nonmetals to form a molecular compound.

$$C(s) + O_2(g) \longrightarrow CO_2(g)$$

- A compound and an element can combine to form another compound.

$$2CO(g) + O_2(g) \longrightarrow 2CO_2(g)$$

- Two compounds may react to form a new compound.

$$CaO(s) + CO_2(g) \longrightarrow CaCO_3(s)$$

*When both reactants of a combination reaction are nonmetal elements, they would combine to produce a molecular compound.* For example, the yellow solid sulfur reacts with oxygen gas to form colorless, gaseous sulfur dioxide (Figure 5.17):

$$S_8(s) + 8O_2(g) \longrightarrow 8SO_2(g)$$

This compound has a pungent odor and is one of the substances that make acid rain.

*A compound and an element may combine to form another compound if one exists with a higher atom:atom ratio.* For example, carbon dioxide has a higher oxygen:carbon atom ratio than does carbon monoxide. Carbon monoxide forms when carbon (coal), hydrocarbons, or other carbon-containing materials are burned in the presence of insufficient oxygen. It is a colorless, odorless, poisonous gas.

$$2C(s) + O_2(g) \longrightarrow 2CO(g)$$

It reacts further with oxygen to form colorless, odorless carbon dioxide (Figure 5.18), which is a component of the atmosphere and of exhaled breath:

$$2C(s) + O_2(g) \longrightarrow 2CO(g)$$

The combination reaction of carbon monoxide and oxygen is shown on a molecular level in Figure 5.19.

**FIGURE 5.17** In a combination reaction, the elements sulfur and oxygen form sulfur dioxide.
©McGraw-Hill Education/Stephen Frisch

CO   CO₂

O₂

**FIGURE 5.19** In the burning of carbon monoxide, oxygen atoms from oxygen molecules combine with carbon monoxide molecules and rearrange to form carbon dioxide molecules.

**FIGURE 5.18** Carbon monoxide burns in air to form carbon dioxide. This is a combination reaction between a compound and an element.
©McGraw-Hill Education/Stephen Frisch

*Two compounds may react to form a new compound.* For example, calcium oxide, a white solid, reacts with carbon dioxide to form white solid calcium carbonate, which is found naturally in limestone, marble, and seashells:

$$CaO(s) + CO_2(g) \longrightarrow CaCO_3(s)$$

Reactions such as this happen during the hardening and aging of cement.

Nonmetals become more reactive the closer they are to the upper-right corner of the periodic table, ignoring the noble gases. The most reactive nonmetal is fluorine, $F_2$.

| EXAMPLE 5.7 | **Combination Reactions** |
|---|---|

When pure calcium metal is exposed to the oxygen in air, a white coating appears on the surface. Predict the product and write a complete, balanced equation for this reaction.

**Solution:**
When two elements react, they form a compound. Since the elements are a metal and a nonmetal, the compound should be ionic. The metal should form a cation, and the nonmetal should form an anion. From their positions in the periodic table, we can predict that the charge on the calcium cation should be 2+ and that the charge on the oxide anion should be 2−. The product should be the compound formed from $Ca^{2+}$ and $O^{2-}$, CaO. From this information, we can write the following word equation:

$$calcium + oxygen \longrightarrow calcium\ oxide$$

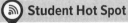

**🔊 Student Hot Spot**

Student data indicate you may struggle with writing balanced equations for combination reactions. Access the SmartBook to view additional Learning Resources on this topic.

**FIGURE 5.20** Calcium reacts with water to form aqueous calcium hydroxide and gaseous hydrogen.
©McGraw-Hill Education/Stephen Frisch

Using formulas, we get the following skeletal equation:

$$Ca(s) + O_2(g) \longrightarrow CaO(s)$$

The oxygen atoms are not balanced, so we place a coefficient of 2 in front of both $Ca(s)$ and $CaO(s)$. This gives us a balanced equation:

$$2Ca(s) + O_2(g) \longrightarrow 2CaO(s)$$

### Consider This 5.7

If an element reacts with a compound to form a new compound and a different element, would this be a combination reaction?

### Practice Problem 5.7

When magnesium metal burns in air, it forms two white compounds. One is the compound expected for the reaction of magnesium with oxygen in air, while the other, magnesium nitride, $Mg_3N_2$, results from a reaction with nitrogen in air. Predict the product of the reaction of magnesium with oxygen. Write complete, balanced equations for the reactions that form both products.

**Further Practice:** Questions 5.59 and 5.60 at the end of the chapter

## Single-Displacement Reactions

In a **single-displacement reaction,** a free element displaces another element from a compound to produce a different compound and a different free element. An example is the thermite reaction between aluminum and iron(III) oxide, used in the production of iron metal for welding (Figure 5.9): In this single-displacement reaction, aluminum metal displaces iron ion from $Fe_2O_3$, producing the new compound $Al_2O_3$ and free iron metal:

$$2Al(s) + Fe_2O_3(s) \longrightarrow Al_2O_3(s) + 2Fe(l)$$

In single-displacement reactions, the displaced element is often a metal, but it can be a nonmetal. Let's consider a common type of single-displacement reaction, one in which a highly reactive metal displaces hydrogen from water. For example, calcium reacts with cold water to form calcium hydroxide and hydrogen gas, as shown in Figure 5.20. Note in the equation for the reaction that calcium displaces only one H atom from each $H_2O$ molecule:

$$Ca(s) + 2H_2O(l) \longrightarrow Ca(OH)_2(aq) + H_2(g)$$

Some metals, such as magnesium, do not react with cold water, but react slowly with steam:

$$Mg(s) + 2H_2O(g) \xrightarrow{\text{heat}} Mg(OH)_2(aq) + H_2(g)$$

Metals that are still less reactive do not react with water at all, but they do displace hydrogen from acids. Cobalt is an example:

$$Co(s) + 2HCl(aq) \longrightarrow CoCl_2(aq) + H_2(g)$$

The reaction of metals with acids is sometimes called corrosion, especially if the result is undesirable. **Corrosion** is a process in which a solid metal slowly deteriorates due to a chemical change. Metals that are even less reactive, such as copper, do not react with either water or acids in a single-displacement reaction.

How do we predict which metals react with water or acids? We can use an **activity series,** a list of the metals in order of their reactivity (e.g., Figure 5.21). *A more active element displaces a less active element from its compounds.* The most active elements, found at the top of the activity series, react with cold water to release $H_2$ gas. The next most active elements react with steam to release $H_2$ gas. All the elements

All single-displacement reactions are oxidation-reduction reactions. In the thermite reaction, aluminum displaces iron from its oxide compound. Aluminum changes its charge from 0 to 3+, while iron changes its charge from 3+ to 0.

Copper, mercury, and gold do not react with acids to release hydrogen gas, but they do react with nitric acid to form the metal nitrate and nitrogen dioxide gas. The reaction for copper is given by the following equation:

$$Cu(s) + 4HNO_3(aq) \longrightarrow$$
$$Cu(NO_3)_2(aq) + 2NO_2(g) + 2H_2O(l)$$

This is another example of an oxidation-reduction reaction.

more active than hydrogen in the series react with acids. None of the elements less active than hydrogen displace hydrogen as $H_2$ gas from any compound. Note that some of the less active metals are commonly found in coins and jewelry.

| EXAMPLE 5.8 | Reaction of Metals with Water and Acids |
| --- | --- |

Antonio is asked to make a magnesium statue to be placed in the middle of a public fountain. Should he take the commission? Why or why not?

**Solution:**

From its position in the activity series, we know that magnesium reacts with steam to release hydrogen gas:

$$Mg(s) + 2H_2O(g) \xrightarrow{heat} Mg(OH)_2(aq) + H_2(g)$$

If splashed with water from the fountain and heated by the sun, the statue will corrode. Antonio should suggest a different metal.

**Consider This 5.8**

If aluminum foil is used to store an acidic food such as lemon slices, is the foil likely to corrode?

**Practice Problem 5.8**

Tomatoes are acidic. Why don't we use iron cans coated with zinc to hold tomato sauce?

**Further Practice:**  Questions 5.65 and 5.66 at the end of the chapter

The displacement of hydrogen from water or acids is just one type of single-displacement reaction. Other elements can also be displaced from their compounds. For example, copper displaces silver. If a piece of copper metal is placed in an aqueous solution of silver nitrate, silver metal is deposited on the surface of the copper (Figure 5.22). The copper is converted to an ionic copper(II) compound in solution:

$$Cu(s) + 2AgNO_3(aq) \longrightarrow Cu(NO_3)_2(aq) + 2Ag(s)$$

As shown in Figure 5.23, atoms on the surface of the copper metal develop a positive charge by losing electrons and go into solution as copper(II) ions. Silver ions in solution, upon contacting the piece of copper metal, lose their positive charge by gaining electrons to form silver atoms that stick to the copper's surface. The anions are not involved in the reaction and remain unchanged in the solution.

| | |
| --- | --- |
| K | |
| Ba | React with |
| Sr | cold water to |
| Ca | give $H_2$ |
| Na | |
| Mg | |
| Al | |
| Mn | React with |
| Zn | steam to |
| Cr | give $H_2$ |
| Fe | |
| Cd | |
| Co | |
| Ni | React with |
| Sn | acids to |
| Pb | give $H_2$ |
| $H_2$ | |
| Sb | |
| Bi | |
| Cu | |
| Ag | Do not react |
| Hg | to give $H_2$ |
| Pd | |
| Pt | |
| Au | |

Increasing activity

**FIGURE 5.21**  In this chemical activity series, the most active elements appear at the top. The notations to the right side indicate the reactivity of these elements with water or acids to release molecular hydrogen gas, $H_2(g)$.

 **ANIMATION:** Oxidation and Reduction Reactions

**FIGURE 5.22**  Copper metal reacts with aqueous silver nitrate to form silver metal and aqueous copper(II) nitrate. The blue color of the solution is due to the formation of $Cu^{2+}(aq)$. A microscopic view shows that the silver metal is deposited as crystals.
©McGraw-Hill Education/Stephen Frisch; (inset): ©Jim Birk

**FIGURE 5.23** Copper atoms are converted to copper(II) ions, and silver ions are converted to silver atoms. Nitrate ions remain unchanged in solution.

A single-displacement reaction was responsible for severe corrosion that occurred in the Statue of Liberty. In the damp environment of New York Harbor, the copper panels on the outside of the statue developed a coating of copper(II) oxide, copper(II) hydroxide, and copper(II) carbonate, while the iron used for the inner framework rusted to form iron(III) oxide. This brought copper metal in contact with iron(III) ions and elemental iron in contact with copper(II) ions. Would either of these combinations undergo a single-displacement reaction? As the National Park Service found out, the answer is yes. The iron was converted to iron(II) ions and the copper(II) ions were converted to copper metal:

$$Fe(s) + CuO(s) \longrightarrow Cu(s) + FeO(s)$$

Consequently, corrosion of the metal framework was accelerated by a factor of 1000, and the statue had to be rebuilt (Figure 5.24).

Will a single-displacement reaction occur with all combinations of metals and ionic compounds? No. Earlier we saw that if we place a piece of copper metal in a solution of silver nitrate, silver metal is deposited and copper goes into solution as copper(II) nitrate. But the opposite does not occur. Nothing happens when we place a piece of silver in a solution of copper(II) nitrate (Figure 5.25). How can we predict if a single-displacement reaction will occur when a metal is added to an ionic compound in solution? We can use the activity series shown in Figure 5.21. The more

**FIGURE 5.24** The Statue of Liberty was closed between 1984 and 1986. It was rebuilt to remedy the effects of corrosion.

©Bettmann/Getty Images

**Before**     **After**

**FIGURE 5.25** Although a single-displacement reaction occurs when a piece of copper metal is placed in a solution of silver nitrate, no reaction occurs when a piece of silver is placed in a solution of copper(II) nitrate. Copper is more active than silver.

©Jim Birk

active metal displaces the less active metal from its ionic compound, but a less active metal cannot displace a more active metal from its ionic compound. Thus, magnesium will displace iron from $FeCl_2$, but tin or copper will not.

---

**EXAMPLE 5.9** | **Displacement of Metal Ions**

Antonio has two pieces of metal, zinc and iron. He also has two aqueous solutions, one of zinc chloride and one of iron(II) chloride. Which metal will corrode when placed in one of the solutions? Which solution? Write a balanced equation for the reaction.

**Solution:**

A metal placed in a solution containing its own ions will not react. Thus, we must consider whether iron metal will displace zinc ions from solution or whether zinc metal will displace iron(II) ions from $FeCl_2$ in solution. To determine which reaction will occur, we consult the activity series. The more active metal will displace ions of the less active metal from solution. Figure 5.21 shows that zinc is more active than iron. Therefore, it will displace $Fe^{2+}$ ions from $FeCl_2$ dissolved in solution. The products of the reaction will be iron metal and zinc chloride. The word equation is

$$\text{zinc} + \text{iron(II) chloride} \longrightarrow \text{zinc chloride} + \text{iron}$$

Substituting the formulas gives us the following skeletal equation:

$$Zn(s) + FeCl_2(aq) \longrightarrow ZnCl_2(aq) + Fe(s)$$

Counting atoms of each element, we see that this is also the balanced equation for the reaction.

**Consider This 5.9**

If a piece of metal is placed in a solution containing two different metal ions, which metal ion will be more likely to convert to its element form?

**Practice Problem 5.9**

Antonio has two other pieces of metal, cobalt and tin. He also has two aqueous solutions containing their ions, one of cobalt(II) sulfate and one of tin(II) sulfate. Which metal will corrode when placed in one of the solutions? Which solution? Write a balanced equation for the single-displacent reaction where corrosion occurs.

**Further Practice:** Questions 5.67 and 5.68 at the end of the chapter

---

## Double-Displacement Reactions

Jennifer mixed a solution of lead(II) acetate with a solution of potassium iodide and formed a yellow solid of lead(II) iodide, leaving potassium acetate in solution:

$$Pb(CH_3CO_2)_2(aq) + 2KI(aq) \longrightarrow PbI_2(s) + 2KCH_3CO_2(aq)$$

Consider a similar reaction. When aqueous sodium sulfide reacts with hydrochloric acid, for example, it forms aqueous sodium chloride and hydrogen sulfide gas:

$$Na_2S(aq) + 2HCl(aq) \longrightarrow 2NaCl(aq) + H_2S(g)$$

When aqueous potassium hydroxide reacts with aqueous nitric acid, it forms water and aqueous potassium nitrate:

$$KOH(aq) + HNO_3(aq) \longrightarrow H_2O(l) + KNO_3(aq)$$

What do these reactions have in common? In each reaction, two compounds exchange parts to form two new compounds. Such a reaction is called a **double-displacement reaction.** In all double-displacement reactions, two compounds exchange ions or elements to form new compounds. What is the *driving force* that causes two compounds to exchange partners? In the first example, called a **precipitation reaction,** one product compound separates from the reaction mixture because it is insoluble. It forms a solid. In the second example, one product compound separates from the reaction mixture as an insoluble gas. In the third example, a stable molecular substance, water, is formed. In each case reactant ions are removed from the solution by forming a new substance—either a solid, a gas, or a stable molecular compound.

Not all combinations of two compounds can undergo a double-displacement reaction. Consider aqueous solutions of the compounds potassium chloride, $KCl$, and nitric acid, $HNO_3$, for example. Because they are electrolytes, they exist as free ions: $K^+(aq)$, $Cl^-(aq)$, $H^+(aq)$, and $NO_3^-(aq)$. Upon mixing, we see no sign of a reaction because no combination of the component ions can form either an insoluble compound or a stable molecular compound.

We will consider three classes of double-displacement reactions: precipitation, gas formation, and acid-base neutralization reactions. For each class we will examine how to determine whether a reaction will occur.

**Precipitation Reactions**    In precipitation reactions, reactants exchange ions to form a precipitate. A **precipitate** is an insoluble ionic compound that does not dissolve in water. *If a possible product is insoluble, a precipitation reaction should occur.* We can predict whether such a compound will form by consulting the solubility rules, some of which are listed in Table 5.4.

Suppose we mix colorless, aqueous solutions of barium chloride and sodium sulfate. As shown in Figure 5.26, this reaction produces a white precipitate, so we know a reaction occurs. To identify the precipitate, we determine which product of double displacement is insoluble in water. After exchanging the cations and anions present in the reactants, we determine that the products are barium sulfate, $BaSO_4$, and sodium chloride, $NaCl$. To determine which of these compounds is the precipitate, we consult the solubility rules.

According to the solubility rules in Table 5.4, most sulfate compounds, sodium compounds, and chloride compounds are water soluble. An exception is barium sulfate, which is insoluble. Sodium chloride is soluble according to the first solubility rule, and

**FIGURE 5.26** When solutions of barium chloride and sodium sulfate mix, a precipitate of barium sulfate forms, leaving sodium chloride in the solution.

©Charles D. Winters/Science Source

**TABLE 5.4** Rules Used to Predict the Solubility of Ionic Compounds

| Ions | Rule |
|---|---|
| $Na^+$, $K^+$, $NH_4^+$ (and other alkali metal ions) | Most compounds of alkali metal and ammonium ions are soluble. |
| $NO_3^-$, $CH_3CO_2^-$ | All nitrates and acetates are soluble. |
| $SO_4^{2-}$ | Most sulfates are soluble. Exceptions are $BaSO_4$, $SrSO_4$, $PbSO_4$, $CaSO_4$, $Hg_2SO_4$, and $Ag_2SO_4$. |
| $Cl^-$, $Br^-$, $I^-$ | Most chlorides, bromides, and iodides are soluble. Exceptions are $AgX$, $Hg_2X_2$, $PbX_2$, and $HgI_2$. ($X = Cl$, $Br$, or $I$.) |
| $Ag^+$ | Silver compounds, except $AgNO_3$ and $AgClO_4$, are insoluble. $AgCH_3CO_2$ is slightly soluble. |
| $O^{2-}$, $OH^-$ | Oxides and hydroxides are insoluble. Exceptions are alkali metal hydroxides, $Ba(OH)_2$, $Sr(OH)_2$, and $Ca(OH)_2$ (somewhat soluble). |
| $S^{2-}$ | Sulfides are insoluble. Exceptions are compounds of $Na^+$, $K^+$, $NH_4^+$, and the alkaline earth metal ions. |
| $CrO_4^{2-}$ | Most chromates are insoluble. Exceptions are compounds of $Na^+$, $K^+$, $NH_4^+$, $Mg^{2+}$, $Ca^{2+}$, $Al^{3+}$, and $Ni^{2+}$. |
| $CO_3^{2-}$, $PO_4^{3-}$, $SO_3^{2-}$, $SiO_3^{2-}$ | Most carbonates, phosphates, sulfites, and silicates are insoluble. Exceptions are compounds of $Na^+$, $K^+$, and $NH_4^+$. |

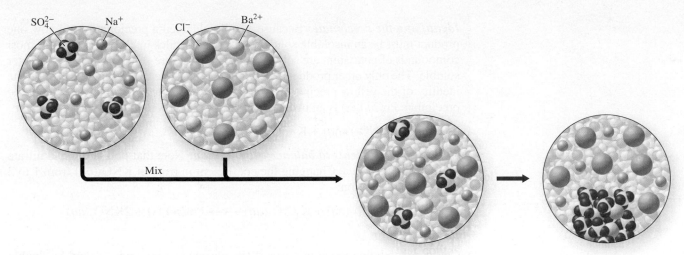

**FIGURE 5.27** In typical double-displacement reactions, one pair of ions combines to form a solid, a gas, or an undissociated molecule, leaving the other pair of ions in solution. In this reaction, barium ions react with the sulfate ions to form the barium sulfate precipitate, $BaSO_4(s)$. Note that the $Na^+(aq)$ and $Cl^-(aq)$ ions undergo no change at all.

therefore remains dissolved in solution. Thus, the precipitate is the insoluble product barium sulfate. From this information, we can write the following skeletal equation:

$$BaCl_2(aq) + Na_2SO_4(aq) \longrightarrow BaSO_4(s) + NaCl(aq)$$

**ANIMATION:** Barium Precipitation

We balance this equation by adding a coefficient of 2 in front of NaCl, giving the following balanced equation:

$$BaCl_2(aq) + Na_2SO_4(aq) \longrightarrow BaSO_4(s) + 2NaCl(aq)$$

Let's also consider this reaction on a molecular level (Figure 5.27). From the figure, we see that mixing the two solutions gives us a solution that contains four kinds of ions. After precipitation of barium sulfate, the solution still contains the two ions, $Na^+(aq)$ and $Cl^-(aq)$, that do not form an insoluble compound. These ions do not undergo a chemical change.

---

**EXAMPLE 5.10** | **Precipitation Reactions**

Janelle finds a recipe for the preparation of the pigment chrome yellow. She dissolves lead(II) nitrate and potassium chromate in separate volumes of water, and then mixes the two solutions. She obtains a yellow precipitate. Identify the precipitate and write a balanced equation for the reaction she carried out.

**Solution:**

***Predicting products and balancing their formulas:*** The starting solutions contain $Pb^{2+}(aq)$, $NO_3^-(aq)$, $K^+(aq)$, and $CrO_4^{2-}(aq)$. We can write the possible product formulas (without physical state symbols for now) by exchanging cations and anions to get the other possible arrangement. The product formulas must be written so that they are balanced with the cation and anion charges cancelling to give a neutral overall charge or zero. In the first product written, the lead ion and chromate are 2+ and 2−, respectively, so the formula is balanced as is, $PbCrO_4$. In the second product written, the potassium ion and nitrate ion are 1+ and 1−, respectively, so that formula is also already balanced, $KNO_3$. In this case, no subscripts are needed to balance the formulas.

$$Pb(NO_3)_2(aq) + K_2CrO_4(aq) \longrightarrow PbCrO_4 + KNO_3$$

©McGraw-Hill Education/Stephen Frisch

*Identifying the precipitate:* Because Janelle obtained a precipitate, we know one product must be an insoluble solid. The solubility rules in Table 5.4 say that most compounds of potassium are soluble and all nitrates are soluble, so $KNO_3$ is water soluble. The only other product is $PbCrO_4$, so it must be the insoluble solid and the identity of the yellow precipitate. We use physical state symbols to show that the precipitate $PbCrO_4(s)$ is an insoluble solid and that $KNO_3(aq)$ is soluble in water:

$$Pb(NO_3)_2(aq) + K_2CrO_4(aq) \longrightarrow PbCrO_4(s) + KNO_3(aq)$$

*Changing coefficients to balance the equation:* Note that potassium and nitrate ions are not balanced. Changing the coefficient in front of $KNO_3(aq)$ from 1 to 2 balances the equation:

$$Pb(NO_3)_2(aq) + K_2CrO_4(aq) \longrightarrow PbCrO_4(s) + 2KNO_3(aq)$$

### Consider This 5.10

Can a precipitation reaction occur if the possible compounds formed by double displacement are both soluble? Creating such an example might help to answer this question.

### Practice Problem 5.10

Jennifer mixes solutions of cadmium(II) nitrate and sodium sulfide and obtains an orange precipitate that is the pigment cadmium orange. Identify the precipitate, and write a balanced equation for the reaction she carried out.

**Further Practice:** Questions 5.77 and 5.78 at the end of the chapter

**Gas-Formation Reactions** In some double-displacement reactions, the formation of an insoluble (or only slightly soluble) gas provides the driving force for a reaction. Thus, *a double-displacement reaction should occur if an insoluble gas would be formed.* Examples of gases that are commonly produced as part of double-displacement reactions are $H_2S(g)$, $CO_2(g)$, and $SO_2(g)$. Let's take a look at the types of compounds that will react to form these gases.

Many sulfide compounds, such as the ZnS used in some television screens, will react with acids to form gaseous hydrogen sulfide:

$$ZnS(s) + 2HCl(aq) \longrightarrow ZnCl_2(aq) + H_2S(g)$$

Sometimes insoluble gases do not form directly from a double-displacement reaction. Instead they form when an unstable product of a double-displacement reaction decomposes. For example, calcium carbonate reacts with hydrochloric acid to form calcium chloride and carbonic acid:

$$CaCO_3(s) + 2HCl(aq) \longrightarrow CaCl_2(aq) + H_2CO_3(aq)$$

Carbonic acid, $H_2CO_3$, is an unstable substance. It readily decomposes to form water and carbon dioxide (Figure 5.28), as described in Table 5.2:

$$H_2CO_3(aq) \longrightarrow H_2O(l) + CO_2(g)$$

The net reaction is

$$CaCO_3(s) + 2HCl(aq) \longrightarrow CaCl_2(aq) + H_2O(l) + CO_2(g)$$

**FIGURE 5.28** When an acid reacts with a metal carbonate, bubbles of carbon dioxide gas form.

©McGraw-Hill Education/Stephen Frisch

Such reactions are responsible for the acid rain damage to marble statues (composed of $CaCO_3$) that Breanna was investigating. Acid rain erodes marble and limestone, pitting their surfaces and diminishing the detail of carvings and castings. In the air, oxides of nitrogen and sulfur, released from motor vehicles

**FIGURE 5.29** A limestone statue was photographed in 1935 *(left)* and again in 1994 *(right)*. The damage resulted from the reaction of acid rain with the calcium carbonate in limestone.
(left): ©NYC Parks Photo Archive/ Fundamental Photographs; (right): ©Kristen Brochmann/Fundamental Photographs

and some industrial processes, form nitric and sulfuric acids, which fall to Earth with the rain. As shown in Figure 5.29, exposure of marble or sandstone statues to acid rain leads to loss of some of the calcium carbonate of which they are made. Many old statues and architectural features have eroded over time because of air pollution.

| EXAMPLE 5.11 | **Gas-Forming Reactions** |

Magnesium sulfite, $MgSO_3$, is used as a bleach to make wood pulp for making paper. A gas is released when this compound is mixed with acid. Identify the gas, and write a balanced chemical equation for the reaction of magnesium sulfite with hydrochloric acid.

**Solution:**

Sulfites react with acid to give sulfur dioxide gas. Initially, we have a double-displacement reaction:

$$MgSO_3(s) + 2HCl(aq) \longrightarrow MgCl_2(aq) + H_2SO_3(aq)$$

In a sufficiently concentrated solution, the sulfurous acid then undergoes a decomposition reaction:

$$H_2SO_3(aq) \longrightarrow H_2O(l) + SO_2(g)$$

The overall reaction forms a metal compound with the anion present in the acid, along with liquid water and sulfur dioxide gas:

$$MgSO_3(s) + 2HCl(aq) \longrightarrow MgCl_2(aq) + H_2O(l) + SO_2(g)$$

**Consider This 5.11**

If a compound contained the $HSO_3^-$ ion instead of the $SO_3^{2-}$ ion, could a gas formation reaction still occur?

**Practice Problem 5.11**

The reaction of barium sulfide with sulfuric acid is used to prepare barium sulfate. Barium sulfate, combined with zinc sulfide and zinc oxide, is used to make the white pigment lithopone. Write a balanced equation for the reaction of barium sulfide with sulfuric acid.

**Further Practice:**  Questions 5.79 and 5.80 at the end of the chapter

Metal oxides can behave like bases when dissolved in water because they react with water to form metal hydroxides. For example, the equation for the reaction of calcium oxide with water is

$$CaO(s) + H_2O(l) \longrightarrow Ca(OH)_2(aq)$$

**Acid-Base Neutralization Reactions**   A **neutralization reaction** is a double-displacement reaction of an acid and a base. Recall from Chapter 3 that acids are compounds that release hydrogen ions, and bases are compounds that neutralize acids. They react with the hydrogen ions released by acids. The most common bases are hydroxide compounds of metals. Normally, *an acid reacts with a base to form an ionic compound and water.* The formation of stable water molecules from hydrogen ions and hydroxide ions is the driving force for neutralization reactions. A common laboratory example of an acid-base neutralization is the reaction between hydrochloric acid and sodium hydroxide:

$$HCl(aq) + NaOH(aq) \longrightarrow NaCl(aq) + H_2O(l)$$

---

**EXAMPLE 5.12**     **Acid-Base Neutralization Reactions**

Upset stomachs are sometimes treated with an antacid that is a suspension of magnesium hydroxide in water. Stomach acid is hydrochloric acid. Write a balanced equation for the reaction that occurs when magnesium hydroxide mixes with stomach acid.

**Solution:**

The reactants are $Mg(OH)_2(s)$ and $HCl(aq)$. A reaction between two compounds is usually a double-displacement reaction. Since we can recognize $Mg(OH)_2$ as a base and $HCl$ as an acid, this must be a neutralization reaction. The products can be predicted by exchanging partners—$Mg^{2+}$ with $Cl^-$ and $H^+$ with $OH^-$—giving $MgCl_2$ and $H_2O$. From the solubility rules, we see that $MgCl_2$ should be soluble. Writing the skeletal equation gives

$$Mg(OH)_2(s) + HCl(aq) \longrightarrow MgCl_2(aq) + H_2O(l)$$

The complete, balanced equation is

$$Mg(OH)_2(s) + 2HCl(aq) \longrightarrow MgCl_2(aq) + 2H_2O(l)$$

**Consider This 5.12**

If a compound contains hydrogen and another nonmetal, but no other elements, is the compound necessarily an acid?

**Practice Problem 5.12**

Calcium oxide is the white powder, lime. When added to water, it makes slaked lime, which is a solution of the base calcium hydroxide. If sulfuric acid is added to slaked lime, calcium sulfate and water form. Write a complete, balanced equation for the reaction between sulfuric acid and calcium hydroxide.

**Further Practice:**  Questions 5.81 and 5.82 at the end of the chapter

## Combustion Reactions

Any reaction that involves oxygen molecules as a reactant and that rapidly produces heat and flame is a **combustion reaction.** We see a flame when a large amount of energy in the form of heat is released very rapidly. The burning of a fuel, such as logs in a fireplace or methane in a gas stove, is a common combustion reaction. Combustion reactions are also used to heat kilns for firing pottery. Many other substances can undergo combustion, as shown in Figure 5.30.

Both metallic and nonmetallic elements can undergo combustion. In such instances, the reactions can also be classified as combination reactions. The following are examples of both combustion and combination reactions:

$$2Mg(s) + O_2(g) \longrightarrow 2MgO(s)$$

$$S(s) + O_2(g) \longrightarrow SO_2(g)$$

Look back at the section on combination reactions and see which of the examples there can also be classified as combustion reactions.

The most common combustion reactions involve hydrocarbons as one of the reactants. *Hydrocarbons* are chemical compounds made only of hydrogen and carbon. Propane, $C_3H_8$, is a hydrocarbon used as a fuel for barbecue grills. The combustion of propane is described by the following equation:

$$C_3H_8(g) + 5O_2(g) \longrightarrow 3CO_2(g) + 4H_2O(g)$$

If insufficient oxygen is present, combustion of a hydrocarbon produces carbon monoxide instead of carbon dioxide. Because carbon monoxide is poisonous, a barbecue grill with charcoal should never be used indoors.

**Nonmetal combustion**

$$S_8(s) + 8O_2(g) \longrightarrow 8SO_2(g)$$

**Metal combustion**

$$2Mg(s) + O_2(g) \longrightarrow 2MgO(s)$$

**Hydrocarbon combustion**

$$CH_4(g) + 2O_2(g) \longrightarrow CO_2(g) + 2H_2O(g)$$

**Hydrocarbon combustion**

$$C_{25}H_{52}(s) + 38O_2(g) \longrightarrow 25CO_2(g) + 26H_2O(g)$$

**FIGURE 5.30** Many substances can undergo combustion reactions. Which of these reactions could also be classified as combination reactions? (top left): ©McGraw-Hill Education/Stephen Frisch; (top right): ©McGraw-Hill Education/Stephen Frisch; (bottom left): ©Sami Sarkis/Getty Images; (bottom right): ©TheDman/iStock/Getty Images

**FIGURE 5.31** In combustion reactions, oxygen combines with other elements in the other reactant to form oxides. In this example, the combustion of methane produces carbon dioxide and water.

> Hydrocarbon compounds that also contain oxygen, such as alcohols and sugars, participate in combustion reactions, producing $CO_2(g)$ and $H_2O(g)$. If reactants contain elements other than carbon, hydrogen, and oxygen, those elements are converted to their oxides.

When plenty of oxygen is available for the reaction, the hydrogen in hydrocarbons forms water vapor, and the carbon forms carbon dioxide, as shown in Figure 5.31.

---

**EXAMPLE 5.13**    **Combustion Reactions**

Antonio uses an oxyacetylene torch to cut some pieces of metal. The heat from the torch comes from the combustion of acetylene, $C_2H_2(g)$. Write a balanced equation for the reaction.

**Solution:**

Acetylene reacts with molecular oxygen to form carbon dioxide and water. The skeletal equation is

$$C_2H_2(g) + O_2(g) \longrightarrow CO_2(g) + H_2O(g)$$

Balancing this equation results in

$$2C_2H_2(g) + 5O_2(g) \longrightarrow 4CO_2(g) + 2H_2O(g)$$

**Consider This 5.13**

If a certain combustion reaction can also be classified as a combination reaction, how many products can be formed?

**Practice Problem 5.13**

Shellac, used as a transparent coating on wood and other materials, is a suspension of a tree sap (resin) in methanol, $CH_3OH(l)$. The mixture is highly combustible. Write an equation to describe the combustion of methanol.

**Further Practice:** Questions 5.85 and 5.86 at the end of the chapter

---

## 5.5 Representing Reactions in Aqueous Solution

A chemical equation is a shorthand way of showing the changes that occur in a chemical reaction. Many of the reactions mentioned in this chapter occur in solution. We have written them as if the reactants and products were discrete units like molecular compounds, but in many cases, they are not. From Chapter 3 recall that electrolytes dissolved in water exist as separated cations and anions. To be accurate in representing reactions that occur in solution, we should represent soluble ionic compounds, as well as strong acids and bases, by the formulas of their component ions. For example, $Pb(NO_3)_2(aq)$ can be represented as

$$Pb^{2+}(aq) + 2NO_3^-(aq)$$

Compounds that are insoluble in water, such as $AgCl(s)$, do not break apart into ions.

Consider the reaction of aqueous solutions of lead(II) nitrate and potassium chromate, both of which are ionic compounds and strong electrolytes. Their reaction is used to prepare the pigment chrome yellow, which is lead(II) chromate, leaving an aqueous solution of potassium nitrate, as shown in Figure 5.32.

Earlier, we wrote the following equation for this reaction:

$$Pb(NO_3)_2(aq) + K_2CrO_4(aq) \longrightarrow PbCrO_4(s) + 2KNO_3(aq)$$

An equation in this form is often called a **molecular equation,** because it represents the substances as if they existed as molecules (or formula units) in solution rather than dissociated into ions. The aqueous physical state ($aq$) indicated for both reactants and one product indicates they are dissolved in water. These soluble ionic compounds exist as separated ions in aqueous solution. The insoluble $PbCrO_4(s)$ product does not exist as separate ions. The following **ionic equation** represents the soluble ionic substances more appropriately:

$$Pb^{2+}(aq) + 2NO_3^-(aq) + 2K^+(aq) + CrO_4^{2-}(aq) \longrightarrow$$
$$PbCrO_4(s) + 2K^+(aq) + 2NO_3^-(aq)$$

Note that $K^+(aq)$ ions and $NO_3^-(aq)$ ions occur on both sides of the equation in equal numbers. They are called **spectator ions** because they do not undergo a change. When these spectator ions are eliminated from both sides of the ionic equation, the reaction is more simply represented as follows:

$$Pb^{2+}(aq) + CrO_4^{2-}(aq) \longrightarrow PbCrO_4(s)$$

Such an equation, called the **net ionic equation,** includes only those substances that are involved in the reaction.

Ionic equations can also be written for some single-displacement reactions. An example is the reaction of copper metal with silver nitrate in aqueous solution (Figures 5.21 and 5.22):

$$Cu(s) + 2AgNO_3(aq) \longrightarrow Cu(NO_3)_2(aq) + 2Ag(s)$$

The soluble ionic compounds are strong electrolytes, so they are written as dissociated ions in the ionic equation:

$$Cu(s) + 2Ag^+(aq) + 2NO_3^-(aq) \longrightarrow Cu^{2+}(aq) + 2NO_3^-(aq) + 2Ag(s)$$

What are the spectator ions? They are those species that undergo no change and therefore are shown on both sides of the ionic equation. In this reaction the spectator ions are only the nitrate ions, $NO_3^-$. The other species undergo a change: copper element changes to copper ion, and the silver ion changes to silver element. The net ionic equation omits the spectator ions and shows only those species that change:

$$Cu(s) + 2Ag^+(aq) \longrightarrow Cu^{2+}(aq) + 2Ag(s)$$

In some aqueous reactions, there are no spectator ions. An example is the reaction of a highly active metal with water:

$$2K(s) + 2H_2O(l) \longrightarrow 2KOH(aq) + H_2(g)$$

Its ionic equation shows no spectator ions:

$$2K(s) + 2H_2O(l) \longrightarrow 2K^+(aq) + 2OH^-(aq) + H_2(g)$$

**FIGURE 5.32**  Mixing solutions of lead nitrate and potassium chromate produces the yellow solid lead(II) chromate and a solution of potassium nitrate.
©McGraw-Hill Education/Stephen Frisch

Metal ions of group IA (1) always form soluble compounds, so they are always spectator ions in double-displacement reactions. Look back at Figure 5.27. Identify the spectator ions in that molecular-level diagram.

When all ions are spectator ions, no reaction occurs.

 **Student Hot Spot**

Student data indicate you may struggle with writing net ionic equations and determining spectator ions. Access the SmartBook to view additional Learning Resources on this topic.

---

| **EXAMPLE 5.14** | **Net Ionic Equations** |

Consider the reaction between colorless aqueous solutions of silver nitrate, $AgNO_3$, and potassium bromide, KBr, which gives a pale yellow precipitate and a colorless solution. The precipitate is used extensively in photographic films and papers. Identify the precipitate and write molecular, ionic, and net ionic equations for the reaction.

**Solution:**

Both compounds are strong electrolytes, so they dissociate into the ions $Ag^+(aq)$, $NO_3^-(aq)$, $K^+(aq)$, and $Br^-(aq)$ in solution. Possible formulas for the precipitate, then, are $AgNO_3$, $KBr$, $AgBr$, and $KNO_3$. The first two choices can be eliminated because we started with solutions of $AgNO_3$ and $KBr$. If they could have precipitated as solids, they would have done so before mixing. That leaves $AgBr$ and $KNO_3$. The solubility rules (Table 5.4) indicate that $KNO_3$ should be soluble and $AgBr$ should be insoluble. Thus, we can write the molecular equation

$$AgNO_3(aq) + KBr(aq) \longrightarrow AgBr(s) + KNO_3(aq)$$

Since $AgNO_3$, $KBr$, and $KNO_3$ are all strong electrolytes in solution, we can write their formulas as the separate ionic formulas, giving the following ionic equation:

$$Ag^+(aq) + NO_3^-(aq) + K^+(aq) + Br^-(aq) \longrightarrow AgBr(s) + K^+(aq) + NO_3^-(aq)$$

Finally, $NO_3^-$ and $K^+$ occur on both sides of the equation, so they are spectator ions. They can be eliminated, giving the following net ionic equation:

$$Ag^+(aq) + Br^-(aq) \longrightarrow AgBr(s)$$

**Consider This 5.14**

If a precipitation reaction forms two insoluble products, are there any spectator ions?

**Practice Problem 5.14**

Although soluble barium compounds are toxic in body fluids, barium sulfate is safely used as a contrast agent for X-ray imaging the gastrointestinal (GI) tract because it is insoluble in water. After a slurry of the compound is ingested, the GI tract becomes visible on an X-ray. Barium sulfate can be prepared by mixing solutions of barium chloride and sodium sulfate. Write balanced molecular, ionic, and net ionic equations for the reaction.

**Further Practice:** Questions 5.103 and 5.104 at the end of the chapter

# CHAPTER REVIEW

## KEY CONCEPTS

- Chemical reactions involve a rearrangement of atoms that converts one set of substances (the reactants) into another set of substances (the products).
  - Molecular-level rearrangements during a reaction often result in observable macroscopic changes such as a transfer of heat, a change in color, or the formation of a solid or gas.
  - Chemical reactions are represented symbolically by chemical equations, with the formulas of the reactants and the products separated by an arrow, and with a coefficient preceding each formula to indicate the relative number of molecules or formula units in the reaction.
  - A chemical equation is balanced when the number of atoms of each component element is the same in the reactants as in the products.
- The products of chemical reactions must be identified experimentally or can be predicted for some classes of reactions.
  - In decomposition reactions, complex substances break down into simpler ones. Patterns of reactivity predict the products of decomposition reactions.
  - Combination reactions make more complex substances from simpler ones. The products of reactions between a metal and a nonmetal are ionic compounds whose formulas can be predicted based on the common charges of ions.

- In single-displacement reactions, an element displaces another element from its compound. An activity series predicts whether a single-displacement reaction will occur.
- In double-displacement reactions, two compounds exchange partners, such as cations and anions. Such reactions occur if the products include an insoluble solid (a precipitate), an insoluble gas, or a stable molecule such as water.
- Combustion is the reaction of a substance with oxygen, often rapidly producing heat and flames. When an excess of oxygen is available, the products of the combustion of hydrocarbons are water vapor and carbon dioxide.
- An ionic equation accurately describes reactions that occur in solution, because it represents soluble electrolytes with the formulas of their component ions.
  - Any ions that occur on both sides of an ionic equation are spectator ions, which don't participate in the chemical reaction.
  - A net ionic equation results when the spectator ions are eliminated from an ionic equation.
  - Net ionic equations focus on the chemical changes that occur.

# KEY TERMS

activity series (5.4)
anhydrous (5.4)
balanced equation (5.3)
chemical equation (5.3)
combination reaction (5.4)
combustion reaction (5.4)

corrosion (5.4)
decomposition reaction (5.4)
double-displacement
  reaction (5.4)
ionic equation (5.5)
molecular equation (5.5)

net ionic equation (5.5)
neutralization reaction (5.4)
precipitate (5.4)
precipitation reaction (5.4)
product (5.1)
reactant (5.1)

single-displacement
  reaction (5.4)
spectator ion (5.5)

# QUESTIONS AND PROBLEMS

The following questions and problems, except those in *Additional Questions* and *Concept Review Questions,* are paired. Questions in a pair focus on the same concept. Answers to selected questions and problems are in Appendix D.

## Matching Definitions with Key Terms

5.1    Match the key terms with the descriptions provided.
(a) a reaction in which a free element displaces another element from a compound to produce a different compound and a different free element
(b) a compound free of molecular water
(c) a form of a chemical equation in which substances are represented as if they existed as molecules (using formula units), even though the substances may exist in solution as ions
(d) a chemical reaction in which a substance breaks down into simpler compounds or elements
(e) a chemical equation that has coefficients in front of each formula to make the number of atoms of each element the same on both sides of the equation
(f) a substance that is converted to another substance during a chemical reaction
(g) an ion that appears on both sides of an ionic equation and is therefore not involved in a chemical change
(h) a reaction that involves molecular oxygen as a reactant and that rapidly produces heat and flames
(i) an insoluble solid deposited from a solution

5.2    Match the key terms with the descriptions provided.
(a) a reaction in which two elements, an element and a compound, or two compounds join to form a new compound

(b) a list of metals in order of activity
(c) an abbreviated representation of a chemical reaction consisting of chemical symbols and formulas
(d) a substance formed from another substance during a chemical reaction
(e) a form of a chemical equation in which ionic compounds are represented as the separated ions
(f) a reaction of an acid and a base to form an ionic compound and water
(g) a reaction in which the reactants in solution exchange ions to form a solid
(h) a form of a chemical equation in which ionic compounds are represented as the separated ions, and spectator ions are eliminated
(i) a reaction in which two compounds exchange ions or elements to form new compounds

## What Is a Chemical Reaction?

5.3    Aluminum metal reacts with oxygen gas in the air to give aluminum oxide solid. Identify the reactants and products in this reaction.

5.4    Nitrogen gas and solid iodine form in an explosive reaction when nitrogen triiodide is detonated. Identify the reactants and products in this reaction.

5.5    Xenon gas reacts with fluorine gas to form xenon tetrafluoride gas. Identify which image represents reactants and then which image represents products.

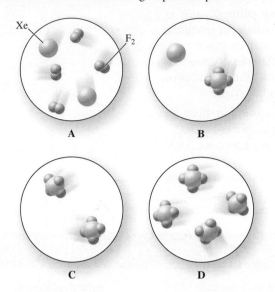

5.6    Ozone gas, $O_3$, reacts with carbon monoxide gas to form oxygen gas and carbon dioxide gas. Identify which image represents reactants and which image represents products.

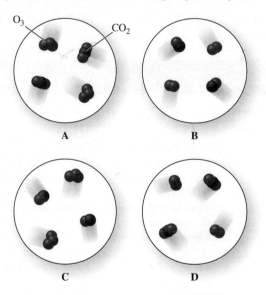

5.7    This molecular-level diagram represents a chemical reaction between nitrogen and hydrogen gases. Is it accurate? If not, what is wrong with it and how would you fix it?

5.8    This molecular-level diagram represents a chemical reaction between hydrogen and oxygen gases. Is it accurate? If not, what is wrong with it and how would you fix it?

5.9    Examine the following molecular-level diagram of a chemical reaction. The product image is incomplete. How would you modify this to show that the law of conservation of mass is obeyed?

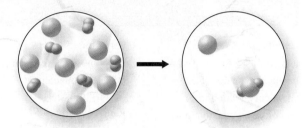

5.10   Examine the following molecular-level diagram of a chemical reaction. The product image is incomplete. How would you modify this to show that the law of conservation of mass is obeyed?

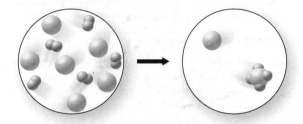

5.11   Sulfur dioxide reacts with oxygen to form sulfur trioxide. Consider the following molecular diagram of a mixture of reactants. Draw a diagram of the products when the reaction is complete.

**5.12**    Nitrogen monoxide reacts with oxygen to form nitrogen dioxide. Consider the following molecular diagram of a mixture of reactants. Draw a diagram of the products when the reaction is complete.

## How Do We Know a Chemical Reaction Occurs?

**5.13**    Examine the following photograph. Identify any evidence that a chemical reaction may be occurring.

©McGraw-Hill Education/Stephen Frisch

**5.14**    Examine the following photograph. Identify any evidence that a chemical reaction may be occurring.

©Charles D. Winters/Science Source

**5.15**    A sample of dry ice (solid $CO_2$) starts to form a white cloud upon standing. Gaseous carbon dioxide is formed. Is this change a chemical reaction? Explain your answer.

**5.16**    A piece of copper metal is placed in a solution of silver nitrate. The solution turns blue, and a black deposit coats the copper metal. Is this change a chemical reaction? Explain your answer.

**5.17**    Examine the following molecular-level diagram. Is a chemical reaction occurring? Explain your answer.

**5.18**    Examine the following molecular-level diagram. Is a chemical reaction occurring? Explain your answer.

**5.19**    Examine the following molecular-level diagram. Is a chemical reaction occurring? Explain your answer.

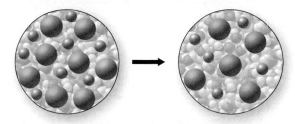

**5.20**    Examine the following molecular-level diagram. Is a chemical reaction occurring? Explain your answer.

## Writing Chemical Equations

**5.21**    What is a chemical equation?

**5.22**    What is the difference between a chemical equation and a chemical reaction?

**5.23**    Consider the following equations. For each, decide whether it represents a chemical reaction or a physical change.
   (a) $CO_2(g) + H_2O(l) \longrightarrow H_2CO_3(aq)$
   (b) $H_2O(s) \longrightarrow H_2O(l)$
   (c) $HOCN(g) \longrightarrow HCNO(g)$

5.24 Consider the following equations. For each, decide whether it represents a chemical reaction or a physical change.

(a) $TiCl_4(aq) + 2H_2O(l) \longrightarrow TiO_2(s) + 4HCl(aq)$

(b) $4Li(s) + O_2(g) \longrightarrow 2Li_2O(s)$

(c) $KBr(s) \longrightarrow K^+(aq) + Br^-(aq)$

5.25 Why must a chemical equation be balanced?

5.26 Why can't subscripts be changed to balance a chemical equation?

5.27 Write complete, balanced equations for each of the following reactions.

(a) When solid sodium hydride, NaH, is added to water, hydrogen gas is released and aqueous sodium hydroxide is formed.

(b) Aluminum metal reacts with chlorine gas to form solid aluminum chloride.

5.28 Write complete, balanced equations for each of the following reactions.

(a) Sodium metal reacts with oxygen gas to form solid sodium oxide.

(b) Solid copper(II) nitrate is heated to produce solid copper(II) oxide, gaseous nitrogen dioxide, and oxygen gas.

5.29 Consider the following molecular-level diagram of a chemical reaction between $N_2(g)$ and $Cl_2(g)$. Write a balanced equation to represent this reaction.

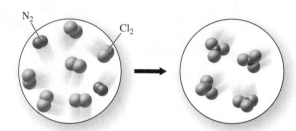

5.30 Consider the following molecular-level diagram of a chemical reaction between $CH_4(g)$ and $O_2(g)$. Write a balanced equation to represent this reaction.

5.31 A piece of magnesium metal is ignited and burns with an intense white flame, leaving behind a white ashlike substance. Consider the following molecular diagrams. Which matches this description of a chemical reaction? Write an equation to represent this reaction.

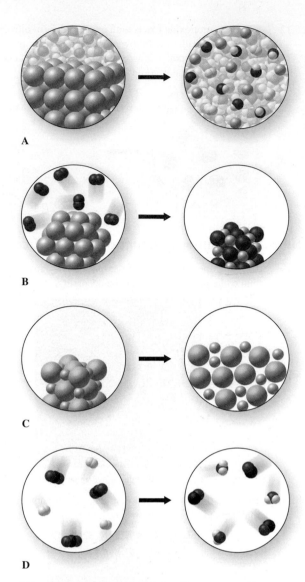

**A**

**B**

**C**

**D**

5.32 When the silver-colored metal sodium reacts with water, it forms a solution of sodium hydroxide and a molecular gas bubbles out of the solution. Consider the molecular diagrams shown in Question 5.31. Which matches this description of a chemical reaction? Write an equation to represent this reaction.

5.33 Consider the incomplete molecular-level diagram of the following reaction:

$$H_2(g) + I_2(g) \longrightarrow 2HI(g)$$

Complete this diagram to show the products expected for this reaction.

**5.34** Consider the incomplete molecular-level diagram of the following reaction:

$$2NO(g) + 2CO(g) \longrightarrow N_2(g) + 2CO_2(g)$$

Complete this diagram to show the products expected for this reaction.

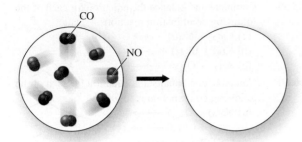

**5.35** Consider the following equation:

$$2NO_2(g) \longrightarrow N_2O_4(g)$$

Draw a molecular-level diagram that shows the reactants and the products for this reaction.

**5.36** Consider the following equation:

$$2F_2(g) + 2H_2O(g) \longrightarrow 4HF(g) + O_2(g)$$

Draw a molecular-level diagram that shows the reactants and the products for this reaction.

**5.37** Balance each of the following skeletal equations:
(a) $Al(s) + Cl_2(g) \longrightarrow AlCl_3(s)$
(b) $Pb(NO_3)_2(aq) + K_2CrO_4(aq) \longrightarrow$
$\qquad\qquad\qquad PbCrO_4(aq) + KNO_3(aq)$
(c) $Li(s) + H_2O(l) \longrightarrow LiOH(aq) + H_2(g)$
(d) $C_6H_{14}(g) + O_2(g) \longrightarrow CO_2(g) + H_2O(g)$

**5.38** Balance each of the following skeletal equations:
(a) $Mg(s) + O_2(g) \longrightarrow MgO(s)$
(b) $Zn(s) + AgNO_3(aq) \longrightarrow Zn(NO_3)_2(aq) + Ag(s)$
(c) $C_2H_2(g) + O_2(g) \longrightarrow CO_2(g) + H_2O(g)$
(d) $C_{12}H_{22}O_{11}(s) + KClO_3(l) \longrightarrow$
$\qquad\qquad\qquad CO_2(g) + H_2O(g) + KCl(s)$

**5.39** Balance each of the following skeletal equations:
(a) $CuCl_2(aq) + AgNO_3(aq) \longrightarrow$
$\qquad\qquad\qquad Cu(NO_3)_2(aq) + AgCl(s)$
(b) $S_8(s) + O_2(g) \longrightarrow SO_2(g)$
(c) $C_3H_8(g) + O_2(g) \longrightarrow CO_2(g) + H_2O(g)$

**5.40** Balance each of the following skeletal equations:
(a) $Al_2O_3(s) + HCl(aq) \longrightarrow AlCl_3(aq) + H_2O(l)$
(b) $PCl_3(l) + AgF(s) \longrightarrow PF_3(g) + AgCl(s)$
(c) $NO_2(g) \longrightarrow NO(g) + O_2(g)$

**5.41** Write a balanced equation that corresponds to the following word equation for a reaction in aqueous solution:

copper + silver nitrate $\longrightarrow$ copper(II) nitrate + silver

**5.42** Write a balanced equation that corresponds to the following word equation for a reaction in aqueous solution:

barium chloride + sulfuric acid $\longrightarrow$

$\qquad\qquad$ barium sulfate + hydrochloric acid

## Predicting Chemical Reactions

**5.43** Describe the identifying characteristics of the following classes of reactions: decomposition, combination, single-displacement, and double-displacement.

**5.44** Give examples of each of the following classes of reactions: decomposition, combination, single-displacement, and double-displacement.

**5.45** Determine which classes of reactions would fit into each of the following categories for the reactants and products. Choices are decomposition, combination, single-displacement, double-displacement, and combustion.

|     | Reactants | Products |
|-----|-----------|----------|
| (a) | Two elements | One compound |
| (b) | One element and one compound | One element and one compound |
| (c) | One compound | Two elements |

**5.46** Determine which classes of reactions would fit into each of the following categories for the reactants and products. Choices are decomposition, combination, single-displacement, double-displacement, and combustion.

|     | Reactants | Products |
|-----|-----------|----------|
| (a) | One compound | One element and one compound |
| (b) | Two compounds | Two compounds |
| (c) | Two compounds | One compound |

**5.47** A solution of sodium chloride is mixed with a solution of lead(II) nitrate. A precipitate of lead(II) chloride results, leaving a solution of sodium nitrate. Into which class(es) does this reaction fit? Choices are decomposition, combination, single-displacement, double-displacement, and combustion.

**5.48** Solid sulfur is ignited and burns in oxygen gas with a blue flame to form sulfur dioxide. Into which class(es) does this reaction fit? Choices are decomposition, combination, single-displacement, double-displacement, and combustion.

**5.49** Consider the following molecular-level diagrams of chemical reactions. Identify each as decomposition, combination, single-displacement, or double-displacement.

(a)

(b)

5.50    Consider the following molecular-level diagrams of chemical reactions. Identify each as decomposition, combination, single-displacement, or double-displacement.

(a)

(b)

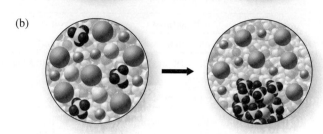

5.51    Balance the following chemical equations. Classify the reactions as decomposition, combination, single-displacement, double-displacement, combustion, or none of these.

(a) $CaCl_2(aq) + Na_2SO_4(aq) \longrightarrow$
$\phantom{xxxxxxxxxxxxx} CaSO_4(s) + NaCl(aq)$

(b) $Ba(s) + HCl(aq) \longrightarrow BaCl_2(aq) + H_2(g)$

(c) $N_2(g) + H_2(g) \longrightarrow NH_3(g)$

(d) $FeO(s) + CO(g) \xrightarrow{heat} Fe(s) + CO_2(g)$

(e) $CaO(s) + H_2O(l) \longrightarrow Ca(OH)_2(aq)$

(f) $Na_2CrO_4(aq) + Pb(NO_3)_2(aq) \longrightarrow$
$\phantom{xxxxxxxxxxxxx} PbCrO_4(s) + NaNO_3(aq)$

(g) $KI(aq) + Cl_2(g) \longrightarrow KCl(aq) + I_2(aq)$

(h) $NaHCO_3(s) \xrightarrow{heat} Na_2CO_3(s) + CO_2(g) + H_2O(g)$

5.52    Balance the following equations and classify the reactions as decomposition, combination, single-displacement, double-displacement, or combustion.

(a) $GaH_3 + N(CH_3)_3 \longrightarrow (CH_3)_3NGaH_3$

(b) $Ca(s) + H_2O(l) \longrightarrow Ca(OH)_2(aq) + H_2(g)$

(c) $N_2(g) + CaC_2(s) \longrightarrow C(s) + CaNCN(s)$

(d) $N_2(g) + Mg(s) \longrightarrow Mg_3N_2(s)$

(e) $NH_4Cl(s) \longrightarrow NH_3(g) + HCl(g)$

(f) $CaO(s) + SO_3(g) \xrightarrow{heat} CaSO_4(s)$

(g) $PCl_5(g) \xrightarrow{heat} PCl_3(g) + Cl_2(g)$

(h) $Ca_3N_2(s) + H_2O(l) \xrightarrow{heat} Ca(OH)_2(aq) + NH_3(g)$

5.53    When heated, nickel(II) carbonate undergoes a decomposition reaction. Write a balanced equation to describe this reaction.

5.54    When heated, platinum(IV) chloride undergoes a decomposition reaction. Write a balanced equation to describe this reaction.

5.55    Complete and balance the equation for each of the following decomposition reactions.

(a) $CaCO_3(s) \xrightarrow{heat}$

(b) $CuSO_4 \cdot 5H_2O(s) \xrightarrow{heat}$

5.56    Complete and balance the equation for each of the following decomposition reactions. [The products in part (b) are elements.]

(a) $Cu(OH)_2(s) \xrightarrow{heat}$

(b) $NaN_3(s) \longrightarrow$

5.57    Magnesium metal reacts with oxygen gas in a combination reaction. Write a balanced equation to describe this reaction.

5.58    Sodium metal reacts with chlorine gas in a combination reaction. Write a balanced equation to describe this reaction.

5.59    Complete and balance the equation for each of the following combination reactions.

(a) $Ca(s) + N_2(g) \longrightarrow$

(b) $K(s) + Br_2(l) \longrightarrow$

(c) $Al(s) + O_2(g) \longrightarrow$

5.60    Complete and balance the equation for each of the following combination reactions.

(a) $Na(s) + O_2(g) \longrightarrow$

(b) $Al(s) + Cl_2(g) \longrightarrow$

(c) $Ca(s) + F_2(g) \longrightarrow$

5.61    Complete and balance the equations for each of the following single-displacement reactions.

(a) $Zn(s) + AgNO_3(aq) \longrightarrow$

(b) $Na(s) + FeCl_2(s) \xrightarrow{heat}$

5.62    Complete and balance each of the following single-displacement reactions.

(a) $Al(s) + CuSO_4(aq) \longrightarrow$

(b) $Cs(s) + H_2O(l) \longrightarrow$

5.63    Zinc metal reacts with a solution of tin(II) chloride in a single-displacement reaction. Write a balanced equation to describe this reaction.

5.64    Magnesium metal reacts with a solution of copper(II) sulfate in a single-displacement reaction. Write a balanced equation to describe this reaction.

5.65    Consider the following metals. Determine which will react with water and which will react with a solution of hydrochloric acid. Write balanced equations to describe any reactions that occur.

(a) Ca        (b) Fe        (c) Cu

5.66    Consider the following metals. Determine which will react with water and which will react with a solution of hydrochloric acid. Write balanced equations to describe any reactions that occur.

(a) Cr        (b) Bi        (c) K

5.67    Which of the following single-displacement reactions will actually occur if the indicated reactants are mixed?

(a) $3Mg(s) + 2AlCl_3(aq) \longrightarrow 3MgCl_2(aq) + 2Al(s)$

(b) $Zn(s) + MgCl_2(aq) \longrightarrow ZnCl_2(aq) + Mg(s)$

(c) $Cu(s) + Pb(NO_3)_2(aq) \longrightarrow Cu(NO_3)_2(aq) + Pb(s)$

(d) $Ni(s) + 2AgNO_3(aq) \longrightarrow Ni(NO_3)_2(aq) + 2Ag(s)$

5.68    Which of the following single-displacement reactions will actually occur if the indicated reactants are mixed?

(a) $Ni(s) + FeCl_2(aq) \longrightarrow NiCl_2(aq) + Fe(s)$

(b) $2Al(s) + 3NiCl_2(aq) \longrightarrow 2AlCl_3(aq) + 3Ni(s)$

(c) $Fe(s) + CuCl_2(aq) \longrightarrow FeCl_2(aq) + Cu(s)$

(d) $3Sn(s) + 2AlCl_3(aq) \longrightarrow 3SnCl_2(aq) + 2Al(s)$

5.69    Which of the following ionic compounds is expected to be soluble in water?

(a) $CuCl_2$

(b) $AgNO_3$

(c) $PbCl_2$

(d) $Cu(OH)_2$

5.70 Which of the following ionic compounds is expected to be insoluble when mixed with water?
(a) $Cr_2S_3$
(b) $Ca(OH)_2$
(c) $BaSO_4$
(d) $(NH_4)_2CO_3$

5.71 Write a balanced equation to describe any precipitation reaction that should occur when the following substances are mixed.
(a) $K_2CO_3(aq)$ and $BaCl_2(aq)$
(b) $CaS(aq)$ and $Hg(NO_3)_2(aq)$
(c) $Pb(NO_3)_2(aq)$ and $K_2SO_4(aq)$

5.72 Write a balanced equation to describe any precipitation reaction that should occur when the following substances are mixed.
(a) $MgSO_4(aq)$ and $BaCl_2(aq)$
(b) $K_2SO_4(aq)$ and $MgCl_2(aq)$
(c) $MgCl_2(aq)$ and $Pb(NO_3)_2(aq)$

5.73 Complete and balance the equation for each of the following double-displacement reactions.
(a) $BaCO_3(s) + H_2SO_4(aq) \longrightarrow$
(b) $CuCl_2(aq) + AgNO_3(aq) \longrightarrow$

5.74 Complete and balance the equation for each of the following double-displacement reactions.
(a) $NaOH(aq) + CuCl_2(aq) \longrightarrow$
(b) $H_2SO_4(aq) + KOH(aq) \longrightarrow$

5.75 Aqueous solutions of sodium sulfate and lead(II) nitrate are mixed and a white solid forms. Identify the solid.

5.76 Aqueous solutions of mercury(I) nitrate, $Hg_2(NO_3)_2$, and calcium iodide, $CaI_2$, are mixed and a yellow solid forms. Identify the solid. [Note that the mercury(I) ion exists as $Hg_2^{2+}$.]

5.77 Aqueous calcium chloride reacts with aqueous potassium carbonate in a double-displacement reaction. Write a balanced equation to describe this reaction.

5.78 Aqueous ammonium chromate reacts with aqueous lead(II) nitrate in a double-displacement reaction. Write a balanced equation to describe this reaction.

5.79 Consider the following double-displacement reactions. For each, identify the driving force that causes the reaction to go to completion.
(a) $Hg(NO_3)_2(aq) + H_2S(aq) \longrightarrow$
$HgS(s) + 2HNO_3(aq)$
(b) $MnS(s) + 2HCl(aq) \longrightarrow MnCl_2(aq) + H_2S(g)$
(c) $Ba(OH)_2(aq) + H_2SO_4(aq) \longrightarrow$
$BaSO_4(s) + 2H_2O(l)$

5.80 Consider the following double-displacement reactions. For each, identify the driving force that causes the reaction to go to completion.
(a) $Al_2O_3(s) + 3H_2SO_4(aq) \longrightarrow$
$Al_2(SO_4)_3(aq) + 3H_2O(l)$
(b) $CaCO_3(s) + H_2SO_4(aq) \longrightarrow$
$CaSO_4(s) + H_2O(l) + CO_2(g)$
(c) $BaCl_2(aq) + Na_2SO_4(aq) \longrightarrow$
$BaSO_4(s) + 2NaCl(aq)$

5.81 Write a balanced equation to describe any acid-base neutralization reaction that might occur when the following substances are mixed.
(a) $H_2S(aq)$ and $Cu(OH)_2(s)$
(b) $CH_4(g)$ and $NaOH(aq)$
(c) $KHSO_4(aq)$ and $KOH(aq)$

5.82 Write a balanced equation to describe any acid-base neutralization reaction that might occur when the following substances are mixed.
(a) $H_2(g)$ and $KOH(aq)$
(b) $H_3PO_4(aq)$ and $NaOH(aq)$
(c) $H_2SO_4(aq)$ and $Fe(OH)_2(s)$

5.83 What substance is always a reactant in a combustion reaction?

5.84 What are the usual products when a hydrocarbon, or a hydrocarbon containing oxygen, react in a combustion reaction?

5.85 Write a formula for the product of the combustion reaction of the following elements with molecular oxygen.
(a) Cs    (b) Pb    (c) Al    (d) $H_2$    (e) C

5.86 Write a formula for the product of the combustion reaction of the following elements with molecular oxygen.
(a) Mg    (b) Fe    (c) Li    (d) $S_8$    (e) $N_2$

5.87 Write formulas for the products of the reactions of the following substances with molecular oxygen.
(a) $CH_4$
(b) CO
(c) Al
(d) $CH_3OH$

5.88 Write formulas for the products of the reactions of the following compounds with molecular oxygen.
(a) Na
(b) P
(c) $C_3H_8$
(d) $HCO_2H$

## Representing Reactions in Aqueous Solution

5.89 Distinguish between an electrolyte and a nonelectrolyte.

5.90 Describe an experimental method for determining whether a solid is an electrolyte or a nonelectrolyte.

5.91 Indicate whether each of the following substances is an electrolyte or a nonelectrolyte.
(a) $NaOH(aq)$
(b) $HCl(aq)$
(c) $C_{12}H_{22}O_{11}(aq)$ (sucrose solution)

5.92 Indicate whether each of the following substances is an electrolyte or a nonelectrolyte.
(a) $CH_3CH_2OH(aq)$
(b) $H_2O(l)$
(c) $NaCl(aq)$

5.93    Consider the following molecular-level diagram of a solution of a compound in water. Is this substance an electrolyte or a nonelectrolyte?

5.94    Consider the following molecular-level diagram of a solution of a compound in water. Is this substance an electrolyte or a nonelectrolyte?

5.95    Indicate whether aqueous solutions of the following substances contain ions. If so, write formulas for the ions in solution.
(a) $C_6H_{12}O_6$ (glucose)
(b) $CH_4$
(c) NaCl

5.96    Indicate whether aqueous solutions of the following substances contain ions. If so, write formulas for the ions in solution.
(a) $I_2$
(b) KI
(c) $CH_3OH$

5.97    Distinguish between a molecular, an ionic, and a net ionic equation.

5.98    Why is it necessary to identify a substance as an electrolyte or a nonelectrolyte when writing a net ionic equation?

5.99    What are spectator ions?

5.100   In which type of equation—molecular, ionic, or net ionic—do we find spectator ions?

5.101   Add the physical state for each substance in the following aqueous reactions. Then write and balance a net ionic equation for each reaction.
(a) $NaCl + Ag_2SO_4 \longrightarrow Na_2SO_4 + AgCl$
(b) $Cu(OH)_2 + HCl \longrightarrow CuCl_2 + H_2O$
(c) $BaCl_2 + Ag_2SO_4 \longrightarrow BaSO_4 + AgCl$

5.102   Add the physical state for each substance in the following aqueous reactions. Then write and balance a net ionic equation for each reaction.
(a) $Cr_2(SO_4)_3 + KOH \longrightarrow Cr(OH)_3 + K_2SO_4$
(b) $K_2CrO_4 + PbCl_2 \longrightarrow KCl + PbCrO_4$
(c) $Na_2SO_3 + H_2SO_4 \longrightarrow Na_2SO_4 + H_2O + SO_2$

5.103   When aqueous solutions of sodium carbonate and calcium bromide mix, a white solid forms.
(a) Identify the solid.
(b) Write a balanced molecular equation for the reaction.
(c) Write a balanced ionic equation for the reaction.
(d) Identify any spectator ions in the reaction.
(e) Write a balanced net ionic equation for the reaction.

5.104   When aqueous solutions of potassium chromate and barium chloride mix, a yellow solid forms.
(a) Identify the solid.
(b) Write a balanced molecular equation for the reaction.
(c) Write a balanced ionic equation for the reaction.
(d) Identify any spectator ions in the reaction.
(e) Write a balanced net ionic equation for the reaction.

5.105   Aqueous calcium chloride reacts with aqueous silver nitrate to form a precipitate of silver chloride and a solution of calcium nitrate. Write a net ionic equation for this reaction.

5.106   Aqueous chromium(III) sulfate reacts with aqueous sodium hydroxide to form a precipitate of chromium(III) hydroxide and a solution of sodium sulfate. Write a net ionic equation for this reaction.

5.107   Consider the following molecular-level diagram for the reaction between Na and $H_2O$ in aqueous solution. Write a net ionic equation for this reaction.

5.108   Consider the following molecular-level diagram for the reaction between Cu and $AgNO_3$ in aqueous solution. Write a net ionic equation for this reaction.

5.109   Write balanced molecular and net ionic equations for the following reactions in aqueous solution.
(a) copper + silver nitrate $\longrightarrow$ copper(II) nitrate + silver
(b) iron(II) oxide + hydrochloric acid $\longrightarrow$ iron(II) chloride + water

5.110   Write balanced molecular and net ionic equations for the following reactions in aqueous solution.
(a) calcium + water $\longrightarrow$ calcium hydroxide + hydrogen
(b) barium chloride + sodium sulfate $\longrightarrow$ barium sulfate + sodium chloride

5.111 Predict whether reactions should occur between aqueous solutions of the following compounds. If so, write balanced molecular and net ionic equations for the reactions. Include physical states.
(a) $Sr(NO_3)_2$ and $H_2SO_4$
(b) $Zn(NO_3)_2$ and $Na_2SO_4$
(c) $CuSO_4$ and $BaS$
(d) $NaHCO_3$ and $CH_3CO_2H$

5.112 Predict whether reactions should occur between aqueous solutions of the following compounds. If so, write balanced molecular and net ionic equations for the reactions. Include physical states.
(a) $(NH_4)_2CO_3$ and $CrCl_3$
(b) $Ba(OH)_2$ and $HCl$
(c) $FeCl_3$ and $NaOH$
(d) $Ca(HCO_3)_2$ and $HNO_3$

## Additional Questions

5.113 Bismuth selenide ($Bi_2Se_3$) is used in semiconductor research. It can be prepared directly from its elements.

$$2Bi + 3Se \longrightarrow Bi_2Se_3$$

Classify the reaction as decomposition, combination, single-displacement, double-displacement, or combustion.

5.114 Heating causes the copper mineral malachite, $CuCO_3 \cdot Cu(OH)_2$, to decompose. Write a balanced equation for the reaction.

5.115 Write a balanced equation to describe any precipitation reaction that might occur when the following substances are mixed.
(a) $ZnSO_4(aq)$ and $Ba(NO_3)_2(aq)$
(b) $Ca(NO_3)_2(aq)$ and $K_3PO_4(aq)$
(c) $ZnSO_4(aq)$ and $BaCl_2(aq)$
(d) $KOH(aq)$ and $MgCl_2(aq)$
(e) $CuSO_4(aq)$ and $BaS(aq)$

5.116 Indicate whether each of the following substances reacts with hydrochloric acid to form water.
(a) $NaOH(aq)$
(b) $Zn(s)$
(c) $CaCO_3(s)$
(d) $Cu(OH)_2(s)$

5.117 Sodium metal reacts with liquid water to form a solution of sodium hydroxide and hydrogen gas. Potassium reacts in a similar manner. Write a balanced equation to represent the reaction of potassium with water.

5.118 The solid compounds $NaCl$ and $Cu(NO_3)_2$ are added to a beaker of water. Describe what happens. Include in your description the identities of any substances that dissolve and any substances that precipitate.

5.119 An aqueous solution contains silver nitrate. Which of the following substances could be added to this solution to precipitate the silver ions as silver chloride: $Cl_2(g)$, $NaCl(s)$, or $CCl_4(l)$?

5.120 A solution contains the ionic compounds $BaCl_2$ and $NaCl$. What ionic compound could be added that would remove the barium ions from the solution but not introduce any new kinds of ions?

5.121 Consider the following net ionic equation:

$$H^+(aq) + OH^-(aq) \longrightarrow H_2O(l)$$

Make a list of three different sets of two chemical compounds, each of which would result in this net ionic equation when they are mixed.

5.122 A solution contains cobalt(II) chloride. Identify three metals that could be placed in the solution to cause the cobalt to be plated out as a metal.

5.123 What reaction, if any, occurs if a piece of copper metal is placed in an aqueous solution containing iron(II) sulfate?

5.124 A solution contains both copper(II) nitrate and lead(II) nitrate. What substance could be added to the solution to precipitate the lead(II) ions, but leave the copper(II) ions in solution? After removal of the lead(II) compound, what could then be added to the solution to precipitate the copper(II) ions?

5.125 Draw molecular-level diagrams to show that a solution of KI is a strong electrolyte and that a solution of CO is a nonelectrolyte.

5.126 Write equations for reactions that could be used to release each of the following elements from a compound that contains the element.
(a) $O_2$
(b) $H_2$
(c) $Cl_2$

5.127 Identify a substance that could be used to release the following metals from a compound that contains the element, using a different substance for each metal.
(a) Cr
(b) Fe
(c) Al
(d) Pb

5.128 Suggest chemical reactions that could be used to identify the contents of four unlabeled bottles containing solutions of the following substances: $CuSO_4$, $Al_2(SO_4)_3$, $Pb(NO_3)_2$, and $FeSO_4$.

5.129 Suggest chemical methods that would separate the following ions or compounds from one another.
(a) $Fe^{3+}$ and $Al^{3+}$
(b) $Ba^{2+}$ and $Mg^{2+}$
(c) $Cl^-$ and $NO_3^-$
(d) $BaSO_4$ and $MgSO_4$

5.130 When copper metal is exposed to the atmosphere over a long period of time, it takes on a green patina (a coating of copper compounds). The green color of copper roofs and the Statue of Liberty are examples. Initially the copper metal reacts with oxygen gas. The product of this reaction reacts in part with water vapor and in part with carbon dioxide in the air. Predict the products and write balanced equations for the three reactions.

5.131 A solution contains $AgNO_3$ and $Ba(NO_3)_2$. What substance could be added to the solution to precipitate $Ag^+$ ions, but leave $Ba^{2+}$ ions in solution? After removal of the $Ag^+$ compound, what could then be added to the solution to precipitate the $Ba^{2+}$ ions?

**5.132** The active ingredient in a well-known antacid containing magnesium is $Mg(OH)_2$. Magnesium hydroxide is insoluble in water, so the product is a mixture of water, flavoring, and other ingredients to suspend the solid $Mg(OH)_2$. When someone suffering from acid indigestion takes this drug, the $Mg(OH)_2$ reacts with the HCl in the stomach. Classify this chemical reaction and write a balanced equation.

**5.133** Describe what you would expect to observe, if anything, when a piece of nickel is placed in a solution of copper(II) nitrate.

**5.134** Describe what you would expect to observe, if anything, when a piece of silver is placed in a solution of copper(II) nitrate.

**5.135** When solutions of hydrochloric acid and sodium hydroxide are mixed, a chemical reaction occurs forming aqueous sodium chloride and water. What would you expect to observe if you ran the reaction in the laboratory?

**5.136** Barium sulfate is used clinically to provide X-ray images of the digestive tract. How might barium sulfate be prepared in the laboratory from a double-displacement reaction?

**5.137** The chemical reaction that occurs in a gas grill is the combustion of propane, $C_3H_8$. Write a balanced equation for this reaction.

**5.138** Some campers and military personnel carry MREs (Meals, Ready to Eat) so that they can prepare a hot meal when a fire cannot be used. The MREs contain a flameless heater in which the heat given off by a chemical reaction is used to cook the food. The chemical reaction that produces the heat is a single-displacement reaction between magnesium and water. Write a balanced chemical equation for this reaction.

**5.139** When a piece of copper is placed in a solution of $AgNO_3$, a single-displacement reaction occurs in which silver metal and copper(II) nitrate form. What happens to the charges on the species involved in this chemical reaction?

**5.140** Barium ions are highly toxic to the human body, yet barium sulfate is ingested to identify gastrointestinal problems through an X-ray. What property of barium sulfate allows it to be ingested without producing aqueous barium ions in the body?

**5.141** Balance the following chemical equations.
(a) $C_4H_{10}(g) + O_2(g) \longrightarrow CO_2(g) + H_2O(g)$
(b) $C_2H_5OH(l) + O_2(g) \longrightarrow CO_2(g) + H_2O(g)$
(c) $HCO_2H(l) + O_2(g) \longrightarrow CO_2(g) + H_2O(g)$
(d) $C_5H_{12}(l) + O_2(g) \longrightarrow CO_2(g) + H_2O(g)$
(e) $CH_3OCH_3(g) + O_2(g) \longrightarrow CO_2(g) + H_2O(g)$

## Concept Review Questions

**5.142** Consider the following chemical reaction:

$$FeO(s) + H_2SO_4(aq) \longrightarrow FeSO_4(aq) + H_2O(l)$$

Identify the products of this reaction.
A. sulfuric acid and iron(II) oxide
B. sulfuric acid and iron(II) sulfate
C. iron(II) oxide and water
D. iron(II) sulfate and water
E. iron(II) oxide and iron(II) sulfate

For each incorrect response, identify each of the listed components as a reactant or a product.

**5.143** All of the following may change during a chemical reaction except
A. the total number of atoms in the system.
B. the temperature of the system.
C. the color of the system.
D. the total number of molecules in the system.
E. the physical state of the system.

For those properties that might change, give an example of a chemical reaction that illustrates this change.

**5.144** Which of the following changes represents a physical change rather than a chemical reaction?
A. When solutions of $AgNO_3$ and NaCl are mixed, a white solid forms.
B. When copper metal is placed in a colorless $AgNO_3$ solution, a grey coating is formed on the surface of the copper, and a blue solution is formed.
C. When water and a green solution of antifreeze are mixed, the resulting solution is a lighter shade of green.
D. When zinc metal is placed in a hydrochloric acid solution, a colorless gas is formed and the zinc dissolves.
E. When a mixture of oxygen and hydrogen gases is ignited, an explosion occurs.

For each incorrect response, write a balanced chemical equation to represent the reaction being described.

**5.145** When the mixture of molecules shown in the diagram below undergoes complete reaction, all of these molecules are converted to products. Which of the following reactions could this represent?

A. $N_2(g) + O_2(g) \longrightarrow 2NO(g)$
B. $N_2(g) + 2Cl_2(g) \longrightarrow N_2Cl_4(g)$
C. $N_2(g) + 2O_2(g) \longrightarrow 2NO_2(g)$
D. $N_2(g) + 3H_2(g) \longrightarrow 2NH_3(g)$
E. $N_2(g) + 3O_2(g) \longrightarrow 2NO_3(g)$

Draw molecular-level images that would represent the reactants in each of the incorrect responses.

**5.146** Consider the following incomplete chemical equation:

$$2Na(s) + Cl_2(g) \longrightarrow$$

Which of the following corresponds to the reaction product coefficient and formula?
A. $NaCl_2(s)$
B. $2NaCl(s)$
C. $Na_2Cl(s)$
D. $2Na_2Cl(s)$
E. $Na_2Cl_2(s)$

For each of the incorrect responses, indicate why it is incorrect.

5.147    Consider the following diagram of a chemical reaction.

Reactants                          Products

This type of reaction can be classified as a
  A. combination reaction.
  B. double-displacement reaction.
  C. single-displacement reaction.
  D. combustion reaction.
  E. decomposition reaction.

If the reaction occurred in the reverse direction, to what class or classes of reaction could it belong?

5.148    Predict the products if any reaction occurs when $Mg(s)$ and $NiCl_2(aq)$ are mixed.
  A. $MgCl(aq)$ and $NiCl(aq)$
  B. $Ni(s)$ and $Cl_2(g)$
  C. $Mg(s)$ and $Ni(s)$
  D. $MgCl_2(aq)$ and $Ni(s)$
  E. No reaction occurs.

Describe what is wrong with the incorrect responses.

5.149    Which of the following pairs of reactants could participate in a combustion reaction? Select all that apply.
  A. $Zn(s) + O_2(g)$
  B. $C(s) + O_2(g)$
  C. $S(s) + H_2(g)$
  D. $O_2(g) + H_2(g)$
  E. $H_2(g) + Cl_2(g)$

For those that would not be classified as a combustion reaction, suggest a change in one reactant that would result in a combustion reaction.

5.150    Select the balanced net ionic equation for the reaction that occurs when aqueous solutions of $Li_2CO_3$ and $CaCl_2$ are mixed.
  A. $2Li^+(aq) + CO_3^{2-}(aq) \longrightarrow Li_2CO_3(aq)$
  B. $2Li^+(aq) + 2Cl^-(aq) \longrightarrow 2LiCl(aq)$
  C. $Li^+(aq) + Cl^-(aq) \longrightarrow LiCl(aq)$
  D. $Ca^{2+}(aq) + CO_3^{2-}(aq) \longrightarrow CaCO_3(s)$
  E. $Li_2CO_3(aq) + CaCl_2(aq) \longrightarrow$
                        $2LiCl(aq) + CaCO_3(s)$

For each incorrect response, indicate what is wrong with that equation.

5.151    Aqueous solutions of $CoBr_2$ and $AgNO_3$ are combined to form a precipitate of $AgBr$. Identify any spectator ions in this reaction.
  A. $Co^{2+}$ and $NO_3^-$
  B. $NO_3^-$ and $Br^-$
  C. $Co^{2+}$ and $Ag^+$
  D. $Br^-$
  E. $NO_3^-$

Explain why the other choices are incorrect.

# Quantities in Chemical Reactions

6.1 The Meaning of a Balanced Equation

6.2 Mole-Mole Conversions

6.3 Mass-Mass Conversions

6.4 Limiting Reactants

6.5 Percent Yield

6.6 Energy Changes

6.7 Heat Changes in Chemical Reactions

Chapter Review

Questions and Problems

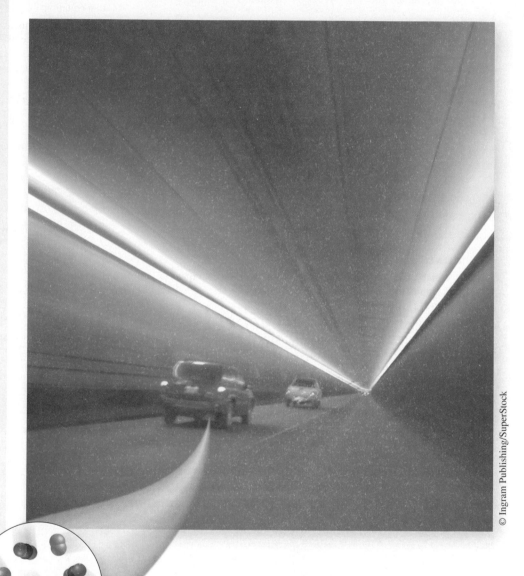

© Ingram Publishing/SuperStock

Imagine the car you will drive five years from now. What will power it? Will it be gasoline? Ethanol? An electric battery? Hydrogen?

Gasoline, a form of petroleum, has been used to power the petroleum-fueled internal-combustion engine (ICE) of automobiles since their invention in the 1860s. Despite steady improvements to the ICE, today's ICE vehicles are only 20%–25% efficient. This means that 75%–80% of the energy in gasoline is wasted. This is not the only disadvantage. The combustion of gasoline in the ICE releases pollutants into the atmosphere, and our reliance on gasoline for transportation places a heavy demand on a limited—and rapidly dwindling—natural resource. Although the auto industry has found ways to significantly reduce many toxic gas emissions, the nontoxic carbon dioxide that is released instead is a major factor in the warming of Earth. It is a contributor to the "greenhouse effect," which may be responsible for the recent increase of Earth's average temperature (often called "global warming").

Because Lily is concerned about the environmental impacts of burning fuels, she drives a car that runs on an alternative fuel, an ethanol blend called E-85. It is 85% ethanol and 15% gasoline. She chose the car because ethanol is a cleaner-burning fuel than gasoline. It can be made from the fermentation of corn or any other carbohydrate. There are concerns, however, that the production of ethanol is worse for the environment than burning regular gasoline.

Joel drives a battery-powered electric vehicle (EV), another alternative to the ICE. Instead of filling up with gasoline, Joel recharges his car battery. His engine is about 60% efficient, has fewer mechanical parts than an ICE, and gives off no tail-pipe emissions. However, the electricity that recharges Joel's car battery comes from a power plant that burns fossil fuel. The combustion process generates carbon dioxide, so there is still the problem of increased $CO_2$ in the atmosphere, although it doesn't come directly from Joel's car. Also, the overall efficiency drops to less than 40% when the loss of energy between the power plant and the wall outlet in Joel's garage is taken into account. The drain on scarce fossil fuel resources is no less than using gasoline.

Lily and Joel have heard that hydrogen-powered cars may soon be available (Figure 6.1). How can hydrogen power a vehicle? When hydrogen reacts with oxygen to form water, a large amount of energy is released. Hydrogen is the major fuel that powered the space shuttle throughout takeoff (Figure 6.2). Hydrogen-powered cars do not burn hydrogen directly; instead, they use a fuel cell, in which the reaction between hydrogen and oxygen is carried out in a controlled way.

A fuel cell works on a different principle than an ICE (Figure 6.3). In an ICE, the fuel is burned and some of the heat energy is converted to work by driving pistons back and forth to run the car. In a hydrogen fuel cell, the hydrogen is not burned. Instead, the cell is separated by a membrane and hydrogen gas runs through one side of the cell, while oxygen runs through the other side. They react together at the membrane, much like a battery. A flow of electrons from hydrogen to oxygen produces an electric current that runs the vehicle's electric motor. In a fuel cell, the hydrogen and oxygen must be continuously delivered to the cell from external storage tanks that can be refilled.

Many people think hydrogen will be the fuel of the future because it is efficient and clean burning. There are, however, problems to overcome. Storing hydrogen gas under high pressure in a vehicle is a safety hazard. Also, hydrogen gas is explosive in air when not controlled. Solutions to these problems have focused on storing the hydrogen as part of a metal hydride compound such as $TiH_3$, or in methanol, $CH_3OH$. Hydrogen can then be obtained from chemical reactions of these compounds as needed. Another problem is that hydrogen is not readily available. It is most easily obtained from reactions of methanol or other fuels, which also form carbon monoxide or carbon dioxide. In this case, no reduction of air pollution occurs.

**FIGURE 6.1** This prototype hydrogen-fueled car has its fuel cell visible in the left side of the opened engine compartment.

Source: US Department of energy

**FIGURE 6.2** Hydrogen is the fuel that launched the space shuttles into space during many operational missions from 1981 to 2011. An external tank carried both the liquid hydrogen and liquid oxygen needed for combustion. The product of this reaction, $H_2O$, is visible as clouds of steam.

Source: NASA

A  Single-cylinder internal-combustion engine (ICE)

B  Hydrogen fuel cell

**FIGURE 6.3** (A) In an internal-combustion engine (ICE), a fuel-air mixture burns inside a combustion chamber, providing the force needed to turn a crankshaft. (B) A hydrogen fuel cell works something like a battery. The electron transfer between hydrogen and oxygen generates a current that runs the vehicle's electric motor.

> Among the principles of green chemistry are the prevention of waste and the maximization of atom economy. These goals can be accomplished by careful calculation of the exact amounts of reactants needed to produce the desired amount of product in a chemical reaction.

As drivers of alternative-fuel vehicles, Lily and Joel are interested in the amount of fuel they need for their vehicles and the energy output provided by the fuel. Whatever the chemical reaction may be, it is often important to know the quantities of the reactants and products and the amount of energy involved in a reaction.

In this chapter we will look at the quantitative aspects of chemical reactions. We will show how to determine the amounts of reactants that are needed and the amounts of products that should be formed in a given reaction. We will investigate the amounts of energy produced in chemical reactions, such as the combustion of fuels, and we will compare the amount of heat released by different fuels. While combustion reactions release energy, some chemical reactions absorb energy. We will look at how to measure the amount of energy released or absorbed during chemical reactions.

### Questions for Consideration

**6.1** What do the coefficients in balanced equations represent?

**6.2** How can we use a balanced equation to relate the number of moles of reactants and products in a chemical reaction?

**6.3** How can we use a balanced equation to relate the mass of reactants and products in a chemical reaction?

**6.4** How do we determine which reactant limits the amount of product that can form?

**6.5** How can we compare the amount of product we actually obtain to the amount we expect to obtain?

**6.6** How can we describe and measure energy changes?

**6.7** How are heat changes involved in chemical reactions?

 **Math Tools Used in This Chapter**

Significant Figures (Math Toolbox 1.2)

Units and Conversions (Math Toolbox 1.3)

Mole Quantities (Math Toolbox 4.1)

# 6.1 The Meaning of a Balanced Equation

As you learned in Chapter 5, a balanced equation represents the identity and physical state of each reactant and product in a chemical reaction. In addition, the coefficients in the equation show the relative amounts of reactants that combine and the relative amounts of products expected to form. Let's consider the combustion reaction of propane, a chemical reaction that occurs when you light a gas grill. The balanced equation for the combustion of propane (Figure 6.4) shows that when 1 propane, $C_3H_8$, molecule reacts, 5 $O_2$ molecules combine with it to form 3 $CO_2$ molecules and 4 $H_2O$ molecules:

$$C_3H_8(g) + 5O_2(g) \longrightarrow 3CO_2(g) + 4H_2O(g)$$

Since the coefficients in this equation represent the *relative* numbers of reactant and product molecules, the equation reveals the amounts of reactants and products at any scale. For example, if 2 molecules of $C_3H_8$ react, they must combine with 10 molecules of $O_2$, producing 6 molecules of $CO_2$ and 8 molecules of $H_2O$. The relative numbers of reactants and products remain the same (1 $C_3H_8$ : 5 $O_2$ : 3 $CO_2$ : 4 $H_2O$).

Suppose 100 molecules of $C_3H_8$ react, or suppose $6.022 \times 10^{23}$ molecules (a mole of molecules) react. How many $O_2$ molecules will combine with each of these quantities? How many molecules of $CO_2$ and $H_2O$ should form in each case? Compare your answers to those shown in Table 6.1.

Because $6.022 \times 10^{23}$ molecules equals 1.000 mol of molecules, we can also say that 1.000 mol of $C_3H_8$ reacts with 5.000 mol of $O_2$ to produce 3.000 mol of $CO_2$ and 4.000 mol of $H_2O$. A balanced equation, therefore, also describes the *relative number of moles* of reactants and products. Table 6.1 summarizes the relationships between reactants and products for the combustion of propane.

If we know the amount of any reactant or product, we can determine the amount of any other reactant or product in the same reaction by using the balanced equation. The process of determining the amounts of substances in a chemical reaction is called **stoichiometry.** In the next sections, we will use stoichiometry to relate amounts of reactants and products using balanced equations.

Propane for barbecue grills is stored under pressure in its liquid state. Air, which is about 20% oxygen, provides the oxygen for the reaction.

**Student Hot Spot**

Student data indicate you may struggle with interpreting chemical equations. Access the SmartBook to view additional Learning Resources on this topic.

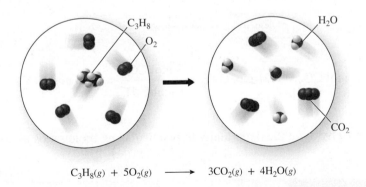

$$C_3H_8(g) + 5O_2(g) \longrightarrow 3CO_2(g) + 4H_2O(g)$$

**FIGURE 6.4** The balanced equation for the combustion of propane shows the relative numbers of propane and oxygen molecules that react and the relative numbers of carbon dioxide and water molecules that are produced.

**TABLE 6.1** Relationships Between Reactants and Products

| $C_3H_8(g)$ + | $5O_2(g)$ $\longrightarrow$ | $3CO_2(g)$ + | $4H_2O(g)$ |
|---|---|---|---|
| 1 molecule | 5 molecules | 3 molecules | 4 molecules |
| 2 molecules | 10 molecules | 6 molecules | 8 molecules |
| 100 molecules | 500 molecules | 300 molecules | 400 molecules |
| $6.022 \times 10^{23}$ molecules | $5 \times (6.022 \times 10^{23})$ molecules | $3 \times (6.022 \times 10^{23})$ molecules | $4 \times (6.022 \times 10^{23})$ molecules |
| 1.000 mol | 5.000 mol | 3.000 mol | 4.000 mol |

## 6.2 Mole-Mole Conversions

If you are having difficulty with mole ratios, review the concept of moles introduced at the beginning of Section 4.2.

A convenient way to relate the moles of a reactant or a product to other reactants or products is to use *mole ratios*. Mole ratios are obtained from the coefficients in a balanced equation. Let's look again at the balanced equation for the combustion of propane:

$$C_3H_8(g) + 5O_2(g) \longrightarrow 3CO_2(g) + 4H_2O(g)$$

The mole ratio of $O_2$ to $C_3H_8$ is 5/1, or $\dfrac{5 \text{ mol } O_2}{1 \text{ mol } C_3H_8}$, because the coefficient in front of $O_2$ is 5 and the coefficient in front of $C_3H_8$ is an implied 1. This mole ratio tells us that 5 mol of $O_2$ are needed for every 1 mol of $C_3H_8$ that reacts. The same equation also shows that the mole ratio of the product $CO_2$ to the reactant $O_2$ is 3/5, or $\dfrac{3 \text{ mol } CO_2}{5 \text{ mol } O_2}$. Mole ratios help us determine the moles of one substance in a reaction when the number of moles of another substance in a reaction is known:

Suppose we want to know the amount of $CO_2$ that will be produced when 9.21 mol of $C_3H_8$ reacts. What mole ratio do we use? From the coefficients in the equation, the mole ratio relating $CO_2$ and $C_3H_8$ can be written in two ways:

$$\frac{3 \text{ mol } CO_2}{1 \text{ mol } C_3H_8} \quad \text{and} \quad \frac{1 \text{ mol } C_3H_8}{3 \text{ mol } CO_2}$$

**MATH TOOLBOX 1.3**

We can use these mole ratios as conversion factors. (See Math Toolbox 1.3 for a review of unit conversions.) We multiply the known moles of $C_3H_8$ by the ratio that will cancel the old units (mol $C_3H_8$) and introduce the new units (mol $CO_2$):

$$\text{mol } CO_2 = 9.21 \text{ mol } C_3H_8 \times \frac{3 \text{ mol } CO_2}{1 \text{ mol } C_3H_8} = 27.6 \text{ mol } CO_2$$

Example 6.1 gives you the opportunity to practice using mole ratios as conversion factors.

---

**EXAMPLE 6.1**     **Mole-Mole Conversions**

If 1.14 mol of $CO_2$ was formed by the combustion of $C_3H_8$, how many moles of $H_2O$ were also formed?

$$C_3H_8(g) + 5O_2(g) \longrightarrow 3CO_2(g) + 4H_2O(g)$$

**Solution:**

We know the number of moles of $CO_2$, and we want to know the moles of $H_2O$:

The relationship we use to convert from moles of $CO_2$ to moles of $H_2O$ is the mole ratio we get from the balanced equation:

First we must ensure that the equation is balanced. Yes it is, as indicated by having the same number of atoms of each element in the reactants and in the products, so the coefficients in the equation give mole relationships between $CO_2$ and $H_2O$, which can be written in two ways:

$$\frac{3 \text{ mol } CO_2}{4 \text{ mol } H_2O} \quad \text{and} \quad \frac{4 \text{ mol } H_2O}{3 \text{ mol } CO_2}$$

The mole ratio to be used as the conversion factor is selected so that multiplying the known moles of $CO_2$ by the ratio will cancel the old units (mol $CO_2$) and introduce the new units (mol $H_2O$):

$$\text{mol } H_2O = 1.14 \, \cancel{\text{mol } CO_2} \times \frac{4 \text{ mol } H_2O}{3 \, \cancel{\text{mol } CO_2}} = 1.52 \text{ mol } H_2O$$

**MATH TOOLBOX 1.3**

Notice that the units cancel properly. We would expect the moles of $H_2O$ to be greater than the moles of $CO_2$, based on the 4:3 ratio in the balanced equation, so this answer makes physical sense.

### Consider This 6.1

If we were interested in the mass of the water formed in the reaction, how would we calculate the grams of water from the moles of water just calculated?

### Practice Problem 6.1

Pure methanol is used as a fuel for race cars in the Indy Racing League and in the Championship Auto Racing Teams. It is used because methanol fires are easier to put out with water than the fires of most other fuels. The balanced equation for the combustion of methanol is

$$2CH_3OH(l) + 3O_2(g) \longrightarrow 2CO_2(g) + 4H_2O(g)$$

Given that 1.00 gal of methanol contains 94.5 mol, how many moles of oxygen gas will react with 1.00 gal of methanol?

**Further Practice:**  Questions 6.11 and 6.12 at the end of the chapter

## 6.3  Mass-Mass Conversions

When we measure the amount of a reactant or product, we do not measure moles because our measuring devices do not count moles. Instead, we typically measure the mass of a reactant or product, but a chemical equation does not directly tell us the *mass* relationship between reactants and products. It tells us the *mole* relationship. This is not a problem, however, because we can convert from mass units (grams) to moles using the molar mass (*MM*) of the substance, as you learned in Chapter 4. Once we convert to moles, we can determine the moles of another substance in the equation using the mole ratio as we did in Section 6.2. Then, if we want to know the mass of the substance, we can convert to grams using its molar mass. The process always involves the same general conversion steps illustrated in Figure 6.5.

**MATH TOOLBOX 4.1**

   To illustrate a mass-mass conversion, let's consider the combination reaction of sodium metal and chlorine gas to form sodium chloride:

$$2Na(s) + Cl_2(g) \longrightarrow 2NaCl(s)$$

Although chemicals are often measured by mass, some substances are more conveniently measured by volume. If the substance is pure and the density is known, then the mass can be determined from the density and volume using the relationship you learned in Chapter 1:

$$\text{Density} = \frac{\text{mass}}{\text{volume}}$$

$$\text{Mass} = \text{density} \times \text{volume}$$

Sometimes reactants are in the gaseous state or are in solution. Calculations involving these types of substances will be discussed in Chapters 9 and 11.

**Student Hot Spot**

Student data indicate you may struggle with mass-mass conversions. Access the SmartBook to view additional Learning Resources on this topic.

**FIGURE 6.5** The process for converting from the mass of one reactant or product to another involves three basic steps.

Suppose we have 9.20 g of sodium metal. How many grams of chlorine gas should react with this amount of sodium in this reaction? Let's follow the general procedure for a mass-mass problem shown in Figure 6.5. Because the balanced equation does not give us a relationship between the grams of Na and the grams of $Cl_2$, we must first convert grams of Na to moles of Na using its molar mass, 22.99 g/mol.

The molar mass of Na gives the conversion factor between grams of Na and moles of Na, which can be written in two ways:

$$\frac{1 \text{ mol Na}}{22.99 \text{ g Na}} \quad \text{and} \quad \frac{22.99 \text{ g Na}}{1 \text{ mol Na}}$$

When we convert from grams to moles, we multiply the known amount in grams by the ratio that will cancel the old units (grams) and introduce the new units (moles):

$$\text{Moles of Na} = 9.20 \text{ g Na} \times \frac{1 \text{ mol Na}}{22.99 \text{ g Na}} = 0.400 \text{ mol Na}$$

Now that we know the moles of the reactant, Na, we can determine the moles of $Cl_2$ using the mole ratio of $Cl_2$ to Na obtained from the balanced equation as the conversion factor:

We multiply the known moles of Na by the ratio that will cancel the old units (mol Na) and introduce the new units (mol $Cl_2$):

$$\text{Moles of } Cl_2 = 0.400 \text{ mol Na} \times \frac{1 \text{ mol } Cl_2}{2 \text{ mol Na}} = 0.200 \text{ mol } Cl_2$$

Once we know the moles of $Cl_2$, all we have left to do is convert from moles to grams using the molar mass of $Cl_2$, 70.90 g/mol.

We multiply 0.200 mol $Cl_2$ by the molar mass ratio that cancels the old units (moles) and introduces the new units (grams):

$$\text{Mass of } Cl_2 = 0.200 \text{ mol } Cl_2 \times \frac{70.90 \text{ g } Cl_2}{1 \text{ mol } Cl_2} = 14.2 \text{ g } Cl_2$$

Notice that the answer has three significant figures because the mole amount of $Cl_2$ has three. (See Math Toolbox 1.2.) When we convert from moles to grams, we always *multiply* by the molar mass.

Alternatively we can do all three conversions consecutively in one calculation:

Grams Na $\xrightarrow{\text{MM of Na}}$ Moles Na $\xrightarrow{\text{mole ratio}}$ Moles Cl$_2$ $\xrightarrow{\text{MM of Cl}_2}$ Grams Cl$_2$

$$\text{Mass of } Cl_2 = 9.20 \text{ g Na} \times \frac{1 \text{ mol Na}}{22.99 \text{ g Na}} \times \frac{1 \text{ mol } Cl_2}{2 \text{ mol Na}} \times \frac{70.90 \text{ g } Cl_2}{1 \text{ mol } Cl_2} = 14.2 \text{ g } Cl_2$$

All mass-mass conversion problems involve the same three-step calculation. Writing the conversion calculation with all steps combined has advantages. It is easier to ensure that the units cancel properly. Also, you are less likely to make mistakes.

This calculation shows that 9.20 g of Na should react with 14.2 g of Cl$_2$. Now let's determine how much NaCl should form. We can use a three-step series of conversions starting with 9.20 g of Na:

Grams Na $\xrightarrow{\text{MM of Na}}$ Moles Na $\xrightarrow{\text{mole ratio}}$ Moles NaCl $\xrightarrow{\text{MM of NaCl}}$ Grams NaCl

$$9.20 \text{ g Na} \times \frac{1 \text{ mol Na}}{22.99 \text{ g Na}} \times \frac{2 \text{ mol NaCl}}{2 \text{ mol Na}} \times \frac{58.44 \text{ g NaCl}}{1 \text{ mol NaCl}} = 23.4 \text{ g NaCl}$$

> The way you round numbers will affect your answer. You'll get one result if you round after each step, but another if you hold the answers to each step in your calculator. Don't be alarmed if your answers are *slightly* different from those given in this textbook. We generally carry excess digits through the entire calculation and round the final answer to the appropriate number of significant figures.

(If we started with the 14.2 g Cl$_2$ calculated earlier, we would have obtained the same answer.)

Alternatively, we can use the law of conservation of mass to determine the mass of NaCl since we know the masses of all but one substance in the reaction, NaCl. The mass of the reactants must equal the mass of the products (Figure 6.6):

$$\text{Mass of reactants = mass of products}$$
$$\text{Mass Na + mass } Cl_2 = \text{mass NaCl}$$
$$9.20 \text{ g Na} + 14.2 \text{ g } Cl_2 = 23.4 \text{ g NaCl}$$

In this example, we could calculate the mass of NaCl in this way because we knew the mass of all the other substances in the reaction. Usually this is not the case, and we have to use the mass-mass conversion procedure to determine the mass of a reactant or product from a known mass of another reactant or product. Example 6.2 gives you practice converting from the mass of one reactant or product to the mass of another.

|  |  |  |
|---|---|---|
| 9.20 g Na | 14.2 g Cl$_2$ | 23.4 g NaCl |
| (0.400 mol) | (0.200 mol) | (0.400 mol) |
| 2Na(*s*) | Cl$_2$(*g*) | 2NaCl(*s*) |

**FIGURE 6.6**  The law of conservation of mass states that the mass of the reactants must equal the mass of the products. Mass is not gained or lost. (The amounts given do not represent the quantities of substances shown.)

## EXAMPLE 6.2      Mass-Mass Conversions

The decomposition of nitroglycerin is an explosive reaction that produces four different gases. What mass of water vapor should be produced when 1.0 g of nitroglycerin decomposes by the following reaction?

$$4C_3H_5O_9N_3(l) \longrightarrow 12CO_2(g) + 6N_2(g) + O_2(g) + 10H_2O(g)$$

**Solution:**

Because we are given grams of nitroglycerin, $C_3H_5O_9N_3$, we must first convert to moles using its molar mass, 227.10 g/mol:

$$\text{Moles of } C_3H_5O_9N_3 = 1.0 \text{ g } C_3H_5O_9N_3 \times \frac{1 \text{ mol } C_3H_5O_9N_3}{227.10 \text{ g } C_3H_5O_9N_3}$$

$$= 0.0044 \text{ mol } C_3H_5O_9N_3$$

Next we use the mole ratio of $H_2O$ to $C_3H_5O_9N_3$ to convert to moles of $H_2O$:

$$\boxed{\text{Moles } C_3H_5O_9N_3} \xrightarrow{\text{mole ratio}} \boxed{\text{Moles } H_2O}$$

$$\text{Moles of } H_2O = 0.0044 \text{ mol } C_3H_5O_9N_3 \times \frac{10 \text{ mol } H_2O}{4 \text{ mol } C_3H_5O_9N_3}$$

$$= 0.011 \text{ mol } H_2O$$

Now that we know the moles of $H_2O$, all we have left to do is convert moles to grams by multiplying by the molar mass of $H_2O$. Note that the units cancel properly.

$$\boxed{\text{Moles } H_2O} \xrightarrow{MM \text{ of } H_2O} \boxed{\text{Grams } H_2O}$$

$$\text{Mass of } H_2O = 0.011 \text{ mol } H_2O \times \frac{18.02 \text{ g } H_2O}{1 \text{ mol } H_2O} = 0.20 \text{ g } H_2O$$

Alternatively, we can do all three conversions consecutively in one calculation:

$$\text{Mass of } H_2O = 1.0 \text{ g } C_3H_5O_9N_3 \times \frac{1 \text{ mol } C_3H_5O_9N_3}{227.10 \text{ g } C_3H_5O_9N_3}$$

$$\times \frac{10 \text{ mol } H_2O}{4 \text{ mol } C_3H_5O_9N_3} \times \frac{18.02 \text{ g } H_2O}{1 \text{ mol } H_2O} = 0.20 \text{ g } H_2O$$

When 1.0 g of nitroglycerin decomposes, 0.20 g of water vapor is produced.

### Consider This 6.2

If we obtained a mass of $H_2O$, or any other product, that was greater than 1.0 g, how would we know that was an incorrect answer?

### Practice Problem 6.2

Hydrogen and oxygen were used as fuel for liftoff during space-shuttle launches. The hydrogen and oxygen were stored in an external tank as compressed liquids. During an average liftoff, 180,000 kg of hydrogen burned as it reacted with oxygen to form water vapor. The balanced equation for this reaction is

$$2H_2(g) + O_2(g) \longrightarrow 2H_2O(g)$$

MATH TOOLBOX 4.1

MATH TOOLBOX 1.3

(a)  What mass of oxygen gas is consumed?
(b)  What mass of water vapor is produced?
(c)  Does the mass of water vapor equal the sum of the masses of the reactants? What law is followed here?

**Further Practice:**  Questions 6.21 and 6.22 at the end of the chapter

## 6.4  Limiting Reactants

When Lily burns ethanol in her car, she relies on oxygen in the air to support the combustion. She doesn't worry about running out of oxygen. A chemist would say that oxygen is in excess for the reaction.

When the space shuttle used hydrogen as a fuel in space, there was no air to provide the oxygen. The shuttle had to take specific amounts of hydrogen and oxygen with it. If too much of one reactant were transported, the extra reactant would have had nothing to react with. It would simply add extra mass to the shuttle. Mass-mass calculations are important in such situations.

In many other reactions, it's important to start with an excess of one reactant. Consider a single-displacement reaction between magnesium metal and aqueous hydrochloric acid:

$$Mg(s) + 2HCl(aq) \longrightarrow MgCl_2(aq) + H_2(g)$$

How do we know if all of each reactant is used up in the reaction? If one of the reactants does not react completely, there will be some of it left over at the end of the reaction. The leftover amount will be mixed with the product, as shown in Figure 6.7. In which of the pictures, Figure 6.7B or C, is magnesium left over? In which is hydrochloric acid left over?

When there is excess magnesium, we see the magnesium metal remaining with the solution. When there is excess hydrochloric acid, no metal remains with the product solution, which also contains excess acid. If we do not want any unreacted magnesium or hydrochloric acid left over, we should mix the reactants in the proper quantities so that their mole ratio is the same as that shown by the coefficients in the balanced equation:

$$\frac{1 \text{ mol Mg}}{2 \text{ mol HCl}}$$

Why don't we use the exact amounts of each reactant that are needed for the reaction? Small errors can always lead to a slight excess of one reactant that might contaminate the product. To ensure that a selected reactant is used completely, we commonly use an excess of the other reactant. For example, silver bromide used in photographic films is prepared by the reaction of potassium bromide and silver nitrate solutions. Potassium bromide is used in excess to ensure that all of the more expensive silver nitrate is used completely.

$HCl(aq)$ can be detected in the product solution with various acid test strips.

    A              B              C

**FIGURE 6.7**  When reactants are not mixed in relative amounts as described by the balanced equation, one reactant does not react completely. (A) Magnesium metal reacts with hydrochloric acid to form magnesium chloride in solution and hydrogen gas. (B) Under this set of conditions, which reactant is in excess when the chemicals are mixed? (C) When the reaction is run again, a different reactant is in excess. Although it cannot be seen, do you know which reactant is in excess this time?
©Jim Birk

A limiting reactant is also known as a limiting reagent.

If the reactants mixed together are *not* present in the exact mole ratio, then we will have too much of one reactant and not enough of another. One reactant reacts completely, and some of the other is left unreacted. The reactant that reacts completely is called the **limiting reactant;** it limits the amount of the other reactant that can react, and it limits the amount of product that can form. A reactant that does not react completely is *in excess*.

---

**EXAMPLE 6.3**    Identifying the Limiting Reactant

Copper metal is added to an aqueous solution of silver nitrate, producing copper(II) nitrate and silver metal. The solution is blue because copper(II) ions in solution produce a blue color. Note that silver metal crystals have formed at the surface of the copper metal wire. Given that this reaction is complete, identify the limiting reactant and the reactant that is in excess.

©McGraw-Hill Education/Stephen Frisch

**Solution:**

If the copper had reacted completely, there would be no solid surface for the silver crystals to reside on and they would fall to the bottom of the beaker. Since the silver is still adhering to the copper wire, we know that some copper is still unreacted. Because we can tell that copper metal is in excess, the silver nitrate dissolved in solution must be the limiting reactant.

**Consider This 6.3**

If you added more silver nitrate to the solution, what would you expect to happen?

**Practice Problem 6.3**

In Example 6.3 you identified the limiting reactant. Describe what you would observe if the other reactant were the limiting reactant.

**Further Practice:** Questions 6.29 and 6.30 at the end of the chapter

---

To gain a better understanding of limiting reactants, let's look at a system that is not a chemical reaction. Suppose Joel wants to build a model solar-powered car. He needs 1 frame, 1 solar cell, 1 electric motor, and 4 wheels (Figure 6.8A). We can write an equation to represent the numbers of parts required to make the car:

1 frame + 1 solar cell + 1 electric motor + 4 wheels = 1 model solar car

To make one or more solar cars, the ratio of cells to motors to frames to wheels is 1:1:1:4. If the numbers of parts are present in the same ratio, then all the parts can be used to make cars. For example, if Joel has 10 cells, 10 electric motors, 10 frames,

**FIGURE 6.8** To make a model solar car, a specific number of each part is required. To make more than one car, the relative number of parts remains the same. If there is a shortage of a particular part, that part limits the number of solar model cars that can be built. It is the limiting part.

and 40 wheels, he can make 10 cars with no parts left over. If the number of parts is different from this ratio, then the number of cars will be limited by at least one of the parts, and Joel will have leftover parts. For example, suppose he has 5 solar cells, 6 motors, 4 frames, and 12 wheels, as shown in Figure 6.8B. How many cars can he make? He has enough frames to make 4 cars. He has more than enough cells and motors, but does he have enough wheels? No, he would need 16 wheels to go with the 4 frames, and he has only 12. So, the number of wheels limits the number of cars to 3. The wheels are the *limiting part*. He has too many of the other parts, so they are in excess (Figure 6.8C). Notice that the terms we are using are similar to those we use in discussing a chemical reaction. The difference is that we are describing parts instead of reactants.

| **EXAMPLE 6.4** | **Limiting Parts** |

Suppose Joel received a shipment of 88 wheels, so he now has a total of 100. He still has 5 solar cells, 6 motors, and 4 frames.

(a) Use the parts ratio described in Figure 6.8A to determine the limiting part(s).
(b) How many cars can Joel make?
(c) Identify which parts are in excess and how many of each will be left over.

### Solution:

(a) The large shipment of wheels provides excess wheels. They are no longer the limiting part. The numbers of cells, motors, and frames needed per car are the same (1:1:1). Since Joel has only 4 frames, and more of each of the other parts, the frame is the limiting part.

(b) The number of limiting parts (frames) determines the number of cars that can be built. Since each car requires 1 frame, a maximum of 4 cars can be built.

(c) Joel has 1 excess cell, 2 excess motors, and 84 excess wheels. They will be left over after 4 complete cars are constructed.

### Consider This 6.4

If Joel wanted to make 25 cars, which parts would he need to order more of?

### Practice Problem 6.4

Suppose Joel then got a shipment of 10 additional frames.

(a) Use the parts ratio described in Figure 6.8A to determine the limiting part(s).
(b) How many cars can Joel make?
(c) Identify which parts are in excess and how many of each will be left over.

**Further Practice:**  Questions 6.31 and 6.32 at the end of the chapter

Now let's look at what occurs at the molecular level in a reaction that has a limiting reactant. Consider the combustion of hydrogen to form water. The molecular-level diagram and balanced equation in Figure 6.9 show that the molecule ratio of $H_2$ to $O_2$ is $\dfrac{2\ H_2 \text{ molecules}}{1\ O_2 \text{ molecule}}$. The *molecule ratio* is the same as the *mole ratio* because both are related to the coefficients in the balanced equation.

In this chemical reaction, the reactant molecules are the "parts" that combine in a specific ratio to make the product $H_2O$ molecules. If the number of $H_2$ molecules mixed is exactly two times the number of $O_2$ molecules, then all the $H_2$ and $O_2$ molecules will react, and no reactants will be left over. If the ratio of $H_2$ to $O_2$ molecules mixed together is *not* exactly $\dfrac{2\ H_2 \text{ molecules}}{1\ O_2 \text{ molecule}}$, then one reactant is the limiting reactant and the other is in excess. For example, suppose 8 molecules of $H_2$ are mixed with 5 molecules of $O_2$, as shown in Figure 6.10. The reactant ratio is not $\dfrac{2\ H_2 \text{ molecules}}{1\ O_2 \text{ molecule}}$, so only one of the reactants is completely consumed. The other is in excess. After the reactants combine in this 2 to 1 ratio, 8 $H_2O$ molecules form. The $H_2$ reactant is the limiting reactant because it was completely used up. It limited the number of

**FIGURE 6.9**  For every 2 $H_2$ molecules that react, 1 $O_2$ molecule reacts with them to form 2 $H_2O$ molecules.

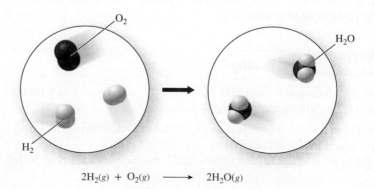

$$2H_2(g) + O_2(g) \longrightarrow 2H_2O(g)$$

Before              After

$$2H_2(g) + O_2(g) \longrightarrow 2H_2O(g)$$

**FIGURE 6.10** When 8 $H_2$ molecules and 5 $O_2$ molecules are combined, all of the $H_2$ reacts and 1 $O_2$ molecule is left over. The limiting reactant is $H_2$, and $O_2$ is in excess.

MATH
TOOLBOX
**1.3**

$H_2O$ molecules that could form. The $O_2$ reactant is left over. It was in excess. When the reaction is complete, the 1 leftover $O_2$ molecule that did not react will also be present in the reaction mixture along with the 8 $H_2O$ molecules that formed.

Suppose there was no "after" picture in Figure 6.10. Could you still determine the limiting reactant? Yes, it could be done in a way similar to the method used to determine the limiting part for building model cars. The molecule ratio in the balanced equation shows the number of molecules of one reactant that is required to react with the given number of molecules of the other reactant. Let's arbitrarily start by determining the number of $H_2$ molecules needed to react with the 5 $O_2$ molecules given in Figure 6.10. We multiply the given number of $O_2$ molecules by the molecule ratio from the balanced equation to get the number of $H_2$ molecules:

$$\text{Molecules of } H_2 = 5 \, \cancel{O_2 \text{ molecules}} \times \frac{2 \, H_2 \text{ molecules}}{1 \, \cancel{O_2 \text{ molecule}}} = 10 \, H_2 \text{ molecules}$$

This calculation tells us that we need 10 $H_2$ molecules to react with the 5 $O_2$ molecules. Since we actually have only 8 $H_2$ molecules, we know that

$$\text{Calculated } H_2 > \text{actual } H_2$$

Since there is not enough $H_2$ to react with all the $O_2$, $H_2$ is the limiting reactant. We come to the same conclusion if we calculate the number of $O_2$ molecules required to react with the 8 $H_2$ molecules:

$$\text{Molecules of } O_2 = 8 \, \cancel{H_2 \text{ molecules}} \times \frac{1 \, O_2 \text{ molecule}}{2 \, \cancel{H_2 \text{ molecules}}} = 4 \, O_2 \text{ molecules}$$

This calculation tells us that we need 4 $O_2$ molecules to react with the 8 $H_2$ molecules. We actually have 5 $O_2$ molecules:

$$\text{Calculated } O_2 < \text{actual } O_2$$

We have 1 extra $O_2$ molecule, so $O_2$ is in excess. The $H_2$ must be the limiting reactant. All 8 $H_2$ molecules will react with 4 of the $O_2$ molecules, and there will be 1 leftover $O_2$ molecule. Does this match the "after" picture in Figure 6.10? It should. It makes no difference which reactant you start with to determine the limiting reactant. Your conclusion will be the same. The following table shows how much is mixed, how much reacts, and the final composition of the mixture.

| | $2H_2(g)$ | + | $O_2(g)$ | $\longrightarrow$ | $2H_2O(g)$ |
|---|---|---|---|---|---|
| Initially mixed | 8 molecules | | 5 molecules | | 0 molecules |
| How much reacts | 8 molecules | | 4 molecules | | |
| Composition of final mixture | 0 molecules | | 1 molecule | | 8 molecules |

The following summarizes the method for determining the limiting reactant.

### Steps for Determining the Limiting Reactant

1. Calculate the amount of one reactant (B) needed to react with the other reactant (A).
2. Compare the calculated amount of B (amount needed) to the actual amount of B that is present initially.
    (a) If calculated B = actual B, there is no limiting reactant. Both A and B will react completely.
    (b) If calculated B > actual B, B is the limiting reactant. Only B will react completely.
    (c) If calculated B < actual B, A is the limiting reactant. Only A will react completely.

> If you calculate the amount of product predicted from the given amount of each reactant, the calculation that gives the *least* amount of product identifies the limiting reactant. Why?

Example 6.5 will give you practice identifying limiting reactants in a chemical reaction.

> The reaction of nitrogen gas with hydrogen gas to produce ammonia is very important. Ammonia is used to produce fertilizer for plants that help feed much of the world population.

### EXAMPLE 6.5    Limiting Reactants—Molecular Level

Consider the combination reaction of nitrogen gas with hydrogen gas to form ammonia gas:

$$N_2(g) + 3H_2(g) \longrightarrow 2NH_3(g)$$

The molecular-level diagram shows the number of $N_2$ and $H_2$ molecules mixed together before the reaction occurs.

N₂
H₂
Before

(a) Identify the limiting reactant, assuming the reaction goes to completion.
(b) Draw the "after" picture. Include the correct number of product molecules and any leftover reactant molecules.
(c) Determine the reactant that is in excess.

**Solution:**

(a) The limiting reactant can be identified by determining the number of molecules of one reactant needed to react with the given number of molecules of the other. Here we will calculate the number of $N_2$ molecules needed to react with the given number of $H_2$ molecules:

MATH
TOOLBOX
**1.3**

$$\text{Molecules of } N_2 = 6 \text{ H}_2 \text{ molecules} \times \frac{1 \text{ N}_2 \text{ molecule}}{3 \text{ H}_2 \text{ molecules}} = 2 \text{ N}_2 \text{ molecules}$$

We need 2 $N_2$ molecules to react with the 6 $H_2$ molecules. We actually have 6 $N_2$ molecules.

$$\text{Calculated } N_2 < \text{actual } N_2$$

We have 4 extra $N_2$ molecules, so $N_2$ is in excess. The $H_2$ must be the limiting reactant.

(b) Your "after" picture should have 4 $NH_3$ molecules and 4 $N_2$ molecules:

$$\text{Molecules of } NH_3 = 6 \; \cancel{H_2 \text{ molecules}} \times \frac{2 \; NH_3 \text{ molecules}}{3 \; \cancel{H_2 \text{ molecules}}}$$

$$= 4 \; NH_3 \text{ molecules}$$

After

(c) Because some $N_2$ molecules are left over, there was too much to react, so it is in excess. The following table summarizes this reaction.

|  | $N_2(g)$ | $+$ | $3H_2(g)$ | $\longrightarrow$ | $2NH_3(g)$ |
|---|---|---|---|---|---|
| Initially mixed | 6 molecules | | 6 molecules | | 0 molecules |
| How much reacts | 2 molecules | | 6 molecules | | |
| Composition of final mixture | 4 molecules | | 0 molecules | | 4 molecules |

The last row in the table shows that $H_2$ reacted completely and some $N_2$ remains, so $H_2$ is the limiting reactant.

## Consider This 6.5

If you wanted no leftover reactant molecules in your "after" picture, how many molecules of the limiting reactant should you add to the "before" picture?

## Practice Problem 6.5

Consider the combination reaction of nitrogen gas with hydrogen gas to form ammonia gas:

$$N_2(g) + 3H_2(g) \longrightarrow 2NH_3(g)$$

The molecular diagram shows the number of $N_2$ and $H_2$ molecules mixed together before the reaction occurs.

Before

(a) Identify the limiting reactant, assuming the reaction goes to completion.
(b) Draw the "after" picture. Include the correct number of product molecules and any leftover reactant molecules.
(c) Determine the reactant that is in excess.

**Further Practice:** Questions 6.35 and 6.36 at the end of the chapter

Now that we have identified limiting reactants on a molecular scale, let's look at the much larger mole scale. The procedure is similar because the molecule ratio and the mole ratio are the same and are given by the balanced equation. Let's look at the reaction of sodium metal and chlorine gas:

$$2Na(s) + Cl_2(g) \longrightarrow 2NaCl(s)$$

Suppose we mix 0.50 mol Na with 0.20 mol $Cl_2$. What is the limiting reactant? We'll follow a similar procedure to what we did when we started with the number of molecules of each reactant. Let's begin by finding the moles of Na needed to react with the 0.20 mol $Cl_2$:

$$\text{Moles of Na} = 0.20 \ \text{mol Cl}_2 \times \frac{2 \ \text{mol Na}}{1 \ \text{mol Cl}_2} = 0.40 \ \text{mol Na}$$

We need 0.40 mol Na, and we actually have 0.50 mol Na:

$$\text{Calculated Na} < \text{actual Na}$$

Na is in excess by 0.10 mol. The $Cl_2$ must be the limiting reactant. We can expect all the $Cl_2$ gas to react and some (0.10 mol) of the Na metal to be left over after the reaction.

We can determine how much NaCl should form by doing a mole-mole calculation. Which reactant do we use? We use the limiting reactant $Cl_2$ because it reacts completely. We multiply the moles of limiting reactant $Cl_2$ by the mole ratio of NaCl to $Cl_2$ to determine the moles of NaCl that can form:

$$\text{Moles of NaCl} = 0.20 \ \text{mol Cl}_2 \times \frac{2 \ \text{mol NaCl}}{1 \ \text{mol Cl}_2} = 0.40 \ \text{mol NaCl}$$

The following table summarizes this reaction.

|  | 2Na(s) | + | Cl$_2$(g) | $\longrightarrow$ | 2NaCl(s) |
|---|---|---|---|---|---|
| Initially mixed | 0.50 mol | | 0.20 mol | | 0.00 mol |
| How much reacts | 0.40 mol | | 0.20 mol | | |
| Composition of final mixture | 0.10 mol | | 0.00 mol | | 0.40 mol |

When 0.50 mol Na is mixed with 0.20 mol $Cl_2$, 0.40 mol Na reacts with 0.20 mol $Cl_2$. When the reaction is complete, 0.40 mol NaCl is produced. Because the Na is in excess by 0.10 mol, that amount, along with the 0.40 mol NaCl that forms, will be present when the reaction is complete.

Example 6.6 will give you practice determining the limiting reactant and the amount of product that can form.

### EXAMPLE 6.6    Limiting Reactants—Mole Scale

Ethylene, $C_2H_4$, undergoes many useful reactions. However, since it's a hydrocarbon, it is flammable and burns in the presence of oxygen according to the equation

$$C_2H_4(g) + 3O_2(g) \longrightarrow 2CO_2(g) + 2H_2O(g)$$

Suppose we mix 0.25 mol $C_2H_4$ with 1.0 mol $O_2$. Identify the limiting reactant and the number of moles of $CO_2$ that can be formed.

**Solution:**

To determine the limiting reactant, first determine if there is enough $C_2H_4$ to react with 1.0 mol $O_2$:

$$\text{Moles of } C_2H_4 = 1.0 \; \cancel{\text{mol } O_2} \times \frac{1 \text{ mol } C_2H_4}{3 \; \cancel{\text{mol } O_2}} = 0.33 \text{ mol } C_2H_4$$

MATH
TOOLBOX
**1.3**

We need 0.33 mol $C_2H_4$, and we actually have only 0.25 mol $C_2H_4$:

$$\text{Calculated } C_2H_4 > \text{actual } C_2H_4$$

Because there is not enough $C_2H_4$ to react with all the $O_2$, $C_2H_4$ is the limiting reactant. (Assure yourself that you would reach the same conclusion if you started the calculation with the given amount of $C_2H_4$ instead of the given amount of $O_2$.) Remember that we must use the amount of limiting reactant to calculate the moles of $CO_2$ product:

$$\text{Moles of } CO_2 = 0.25 \; \cancel{\text{mol } C_2H_4} \times \frac{2 \text{ mol } CO_2}{1 \; \cancel{\text{mol } C_2H_4}} = 0.50 \text{ mol } CO_2$$

We also use the amount of the limiting reactant to calculate the amount of $O_2$ that will react:

$$\text{Moles of } O_2 \text{ that react} = 0.25 \; \cancel{\text{mol } C_2H_4} \times \frac{3 \text{ mol } O_2}{1 \; \cancel{\text{mol } C_2H_4}} = 0.75 \text{ mol } O_2$$

This reaction is summarized in the following table.

| | $C_2H_4(g)$ | + | $3O_2(g)$ | $\longrightarrow$ | $2CO_2(g)$ | + | $2H_2O(g)$ |
|---|---|---|---|---|---|---|---|
| Initially mixed | 0.25 mol | | 1.00 mol | | 0.00 mol | | 0.00 mol |
| How much reacts | 0.25 mol | | 0.75 mol | | | | |
| Composition of final mixture | 0.00 mol | | 0.25 mol | | 0.50 mol | | 0.50 mol |

**Consider This 6.6**

If we mixed 0.50 mol $O_2$ with 0.25 mol $C_2H_4$ in the same reaction, which would be the limiting reactant?

**Practice Problem 6.6**

In the double-displacement reaction of aqueous silver nitrate and magnesium chloride, silver chloride precipitates and aqueous magnesium nitrate forms:

$$2AgNO_3(aq) + MgCl_2(aq) \longrightarrow 2AgCl(s) + Mg(NO_3)_2(aq)$$

Suppose we mix 1.0 mol of silver nitrate with 1.0 mol of magnesium chloride. Identify the limiting reactant and the number of moles of solid AgCl that can form.

**Further Practice:** Questions 6.41 and 6.42 at the end of the chapter

**Student Hot Spot**

Student data indicate you may struggle with determining the mass of product when there is a limiting reactant. Access the SmartBook to view additional Learning Resources on this topic.

Generally we measure amounts of reactants in units of grams. Given grams of reactants, we must first convert from grams to moles for each reactant. Example 6.7 shows how to determine the limiting reactant when grams of reactants are given.

Aluminum reacts with bromine rather spectacularly to produce aluminum bromide.
©Tom Pantages

MATH TOOLBOX 4.1

Although the mass ratio of $Br_2$ to Al mixed together is 45 g/30 g, or 3/2, this is not the mole ratio. *When given the mass of reactants, we must always first convert to moles before determining the amount of another reaction species.*

You may have been able to see merely by looking at the balanced equation that $Br_2$ is the limiting reactant. The balanced equation shows that the number of moles of $Br_2$ must be greater than that of Al, but the number of moles of $Br_2$ we have is less than that of Al.

| EXAMPLE 6.7 | Limiting Reactants When Given the Mass of Reactants |
|---|---|

When liquid bromine is mixed with aluminum metal, a combination reaction occurs, forming aluminum bromide:

$$2Al(s) + 3Br_2(l) \longrightarrow 2AlBr_3(s)$$

Determine the limiting reactant and mass of $AlBr_3$ produced when 45.0 g of $Br_2$ is added to 30.0 g of Al.

**Solution:**

Because we are given the mass of reactants, we must first convert from grams to moles for each reactant using their molar masses:

$$\text{Moles of Al} = 30.0 \text{ g Al} \times \frac{1 \text{ mol Al}}{26.98 \text{ g Al}} = 1.11 \text{ mol Al}$$

$$\text{Moles of } Br_2 = 45.0 \text{ g } Br_2 \times \frac{1 \text{ mol } Br_2}{159.89 \text{ g } Br_2} = 0.281 \text{ mol } Br_2$$

Next we determine the number of moles of one reactant needed to react with the other. Here we will determine the moles of $Br_2$ needed to react with 1.11 mol Al:

$$\text{Moles of } Br_2 = 1.11 \text{ mol Al} \times \frac{3 \text{ mol } Br_2}{2 \text{ mol Al}} = 1.67 \text{ mol } Br_2$$

We need 1.67 mol $Br_2$, but we actually have only 0.281 mol $Br_2$:

$$\text{Calculated } Br_2 > \text{actual } Br_2$$

Because $Br_2$ is in short supply, it is the limiting reactant. (Convince yourself that you would come to the same conclusion if you started the calculation with the given amount of $Br_2$ instead of the given amount of Al.)

To determine the mass of product, we can do a mass-mass conversion starting with the given grams of $Br_2$. Or, we can start with the moles of $Br_2$ calculated from the grams of $Br_2$, convert moles of $Br_2$ to moles of $AlBr_3$, and then convert moles of $AlBr_3$ to grams of $AlBr_3$:

$$\boxed{\text{Moles } Br_2} \xrightarrow{\text{mole ratio}} \boxed{\text{Moles } AlBr_3} \xrightarrow{MM \text{ of } AlBr_3} \boxed{\text{Grams } AlBr_3}$$

$$\text{Mass of } AlBr_2 = 0.281 \text{ mol } Br_2 \times \frac{2 \text{ mol } AlBr_3}{3 \text{ mol } Br_2} \times \frac{266.7 \text{ g } AlBr_3}{1 \text{ mol } AlBr_3}$$

$$= 50.0 \text{ g } AlBr_3$$

The maximum amount of $AlBr_3$ that can be produced is 50.0 g.

**Consider This 6.7**

If we doubled the mass of Al, how would the mass of $AlBr_3$ produced change?

**Practice Problem 6.7**

Sodium carbonate reacts with hydrochloric acid to produce aqueous sodium chloride, water, and carbon dioxide gas. The balanced equation is

$$Na_2CO_3(s) + 2HCl(aq) \longrightarrow 2NaCl(aq) + H_2O(l) + CO_2(g)$$

Determine the limiting reactant and mass of $CO_2$ gas produced when 11.0 g of $Na_2CO_3$ is added to a solution that contains 11.0 g of HCl.

©McGraw-Hill Education/Stephen Frisch

**Further Practice:** Questions 6.43 and 6.44 at the end of the chapter

## 6.5 Percent Yield

When the manufacturers of E-85 produce ethanol to power Lily's car, they calculate the amount of ethanol product that *should* form from a reaction. This is the **theoretical yield,** the maximum amount of product that can be obtained from the given amounts of reactants. In Example 6.7 for instance, we calculated the theoretical yield of $AlBr_3$ as 50.0 g.

When chemists do experiments and when companies manufacture fuels, they seldom obtain the amount of product predicted by the calculation. They fall short of the theoretical yield for any number of reasons. Reactants and products can be lost due to spillage or during transfer from one container to another (Figure 6.11); and alternative reactions (side reactions) can occur that produce different products. The amount of product we measure in the laboratory is the **actual yield.** The actual yield is usually less than the theoretical yield; however, it can appear greater if a solid is weighed when still wet or if a product is contaminated.

The **percent yield** describes how much of a product is actually formed in comparison to how much should have been formed. It is the actual yield expressed as a percentage of the theoretical yield:

$$\text{Percent yield} = \frac{\text{actual yield}}{\text{theoretical yield}} \times 100\%$$

The actual yield and theoretical yield can be expressed in any units, as long as those units are the same. Usually yields are reported in mass units, but yields in units of moles or molecules are acceptable.

Suppose the reaction of aluminum metal and liquid bromine described in Example 6.7 produced only 26.8 g $AlBr_3$. What is the percent yield? Since we calculated the amount of $AlBr_3$ that should form as 50.0 g, we know the theoretical yield. The percent yield is the ratio of the actual yield to the theoretical yield multiplied by 100:

$$\text{Percent yield} = \frac{26.8 \text{ g}}{50.0 \text{ g}} \times 100\% = 53.6\%$$

**FIGURE 6.11** The actual yield is usually less than the theoretical yield due to loss during reaction steps, purification, and transfer processes, or due to side reactions.

©McGraw-Hill Education/Stephen Frisch

©Richard Megna/Fundamental Photographs

**MATH TOOLBOX 1.3**

Example 6.8 will help you better understand percent yield, theoretical yield, and actual yield.

---

**EXAMPLE 6.8**    **Percent Yield, Theoretical Yield, and Actual Yield**

Sodium metal reacts with water in a single-displacement reaction to produce aqueous sodium hydroxide and hydrogen gas. The balanced equation is

$$2Na(s) + 2H_2O(l) \longrightarrow 2NaOH(aq) + H_2(g)$$

When 0.50 mol Na is placed in water, all the sodium metal reacts, and the hydrogen gas produced is isolated. It is determined that 0.21 mol $H_2$ has been produced. What is the percent yield of $H_2$?

**Solution:**

To calculate percent yield, we divide the actual yield by the theoretical yield, and then multiply by 100:

$$\text{Percent yield} = \frac{\text{actual yield}}{\text{theoretical yield}} \times 100\%$$

The actual yield is the number of moles of $H_2$ that was isolated once the reaction occurred (0.21 mol). The theoretical yield must be calculated from the amount of limiting reactant that reacted. In this case we can assume the Na is the limiting reactant (not the water) because all the Na reacted. We calculate the theoretical yield by calculating the maximum number of moles of $H_2$ that should be produced from the moles of limiting reactant Na:

$$\text{Moles of } H_2 = 0.50 \text{ mol Na} \times \frac{1 \text{ mol } H_2}{2 \text{ mol Na}} = 0.25 \text{ mol } H_2 \text{ (theoretical yield)}$$

We substitute the values of actual and theoretical yield into the percent yield equation and solve for percent yield:

$$\text{Percent yield} = \frac{0.21 \text{ mol } H_2}{0.25 \text{ mol } H_2} \times 100\% = 84\%$$

**Consider This 6.8**

If we doubled the amounts of both reactants, predict the moles of $H_2$ produced assuming the same percent yield you just calculated.

**Practice Problem 6.8**

Toxic carbon monoxide is a by-product of the preparation of hydrogen fuel from methanol for use in hydrogen-powered vehicles. When carbon monoxide reacts with oxygen gas, the greenhouse gas carbon dioxide is produced:

$$2CO(g) + O_2(g) \longrightarrow 2CO_2(g)$$

When 5.0 mol CO is mixed with excess $O_2$, the reaction occurs to give 3.4 mol $CO_2$. What is the percent yield of $CO_2$?

**Further Practice:** Questions 6.55 and 6.56 at the end of the chapter

---

If we know the expected percent yield for a reaction, we can predict the amount of product that can realistically be made. Such calculations are important in chemical and pharmaceutical industries where percent yield is a significant factor in determining profits.

# 6.6 Energy Changes

When hydrogen and oxygen react to produce water, energy is released. In a hydrogen fuel cell, most of this energy is converted to electric energy, which can be used to run a vehicle. Suppose we combine the same amounts of hydrogen and oxygen in a combustion reaction, not in a fuel cell. Would the same amount of energy be released by the reaction? The answer is yes, but the energy would not be in such a usable form. Some would be lost as *heat* to the surroundings, causing the temperature of the environment to increase. This is what happens when we burn other fuels, such as propane in a gas grill or wood in a bonfire. Whether the energy is converted to useful work or is dissipated into the environment as heat, the amount of energy released is the same. So far in this chapter, we have looked only at the amounts of reactants and products in a chemical reaction. Here we will investigate the amount of energy absorbed or released by a chemical reaction.

> See Section 1.3 for an introduction to energy.

## Law of Conservation of Energy

As described in Chapter 1, energy can take many forms, and it can be converted from one form to another. Energy can also be transferred as heat. In either case, the amount of energy remains the same. The **law of conservation of energy** expresses this principle: Energy can be converted or transferred, but it cannot be created or destroyed. Thus, when we "use" energy, we are not using it up. We are converting it to a different form of energy. However, we may be converting it into forms that are not useful, such as heat lost to the surrounding environment. When gasoline is burned in an engine, the chemical energy in the gasoline is converted to mechanical energy and heat energy (Figure 6.12). The mechanical energy does useful work, but the heat energy is generally lost to the atmosphere. If gasoline is burned in a barrel, most of the energy would be lost as heat. In order to make the energy possessed by an object most useful to us, we try to minimize the amount of energy lost as heat.

*Efficiency* is a measure of the amount of useful work that is achieved from an energy conversion. No conversion is totally efficient, as shown in Figure 6.13. The process of conversion always generates some heat that is lost even before useful work can be done. The efficiency of a conversion is the percentage that ends up in the form we want. Consider, for example, the series of conversions from the chemical energy stored in coal to the light energy we see when we turn on a desk lamp (Figure 6.14). When coal is burned in a steam boiler, about 88% of its chemical energy is converted to heat energy, which can be used to heat water and form steam. A steam turbine at the power plant can convert about 46% of the heat energy in the steam to mechanical energy, and an electric generator can convert about 98% of the mechanical energy to electric energy. Finally, when the electric energy is used to operate an incandescent lightbulb, only about 4% is converted to

What type of lightbulbs do you have in your room? Fluorescents? Incandescents? LEDs? The incandescent lightbulbs used regularly in the past are being phased out by the government because they emit large amounts of energy as heat. The now popular fluorescent and LED bulbs generate light by processes that are 4 to 10 times more efficient than incandescent bulbs. A 60-watt incandescent lightbulb can be replaced by a 15-watt fluorescent or an 8-watt LED, and the same amount of light will be emitted. Another benefit is that the lifetimes of fluorescent and LED lightbulbs are significantly longer, so you don't have to replace them as often.

**FIGURE 6.12** (A) When gasoline burns in an ICE, about 32% of the energy released is converted to work and 68% is immediately lost to the surrounding environment as heat. (B) When gasoline is burned in a barrel, 100% of the energy released is immediately lost as heat. The total energy released is the same.

**FIGURE 6.13** The conversion of one kind of energy to another is usually not completely efficient, because some energy is converted into undesired forms. Some typical conversion efficiencies are shown.

**FIGURE 6.14** From power plant to desk lamp, the overall conversion efficiency is only 1.6%. The other 98.4% of coal's energy is converted to heat energy before ever performing the desired function.

light, while the rest is lost to the surroundings as heat. In the overall process of burning coal to operate a lightbulb, only 1.6% of the chemical energy ends up as light energy. The rest has been lost along the way because of inefficient conversions. Most of the "lost" energy ends up as heat dissipated to the air or into the bodies of water used to cool power plants.

## Energy Changes That Accompany Chemical Reactions

Lily wonders how the alternative fuel she uses provides the energy that moves her car from one place to another. Where does the energy come from? It comes from chemical energy that is stored as potential energy in the fuel. It is stored in the *bonds* that hold atoms together in molecules. Some substances release some of their potential energy when they react to form substances that store less potential energy. Ethanol, for example, stores a large amount of potential energy in the bonds that hold its atoms together. Some of that energy is released as heat when it reacts with oxygen. A reaction that releases energy is called an **exothermic reaction.**

Not all reactions release energy. Some require a continuous energy input. A reaction that absorbs energy from the surroundings is an **endothermic reaction.** In

**FIGURE 6.15** An exothermic reaction is thought of as being "downhill" because the products contain less stored energy than the reactants. An endothermic reaction is thought of as being "uphill" because the products contain more potential energy.

an endothermic reaction, the products have more potential energy than the reactants. Figure 6.15 shows *reaction profiles,* which describe the overall energy changes that accompany exothermic and endothermic reactions. As shown by the endothermic reaction profile, the energy of the reactants is less than the energy of the products. The difference is the energy absorbed as the reaction takes place. In the exothermic reaction profile, the reactants contain more energy than the products. The difference is the energy released during the reaction. These reaction profiles also show us that if a reaction is exothermic, it will become endothermic if it is run in the reverse direction. And if a reaction is endothermic, its reverse reaction is exothermic.

Physical changes, such as phase changes, can also be endothermic or exothermic. Sweating is a natural process that helps cool our bodies when the water in sweat evaporates. Evaporation is an endothermic process, so it absorbs heat from our skin. Our skin cools, helping to maintain normal body temperature.

---

**EXAMPLE 6.9** | **Endothermic and Exothermic Reactions**

The following is the reaction used in a hydrogen fuel cell to produce electricity:

$$2H_2(g) + O_2(g) \longrightarrow 2H_2O(g)$$

(a) Is this reaction endothermic or exothermic?
(b) The reverse of this reaction, the formation of hydrogen and oxygen from water, can be accomplished by running electricity through water. Is this reaction endothermic or exothermic? Explain.

**Solution:**

(a) The reaction releases the energy that is converted to useful energy (electricity) by the fuel cell. Since energy is released by the reaction, it is exothermic.
(b) The reverse of this reaction requires a continuous energy input (electricity) to convert from reactants to products, so it is an endothermic reaction. If a reaction is exothermic, its reverse reaction is endothermic.

**Consider This 6.9**

Which is higher in potential energy: A mixture of $H_2$ and $O_2$ gases or pure $H_2O$ gas?

**Practice Problem 6.9**

When a car battery does not work, it can usually be recharged so that it is useful again. Recharging a battery involves running the reaction in reverse. Is this reaction an endothermic or exothermic process? Explain.

**Further Practice:** Questions 6.67 and 6.68 at the end of the chapter

**Nutrition Facts**

Serving Size:          1 cup (54g/1.9 oz.)
Servings Per Container:        About 9

**Amount Per Serving**

| **Calories** 190 | **Calories from Fat** 10 |
|---|---|
| | % Daily Value** |
| **Total Fat** 1g* | 2% |
| Saturated Fat 0g | 0% |
| Trans Fat 0g | |
| **Cholesterol** 0mg | 0% |
| **Sodium** 0mg | 0% |
| **Potassium** 180mg | 5% |
| **Total Carbohydrate** 45g | 15% |
| Dietary Fiber 6g | 24% |
| Soluble Fiber 1g | |
| Insoluble Fiber 5g | |
| Sugars 7g | |
| Other Carbohydrates 32g | |
| **Protein** 5g | |

| Vitamin A 0% | • | Vitamin C 0% |
|---|---|---|
| Calcium 0% | • | Iron 8% |

\* Amount in cereal. One half cup of fat free milk
contributes an additional 40 calories, 65mg sodium,
6g total carbohydrate (6g sugars), and 4g protein.
\*\* Percent Daily Values are based on a 2,000 calorie diet.
Your daily values may be higher or lower depending on
your calorie needs.

| | Calories: | 2,000 | 2,500 |
|---|---|---|---|
| Total Fat | Less Than | 65g | 80g |
| Sat. Fat | Less Than | 20g | 25g |
| Cholesterol | Less Than | 300mg | 300mg |
| Sodium | Less Than | 2,400mg | 2,400mg |
| Potassium | | 3,500mg | 3,500mg |
| Total Carbohydrate | | 300g | 375g |
| Dietary Fiber | | 25g | 30g |
| Protein | | 50g | 65g |

Calories per gram:
    Fat 9 • Carbohydrate 4 • Protein 4

**INGREDIENTS:** Organic Whole Grain Wheat,
Organic Evaporated Cane Juice, Natural Flavor.

**FIGURE 6.16** A nutritional label gives the energy content in Calories (or kilocalories) per serving. Also listed are grams of fat, carbohydrates, and proteins. This label also lists the Calories from fat.

©Brian Moeskau/Moeskau Photography

## Quantities of Heat

For a chemical reaction, the energy change that is easiest to measure is the *heat* absorbed or released. In order to understand how we measure heat changes, we must first learn more about heat and how it is related to temperature changes. **Heat** is energy that is transferred between two objects because of a difference in their temperatures. Consider the heat associated with some physical changes. If we place warm pancakes on a plate, for example, heat is transferred from the pancakes to the plate. The pancakes lose heat and get cooler, while the plate gains heat and gets hotter until the temperature of the pancakes and the temperature of the plate are the same. Heat is always transferred from the hotter to the colder object. When two objects that are in contact become equal in temperature, they have reached a state of *thermal equilibrium.*

**Units of Energy**    Many packaged foods are labeled with the energy content. For example, a small box of raisins contains 130 Calories (Cal with a capital *C*). How much energy is this? Nutritionists and chemists use related, but not identical, units to measure energy. Chemists measure energy in units of *joules* (J) or *calories* (cal with a lowercase *c*). One calorie is the amount of heat energy needed to raise the temperature of 1 g of water by 1°C. A joule is smaller than a calorie:

$$4.184 \text{ J} = 1 \text{ cal}$$

A *kilojoule* (kJ), or 1000 J, is approximately the amount of energy that is emitted when a wooden kitchen match burns completely. The calorie used in chemistry should not be confused with the *Calorie* used by nutritionists (Figure 6.16), which is actually a kilocalorie (kcal), or 1000 cal.

---

| **EXAMPLE 6.10** | **Units of Energy** |
|---|---|

A can of cola from the United States contains 180 Cal, and a can containing the same amount of cola from Australia contains 900 J. Which contains more energy? Which is the diet cola?

**Solution:**

To answer this question, we must compare the energy in the two drinks using the same units. Thus, we must convert one value into the same unit as the other value. We will convert energy in Calories to energy in joules:

MATH
TOOLBOX
**1.3**

We have no direct relationship between these two quantities, so we have to break this down into two steps. First, note that the relationships between these two quantities are 1 Cal = 1000 cal and 1 cal = 4.184 J:

$$\text{Energy in J} = 180 \text{ Cal} \times \frac{1000 \text{ cal}}{1 \text{ Cal}} \times \frac{4.184 \text{ J}}{1 \text{ cal}} = 7.5 \times 10^5 \text{ J}$$

This is considerably more energy than the 900 J found in the Australian cola, which must be the diet cola. Another way to solve the problem would be to convert 900 J to Calories and make the comparison.

### Consider This 6.10

If we wanted to know the energy value of the soda in units of joules per gram (J/g), how would we calculate that and what additional information would we need?

### Practice Problem 6.10

How many joules of energy are in a candy bar that contains 235 Cal?

**Further Practice:** Questions 6.79 and 6.80 at the end of the chapter

**Specific Heat** When heat is added to a substance, the temperature of that substance increases. The amount of heat that must be added to 1 g of a substance to raise its temperature by 1°C is a unique property of that substance, called its **specific heat.** Specific heat values are expressed in units of joules per gram per degree Celsius [J/(g °C)] or calories per gram per degree Celsius [cal/(g °C)]. The specific heat of liquid water is 1.000 cal/(g °C) or 4.184 J/(g °C). This means that if we have a 1.00-g sample of water and we want to increase its temperature by 1.00°C, we must add 1.00 cal or 4.18 J of heat. Similarly, if we have a 2.00-g sample of water and want to increase its temperature by 1.00°C, we must add 2.00 cal or 8.37 J of heat. The more of that substance we have, the more heat we need to add for the same temperature change. How much energy will increase the temperature of a 1.00-g sample by 10.0°C? You need 10.0 times as much heat for a 10.0°C increase than for a 1.00°C increase; you need 10.0 cal or 41.8 J. Some representative values of specific heat for various substances are presented in Table 6.2.

When you eat pizza fresh from the oven, why do you burn the top of your mouth and not your tongue? One factor is that the specific heat of cheese is greater than that of the crust. The cheese transfers a greater amount of heat for a given temperature change. Another factor is the greater ability of the cheese to conduct heat. It transfers heat faster.

**TABLE 6.2** Specific Heats of Some Substances

| Substance | Specific Heat J/(g °C) | Specific Heat cal/(g °C) | Substance | Specific Heat J/(g °C) | Specific Heat cal/(g °C) |
|---|---|---|---|---|---|
| aluminum(s) | 0.895 | 0.214 | water(s) | 2.027 | 0.484 |
| carbon (diamond) | 0.508 | 0.121 | water(l) | 4.184 | 1.000 |
| carbon (graphite) | 0.708 | 0.169 | water(g) | 2.015 | 0.482 |
| calcium(s) | 0.656 | 0.157 | asphalt | 0.920 | 0.220 |
| chromium(s) | 0.450 | 0.108 | bone | 0.440 | 0.105 |
| copper(s) | 0.377 | 0.0900 | brick | 0.84 | 0.20 |
| gold(s) | 0.129 | 0.0310 | cheddar cheese | 2.60 | 0.621 |
| iodine(s) | 0.214 | 0.0510 | concrete | 0.88 | 0.21 |
| iron(s) | 0.448 | 0.107 | glass | 0.84 | 0.20 |
| lead(s) | 0.129 | 0.0310 | granite | 0.79 | 0.19 |
| mercury(l) | 0.140 | 0.0335 | marble | 0.86 | 0.21 |
| silver(s) | 0.234 | 0.0560 | olive oil | 1.79 | 0.428 |
| tin(s) | 0.222 | 0.0530 | sand | 0.835 | 0.200 |
| uranium(s) | 0.117 | 0.0280 | strawberries | 3.89 | 0.930 |
| | | | wax | 2.89 | 0.69 |

The amount of heat absorbed by a substance when it increases in temperature equals the amount of heat released when it cools back down to the same temperature. The amount of heat transferred to or from a substance is related to the mass, the specific heat, and the change, or difference, in temperature.

---

### EXAMPLE 6.11    Factors Affecting Heat

Samples of different metals having the same mass are heated to the same temperature and are then allowed to cool in wax. When all have cooled to the same temperature, they have melted through different amounts of the wax. Which of the metals likely has the greatest specific heat?

**Solution:**

From the figure, we see that metal B melted the wax to the lowest level. It did so because it released the greatest amount of heat. Consider the factors that determine the amount of heat transferred. Greater heat could arise from a greater mass, greater specific heat, or greater temperature change. We started with samples of the same mass, and they all underwent the same temperature change, so these are not factors. Therefore, the specific heat must be greater for metal B than for the other metals.

**Consider This 6.11**

If four pieces of the same metal but different masses were heated to the same temperature and then allowed to cool in wax, which metal piece should melt the wax the most? The one with the greatest mass or the least mass, or would all melt the same amount of wax?

**Practice Problem 6.11**

A drop of boiling water spilled on your hand stings, but a cup of boiling water spilled on your hand burns. What factor explains the difference?

**Further Practice:** Questions 6.81 and 6.82 at the end of the chapter

---

If we know the specific heat of an object, its mass, and its change in temperature, we can determine the amount of heat absorbed or released. We can calculate the heat ($q$) associated with a temperature change by multiplying the mass in grams $m$ of a substance by its specific heat $C$ and by the temperature change $\Delta T$ (which is the difference between the final temperature and the initial temperature, $T_f - T_i$):

$$q = m \times C \times \Delta T$$

Heat    Mass    Specific heat    Temperature change

When the temperature of a substance increases ($\Delta T$ is positive), $q$ has a positive value; heat is absorbed by the substance and it gets hotter. When the temperature decreases ($\Delta T$ is negative), $q$ is negative; heat is released by the substance and it cools down.

We can't measure heat directly. There is no heat-measuring device. Instead, we measure the initial and final temperatures of an object and calculate heat from the specific heat, mass, and temperature change. Example 6.12 demonstrates this.

---

**EXAMPLE 6.12    Heat and Temperature Change**

A simple solar energy heater is a glass-topped box containing rocks. Sunlight heats the rocks during the day, and air is blown over them at night to heat a house. The box contains $7.5 \times 10^4$ g of rocks, which have a specific heat of 0.49 J/(g °C). How much additional heat can the rocks store when they warm from a nighttime temperature of 18°C (65°F) to a daytime temperature of 43°C (110°F)?

**Solution:**

The temperature of the rocks increases, so they absorb heat, and we expect the calculated value of $q$ to be positive. To calculate heat we use the following equation:

$$q = m \times C \times \Delta T$$

The mass $m$ and specific heat $C$ values are given. The temperature change $\Delta T$ is not given, but we can calculate $\Delta T$ from the final and initial temperatures:

$$\Delta T = T_f - T_i = 43°C - 18°C = 25°C$$

Substituting values of mass, specific heat, and temperature change into the equation gives the heat change for the rocks:

$$q = m \times C \times \Delta T$$

$$= 7.5 \times 10^4 \, \cancel{g} \times 0.49 \, \frac{J}{\cancel{g} \, °\cancel{C}} \times 25 \, °\cancel{C}$$

$$= 7.5 \times 10^4 \times 0.49 \, J \times 25$$

$$= 9.2 \times 10^5 \, J \quad \text{or} \quad 9.2 \times 10^2 \, kJ$$

Note that $q$ is positive as we predicted, which means that the rocks have absorbed heat.

**Consider This 6.12**

If more rocks were added to the solar heater, would the amount of heat absorbed change?

**Practice Problem 6.12**

The specific heat of aluminum is 0.895 J/(g °C). If 156 g of aluminum at 75.0°C is cooled to 25.5°C, how much heat is transferred? What is the sign of $q$, and what is its significance?

**Further Practice:**  Questions 6.85 and 6.86 at the end of the chapter

---

More sophisticated solar energy systems use silicon semiconductor panels that convert sunlight into electricity. Use of solar energy for residential homes is becoming more popular as the cost of solar panels continues to decrease.

©Elenathewise/Getty Images

---

We have calculated heat from known values of mass, specific heat, and temperature change. Since the heat equation uses just these four quantities, any one can be determined if the other three are known.

In the examples we have looked at so far, we have focused on the amount of heat absorbed or released by an object such as a rock or a metal. Sometimes, however, we are interested not in an object, but in a process such as a chemical reaction. We call whatever we are interested in—whether an object or a process—the *system*. When a system absorbs heat, where does the heat come from? When a system releases heat, where does the heat go? We know from the law of conservation of

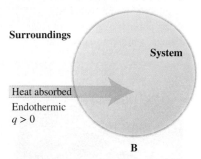

**FIGURE 6.17** (A) When the system releases heat to the surroundings, $q$ is a negative value ($q < 0$). (B) When the system absorbs heat from the surroundings, $q$ is a positive value ($q > 0$).

energy that energy is not created or destroyed. Heat absorbed by a system comes from its *surroundings,* which might be the air around it—or if in solution—the water around it. Heat released by a system is transferred to its surroundings.

Consider a car engine that heats up due to the combustion reaction in the cylinder. If we describe the system as the combustion reaction in the cylinder, is the value of $q$ for the system positive or negative? We can't take the temperature of the combustion reaction, but we can tell that the surroundings of the combustion reaction (mostly the engine) increases in temperature. The car's engine must be absorbing heat. From the law of conservation of energy, we know that the heat is coming from the combustion reaction. The combustion reaction is releasing heat, so its value of $q$ is negative. Figure 6.17 shows the relationship between a system, its surroundings, and heat transfer.

For some substances such as water, it's straightforward to calculate the amount of heat absorbed or released when it undergoes a process involving heat transfer. We know the specific heat, and mass is easily measured. The initial and final temperatures of the water are easily measured, too. But the calculation is more difficult for an object involved in heat transfer, such as a metal pipe. We might not know its specific heat, and we can't directly measure the temperature change of the pipe. Instead, we must design a heat transfer process that will let us determine the amount of heat transferred from the pipe to its surroundings, such as a known amount of water. In this case, we call the piece of pipe the system because it's what we are interested in. If we ensure that heat does not escape to anything but the water, then the water is the surroundings. There is no net loss of heat, so the heat change of the system plus the heat change of the surroundings equals zero:

$$q_{\text{system}} + q_{\text{surroundings}} = 0$$

In this example

$$q_{\text{pipe}} + q_{\text{water}} = 0$$

Suppose a 92.0-g piece of pipe is heated and then placed in an insulated container filled with 100.0 g of water at 25.00°C (Figure 6.18). The final temperature of the mixture is 29.45°C. If we assume no heat is lost from the water to the container, we can calculate the amount of heat the pipe loses to the water. It is the same as the amount of heat the water gains, which equals the mass of the water multiplied by the specific heat of water and its temperature change:

$$q_{\text{water}} = m \times C \times \Delta T$$

$$= 100.0 \text{ g} \times 4.184 \frac{\text{J}}{\text{g °C}} \times (29.45 \text{ °C} - 25.00 \text{ °C})$$

$$= 1860 \text{ J}$$

From the law of conservation of energy, the heat change of the pipe plus the heat change of the water equals zero:

$$q_{\text{pipe}} + q_{\text{water}} = 0$$

The heat change for the pipe, then, is the same value, but with the opposite sign:

$$q_{\text{pipe}} = -q_{\text{water}} = -1860 \text{ J}$$

The process just described is commonly used to determine the heat change for an object for which we do not know the specific heat, or if the initial temperature of the object is difficult to measure. The insulated container is commonly called a *calorimeter*. The measurement of heat transfer using a calorimeter is called *calorimetry*.

**FIGURE 6.18** The amount of heat released by an object can be determined by measuring the amount of heat absorbed by the water, assuming the thermometer and stirrer absorb no heat. This experimental process is called *calorimetry.* The insulated container is called a *calorimeter.*

| EXAMPLE 6.13 | **Calorimetry** |

A brick is placed in an insulated calorimeter containing 5.000 kg of water. The temperature of the water decreases from 25.0°C to 19.4°C when thermal equilibrium is reached.

(a) Was the initial temperature of the brick greater than or less than the initial temperature of the water? Explain.
(b) What is the heat change $q$ of the brick?

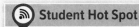

**Student Hot Spot**

Student data indicate you may struggle with calorimetery calculations. Access the SmartBook to view additional Learning Resources on this topic.

**Solution:**

(a) The temperature of the water decreased, so the temperature of the brick must have increased. Since they both have the same temperature at thermal equilibrium, the brick must have started out at a lower temperature than the initial temperature of the water.

(b) We don't know the mass or temperature of the brick, so we can't determine the heat change $q$ for the brick directly. However, we do know the mass, specific heat, and temperature change of the water. The water is the brick's surroundings, so if we determine the heat change of the water, we can use the law of conservation of energy to determine the heat change of the brick. To determine the heat change of the water, we use the equation

$$q_{water} = m \times C \times \Delta T$$

The temperature change $\Delta T$ is the difference between the final temperature and the initial temperature:

$$\Delta T = T_f - T_i$$
$$= 19.4°C - 25.0°C$$
$$= -5.6°C$$

The mass of the water must be in units of grams, so we must convert kilograms to grams:

$$\text{Mass in grams} = 5.000 \text{ kg} \times \frac{1000 \text{ g}}{1 \text{ kg}} = 5.000 \times 10^3 \text{ g}$$

Substituting values for the mass of the water, specific heat of the water, and the temperature change of the water, we get

$$q_{water} = 5.000 \times 10^3 \text{ g} \times 4.184 \frac{J}{g \, °C} \times (-5.6 \, °C)$$

$$= -1.2 \times 10^5 \text{ J}$$

The negative value of $q_{water}$ means that the water released heat. From the law of conservation of energy, we know that the brick absorbed the same amount of heat that the water released. The heat change of the brick plus the heat change of the water must sum to zero. This means that the heat change of the water is equal and opposite in sign to the heat change of the brick:

$$q_{brick} + q_{water} = 0$$
$$q_{brick} = -q_{water} = +1.2 \times 10^5 \text{ J}$$

**Consider This 6.13**

If we use a greater mass of water, the temperature of the water drops to only 21.0°C at thermal equilibrium. Would the heat released from the brick increase, decrease, or remain the same?

**Practice Problem 6.13**

A sample of a metal alloy is heated and then placed in 125.0 g of water held in a calorimeter at 22.5°C. The final temperature of the water is 29.0°C. Assume heat exchange only occurs between the water and the alloy.

(a) Was the initial temperature of the alloy greater than or less than the initial temperature of the water? Explain.
(b) What is the heat change of the alloy?

**Further Practice:** Questions 6.89 and 6.90 at the end of the chapter

## 6.7 Heat Changes in Chemical Reactions

Lily wonders how the amount of heat released by combustion of a fuel such as ethanol is determined. A common way is to use a special type of calorimeter called a *bomb calorimeter* (Figure 6.19). It's similar to the calorimeter used for objects like pipes and bricks, but the reaction is kept separate from the water in a compartment that contains an ignition wire. A specific amount of the reactant is placed in the compartment. A small amount of electricity sent through the ignition wire gets the reaction going. Since all combustion reactions are exothermic, the reaction releases heat that the water absorbs.

By measuring the temperature change of the water, we can calculate the heat change of the water, $q_{water}$, using its specific heat. Assuming no heat is absorbed by the calorimeter, we can determine the heat change for the reaction, $q_{reaction}$, using the law of conservation of energy:

$$q_{reaction} + q_{water} = 0$$

$$q_{reaction} = -q_{water}$$

For example, if 5.0 g of octane burns in a calorimeter, and the surrounding water absorbs 240 kJ of heat ($q_{water} = +240$ kJ), then the combustion reaction must have released 240 kJ of heat:

$$q_{reaction} = -q_{water} = -240 \text{ kJ}$$

A bomb calorimeter can be used to determine the Calorie (energy) content of foods. Foods are fuels. They release their energy slowly when oxidized in the body or rapidly when burned in a calorimeter. Either way, the amount of energy is the same.

**FIGURE 6.19** A bomb calorimeter is commonly used to measure the amount of heat released in combustion reactions. The water absorbs the heat released by the reaction.

The amount of energy released depends on the amount of substance burned. A few drops of a substance will produce much less heat than a gallon of that substance. In combustion reactions, it's common to report the heat change in terms of the amount of the substance that reacts with oxygen. One common way to report heat changes is in units of *kilojoules per gram* (kJ/g). We just calculated the heat change for burning 5.0 g of octane. The heat change *per gram* is one-fifth of the heat change when 5.0 g burns:

$$\frac{-240 \text{ kJ}}{5.0 \text{ g}} = -48 \text{ kJ/g}$$

Another way to report heat changes in chemical reactions is in units of *kilojoules per mole* (kJ/mol). Consider the combustion of propane gas, $C_3H_8$, used to fuel a gas grill:

$$C_3H_8(g) + 5O_2(g) \longrightarrow 3CO_2(g) + 4H_2O(g)$$

When 1 mol of propane burns, 2220 kJ of heat is released. The heat change for the combustion of propane, then, is −2220 kJ per mol, or

$$q_{\text{reaction}} = -2220 \text{ kJ/mol}$$

This ratio is described as the heat of reaction, and it allows us to calculate the heat change when any molar amount of propane reacts. For example, if 2.00 mol of propane reacts, twice as much heat is released. This can be calculated by multiplying the 2.00 mol propane by the heat change per mole:

$$\text{Heat change} = 2.00 \text{ mol } C_3H_8 \times \frac{-2220 \text{ kJ}}{1 \text{ mol } C_3H_8} = -4440 \text{ kJ}$$

MATH TOOLBOX **1.3**

**Student Hot Spot**

Student data indicate you may struggle with heat changes in chemical reactions. Access the SmartBook to view additional Learning Resources on this topic.

---

### EXAMPLE 6.14   Heat Changes in Chemical Reactions

The ethanol, $CH_3CH_2OH$, in Lily's car burns with excess oxygen according to the following equation:

$$CH_3CH_2OH(l) + 3O_2(g) \longrightarrow 2CO_2(g) + 3H_2O(l)$$

The heat change $q$ for this reaction is −1367 kJ per mole of $CH_3CH_2OH$ that reacts.

(a)  Is this reaction endothermic or exothermic?
(b)  What is the heat change when 0.200 mol of ethanol burns?

**Solution:**

(a)  The value of $q$ is negative, so the reaction is exothermic; energy is released by the reaction.
(b)  To calculate the heat change when 0.200 mol $CH_3CH_2OH$ reacts, multiply 0.200 mol $CH_3CH_2OH$ by the heat change per mole of $CH_3CH_2OH$:

$$\text{Heat change} = 0.200 \text{ mol } CH_3CH_2OH \times \frac{-1367 \text{ kJ}}{1 \text{ mol } CH_3CH_2OH} = -273 \text{ kJ}$$

MATH TOOLBOX **1.3**

When 0.200 mol $CH_3CH_2OH$ reacts, 273 kJ of heat is *released* to the surroundings.

**Consider This 6.14**

If only 0.1000 mol $CH_3CH_2OH$ burns by this reaction, how much heat would be released to the surroundings?

**Practice Problem 6.14**

The balanced equation for the combination reaction of hydrogen gas and solid iodine is

$$H_2(g) + I_2(s) \longrightarrow 2HI(g)$$

The heat change $q$ for this reaction is +53.00 kJ per mole of $I_2$ that reacts.

(a) Is this reaction endothermic or exothermic?
(b) What is the heat change when 2.50 mol of $I_2$ reacts?

**Further Practice:** Questions 6.99 and 6.100 at the end of the chapter

The combustion of ethanol releases less energy than the combustion of the hydrocarbons, like octane, that are found in gasoline. Thus, ethanol, although being used as a gasoline additive or replacement, is a less efficient fuel than gasoline. On the other hand, fuels derived from plant materials are renewable, consistent with the principles of green chemistry.

Foods and fuels are usually sold by volume or weight, so the kJ/g ratio is often useful. A mole of liquid octane, $C_8H_{18}$, for example, has a much greater mass and volume than a mole of liquid propane, $C_3H_8$. Table 6.3 shows the energy changes that correspond to the combustion of 1 g of a variety of fuels. Note that the energy released *per mole* generally increases with the number of carbon atoms in the molecular formula. However, the energy released *per gram* is relatively constant for compounds that contain only carbon and hydrogen. The energy released per gram is lower for methanol and ethanol, which also contain oxygen. Hydrogen provides the most energy per gram, which is one reason why it offers promise as a fuel of the future.

Fuels aren't the only substances that can be described in terms of energy content and mass. The energy content of foods is also reported in energy units per gram, and most of the world reports the energy content of foods in kilojoules. In the United States, the preferred unit is Cal/g, which is the same as kcal/g. The energy content of foods comes primarily from proteins, fats, and carbohydrates. The energy values of these components are listed in Table 6.4. Note that fats have a much higher energy value per gram than proteins or carbohydrates. This is one reason why many weight watchers limit the amount of fat in their diet.

Notice in Table 6.3 that a relatively small amount of energy is released when TNT (trinitrotoluene) burns. Other fuels release much more energy per mole or per gram. Is this surprising? The explosive nature of TNT has to do with its detonation *velocity*. The *rate* at which the reaction occurs is greater than for other fuels.

**TABLE 6.3** Heat Changes for the Combustion of Fuels

Substance $+ \ xO_2(g) \longrightarrow yCO_2(g) + zH_2O(l)$

| Fuel | Formula | Heat Change kJ/mol | Heat Change kJ/g |
|------|---------|------|------|
| hydrogen | $H_2(g)$ | −286 | −143 |
| methanol | $CH_3OH(l)$ | −726 | −23 |
| methane | $CH_4(g)$ | −891 | −56 |
| ethanol | $CH_3CH_2OH(l)$ | −1367 | −30 |
| acetylene | $C_2H_2(g)$ | −1301 | −50 |
| ethylene | $C_2H_4(g)$ | −1411 | −50 |
| ethane | $C_2H_6(g)$ | −1561 | −52 |
| propane | $C_3H_8(g)$ | −2219 | −50 |
| butane | $C_4H_{10}(l)$ | −2878 | −50 |
| octane | $C_8H_{18}(l)$ | −5471 | −48 |
| wood | — | — | −15 |
| TNT | $C_7H_5(NO_2)_3(s)$ | −3434 | −15 |

**TABLE 6.4** Energy Values of Food Components

| Food Component | kcal/g (Cal/g) | kJ/g |
|---|---|---|
| proteins | 4 | 17 |
| fats | 9 | 38 |
| carbohydrates | 4 | 17 |

The data in Table 6.4 are used to determine the energy content of foods as presented in most nutritional labels (Figure 6.16). Instead of burning every food in a calorimeter, food industry scientists determine the grams of protein, fat, and carbohydrate present in one serving. The mass of each is multiplied by their respective energy values. The calories from proteins, fats, and carbohydrates are then added to get the total energy content. A typical food label also lists the number of calories from fat.

# CHAPTER REVIEW

## KEY CONCEPTS

- We use balanced equations to determine quantities of reactants and products involved in chemical reactions. The coefficients tell us the relative numbers of moles of reactants that combine and products that form.
  - We combine coefficients to give mole ratios (or molecule ratios) that allow us to calculate moles of one reaction species from another one that is known.
  - If the mass in grams is given or required, conversion between grams and moles using molar mass must be added to the calculation.
- When reactants are not present in exactly the right amounts for complete reaction of all reactants, only the limiting reactant reacts completely.
  - Because the limiting reactant is in short supply, its amount limits the amount of product that can form. This is why we use the amount of the limiting reactant to calculate the theoretical yield—the amount of product that is predicted.
  - If the reaction is done in the laboratory, the amount of product actually obtained can be compared to the theoretical yield by calculating a percent yield.
- Energy changes accompany chemical and physical changes.
  - Endothermic reactions absorb heat from the surroundings, and exothermic reactions release heat to the surroundings.
  - The specific heat value for a substance is the amount of heat needed to raise the temperature of 1 g of a substance by 1°C.
  - A substance's specific heat value, $C$, allows us to relate the heat change of a substance, $q$, its mass, $m$, and temperature change, $\Delta T$, by the equation:

$$q = m \times C \times \Delta T$$

- Calorimetry is used to determine the heat change for a system (usually an object or a chemical reaction).
  - Calorimetry usually involves calculating the heat change of the surroundings ($q_{surroundings}$) from its measured temperature change. The law of conservation of energy is then used to relate the heat change of the system to the heat change of the surroundings: $q_{system} = -q_{surroundings}$
  - Heat changes for chemical reactions are usually reported in units of kJ/mol or kJ/g of a specific reactant.

## KEY RELATIONSHIPS

| Relationship | Equation |
|---|---|
| The percent yield for a reaction is a ratio of the amount of product actually obtained in a chemical reaction to the theoretical yield, the amount calculated from the limiting reactant. | $\text{Percent yield} = \dfrac{\text{actual yield}}{\text{theoretical yield}} \times 100\%$ |
| The heat change for an object is equal to the product of its mass, its specific heat, and its temperature change. The temperature change $\Delta T$ is the difference between the final temperature and the initial temperature. | $q = m \times C \times \Delta T$ |

## KEY TERMS

actual yield (6.5)

endothermic reaction (6.6)

exothermic reaction (6.6)

heat (6.6)

law of conservation of
  energy (6.6)

limiting reactant (6.4)

percent yield (6.5)

specific heat (6.6)

theoretical yield (6.5)

## QUESTIONS AND PROBLEMS

The following questions and problems, except those in *Additional Questions* and *Concept Review Questions,* are paired. Questions in a pair focus on the same concept. Answers to selected questions and problems are in Appendix D.

### Matching Definitions with Key Terms

6.1    Match the key terms with the descriptions provided.
  (a) the process of determining the amounts of reactants and products in a chemical reaction
  (b) the energy that is transferred between objects due to a difference in their temperatures
  (c) a chemical change that absorbs heat
  (d) the amount of heat required to raise the temperature of 1 g of a substance by 1°C
  (e) the amount of product actually obtained in the laboratory from a reaction

6.2    Match the key terms with the descriptions provided.
  (a) the reactant that is completely used up in a reaction and determines the amount of product that can form
  (b) a chemical change that releases heat
  (c) the maximum amount of product that can be obtained in a chemical reaction from known amounts of reactants; the amount of product calculated by assuming all of the limiting reactant is consumed
  (d) the principle that states that the total amount of energy remains the same no matter how a physical or chemical change occurs
  (e) the amount of a product actually obtained as a proportion of the amount expected, expressed as a percentage

### The Meaning of a Balanced Equation

6.3    When 1 molecule of glucose, $C_6H_{12}O_6$, is metabolized in our bodies, it combines with 6 $O_2$ molecules to form 6 $CO_2$ molecules and 6 $H_2O$ molecules.
  (a) Write a balanced equation for this reaction.
  (b) How many $CO_2$ molecules are formed when 12 glucose molecules react?
  (c) If 30 $O_2$ molecules react, how many glucose molecules also react?

6.4    When 2 molecules of methanol, $CH_3OH$, react with oxygen, they combine with 3 $O_2$ molecules to form 2 $CO_2$ molecules and 4 $H_2O$ molecules.
  (a) Write a balanced equation for this reaction.
  (b) How many $H_2O$ molecules are formed when 10 methanol molecules react?
  (c) If 30 $O_2$ molecules react, how many methanol molecules react?

6.5    What do the coefficients in a balanced equation represent?

6.6    Why is a balanced equation important when determining the amount of product that can be produced from a given amount of reactant?

6.7    The balanced equation for the dissolving of magnesium nitrate in water is

$$Mg(NO_3)_2(s) \xrightarrow{\;H_2O\;} Mg^{2+}(aq) + 2NO_3^-(aq)$$

  (a) How many $Mg^{2+}(aq)$ and $NO_3^-(aq)$ ions form for each $Mg(NO_3)_2$ formula unit that dissolves?
  (b) How many moles of $Mg^{2+}$ and $NO_3^-$ form for each mole of $Mg(NO_3)_2$ that dissolves?

**6.8** The balanced equation for the dissolving of sodium phosphate in water is

$$Na_3PO_4(s) \xrightarrow{H_2O} 3Na^+(aq) + PO_4^{3-}(aq)$$

(a) How many $Na^+$ and $PO_4^{3-}$ ions form for each $Na_3PO_4$ formula unit that dissolves?

(b) How many moles of $Na^+$ and $PO_4^{3-}$ form for each mole of $Na_3PO_4$ that dissolves?

**6.9** The molecular-level diagram shows the reactants (before the reaction) and the products (after the reaction). Which of the following is the best balanced equation to represent this reaction? For the choices you did not select, explain why they are incorrect.

(a) $2A + 4B \longrightarrow 4AB$
(b) $A_2 + B_2 \longrightarrow AB_2$
(c) $2A_2 + 4B_2 \longrightarrow 4AB_2$
(d) $A_2 + 2B_2 \longrightarrow 2AB_2$

**6.10** The molecular-level diagram shows the reactants (before the reaction) and the products (after the reaction). Which of the following is the best balanced equation to represent this reaction on any scale? For the choices you did not select, explain why they are incorrect.

(a) $A_2 + B_2 \longrightarrow AB_3$
(b) $A_2 + 2B_2 \longrightarrow 2AB_3$
(c) $A_2 + 3B_2 \longrightarrow 2AB_3$
(d) $3A_2 + 9B_2 \longrightarrow 6AB_3$

## Mole-Mole Conversions

**6.11** Benzene, $C_6H_6$, is a known carcinogen that burns in air according to the following equation:

$$2C_6H_6(l) + 15O_2(g) \longrightarrow 12CO_2(g) + 6H_2O(g)$$

(a) What is the mole ratio of $O_2$ to $C_6H_6$?
(b) How many moles of $O_2$ are required to react with each mole of $C_6H_6$?
(c) How many moles of $O_2$ are required to react with 0.38 mol of $C_6H_6$?

**6.12** Aluminum reacts with chlorine gas according to the following equation:

$$2Al(s) + 3Cl_2(g) \longrightarrow 2AlCl_3(s)$$

(a) What is the mole ratio of $Cl_2$ to $Al$?
(b) How many moles of $Cl_2$ are required to react with each mole of $Al$?
(c) How many moles of $Cl_2$ are required to react with 0.085 mol of $Al$?

**6.13** Suppose you have a piece of aluminum that you want to react completely by the following single-displacement reaction:

$$2Al(s) + 6HNO_3(aq) \longrightarrow 2Al(NO_3)_3(aq) + 3H_2(g)$$

(a) What mole ratio would you use in the following equation to determine the number of moles of $HNO_3$ needed to react with a known amount of $Al$?

$$\text{mol Al} \times \text{————} = \text{mol } HNO_3$$

(b) If you add more than enough nitric acid so that all the aluminum reacts, what mole ratio would you use in the following equation to determine the moles of $H_2$ produced?

$$\text{mol Al} \times \text{————} = \text{mol } H_2$$

(c) Suppose you know the number of moles of $H_2$ product formed and you want to know the number of moles of $Al$ that reacted. What mole ratio would you use in the following equation?

$$\text{mol } H_2 \times \text{————} = \text{mol Al}$$

**6.14** Suppose you want to convert iron ore to a specific amount of pure iron using the following reaction:

$$Fe_3O_4(s) + 4CO(g) \longrightarrow 3Fe(s) + 4CO_2(g)$$

(a) What mole ratio would you use in the following equation to determine the number of moles of $CO$ needed to react with a known amount of $Fe_3O_4$?

$$\text{mol } Fe_3O_4 \times \text{————} = \text{mol CO}$$

(b) If you add more than enough $CO$ so that all the $Fe_3O_4$ reacts, what mole ratio would you use in the following equation to determine the moles of $CO_2$ produced?

$$\text{mol } Fe_3O_4 \times \text{————} = \text{mol } CO_2$$

(c) Suppose you know the number of moles of $Fe$ product formed and you want to know the number of moles of $CO$ that reacted. What mole ratio would you use in the following equation?

$$\text{mol Fe} \times \text{————} = \text{mol CO}$$

**6.15** Consider the combustion reaction of propane, used in barbecue grills:

$$C_3H_8(g) + 5O_2(g) \longrightarrow 3CO_2(g) + 4H_2O(g)$$

Which of the following is conserved in this reaction?
(a) moles of molecules
(b) moles of atoms
(c) atoms
(d) mass
(e) Which of your answers to (a) through (d) are true for any reaction?

6.16 Consider the combination reaction of hydrogen and iodine:

$$H_2(g) + I_2(s) \longrightarrow 2HI(g)$$

Which of the following is conserved in this reaction?
(a) moles of molecules
(b) moles of atoms
(c) atoms
(d) mass
(e) Which of your answers to (a) through (d) are true for any reaction?

6.17 The balanced equation for the decomposition of TNT, $C_7H_5(NO_2)_3$, is

$$2C_7H_5(NO_2)_3(s) \longrightarrow 7C(s) + 7CO(g) + 3N_2(g) + 5H_2O(g)$$

(a) Write the mole ratios that relate moles of each product to the reactant $C_7H_5(NO_2)_3$.
(b) How many moles of each product would form if 1.00 mol of $C_7H_5(NO_2)_3$ reacts?
(c) How many moles of each product would form if 6.25 mol of $C_7H_5(NO_2)_3$ react?

6.18 The balanced equation for the decomposition of ammonium nitrate is

$$2NH_4NO_3(s) \longrightarrow 2N_2(g) + O_2(g) + 4H_2O(g)$$

(a) Write the mole ratios that relate moles of each product to the reactant $NH_4NO_3$.
(b) How many moles of each product would form if 1.00 mol of $NH_4NO_3$ reacts?
(c) How many moles of each product would form if 12.4 mol of $NH_4NO_3$ react?

## Mass-Mass Conversions

6.19 When copper(II) sulfate pentahydrate, $CuSO_4 \cdot 5H_2O$, is heated, it decomposes to the dehydrated form. The waters of hydration are released from the solid crystal and form water vapor. The hydrated form is medium blue, and the dehydrated solid is light blue. The balanced equation is

$$CuSO_4 \cdot 5H_2O(s) \xrightarrow{\text{heat}} CuSO_4(s) + 5H_2O(g)$$

©Brian Moeskau/Moeskau Photography

(a) What is the molar mass of $CuSO_4 \cdot 5H_2O$?
(b) What is the molar mass of $CuSO_4$?
(c) If 1.00 g $CuSO_4 \cdot 5H_2O$ is decomposed to $CuSO_4$, predict the mass of the remaining light blue solid.

6.20 When the reddish-brown mercury(II) oxide, HgO, is heated, it decomposes to its elements, liquid mercury metal and oxygen gas:

$$2HgO(s) \xrightarrow{\text{heat}} 2Hg(l) + O_2(g)$$

©McGraw-Hill Education/
Stephen Frisch

(a) What is the molar mass of HgO?
(b) What is the molar mass of Hg?
(c) If 2.00 g HgO is decomposed to Hg, predict the mass of the pure Hg metal produced.

6.21 When active metals such as sodium are exposed to air, they quickly form a coating of metal oxide. The balanced equation for the reaction of sodium metal with oxygen gas is

$$4Na(s) + O_2(g) \longrightarrow 2Na_2O(s)$$

Suppose a piece of sodium metal gains 2.05 g of mass after being exposed to air. Assume that this gain can be attributed to its reaction with oxygen.
(a) What mass of $O_2$ reacted with the Na?
(b) What mass of Na reacted?
(c) What mass of $Na_2O$ formed?

6.22 When aluminum metal is exposed to oxygen gas, a coating of aluminum oxide forms on the surface of the aluminum. The balanced equation for the reaction of aluminum metal with oxygen gas is

$$4Al(s) + 3O_2(g) \longrightarrow 2Al_2O_3(s)$$

Suppose a sheet of pure aluminum gains 0.0900 g of mass when exposed to air. Assume that this gain can be attributed to its reaction with oxygen.
(a) What mass of $O_2$ reacted with the Al?
(b) What mass of Al reacted?
(c) What mass of $Al_2O_3$ formed?

6.23 Diiodine pentoxide is used in respirators to remove carbon monoxide from air:

$$I_2O_5(s) + 5CO(g) \longrightarrow I_2(s) + 5CO_2(g)$$

(a) What mass of carbon monoxide could be removed from air by a respirator that contains 50.0 g of diiodine pentoxide?
(b) What mass of $I_2$ would remain in the respirator?

6.24 Ethanol, used in alcoholic beverages, can be produced by the fermentation of sucrose, which is found in sugar cane and other plants. The balanced equation for the fermentation process is

$$C_{12}H_{22}O_{11}(s) + H_2O(l) \longrightarrow 4C_2H_5OH(l) + 4CO_2(g)$$

(a) What mass of ethanol, $C_2H_5OH$, would be produced when 2.50 g sucrose reacts by this process?
(b) What mass of $CO_2$ would also be produced?

6.25 For each of the following *unbalanced* equations, balance the equation and then determine how many grams of the second reactant would be required to react completely with 0.600 g of the first reactant.
(a) $Cr(s) + Cl_2(g) \longrightarrow CrCl_3(s)$
(b) $RbO_2(s) + H_2O(l) \longrightarrow O_2(g) + RbOH(s)$
(c) $C_5H_{12}(g) + O_2(g) \longrightarrow CO_2(g) + H_2O(g)$
(d) $Li(s) + Cl_2(g) \longrightarrow LiCl(s)$

6.26 For each of the following *unbalanced* equations, balance the equation and then determine how many grams of the first reactant would be required to react completely to give 1.70 g of the product indicated in boldface.
(a) $C_6H_{12}(l) + O_2(g) \longrightarrow \mathbf{CO_2}(g) + H_2O(g)$
(b) $Ni_3(s) \longrightarrow N_2(g) + \mathbf{I_2}(s)$
(c) $Na(s) + O_2(g) \longrightarrow \mathbf{Na_2O}(s)$
(d) $Ca(s) + HCl(aq) \longrightarrow CaCl_2(aq) + \mathbf{H_2}(g)$

6.27 Calcium carbonate, $CaCO_3$, is often used in commercial antacids. It acts to reduce the acidity in the stomach by neutralizing stomach acid, which is mostly HCl, by the following reaction:

$$CaCO_3(s) + 2HCl(aq) \longrightarrow CaCl_2(aq) + CO_2(g) + H_2O(l)$$

What mass of $CaCO_3$ is needed to neutralize 0.020 mol HCl?

6.28 Sulfuric acid is commonly used as an electrolyte in car batteries. Suppose you spill some on your garage floor. Before cleaning it up, you wisely decide to neutralize it with sodium bicarbonate (baking soda) from your kitchen. The reaction of sodium bicarbonate and sulfuric acid is

$$2NaHCO_3(s) + H_2SO_4(aq) \longrightarrow$$
$$Na_2SO_4(aq) + 2CO_2(g) + 2H_2O(l)$$

You estimate that your acid spill contains about 2.0 mol $H_2SO_4$. What mass of $NaHCO_3$ do you need to neutralize the acid?

## Limiting Reactants

6.29 When a fuel is burned in the air, what is typically the limiting reactant?

6.30 An acidic solution is added to a basic solution. If the resulting solution is basic, is the acid or the base the limiting reactant?

6.31 Suppose you are making turkey sandwiches for your friends. You have lots of friends, so you want to make as many sandwiches as possible. Because you are on a budget, you only have 24 pieces of bread and 15 pieces of turkey. Each sandwich consists of 2 pieces of bread and 1 piece of turkey.
(a) How many sandwiches can you make?
(b) What is the limiting part?
(c) What part is left over? How many are left over?

6.32 Your little sister is having a party and you are in charge of the party goody bags. You decide that each bag should have 3 candy bars, 1 bottle of nail polish, and 2 pieces of bubble gum. You bought an 18-pack of candy bars, 12 bottles of nail polish, and 20 pieces of bubble gum.
(a) What is the ratio of candy bars to bottles of nail polish to pieces of bubble gum in the goody bags?
(b) What is the limiting "goody"?
(c) Assuming you have an unlimited supply of bags, how many goody bags can you make?
(d) How many of each type of goody will be left over?

6.33 The balanced equation for the reaction of nitrogen and fluorine gas to form nitrogen trifluoride is

$$N_2(g) + 3F_2(g) \longrightarrow 2NF_3(g)$$

What is the limiting reactant when each of the following sets of quantities of reactants is mixed?
(a) 9 $N_2$ molecules and 9 $F_2$ molecules
(b) 5 $N_2$ molecules and 20 $F_2$ molecules
(c) 6 $N_2$ molecules and 18 $F_2$ molecules

6.34 The balanced equation for the reaction of chromium metal and chlorine gas is

$$2Cr(s) + 3Cl_2(g) \longrightarrow 2CrCl_3(s)$$

What is the limiting reactant when each of the following sets of quantities of reactants is mixed?
(a) 4 Cr atoms and 6 $Cl_2$ molecules
(b) 6 Cr atoms and 10 $Cl_2$ molecules
(c) 12 Cr atoms and 20 $Cl_2$ molecules

6.35 The molecular-level diagram shows a mixture of reactant molecules (three $O_2$ molecules and eight $H_2$ molecules) for the following reaction:

$$2H_2(g) + O_2(g) \longrightarrow 2H_2O(g)$$

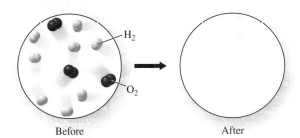

Before          After

(a) Draw what the mixture should look like when the reaction is complete.
(b) What is the limiting reactant?
(c) Which reactant is left over?

6.36 The molecular-level diagram shows a mixture of reactant molecules for the following reaction:

$$2C_2H_2(g) + 5O_2(g) \longrightarrow 4CO_2(g) + 2H_2O(g)$$

Before          After

(a) Draw what the mixture should look like when the reaction is complete.
(b) What is the limiting reactant
(c) Which reactant is left over?

6.37 Use the balanced equation for the combustion of ethane to complete the table.

| $2C_2H_6(g) + 7O_2(g) \longrightarrow 4CO_2(g) + 6H_2O(g)$ | | | |
|---|---|---|---|
| **Initially mixed** | 6 molecules | 18 molecules | 0 molecules | 0 molecules |
| **How much reacts** | | | — | — |
| **Composition of final mixture** | | | | |

6.38 Use the balanced equation for the combustion of ethane to complete the table.

| $2C_2H_6(g) + 7O_2(g) \longrightarrow 4CO_2(g) + 6H_2O(g)$ | | | |
|---|---|---|---|
| **Initially mixed** | 10 molecules | 20 molecules | 0 molecules | 0 molecules |
| **How much reacts** | | | — | — |
| **Composition of final mixture** | | | | |

6.39 Use the balanced equation for the combustion of butane to complete the table.

| $2C_4H_{10}(g) + 13O_2(g) \longrightarrow 8CO_2(g) + 10H_2O(g)$ | | | |
|---|---|---|---|
| **Initially mixed** | 3.10 mol | 13.0 mol | 0.00 mol | 0.00 mol |
| **How much reacts** | | | — | — |
| **Composition of final mixture** | | | | |

6.40 Use the balanced equation for the combustion of butane to complete the table.

| $2C_4H_{10}(g) + 13O_2(g) \longrightarrow 8CO_2(g) + 10H_2O(g)$ | | | |
|---|---|---|---|
| **Initially mixed** | 0.500 mol | 3.50 mol | 0.00 mol | 0.00 mol |
| **How much reacts** | | | — | — |
| **Composition of final mixture** | | | | |

6.41 The balanced equation for the reaction of phosphorus and oxygen gas to form tetraphosphorus decoxide is

$$P_4(g) + 5O_2(g) \longrightarrow P_4O_{10}(s)$$

What is the limiting reactant when each of the following sets of quantities of reactants is mixed?
(a) 0.50 mol $P_4$ and 5.0 mol $O_2$
(b) 0.20 mol $P_4$ and 1.0 mol $O_2$
(c) 0.25 mol $P_4$ and 0.75 mol $O_2$

6.42 Tetraphosphorus trisulfide, $P_4S_3$, is used in the heads of wooden matches. This material is manufactured by the heating of a mixture of red phosphorus and sulfur.

$$8P_4(s) + 3S_8(s) \longrightarrow 8P_4S_3(s)$$

What is the limiting reactant when each of the following sets of quantities of reactants is mixed?
(a) 4.4 mol $P_4$ and 1.1 mol $S_8$
(b) 0.20 mol $P_4$ and 0.10 mol $S_8$
(c) 0.75 mol $P_4$ and 0.28 mol $S_8$

6.43 The balanced equation for the reaction of nitrogen and fluorine to form nitrogen trifluoride is

$$N_2(g) + 3F_2(g) \longrightarrow 2NF_3(g)$$

Assume that 10.0 g of $N_2$ is mixed with 10.0 g of $F_2$.
(a) What is the limiting reactant?
(b) What is the maximum amount of $NF_3$, in grams, that can be produced?

6.44 The balanced equation for the reaction of aluminum metal and chlorine gas is

$$2Al(s) + 3Cl_2(g) \longrightarrow 2AlCl_3(s)$$

Assume that 0.40 g Al is mixed with 0.60 g $Cl_2$.
(a) What is the limiting reactant?
(b) What is the maximum amount of $AlCl_3$, in grams, that can be produced?

6.45 Sodium bicarbonate reacts with hydrochloric acid in a gas-forming reaction to produce aqueous sodium chloride, water, and carbon dioxide gas:

$$NaHCO_3(s) + HCl(aq) \longrightarrow$$
$$NaCl(aq) + H_2O(l) + CO_2(g)$$

Determine the limiting reactant and mass of $CO_2$ gas produced when 8.0 g of $NaHCO_3$ is added to a solution that contains 4.0 g of HCl.

6.46 Sodium metal reacts with water in the following single-displacement reaction:

$$2Na(s) + 2H_2O(l) \longrightarrow 2NaOH(aq) + H_2(g)$$

Determine the limiting reactant and mass of $H_2$ gas produced when 2.0 g of Na is added to 10.0 g $H_2O$.

**6.47** Use the balanced equation for the combustion of ethane to complete the table.

| $2C_2H_6(g) + 7O_2(g) \longrightarrow 4CO_2(g) + 6H_2O(g)$ | | | | |
|---|---|---|---|---|
| Initially mixed | 0.260 g | 1.00 g | 0.00 g | 0.00 g |
| How much reacts | | | — | — |
| Composition of final mixture | | | | |

**6.48** Use the balanced equation for the combustion of butane to complete the table.

| $2C_4H_{10}(g) + 13O_2(g) \longrightarrow 8CO_2(g) + 10H_2O(g)$ | | | | |
|---|---|---|---|---|
| Initially mixed | 0.130 g | 0.580 g | 0.00 g | 0.00 g |
| How much reacts | | | — | — |
| Composition of final mixture | | | | |

**6.49** An aqueous solution containing 10.0 g NaOH is added to an aqueous solution containing 10.0 g $HNO_3$.
(a) Write a balanced equation for the acid-base reaction that should occur.
(b) Identify the limiting reactant.
(c) Will the solution be acidic or basic when the reaction is complete?

**6.50** An aqueous solution containing 10.0 g NaOH is added to an aqueous solution containing 10.0 g $H_3PO_4$.
(a) Write a balanced equation for the acid-base reaction that should occur.
(b) Identify the limiting reactant.
(c) Will the solution be acidic or basic when the reaction is complete?

## Percent Yield

**6.51** In a process for obtaining pure copper from copper ore, 55.6 g of copper metal was obtained. Is this the theoretical yield or actual yield?

**6.52** After determining the limiting reactant in a reaction, you calculate the amount of product that can form from the limiting reactant. Is this the theoretical yield or the actual yield?

**6.53** A student carried out the following precipitation reaction:

$$BaCl_2(aq) + Na_2CO_3(aq) \longrightarrow BaCO_3(s) + 2NaCl(aq)$$

After no more precipitate appeared to form, he filtered the barium carbonate solid and then immediately weighed it. Using this mass, he calculated the percent yield as 105%. Is this possible? Explain.

**6.54** What factors may contribute to a percent yield less than 100%?

**6.55** A student was synthesizing aspirin in the laboratory. Using the amount of limiting reactant, she calculated the mass of aspirin that should form as 8.95 g. When she weighed her aspirin product on the balance, its mass was 7.44 g.
(a) What is the actual yield of aspirin?
(b) What is the theoretical yield of aspirin?
(c) Calculate the percent yield for this synthesis.

**6.56** A student added zinc metal to a copper(II) nitrate solution to obtain copper metal by a single-displacement reaction. From the known moles of copper(II) nitrate in solution, she calculated the expected mass of copper metal as 6.7 g. When she weighed the copper on a balance, the mass was 5.7 g.
(a) What is the actual yield of copper metal?
(b) What is the theoretical yield of copper metal?
(c) Calculate the percent yield.

**6.57** The combination reaction of sodium metal and chlorine gas to form sodium chloride is represented by the following balanced equation:

$$2Na(s) + Cl_2(g) \longrightarrow 2NaCl(s)$$

If 5.00 g of sodium metal reacts completely with excess chlorine gas, and 11.5 g NaCl is actually obtained, what is the percent yield of sodium chloride?

**6.58** Ammonia is synthesized commercially from nitrogen gas and hydrogen gas for the production of fertilizers:

$$N_2(g) + 3H_2(g) \longrightarrow 2NH_3(g)$$

If 100.0 g of nitrogen reacts completely with excess hydrogen, and 34.0 g of $NH_3$ are obtained, what is the percent yield of ammonia?

**6.59** When $I_2$ is mixed with excess $H_2$, 0.80 mol HI is isolated in the laboratory. If this reaction occurs to give an 85% yield, how many moles of $I_2$ reacted? (Hint: Write a balanced equation.)

**6.60** The reaction of lithium metal and water to form lithium hydroxide and hydrogen gas is represented by the following balanced equation:

$$2Li(s) + 2H_2O(l) \longrightarrow 2LiOH(aq) + H_2(g)$$

When Li is mixed with excess water, 0.30 mol $H_2$ gas is isolated in the laboratory. If this reaction occurs to give an 85% yield of $H_2$, how many moles of Li reacted?

**6.61** The combination reaction of magnesium metal and bromine to form magnesium bromide is represented by the following balanced equation:

$$Mg(s) + Br_2(l) \longrightarrow MgBr_2(l)$$

If 1.0 mol of Mg is mixed with 2.0 mol of $Br_2$, and 0.84 mol of $MgBr_2$ is obtained, what is the percent yield for the reaction?

**6.62** The combination reaction of sodium metal and nitrogen gas to form sodium nitride is represented by the following balanced equation:

$$6Na(s) + N_2(g) \longrightarrow 2Na_3N(s)$$

If 0.30 mol Na is mixed with 0.60 mol $N_2$, and 0.092 mol $Na_3N$ is obtained, what is the percent yield for the reaction?

## Energy Changes

6.63 If energy cannot be created or destroyed, what happens to the energy of a ball as it rolls down a hill and rests at the bottom?

6.64 If energy cannot be created, what is the source of the energy that is released when gasoline is burned?

6.65 Consider the reaction of hydrogen and oxygen in a hydrogen-fuel vehicle. Describe the transformation of energy in this process in terms of potential and kinetic energy.

6.66 Consider propane fuel being burned in a propane-fueled barbecue grill. Describe the transformation of energy in this process in terms of potential and kinetic energy.

6.67 When the solids ammonium thiocyanate and barium hydroxide are mixed in a beaker, a solution forms, and the temperature of the solution drops to $-5°C$. Is this an endothermic or exothermic reaction? Explain.

6.68 When baking soda is added to a solution of sulfuric acid, the container becomes warmer to the touch. Is this an endothermic or exothermic reaction? Explain.

6.69 A doctor sprays a liquid on your skin to numb it before injecting a painkiller. You notice that the liquid evaporates quickly. You also notice that your skin feels cold. Explain why your skin feels cold.

6.70 Explain why sweating is a natural way for your body to cool itself.

6.71 What is the relationship between the potential energy of the reactants and products in an exothermic reaction?

6.72 If the reactants have a lower potential energy than the products, is the reaction exothermic or endothermic?

6.73 Convert an energy of 526 cal to units of joules.

6.74 Convert an energy of 225 cal to units of joules.

6.75 Convert an energy of 145 kJ to units of calories.

6.76 Convert an energy of 0.675 kJ to units of calories.

6.77 Convert an energy of 876 J to units of Calories.

6.78 Convert an energy of 2430 J to units of Calories.

6.79 Sodium cyclamate, $NaC_6H_{12}NSO_3$, was a popular nonsugar sweetener until it was banned by the Food and Drug Administration. Its sweetness is about 30 times that of sugar (or sucrose, $C_{12}H_{22}O_{11}$). The energy that can be supplied by ingestion of each compound is 16.03 kJ/g for sodium cyclamate and 16.49 kJ/g for sucrose. What is the energy savings in Calories resulting from using 1.00 g of sodium cyclamate in place of 30.0 g of sucrose?

6.80 In some areas, gasoline is formulated to contain 15% ethanol during some seasons. If gasoline releases 11.4 kcal/g upon combustion, but ethanol releases only 7.12 kcal/g, how much less energy in kJ/g is provided by the mixture than by pure gasoline?

6.81 If the same amount of heat is added to 50.0 g each of the metals, aluminum, copper, and lead, all at the same initial temperature, which metal will have the highest final temperature? Refer to Table 6.2 for specific heat values.

6.82 We add the same amount of heat to three different samples of the same liquid substance. The temperature of the first sample increases by the greatest amount. How can we explain the different temperature changes in each sample?

6.83 How much heat has to be added to 528 g of copper at 22.3°C to raise the temperature of the copper to 49.8°C? (See Table 6.2.)

6.84 How much heat has to be added to 235 g of iron at 25.0°C to raise the temperature of the iron to 250.0°C? (See Table 6.2.)

6.85 What is the heat change when 1.25 g of water vapor (steam) at 185.3°C is cooled to 102.1°C? The specific heat of steam is 2.02 J/(g °C).

6.86 What is the heat change when 55.0 g of water cools from 60.0°C to 25.5°C?

6.87 What are some characteristics of a good calorimeter?

6.88 What information can be obtained from a calorimetry experiment?

6.89 A rock is added to 60.0 g of water. The initial temperature of the water is 25.0°C. The final temperature of the water and the rock is 30.1°C.
   (a) Did the rock gain or lose heat?
   (b) What is the heat change of the rock?

6.90 A marble is added to 40.0 g of water. The initial temperature of the water is 25.0°C. The final temperature of the water and the marble is 22.2°C.
   (a) Did the marble gain or lose heat?
   (b) What is the heat change of the marble?

6.91 A piece of copper is placed in 75.0 g of water held in a calorimeter at 45.0°C. The final temperature of the water is 36.2°C.
   (a) Is the initial temperature of the copper greater than or less than the initial temperature of the water?
   (b) Did the copper gain or lose heat?
   (c) What is the heat change of the copper?

6.92 A scalpel is heated in an autoclave and then dropped into 145.0 g of water held in a calorimeter at 22.0°C. The final temperature of the water is 33.1°C.
   (a) Did the scalpel gain or lose heat?
   (b) Did the water gain or lose heat?
   (c) What is the heat change of the scalpel?

6.93 What is the final temperature when 70.0 g of water at 80.0°C is mixed with 30.0 g of water at 25.0°C?

6.94 What is the final temperature when a 100.0-g sample of copper pellets at 100.0°C is added to a calorimeter containing 140.0 g of water at 25.0°C? Refer to Table 6.2 for specific heat values.

## Heat Changes in Chemical Reactions

6.95 Why can't you measure the temperature change of a chemical reaction?

6.96 Why do you have to first determine the heat change of the surroundings for a chemical reaction?

6.97 You carry out a chemical reaction in a calorimeter to determine the heat change for the reaction. Identify the system and surroundings.

**6.98**  When a chemical reaction occurs in a calorimeter containing water, and the temperature of the water decreases, is the reaction endothermic or exothermic?

**6.99**  A mixture of carbon monoxide, hydrogen, and oxygen gases, known as *water gas,* can be used as an industrial fuel. The reaction of these gases is

$$CO(g) + H_2(g) + O_2(g) \longrightarrow CO_2(g) + H_2O(g)$$

This reaction releases 525 kJ of heat for every mole of carbon monoxide that reacts.
   (a) Is this reaction endothermic or exothermic?
   (b) What is the energy change for this reaction in units of kJ/mol of carbon monoxide?

**6.100**  The reaction that produces the *water gas* mixture, described in Question 6.99, is

$$C(s) + H_2O(g) \longrightarrow CO(g) + H_2(g)$$

This reaction requires an input of 131 kJ of heat for every mole of carbon that reacts.
   (a) Is this reaction endothermic or exothermic?
   (b) What is the energy change for this reaction in units of kJ/mol of carbon?

**6.101**  When a 6.00-g sample of coal is burned, it releases enough heat to raise the temperature of 2010 g of water from 24.0°C to 41.5°C.
   (a) How much heat did the coal release as it burned?
   (b) Calculate the heat of combustion of coal in units of kJ/g.

**6.102**  When a 4.00-g sample of magnesium metal is burned, it produces enough heat to raise the temperature of 2010 g of water from 24.00°C to 33.10°C.
   (a) How much heat did the magnesium release as it burned?
   (b) Calculate the heat of combustion of magnesium in units of kJ/g.

**6.103**  A 2.00-g peanut is burned in a bomb calorimeter containing 1200 g of water. The temperature of the water increases from 25.00°C to 30.25°C.
   (a) How much heat, in joules, did the peanut release as it burned?
   (b) Calculate the heat content in units of calories and Calories.
   (c) Calculate the energy value in units of Cal/g.

**6.104**  A 5.00-g piece of fat is burned in a bomb calorimeter containing 4050 g of water. The temperature of the water increases by 12.4°C.
   (a) How much heat, in joules, did the fat release as it burned?
   (b) Calculate the heat content in units of calories and Calories.
   (c) Calculate the energy value in units of Cal/g.
   (d) Calculate the heat content in units of Cal/mol. Assume the fat is all tristearin, a typical fat with the molecular formula $C_{57}H_{110}O_6$.

**6.105**  The heat change that accompanies the formation of 1.00 mol of carbon dioxide from its elements is $-393.7$ kJ/mol. What heat change accompanies the formation of 0.650 mol $CO_2$?

**6.106**  The decomposition of hydrogen peroxide to form water and oxygen gas releases 196.6 kJ per mole of hydrogen peroxide. This reaction occurs when hydrogen peroxide is placed on a cut to sterilize it. How much heat is released when 5.25 mol $H_2O_2$ decompose?

## Additional Questions

**6.107**  For each ionic compound listed, determine the number of moles of ions that exist in solution when 1.5 mol of the ionic solid dissolves in water.
   (a) $MgCl_2$
   (b) $Al(NO_3)_3$
   (c) $NH_4NO_3$

**6.108**  Certain drain cleaners are a mixture of sodium hydroxide and powdered aluminum. When dissolved in water, the sodium hydroxide reacts with the aluminum and the water to produce hydrogen gas.

$$2Al(s) + 2NaOH(aq) + 6H_2O(l) \longrightarrow 2NaAl(OH)_4(aq) + 3H_2(g)$$

The sodium hydroxide helps dissolve grease, and the hydrogen gas provides a mixing and scrubbing action. What mass of hydrogen gas would be formed from a reaction of 2.48 g Al and 4.76 g NaOH in water?

**6.109**  When silver nitrate is added to an aqueous solution of calcium chloride, a precipitation reaction occurs that removes the chloride ions from solution.

$$2AgNO_3(s) + CaCl_2(aq) \longrightarrow 2AgCl(s) + Ca(NO_3)_2(aq)$$

   (a) If a solution contains 10.0 g $CaCl_2$, what mass of $AgNO_3$ should be added to remove all of the chloride ions from solution?
   (b) When enough $AgNO_3$ is added so that all 10.0 g of $CaCl_2$ react, what mass of the AgCl precipitate should form?

**6.110**  Sometimes it is more convenient to measure the amount of a liquid reactant in volume units instead of mass units. Given that 1.00 gal is 3.79 L and the density of methanol is 0.793 g/mL, what mass of oxygen is needed to react with 1.00 gal of methanol? The balanced equation is

$$2CH_3OH(l) + 3O_2(g) \longrightarrow 2CO_2(g) + 4H_2O(g)$$

**6.111**  The balanced equation for the combustion of octane is

$$2C_8H_{18}(l) + 25O_2(g) \longrightarrow 16CO_2(g) + 18H_2O(g)$$

What mass of oxygen is needed to react with 1.00 gal of octane? (A gallon is 3.79 L, and the density of octane is 0.703 g/mL.)

**6.112**  When lead(II) nitrate is added to an aqueous solution of sodium iodide, a precipitation reaction occurs that removes the iodide ion from solution.

$$Pb(NO_3)_2(s) + 2NaI(aq) \longrightarrow 2PbI_2(s) + 2NaNO_3(aq)$$

   (a) If a solution contains 7.00 g NaI, what mass of $Pb(NO_3)_2$ should be added to remove all of the iodide ion from solution?
   (b) When enough $Pb(NO_3)_2$ is added so that all 7.00 g of NaI reacts, what mass of $PbI_2$ should precipitate?

6.113 A 5.0-g sample of greenish-yellow $Cl_2$ gas is added to a 10.0-g sample of gray potassium metal to form a white solid in a sealed container.
(a) Write a balanced equation for the combination reaction that should occur.
(b) Identify the limiting reactant.
(c) Predict the appearance of the substances left in the reaction container once the reaction is complete.
(d) Calculate the mass of product that should form.
(e) What mass of excess reactant should be mixed with the product after the reaction is complete?

6.114 A 5.00-g piece of copper metal is added to a colorless aqueous solution that contains 20.0 g of $Zn(NO_3)_2$. The products that form are a water-soluble blue ionic compound and a gray metal.
(a) Write a balanced equation for the single-displacement reaction that should occur.
(b) Identify the limiting reactant.
(c) Predict the appearance of the substances left in the reaction container once the reaction is complete.

6.115 The equation for the combustion of hydrogen is

$$2H_2(g) + O_2(g) \longrightarrow 2H_2O(g)$$

(a) What is the mole ratio of $O_2$ to $H_2$?
(b) If 1.0 g of $H_2$ reacts, what mass of $O_2$ will react with it, and what mass of $H_2O$ should form?
(c) When 1.0 g $H_2$ is mixed with 4.0 g $O_2$, what is the theoretical yield of $H_2O$?

6.116 The decomposition reaction of calcium carbonate is represented by the following balanced equation:

$$CaCO_3(s) \xrightarrow{\text{heat}} CaO(s) + CO_2(g)$$

After a 15.8-g sample of calcium carbonate was heated in an open container to cause decomposition, the mass of the remaining solid was determined to be 9.10 g. The student is unsure if the reaction is complete, so the solid could contain unreacted $CaCO_3$.
(a) Determine the theoretical yield of CaO and the theoretical yield of $CO_2$.
(b) Determine the percent yield of CaO and the percent yield of $CO_2$.
(c) Why are the percent yields for CaO and $CO_2$ different?

6.117 What is the temperature of a 125-g sample of tin, initially at 26.2°C, if 155 J of heat is added to the tin? See Table 6.2.

6.118 Calculate the temperature change that results when 2.15 kJ of heat is supplied to 45.0 g of aluminum. See Table 6.2.

6.119 If 20.0 g of an unknown metal cools from 50.0°C to 25.5°C and loses 220.6 J of heat, what is the specific heat of the metal in joules per gram per degree Celsius? Refer to Table 6.2 to determine the possible identity of the metal.

6.120 What is the specific heat of an unknown metal if 658 J of heat increases the temperature of 1.00 kg of the metal by 5.10°C? Refer to Table 6.2 to determine the possible identity of the metal.

6.121 When 0.250 mol of glucose is broken down in your body, 701 kJ of energy is released. What is the energy change in units of kJ/mol and Cal/mol?

6.122 A cold object at 5°C is placed in an insulated cup of water at 25°C. Determine whether or not each of the following will occur.
(a) The increase in temperature of the object will be equal to the decrease in temperature of the water.
(b) Only the object will change temperature.
(c) Only the water will change temperature.
(d) The heat gain by the object will equal the heat lost by the water.
(e) The final temperature will be somewhere between 5°C and 25°C.

6.123 If 20.0 g of lead at 24.5°C is added to 105 g of water at 55.0°C, what will be the final temperature of the mixture? See Table 6.2.

6.124 Methylhydrazine, $CH_6N_2$, is commonly used as a liquid rocket fuel. The heat of combustion of methylhydrazine is $-1.30 \times 10^3$ kJ/mol. How much heat is released when 100.0 g of methylhydrazine is burned?

6.125 A 175-g pipe is heated to 78.24°C. The pipe is then placed in 100.0 g of water held in a calorimeter at 25.00°C. The final temperature of the mixture is 33.43°C. The specific heat of water is 4.184 J/(g °C). Assuming no heat is transferred out of the water, what is the specific heat of the pipe? Using Table 6.2, determine what the pipe could be made of.

6.126 A 150.0-g sample of copper is heated to 89.3°C. The copper is then placed in 125.0 g of water held in a calorimeter at 22.5°C. The final temperature of the mixture is 29.0°C. Assuming no heat is lost from the water, what is the specific heat of copper?

6.127 How many moles of aqueous magnesium ions and chloride ions are formed when 0.250 mol of magnesium chloride dissolves in water?

6.128 How many moles of aqueous sodium ions and sulfide ions are formed when 1.50 mol of sodium sulfide dissolves in water?

6.129 How many moles of aqueous potassium ions and sulfate ions are formed when 50.0 g of $K_2SO_4$ dissolves in water?

6.130 How many moles of aqueous aluminum ions and nitrate ions are formed when 145.0 g of $Al(NO_3)_3$ dissolves in water?

6.131 The production of ethanol for use as a fuel can be accomplished through fermentation of glucose, a type of sugar:

$$C_6H_{12}O_6(s) \longrightarrow 2CH_3CH_2OH(l) + 2CO_2(g)$$

What mass of ethanol can be produced from the fermentation of 5.0 kg of glucose?

6.132 The deployment of airbags during a car accident results from a chemical reaction that, until the 1990s, involved sodium azide, $NaN_3$. Sodium azide has since been replaced by less toxic substances that react in a similar way. In older airbags a crash triggers the decomposition of $NaN_3$ to form elemental sodium and nitrogen gas, which fills the airbag. How many moles of $N_2$ gas are produced when 150.0 g $NaN_3$ decomposes?

6.133    When mixed, solutions of barium chloride, $BaCl_2$, and potassium chromate, $K_2CrO_4$, form a yellow precipitate of barium chromate, $BaCrO_4$. The balanced equation is

$$BaCl_2(aq) + K_2CrO_4(aq) \longrightarrow BaCrO_4(s) + 2KCl(aq)$$

How many grams of barium chromate are expected when a solution containing 0.45 mol $BaCl_2$ is mixed with a solution containing 0.20 mol $K_2CrO_4$?

6.134    When mixed, solutions of silver nitrate, $AgNO_3$, and sodium sulfate, $Na_2SO_4$, form a precipitate of silver sulfate, $Ag_2SO_4$. The balanced equation is

$$2AgNO_3(aq) + Na_2SO_4(aq) \longrightarrow Ag_2SO_4(s) + 2NaNO_3(aq)$$

How many grams of silver sulfate are expected when a solution containing 0.20 mol $AgNO_3$ is mixed with a solution containing 0.15 mol $Na_2SO_4$?

6.135    Sodium metal reacts with liquid bromine to form sodium bromide. When 20.0 g of sodium is placed in 100.0 g of bromine, how many moles of sodium are expected to react?

6.136    When magnesium is heated in air, it reacts with oxygen to form magnesium oxide. Suppose a piece of magnesium reacts with oxygen and the mass of the solid increases by 1.75 g. How many grams of magnesium reacted?

## Concept Review Questions

6.137    Use the following balanced equation for the reaction of titanium metal and oxygen gas to identify the false statement.

$$4Ti(s) + 3O_2(g) \longrightarrow 2Ti_2O_3(s)$$

A. 1 mole of $Ti_2O_3$ should be produced when 2 moles of Ti react.

B. 2 grams of $Ti_2O_3$ should be produced when 4 grams of Ti react.

C. The mass of $Ti_2O_3$ produced should be equal to the sum of the masses of Ti and $O_2$ that react.

D. The moles of $Ti_2O_3$ produced should be less than the sum of the moles of Ti and $O_2$ that react.

E. The number of atoms in the product should be equal to the number of atoms in the reactants.

Change the false statement to a true statement.

6.138    Which of the following is not always conserved in a chemical reaction?

A. number of molecules
B. moles of atoms
C. number of atoms
D. mass
D. All of these are always conserved.

Give an example of a chemical reaction that illustrates the answer.

6.139    The dissolving of magnesium nitrate can be represented by the following equation:

$$Mg(NO_3)_2(s) \longrightarrow Mg^{2+}(aq) + 2NO_3^-(aq)$$

How many moles of ions would be in a solution if 2.5 mol of $Mg(NO_3)_2$ dissolves?

A. 2.5 mol ions
B. 5.0 mol ions
C. 7.5 mol ions
D. 20 mol ions
E. 0.83 mol ions

Determine the number of moles of $Mg(NO_3)_2$ that would dissolve to form the number of ions in each incorrect answer.

6.140    Which of the following is not the amount of the indicated product formed when 1.0 mol of $C_3H_5N_3O_9$ decomposes?

$$4C_3H_5N_3O_9(l) \longrightarrow 6N_2(g) + 12CO_2(g) + 10H_2O(g) + O_2(g)$$

A. 1.5 mol $N_2$
B. 3.0 mol $CO_2$
C. 2.5 mol $H_2O$
D. 1.0 mol $O_2$
E. All of these are correct.

For the incorrect value, determine the correct value for that product.

6.141    The molecular-level diagram shows 4 $H_2$ molecules and 3 $O_2$ molecules prior to reaction. Given the balanced equation for the reaction, identify the limiting reactant and describe the number of $H_2O$ molecules that should be produced.

$$2H_2(g) + O_2(g) \longrightarrow 2H_2O(g)$$

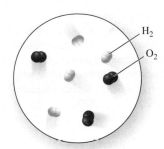

A. $H_2$ is the limiting reactant; 2 $H_2O$ molecules should be produced.

B. $H_2$ is the limiting reactant; 4 $H_2O$ molecules should be produced.

C. $O_2$ is the limiting reactant; 6 $H_2O$ molecules should be produced.

D. $O_2$ is the limiting reactant; 2 $H_2O$ molecules should be produced.

E. There is no limiting reactant.

How many additional limiting reactant molecules are needed to react with the amount of reactant that is in excess?

6.142   When calculating percent yield for a reaction, the theoretical yield is:
A. always 1.0 g.
B. calculated from the amount of limiting reactant.
C. equal to the amount of limiting reactant.
D. equal to the amount of leftover reactant.
E. the amount of product obtained in the lab.
How is *theoretical yield* used when calculating percent yield for a reaction?

6.143   If 10.0 mol of $N_2$ and 10.0 mol of $H_2$ are mixed, and the reaction shown goes to completion, which reactant is in excess and how much of it remains unreacted?

$$N_2(g) + 3H_2(g) \longrightarrow 2NH_3(g)$$

A. $H_2$ is in excess, and 1.0 mol of it should remain unreacted.
B. $H_2$ is the excess reactant, and 7.0 mol of it should remain unreacted.
C. $N_2$ is in excess, and 1.0 mol of it should remain unreacted.
D. $N_2$ is in excess, and 6.7 mol of it should remain unreacted.
E. There is no limiting reactant, and both reactants should react completely.

For each incorrect answer, determine the amounts of reactants that would need to be mixed to make those true statements.

6.144   Which of the following best describes an endothermic chemical reaction?
A. Heat is released by the reaction and transferred to the surroundings.
B. The potential energy of the reactants is higher than the energy of the products.
C. The temperature of the surroundings decreases.
D. The process can only run in reverse.
E. Mass is converted into energy.

For each incorrect answer, explain why it is not a description of an endothermic reaction.

6.145   What is the heat change (in kJ) when a 36.0 g sample of water ($C = 4.184$ J/g°C) is cooled from 80.0°C to 25.0°C? Is heat absorbed or released by the water?
A. 8.28 kJ; heat is absorbed.
B. −8.28 kJ; heat is released.
C. 8.28 kJ; heat is released.
D. −8.28 kJ; heat is absorbed.
E. None of these are correct.
Which of these answers would be correct if the question asked for the heat change of the surroundings?

6.146   If the same amount of heat were added to each of the following samples, which would undergo the greatest temperature change, $\Delta T$?
A. 1.0 g olive oil, specific heat is 1.79 J/(g°C)
B. 5.0 g olive oil, specific heat is 1.79 J/(g°C)
C. 1.0 g water, specific heat is 4.184 J/(g°C)
D. 5.0 g water, specific heat is 4.184 J/(g°C)
E. These would all undergo the same temperature change.

Use the terms proportional or inversely proportional to describe the relationship between temperature change and each of the following: mass, specific heat, and heat change.

# Electron Structure
# of the Atom

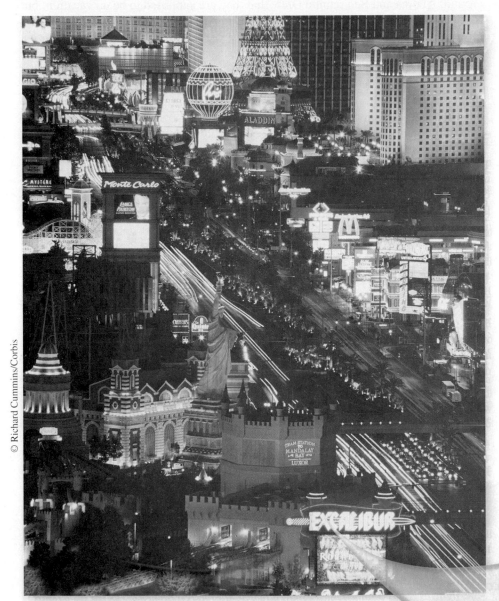

© Richard Cummins/Corbis

**7.1 Electromagnetic Radiation and Energy**

**7.2 The Bohr Model of the Hydrogen Atom**

**7.3 The Modern Model of the Atom**

**7.4 Periodicity of Electron Configurations**

**7.5 Valence Electrons for the Main-Group Elements**

**7.6 Electron Configurations for Ions**

**7.7 Periodic Properties of Atoms**

**Chapter Review**

**Questions and Problems**

ndrea, Ben, and Drew often meet outside of class to go over homework problems and to study for quizzes and exams. After many weeks of hard work, they agree they need to get away from chemistry for a while, so Andrea suggests a weekend trip by car to Las Vegas. They don't suspect that they will encounter chemistry at every turn.

As they approach Las Vegas, the bright lights of the city's signs and advertising displays tell them that their destination is near. Drew, the most studious of the three, remarks that—while the bright lights of Las Vegas are dazzling in their array of colors—good chemistry students should be just as appreciative of the light that is all around. Andrea and Ben remind Drew that they are supposed to be on vacation, but a discussion about the Sun begins anyway. Drew tells the others that the Sun is the most important source of light and energy. It provides the light that allows us to see. The light from the Sun provides the energy that plants use to convert carbon dioxide and water into the sugar glucose. As sunlight hits our atmosphere and our planet's surface, some of its energy converts to heat energy that warms the planet.

Drew reminds his friends of the rainbow they saw after a brief rain on their way to Las Vegas. Sunlight is composed of all the possible colors of light. The mixture results in a continuum of colors that we see as white light. A rainbow is a separation of the colors in sunlight. It appears when sunlight passes through water droplets in the atmosphere. The water acts as a prism to separate the colors that make up sunlight. Light from an incandescent lightbulb is also white light. Shine it through a prism and a rainbowlike separation of colors appears (Figure 7.1).

During their first evening in Las Vegas, the friends watch a spectacular fireworks show. Ben wonders how all the different colors and designs are created. Fireworks generally consist of two components: the compounds that produce colored light and the explosives. Makers of fireworks use different chemical compounds to get different colors (Figure 7.2). For example, they use strontium nitrate or strontium carbonate to produce red light, and barium nitrate or barium chlorate to make green light. Copper compounds produce bluish-green light. Magnesium and aluminum metals are generally used to produce white fireworks. A potassium nitrate and sulfur mixture produces the white smoke that often accompanies a grand finale.

The explosives in fireworks provide the energy that propels the compounds into the air. Explosions are exothermic reactions that release energy to the surroundings. Some of the energy released changes into kinetic energy, causing the fireworks' components to move quickly through the air. More important, the explosion causes

**FIGURE 7.1** White light is composed of different colors that can be separated by a prism. A rainbow results from the separation of white sunlight by raindrops.
©PhotoLink/Getty Images

**FIGURE 7.2** Commercial fireworks used by private companies or sold to the public are composed of two main components: explosive compounds and metals or compounds that will emit a specific type of light. Shown here are finely ground ionic compounds and metals that are used to produce light and special effects. Pure metals, such as magnesium, are often used to produce the white sparkles seen in firework shows.
(left): ©Flickr RF/Getty Images; (right): ©Tom Pantages

the compounds to release light. Some of the energy from the explosion is absorbed by ionic compounds or metal ions, which then release energy in the form of light. The heat from a simple flame will also release light. For example, if we hold a small sample of strontium nitrate, $Sr(NO_3)_2$, in a flame, the flame changes to a red color (Figure 7.3). The red color of the flame is the same as we see in strontium nitrate fireworks.

Ben, Andrea, and Drew walk past the casinos and hotels of Las Vegas, admiring the brilliant displays of neon lights and talking about how they work. The "neon" lights of Las Vegas duplicate all the colors of the rainbow, but the source is not white light. If the light from a neon light passes through a prism, only a few of all possible colors appear. Neon light, then, is different from white light; it does not contain a continuum of colors. Ben says the colored light from neon signs is produced in much the same way as the colors of fireworks, but the energy absorbed by the atoms of gas inside the tube comes from electricity, not from a flame or an explosion.

The three students decide to attend a concert in Las Vegas. As they enter the auditorium, a laser light show is in progress. Dozens of beams of colored light move to music in patterns across the stage. Andrea notices that the lines are very narrow and intense, much different from the light given off by the fireworks show and the neon lights. Laser light, she remembers from chemistry class, consists of only one color of light that travels in a narrow and intensely colored beam. Different lasers produce different colors of light (Figure 7.4). Because visible laser light is fun to look at, lasers are great for entertainment, but they have practical uses, too. In surgery the narrow beam can be used to make tiny incisions in a very small area, such as in the cornea of the eye during lasik surgery. Lasers have also been found effective in collapsing varicose veins without the pain and swelling that accompanies traditional surgical methods.

Whether we're looking at neon lights, fireworks, or laser lights, we see light that is constantly being released by atoms. Learning about light and how it interacts with atoms will help you better understand the arrangement and behavior of electrons in atoms. It will also help you relate the microscopic and macroscopic properties of the elements to their positions in the periodic table.

## Questions for Consideration

**7.1** What is light and how can we describe it?

**7.2** How can we describe the behavior of the electron in a hydrogen atom?

**7.3** How can we describe electrons in all atoms?

**7.4** How does the arrangement of electrons in atoms relate to the arrangement of elements in the periodic table?

**7.5** Which electrons are chemically important in an atom?

**7.6** How does the electron arrangement in ions differ from that in atoms?

**7.7** What properties of atoms are related to their electron arrangements?

## Math Tools Used in This Chapter

Units and Conversions (Math Toolbox 1.3)

## 7.1 Electromagnetic Radiation and Energy

Light is a form of energy called **electromagnetic radiation**, or *radiant energy.* All types of electromagnetic radiation travel through space as oscillating waves, and—as their name suggests—they are synchronized electric and magnetic fields. In a vacuum, they all move at the same speed, the *speed of light,* $3.0 \times 10^8$ m/s. When they move through another medium, such as air or glass, the speed is slower. The light we see with our eyes, visible light, is only a small fraction of the wide range of electromagnetic radiation. Other forms include X-rays, ultraviolet light, infrared light, microwaves, and radio waves.

**ANIMATION:** Flame Tests

**FIGURE 7.3** When a sample of $Sr(NO_3)_2$ is placed in a flame, the flame turns red. This compound is used to make red fireworks.
©McGraw-Hill Education/Stephen Frisch

At one time, neon lights contained only the noble gas neon. They glowed a red-orange. Today, we use *neon light* as a general term to describe tubes filled with other noble gases or a mixture of them. The color we see depends on the gases inside, typically pale orange from helium, lavender from argon, gray-green from krypton, and gray or blue-gray from xenon. The color can also be changed by coating the inside of the glass tube with a solid called a phosphor.

**ANIMATION:** Discharge Tubes

**FIGURE 7.4** Laser lights can be entertaining. They can also be used for practical purposes such as surgery.
(top): ©iStockphoto/Getty Images;
(bottom): ©Corbis/SuperStock

## Properties of Electromagnetic Radiation

What makes different kinds of electromagnetic radiation different? One variable property is their wavelengths. As shown in Figure 7.5, the **wavelength** (symbolized by the Greek letter lambda, $\lambda$) is the distance between two corresponding points on a wave. They can be two crests, two troughs, or any other two corresponding points along the wave. The wavelengths are characteristics of different kinds of electromagnetic radiation. The shorter wavelengths are the gamma rays and X-rays. The longer wavelengths are microwaves and radio waves.

Another property is frequency. To understand it, imagine that you are standing on a boat dock on a calm day. The waves hit the dock at regular, slow intervals. Now imagine that you are standing on that same dock on a windy day. The waves hit the dock much more frequently. These waves have a higher *frequency* than the waves on a calm day. For light, the **frequency** (symbolized by the Greek letter nu, $\nu$) is a measure of the number of wave cycles that move through a point in space in 1 second. It is described in units of 1/s ($s^{-1}$) or **hertz (Hz).**

Both wavelength and frequency can be used to characterize electromagnetic radiation. The *electromagnetic spectrum* in Figure 7.6 shows the different types of electromagnetic radiation, listed from shortest wavelength (gamma rays) to longest wavelength (TV and radio waves) in units of nanometers (nm). Notice that as wavelength increases, frequency decreases. As one goes up, the other one goes down. That is, *wavelength and frequency are inversely proportional.* (For more information on direct and inverse relationships, see Math Toolbox 9.2.) Compare the two

> Although we describe frequency as wave cycles per second, the units of frequency are usually written as $s^{-1}$ or 1/s, pronounced "inverse seconds" or "per second." Wave cycles are not considered units, so there is no numerator unit; the denominator unit is seconds.

**FIGURE 7.5** Wavelength ($\lambda$) is the distance between two corresponding points on a wave. The same value would be obtained if the measurement were made between two crests, two troughs, or between any other two corresponding points along the wave.

**FIGURE 7.6** The electromagnetic spectrum includes all the types of electromagnetic radiation, shown here in order from shortest wavelength to longest wavelength. As wavelength increases, frequency decreases. The visible region of the electromagnetic spectrum is the region between the ultraviolet and the infrared. It represents a small portion of the electromagnetic spectrum.

Infrared light has s wavelength ($\lambda$) of 1000 nm and a frequency ($\nu$) of $3 \times 10^{14}$/s

Green light has s wavelength ($\lambda$) of 500 nm and a frequency ($\nu$) of $6 \times 10^{14}$/s

**FIGURE 7.7** The wavelength of the infrared (IR) light wave is twice that of the green light. The frequency of the infrared light wave is one-half that of the green. Wavelength and frequency are inversely proportional. (Although the IR laser wave is shown in red, IR light is not visible to humans.)

Although we cannot see infrared radiation, we can sometimes feel its effects as heat. All objects emit infrared radiation. The wavelength at which an object radiates most intensely depends on its temperature. The cooler the object, the longer the wavelength. We can detect infrared radiation using special cameras and film that detect differences in temperature. In this thermal image of a human brain, the green and yellow areas are the warmest and the blue and black areas are the coldest.

©Janis Christie/Getty Images

waves shown in Figure 7.7. Which has the higher frequency? The one with the shorter wavelength.

Our eyes can detect light only in the visible region of the electromagnetic spectrum. This region is composed of the different colors of light, with red having the longest wavelength and violet the shortest. Atoms absorb light. This is why grass appears green to us. When light from the Sun shines on grass, the atoms and molecules in the grass absorb *all* the colors of light *except* green. The green wavelengths not absorbed by the grass reflect back to our eyes, and we see green. A white object absorbs little or no visible light. A black object absorbs most or all visible light. These properties of atoms and light can fool our eyes sometimes. Suppose you park your red car at night in a parking lot lit by sodium vapor lamps. They emit only yellow light. Since no red light shines on the car, no red light can reflect back. Your car will look a different color!

Light can be characterized by its wavelength or its frequency, and also by its energy. A unit of light energy can be thought of as a "packet" of energy called a **photon.** The energy, *E,* of a photon depends on the wavelength or frequency. Electromagnetic radiation of long wavelength and low frequency, such as radio waves, is composed of low-energy photons. Short-wavelength radiation with high frequencies is composed of high-energy photons. *Photon energy is directly proportional to frequency.* As one gets larger, so does the other. Photon energy is *inversely proportional* to wavelength. As one increases, the other decreases.

A laser such as that used to create a light show produces *monochromatic light,* which is light of only one wavelength and composed of photons of identical energy. One reason that a beam of laser light does not spread out is that the waves are synchronized. They move together so that the peaks are perfectly lined up. Common lasers emit light in the visible or the infrared region of the electromagnetic spectrum. In Example 7.1, we will compare the wavelength, frequency, and photon energy of different lasers and of different types of electromagnetic radiation.

In addition to surgery, the uses of lasers include drilling out cavities in teeth, playing compact disks (CDs) and digital video disks (DVDs), scanning supermarket prices, reading and writing computer media, cutting and welding of metals, measuring distances, aiming weapons, and removing birthmarks and tattoos.

| EXAMPLE 7.1 | **Relative Values of Wavelength, Frequency, and Photon Energy** |
|---|---|

Andrea, Ben, and Drew notice that their chemistry teacher uses two different laser pointers. One emits red light and the other emits green light.

(a) Which laser light has the longer wavelength?
(b) Which has the higher frequency?
(c) Which is composed of photons of higher energy?

Use a mnemonic to remember wavelengths and colors of visible light: red, orange, yellow, green, blue, indigo, and violet. From longest to shortest wavelength, the first letters of the colors spell a man's name: ROY G. BIV.

Red (700 nm)
Orange (650 nm)
Yellow (580 nm)
Green (500 nm)
Blue (480 nm)
Indigo (430 nm)
Violet (400 nm)

**Solution:**

(a) Relative wavelengths are shown in the electromagnetic spectrum in Figure 7.6. Red light is farther to the right with the longer wavelength. Green light has the shorter wavelength. If you do not have an electromagnetic spectrum available, the mnemonic ROYGBIV provides the order, starting with red (R), the color of light with the longest wavelength. Green (G) light falls in the middle of the visible region, with a shorter wavelength. The shortest is violet (V).

(b) Frequency and wavelength are inversely proportional. When frequency is relatively large, wavelength is relatively small. Since green light has the shorter wavelength, it has the higher frequency.

(c) Photon energy is proportional to frequency. Light with a high frequency has a relatively greater photon energy. Since green light has the higher frequency, it also has the greater photon energy. You can also infer relative photon energies by comparing wavelengths. Photon energy is inversely related to wavelength; so green light—with the shorter wavelength—has the higher photon energy.

**Consider This 7.1**

What color laser would provide higher-energy photons than green or red?

**Practice Problem 7.1**

The Sun emits a wide range of electromagnetic radiation. The light that gets to Earth is mostly visible light, infrared light, and ultraviolet light. We can see the visible light with our eyes. We can feel the infrared light as heat. It warms Earth and the atmosphere. We use sunscreen to protect our skin from the harmful effects of ultraviolet light.

(a) Which type of light has the longest wavelength: visible, infrared, or ultraviolet (UV)?

(b) Which has the highest frequency?

(c) Which is composed of the highest-energy photons?

**Further Practice:**   Questions 7.13 and 7.14 at the end of the chapter

---

Mathematical relationships allow us to calculate the wavelength ($\lambda$), frequency ($\nu$), and photon energy ($E_{photon}$) of a specific type of electromagnetic radiation. The product of frequency and wavelength is a constant, $c$, which is the speed of light, $3.00 \times 10^8$ m/s:

$$\nu\lambda = c$$

This equation shows the inverse relationship between wavelength and frequency. Since the speed of light $c$ is a constant, when frequency is large, wavelength must be small; when wavelength is large, frequency must be small. If we know the value of either wavelength or frequency, we can calculate the unknown value by rearranging the equation:

$$\nu = \frac{c}{\lambda} \quad \text{or} \quad \lambda = \frac{c}{\nu}$$

When calculating frequency from wavelength, wavelength must be in units of meters, so it will cancel properly with the units of the speed of light. Because wavelengths are commonly expressed in units of nanometers (1 nm = $10^{-9}$ m or $10^9$ nm = 1 m), conversion to meters is required. When calculating wavelength from frequency, note that units cancel to give units of meters.

To solve for frequency $\nu$, we divide both sides of the equation $\nu\lambda = c$ by wavelength $\lambda$:

$$\frac{\nu\cancel{\lambda}}{\cancel{\lambda}} = \frac{c}{\lambda}$$

Now we have an expression that solves for frequency:

$$\nu = \frac{c}{\lambda}$$

The mathematical relationship between photon energy and frequency is described by the following equation where $h$, Planck's constant, is $6.626 \times 10^{-34}$ J · s:

$$E_{\text{photon}} = h\nu$$

The equation shows the directly proportional relationship between photon energy and frequency. The inverse relationship between photon energy and wavelength is given by the following equation where $h$ is Planck's constant and $c$ is the speed of light:

$$E_{\text{photon}} = \frac{hc}{\lambda}$$

---

**EXAMPLE 7.2** | **Calculating Wavelength, Frequency, and Photon Energy**

Infrared light emitted from a TV remote control or from a wireless computer mouse has a wavelength of 805 nm.

(a) What is the frequency of the light?
(b) What is the energy of its photons?

**Solution:**

(a) Recall the relationship between wavelength, $\lambda$, and frequency, $\nu$:

$$\nu\lambda = c$$

where $c$ is the speed of light, $3.00 \times 10^8$ m/s. We know wavelength, so we can rearrange the equation to solve for frequency:

$$\nu = \frac{c}{\lambda}$$

Wavelength is given in units of nanometers, so we must convert to meters to cancel units with the units of speed of light (m/s). (For a review of unit conversions, see Math Toolbox 1.3.) The relationship between nanometers and meters is

 **MATH TOOLBOX 1.3**

$$1 \text{ m} = 10^9 \text{ nm}$$

When we convert from nanometers to meters, we multiply the known value in nanometers by the conversion factor that has the nanometer units in the denominator to cancel units properly:

$$805 \ \cancel{\text{nm}} \times \frac{1 \text{ m}}{10^9 \ \cancel{\text{nm}}} = 8.05 \times 10^{-7} \text{ m}$$

$$\lambda = 8.05 \times 10^{-7} \text{ m}$$

Now that we have the appropriate units for wavelength, we can divide the speed of light by its value to get frequency:

$$\nu = \frac{c}{\lambda}$$

$$= \frac{3.00 \times 10^8 \ \cancel{\text{m}} \text{/s}}{8.05 \times 10^{-7} \ \cancel{\text{m}}} = 3.73 \times 10^{14} \text{ /s} \quad \text{or} \quad 3.73 \times 10^{14} \text{ Hz}$$

(b) The energy of a photon can be calculated from frequency using the following equation:

$$E_{photon} = h\nu$$

where $h$ is Planck's constant, $6.626 \times 10^{-34}$ J · s. Since we have just calculated frequency, we can substitute values of $h$ and $\nu$ into the equation:

$$E_{photon} = 6.626 \times 10^{-34}\,J\cdot s \times 3.73 \times 10^{14}\,\frac{1}{s} = 2.47 \times 10^{-19}\,J$$

Units of seconds cancel, giving units of joules, which is a unit of energy. The energy of a single photon of this light is $2.47 \times 10^{-19}$ J.

### Consider This 7.2

If a 60-milliwatt TV remote emits a total energy of light of $6.0 \times 10^4$ J every second with photons of an energy of $2.47 \times 10^{-19}$ J/photon, what is the best estimate of the number of photons emitted from the TV remote in 1 second: 100 photons, $10^6$ photons, $10^{23}$ photons?

### Practice Problem 7.2

Microwaves are to the right of infrared radiation in the electromagnetic spectrum shown in Figure 7.6, so their wavelengths are longer. The microwaves used to cook food have wavelengths that are about 11 to 12 cm long! If the wavelength of light of a microwave beam is 11.5 cm, what are (a) the frequency of the radiation and (b) the energy of its photons?

**Further Practice:** Questions 7.17 and 7.18 at the end of the chapter

## Atomic Spectra

When visible light passes through a prism, its components separate into a sequence of colors called a *spectrum*. Sunlight or light from an incandescent bulb produces a **continuous spectrum,** which contains all the wavelengths of light in the visible region (Figure 7.8A). A light source that produces a continuous spectrum is called *white light.*

What happens when we pass colored light, such as that from a neon sign, through a prism? Unlike white light, colored light is not composed of all the wavelengths. Its spectrum is not continuous. A true neon light, for example, consists of a tube containing only neon atoms under very low pressure. When an electric current moves through the gas, the neon atoms absorb energy and then emit energy as light. In other words, electric energy is converted to light energy. If we pass this light through a slit and then through a prism, the result is a set of distinct colored lines, each representing a single wavelength of light. The pattern of these lines is a **line spectrum.** The line spectra for neon and some other elements are shown in Figure 7.8B.

When an element is heated or given an electric charge, the energy absorbed can be transformed into light energy. The light an element produces in this way contains only specific colors, as shown by its line spectrum. The line spectrum for each element is unique, so it can be used to identify the element. It is like the element's "fingerprint." In Section 7.2 we see how the line spectrum for hydrogen helps to explain the electron structure of the hydrogen atom.

Measurement of line spectra can be used to identify unknown materials. Different types of *spectroscopy* are used in science, medicine, industry, and forensics.

**FIGURE 7.8** (A) When white light from sunlight or an incandescent lightbulb passes through a prism, a continuous spectrum is produced. (B) The colored light emitted by a substance produces a line spectrum that is unique to that substance. The line spectra for mercury, strontium, and neon are shown here.

## 7.2 The Bohr Model of the Hydrogen Atom

How can we explain the different colors of light that Andrea, Ben, and Drew saw in Las Vegas? How do neon atoms produce the distinctive red-orange color of a true neon light? In Chapter 2 you learned about the basic structure of the atom. An atom consists of a relatively small, positively charged nucleus made of protons and neutrons. Although the nucleus occupies little space in an atom, it accounts for the majority of its mass. The electrons occupy a relatively large volume of space in the atom surrounding the nucleus.

Recall from Chapter 2 that the nuclear model of the atom was proposed by Rutherford as a result of his gold foil experiment. When Rutherford proposed his theory in 1911, many scientists, especially J. J. Thomson, found it difficult to accept. Although the evidence was compelling, Rutherford's model didn't follow the basic laws of physics known at the time. How could negative and positive charges—which normally attract each other—remain separate? What prevents electrons from tumbling toward the nucleus?

A year after Rutherford published his nuclear model of the atom, a young man named Niels Bohr came to work with Rutherford. Bohr had just completed his doctoral thesis in physics. One of his first assignments with Rutherford was to explain the stability of the arrangement of the electrons in the nuclear model of the atom. One possible explanation that Bohr proposed is described as a planetary model, in which electrons orbit the nucleus similar to the way planets orbit the Sun. Bohr based his ideas on work done by a German physicist, Max Planck, who thought that energy produced by atoms is **quantized;** that is, it can have only certain values. Bohr suggested that the energies of electrons in an atom are quantized. They can have only certain values.

In his model of a hydrogen atom, Bohr proposed that hydrogen's single electron revolves about the nucleus in one of several possible, fixed orbits, each with a specific radius and energy (Figure 7.9). He labeled each orbit with an integer value of $n$,

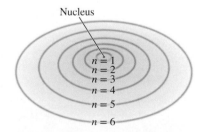

**FIGURE 7.9** In the Bohr model of the hydrogen atom, the nucleus is at the center of the atom, and the electron is located in one of the circular orbits ($n = 1$ through $n = 6$), each with a specific energy and radius.

The quantized behavior of an electron in an atom is similar to the notes played on a keyboard. You can play any of the notes that correspond to keys, but you can't play a note between two adjacent keys. The notes you play are quantized because only certain notes are possible. The difference between notes is also quantized.

**ANIMATION:** Emission in an Excited Hydrogen Atom

Why don't we see more than four colored lines in the hydrogen spectrum?

**Student Hot Spot**

Student data indicate you may struggle with the Bohr model of the hydrogen atom. Access the SmartBook to view additional Learning Resources on this topic.

with $n = 1$ being the orbit closest to the nucleus. He explained that when the electron is in an orbit close to the nucleus, it has a relatively low energy. The electron can absorb energy and move to a higher-energy orbit that is farther from the nucleus. When the electron drops down to a lower orbit, energy is released. The possible energies of the electron are restricted to those of allowed orbits. The amount of energy that can be absorbed or released is restricted to the possible transitions between the energy levels defined by the orbits. Since the energy of an electron in an atom is quantized, its energy changes are quantized, and the energy absorbed or released is quantized, Bohr said.

The scientific community accepted Bohr's model because it explained the hydrogen line spectrum. As shown in Figure 7.10, the four lines in the line spectrum correlate with four of the electron transitions that Bohr proposed. Each line in the spectrum has a specific wavelength, frequency, and photon energy. Bohr thought that the transition or movement of an electron from a higher-energy orbit to a lower-energy orbit releases energy. A transition to the $n = 2$ orbit produces an emission of visible light. The possible transitions correspond to lines in the hydrogen line spectrum: $n = 6$ to $n = 2$ (violet); $n = 5$ to $n = 2$ (blue); $n = 4$ to $n = 2$ (green); and $n = 3$ to $n = 2$ (red). In Bohr's model, transitions to the $n = 1$ orbit are higher-energy transitions that result in the emission of ultraviolet light. Transitions to the $n = 3$ orbit are lower-energy transitions that result in the emission of infrared light.

Bohr said that when an electron moves from a higher-energy orbit to one that is lower in energy, an electron releases an amount of energy, $\Delta E$, equal to the *difference* in energy between the two orbits:

$$\Delta E = E_{\text{final}} - E_{\text{initial}}$$

The value of $\Delta E$ is negative for a downward transition because energy is released as a photon. The energy of the photon, $E_{\text{photon}}$, equals the absolute value of the energy change of the electron. We express this relationship using lines around $\Delta E$ to show that we take the absolute value of $\Delta E$, which is its positive value:

$$E_{\text{photon}} = |\Delta E|$$

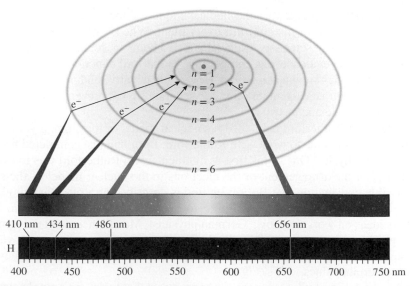

**FIGURE 7.10** Four distinct lines appear in the line spectrum of hydrogen. In the Bohr model, the difference in energy between certain orbits corresponds to the photon energies of the lines that appear in the hydrogen line spectrum. As the electron moves from a higher-energy orbit to a lower-energy orbit, energy is released as light.

| EXAMPLE 7.3 | **The Bohr Model and the Hydrogen Line Spectrum** |

Use the Bohr model of the hydrogen atom and the hydrogen line spectrum in Figure 7.10 to answer the following questions.

(a)  What is the photon energy of the light that produces the red line in the hydrogen spectrum?

(b)  Of the four lines in the hydrogen line spectrum, which has the highest photon energy?

**Solution:**

(a)  In the hydrogen line spectrum in Figure 7.10, the wavelength of red light is given as 656 nm. Knowing the wavelength, $\lambda$, we can find $E_{photon}$ using the equation that relates wavelength to photon energy:

$$E_{photon} = \frac{hc}{\lambda}$$

We must convert the wavelength units from nanometers to meters so units cancel, just as we did in Example 7.2:

$$656 \ \cancel{nm} \times \frac{1 \text{ m}}{10^9 \ \cancel{nm}} = 6.56 \times 10^{-7} \text{ m}$$

Now we substitute Planck's constant, the speed of light, and the wavelength into the equation for photon energy:

$$E_{photon} = \frac{6.626 \times 10^{-34} \text{ J} \cdot \cancel{s} \times 3.00 \times 10^8 \ \frac{\cancel{m}}{\cancel{s}}}{6.56 \times 10^{-7} \ \cancel{m}} = 3.03 \times 10^{-19} \text{ J}$$

(b)  From the equation that relates photon energy to wavelength, we see that photon energy is *inversely* proportional to wavelength—the shorter the wavelength, the higher the photon energy. In Figure 7.10 we see that the violet line in the hydrogen spectrum has the shortest wavelength, so it has the highest photon energy. The highest photon energy, $E_{photon}$, corresponds to the largest electron transition $\Delta E$, which is the difference in energy between the orbits. The $n = 6$ orbit is further in energy from the $n = 2$ orbit than are the other orbits, so the transition from $n = 6$ to $n = 2$ emits light of highest energy.

**Consider This 7.3**

There are electron transitions in the hydrogen atom that emit light we cannot see. One of these transitions is from $n = 7$ to $n = 2$. If an electron undergoes this transition, what type of radiation should be emitted?

**Practice Problem 7.3**

Use the Bohr model of the hydrogen atom and the hydrogen line spectrum in Figure 7.10 to answer the following questions.

(a)  What is the photon energy of the light that produces the violet line in the hydrogen spectrum?

(b)  Of the four lines in the hydrogen line spectrum, which has the lowest photon energy?

**Further Practice:**  Questions 7.35 and 7.36 at the end of the chapter

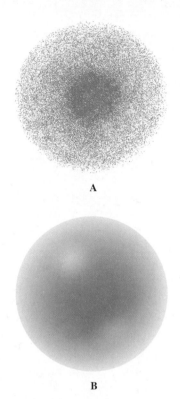

**A**

**B**

**FIGURE 7.11** (A) This probability map is for the hydrogen electron in its lowest-energy state. The electron is more likely to be found in the darker regions. (B) The size of this orbital is defined by an enclosed region where the likelihood of finding an electron is 95%.

While the Bohr model contributed to a better understanding of the atom, it fell short in the long run because it explained only the hydrogen atom. It did not work for atoms that contained more than one electron. In Section 7.3 we will modify the Bohr model to describe our modern concept of the atom.

## 7.3 The Modern Model of the Atom

By the 1920s scientists started to work on a new model of the atom that could account for the line spectra of other elements. Erwin Schrödinger, from Austria, developed a mathematical model that seemed to work for all atoms. Schrödinger's model of the atom was similar to Bohr's model in that the energies of electrons were quantized. They could have only certain energies. The new model, however, describes electrons as occupying *orbitals*, not orbits. An **orbital** is a three-dimensional region in space where the electron is likely to be found, not a circular pathway. The mathematical model was based on the probability of finding an electron in a region outside the nucleus. From it chemists created a probability map to show where an electron is likely to be found. The diagram in Figure 7.11A is the probability map for a hydrogen electron in its lowest-energy state. The darker regions represent areas where the electron is most likely to be found. Because the likelihood of finding an electron in the outer regions of the probability map is small, it is customary to show an enclosed region in space where the electron is likely to be found 95% of the time (Figure 7.11B). We will use orbital representations without dots for the rest of this chapter.

In the modern model of the atom, orbitals of similar size are considered to be in the same **principal energy level.** The first six principal energy levels for the electron in a hydrogen atom are shown in Figure 7.12. In the hydrogen atom, the energies of the orbitals are restricted to the energies of the principal energy levels.

Orbitals come in different shapes and sizes. There are four types of occupied orbitals in atoms in their lowest-energy states, which we label with the letters *s, p, d,* and *f.* The general shapes of these orbitals are shown in Figure 7.13. These orbitals are distributed into the principal energy levels.

Orbitals at lower-energy levels are smaller. Orbitals at the higher-energy levels are larger and extend farther from the nucleus. The first principal energy level ($n = 1$) consists of a single *s* orbital. We call this *s* orbital a 1*s* orbital because it is in the $n = 1$ principal energy level and it has the *s* orbital shape:

1*s*

Principal energy level          Orbital shape

Energy-level diagram

Bohr diagram

**FIGURE 7.12** In the modern model of the atom, principal energy levels describe the allowed energies of the electrons in orbitals. In the hydrogen atom, the allowed energies for the electron are the same as the energies of the orbits in the Bohr model.

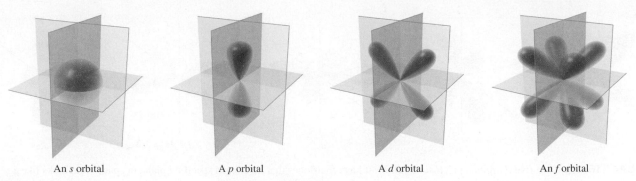

An *s* orbital         A *p* orbital         A *d* orbital         An *f* orbital

**FIGURE 7.13**  The *s, p, d,* and *f* orbitals differ in shape and number of lobes.

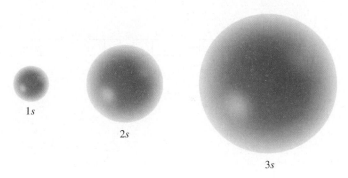

1*s*

2*s*

3*s*

**FIGURE 7.14**  The size of the orbital increases as the principal energy level increases.

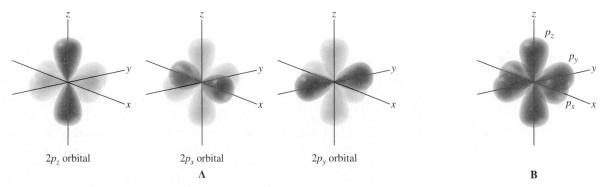

2$p_z$ orbital          2$p_x$ orbital          2$p_y$ orbital

**A**                                                **B**

**FIGURE 7.15**  (A) A *p* sublevel is composed of three *p* orbitals. They have the same shape, but they point in different directions in space. They lie perpendicular to one another, so we can imagine them lying along perpendicular lines that we call *x, y,* and *z* axes. The $p_z$ orbital lies along the *z* axis, the $p_x$ orbital lies along the *x* axis, and the $p_y$ orbital lies along the *y* axis. (B) The three *p* orbitals coexist with their centers at the nucleus.

The second principal energy level ($n = 2$) consists of *s* and *p* orbitals called the 2*s* and 2*p* orbitals. The 2*s* orbital is like the 1*s* orbital, but it is larger  (Figure 7.14). A single *p* orbital, as depicted in Figure 7.13, has a figure-eight or dumbbell shape. In their respective principal energy levels, *p* orbitals always come in sets of three (Figure 7.15). The three 2*p* orbitals lie with their centers overlapping at the nucleus and perpendicular to one another. We can think of each two-lobed orbital as directed along one of the *x, y,* or *z* axes that are each perpendicular to one another. We label these orbitals accordingly as $p_x$, $p_y$, and $p_z$ orbitals.

Because there are two types of orbitals in the second principal energy level, we say that there are two sublevels: the 2*s* sublevel and the 2*p* sublevel. A **sublevel** consists of just one type of orbital at a specific principal energy level. The 2*s* sublevel consists of a single 2*s* orbital, and the 2*p* sublevel consists of three 2*p* orbitals.

**ANIMATION:** Atomic Line Spectra

$d_{yz}$          $d_{xz}$          $d_{xy}$          $d_{x^2-y^2}$          $d_{z^2}$

**FIGURE 7.16**  There are five *d* orbitals in a *d* sublevel. Most have four lobes that lie along a specified plane in space. One looks like a *p* orbital with a donut shape around its center.

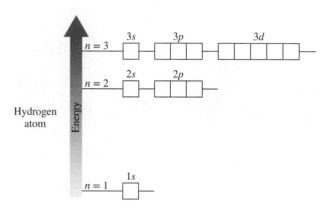

**FIGURE 7.17**  The orbital diagram for the hydrogen atom shows the sublevels and orbitals that can exist at each principal energy level. Each box represents an orbital, and groups of boxes represent sublevels. For the hydrogen atom only, the sublevels within a principal energy level all have the same energy. Here we show only the first three principal energy levels for hydrogen.

At the third principal energy level ($n = 3$), three sublevels are available: the 3*s*, 3*p*, and 3*d* sublevels. The 3*s* sublevel consists of a single 3*s* orbital, and the 3*p* sublevel consists of a set of three 3*p* orbitals. The 3*d* sublevel consists of five *d* orbitals (Figure 7.16). The fourth principal energy level ($n = 4$) consists of four sublevels: the 4*s*, 4*p*, 4*d*, and 4*f* sublevels. The 4*f* sublevel consists of seven 4*f* orbitals.

A convenient way to show the distribution of sublevels and orbitals in an atom is with an orbital diagram. The one shown in Figure 7.17 is specifically for the hydrogen atom. The boxes represent orbitals, and the groups of boxes represent sublevels.

In the modern model of the hydrogen atom, the energy levels have the same energies as the orbits described in the Bohr model. The hydrogen line spectrum is produced by the movement of electrons in an atom from orbitals in higher principal energy levels to orbitals in lower principal energy levels. The hydrogen atom line spectrum is simple to explain because the orbitals in the same principal energy level have the same energy. This is not the case for atoms with more than one electron.

## Orbital Diagrams for Multielectron Atoms

Sublevels in the same principal energy level have the same energy only in atoms with one electron. In atoms with more than one electron, the interaction among electrons causes the sublevels within the same principal energy level to have different energies. The result is an orbital diagram like that shown in Figure 7.18. Notice that, within a given principal energy level, the *p* orbitals are higher in energy than the *s* orbitals, and the *d* orbitals are higher in energy than the *p* orbitals. The 3*d* orbitals are so high that they are slightly higher in energy than the 4*s* orbitals.

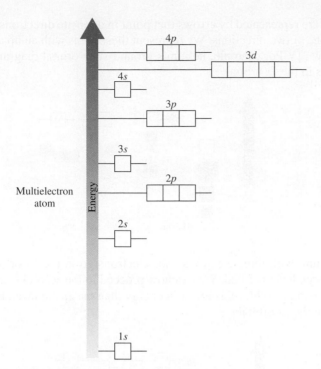

**FIGURE 7.18** The orbital diagram for a multielectron atom shows the same sublevels and orbitals as for the hydrogen atom. However, in atoms that contain more than one electron, the sublevels within a principal energy level have different energies. Note that the energy of the 3*d* sublevel is higher than the energy of the 4*s* sublevel.

How are electrons in atoms arranged into these principal energy levels, sublevels, and orbitals? Can we predict the electron arrangement for a particular atom? We can determine the arrangement of electrons for an atom in its **ground state,** the lowest-energy state of the atom. Consider the orbital diagram that represents how electrons are arranged in a carbon atom in its ground state:

The boxes represent orbitals, and the arrows represent electrons. Can you come up with some rules for the arrangement of electrons? What is the maximum number of electrons per orbital? Why do some orbitals contain only one electron?

When we create an orbital diagram for a particular atom, we fill in electrons starting with a bare nucleus and add one electron at a time. (This is not how atoms are created in nature.) You may have noticed that the lowest-energy orbitals have two electrons in them, and some of the higher-energy orbitals have only one or none. This is a result of the **aufbau principle** (from the German *aufbauen,* "to build up"), which states that *electrons fill orbitals starting with the lowest-energy orbitals.* For the hydrogen atom in its ground state, its electron is in the 1*s* orbital. You may have also noted that no more than two electrons occupy each orbital. This is the basis for the **Pauli exclusion principle,** which states that *a maximum of two electrons can occupy each orbital, and they must have opposite spins.* Electrons in the

An electron behaves as though it were spinning on its axis, like a bicycle wheel. It can spin in only two directions, just like a bicycle wheel can spin only forward or backward. Two electrons in a given orbital must have opposite spin directions. This spin property of electrons can be detected in a magnetic field, using a technique known as electron paramagnetic resonance (EPR) or electron spin resonance (ESR).

same orbital are represented by arrows that point in opposite directions. An electron can spin in one of two directions. We represent these spins with an up arrow ($\uparrow$) and a down arrow ($\downarrow$). For example, helium's ground-state orbital diagram consists of two electrons in the $1s$ orbital:

Helium

For lithium, with three electrons, two electrons go in the $1s$ orbital. The first principal energy level is filled. We can now proceed to the second principal energy level. Because the $2s$ sublevel is lower in energy than the $2p$ sublevel, the third electron goes into the $2s$ orbital:

Lithium

Boron, with five electrons, has an orbital diagram with two electrons in the $1s$ orbital, two electrons in the $2s$ orbital, and one electron in one of the $2p$ orbitals. It makes no difference which $p$ orbital we place the electron in because they are equal in energy. The orbital diagram for boron is

Boron

What happens when there is more than one electron in a sublevel that contains more than one orbital, such as a $p$ sublevel? Are electrons paired up right away, or are they kept in separate orbitals until pairing is required? Did you notice in carbon's orbital diagram that the electrons in the $p$ sublevel are unpaired in separate $p$ orbitals? This brings us to the third rule, **Hund's rule:** *Electrons are distributed into orbitals of identical energy (same sublevel) in such a way as to give the maximum number of unpaired electrons.* Hund's rule is a consequence of the fact that negatively charged electrons repel each other. It follows that electrons should be found in separate orbitals if no energy is required to get there. What would the orbital diagrams for nitrogen and oxygen look like? How many unpaired electrons would these atoms have?

It is common to show a simpler orbital diagram, where all the orbitals are represented on one line, with sublevels labeled. We must remember that the energy of the sublevels increases from left to right. For example, we can represent the orbital diagram for carbon as

Notice the correspondence between this orbital diagram and the simpler one shown in the adjacent paragraph.

Carbon

---

**EXAMPLE 7.4** | **Orbital Filling Rules for Ground-State Electron Diagrams**

Identify the ground-state orbital diagrams that are incorrect. For those that are incorrect, identify the filling rule that is violated.

(a) Be

(b) O

(c) Mg

**Solution:**

(a) The orbital diagram is correct for beryllium.

(b) The orbital diagram for oxygen is not correct. Hund's rule is not obeyed. The orbitals in the 2p sublevel have the same energy. We must place electrons into individual 2p orbitals singly until an electron occupies all three orbitals before starting to pair electrons. Since there are four electrons in the 2p sublevel, two should be paired and two should be unpaired:

O

(c) The orbital diagram for magnesium is not correct. The two electrons in the 3s orbital have the same spin, so the Pauli exclusion principle is not followed. The arrows in the 3s orbital should be pointing in opposite directions to represent electrons with opposite spin:

Mg

**Consider This 7.4**

In the orbital diagram for oxygen, does it matter which p orbital contains the paired electrons?

**Practice Problem 7.4**

Identify the orbital diagrams that are incorrect. For those that are incorrect, identify the filling rule that is violated.

(a) N

(b) Ne

(c) Na

**Further Practice:** Questions 7.45 and 7.46 at the end of the chapter

## Electron Configurations

Although orbital diagrams are useful in describing the arrangement of electrons in an atom, they take up a lot of space. Fortunately, there is a shorter way to convey the same information, called an **electron configuration,** which shows the distribution of electrons among sublevels. When writing electron configurations, we write the number of the principal energy level followed by the symbol for the sublevel. We then add a superscript to each sublevel symbol to indicate the number of electrons in that sublevel. Consider the arrangement of electrons in a carbon atom, as shown by its orbital diagram:

The lowest-energy sublevel, 1s, contains two electrons. This is represented as

$$1s^2$$

The next higher-energy sublevel, 2s, also contains two electrons:

$$2s^2$$

The last sublevel that contains electrons is the 2p sublevel. We do not have to show the number of electrons in each p orbital. We write the electron configuration to show the total number of electrons in the 2p sublevel:

$$2p^2$$

Combining the sublevels, we get the electron configuration for carbon:

$$C \quad 1s^2 2s^2 2p^2$$

Electron configurations are convenient, but they do not typically show how electrons are arranged in orbitals within the same sublevel. The electron configurations for the first 10 elements are shown in Figure 7.19, along with their orbital diagrams.

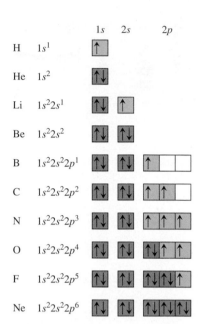

**FIGURE 7.19** Orbital diagrams give information about how electrons are arranged within a sublevel. The orbital diagrams and corresponding electron configurations are shown for the first 10 elements.

---

**EXAMPLE 7.5**     **Writing Electron Configurations**

Use the orbital diagram in Figure 7.18 to write the electron configurations for the following:

(a)  Na

(b)  Ar

(c)  Br

**Solution:**

(a)  The atomic number of sodium is 11, so it has 11 protons and 11 electrons. We distribute electrons into the orbitals in Figure 7.18, starting with the lowest-energy orbital, 1s. By putting a maximum of 2 electrons in each orbital, we get the electron configuration

$$1s^2 2s^2 2p^6 3s^1$$

(b)  The atomic number of argon is 18. It has 18 electrons, which is enough to fill sublevels through 3p. The electron configuration is

$$1s^2 2s^2 2p^6 3s^2 3p^6$$

(c)  From the periodic table we see that bromine has 35 electrons. Filling these electrons into the orbital diagram from lowest- to highest-energy sublevels, we get the following electron configuration:

$$1s^2 2s^2 2p^6 3s^2 3p^6 4s^2 3d^{10} 4p^5$$

Electrons fill the $4s$ orbital before filling the $3d$ orbitals because the $4s$ sublevel is lower in energy than the $3d$.

**Consider This 7.5**

What part of the electron configuration for Na corresponds to the electron configuration for Ne?

**Practice Problem 7.5**

Use the orbital diagram in Figure 7.18 to write the electron configurations for the following:

(a) Al
(b) Sc
(c) K

**Further Practice:**  Questions 7.47 and 7.48 at the end of the chapter

Writing electron configurations for the first 36 elements is easy if you have an orbital diagram like Figure 7.18. In Section 7.4 we will see how the arrangement of elements in the periodic table relates to the electron configurations of the elements. Once you understand the patterns, you won't need an orbital diagram to write electron configurations. You will only need a periodic table.

## 7.4  Periodicity of Electron Configurations

The neon in the signs that Andrea, Ben, and Drew saw in Las Vegas emits light when charged with electricity, but it is *inert*. That is, it does not react chemically with other elements. The inert character of neon and the other noble gases relates to their similar electron configurations. In Chapter 2 you saw that elements in the same group of the periodic table have similar properties. Here you will see that elements in the same group also have similar electron configurations.

Consider the alkali metals, the group IA (1) elements. The electron configurations for Li, Na, K, and Rb are

| | |
|---|---|
| Li | $1s^2 2s^1$ |
| Na | $1s^2 2s^2 2p^6 3s^1$ |
| K | $1s^2 2s^2 2p^6 3s^2 3p^6 4s^1$ |
| Rb | $1s^2 2s^2 2p^6 3s^2 3p^6 4s^2 3d^{10} 4p^6 5s^1$ |

The alkali metals are all soft, reactive metals. They are so reactive that they are not found in nature as metals, but only as ionic compounds or dissolved ions. Since the metals react with oxygen in air and with water, they must be stored in closed containers.

What patterns do you see? These group IA (1) elements all have one electron in the last-occupied sublevel, the $s$ sublevel. Each ends in $ns^1$, where $n$ is the principal energy level.

The group IIA (2) elements have a similar pattern. Each ends in $ns^2$:

| | |
|---|---|
| Be | $1s^2 2s^2$ |
| Mg | $1s^2 2s^2 2p^6 3s^2$ |
| Ca | $1s^2 2s^2 2p^6 3s^2 3p^6 4s^2$ |
| Sr | $1s^2 2s^2 2p^6 3s^2 3p^6 4s^2 3d^{10} 4p^6 5s^2$ |

The periodic table can be subdivided into different blocks (Figure 7.20). These blocks correspond to the highest energy electrons—those that are in the last-occupied sublevel when we write electron configurations. The group IA (1) and IIA (2) elements are in the *s block*. Helium is also an *s* block element.

**FIGURE 7.20** The periodic table can be subdivided into $s$, $p$, $d$, and $f$ blocks that correspond to the last-occupied sublevel for the element in that block.

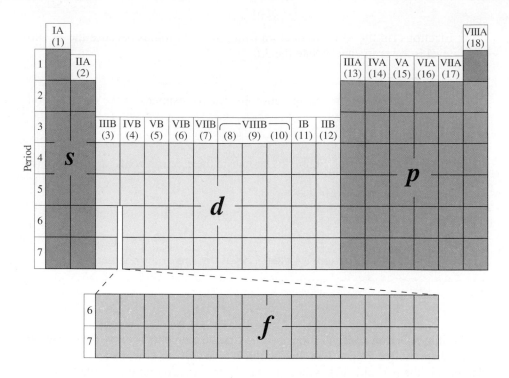

Next let's consider a few of the halogens, group VIIA (17). What pattern do you see in the electron configurations of F, Cl, and Br?

F        $1s^2 2s^2 2p^5$
Cl       $1s^2 2s^2 2p^6 3s^2 3p^5$
Br       $1s^2 2s^2 2p^6 3s^2 3p^6 4s^2 3d^{10} 4p^5$

Helium has its electrons in the $1s$ sublevel. It cannot have an electron in a $p$ sublevel, like the rest of the noble gases, because it has only two electrons and there is no $1p$ orbital.

For the halogens, five electrons occupy the highest-energy sublevel, the $p$ sublevel. Each ends in $np^5$. The noble gases, group VIIIA (18)—each with one more electron than the halogen it follows—have six electrons in the last-occupied $p$ sublevel. Each ends in $np^6$:

Ne       $1s^2 2s^2 2p^6$
Ar       $1s^2 2s^2 2p^6 3s^2 3p^6$
Kr       $1s^2 2s^2 2p^6 3s^2 3p^6 4s^2 3d^{10} 4p^6$

The halogens and noble gases, along with the other elements in groups IIIA (13) through VIIIA (18), are in the *p block* of the periodic table because their highest-energy electron is in a $p$ sublevel (Figure 7.20). Look at the elements in group VIA (16). How do their electron configurations end? Do they all end with the same number of electrons in the last-occupied sublevel?

The transition metals are hard, with high melting points and boiling points. They are used extensively to manufacture metallic objects where hardness, strength, low reactivity, or electrical conductivity is an important property.

Next let's consider the transition metals. As shown in Figure 7.20, the transition metals are in the *d block*. Their highest-energy electrons are in $d$ orbitals. Most of the time, the transition metals follow a similar pattern as the main-group elements in the $s$ and $p$ blocks. For example, the group IIIB (3) transition metals (Sc and Y) have one electron in the last-occupied $d$ sublevel ($nd^1$). The group IIB (12) transition metals (Zn and Cd) have 10 electrons ($nd^{10}$).

Let's use these periodic trends to write electron configurations for elements. As atomic numbers increase, the number of electrons increases. So, if we follow the atomic symbols of the elements in the periodic table, we should be able to write an electron configuration for any element as we go. As we enter a new $s$, $p$, or $d$ block,

**FIGURE 7.21** The periodic table helps us determine the electron configuration for an element. The principal energy level number $n$ written before the sublevel letter is obtained from the period number. For the $s$ and $p$ blocks, $n$ is the period number (row number). For the $d$ block, $n$ is one less than the period number. For the $f$ block, $n$ is two less than the period number.

we know to change to a new sublevel. The principal energy level number $n$ written before the sublevel letter is obtained from the period number. For the $s$ and $p$ blocks, $n$ is the period number (row number). For the $d$ block, $n$ is one less than the period number. Figure 7.21 helps to show the relationship between sublevels and the period numbers.

Let's write the electron configuration for phosphorus using the periodic table. Phosphorus is a main-group element in the $p$ block, so we expect its highest-energy electrons to be $p$ electrons. Following the arrows in the first periodic table in Figure 7.22, we start in period 1. There is only an $s$ block in period 1, so only the $1s$ orbital is filled. We write

$$1s^2 \quad \text{(period 1)}$$

In the next row, period 2, there are two columns in the $s$ block and six columns in the $p$ block giving

$$2s^2 2p^6 \quad \text{(period 2)}$$

In the third row, period 3, there are two columns in the $s$ block. In the $p$ block, we fill in only three electrons because phosphorus is in the third column of the $p$ block:

$$3s^2 3p^3 \quad \text{(period 3)}$$

Combining the configurations from each period, we get the electron configuration for phosphorus:

$$\text{P} \quad 1s^2 2s^2 2p^6 3s^2 3p^3$$

The superscripts in the electron configuration for phosphorus should add to 15, the number of electrons in a phosphorus atom.

This image shows the filling order for writing electron configurations. You may have used a diagram like this in a previous course. However, if you follow the arrows, you will see how the order follows the periodic table shown in Figure 7.21. Thus, a device like this is unnecessary because the periodic table provides the information you need.

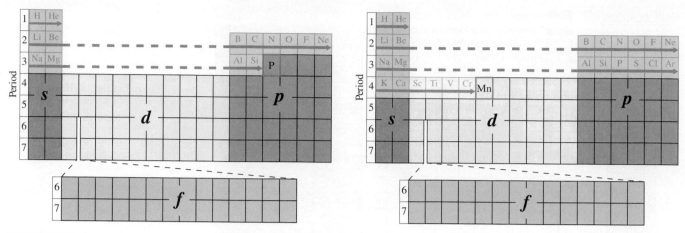

**FIGURE 7.22** To determine the electron configuration for an element, follow the arrows to obtain its electron configuration, filling in one electron for each box (element symbol). When writing the electron configuration for elements past the third period, remember to label the *d* sublevel with a number one less than the period number.

Next let's write the electron configuration for manganese, Mn. We follow the arrows in the second periodic table in Figure 7.22 from hydrogen to manganese. Manganese is a transition metal in the *d* block, so we expect its highest-energy electrons to be *d* electrons. Manganese is in period 4, so the last-occupied *d* orbitals are the $(4 − 1)d$ or $3d$ orbitals. To write the electron configuration, we follow the same general procedure for manganese as we did for phosphorus. In period 1 we fill the first principal energy level:

$$1s^2 \quad \text{(period 1)}$$

In period 2 we fill the *s* and *p* sublevels in the second principal energy level:

$$2s^2 2p^6 \quad \text{(period 2)}$$

In period 3 we fill the *s* and *p* sublevels in the third principal energy level:

$$3s^2 3p^6 \quad \text{(period 3)}$$

In period 4 we first fill the $4s$ sublevel; then we continue to the *d* block, which is the $3d$ sublevel (the period number − 1). Manganese is in the fifth column of the *d* block, so there are five $3d$ electrons. The electron configuration for the *s* and *d* block electrons from period 4 is

$$4s^2 3d^5 \quad \text{(period 4)}$$

Combining the configurations from each period, we get the electron configuration for manganese:

$$\text{Mn} \quad 1s^2 2s^2 2p^6 3s^2 3p^6 4s^2 3d^5$$

Do the superscripts total to the number of electrons in a manganese atom?

Example 7.6 will give you practice at writing electron configurations using the periodic table.

| EXAMPLE 7.6 | **Writing Electron Configurations Using the Periodic Table** |
|---|---|

Use a periodic table to write the electron configurations for the following elements.

(a) Si
(b) Zr
(c) Br

**Solution:**

(a) Silicon, Si, is in the second column of the $p$ block, and it is in period 3, so its electron configuration will end in $3p^2$. The rest of its electron configuration is obtained by following the periodic table in order of increasing atomic number starting with H and He in period 1. We get $1s^2$ for period 1, $2s^2 2p^6$ for period 2, and $3s^2 3p^2$ for period 3. Combining these we get the electron configuration for Si:

$$\text{Si} \quad 1s^2 2s^2 2p^6 3s^2 3p^2$$

The superscripts add to 14, the number of electrons in a silicon atom.

(b) Zirconium, Zr, is in the second column of the $d$ block, and it is in period 5. Its electron configuration will end in $4d^2$ (not $5d^2$) because the principal energy level number for the $d$-block elements is always the period number minus one $(5 - 1 = 4)$. The electron configuration is obtained in the same way as for Si. For periods 1, 2, and 3, we get the first part of the electron configuration: $1s^2 2s^2 2p^6 3s^2 3p^6$. In period 4, we add the $s$-block electrons, $4s^2$; the $d$-block electrons, $3d^{10}$; and the $p$-block electrons, $4p^6$. In period 5, we add the $s$-block electrons, $5s^2$, and the two $d$-block electrons, $4d^2$. Combining the sublevels from all five periods, we get the electron configuration for Zr:

$$\text{Zr} \quad 1s^2 2s^2 2p^6 3s^2 3p^6 4s^2 3d^{10} 4p^6 5s^2 4d^2$$

The superscripts add to 40, the number of electrons in a zirconium atom.

(c) Bromine, Br, is in the fifth column of the $p$ block, and it is in period 4. Its electron configuration will end in $4p^5$. For periods 1, 2, and 3, we get the first part of the electron configuration: $1s^2 2s^2 2p^6 3s^2 3p^6$. In period 4, we add the $s$-block electrons, $4s^2$; the $d$-block electrons, $3d^{10}$; and the $p$-block electrons, $4p^5$. Combining the sublevels from all four periods, we get the electron configuration for bromine:

$$\text{Br} \quad 1s^2 2s^2 2p^6 3s^2 3p^6 4s^2 3d^{10} 4p^5$$

**Consider This 7.6**

How does the electron configuration of the noble gas Kr differ from that of Zr?

**Practice Problem 7.6**

Use a periodic table to write the electron configurations for the following elements.

(a) S
(b) Rb
(c) Cd

**Further Practice:** Questions 7.61 and 7.62 at the end of the chapter

For elements with lengthy electron configurations, chemists often use *abbreviated electron configurations*. Notice that argon and calcium differ only by the $4s^2$ electrons at the end of the electron configuration for Ca:

$$\text{Ar} \qquad 1s^2 2s^2 2p^6 3s^2 3p^6$$
$$\text{Ca} \qquad 1s^2 2s^2 2p^6 3s^2 3p^6 4s^2$$

Instead of writing the complete electron configuration for Ca, we will use the symbol [Ar] to represent the part of calcium's electron configuration that corresponds to argon. The abbreviated electron configuration for Ca substitutes [Ar] for the first part of its electron configuration:

$$\text{Ca} \qquad [\text{Ar}]4s^2$$

By convention, electron configurations are only abbreviated using noble-gas configurations. To write an abbreviated electron configuration for an element, locate its position on the periodic table and find the closest noble gas in the period above the element. Use this noble gas in the abbreviation. Once you have written the noble-gas symbol in brackets, add the additional electron configuration needed to reach the element on the periodic table.

What is the abbreviated electron configuration for bromine, Br? The closest noble gas above bromine is argon. We enclose the symbol for argon in brackets to represent its electron configuration. Argon is at the end of period 3, so we add the electrons that correspond to the blocks we must pass through in period 4 to get to bromine, just as we do when we write the complete electron configuration. The abbreviated electron configuration is

$$\text{Br} \quad [\text{Ar}]4s^2 3d^{10} 4p^5$$

We have not yet mentioned electron configurations that include the $f$ orbitals. Cerium (58), the first element of the lanthanide series, is the first element in the $f$-block series (Figure 7.21). It is enough to note here that the lanthanide and actinide elements are in the $f$ block because their highest-energy electron is in an $f$ sublevel. There are 14 lanthanides, this number corresponding to the number of electrons that can fill the seven $4f$ orbitals. Similarly, the actinide series (elements 90 to 103) corresponds to the filling of the $5f$ orbitals. The lanthanides and actinides are rarely encountered, and many irregularities occur in their electron configurations.

## 7.5  Valence Electrons for the Main-Group Elements

There are a few minor exceptions to the normal filling order for electron configurations. Notably, the lanthanides and actinides vary from a predictable pattern. Also some of the transition metal elements in group VIB (6) and group IB (11) have electron configurations different from what the filling rules predict. The deviations result from the extra stability of half-filled and completely filled $d$ sublevels. Can you find any exceptions for period 4 elements in Figure 7.23?

Elements in the same group have the same number of electrons in their last-occupied sublevel and principal energy level. The last-occupied principal energy level is called the **valence level,** or *valence shell*. It is highest in energy and contains orbitals that are larger than orbitals in lower principal energy levels. An electron that occupies the valence level is a **valence electron.** An electron in a principal energy level below the valence level is called a **core electron,** or inner electron. The periodic table in Figure 7.23 shows valence electron configurations for all the elements. The similar properties of same-group elements, discussed in Chapter 2 and in Section 7.7, relate to the number of valence electrons in the atoms. Let's look closely at the valence electrons of the main-group elements.

One way to determine the number of valence electrons for an element is to look at its electron configuration or its abbreviated electron configuration. For example, the electron configuration for silicon is

$$\text{Si} \quad 1s^2 2s^2 2p^6 3s^2 3p^2 \quad \text{or} \quad [\text{Ne}]3s^2 3p^2$$

The valence level is the $n = 3$ principal energy level, which contains the $3s$ and $3p$ sublevels. From the electron configuration, we can see that there are a total of four electrons in the $n = 3$ principal energy level. Thus, silicon has four valence electrons. Note that the valence electrons in silicon are the electrons that are written after the electron configuration for neon. Electrons that make up a noble-gas electron configuration are core electrons held very strongly by the nucleus. For some elements, however, some of the core electrons are *not* part of a noble-gas electron configuration. For example, the abbreviated electron configuration for bromine is

$$\text{Br} \quad [\text{Ar}]4s^2 3d^{10} 4p^5$$

The $d$-sublevel electrons are sometimes counted as valence electrons for transition metals, but the behavior of these electrons is less predictable than the valence electrons for main-group elements.

The valence level is the fourth principal energy level ($n = 4$), which contains seven valence electrons. The $3d$ electrons are not valence electrons since they are in the third principal energy level ($n = 3$). *For any main-group element, the number of*

**FIGURE 7.23** This periodic table shows the valence electron configurations for all the elements. The last-occupied sublevel for each element corresponds to the *s, p, d,* or *f* block designated in the periodic table.

*valence electrons is equal to the number of electrons in the highest-energy s and p sublevels.*

A more convenient way to determine the valence level and the number of valence electrons is to use the periodic table. In Section 7.4 we saw how the periodic table predicts an element's electron configuration. The principal energy level number is the same as the period number for the *s* and *p* sublevels. Now we know that the valence electrons occupy the highest-energy *s* and *p* sublevels. This tells us that the period number of the element equals the valence level number. The number of valence electrons can be determined from the Roman numeral group number for main group elements. Except for helium, the Roman numeral group number corresponds to the number of valence electrons because the group number is a sum of the *s* and *p* valence electrons. Silicon, for example, is in group IVA (14); the Roman numeral IV corresponds to four valence electrons, two *s* electrons and two *p* electrons. Bromine, in group VIIA (17), has seven valence electrons, as do all the other elements, the halogens, in that group. The periodic table shows that *elements in the same group have the same number of valence electrons.*

Example 7.7 will give you practice determining the number of valence electrons for main-group elements.

---

**EXAMPLE 7.7**    **Number of Valence Electrons**

For each of the following elements, determine the number of valence electrons.

(a) nitrogen
(b) potassium
(c) oxygen

**Solution:**

(a) The electron configuration for nitrogen, N, is

$$1s^2 2s^2 2p^3$$

The valence level has the highest principal energy level number, $n = 2$. There are five electrons in the valence level (two electrons in the $2s$ sublevel and three electrons in the $2p$ sublevel), so nitrogen has five valence electrons. We can also use the periodic table to determine the number of valence electrons. Nitrogen is in group VA (15). The Roman numeral group number (V) corresponds to a nitrogen atom's five valence electrons. Using the periodic table to determine valence electrons is a faster method, so we will use it for parts (b) and (c).

(b) Potassium, K, is in group IA, so it has just one valence electron, the $4s^1$ valence electron.

(c) Oxygen, O, is in group VIA (16), so it has six valence electrons, the $2s^2 2p^4$ valence electrons.

**Consider This 7.7**

If an unknown main-group element is known to have four valence electrons, identify the group of elements that it's in.

**Practice Problem 7.7**

For each of the following elements, determine the number of valence electrons.

(a) magnesium
(b) carbon
(c) lead

**Further Practice:** Questions 7.77 and 7.78 at the end of the chapter

---

## 7.6 Electron Configurations for Ions

The electron configurations we have written so far are for atoms. Because atoms have no charge, the number of electrons equals the number of protons, which is the atomic number. Ionic substances, such as those that create the colors in fireworks, consist of charged ions, not neutral atoms. Ions are charged because their number of electrons is greater than or less than their number of protons. To write electron configurations for ions, we adjust the number of valence electrons to account for the charge.

Cations have fewer electrons than protons. To write their electron configurations, we remove valence electrons from the atom's electron configuration. Consider calcium as an example. A calcium atom has the following electron configuration:

$$Ca \quad 1s^2 2s^2 2p^6 3s^2 3p^6 4s^2$$

When a $Ca^{2+}$ ion forms, two electrons are lost from the valence $4s$ sublevel. We write the electron configuration for a $Ca^{2+}$ ion as

$$Ca^{2+} \quad 1s^2 2s^2 2p^6 3s^2 3p^6$$

Anions have more electrons than protons. To write their electron configurations, we add electrons to the valence level of the atom's electron configuration. Consider chlorine as an example. A chlorine atom has the following electron configuration:

$$Cl \quad 1s^2 2s^2 2p^6 3s^2 3p^5$$

When a $Cl^-$ ion forms from a chlorine atom, one electron is gained. It is added to the valence $3p$ sublevel. The electron configuration for a $Cl^-$ ion is

$$Cl^- \quad 1s^2 2s^2 2p^6 3s^2 3p^6$$

The electron configurations for $Ca^{2+}$ and $Cl^-$ are identical, and they are the same as for an argon atom. The $Ca^{2+}$ cation, the $Cl^-$ anion, and the Ar atom are **isoelectronic** because they have the same number of electrons. Even so, the three are different because they have different numbers of protons. Can you think of other common ions that are isoelectronic with these three?

---

| **EXAMPLE 7.8** | **Electron Configurations for Ions** |
|---|---|

Write the abbreviated electron configuration for each of the following ions. For each ion identify another ion that is isoelectronic with it.

(a)  $S^{2-}$
(b)  $Sr^{2+}$
(c)  $Al^{3+}$

**Solution:**

(a)  To write the abbreviated electron configuration for the sulfide ion, $S^{2-}$, start with the abbreviated electron configuration for the sulfur atom:

$$S \quad [Ne]3s^2 3p^4$$

The $S^{2-}$ anion has a 2− charge, so we add two electrons to the highest-energy valence sublevel (the $3p$ sublevel). The electron configuration for $S^{2-}$ is

$$S^{2-} \quad [Ne]3s^2 3p^6 \text{ or } [Ar]$$

Another ion that is isoelectronic with $S^{2-}$ must have the same number of electrons and the same electron configuration. The electron configuration for $S^{2-}$ is the same as that of the noble gas argon, so ions that are isoelectronic with Ar are also isoelectronic with $S^{2-}$. Examples include $P^{3-}$, $Cl^-$, $K^+$, and $Ca^{2+}$.

(b)  To write the abbreviated electron configuration for the strontium ion, $Sr^{2+}$, start with the abbreviated electron configuration for the strontium atom:

$$Sr \quad [Kr]5s^2$$

The $Sr^{2+}$ cation has a 2+ charge, so we remove two electrons from the highest-energy valence sublevel (the $5s$ sublevel). The electron configuration for $Sr^{2+}$ is

$$Sr^{2+} \quad [Kr]$$

Ions that are isoelectronic with $Sr^{2+}$ have the same electron configuration as the noble gas krypton. Ions that are isoelectronic with Kr are also isoelectronic with $Sr^{2+}$. Examples are $Se^{2-}$, $Br^-$, and $Rb^+$. Some transition metals are also isoelectronic with $Sr^{2+}$. An example is $Zr^{4+}$.

(c) To write the abbreviated electron configuration for the $Al^{3+}$ ion, start with the abbreviated electron configuration for the aluminum atom:

$$Al \quad [Ne]3s^23p^1$$

The $Al^{3+}$ cation has a 3+ charge, so we must remove three electrons from the $n = 3$ valence level. The electron configuration for $Al^{3+}$ is the same as that for neon.

$$Al^{3+} \quad [Ne]$$

The $Al^{3+}$ ion has 10 electrons like neon. Other ions with 10 electrons include $O^{2-}$, $F^-$, $Na^+$, and $Mg^{2+}$.

**Consider This 7.8**

What are some ways that isoelectronic ions differ?

**Practice Problem 7.8**

Write the abbreviated electron configuration for each of the following ions. For each ion, identify another ion that is isoelectronic with it.

(a) $Br^-$

(b) $N^{3-}$

(c) $K^+$

**Further Practice:** Questions 7.81 and 7.82 at the end of the chapter

## 7.7  Periodic Properties of Atoms

The electronic structure of atoms explains what happens when light interacts with matter. When atoms absorb energy, electrons move into orbitals at higher energy levels. When the electrons drop to lower energy positions, energy is released in the form of electromagnetic radiation that is sometimes in the visible region of the spectrum. These transitions of electrons account for the colors of light that Andrea, Ben, and Drew saw in the fireworks, neon lights, and lasers in Las Vegas.

In this section we'll discuss how electron arrangements relate to other properties of elements, including reactivity, tendency to lose electrons, and atomic size.

### Chemical Reactivity and Electron Configurations

Recall from Chapter 5 the activity series that predicts the relative reactivity of metals. Notice the elements that are at the top of the activity series, as shown in Figure 7.24. Can you see the pattern? Most are alkali metals, group IA (1), and alkaline earth metals, group IIA (2). They are the most reactive metals. These metals are so reactive that they are not found in nature as pure elements. They are usually found as ions combined with nonmetals in compounds.

The alkali metals, in the first group in the periodic table, are commonly found in nature as oxides. Their stable compounds with oxygen have the general formula $M_2O$, where M represents an alkali metal:

$$Li_2O \quad Na_2O \quad K_2O \quad Rb_2O \quad Cs_2O$$

When the alkali metals combine with oxygen or with other elements in compounds, they are 1+ ions. The electron configurations for the alkali metals explain this similarity:

$$Li \qquad 1s^22s^1 = [He]2s^1$$
$$Na \qquad 1s^22s^22p^63s^1 = [Ne]3s^1$$

**FIGURE 7.24** In this chemical activity series, the most active elements appear at the top.

K     $1s^2 2s^2 2p^6 3s^2 3p^6 4s^1 = [Ar]4s^1$

Rb    $1s^2 2s^2 2p^6 3s^2 3p^6 4s^2 3d^{10} 4p^6 5s^1 = [Kr]5s^1$

Cs    $1s^2 2s^2 2p^6 3s^2 3p^6 4s^2 3d^{10} 4p^6 5s^2 4d^{10} 5p^6 6s^1 = [Xe]6s^1$

What do these elements have in common? All have a single valence electron in an *s* orbital. For each, the loss of the valence *s* electron gives the electron configuration of the noble gas that precedes it in the periodic table.

When the alkali metals form 1+ ions, they lose their only valence electron to get an electron configuration of a noble gas:

Li$^+$    $1s^2 = [He]$

Na$^+$    $1s^2 2s^2 2p^6 = [Ne]$

K$^+$     $1s^2 2s^2 2p^6 3s^2 3p^6 = [Ar]$

Rb$^+$    $1s^2 2s^2 2p^6 3s^2 3p^6 4s^2 3d^{10} 4p^6 = [Kr]$

Cs$^+$    $1s^2 2s^2 2p^6 3s^2 3p^6 4s^2 3d^{10} 4p^6 5s^2 4d^{10} 5p^6 = [Xe]$

Because francium is very rare and highly radioactive, we won't discuss it here.

*Because valence electrons are farthest from the nucleus, they're the ones that participate in chemical reactions.* Chemical reactions of the elements involve changes in the electron configurations—that is, addition, removal, or sharing of electrons—to give more stable configurations. Usually, a change in the number of electrons results in a noble-gas electron configuration:

**ANIMATION:** Properties of Alkali and Alkaline Earth Materials

He    $1s^2$

Ne    $[He]2s^2 2p^6$

Ar    $[Ne]3s^2 3p^6$

Kr    $[Ar]3d^{10} 4s^2 4p^6$

Xe    $[Kr]4d^{10} 5s^2 5p^6$

What do the electron configurations of the noble gases have in common? Their valence *s* and *p* sublevels are full. Because very few compounds of the noble gases are known (and must be synthesized under very special conditions), we can conclude that the filled *s* and *p* sublevels give exceptional stability to any atom or ion.

The alkaline earth metals are also quite reactive, although not as reactive as the alkali metals. The abbreviated electron configurations for the alkaline earth metals are

Be    $[He]2s^2$

Mg    $[Ne]3s^2$

Ca    $[Ar]4s^2$

Sr    $[Kr]5s^2$

Ba    $[Xe]6s^2$

These elements are commonly found in nature as 2+ ions combined with oxygen or with halogens. The ions of the alkaline earth metals have noble-gas electron configurations:

Be$^{2+}$    $[He]$

Mg$^{2+}$    $[Ne]$

Ca$^{2+}$    $[Ar]$

Sr$^{2+}$    $[Kr]$

Ba$^{2+}$    $[Xe]$

Lithium          Sodium          Potassium          Rubidium          Cesium

**FIGURE 7.25** All the alkali metals react with water. Their reactivity increases going down the group in the periodic table.
(lithium): ©Chemistry at Work/Videodisk/Videodiscovery, Inc.; (sodium): ©Richard Megna/Fundamental Photographs; (potassium): ©Richard Megna/Fundamental Photographs; (rubidium, cesium): ©Chemistry at Work/Videodisk/Videodiscovery, Inc.

We've considered the metals on the far left of the periodic table. Let's now consider the halogens on the right, next to the noble gases. The halogens are highly reactive nonmetals, so they are rarely found in nature as pure elements. They are found as 1− ions combined with metal ions, such as in NaCl or $CaF_2$. The chemical behavior of the halogen elements is based on *gaining* one electron. How are their electron configurations similar?

F      $[He]2s^2 2p^5$

Cl     $[Ne]3s^2 3p^5$

Br     $[Ar]3d^{10}4s^2 4p^5$

I      $[Kr]4d^{10}5s^2 5p^5$

What ions do these elements form? How can we explain their ion formation in terms of forming the more stable electron configuration of a noble gas?

The modern model of the atom, with its description of electronic structure, provides an explanation for the similar properties of elements within the same group of the periodic table. However, elements in the same group do not have exactly the same properties. Many properties vary in a regular fashion moving down a group. As an example, let's look more closely at the chemical reactivity and electron configurations of the alkali metals.

If our students on their Las Vegas trip threw chunks of the alkali metals into one of the spectacular water fountains in Las Vegas, they'd observe some evidence of chemical reactions—some of them quite spectacular. *Of course, the students would not actually do this because the reactions are very dangerous.* The reactions of the alkali metals with water are shown in Figure 7.25.

Beginning with lithium, Ben, Andrea, and Drew might notice the chunk fizzle a little, a sign that a gas is being produced. The gas is hydrogen, which is produced by a single-displacement reaction. Sodium, the next alkali metal down the group in the periodic table, reacts more vigorously. Things get even more interesting when potassium meets water. It reacts violently, and the hydrogen usually ignites to give a flame. Rubidium reacts even more explosively. Cesium's reaction is so violent that it blows out the side of the container. Rubidium and cesium are both so reactive that they will react quickly with the oxygen in the air. Scientists must work with them in an unreactive argon environment.

The reactivity of the alkali metals increases as we go down the group in the periodic table. This general trend is also observed with the alkaline earth metals, group IIA (2), which are less reactive than the alkali metals. How can we explain the variation in reactivity within a group? Next, we will learn about atomic properties that help explain these trends. We will also see why properties change gradually as we move across the periodic table from left to right.

## Ionization Energy

When metal elements react, they lose valence electrons and form cations. *The more easily metal atoms give up their valence electrons, the more reactive they are.* One important factor is **ionization energy** (*IE*), which is a measure of the energy required to remove a valence electron from a gaseous atom to form a gaseous ion. It is usually expressed as the energy required to remove an electron from each atom in exactly 1 mol $(6.022 \times 10^{23})$ of atoms. For example, consider the ionization process for lithium, which can be represented with the following equation:

$$\text{Li}(g) \longrightarrow \text{Li}^+(g) + \text{e}^- \quad IE = 520 \text{ kJ/mol}$$

In other words, 520 kJ of energy would be required to remove the outermost electron from each atom in 1 mol of lithium. Ionization energy is related to electron arrangement and helps explain differences in the reactivities of metals. *In general, atoms with low ionization energies do not bind their valence electrons very tightly, so they are very reactive.*

**Trends in Ionization Energies** The ionization energies of the main-group elements are given in Figure 7.26. What patterns can you find? Notice that the ionization energies of the alkali and alkaline earth metals are relatively low. If we look specifically at elements in the same group, we can see that their ionization energies decrease as we move down a group. This happens because the valence electron that's removed upon ionization is farther from the nucleus as we move down a group. Electrons farther from the nucleus are easier to remove. They are not as strongly attracted to the positive charge of the nucleus (Figure 7.27). In addition, valence electrons are shielded from the attraction of protons by core electrons closer to the nucleus. As the number of core sublevels increases, so does the effect of shielding. This explains why elements such as cesium are so reactive.

You may have observed another pattern in ionization energies from Figure 7.26. They increase from left to right on the periodic table. What is different about atoms in the same period that makes ionization energy increase? The increase in the number of electrons does not explain this trend. As we move across a period from left to right, electrons are added to the same principal energy level. This has a negligible

You might wonder what happens when an atom gains an electron to become an anion. Most elements release energy when they gain an electron, but the pattern is less predictable, so we will not discuss it in this text.

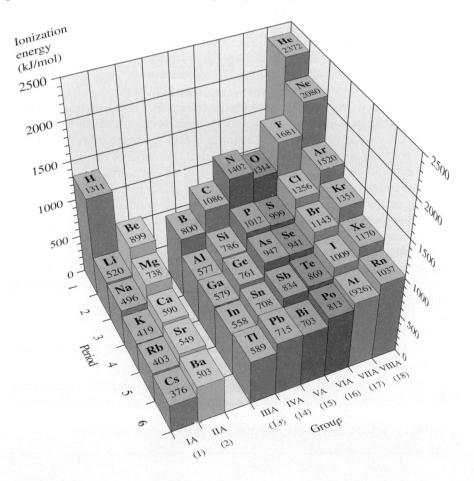

**FIGURE 7.26** Ionization energies for the main-group elements are given in units of kJ/mol. They are the amount of energy required to remove one electron from each atom in 1 mol of atoms of the element when in the gaseous state.

**FIGURE 7.27** A valence electron in a lower principal energy level is closer to the nucleus than a valence electron in a higher principal energy level. The attraction between the nucleus and the valence electron is greater for smaller atoms, resulting in a greater ionization energy.

In smaller atoms, electrons are closer to the nucleus, and therefore strongly attracted by the positive nuclear charge.

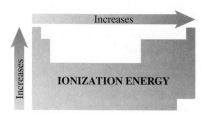

**FIGURE 7.28** Ionization energy tends to increase going up the periodic table and from left to right.

effect on ionization energy. However, an increase in the number of protons in the nucleus as we move across the periodic table causes ionization energy to increase because of the additional positive charge. This increase in nuclear charge pulls the electrons in the valence level closer to the nucleus and makes removal of a valence electron more difficult. The general tendency is for ionization energy to increase from bottom to top and from left to right across the periodic table (Figure 7.28).

---

**EXAMPLE 7.9** | **Trends in Ionization Energy**

Without looking at Figure 7.26, use periodic trends in ionization energy to answer the following questions. Explain your reasoning.

(a) Which element, carbon or fluorine, should have the greater ionization energy? Why?

(b) Which element, strontium or magnesium, should have the greater ionization energy? Why?

**Solution:**

(a) Carbon and fluorine are in the same period in the periodic table, with fluorine farther to the right. Ionization energy increases from left to right across a period, with exceptions for the first and fourth electrons in the *p* orbitals. Since neither of these elements is one of the exceptions, fluorine should have the greater ionization energy.

(b) Strontium and magnesium are in the same group in the periodic table, with magnesium closer to the top. Ionization energy increases up a group, so magnesium should have the greater ionization energy.

**Consider This 7.9**

If you were asked which element requires the greatest amount of energy to remove a valence electron, would your answers in (a) and (b) be the same or different than when asked about greatest ionization energy?

**Practice Problem 7.9**

Without looking at Figure 7.26, use periodic trends in ionization energy to answer the following questions. Explain your reasoning.

(a) Which element, potassium or lithium, should have the greater ionization energy?

(b) Which element, silicon or chlorine, should have the greater ionization energy?

**Further Practice:** Questions 7.89 and 7.90 at the end of the chapter

**Successive Ionization Energies**   In this section we have focused on the ionization energy that removes one electron from a neutral atom. This is sometimes referred to as the first ionization energy, $IE_1$. Suppose we remove a second electron. Will its removal require the same amount of energy as the first? For example, consider a lithium atom, which has three electrons. The first ionization energy of lithium corresponds to the removal of the first electron. To remove an electron from 1 mol of lithium atoms requires 520 kJ of energy. We represent the process involved for the first ionization energy, $IE_1$, with the following equation:

$$\text{Li}(g) \longrightarrow \text{Li}^+(g) + e^- \quad IE_1 = 520 \text{ kJ/mol}$$

To remove an electron from 1 mol of $\text{Li}^+$ ions requires 7298 kJ. This is the second ionization energy, $IE_2$, and it is represented with the following equation:

$$\text{Li}^+(g) \longrightarrow \text{Li}^{2+}(g) + e^- \quad IE_2 = 7298 \text{ kJ/mol}$$

Removal of an electron from 1 mol of $\text{Li}^{2+}$ ions requires 11,815 kJ, which is the third ionization energy, $IE_3$.

$$\text{Li}^{2+}(g) \longrightarrow \text{Li}^{3+}(g) + e^- \quad IE_3 = 11,815 \text{ kJ/mol}$$

Ionization energy increases with each successive removal of an electron. How can we explain this trend? Remember that the second electron is removed from a 1+ charged ion. It is harder to remove because there are fewer electrons, so the ones that remain feel the attraction of the positively charged nucleus more strongly. The same is true for further removal of electrons: $(IE_3 > IE_2 > IE_1)$. The electrons are held more strongly as the positive charge on the ion increases.

The increase in ionization energy with successive removal of electrons is not gradual. For lithium, the second ionization energy is more than 10 times greater than the first! How does the electron configuration of lithium explain this?

$$\text{Li} \quad 1s^2 2s^1$$

The first electron to be removed is a $2s$ valence electron. The second electron to be removed is a $1s$ core electron. The nucleus holds core electrons much more tightly than valence electrons. The large second and third ionization energies for lithium show why lithium does not form $\text{Li}^{2+}$ or $\text{Li}^{3+}$ ions except under extreme conditions.

Table 7.1 shows the successive ionization energies for the elements in period 2. Removal of core electrons (to the right of the stair-step line) requires much more energy than removal of valence electrons (to the left of the line).

**TABLE 7.1** Successive Ionization Energies for Period 2 Elements

| Element | $IE_1$ | $IE_2$ | $IE_3$ | $IE_4$ | $IE_5$ | $IE_6$ | $IE_7$ | $IE_8$ | $IE_9$ | $IE_{10}$ |
|---------|--------|--------|--------|--------|--------|--------|--------|--------|--------|-----------|
| Li | 520 | 7298 | 11,815 | | | | | | | |
| Be | 899 | 1757 | 14,849 | 21,006 | | Core electrons | | | | |
| B | 800 | 2427 | 3660 | 25,026 | 32,827 | | | | | |
| C | 1086 | 2353 | 4620 | 6222 | 37,830 | 47,277 | | | | |
| N | 1402 | 2856 | 4582 | 7475 | 9445 | 53,266 | 64,360 | | | |
| O | 1314 | 3388 | 5300 | 7469 | 10,989 | 13,326 | 71,334 | 84,078 | | |
| F | 1681 | 3374 | 6050 | 8408 | 11,023 | 15,164 | 17,868 | 92,038 | 106,434 | |
| Ne | 2080 | 3952 | 6122 | 9370 | 12,178 | 15,238 | 19,999 | 23,069 | 115,379 | 131,431 |

Note: Ionization energies are given in kJ/mol.

Cl—Cl

**FIGURE 7.29** To determine the size of an atom, scientists measure the distance between the centers of two bonded atoms. For identical atoms, the atomic radius is one-half this distance.

## Atomic Size

Another trend that shows a pattern throughout the periodic table is the size of atoms. The size of an atom is often described in terms of its **atomic radius,** the distance from the center of the nucleus to the outer edge of the atom. The radius of an atom is often reported in units of picometers (1 pm = $10^{-12}$ m). Because the outer edge of an atom is not sharply defined, scientists measure the distance between the centers of two identical bonded atoms to determine atomic radii. The radius is one-half this distance (Figure 7.29).

Atomic radii for the main-group elements are shown in Figure 7.30. Do you observe any patterns? Notice that the atomic radius generally increases as we move down a group. How might you explain this trend? The electron configurations for the elements in a group provide the answer. For example, recall the electron configurations for the alkali metals, group IA (1):

| | | | |
|---|---|---|---|
| Li | [He]$2s^1$ | Rb | [Kr]$5s^1$ |
| Na | [Ne]$3s^1$ | Cs | [Xe]$6s^1$ |
| K | [Ar]$4s^1$ | | |

Lithium's valence electron is in the $2s$ orbital; the valence level is $n = 2$, very close to the nucleus. At the bottom of the group is cesium. Its valence level is $n = 6$, much farther from the nucleus. As we go from top to bottom in a group, the atomic radius increases because the valence electrons are in larger orbitals.

As we move from left to right across a period, atomic size generally decreases. You might think that size should increase because the number of electrons increases, but the electrons are added to the same principal energy level while the positive charge in the nucleus increases. As we move from left to right across a period, the increased positive charge in the nucleus becomes more effective at drawing the valence electrons closer to the center of the atom, causing atoms to the right on the periodic table to be smaller in size. Trends in atomic size are shown in Figure 7.31.

**FIGURE 7.30** This periodic table shows the relative atomic size of the main-group elements, along with their atomic radii in units of picometers. (Except for the noble gases, the atomic radius is determined as half the distance between two bonded atoms.)

## EXAMPLE 7.10 | Relative Sizes of Atoms

(a) Which atom is larger, carbon or fluorine? Explain.
(b) Which atom is larger, strontium or magnesium? Explain.

**Solution:**

(a) As we go from left to right in a period, size decreases because the increase in positive nuclear charge pulls the valence electrons closer to the nucleus. Carbon atoms are larger than fluorine atoms.
(b) When we move down a group on the periodic table, size increases because the valence electrons are in principal energy levels farther from the nucleus. Strontium atoms are bigger than magnesium atoms.

**Consider This 7.10**

If you were asked to compare atomic sizes of carbon and phosphorus, what trend in Figure 7.30 helps you justify that phosphorus atoms are larger than carbon atoms without referring to atomic radii values?

**Practice Problem 7.10**

(a) Which atom is larger, potassium or lithium? Explain.
(b) Which atom is larger, silicon or chlorine? Explain.

**Further Practice:** Questions 7.101 and 7.102 at the end of the chapter

**FIGURE 7.31** Atomic size increases down a group and from right to left in a period.

## Sizes of Ions

When ions form from neutral atoms, the radii of the atoms change. Consider what happens when lithium ionizes (Figure 7.32A). A neutral atom of lithium has 3 electrons and 3 protons. The ion has 2 electrons and 3 protons. The nuclear charge remains the same, while the number of electrons decreases. In the ion, the protons draw the remaining electrons closer to the nucleus. Consequently, the $Li^+$ ion is smaller than the neutral atom.

The opposite happens with anions. For example, consider the formation of the fluoride ion (Figure 7.32B). The neutral atom has 9 electrons and 9 protons. The $F^-$ ion has 10 electrons and 9 protons. Because there are more electrons, the protons cannot hold the electrons as close to the nucleus. Thus, the ion is larger than the neutral atom. A summary of the sizes of many main-group ions is given in Figure 7.33.

There are so many factors involved in determining the sizes of ions that no single trend spans the entire periodic table. However, for a series of ions with the same number of electrons—*an isoelectronic series*—consider the following ions:

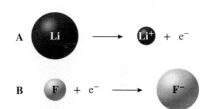

**FIGURE 7.32** (A) A cation is smaller than the neutral atom from which it forms. The nucleus holds the cation's fewer electrons more tightly. (B) An anion is larger than the neutral atom from which it forms. The nucleus holds the anion's greater number of electrons less tightly.

| Ion | | Radius (pm) | Electron Configuration | Protons |
|-----|---|-------------|------------------------|---------|
| $S^{2-}$ | | 184 | $1s^2 2s^2 2p^6 3s^2 3p^6$ | 16 |
| $Cl^-$ | | 181 | $1s^2 2s^2 2p^6 3s^2 3p^6$ | 17 |
| $K^+$ | | 133 | $1s^2 2s^2 2p^6 3s^2 3p^6$ | 19 |
| $Ca^{2+}$ | | 99 | $1s^2 2s^2 2p^6 3s^2 3p^6$ | 20 |
| $Sc^{3+}$ | | 81 | $1s^2 2s^2 2p^6 3s^2 3p^6$ | 21 |

Note the relative sizes of the $Na^+$ and $Cl^-$ ions in the NaCl structure. Why is $Na^+$ smaller than $Cl^-$?

$Cl^-$
$Na^+$

**FIGURE 7.33** Shown here are the ionic radii for ions of the main-group elements. Note the trends down a group and for ions that are isoelectronic.

All of the ions in this series have the same number of electrons and the same electron configuration. However, they have different numbers of protons (atomic numbers). *For any isoelectronic series, as the number of protons increases, the ion size decreases.*

---

**EXAMPLE 7.11**　　**Sizes of Ions in an Isoelectronic Series**

Arrange the following isoelectronic ions in order of increasing radius: $Al^{3+}$, $F^-$, $Mg^{2+}$, $N^{3-}$, $Na^+$, $O^{2-}$.

**Solution:**

Each of these ions has 10 electrons. The size of isoelectronic ions increases with decreasing nuclear charge (the number of protons). Thus, $N^{3-}$, with a nuclear charge of 7+, should be the largest ion, and $Al^{3+}$, with a nuclear charge of 13+, should be the smallest. The order is $Al^{3+} < Mg^{2+} < Na^+ < F^- < O^{2-} < N^{3-}$.

**Consider This 7.11**

How does the size of each ion in the example compare to the size of a neutral atom of that element?

**Practice Problem 7.11**

Arrange the following isoelectronic ions in order of increasing radius: $Br^-$, $Se^{2-}$, $Rb^+$, $Sr^{2+}$, $Y^{3+}$.

**Further Practice:** Questions 7.107 and 7.108 at the end of the chapter

# CHAPTER REVIEW

## KEY CONCEPTS

- Scientists interested in light led the way to our current understanding of atoms—specifically, how electrons are arranged in atoms.
  - Light can be described by its wavelength, frequency, or photon energy. The light we see is a small portion of the electromagnetic spectrum, the visible region.
  - When an element is heated or given an electric charge, the energy absorbed by its atoms is emitted as light energy. The light an element produces in this way contains only specific colors, as shown by its line spectrum.
- Bohr's model of the hydrogen atom helped scientists understand the quantized nature of electrons—that they have specific allowed energies.
  - Bohr showed that the four lines in the hydrogen spectrum result from electron transitions from higher-energy orbits ($n = 3$, 4, 5, and 6) to the $n = 2$ orbit.
  - Bohr's model only explained the hydrogen atom spectrum; however, the fundamental ideas led to the modern model of the atom.
- The idea that the energies of electrons are quantized is retained in the modern model of the atom, where electrons are described as being in specific principal energy levels, sublevels, and orbitals.
  - An orbital is a three-dimensional region in space where an electron is likely to be found.
  - Sublevels contain sets of orbitals with the same letter designations: s, p, d, or f.
  - The s sublevel contains one orbital, the p sublevel contains three orbitals, the d sublevel contains five orbitals, and the f sublevel contains seven orbitals.
  - Orbital diagrams show the relative energies of the orbitals, and electrons occupy those orbitals following the aufbau principle, the Pauli exclusion principle, and Hund's rule.
- Electron configurations are shorthand notations that show the distribution of electrons in atoms.
  - The periodic table can be divided into s, p, d, and f blocks and is a useful tool for writing electron configurations.
  - Valence electrons are the outer electrons that are held less tightly by the nucleus than the core electrons. Valence electrons for main-group elements are the electrons in the highest principal energy level (highest $n$ value); they are the electrons in the last-occupied s and p sublevels, so elements in the same group have the same number of valence electrons.
  - Electron configurations for ions are written by adjusting the number of valence electrons.
- We can explain many of the trends in chemical and physical properties of elements by looking at specific atomic properties such as ionization energy and atomic size.
  - Ionization energy ($IE$) is a measure of the amount of energy required to remove the highest-energy valence electron from an atom. There is a periodic trend for ionization energy values: they increase going up a group and from left to right within a period.
  - Because successive electrons are more difficult to remove, successive ionization energies increase in the order $IE_1 < IE_2 < IE_3$, etc.
  - The periodic trend for atomic size is opposite that of ionization energy: atomic size increases going down a group and from right to left within a period.
  - Cations are smaller than their neutral atoms, and anions are larger than their neutral atoms. Because isoelectronic ions differ only in the number of protons, size decreases as the number of protons increases.

## KEY RELATIONSHIPS

| Relationship | Equation |
|---|---|
| The frequency times the wavelength is equal to a constant, the speed of light. Frequency and wavelength are inversely proportional. | $\nu\lambda = c$ |
| The energy of a photon is equal to Planck's constant times the light frequency. Photon energy is proportional to frequency. | $E_{\text{photon}} = h\nu$ |
| The energy of a photon is equal to Planck's constant times the speed of light divided by wavelength. The energy of a photon is inversely proportional to wavelength. | $E_{\text{photon}} = \dfrac{hc}{\lambda}$ |
| The change in energy of an electron is equal to the difference between the final energy level and the initial energy level occupied by the electron. | $\Delta E = E_{\text{final}} - E_{\text{initial}}$ |
| The energy of a photon released due to an electron transition is equal to the absolute value of the change in energy of the electron, or the difference in energy between two energy levels. | $E_{\text{photon}} = |\Delta E|$ |

## KEY TERMS

atomic radius (7.7)

aufbau principle (7.3)

continuous spectrum (7.1)

core electron (7.5)

electromagnetic radiation (7.1)

electron configuration (7.3)

frequency (7.1)

ground state (7.3)

hertz (Hz) (7.1)

Hund's rule (7.3)

ionization energy (7.7)

isoelectronic (7.6)

line spectrum (7.1)

orbital (7.3)

Pauli exclusion principle (7.3)

photon (7.1)

principal energy level (7.3)

quantized (7.2)

sublevel (7.3)

valence electron (7.5)

valence level (7.5)

wavelength (7.1)

## QUESTIONS AND PROBLEMS

The following questions and problems, except those in *Additional Questions* and *Concept Review Questions,* are paired. Questions in a pair focus on the same concept. Answers to selected questions and problems are in Appendix D.

### Matching Definitions with Key Terms

7.1 Match the key terms with the descriptions provided.
(a) energy that travels through space at the speed of light as oscillating waves
(b) the number of wave cycles passing a stationary point in 1 second
(c) the amount of energy required to remove a valence electron from an atom or ion in the gaseous state
(d) a rule stating that electrons are distributed into orbitals of the same energy in such a way as to give the maximum number of unpaired electrons
(e) a description of the distribution of electrons in an atom's sublevels
(f) an inner electron that is not a valence electron
(g) a region in space where an electron is likely to be found
(h) a spectrum that consists of all the wavelengths of visible light
(i) having the same number of electrons

7.2 Match the key terms with the descriptions provided.
(a) the distance between two corresponding points on a wave
(b) the smallest packet of energy of a specific type of electromagnetic radiation
(c) a measure of the size of an atom, measured from the nucleus to the outer edge of the atom
(d) a rule stating that a maximum of two electrons can occupy an orbital and that the electrons must have opposite spin
(e) an electron that occupies the highest principal energy level
(f) the lowest-energy electron configuration
(g) an energy level consisting of orbitals of the same type and energy
(h) a spectrum that consists of only certain wavelengths of light; a "fingerprint" for an element
(i) having only specific allowed values

## Electromagnetic Radiation and Energy

7.3    List three types of electromagnetic radiation that have longer wavelengths than visible light.

7.4    List three types of electromagnetic radiation that have higher frequencies than visible light.

7.5    Draw a picture of two waves, one with twice the frequency of the other. Label the wave with the higher frequency.

7.6    Draw a picture of two waves, one with three times the wavelength of the other. Label the wave with the longer wavelength.

7.7    List the following colors of visible light from shortest wavelength to longest wavelength: orange, blue, red, yellow.

7.8    List the following colors of visible light from lowest frequency to highest frequency: orange, green, violet, yellow.

7.9    What does it mean when we say that wavelength is inversely proportional to frequency?

7.10   What does it mean when we say that photon energy is proportional to frequency?

7.11   What type of radiation has wavelengths just slightly longer than red light?

7.12   What type of radiation has wavelengths just slightly shorter than violet light?

7.13   Which type of electromagnetic radiation is composed of the highest-energy photons? Has the longest wavelength? Has the highest frequency?

7.14   Which color of light in the visible region is composed of the lowest-energy photons? Has the shortest wavelength? Has the lowest frequency?

7.15   Calculate the frequency of light that has a wavelength of 75.0 nm. What type of radiation is this?

7.16   Calculate the wavelength of light that has a frequency of $5.00 \times 10^{14} \ \text{s}^{-1}$. What type of radiation is this?

7.17   What is the photon energy of light with a wavelength of 465 nm? What type of radiation is this?

7.18   What is the frequency of light that is composed of photons that each has an energy of $1.99 \times 10^{-25}$ J? What type of radiation is this?

7.19   What type of light gives a continuous spectrum?

7.20   Should colored light from a heated ionic compound produce a continuous spectrum or a line spectrum?

7.21   Do any two elements produce the same line spectrum?

7.22   The line spectrum for neon is shown in Figure 7.8B. Are there any conditions when pure neon would not produce the same set of lines?

## The Bohr Model of the Hydrogen Atom

7.23   Bohr proposed that the electron in a hydrogen atom circles the nucleus in orbits. In his model, could electrons exist between orbits? Explain.

7.24   In the Bohr model, could the electron ever fall below the $n = 1$ orbit? Explain.

7.25   In the Bohr model, how does the electron move to a higher-energy orbit or to a lower-energy orbit?

7.26   In the Bohr model, is light emitted or absorbed when an electron moves from a higher-energy orbit to a lower-energy orbit?

7.27   In the Bohr model, how many photons are emitted when an electron moves directly from the $n = 5$ to the $n = 2$ orbit?

7.28   In the Bohr model, how many photons are emitted when an electron moves from the $n = 5$ to the $n = 3$ orbit, and then from the $n = 3$ to the $n = 2$ orbit?

7.29   In the Bohr model, which of the following electron transitions in a hydrogen atom results in the emission of the highest-energy photon?
       $n = 6$ to $n = 2$
       $n = 5$ to $n = 3$

7.30   In the Bohr model, which of the following electron transitions in a hydrogen atom results in the emission of the highest-energy photon?
       $n = 3$ to $n = 2$
       $n = 4$ to $n = 3$

7.31   In the Bohr model, which of the following electron transitions in a hydrogen atom results in emission of the longest wavelength of light?
       $n = 6$ to $n = 3$
       $n = 5$ to $n = 3$

7.32   In the Bohr model, which of the following electron transitions in a hydrogen atom results in emission of the longest wavelength of light?
       $n = 3$ to $n = 2$
       $n = 4$ to $n = 3$

7.33   Explain how the Bohr model accounts for the four colored lines in the hydrogen line spectrum.

7.34   Which electron transition in the Bohr hydrogen atom corresponds to each of the colored lines in the hydrogen line spectrum (red, green, blue, violet)?

7.35   The wavelength of the blue light in the hydrogen line spectrum is 434.1 nm. What is the photon energy of the blue light emitted by a hydrogen atom?

7.36   The wavelength of the violet light emitted from a hydrogen atom is 410.1 nm. This light is a result of electron transitions from the $n = 6$ to the $n = 2$ energy levels. How much higher in energy is the $n = 6$ energy level than the $n = 2$ energy level?

## The Modern Model of the Atom

7.37   How do the *orbitals* described in the modern model of the atom differ from the *orbits* described in the Bohr model?

7.38   Draw a picture of an $n = 1$ Bohr orbit and a $1s$ orbital. Why is it more difficult to draw a $1s$ orbital?

7.39   Which of the images represents a $p$ orbital?

A          B          C          D

**7.40**   Which of the images in Question 7.39 represents an *s* orbital?

**7.41**   Draw a picture of a 2*p* orbital and a 3*p* orbital. How do they differ?

**7.42**   Draw a picture of a typical 3*d* orbital and a typical 4*d* orbital. What is the difference between the 3*d* orbital and the 4*d* orbital?

**7.43**   How many orbitals are in each of the following sublevels?
(a) 4*s*          (c) 3*d*          (e) 2*s*
(b) 2*p*          (d) 4*f*          (f) 4*p*

**7.44**   Which sublevels are in each of the following principal energy levels?
(a) $n = 1$     (b) $n = 2$     (c) $n = 3$     (d) $n = 4$

**7.45**   Complete the following orbital diagrams for the ground-state electron arrangement for silicon, boron, and phosphorus. Be sure to follow the three orbital filling rules: the aufbau principle, the Pauli exclusion principle, and Hund's rule.

|     | 1*s* | 2*s* | 2*p* | 3*s* | 3*p* |
|-----|------|------|------|------|------|
| Si  | □    | □    | □□□  | □    | □□□  |
| Li  | □    | □    | □□□  | □    | □□□  |
| P   | □    | □    | □□□  | □    | □□□  |

**7.46**   Complete the following orbital diagrams for the ground-state electron arrangement for aluminum, chlorine, and sulfur. Be sure to follow the three orbital filling rules: the aufbau principle, the Pauli exclusion principle, and Hund's rule.

|     | 1*s* | 2*s* | 2*p* | 3*s* | 3*p* |
|-----|------|------|------|------|------|
| Al  | □    | □    | □□□  | □    | □□□  |
| Cl  | □    | □    | □□□  | □    | □□□  |
| S   | □    | □    | □□□  | □    | □□□  |

**7.47**   Write the electron configuration for each of the following elements.
(a) silicon
(b) lithium
(c) magnesium

**7.48**   Write the electron configuration for each of the following elements.
(a) calcium
(b) neon
(c) vanadium

**7.49**   What is the maximum number of electrons that can fill a 2*p* orbital?

**7.50**   What is the maximum number of electrons that can fill a 2*p* sublevel?

**7.51**   What is the maximum number of electrons that can fill the second principal energy level?

**7.52**   What is the maximum number of electrons that can fill the first principal energy level?

**7.53**   How many unpaired electrons are in a phosphorus atom?

**7.54**   How many unpaired electrons are in a chlorine atom?

## Periodicity of Electron Configurations

**7.55**   Which elements have five electrons in the highest-energy *p* sublevel in their ground state?

**7.56**   Which elements have only one electron in their highest-energy *s* orbital in their ground state?

**7.57**   Which elements have a partially filled *d* sublevel?

**7.58**   Which elements have a partially filled *f* sublevel?

**7.59**   Which element has two electrons in the 3*p* sublevel?

**7.60**   Which element has three electrons in its 3*d* sublevel?

**7.61**   Write the complete electron configurations for atoms of the following elements.
(a) Na          (b) Mn          (c) Se

**7.62**   Write the complete electron configurations for atoms of the following elements.
(a) Sc          (b) As          (c) Ba

**7.63**   What is wrong with the following electron configuration for bromine, Br?
$$1s^2 2s^2 2p^6 3s^2 3p^6 4s^2 4d^{10} 4p^6$$

**7.64**   What is wrong with the following electron configuration for rubidium, Rb?
$$1s^2 2s^2 2p^6 3s^2 3p^6 4s^2 3d^{10} 3p^6 4s^2$$

**7.65**   Identify the elements that have the following abbreviated electron configurations.
(a) [Ne] $3s^2 3p^5$
(b) [Ar] $4s^2 3d^7$
(c) [Xe] $6s^1$

**7.66**   Identify the elements that have the following abbreviated electron configurations.
(a) [Ar] $4s^1$
(b) [Xe] $6s^2 4f^{14} 5d^{10}$
(c) [Kr] $5s^2 4d^{10} 5p^5$

**7.67**   Write the abbreviated electron configuration for each of the following elements.
(a) Na          (b) Mn          (c) Se

**7.68**   Write the abbreviated electron configuration for each of the following elements.
(a) Sc          (b) As          (c) Ba

**7.69**   Which element has two filled *p* sublevels and five electrons in one of its *p* sublevels?

**7.70**   Which element has four filled *s* sublevels and no *d* electrons?

## Valence Electrons for the Main-Group Elements

**7.71**   How do you identify the valence electrons from an electron configuration?

**7.72**   How do you identify valence electrons from the periodic table?

**7.73**   Are the valence electrons always all of the electrons that follow the noble-gas electron configuration in an abbreviated electron configuration? Explain.

**7.74**   For the element gallium, Ga, are the 3*d* electrons classified as valence electrons or core electrons? Explain.

7.75    Explain why the number of valence electrons for a main-group element is equal to its Roman numeral group number.

7.76    Which elements have the same number of valence electrons as selenium?

7.77    Determine the valence level and the number of valence electrons for each of the following elements.
(a) Al        (b) S        (c) As

7.78    Determine the valence level and the number of valence electrons for each of the following elements.
(a) Sr        (b) Ga        (c) I

## Electron Configurations for Ions

7.79    How does the electron configuration for a cation differ from that of its neutral atom? How is it similar?

7.80    How does the electron configuration for an anion differ from that of its neutral atom? How is it similar?

7.81    Write the complete and abbreviated electron configuration for each of the following ions. For each ion, identify another ion that is isoelectronic with it.
(a) $Mg^{2+}$        (b) $O^{2-}$        (c) $Ga^{3+}$

7.82    Write the complete and abbreviated electron configuration for each of the following ions. For each ion, identify another ion that is isoelectronic with it.
(a) $Na^+$        (b) $P^{3-}$        (c) $H^-$

7.83    Identify at least three ions that are isoelectronic with neon.

7.84    Identify at least three ions that are isoelectronic with krypton.

## Periodic Properties of Atoms

7.85    Explain, in terms of ionization energies, why potassium is more reactive than sodium.

7.86    Explain, in terms of ionization energies, why strontium is more reactive than calcium.

7.87    Explain, in terms of ionization energies, why potassium is more reactive than calcium.

7.88    Explain, in terms of ionization energies, why the alkali and alkaline earth metals are not used to make coins or jewelry.

7.89    Rank the following elements in order of increasing ionization energy: P, O, S.

7.90    Rank the following elements in order of increasing ionization energy: Mg, P, O.

7.91    Use electron configurations to explain why more energy is required to remove an electron from a lithium atom than from a sodium atom.

7.92    Use electron configurations to explain why more energy is required to remove an electron from a bromine atom than from an iodine atom.

7.93    Use electron configurations to explain why the ionization energy for fluorine is greater than that for oxygen.

7.94    Use electron configurations to explain why the ionization energy for calcium is greater than that for potassium.

7.95    Write balanced equations that represent the processes that correspond to the first and second ionization energies for magnesium.

7.96    Which ionization energy ($IE_1$, $IE_2$, or $IE_3$) corresponds to the following process?
$$Al^{2+}(g) \longrightarrow Al^{3+}(g) + e^-$$

7.97    Which do you expect to have the higher third ionization energy, $IE_3$, magnesium or aluminum? Explain.

7.98    Which do you expect to have the higher second ionization energy, $IE_2$, potassium or calcium? Explain.

7.99    Explain, in terms of ionization energies, why the following ions do not form under normal conditions.
(a) $F^+$        (b) $Na^{2+}$        (c) $Ne^+$

7.100    Explain, in terms of ionization energies, why the following ions do not form under normal conditions.
(a) $Ba^{3+}$        (b) $O^{2+}$        (c) $Cl^+$

7.101    Rank the following elements in order of increasing atomic size: P, O, S.

7.102    Rank the following elements in order of increasing atomic size: Mg, P, O.

7.103    Why are chlorine atoms smaller than sulfur atoms?

7.104    Why are krypton atoms larger than argon atoms?

7.105    For each pair, identify the larger atom or ion.
(a) Mg or $Mg^{2+}$        (b) P or $P^{3-}$

7.106    For each pair, identify the larger atom or ion.
(a) F or $F^-$        (b) Sr or $Sr^{2+}$

7.107    Which is the larger ion, $K^+$ or $Ca^{2+}$? Explain.

7.108    Which is the larger ion, $S^{2-}$ or $Cl^-$? Explain.

## Additional Questions

7.109    In the modern model of the atom, we use orbital diagrams to describe the electron structure of the atom. How is the orbital diagram for the hydrogen atom different from the orbital diagram for all other atoms?

7.110    Consider the electron configurations of the noble gases. Which valence sublevels are completely filled for each noble gas?

7.111    Write the abbreviated electron configuration for each of the following elements.
(a) Bi        (b) Rn        (c) Ra

7.112    When writing electron configurations for elements in period 4, why are electrons placed in the $4s$ sublevel before the $3d$ sublevel?

7.113    How many $p$ orbitals are filled with electrons in a xenon atom?

7.114    How many $s$ orbitals are filled with electrons in a zirconium atom?

7.115    How many $d$ orbitals are filled with electrons in a cadmium atom?

7.116    Is it better to describe the noble gases as having eight valence electrons or zero valence electrons?

7.117    Why is the periodic trend in atomic size opposite to the trend in ionization energy?

7.118    Describe the changes in the sizes of the copper and silver atoms during the following single-displacement reaction.
$$Cu(s) + 2AgNO_3(aq) \longrightarrow Cu(NO_3)_2(aq) + 2Ag(s)$$

7.119 Describe the changes in the sizes of the sodium and chlorine atoms during the following combination reaction.

$$2Na(s) + Cl_2(g) \longrightarrow 2NaCl(s)$$

7.120 How would you explain light *absorption* by an atom in terms of electron transitions?

7.121 These photos show neon and krypton as clear gases in flasks. When electricity is passed through different tubes that contain these gases under low pressure, they glow different colors. What do the colors reveal about the wavelength, frequency, and energy of light that is absorbed, and then released, to cause the tubes to glow?

| Neon | Krypton |
|---|---|
| ©McGraw-Hill Education/ Stephen Frisch | ©McGraw-Hill Education/ Stephen Frisch |

7.122 The aufbau principle, the Pauli exclusion principle, and Hund's rule all describe some aspect of how electrons are distributed in electron sublevels. What do all these principles have in common?

7.123 Indicate which principle is violated in each of the following incorrect ground-state electron configuration diagrams, and correct the diagrams.

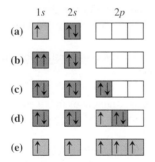

7.124 Which element has the following ground-state electron configuration: $1s^2 2s^2 2p^6 3s^2 3p^4$?

7.125 Classify an element having each of the following ground-state electron configurations as an alkali metal, alkaline earth metal, nonmetal, halogen, transition metal, or noble gas.
   (a) $[Ne]3s^1$
   (b) $[Ne]3s^2 3p^3$
   (c) $[Ar]4s^2 3d^{10} 4p^5$
   (d) $[Kr]5s^2 4d^1$
   (e) $[Kr]5s^2 4d^{10} 5p^6$

7.126 How many valence electrons are found in the halogens, the alkali metals, and the alkaline earth metals?

7.127 Which of the following will be the same for the ions $Rb^+$, $Sr^{2+}$, $Br^-$, and $Se^{2-}$: number of protons, number of

neutrons, number of electrons, atomic mass, electron configuration?

7.128 The electron configuration of a vanadium ion is $[Ar]3d^3$. What is the charge on this vanadium ion?

7.129 Compare strontium with rubidium in terms of the following properties: atomic radius, number of valence electrons, ionization energy.

7.130 Place the following elements, Mg, P, Cl, Al, and Na, in order of increasing size.

7.131 It takes energy to ionize any atom. It takes progressively more and more energy for each successive electron that is removed, and the increase is somewhat smooth. For the removal of which electron does boron exhibit a sudden marked increase in its ionization energy?

## Concept Review Questions

7.132 Which of the following statements is correct?
   A. Green light has a longer wavelength than orange light.
   B. Green light has a higher frequency than blue light.
   C. Blue light consists of higher-energy photons than red light.
   D. Photon energy and frequency are inversely proportional.
   E. Wavelength and frequency are directly proportional.
   Modify each false statement to make it a true statement.

7.133 Use the energy-level diagram shown to help you determine which of the following electron transitions in the hydrogen atom will result in emission of light with the longest wavelength.

Energy-level diagram

   A. $n = 1$ to $n = 2$
   B. $n = 1$ to $n = 3$
   C. $n = 2$ to $n = 1$
   D. $n = 3$ to $n = 1$
   E. All of these will result in emission of light with the same wavelength.
   Which of the transitions would result in emission of light with the greatest photon energy?

7.134 How many unpaired electrons are in a single atom of phosphorus when in its ground state?
   A. 0            D. 3
   B. 1            E. 5
   C. 2
   For each incorrect answer, identify an element that should have the given number of unpaired electrons in its ground-state electron configuration.

7.135 Which of the following statements best describes a *p* orbital?

A. a two-lobed two-dimensional pathway that an electron follows
B. a two-lobed three-dimensional region in space
C. a four-lobed two-dimensional region in space
D. a spherical three-dimensional region in space
E. a circular pathway that an electron follows

Which of the other choices is a description of an *s* orbital, and which is a description of a *d* orbital?

7.136 Which element has a ground-state electron configuration with two electrons in its 4*d* sublevel?

A. Ge
B. Sn
C. Mo
D. Ti
E. Zr

For each of the other elements, identify the sublevel that is partially filled with electrons.

7.137 Which of the following is the correct abbreviated ground-state electron configuration for germanium, Ge?

A. $[Ne]3s^2 3p^2$
B. $[Ne]4s^2 3d^6 4p^6$
C. $[Ar]4s^2 4d^{10} 4p^2$
D. $[Ar]4s^2 4d^8 4p^4$
E. $[Ar]4s^2 3d^{10} 4p^2$

For each incorrect answer, determine if it is a correct ground-state electron configuration for any other element, and if so, identify the element.

7.138 Which of the following is a true statement about valence electrons in atoms of main-group elements?

A. They are in the lowest-energy principal energy level.
B. They require the most amount of energy to remove.
C. Their ionization energies are greater than that for core electrons.
D. They are removed when anions form from neutral atoms.
E. The valence electrons are the highest-energy *s* and *p* electrons.

Change each false statement to a true statement.

7.139 Which of the following is a true statement when comparing the atomic sizes of fluorine and oxygen?

A. O atoms are smaller than F atoms because O atoms have a greater positive nuclear charge.
B. F atoms are smaller than O atoms because F atoms have a greater positive nuclear charge.
C. O atoms are smaller than F atoms because O atoms have fewer electrons.
D. F atoms are smaller than O atoms because F atoms have fewer electrons.
E. O atoms and F atoms are the same size because their valence electrons are in the same kind of orbitals.

How does your answer also help explain the relative first ionization energy values for these two elements?

7.140 When the elements sodium and chlorine react, they become ions and form the ionic compound NaCl:

$$2Na(s) + Cl_2(g) \longrightarrow 2NaCl(s)$$

Which decreases in size as the reactant atoms convert to ions in this reaction?

A. Na
B. Cl
C. both Na and Cl
D. neither Na nor Cl

Which increases in size during the reaction?

7.141 Which elements require the most energy to remove a valence electron?

A. nonmetals in the lower right area of the periodic table
B. nonmetals in the upper right area of the periodic table
C. metals in the lower left area of the periodic table
D. metals in the upper left area of the periodic table
E. transition metals in the center of the periodic table

How does this property affect the chemical reactivity of those elements?

7.142 In general, how do ionization energy trends relate to the activity series used to predict reactivity in single-displacement reactions?

A. More active metals have lower ionization energy values.
B. More active metals have higher ionization energy values.
C. Activity is greatest for those metals with intermediate ionization energies.
D. Activity is greatest for those metals with low second ionization energies ($IE_2$).
E. There is no relationship between ionization energy and the activity of a metal.

In general, where are the most active metals located on the periodic table?

# Chemical Bonding

8.1 Types of Bonds

8.2 Ionic Bonding

8.3 Covalent Bonding

8.4 Bonding in Carbon
Compounds

8.5 Shapes of Molecules

Chapter Review

Questions and Problems

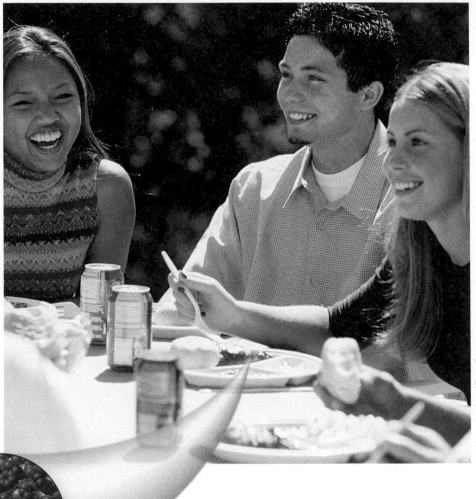

©SW Productions/Getty Images

Michael stops by the snack bar and picks up a hamburger, fries, and a soft drink. He takes his food to a park near campus, where he meets Ashley and Amanda for lunch. Ashley brought a salad from home, made of a variety of vegetables, including morel mushrooms that her family had picked in the woods the previous weekend. Amanda has a tuna sandwich on whole wheat bread.

As Michael slathers ketchup on his quarter-pound beef patty, his eating habits draw criticism from Ashley, a serious vegetarian. Michael defends his choices with a laugh, saying he needs the protein in meat if he is going to lift weights at the gym. Ashley argues that her mushrooms also contain protein and she has no problems lifting weights at the gym. Amanda interrupts, pointing out that it's not the protein itself that is necessary for good nutrition, but the amino acids that proteins are made from. The human body contains enzymes that can break down proteins by breaking bonds between atoms that hold protein molecules together. The body then reassembles the resulting amino acids into human proteins. Other enzymes process carbohydrates and fats, also needed in a balanced diet.

Humans must eat food that contains carbon, along with a number of other elements, in order to live, grow, develop, and stay healthy. But why do we have to consume carbon-based food? Carbon atoms form the backbone of most of the molecules that are in our bodies, as well as in the plants and animals that we eat. Carbon is also found in some nonliving things—including rocks, ocean water, and the atmosphere—as well as in coal, oil, and natural gas, which are the remains of plants and animals that lived millions of years ago. Carbon atoms are used over and over again. They move between and among organisms and the environment in a continuous cycle, called the carbon cycle (Figure 8.1).

To see how the carbon cycle works, let's trace the possible history of a carbon atom in a mushroom in Ashley's salad. This carbon atom has been around for a long time, but not always as a part of a molecule in the mushroom. Eons ago it was part of a carbon dioxide molecule in the air. It was taken up by a leaf of a tree that, through the process of photosynthesis, incorporated the carbon atom along with hydrogen from water into a glucose molecule. After the tree died and decomposed, it sank into the swamp and formed part of a layer of peat, partially carbonized vegetable matter. Over time the area dried and a river deposited layers of sediment on top of it, burying the peat and subjecting it to heat and pressure over millions of years. The carbon atom then became a part of a layer of coal.

Glycine is an example of an amino acid that the body uses as structural components of proteins. Various chemical groups can replace the highlighted hydrogen atom to make a different amino acid.

$$CO_2H$$
$$H_2N-\overset{|}{\underset{|}{C}}-H$$
$$H$$

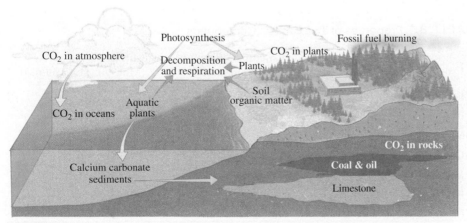

**FIGURE 8.1** The movement of carbon around our planet is summarized by the carbon cycle. Some of the carbon transfer processes are rapid, while others take millions of years.

The coal was mined and burned to produce heat causing the carbon atom to combine with oxygen. It returned to the atmosphere as yet another carbon dioxide molecule before being taken up by another tree, which subsequently died and began decomposing. In the rotting log, a mushroom spore began to grow. It absorbed the carbon atom and incorporated it into an amino acid. Inside the cells of the mushroom, many such amino acids were joined to form a protein molecule. The protein ended up in Ashley's salad.

Michael says that there is more to the story. Carbon is all around us. Large amounts of carbon dioxide dissolve into the oceans, where the gas is converted to calcium carbonate, $CaCO_3$, in coral and in the shells of other marine organisms. Calcium carbonate is also found in rocks such as limestone and marble. Interesting formations of limestone build up in caves, where the mineral dissolves under acidic conditions and then redeposits as stalactites and stalagmites when the water evaporates. Michael realizes that calcium carbonate is the active ingredient in an antacid tablet he takes to relieve his upset stomach after lunch.

In this chapter we will examine the forces that hold atoms together in compounds and molecular elements. These forces are *chemical bonds*. Understanding chemical bonding helps us explain how compounds form and behave. In this chapter we will see how chemical bonds and the shapes of molecules explain the physical and chemical properties of many compounds, including those that contain carbon and those that don't. Before we explore these topics, consider the following questions.

### Questions for Consideration

**8.1** How can we classify the types of bonding of different compounds?
**8.2** What is the nature of the bonding in ionic compounds?
**8.3** What is the nature of the bonding in molecular compounds?
**8.4** What types of bonds are found in carbon compounds?
**8.5** How do we predict and explain the shapes of molecular compounds and polyatomic ions?

## 8.1 Types of Bonds

In our review of the carbon cycle, we mentioned two carbon compounds, $CaCO_3$ in limestone and $CO_2$ in the air. Recall from Chapter 3 that calcium carbonate, $CaCO_3$, is classified as an ionic compound because it is composed of a metal cation and a polyatomic anion. Carbon dioxide, $CO_2$, is classified as a molecular compound because it is composed of nonmetals. Consider the significantly different physical and chemical properties of these two compounds, as summarized in Table 8.1.

**TABLE 8.1** Properties of Two Carbon Compounds

| Property | Calcium Carbonate (Ionic compound) | Carbon Dioxide (Molecular compound) |
| --- | --- | --- |
| Formula | $CaCO_3$ | $CO_2$ |
| Physical state | White solid | Colorless gas |
| Molar mass (g/mol) | 100.1 | 44.01 |
| Density (g/mL) | 2.71 | 0.00198 |
| Melting point (°C) | 1339 (at high pressure) | −56.6 (at 5.11 atm) |
| Boiling point (°C) | Decomposes | Sublimes at −78.6 |
| Electrical conductivity as solid | Very low | Very low |
| Electrical conductivity as liquid | High | Very low |
| Dissolves in | Acids | $H_2O$, $CCl_4$ |

**TABLE 8.2** General Properties of Ionic and Covalent Substances

| Ionic | Covalent |
|---|---|
| Crystalline solids | Gases, liquids, or solids |
| Hard and brittle solids | Brittle and weak solids, or soft and waxy solids |
| Very high melting point | Low melting point |
| Very high boiling point | Low boiling point |
| Good electrical conductor when molten or in solution | Poor conductor of electricity and heat |
| Often soluble in water but not in carbon tetrachloride | Often soluble in carbon tetrachloride but not in water |

Why do these compounds have such different properties? To a considerable extent, the differences can be attributed to chemical bonds. A **chemical bond** is a force that holds atoms together in a molecule or compound. Calcium carbonate is an ionic compound made of $Ca^{2+}$ ions and $CO_3^{2-}$ ions. Ionic compounds are held together by *ionic bonds*. Carbon dioxide is a molecular compound made of $CO_2$ molecules. In each molecule, *covalent bonds* connect the carbon and oxygen atoms. Molecular substances are also called *covalent substances*.

The nature of these bonds results in the differences in properties observed for the two classes of compounds. General properties of ionic and covalent compounds are summarized in Table 8.2. For example, consider the difference in solubility as described in the last entry of the table. Ionic compounds are often soluble in water. However, they do not dissolve in carbon tetrachloride, $CCl_4$, a solvent with properties different from water. (One of these properties is *polarity,* which you will learn more about later in this chapter.) In contrast, covalent substances tend to be soluble in carbon tetrachloride but not water. Structure and bonding within substances affects their solubility in different solvents.

**ANIMATION:** Ionic vs Covalent Bonding

## Ionic and Covalent Bonding

Electron configurations help explain how and why atoms form different types of bonds in compounds that exhibit such different properties. In bonding, atoms tend to lose, gain, or share *valence* electrons to achieve a more stable electron configuration similar to that of a noble gas. In the simplest type of **ionic bonding,** one or more electrons are transferred from a metal to a nonmetal, forming a positively charged metal cation and a negatively charged nonmetal anion, each with a more stable electron configuration. For example, the ionic compound lithium fluoride, LiF, is composed of $Li^+$ and $F^-$, each with a noble gas electron configuration. The ions are held together by *electrostatic forces,* the attractions between the opposite charges of ions. The attractions are offset in part by repulsions between ions of like charge. Because each ion is surrounded by and attracted to several ions of opposite charge (Figure 8.2A), a lot of energy is required to separate the ions from one another. The strong ionic bonds explain the high melting points of ionic solids and the hard, crystalline form that many ionic compounds take.

In **covalent bonding,** electrons are shared between atoms in pairs, forming molecules in which each atom has a noble-gas electron configuration. In $CO_2$, each carbon atom is strongly covalently bonded to two oxygen atoms. However, the molecules are not bonded to other molecules (Figure 8.2B). This explains why molecular compounds generally have lower melting and boiling points than ionic compounds and usually occur naturally as gases, liquids, or soft solids. Covalent bonding generally occurs between nonmetals in molecular elements, molecular compounds, and polyatomic ions.

A $CaCO_3$

B $CO_2$

**FIGURE 8.2** (A) Calcium carbonate, $CaCO_3$, consists of arrays of calcium cations and carbonate anions held together by ionic bonds, attractions between their opposite charges. (Note that the atom labels indicate the elements but do not specify charge.) In each carbonate ion, $CO_3^{2-}$, the carbon atom is covalently bonded to three oxygen atoms. (B) The atoms in each $CO_2$ molecule are held together by covalent bonds. Molecules of carbon dioxide are not bonded to one another.

Notice in Figure 8.2A that calcium carbonate contains calcium ions and carbonate ions. The carbonate ions are polyatomic ions each consisting of one carbon atom bonded to three oxygen atoms. These are bonds between nonmetals, so they are covalent bonds where electrons are shared. Thus, calcium carbonate contains both ionic and covalent bonds.

---

### EXAMPLE 8.1    Types of Bonding

Identify the type of bonding in each of the following substances.

(a) $CaCl_2$
(b) $NO_2$
(c) $NaNO_3$

**Solution:**

(a) $CaCl_2$ contains a metal and a nonmetal, so the bonding is ionic.
(b) $NO_2$ contains two nonmetals, so the bonding is covalent.
(c) $NaNO_3$ has a metal and two nonmetals. Because the nonmetals form a polyatomic anion, $NO_3^-$, the bonding between nitrogen and oxygen is covalent. The bonding between the $Na^+$ cation and the $NO_3^-$ polyatomic anion is ionic.

**Consider This 8.1**

If a nonmetal were bonded to a metalloid, what type of bonding would be expected?

**Practice Problem 8.1**

Identify the type of bonding in each of the following substances.

(a) $NaF$
(b) $ClO_2$
(c) $FeSO_4$

**Further Practice:** Questions 8.9 and 8.10 at the end of the chapter

---

Consider the following groups of chemical substances organized by type of bonding:

$$CaCl_2, K_2O, CsF \qquad CO, NO, H_2O \qquad H_2, N_2, O_3$$

How would you classify the bonding in these groups? The first group consists of compounds made of a metal and a nonmetal. The second group consists of compounds made of nonmetals. The third group consists of diatomic and triatomic elements, each containing identical elements. Based on what you've read so far, you might classify the bonding in the first group as ionic and in the second and third groups as covalent. You would be right, but there is more to the story. Ionic and covalent bonding represent two extremes of a continuous range of bonding possibilities. Ionic bonding assumes a complete transfer of electrons from one atom to another. It generally applies to compounds formed between metallic and nonmetallic elements, such as CsF. Covalent bonding best applies to compounds of two nonmetals. A pair of electrons may be shared equally by the two bonded atoms, as in $Cl_2$ and $H_2$. However, a covalent bond between two different nonmetals usually does *not* result in an *equal* sharing of electrons. This is the case in compounds like CO or $H_2O$. Instead they bond through a partial transfer of electrons from one atom to the other, with *unequal sharing* of electrons. The resulting bond has some characteristics of an ionic bond and some characteristics of a covalent bond.

A whole range of bonding types has this mixed ionic and covalent character. For convenience the bonds in a compound are usually classified as the dominant type. For example, the bonding in NaCl is considered ionic, and the bonding in $CO_2$ or HF is called covalent, although each has some of the other's character.

## Polar and Nonpolar Covalent Bonds

Let's examine the unequal sharing of electrons in a covalent bond such as that in hydrogen fluoride. On the average, the electron pair is closer to the fluorine atom than to the hydrogen atom. Although the bonding in HF is classified as covalent, it does have some ionic character because the electron from the hydrogen atom has been partially transferred to the fluorine atom. The amount of ionic character in a principally covalent bond can be judged from the polarity of the bond. The **polarity** of a bond is the degree of transfer of bonding electrons from one atom to another. Bonding patterns fall on a continuum from complete transfer and no sharing of electrons at one extreme to no transfer and equal sharing of electrons at the other extreme. The continuum is shown in Figure 8.3.

As Figure 8.3 shows, covalent bonds can be subdivided into two general types. In a **nonpolar covalent bond,** electrons are equally shared and charge is evenly distributed over the two bonded atoms. In a **polar covalent bond,** the electrons tend to be found closer, on the average, to one of the atoms than the other. This unequal sharing leads to the development of partial charges on the two atoms, as in the compound HF (Figure 8.4). The bonding electrons are closer to the fluorine atom than the hydrogen atom, so fluorine has a partial negative charge ($\delta-$). The hydrogen atom has the bonding electrons farther away, so it develops a partial positive charge ($\delta+$).

## Electronegativity

How do we know whether a covalent bond is polar or nonpolar? In the early 1930s, Linus Pauling developed the idea that charges could be separated in a bond. He examined the difference in properties between the covalent bonds formed between identical elements (such as in $H_2$ and $F_2$) and those formed between different elements (such as in HF). The bonds between different elements are stronger. Pauling proposed that a partial transfer of electrons, resulting in partial ionic character, would explain the extra stability. The partial transfer occurs because one atom attracts the shared electrons more strongly than the other does. Pauling called the ability of an atom to attract bonding electrons **electronegativity.** If two atoms have the same electronegativity, the electrons are shared equally, and the bond is nonpolar covalent. The greater the difference in electronegativity, the more ionic character the bond has, the greater the partial ionic charges on the atoms, and the more polar the bond.

By examining the stability of various bonds, Pauling devised a scale of electronegativity values, with fluorine at the highest value (4.0) and francium at the lowest (0.7). Electronegativity values for most of the elements are given in Figure 8.5. The higher the electronegativity value of an element, the greater its attraction for electrons.

 **ANIMATION:** Electronegativity

Linus Pauling was a chemistry professor at the California Institute of Technology. He received two Nobel Prizes. The first, the Chemistry Prize in 1954, was for his work on the chemical bond. His second, the Peace Prize in 1963, acknowledged his efforts to stop nuclear weapons testing.

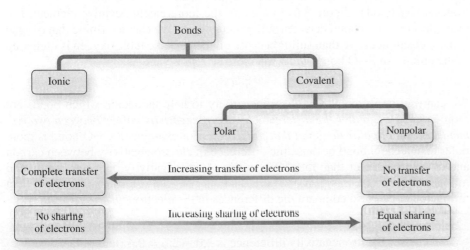

**FIGURE 8.3** Bonds can be classified by the extent of transfer or sharing of electrons.

$\delta+$  $\delta-$

**FIGURE 8.4**  In HF, the shared electrons are closer, on the average, to the fluorine atom than to the hydrogen atom. Consequently, fluorine develops a partial negative charge ($\delta-$) and hydrogen develops a partial positive charge ($\delta+$). The Greek letter delta ($\delta$) is used to indicate partial charge and is followed by the sign of the charge.

**FIGURE 8.5** Electronegativity values
of the elements.

Notice that electronegativity
values for He, Ne, Ar, and Rn are
not shown. These elements are not
known to form any bonds under
normal conditions.

Electronegativity values show periodic trends. The nonmetals tend to gain electrons, so their electronegativity values are higher. The metals tend to lose electrons, so their electronegativity values are lower. Fluorine, with the highest electronegativity value, is the most reactive of the nonmetals. Francium, with the lowest electronegativity value, is the most reactive of the metals. Electronegativity values tend to increase from bottom to top of a group and from left to right across a period, as illustrated in Figure 8.6. Note that hydrogen does not fit the general trend. Its electronegativity value is between that of boron and carbon. This information is useful when predicting relative electronegativity values for elements in compounds containing hydrogen.

In a covalent bond the more electronegative element has a partial negative charge ($\delta-$) and the less electronegative element has the partial positive charge ($\delta+$). If you have a table of electronegativity values like Figure 8.5, you can reference the values to determine where to place the partial charges. However, you should try to use electronegativity trends (Figure 8.6) to identify the more electronegative element. For example, consider a bond between sulfur and oxygen. We can determine that oxygen is more electronegative than sulfur because on the periodic table oxygen is higher up in the group. An S—O bond would therefore be represented as

$$^{\delta+}S-O^{\delta-}$$

We can use the difference in electronegativity to help us decide which bonds are more polar. *The greater the difference in electronegativity values between two elements, the more polar the bond that joins them.* For example, a C—O bond is more polar than a C—N bond because the difference in electronegativity between carbon and oxygen is greater than the difference in electronegativity between carbon and nitrogen. We can see this from the periodic table: oxygen is farther to the right than nitrogen. We can also compare the differences in the electronegativity values:

C—O   Electronegativity difference is 3.4 − 2.6 = 0.8 (more polar)
C—N   Electronegativity difference is 3.0 − 2.6 = 0.4 (less polar)

**FIGURE 8.6** Trends in the values of
electronegativity of the elements.

| EXAMPLE 8.2 | Polarity of Bonds |
|---|---|

Which of the following molecules contain polar bonds? If a bond is polar, which atom has the partial negative charge?

(a) $CO_2$     (b) $O_3$     (c) $NCl_3$     (d) $CH_4$

Which bond is the most polar? (In the formulas of compounds, the first element is the central atom.)

**Solution:**

To determine whether the bonds in a molecule are polar, we first find the electronegativity values of the bonded atoms shown in Figure 8.5. If the values are identical, the bond is nonpolar. If the values are different, the bond is polar, and the atom with the greater electronegativity has the partial negative charge. We can also use the electronegativity trends shown in Figure 8.6.

(a)  The electronegativity of carbon is 2.6. The value for oxygen is 3.4. The bond is polar, with the partial negative charge on the oxygen atom.
(b)  The bonds are between three oxygen atoms, so there is no electronegativity difference and the bonds are nonpolar.
(c)  Nitrogen and chlorine atoms have electronegativities of 3.0 and 3.2, respectively. The bond is polar, with the partial negative charge on the chlorine atoms.
(d)  The bonds are between carbon and hydrogen. The electronegativity of carbon is 2.6, and the value for hydrogen is 2.2. The bond is polar, with the partial negative charge on the carbon atom.

The most polar bond will be the one with the greatest difference in electronegativities. The difference for C−O is $3.4 - 2.6 = 0.8$. The difference for C−H is $2.6 - 2.2 = 0.4$. Thus, the C−O bond is more polar than the C−H bond. The difference for N−Cl is $3.2 - 3.0 = 0.2$, so this bond is less polar than the C−O or C−H bond. These results are summarized as follows in order of increasing polarity.

| Bond | Difference in Electronegativity Values | Bond Type |
|---|---|---|
| O−O | $3.5 - 3.5 = 0$ | Nonpolar covalent |
| N−Cl | $3.2 - 3.0 = 0.2$ | Polar covalent |
| C−H | $2.6 - 2.2 = 0.4$ | Polar covalent |
| C−O | $3.4 - 2.6 = 0.8$ (most polar) | Polar covalent |

**Consider This 8.2**

If a table of electronegativities is not available, how could you decide on the polarity of the bonds in a molecule?

**Practice Problem 8.2**

Which of the following molecules contain polar bonds? If a bond is polar, which atom has the partial negative charge?

(a) $SO_2$        (b) $N_2$        (c) $PH_3$        (d) $CCl_4$

Which bond is the most polar? (In the formulas of compounds, the first element is the central atom.)

**Further Practice:** Questions 8.23 and 8.24 at the end of the chapter

# 8.2 Ionic Bonding

Nonpolar covalent bonds lie at one end of the continuum shown in Figure 8.3. Ionic bonds lie at the other. They are familiar to you already in compounds such as sodium chloride, the table salt that Michael likes on his fries.

The formation of ions and ionic bonds relates to the electron configurations of the elements. We can see the relationship in the periodic table. Each element immediately *following* a noble gas is a metal, which has a strong tendency to *lose* electrons and thereby achieve a stable noble-gas electron configuration. When giving up electrons to gain a noble-gas configuration, the metal develops a *positive* electrical charge, forming a *cation*. An example is the metal sodium, which follows the noble gas neon on the periodic table. When a sodium atom forms $Na^+$, it loses an electron to achieve neon's electron configuration. Each element immediately *preceding* a noble gas in the periodic table is a nonmetal, which has a strong tendency to *gain* electrons to achieve a noble-gas configuration. In so doing, it develops a *negative* electrical charge, forming an *anion*. An example is the nonmetal chlorine, which precedes the noble gas argon on the periodic table. When a chlorine atom forms $Cl^-$, it gains an electron to achieve argon's electron configuration.

## Lewis Symbols

Gilbert Newton Lewis was a chemistry professor at the University of California, Berkeley. He proposed the concept of covalent bonding in 1916.

The electrons involved in bonding are the valence electrons (Section 7.5). A convenient way to show the valence electrons is with a **Lewis symbol** or an *electron dot symbol,* in which dots placed around an element's symbol represent valence electrons. The dots are placed *singly* on the four sides of the elemental symbol, in any order, and then paired as necessary. An element with a noble-gas configuration is surrounded by four pairs of dots representing eight electrons, an *octet.*

For the main-group elements, the valence electrons are located in the highest-energy *ns* and *np* orbitals, where *n* designates the highest principal energy level for that element.

Table 8.3 shows the Lewis symbols for the period 2 elements. They follow a progression starting with one valence electron for Li and ending with eight for Ne. Try drawing the Lewis symbols for the period 3 elements, Na through Ar. What pattern develops? It should look the same, starting with one valence electron for the group IA (1) elements and ending with eight in group VIIIA (18). All the noble-gas elements have eight valence electrons occurring in four pairs, except for helium, which only has two.

| **EXAMPLE 8.3** | **Lewis Symbols of Atoms** |
| --- | --- |

The electron configuration of sulfur is $1s^2 2s^2 2p^6 3s^2 3p^4$. Write the Lewis symbol for the sulfur atom.

**Solution:**

The valence electrons are those designated $3s^2 3p^4$. Sulfur has six valence electrons, so it has two electron pairs and two unpaired electrons: $:\!\dot{\underset{\cdot\cdot}{S}}\!\cdot$

**Consider This 8.3**

How would the Lewis symbol change if the atom were selenium instead of sulfur?

**Practice Problem 8.3**

Write the Lewis symbol for the silicon atom.

**Further Practice:**  Questions 8.29 and 8.30 at the end of the chapter

**TABLE 8.3** Lewis Symbols of the Period 2 Elements

| | Group | | | | | | | |
| --- | --- | --- | --- | --- | --- | --- | --- | --- |
| | **IA (1)** | **IIA (2)** | **IIIA (13)** | **IVA (14)** | **VA (15)** | **VIA (16)** | **VIIA (17)** | **VIIIA (18)** |
| Electron configuration | $1s^2 2s^1$ | $1s^2 2s^2$ | $1s^2 2s^2 2p^1$ | $1s^2 2s^2 2p^2$ | $1s^2 2s^2 2p^3$ | $1s^2 2s^2 2p^4$ | $1s^2 2s^2 2p^5$ | $1s^2 2s^2 2p^6$ |
| Valence electrons | 1 | 2 | 3 | 4 | 5 | 6 | 7 | 8 |
| Lewis symbol | Li· | ·Be· | $\dot{B}\cdot$ | $\cdot\dot{C}\cdot$ | $:\!\dot{N}\!\cdot$ | $:\!\dot{\underset{\cdot\cdot}{O}}\!\cdot$ | $:\!\dot{\underset{\cdot\cdot}{F}}\!\cdot$ | $:\!\overset{\cdot\cdot}{\underset{\cdot\cdot}{Ne}}\!:$ |

The Lewis symbols for hydrogen and helium in period 1 do not fit the pattern. Since their valence levels can hold only two electrons, only one side of the elemental symbol is used. Hydrogen is represented as H·, which corresponds to the electron configuration $1s^1$. Helium's Lewis symbol is He: with two valence electrons, representing the filled level $1s^2$. The hydride ion, H:⁻, formed by adding one electron to the hydrogen atom, is represented as H:⁻. The hydrogen ion, H⁺, is formed by removing one electron from the hydrogen atom. Since it has no remaining electrons, its Lewis symbol H⁺ has no dots.

The loss of electrons from a metal and the gain of electrons by a nonmetal create the simplest kind of ionic bond. The transfer must happen at the same time. A metal can lose electrons to form a cation only when a nonmetal is available to gain them. The nonmetal then forms an anion with a stable octet of valence electrons. The following equations use Lewis symbols to illustrate this process in the formation of sodium chloride:

$$\text{Na·} \longrightarrow \text{Na}^+ + e^-$$

$$:\ddot{\text{Cl}}· + e^- \longrightarrow :\ddot{\text{Cl}}:^-$$

$$\text{Na}^+ + :\ddot{\text{Cl}}:^- \longrightarrow (\text{Na}^+)(:\ddot{\text{Cl}}:^-) \quad \text{or} \quad \text{NaCl}$$

The sodium atom loses one electron to form a sodium ion, which has a noble-gas configuration. This electron is transferred to a chlorine atom, which becomes a chloride ion; this ion also has a noble-gas configuration. The ions associate to form sodium chloride, a compound held together by attractions between the opposite charges of the ions.

Because sodium needs to lose one electron and chlorine needs to gain one, the atoms combine in a 1:1 ratio, which leads to electrical neutrality. Other ions may combine in different ratios. Consider, for example, the formation of aluminum chloride:

$$·\dot{\text{Al}}· \longrightarrow \text{Al}^{3+} + 3e^-$$

$$3:\ddot{\text{Cl}}· + 3e^- \longrightarrow 3:\ddot{\text{Cl}}:^-$$

$$\text{Al}^{3+} + 3:\ddot{\text{Cl}}:^- \longrightarrow (\text{Al}^{3+})(:\ddot{\text{Cl}}:^-)_3 \quad \text{or} \quad \text{AlCl}_3$$

Aluminum loses three electrons to achieve a noble-gas electron configuration. Since it takes only one electron to complete the chlorine atom's octet, three chlorine atoms are required to use the three electrons given up by the aluminum atom. We now have an aluminum ion and three chloride ions, which associate in an electrically neutral 1:3 ratio to form ionic aluminum chloride, $\text{AlCl}_3$. This ratio is maintained by an array of aluminum cations and chloride anions arranged in a regular pattern.

Transition metals are not usually represented with Lewis symbols since they may have as many as 10 $d$ electrons, which would make the symbols cumbersome.

If necessary, review the charges of common ions in Section 3.2.

When an ionic compound forms from its elements, quintillions ($10^{18}$) of cations and quintillions of anions form and bond to give a single ionic crystal.

## EXAMPLE 8.4 | Lewis Symbols of Ions

Write the Lewis symbols for the magnesium and sulfide ions. Use the Lewis symbols and equations to show the process of magnesium sulfide formation.

### Solution:

Magnesium can lose two valence electrons to achieve a noble-gas configuration, giving the $\text{Mg}^{2+}$ cation with no remaining valence electrons:

$$·\text{Mg}· \longrightarrow \text{Mg}^{2+} + 2e^-$$

Sulfur has six valence electrons—two electron pairs and two unpaired electrons. It gains two electrons to form the $\text{S}^{2-}$ anion:

$$:\dot{\text{S}}· + 2e^- \longrightarrow :\ddot{\text{S}}:^{2-}$$

These ions associate in a 1:1 ratio to form MgS:

$$\text{Mg}^{2+} + :\ddot{\text{S}}:^{2-} \longrightarrow (\text{Mg}^{2+})(:\ddot{\text{S}}:^{2-}) \quad \text{or} \quad \text{MgS}$$

**FIGURE 8.7** In an ionic crystal, ions of like charge repel one another, while ions of opposite charge attract. The regular pattern of their arrangement ensures that attractions are stronger than repulsions.

**FIGURE 8.8** The sodium chloride crystal is represented at the molecular level in two ways. In each, look for the arrangement of one ion surrounded by six ions of opposite charge. (A) A space-filling model shows the relative sizes of the sodium and chloride ions accurately, but masks the details of the crystal structure. (B) A ball-and-stick model reveals the inside of the crystal, but does not show relative sizes and distances correctly. (C) A photograph of some magnified NaCl crystals shows how the shape of the crystals corresponds with the arrangement of ions within the crystals.
©Jim Birk

---

### Consider This 8.4

If the sulfide ion combined with a 1+ cation, such as K$^+$, instead of a 2+ cation, how would the Lewis symbol for the compound differ?

### Practice Problem 8.4

Write the Lewis symbols for the beryllium and nitride ions. Use the Lewis symbols and equations to show the process of beryllium nitride formation.

**Further Practice:** Questions 8.31 and 8.32 at the end of the chapter

---

## Structures of Ionic Crystals

The bonds that hold ions together in ionic compounds result from attractions between oppositely charged ions. If like-charged ions are close to one another, their repulsions partially offset these attractive forces, as shown in Figure 8.7. How do the ions fit together to make a stable ionic compound? An ion, which may be represented as a charged sphere, exerts a force equally in all directions, so ions of opposite charge surround it. The pattern is a **crystal lattice,** and the result is an **ionic crystal,** in which the ions are arranged in a regular geometric pattern that maximizes the attractive forces and minimizes the repulsive forces.

The charges and sizes of the ions largely determine the characteristic patterns of ionic crystals. For example, many ionic salts with a 1:1 ratio of ions have the same structure as sodium chloride, with each ion surrounded by six ions of opposite charge (Figure 8.8).

For a crystal lattice to form, cations and anions must come into contact. If the cations are smaller or larger relative to the anions, a different structure develops. For example, if a salt contains a large cation, such as a cesium ion, each is surrounded by eight similarly sized ions of opposite charge. The crystal lattice for cesium chloride is shown in Figure 8.9. For other compositions, such as salts with 1:2 or 2:1 ratios of ions, different patterns form. A typical structure shown in Figure 8.10 is that of CaF$_2$, found in the mineral fluorite.

**FIGURE 8.9** The cesium chloride structure is adopted when cations and anions are about the same size. (The image shown for CsCl displays only a portion of the structure for this ionic compound.)

**FIGURE 8.10** Ionic compounds that have twice as many of one ion as the other often adopt the same structure as CaF$_2$, found in the mineral fluorite, used to make camera lenses for taking ultraviolet images.
©McGraw-Hill Education/Doug Sherman

**FIGURE 8.11** When a crystal such as NaCl is struck with sufficient force, layers of ions shift. The attractive forces between ions of opposite charge move out of alignment. Layers of like charge align, and they are pushed apart by repulsive forces. The crystal splits along these lines.
©Charles D. Winters/Science Source

Ionic bonds are very strong because of the large number of interactions between oppositely charged ions. It takes a lot of energy to separate an ion from the solid structure. Because melting and boiling require the separation of ions, ionic compounds have high melting and boiling points.

Strong attractive forces also make ionic crystals hard and brittle. If struck forcefully, ionic crystals shatter. They break because the force shifts the alignment of ions from the stable interaction of opposite charges to an unstable alignment of like charges (Figure 8.11). Along a shifted layer, strong repulsive forces replace strong attractive forces, and the crystal cracks.

The crystalline structure of ionic compounds also explains why solid ionic salts are poor conductors of electricity. The ions are not mobile but are held in place in the crystal lattice. However, melting greatly increases the electrical conductivity of ionic salts, because the ions in the liquid are free to move around.

## 8.3 Covalent Bonding

When two nonmetals form a compound, as in the carbon dioxide that we exhale, the bonds within the molecule are covalent. Since the atoms do not give up electrons, they cannot achieve a noble-gas configuration, but they can come close by sharing electrons. This compromise allows them to approximate a noble-gas configuration, with the shared electrons being a part of the electron configuration of both atoms.

Covalent bonds involve localized attractive forces between atoms that are held together tightly in molecules. A covalent bond is strong because each shared electron interacts simultaneously with two nuclei. Electrons and nuclei attract one another because of their opposite charges. At the same time, electrons repel one another and nuclei repel each other because of their like charges. When the electrons are located primarily between the two nuclei, the attractions are maximized and the repulsions are minimized, resulting in a covalent bond, as shown in Figure 8.12.

Although the attractions (bonds) between atoms within the molecule are strong, the molecules are not bonded to other molecules. Consequently, covalent compounds have low melting and boiling points, and they often exist as gases or liquids. As solids, they are brittle or soft, because their molecules are not held together strongly. For example, in solid carbon dioxide (dry ice), the individual molecules remain intact and are identifiable in the molecular model (Figure 8.13).

### The Octet Rule

The number of bonds an element can form depends on its tendency to achieve a noble-gas configuration with an octet of eight valence electrons. Review the Lewis

**FIGURE 8.12** A covalent bond results when attractive forces between opposite charges are greater than repulsive forces between like charges.

**FIGURE 8.13** In solid carbon dioxide, $CO_2$, molecules pack together in a regular array, but they retain their identity and can be removed from the crystal easily.
©Jim Birk

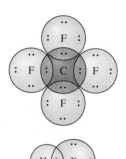

**FIGURE 8.14** If shared electrons are counted for both atoms in a bond, each atom in $CF_4$ has a noble-gas configuration. In HF, the fluorine atom has an octet, but the hydrogen atom has only two electrons.

**FIGURE 8.15** We think of the formation of hydrogen molecules as two H atoms coming together to share a pair of electrons. The hydrogen atoms do not combine unless they get close enough together. Hydrogen molecules form when orbitals overlap, creating a region in space where both of their electrons can be found.

symbols of the elements carbon through fluorine in Table 8.3. How many bonds do you think each atom will make? Do your predictions result in an electron configuration containing an octet of valence electrons?

The tendency of an atom to achieve an electron configuration having eight valence electrons is known as the **octet rule.** Thus, fluorine, with seven valence electrons ($:\ddot{F}\cdot$), normally forms one covalent bond, since it needs one electron to reach a noble-gas configuration. Carbon, with four valence electrons ($\cdot\dot{C}\cdot$), must gain four electrons to reach a noble-gas configuration, so it usually forms four bonds. When these two elements combine in a 1:4 ratio, carbon forms four bonds and each fluorine forms one bond, giving the covalent compound $CF_4$. As shown in Figure 8.14, if we count shared electrons for both atoms, we find eight electrons around the carbon atom and around each fluorine atom, the same number as in the noble gas neon. Fluorine also forms a covalent compound by sharing electrons with hydrogen. Because they each need one electron to reach a noble-gas configuration, they combine in a 1:1 ratio to form HF (Figure 8.14). While carbon and fluorine form bonds to reach the same electron configuration as neon, hydrogen forms one bond to reach the same configuration as helium. Hydrogen, with only a $1s$ orbital in its valence level, has two electrons when it forms a covalent bond.

The Lewis symbols for the nonmetals in period 2 show the number of bonds each typically forms:

$$\cdot\dot{C}\cdot \quad :\dot{N}\cdot \quad :\ddot{O}\cdot \quad :\ddot{F}\cdot \quad :\ddot{Ne}:$$
$$4 \qquad 3 \qquad 2 \qquad 1 \qquad 0$$

To reach an octet, carbon needs four electrons and generally forms four bonds. Nitrogen needs three electrons, so it typically forms three bonds. Oxygen forms two bonds, and fluorine forms one bond. Since neon has a valence level filled with electrons, it forms no bonds.

## Lewis Structures for the Diatomic Elements

In some cases the only way to achieve an octet of electrons around each atom in a diatomic molecule is to share more than two electrons. Depending on the number of electrons the atoms share, covalent bonds are classified as single, double, or triple.

**Single Covalent Bonds**    A covalent bond that consists of a pair of electrons shared by two atoms is a **single bond.** Each atom has a half-filled orbital in the valence level, and the orbitals overlap to allow the electron pair to be a part of both atoms. For example, in the formation of the diatomic hydrogen molecule, each hydrogen atom has an electron in the $1s$ orbital. When the atoms come together, as shown in Figure 8.15, the orbitals overlap and the atoms share an electron pair:

$$H\cdot + \cdot H \longrightarrow H:H$$

This representation of a hydrogen molecule is a **Lewis structure** (or *electron dot formula,* or *Lewis formula*), in which the atoms are shown separately and the

**FIGURE 8.16** All the halogen elements exist as diatomic molecules. Each atom has an octet of electrons, as shown by the circles, because they share one electron pair between the bonded atoms. The shared electron pair is a single covalent bond, shown here in the overlapping circles. The electron pairs that are not in the overlapping circles are the unshared (nonbonding) electrons.

valence electrons are represented by dots. Lewis structures are sometimes simplified by showing each bond not as a pair of dots, but as a line. For $H_2$, the Lewis structure could be either H:H or H—H.

The halogens, each with seven valence electrons, also share one pair of electrons in diatomic molecules. Lewis structures that show the distribution of electrons in the halogen molecules are given in Figure 8.16. Each atom in these structures has an octet because the shared electrons are counted for each atom. The covalent bond in each of these molecules is a single bond, since it arises from the sharing of a single electron pair. There are also three pairs of unshared (nonbonding) electrons on each atom of each molecule.

**Double and Triple Bonds**   Consider the $O_2$ molecule. Can a single covalent bond be formed between two oxygen atoms and satisfy the octet rule? With only a single bond $O_2$ would have one of these Lewis structures:

$$:\ddot{O}:\ddot{O}: \qquad :\ddot{O}:\ddot{O}: \qquad :\ddot{O}:\ddot{O}:$$

Do any of these satisfy the octet rule? No. Can you think of a way to resolve the problem?

Some combinations of atoms do not have enough electrons to satisfy the octet rule with a single bond. To achieve an octet, they must share more than one pair of electrons. In $O_2$, for example, oxygen atoms, each with six valence electrons, must share two pairs of electrons to achieve an octet, while in $N_2$ the nitrogen atoms, each with only five valence electrons, must share three pairs (Figure 8.17). This bonding pattern reflects the number of bonds normally formed by these elements in order to reach an octet. Oxygen usually forms two bonds, and nitrogen normally forms three.

The sharing of two pairs of electrons is a **double bond.** The sharing of three pairs of electrons is a **triple bond.** In Lewis structures, two pairs of dots or two parallel lines represent a double bond. Three pairs of dots or three parallel lines represent a triple bond.

The length and strength of single, double, and triple bonds differ. Normally, breaking double bonds requires more energy than breaking single bonds. In turn, breaking triple bonds requires more energy than breaking double bonds. The energy required to break bonds provides a measure of *bond strength*. In general, triple bonds are stronger than double bonds, and double bonds are stronger than single bonds. Furthermore, the *bond length* (the distance between the centers of the bonded atoms) is greater for single than for double bonds and greater for double than triple bonds.

## Valence Electrons and Number of Bonds

Atoms combine in different ways to achieve octets. Oxygen, for example, typically forms two bonds, but it can do so in different ways. When oxygen atoms combine in molecular oxygen, $O_2$, they form a double bond:

$$:\dot{O}\cdot + \cdot\dot{O}: \longrightarrow :\ddot{O}::\ddot{O}: \quad \text{or} \quad :\ddot{O}=\ddot{O}:$$

But in combining with hydrogen atoms to form water, $H_2O$, oxygen may form two single bonds—one with each of two hydrogen atoms:

$$H\cdot + \cdot\ddot{O}\cdot + \cdot H \longrightarrow H:\ddot{O}:H \quad \text{or} \quad H-\ddot{O}-H$$

**FIGURE 8.17** The diatomic oxygen molecule has a double bond, formed by the sharing of two electron pairs between the oxygen atoms. In the diatomic nitrogen molecule, each atom shares three electron pairs with the other atom, forming a triple bond. The shared bonding electron pairs are shown in the areas where the circles overlap.

Lewis structures show the numbers of shared and unshared valence electrons. They are not meant to represent the three-dimensional shapes of molecules.

**FIGURE 8.18** Phosphorus forms a molecule containing four atoms. Molecular sulfur contains eight atoms. Single bonds connect the atoms in each molecule.

In a different compound, two oxygen atoms join with a single bond, with two hydrogen atoms supplying the additional electrons, to form hydrogen peroxide, $H_2O_2$:

$$H\cdot \; + \; \cdot\ddot{O}\cdot \; + \; \cdot\ddot{O}\cdot \; + \; \cdot H \; \longrightarrow \; H\!:\!\ddot{O}\!:\!\ddot{O}\!:\!H \quad \text{or} \quad H-\ddot{O}-\ddot{O}-H$$

Similarly a nitrogen atom, with five valence electrons, shares three electron pairs with another nitrogen atom to form molecular nitrogen, $N_2$:

$$:\!\dot{N}\!\cdot \; + \; \cdot\dot{N}\!: \; \longrightarrow \; :\!N\!:\!:\!:\!N\!: \quad \text{or} \quad :\!N\!\equiv\!N\!:$$

Nitrogen forms single bonds with three hydrogen atoms in ammonia, $NH_3$:

$$H\cdot \; + \; H\cdot \; + \; H\cdot \; + \; :\!\dot{N}\!\cdot \; \longrightarrow \; \begin{matrix} & H \\ :\!\ddot{N}\!:\!H \end{matrix} \quad \text{or} \quad \begin{matrix} & H \\ & | \\ :\!N\!-\!H \\ & | \\ & H \end{matrix}$$

Like oxygen, nitrogen can bond to itself with a single bond, if another element is also present, as in hydrazine, $N_2H_4$:

$$H\cdot + H\cdot + \cdot\dot{N}\cdot + \cdot\dot{N}\cdot + H\cdot + H\cdot \longrightarrow \begin{matrix} H\;H \\ H\!:\!\ddot{N}\!:\!\ddot{N}\!:\!H \end{matrix} \quad \text{or} \quad \begin{matrix} H\;\;H \\ |\;\;\; | \\ H\!-\!\ddot{N}\!-\!\ddot{N}\!-\!H \end{matrix}$$

Instead of forming double or triple bonds, some elements form molecules containing more than two atoms, each connected to others with single bonds. Phosphorus, which needs to form three bonds to complete its octet, can bond to three other phosphorus atoms, resulting in a $P_4$ molecule with each phosphorus atom at the corner of a tetrahedron. In contrast, sulfur needs to form two bonds to complete its octet. It can do this in several ways, but the most common is to form a ring of eight sulfur atoms, giving $S_8$. The common forms of the elements phosphorus and sulfur are shown in Figure 8.18.

---

**EXAMPLE 8.5**   **Lewis Structure of a Compound**

Determine the chemical formula of a simple compound that follows the octet rule and is formed from sulfur and chlorine atoms. Use a Lewis structure to describe the bonding in this compound.

**Solution:**

Sulfur has six valence electrons. It can fill its octet by sharing two electrons. Thus, a sulfur atom tends to form two single bonds or one double bond. Chlorine has seven valence electrons and needs to share one by forming a single bond. Thus, we would expect that two chlorine atoms could combine with one sulfur atom to form $SCl_2$, with single bonds between the atoms:

$$:\!\ddot{\underset{..}{C}l}\!:\!\ddot{\underset{..}{S}}\!:\!\ddot{\underset{..}{C}l}\!: \quad \text{or} \quad :\!\ddot{\underset{..}{C}l}\!-\!\ddot{\underset{..}{S}}\!-\!\ddot{\underset{..}{C}l}\!:$$

Notice that each atom has a share of an octet of electrons and that the total number of electrons equals the sum of the valence electrons of the component atoms.

**Consider This 8.5**

If chlorine formed a compound with phosphorus instead of sulfur, how would the Lewis structure differ?

**Practice Problem 8.5**

Determine the formula of a simple compound that follows the octet rule and is formed from nitrogen and fluorine atoms. Use a Lewis structure to describe the bonding in this compound.

**Further Practice:** Questions 8.51 and 8.52 at the end of the chapter

## Structures of Covalent Molecules

**Drawing Lewis Structures** Let's look more closely at how to develop Lewis structures for common molecules and ions. Consider the following small molecules containing carbon, each of which can be found in nature as part of the carbon cycle:

$$
\begin{array}{c}
\text{H} \\
| \\
\text{H}-\overset{|}{\underset{|}{\text{C}}}-\text{H} \\
\text{H}
\end{array}
\quad
\begin{array}{c}
:\ddot{\text{C}}\text{l}: \\
| \\
\text{H}-\overset{|}{\underset{|}{\text{C}}}-\text{H} \\
:\ddot{\text{C}}\text{l}:
\end{array}
\quad
\begin{array}{c}
\text{H} \\
| \\
\text{H}-\text{C}=\ddot{\text{O}}:
\end{array}
\quad
:\ddot{\text{O}}=\text{C}=\ddot{\text{O}}:
\quad
:\text{C}\equiv\text{O}:
\quad
\text{H}-\text{C}\equiv\text{N}:
$$

Does each atom in these molecules follow the octet rule? You already know that hydrogen does not. It gets a share of only two electrons. What about the other atoms? Does each molecule have the same number of valence electrons as that from the atoms that compose it?

When drawing Lewis structures, it is helpful to follow a set of steps that will efficiently result in a structure with an octet on each atom, and with the correct number of total valence electrons. Let's use chloromethane, $CH_3Cl$, to show how we follow these steps to arrive at a correct Lewis structure.

**Step 1. Draw an atomic skeleton.** From the formula, $CH_3Cl$, we can conclude that hydrogen cannot be the central atom. It is present in the greatest number, and it can only form one bond. Chlorine has seven valence electrons and generally forms only one bond. Carbon, with four valence electrons, can form four bonds, so it should be the central atom.

$$
\begin{array}{c}
\text{H} \\
\\
\text{H} \quad \text{C} \quad \text{Cl} \\
\\
\text{H}
\end{array}
$$

**Step 2. Sum the valence electrons.** In a molecule of chloromethane, the carbon atom has 4 valence electrons, the chlorine atom has 7, and each of the three hydrogen atoms has 1, giving *a total of 14 valence electrons.*

**Step 3. Place two electrons, a single bond, between each pair of bonded atoms.**

$$
\begin{array}{c}
\text{H} \\
| \\
\text{H}-\overset{|}{\underset{|}{\text{C}}}-\text{Cl} \\
\text{H}
\end{array}
$$

**Step 4. Add remaining electrons to complete the octet of each outer atom and then to the central atom if necessary and if there are electrons available.** The shared bonding electrons in $CH_3Cl$ contribute a total of eight valence electrons. The hydrogen atoms are satisfied with their two bonding electrons (same number of electrons as helium). Completing the octet around chlorine with 6 unshared electrons gives a total of 14 valence electrons (the number we should have, as determined in step 2). The central carbon atom has an octet already, with four pairs of shared bonding electrons. *The structure is complete because all of the atoms have octets (except hydrogen, which needs only two) and the structure has the correct total number of valence electrons.*

$$
\begin{array}{c}
\text{H} \\
| \\
\text{H}-\overset{|}{\underset{|}{\text{C}}}-\ddot{\text{C}}\text{l}: \\
\text{H}
\end{array}
$$

**Step 5. If needed to satisfy the octet rule, shift unshared electron pairs to bonding positions to form double or triple bonds.** This step is needed only if the central atom does not have an octet, but all of the allowed electrons have been added to the

Electrons are *paired* in Lewis structures as bonding (shared) electron pairs and as nonbonding (unshared) electron pairs.

Can you predict the central atom in a molecule of $POCl_3$? Phosphorus is the central atom because it usually forms the greatest number of bonds and is the least electronegative.

Elements in the third and higher periods, such as sulfur, phosphorus, and the heavier halogens and noble gases, can form compounds with 10 or even 12 electrons surrounding the central atom. An example is $SF_4$, with 10 electrons around the sulfur:

$$
\begin{array}{c}
:\ddot{\text{F}} \qquad \ddot{\text{F}}: \\
\diagdown \quad \diagup \\
\ddot{\text{S}} \\
\diagup \quad \diagdown \\
:\ddot{\text{F}} \qquad \ddot{\text{F}}:
\end{array}
$$

structure. In this example, this step wasn't needed because the central atom has an octet of electrons. As we will see in Example 8.6, double bonds are required to complete the octet of each atom in the $CO_2$ molecule.

### Steps for Determining Lewis Structures

1. *Draw an atomic skeleton.* Place the symbols of all the elements in the compound in the correct locations with respect to one another. To draw correct structures, you must know which atoms are connected by bonds.
   - The arrangement of atoms is usually symmetrical.
   - In a molecule of two different elements, the one with the greater number of atoms usually surrounds the one with the lesser number of atoms. If it is still not obvious, the central atom is usually the atom able to form the most bonds.
   - The central atom, the one surrounded by the other atoms, tends to be the one that is less electronegative and is present in the least quantity. This atom usually forms the greater number of bonds and is found farther toward the bottom left side of the periodic table.
   - Hydrogen atoms are generally on the outside of the molecule.
   - The chemical formula may give clues about the arrangement of atoms.
2. *Sum the valence electrons from each atom to get the total number of valence electrons.* If the covalent species is a polyatomic ion, be sure to take into account any ionic charge, subtracting electrons for positive charge and adding electrons for negative charge.
3. *Place two electrons, a single bond, between each pair of bonded atoms.*
4. *If you have not placed all the valence electrons in the formula, add any remaining electrons, as unshared electron pairs, to satisfy the octet rule.*
   - Add pairs of electrons first to complete the octet of atoms surrounding the central atom. Then add any remaining electrons in pairs to the central atom.
   - When placing electrons around the atoms, recall that hydrogen has only 2 electrons when bonded, and period 2 elements have 8. Elements from periods 3 to 7 usually have 8 electrons, but 10 or 12 are also possible.
5. *If necessary to satisfy the octet rule, shift unshared electron pairs from nonbonded positions on atoms with completed octets to positions between atoms to make double or triple bonds.*

### EXAMPLE 8.6    Lewis Structures of Molecules

Draw a Lewis structure for carbon dioxide, $CO_2$, the primary carbon compound found in the atmosphere.

**Solution:**

***Step 1.*** The atomic skeleton of carbon dioxide has oxygen atoms surrounding the carbon atom:

$$O \quad C \quad O$$

***Step 2.*** The sum of valence electrons in carbon dioxide is 16, 4 from carbon and 6 from each oxygen.

***Step 3.*** Place a bond between the carbon atom and each oxygen atom, but not between the oxygen atoms, using four electrons:

$$O-C-O$$

***Step 4.*** There are 12 remaining valence electrons. Add pairs of electrons to satisfy the octets of the oxygen atoms:

$$:\ddot{O}-C-\ddot{O}:$$

***Step 5.*** This uses all 16 valence electrons. Is this a satisfactory Lewis structure? No, because the carbon atom does not have an octet of electrons. We have to shift a pair of electrons from one of the oxygen atoms to the bond, making a double bond:

$$:\ddot{O}=C-\ddot{O}:$$

How many electrons surround the carbon atom now? There are 6. Since this still does not satisfy the octet rule for carbon, we have to shift an electron pair from the other oxygen atom to its bond, making another double bond:

$$:\ddot{O}=C=\ddot{O}:$$

Count the electrons now to assure yourself that this structure satisfies the octet rule for each atom.

### Consider This 8.6

How would the Lewis structure for $CS_2$ differ from the Lewis structure for $CO_2$?

### Practice Problem 8.6

Formaldehyde, $CH_2O$, is present in smoke from burning wood and coal, such as from the grill used to cook Michael's hamburger. It is toxic and often used as a tissue preservative. A formaldehyde molecule has the two hydrogen atoms and the oxygen atom attached to the carbon atom. Draw a Lewis structure for the formaldehyde molecule.

**Further Practice:** Questions 8.53 and 8.54 at the end of the chapter

How do we know when a Lewis structure has one or more double bonds? After step 4 of the procedure, if the central atom lacks two electrons, it must form a double bond with another atom. If it lacks four electrons, two double bonds may form, as in $CO_2$. Another possibility is the formation of a triple bond. For example, the Lewis structure for hydrogen cyanide, $H:C:::N:$ or $H-C\equiv N:$, has a triple bond between carbon and nitrogen.

*When drawing Lewis structures for polyatomic ions*, we follow the same steps but we must take into account the charge when summing the total number of valence electrons in step 2. For anions, we add an electron for each negative charge. For cations, we subtract electrons for each positive charge. For example, the cyanide ion, $CN^-$, has a total of 10 valence electrons. The carbon atom contributes 4, the nitrogen atom contributes 5, and the $-1$ charge indicates 1 extra electron. Following the steps for drawing Lewis structures, we get

$$[:C\equiv N:]^-$$

For polyatomic ions, brackets are usually used to show that the charge is spread out over the entire ion, and not localized on one atom.

*When creating Lewis structures for oxoacids,* we must be careful when drawing the skeletal structure in step 1. In oxoacids, acidic hydrogen atoms are bonded to the oxygen atoms, not to the central atom. For example, the skeletal structure obtained for $HNO_2$ after step 3 is

$$H-O-N-O$$

The structure after completing the octets on the oxygen atoms with the remaining valence electrons is

$$H-\ddot{O}-N-\ddot{O}:$$

There are 16 electrons shown on the structure; however, there are a total of 18 electrons contributed from all the atoms. Therefore, we add 2 electrons to the nitrogen atom:

$$H-\overset{..}{\underset{..}{O}}-\overset{..}{N}-\overset{..}{\underset{..}{O}}:$$

Because the nitrogen atom still does not have an octet, we move 2 unshared electrons from an oxygen atom to make a double bond (step 5). If we make the double bond to the oxygen that is not bonded to hydrogen, then both oxygen atoms will have two bonds:

$$H-\overset{..}{\underset{..}{O}}-\overset{..}{N}=\overset{..}{\underset{..}{O}}:$$

The structure is now complete with octets on each atom and the total number of valence electrons equaling 18. If we instead drew the double bond to the other oxygen atom, it would have three bonds to it, while the other oxygen would have only one:

$$H-\overset{..}{\underset{..}{O}}=\overset{..}{N}-\overset{..}{\underset{..}{O}}:$$   (Incorrect structure)

This structure is not as good because the oxygen atoms do not each have two bonds to them, as is preferred and is possible in the $HNO_2$ molecule.

*Polyatomic ions and molecules with the same number of valence electrons have similar Lewis structures.* For example, $NO_2^-$ and $SO_2$ both have 18 valence electrons. Notice the similarity in their Lewis structures:

$$\left[:\overset{..}{O}=\overset{..}{N}-\overset{..}{\underset{..}{O}}:\right]^- \quad \text{and} \quad :\overset{..}{O}=\overset{..}{S}-\overset{..}{\underset{..}{O}}:$$

Two other examples are the carbonate ion, $CO_3^{2-}$, and sulfur trioxide, $SO_3$, which have 24 electrons and similar Lewis structures:

$$\left[:\overset{..}{O}=\overset{\overset{\textstyle :\overset{..}{O}:}{|}}{\underset{..}{C}}-\overset{..}{\underset{..}{O}}:\right]^{2-} \quad \text{and} \quad :\overset{..}{O}=\overset{\overset{\textstyle :\overset{..}{O}:}{|}}{S}-\overset{..}{\underset{..}{O}}:$$

**Resonance**   Sometimes Lewis structures do not give accurate pictures of the bonding as it is known to exist in nature. Consider the ozone molecule, $O_3$, for example. The Lewis structure of this molecule can be drawn in two ways, each having one double bond and one single bond:

$$:\overset{..}{O}=\overset{..}{O}-\overset{..}{\underset{..}{O}}: \quad \text{or} \quad :\overset{..}{\underset{..}{O}}-\overset{..}{O}=\overset{..}{\underset{..}{O}}:$$

Do these formulas satisfy the octet rule for each atom? Yes, but neither of them accurately explains experimental evidence about the oxygen-oxygen bonds in ozone. Measurements on other molecules show that the normal length of an oxygen-oxygen single bond is 148 pm. The usual length of an oxygen-oxygen double bond is 121 pm, significantly shorter than a single bond. But in $O_3$, both bonds are the same length, 126 pm, a measurement that falls between those of single and double bonds. How can we explain this kind of bond?

The concept of *resonance* allows us to represent this kind of bonding while retaining the utility and simplicity of Lewis structures. According to this concept, the electron arrangement in molecules like ozone is represented not by a single Lewis structure but by two or more, each illustrating a different aspect of the true arrangement of electrons. The actual molecule is a composite of the structures drawn and is called a **resonance hybrid.**

The term *resonance* may suggest a resonating, or oscillating, movement between the different structures, but this is not the case. The true structure is not one of the Lewis structures some of the time and the other the rest of the time. The real structure always has some of the characteristics of each contributing structure. It is an "average" of the contributions.

Using Lewis structures to represent the bonding in molecules like ozone is similar to using a photograph to represent the structure of a seashell. The photograph of a seashell is not the seashell but only a representation of it, in the same way that a Lewis structure is only a representation of a molecule. Figure 8.19 shows two aspects of a seashell. The shell is best represented by a composite of the two photos.

To represent a resonance hybrid, we draw the contributing Lewis structures in the usual way and connect them with double-headed arrows. The two resonance structures for ozone, for instance, are drawn as follows:

$$:\ddot{O}=\ddot{O}-\ddot{O}: \longleftrightarrow :\ddot{O}-\ddot{O}=\ddot{O}:$$

The composite structure of these two resonance forms is sometimes depicted as a single structure with dashed or dotted lines to show the sharing of the double-bond character over two sets of bonded atoms:

$$O \dot= O \dot= O$$

Such a structure emphasizes the delocalized nature of two of the electrons in the double bond, with resonance forms differing only in the locations of these electrons. As shown, both contributing resonance structures of $O_3$ have the same atomic arrangement; only the location of the double bond differs.

Many oxoanions and oxoacids also exhibit resonance. For example, nitric acid, $HNO_3$, has two resonance strsuctures:

$$H-\ddot{O}-N=\ddot{O} \quad \longleftrightarrow \quad H-\ddot{O}-N-\ddot{O}:$$

Although we might expect nitric acid to have three resonance forms—with a double bond between the nitrogen and each of the oxygen atoms—it has only two. As mentioned earlier in this chapter, oxoacids do not have hydrogen atoms bonded to oxygen atoms that have double bonds to other elements. This example also emphasizes that resonance forms differ only in the location of the double bond. It is not legitimate to change the relative positions of the atoms. If we moved one (such as the hydrogen), we might represent an entirely different molecule, or we might show the molecule rotated in space, not a different resonance form.

The most molecules that exhibit resonance have a central atom bonded to two (or more) other identical atoms. One bond is double and the other (or others) is single. The double bond can be exchanged with any single bond to give a valid resonance form. In the following example, we'll examine the resonance forms of the carbonate ion, the primary geological form of carbon, to illustrate these points.

**FIGURE 8.19** Two photographs of a seashell, neither of which accurately represents the true structure of the shell. The two views together are a better representation.
©Jim Birk

When chemists say electrons are "localized," they mean that the electrons tend to be located in a small region of space primarily between two nuclei. In contrast, "delocalized" electrons are associated with three or more nuclei.

**EXAMPLE 8.7** | **Resonance Structures**

Carbon occurs in nature in the form of carbonate ion in limestone and seashells. Draw the resonance hybrids of the Lewis structure for the carbonate ion, $CO_3^{2-}$.

**Solution:**
Following the steps in the procedure outlined earlier, we obtain the first Lewis structure:

$$\left[ :\ddot{O}=\overset{\ddot{O}:}{\underset{}{C}}-\ddot{O}: \right]^{2-}$$

It is possible to show the double bond with either of the other two oxygen atoms:

$$\left[ :\ddot{O}-\overset{\ddot{O}:}{\underset{}{C}}=\ddot{O}: \right]^{2-} \quad or \quad \left[ :\ddot{O}-\overset{\ddot{O}}{\underset{}{C}}-\ddot{O}: \right]^{2-}$$

The actual structure is a resonance hybrid of these three structures:

$$\left[\begin{array}{c} :\!\ddot{O}\!: \\ | \\ :\!\ddot{O}\!=\!C\!-\!\ddot{O}\!: \end{array}\right]^{2-} \longleftrightarrow \left[\begin{array}{c} :\!\ddot{O}\!: \\ | \\ :\!\ddot{O}\!-\!C\!=\!\ddot{O}\!: \end{array}\right]^{2-} \longleftrightarrow \left[\begin{array}{c} :\!\ddot{O} \\ \| \\ :\!\ddot{O}\!-\!C\!-\!\ddot{O}\!: \end{array}\right]^{2-}$$

### Consider This 8.7

If a molecule had two hydrogen atoms attached to a central atom using single bonds and one oxygen attached to the central atom using a double bond, would the molecule exhibit resonance?

### Practice Problem 8.7

Before 1842 the most common anesthetics were whiskey and a blow on the head. After that, nitrous oxide, also known as laughing gas, came into use. Along with carbon dioxide, nitrous oxide has been implicated in global warming. It is $N_2O$, with its atoms arranged as NNO. Draw the resonance structures of the resonance hybrid for nitrous oxide.

**Further Practice:** Questions 8.63 and 8.64 at the end of the chapter

## Exceptions to the Octet Rule

Although most molecules and polyatomic ions satisfy the octet rule, many do not. Exceptions generally fall into three categories: odd-electron molecules, incomplete octets, and expanded valence levels.

In a molecule that contains an odd number of valence electrons, one electron must remain unpaired, so one of the atoms cannot have an octet. Generally the atom with an incomplete octet is the one with the lower electronegativity. An example is nitrogen monoxide, NO, which is an electron-deficient molecule with the following Lewis structure:

$$\cdot\ddot{N}\!=\!\ddot{O}\!: \qquad \text{(odd-electron molecule)}$$

An unpaired electron makes this a reactive molecule, because it will react with any other substance that can readily supply an electron. It may react, for example, with chlorine to form the more stable nitrosyl chloride:

$$:\!\ddot{C}\!l\!-\!\ddot{N}\!=\!\ddot{O}\!: \qquad \text{(odd-electron molecule)}$$

Color is characteristic of substances composed of odd-electron molecules. Nitrogen monoxide is deep blue in the liquid and solid states, although most other nitrogen oxides are colorless. Another nitrogen oxide with an odd number of electrons is the reddish-brown compound nitrogen dioxide.

Some atoms, notably boron, participate in covalent bonding but do not have enough valence electrons to form an octet. Boron has only three valence electrons, which form three covalent bonds, as in the very reactive $BH_3$ and $BF_3$ molecules:

$$\begin{array}{cc} \begin{array}{c} H \\ | \\ H\!-\!B\!-\!H \end{array} & \begin{array}{c} :\!\ddot{F}\!: \\ | \\ :\!\ddot{F}\!-\!B\!-\!\ddot{F}\!: \end{array} \end{array} \qquad \text{(molecules with incomplete octets)}$$

Some compounds have more than eight electrons around the central atom, with what is called an expanded valence level. Examples are $SF_6$ and $XeF_4$:

$$\begin{array}{cc} \begin{array}{c} :\!\ddot{F}\!: \quad :\!\ddot{F}\!: \\ \diagdown \; \diagup \\ :\!\ddot{F}\!-\!S\!-\!\ddot{F}\!: \\ \diagup \; \diagdown \\ :\!\ddot{F}\!: \quad :\!\ddot{F}\!: \end{array} & \begin{array}{c} :\!\ddot{F}\!: \quad :\!\ddot{F}\!: \\ \diagdown \; \diagup \\ :\!\ddot{Xe}\!: \\ \diagup \; \diagdown \\ :\!\ddot{F}\!: \quad :\!\ddot{F}\!: \end{array} \end{array} \qquad \text{(molecules with expanded valence levels)}$$

Although these are interesting exceptions, we will focus on compounds in which the central atom obeys the octet rule.

In Chapter 7 we saw that color in elements arises from the transition of an electron from one orbital to another having a different energy. The same can happen in molecules when there is a half-filled orbital with only one electron. The transfer of electrons from another orbital of different energy to the half-filled orbital results in the absorption or emission of light energy. This causes the substance to be colored. Color also appears in some substances that contain double bonds.

## 8.4 Bonding in Carbon Compounds

Carbon is part of more different chemical compounds than any other element except hydrogen. Carbon forms the backbone of nearly every molecule that is involved in life processes in plants and animals on Earth. Why does carbon have this role, rather than some other element such as boron, nitrogen, or oxygen? In part the answer lies in carbon's four valence electrons and its ability to form four covalent bonds. Boron, with three valence electrons, and nitrogen, with five, can form only three covalent bonds. Oxygen forms only two. Carbon is versatile in its bonding. It can form single, double, or triple bonds to other carbon atoms or to atoms of other elements:

$$ -\overset{\displaystyle |}{\underset{\displaystyle |}{C}}-  \qquad \overset{\diagdown}{\underset{\diagup}{C}}=  \qquad =C=  \qquad -C\equiv $$

In addition, carbon-carbon bonds are very strong, so molecules formed by carbon are very stable. Long chains of carbon atoms are found in nature and in synthetic polymers, which contain very large molecules. Despite their size, all can be derived from the simplest of the carbon compounds, the hydrocarbons.

### Hydrocarbons

A compound of hydrogen and carbon is a **hydrocarbon.** Hydrocarbons can be categorized into several classes and subclasses (Figure 8.20). Methane, $CH_4$, the simplest hydrocarbon, has one carbon atom bonded to four hydrogen atoms (Figure 8.21). The next larger hydrocarbon is ethane, $C_2H_6$, in which each carbon atom is attached to the other carbon atom and three hydrogen atoms. Hydrocarbons can also be found in branched arrangements, where some carbon atoms are attached to three or four other carbon atoms. Any one of these hydrocarbons, which contain only carbon-carbon single bonds, is called an **alkane** (Figure 8.21).

A hydrocarbon that contains a carbon-carbon double bond is an **alkene.** The simplest is ethene, or ethylene, $C_2H_4$, which has a double bond between the carbon atoms:

$$ \underset{H}{\overset{H}{\diagdown}} C = C \overset{H}{\underset{H}{\diagup}}  \qquad \text{or} \qquad H_2C = CH_2 $$

The hydrocarbons and their derivatives are so numerous that they form the subject of an entire branch of chemistry, organic chemistry, which we will explore in more detail in Chapter 16.

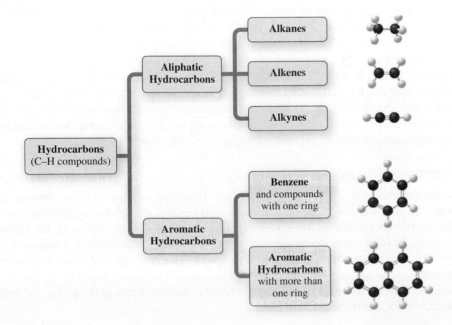

**FIGURE 8.20** Classes of hydrocarbons.

**FIGURE 8.21** Alkanes contain only single bonds. Additional structures could be drawn for $C_8H_{18}$ and $C_5H_{12}$.

Benzene                                            Naphthalene

**FIGURE 8.22** Benzene and naphthalene are aromatic hydrocarbons. Some of their electrons are delocalized in six-membered rings of carbon atoms. Each resonance form has alternating single and double bonds.

An **alkyne** is a hydrocarbon that contains a carbon-carbon triple bond. The simplest example is ethyne, also known as acetylene, $H-C\equiv C-H$, used in the oxyacetylene welding torch.

Large alkanes such as octane, $C_8H_{18}$, are quite common. However, formulas such as $C_8H_{18}$ don't give much information about how the atoms are connected. In all hydrocarbons, the hydrogen atoms are attached to carbon atoms, while the carbon atoms can be attached to one another in a variety of ways. Molecules with the same composition but different arrangements of atoms are called *isomers*. Octane, $C_8H_{18}$, has 15 isomers. The number of isomers increases rapidly as the number of carbons increases. For example, 75 isomers of $C_{10}H_{22}$ and 366,319 isomers of $C_{20}H_{42}$ are possible.

Any of these types is an **aliphatic hydrocarbon,** a class in which the bonds are all localized single, double, or triple bonds. There are no resonance forms. Another class of hydrocarbons has carbon atoms arranged in a six-atom ring with alternating single and double bonds—or, more accurately, delocalized bonds. Members of this class are represented by resonance structures. Such a compound is an **aromatic hydrocarbon.** Examples are benzene, $C_6H_6$, and naphthalene, $C_{10}H_8$, shown in Figure 8.22.

## Functional Groups

Many other carbon compounds—including those that are important to the maintenance of life, such as proteins, enzymes, and genetic material—are derivatives of the hydrocarbons; that is, their structures are based on the structures of the hydrocarbons. In such compounds, other groups of atoms are substituted for one or more hydrogen atoms on the hydrocarbon framework. The properties of these derivatives depend on the atoms involved in the substitution. The group that is introduced is called a **functional group.** It is this part of the molecule that gives a class of compounds its characteristic properties.

**TABLE 8.4** Functional Groups in Hydrocarbons

| Class | Functional Group | Example | Structural Formula |
|---|---|---|---|
| alcohol | $-OH$ | methyl alcohol | $H_3C-OH$ |
| ether | $-O-$ | dimethyl ether | $H_3C-O-CH_3$ |
| aldehyde | $\overset{\displaystyle O}{\overset{\|}{-C-H}}$ | acetaldehyde | $H_3C-\overset{\displaystyle O}{\overset{\|}{C}}-H$ |
| ketone | $\overset{\displaystyle O}{\overset{\|}{-C-}}$ | acetone | $H_3C-\overset{\displaystyle O}{\overset{\|}{C}}-CH_3$ |
| carboxylic acid | $\overset{\displaystyle O}{\overset{\|}{-C-OH}}$ | acetic acid | $H_3C-\overset{\displaystyle O}{\overset{\|}{C}}-OH$ |
| ester | $\overset{\displaystyle O}{\overset{\|}{-C-O-}}$ | ethyl acetate | $H_3C-\overset{\displaystyle O}{\overset{\|}{C}}-O-C_2H_5$ |
| amine | $\overset{\displaystyle |}{-N-}$ | methyl amine | $H_3C-\overset{\displaystyle H}{\overset{\|}{N}}-H$ |

Note: Nonbonding electrons are not shown.

For example, if a hydroxyl group ($-OH$) replaces a hydrogen atom in the formula of an alkane, the new formula represents an **alcohol.** Methyl alcohol (or methanol) is $CH_3OH$, derived from the formula for methane, $CH_4$. Substituting a hydroxyl group for a hydrogen in the formula for ethane, $CH_3CH_3$, gives the formula for ethyl alcohol (or ethanol), $CH_3CH_2OH$. The most common functional groups and examples of organic molecules in each class are given in Table 8.4.

Molecules can contain more than one functional group. For example, the proteins in the mushrooms in Ashley's salad are chains of amino acids, which have both amine and carboxylic acid functional groups. An example is glycine, $H_2NCH_2CO_2H$:

$$H-\overset{\displaystyle H}{\underset{\displaystyle H}{N}}-\overset{\displaystyle H}{\underset{\displaystyle H}{C}}-\overset{\displaystyle :O:}{C}-\overset{..}{\underset{..}{O}}-H$$

Note that in the structure of glycine, each carbon atom has four pairs of bonding electrons, the nitrogen atom has three pairs of bonding electrons, and each oxygen atom has two pairs of bonding electrons. Remember that these are the numbers of bonds these atoms have, when possible, to obtain their octets. Drawing Lewis structures for organic molecules can be done by drawing the predicted number of bonds to carbon, nitrogen, oxygen, and hydrogen, and then filling in the unshared electrons when needed to complete the octets. Example 8.8 illustrates this process.

---

**EXAMPLE 8.8    Functional Groups**

Draw the Lewis structure of an ether that has the formula $C_3H_8O$.

**Solution:**

As shown in Table 8.4, in an ether, the oxygen atom must be located between two of the carbon atoms:

$$C-C-O-C$$

The hydrogen atoms are then bonded to carbon atoms, since the oxygen atom already has its normal two bonds:

$$H-\overset{\overset{\displaystyle H}{|}}{\underset{\underset{\displaystyle H}{|}}{C}}-\overset{\overset{\displaystyle H}{|}}{\underset{\underset{\displaystyle H}{|}}{C}}-O-\overset{\overset{\displaystyle H}{|}}{\underset{\underset{\displaystyle H}{|}}{C}}-H$$

Placing the oxygen atom after the first carbon atom gives the same structure, simply flipped on the page left to right. We complete the structure by adding two pairs of nonbonding electrons to the oxygen to complete its octet:

$$H-\overset{\overset{\displaystyle H}{|}}{\underset{\underset{\displaystyle H}{|}}{C}}-\overset{\overset{\displaystyle H}{|}}{\underset{\underset{\displaystyle H}{|}}{C}}-\overset{..}{\underset{..}{O}}-\overset{\overset{\displaystyle H}{|}}{\underset{\underset{\displaystyle H}{|}}{C}}-H$$

**Consider This 8.8**
Could an alcohol have the same molecular formula, $C_3H_8O$?

**Practice Problem 8.8**
Draw the Lewis structure of a ketone that has the formula $C_3H_6O$.

**Further Practice:** Questions 8.81 and 8.82 at the end of the chapter

## 8.5 Shapes of Molecules

Michael's and Ashley's lunch foods supply some of the carbon-based molecules their bodies need to sustain their lives. But why did they choose hamburgers and salad instead of rice cakes? They would probably say they like the flavor and aroma of their favorite foods, but how do the human senses of taste and smell actually work? Although these senses have a chemical basis, determining how they operate has not been simple. It has, however, become clear that there are receptor sites in the nose and tongue where molecules can fit. The receptors are fashioned from the proteins that make up the membranes of cells. The proteins are interlinked in various ways, creating pockets in the structure of the cell membrane. The pockets act as receptors for other molecules if two conditions are met. First, the molecules must be the right size and shape to fit into them. Second, the molecules must interact in a certain way with atoms in the membrane protein, causing a change that can trigger a nerve impulse. It's the nerve impulse that the brain interprets as, "Wow, that smells great!" or "Ugh! I can't eat that!"

Recognizable tastes fall into several categories. One is sweet, and it is highly attractive to humans for good reason. Lactose is a sweet substance found in mothers' milk. The sweet taste encourages babies to consume more. Other sweet substances include glucose and sucrose, which occur naturally in plants. While it has been difficult to relate the structures of molecules to their tastes, a common feature of sweet molecules is the size of a particular portion of the molecule. This part of a sweet molecule presumably fits into the receptor site. Sweet molecules commonly have an —H (or —OH) separated from an —O— atom by about 300 pm, as illustrated in Figure 8.23.

Although humans can distinguish thousands of odors, one theory proposes that there are only seven primary odors, each associated with a different type of receptor in the nose. When a gaseous molecule enters the nose and interacts with a particular receptor, a nerve impulse is generated and sent to the brain. A molecule that can fit into more than one type of receptor triggers multiple signals,

The 2004 Nobel Prize in Physiology or Medicine was given to Richard Axel of Columbia University and Linda Buck of the Fred Hutchinson Cancer Center in Seattle for their discoveries of oderant receptors and the organization of the olfactory system.

~300 pm

**FIGURE 8.23** Glucose is sweet because it contains —H and —OH groups that fit into a taste receptor site on the tongue.

| Molecule in Receptor Cavity | | | | | | | |
|---|---|---|---|---|---|---|---|
| **Molecular Characteristics** | Disk shape | Spherical shape | Rod shape | Wedge shape | Disk and tail shape | Attraction to negative centers | Attraction to positive centers |
| **Primary Odor** | Musky | Camphoric | Ethereal | Pepperminty | Floral | Pungent | Putrid |
| **Chemical Example** | Xylene | Camphor | Diethyl ether | Menthol | α-Amyl pyridine | Formic acid | Hydrogen sulfide |
| **Common Examples** | Musk perfume/ aftershave | Mothballs | Pears | Mint gum/ mouthwash | Roses | Vinegar | Rotten eggs |

**FIGURE 8.24** The primary odors arise from interactions between molecules and shaped receptors. However, two of the receptors involve not shape, but interactions with charge separations in molecules.

generating a composite odor. The proposed receptor cavities and some molecules that give each of the seven primary odors are shown in Figure 8.24. Which combination of primary odors do you think gives the aroma of Michael's hamburger or Ashley's mushrooms?

## The Valence-Shell Electron-Pair Repulsion Theory

As the senses of taste and smell demonstrate, the observable properties of substances derive from the three-dimensional shapes of their molecules. Shape, in turn, is a function of the arrangement of atoms in a molecule. The relative locations of electron pairs around a central atom play a significant role in determining a molecule's three-dimensional shape. Negatively charged electrons repel one another, so electron pairs stay as far apart as possible. The tendency of electron pairs to adjust the orientation of their orbitals to maximize the distance between them is the basis of the **valence-shell electron-pair repulsion (VSEPR) theory.** (*Valence shell* is another name for *valence level*.)

To predict the geometric shape of a molecule or an ion using VSEPR theory, we first need to know how electrons are arranged in a molecule. In particular we need to know the number of unshared electron pairs and single, double, or triple bonds to atoms surrounding the central atom, which we will call *electron domains*. We can get this information by examining the Lewis structure. For example, $CH_4$ has four single bonds to atoms and no unshared electron pairs surrounding the carbon atom, so there are four electron domains. In the $CO_2$ molecule, carbon has only two double bonds to oxygen atoms, as shown in the Lewis structure $:\ddot{O}=C=\ddot{O}:$, and it has no unshared pairs of electron. There are just two electron domains on carbon. The nitrogen in the $NH_3$ molecule has three bonds to atoms and one unshared electron pair, so it has four electron domains. The electron domains (sets of bonding electrons to atoms and unshared electron pairs) are arranged around the central atom as far apart as possible. The result is a shape characterized by the **bond angle** between the central atom and the atoms bonded to it. What shapes and angles would you predict for molecules having two, three, and four electron domains? How would you make these predictions based on their repulsions?

To see what angles to predict, let's examine the shape commonly found in hydrocarbons and many other molecules that arises from four electron pairs or atoms. The Lewis structures look flat, as in $CH_4$:

**ANIMATION:** VSEPR Theory and the Shapes of Molecules

A Lewis structure does not show the shape of a molecule, but it does help us to determine molecular shape.

Since it is hard to see how atoms are arranged in a two-dimensional line drawing, chemists often use a system of solid lines, dashed lines, and wedges to depict the three-dimensional structure of molecules. Solid lines indicate bonds in the plane of the paper. Dashed lines are bonds that point back from the plane of the paper. Wedges depict bonds that come forward from the plane of the paper:

**FIGURE 8.25** Macroscopic objects that are fastened to a common center adopt the same structures as molecules because, like electron domains, they stay as far apart as possible.
©McGraw-Hill Education/Stephen Frisch

This structure looks as if it would have bond angles of 90°:

$$H-\overset{\displaystyle H}{\underset{\displaystyle H}{C}}-H \quad 90° \quad \text{(incorrect bond angle)}$$

However, this does not accurately represent the three-dimensional structure of the $CH_4$ molecule, which is not flat. If we move atoms to correct the flatness, we change the bond angles. The larger the angles, the farther away from one another the atoms are. Imagine the "flat" $CH_4$ molecule in a plane perpendicular to the plane of this page. If we move only the carbon atom relative to the hydrogen atoms the H—C—H angles actually get smaller. The only way to make the angles larger is to move two hydrogen atoms up and two down:

Before                              After

How large can the angles get if we move two up and two down? When the angles all become equal, they have values of 109.5°, giving a tetrahedral shape. This structure is usually shown in a rotated form:

$$H-\overset{\displaystyle H}{\underset{\displaystyle H}{C}}\cdots H \quad 109.5°$$

Table 8.5 shows the geometries that achieve the maximum distance between combinations of two, three, or four electron domains (atoms or unshared electron pairs) on the central atom. The geometries shown in the table are adopted by any set of objects that are attached at a common point and that stay as far apart as possible. For example, the balloons shown in Figure 8.25 adopt the same geometries as electron domains around a central atom. These structures, called *parent structures,* help to predict a molecule's shape.

**TABLE 8.5** Geometric Structures Arising from Different Numbers of Atoms

| Number of Electron Domains (Atoms or Electron Pairs) | Parent Structure | Geometric Arrangement of Atoms or Electron Pairs |
|:---:|:---:|:---:|
| 2 | Linear | 180° |
| 3 | Trigonal planar | 120° |
| 4 | Tetrahedral | 109.5° |

## EXAMPLE 8.9 | VSEPR Parent Structures

What is the parent structure for $H_2O$, in which oxygen is bonded to two other atoms and has two unshared pairs of electrons?

**Solution:**

Because the parent structure depends on the number of atoms and unshared electron pairs around the central atoms, we must first draw the Lewis structure for $H_2O$:

$$H-\overset{\cdot\cdot}{\underset{\cdot\cdot}{O}}-H$$

There are two single bonds to atoms and two pairs of unshared electrons around the oxygen atom, giving a total of four electron domains. From Table 8.5, we see that the parent structure is tetrahedral with angles of 109.5°. The geometric arrangement of atoms and unshared electron pairs is

**Consider This 8.9**

If the oxygen atom were bonded to three hydrogen atoms with one unshared pair of electrons, as in $H_3O^+$, what would be the parent structure?

**Practice Problem 8.9**

What is the parent structure of the $NO_2$ ion, in which nitrogen is bonded to two other atoms and has one unshared pair of electrons?

**Further Practice:** Questions 8.89 and 8.90 at the end of the chapter

**📶 Student Hot Spot**

Student data indicate you may struggle with determining the shape of a molecule from a Lewis structure. Access the SmartBook to view additional Learning Resources on this topic.

Consider the Lewis structures for several molecules and ions having the general composition $AX_3$, shown in Figure 8.26. Some have a trigonal planar shape, and some have a trigonal pyramidal shape. Why do they differ? If the central atom contains any unshared pairs of electrons, the molecular shape is not that of the parent structure, but is derived from it. The molecular shape we use to describe a molecule is based on the arrangement of only the groups of *bonding electrons* (or atoms) within the parent structure, and not the unshared pairs of electrons; the shape is a

**FIGURE 8.26** Some molecules and ions with the formula $AX_3$ have a trigonal planar shape, while others have a trigonal pyramidal shape.

The bond angles in real molecules often differ a little from the values predicted by the VSEPR theory. Differences occur because unshared pairs of electrons occupy more space than bonded pairs. The bonded pairs are forced together somewhat, resulting in slightly smaller bond angles than predicted. For example, ammonia, which has one unshared pair of electrons, has a bond angle of 107° rather than the predicted 109.5°. Water has two unshared pairs of electrons and a bond angle of about 105°. A similar effect is caused by a double bond, which occupies more space than a single bond.

**ANIMATION:** VSEPR

**TABLE 8.6** Arrangement of Electron Domains and Molecular Shapes

| Electron Domains | Number of Bonded Atoms | Number of Unshared Pairs | Molecular Shape | Examples |
|---|---|---|---|---|
| **Parent Structure: Linear** | | | | |
| 2 | 2 | 0 | Linear<br>X — A — X | (AX$_2$)<br>BeCl$_2$, CO$_2$, HCN |
| **Parent Structure: Trigonal Planar** | | | | |
| 3 | 3 | 0 | Trigonal planar | (AX$_3$)<br>BF$_3$, BH$_3$, SO$_3$, NO$_3^-$ |
| 3 | 2 | 1 | Bent | (AX$_2$)<br>SO$_2$, NO$_2^-$ |
| **Parent Structure: Tetrahedral** | | | | |
| 4 | 4 | 0 | Tetrahedral | (AX$_4$)<br>CH$_4$, CH$_2$Cl$_2$, SiCl$_4$, POCl |
| 4 | 3 | 1 | Trigonal pyramidal | (AX$_3$)<br>NH$_3$, PF$_3$, NH$_2$Cl |
| 4 | 2 | 2 | Bent | (AX$_2$)<br>H$_2$O, OF$_2$, BrO$_2^-$, SCl$_2$ |

**FIGURE 8.27** The four electron domains in water are arranged around the oxygen atom in orbitals that point approximately to the corners of a tetrahedron. Hydrogen atoms occupy only two of the positions, so the molecular shape is described as bent, not tetrahedral.

description of the locations of the atoms in the molecule. Consider, as an example, the water molecule from Example 8.9, which has two atoms and two unshared pairs of electrons around the central atom. The four electron pairs in H$_2$O arrange themselves tetrahedrally around the oxygen atom, as described in Table 8.6 and shown in Figure 8.27. But because we consider only the two bonding pairs when we describe the molecular shape, we do not say that the water molecule shape is tetrahedral. Rather, we describe it as bent. All possible shapes for molecules that have two to four electron domains are summarized in Table 8.6. The bond angles for these structures are approximately those of the parent structures.

Notice in Table 8.6 that CO$_2$ is linear, but SO$_2$—with the same number of atoms—is bent. We can understand this difference using VSEPR theory. Examine the Lewis structures of these molecules:

$$:\overset{..}{O}=C=\overset{..}{O}: \qquad :\overset{..}{O}=\overset{..}{S}-\overset{..}{\underset{..}{O}}:$$

What is different about them? The carbon in $CO_2$ has two bonded atoms and no unshared electron pairs on the central atom, so it has two electron domains and a linear structure with a bond angle of 180°. In contrast, sulfur in $SO_2$ has two bonded atoms and one unshared electron pair on the central atom. The unshared pair causes the molecule to be bent with a bond angle of approximately 120°. Models of these molecules are shown in Figure 8.28.

To predict the shape of a molecule according to VSEPR theory, use the following four steps.

**FIGURE 8.28** Carbon dioxide is linear, whereas sulfur dioxide is bent, because there is an unshared electron pair on the sulfur atom.

### Steps for Predicting the Shape of a Molecule

1. Draw a Lewis stucture.
2. Count the number of atoms bonded to the central atom, and count unshared electron pairs on the central atom.
3. Add the number of atoms and the number of unshared electron pairs around the central atom. The total indicates the number of electron domains and thus the parent structure.
4. The molecular shape is derived from the parent structure by considering only the positions in the structure occupied by bonded atoms.

### EXAMPLE 8.10 | Shapes of Molecules and Ions

The carbonate ion, $CO_3^{2-}$, is found in limestone and in Michael's antacid tablets. Predict its shape. What are the approximate O—C—O bond angles?

**Solution:**

1. First, we draw a Lewis structure, following the procedure from Section 8.3:

$$\left[ \begin{array}{c} :\ddot{O}: \\ | \\ :\ddot{O}=\overset{}{C}-\ddot{O}: \end{array} \right]^{2-}$$

2. There are two other resonance forms, which have the double bond in other positions, but we need obtain only one Lewis structure to determine the shape.
3. Next, count the bonded atoms and unshared electron pairs on the central atom. The carbon atom has three bonded oxygen atoms and no unshared pairs of electrons around it.
4. The total number of electron domains (bonded atoms and unshared electron pairs) is three, so the parent structure is trigonal planar.
5. Since there are no unshared electron pairs around carbon, the shape of this ion is the same as the parent structure: trigonal planar. The bond angle in a trigonal planar structure is 120°.

**Consider This 8.10**

If the carbon had three atoms bonded to it but also had an unshared electron pair, would the bond angle be the same as the O—C—O bond angle in the carbonate ion?

**Practice Problem 8.10**

What is the shape of the nitrite ion, $NO_2^-$? What is the O—N—O bond angle?

**Further Practice:** Questions 8.91 and 8.92 at the end of the chapter

For molecules with more than one central atom, we can repeat the procedure for each central atom, a process that allows us to predict the structures of molecules as large and complex as proteins. Let's consider one amino acid, glycine, which might be found in a protein. It has a sweet taste and is found in gelatin and other animal products; it was probably in Michael's hamburger. Its formula is $H_2NCH_2CO_2H$. How do we predict its structure?

We start, as before, by writing a Lewis structure for the molecule:

$$H-\ddot{N}-\underset{\underset{H}{|}}{\overset{\overset{H}{|}}{C}}-\underset{\ddot{\ddot{O}}}{\overset{\overset{\ddot{O}\!:}{\|}}{C}}-\ddot{O}-H$$

This molecule contains four atoms that act as central atoms: the nitrogen atom, both carbon atoms, and one of the oxygen atoms.

- The nitrogen atom has three atoms bonded to it and one unshared electron pair, so it has four electron domains. Thus, the parent structure around this nitrogen atom is tetrahedral, giving bond angles of approximately 109.5°. Because there is one unshared pair of electrons, the shape around the nitrogen atom is trigonal pyramidal.
- The first carbon atom is bonded to four other atoms (N, two H, and C) and has no unshared electron pairs, so the shape around it is tetrahedral, with bond angles of about 109.5°.
- The second carbon atom is bonded to three other atoms (C and two O) and has no unshared electron pairs, so the shape around it is trigonal planar, with bond angles of around 120°.
- Finally one of the oxygen atoms has two atoms (C and H) bonded to it and has two unshared electron pairs. The parent structure around it is tetrahedral, but with two unshared pairs, the shape is bent. The bond angle will be around 109.5°.

Verify each of these predictions by examining the molecular model of glycine in Figure 8.29.

Molecular shapes are important in living systems. For example, in humans, red blood cells carry oxygen through the bloodstream to all body cells. They can act as delivery systems because they contain hemoglobin, a large protein molecule that contains four heme groups, one of which is shown in Figure 8.30A. Oxygen molecules bind to the iron ions in heme, as shown in Figure 8.30B. Notice the bent $Fe-O=O$ arrangement, which is consistent with the two unshared pairs of electrons on each oxygen atom:

$$:\!\ddot{O}=\ddot{O}\!:$$

Another part of the hemoglobin protein is an amino acid called histidine. The histidine group projects down near the top of the heme group, just fitting into the space around the oxygen molecule (Figure 8.30C).

**FIGURE 8.29** VSEPR theory predicts the structure of glycine.

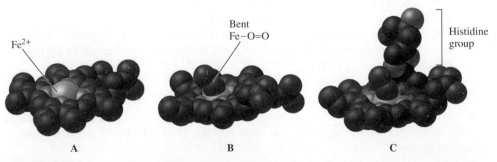

**FIGURE 8.30** (A) The heme group in hemoglobin consists of an iron(II) ion bonded to four nitrogen atoms, each surrounded by a planar arrangement of carbon atoms. (The hydrogen atoms are not shown for clarity.) (B) When oxygen binds to heme, the $Fe-O=O$ arrangement is bent, not linear, since the electrons surrounding each oxygen atom are located at the corners of a trigonal plane. (C) A histidine group in the hemoglobin protein fits into the space next to the oxygen molecule.

FIGURE 8.31 (A) When carbon monoxide binds to the iron in heme, the Fe—C≡O arrangement is expected to be linear. (B) However, in hemoglobin, a histidine group is close to the heme group, forcing the carbon monoxide to assume a bent arrangement. This makes the iron-carbon bond weaker than if the arrangement remained linear.

This system works well as long as carbon monoxide, a highly poisonous gas, does not get into the bloodstream. If it does, it interferes with the distribution of oxygen molecules to cells. As shown in Figure 8.31A, the arrangement of CO bonded to iron in hemoglobin is linear, since the carbon in carbon monoxide has only one unshared electron pair:

$$:C\equiv O:$$

In this form, the carbon monoxide binds to the iron 25,000 times more strongly than oxygen does. Under these conditions, any exposure to carbon monoxide would completely negate the ability of heme to bind to oxygen, and there would be no possibility of surviving such exposure.

However, nature has provided some protection. As shown in Figure 8.31B, the histidine group forces the carbon monoxide to bind at an angle. In this form, carbon monoxide binds only 200 times more strongly than oxygen. When hemoglobin binds to carbon monoxide, the red blood cells can no longer transport oxygen to the cells, but continued exposure to oxygen can replace the carbon monoxide, counteracting the toxic effect.

## Polarity of Molecules

In Section 8.1 you learned that bonds are polar if the bonded atoms share electrons unequally, resulting in partial charges on the atoms. Molecules with polar bonds can be polar molecules under certain conditions. Here we will look at how to determine if a molecule is polar.

In a diatomic molecule composed of atoms having different electronegativities, the polarity of the molecule lies along the plane of the bond. Since there are only two atoms, one of them must be partially positive ($\delta+$) and the other partially negative ($\delta-$). This can be represented by an arrow, with the + sign at the position of partial positive charge ($\delta+$) and the arrowhead at the position of partial negative charge ($\delta-$):

$$\delta+ \ \delta-$$
$$H-F$$

The direction of the polarity is evident in the case of a diatomic molecule with only one bond, but polyatomic molecules with more than one bond are more complex. The polarity of the bonds and the polarity of the molecule may be different.

A polyatomic molecule that has nonpolar bonds cannot be polar. For example, in $NBr_3$, nitrogen and bromine have the same electronegativity, so each bond is nonpolar and the molecule must be nonpolar. But a polyatomic molecule with polar

**ANIMATION:** Influence of Shape on Polarity

Carbon-hydrogen bonds are almost nonpolar, so hydrocarbons are generally considered to be essentially nonpolar. Because of the geometry of hydrocarbons, the slight polarity of the bonds cancels out.

**FIGURE 8.32** The polar molecule $H_2O$ aligns with the electric field in a capacitor. Notice that the negative side of each $H_2O$ molecule is attracted to the positively charged plate, and the positive side of each $H_2O$ molecule is attracted to the negatively charged plate. The nonpolar $CO_2$ molecules are unaffected by the electric field because they do not have positive and negative sides.

**FIGURE 8.33** In symmetrical molecules with identical atoms surrounding a central atom, such as $SO_3$ and $CCl_4$, the bond polarities cancel, resulting in *nonpolar molecules*.

bonds may be polar or nonpolar, depending on its geometry. We can determine this experimentally by placing a sample of the substance in an electric field, such as between the electrostatically charged plates of a capacitor. The greater the polarity of the substance, the more voltage it takes to charge the plates of the capacitor to the same extent. As shown in Figure 8.32, if the molecules are polar, they align with the electric field. If they are nonpolar, they retain random orientations in the electric field.

The polarity of a bond can be canceled by the polarity of another bond in the opposite direction. Consider a molecule of gaseous beryllium chloride (linear $BeCl_2$), for example. Although it has two polar Be—Cl bonds, it is nonpolar. The bonds in $BeCl_2$ are 180° apart, and the polarities of the two bonds point in opposite directions. Thus, the charge separation places partial negative charges at the two chlorine positions and a partial positive charge at the beryllium position:

$$\delta - \quad \delta + \quad \delta -$$
$$Cl - Be - Cl$$
$$\longleftarrow \; + \; \longrightarrow$$

No net polarity

The polarities of the two bonds are exactly equal in magnitude and opposite in direction. They cancel each other out, making the molecule nonpolar.

Any molecule in which the relationship of the bonded atoms is completely symmetrical—linear, trigonal planar, or tetrahedral (Table 8.6)—experiences this sort of cancellation and is nonpolar. Thus, cancellation of polarity occurs in trigonal planar molecules such as $SO_3$ and tetrahedral molecules such as $CCl_4$ (Figure 8.33). In such molecules, the central atom is surrounded by identical atoms. However, if the atoms surrounding the central atom are not identical, the molecule's shape is not completely symmetrical. Such molecules are polar if the bonds are polar. Chloroform, $CHCl_3$, is polar, as are chloromethane, $CH_3Cl$, and dichloromethane, $CH_2Cl_2$ (Figure 8.34). In these molecules, the atoms surrounding the carbon are not identical, so the polarities of the bonds do not cancel, resulting in a net polarity for the molecule.

In summary, we must consider two factors when deciding whether a molecule is polar: the bond polarity and the molecular shape.

- If the bonds are polar *and* the molecular shape is *not* symmetrical, then the molecule is polar.
- If all the bonds are nonpolar *or* if the molecular shape is symmetrical, the molecule is nonpolar.

**ANIMATION:** Polarity of Molecules

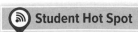 **Student Hot Spot**

Student data indicate you may struggle with molecular polarity. Access the SmartBook to view additional Learning Resources on this topic.

| **EXAMPLE 8.11** | **Polarity of Molecules** |

Predict whether $H_2S$ is a polar or a nonpolar molecule.

**Solution:**

The molecule $H_2S$ has polar bonds because sulfur and hydrogen have different electronegativities (see Figure 8.5). To determine the molecular shape, we start by drawing its Lewis structure:

$$H-\ddot{S}-H$$

There are four electron domains around the sulfur, so the parent structure is tetrahedral. The hydrogen atoms occupy two of the four positions around the sulfur so the molecule is a bent shape:

Because the bonds are polar and the molecule is not completely symmetrical, $H_2S$ is polar.

**Consider This 8.11**

If a molecule with two hydrogens bonded to a central atom were linear (such as $BeH_2$), would it be polar?

**Practice Problem 8.11**

Predict whether $C_2H_6$, $NO_2$, $CO_2$, $SO_2$, and $SO_3$ are polar or nonpolar molecules.

**Further Practice:** Questions 8.111 and 8.112 at the end of the chapter

**FIGURE 8.34** In asymmetric molecules like these, the atoms surrounding the central atom are not identical and have different bond polarities. The side of each molecule with the chlorine atoms is more negative than the side with hydrogens, so they are *polar molecules*.

Polarity determines many interactions between molecules. For example, nonpolar molecules exist as gases at lower temperatures than do polar molecules of similar size. Polar liquids freeze at higher temperatures than do nonpolar liquids of similar size. Ionic salts and polar liquids dissolve better in polar liquids than in nonpolar liquids. On the other hand, nonpolar liquids dissolve better in other nonpolar liquids than in polar liquids. Oil, which is composed of largely nonpolar hydrocarbons, does not dissolve in polar water (Figure 8.35).

**FIGURE 8.35** Nonpolar oil in the top layer of the cylinder does not dissolve in polar water in the bottom layer.
©David A. Tietz/Editorial Image, LLC

# CHAPTER REVIEW

## KEY CONCEPTS

- A chemical bond is the force that holds atoms together in a molecule or compound.
  - Different types of chemical bonds give substances different properties.
- Electronegativity is a measure of the tendency for an atom to attract electrons in a bond.
  - The electrons in a covalent bond are not always shared equally, in which case the covalent bonds are polar.

- Chemical bonds tend to form in conformance with the octet rule.
  - Atoms tend to lose, gain, or share electrons to achieve a more stable electron configuration when bonding to form a compound.
  - According to the octet rule, atoms tend to be associated with eight valence electrons.
  - To obtain the noble-gas configuration of helium, hydrogen usually has two electrons when bonded. There are some other exceptions to the octet rule.
- Ionic bonds form either through sharing of electrons or through transfer of electrons.
  - Ionic bonding involves the complete transfer of one or more electrons from a metal to a nonmetal, often giving both elements a noble-gas configuration.
  - In ionic compounds, electrostatic forces hold the ions together in crystal lattices.
- Covalent bonds form through sharing electrons between atoms.
  - Lewis structures represent the arrangement of bonding and unshared valence electrons in molecules and ions.
  - Two atoms can share one electron pair to form a single bond, two electron pairs to form a double bond, or three electron pairs to form a triple bond.
  - When there is more than one way to draw a Lewis structure, the molecule exhibits resonance. The actual bonding is a composite of all the resonance structures, and two or more Lewis structures are needed to show the arrangement of electrons in a covalent molecule or ion.
- Carbon atoms bond together readily to form chains of atoms that are surrounded by hydrogen atoms to form the hydrocarbons.
  - Hydrocarbons can be classified as aliphatic or aromatic.
  - Aliphatic hydrocarbons can be further classified as alkanes, alkenes, or alkynes, depending on the types of bonds between carbon atoms.
  - Aromatic hydrocarbons contain rings of six carbon atoms with delocalized bonds.
  - Many other carbon compounds are formed from hydrocarbons by the inclusion of functional groups.
- VSEPR theory, which assumes that electron domains stay as far apart as possible, predicts the shapes of molecules.
  - The total number of bonded atoms and of unshared electron pairs on the central atom determines the parent structure.
  - The molecular shape is determined by the position of atoms around the central atom, which is affected by the presence of unshared electron pairs.
  - A nonpolar molecule has either nonpolar bonds or a symmetrical structure.
  - A molecule is polar if the bonds are polar and the molecule is not symmetrical.

# KEY TERMS

alcohol (8.4)

aliphatic hydrocarbon (8.4)

alkane (8.4)

alkene (8.4)

alkyne (8.4)

aromatic hydrocarbon (8.4)

bond angle (8.5)

chemical bond (8.1)

covalent bonding (8.1)

crystal lattice (8.2)

double bond (8.3)

electronegativity (8.1)

functional group (8.4)

hydrocarbon (8.4)

ionic bonding (8.1)

ionic crystal (8.2)

Lewis structure (electron dot structure) (8.3)

Lewis symbol (electron dot symbol) (8.2)

nonpolar covalent bond (8.1)

octet rule (8.3)

polar covalent bond (8.1)

polarity (8.1)

resonance hybrid (8.3)

single bond (8.3)

triple bond (8.3)

valence-shell electron-pair repulsion (VSEPR) theory (8.5)

# QUESTIONS AND PROBLEMS

The following questions and problems, except those in *Additional Questions* and *Concept Review Questions,* are paired. Questions in a pair focus on the same concept. Answers to selected questions and problems are in Appendix D.

## Matching Definitions with Key Terms

8.1     Match the key terms with the descriptions provided.
(a) a covalent bond that involves the sharing of one pair of electrons
(b) a hydrocarbon containing only carbon-carbon single bonds
(c) bonding between two atoms resulting from the sharing of electrons
(d) a solid structure in which the ions are arranged in a regular repeating pattern
(e) a rule stating that atoms tend to gain, lose, or share electrons to achieve an electron configuration with eight electrons in the valence shell
(f) a bond resulting from the unequal sharing of electron pairs
(g) a hydrocarbon containing carbon-carbon triple bonds
(h) a measure of the ability of an atom to attract electrons within a bond to itself
(i) a covalent bond that involves the sharing of three pairs of electrons
(j) the repeating pattern of atoms or ions in a crystal
(k) representation of a compound consisting of the symbols of the component elements, each surrounded by dots representing shared and unshared electrons
(l) force that holds atoms together in a molecule or compound

8.2     Match the key terms with the descriptions provided.
(a) bonding between cations and anions resulting from electrostatic attractions of opposite charges
(b) a covalent bond that involves the sharing of two pairs of electrons
(c) a group of atoms substituted for hydrogen in the formula of a hydrocarbon that gives the compound its characteristic properties
(d) representation of an atom consisting of the symbol for the element surrounded by a number of dots equal to the number of valence electrons
(e) the angle between the two lines defined by a central atom attached to two surrounding atoms
(f) a bond resulting from the equal sharing of electron pairs
(g) a hydrocarbon containing carbon-carbon double bonds
(h) the separation of electronic charge within a bond (or molecule)
(i) a class of chemical compounds containing only carbon and hydrogen
(j) an average or composite Lewis structure derived from two or more valid Lewis structures that closely represents the bonding in a molecule
(k) a hydrocarbon derivative containing the −OH group
(l) the shapes of molecules result from the tendency for electron pairs to maximize the distance between them to minimize repulsions

## Types of Bonds

8.3     What is a chemical bond?

8.4     Describe the difference between ionic and covalent bonding.

8.5     Which type of elements are most likely to form compounds using ionic bonding?

8.6     Which type of elements are most likely to form compounds using covalent bonding?

8.7     Which of the following compounds are likely to have covalent bonds?
(a) HF                    (c) $NCl_3$                (e) $CF_4$
(b) NaF                   (d) $MgBr_2$

8.8     Which of the following compounds are likely to have ionic bonds?
(a) CsCl                  (c) $OCl_2$                (e) $IF_5$
(b) $CaCl_2$             (d) $NBr_3$

8.9     Identify the bonding in each of the following substances as ionic or covalent.
(a) $CaF_2$             (c) $CO_2$
(b) $Cl_2$              (d) $FeCl_3$

8.10    Identify the bonding in each of the following substances as ionic or covalent.
(a) NaCl                  (c) $CH_4$
(b) $H_2$               (d) $SiO_2$

8.11    Which of the following images represents an ionic compound? (Note that the atom labels indicate the elements but do not necessarily specify charge.)

A                         B

C               D               E

8.12  Which of the following images represents a covalent compound?

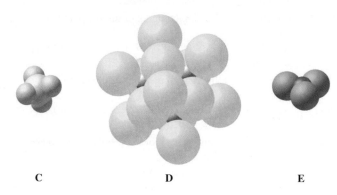

A                         B

C            D            E

8.13  Which of the following compounds are likely to occur as gases at room temperature?
(a) $CH_4$          (d) KBr
(b) $Ca_3N_2$       (e) HCl
(c) $SF_4$

8.14  Which of the following compounds are likely to occur as solids at room temperature?
(a) BrCl          (d) CaO
(b) $CaCO_3$      (e) NO
(c) $SO_2$

8.15  Predict whether each of the following substances is likely to have a relatively high or low boiling point.
(a) $TlCl_3$       (d) CaO
(b) CsCl          (e) $O_2$
(c) $CO_2$

8.16  Predict whether each of the following substances is likely to have a relatively high or low melting point.
(a) $P_4$          (d) AlN
(b) $CaCl_2$       (e) $CaC_2$
(c) $CCl_4$

8.17  Describe how electronegativity values change going down a group in the periodic table.

8.18  Compare the electronegativity of metallic and nonmetallic elements.

8.19  What kinds of bonds are always nonpolar?

8.20  Describe how to decide whether a bond is polar.

8.21  Using periodic trends, arrange the following atoms in order of increasing electronegativity.
(a) Br, Cl, F, N, O       (b) C, F, H, N, O

8.22  Using periodic trends, arrange the following atoms in order of increasing electronegativity.
(a) B, C, H, Se, Si       (b) C, Ca, Cl, Cs, Cu

8.23  Decide which bond in each pair is polar and explain your reasoning. Place $\delta+$ and $\delta-$ superscripts on the element symbols for each polar bond to indicate the direction of the polarity.
(a) H—F and H—H          (c) H—H and B—H
(b) Cl—I and Cl—Cl

8.24  Decide which bond in each pair is polar and explain your reasoning. Place $\delta+$ and $\delta-$ superscripts on the element symbols for each polar bond to indicate the direction of the polarity.
(a) F—F and C—F          (c) O—O and N—O
(b) O—H and H—H

8.25  Arrange the following bonds in order of increasing polarity.
(a) O—H, C—H, H—H, F—H
(b) O—Cl, C—Cl, H—Cl, F—Cl

8.26  Arrange the following bonds in order of increasing polarity.
(a) H—F, F—F, H—H, H—I
(b) B—F, O—F, C—F, H—F

## Ionic Bonding

8.27  What information can be determined from a Lewis symbol?

8.28  What is the maximum number of valence electrons represented in the Lewis symbols of the main-group elements?

8.29  Draw Lewis symbols showing the valence electrons of the following atoms.
(a) C          (c) Se          (e) Cs
(b) I          (d) Sr          (f) Ar

8.30  Draw Lewis symbols showing the valence electrons of the following atoms.
(a) Kr          (c) F          (e) Ba
(b) Sb          (d) In          (f) N

8.31  Draw Lewis symbols showing the valence electrons of the following ions.
(a) $Cl^-$       (c) $S^{2-}$       (e) $B^{3+}$
(b) $Sc^{3+}$    (d) $Ba^{2+}$

8.32  Draw Lewis symbols showing the valence electrons of the following ions.
(a) $N^{3-}$     (c) $Br^-$       (e) $Al^{3+}$
(b) $C^{4-}$     (d) $Mg^{2+}$

8.33  Write a formula for each of the following ionic salts using the Lewis symbols of the ions.
(a) LiCl          (b) $BaCl_2$          (c) BaS

8.34  Write a formula for each of the following ionic salts using the Lewis symbols of the ions.
(a) $Na_2O$       (b) LiOH          (c) CsF

8.35  Explain why $K^+$ is known in nature, but $K^{2+}$ is not.

8.36  Explain why $Cl^-$ is known in nature, but $Cl^{2-}$ is not.

8.37  Explain why sodium fluoride has the composition represented by the formula NaF.

8.38  Explain why calcium oxide has the composition represented by the formula CaO.

8.39  What is the difference between an ionic crystal and a crystal lattice?

8.40    What holds ions together in a crystal lattice?

8.41    Describe the sodium chloride structure shown in Figure 8.8. How many chloride ions surround each sodium ion?

8.42    Describe the cesium chloride structure shown in Figure 8.9. How many chloride ions surround each cesium ion?

8.43    Why does $CaF_2$ have a different crystal structure than NaCl?

8.44    Would you expect LiCl to have the same crystal structure as NaCl or CsCl? Explain your selection.

## Covalent Bonding

8.45    Draw the Lewis structures for $O_2$ and $F_2$.
    (a) How many valence electrons are around each atom in each of these structures?
    (b) Describe the bonding in each as a single, double, or triple covalent bond.

8.46    Draw the Lewis structures for $I_2$ and $N_2$.
    (a) How many valence electrons are around each atom in each of these structures?
    (b) Describe the bonding in each as a single, double, or triple covalent bond.

8.47    Why does hydrogen exist as a diatomic molecule?

8.48    How many electrons does each hydrogen have in the $H_2$ molecule?

8.49    How many single bonds are typically formed by the following atoms?
    (a) H    (b) N    (c) F    (d) Ne

8.50    How many single bonds are typically formed by the following atoms?
    (a) B    (b) O    (c) I    (d) C

8.51    Identify a main-group element (X) could form each of the following compounds.
    (a) $:\!\ddot{C}l\!:\!X\!:\!\ddot{C}l\!:$    (b) $:\!\ddot{F}\!:\!X\!:\!\ddot{F}\!:$    (c) $:\!\ddot{C}l\!:\!X\!:\!\ddot{C}l\!:$

8.52    Identify a main-group element (X) could form each of the following compounds.
    (a) $:\!\ddot{O}\!=\!X\!=\!\ddot{O}\!:$    (b) $:\!\ddot{Br}\!:\!X\!:\!\ddot{Br}\!:$    (c) $:\!X\!-\!X\!:$

8.53    Draw a Lewis structure for each of the following:
    (a) HCN    (c) $C_2H_2$    (e) $C_2H_6$
    (b) $H_3CCN$    (d) $C_2H_4$

8.54    Draw a Lewis structure for each of the following:
    (a) $NH_2OH$    (c) $C_2H_3Cl$    (e) HOCl
    (b) $CCl_4$    (d) $C_2Br_2$

8.55    Draw a Lewis structure for each of the following:
    (a) $NO_3^-$    (c) $SO_3^{2-}$    (e) $NO^+$
    (b) $SO_4^{2-}$    (d) $NO_2^-$

8.56    Draw a Lewis structure for each of the following:
    (a) $NH_4^+$    (c) $(CH_3)_2CO$    (e) $CH_3NO_2$
    (b) $H_2CO$    (d) $OH^-$

8.57    Determine whether each of the following has the same number of bonding electrons as the cyanide ion, $CN^-$.
    (a) $O_2$    (c) CO    (e) $NH_3$
    (b) $NO^+$    (d) $N_2$

8.58    Determine whether each of the following has the same number of bonding electrons as the sulfur dioxide molecule, $SO_2$.
    (a) $NO_2^-$    (c) $CO_2$    (e) $N_2$
    (b) $OF_2$    (d) $O_3$

8.59    When is it necessary to use the concept of resonance?

8.60    How is the concept of resonance consistent with the octet rule?

8.61    Indicate whether or not each of the following molecules or ions exhibits resonance.
    (a) $O_2$    (c) $SO_2$    (e) $SO_3^{2-}$
    (b) $H_2O$    (d) $NO_2$

8.62    Indicate whether or not each of the following molecules or ions exhibits resonance.
    (a) $N_2$    (c) $ClO_2^-$    (e) $SO_3$
    (b) $OF_2$    (d) HCN

8.63    Draw a Lewis structure, including the resonance forms, for each of the following molecules or ions.
    (a) $NO_2^-$    (c) $CH_3CO_2^-$    (e) $HNO_3$
    (b) $SO_3$    (d) $CO_3^{2-}$

8.64    Draw a Lewis structure, including the resonance forms, for each of the following molecules or ions.
    (a) $CS_2$    (c) $HNO_2$    (e) $SO_3^{2-}$
    (b) $NCO^-$    (d) $SO_4^{2-}$

8.65    In HF, the hydrogen atom shares two electrons with the fluorine atom, but has no unshared electron pairs. Discuss why this observation is consistent with the principle behind the octet rule.

8.66    Describe the bonding in $S_2Cl_2$. The atoms are connected in the order, Cl S S Cl. Why does this molecule not exist as SCl?

8.67    Decide whether the indicated atom obeys the octet rule. If not, indicate how the octet rule is broken.
    (a) O in $H_2O$    (c) F in $SF_4$
    (b) S in $SF_4$    (d) S in $SF_2$

8.68    Decide whether the indicated atom obeys the octet rule. If not, indicate how the octet rule is broken.
    (a) B in $BCl_3$    (c) F in $F_2$
    (b) Cl in $ClF_2^-$ ion    (d) S in $SF_6$

8.69    An atom in each of the following molecules does not obey the octet rule. Decide which atom violates the rule and explain the nature of the violation.
    (a) $SF_4$    (c) $XeF_6$
    (b) $BH_3$    (d) $ClO_2$

8.70    An atom in each of the following molecules does not obey the octet rule. Decide which atom violates the rule and explain the nature of the violation.
    (a) $NO_2$    (c) $BeCl_2$
    (b) $XeF_2$    (d) $ICl_3$

## Bonding in Carbon Compounds

8.71    What are the different classes of hydrocarbons?

8.72    How many bonds does carbon normally form? What are the different ways it can form this many bonds?

8.73    Draw the Lewis structure of benzene, $C_6H_6$, a cyclic compound.

8.74    Draw the Lewis structure of ethylene, $C_2H_4$.

8.75 Identify the class of organic substance for each of the following molecules.

(a) CH₃—OH  (d) H₃C—CH=CH₂

(b) H₃C—CH₃  (e) H₃C—C̈—CH₃ (with O double-bonded above C)

(c) H₃C—O—CH₃

8.76 Identify the class of organic substance for each of the following molecules.

(a) H₃C—NH₂  (d) H₃C—C̈—OH (with O double-bonded above C)

(b) H₂C=CH₂  (e) C₆H₆

(c) H₃C—C̈—H (with O double-bonded above C)

8.77 Identify the class of organic substance for each of the following molecules.

(a) CH₃CH₂OH  (c) CH₃CH₂CHO

(b) CH₃C≡CH  (d) CH₃CH₂OCH₂CH₃

8.78 Identify the class of organic substance for each of the following molecules.

(a) (CH₃)₂NH  (c) CH₃CH₂CO₂H

(b) C₂H₂  (d) CH₃CH₂COCH₃

8.79 Identify the class of organic substance for each of the following molecules.

(a)          (b)

(c)

8.80 Identify the class of organic substance for each of the following molecules.

(a)          (b)

(c)

8.81 Draw the Lewis structure for an aldehyde that has the molecular formula C₄H₈O.

8.82 Draw the Lewis structure for a ketone that has the molecular formula C₄H₈O.

## Shapes of Molecules

8.83 How can VSEPR theory be used to predict molecular shapes?

8.84 Why are unshared pairs of electrons on a central atom not considered to be part of the molecular shape?

8.85 Why is it important to draw Lewis structures before predicting molecular shapes?

8.86 Explain how nonbonding pairs of electrons influence molecular shape.

8.87 Draw each of the following geometric arrangements.

(a) tetrahedral  (c) bent

(b) trigonal planar  (d) trigonal pyramidal

8.88 In which of the following molecular shapes would you expect to find one or more unshared pairs of electrons on the central atom?

(a) bent  (c) trigonal planar

(b) tetrahedral  (d) trigonal pyramidal

8.89 Predict the parent structures of the following molecules.

(a) BeCl₂  (c) SCl₂  (e) H₂Te  (g) BBr₃

(b) PH₃  (d) SO₂  (f) SiH₄  (h) H₂O

8.90 Predict the parent structures of the following molecules.

(a) OCS  (c) FCN  (e) PF₃  (g) NO₂

(b) FNO  (d) HN₃  (f) SF₂

8.91 Predict the shapes and give approximate bond angles for the following molecules.

(a) BeCl₂  (c) SCl₂  (e) H₂Te  (g) BBr₃

(b) PH₃  (d) SO₂  (f) SiH₄  (h) H₂O

8.92 Predict the shapes and give approximate bond angles for the following molecules.

(a) OCS  (c) FCN  (e) PF₃  (g) NO₂

(b) FNO  (d) HN₃  (f) SF₂

8.93 Predict the bond angles in the following molecules.

(a) NH₃  (c) HCl  (e) BF₃  (g) PCl₃

(b) H₂O  (d) HCN  (f) H₂CO

8.94 Predict the bond angles in the following molecules.

(a) NF₃  (c) CBr₄  (e) PH₃

(b) SO₂Cl₂  (d) F₂CO  (f) HOCl

8.95 Give an example of a molecule or polyatomic ion that has the following features.

(a) three bonded atoms, no unshared electrons on the central atom

(b) three bonded atoms, one unshared pair of electrons on the central atom

(c) two bonded atoms, two unshared pairs of electrons on the central atom

8.96 Give an example of a molecule or polyatomic ion that has the following features.

(a) four bonded atoms, no unshared electrons on the central atom

(b) two bonded atoms, no unshared electrons on the central atom

(c) two bonded atoms, one unshared pair of electrons on the central atom

8.97 Identify a molecule or ion that could have the following molecular shapes.

    (a)        (b)        (c)        (d)

8.98 Identify a molecule or ion that could have the following molecular shapes.

    (a)        (b)        (c)        (d)

8.99 Is this the shape of $NO_3^-$ or $ClO_3^-$?

8.100 Is this the shape of $SCl_2$ or $BeCl_2$?

8.101 Which of the following molecules or ions have unshared electron pairs on the central atom?

   **A**       **B**       **C**       **D**

8.102 Which of the following molecules or ions have unshared electron pairs on the central atom?

   **A**       **B**       **C**       **D**

8.103 Hydrazine, $N_2H_4$, is a colorless, oily liquid that fumes in air and has an odor much like that of ammonia. It is used as a rocket fuel. The order of its atoms is $H_2NNH_2$. How many unshared pairs of electrons are on each nitrogen atom?

8.104 Oxalic acid, $H_2C_2O_4$, a poisonous colorless solid, is found in some vegetables such as spinach and rhubarb. It is present in concentrations well below the toxic limit, so you can't use this as a reason to refuse a helping of spinach. The order of atoms in a molecule of oxalic acid is $HO_2CCO_2H$. How many unshared pairs of electrons are on the carbon and oxygen atoms?

8.105 Chloropicrin, $Cl_3CNO_2$, is an insecticide that has been used against insects that attack cereals and grains. It is a liquid with an intense odor. Predict the Cl—C—Cl, Cl—C—N, C—N—O, and O—N—O bond angles in a molecule of chloropicrin.

8.106 Fuel cells are used in many areas, such as the aerospace industry, where energy efficiency is more important than high power output. One fuel cell uses methanol, $CH_3OH$, as a fuel. Predict the H—C—H, H—C—O, and C—O—H bond angles.

8.107 Distinguish between bond polarity and molecular polarity.

8.108 Why does molecular polarity depend not only on bond polarity but also on the geometry of the molecule?

8.109 Explain how carbon tetrachloride can have polar bonds but still be a nonpolar molecule.

8.110 Explain why hydrocarbons are all essentially nonpolar substances.

8.111 Are the following molecules polar or nonpolar?
(a) HI
(b) $CHF_3$
(c) $SO_2Cl_2$
(d) $PF_3$

8.112 Are the following molecules polar or nonpolar?
(a) $BF_3$               (c) $CH_2Cl_2$

8.113 For each pair of molecules decide which molecule is polar and explain why it is polar while the other is not.
(a) $SO_2$, $CO_2$       (c) $SeCl_2$, $BeCl_2$
(b) $SO_2$, $SO_3$       (d) $CH_4$, $CH_3I$

8.114 Explain why the first molecule of each pair is polar and the second is not.
(a) $CH_2Cl_2$, $CCl_4$     (c) $BF_2Cl$, $BF_3$
(b) $PF_3$, $BF_3$         (d) $SO_2$, $O_3$

8.115 Which of the following molecules would align with an electric field if placed in a capacitor? Assume the bonds are polar.

   **A**       **B**       **C**       **D**

8.116 Which of the following molecules would align with an electric field if placed in a capacitor? Assume the bonds are polar.

   **A**       **B**       **C**       **D**

8.117 Which molecule, $CF_4$ or $CCl_2F_2$, is most likely to be soluble in water?

8.118 Which substance, $SO_2$ or $CO_2$, is likely to be more soluble in water?

8.119 Which of these molecules is polar? Assume the bonds are polar.

   **A**       **B**

8.120 Which of these molecules is polar? Assume the bonds are polar.

   **A**       **B**

## Additional Questions

**8.121** Draw Lewis symbols showing the valence electrons of the following atoms.
  (a) Br      (c) S      (e) Be
  (b) Pb      (d) Ca     (f) Xe

**8.122** Draw Lewis symbols showing the valence electrons of the following ions.
  (a) $P^{3-}$    (c) $Se^{2-}$   (e) $C^{4-}$
  (b) $In^{3+}$   (d) $Be^{2+}$

**8.123** Arrange the following atoms in order of decreasing electronegativity: Br, Cl, F, I.

**8.124** Which of the following substances is likely to be ionic?
  (a) $Cl_2$    (c) $BH_3$    (e) $SiO_2$
  (b) $Na_2O$   (d) $ICl$

**8.125** Classify each of the following substances according to the primary type of bonding.
  (a) HCN       (c) $P_4$     (e) $MnCl_2$
  (b) $AuCl_3$  (d) $SO_2$

**8.126** Draw a Lewis structure for each of the following.
  (a) $HClO_2$   (c) $HClO_4$   (e) $ClO_3^-$
  (b) $HClO_3$   (d) $BrF_3$

**8.127** Draw a Lewis structure for each of the following molecules.
  (a) $N_2H_2$   (c) $AsF_3$    (e) CO
  (b) $CS_2$     (d) $CO_2$

**8.128** Draw a Lewis structure, including the resonance forms, for each of the following molecules or ions.
  (a) $OCN^-$    (c) $ClO_2^-$   (e) $H_2CO_3$
  (b) $N_3^-$    (d) $PO_4^{3-}$

**8.129** Draw the best Lewis structure for each of the following. Include any necessary resonance forms.
  (a) $ClO_4^-$   (c) $NCO^-$    (e) $BF_3$
  (b) $NO_2^-$    (d) $HCO_2^-$

**8.130** Gaseous aluminum chloride exists as a dimer, $Al_2Cl_6$, with two chlorines bridging between the two aluminum atoms: $Cl_2AlCl_2AlCl_2$. Draw a Lewis structure for $Al_2Cl_6$ and for $AlCl_3$. Discuss a reason why $AlCl_3$ forms $Al_2Cl_6$.

**8.131** Describe the molecular shape of each of the following.
  (a) $SiCl_4$   (d) $IO_3^-$   (g) $GeH_4$   (j) $ClO_2^-$
  (b) $GaCl_3$   (e) $PCl_4^+$  (h) $SOCl_2$
  (c) $NCl_2^+$  (f) $OF_2$     (i) $Br_2O$

**8.132** Describe the structure and bonding in sulfuric acid, $H_2SO_4$, and in its two ions, $HSO_4^-$ and $SO_4^{2-}$.

**8.133** Decide which of each pair of gaseous molecules is polar, and explain why that molecule is polar while the other is not.
  (a) $BeCl_2$, $OCl_2$       (c) $BCl_3$, $AsCl_3$
  (b) $PH_3$, $BH_3$          (d) $SiH_4$, $NH_3$

**8.134** Which of the following are nonpolar molecules, even though they have polar bonds?
  (a) $CH_2Cl_2$   (c) $SCl_2$    (e) $PBr_3$
  (b) $SiCl_4$     (d) $ClO_2$

**8.135** For each pair of molecules decide which molecule is most polar and explain your answer.
  (a) $CCl_4$, $CH_2Cl_2$      (c) $NF_3$, $NH_3$
  (b) $CH_3F$, $CH_3Br$        (d) $OF_2$, $H_2O$

**8.136** Based on predicted bond angles from VSEPR theory, why is the ozone molecule, $O_3$, not a cyclic molecule?

**8.137** Predict whether each of the following substances is likely to have a relatively high or low boiling point.
  (a) $N_2$      (d) $MgBr_2$
  (b) $NO_2$     (e) $NH_3$
  (c) KCl

**8.138** Predict whether each of the following substances is likely to have a relatively high or low melting point.
  (a) $SO_2$     (d) $Na_2SO_4$
  (b) $BaCl_2$   (e) $CH_4$
  (c) $F_2$

**8.139** Identify the bonding in each of the following substances as ionic, covalent, or both.
  (a) $Br_2$
  (b) $Na_2SO_4$
  (c) KOH

**8.140** Identify the bonding in each of the following substances as ionic, covalent, or both.
  (a) $Mg(NO_3)_2$
  (b) $SO_3$
  (c) $K_3PO_4$

**8.141** There are two different alcohols with the formula $C_3H_7OH$. Draw Lewis structures for the two alcohols.

**8.142** The proteins in our bodies are built from small organic molecules called amino acids. An amino acid is a class of organic compound that contains both an amine functional group and a carboxylic acid functional group. Draw the Lewis structure for an amino acid that has one $-CH_2-$ sandwiched between $-NH_2$ and $-CO_2H$ groups.

**8.143** How many single, double, or triple bonds are in an HCN molecule? How many unshared electrons are present?

**8.144** Plastic food storage containers are often made of polypropylene. The compound used to make polypropylene has the formula $CH_3CHCH_2$. How many single, double, or triple bonds are present in this molecule?

**8.145** Draw the Lewis structure for a ketone containing four carbon atoms and eight hydrogen atoms.

**8.146** Compare the molecular shape around each carbon atom and $H-C-C$ bond angles in $C_2H_6$, $C_2H_4$, and $C_2H_2$. How are these compounds classified?

## Concept Review Questions

**8.147** Which compound contains both covalent and ionic bonds?
  A. $NaHSO_4$
  B. NaF
  C. HF
  D. NaH
  E. $SO_2Cl_2$
  Identify the type of bond in each of the other compounds.

8.148    The bonds in $O_3$ are expected to be
   A. ionic because of a large difference in electronegativity.
   B. nonpolar covalent because of no difference in electronegativity.
   C. polar covalent because of no difference in electronegativity.
   D. polar covalent because of a moderate difference in electronegativity.
   E. nonpolar covalent because of a large difference in electronegativity.

   Describe why you did not choose the other responses.

8.149    Which of the following is a true statement about ionic crystals?
   A. The strong electrostatic forces between ions results in low melting points.
   B. The strong attractive forces between ions cause the crystal to be malleable.
   C. The static arrangement of ions in an ionic crystal causes ionic compounds to be poor electrical conductors.
   D. The arrangement of ions in an ionic crystal maximizes repulsive forces.
   E. The arrangement of ions in an ionic crystal minimizes attractive forces.

   For each false statement, modify it so that it's a true statement.

8.150    Which of the following bonds is most polar?
   A. C−F
   B. C−Cl
   C. C−S
   D. C−H
   E. F−F

   For each incorrect answer that is a polar bond, identify the atom that has a partial negative charge.

8.151    Which of the following always violates the octet rule?
   A. a molecule or ion with sulfur as the central atom
   B. a molecule with a central atom from the third period
   C. a molecule with an odd number of electrons
   D. a molecule with carbon as the central atom
   E. a molecule containing multiple bonds

   For each of the responses not selected, provide an example of a compound that illustrates why that response is not correct.

8.152    Identify the main-group element X that could form this compound:

$$:\overset{\cdot\cdot}{\underset{}{Cl}}:$$
$$|$$
$$:\overset{\cdot\cdot}{\underset{\cdot\cdot}{Cl}}-X-\overset{\cdot\cdot}{\underset{\cdot\cdot}{Cl}}:$$

   A. C
   B. O
   C. N
   D. F
   E. B

   Explain why each of the other elements could not form this compound.

8.153    Which of the following has a Lewis structure most like that of $SO_3^{2-}$?
   A. $SO_4^{2-}$
   B. $NO_3^-$
   C. $SO_3$
   D. $NF_3$
   E. $CO_3^{2-}$

   Why does each of the other species have a different Lewis structure?

8.154    Which of the following has a double bond?
   A. $N_2H_4$
   B. $C_2H_2$
   C. $C_2H_4$
   D. $C_2H_6$
   E. $N_2$

   For each incorrect answer, describe the bonding between carbon atoms or nitrogen atoms.

8.155    Which of the following statements about resonance is true?
   A. Resonance is needed when a single Lewis structure does not provide an adequate representation of the bonding.
   B. Resonance describes the periodic vibration of electrons.
   C. Resonance describes a less-stable configuration than does any one contributing resonance structures.
   D. Resonance describes the bonding as oscillating between the contributing resonance structures.
   E. The contributing resonance structures differ in the arrangement of atoms.

   Explain what is wrong with the incorrect statements.

8.156    Which of the following molecules is an ether?
   A. $H_3C-CH_2-OH$

   B. $H_3C-\overset{\overset{\displaystyle O}{\|}}{C}-H$

   C. $H_3C-\overset{\overset{\displaystyle O}{\|}}{C}-CH_3$

   D. $H_3C-\overset{\overset{\displaystyle O}{\|}}{C}-OH$

   E. $H_3C-O-CH_3$

   Classify each of the other molecules.

8.157    Which of the following molecules has a bent molecular shape with bond angles of 120°?
   A. $H_2O$
   B. $SO_2$
   C. $SF_2$
   D. $CO_2$
   E. $H_2CO$

   For each molecule that is not correct, describe its parent structure, bond angles, and molecular shape.

8.158    Which of the following molecules is polar?
   A. $CO_2$
   B. $SO_2$
   C. $SO_3$
   D. $BH_3$
   E. $O_2$

   Explain why the other molecules are not polar.

# The Gaseous State

9.1 The Behavior of Gases

9.2 Factors That Affect the Properties of Gases

9.3 The Ideal Gas Law

9.4 Kinetic-Molecular Theory of Gases

9.5 Gases and Chemical Reactions

Math Toolbox 9.1 Graphing

Math Toolbox 9.2 Solving Simple Algebraic Equations

Chapter Review

Questions and Problems

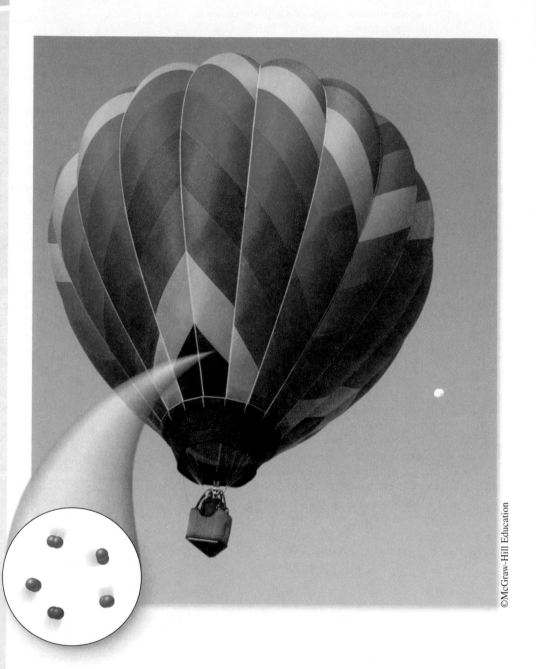

Joel is a student in an atmospheric science course. For a class project, he and his lab team are launching a weather balloon. Their balloon is filled with helium and made of a latex material that expands as external pressure decreases. The equipment attached to the balloon includes a number of instruments that Joel and his group will use to monitor atmospheric pressure and temperature as the balloon rises. Joel's project is part of a larger scientific investigation. His launch time is scheduled to synchronize with the release of hundreds of other similar balloons from stations all over the planet (Figure 9.1). After that, the team will work throughout the day to collect and analyze data. Come evening, they will have earned a rest and a small celebration. They plan to get together at Joel's house for burgers cooked on his gas grill.

But first comes the launch. Joel and his team know what to expect. After the release, their balloon will ascend through the lowest layer of Earth's atmosphere. The region from the surface to about 11 km (6.8 mi) above it is the *troposphere* (Figure 9.2). From Earth's surface to the outer edge of the troposphere, the pressure drops by around 80%. There is also a steady decrease in temperature, about 7°C (or 7 K) for every kilometer increase in altitude. Although Joel and the other observers can't see it, the balloon expands as it rises through the atmosphere.

Beyond the troposphere lies the *stratosphere*, a region that extends to about 50 km (31 mi) in altitude above Earth's surface. As the balloon enters it, pressure continues to drop, and the balloon expands even more. Unfortunately, a weather balloon cannot withstand the atmospheric conditions in the upper stratosphere. When it reaches about 30 km (19 mi), the latex stretches beyond its capacity and the balloon bursts. As the data-gathering equipment and transmitter fall, a small parachute opens to ease the instruments' landing. Joel and his group hope they can recover their equipment.

The instruments attached to Joel's weather balloon transmit data about temperature, relative humidity, and pressure. These characteristics of the atmosphere are associated with the behavior of gases. Of the three physical states of matter—gaseous, liquid, and solid—the gaseous state is the simplest, because many properties of gases are the same, regardless of the chemical makeup of the particular gas.

The atmosphere is a natural resource that we too often take for granted. Gases in the atmosphere are necessary to our life and well-being. Both oxygen gas and carbon dioxide gas, for example, are vital to the maintenance of life. Plants produce oxygen that can be consumed by animals. Carbon dioxide plays a complementary

**FIGURE 9.1** Weather balloons are released simultaneously from hundreds of locations around the planet to forecast and report weather conditions.
©David R. Frazier

List all the ways you can change the size of a balloon containing air. Keep your list handy as we discuss the behavior of gases in this chapter.

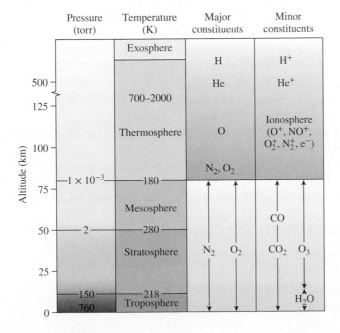

**FIGURE 9.2** The troposphere is a small portion of the atmosphere surrounding Earth. Notice what happens to temperature and pressure as the distance from the surface increases.

**TABLE 9.1** Volume Percent of Gases in the Atmosphere

| Gas | Volume Percent | Gas | Volume Percent | Gas | Volume Percent |
|---|---|---|---|---|---|
| $N_2$ | 78.09 | $CH_4$ | 0.00015 | $O_3$ | 0.000002 |
| $O_2$ | 20.94 | Kr | 0.0001 | $NH_3$ | 0.000001 |
| Ar | 0.93 | $H_2$ | 0.00005 | $NO_2$ | 0.0000001 |
| $CO_2$ | 0.032 | $N_2O$ | 0.000025 | $SO_2$ | 0.00000002 |
| Ne | 0.0018 | CO | 0.00001 | $H_2O$ | Varies |
| He | 0.00052 | Xe | 0.000008 | | |

Carbon dioxide helps to regulate breathing. A common malfunction of the breathing mechanism is hiccups. One cure involves breathing into a paper bag, which builds up the level of $CO_2$ in the bloodstream, stimulating normal breathing. The hiccups disappear. High levels of carbon dioxide in the body are not beneficial, however. Breathing air that contains more than 10% $CO_2$ causes humans to lose consciousness, and continued exposure to high $CO_2$ levels can lead to respiratory failure and death.

If you live in a metropolitan area, air around your city contains many other gases that make up smog. These gases include $NO_2$ and $SO_2$. Small particles like dust also contribute to smog.

The volume percent of water in air is different from relative humidity, which is also a percentage. Relative humidity is a measure of the mass of water vapor present relative to the mass of water needed to saturate the air at a given temperature.

role; it is exhaled by animals and used as a raw material for food production in plants. Although we exhale carbon dioxide as a waste product, it is important to us. Its production during the breakdown of nutrients in the body helps to maintain proper blood acidity, which is essential for good health.

The atmosphere is estimated to have a total mass of about $5.2 \times 10^{18}$ kg, only about 0.03% of the mass of Earth. Nevertheless the gases in the atmosphere are attracted to Earth by gravity. The levels of the atmosphere differ in their composition (Figure 9.2). The troposphere contains most of the atmosphere's gas molecules, primarily $N_2$, $O_2$, $CO_2$, $H_2O$, and Ar. The stratosphere contains, in addition to other gases, some ozone, $O_3$. Ozone is toxic to humans and animals when they breathe it, and it is responsible in part for photochemical smog at low altitudes. In the stratosphere, however, ozone absorbs harmful ultraviolet radiation from the Sun before it reaches Earth. We'll discuss more about ozone in Chapter 12.

Table 9.1 shows the composition of the atmosphere at ground level. The principal components are nitrogen and oxygen, which make up 99% of the substances in air. Air pollution or humidity may cause the composition to differ from that given in the table. The amount of water vapor in the air varies considerably from one location to another, from as high as 5% of the total volume in hot humid areas such as the tropics to as low as 0.01% in cold areas such as polar regions and in dry areas such as deserts.

In this chapter we will examine the physical properties of gases, especially how they behave when subjected to pressure or temperature changes. This behavior is described by several laws, called the gas laws, which allow us to predict how gases respond to changes in their environment. To explain the gas laws, we will develop a model for the behavior of gases at an atomic or molecular level. As you read this chapter, consider the following questions.

## Questions for Consideration

**9.1** What are some general properties of gases?

**9.2** How does the behavior of gases vary with changes in pressure, temperature, volume, and number of molecules (atoms)?

**9.3** What are the mathematical relationships between volume, pressure, temperature, and amount of gas?

**9.4** What is the theory that explains the behavior of gases in terms of atomic or molecular motion?

**9.5** How can quantities of gases in chemical reactions be calculated?

 **Math Tools Used in This Chapter**

Units and Conversions (Math Toolbox 1.3)

Mole Quantities (Math Toolbox 4.1)

Graphing (Math Toolbox 9.1)

Solving Simple Algebraic Equations (Math Toolbox 9.2)

# 9.1 The Behavior of Gases

Joel's weather balloon responds to changing conditions because of the unique properties of gases. The gases found in the atmosphere—whether major components like $O_2$ and $N_2$, or minor ones like CO and Ar—share some common properties. Among other things, gases can be described in terms of pressure, volume, temperature, and the amount of the gas (number of particles). If one of these properties changes, one or more of the others will change also.

Gases can also be described in terms of density. Recall from Chapter 1 that density is mass per unit volume. Compared to liquids and solids, all gases have relatively low densities. Their densities explain, in part, why they exist above Earth and bubble up through liquids. The density of oxygen, for example, is 0.0013 g/mL under normal conditions of temperature (20°C) and atmospheric pressure (sea level). By comparison, the density of liquid water is 1.0 g/mL. Gases can be compressed into a container, such as the fuel tank of Joel's gas grill, if we exert a force on them. In Joel's grill, propane is compressed to the liquid state. When gases are stored under pressure, their densities increase.

Such properties are easy to observe, but what is going on with gas particles that might explain some of the changes we can see and measure? The properties of gas particles—whether atoms or molecules—explain why, among other things, gases have low densities and are compressible.

- *Gases consist of particles that are relatively far apart (as compared to solids and liquids).* The large spaces between particles explain why gases have much lower densities than liquids and solids. Because gases have so much empty space, it is easy to compress them to smaller volumes by applying external pressure.
- *Gas particles move about rapidly.* For example, the average velocity of oxygen gas molecules at 20°C and normal atmospheric pressure is 0.44 km/s (980 mi/h).
- *Gas particles have little effect on one another unless they collide.* Gas particles have only a small attraction for one another at normal temperatures and pressures, so they move about freely in all directions. When they collide, gas particles simply bounce off of each other.
- *Gases expand to fill their containers.* Gases take the volume and shape of their containers as a consequence of their random motion.

## Temperature and Density

Now let's apply the properties of gas particles to some other observations about gas behavior. All gases expand if heated and contract if cooled. This is one of the properties that cause winds in our atmosphere. Solar heating causes gases in the atmosphere to expand, creating currents that influence weather conditions. Why do gases expand when heated? The additional heat increases the kinetic energy of gas particles, making them move faster. The gas particles move farther away from each other because of the faster motion. Consider the representations for gas particles at different temperatures shown in Figure 9.3. At the warmer temperature, the gas particles are farther apart. In an open system, where gases are free to expand or contract, the number of gas particles changes within a given volume if the temperature changes. At warmer temperatures gas particles move farther apart, so there are fewer gas particles in a given volume. Consequently, warm gases have lower densities. In the atmosphere, gases at the surface are warmer than those above them. The warmer gas particles with lower densities rise, displacing some of the gas particles at higher altitudes. This movement of warm and cold air creates the currents that cause winds.

To further explore the properties of gases, let's consider a hot-air balloon (Figure 9.4). When the air heats up inside the filled balloon, the gas particles move faster and farther apart. The gas expands, but the balloon's volume is fixed by its inelastic fabric. The only way for the gas to expand is for some of it to escape through the opening at the bottom of the balloon. With fewer gas particles that

What is in the space between gas molecules?

Cool gas

Warm gas

**FIGURE 9.3** Gases expand when heated. At a higher temperature, gas molecules are farther apart. Since density is the ratio of mass to volume, warm gases in an open system have less mass in the same volume, so their density is less than for the same gas at a lower temperature.

You may notice the effect of different air densities if you take a shower in a cool room. The steam rises to the ceiling because the warm water vapor is less dense than the cooler gas that is higher in the room.

**FIGURE 9.4** When the gas in a hot-air balloon is heated, the gas expands, filling the balloon. The opening in the bottom allows air to move in and out of the balloon. When the balloon is completely filled, the balloon can no longer expand. As the gas in the balloon is heated further, some gas particles escape, decreasing the density of the balloon. Eventually the air in the balloon becomes less dense than the surrounding air. The balloon rises.
©McGraw-Hill Education

Which finger hurts? Use the discussion of pressure to explain your answer.

©Brian Moeskau/Moeskau Photography

are now farther apart, the density of the air inside the balloon decreases. Since substances with lower densities float on those of higher densities, the balloon rises.

## Pressure

Changes in pressure and amount of gas also affect the behavior of gases. If you've ever put air into a tire that looks low or flat, you've probably noticed that the tire expands as air is forced into it. The increase in volume of the tire results from an increase in the number of gas particles. A pressure gauge indicates when the gas inside the tire exerts enough pressure so the car can be driven safely (Figure 9.5).

As another example, consider an exercise ball used in fitness training (Figure 9.6). The gas particles within the ball exert a pressure that supports the body's weight.

When we blow up a balloon, it inflates because of the increasing number of gas particles colliding against its wall. Gas particles are in constant motion and continuously bounce off the walls of the container. Each time a particle hits the wall, it exerts a force against the container (Figure 9.7). **Pressure** (*P*) is the amount of force applied per unit area:

$$\text{Pressure } (P) = \frac{\text{force}}{\text{area}}$$

**FIGURE 9.5** When a flat tire is inflated, gas particles fill the tire and push against its sides. The proper amount of pressure allows the car to be driven safely.
©Brian Moeskau/Moeskau Photography

**FIGURE 9.6** In an exercise ball, air pressure is great enough to support the weight of an adult body.
©Stockbyte/PunchStock/Getty Images

We can describe pressure as the force of gas particles hitting the walls of the container divided by the surface area of the container:

$$\text{Pressure } (P) = \frac{\text{force of gas particles}}{\text{area of container}}$$

To distinguish between force and pressure, consider hammering a nail into a piece of wood. If we exert the right amount of force with the hammer on the nail as shown in Figure 9.8A, the nail will penetrate the board. However, if we turn the nail upside down, it will not go into the board, even if we exert the same force (Figure 9.8B). The pressure exerted is significantly reduced when the nail is upside down because the area of the nail head is larger than the area of the tip.

A moderate force can exert a tremendous pressure if the force is applied to a small area. For example, suppose a 115-lb person steps on an upright nail. If the nail point has an area of $0.0100 \text{ in}^2$, the pressure generated is

$$\text{Pressure } (P) = \frac{\text{force}}{\text{area}} = \frac{115 \text{ lb}}{0.0100 \text{ in}^2} = 11{,}500 \text{ lb/in}^2$$

That's enough pressure to push the nail into the person's foot.

Like a hammer pounding a nail, gases exert a force on the walls of a container. Consider the images shown in Figure 9.9. In Figure 9.9A, fewer gas atoms are present in the same size container. Fewer atoms exert less force per unit area on the walls of the container because there are fewer collisions per unit time. The result is a lower pressure. More atoms, as shown in Figure 9.9B, exert more force because there are more collisions per unit time.

We can relate the pressure that gases exert on the wall of a container to blood pressure. When the heart beats, blood exerts pressure on artery walls in the circulatory system. Blood pressure is typically measured using an inflatable cuff wrapped around the upper arm. As the cuff inflates, gas particles within it exert pressure on the blood vessels in the arm, temporarily cutting off circulation. As air is released from the cuff, blood begins to flow through the arm again, and its pressure can be measured.

Pressure can be expressed in several ways. The one that we encounter most often in our daily lives is pounds per square inch (lb/in² or psi). Many automobile tires are inflated to a pressure of 30 psi. Bicycle tires are inflated to pressures between 60 and 100 psi. The pressure exerted by air molecules in the atmosphere is much smaller than what we create inside tires—only about 14.7 psi. However, the pressure of the atmosphere is significant, as Figure 9.10 shows. A vacuum pump attached to the can removes air from the can. The pressure of the gases outside the evacuated can is great enough to crush it.

We encounter another measure of pressure in our daily weather reports. For example, a forecaster might say, "Today's barometric pressure is 29.95 inches and falling." How can inches describe pressure? A device called a **barometer** is used to measure atmospheric pressure. The original barometer was a tube closed at one end, filled with mercury, and then inverted into a container of mercury. The height of the mercury column rises or falls so that the pressure caused by the weight of the

**FIGURE 9.7** The packed structure on the right represents the rigid walls of a container made of glass ($SiO_2$). Gas particles within the container are in constant motion, exerting a force on the glass wall when they hit it.

A    B

**FIGURE 9.8** In both cases the hammer exerts the same force on the nail. When the nail is upside down, the force is spread over a larger area.

**FIGURE 9.10** When the air inside a metal can is evacuated with a vacuum pump, the pressure of the outside air crushes the can.
©McGraw-Hill Education/Stephen Frisch

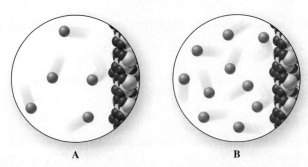

A        B

**FIGURE 9.9** (A) Gas particles exert pressure on the wall of the glass container. (B) With a greater number of gas particles, pressure increases because of the increased number of collisions over a given area.

Vacuum above
mercury column

Pressure due to
mercury column

Atmospheric
pressure, $P_{atm}$

$\Delta h = 760$ mm Hg
(29.9 in)

Dish filled with mercury

**FIGURE 9.11** Mercury barometers can be used to measure atmospheric pressure. The pressure of the atmosphere on the mercury in the basin forces the column of mercury to a height ($\Delta h$) that indicates the atmospheric pressure.

mercury column equals the pressure exerted by the atmosphere on the mercury pool, as shown in Figure 9.11. Thus, changes in atmospheric pressure cause changes in the height of the mercury column. The greater the height of the mercury in the column, the higher the atmospheric pressure. Although modern barometers are designed differently, units based on the height of the mercury column in a barometer, measured in inches, are commonly used to express atmospheric pressures in the United States.

The pressure of the atmosphere at sea level under normal weather conditions corresponds to a column of mercury that is 29.9 in (76.0 cm) tall. This pressure is the basis for a unit, the standard *atmosphere* (atm), which relates to barometric units:

$$1 \text{ atm} = 29.9 \text{ in Hg} = 76 \text{ cm Hg} = 760 \text{ mm Hg}$$

The pressures of gases in chemical systems are often fractions of 1 atm, so they are commonly measured in units of mm Hg, which has been given the name *torr*. Since 1 torr is the same as 1 mm Hg, the torr is related to the unit of atmospheres:

$$1 \text{ atm} = 760 \text{ torr} = 760 \text{ mm Hg}$$

In this book we will generally use units of either atm or torr, since they are most useful for the kinds of measurements chemists make. However, we sometimes need to convert to other units. The official SI unit of pressure is the *pascal* (Pa). This unit is related to the pressure in atmospheres:

$$1 \text{ atm} = 101,325 \text{ Pa}$$

Atmospheric pressure in weather reports broadcast in countries that use the metric system is expressed in hectapascals or kilopascals. Pounds per square inch can also be used to relate to atmospheres:

$$1 \text{ atm} = 14.7 \text{ lb/in}^2$$

Conversions among these units are most easily carried out by using their relationship to a standard atmosphere, as Example 9.1 demonstrates.

---

**EXAMPLE 9.1**  |  **Converting Pressure Units**

Express the pressure 735 torr in units of atmospheres and pascals.

**Solution:**

We want to convert pressure units from torr to atmospheres:

Pressure in torr $\xrightarrow{?}$ Pressure in atm

The relationship between torr and atm is

$$1 \text{ atm} = 760 \text{ torr}$$

This relationship can be used as a conversion factor as described in Math Toolbox 1.3:

$$P_{atm} = 735 \text{ torr} \times \frac{1 \text{ atm}}{760 \text{ torr}} = 0.967 \text{ atm}$$

We can convert to pascals by using the following conversion factor:

$$1 \text{ atm} = 101,325 \text{ Pa}$$

$$P_{Pa} = 0.967 \text{ atm} \times \frac{101,325 \text{ Pa}}{1 \text{ atm}} = 9.80 \times 10^4 \text{ Pa}$$

**Consider This 9.1**

At an elevation of 5431 feet, Denver's average atmospheric pressure is around 0.822 atm. What is this pressure expressed in units of torr? Would you expect to observe an atmospheric pressure of 735 torr closer to sea level or closer to Denver?

MATH
TOOLBOX
**1.3**

The unit *torr* is used to credit the scientist, Evangelista Torricelli (1608–1647), who discovered the principle associated with the barometer. At the time of his discovery, one of his supporters, Pascal, took a barometer to the top of a mountain in France, measured the pressure at the top, and compared it to the pressure reading taken from a barometer at the bottom. As expected, the pressure on the mountaintop was less than the pressure at the bottom because the density of the atmosphere is lower at higher altitudes.

## 9.2 Factors That Affect the Properties of Gases

Joel's weather balloon encounters changes in temperature, pressure, and volume as it rises from the surface of Earth to higher altitudes. Each of these factors affects the behavior of a fixed amount of gas. Another factor is the number of gas particles present. In this section we'll describe the relationships among these factors that allow us to make quantitative predictions about gases.

### Volume and Pressure

Joel and his team measure changes in atmospheric pressure as their weather balloon rises. The pressure change affects the volume of the balloon. Notice in Figure 9.12 that a balloon expands when atmospheric pressure decreases. Why does this occur? To maintain equal pressures inside and outside the balloon, the gas particles spread out. The only way they can do this in an enclosed balloon is if the volume of the balloon is increased.

As another example, what happens to the size of small gas bubbles released at the bottom of a large fish tank (Figure 9.13)? If you've ever jumped into deep water in a pool or lake, you've felt an increase in pressure below the surface of the water. Bubbles in a deep fish tank experience the same phenomenon. As a bubble rises, the pressure on it decreases, since there is not as much water pressing down on it. The size of the bubble therefore increases as it rises.

The effect of pressure on gas volume can also be observed in a piston as shown in Figure 9.14. In this system a fixed amount of gas is sealed inside. When the piston is pushed down, external pressure is applied to the gas within the container. Once the piston is still, the pressure exerted by the piston is equal to the pressure on the inside, so the pressure the trapped gas exerts on the inside walls of the container has increased.

If the volume and pressure are measured as the gas is compressed, these quantities can be plotted on a graph as shown in Figure 9.15. From this graph, can you determine a relationship between volume and pressure? What happens to the volume

**FIGURE 9.12** As the weather balloon in Figure 9.1 ascends, it expands to maintain equal pressures inside and outside the balloon. Because of decreased atmospheric pressure, the pressure of the gas within the balloon decreases as the balloon expands and the volume of the balloon increases. The pressures inside and outside the balloon then become equal. Eventually the balloon expands so much that it bursts.

©David R. Frazier

**FIGURE 9.13** Bubbles in a fish tank increase in size as they move toward the surface of the water. Bubbles on the bottom experience a greater pressure from the surrounding water molecules than those at the top.

©Juniors Bildarchiv GmbH/Alamy Stock Photo

**B**

**FIGURE 9.14** (A) Gas atoms in a cylinder with a movable piston. (B) When the piston moves down, the volume decreases and atoms move closer together, exerting a greater pressure on the walls of the cylinder.

**FIGURE 9.15** This graph shows the relationship of volume and pressure for a gas at constant temperature. What happens to volume when pressure increases?

when the pressure increases? What happens to the pressure if the volume increases? What happens to particles inside a container as pressure and volume vary along the curve? Use your interpretation of the graph to answer the questions in Example 9.2.

| EXAMPLE 9.2 | **Graphical Relationship of Volume and Pressure for a Gas** |
|---|---|

The piston shown in the figure represents starting conditions for a helium gas sample. Suppose the volume and pressure correspond to point *A* on the graph. If the pressure increases by a factor of 2, what point along the curve corresponds to the new volume and pressure conditions?

**MATH TOOLBOX**

**9.1**

**Solution:**

Inspection of the graph indicates that when the pressure doubles, the volume is halved. Point *A*, the starting conditions of the gas in the piston, is about 0.8 atm in pressure and approximately 7 L in volume. If the pressure is doubled to 1.6 atm, the volume falls to about 3.5 L. This corresponds to point *C*.

**Consider This 9.2**

With the initial pressure and volume at point *A*, what happens to pressure and volume if conditions change to point *B*?

**Practice Problem 9.2**

The atomic-level image in the figure represents the atoms present in a microscopic volume under the starting conditions defined by point *A*. Draw a picture to represent the atoms in the same microscopic volume at point *C*.

**Further Practice:**  Questions 9.35 and 9.36 at the end of the chapter

---

The relationship you just examined is known as **Boyle's law:** *For a given mass of gas at constant temperature, volume varies inversely with pressure.* That is, as pressure increases, volume decreases when temperature is constant. Or, looking at the relationship another way, as volume increases, pressure decreases. To understand this relationship, consider how gas molecules behave inside a container. They are constantly moving and bouncing off the container walls. The impacts exert force and thus pressure on the inside walls. When an external pressure compresses the gas to a smaller volume, the molecules do not have to travel as far to hit the walls, so they collide with it more often. The more frequent the collisions, the more force is exerted on the walls, and the higher the resulting pressure.

If we look at the data in Figure 9.15 more quantitatively, we can see that doubling the pressure halves the volume, tripling the pressure cuts the volume to one-third its original value, and so on. We can express this relationship mathematically, using Boyle's law. The volume occupied by a gas is inversely proportional to its pressure, with *V* as volume and *P* as pressure:

$$V \propto \frac{1}{P}$$

We can also use a constant to describe an equivalent relationship:

$$V = \text{constant} \times \frac{1}{P} \quad \text{or} \quad V = \frac{\text{constant}}{P}$$

To illustrate this inverse relationship, the data shown in Figure 9.15 can be graphed in a different way. If we plot the volume versus the inverse of the pressure (1/*P*), we get a straight line as shown in Figure 9.16.

In 1622, Robert Boyle, an English chemist, discovered the relationship between pressure and volume in a gas.

**ANIMATION:** Boyle's Law Experiment

When variables are proportionally related, the symbol ∝ is used to express the mathematical relationship. For example, if *x* is directly proportional to *y*, we can express the relationship as *x* ∝ *y*. When two variables are inversely proportional, we can use a similar expression, except we must show the inverse of one of the variables:

$$x \propto \frac{1}{y} \quad \text{or} \quad y \propto \frac{1}{x}$$

For more information, see Math Toolbox 9.1.

**FIGURE 9.16**  Plotting volume against the inverse of pressure (1/*P*) gives a straight line. For more on this relationship, see Math Toolbox 9.1.

The equation that relates volume and pressure can be rearranged:

$$PV = \text{constant}$$

The constant has the same value for all values of pressure and volume, provided the amount (in terms of mass or moles) of gas and the temperature remain the same. Look again at Figure 9.15. Notice that multiplying volume by pressure at two different points gives the same value. That is, at two different sets of $P$ and $V$, the values of their products equal a constant:

$$P_1V_1 = \text{constant} \qquad P_2V_2 = \text{constant}$$

The subscripts refer to any two points on a plot of volume and pressure. If both $P_1V_1$ and $P_2V_2$ equal a constant, they must also equal each other:

$$P_1V_1 = P_2V_2$$

As long as the temperature and mass (or moles) of the gas remain constant, Boyle's law holds for many gases under normal atmospheric conditions. A gas that behaves according to predicted linear relationships is called an **ideal gas.**

**MATH TOOLBOX 9.2**

The equation just given can be used for many pressure-volume calculations. If three of the four values are known, the other can be calculated, as Example 9.3 demonstrates. (See Math Toolbox 9.2 for more information on rearranging simple algebraic expressions.)

---

**EXAMPLE 9.3** | **Pressure-Volume Relationship**

A balloon contains 512 mL of helium when filled at 1.00 atm. What would be the volume of the balloon if it were subjected to 2.50 atm of pressure?

**Solution:**

Before we use Boyle's law to calculate the final volume, let's consider the problem qualitatively. What do you expect to happen to the balloon if we apply a pressure that's greater than the original pressure? Will the volume increase or decrease? By what factor do you expect the volume to change? We would expect the volume to decrease with an increase in the pressure because pressure and volume are inversely proportional. Since pressure increases by a factor of 2.50, we would expect volume to decrease by a factor of 2.50. Let's now apply Boyle's law to calculate the final pressure. $P_1$ is 1.00 atm, $P_2$ is 2.50 atm, $V_1$ is 512 mL, and $V_2$ is unknown.

| Initial Conditions | Final Conditions |
|---|---|
| $P_1 = 1.00$ atm | $P_2 = 2.50$ atm |
| $V_1 = 512$ mL | $V_2 = ?$ |

We can use the following relationship to determine the new volume:

$$P_1V_1 = P_2V_2$$

This equation can be rearranged to solve for the unknown volume:

$$V_2 = \frac{P_1V_1}{P_2}$$

Now we can substitute $P_1$, $V_1$, and $P_2$ into the expression and solve for $V_2$:

$$V_2 = \frac{(512 \text{ mL})(1.00 \text{ atm})}{(2.50 \text{ atm})} = 205 \text{ mL}$$

---

The top photograph of an unopened bag of potato chips was taken at an elevation of 7000 ft (or 2120 m), while the bottom photograph of the same bag was taken at 1220 ft (or 370 m). Why is the bag puffier at higher elevations?

## Consider This 9.3

A common error in calculations such as this is switching the values of pressure. Always ask yourself, "Does this answer make sense?" Suppose you got an answer of 1280 mL. Explain why this answer doesn't make sense.

## Practice Problem 9.3

What pressure is needed to compress 455 mL of oxygen gas at 2.50 atm to a volume of 282 mL?

**Further Practice:** Questions 9.37 and 9.38 at the end of the chapter

## Volume and Temperature

When an inflated balloon is placed in hot sunlight, it gets larger. It may even stretch beyond its limits and burst. When an inflated balloon is placed in a bath of very cold liquid nitrogen, it deflates to a very small volume, but it regains its original volume if allowed to warm to room temperature again, as shown in Figure 9.17. Suppose the volume of a confined gas is carefully measured at constant pressure as the temperature is systematically changed (Figure 9.18). From the graph, can you determine a relationship between volume and temperature? What happens to the volume when the temperature increases? Is volume directly proportional or inversely proportional to temperature? What happens to the particles inside the container as volume and temperature vary along the plot? Use the graph to answer the questions in Example 9.4.

**FIGURE 9.17** When an air-filled balloon is cooled to 77 K in liquid nitrogen, the volume of the air inside it decreases drastically. When the balloon warms back up to room temperature, the gas regains its original volume.
©McGraw-Hill Education/Charles D. Winters

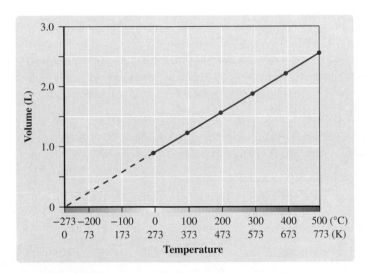

**FIGURE 9.18** What happens to the volume of an ideal gas as temperature increases?

Recall that the temperature in kelvins is equal to the Celsius temperature plus 273.15:

$$T_{\text{kelvin}} = T_{\text{Celsius}} + 273.15$$

## EXAMPLE 9.4     Graphical Relationship of Volume and Temperature for a Gas

The piston shown in the figure on the next page represents starting conditions for a helium gas sample. Pressure is held constant in this container by allowing the piston to move freely as temperature changes. Suppose the initial volume and temperature correspond to point *A* on the graph. If the temperature in kelvins increases by a factor of 2, what point along the line corresponds to the new volume and temperature conditions?

 **Student Hot Spot**

Student data indicate you may struggle with volume-temperature relationships. Access the SmartBook to view additional Learning Resources on this topic.

MATH
TOOLBOX
**9.1**

**Solution:**

The graph shows that volume increases as temperature increases. With the Celsius temperature scale, we have to consider negative values of temperature. The Kelvin temperature scale allows us to look at absolute values of temperature, avoiding negative numbers. When the temperature is 373 K, the volume is about 1.0 L. If the temperature doubles to 746 K, the volume increases to about 2.0 L. This corresponds to point C.

**Consider This 9.4**

With the initial volume and temperature at point A, what happens to volume and temperature if conditions change to point B?

**Practice Problem 9.4**

The atomic-level image in the figure represents the atoms present in a microscopic volume under the starting conditions defined by point A. Draw a picture to represent the atoms in the same microscopic volume at point C.

**Further Practice:** Questions 9.47 and 9.48 at the end of the chapter

Jacques Charles and Joseph Gay-Lussac, both French scientists, independently discovered the relationship between temperature and volume. Charles made his discovery in 1787, and Gay-Lussac completed his study in 1802. Both were interested in hot-air balloons. In 1804 Gay-Lussac took a hot-air balloon up to 23,000 ft (about 7000 m) to study the atmosphere.

Jacques Charles and Joseph Louis Gay-Lussac investigated the effect of temperature on gas volume. They found that a fixed amount of gas at a fixed pressure increases in volume linearly with temperature, as shown in Figure 9.18. The relationship between temperature and volume is known as **Charles's law:** *For a given mass of gas at constant pressure, volume is directly proportional to temperature on an absolute (Kelvin) scale.* In other words, if temperature increases, volume increases. In their calculations, Charles and Gay-Lussac did not see a direct relationship between volume and degrees Celsius. An adjustment of the temperature in degrees Celsius to kelvins avoids negative temperatures and provides a straightforward mathematical relationship between volume and temperature. This can be expressed mathematically with the expression

$$V \propto T$$

ANIMATION: Charles's Law

We can use a constant to describe an equivalent relationship:

$$V = \text{constant} \times T \quad \text{or} \quad \frac{V}{T} = \text{constant}$$

Figure 9.18 shows that dividing volume by temperature at two different points gives the same value. That is, at two different sets of $V$ and $T$ values, their ratios equal a constant (when the pressure is constant):

$$\frac{V_1}{T_1} = \text{constant} \qquad \frac{V_2}{T_2} = \text{constant}$$

The subscripts refer to any two points on a plot of volume and temperature. If both ratios equal a constant, they must also equal each other:

$$\frac{V_1}{T_1} = \frac{V_2}{T_2}$$

This equation can be used to calculate the change in the volume occupied by a fixed amount of gas at constant pressure resulting from a change in the absolute temperature of the gas. It also allows us to calculate the temperature change required to achieve a specific change in volume. If any three values are known, the fourth can be calculated. (See Math Toolbox 9.2 for more information on rearranging simple algebraic expressions.)

> At absolute zero, 0 K, all motion would cease and the volume of an ideal gas would be zero. This temperature has not been achieved. At normal pressures, gases condense or solidify as absolute zero is approached.

> It is essential in these calculations that the temperature is expressed in units of kelvins.

## EXAMPLE 9.5  Volume-Temperature Relationship

If a sample of chlorine gas occupies 50.0 mL at 100.0°C, what is its volume at 25.0°C at constant pressure?

### Solution:

Before we use Charles's law to calculate the final volume, let's consider the problem qualitatively. What do you expect to happen to the volume of the balloon when the temperature decreases? Since the gas particles slow down at the lower temperature, their kinetic energy decreases, and the volume of the balloon decreases to maintain a constant pressure. To calculate the final volume we use Charles's law, remembering that we must convert to the Kelvin temperature scale. $V_1$ is 50.0 mL, $V_2$ is unknown, $T_1$ is 100.0°C + 273.15 = 373.2 K, and $T_2$ is 25.0°C + 273.15 = 298.2 K.

| Initial Conditions | Final Conditions |
|---|---|
| $V_1 = 50.0$ mL | $V_2 = ?$ |
| $T_1 = 373.2$ K | $T_2 = 298.2$ K |

We can use the following relationship to determine the new volume:

$$\frac{V_1}{T_1} = \frac{V_2}{T_2}$$

This equation can be rearranged algebraically to solve for the unknown volume:

$$V_2 = \frac{V_1 T_2}{T_1}$$

MATH
TOOLBOX
**9.2**

Now we can substitute $T_1$, $V_1$, and $T_2$ into the expression and solve for $V_2$:

$$V_2 = \frac{(50.0 \text{ mL})(298.2 \cancel{K})}{(373.2 \cancel{K})} = 40.0 \text{ mL}$$

As usual, we should ask if the answer makes sense. It does, based on the prediction that $V_2$ should be less than $V_1$ when the temperature decreases.

**Consider This 9.5**

What if the temperature drops by another 75.0°C from 25.0°C to −50.0°C, and a negative value is calculated for the new volume? How would such an answer that doesn't make physical sense be obtained? What is the correct value for the new volume?

**Practice Problem 9.5**

A sample of carbon monoxide gas occupies 150.0 mL at 25.0°C. It is then cooled at constant pressure until it occupies 100.0 mL. What is the new temperature in degrees Celsius?

**Further Practice:** Questions 9.49 and 9.50 at the end of the chapter

## Volume, Pressure, and Temperature

The volume of Joel's weather balloon increases as it ascends to higher altitudes. Can the gas laws we've discussed so far explain this change? To answer the question, we need to consider how pressure, temperature, and volume relate. Recall that at constant temperature, volume is inversely proportional to pressure:

$$V \propto \frac{1}{P}$$

At constant pressure, volume is directly proportional to temperature:

$$V \propto T$$

The interdependence of these variables can be summarized by the **combined gas law,** which states that for a constant amount of gas, volume is proportional to absolute temperature divided by pressure:

$$V \propto \frac{T}{P}$$

Because of the proportionality relationship, we can use a constant to relate volume to temperature and pressure:

$$V = \text{constant} \times \frac{T}{P} \quad \text{or} \quad \frac{PV}{T} = \text{constant}$$

The relationship between $P$, $V$, and $T$ holds for two sets of conditions (when the amount of gas is constant):

$$\frac{P_1V_1}{T_1} = \text{constant} \qquad \frac{P_2V_2}{T_2} = \text{constant}$$

As usual, for any two sets of conditions, if the quantities equal a constant, they must also equal each other:

$$\frac{P_1V_1}{T_1} = \frac{P_2V_2}{T_2}$$

**🔊 Student Hot Spot**

Student data indicate you may struggle with pressure-volume-temperature relationships. Access the SmartBook to view additional Learning Resources on this topic.

Now that we have a relationship between pressure, temperature, and volume, let's see if we can explain what happens to Joel's weather balloon. Recall that temperature decreases with an increase in altitude. If we assume that the temperature at the surface is about 300 K (approximately room temperature), we see in Figure 9.2 that the temperature decreases to around 218 K (−55°C) at the upper limit of the troposphere. If pressure were held constant, we'd expect the volume of the gas inside the weather balloon to decrease with the drop in temperature. However, pressure is not constant. Figure 9.2 shows that it decreases from 760 to 150 torr as the balloon rises its first 11 km.

We can use the combined gas law to rationalize the observed change in volume for the weather balloon. Rearranging this gas law and solving for $V_2$ gives

MATH
TOOLBOX
**9.2**

$$V_2 = V_1 \times \frac{P_1}{P_2} \times \frac{T_2}{T_1}$$

We can substitute these quantities into the combined gas law and calculate how the volume changes with respect to the original volume:

$$V_2 = V_1 \times \frac{760 \text{ torr}}{150 \text{ torr}} \times \frac{218 \text{ K}}{300 \text{ K}} = 3.7 V_1$$

The volume of the balloon should increase by a factor of 3.7.

The combined gas law results from merging Boyle's law and Charles's law. If temperature is held constant ($T_2 = T_1$), the temperature terms cancel, giving Boyle's law:

$$\frac{P_1 V_1}{T_1} = \frac{P_2 V_2}{T_2}$$

$$\frac{P_1 V_1}{T_1} = \frac{P_2 V_2}{T_2}$$

$$P_1 V_1 = P_2 V_2$$

If pressure is held constant ($P_2 = P_1$), the pressure terms cancel, giving Charles's law:

$$\frac{P_1 V_1}{T_1} = \frac{P_1 V_2}{T_2}$$

$$\frac{V_1}{T_1} = \frac{V_2}{T_2}$$

The combined gas law predicts what happens to one of the gas properties—$P$, $V$, or $T$—when one or both of the other properties change. The equation can be rearranged to any form necessary to solve for a given unknown variable, provided the other variables are known. Information on rearranging simple algebraic expressions can be found in Math Toolbox 9.2.

> We can look at this problem in another way. The pressure decreases by a factor of 5 (760/150) and temperature decreases by only a factor of 1.4 (300/218). The change in pressure is greater than the change in temperature, so pressure plays a more important role in explaining the change in volume of the weather balloon.

---

**EXAMPLE 9.6**     **Combined Gas Law**

If a sample of argon gas occupies 2.50 L at 100.0°C and 5.00 atm, what volume will it occupy at 0.0°C and 1.00 atm?

**Solution:**

Volume, pressure, and temperature change in this process, so we will use the combined gas law. $V_1$ is 2.50 L, $V_2$ is unknown, $P_1$ is 5.00 atm, $P_2$ is 1.00 atm, $T_1$ is $100.0°C + 273.15 = 373.2$ K, and $T_2$ is $0.0°C + 273.15 = 273.2$ K.

| Initial Conditions | Final Conditions |
|---|---|
| $V_1 = 2.50$ L | $V_2 = ?$ |
| $P_1 = 5.00$ atm | $P_2 = 1.00$ atm |
| $T_1 = 373.2$ K | $T_2 = 273.2$ K |

The combined gas law is

$$\frac{P_1 V_1}{T_1} = \frac{P_2 V_2}{T_2}$$

This equation can be rearranged to solve for the unknown volume:

$$V_2 = \frac{P_1 V_1}{T_1} \times \frac{T_2}{P_2}$$

$$= \frac{(5.00 \text{ atm}) (2.50 \text{ L})}{373.2 \text{ K}} \times \frac{273.2 \text{ K}}{1.00 \text{ atm}}$$

$$= 9.15 \text{ L}$$

### Consider This 9.6

With a decrease in temperature, volume is expected to decrease at constant pressure. What explains why there is a dramatic increase in volume from 2.50 L to 9.15 L?

### Practice Problem 9.6

A sample of hydrogen gas occupies 1.25 L at 80.0°C and 2.75 atm. What volume will it occupy at 185°C and 5.00 atm?

**Further Practice:** Questions 9.63 and 9.64 at the end of the chapter

**FIGURE 9.19** When hydrogen gas reacts with oxygen gas to form water vapor at constant temperature and pressure, the volume ratios of reactants and products are the same as the mole ratios given by the chemical equation.

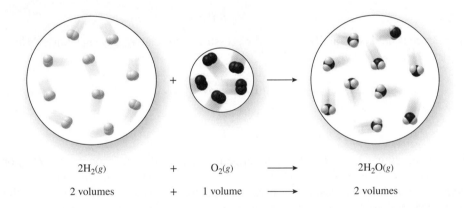

| $2H_2(g)$ | + | $O_2(g)$ | ⟶ | $2H_2O(g)$ |
|-----------|---|----------|---|------------|
| 2 volumes | + | 1 volume | ⟶ | 2 volumes |

## Gay-Lussac's Law of Combining Volumes

During his investigations of gases and balloons, Gay-Lussac also examined gaseous chemical reactions. In 1808 he measured the volumes occupied by gases before and after reactions. He found that—under conditions of constant pressure and temperature—the volumes of gaseous reactants and gaseous products always relate in a ratio of small whole numbers. For example, as shown in Figure 9.19, 2 volumes of hydrogen gas react with 1 volume of oxygen gas to produce 2 volumes of water vapor. If temperature and pressure conditions remain constant, the volume ratio of reactants and products is the same as the mole ratio, 2:1:2. **Gay-Lussac's law of combining volumes** states that gases combine in simple whole-number volume proportions at constant temperature and pressure. Although Gay-Lussac did not understand the reasons for the ratios, we can now see that the volume proportions of reacting gases are the same as the proportions of reacting molecules.

## Avogadro's Hypothesis

Amadeo Avogadro, an Italian physicist, developed his hypothesis in 1811.

Consider what happened when Joel filled his weather balloon with helium. When he pumped in the first measured amount, the balloon expanded to a small volume. A second addition of the same amount expanded the balloon to about twice that volume, assuming the number of gas particles added was the same each time. The volume, then, depends on the amount of gas blown into the balloon.

Amadeo Avogadro recognized the implications of Gay-Lussac's law of combining volumes. He hypothesized that the volume occupied by a gas at a given temperature and pressure is proportional to the number of gas particles and thus to the moles of the gas. From Figure 9.20 we can see that as the moles of gas particles ($n$) increases at constant temperature and pressure, the volume occupied by the gas also increases:

$$V \propto n$$

We can use a constant to describe an equivalent relationship between volume and amount of gas:

$$V = \text{constant} \times n \qquad \text{or} \qquad \frac{V}{n} = \text{constant}$$

Figure 9.20 reveals that dividing volume by moles of gas at two different points gives the same value. That is, at two different sets of $V$ and $n$ values, their ratios are equal to a constant (when pressure and temperature are constant):

$$\frac{V_1}{n_1} = \text{constant} \qquad\qquad \frac{V_2}{n_2} = \text{constant}$$

The subscripts refer to any two points on a plot of volume and moles of gas. If both ratios are equal to a constant, they must also equal each other:

$$\frac{V_1}{n_1} = \frac{V_2}{n_2}$$

This equation can be used to calculate the change in the volume of a gas when the amount of gas changes at constant pressure and temperature. Why would the volume change in this manner?

**MATH TOOLBOX**
**9.1**

## EXAMPLE 9.7    Volume and Amount of Gas

If a 10.0-L balloon contains 0.80 mol of a gas, what will be the volume of a balloon that contains 0.20 mol of the gas if temperature and pressure remain constant?

**Solution:**

Before we calculate the final volume, let's consider the problem qualitatively. What do you expect to happen to the volume of the balloon when the number of gas particles decreases? Since there are fewer gas particles, the volume of the balloon must decrease to maintain a constant pressure. To calculate the final volume we can use the following relationship:

$$\frac{V_1}{n_1} = \frac{V_2}{n_2}$$

$V_1$ is 10.0 L, $V_2$ is unknown, $n_1$ is 0.80 mol, and $n_2$ is 0.20 mol.

| Initial Conditions | Final Conditions |
|---|---|
| $V_1 = 10.0$ L | $V_2 = ?$ |
| $n_1 = 0.80$ mol | $n_2 = 0.20$ mol |

**MATH TOOLBOX 9.2**

Rearranging the equation to solve for the unknown volume, we get

$$V_2 = \frac{V_1 n_2}{n_1}$$

Now we can substitute $n_1$, $V_1$, and $n_2$ into the expression and solve for $V_2$:

$$V_2 = \frac{(10.00 \text{ L}) (0.20 \text{ mol})}{0.80 \text{ mol}} = 2.5 \text{ L}$$

Does this answer make sense? It does because we expected the volume to decrease when the amount of gas decreased.

### Consider This 9.7

What is the ratio of volume to moles under initial and final conditions? Does it make sense that the values are the same?

### Practice Problem 9.7

If a balloon initially contains 0.35 mol of gas and occupies 8.2 L, what is the volume of the balloon when 1.2 mol of gas is present if temperature and pressure remain constant?

**Further Practice:**  Questions 9.73 and 9.74 at the end of the chapter

**Avogadro's hypothesis** states that at a given pressure and temperature, equal volumes of all gases contain equal numbers of moles (or particles). In other words, 1 mol of $O_2$ gas has the same volume as 1 mol of $CH_4$ gas, if they are both at the same temperature and pressure. At 0°C and 1 atm, 1 mol of any ideal gas occupies a volume of 22.414 L, regardless of the gas. This volume is called the **molar volume.** The temperature and pressure conditions of 0°C and 1 atm are called **standard temperature and pressure (STP).** The volume at STP can be calculated from the number of moles and the molar volume, 22.414 L/mol.

---

### EXAMPLE 9.8      Avogadro's Hypothesis

What volume is occupied by 3.00 mol of argon gas at STP?

**Solution:**

At standard temperature and pressure, 1 mol of gas occupies 22.414 L, the molar volume of a gas. We can use this relationship as a conversion factor to determine the volume occupied by 3.00 mol of gas:

**MATH TOOLBOX 1.3**

$$V = 3.00 \text{ mol} \times \frac{22.414 \text{ L}}{1 \text{ mol}} = 67.2 \text{ L}$$

Note that the molar volume relationship (22.414 L/mol) holds for a gas at 0°C and 1 atm only.

**Consider This 9.8**

What volume is occupied by 3.00 mol of neon gas at STP?

**Practice Problem 9.8**

What volume will 7.50 mol of chlorine gas occupy at STP?

**Further Practice:** Questions 9.61 and 9.62 at the end of the chapter

We can calculate the volume occupied by a gas under any other conditions of temperature and pressure from the volume of a sample at STP by using the combined gas law.

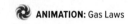 **ANIMATION:** Gas Laws

| EXAMPLE 9.9 | **Determining the Volume of a Gas at Non-STP Conditions** |
|---|---|

What volume would 1.00 mol of an ideal gas occupy at 25.0°C and 1.00 atm?

**Solution:**

At standard temperature and pressure, 1 mol of gas occupies 22.414 L, the molar volume of a gas. This gas is at standard pressure (1.00 atm) but not at standard temperature (0°C). However, since we know the molar volume of a gas at STP, we can use the combined gas law to determine the volume at non-STP conditions.

Before we use the combined gas law to calculate the final volume, let's consider the problem qualitatively. What do you expect to happen to the volume with an increase in temperature and a pressure that does not change? Since the gas particles speed up at higher temperatures, the volume should increase. $V_1$ is the molar volume, 22.414 L; $V_2$ is unknown; $P_1$ is the standard pressure, 1.00 atm; $P_2$ is 1.00 atm; $T_1$ is the standard temperature, 0.00°C + 273.15 = 273.15 K; and $T_2$ is 25.0°C + 273.15 = 298.2 K.

| **Initial Conditions** | **Final Conditions** |
|---|---|
| $V_1 = 22.414$ L | $V_2 = ?$ |
| $P_1 = 1.00$ atm | $P_2 = 1.00$ atm |
| $T_1 = 273.15$ K | $T_2 = 298.2$ K |

The combined gas law is

$$\frac{P_1 V_1}{T_1} = \frac{P_2 V_2}{T_2}$$

This equation can be rearranged to solve for the unknown volume:

$$V_2 = \frac{P_1 V_1}{T_1} \times \frac{T_2}{P_2}$$

 MATH TOOLBOX **9.2**

$$= \frac{(1.00 \text{ atm})(22.414 \text{ L})}{273.15 \text{ K}} \times \frac{298.2 \text{ K}}{1.00 \text{ atm}}$$

$$= 24.5 \text{ L}$$

Notice that since $P_1 = P_2$, the pressures cancel in the equation, and pressure could have been omitted from the equation before values were inserted. As in previous examples, we ask if the answer makes sense. It does, because an increase in temperature, with pressure staying the same, results in an increase in volume.

## 9.3 The Ideal Gas Law

So far we've used the relationships among pressure ($P$), volume ($V$), absolute temperature ($T$), and number of moles ($n$) to determine the consequences of changing one of them. Suppose we want to describe a gas under a specific set of static conditions. The relationships among the individual gas laws can be manipulated to give us a general equation that relates volume, pressure, temperature, and amount of gas. Let's review the relationships we've discussed so far.

Boyle's law: $V \propto \dfrac{1}{P}$    at constant $n$, $T$

Charles's law: $V \propto T$    at constant $n$, $P$

Avogadro's hypothesis: $V \propto n$    at constant $P$, $T$

These three equations can be summarized in the form of a single relationship. Since volume is proportional to each of the three factors, it must be simultaneously proportional to all of them:

$$V \propto \frac{nT}{P}$$

Since these variables are proportionally related, we can use a constant to express an equivalent relationship:

$$V = \text{constant} \times \frac{nT}{P}$$

Because it relates all the terms from the individual gas laws, we have a special name for the constant in this expression, the **ideal gas constant,** and we use $R$ to represent it. Substituting the ideal gas constant $R$ into the expression, we obtain the following equation:

$$V = \frac{RnT}{P}$$

 **MATH TOOLBOX 9.2**

Rearranging yields an equation known as the **ideal gas law:**

$$PV = nRT$$

An ideal gas is a gas that follows behavior predicted by the ideal gas law. Using the observation that 1 mol of an ideal gas at STP occupies 22.414 L, we can calculate the value of $R$ as follows:

$$R = \frac{PV}{nT}$$

$$= \frac{(1.000 \text{ atm})(22.414 \text{ L})}{(1.000 \text{ mol})(273.15 \text{ K})}$$

$$= 0.08206 \frac{\text{L} \cdot \text{atm}}{\text{mol} \cdot \text{K}}$$

This constant is valid only if we measure volume in liters, pressure in atmospheres, and temperature in kelvins. Measurements in other units must be converted to use this constant.

The ideal gas law includes all the information summarized by Boyle's law, Charles's law, and Avogadro's hypothesis. Like them, the ideal gas law is valid only for an ideal gas. It indicates that the volume occupied by an ideal gas is directly proportional to the absolute temperature and to the number of moles of gas molecules. It is inversely proportional to pressure. A single equation that relates all the variables is a useful tool, because given any three values and the value of $R$, we can calculate the unknown quantity.

## Calculations with the Ideal Gas Law

To solve a gas law problem, we must first determine what is going on in the gas system and then select the appropriate mathematical relationship. For changes in volume, pressure, and temperature, we generally use the combined gas law. In contrast, the ideal gas law allows us to determine any one of the quantities volume, pressure, temperature, or moles of the gas given the other three. If the identity of the gas is known, we can also calculate its mass using its molar mass. The ideal gas law can also be used to determine the density of a gas. Can you see how to rearrange the ideal gas law to solve for each quantity? If not, review Math Toolbox 9.2.

**Moles of a Gas**    If we are given the $P$, $V$, and $T$ conditions for a sample of a gas, we can calculate the number of moles of the gas, using the ideal gas law. We need only rearrange the equation $PV = nRT$ to solve for the number of moles. Remember that all the other quantities must be in units that match the value used for the ideal gas constant.

Small gas particles like helium and hydrogen behave most ideally because they have very weak forces of attraction between the atoms or molecules. When there are stronger forces of attraction or repulsion between gas particles, their properties deviate from predictions made by the ideal gas law. This deviation also occurs at very high pressures and low temperatures. When deviations from ideal gas behavior are observed, the gas is called a *real gas*. We will discuss forces of attraction between atoms and molecules in greater detail in Chapter 10.

 MATH TOOLBOX **9.2**

---

| EXAMPLE 9.10 | Calculating Moles of a Gas |
|---|---|

The volume of a propane cylinder used in Joel's gas grill is 0.960 L. When filled, the cylinder contains liquid propane stored under pressure. When the cylinder is "empty," it contains propane gas molecules at atmospheric pressure and temperature. How many moles of propane gas remain in a cylinder when it is empty if the surrounding atmospheric conditions are 25.0°C and 745 torr?

**Solution:**

First, we must convert to the desired units of $T$ and $P$:

$$T = 25.0°C + 273.15 = 298.2 \text{ K}$$

$$P = 745 \text{ torr} \times \frac{1 \text{ atm}}{760 \text{ torr}} = 0.980 \text{ atm}$$

Next, we rearrange the ideal gas law and substitute the appropriate values to calculate the number of moles:

$$PV = nRT$$

$$n = \frac{PV}{RT}$$

$$= \frac{(0.980 \text{ atm}) (0.960 \text{ L})}{\left( 0.08206 \dfrac{\text{L} \cdot \text{atm}}{\text{mol} \cdot \text{K}} \right) (298.2 \text{ K})}$$

$$= 0.0384 \text{ mol}$$

 MATH TOOLBOX **9.2**

As a final step, we should ask if this answer seems reasonable. At standard temperature (273.15 K) and pressure (1.00 atm), 1 mol of an ideal gas occupies 22.414 L, the molar volume. In this case the temperature is 25° higher than standard, the pressure is close to standard pressure, and the volume is significantly less than the molar volume of a gas. We would expect significantly fewer moles of gas present in the cylinder under these conditions, and that's what our calculation shows.

### Consider This 9.10

A common error in ideal gas law calculations involves forgetting to convert measurements to units that correspond to those in the ideal gas constant. In this calculation forgetting to convert pressure from torr to atm and temperature from °C to kelvins yields a value of 349. Do units cancel appropriately? Does this quantity make physical sense?

### Practice Problem 9.10

The volume of an oxygen cylinder used as a portable breathing supply is 2.025 L. When the cylinder is empty at 29.2°C, it has a pressure of 723 torr. How many moles of oxygen gas remain in the cylinder?

**Further Practice:** Questions 9.83 and 9.84 at the end of the chapter

---

**Mass of a Gas**  Once the number of moles of gas is determined from the ideal gas law, we can calculate the mass of the gas if we know its identity. Knowing the number of moles, we can use the molar mass of the gas to convert moles to grams.

---

### EXAMPLE 9.11 | Calculating the Mass of a Gas

What mass of propane, $C_3H_8$, is left in the cylinder described in Example 9.10?

**Solution:**

Recall that molar mass provides a relationship between moles of a substance and grams of that substance. The molar mass of propane is 44.10 g/mol. The mass of 0.0384 mol of propane can be determined by multiplying moles by the molar mass:

$$\text{Mass} = 0.0384 \ \cancel{\text{mol } C_3H_8} \times \frac{44.10 \text{ g } C_3H_8}{1 \ \cancel{\text{mol } C_3H_8}} = 1.69 \text{ g } C_3H_8$$

### Consider This 9.11

Would an "empty" butane, $C_4H_{10}$, cylinder under the same conditions contain the same moles and mass as the propane cylinder?

### Practice Problem 9.11

The volume of an oxygen cylinder used as a portable breathing supply is 1.85 L. What mass of oxygen gas remains in the cylinder when it is empty if the pressure is 755 torr and the temperature is 18.1°C?

**Further Practice:** Questions 9.93 and 9.94 at the end of the chapter

---

**Density of a Gas**  If we know the mass and volume of a gas, we can determine its density. The propane described in Example 9.11 has a mass of 1.69 g. Recall from Example 9.10 that the volume of this gas is 0.960 L. We can now calculate the density:

$$\text{Density} = \frac{\text{mass}}{\text{volume}}$$

$$= \frac{1.69 \text{ g}}{0.960 \text{ L}} = 1.76 \text{ g/L}$$

The densities of gaseous substances are much more dependent upon conditions of temperature and pressure than the densities of liquids and solids. As shown earlier in Figure 9.14, a volume reduction causes gas particles to move closer together. The greater mass-to-volume ratio results in a greater gas density. What would be the new density of the propane we just described if its volume was reduced to one-half its original volume?

$$\text{Density} = \frac{\text{mass}}{\text{volume}} = \frac{1.69 \text{ g}}{0.5 \times 0.960 \text{ L}} = \frac{1.69 \text{ g}}{0.480 \text{ L}} = 3.52 \text{ g/L}$$

Note that the density increased by a factor of 2.

Under conditions of constant temperature and pressure, the density of a gas is independent of sample size. If we know the temperature, pressure, and identity of a gas, we can calculate its density. One way to approach this type of problem is to assume a 1.00-L sample. Density is grams divided by liters, so we must determine the mass of the gas in this sample. We can do this using the $PV = nRT$ equation to solve for moles ($n$), and then solve for grams using molar mass. As an example, let's calculate the density of chlorine gas at a temperature of 373 K and a pressure of 1.00 atm:

$$PV = nRT$$

$$n = \frac{PV}{RT} = \frac{(1.00 \text{ atm})(1.00 \text{ L})}{\left(0.08206 \dfrac{\text{L} \cdot \text{atm}}{\text{K} \cdot \text{mol}}\right)(373 \text{ K})} = 0.0327 \text{ mol Cl}_2$$

We use the molar mass of $Cl_2$ to convert moles of $Cl_2$ to grams:

$$\text{Mass of Cl}_2 = 0.0327 \text{ mol Cl}_2 \times \frac{70.90 \text{ g Cl}_2}{1 \text{ mol Cl}_2} = 2.32 \text{ g Cl}_2$$

Lastly we combine the mass and volume of the gas to determine the density:

$$\text{Density} = \frac{\text{mass}}{\text{volume}} = \frac{2.32 \text{ g}}{1.00 \text{ L}} = 2.32 \text{ g/L}$$

Note that the density of the gas is the same numerical value as the mass of the sample because the sample volume is 1.00 L.

How would the mass, volume, and density of a gas change if the sample size under the same conditions were 2.00 L instead of 1.00 L? The mass and volume would both double, but the density would not change.

How would the density of HCl compare to that of $Cl_2$ under the same conditions of temperature and pressure? The molar mass of HCl is about half that of $Cl_2$, so the mass of the same volume sample would be about half that of the $Cl_2$ sample. The resulting density of HCl would therefore be about half that of $Cl_2$.

**Determining the Molar Mass from Gas Density**   If we are given the density of an unknown gas at a specific temperature and pressure, it is possible to predict the identity of the gas. This is because the ideal gas law provides a way to convert density units (g/L) to molar mass units (g/mol). If we can find molar mass, we can make a good guess at the gas identity. We just have to convert liters to moles using the $PV = nRT$ equation. As an example let's consider a gas that has a density of 0.654 g/L at a temperature of 298 K and a pressure of 1.00 atm. What is the molar mass of the gas? The following shows the relationship between the units for density and molar mass:

$$\text{Density} = \frac{\text{mass (g)}}{\text{volume}} \qquad \text{Molar mass} = \frac{\text{mass (g)}}{\text{mole}}$$

The density given tells us the mass of gas in 1.00 L. Using the ideal gas law, we can calculate the number of moles in a 1.00-L volume:

$$PV = nRT$$

$$n = \frac{PV}{RT} = \frac{(1.00 \text{ atm})(1.00 \text{ L})}{\left(0.08206 \dfrac{\text{L} \cdot \text{atm}}{\text{K} \cdot \text{mol}}\right)(298 \text{ K})} = 0.0409 \text{ mol gas}$$

Substituting the mass and the number of moles of gas that occupy the 1.00-L volume into the molar mass equation gives

$$\text{Molar mass} = \frac{\text{mass (g)}}{\text{moles}} = \frac{0.654 \text{ g}}{0.0409 \text{ mol}} = 16.0 \text{ g/mol}$$

What substance has a molar mass of 16.0 g/mol and is a gas at 298 K (room temperature)? The molar mass of methane, $CH_4$, is 16.04 g/mol, and is a gas at room temperature, so this is likely the identity of the gas. The next closest possibility is $NH_3$ at 17.03 g/mol, then $H_2O$ at 18.02 g/mol. (Oxygen atoms have a molar mass of 16.00 g/mol; however, naturally occurring oxygen is $O_2$ with a molar mass of 32.00 g/mol.)

## Dalton's Law of Partial Pressures

If we have a mixture of gases such as those in Earth's atmosphere, we sometimes want to know the properties of one or more of the individual substances in the mixture. The air we breathe consists of many gases, mostly nitrogen and oxygen. On humid days it contains more water vapor than on dry days. John Dalton studied mixtures of dry air and water vapor. He discovered that the total pressure of the mixture increased incrementally with the amount of water vapor. The consequence of his discovery is **Dalton's law of partial pressures:** *Gases in a mixture behave independently and exert the same pressure they would exert if they were in a container alone.* Thus, the total pressure exerted by a mixture of gases is the sum of the *partial pressures* of the component gases:

$$P_{\text{total}} = P_A + P_B + P_C + \cdots$$

where A, B, and C represent the gases in the mixture. For example, the total pressure of the system studied by Dalton is given by the equation:

$$P_{\text{total}} = P_{\text{dry air}} + P_{\text{water}}$$

The partial pressure of each gas in a mixture is proportional to the number of molecules of that gas (Figure 9.21). Mixtures of gases have the same properties as pure gases, provided that all gases in the mixtures are ideal gases and do not react chemically with each other. We often deal with mixtures of gases when we run chemical reactions. Even when a reaction produces only a single gaseous product, the procedure used to collect it may introduce another gas. For example, a technique often used to collect gases from a chemical reaction is the displacement of water from an inverted bottle, as shown in Figure 9.22. This technique involves bubbling the gas through a tube into a bottle filled with water until the gas pushes all the liquid water from the bottle.

**FIGURE 9.21** The partial pressure of each gas in a mixture is in proportion to its number of molecules. Visible in this sample of air at 5.00 atm pressure are 3 $O_2$ molecules and 12 $N_2$ molecules. Thus, $\frac{3}{15}$ or 20% of the total number of molecules are $O_2$ molecules, and $\frac{12}{15}$ or 80% of the molecules are $N_2$ molecules. The partial pressure of $O_2$ is then 20% of the total pressure, or 1.00 atm. The partial pressure of $N_2$ is 80% of the total pressure, or 4.00 atm.

**FIGURE 9.22** Gaseous products can be isolated by displacing water in a closed container. Because water vaporizes to some extent at room temperature, some water vapor will be present with the gaseous product.

**TABLE 9.2** Vapor Pressure of Water at Various Temperatures

| Temperature (°C) | Vapor Pressure (torr) | Temperature (°C) | Vapor Pressure (torr) |
|---|---|---|---|
| 0 | 4.6 | 28 | 28.3 |
| 5 | 6.5 | 29 | 30.0 |
| 10 | 9.2 | 30 | 31.8 |
| 15 | 12.8 | 35 | 42.2 |
| 16 | 13.6 | 40 | 55.3 |
| 17 | 14.5 | 45 | 71.9 |
| 18 | 15.5 | 50 | 92.5 |
| 19 | 16.5 | 60 | 149.4 |
| 20 | 17.5 | 70 | 233.7 |
| 21 | 18.6 | 80 | 355.1 |
| 22 | 19.8 | 90 | 525.8 |
| 23 | 21.1 | 100 | 760.0 |
| 24 | 22.4 | 110 | 1,074.6 |
| 25 | 23.8 | 150 | 3,570.5 |
| 26 | 25.2 | 200 | 11,659.2 |
| 27 | 26.7 | 300 | 64,432.8 |

For values at temperatures between the listed entries, vapor pressure can be calculated from the following equation, valid for water in the temperature range of 10°C to 30°C:

$$P_{vapor} = 4.1243 + 0.43562t + 0.0027016t^2 + 0.00045158t^3$$

where $P_{vapor}$ is the vapor pressure in torr and $t$ is the temperature in degrees Celsius.

For example, heating molten potassium bromate can produce oxygen gas:

$$2KBrO_3(l) \xrightarrow{heat} 2KBr(s) + 3O_2(g)$$

Since oxygen gas is not very soluble in water and does not react with it, this technique should collect all the gas in a pure state. Unfortunately it does not. The collected gas is a mixture of the oxygen generated by the reaction and some water vapor formed by evaporation. The amount of water vapor contained in the gas is most readily measured by the pressure it exerts at that temperature, called the *vapor pressure*. Vapor pressures for water are given in Table 9.2. How can we use this table to calculate the partial pressure of a collected gas?

To determine the pressure exerted by a gas collected in this fashion at a specific temperature, we merely subtract the vapor pressure of water from the total pressure of the mixture. This calculation is an application of Dalton's law of partial pressures. The total pressure is atmospheric pressure in this case, since the pressure inside the bottle is the same as that outside the bottle. The vapor pressure of water, found in Table 9.2, is subtracted from the total pressure to give the partial pressure of the gas. From the partial pressure of the gas and its volume and temperature, the ideal gas law can be used to calculate the moles of gas collected.

**ANIMATION:** Collecting a Gas Over Water

When collecting oxygen gas, why do we fill the bottle with water and not air?

**Student Hot Spot**

Student data indicate you may struggle with application of Dalton's law. Access the SmartBook to view additional Learning Resources on this topic.

Scientists use the term *vapor* to describe the gas phase of a substance that is normally a liquid or solid.

**EXAMPLE 9.12** | **Dalton's Law of Partial Pressures**

Suppose we produce gaseous oxygen by the following reaction:

$$2KClO_3(s) \xrightarrow{heat} 2KCl(s) + 3O_2(g)$$

If we collect 1.50 L of $O_2$ over water at 27.0°C (300.2 K) and 758 torr using an apparatus like that shown in Figure 9.22, how many moles of $O_2$ are produced?

**Solution:**

First, we obtain the partial pressure of water at 27.0°C from Table 9.2. Then we calculate the pressure of the oxygen gas using a rearrangement of Dalton's law:

$$P_{total} = P_{oxygen} + P_{water}$$
$$P_{oxygen} = P_{total} - P_{water}$$
$$= 758 \text{ torr} - 26.7 \text{ torr} = 731 \text{ torr}$$

To use the ideal gas law we must convert the pressure in units of torr to atm:

$$P_{oxygen} = 731 \text{ torr} \times \frac{1 \text{ atm}}{760 \text{ torr}} = 0.962 \text{ atm}$$

Finally we can calculate the number of moles of oxygen using the ideal gas equation:

$$PV = nRT$$

This equation can be rearranged and solved for $n$:

$$n = \frac{PV}{RT}$$

$$= \frac{(0.962 \text{ atm})(1.50 \text{ L})}{\left(0.08206 \dfrac{\text{L} \cdot \text{atm}}{\text{mol} \cdot \text{K}}\right)(300.2 \text{ K})}$$

$$= 0.0586 \text{ mol O}_2$$

**Consider This 9.12**

If the correction for the partial pressure of water weren't made, what would that do to the calculated moles of oxygen produced?

**Practice Problem 9.12**

Suppose 2.25 L of $H_2$ gas is collected over water at 18.0°C and 722.8 torr. How many moles of $H_2$ are produced in this reaction?

**Further Practice:** Questions 9.97 and 9.98 at the end of the chapter

## 9.4 Kinetic-Molecular Theory of Gases

In Section 9.2 we described an ideal gas as a gas that behaves according to predicted proportional relationships. That is, it obeys the gas laws. The **kinetic-molecular theory of gases** is a model that explains experimental observations about gases under normal temperature and pressure conditions that we encounter in our environment.

### Postulates of Kinetic-Molecular Theory

Five postulates of the kinetic-molecular theory of gases allow us to describe and predict ideal gas behavior.

1. *Gases are composed of small and widely separated particles* (molecules or atoms). Because the particles are widely separated, the actual volume of the particles is very small compared to the total volume occupied by the gas, which is mostly empty space. This postulate correctly predicts that the volume occupied by a gas is much larger than that of a liquid or a solid, in which the particles are much closer together. Because gas particles are so widely separated, gases have relatively low densities compared to liquids and solids. They are also highly compressible, since the particles can easily be squeezed much closer together and remain a gas.

To explain some properties of gases, such as why they are easy to compress, some size comparisons are useful. The volume of a helium atom is known to be $1.15 \times 10^{-26}$ L. The actual volume of 1 mol of helium atoms is $6.92 \times 10^{-3}$ L. However, the total volume occupied by 1 mol of gaseous helium at STP is 22.4 L. Thus, only 0.03% of the volume occupied by the gas is actually taken up by the atoms.

2. *Particles of a gas behave independently of one another.* Because gas particles are far apart, they move independently of one another unless they collide. That is, no forces of attraction or repulsion operate between and among gas particles. This postulate explains Dalton's law of partial pressures. If gas particles move independently, the presence of more than one kind is irrelevant to the total pressure exerted by the mixture of gases. Only the total number of particles is important in determining the pressure. (This postulate is not valid at very high pressures or very low temperatures.)

3. *Each particle in a gas is in rapid, straight-line motion until it collides with another molecule or with its container.* When collisions do occur, the collisions are perfectly elastic. That is, energy may be transferred from one particle to another, but there is no net loss of energy. This postulate also explains why gases fill their containers.

4. *The pressure of a gas arises from the sum of the collisions of the particles with the walls of the container.* This postulate explains Boyle's law, which states that pressure is inversely proportional to volume at constant temperature. For the same amount of gas, the smaller the volume of the container, the more collisions occur per unit area. The average distance traveled by a gas particle before a collision is less in a smaller volume. More collisions in a given area produce greater pressure. This postulate also predicts that pressure should be proportional to the number of moles of gas particles. The more gas particles, the greater the frequency of collisions with the walls, so the greater the pressure.

5. *The average kinetic energy of gas particles depends only on the absolute temperature.* What is average kinetic energy? To answer this question we must examine the velocity, or speed, of gas particles. The relationship between the velocity ($v$) and kinetic energy ($KE$) for a given gas particle is given by the equation

$$KE = \tfrac{1}{2} mv^2$$

where $m$ is the mass of a gas particle. In Figure 9.23 look at the line that corresponds to gas particle velocities at low temperature. Notice that it isn't a single velocity, but a distribution of velocities. To deal with this distribution, it's convenient to describe the average kinetic energy for all the gas particles using the equation

$$KE_{av} = \tfrac{1}{2} m(v_{av})^2$$

where the "av" subscript stands for the average, or mean, of both kinetic energy and velocity. The average velocity is observed for some, but not all, of the gas particles at a given temperature. The rest have velocities either greater than or

The average velocity of nitrogen molecules is about 500 m/s at room temperature. If gas molecules move at such high speeds, why does it seem to take a long time for smells to permeate a room?

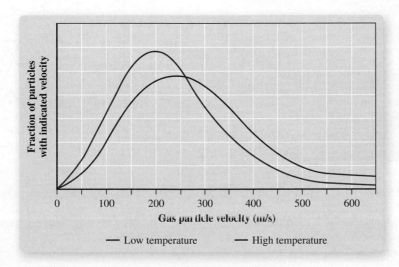

**FIGURE 9.23**  Gas particles move at various speeds when the temperature changes. At a given temperature, the speed of molecules varies across a definable range. At the higher temperature, there is a larger fraction of particles moving at greater velocities.

less than the average value. The distribution of velocities means that the gas particles also have a distribution of kinetic energies. Notice in Figure 9.23 that when the temperature increases, the distribution of velocities shifts. The average velocity is greater for the gas particles at higher temperatures. If the particles move faster, they strike the walls more frequently and with greater energy, so the pressure increases as temperature increases.

What does the equation $KE_{av} = \frac{1}{2}m(\nu_{av})^2$ tell us about the relative average molecular velocities of different gases at the same temperature? Remember that at a given temperature average kinetic energy, $KE_{av}$, is constant regardless of the gas identity. This equation shows an *inverse* relationship between mass and velocity when temperature (and therefore kinetic energy) is constant: the larger the mass the lower the velocity. When comparing two gases of different molar mass at the same temperature, the smaller mass particles have greater average velocities and move faster than the larger mass particles. Which has the greater average molecular velocity at 25°C: helium atoms or neon atoms?

## Diffusion and Effusion

Kinetic-molecular theory explains much of the behavior of gases. Our everyday experiences tell us that gas particles are in motion. When Joel's friends get close to his house, they can smell the meat he has already begun to cook on his grill. They smell the food because gas particles disperse through the air to their noses. **Diffusion** is the movement of gas particles from regions of high concentration to regions of low concentration (Figure 9.24). Substances can be smelled from long distances because the gaseous molecules diffuse rapidly through the atmosphere.

Although particles of different gases at the same temperature have the same average kinetic energy, gases with different molecular masses have different average velocities. Recall the relationship between velocity, molecular mass, and kinetic energy for a collection of gas particles:

$$KE_{av} = \frac{1}{2}m(\nu_{av})^2$$

This equation shows that mass and velocity both influence kinetic energy. For the particles of two gases with different molar masses to have the same average kinetic energy, they must move through space at different average velocities. To compensate for a greater molar mass, heavy gas particles move at lower speeds. Light molecules or atoms move at much higher speeds.

Graham's law, formulated by the Scottish chemist, Thomas Graham, makes use of the equation for average kinetic energy. Two gases (labeled 1 and 2) at the same temperature have the same kinetic energy:

$$\frac{1}{2}m_1(\nu_1)^2 = \frac{1}{2}m_2(\nu_2)^2$$

Eliminating $\frac{1}{2}$ on both sides of the equation gives:

$$m_1(\nu_1)^2 = m_2(\nu_2)^2$$

This equation can be rearranged to compare the velocities of the two gases:

$$\frac{\nu_1}{\nu_2} = \sqrt{\frac{m_2}{m_1}}$$

Measurement of the relative rates of diffusion of two gases can be used to determine their relative molecular masses or molar masses.

### 🔊 Student Hot Spot

Student data indicate you may struggle with relative rates of diffusion. Access the SmartBook to view additional Learning Resources on this topic.

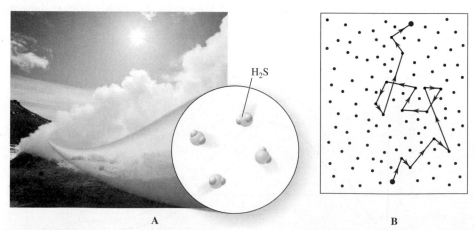

**FIGURE 9.24** (A) The hydrogen sulfide gas that comes out of a "sulfur vent" disperses through the air. Since gas particles are in constant motion, they can diffuse through space. (B) Although gas particles move rapidly, they encounter other gas particles along the way, as shown for a typical pathway for a gas particle.
©Justin Lewis/Iconica/Getty Images

For example, consider the experiment shown in Figure 9.25. In this reaction $NH_3$ gas reacts with HCl gas to produce solid $NH_4Cl$. The solid forms closer to the point where HCl is released in the tube because HCl molecules do not diffuse as fast as the lighter $NH_3$ molecules.

In addition to diffusion, we observe evidence of the motion of gas particles in another way. Consider Joel's weather balloon when it is filled with helium. If he fails to release it and leaves it sitting on the launch pad, it loses its buoyancy and eventually shrinks to a fraction of its inflated size. The escape of helium atoms through the natural holes in the structure of the latex explains this result. The passage of a gas through a small opening into a vacuum is called **effusion** (Figure 9.26). As with diffusion, small particles can effuse faster than large ones.

## 9.5 Gases and Chemical Reactions

Chemical reactions produce many of the gases in our atmosphere. For example, the burning of propane in Joel's grill yields carbon dioxide gas and water vapor. Oxygen in the air is used to burn the propane. Because gases are involved in many chemical reactions, it's important to understand how to calculate the quantities of gaseous reactants and products.

### Product Volume from Reactant Volume

Recall that Gay-Lussac found that volumes of gases in reactions occur in small whole-number ratios. The ratios correspond to the coefficients in the balanced equation for the reaction. Consider the reaction of hydrogen with oxygen to form water vapor:

$$2H_2(g) + O_2(g) \longrightarrow 2H_2O(g)$$

If 2.38 L of $H_2$ at 25.0°C and 720 torr is allowed to react completely with 1.19 L of $O_2$ under the same conditions of temperature and pressure, we expect to form exactly 2.38 L of gaseous $H_2O$. This example shows how we can determine the volume of one gaseous substance from the volume of another in a chemical reaction. When temperature and pressure are constant, the volume-to-volume relationship, as described by Gay-Lussac's law of combining volumes, is the same as the mole-to-mole ratios given by the coefficients in the chemical equation. In our example, the ratio of the volumes is 2 $H_2$ : 1 $O_2$ : 2 $H_2O$.

Such simple conversions work only if all the substances are gases at the same pressure and temperature. If temperature and pressure vary, the volume-to-mole relationship is given by the ideal gas law, and the mole-to-mole relationship is provided by the balanced chemical equation:

$$\boxed{\text{Volume A}} \xrightarrow{PV = nRT} \boxed{\text{mol A}} \xrightarrow{\text{mole ratio}} \boxed{\text{mol B}} \xrightarrow{PV = nRT} \boxed{\text{Volume B}}$$

Example 9.13 shows how to perform a calculation of this kind.

**EXAMPLE 9.13** | **Product Volume from Reactant Volume**

A sample of hydrogen gas has a volume of 8.00 L at a pressure of 5.00 atm and a temperature of 25.0°C (298.2 K). What volume of gaseous water is produced by the following reaction at 150.0°C (423.2 K) and 0.947 atm, if all of the hydrogen gas reacts with copper(II) oxide?

$$CuO(s) + H_2(g) \longrightarrow Cu(s) + H_2O(g)$$

**FIGURE 9.25** Ammonia ($NH_3$) and hydrogen chloride (HCl) gases react to form solid ammonium chloride, $NH_4Cl$:

$$NH_3(g) + HCl(g) \longrightarrow NH_4Cl(s)$$

The reactants have to meet before a reaction can occur. Notice that the solid does not form in the middle of the tube. $NH_3$ is a lighter molecule, so it travels faster than HCl. (Note that HCl and $NH_3$ are colorless gases.)

**FIGURE 9.26** Gas particles pass through small openings into a vacuum. This process is called *effusion*.

**Solution:**

Because the conditions for the products and reactants are different in terms of temperature and pressure, we cannot use Gay-Lussac's law of combining volumes to solve this problem. We'll have to use the following sequence of calculations:

We are given gas volume, not moles, so we must first calculate the moles of hydrogen gas from the ideal gas law:

**MATH TOOLBOX 9.2**

$$PV = n_{H_2}RT$$

$$n_{H_2} = \frac{PV}{RT}$$

$$= \frac{(5.00 \text{ atm})(8.00 \text{ L})}{\left(0.08206 \dfrac{\text{L} \cdot \text{atm}}{\text{mol} \cdot \text{K}}\right)(298.2 \text{ K})}$$

$$= 1.63 \text{ mol H}_2$$

Next, we use the coefficients from the balanced equation to calculate the moles of gaseous water:

**MATH TOOLBOX 1.3**

$$n_{H_2O} = 1.63 \text{ mol H}_2 \times \frac{1 \text{ mol H}_2O}{1 \text{ mol H}_2} = 1.63 \text{ mol H}_2O$$

Finally, we calculate the volume of water vapor produced, using the ideal gas law solved for *V*:

$$V_{H_2O} = \frac{n_{H_2O}RT}{P}$$

$$= \frac{1.63 \text{ mol} \times \left(0.08206 \dfrac{\text{L} \cdot \text{atm}}{\text{mol} \cdot \text{K}}\right) \times 423.2 \text{ K}}{0.947 \text{ atm}}$$

$$= 59.8 \text{ L H}_2O$$

**Consider This 9.13**

What is the density of the water vapor produced under these conditions?

**Practice Problem 9.13**

A sample of hydrogen gas has a volume of 7.49 L at a pressure of 22.0 atm and a temperature of 32.0°C. What volume of gaseous water is produced by the following reaction at 125.0°C and 0.975 atm, if all of the hydrogen gas reacts with iron(III) oxide?

$$Fe_2O_3(s) + 3H_2(g) \longrightarrow 2Fe(s) + 3H_2O(g)$$

**Further Practice:** Questions 9.117 and 9.118 at the end of the chapter

## Moles and Mass from Volume

Many chemical reactions have reactants or products in different physical states. For example, solid sodium metal reacts with chlorine gas to produce solid sodium chloride:

$$2Na(s) + Cl_2(g) \longrightarrow 2NaCl(s)$$

If we know the mass of $Cl_2$, we can calculate the mass of NaCl produced using the procedure described in Section 6.3:

For a review of determining quantities in chemical reactions, see Chapter 6.

$$\boxed{g\ Cl_2} \xrightarrow{\text{molar mass}} \boxed{mol\ Cl_2} \xrightarrow{\text{mole ratio}} \boxed{mol\ NaCl} \xrightarrow{\text{molar mass}} \boxed{g\ NaCl}$$

Suppose we want to know how much NaCl can be prepared from a known *volume* of $Cl_2$ under given conditions of temperature and pressure. Knowing the volume of $Cl_2$ lets us calculate the number of moles of this gaseous reactant using the ideal gas law. This result is used with the coefficients from the balanced chemical equation to calculate the moles of NaCl. The mass of the product can, in turn, be obtained from the number of moles of product and its molar mass. Slight modification of the first step of the previous problem-solving scheme allows us to calculate the mass of NaCl from the volume of $Cl_2$ gas:

$$\boxed{\text{Volume } Cl_2} \xrightarrow{PV = nRT} \boxed{mol\ Cl_2} \xrightarrow{\text{mole ratio}} \boxed{mol\ NaCl} \xrightarrow{\text{molar mass}} \boxed{g\ NaCl}$$

---

### EXAMPLE 9.14 | Mass from Volume

How many moles of calcium carbonate form if 3.45 L of $CO_2$, measured at 45.0°C (318.2 K) and 1.37 atm, react with excess CaO? How many grams? The balanced equation is as follows:

$$CaO(s) + CO_2(g) \longrightarrow CaCO_3(s)$$

**Solution:**

We will have to perform the following series of calculations to solve this problem:

$$\boxed{\text{Volume } CO_2} \xrightarrow{PV = nRT} \boxed{mol\ CO_2} \xrightarrow{\text{mole ratio}} \boxed{mol\ CaCO_3} \xrightarrow{\text{molar mass}} \boxed{g\ CaCO_3}$$

First, we use the ideal gas law solved for $n$ to calculate the moles of $CO_2$:

MATH TOOLBOX **9.2**

$$n_{CO_2} = \frac{(1.37\ \text{atm})(3.45\ L)}{\left(0.08206\ \dfrac{L \cdot \text{atm}}{\text{mol} \cdot K}\right)(318.2\ K)}$$

$$= 0.181\ \text{mol } CO_2$$

We calculate the number of moles of calcium carbonate from the number of moles of carbon dioxide by using coefficients in the balanced equation:

$$\boxed{mol\ CO_2} \xrightarrow{\text{mole ratio}} \boxed{mol\ CaCO_3}$$

$$n_{CaCO_3} = 0.181\ \text{mol } CO_2 \times \frac{1\ \text{mol } CaCO_3}{1\ \text{mol } CO_2} = 0.181\ \text{mol } CaCO_3$$

MATH TOOLBOX **1.3**

Then we obtain the mass of calcium carbonate from its molar mass, 100.1 g/mol:

$$\boxed{mol\ CaCO_3} \xrightarrow{\text{molar mass}} \boxed{\text{Mass } CaCO_3}$$

$$\text{Mass of } CaCO_3 = 0.181\ \text{mol } CaCO_3 \times \frac{100.1\ \text{g } CaCO_3}{1\ \text{mol } CaCO_3} = 18.1\ \text{g } CaCO_3$$

**Consider This 9.14**

What masses of CaO and $CO_2$ must react to produce this amount of $CaCO_3$?

**Practice Problem 9.14**

How many moles of $H_2SO_4$ form if 2.38 L of $SO_3$ gas, measured at 65.0°C and 1.05 atm, react with sufficient water? How many grams? The balanced equation is as follows:

$$SO_3(g) + H_2O(l) \longrightarrow H_2SO_4(l)$$

**Further Practice:** Questions 9.119 and 9.120 at the end of the chapter

When determining the amount of a gaseous reactant or product in a chemical reaction from the mass of another reactant or product, the reverse of the process described in Example 9.14 can be followed.

 ## Math Toolbox 9.1   Graphing

### Relationships Between Variables

In a controlled scientific experiment involving measured quantities, values for one variable (the dependent variable) are determined as another variable changes (the independent variable). We often find one of two basic relationships between the variables. In a *direct* relationship, both variables increase (or decrease) together. That is, as one increases, so does the other. In an *inverse* relationship, one variable decreases as the other increases.

Large amounts of data often reveal proportional relationships between two variables that can be represented using a single algebraic equation. *Proportional* means the two variables are related by a constant factor, which we can represent in an algebraic equation as $k$. Proportional relationships are not always obvious from raw data. Determining a relationship—and consequently an equation that relates proportional variables—often requires constructing an appropriate graph. For example, consider the following data of the volume of a gas at different absolute temperatures:

| Volume (L) | Temperature (K) |
|---|---|
| 2.20 | 300 |
| 2.37 | 325 |
| 2.59 | 350 |
| 2.95 | 400 |
| 3.30 | 450 |
| 3.71 | 500 |
| 4.37 | 600 |
| 5.17 | 700 |

Although the data show that volume increases with absolute temperature (a direct relationship), they do not reveal whether an algebraic proportion exists between the two variables. We have to construct a graph of the data to see if a proportional relationship exists.

### Graphing

If you are creating a graph by hand, the following steps will generally lead to a graph that presents data in a clear and easily interpreted manner.

1. Carefully examine the data to be plotted. Organize the data in a table to facilitate transfer to the graph.
2. Identify and label the axes. The *dependent variable* is the variable measured experimentally. Plot it on the vertical axis (the $y$ axis). The *independent variable* is the variable controlled and varied by the experimenter. Plot it on the horizontal axis (the $x$ axis). Clearly label each axis with the name of the variable and the units. Put the units either in parentheses following the name of the variable or separated from the variable name by a comma: volume (L) or volume, L.
3. Decide on the scales and limits of values to be plotted along each axis. The graph should nearly fill the graph paper. Adjust the scale so the data can be entered easily on the graph and subsequently read off the graph with about the same precision as they were measured. The value of each major division on an axis is usually selected so that minor divisions represent workable numbers of units. To avoid using a small area of the graph, it may not always be desirable for the scale of each axis to begin at zero. Increments on the $x$ axis may differ from increments on the $y$ axis.
4. After selecting the scales to be plotted along each axis, clearly mark the value of each major division along each axis.
5. Place dots on the graph at appropriate intersecting $y$ and $x$ values to represent each data point. If necessary to identify the plotted points, draw a small circle around each with a diameter of 1 to 2 mm. If you plot several sets of data on the same graph, use different geometric figures, such as squares or triangles, to outline the other sets of data points.
6. Draw a smooth line through the data set. Because of experimental error, not all points will fall precisely on the line, but

**Math Toolbox 9.1 (continued)**

it is generally assumed that the dependent variable actually varies smoothly with the independent variable. For that reason, do *not* connect the points dot-to-dot with separate straight-line segments. If the data appear to follow a straight line, use a straight edge to draw the line. If the data follow a curve, use a graphing aid such as a French curve. Label the completed graph with an appropriate identifying title in a clear space near the top.

### EXAMPLE 9.15  Creating a Graph

Create a graph of the volume and temperature data given in the earlier table.

**Solution:**

We will follow the steps just outlined.

1. The data are already organized in a table.
2. Assign volume with units of liters (the dependent variable) to the $y$ axis and temperature with units of kelvins (the independent variable) to the $x$ axis since temperature is the variable we are controlling.
3. Suppose we have a piece of graph paper that is four major divisions high and six major divisions wide. The $y$ axis then has four major divisions, and we need to cover 5.17 L − 2.20 L = 2.97 L. To keep even units for each minor division, assign 1.0 unit per major division. On the $x$ axis, we have six major divisions to cover 700 K − 300 K = 400 K. We assign 100 K per major division.
4. The divisions on the $y$ axis are assigned values from 2 to 6, and the divisions on the $x$ axis are assigned values from 200 to 700.
5. Each data point is placed on the graph. We will use dots large enough to identify without being encircled.
6. A straight line can be drawn through all the data points. Following these steps yields the following graph:

**Practice Problem 9.15**

The population of Earth increased over an 85-year period, as shown in the table. Draw a graph to show the relationship between population and time. Do the data conform to a straight line?

| Population (in billions) | Year |
|---|---|
| 1.80 | 1920 |
| 2.00 | 1930 |
| 2.25 | 1940 |
| 2.52 | 1950 |
| 2.69 | 1955 |
| 3.00 | 1960 |
| 3.29 | 1965 |
| 3.63 | 1970 |
| 4.07 | 1975 |
| 4.44 | 1980 |
| 4.84 | 1985 |
| 5.28 | 1990 |
| 5.69 | 1995 |
| 6.09 | 2000 |
| 6.46 | 2005 |

**Further Practice:**  Questions 9.5 and 9.6 at the end of the chapter

### Proportional and Reciprocal Relationships

In general, a straight line through the data points shows that the two variables are directly proportional if the line passes through the origin (0,0). An extended graph would show that volume and temperature are directly proportional. Their relationship can be represented as $y \propto x$ or $y = kx$. (The slope of the line equals $k$.)

Consider another set of data:

| Volume (L) | Pressure (atm) |
|---|---|
| 2.20 | 1.00 |
| 2.32 | 0.95 |
| 2.59 | 0.85 |
| 2.93 | 0.75 |
| 3.28 | 0.67 |
| 3.67 | 0.60 |
| 4.40 | 0.50 |
| 5.12 | 0.43 |

**Math Toolbox 9.1 (continued)**

These data show that volume increases as pressure decreases, although we cannot tell if the relationship is proportional. Following the steps for drawing a graph, we obtain the following plot:

A smooth line through the data points forms a curve. The data reveal an inverse relationship between volume and pressure; that is, volume decreases as pressure increases. However, we don't have a straight line, so it is not easy to see if the relationship is proportional. When scientists encounter such data, they often try to manipulate the values mathematically to see if a straight line can be plotted.

Let's see what happens if we take the reciprocal of pressure. (The reciprocal is 1 divided by the quantity we are interested in—in this case, 1/pressure.)

| Volume (L) | 1/Pressure |
|------------|------------|
| 2.20       | 1.00       |
| 2.32       | 1.05       |
| 2.59       | 1.18       |
| 2.93       | 1.33       |
| 3.28       | 1.49       |
| 3.67       | 1.67       |
| 4.40       | 2.00       |
| 5.12       | 2.33       |

These data show that as 1/pressure increases, the volume also increases. However, we cannot tell if the data are proportional until we create a graph.

In general, if a graph of one variable versus the reciprocal of another variable yields a straight line that would pass through the origin, then the two variables are inversely (or reciprocally) proportional. That is,

$$x \propto \frac{1}{y} \quad \text{and} \quad y \propto \frac{1}{x}$$

The general equation is $y = k(1/x)$. (Find $k$ by determining the slope of the line.)

We can obtain values of the dependent variable at any value of the independent variable found on the graph, even if we did not make a measurement at that value. We can also determine the value of the dependent variable that would occur for a desired value of the independent variable. This is illustrated for the gas volume versus 1/pressure graph in Example 9.16.

### EXAMPLE 9.16   Reading Data from a Graph

We want to know the value of pressure when the volume is 3.0 L, using the graph of volume versus 1/pressure.

**Solution:**

We find this volume on the graph and read the value of 1/pressure to be 1.35 from the position of the straight line. Taking the inverse of this quantity, we get a pressure of 0.741 atm:

$$\frac{1}{\text{Pressure}} = 1.35$$

$$\text{Pressure} = \frac{1}{1.35} = 0.741 \text{ atm}$$

**Practice Problem 9.16**

We can also determine the volume at any given pressure. Use the graph to find the volume when the pressure is 0.550 atm.

**Further Practice:** Questions 9.9 and 9.10 at the end of the chapter

# Math Toolbox 9.2    Solving Simple Algebraic Equations

Algebraic expressions represent many chemical principles, so understanding the principles requires solving and manipulating such equations. An algebraic equation is a simple statement of equality. For example, the equality $9x + 12x = 63$ is confirmed when $x = 3$:

$$(9 \times 3) + (12 \times 3) = 63$$

## Manipulating Equations

We can manipulate an equation in any way that does not destroy the equality. The usual purpose is to obtain a value for an unknown quantity. Operations that will maintain the equality are

- adding the same number to both sides of the equation
- subtracting the same number from both sides of the equation
- multiplying or dividing both sides of the equation by the same number
- raising both sides of the equation to the same power

Consider the equation $16x - 32 = 16$. To solve for $x$, we first add 32 to both sides of the equation:

$$16x - 32 + 32 = 16 + 32$$
$$16x = 48$$

We then divide both sides of the equation by 16:

$$\frac{16x}{16} = \frac{48}{16}$$
$$x = 3$$

As a second example, consider the equation $\frac{1}{4}x + 4 = 12$. We subtract 4 from both sides of the equation:

$$\frac{1}{4}x + 4 - 4 = 12 - 4$$
$$\frac{1}{4}x = 8$$

Then we multiply each side of the equation by 4:

$$4 \times \frac{1}{4}x = 4 \times 8$$
$$x = 32$$

Now consider $4x = 15 + x$. To solve for $x$, we begin by moving all of the terms that contain $x$ to one side of the equation. Subtracting $x$ from both sides of the equation will accomplish this:

$$4x - x = 15 + x - x$$
$$3x = 15$$

Dividing both sides by 3, we solve for $x$:

$$\frac{3x}{3} = \frac{15}{3}$$
$$x = 5$$

For additional practice, work Questions 9.11 and 9.12 at the end of the chapter.

## Multiple Variables

So far we have considered equations that contain only one variable (unknown). Many chemistry relationships involve more than one variable. For example, to convert temperatures between Celsius ($T_C$) and Fahrenheit ($T_F$) scales, we can use the equation

$$T_C = \tfrac{5}{9}(T_F - 32)$$

Using this equation, we can calculate one of the variables if we know the value of the other, as illustrated in Example 9.17.

### EXAMPLE 9.17  Converting Temperature Scales

The temperature in a lab is 72.5°F. What is the temperature in degrees Celsius?

**Solution:**
We begin by assigning the quantity we know to the correct variable.

| Given Quantity | Unknown Quantity |
|---|---|
| $T_F = 72.5°F$ | $T_C = ?$ |

To calculate the temperature in degrees Celsius, substitute the known value for $T_F$ and solve the equation for $T_C$:

$$T_C = \tfrac{5}{9}(T_F - 32) = \tfrac{5}{9}(72.5°F - 32)$$
$$= \tfrac{5}{9}(40.5°F) = 22.5°C$$

### Practice Problem 9.17

The temperature of the human body measured orally is normally 36.8°C. What is this temperature in degrees Fahrenheit?

**Further Practice:** Questions 9.13 and 9.14 at the end of the chapter

## Rearranging Equations

It is often useful to rearrange an equation to isolate the variable of interest on one side. For example, the relationship $PV = nRT$ contains five variables. Suppose we know $P$, $V$, $R$, and $T$ and want to solve for $n$. We can rearrange the equation by dividing both sides by $R$ and $T$. This step removes them from the right side of the equation, leaving an equality statement for $n$ alone:

$$\frac{PV}{RT} = \frac{n\cancel{R}\cancel{T}}{\cancel{R}\cancel{T}} = n$$

The equation is usually rearranged so that the unknown variable is on the left of the equality:

$$n = \frac{PV}{RT}$$

We can use this relationship to obtain the value of $n$ when the other variables are known, as shown in Example 9.18. Units have been omitted from these problems since the focus is on practicing how to rearrange equations.

**Math Toolbox 9.2 (continued)**

**EXAMPLE 9.18   Rearranging and Solving Equations**

Given that $PV = nRT$, what is the value of $n$ if $P$ is 1.02, $V$ is 0.968, $R$ is 0.08206, and $T$ is 298.15?

**Solution:**

It is often useful to organize the data in some way.

| Given Quantities | Unknown Quantity |
|---|---|
| $P = 1.02$ | $n = ?$ |
| $V = 0.968$ | |
| $R = 0.08206$ | |
| $T = 298.15$ | |

Then substitute the known quantities into the equation and calculate the unknown term:

$$n = \frac{1.02 \times 0.968}{0.08206 \times 298.15} = 0.0404$$

**Practice Problem 9.18**

Given that $PV = nRT$, calculate $T$ when $P$ is 2.25, $V$ is 12.4, $n$ is 1.43, and $R$ is 0.08206.

**Further Practice:** Questions 9.15 and 9.16 at the end of the chapter

# CHAPTER REVIEW

## KEY CONCEPTS

- Several properties—including pressure, volume, temperature, and amount—describe the behavior of gases.
  - Pressure is force per unit area.
  - Gas pressure results from collisions of gas particles with the walls of the container.
  - Under conditions of constant temperature, the volume of a gas decreases when pressure increases (Boyle's law). Volume is inversely proportional to pressure at constant temperature.
  - Under conditions of constant pressure, the volume of a gas increases when temperature increases (Charles's law). Volume is directly proportional to pressure at constant temperature.
  - When not in a rigid container, gases expand when heated and contract when cooled, so warm gases have lower densities.
  - When more gas is added to a container that is allowed to expand, the volume increases due to the presence of more gas particles (Avogadro's hypothesis).
  - The combined gas law relates temperature, pressure, and volume for a fixed amount (in moles) of gas.
- Pressure, volume, temperature, and amount of gas relate mathematically according to the ideal gas law ($PV = nRT$).
  - Given the ideal gas constant and three of the factors, the fourth can be calculated.
  - The density (mass/volume) of a gas can also be calculated from the ideal gas law when given conditions of temperature and pressure.
  - Gases in a mixture behave independently and exert the same pressure they would exert if they were in a container alone. The total pressure of the gas mixture is the sum of each gas pressure (Dalton's law of partial pressures).

- The five postulates of the kinetic-molecular theory explain gas behavior at the particulate level.
  - Gases are composed of small and widely separate particles (molecules or atoms).
  - Particles of a gas behave independently of one another.
  - Each particle in a gas is in rapid, straight-line motion until it collides with another particle or with its container.
  - The pressure of a gas arises from the sum of the collisions of the particles with the walls of the container.
  - The average kinetic energy of gas particles depends only on the absolute temperature.
  - In addition to the gas laws, these postulates of kinetic-molecular theory also account for other properties of gases such as diffusion (the movement of gas particles from an area of high concentration to low concentration) and effusion (the movement of gases through a small opening in a container into a vacuum).
- The ideal gas law can be used in conjunction with reaction calculations discussed in previous chapters to calculate quantities of gases involved in chemical reactions.

## KEY RELATIONSHIPS

| Relationship | Equation |
|---|---|
| Pressure can be expressed in different units. | 1 atm = 101,325 Pa = 760 mm Hg = 760 torr = 14.7 lb/in$^2$ |
| Pressure is inversely proportional to volume at constant temperature and moles (Boyle's law). | $P_1V_1 = P_2V_2$ (constant $T$ and $n$) |
| Temperature is proportional to volume at constant pressure and moles (Charles's law). | $\dfrac{V_1}{T_1} = \dfrac{V_2}{T_2}$ (constant $P$ and $n$) |
| Volume is proportional to temperature divided by the pressure if the amount of gas is constant (combined gas law). | $\dfrac{P_1V_1}{T_1} = \dfrac{P_2V_2}{T_2}$ (constant $n$) |
| Volume is proportional to the amount of gas (moles) at constant temperature and pressure (Avogadro's hypothesis). | $\dfrac{V_1}{n_1} = \dfrac{V_2}{n_2}$ (constant $T$ and $P$) |
| The amount of gas (moles), and its pressure, volume, and temperature are related by the ideal gas law. | $PV = nRT$ |
| For a mixture of gases, the sum of the individual pressures is equal to the total pressure (Dalton's law of partial pressures). | $P_{\text{total}} = P_A + P_B + P_C + \cdots$ |
| The average kinetic energy of gas particles is related to their mass and average velocity. | $KE_{av} = \dfrac{1}{2}m(v_{av})^2$ |

## KEY TERMS

Avogadro's hypothesis (9.2)
barometer (9.1)
Boyle's law (9.2)
Charles's law (9.2)
combined gas law (9.2)

Dalton's law of partial
   pressures (9.3)
diffusion (9.4)
effusion (9.4)
Gay-Lussac's law of
   combining volumes (9.2)

ideal gas (9.2)
ideal gas constant, $R$ (9.3)
ideal gas law (9.3)
kinetic-molecular theory of
   gases (9.4)

molar volume (9.2)
pressure (9.1)
standard temperature and
   pressure (STP) (9.2)

# QUESTIONS AND PROBLEMS

The following questions and problems, except those in *Additional Questions* and *Concept Review Questions*, are paired. Questions in a pair focus on the same concept. Answers to selected questions and problems are in Appendix D.

## Matching Definitions with Key Terms

9.1 Match the key terms with the descriptions provided.
   (a) the movement of gas particles through a small opening into a vacuum
   (b) law stating that at constant temperature, the volume occupied by a fixed amount of a gas is inversely proportional to its pressure
   (c) law that describes the relationship between initial and final conditions of pressure, volume, and temperature for a fixed amount of a gas
   (d) a gas that follows predicted behavior, as described by the ideal gas law
   (e) the amount of force applied per unit area
   (f) law stating that gases in a mixture behave independently and exert the same pressure they would if they were in the container alone
   (g) the volume occupied by 1 mol of a gas, which equals 22.414 L at STP for an ideal gas
   (h) a constant used in the ideal gas law that relates pressure, volume, amount of gas, and temperature

9.2 Match the key terms with the descriptions provided.
   (a) the movement of gas particles through the atmosphere from high concentration to low
   (b) law stating that at constant pressure, the volume occupied by a fixed amount of gas is directly proportional to its absolute temperature
   (c) law stating that in chemical reactions occurring at constant pressure and temperature, the volumes of gaseous reactants and products are in ratios of small whole numbers
   (d) law stating the relationship between pressure, volume, temperature, and amount of an ideal gas
   (e) a closed tube filled with mercury and inverted into a pool of mercury; instrument used to measure atmospheric pressure
   (f) a model of ideal gas behavior; assumes that gas particles occupy no volume and that there are no intermolecular forces between them
   (g) 0°C and 1 atm
   (h) proposal that equal volumes of all gases contain equal numbers of particles (at constant temperature and pressure)

## Math Toolbox Questions

9.3 A student measured the following data for the solubility of a compound in water at various temperatures. The student prepared a graph of these data. List all the incorrect features of this graph.

| Solubility | Temperature |
| --- | --- |
| 12.8 | 17 |
| 21.0 | 23 |
| 25.0 | 31 |
| 30.5 | 39 |
| 37.0 | 43 |
| 39.0 | 51 |

9.4 Use the data from Question 9.3 to prepare a correctly designed graph.

9.5 A series of organic compounds called the alkanes has the general formula $C_nH_{2n+2}$. Plot the boiling point of the alkanes versus the number of carbon atoms in the alkane. Is there a pattern? Are the data linear?

| Alkane | Boiling Point (°C) |
| --- | --- |
| $CH_4$ | −161 |
| $C_2H_6$ | −89 |
| $C_3H_8$ | −44 |
| $C_4H_{10}$ | −0.5 |
| $C_5H_{12}$ | 36 |
| $C_6H_{14}$ | 68 |
| $C_7H_{16}$ | 98 |
| $C_8H_{18}$ | 126 |
| $C_9H_{20}$ | 151 |
| $C_{10}H_{22}$ | 174 |

9.6    Plot the solubility of glycine, an amino acid, in water as a function of temperature. Write an equation to describe this relationship and determine the slope and intercept.

| Solubility (g glycine/100 g water) | Temperature (°C) |
|---|---|
| 14.2 | 0 |
| 25.0 | 25 |
| 39.1 | 50 |
| 54.4 | 75 |
| 67.2 | 100 |

9.7    Boyle used a U-tube to investigate gas properties. As shown in the figure, a gas was trapped in the closed arm of the U-tube at 29.9 in Hg, and the pressure was varied by adding mercury to the open arm. The total pressure exerted on the gas is the sum of the atmospheric pressure (29.9 in Hg) and the pressure due to the addition of mercury as measured by the difference in mercury height. Boyle recorded the following data:

| Length of Gas Column (in) | Difference Between Mercury Levels (in), $\Delta h$ |
|---|---|
| 48 | 0.0 |
| 44 | 2.8 |
| 40 | 6.2 |
| 36 | 10.1 |
| 32 | 15.1 |
| 28 | 21.2 |
| 24 | 29.7 |
| 22 | 35.0 |
| 20 | 41.6 |
| 18 | 48.8 |
| 16 | 58.2 |
| 14 | 71.3 |
| 12 | 88.5 |

Graph these data. What does the graph show about the relationship between volume and pressure?

9.8    For the data in Question 9.7, graph the second column against the inverse of the data in the first column. What does the graph show about the relationship between volume and pressure?

9.9    The vapor pressure of water varies with temperature as shown in the graph. What is the value of the vapor pressure at 22°C? At 38°C?

**9.10**   Plot the velocity of sound at 25°C as a function of the density of the liquid medium through which it is passing. Is there a relationship between these variables? What is the value of the velocity for a liquid with a density of 0.92 g/mL?

| Liquid | Velocity of Sound (m/s) | Density (g/mL) |
|---|---|---|
| ethyl ether | 985 | 0.713 |
| acetone | 1174 | 0.79 |
| ethanol | 1207 | 0.79 |
| methanol | 1103 | 0.791 |
| kerosene | 1324 | 0.81 |
| benzene | 1295 | 0.870 |
| turpentine | 1255 | 0.88 |
| castor oil | 1477 | 0.969 |
| distilled water | 1497 | 0.998 |
| seawater | 1531 | 1.025 |
| ethylene glycol | 1658 | 1.113 |
| glycerol | 1904 | 1.26 |

**9.11**   Solve the following equations for $x$.
(a) $5x + 1 = 3$
(b) $0.412/x = 2.00$
(c) $x^2 = 32 - x^2$
(d) $2x = 6 - x$

**9.12**   Solve the following equations for $x$.
(a) $14x + 16 = 44$
(b) $\frac{1}{3}x + 3 = 5$
(c) $12x = 4(12 - x)$
(d) $3x + 15 = 4x + 12$

**9.13**   Convert the following temperatures from degrees Fahrenheit to degrees Celsius.
(a) 212°F
(b) 80.0°F
(c) 32.0°F
(d) −40.0°F

**9.14**   Convert the following temperatures from degrees Celsius to degrees Fahrenheit.
(a) 37.0°C
(b) 212°C
(c) 100.0°C
(d) −40.0°C

**9.15**   Given that $PV = nRT$, solve for the unknown quantity.

|  | $P$ | $V$ | $n$ | $R$ | $T$ |
|---|---|---|---|---|---|
| (a) | 1.00 | 1.00 | 0.500 | 0.08206 | ? |
| (b) | 0.750 | 3.00 | ? | 0.08206 | 237 |
| (c) | 3.25 | ? | 1.50 | 0.08206 | 455 |
| (d) | ? | 15.0 | 2.67 | 0.08206 | 322 |

**9.16**   Given that $PV = nRT$, solve for the unknown quantity.

|  | $P$ | $V$ | $n$ | $R$ | $T$ |
|---|---|---|---|---|---|
| (a) | 3.55 | 1.75 | 0.205 | 0.08206 | ? |
| (b) | 1.00 | 22.5 | ? | 0.08206 | 298 |
| (c) | 0.125 | ? | 3.00 | 0.08206 | 535 |
| (d) | ? | 6.25 | 1.57 | 0.08206 | 343 |

## The Behavior of Gases

**9.17**   What are some general properties of gases?

**9.18**   In general, how do the properties of gases differ from the properties of liquids and solids?

**9.19**   How does the density of warm air differ from the density of cooler air?

**9.20**   Why does warm air rise?

**9.21**   The figure shows atoms of a gas at a particular temperature. In the blank circle, show the arrangement of atoms when the temperature decreases and pressure remains constant.

Before                                        After

9.22 The figure shows atoms of a gas at a particular temperature. Students were asked to select images that show what happens when the temperature increases and pressure remains constant. Many students selected the images shown. What is wrong with each of images (a) to (d)?

Before

(a)                    (b)

(c)                    (d)

9.23 What is gas pressure?

9.24 Why do gases exert pressure on the walls of their container?

9.25 How is pressure measured?

9.26 (a) What are the common units of pressure? (b) How are they related?

9.27 The figure shows atoms of a gas at a particular pressure. In the blank circle show the arrangement of atoms when the volume increases and temperature remains constant.

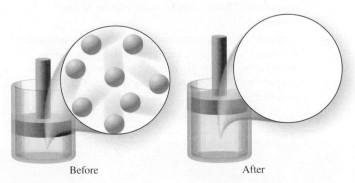

Before          After

9.28 The figure shows atoms of a gas at a particular temperature. Students were asked to select images that show what happens when the volume increases and temperature remains constant. Many students selected the images shown. What is wrong with each of images (a) to (d)?

Before

(a)                    (b)

(c)                    (d)

9.29 Perform the following pressure conversions.
(a) 745 torr to atm
(b) 1.23 atm to torr
(c) 90.1 mm Hg to atm
(d) 0.643 kPa to Pa
(e) $1.35 \times 10^5$ Pa to mm Hg
(f) $7.51 \times 10^4$ Pa to torr
(g) 798 torr to Pa
(h) 29.3 cm Hg to mm Hg

9.30 Perform the following pressure conversions.
(a) 1.15 atm to torr
(b) 968 torr to atm
(c) $2.50 \times 10^5$ Pa to atm
(d) 695 mm Hg to torr
(e) 0.953 atm to Pa
(f) 653 torr to mm Hg
(g) 1545 mm Hg to Pa
(h) 3.73 kPa to atm

9.31 If the barometric pressure in Denver, Colorado, is 30.24 inches of mercury (in Hg), what is the pressure in units of pascals? (Recall that 2.54 cm = 1 in.)

9.32 If the barometric pressure in Vancouver, Canada, is 101.19 kPa, what is the pressure in units of inches of mercury (in Hg)? (Recall that 2.54 cm = 1 in.)

## Factors That Affect the Properties of Gases

9.33 What does Boyle's law tell us about the effect of pressure on the volume of a gas?

9.34 What is Boyle's law in mathematical terms?

9.35 Consider a gas in a container with a movable piston. What happens to the gas particles if the piston is moved to decrease the volume of the container? What happens to the pressure if the piston is moved to give one-half of the original volume of the cylinder? In the blank circle, show the arrangement of atoms when the volume decreases by one-half.

Before    After

9.36 Consider a gas in a container with a movable piston. What happens to the gas particles if the piston is moved to increase the volume of the container? What happens to the pressure if the piston is moved to triple the original volume of the cylinder? In the blank circle, show the arrangement of atoms when the volume is tripled.

Before    After

9.37 Given a fixed quantity of a gas at constant temperature, calculate the new volume the gas would occupy if the pressure were changed as shown in the following table.

|  | Initial Volume | Initial Pressure | Final Pressure | Final Volume |
|---|---|---|---|---|
| (a) | 6.00 L | 3.00 atm | 5.00 atm | ? |
| (b) | 40.0 mL | 60.0 torr | 90.0 torr | ? |
| (c) | 2.50 mL | 40.0 torr | 255 torr | ? |

9.38 Given a fixed quantity of a gas at constant temperature, calculate the new volume the gas would occupy if the pressure were changed as shown in the following table.

|  | Initial Volume | Initial Pressure | Final Pressure | Final Volume |
|---|---|---|---|---|
| (a) | 2.50 L | 2.50 atm | 725 atm | ? |
| (b) | 6.25 mL | 825 torr | 456 torr | ? |
| (c) | 450 mL | 50.0 torr | 30.0 torr | ? |

9.39 Given a fixed quantity of a gas at constant temperature, calculate the new pressure the gas would exert if the volume were changed as shown in the following table.

|  | Initial Pressure | Initial Volume | Final Volume | Final Pressure |
|---|---|---|---|---|
| (a) | 602 torr | 405 mL | 1512 mL | ? |
| (b) | 0.00100 torr | 1.50 L | 15.0 mL | ? |
| (c) | 0.832 atm | 805 L | 37.5 L | ? |

9.40 Given a fixed quantity of a gas at constant temperature, calculate the new pressure the gas would exert if the volume were changed as shown in the following table.

|  | Initial Pressure | Initial Volume | Final Volume | Final Pressure |
|---|---|---|---|---|
| (a) | 845 torr | 155 mL | 1.55 L | ? |
| (b) | 5.30 atm | 2.85 L | 4.50 L | ? |
| (c) | 755 torr | 2.00 L | 5500 mL | ? |

9.41 What pressure is required to compress 925 L of $N_2$ at 1.25 atm into a container whose volume is 6.35 L?

9.42 What will be the final volume of 186 mL of $Cl_2$ if the pressure increases from 0.945 atm to 1.76 atm?

9.43 What volume of $H_2$ must be collected at 725 torr to have 0.550 L of the gas at 1.00 atm?

9.44 A 2.50-L flask is filled with air at 2.00 atm. What size flask is needed to hold this much air at 735 torr?

9.45 What does Charles's law tell us about the effect of temperature on the volume of a gas?

9.46 What is Charles's law in mathematical terms?

9.47 Consider a gas in a container that can adjust its volume to maintain constant pressure. Suppose the gas is heated. What happens to the gas particles with the increase in temperature? What happens to the volume of the container?

9.48 Consider a gas in a container that can adjust its volume to maintain constant pressure. Suppose the gas is cooled. What happens to the gas particles with the decrease in temperature? What happens to the volume of the container?

**9.49** For a fixed amount of gas held at constant pressure, calculate the new volume the gas would occupy if the temperature were changed as shown in the following table.

|     | Initial Volume | Initial Temperature | Final Temperature | Final Volume |
| --- | --- | --- | --- | --- |
| (a) | 6.00 L | 30.0°C | 0.0°C | ? |
| (b) | 212 mL | −60.0°C | 401.0°C | ? |
| (c) | 47.5 L | 212 K | 337 K | ? |

**9.50** For a fixed amount of gas held at constant pressure, calculate the new volume the gas would occupy if the temperature were changed as shown in the following table.

|     | Initial Volume | Initial Temperature | Final Temperature | Final Volume |
| --- | --- | --- | --- | --- |
| (a) | 224 L | 0.0°C | 100.0°C | ? |
| (b) | 152 mL | 45 K | 450 K | ? |
| (c) | 156 mL | 45°C | 450°C | ? |

**9.51** For a fixed amount of gas held at constant pressure, calculate the temperature in degrees Celsius to which the gas would have to be changed to achieve the change in volume shown in the following table.

|     | Initial Temperature | Initial Volume | Final Volume | Final Temperature |
| --- | --- | --- | --- | --- |
| (a) | 0.0°C | 70.0 mL | 140.0 mL | ? |
| (b) | −37°C | 2.55 L | 85 mL | ? |
| (c) | 165 K | 87.5 L | 135 L | ? |

**9.52** For a fixed amount of gas held at constant pressure, calculate the temperature in degrees Celsius to which the gas would have to be changed to achieve the change in volume shown in the following table.

|     | Initial Temperature | Initial Volume | Final Volume | Final Temperature |
| --- | --- | --- | --- | --- |
| (a) | 100.0°C | 250.0 mL | 100.0 mL | ? |
| (b) | 27.5°C | 125 mL | 148 mL | ? |
| (c) | 300 K | 13.7 L | 57.2 L | ? |

**9.53** A bubble of air has a volume of 0.150 mL at 24.2°C. If the pressure remains constant, what is the volume of the bubble at 62.5°C?

**9.54** If 79.0 L of helium at 27.0°C is compressed to 32.0 L at constant pressure, what is the new temperature?

**9.55** Suppose a steel tank is filled with a gas at sea level. This tank keeps the volume constant. If the tank is placed in the bed of a truck and driven up a mountain road to an elevation of 10,000 ft, what happens to the gas particles?

**9.56** Suppose a flexible container is filled with a gas at an elevation of 9000 ft. What will happen to this gas if the container is transported to sea level?

**9.57** Assume that the volume of a fixed amount of gas in a rigid container does not change. Calculate the pressure the gas would exert if the temperature were changed as shown in the following table.

|     | Initial Pressure | Initial Temperature | Final Temperature | Final Pressure |
| --- | --- | --- | --- | --- |
| (a) | 302 torr | 0.0°C | 105.0°C | ? |
| (b) | 735 torr | 25.0°C | 0.0°C | ? |
| (c) | 3.25 atm | 273 K | 373 K | ? |

**9.58** Assume that the volume of a fixed amount of gas in a rigid container does not change. Calculate the pressure the gas would exert if the temperature were changed as shown in the following table.

|     | Initial Pressure | Initial Temperature | Final Temperature | Final Pressure |
| --- | --- | --- | --- | --- |
| (a) | 255 torr | 225 K | 315 K | ? |
| (b) | 895 torr | 25°C | 206 K | ? |
| (c) | 2.74 atm | 150°C | 23°C | ? |

**9.59** Assume that the volume of a fixed amount of gas in a rigid container does not change. Calculate the temperature in degrees Celsius to which the gas would have to be changed to achieve the change in pressure shown in the following table.

|     | Initial Temperature | Initial Pressure | Final Pressure | Final Temperature |
| --- | --- | --- | --- | --- |
| (a) | 30.0°C | 1525 torr | 915 torr | ? |
| (b) | 250.0°C | 0.70 atm | 1042 torr | ? |
| (c) | 355 K | 500.0 torr | 1000.0 torr | ? |

**9.60** Assume that the volume of a fixed amount of gas in a rigid container does not change. Calculate the temperature in degrees Celsius to which the gas would have to be changed to achieve the change in pressure shown in the following table.

|     | Initial Temperature | Initial Pressure | Final Pressure | Final Temperature |
| --- | --- | --- | --- | --- |
| (a) | 25.0°C | 243 torr | 735 torr | ? |
| (b) | 205 K | 2.35 atm | 1.20 atm | ? |
| (c) | 375 K | 875 torr | 0.85 atm | ? |

**9.61** A steel tank contains acetylene gas at a pressure of 7.25 atm at 18.5°C. What is the pressure at 37.2°C?

**9.62** A steel tank contains gas at a pressure of 6.75 atm at 25.0°C. At what temperature will the pressure decrease to 1.25 atm?

9.63 For a gas under a given initial set of conditions, calculate the final value for the variable indicated if the other two variables change as described in the following table.

|  | (a) | (b) | (c) |
|---|---|---|---|
| **Initial volume** | 2.50 L | 125 L | 455 mL |
| **Initial pressure** | 0.50 atm | 0.250 atm | 200.0 torr |
| **Initial temperature** | 20.0°C | 25°C | 300 K |
| **Final volume** | ? | 62.0 L | 200.0 mL |
| **Final pressure** | 760.0 torr | 100.0 torr | ? |
| **Final temperature** | 0.0°C | ? | 327°C |

9.64 For a gas under a given initial set of conditions, calculate the final value for the variable indicated if the other two variables change as described in the following table.

|  | (a) | (b) | (c) |
|---|---|---|---|
| **Initial volume** | 601 mL | 237 mL | 1.12 L |
| **Initial pressure** | 900.0 torr | 75.0 atm | 760.0 torr |
| **Initial temperature** | −10.0°C | 147 K | 0.0°C |
| **Final volume** | 1200.0 mL | 474 mL | ? |
| **Final pressure** | ? | 150.0 atm | 700.0 torr |
| **Final temperature** | 0.0°C | ? | 25.0°C |

9.65 What volume of $O_2$ at STP can be pumped into a 0.500-L tank at 24.5°C to give a pressure of 3.50 atm?

9.66 If 22.0 L of $N_2$ at 25.0°C and 725 torr are heated to 134°C and compressed to 4.50 L, what is the new pressure?

9.67 What is Gay-Lussac's law?

9.68 What is the relationship between the volume of a gaseous reactant and the volume of a gaseous product when temperature and pressure are constant?

9.69 What is the molar volume of neon gas at standard temperature and pressure (STP)?

9.70 What is the volume occupied by 1.00 mol of $H_2$ gas at STP?

9.71 Given the following volumes of gases at STP, calculate the number of moles of each gas and the mass of the gas.
(a) 8.62 L $CH_4$ (b) 350.0 mL Xe (c) 48.1 L CO

9.72 Given the following volumes of gases at STP, calculate the number of moles of each gas and the mass of the gas.
(a) 135 mL $H_2$ (b) 8.96 L $N_2$ (c) 0.75 L He

9.73 The two balloons shown each have a volume of 1.5 L. They are tethered together and are at the same temperature and pressure. Which balloon has the greater number of gas particles? Which balloon has the greater mass? Which balloon has the greater density?

9.74 The two balloons shown each have a volume of 1.5 L. They are tethered together and are at the same temperature and pressure. Which balloon has the greater number of gas particles? Which balloon has the greater mass? Which balloon has the greater density?

9.75 Given the following amounts of gases, calculate the number of moles of each gas. Calculate the volume each amount of gas would occupy at STP.
(a) 5.8 g $NH_3$ (b) 48 g $O_2$ (c) 10.8 g He

9.76 Given the following amounts of gases, calculate the number of moles of each gas. Calculate the volume each amount of gas would occupy at STP.
(a) 2.2 g $CO_2$ (b) 5.6 g $N_2$ (c) 145 g Ar

9.77 Assuming all the gas is removed from the tank when filling balloons, how many 1.0-L balloons can be filled from this tank at STP?

425 g He

9.78 Assuming all the gas is removed from the tank when filling balloons, how many 0.75-L balloons can be filled from this tank at STP?

370 g $CO_2$

9.79 What will be the volume at STP of $O_2$ gas that occupies 12.0 L at 25.0°C assuming constant pressure?

9.80 What will be the volume of 1.50 mol $H_2$ gas at 30.0°C and 0.975 atm?

## The Ideal Gas Law

9.81 What is an ideal gas?

9.82 (a) State the ideal gas law. (b) What is it used for?

9.83 Given the following amounts of gases, calculate the number of moles of each gas. Calculate the volume each amount of gas would occupy at 100.0°C and 15.0 atm.
(a) 5.8 g $NH_3$ (b) 48 g $O_2$ (c) 10.8 g He

9.84 Given the following amounts of gases, calculate the number of moles of each gas. Calculate the volume each amount of gas would occupy at 75.0°C and 3.5 atm.
(a) 2.2 g $CO_2$ (b) 5.6 g $N_2$ (c) 7.5 g Ar

9.85 Given the following volumes of gases at 87.5°C and 722 torr, calculate the number of moles of each gas and the mass of the gas.
(a) 7.62 L $CH_4$ (b) 135 mL $H_2$ (c) 8.96 L $N_2$

9.86 Given the following volumes of gases at 54.0°C and 790 torr, calculate the number of moles of each gas and the mass of the gas.
(a) 150.0 mL Xe (b) 38.1 L CO (c) 2.5 L $O_2$

9.87 If two balloons at the same temperature and pressure contain equal volumes of $CO_2$ and He, why does the balloon containing $CO_2$ stay on the ground and the helium balloon float?

©Tom Pantages

9.88 These two balloons contain the same number of moles of $N_2$ gas but at different temperatures. Which balloon is at a higher temperature? How do these balloons differ in their densities? Draw representations of gas particles to show how the molecular-level images of $N_2$ molecules differ in the two balloons.

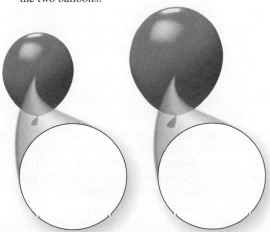

©Fuse/Getty Images

9.89 Calculate the density in grams per liter of the following gases at STP.
(a) $NH_3$ (b) $N_2$ (c) $N_2O$

9.90 Calculate the density in grams per liter of the following gases at STP.
(a) NO (b) $NO_2$ (c) $O_2$

9.91 Calculate the density in grams per liter of the following gases at 25.0°C and 735 torr.
(a) $NH_3$ (b) $N_2$ (c) $N_2O$

9.92 Calculate the density in grams per liter of the following gases at 25.0°C and 735 torr.
(a) NO (b) $NO_2$ (c) $O_2$

9.93 If 5.00-L containers are filled to a pressure of 840 torr at 50.0°C with the following gases, calculate the mass of each gas in its container.
(a) $H_2$ (b) $CH_4$ (c) $SO_2$

9.94 If 2.50-L containers are filled to a pressure of 650 torr at 25.0°C with the following gases, calculate the mass of each gas in its container.
(a) $O_2$ (b) $CO_2$ (c) He

9.95 What is Dalton's law of partial pressures?

9.96 How is Dalton's law of partial pressures used to determine the pressure and moles of a sample of gas that is collected over water?

9.97 A sample of oxygen gas saturated with water vapor at 22°C exerts a total pressure of 728 torr. What is the partial pressure of oxygen in the flask? The vapor pressure of water at 22°C is 20.0 torr.

9.98 A total of 0.400 L of hydrogen gas is collected over water at 18°C. The total pressure is 742 torr. If the vapor pressure of water at 18°C is 15.5 torr, what is the partial pressure of hydrogen?

9.99 A tank contains 78.0 g of $N_2$ and 42.0 g of Ne at a total pressure of 3.75 atm and a temperature of 50.0°C. Calculate the following quantities.
(a) moles of $N_2$
(b) moles of Ne
(c) partial pressure of $N_2$
(d) partial pressure of Ne

9.100 A tank contains 150.0 g of $CO_2$ and 24.0 g of $O_2$ at a total pressure of 4.25 atm and a temperature of 25.0°C. Calculate the following quantities.
(a) moles of $CO_2$
(b) moles of $O_2$
(c) partial pressure of $CO_2$
(d) partial pressure of $O_2$

9.101 Given the following densities (in grams per liter) of gases at STP, calculate the molar mass of each gas.
(a) 1.785
(b) 1.340
(c) 2.052
(d) 0.905
(e) 0.714

9.102 Given the following densities (in grams per liter) of gases at 27.0°C and 745 torr, calculate the molar mass of each gas.
(a) 2.436 (d) 3.450
(b) 0.842 (e) 1.706
(c) 1.325

## Kinetic-Molecular Theory of Gases

9.103   State the five postulates of the kinetic-molecular theory.

9.104   How is kinetic energy related to gas pressure?

9.105   Explain in molecular terms why the pressure of a gas increases with increasing temperature when the volume is held constant.

9.106   Joel fills a weather balloon with helium gas in a room at 22°C. He takes the balloon outside where the pressure is the same, but the temperature is 0°C. Draw pictures of particles to show why, at constant pressure, the volume occupied by the helium gas decreases as temperature decreases.

9.107   At constant temperature, the pressure of a gas is inversely proportional to its volume. Explain this relationship in molecular terms.

9.108   Explain why the pressure of a gas sample is directly proportional to the moles of gas particles in the sample.

9.109   At the same temperature, rank the following substances in order of increasing average velocity: $CO_2$, He, $H_2$, $CH_4$.

9.110   At the same temperature, rank the following substances in order of increasing average kinetic energy: $CO_2$, He, $H_2$, $CH_4$.

9.111   What is the relationship between rate of diffusion of a gas and its molar mass?

9.112   What is the relationship between rate of effusion of a gas and its molar mass?

9.113   If equal amounts of helium and neon are placed in a porous container, which gas will escape faster?

9.114   If equal amounts of xenon and argon are placed in a porous container, which gas will escape faster?

## Gases and Chemical Reactions

9.115   What volume of hydrogen is required to react with 12 L of oxygen under the same conditions?

$$2H_2(g) + O_2(g) \longrightarrow 2H_2O(g)$$

9.116   What volume of nitrogen is required to react with 9 L of hydrogen under the same conditions?

$$3H_2(g) + N_2(g) \longrightarrow 2NH_3(g)$$

9.117   Hexane burns according to the following equation:

$$2C_6H_{14}(g) + 19O_2(g) \longrightarrow 12CO_2(g) + 14H_2O(g)$$

What volume of $CO_2$ forms when 8.00 L of hexane burn, assuming the two volumes are measured under the same conditions? What volume of oxygen will be needed?

9.118   When nitric acid is synthesized from ammonia, the first step in the process is

$$4NH_3(g) + 5O_2(g) \longrightarrow 4NO(g) + 6H_2O(g)$$

What volume of NO at STP can be formed from 1250 L of $NH_3$ at 325°C and 4.25 atm, assuming temperature and pressure conditions remain constant?

9.119   Nitrous oxide can be formed by the thermal decomposition of ammonium nitrate:

$$NH_4NO_3(s) \xrightarrow{\text{heat}} N_2O(g) + 2H_2O(g)$$

What mass of ammonium nitrate is required to produce 145 L of $N_2O$ at 2850 torr and 42°C?

9.120   Nitric oxide is produced in the reaction between copper metal and nitric acid:

$$3Cu(s) + 8HNO_3(aq) \longrightarrow$$
$$3Cu(NO_3)_2(aq) + 4H_2O(l) + 2NO(g)$$

What mass of copper is required to produce 15.0 L of NO at 725 torr and 20.0°C?

## Additional Questions

9.121   Why does a weather balloon change in volume when external pressure changes?

9.122   Why do gases expand when they are heated?

9.123   How can a hot-air balloon float in air?

9.124   Why are gas bubbles smaller at the bottom of a fish tank?

9.125   What would have to be the surface temperature to observe no change in volume in a weather balloon once it reaches the outer edge of the troposphere? Assume the pressure at the surface is 760 torr and changes to 150 torr. The temperature at the outer edge of the troposphere is about 218 K.

9.126   If gas molecules move at high speeds, why does it seem to take a long time for smells to permeate a room?

9.127   In macroscopic and molecular-level terms, describe what will happen to this balloon if temperature increases while pressure remains constant. What will happen if temperature decreases at constant pressure? What will happen if the balloon is taken up to an elevation of 10,000 feet?

©Jules Frazier/Getty Images

9.128   In older air bags used in automobiles, the nitrogen gas that fills the bags was produced by the decomposition of sodium azide, $NaN_3$:

$$2NaN_3(s) \longrightarrow 2Na(s) + 3N_2(g)$$

What mass of sodium azide is needed to fill a 2.50-L air bag with nitrogen gas at a pressure of 1140 torr and 25°C?

9.129   Oxygen can be produced by the thermal decomposition of mercury(II) oxide:

$$2HgO(s) \xrightarrow{\text{heat}} 2Hg(l) + O_2(g)$$

What volume of $O_2$ is produced at 50.0°C and 0.947 atm by the decomposition of 27.0 g of HgO?

9.130   A sample of argon measuring 3.00 L at 27.0°C and 765 torr was bubbled through and collected over water. The pressure of the new system is 778 torr at 15.0°C. What is the volume occupied by the gas? The vapor pressure of water at 15.0°C is 13 torr.

9.131   Consider the combustion of butene:

$$C_4H_8(g) + 6O_2(g) \longrightarrow 4CO_2(g) + 4H_2O(g)$$

What volume of butene at 188°C and 2.50 atm can be burned with 12.0 L of $O_2$ at 745 torr and 25.0°C?

9.132   A 2.00-L tank contains a mixture of 72.0 g $N_2$ and 66.0 g Ar. Calculate the total pressure of gas in the tank at 45.0°C. Calculate the partial pressure of each gas.

9.133   Copper(II) oxide can be reduced to copper metal by heating in a stream of hydrogen gas:

$$CuO(s) + H_2(g) \xrightarrow{\text{heat}} Cu(s) + H_2O(g)$$

What volume of hydrogen at 27.0°C and 722 torr would be required to react with 85.0 g of CuO?

9.134   The given images represent a collection of gas molecules at two different pressures. Which image shows molecules at the greatest pressure? What changes in conditions could explain the changes in this gas sample at a molecular level?

Before                    After

9.135   If all volumes of gases are measured under the same conditions, what volume of the product will be formed from reaction of 5.00 L of $H_2$ in each of the following reactions, assuming they go to completion?
    (a) $S(s) + H_2(g) \longrightarrow H_2S(g)$
    (b) $N_2(g) + 3H_2(g) \longrightarrow 2NH_3(g)$
    (c) $C_6H_6(g) + 3H_2(g) \longrightarrow C_6H_{12}(g)$

9.136   A cement company uses $1.00 \times 10^5$ kg of limestone daily. The limestone decomposes upon heating, forming lime and carbon dioxide:

$$CaCO_3(s) \xrightarrow{\text{heat}} CaO(s) + CO_2(g)$$

What volume of $CO_2$ at 735 torr and 25.0°C is released into the atmosphere daily by this company?

9.137   Calcium carbide, $CaC_2$, is made by heating lime, CaO, with carbon. The lime is made by heating limestone, $CaCO_3$. Acetylene, $C_2H_2$, can be made from the calcium carbide by reacting it with water:

$$CaC_2(s) + 2H_2O(l) \longrightarrow Ca(OH)_2(aq) + C_2H_2(g)$$

If 1.25 L of acetylene at 0.750 atm can be made from 5.00 g of limestone, what is the temperature of the gas?

9.138   Oxygen gas can be generated by heating potassium chlorate:

$$2KClO_3(s) \xrightarrow{\text{heat}} 2KCl(s) + 3O_2(g)$$

What volume of oxygen gas, collected by displacement of water and measured at 735.0 torr and 70.0°C, will be formed by the decomposition of 13.5 g of potassium chlorate?

9.139   (a) How is the volume occupied by a gas related to the number of moles of that gas?
    (b) What other variable(s) must be held constant to show experimentally that this relationship is valid?

9.140   The density of air at 760.0 torr and 25.0°C is 1.186 g/L.
    (a) Calculate the average molar mass of air.
    (b) From this value, and assuming that air contains only molecular nitrogen and molecular oxygen gases, calculate the mass percent of $N_2$ and of $O_2$ in air.

9.141   A sample of an unknown liquid weighing 0.495 g is collected as vapor in a 127-mL flask at 98°C. The pressure of the vapor in the flask is then measured and found to be 691 torr. What is the molar mass of the liquid?

9.142   Airplanes are pressurized for the comfort and safety of the passengers and flight crew. Generally, the air pressure in the cabin is equivalent to around 3000 ft (915 m), which is a pressure of 686 torr, about 90% of that at sea level. A pressurization equivalent to 8000 ft (2500 m) would still be safe, with a pressure of 570 torr, about 74% of that at sea level. What would happen to the passengers if the aircraft was not pressurized and flew at 30,000 ft (9150 m)?

9.143   The explosion of nitroglycerin can be described by the following equation:

$$4C_3H_5(ONO_2)_3(l) \longrightarrow$$
$$12CO_2(g) + 10H_2O(g) + 6N_2(g) + O_2(g)$$

What is the total volume of gases produced at 2.00 atm and 275°C from 1.00 g of nitroglycerin?

9.144   Why do air bubbles get larger when they rise toward the surface of water?

9.145   Predict if and how each of the following will change when the *volume* of an oxygen gas sample is reduced at *constant temperature*.
    (a) average kinetic energy
    (b) average molecular velocity
    (c) pressure

9.146   Predict if and how each of the following will change when the *temperature* of an oxygen gas sample is reduced at *constant volume* (rigid container).
    (a) average kinetic energy
    (b) average molecular velocity
    (c) pressure

9.147   Predict if and how each of the following will change when the *temperature* of an oxygen gas sample in a perfectly elastic container is reduced at *constant pressure*.
    (a) average kinetic energy
    (b) average molecular velocity
    (c) volume

9.148   For each of the following gases, determine the moles and mass contained in a 20.0-L volume at a pressure of 389.0 torr at a temperature of 300.0 K.
    (a) neon
    (b) hydrogen
    (c) carbon dioxide

9.149   Hydrogen gas has a density of 0.090 g/L under a certain set of conditions of temperature and pressure. Under the same set of conditions, an unknown gaseous substance was found to have a density of 0.716 g/L, which is about 8.0 times the density of hydrogen. Which of the gases $O_2$, $CH_4$, or Ne is most likely to be the unknown gas?

9.150    Explain each of the following phenomena in terms of gas laws:
  (a) When flying in an airplane, you drink half a bottle of water and then seal it. When the plane lands, you notice that the bottle is crushed.
  (b) Unopened potato chip bags are puffier at 8000 ft elevation than at sea level.
  (c) On a hot summer day in Arizona, soccer balls have been known to explode when left in the trunk of a parked car.
  (d) The air in a hot-air balloon is less dense than in a balloon at colder temperatures.

9.151    Describe three ways to increase the pressure of a contained gas.

9.152    The figure shows oxygen gas being bubbled through water and collected above the water, along with some water vapor. Assume the gas is collected at 50.0°C.

  (a) Use Table 9.2 in your text to determine the partial pressure of the water vapor, $P_{H_2O}$.
  (b) Calculate the partial pressure of oxygen, $P_{O_2}$, using the ratio of the $H_2O$ and $O_2$ gases in the molecular-level image and Dalton's law of partial pressures.
  (c) What is the total pressure of the gas collected, $P_{total}$?
  (d) If the volume of the collected gas is 120.0 mL, how many moles of oxygen were collected?
  (e) If the balanced equation for the reaction that produced the oxygen is

$$2KClO_3(s) \xrightarrow{\text{heat}} 2KCl(s) + 3O_2(g),$$

  how many moles of $KClO_3$ reacted?

9.153    Although the planet Venus is similar in size and composition to Earth, the atmosphere at its surface is not conducive to life. The average temperature is about 740 K, the atmospheric pressure is about 91 atm, and the atmosphere is mostly carbon dioxide.
  (a) Calculate the moles of atmospheric gas in 1.0 L at the surface of Venus.
  (b) Calculate the mass of atmospheric gas (assume it all to be carbon dioxide) in 1.0 L.
  (c) Calculate the density of Venus's atmosphere at its surface in units of both g/L and g/mL.

## Concept Review Questions

9.154    Which of the following statements is correct?
  A. The densities of gases are greater than the densities of liquids and solids.
  B. Gases consist of particles that are relatively close together compared to liquids and solids.
  C. Gas pressure arises from the force exerted by gas particles per unit area of the container.
  D. Warm air is more dense than cold air.
  E. When the temperature of a gas increases, the density of the gas increases.

Modify each incorrect statement to make it a true statement.

9.155    Which of the following is the greatest pressure?
  A. 12.1 lb/in$^2$
  B. 5.0 atm
  C. 34.1 in Hg
  D. 760 torr
  E. 64,000 Pa

Arrange these quantities in order of increasing pressure.

9.156    In January, a balloon is filled to 1.50 L at 25°C inside, then taken outside where the temperature is −20.0°C. What will happen to the balloon when it is taken outside?
  A. The balloon will compress because the low heat will speed the molecules up.
  B. Nothing happens to the balloon because temperature has no effect on the behavior of gases.
  C. The balloon expands because the molecules get bigger.
  D. The balloon expands because the molecules move faster.
  E. The balloon will compress because the molecules will slow down.

Modify each incorrect statement to make it a true statement.

9.157 The "Before" figure shows atoms of a gas at a particular temperature. If the volume decreases while temperature remains constant, which of the images corresponds to the final condition of the gas?

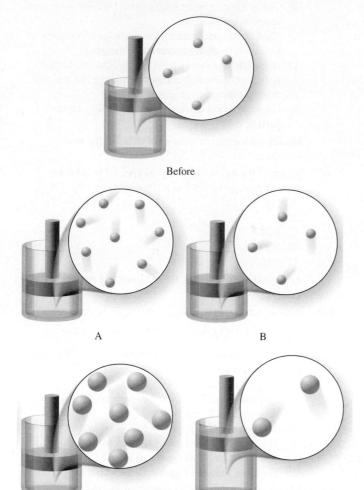

Before

A

B

C

D

E

Explain what is wrong with the incorrect images.

9.158 Which of the following statements is correct?
A. The volume of a gas is directly proportional to pressure at constant temperature.
B. The volume of a gas is inversely proportional to temperature at constant pressure.
C. The volume of a gas is inversely proportional to the pressure at constant temperature.
D. Equal volumes of different gases will contain different moles of gas depending on the molar masses of the gases, at a given temperature and pressure.
E. The volume of a gas is inversely proportional to the moles of gas at constant temperature and pressure.

Modify each incorrect statement to make it a true statement.

9.159 A sample of helium gas at 25°C and 745 torr occupies a volume of 3.84 L. Which of the following mathematical operations would correctly determine the number of moles of helium?

A. $n = \dfrac{\left(0.08206\ \dfrac{L \cdot atm}{mol \cdot K}\right)(25°C)}{\left(745\ torr \times \dfrac{1\ atm}{760\ torr}\right)(3.84\ L)}$

B. $n = \dfrac{(745\ torr)(3.84\ L)}{\left(0.08206\ \dfrac{mol \cdot K}{L \cdot atm}\right)(25°C)}$

C. $n = \dfrac{\left(745\ torr \times \dfrac{1\ atm}{760\ torr}\right)(3.84\ L)}{\left(0.08206\ \dfrac{L \cdot atm}{mol \cdot K}\right)(298\ K)}$

D. $n = \dfrac{\left(0.08206\ \dfrac{L \cdot atm}{mol \cdot K}\right)\left(745\ torr \times \dfrac{1\ atm}{760\ torr}\right)}{(3.84\ L)(298\ K)}$

E. $n = \dfrac{(760\ atm)(22.414\ L)}{\left(0.08206\ \dfrac{L \cdot atm}{mol \cdot K}\right)(273.15\ K)}$

Explain what is wrong with the incorrect responses.

9.160   If 1.25 mol of $N_2$ at 25.0°C and 755 torr is heated to 50.0°C and the pressure changes to 978 torr, what is expected to be observed?

   A. The volume will double because the temperature doubles and pressure decreases.
   B. The initial volume is 2.58 L.
   C. The volume is expected to increase because temperature changes by a larger factor than pressure.
   D. The final volume is 25.8 L.
   E. The volume of any gas is 22.4 L regardless of the amount of gas or conditions of temperature and pressure.

   Modify each incorrect statement to make it a true statement.

9.161   If five 2.50-L containers are each filled to a pressure of 1.12 atm at 25°C with one of the following gases, which gas will have the greatest density?

   A. $N_2$
   B. $O_2$
   C. $CO_2$
   D. He
   E. $SO_2$

   Calculate the density of each gas and arrange them in order of increasing density.

9.162   A 1.5-L container of He and a 1.5-L container of $O_2$ are both at 22°C and 760 torr. Which of the following statements about these gas samples is true?

   A. The He and $O_2$ molecules have the same mass.
   B. There are more He moles than $O_2$ moles.
   C. The average kinetic energy of the He molecules and $O_2$ molecules is the same.
   D. The $O_2$ molecules on average are moving faster than the He molecules.
   E. The volume of the $O_2$ container is greater than the volume of the He container.

   Modify each incorrect statement to make it a true statement.

9.163   Which of the following gases occupies the greatest volume at STP?

   A. 46.2 g $N_2$
   B. 24.3 g $H_2$
   C. 54.1 g $CO_2$
   D. 39.4 g Ar
   E. 21.5 g $NH_3$

   Calculate the volume of each gas and arrange them in order of increasing volume.

9.164   Which of the following statements is correct according to kinetic-molecular theory?

   A. The average kinetic energy of gas particles depends on the identity of the gas.
   B. The average velocity of gas particles decreases when temperature increases.
   C. The volume occupied by gas particles is comparable to the total volume occupied by the gas.
   D. A gas particle moves in rapid, straight-line motion until it collides with another particle or the container, then it stops.
   E. Gases are composed of small and widely separated particles.

   Modify each incorrect statement to make it a true statement.

9.165   Which of the following gases is expected to have the greatest average velocity at STP?

   A. $N_2$
   B. $H_2$
   C. $SO_2$
   D. Ne
   E. $NH_3$

   Arrange these gases in order of increasing average velocity.

9.166   Butane burns with oxygen according to the equation:

   $$2C_4H_{10}(g) + 13O_2(g) \longrightarrow 8CO_2(g) + 10H_2O(g)$$

   If 5.00 L of butane is allowed to burn and all of the gases are under the same conditions, which of the following statements is correct?

   A. The volume of $O_2$ required to completely react with the butane is 13.0 L.
   B. The volume of $CO_2$ produced is 20.0 L.
   C. The total volume of gases is expected to decrease.
   D. The volume of $H_2O$ produced is 1.00 L.
   E. The final volume of gases produced is 18.0 L.

   Modify each incorrect statement to make it a true statement.

9.167   Oxygen can be produced by heating potassium chlorate according to the reaction:

   $$2KClO_3(s) \longrightarrow 2KCl(s) + 3O_2(g)$$

   If 7.5 g $KClO_3$ completely reacts and the oxygen gas is collected over water at 29.0°C and 765 torr, which of the following statements is correct?

   A. 11.25 mol of oxygen is expected to be produced.
   B. The partial pressure of water at this temperature is 765 torr.
   C. The partial pressure of oxygen is 735 torr.
   D. 122.55 g of $KClO_3$ reacts.
   E. 0.0918 mol of $KClO_3$ reacts.

   Modify each incorrect statement to make it a true statement.

# The Liquid and Solid States

**10.1** Changes of State

**10.2** Intermolecular Forces

**10.3** Properties of Liquids

**10.4** Properties of Solids

Chapter Review

Questions and Problems

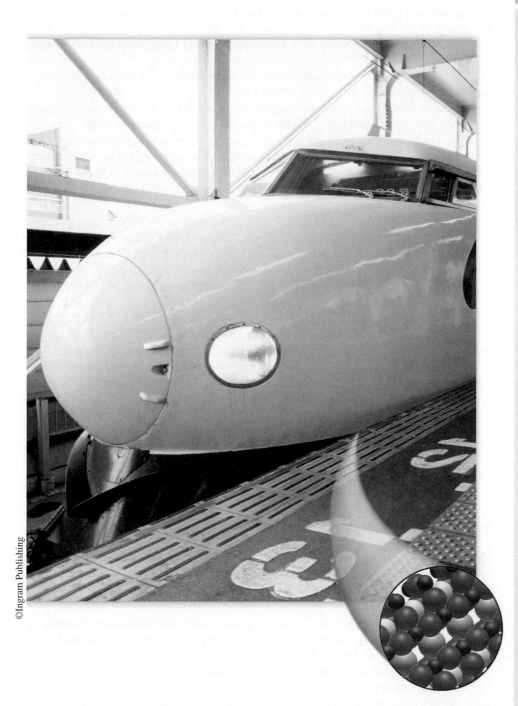

After her chemistry class on Friday, Nicole left for the weekend to attend a family reunion. At the reunion she talked with her mother, grandfather, and great-grandfather about how much she loved chemistry. Her only complaint was that her professor wouldn't allow her to use the chemistry app on her new smart phone on exams. She thought she could do better if she didn't have to memorize all those facts and equations. Her mother said that it was similar when she went to college. Students weren't allowed to use their electronic calculators on exams. Her grandfather said that in his day, they weren't allowed to bring slide rules, because their professors feared that math skills would suffer. Her great-grandfather laughed. "I remember going to the blackboard for tests and using chalk to show my work," he said, "and on some tests, even those were forbidden. We had to solve all the problems in our heads!"

The theme of each of the stories was the same. Instructors worried that "modern" technology would interfere with their students' learning. In each case, the innovation came as a result of a new material or a new use for an old one. For example, the first chalkboards of the mid-1800s were made of slate from quarries. They replaced an older technology: wooden boards coated with black, gritty paint to make them easier to write on. Those materials came from natural resources. More recent technologies, such as the LCD (liquid-crystal display) screen used in Nicole's smart phone, are made from modern synthetic materials, including polymers and semiconductors.

Materials have been important to humans throughout history, so much so that several historical eras are named for the materials most used at the time. During the Stone Age, for example, tools were made from rocks, minerals, or bone that could be chipped into useful shapes. A popular material was the hard mineral flint, a form of quartz. Around 8000 BC, naturally occurring copper was beaten into shape and used in tools. About 5000 BC, methods for removing copper from its ores were discovered. Around 3500 BC, it was discovered that copper could be made harder by melting it with tin, creating bronze. Bronze tools were much better than copper ones, and we call the era the Bronze Age.

Around 1500 BC, iron was recovered from its ores by heating them with charcoal, a discovery probably made by accident. During the Iron Age, tools were made from impure iron by heating, pounding, and reheating the metal, which hardened it. Conversion of a solid to a liquid, modifying the composition of the liquid, and conversion back to a solid are common techniques still used to form new materials.

In modern times, sometimes called the Age of Materials, we have added plastics and many ceramics—along with new metals—to our repertoire of materials that can be used to make tools. For example, superconductors can be used with rapid commuter trains, as shown in the opening photograph for this chapter. Materials science is concerned with the synthesis, processing, composition, structure, properties, and performance of new materials.

Many modern materials are *alloys* (mixtures of metals) with useful properties. One widely used alloy is Nitinol (Figure 10.1), named for its composition and its discoverers, *Ni*ckel *Ti*tanium *Na*val *O*rdnance *L*aboratory. This "shape-memory" metal is one of several recently developed "smart" materials. If its shape changes, mild heating returns it to its original form. Shape-memory metals are used for self-repairing eyeglass frames as well as life-saving medical devices and a whimsical shirt that rolls up its own sleeves when its environment gets hot. Numerous other materials have been used to make implants for the human body, including heart valves, blood vessels, pacemakers, dental implants, and joint replacements (Figure 10.2).

Another new type of material is a class of liquids called ferrofluids. They are used in stereo-speaker drivers as coolants and damping fluids that prevent unwanted vibrations. They work for these purposes because they have the fluid properties of liquids, but they respond to a magnetic field as if they were solids. To prepare ferrofluids, very small particles of a magnetic material like magnetite, $Fe_3O_4$, are suspended in oil. A long-chain carboxylic acid such as oleic acid, $C_{17}H_{33}CO_2H$, is

**FIGURE 10.1** Nitinol, a shape-memory metal, is used for eyeglass frames that repair themselves when bent out of shape.
©Tom Pantages

A                                   B                                   C

**FIGURE 10.2** (A) A hip joint that has been degraded by arthritis can be repaired by replacing the bone in both parts of the joint with (B) an implant. (C) The replacement works better when coated with synthetic hydroxyapatite, a material found naturally in human bone.
(A), (C): ©Image Source/Getty Images; (B): ©PhotoDisc/Getty Images

added to keep the magnetite particles from precipitating from the mixture. The polar ends of the oleic acid molecules interact with the magnetite particles, while the nonpolar ends interact with the oil, keeping the magnetite particles apart. In the absence of a magnetic field, ferrofluids pour just like the oil they are made from. However, in the presence of a magnetic field, the oil thickens, and the magnetite particles align with the magnetic field. As shown in Figure 10.3, the oil forms spikes in the pattern of the magnetic field. Ferrofluids are used in applications such as magnetic brakes. The oil is suspended between two plates, one of which rotates. When a magnetic field is applied, the oil thickens, and the plate can no longer turn.

Although many new materials are mixtures of substances, we can learn a lot about their properties by studying the properties of their pure components. These new materials are examples of solids and liquids, so it is important to understand the properties of these states of matter. In this chapter, we will examine the properties and structures of the solid and liquid states. In addition we will investigate the transitions between these states of matter.

## Questions for Consideration

**10.1** How do the properties of liquids and solids differ, and what happens when substances undergo a change of state?

**10.2** Why do atoms or molecules hold together as liquids or solids instead of existing only as gases?

**10.3** What are some unique properties of liquids?

**10.4** How do atoms or ions fit together to make solids, and how does their arrangement affect their properties?

**FIGURE 10.3** In a magnetic field, a ferrofluid thickens and forms spikes in the pattern of the magnetic field, much like iron filings form a pattern when sprinkled near a magnet.
©Jonathan Mitchell/Photoshot/Newscom

 **Math Tools Used in This Chapter**

Units and Conversions (Math Toolbox 1.3)

Mole Quantities (Math Toolbox 4.1)

Graphing (Math Toolbox 9.1)

**ANIMATION:** Three States of Matter

## 10.1 Changes of State

In Chapter 1 we discussed the physical states of matter: solid, liquid, and gas. The general properties of these three states are summarized in Table 10.1. To understand their differences, try drawing pictures of atoms or molecules in each of the three states. Then compare your drawings to those shown in Figure 10.4.

Your drawings should show a difference in the distance between the particles in the different states of matter. Relatively strong forces hold particles in fixed positions, close together in solids. The particles cannot move readily, so a solid is rigid, retains its shape, and is not easily compressed. A liquid, on the other hand, generally has its molecules slightly farther apart. Attractive forces are not strong enough to hold the particles rigidly in place, so they can slip past one another and move

Although particles in solids move with difficulty, they are not completely still. They vibrate around an average position. These vibrations increase in frequency as a solid material is heated.

In most substances the particles are closer together in the solid state than in the liquid state, so the solid is denser. Water is an exception. Because of the unique arrangement of its molecules, ice has open spaces in its structure. Thus, on the average, water molecules fill more space in the liquid than in the solid state. How does this observation explain why ice floats on liquid water?

**TABLE 10.1** General Properties of the States of Matter

| Solid | Liquid | Gas |
|---|---|---|
| Fixed shape | No fixed shape | No fixed shape |
| Shape not set by container | Takes shape of filled portion of the container | Takes shape of container |
| Shape remains rigid | Can be poured | Fills container |
| Particles fixed in place but vibrate around a fixed position | Particles move past one another | Particles move through space |
| Little or no volume change under moderate pressure | Can be compressed slightly by moderate pressure | Compressed under moderate pressure |
| Little free space between particles | Some free space between particles | Particles are widely separated with much free space |

Solid Cu     Liquid $H_2O$     Gaseous $NO_2$

**FIGURE 10.4** The macroscopic properties of solids, liquids, and gases result from the distance that separates their atoms or molecules and from how freely these particles can move.

(left, middle): ©Brian Moeskau/Moeskau Photography; (right): ©McGraw-Hill Education/Stephen Frisch around. Thus, a sample of a liquid takes the shape of its container; and because there is some space between particles, a liquid can be compressed slightly. The relative incompressibility of liquids is useful in liquid hydraulic systems such as those used in automobile brakes.

All substances can exist in any of the three physical states under certain conditions, which may be extreme. For example, nickel, a component of Nitinol, is a solid unless heated above 1455°C. Then it becomes liquid. Figure 10.5 shows liquid aluminum being poured at high temperatures. Metals can change to the gaseous state if heated even hotter. Nickel boils at 2800°C. *Vaporization* of metals is applied in the materials science technique of vapor deposition. In this technique, gaseous metal atoms deposit on the surface of interest, sometimes in very thin layers.

Water is the only common substance that exists in all three physical states under normal atmospheric conditions. Liquid water in underground pools is expelled from a geyser as steam. An iceberg floats in water and gradually melts to produce more liquid water. When an iceberg floats in liquid water, invisible water vapor is also present in the air, so three physical states—gas, liquid, and solid—are present (Figure 10.6). Transitions between these states are *changes of state* or *phase changes*.

Contrary to popular science fiction, vaporization does not mean blasting something into oblivion with a ray gun. The term *vapor* refers to the gaseous state of a substance that normally exists as a solid or a liquid. Vaporization is a synonym for evaporation.

**ANIMATION:** Unique Properties of Water

## Liquid-Gas Phase Changes

At any given temperature, molecules in a liquid have a range of kinetic energies. Some move faster and some move slower than the average. What happens if a molecule with a high kinetic energy happens to be close to the liquid's surface? If it is moving in the right direction, it may escape into the gas state. This process is called **evaporation** or vaporization (Figure 10.7).

Because the molecules of highest kinetic energy are the ones most likely to escape from the liquid, the average kinetic energy of the remaining liquid molecules decreases as evaporation proceeds. As we saw in Chapter 9, kinetic energy is proportional to absolute temperature, so unless energy is supplied, evaporation decreases the temperature of the liquid. To evaporate at a constant temperature, a liquid must absorb heat from its surroundings. The process is endothermic.

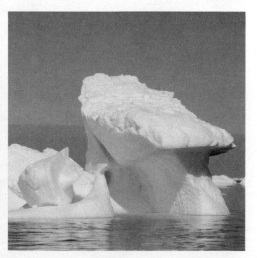

**FIGURE 10.5** When heated above 2467°C, aluminum melts and can be poured just like any other liquid.

©Chris Knapton/Alamy Stock Photo

**FIGURE 10.6** Water exists in all three physical states under conditions found on the surface of Earth. Water in the gaseous state, called water vapor, is not visible. Clouds (and steam) are actually tiny droplets of liquid water.

(left): ©Emanuele Taroni/Getty Images; (right): ©Lissa Harrison

Evaporation

**FIGURE 10.7** At the surface of a liquid, some molecules have sufficient energy to escape into the gas state.

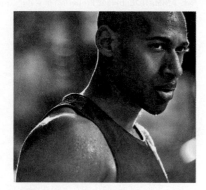

**FIGURE 10.8** Evaporation of a liquid, such as a runner's sweat, is endothermic, so it causes cooling.
©Patrik Giardino/Corbis/Getty Images

The reverse of evaporation, condensation, is an exothermic process. Gas particles must release energy to condense into the lower-energy liquid state.

The cooling effect of evaporation is significant in many processes. Some examples include cooling by perspiration (Figure 10.8), evaporative cooling (swamp cooling) in homes, cooling car interiors with liquid sprays, and the use of canvas water bags for desert travel. Have you ever noticed feeling cold when leaving a shower or swimming pool? That's evaporative cooling, and it can be used to treat a sick baby running a high fever. A sponge bath lowers body temperature because the evaporation of water cools the skin. The wind chill factor reported in weather forecasts is another measure of evaporative cooling. Moving air evaporates water from the skin faster than still air, so the air feels colder when the weather is windy. The increased cooling is sensed by the skin as a lower air temperature.

The extent of evaporation depends on how the liquid is contained. In an open container, molecules that are more energetic move away from the liquid's surface into the surrounding atmosphere in the gaseous state. Since the container is open, the process continues until all of the liquid has evaporated. Some of the evaporated molecules may return to the liquid state if they happen to move toward the liquid rather than away from it, but most of the evaporated molecules stay in the gaseous state moving farther away. The average energy of the liquid remains constant, because the liquid absorbs heat from its surroundings, maintaining a constant temperature. However, if the container is insulated, heat cannot be absorbed from the surroundings as easily, so the liquid cools as evaporation proceeds. As a result, the process slows, because the average energy of the molecules decreases.

What happens if we place a pure liquid in a closed container (Figure 10.9)? At first only evaporation occurs, since there are no vapor molecules. As evaporation continues, molecules begin to accumulate in the gaseous state. They cannot move completely away from the liquid, and many move toward it. When they hit the liquid, they reenter the liquid state, a process called **condensation.** Evaporation and condensation continue at rates that depend on two factors: temperature and the concentration of molecules available to undergo each process. Ultimately the rate of condensation equals the rate of evaporation, and the concentration of molecules in the gaseous state becomes constant. This situation, in which opposing processes occur at equal rates, is called a state of dynamic **equilibrium.** It can be represented by the equation

$$\text{liquid} \underset{\text{condensation}}{\overset{\text{evaporation}}{\rightleftharpoons}} \text{gas}$$

A double arrow represents a state of equilibrium involving a reversible process.

**FIGURE 10.9** In a closed container the number of gas molecules increases until equilibrium is attained. In this state of equilibrium, the rate of liquid molecules evaporating equals the rate of gas molecules condensing. Thus, the number of gas molecules remains constant. (Note that original air molecules are not shown.)

At equilibrium the rate of evaporation equals the rate of condensation, so the concentration of the substance in the gaseous state is constant, although evaporation and condensation continue to occur. The concentration of gas can be measured by the pressure exerted by the gas molecules above the liquid. At equilibrium this pressure is called the **vapor pressure.** Vapor pressure increases as temperature increases. At higher temperatures, the average kinetic energy of the liquid molecules is greater, so the molecules can escape from the liquid more easily. The concentration of the substance in the gas state at equilibrium is therefore greater at higher temperatures.

When the vapor pressure equals the external atmospheric pressure, the liquid begins to boil. The temperature of a pure, boiling liquid remains constant until all the liquid has vaporized. Adding more heat to the liquid increases the rate of evaporation, not the temperature. The temperature at which boiling occurs—at which the vapor pressure equals the external pressure of the atmosphere—is called the **boiling point.** The boiling point for a given liquid depends on the atmospheric pressure. Figure 10.10 shows that boiling points, which are the temperatures corresponding to points on the curves, increase as the pressure rises. For example, if the pressure is 2.0 atm, the boiling point of water is 120°C. When the atmospheric pressure is exactly 1 atm, the temperature at which boiling occurs is called the **normal boiling point.**

An atmospheric pressure of 1 atm is normal only at sea level, however. At higher altitudes atmospheric pressure is less than 1 atm, and liquids boil at lower temperatures. The normal boiling point of water, for example, is 100.0°C. At about 2750 m (9000 ft) above sea level, such as at the Tuolumne Meadows campground in Yosemite National Park, the atmospheric pressure is about 0.72 atm, and water boils at 91°C. It takes almost twice as long to cook an egg in boiling water at this elevation than at sea level, because it must cook at a lower temperature. Water cannot be heated above its boiling point because the added energy is used to evaporate the water. However, a liquid held under a pressure higher than atmospheric pressure can be heated above its normal boiling point. A pressure cooker has a sealed lid with a pressure valve, so the vapor pressure can rise above atmospheric pressure. Since water boils at a higher temperature under such conditions, the food cooks more rapidly.

 **ANIMATION:** Equilibrium Vapor Pressure

Recall from Chapter 9 that the pressure of the atmosphere at sea level under normal weather conditions corresponds to a column of mercury that is 29.9 in (76.0 cm) tall. This pressure is the basis for a unit, the standard *atmosphere* (atm), which equals 14.7 lb/in$^2$.

 **MATH TOOLBOX 9.1**

**FIGURE 10.10** The vapor pressure of all substances increases as the temperature increases. The temperature at which the vapor pressure is 1 atm is the normal boiling point. All temperatures corresponding to points on the curves are boiling points for the corresponding pressures.

| **EXAMPLE 10.1** | **Vapor Pressure and Boiling Point** |

Consider the following vapor pressure curve for acetone, which is used in nail polish removers.

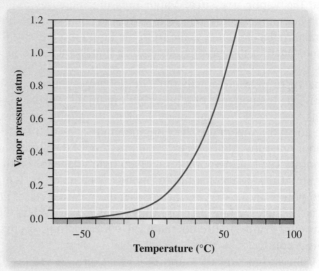

What is the normal boiling point of acetone? What is its boiling point at 0.50 atm?

**Solution:**

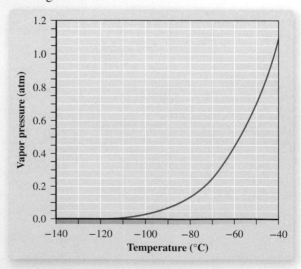

The normal boiling point is the temperature at which the vapor pressure is exactly 1 atm. (If you need help interpreting graphs, see Math Toolbox 9.1.) If we locate the point on the curve at which the pressure is 1 atm and then follow that point down to the temperature axis, we read a temperature of 56°C. The boiling point at 0.50 atm is found in the same way. We locate the point on the curve that has a vapor pressure of 0.50 atm. We follow this point down to the temperature axis and find a value of 36°C.

**Consider This 10.1**

If, for a different substance, the vapor pressure values measured between 0°C and 50°C were all lower at each temperature, would the normal boiling point be higher or lower?

**Practice Problem 10.1**

Consider the following vapor pressure curve for propane, a compound that is used as a fuel in barbecue grills.

What is the normal boiling point of propane? What is its boiling point at 0.40 atm?

**Further Practice:**  Questions 10.13 and 10.14 at the end of the chapter

## Liquid-Solid Phase Changes

The average kinetic energy of liquid molecules decreases as liquids cool. When the kinetic energy of the liquid molecules falls low enough, the molecules become fixed in position in the solid state (although still vibrating), and the liquid freezes. The freezing of a liquid into the solid state occurs at a temperature called the **freezing point.** At this temperature the solid and liquid coexist in a state of equilibrium (Figure 10.11). The temperature is called the **normal freezing point** when this state of equilibrium is reached under a pressure of exactly 1 atm. The temperature remains constant from the time a liquid starts freezing until all the liquid has solidified. After that, release of more heat causes a further decrease in temperature of the solid. Freezing is an exothermic process. Citrus growers use this principle to protect their crops from freezing during cold nights. They spray water on the trees. The freezing of the sprayed water on the surface of the plant releases enough heat to keep the water inside the plant cells from freezing.

Melting is the reverse of freezing. Both processes involve the solid and liquid phases:

$$\text{solid} \underset{\text{freezing}}{\overset{\text{melting}}{\rightleftharpoons}} \text{liquid}$$

Thus, the **melting point** of a solid and the freezing point of its liquid state are the same temperature. When a molecular substance melts, the molecules retain their identity. When an ionic substance melts, the ions become more randomly arranged (Figure 10.12). Energy must be absorbed to separate the particles during melting, so it is an endothermic process.

**FIGURE 10.11**  At the freezing point (or melting point), the solid and liquid forms of a substance are in a state of equilibrium.

©John A.Rizzo/Getty Images

## Solid-Gas Phase Changes

Solids, like liquids, have a characteristic vapor pressure. The vapor pressures of most solids are rather low, much lower than those of their liquids. However, a few have high vapor pressures and can change directly from the solid to the gaseous state without going through the liquid state. The evaporation of a solid is called **sublimation:**

$$\text{solid} \underset{\text{deposition}}{\overset{\text{sublimation}}{\rightleftharpoons}} \text{gas}$$

A

B

**FIGURE 10.12**  (A) When a covalent substance such as water melts, the molecules retain their identity and composition. (B) When an ionic substance such as sodium chloride melts, the ions assume a more random arrangement.

**FIGURE 10.13** When solid iodine is heated, it sublimes into the gaseous state. It returns to the solid state on the cold surface of the upper tube filled with ice.
©McGraw-Hill Education/Richard Megne

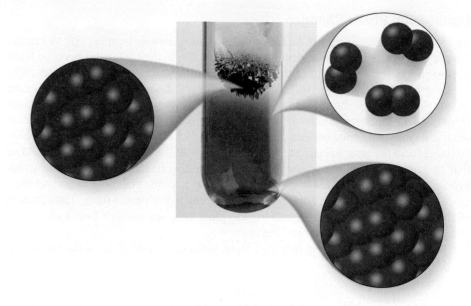

Is sublimation exothermic or endothermic? Is deposition exothermic or endothermic?

An example is dry ice (solid $CO_2$). It sublimes from solid to gas, disappearing into the air. Other common substances that sublime are mothballs and iodine (Figure 10.13). Have you ever noticed how ice cubes get smaller when kept in a freezer for a long time? Solid water sublimes at temperatures below 0°C. It evaporates because it is in equilibrium with the gaseous state, and no liquid state can exist. Wet clothing hung outside in freezing weather dries slowly for this reason.

The reverse of sublimation is **deposition.** Deposition is a change from gas to solid without going through the liquid state. The formation of snow is a deposition process. The deposition of water vapor in the atmosphere releases heat, so the air temperature rises a degree or two when it begins to snow.

---

**EXAMPLE 10.2** | **Changes in the State of Matter**

Identify the process shown in the following diagram. Is the process endothermic or exothermic?

**Solution:**

We can identify the starting material as a liquid, based on the small amount of space between particles and the random arrangement of particles. In the ending material, there is a large distance between particles, so we can identify it as a gas. A liquid-to-gas transformation is evaporation. To go into the gas phase, the atoms must absorb energy. Evaporation is endothermic.

**Consider This 10.2**

If the starting material were a solid, would the process be exothermic or endothermic?

**Practice Problem 10.2**

Identify the process shown in the following diagram. Is the process endothermic or exothermic?

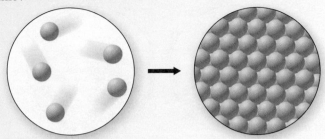

**Further Practice:**   Questions 10.17 and 10.18 at the end of the chapter

## Cooling and Heating Curves

Consider a substance in the gaseous state. When cooled at constant pressure, its volume decreases, but it remains as a gas up to a point. When will it convert to a liquid? When the temperature falls enough that the pressure of the gas matches the vapor pressure of the liquid, the gas starts to condense. This happens at the boiling point of the liquid. If we continue to remove heat, more of the gas condenses to the liquid state. When gas and liquid are in equilibrium, removal of more heat does not decrease the temperature. It results only in further condensation. The first plateau in the *cooling curve* shown in Figure 10.14A illustrates this process. Once all the gas condenses, removal of heat decreases the temperature of the liquid. The temperature drop halts again at the freezing point, shown as the second plateau in Figure 10.14A. Once all the liquid freezes to solid, further removal of heat lowers the temperature of the solid. If we reverse the process and add heat to a solid, we proceed along a heating curve (Figure 10.14B), with plateaus also observed when the substance undergoes changes of state.

In the cooling curves for some substances, we may observe a liquid becoming colder than the freezing point, a phenomenon called *supercooling*. Solidification requires some matrix around which a solid can form—perhaps imperfections in a container wall, pieces of dust, or other solid particles. Thus, in a very clean system, solidification may be delayed. Usually, after supercooling a degree or two, the substance begins to crystallize, and the temperature returns to the freezing point. Supercooling is common in substances such as glass, tars, rubbers, and plastics.

A   Cooling curve

B   Heating curve

**FIGURE 10.14**  (A) When heat is removed at a constant rate from a gas, it cools until it reaches the boiling point of the liquid. There the temperature remains constant until the gas completely condenses to liquid. The temperature then drops again until it reaches the melting point of the solid. The temperature remains constant as the liquid converts to the solid state. Then the temperature drops once more. These changes are summarized in a cooling curve. (B) The same changes in reverse order occur when heat is added to a solid, as shown in a heating curve.

## EXAMPLE 10.3 | Cooling and Heating Curves

The following diagrams represent two different changes of state.

(a)                    (b)

For each change of state, although heat is being added, the temperature does not change. Identify where each change of state would be found on the following heating curve:

**Solution:**

(a) This molecular diagram shows the liquid and gaseous states coexisting. Since the temperature does not change during the process, this corresponds to any point on the *DE* plateau in the heating curve.

(b) This molecular diagram shows solid and liquid together, which corresponds to any point on the *BC* plateau in the heating curve.

**Consider This 10.3**

If heat were being removed from a substance, how would the graph and diagrams from the example differ?

**Practice Problem 10.3**

The following diagrams represent the physical states of a substance.

(a)                    (b)                    (c)

Identify where each state would predominate on the heating curve given in the example.

**Further Practice:** Questions 10.31 and 10.32 at the end of the chapter

## Energy Changes

Consider a piece of ice at a temperature of −40.0°C. If we heat the ice at a constant pressure of 1 atm, it will go through the temperature and phase changes shown by the heating curve in Figure 10.14. The temperature will rise until the ice reaches 0.0°C, at which point it will start to melt. The temperature will stay constant until the ice has melted completely, because the added energy is being used to overcome the forces that hold the water molecules together in ice. When the ice has melted, the temperature will start to rise again as determined by the specific heat of water. When the water reaches 100.0°C, it will start to boil; it will stay at 100.0°C until it has completely vaporized. Continued heating will then increase the temperature of the steam to a temperature such as 120.0°C.

The energy required to change the temperature of any physical state of matter is determined by the specific heat of that state. As described in Chapter 6, the heat absorbed or released by a substance when it changes temperature is given by the equation

$$q = m \times C \times \Delta T$$

In this equation, $m$ is the mass, $\Delta T$ is the temperature change, and $C$ is the specific heat. The values of the specific heat for the physical states of water are 2.03 J/(g °C) for ice, 4.18 J/(g °C) for liquid water, and 2.02 J/(g °C) for gaseous water.

Even though each change of state takes place at a constant temperature, an energy change occurs. The solid form of a substance contains less energy than the liquid form, and the liquid contains less energy than the gas. The change in energy for each process is called the heat of that process ($q$). For example, the *molar heat of fusion* is the energy required to melt 1 mol of a substance. The molar heat of fusion of ice, $H_2O(s)$, is 6.01 kJ/mol, or $6.01 \times 10^3$ J/mol. To melt 5.00 mol of ice, $5.00 \times 6.01 \times 10^3$ J of energy must be added. The *molar heat of vaporization* is the energy required to evaporate 1 mol of a liquid. For liquid water, the molar heat of vaporization is 40.7 kJ/mol, or $4.07 \times 10^4$ J/mol.

It is possible to calculate the heat required to carry out the entire process by which ice is transformed to steam, or any other process involving temperature changes and changes in physical state. We can represent the process as follows:

The energy change for the entire process is the sum of the heat used for the individual steps in the process.

$$q = q_{\text{heat ice}} + q_{\text{fusion}} + q_{\text{heat water}} + q_{\text{vaporization}} + q_{\text{heat steam}}$$

Suppose we start with 1.000 mol (18.02 g) of ice at −40.0°C. To calculate how much heat is required to raise the temperature of ice to 0.0°C, the point at which it will start to melt, we use the following equation:

$$q_{\text{heat ice}} = m \times C \times \Delta T$$

What is happening during the horizontal portions of the heating curve for water at 0.0°C and 100.0°C? What happens to the heat that is being added during these two times?

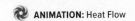

The units for molar heat of fusion are J/mol. There is no temperature component to this quantity because the temperature of the ice does not change as it melts. Recall that the energy absorbed is used to melt the ice, not increase its temperature.

MATH TOOLBOX 1.3

The mass is 18.02 g, the specific heat is 2.03 J/(g °C), and the temperature change is 40.0°C:

$$q_{\text{heat ice}} = 18.02\,\cancel{g} \times \frac{2.03\text{ J}}{\cancel{g}\,°\cancel{C}} \times 40.0°\cancel{C} = 1.46 \times 10^3\text{ J}$$

The units cancel giving joules, a unit of energy.

The molar heat of fusion for the melting of ice is 6010 J/mol. To obtain the heat change, we multiply the molar heat of fusion by the number of moles, $n$:

$$q_{\text{fusion}} = n \times \text{molar heat of fusion}$$

$$q_{\text{fusion}} = 1.000\,\cancel{\text{mol}} \times \frac{6.01 \times 10^3\text{ J}}{\cancel{\text{mol}}} = 6.01 \times 10^3\text{ J}$$

As before, units cancel giving joules, a unit of energy.

The temperature of the water, now liquid, is then raised by 100.0°C. The heat required for this part of the process is calculated from the specific heat of liquid water:

$$q_{\text{heat water}} = m \times C \times \Delta T$$

specific heat of water
Water at 0.0°C → Water at 100.0°C

$$q_{\text{heat water}} = 18.02\,\cancel{g} \times \frac{4.18\text{ J}}{\cancel{g}\,°\cancel{C}} \times 100.0°\cancel{C} = 7.53 \times 10^3\text{ J}$$

The evaporation of water, which takes place as the liquid boils, requires $4.07 \times 10^4$ J/mol, the molar heat of vaporization. To obtain the heat change, we multiply the molar heat of vaporization by the number of moles, $n$:

$$q_{\text{vaporization}} = n \times \text{molar heat of vaporization}$$

heat of vaporization
Water at 100.0°C → Steam at 100.0°C

$$q_{\text{vaporization}} = 1.000\,\cancel{\text{mol}} \times \frac{4.07 \times 10^4\text{ J}}{\cancel{\text{mol}}} = 4.07 \times 10^4\text{ J}$$

The steam is then heated further to 120.0°C, a temperature change of 20.0°C. The change is determined from the specific heat of steam, 2.02 J/(g °C):

$$q_{\text{heat steam}} = m \times C \times \Delta T$$

specific heat of steam
Steam at 100.0°C → Steam at 120.0°C

$$q_{\text{heat steam}} = 18.02\,\cancel{g} \times \frac{2.02\text{ J}}{\cancel{g}\,°\cancel{C}} \times 20.0°\cancel{C} = 7.28 \times 10^2\text{ J}$$

The total heat required is the sum of the heat changes for each step of the process:

$$q = (1.46 \times 10^3\text{ J}) + (6.01 \times 10^3\text{ J}) + (7.53 \times 10^3\text{ J})$$
$$+ (4.07 \times 10^4\text{ J}) + (7.28 \times 10^2\text{ J})$$
$$= 5.64 \times 10^4\text{ J} = 56.4\text{ kJ}$$

**Student Hot Spot**

Student data indicate you may struggle with energy calculations involving changes in physical state. Access the SmartBook to view additional Learning Resources on this topic.

## EXAMPLE 10.4    Energy for Phase Changes

Calculate the heat absorbed when 46.0 g of ice at −10.0°C is converted to liquid water at 85.0°C. The specific heat of ice is 2.03 J/(g °C), the molar heat of fusion of ice is 6010 J/mol, and the specific heat of water is 4.18 J/(g °C).

**Solution:**

The heat is calculated by considering the three-step process of warming ice from −10.0°C to 0.0°C, melting the ice, and heating liquid water to 85.0°C:

To calculate the heat required to warm ice to 0.0°C, the point at which it will begin to melt, we use the equation

$$q_{\text{heat ice}} = m \times C \times \Delta T$$

The mass is 46.0 g, the specific heat is 2.03 J/(g °C), and the temperature change is 10.0°C:

$$q_{\text{heat ice}} = 46.0\,\cancel{g} \times \frac{2.03\ \text{J}}{\cancel{g}\,°\cancel{C}} \times 10.0\,°\cancel{C} = 934\ \text{J}$$

The heat of fusion of ice is 6010 J/mol. The units indicate that the quantity of ice must be expressed in units of moles:

MATH
TOOLBOX
**4.1**

$$46.0\ \text{g}\,\cancel{H_2O} \times \frac{1\ \text{mol}\ H_2O}{18.02\ \text{g}\,\cancel{H_2O}} = 2.55\ \text{mol}\ H_2O$$

To obtain the heat change for melting ice, we multiply the heat of fusion for ice by the moles, $n$:

$$q_{\text{fusion}} = n \times \text{molar heat of fusion}$$

$$q_{\text{fusion}} = 2.55\,\cancel{\text{mol}} \times \frac{6.01 \times 10^3\ \text{J}}{\cancel{\text{mol}}} = 1.53 \times 10^4\ \text{J}$$

The heat required to warm the water, now a liquid, to 85.0°C is calculated from the mass, the specific heat of water, and the change in temperature:

$$q_{\text{heat ice}} = m \times C \times \Delta T$$

$$q_{\text{heat water}} = 46.0\,\cancel{g} \times \frac{4.18\ \text{J}}{\cancel{g}\,°\cancel{C}} \times 85.0\,°\cancel{C} = 1.63 \times 10^4\ \text{J}$$

The total heat absorbed is the sum of the heat changes for each step of the process:

$$q = 934 \text{ J} + (1.53 \times 10^4 \text{ J}) + (1.63 \times 10^4 \text{ J})$$
$$= 3.25 \times 10^4 \text{ J} = 32.5 \text{ kJ}$$

**Consider This 10.4**

If we started with ice at −20.0°C instead of −10.0°C, which energy terms would change?

**Practice Problem 10.4**

Calculate the heat absorbed when 125 g of water at 72.0°C is converted to steam at 186.4°C. The specific heat of water is 4.18 J/(g °C), the heat of vaporization of water is $4.07 \times 10^4$ J/mol, and the specific heat of steam is 2.02 J/(g °C).

**Further Practice:** Questions 10.37 and 10.38 at the end of the chapter

## 10.2 Intermolecular Forces

Why do some samples of matter exist under ordinary conditions as gases, while others are liquids or solids? The existence of the liquid and solid states indicates that some attractive force must pull molecules together. An attractive force that operates *between* molecules is called an **intermolecular force.** In contrast, forces *within* molecules, called bonds, hold atoms together in molecules. Intermolecular forces are much weaker than bonding forces. The energies of intermolecular forces are usually in the range of 0.05 to 40 kJ/mol, while those of most bonds are in the range of 200 to 900 kJ/mol.

What is the origin of these intermolecular forces? They arise from the interaction of positive and negative charges. We will examine three such forces: the *London dispersion force,* the *dipole-dipole force,* and a special case of dipole-dipole attraction, the *hydrogen bond.* Characteristics of these forces, as well as chemical bonds, are summarized in Table 10.2. We will discuss each intermolecular force in more detail in this section.

**TABLE 10.2**  Intermolecular Forces and Bonds

| Type of Force | Type of Interaction | Occurrence |
|---|---|---|
| London dispersion force | A temporary dipole in one molecule induces the formation of a temporary dipole in a nearby molecule and is attracted to it. | All atoms and molecules |
| Dipole-dipole force | Polar molecules (permanent dipoles) attract one another. | Polar molecules |
| Hydrogen-bonding force | An attraction between a hydrogen atom covalently bonded to O, N, or F in one molecule and the O, N, or F atom of another polar molecule. | Polar molecules containing unpaired electrons and a hydrogen bonded to nitrogen, oxygen, or fluorine |
| Covalent bond | Nuclei of two atoms attract the electrons shared between them. | Nonmetal-nonmetal compounds |
| Ionic bond | Cations and anions attract one another. | Metal-nonmetal compounds |

## London Dispersion Forces

Electrons moving around an atom or nonpolar molecule are not always distributed symmetrically. If the electrons happen to be more on one side at an instant in time, that side will be more negative than normal, and the other side will be more positive. This situation leads to the formation of a temporary dipole, or **instantaneous dipole,** which—like any dipole—contains partial charges (as discussed in Chapter 8). The positive end of the dipole exerts an attractive force on nearby electrons, causing an adjacent atom to develop into another temporary dipole, called an **induced dipole.** This effect passes on to more atoms, resulting in a kind of electron "choreography" in which the movements of the electrons in nearby atoms correlate (Figure 10.15). The attraction between these temporary dipoles is called a **London dispersion force.** It occurs between all atoms and molecules, and it is the only intermolecular force at work in nonpolar substances. Although London dispersion forces are relatively weak, they are strong enough to cause substances normally found as gases, such as neon and methane, to liquefy at high pressures or low temperatures.

London dispersion forces tend to be stronger the larger the atom or molecule. For example, $I_2$ molecules interact with each other more strongly than $Br_2$ molecules, which interact more strongly than $Cl_2$ molecules. This explains why $Cl_2$ is a gas, $Br_2$ is a liquid, and $I_2$ is a solid at room temperature (Figure 10.16). The size of the electron cloud surrounding the atom or molecule, which depends on the number of electrons, causes this trend. The larger the electron cloud, the easier it is to distort it, resulting in stronger London dispersion forces. As the molecules get larger, their intermolecular forces get stronger and this leads to substances existing in liquid or solid states.

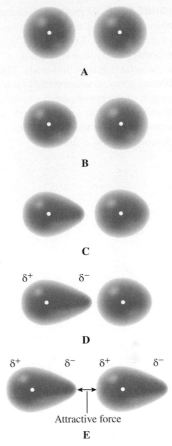

**FIGURE 10.15** London dispersion forces arise from attractions between temporary dipoles. (A) Ordinarily, electrons distribute symmetrically in atoms or nonpolar molecules. (B) An electron cloud may become distorted as electrons travel around the nucleus. (C) If this distortion gets large enough, it may affect adjacent atoms or molecules. (D) The distortion of the electron cloud causes the formation of a temporary or instantaneous dipole. (E) This temporary dipole induces the formation of a dipole in the electron cloud of a nearby atom or molecule, resulting in attractive forces between the instantaneous dipole and the induced dipole. This attraction constitutes the London dispersion force.

---

| **EXAMPLE 10.5** | **London Dispersion Forces** |
| --- | --- |

Which has the stronger London dispersion forces, argon or xenon?

**Solution:**

Since argon has 18 electrons and xenon has 54 electrons, the electron cloud of xenon is larger than that of argon. The larger electron cloud is easier to distort, so xenon has stronger London dispersion forces.

**Consider This 10.5**

If we compare London dispersion forces of all the noble gases, which one would have the weakest London dispersion forces?

**Practice Problem 10.5**

Which has the stronger London dispersion forces, $CH_4$ or $SiH_4$?

**Further Practice:** Questions 10.45 and 10.46 at the end of the chapter

---

In very large molecules, London dispersion forces can get quite strong. Examples can be found in liquid crystals, such as the display used in Nicole's handheld computer discussed in the introduction to this chapter. Liquid crystals can flow like liquids, but have the structural order of solids. Their liquid-crystalline state occurs during melting as a transition between the solid state and the true liquid state. Most liquid-crystal films change color when the molecules change their alignment. The color change can be controlled by temperature, mechanical stress, magnetic fields, or electric fields.

London dispersion forces are named after Fritz London, a German-American physicist who first proposed an explanation for them in 1930.

**FIGURE 10.16** At room temperature, chlorine is a pale green gas, bromine is a reddish-brown liquid that evaporates easily, and iodine is a purple solid that sublimes easily. The different physical states relate to the strength of the intermolecular forces, which in turn relate to the sizes of the electron clouds in these molecules.
©Sciencephotos/Alamy Stock Photo

$Cl_2$          $Br_2$          $I_2$

Why do these molecules behave in this way? Liquid-crystalline substances consist of long, narrow, rigid organic molecules. One example is the molecule 4-methoxybenzylidene-4′-$n$-butylaniline, which behaves as a liquid crystal over the temperature range of 21°C to 47°C:

In liquid crystals, molecules align parallel to one another, as shown in Figure 10.17. The alignment occurs because strong London dispersion forces operate among large molecules. When an electric field is applied to a film of some liquid crystals, the molecular alignment changes and parts of the film turn black, forming the shapes we see on an LCD screen. Some liquid crystals reflect different colors of light under different conditions, such as changing temperatures.

A                                           B

**FIGURE 10.17** (A) A liquid crystal is magnified 320 times, showing layers of parallel molecules. (B) At a higher magnification, a scanning tunneling microscope image shows individual octadecanol molecules aligned with one another, typical of long organic molecules.

(A): ©Oleg D. Lavrentovich; (B): ©B.L. Ramakrishna/Interactive Nano-Visualization for Science and Engineering Education (IN-VSEE) an NSF funded project at ASU

## Dipole-Dipole Forces

In Section 8.4 we discussed the separation of charge within molecules that causes polarity. The attractions among polar molecules result in a second type of intermolecular attraction, the **dipole-dipole force.** With polar molecules, the partially positive end of one molecule attracts the partially negative ends of other molecules (Figure 10.18). In general, attractions between polar molecules are stronger than those between nonpolar molecules of similar size. Consider the boiling points of two molecular substances, $O_2$ and NO. The boiling points of these substances are $-183°C$ and $-152°C$, respectively. Why are these differences observed? Both have a double bond between their atoms, and they differ by only one electron. Their electron clouds are similar in size, but the $O_2$ molecule is nonpolar and the NO molecule is polar. The polar molecule has greater intermolecular forces than the nonpolar molecule, because it has both dipole-dipole and London forces. With stronger intermolecular forces, more energy is required to separate the liquid-state molecules and convert them to gas-state molecules. The increased energy demand causes the boiling point of NO to be higher than that of $O_2$.

**Dipole-dipole attractions**

**FIGURE 10.18** Polar molecules like sulfur dioxide attract each other when the positive end of one molecule aligns with the negative end of another molecule.

---

### EXAMPLE 10.6 | Dipole-Dipole Forces

Which of the following molecules experience dipole-dipole forces in the liquid state?

(a) $SO_3$    (b) $CO_2$    (c) $PCl_3$    (d) $NO_2$

**Solution:**

Recall from Chapter 8 that two conditions are necessary for a molecule to be polar—which is, in turn, necessary for the molecule to experience dipole-dipole forces. First, the molecule cannot be symmetrical. Second, there must be an electronegativity difference between the central atom and the atoms surrounding it. If necessary, review Chapter 8 to determine the shapes of these molecules starting with their Lewis formulas.

(a) The $SO_3$ molecule is trigonal planar, so it is symmetrical.

 No net dipole

Despite the difference in electronegativity between sulfur and oxygen, the symmetry of the molecule causes the bond polarities to cancel and make this molecule nonpolar. Consequently, it cannot experience dipole-dipole forces.

(b) The $CO_2$ molecule is linear, so it is symmetrical.

 No net dipole

It is nonpolar and does not experience dipole-dipole forces.

(c) The $PCl_3$ molecule is trigonal pyramidal, so it is not symmetrical.

 Net dipole

There is a difference in electronegativity between phosphorus and chlorine, so the bonds are polar. Therefore, the molecule is polar and will experience dipole-dipole forces.

(d) The $NO_2$ molecule is bent, so it is not symmetrical.

 Net dipole

The electronegativities of nitrogen and oxygen are different, so the bonds are polar. The molecule is polar and will experience dipole-dipole forces.

**Consider This 10.6**

If the molecule $NO_2$ were linear instead of bent, would it experience dipole-dipole forces in the liquid state?

**Practice Problem 10.6**

Which of the following molecules experience dipole-dipole forces in the liquid state?

(a) $SCl_2$      (b) CO      (c) $NH_3$      (d) $CCl_4$

**Further Practice:** Questions 10.47 and 10.48 at the end of the chapter

## Hydrogen Bonding

**MATH TOOLBOX 9.1**

The boiling points of different substances vary widely. Consider the boiling points of some elements and nonpolar molecular compounds shown in Figure 10.19. The figure reveals that—in a related series of substances—the boiling point increases as the atoms or molecules get larger. The same is true of polar molecules. The larger the molecule, the higher its boiling point, because the London dispersion forces are greater.

Figure 10.20, however, reveals an irregularity. As expected, the boiling points of the nonpolar hydrides of group IVA (14) elements in the periodic table (beginning with $CH_4$) increase as the size of the electron cloud increases. The same occurs with most of the polar hydrides of elements from groups VA (15), VIA (16), and VIIA (17). However, $NH_3$, $H_2O$, and HF have unexpectedly high boiling points, and many of their other properties are unusual as well. What is special about these molecules? Their intermolecular forces must be unusually strong.

Especially strong dipole-dipole forces exist between polar molecules that contain hydrogen attached to a small, highly electronegative element (specifically

> Hydrogen bonding in pure substances generally occurs where there are H–F, H–O or H–N bonds within the molecules. Hydrogen bonding is so strong that some electron sharing occurs between molecules, but not as much as in true covalent bonds within molecules.

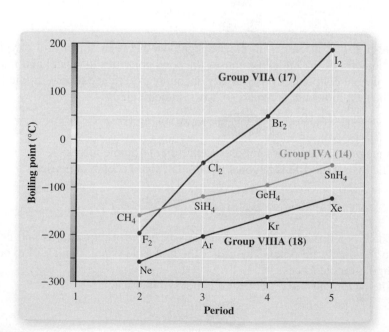

**FIGURE 10.19** Within groups of the periodic table, the boiling points of nonpolar substances increase with the size of the molecule.

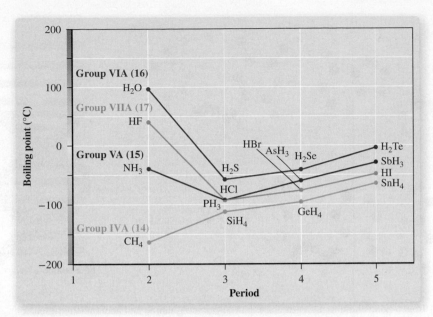

**FIGURE 10.20** Boiling points generally increase with the size of the molecule, but some relatively small molecules have high boiling points, due to hydrogen bonding.

nitrogen, oxygen, or fluorine). Such a force is called a **hydrogen bond,** although there is no bond between molecules in the usual sense. The hydrogen bond is actually a special type of dipole-dipole force. The strength of most hydrogen bonding is in the range of 20 to 40 kJ/mol, although a few examples have energies as high as 150 kJ/mol. This is large for intermolecular forces, but still smaller than the forces in covalent bonds. For example, the strength of the covalent H—H bond in $H_2$ is 436 kJ/mol.

Hydrogen bonding occurs in specific directions. The attraction between molecules occurs with a hydrogen atom covalently bonded to a highly electronegative atom in one molecule pointing at an unshared electron pair on a highly electronegative atom in an adjacent molecule. Some examples are shown in Figure 10.21. The

In rare cases, hydrogen bonding has also been observed between hydrogen atoms and atoms other than N, O, or F. Because this is not common, it will not be considered in this text.

Water        Ammonia in water        Ammonia

Hydrogen fluoride

**FIGURE 10.21** Here three dots represent hydrogen bonding. These forces are experienced between a hydrogen atom on one molecule aligned with an unshared pair of electrons on a highly electronegative atom on another molecule. The two molecules may be alike or different.

DNA double helix

Adenine-thymine hydrogen bonds

Cytosine-guanine hydrogen bonds

A

B

**FIGURE 10.22** (A) In DNA, hydrogen bonds hold together the base pairs cytosine-guanine and adenine-thymine. (B) These bonds form the "rungs" in the spiral ladder that is the DNA double helix.

hydrogen is closer to the atom to which it is covalently bonded than to the other molecule. Hydrogen bonding is an important force in living systems, where it maintains much of the structure of proteins and nucleic acids, as shown in Figure 10.22. The hydrogen bonding stabilizes molecular shapes, thereby protecting the ability of such molecules to carry out their biological functions.

---

### EXAMPLE 10.7 | Hydrogen Bonding

Identify the molecules from the following list that experience hydrogen bonding with other molecules in the pure liquid state: $CH_3F$, $CH_3CH_2OH$, and $CH_3-O-CH_3$.

**Solution:**

For a molecule in a pure liquid to experience hydrogen bonding with other molecules, it must contain a nitrogen, oxygen, or fluorine atom, and a hydrogen atom must be bonded to one of them. The $CH_3F$ molecule contains hydrogen and fluorine, but hydrogen is not bonded to fluorine, so it does not experience hydrogen bonding. The $CH_3CH_2OH$ molecule does contain hydrogen bonded to oxygen, so it will experience hydrogen bonding. The $CH_3-O-CH_3$ molecule contains hydrogen and oxygen, but hydrogen is not bonded to oxygen. It does not experience hydrogen bonding.

**Consider This 10.7**

Which hydrogen atoms in methylamine, $CH_3NH_2$, participate in hydrogen bonding with other molecules when in the pure liquid state?

**Practice Problem 10.7**

Identify the molecules from the following list that experience hydrogen bonding in the pure liquid state: $N(CH_3)_3$, $CH_3CO_2H$, and $HOCl$.

**Further Practice:** Questions 10.53 and 10.54 at the end of the chapter

## Trends in Intermolecular Forces

Intermolecular forces vary in strength. Assuming the molecules are similar in size, the magnitudes of intermolecular forces generally compare in the following way:

molecules experiencing     molecules also      molecules also

London dispersion   <   experiencing   <   experiencing

forces only        dipole-dipole forces    hydrogen bonding

This relative order may not hold if we compare the London dispersion forces in very large molecules with the dipole-dipole forces in small molecules.

How do we know which intermolecular forces to expect in a given substance? London dispersion forces occur in all substances. If the substance is composed of polar molecules, then it also has dipole-dipole forces. If the polar substance consists of molecules containing H–F, H–O, or H–N bonds, then it also has hydrogen bonding forces. These relationships are summarized in Figure 10.23.

Because many properties depend on the total strength of all the intermolecular forces in a substance, it is useful to compare the intermolecular force strengths of substances. The following steps are a helpful guide for comparing the total intermolecular force strengths of different substances:

1. First determine if a substance is polar and has hydrogen-bonding forces. If so, then it likely has the stronger intermolecular forces. For example, $NH_3$ has stronger intermolecular forces than $NF_3$ because $NH_3$ has hydrogen-bonding forces and $NF_3$ does not. An exception to this is if we are comparing substances of dramatically different molar masses such as $CCl_4$ (153.8 g/mol) and HF (20.01 g/mol); in this case it turns out that $CCl_4$ has the stronger forces.

2. If we are comparing polar and/or nonpolar substances with no hydrogen bonding and somewhat different molar mass, then the substance with the greater molar mass (and therefore larger electron clouds) will have the stronger total intermolecular forces due to the stronger London dispersion forces. For example, $CCl_4$ has stronger intermolecular forces than $CHCl_3$ even though $CHCl_3$ has dipole-dipole forces. This is because $CCl_4$ has significantly stronger London dispersion forces than $CHCl_3$.

3. If we are comparing substances with about the same molar mass, and one or both is polar, then the most polar substance should have the stronger intermolecular force strength due to the greater dipole-dipole forces. For example, CO and $N_2$ have about the same molar mass and therefore about the same strength of London dispersion forces. The additional dipole-dipole forces between the polar CO molecules cause CO to have the stronger total intermolecular forces.

It is important to note that these are only guidelines for predicting relative strengths of intermolecular forces. A more accurate way is to compare boiling points. The substance with the higher boiling point has the stronger total intermolecular forces.

**FIGURE 10.23** All molecules experience London dispersion forces. The subset of polar molecules also exhibits dipole-dipole forces. Only polar molecules that contain H–F, H–O, or H–N bonds are also subject to hydrogen-bonding forces.

---

**EXAMPLE 10.8**   **Trends in Intermolecular Forces**

List the following molecules in order of increasing strength of their intermolecular forces: $F_2$, HCl, and $H_2O_2$.

**Solution:**

Of these three substances, only $H_2O_2$ is polar and capable of hydrogen bonding. Because the other substances are not drastically different in molar mass, we can predict that $H_2O_2$ has the strongest intermolecular forces. The other two substances have almost the same molar mass (and therefore number of electrons), so we cannot just compare London dispersion force strength. HCl is polar, so it has stronger forces than $F_2$ due to its additional dipole-dipole forces. The order of increasing intermolecular force strength is: $F_2 < HCl < H_2O_2$.

**Consider This 10.8**

If we wanted to compare the intermolecular forces in HCl and $H_2S$, what information would we have to consider to make the comparison?

**Practice Problem 10.8**

Indicate which member of each pair of molecules has the stronger intermolecular forces and explain your reasoning.

(a)  $N_2$ and $Cl_2$    (b)  $PH_3$ and $NH_3$    (c)  $SO_2$ and $CO_2$    (d)  $H_2S$ and $H_2Te$

**Further Practice:**  Questions 10.63 and 10.64 at the end of the chapter

---

Consider the substances methyl alcohol, $CH_3OH$, and methyl mercaptan, $CH_3SH$. How should their boiling points compare? Since boiling points increase with increasing intermolecular forces, we must decide which intermolecular forces are present in each substance by first identifying each substance as nonpolar or polar. Each has a methyl group, $-CH_3$, and a hydrogen atom bonded to an atom that contains two unshared electron pairs (O or S), so each is bent and polar. In $CH_3OH$, the hydrogen atom is bonded to an oxygen atom, but in $CH_3SH$, a hydrogen atom is not bonded to F, O, or N. Thus, $CH_3OH$ experiences hydrogen-bonding forces, while $CH_3SH$ experiences only dipole-dipole forces. (Both also experience London forces.) Since hydrogen-bonding forces are stronger than dipole-dipole forces and London forces, we expect $CH_3OH$ to have the higher boiling point. Measurements indicate boiling points of 64.7°C for $CH_3OH$ and 5.95°C for $CH_3SH$, in agreement with our predictions.

 **Student Hot Spot**

Student data indicate you may struggle with how intermolecular forces relate to boiling points. Access the SmartBook to view additional Learning Resources on this topic.

---

**EXAMPLE 10.9**   **Trends in Boiling Points**

Arrange the following liquids in order of increasing boiling points: $CH_4$, $GeH_4$, and $SiH_4$.

**Solution:**

None of these molecules is polar, so we must consider only the magnitude of the London dispersion forces in each liquid. These forces increase as the molecule becomes larger, due to a more easily distorted electron cloud. As intermolecular forces increase, the energy required to evaporate the liquids increases, and the temperature needed to supply the energy increases. Thus, we predict the order of boiling points as $CH_4 < SiH_4 < GeH_4$. This agrees with measured values of the boiling points: $CH_4$ (−164°C) < $SiH_4$ (−112°C) < $GeH_4$ (−89°C).

**Consider This 10.9**

If we wanted to compare the melting points of these three compounds in their solid state, would the trend differ, and if so, how?

**Practice Problem 10.9**

Arrange the following liquids in order of increasing boiling points: $CHBr_3$, $CHCl_3$, and $CHI_3$.

**Further Practice:** Questions 10.67 and 10.68 at the end of the chapter

---

Evaporating a liquid requires overcoming the intermolecular attractions among its molecules. At a constant temperature, vapor pressure decreases with the increasing strength of the intermolecular forces. Accordingly we would expect that vapor pressure would be lowest for compounds that have hydrogen bonding. Polar molecules should have the next highest vapor pressures. Nonpolar molecules of similar size should have the highest values. Among nonpolar molecules, those with large electron clouds, such as $C_6H_6$ and $CCl_4$, should have lower vapor pressures than those with small electron clouds, such as $CH_4$.

---

| EXAMPLE 10.10 | Trends in Vapor Pressure |
|---|---|

Explain why the vapor pressure of water is about 0.053 atm at 35°C, while that of diethyl ether, $CH_3CH_2-O-CH_2CH_3$, is about 1 atm at the same temperature.

**Solution:**

Both molecules are polar and so undergo dipole-dipole interactions. However, diethyl ether, unlike water, does not have hydrogen bonding. Since hydrogen bonding is stronger than dipole-dipole forces, the intermolecular forces are stronger in water than in diethyl ether. Thus, at a given temperature, more diethyl ether evaporates, giving a higher vapor pressure.

**Consider This 10.10**

If we had two compounds of similar molar mass, one of which was polar and the other was nonpolar, which would have the higher vapor pressure at a given temperature?

**Practice Problem 10.10**

Which substance, $O_3(l)$ or $O_2(l)$, has the higher vapor pressure at a given temperature? Explain your reasoning.

**Further Practice:** Questions 10.75 and 10.76 at the end of the chapter

---

Trends in vapor pressure are opposite those for the melting point and boiling point, but the reasons are the same. Let's consider $CH_3OH$ and $CH_3SH$ again. The boiling points are 64.7°C for $CH_3OH$ and 5.95°C for $CH_3SH$, consistent with the stronger intermolecular forces in $CH_3OH$. The stronger the intermolecular forces, the higher the temperature needed to overcome them and evaporate the liquid. At a given temperature, then, the substance with the stronger intermolecular forces has the lower concentration of molecules in the gaseous state, and therefore the lower vapor pressure. Observed data are consistent with this conclusion. The values of the vapor pressure at 0°C are 0.039 atm for $CH_3OH$ and 0.626 atm for $CH_3SH$.

---

### EXAMPLE 10.11 | Trends in Molecular Properties

Consider the substances $CH_4$ and $NH_3$. Which has the stronger intermolecular forces? Which has the higher melting point? Boiling point? Vapor pressure?

**Solution:**

$CH_4$ is nonpolar, while $NH_3$ can hydrogen-bond. Thus, $NH_3$ has the stronger intermolecular forces. It takes more energy to overcome the intermolecular forces in $NH_3$ than in $CH_4$, so a higher temperature will be required to melt or boil $NH_3$. Because of the lower energy required for evaporation, at any given temperature $CH_4$ will have a higher vapor pressure than $NH_3$.

**Consider This 10.11**

How would you explain that $H_2O$ has a lower vapor pressure than $CH_3OH$?

**Practice Problem 10.11**

Consider the substances CO and HF. Which has the stronger intermolecular forces? Which has the higher melting point? Boiling point? Vapor pressure?

**Further Practice:** Questions 10.77 and 10.78 at the end of the chapter

**FIGURE 10.24** (A) When the water in a full bottle freezes, it expands beyond the capacity of the bottle. (B) Freezing water expands in cracks, splitting rocks and paved roads.

(A): ©Martyn F. Chillmaid/Science Source;
(B): ©Arthur S. Aubry/Getty Images

## 10.3 Properties of Liquids

The properties of liquids are related to the distance between particles (usually molecules) and to intermolecular forces. Particles in a liquid are much closer together than particles in a gas. However, unlike particles in a solid, liquid particles are not fixed in position. They move about and collide with one another. Because of their intermolecular forces, liquid particles move shorter distances before they collide than do the particles in a gas. They have less independence than gas particles, but more than solid particles.

### Density

The densities of the states of matter are related to the distance between particles. Generally, liquids are denser than gases by a factor of about 1000 and nearly as dense (90%–95%) as solids. As you might expect, most substances are denser when solid than when liquid, because their molecules or atoms are closer together. For example, liquid aluminum at 659°C has a density of 2.38 g/mL, while the density of solid aluminum at 25°C is 2.70 g/mL. Water, however, is an exception. Liquid water has its greatest density, 1.0000 g/mL, at 4°C. At 0°C, solid water has a density of 0.9167 g/mL and liquid water has a density of 0.9998 g/mL. The implication of this density comparison is that water expands when it freezes (Figure 10.24). As a result, ice floats on water, allowing plant and animal life to survive in lakes and streams during the winter. If ice sank, entire lakes would freeze and probably never thaw completely. The expansion of freezing water is responsible for much weathering action. When water freezes in cracks in rock, for example, the rock may be broken up, a process that helps create soil. We see a similar effect in frozen roadbeds, engine blocks, and glass containers, so this action is not always beneficial.

Why does water expand on freezing while most other liquids contract? The answer lies in the arrangement of its molecules. In ice, hydrogen bonding fixes water molecules into a *hexagonal* arrangement. The centers of the hexagons line up to form open channels, leaving a lot of empty space in the solid. In liquid water, the rigid structure is relaxed, and water molecules fill some of the unused space (Figure 10.25).

**FIGURE 10.25**  When ice melts, the open spaces in the ice structure fill with water molecules, so the liquid occupies a smaller volume.
©John A. Rizzo/Getty Images

## Viscosity

Liquids, like gases, are fluids. A **fluid** is any substance that can flow. Most liquids—such as water, alcohol, and gasoline—flow readily, although not as easily as gases. Solids, in contrast, cannot be observed to flow at all over normal time periods, such as days or weeks. The resistance of a substance to flow is its **viscosity.** In general, the viscosity of liquids is low, closer to the low viscosity of gases than to the high viscosity of solids.

However, the viscosities of different liquids vary, generally increasing with the magnitude of their intermolecular forces and with molecular size. At the low end of the range is liquid helium-3 below a temperature of 2.1 K (−271°C). It has such a low viscosity that when it is placed in an open beaker, it actually flows up the sides and out of the beaker. At the other end of the spectrum are such highly viscous liquids as oils (castor oil is about 1000 times more viscous than water), honey, syrup, molasses, and hot tar, a material used to pave roads. The viscosity of a liquid generally decreases as the liquid is heated, because heat energy partially overcomes intermolecular forces. Ferrofluids, discussed in the introduction, normally have the viscosity of oil, but they become more viscous when placed in a magnetic field (Figure 10.3).

The cliché, slow as molasses in January, used to describe high viscosity, may be misleading. On January 15, 1919, a large vat of molasses burst when the temperature reached an unseasonable 43°F, sending over 2 million gallons of molasses down the streets of Boston. The 8-ft wave knocked down buildings and traveled so fast that people in its path couldn't outrun it. The flood killed 21 people and injured many more.

---

**EXAMPLE 10.12** | **Viscosity**

Gasoline contains hydrocarbons such as octane, $C_8H_{18}$. Motor oils contain much larger molecules, such as $C_{18}H_{38}$. Which is the more viscous liquid?

**Solution:**

Viscosity is greater for molecules that have stronger intermolecular forces. Both $C_8H_{18}$ and $C_{18}H_{38}$ are nonpolar molecules, so they have only London dispersion forces. Because $C_{18}H_{38}$ is the larger molecule, it should have the stronger intermolecular forces. Thus, motor oil is more viscous than gasoline.

**Consider This 10.12**

If one substance is polar and the other is not, and they are both about the same size, which would be expected to have the greater viscosity?

**Practice Problem 10.12**

The Society of Automotive Engineers rates the viscosity of motor oils. Some typical ratings are SAE 40 and SAE 15, where the number is a measure of the relative viscosity. Higher numbers are given to oils that are more viscous. Oils suitable for use during winter have a W placed after the number in the viscosity rating. Which of these oils (SAE 40 or SAE 15) could be used during the winter and thus have a W in its rating?

**Further Practice:** Questions 10.83 and 10.84 at the end of the chapter

## Surface Tension

Small droplets of liquids tend to have spherical shapes (Figure 10.26). Water beads up when placed on a waxed surface. Soap bubbles and raindrops are spherical. Why does this happen? As you might expect, it has to do with intermolecular forces. Molecules within a liquid experience forces of attraction equally in all directions, as shown in Figure 10.27. However, there are no forces above the surface of the liquid, but there are normal forces below. Because of this imbalance, forces of attraction tend to pull molecules toward the interior of the liquid. The attraction minimizes the number of molecules at the surface, and thus minimizes the surface area. For a given volume, a sphere is the shape with the smallest surface area.

In order for its surface to be deformed from a spherical shape, the liquid must increase its surface area. To do this, molecules must overcome intermolecular forces in order to move from the interior of the liquid to the surface. The energy required to do this is called **surface tension,** the amount of work necessary to increase the surface area of a liquid by a unit amount. Surface tension causes a liquid surface to behave like a stretched membrane. The greater the intermolecular forces in a liquid, the greater the surface tension. Surface tension decreases somewhat as temperature increases, because intermolecular forces are overcome to some extent by greater kinetic energy at higher temperatures.

**FIGURE 10.26** Water droplets have a spherical shape that minimizes their surface area.

(top): ©Ingram Publishing/age Fotostock; (bottom): ©vonviper/iStock/Getty Images

 **ANIMATION:** Surface Tension

**FIGURE 10.27** Unbalanced intermolecular forces on the surface of a liquid pull molecules back toward the interior. Molecules below the surface experience forces of attraction that are about equal in all directions.
©Jim Birk

The surface tension of water is great enough to float an insect (Figure 10.28). The force of the insect's mass pushing down is less than the amount of energy needed to overcome intermolecular forces and cause an increase in the water's surface area. The same is true of a steel needle laid carefully on the surface of water. However, if a detergent is added to the water, both the insect and the needle will sink because the surface tension is lowered and the surface is easier to deform. The detergent molecules interfere with the water-water attractive interactions, making it easier to move water molecules apart. This effect is put to use in laundry detergents. They work in part because they reduce the surface tension, so the water can wet the clothing better, allowing the detergent to surround the dirt particles.

We observe another result of intermolecular forces when a capillary tube is placed in a liquid. As shown in Figure 10.29, water rises in a glass capillary tube, a phenomenon called *capillary action.* In a tube, water-glass interactions are stronger than the intermolecular forces in water alone, so water rises. The extent of the rise depends on the size of the capillary. Water rises until the water-glass interactions are just offset by the pull of gravity on the column of water. Capillary action is partially responsible for the rise of sap in trees and for a towel's absorption of water.

Intermolecular forces also produce the **meniscus,** the curved surface of a liquid in a container. The meniscus may be either concave or convex, depending on whether the intermolecular forces among liquid molecules are greater than or less than the forces of attraction between the liquid and the container walls. As Figure 10.30 shows, the surface is concave for water because the water-glass interactions are stronger than the water-water interactions. In contrast, the intermolecular forces in mercury are stronger than mercury-glass interactions, so mercury tends to withdraw into itself and repel the glass, forming a convex surface.

Detergents are in a class of compounds called surfactants, which modify surface tension.

**FIGURE 10.28** Water striders can walk across the surface of water because their weight is insufficient to overcome intermolecular forces among water molecules. However, if a detergent is added, surface tension is lessened, and the insect sinks. If it does end up under water, it has to exert considerable force to climb back out. You can examine a similar effect by carefully floating a needle on water, and then adding a drop of dish detergent.
Source: Rosalie LaRue/NPS

**FIGURE 10.29** (A) Water rises in tubes due to capillary action. A hemocrit (capillary tube) is filled with blood using capillary action. (B) The drying action of a towel is due in part to capillary action.
(A): ©SPL/Science Source; (B): ©Keith Brofsky/Getty Images

**FIGURE 10.30** The meniscus for water (left) is concave because the interactions between the silicon dioxide in the glass and the water molecules are stronger than the interactions among water molecules. The meniscus for mercury (right) is convex because the interactions among mercury atoms are stronger than the interactions between mercury and silicon dioxide.
©Charles D. Winters/Science Source

$H_2O$    $Hg$

---

**EXAMPLE 10.13** | **Surface Tension**

Water spreads out on a glass surface but beads up on a waxed surface. Explain the difference.

**Solution:**

The intermolecular forces among water molecules are less than the interactions between water molecules and glass, so the water spreads out on glass. However, on a waxed surface, the opposite is true. The surface water molecules are attracted to the internal water molecules more strongly than to the wax molecules. The water drops form a spherical shape that minimizes their surface area.

**Consider This 10.13**

If a drop of water and a drop of alcohol are placed on a piece of waxed paper, the alcohol spreads out more than the water. Which has the greater surface tension?

**Practice Problem 10.13**

Scientists want to design a new material that can be coated on a car body so the surface of the car can be wet with water without having the water form beads. What properties of the new material should they be concerned about?

**Further Practice:** Questions 10.85 and 10.86 at the end of the chapter

---

Glass panes in old churches in Europe are often thicker at the bottom than at the top. For some time, this was considered evidence that over long periods of time, glass actually flows, a characteristic of liquids. However, investigation of traditional glass-making methods revealed that glass panes were usually not of uniform thickness. It is more likely that the thicker and heavier end of a glass pane was mounted in the bottom of the window frame.

🌀 **ANIMATION:** Packing Spheres—Cubic Close

**FIGURE 10.31** (A) Quartz, a crystalline solid, has faces that are planes at characteristic angles. (B) Glass, an amorphous solid of nearly the same composition, does not fracture along characteristic planes.
(A): ©Siede Preis/Getty Images;
(B): ©Fotosearch RF/Glow Images

## 10.4 Properties of Solids

The most obvious properties of solids are their definite sizes and rigid shapes. Solids are not very compressible, and they usually do not flow. Attractive forces hold the particles (ions, atoms, or molecules) in solids together in a rigid, three-dimensional array. If the temperature of a liquid drops rapidly, the particles may solidify in a partially disordered state, particularly if they are long molecules that can get tangled together. The resulting solid is said to be an **amorphous solid.** Its particles are arranged somewhat randomly, and it lacks a regular form. Glass and most other polymers, such as rubber and plastics, belong in this class. If, in contrast, a liquid solidifies slowly, allowing the array of particles to become well ordered, the result is a **crystalline solid.** Most solids are of this type. Such solids can also be precipitated from solutions containing ionic compounds. The symmetrical arrangement of planar faces typical of a crystalline solid is shown in Figure 10.31. In contrast, amorphous solids rarely show regular planar faces.

### Crystals and Crystal Lattices

A **crystal** is an orderly, repeating, three-dimensional assembly of fundamental particles (atoms, molecules, or ions). The pattern formed by these arrays is the **crystal lattice** or *crystal structure.* One layer in the crystal lattice of a typical metal is

A                                         B

shown in Figure 10.32. Notice the orderly arrangement of atoms. It arises from a tendency of atoms to use space efficiently—that is, to be as close to one another as possible. The arrangement shown is called *close packing* because it is the most efficient way to use space. Six atoms surround and touch each atom in this structure. This arrangement is found at a macroscopic level, both in nature and in human endeavors, as shown in Figure 10.33.

In a three-dimensional crystal, more than one layer of atoms must be considered, but each layer has a similar orderly arrangement. The arrangement of atoms in many metals is the same as the arrangement of fruit stacked in grocery stores (Figure 10.34). In these structures, one layer of close-packed atoms is placed on another layer, with the atoms in the second layer fitting into depressions between atoms in the first. Successive layers fit together in a similar fashion.

Use marbles placed in a box to examine the close-packed structure in three dimensions. Arrange the first layer to place the marbles as close to one another as possible, with six marbles surrounding each individual. Then create a second layer by placing marbles over the depressions in the first layer. Continue for as many layers as you wish to examine.

**FIGURE 10.32** A scanning tunneling microscope image of the surface of a copper crystal shows how the atoms pack closely together in an arrangement called close packing or cubic close packing. Count the number of atoms that surround and touch a single atom.
©Drs. Ali Yazdani & Daniel J. Hornbaker/ Science Source

**FIGURE 10.33** Close packing occurs in many places, both naturally and in human-made objects. From the left: pineapple, wasp nest, bubble wrap, and dimples in a golf ball.
(left): ©Aaron Roeth Photography; (middle, left): ©Jim Birk; (middle, right), (right): ©Ryan McVay/Getty Images

**FIGURE 10.34** Grocers stack fruit in the same close-packed arrangement as atoms in metals.
(left): ©renaschild/iStock/Getty Images

## Types of Crystalline Solids

Solids show a wide range of properties, such as melting point, hardness, and malleability. For most solids, the types of forces holding the fundamental particles together explain these properties and correlate with them. The useful classification system presented in Table 10.3 organizes solids into four groups: *metallic, ionic, molecular* (which may be either polar or nonpolar), and *network*. The table describes some of the properties of these classes and gives examples of each.

Consider two solids, the first of which conducts heat, while the second does not. Both solids are soft, but only the first is malleable. The first conducts electricity, but the second is an electrical insulator. How do we classify these solids? Only metallic solids are good heat and electrical conductors, so the first solid must be a metal. We could place the other solid in any of the other classes if we considered only these two properties. However, only molecular solids (along with a few metals, which this solid is not) are soft. So, the second solid must be a molecular solid.

We must consider several properties to classify any solid. It is difficult to predict whether a covalent substance will form a molecular solid or a network solid, but we can distinguish between them if we know their properties.

---

**EXAMPLE 10.14** | **Types of Crystalline Solids**

A crystalline solid is hard, breaks when it is struck with a hammer, and is an electrical insulator; but it conducts electricity when it is melted. What type of solid is it?

**Solution:**

A hard solid could be metallic, ionic, or network. The observation that it does not conduct electricity as a solid is consistent with either an ionic or a network solid. Since it conducts electricity when melted, it must be an ionic solid.

**Consider This 10.14**

If a solid is soft, what categories can be eliminated when classifying it?

**Practice Problem 10.14**

A crystalline solid is very hard, does not conduct electricity when solid or when melted, and has a high melting point. What type of solid is it?

**Further Practice:** Questions 10.101 and 10.102 at the end of the chapter

---

**TABLE 10.3** Classification of Solids

| Type of Solid | Fundamental Particles | Attractive Forces | Properties | Examples |
|---|---|---|---|---|
| Metallic | Atoms | Attractions between nuclei and delocalized valence electrons | Low melting point and soft, or high melting point and hard; good heat and electrical conductors; malleable and ductile | Hg, Na, Fe, Au, Ag, Cu, Cr, alloys such as Nitinol |
| Ionic | Cations and anions | Ionic bonds | High melting point, hard, brittle; nonconductors when solid; electrical conductors when melted | NaCl, CsCl, $CaF_2$, $CaCO_3$, $KNO_3$, $Ca_5(PO_4)_3OH$, $YBa_2Cu_3O_7$ |
| Molecular | Polar molecules | Dipole-dipole forces | Low to moderate melting point, variable hardness, may be brittle; nonconductors | $H_2O$, $NH_3$, $SO_2$, $CH_3CO_2H$, $CH_2Cl_2$ |
| | Nonpolar molecules | London dispersion forces | Low melting point, soft; poor heat conductors; electrical insulators | $CH_4$, $O_2$, $CCl_4$, $CO_2$, $C_6H_6$, $I_2$, $S_8$ |
| Network | Atoms | Covalent bonds | Very high melting point, very hard, somewhat brittle; nonconductors or semiconductors | C (diamond), SiC (carborundum), $SiO_2$ (silica), $Al_2O_3$ (alundum), BN (boron nitride) |

**Metallic Solids**   In metals, valence electrons move freely through all parts of the metal. This explains why metals are good conductors of electricity. The bonds between atoms result from attractions between the shared valence electrons and the metal nuclei surrounded by their core electrons, which remain fixed in position. These attractions are spread out over the entire piece of metal. Because the attractions are not localized, it is easy to move atoms by applying force, making metals malleable and ductile. Depending on the strength of the attractions, metals can range from soft to hard and can melt at low to high temperatures. The soft, low-melting-point metals are primarily the group IA (1) elements. The properties of metals can be changed by adding impurities. Remember the bronze that gave its name to an entire age in human history? Mixing copper and tin makes bronze, which has different properties than either of the two pure metals.

Mixing a metal with one or more additional metallic or nonmetallic elements forms an **alloy.** Alloys have properties different from those of their parent elements. For example, as shown in Figure 10.35, zinc added to copper makes the alloy brass. In brass, zinc atoms replace some of the positions in the metal structure formerly occupied by copper atoms. Brass is more resistant to tarnish and takes a better polish than copper, so it is commonly used in plumbing fixtures.

Pure iron is relatively soft and malleable. In steel, carbon atoms fit into holes in the original iron structure. As more carbon is added, the steel becomes harder and stronger. Other elements, such as chromium, vanadium, nickel, molybdenum, or tungsten, can also be added to iron to make steels with various properties. Some typical alloys, including stainless steel, are described in Table 10.4.

In the introduction, we discussed Nitinol, an alloy with peculiar properties. This alloy containing nickel and titanium can "remember" its original shape. How? This property is related to the structure of the metal (Figure 10.36). Actually, the metal has two slightly different structures, one at low temperature and another at high temperature. The two structures change from one to another by relatively minor shifts of the planes of atoms. When the metal is bent at low temperature, stress is placed on the planes of atoms. When the metal is then heated, the stress is relieved by shifts of atoms to the high-temperature structure, giving back the original shape. Upon cooling, the original shape is maintained. If the metal is bent when heated at a high temperature, the new shape will be retained when the metal is cooled.

Cu

Zn coating

Brass alloy

**FIGURE 10.35**   When copper (left) is chemically treated with zinc metal, it gets a coating of zinc (center). If this mixture is heated, the zinc atoms move into the crystal structure of the copper, giving the alloy brass (right). In brass, zinc atoms occupy some of the positions formerly occupied by copper atoms.
©Brian Moeskau/Moeskau Photography

**TABLE 10.4**  Examples of Alloys

| Alloy | Metals | Properties | Uses |
|---|---|---|---|
| Brass | 67% Cu, 33% Zn | Ductile, takes a polish | Hardware |
| Dental amalgam | 70% Ag, 25% Pb, 3% Cu, 2% Hg | Easily worked | Dental fillings |
| 14-karat gold | 58% Au, with various amounts of Ag, Cu, and others | Harder than pure gold | Jewelry |
| Monel metal | 67% Ni, 30% Cu, with various amounts of Fe, Mn, Al or Ti | Resistant to corrosion, bright finish | Kitchen fixtures |
| Nichrome | 60% Ni, 40% Cr | High melting point, low electrical conductivity | Heating coils in stoves, toasters, ovens |
| Nitinol | 49% Ni, 51% Ti | Shape memory | Medical and dental implants, eyeglass frames |
| Pewter | 85% Sn, 7% Cu, 6% Bi, 2% Sb | Easily cast into shapes | Utensils, cups |
| Plumber's solder | 67% Pb, 33% Sn | Low melting point, soft | Soldering pipe joints |
| Stainless steel | 80.6% Fe, 0.4% C, 18% Cr, 1% Ni | Resistant to corrosion | Tableware |
| Sterling silver | 92.5% Ag, 7.5% Cu | Bright finish | Tableware |

A                                                                      B

**FIGURE 10.36**  (A) Nitinol can change between this and another closely related structure as its temperature changes. (B) An expanded view of a slice through the Nitinol structures shows the small change in position of the atoms that converts one structure to the other.

**Ionic Solids**    Recall from Chapter 8 that ionic solids contain cations and anions arranged in crystalline lattices. The forces holding the particles together are cation-anion attractive electrostatic forces, offset somewhat by cation-cation and anion-anion repulsive electrostatic forces. Because the attractive electrostatic forces, called ionic bonds, are strong, ionic crystals have high melting points. It is difficult to displace ions, so the crystals tend to be hard. They are also brittle. Although displacement is difficult, only a small amount is necessary to break a crystal. Ionic solids are not electrical conductors because the ions are fixed in position. However, melted or dissolved ionic solids, in which the ions are free to move and carry electrical charge, are good conductors.

Many ionic crystal structures resemble the close-packed arrangements we saw for metals. Close-packed structures use only 74% of the available volume, so there are holes left within them. We saw that metals can form alloys by placing other atoms in holes in the metal structure. Ionic crystals are analogous to alloys. In ionic crystals, the larger ions (usually the anions) occupy close-packed positions, and the smaller ions (usually cations) fit into holes appropriate to their size. Some ionic crystals formed in this way are shown in Figure 10.37.

Some ionic solids can be classified as *superconducting* materials. They offer no resistance to the conduction of an electric current, and they repel magnetic fields. Early superconducting materials worked only at very low temperatures, such as when cooled by liquid helium to below 4 K ($-269°C$), so they had no practical applications. But materials prepared recently are superconductors at higher temperatures. For example,

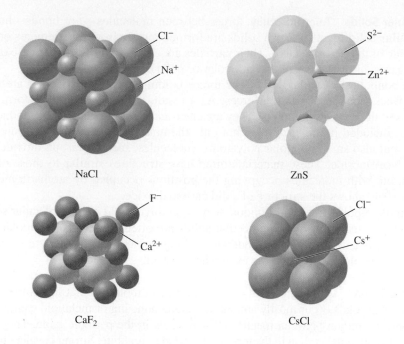

**FIGURE 10.37** Many ionic crystals form by an efficient packing of anions, with cations occupying holes in the resulting structure.

as shown in Figure 10.38A, a magnet floats (levitates) above a disk of superconducting material cooled to liquid nitrogen temperatures, about 77 K ($-196°C$). Such superconductors are used in magnetic resonance imaging (MRI) equipment in hospitals. Extremely fast trains are being tested that can levitate above superconducting tracks (Figure 10.38B). Although the nature of superconducting materials is not thoroughly understood, it appears that the crystal structure of the solid material plays a role. The layered structure shown in Figure 10.39 for the superconductor $YBa_2Cu_3O_7$ seems to be a feature of all superconducting materials discovered so far.

A

B

**FIGURE 10.38** (A) A magnet levitates above a disk of superconducting material. (B) A Japanese train levitates above superconducting tracks.
(A): ©Fuse/Getty Images; (B): ©Martin Bond/Science Source

$YBa_2Cu_3O_7$

**FIGURE 10.39** High-temperature superconductors are ionic solids that form layered structures. In this structure, alternating copper and oxide ions form cubes, barium ions fit in holes in the center of the cubes, and yttrium ions fit into the spaces between layers created by the other ions.

**FIGURE 10.40** In solid carbon dioxide (dry ice), $CO_2$ molecules are arranged in a close-packed pattern similar to that of the atoms in gold metal (Figure 10.32). Molecules of $CO_2$ are arranged at angles relative to one another, providing a better fit into the available space.

**Molecular Solids** Intermolecular forces between molecules—not bonds—hold a **molecular solid** together. Such solids are formed mostly from the elements on the right side of the periodic table. Their particles are either atoms, as in the case of the noble gases, or molecules (either nonpolar or polar).

In nonpolar molecular solids, the forces holding molecules together are relatively weak London dispersion forces. As a result, these solids tend to form soft crystals with low melting points. They are electrical insulators and poor conductors of heat. Included in this group are not only the noble gases and the diatomic elements, but also small nonpolar polyatomic molecules, such as carbon dioxide and sulfur hexafluoride. These materials often have structures similar to those of the metals, but with molecules occupying the positions occupied by atoms in metals. Figure 10.40 shows the structure of solid carbon dioxide.

Dipole-dipole forces and London dispersion forces hold polar molecular solids together. Consequently, polar molecular solids are often harder than nonpolar molecular solids and have low to moderate melting points. Because they contain no ions, they are electrical insulators even when molten.

**Network Solids** A **network solid** consists of a giant molecule that forms the entire crystal. Such solids occur mostly among the semiconducting or metalloid elements in the transition region between metals and nonmetals in the periodic table. They also form from elemental carbon in the form of diamond or graphite. Strong covalent bonds connect the atoms in network solids. The crystals of network solids have very stable, three-dimensional structures because their covalent bonds point in specific directions.

Many network solids have a diamond-like structure (Figure 10.41) or some modification of it, in which each atom is bonded to four other atoms in a tetrahedral arrangement. A typical modification is that in the silicon dioxide crystal, in which oxygen atoms form bridges between the silicon atoms (Figure 10.42). In most network solids, the electrons are localized in covalent bonds, making them poor electrical conductors. Some are semiconductors, with a very slight electrical conductivity; they are used to make computer chips. Because their bonds tend to be strong, network solids tend to have high melting points and to be very hard. Among them are the hardest substances known. They are often used in cutting tools.

A different structure is found in the graphite form of carbon. It consists of a two-dimensional network structure of layers of carbon atoms connected by covalent bonds. Relatively weak London dispersion forces hold the layers together (Figure 10.43). Layers

**FIGURE 10.41** In diamond, used in jewelry and polishing agents, each carbon atom is bonded to four other carbon atoms. These four carbon atoms are arranged at the corners of a tetrahedron. The network arrangement of carbon atoms contributes to the strength and stability of diamond.
©Steve Hamblin/Alamy Stock Photos

**FIGURE 10.42** In silicon dioxide, used in sandpaper, each silicon atom is bonded to four oxygen atoms in a tetrahedral arrangement. Each oxygen atom is bonded to two silicon atoms. This gives one silicon atom for every two oxygen atoms and an empirical formula of $SiO_2$.
©C SquaredStudios/Getty Images

**FIGURE 10.43** In graphite, each carbon atom is covalently bonded to three other carbon atoms in the same layer. London dispersion forces hold layers of carbon atoms together in graphite, which is used in pencil lead. The forces are relatively weak, so layers slide easily over one another. (Note that the distance between layers is exaggerated in this image.)
©Achim Prill/123RF

of atoms in this structure can slide easily over one another, so graphite is a good lubricant and is used in pencil leads. The electrons in graphite move within the layers readily—in contrast to most network solids—so it is a good electrical conductor in the direction of its layers. For this reason, graphite is often used as an electrode in batteries (see Chapter 14).

| EXAMPLE 10.15 | Structures of Solids |
|---|---|

Identify the type of solid shown by the following molecular-level image. What types of forces hold the particles together in this solid?

**Solution:**

It is possible to identify individual molecules, each containing one nitrogen atom and three hydrogen atoms. This composition corresponds to ammonia, $NH_3$. Because the structure contains molecules, it must be a molecular solid. Covalent bonds hold together the atoms of the ammonia molecule. The molecules are attracted to one another by hydrogen-bonding forces. We know they are at work because hydrogen is attached to nitrogen, one of the elements that participate in hydrogen bonding.

**Consider This 10.15**

If individual molecules are not present in the structure of a solid, what types of solids are possible?

**Practice Problem 10.15**

Identify the type of solid shown by the following molecular-level image. What types of forces hold the particles together in this solid?

**Further Practice:** Questions 10.115 and 10.116 at the end of the chapter

# CHAPTER REVIEW

## KEY CONCEPTS

- The transitions among the three states of matter are called changes of state.
  - Changes of state occur with changes in pressure and temperature.
  - Liquids change to gases by the process of evaporation which is endothermic; the reverse process is called condensation and is exothermic. The pressure of the gas in equilibrium with a liquid is called the vapor pressure. The temperature at which the vapor pressure is 1 atm is called the normal boiling point.
  - Liquids change to solids by the process of freezing which is exothermic, and the reverse process is called melting which is exothermic. When the solid is in equilibrium with the liquid at 1 atm pressure, the temperature is the normal freezing point.
  - Cooling and heating curves describe the changes in physical state that occur with temperature changes at constant pressure.
  - A substance's specific heat value, $C$, allows us to relate the heat change of a substance, $q$, its mass, $m$, and temperature change, $\Delta T$, by the equation: $q = m \times C \times \Delta T$. The heat required for a change of state is the product of the moles of substance and the molar heat of fusion or molar heat of vaporization.
- Substances exist in the liquid and solid states because of intermolecular forces that hold atoms or molecules together.
  - All substances experience London dispersion forces when in the liquid or solid states. London dispersion forces result from instantaneous dipoles causing attractions between molecules. The strength of London dispersion forces increases as molecular size (or molar mass) increases.
  - Dipole-dipole forces occur between polar molecules. When comparing substances of very similar molar mass (and therefore similar London dispersion force strength), the polar substance should have the stronger total intermolecular forces.
  - Hydrogen bonding is a particularly strong type of dipole-dipole force that occurs between molecules that contain hydrogen that is bonded to a small-sized electronegative atom (N, O, or F) within that molecule.
- The properties of liquids depend on the strength of intermolecular forces.
  - Viscosity, or resistance to flow, generally increases with intermolecular force strength and molecular size.
  - A unique property of liquids is surface tension, which is the work necessary to expand the surface of a liquid.
  - Vapor pressure is the partial pressure of gas molecules above a liquid or solid when the two states are in equilibrium. Vapor pressure decreases when intermolecular forces increase.
  - Many properties of liquid water and ice can be explained by the existence of strong hydrogen-bonding forces.
- Most solids exist in a crystalline form.
  - Crystals contain well-ordered and symmetrical arrangements of planar faces, which are reflected in a similarly ordered arrangement of atoms, molecules, or ions in a crystal lattice.
  - Metals tend to form a structure that gives the most efficient packing of atoms, called a close-packed structure.
  - Alloys form from metals by substituting atoms of one metal for atoms of another, or by placing other atoms into holes in a metal structure.
  - Many ionic salts adopt close-packed structures like those of metals.
  - Crystalline solids can be classified as metallic, ionic, molecular, or network solids. The classification is based on the types of forces holding the particles together in the solid state.
  - The structures and the types of forces in a solid explain its properties.

# KEY RELATIONSHIPS

| Relationship | Equation |
|---|---|
| The heat required to raise the temperature of a substance is directly related to the mass, specific heat, and temperature change. | $q = m \times C \times \Delta T$ |
| The heat required for a change of state of a substance is the product of the number of moles and the molar heat of the change of state. | $q_{\text{fusion}} = n \times \text{heat of fusion}$ <br> $q_{\text{vaporization}} = n \times \text{heat of vaporization}$ |

# KEY TERMS

| | | | |
|---|---|---|---|
| alloy (10.4) | deposition (10.1) | induced dipole (10.2) | network solid (10.4) |
| amorphous solid (10.4) | dipole-dipole force (10.2) | instantaneous dipole (10.2) | normal boiling point (10.1) |
| boiling point (10.1) | equilibrium (10.1) | intermolecular force (10.2) | normal freezing point (10.1) |
| condensation (10.1) | evaporation (10.1) | London dispersion force (10.2) | sublimation (10.1) |
| crystal (10.4) | fluid (10.3) | melting point (10.1) | surface tension (10.3) |
| crystal lattice (10.4) | freezing point (10.1) | meniscus (10.3) | vapor pressure (10.1) |
| crystalline solid (10.4) | hydrogen bond (10.2) | molecular solid (10.4) | viscosity (10.3) |

# QUESTIONS AND PROBLEMS

The following questions and problems, except those in *Additional Questions* and *Concept Review Questions,* are paired. Questions in a pair focus on the same concept. Answers to selected questions and problems are in Appendix D.

## Matching Definitions with Key Terms

**10.1**  Match the key terms with the descriptions provided.
  (a)  the partial pressure of gas molecules above a liquid (or solid) when the two states are in equilibrium
  (b)  the temperature at which the liquid and solid states of a substance are in equilibrium (identical to the freezing point)
  (c)  a state of balance between opposing processes
  (d)  the process by which atoms or molecules pass from the liquid state to the gaseous state
  (e)  a temporary dipole formed by the attraction of the electrons in a molecule or atom to a nearby dipole
  (f)  a mixture of a metal with one or more additional metallic or nonmetallic elements
  (g)  the temperature at which the vapor pressure of a liquid equals the external pressure
  (h)  vaporization of a solid
  (i)  a solid made up of discrete molecules held together by intermolecular forces
  (j)  the forces of attraction that occur between molecules
  (k)  the attractive forces that result from formation of temporary dipoles in molecules
  (l)  the temperature at which the liquid and solid states of a substance are in equilibrium at a pressure of 1 atm
  (m)  a solid having a shape bounded by planes intersecting at characteristic angles

**10.2**  Match the key terms with the descriptions provided.
  (a)  resistance to flow
  (b)  especially strong dipole-dipole forces between polar molecules that contain hydrogen attached to a highly electronegative element
  (c)  the transition of gas-state atoms or molecules to the solid state
  (d)  a substance that can flow (a gas or a liquid)
  (e)  the geometric arrangement of points occupied by particles in a crystalline solid
  (f)  the curved upper surface of a liquid column
  (g)  the temperature at which the liquid and solid states of a substance are in equilibrium (identical to the melting point)
  (h)  a solid in which all atoms are held together by covalent bonds to make a giant molecule
  (i)  a substance in which the particles are arranged in a regular, repeating geometric structure
  (j)  the temperature at which the vapor pressure of a liquid equals the standard atmospheric pressure of 1 atm
  (k)  a temporary dipole formed by the movement of electrons within a molecule
  (l)  the work necessary to expand the surface of a liquid; the property of a liquid surface that causes it to behave like a stretched membrane
  (m)  the transition of gas-state atoms or molecules to the liquid state

## Changes of State

10.3     Compare and contrast liquids and gases at a molecular level.

10.4     Compare and contrast liquids and solids at a molecular level.

10.5     Why does a liquid take the shape of its container, but a solid does not?

10.6     Why does a gas fill its container, but a liquid does not?

10.7     Consider the following picture of a substance. What is its physical state? Explain how you reached your conclusion.

©VasilySmirnov/iStock/Getty Images

10.8     Consider the following picture of a substance in a container. What is its physical state? Explain how you reached your conclusion.

©Jim Birk

10.9     Consider the following atomic-level diagram of a substance. What is its physical state? Explain how you reached your conclusion.

10.10    Consider the following atomic-level diagram of a substance. What is its physical state? Explain how you reached your conclusion.

10.11    Identify each of the phases of water shown in the following image. What terms are used to identify each of the phase changes indicated by the arrows?

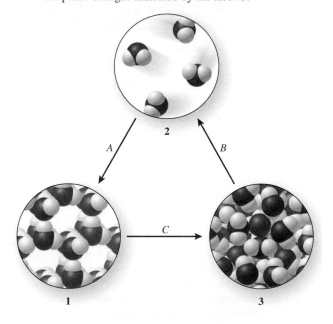

10.12 Identify each of the phases of water shown in the following image. What terms are used to identify each of the phase changes indicated by the arrows?

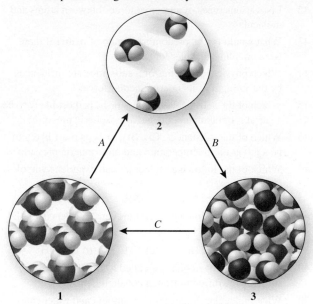

10.13 How is the vapor pressure of a liquid related to its boiling point?

10.14 Why do liquids boil at lower temperatures in the mountains?

10.15 Is evaporation exothermic or endothermic? Describe how energy is involved in this process.

10.16 Is condensation exothermic or endothermic? Describe how energy is involved in this process.

10.17 Draw the phase in the empty circle to illustrate the process of condensation.

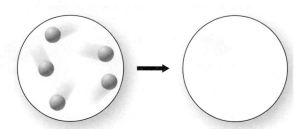

10.18 Draw the phase in the empty circle to illustrate the process of melting.

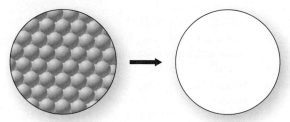

10.19 Identify the phase change shown in the molecular diagram and indicate whether it is exothermic or endothermic.

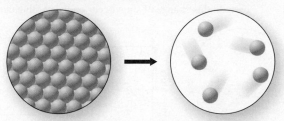

10.20 Identify the phase change shown in the molecular diagram and indicate whether it is exothermic or endothermic.

10.21 Explain why a gas converts to a liquid when it is cooled sufficiently.

10.22 Explain why a liquid converts to a solid when it is cooled sufficiently.

10.23 Why can the evaporation of water be used to cool a house?

10.24 Skin swabbed with alcohol is cooled below room temperature, even though the alcohol is at room temperature. Explain this observation.

10.25 Why does pasta take longer to cook in boiling water at a mountaintop campsite than at the beach?

10.26 Would the boiling point of water be above or below 100°C in Death Valley, which is at an altitude of 86 m below sea level?

10.27 Provide a molecular explanation for the increase in vapor pressure with increasing temperature.

10.28 Explain the process of evaporation in terms of molecular behavior in the liquid state.

10.29 Gallium has a melting point of 30°C and a boiling point of 1983°C. What is the physical state of gallium at 100°C? At 15°C?

10.30 Hydrogen has a melting point of −259°C and a boiling point of −253°C. What is the physical state of hydrogen at −263°C? At −255°C?

10.31 The pictured change of state occurs at constant temperature, even though heat is being removed. Where would this change of state occur on the cooling curve?

10.32 The pictured change of state occurs at constant temperature, even though heat is being removed. Where would this change of state occur on the cooling curve?

10.33 Calculate the amount of heat required for the evaporation of 105.0 g of water at 100.0°C.

10.34 Calculate the amount of heat required when 1.85 mol of ice is melted to liquid water at 0.0°C.

10.35 Calculate the amount of heat required when 15.0 g of water at 52.5°C is converted to steam at 238.2°C.

10.36 What is the amount of heat required to convert 105 g of ice at −15.2°C to liquid water at 35.6°C?

10.37 Calculate the heat absorbed when 542 g of ice at −15.0°C melts and then the water is converted to steam at 145°C. The specific heat of ice is 2.03 J/(g °C), the heat of fusion of ice is $6.01 \times 10^3$ J/mol, the specific heat of water is 4.18 J/(g °C), the heat of vaporization of water is $4.07 \times 10^4$ J/mol, and the specific heat of steam is 2.02 J/(g °C).

10.38 Calculate the heat released when 32.0 g of steam at 115.0°C condenses and then the water is converted to ice at −15.0°C. (See Question 10.37 for heat constants that may be useful in answering this question.)

10.39 Calculate the heat absorbed when 125 g of liquid ethanol at 25.0°C is converted to gaseous ethanol at 96.0°C. The boiling point of ethanol is 78.4°C, the specific heat of liquid ethanol is 2.44 J/(g °C), the heat of vaporization of ethanol is $3.86 \times 10^4$ J/mol, and the specific heat of gaseous ethanol is 1.42 J/(g °C).

10.40 Calculate the heat released when 84.6 g of gaseous ethanol at 90.0°C is converted to liquid ethanol at 54.2°C. The boiling point of ethanol is 78.4°C. (See Question 10.39 for heat constants that may be useful in answering this question.)

## Intermolecular Forces

10.41 What is meant by the term *intermolecular force*?

10.42 List various types of attractive forces between atoms and molecules.

10.43 What would be the predominant state of matter if there were no intermolecular forces?

10.44 Which physical properties of a substance are influenced by the strength of its intermolecular forces?

10.45 Which of the substances, Ar, Hg, or $I_2$, is most likely to be a gas at room temperature and atmospheric pressure?

10.46 Which of the substances, $Cl_2$, $Br_2$, or $I_2$, is most likely to be a gas at room temperature and atmospheric pressure?

10.47 Which of the following molecules can experience dipole-dipole forces?
(a) $CO_2$  (b) NO  (c) $NF_3$  (d) $CH_3Cl$

10.48 Which of the following molecules can experience dipole-dipole forces?
(a) $OF_2$  (b) $Br_2$  (c) $NO_2$  (d) $CH_3OH$

10.49 Should a hydrogen bond be called a "bond"? How is it similar to or different from a covalent bond?

10.50 What molecular properties of ammonia and water cause similarities in their physical properties?

10.51 Indicate which of the following molecules has hydrogen bonding when the compound is in the pure liquid state.

A                  B                  C

10.52 Indicate which of the following molecules has hydrogen bonding when the compound is in the pure liquid state.

A                  B                  C

10.53 Which of the following pure substances can participate in hydrogen bonding?
(a) $H_2O$  (b) $NH_3$  (c) $H_2Se$

10.54 Which of the following pure substances can participate in hydrogen bonding?
(a) $PH_3$  (b) $HCO_2H$  (c) HCl

10.55 Which of the following substances can form hydrogen bonds with water molecules?
(a) $SCl_2$  (b) HF  (c) $BeH_2$

10.56 Which of the following substances can form hydrogen bonds with water molecules?
(a) $CH_4$  (b) $CH_3OH$  (c) $CH_3CH_2NH_2$

10.57 Why do different liquids have different vapor pressures?

10.58 Why do different solids have different melting points?

10.59 Why are intermolecular forces greater for polar molecules than for nonpolar molecules that are about the same size?

10.60 Why are intermolecular forces greater for large molecules than for small molecules?

10.61   Under what circumstances might nonpolar molecules have intermolecular forces greater than those of polar molecules?

10.62   The intermolecular forces in water, $H_2O$, are greater than those in dihydrogen sulfide, $H_2S$. Explain this observation.

10.63   List all of the intermolecular forces in each of the following substances.
(a) $C_6H_6$          (b) $NH_3$          (c) $CS_2$          (d) $CHCl_3$

10.64   List all of the intermolecular forces in each of the following substances.
(a) $CH_3NH_2$          (b) $SO_2$          (c) $CO_2$          (d) $NO_2$

10.65   Describe the interactions that must be overcome to melt or boil the following substances.
(a) krypton                    (c) methane
(b) carbon monoxide            (d) ammonia

10.66   Describe the interactions that must be overcome to melt or boil the following substances.
(a) lithium bromide            (c) sulfur dioxide
(b) carbon dioxide             (d) carbon tetrachloride

10.67   Arrange the following substances in order of increasing boiling point: $CH_3CH_2OH$, $CH_3OH$, and $CH_3CH_2CH_2OH$.

10.68   Arrange the following substances in order of increasing boiling point: $AsH_3$, $NH_3$, and $PH_3$.

10.69   A substance composed of which of the following molecules will have the stronger intermolecular forces?

$CO_2$          $NO_2$

10.70   A substance composed of which of the following molecules will have the stronger intermolecular forces?

$C_2H_6$          $CH_4$

10.71   The boiling point of He is 4 K, while that of $H_2$ is 20 K. Explain why hydrogen boils at a higher temperature than helium.

10.72   The melting point of $CF_4$ is −184°C, while that of $CCl_4$ is −23°C. Explain the difference in melting points.

10.73   The boiling point of $F_2$ is −188°C, while that of HCl is −85°C. Explain the difference in boiling points.

10.74   The boiling point of $N_2O$ is −88°C, while that of $O_3$ is −112°C. Explain the difference in boiling points.

10.75   Which substance in each pair has the higher equilibrium vapor pressure at a given temperature?
(a) $CH_3CH_2OH$ or $CH_3OH$          (c) $NH_3$ or $H_2O$
(b) $CH_3OH$ or $H_2O$

10.76   Which substance in each pair has the higher equilibrium vapor pressure at a given temperature?
(a) $Cl_2$ or $Br_2$                    (c) $CH_4$ or $NH_3$
(b) $PH_3$ or $AsH_3$

10.77   Arrange the following substances in order of increasing strength of intermolecular forces: $H_2O$, He, $I_2$, and $N_2$.

10.78   Arrange the following substances in order of increasing strength of intermolecular forces: $NH_3$, Ne, $O_2$, and $H_2$.

10.79   Explain why acetone boils at 56.2°C, while $H_2O$ boils at 100°C. The structure of acetone is

$$\underset{H_3C}{}\overset{\overset{\displaystyle O}{\|}}{C}\underset{}{}CH_3$$

10.80   Explain why the equilibrium vapor pressure of $H_2O$ is 72 torr at 45°C, while that of $CS_2$ is 760 torr.

## Properties of Liquids

10.81   Which properties of liquids are different from those of solids?

10.82   Which properties of liquids are different from those of gases?

10.83   What is viscosity? Discuss properties of liquids that are controlled by viscosity.

10.84   Why does the viscosity of a liquid decrease as the temperature increases?

10.85   What is surface tension? Discuss properties of liquids that are controlled by surface tension.

10.86   Why does the surface tension of water decrease when soap is added to it?

10.87   Why do liquids tend to assume spherical shapes?

10.88   Why do many liquids rise in a glass tube?

10.89   How would living things be affected if ice were more dense than liquid water?

10.90   In what ways is water different from other liquids? Why is it different?

## Properties of Solids

10.91   What is a crystal? Draw a picture of a crystalline substance.

10.92   What is a crystal lattice? Draw a picture of what a crystal lattice might look like.

10.93   Why would we want to know the crystal structure of a solid?

10.94   Why do crystalline solids typically have symmetrical planar faces?

10.95   What is a common structure of metals?

10.96   What tendency exhibited by metal atoms is also found in macroscopic nature?

10.97   Some solids, such as tar, do flow. What is different about these solids?

10.98   Distinguish between amorphous and crystalline solids.

10.99   Why is it useful to classify solids?

10.100  What are the four major types of crystalline solids?

10.101  What type of crystalline solid has a high melting point, conducts electricity when molten but not as a solid, and is brittle?

10.102  What type of crystalline solid conducts electricity as a solid and when molten, is ductile, and has a lustrous appearance?

10.103  Explain why sodium is malleable, but sodium chloride is brittle.

10.104  Explain why solid $H_2O$ is hard, but solid $I_2$ is soft.

10.105  Compare the structures of metals and the structures of ionic salts.

10.106  How are the structures of ionic crystals related to those of metallic crystals? How are they different?

10.107  Distinguish between the structures of molecular and ionic solids. Give examples of each and contrast their properties.

10.108  Distinguish between the structures of metallic and ionic solids. Give examples of each and contrast their properties.

10.109  What forces hold the following substances together in the solid state?
(a) diamond     (b) $CO_2$     (c) $H_2O$     (d) $SiO_2$

10.110  What forces hold the following substances together in the solid state?
(a) graphite     (b) $CH_4$     (c) $NH_3$     (d) Fe

10.111  Predict what type of solid (metallic, ionic, molecular, or network) each of the following substances should form.
(a) $CaCl_2$     (b) $N_2$     (c) Ti     (d) BN     (e) $SO_3$

10.112  Predict what type of solid (metallic, ionic, molecular, or network) each of the following substances should form.
(a) Cu     (b) $P_4$     (c) $SiO_2$     (d) $NH_3$     (e) CuO

10.113  Indicate whether each of the following solids is a molecular solid.
(a) glass, $SiO_2$     (d) Fe
(b) ice, $H_2O$     (e) dry ice, $CO_2$
(c) $CaCl_2$

10.114  Indicate whether each of the following solids is a network solid.
(a) SiC     (d) Mn
(b) $NH_3$     (e) $CCl_4$
(c) $NiBr_2$

10.115  Identify the type of solid shown by the following image.

Cl$_2$

10.116  Identify the type of solid shown by the following image.

O
C

10.117  Consider the melting points of the following solids. Which of these solids is a molecular solid?
$SiO_2$     1610°C
$SiF_4$     −90°C
SiC     2700°C

10.118  Consider the melting points of the following solids. Which of these solids is a network solid?
$CO_2$     −57°C
$GeBr_4$     26°C
BN     3000°C

## Additional Questions

10.119  Explain why wet laundry hung out in freezing weather will eventually dry.

10.120  Arrange the following substances in order of increasing boiling point: $Br_2$, $Cl_2$, and $I_2$.

10.121  Match the compounds in the first list with the boiling points in the second list:
$CH_4$     0°C
$C_2H_6$     −42°C
$C_3H_8$     −88°C
$C_4H_{10}$     −62°C
Explain the trend that you have described.

10.122  When a vacuum pump is connected to a container of water, the water begins to boil. Explain this phenomenon.

10.123  Why does liquid water spread out when poured on a flat glass plate, while liquid mercury forms spherical drops?

10.124  Arrange the following substances in order of increasing melting points: $CH_4$, $GeH_4$, $SiH_4$, and $SnH_4$.

10.125  Indicate whether or not each of the following solids is an ionic solid.
(a) glass, $SiO_2$     (d) Fe
(b) ice, $H_2O$     (e) dry ice, $CO_2$
(c) $CaCl_2$

10.126  Explain how an increase in the strength of intermolecular forces affects each of the following properties of a substance.
(a) vapor pressure of a solid
(b) hardness of a molecular solid
(c) boiling point of a liquid
(d) viscosity of a liquid
(e) melting point of a molecular solid

10.127  Which substance in each pair has the higher vapor pressure at the same temperature?
(a) $Cl_2$ or $I_2$     (c) $PF_3$ or $NF_3$
(b) $CH_3OH$ or $CH_3SH$

10.128  Explain why diamond is a good insulator, while graphite is a good conductor of electricity.

10.129  Compare the following substances with respect to electrical conductivity, hardness, melting point, and vapor pressure at room temperature: NaCl, SiC, $SiCl_4$, and Fe.

10.130  Which characteristic of molecules changes in magnitude from one state to another and is largely responsible for whether a substance is a gas, a liquid, or a solid?

10.131  List the names for all possible phase changes.

10.132  Indicate which phase changes are exothermic and which are endothermic.

10.133 Draw a picture in the empty circle to illustrate what happens when solid chlorine, $Cl_2$, melts.

10.134 Predict the order of increasing intermolecular force strength for the following and explain your reasoning: HF, HCl, and HBr.

10.135 Which of the following substances have dipole-dipole forces when in the liquid state: $CH_4$, $CH_3Cl$, $CH_2Cl_2$, $CCl_4$, $NCl_3$?

10.136 How do vapor pressure, viscosity, and surface tension change when the temperature of a liquid is increased?

10.137 The rating given to motor oils such as 5W and 10W describe the oil's relative viscosity determined by measuring flow rates through a tiny hole at a specified temperature. Oils with the higher ratings have the greater viscosities.
  (a) Assuming that only London dispersion forces are present in motor oils, predict which motor oil, 5W or 10W, is composed of hydrocarbons of the higher molar mass.
  (b) The viscosity of a motor oil is important to its function and energy efficiency. Explain why motor oils with a low viscosity rating are commonly used in very cold climates.

10.138 Given that the boiling points of ethanol and diethyl ether are 78.1°C and 34.6°C, respectively, which should have the higher vapor pressure at 20.0°C?

10.139 If you are slowly increasing the temperature of a mixture of methanol, $CH_3OH$, and ethanol, $CH_3CH_2OH$, which should boil off first?

10.140 Describe the intermolecular forces that hold bromine molecules together in the liquid state between the temperatures of −7.2°C and 58.8°C, and in the solid state below −7.2°C.

10.141 How would the world be different if hydrogen bonding were not as strong as it is and the boiling point of $H_2O$ were close to that of $H_2S$ (−60.7°C)?

10.142 Predict the order of increasing boiling point for the following substances: $SO_3$, $SO_2$, $CO_2$, $CH_3OH$.

10.143 Compare the structures of the two isomers of 1,2-dichloroethene. These are different compounds (because rotation does not occur about the double bond) with differing boiling points: 47.5°C and 60.3°C. Match the boiling points to the compound structures shown:

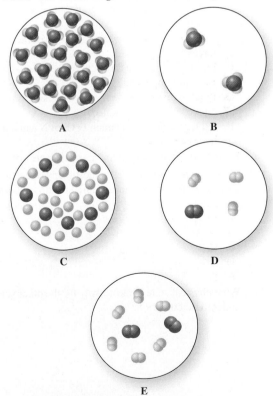

Trans          Cis

10.144 The boiling point of 1-propanol, $CH_3CH_2CH_2OH$, is 39.8°C, and the boiling point of ethylene glycol, $HOCH_2CH_2OH$ is 245°C. Explain this large difference in boiling point in terms of structure and intermolecular forces in these substances.

## Concept Review

10.145 A sample of liquid ammonia is completely vaporized in a closed container as shown:

Which of the following diagrams best represents what you would "see" in the magnified view of the vapor?

Explain why the other images are incorrect.

10.146   Select the pair of substances that lists the one with the higher equilibrium vapor pressure first.
   A. $H_2Se$, $H_2S$
   B. HF, HBr
   C. $CH_3OH$, $CH_3SH$
   D. $SiH_4$, $CH_4$
   E. $F_2$, $Br_2$
   Explain why the substances in each pair have their respective lower and higher vapor pressures.

10.147   Which of the following molecules experience dipole-dipole forces in the liquid state?
   A. $PF_3$
   B. $CO_2$
   C. $CF_4$
   D. $SO_3$
   E. $H_2$
   What type of intermolecular forces would be experienced by the other molecules in the liquid state?

10.148   Which of the following are attractive forces that act between molecules of $H_2O$?
   A. London dispersion forces, dipole-dipole forces, and hydrogen-bonding forces
   B. dipole-dipole forces and hydrogen-bonding forces
   C. hydrogen-bonding forces only
   D. covalent bonds
   E. ionic bonds
   Explain why the other responses are incorrect.

10.149   Which one of the following statements regarding intermolecular forces is correct?
   A. An intermolecular force is an attractive force that operates within a molecule.
   B. Bonding forces are weaker than intermolecular forces.
   C. London dispersion forces occur between all molecules in liquids and solids.
   D. London dispersion forces are the result of permanent dipoles in atoms or molecules.
   E. All molecules that contain F, O, or N can undergo hydrogen bonding.
   Rewrite the incorrect statements so they are correct.

10.150   Which of these substances is most likely to occur in the gaseous state at room temperature and atmospheric pressure?
   A. $CCl_4$
   B. NaBr
   C. $H_2S$
   D. $Br_2$
   E. Fe
   What physical state is most likely for the other substances under these conditions?

10.151   Which of the following statements regarding the liquid state is correct?
   A. Molecules with strong intermolecular forces have low surface tension.
   B. Surface tension increases as temperature increases.
   C. The lower the viscosity of a liquid, the stronger its intermolecular forces.
   D. As temperature increases, the viscosity of a liquid increases.
   E. Water rises in a glass capillary tube due to attractions between water molecules and the molecules of the glass.
   Rewrite the incorrect statements to make them correct.

10.152   Which of the following statements regarding the solid state is correct?
   A. Solids are highly compressible.
   B. Solids have rigid shapes that adjust to the shapes of their containers.
   C. Substances that solidify in a very orderly manner form amorphous solids.
   D. Most polymers, such as rubbers and plastics, are crystalline.
   E. Solids normally do not flow.
   Rewrite the incorrect statements to make them correct.

10.153   Which of the following solids is most likely a molecular solid, based on their composition and their melting points?
   A. Ni, 1455°C
   B. $SiO_2$, 1610°C
   C. $GeI_4$, 144°C
   D. $NiCl_2$, 1001°C
   E. BN, 3000°C
   Classify the other solids as network, ionic, or metallic.

10.154   Solids that are characterized by low melting points, softness, and low electrical conductivities are classified as _____ solids.
   A. ionic
   B. molecular
   C. metallic
   D. network
   E. both ionic and molecular
   Describe these properties for the other types of solid.

# Solutions

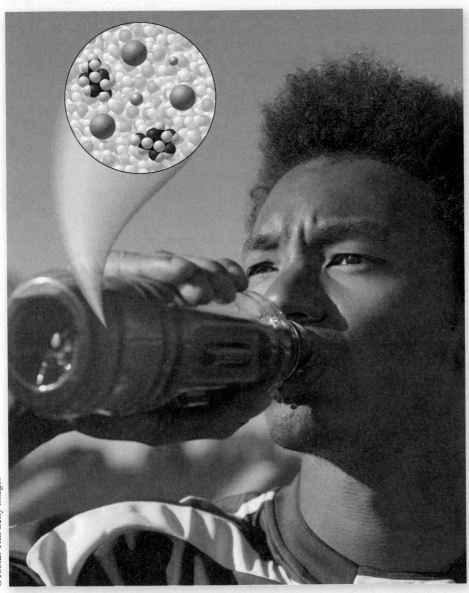

11.1 The Composition of
Solutions

11.2 The Solution Process

11.3 Factors That Affect
Solubility

11.4 Measuring Concentrations
of Solutions

11.5 Quantities for Reactions
That Occur in Aqueous
Solution

11.6 Colligative Properties

Chapter Review

Questions and Problems

©Tobias Titz/Getty Images

Make a list of all the foods and drinks you consume in a week. Which of them are solutions?

**FIGURE 11.1** Some drugs are delivered orally in the form of tablets or capsules. They are often coated or encapsulated to prevent digestion in the stomach and to allow for time-release of the active ingredient. Other drugs are delivered by syringe or from an IV bag.
(left): ©Don Wilkie/Getty Images; (right): ©RandyAllbritton/Getty Images

**FIGURE 11.2** A variety of antacids is available for people who suffer from indigestion. The active ingredient in some is a base such as $Mg(OH)_2$ that gets to the stomach without injuring the tissues of the mouth and esophagus.
©BrianMoeskau/Moeskau Photography

A solution can contain more than one solute. Aqueous solutions in the human body such as blood, sweat, and tears contain many solutes.

Megan and Derek participate in numerous athletic activities on their college campus. They run or cycle every day, play several team sports, and work out on the free weights at the gym. To maintain their physical fitness they carefully select the foods they consume, and they supplement their diets with protein shakes and vitamins. They also drink sports drinks when they exercise to help replenish lost electrolytes. In their chemistry course, they are studying solubility: how well one substance dissolves in another. It doesn't take them long to recognize the importance of this topic to the transport systems that carry nutrients, drugs, and other materials to every cell in the human body.

Anything we ingest is transported through the body in specific ways that depend on its solubility. For example, a drug taken orally must withstand the acidic conditions of the stomach. Some drugs cannot, so they must be administered by injection. A drug taken by injection into fatty tissue must be able to work its way into the circulatory system. If it cannot, it may need to be administered directly into the bloodstream via injection or intravenous (IV) delivery. Because the primary component of IV delivery is water, drugs administered in this way must be water soluble (Figure 11.1).

For a chemistry assignment, Megan and Derek studied the active ingredients in over-the-counter medications used to treat acid indigestion (Figure 11.2). Because antacids neutralize stomach acid, they assumed that all antacids would contain a base. An acid provides $H^+$ ions in solution and a base often produces $OH^-$ ions. They combine to form water. Megan and Derek found that the bases most commonly used in a chemistry lab, sodium hydroxide and potassium hydroxide, were not present in antacid medications. Instead, they discovered that magnesium hydroxide and calcium hydroxide were commonly used. They wondered why NaOH and KOH were not used, but $Mg(OH)_2$ and $Ca(OH)_2$ were. Derek looked up the solubility rules for hydroxides. He discovered that $Mg(OH)_2$ and $Ca(OH)_2$ are insoluble in water, but NaOH and KOH are soluble. Megan wondered if solubility might explain why some bases are used in antacids and others are not.

Consider the antacid magnesium hydroxide, which is sold as a suspension in water. It travels down the esophagus and into the stomach without causing damage along the way. Because $Mg(OH)_2$ is insoluble in water, it does not provide very much hydroxide ion until it is in an acidic environment, as in the stomach. Once there, $Mg(OH)_2$ can safely neutralize excess acid. NaOH would not perform in this way. It's very soluble in water and can, therefore, dissolve in saliva in the mouth. It would damage the lining of the mouth and the esophagus before ever getting to the site of action, the stomach. Magnesium hydroxide is available as a liquid, but the $Mg(OH)_2$ is not dissolved. It is suspended as tiny particles in the liquid carrier. Many other antacids are available as chewable tablets. When chewed, they form a suspension in the mouth similar in consistency to the magnesium hydroxide suspension.

Because our bodies are 70% water, anything we consume interacts with water in some way. As Megan and Derek discovered in their study of antacids, some substances easily dissolve in water and others do not. Those that don't may remain in the body for longer periods of time. For example, vitamin C and the various B vitamins are water soluble. A person must consume them on a regular basis to maintain good physical health because they are readily excreted in the urine. Other vitamins—including A, D, E, and K—are fat soluble; they are stored in fatty tissue for long periods. These vitamins should not be consumed in large quantities, because they can build up to toxic levels in body tissues.

Although we usually think of solutions as liquids, solids and gases can also form solutions. In the medical field, solid solutions of metals—commonly called alloys—are important in dentistry (braces), optometry (glasses), and surgery (metal implants). The air we breathe is a gaseous solution, mostly of nitrogen and oxygen.

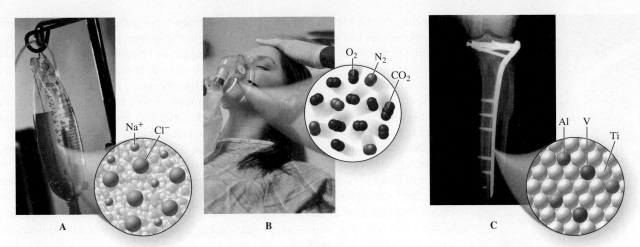

**FIGURE 11.3** (A) A saline IV bag contains a solution of NaCl in water. (B) Gases can also form solutions. Hospital patients can wear masks through which higher oxygen concentrations are provided. Carbon dioxide is also present in the solution because the patient exhales into the mask. (C) Solutions can also be solids. Screws used in orthopedic surgery are often made of an alloy (metal solution) of titanium (90%), aluminum (6%), and vanadium (4%).

(A): ©Photodisc/Getty Images; (B): ©M. Constantini/PhotoAlto; (C): ©Thinkstock Images/Getty Images

The oxygen masks used in hospitals (Figure 11.3) provide higher concentrations of oxygen to patients with respiratory problems. The oxygen tanks that are often seen at sporting events enrich the supply of oxygen that gets to the lungs and help athletes recover more quickly from strenuous activity.

This chapter examines solutions: their composition, how they form, and how dissolved substances affect the properties of solvents. Consider the following questions.

## Questions for Consideration

**11.1** What is a solution?

**11.2** How do substances dissolve to make a solution?

**11.3** What determines whether one substance will dissolve in another?

**11.4** How can we measure the concentration of a solution?

**11.5** In a chemical reaction, how can we relate the amount of one reactant or product in solution to the amount of another?

**11.6** What properties of solutions depend only on the concentration of dissolved particles, not on the identity of the solute?

 **Math Tools Used in This Chapter**

Units and Conversions (Math Toolbox 1.3)
Mole Quantities (Math Toolbox 4.1)

## 11.1 The Composition of Solutions

Recall from Chapter 1 that a **solution** is a homogeneous mixture with uniform composition throughout. Although solutions can be mixtures of solids, liquids, or gases, in this chapter we'll focus on solutions made of a substance dissolved in a liquid. The substance being dissolved is the **solute** (usually present in the lesser amount), and the substance doing the dissolving is the **solvent** (usually present in the greater amount). A solution in which the solvent is water is an **aqueous solution.** Figure 11.4 shows the preparation of an aqueous solution of sodium chloride.

Sometimes it is difficult to distinguish between solvent and solute. The solvent is usually present in the largest amount but this is not always the case. For example, ethanol, $CH_3CH_2OH$, and water can dissolve in each other in all proportions (they are completely miscible), so the relative amounts do not differentiate the solvent from the solute. In such cases, water is considered the solvent, even when the other substance is present in greater quantity.

**FIGURE 11.4** In a sodium chloride solution, the sodium chloride is considered the solute because it dissolves and is present in a lesser amount. Water is the solvent because it does the dissolving and is present in a greater amount.
©McGraw-Hill Education/Jill Braaten

Athletes like Megan and Derek often consume sports drinks to replenish water and electrolytes lost during physical exertion. In the human body, electrolytes serve several purposes. In some cases they participate in chemical reactions. In other cases chemical changes do not occur, but the electrolyte is essential nonetheless. For example, electrolytes help maintain water balance between the interior of cells and the fluid that surrounds them. The electrical properties of electrolytes are also important in nerve function. Imbalances in electrolytes can cause a variety of conditions ranging from muscle cramps to cardiovascular failure.

*Electrolyte solutions* contain a solute that dissociates or ionizes in a solvent, producing ions. The presence of ions allows electric current to pass through the solution. Examples of electrolytes are soluble ionic compounds such as NaCl, strong acids, and strong bases (Figure 11.5). Acids provide $H^+$ ions in aqueous solution. Bases neutralize acids and often provide $OH^-$ ions in aqueous solution. Examples of the dissociation or ionization of salts, acids, and bases in water can be described using the following equations:

$$NaCl(s) \xrightarrow{H_2O(l)} Na^+(aq) + Cl^-(aq)$$

$$HCl(g) \xrightarrow{H_2O(l)} H^+(aq) + Cl^-(aq)$$

$$NaOH(s) \xrightarrow{H_2O(l)} Na^+(aq) + OH^-(aq)$$

How can we determine if an ionic compound will dissociate in water to give ions? How readily or completely a solute dissolves in a solvent is its *solubility*. Unfortunately the solubilities for ionic compounds do not follow a predictable

**FIGURE 11.5** (A) Soluble ionic compounds like NaCl, (B) strong acids like HCl, and (C) strong bases like NaOH dissociate completely in water to produce ions. They are examples of strong electrolytes.
(A): ©McGraw-Hill Education/Jill Braaten; (B): ©Brian Moeskau/Moeskau Photography; (C): ©Charles D. Winters

**TABLE 11.1** Rules Used to Predict the Solubility of Ionic Salts

| Ions | Rule |
| --- | --- |
| $Na^+$, $K^+$, $NH_4^+$ (and other alkali metal ions) | Most compounds of alkali metal and ammonium ions are soluble. |
| $NO_3^-$, $CH_3CO_2^-$ | All nitrates and acetates are soluble. |
| $SO_4^{2-}$ | Most sulfates are soluble. Exceptions are $BaSO_4$, $SrSO_4$, $PbSO_4$, $CaSO_4$, $Hg_2SO_4$, and $Ag_2SO_4$. |
| $Cl^-$, $Br^-$, $I^-$ | Most chlorides, bromides, and iodides are soluble. Exceptions are $AgX$, $Hg_2X_2$, $PbX_2$, and $HgI_2$. ($X$ = Cl, Br, or I.) |
| $Ag^+$ | Silver compounds, except $AgNO_3$ and $AgClO_4$, are insoluble. $AgCH_3CO_2$ is slightly soluble. |
| $O^{2-}$, $OH^-$ | Oxides and hydroxides are insoluble. Exceptions are alkali metal hydroxides, $Ba(OH)_2$, $Sr(OH)_2$, and $Ca(OH)_2$ (somewhat soluble). |
| $S^{2-}$ | Sulfides are insoluble. Exceptions are compounds of $Na^+$, $K^+$, $NH_4^+$, and the alkaline earth metal ions. |
| $CrO_4^{2-}$ | Most chromates are insoluble. Exceptions are compounds of $Na^+$, $K^+$, $NH_4^+$, $Mg^{2+}$, $Ca^{2+}$, $Al^{3+}$, and $Ni^{2+}$. |
| $CO_3^{2-}$, $PO_4^{3-}$, $SO_3^{2-}$, $SiO_3^{2-}$ | Most carbonates, phosphates, sulfites, and silicates are insoluble. Exceptions are compounds of $Na^+$, $K^+$, and $NH_4^+$. |

pattern. They must be recalled from experimental data. The solubility rules derived from data are summarized in Table 11.1. In general, a soluble ionic compound dissociates into ions when dissolved in water. Insoluble compounds remain mostly undissociated in their ionic, crystalline structure. One example is the $Mg(OH)_2$ in liquid antacids (Figure 11.6). Table 11.1 shows that most hydroxides are insoluble, with some exceptions. Since $Mg(OH)_2$ is not one of the exceptions, we can assume it is insoluble—thus, it's safe to use as an antacid.

In sports drinks, sodium chloride helps reestablish electrolyte balance. The drinks also contain simple sugars, such as fructose or glucose. Solutions of sodium chloride and glucose vary in their ability to conduct electricity because sodium chloride dissociates into ions and glucose does not. Recall from Chapter 3 that we can determine whether a compound has dissociated into ions in solution if the solution conducts electricity. A solute that dissociates completely into ions in aqueous solution is a *strong electrolyte* (Figure 11.7). A *weak electrolyte* dissociates only partially into ions. A *nonelectrolyte* does not dissociate into ions.

*Nonelectrolyte* substances retain their molecular structure when dissolved. Glucose and fructose remain as intact molecules when dissolved in water. Both have the same formula, $C_6H_{12}O_6$, but they differ in how their atoms are arranged.

Recall from Chapter 3 that the term *dissociation* describes the process of dissolving for compounds that consist of ions. When a compound that is not ionic provides ions when dissolved in water, the process is called *ionization*.

Other electrolytes in sports drinks include potassium citrate, magnesium chloride, calcium chloride, and potassium phosphate.

Weak electrolytes are usually weak acids and bases.

$Mg(OH)_2$ crystals

**FIGURE 11.6** This liquid antacid is a suspension of magnesium hydroxide in water. Because $Mg(OH)_2$ is insoluble, it does not dissociate into ions in water.
©Brian Moeskau/Moeskau Photography

**FIGURE 11.7** Soluble ionic compounds, strong acids, and strong bases dissociate completely into ions. They are considered to be strong electrolytes because their aqueous solutions are good conductors of electricity.
©McGraw-Hill Education/Stephen Frisch

HNO₃(*aq*)

HCl(*aq*)

NaCl(*aq*)

NaOH(*aq*)

They each retain their molecular structure when dissolved in water, in the same way as does the simpler molecular compound, hydrogen peroxide, $H_2O_2$ (Figure 11.8). A 3% solution of this compound is often used to sterilize cuts and scrapes.

Another nonelectrolyte compound, ethanol, $CH_3CH_2OH$, behaves the same way. It retains its molecular structure in aqueous solutions. We can describe the mixing of ethanol with water using the following equation:

$$CH_3CH_2OH(l) \xrightarrow{H_2O(l)} CH_3CH_2OH(aq)$$

Ethanol is completely **miscible** with water. That is, water and ethanol can be mixed in any proportion to give a homogeneous solution. We'll talk more about miscibility in Section 11.4.

To summarize, when we determine the composition of a solution we must first classify the substance as an electrolyte or a nonelectrolyte. In aqueuous solution, acids, bases, and ionic compounds form separate ions, and most molecular compounds retain their molecular structure. The following examples will give you practice.

$H_2O_2$(*aq*)

**FIGURE 11.8** When soluble nonelectrolytes such as hydrogen peroxide, $H_2O_2$, dissolve in water, they retain their molecular structure.
©Brian Moeskau/Moeskau Photography

| EXAMPLE 11.1 | Composition of Solutions |

What ions, atoms, or molecules are present in solution when the following substances mix with water? Write a balanced equation to represent the process of dissolving each substance in water.

(a)  $HNO_3(l)$      (b)  $Ba(OH)_2(s)$      (c)  $C_{12}H_{22}O_{11}(s)$

**Solution:**

(a)  Nitric acid, $HNO_3$, is a strong acid that ionizes completely in water to give $H^+(aq)$ and $NO_3^-(aq)$ ions. Water molecules are also present. The equation for dissolving $HNO_3$ is

$$HNO_3(l) \xrightarrow{H_2O(l)} H^+(aq) + NO_3^-(aq)$$

(b) From Table 11.1 we know that ionic compounds containing $OH^-$ ions are insoluble, with barium hydroxide, $Ba(OH)_2$, being an exception. Barium hydroxide is an electrolyte that dissociates in water to give $Ba^{2+}(aq)$ and $OH^-(aq)$ ions in a 1 to 2 ratio. The equation for dissolving $Ba(OH)_2$ is

$$Ba(OH)_2(s) \xrightarrow{H_2O(l)} Ba^{2+}(aq) + 2OH^-(aq)$$

(c) Sucrose, $C_{12}H_{22}O_{11}$, is a molecular compound that does not dissociate into ions when it dissolves. Intact molecules of $C_{12}H_{22}O_{11}$ exist in the solution. The equation for dissolving $C_{12}H_{22}O_{11}$ is

$$C_{12}H_{22}O_{11}(s) \xrightarrow{H_2O(l)} C_{12}H_{22}O_{11}(aq)$$

**Consider This 11.1**

What would you expect to observe if $BaSO_4$ were mixed with water?

**Practice Problem 11.1**

What ions, atoms, or molecules are present in solution when the following substances are mixed with water? Write a balanced equation to represent the process of dissolving each substance in water.

(a) $ZnCl_2(s)$     (b) $CH_3OH(l)$     (c) $KNO_3(s)$

**Further Practice:** Questions 11.13 and 11.14 at the end of the chapter

For a review of how to classify compounds as ionic or molecular, consult Section 3.1.

 **Student Hot Spot**

Student data indicate you may struggle with representing a dissolving process with a balanced equation. Access the SmartBook to view additional Learning Resources on this topic.

## 11.2 The Solution Process

A typical sports drink contains carbohydrates, usually sucrose and other sugars; ionic compounds, like sodium chloride and potassium chloride; and fruit juice for flavor. How do these substances form a solution? Let's begin by examining the arrangement of particles in one of the solutes, NaCl, and in the solvent, water. Notice the orderly arrangement of particles in the solute and in the solvent before mixing, as shown in Figure 11.9A. Ionic bonds hold the ions of NaCl together. London dispersion forces and hydrogen bonds hold the molecules of pure water together. (Hydrogen bonding is the stronger of the two intermolecular forces in water.) After mixing (Figure 11.9C), when a solution has formed, $Na^+$ and $Cl^-$ ions distribute uniformly through the solvent, water. The formation of the solution must involve, therefore, the destruction of the ionic bonds in NaCl and some of the intermolecular forces in water (Figure 11.9B). Within the solution, new forces of attraction form between the particles that make up the solute and the solvent molecules.

The force of attraction between an ion and a polar molecule is called an **ion-dipole force** (Figure 11.10). In a sodium chloride solution, several water molecules surround each ion, creating multiple ion-dipole forces. The process by which a solute particle is surrounded by solvent water molecules in a solution is called *hydration.* (If the solvent is not water, the process is called *solvation.*) The hydrated ions are surrounded by water molecules oriented so that the partially negative polar ends of the water molecules are closest to the cations and the partially positive polar ends are closest to the anions.

When an ionic compound dissolves in water, the process can be summarized as follows:

- Ionic bonds in the solute break.
- Hydrogen bonds between water molecules break.
- Ion-dipole forces form between ions and water molecules.

As discussed in Chapter 6, changes in energy accompany all physical and chemical processes. In the formation of a solution, energy must be supplied to break the intermolecular forces and/or bonds in both the solute and solvent. It comes, at least in part,

 **ANIMATION:** Dissolving Table Salt

Pure ionic compounds require high temperatures to break their bonds. What provides the energy to break ionic bonds when these compounds dissolve?

**ANIMATION:** The Hydration Process

A                    B                    C

Ionic bonds     Some hydrogen
in NaCl must    bonds in water
be broken       must be broken

**FIGURE 11.9**  (A) In solid sodium chloride, ions are held together by ionic bonds. Water molecules are held together by intermolecular forces, especially hydrogen bonds. (B) The bonds in sodium chloride and some of the hydrogen bonds in water must be broken to form the aqueous solution. (C) In aqueous solution, $Na^+$ and $Cl^-$ ions distribute throughout.

(all): ©Brian Moeskau/Moeskau Photography

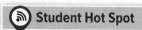
**Student Hot Spot**

Student data indicate you may struggle with understanding the changes in attractive forces when an ionic compound dissolves. Access the SmartBook to view additional Learning Resources on this topic.

**FIGURE 11.10**  When a substance dissolves in water to produce ions, the ions are attracted to water molecules by ion-dipole forces. Notice how the partially negative oxygen atoms on the water molecules arrange themselves around the positive sodium ion. The partially positive hydrogen atoms on the water molecules are attracted to the negative chloride ion.

from the new intermolecular forces that form between solute and solvent particles. The difference between the energy required for the separation of solute and solvent and the energy released upon formation of ion-dipole interactions is the *heat of solution.* The energy changes involved in the solution process are graphically summarized in Figure 11.11, using the formation of two different solutions as examples. When the solvation process provides more energy than is needed to separate the pure solute and solvent particles (Figure 11.11A), the heat of solution is negative, and the overall dissolving process is exothermic. In this case the container in which the solution is made may feel warm because the solution process releases energy to the surroundings.

When the separation processes require more energy than the solvation process releases (Figure 11.11B), the heat of solution is positive, and the overall dissolving process is endothermic. The container in which the solution is made may feel cool to the touch because the solution process absorbs energy from the surroundings.

When a solution forms between a nonpolar solute and a nonpolar solvent, different attractive forces between the molecules must be broken. For example, consider a solution in which iodine, $I_2$, dissolves in carbon tetrachloride, $CCl_4$ (Figure 11.12). London dispersion forces are broken between the solute molecules and between the solvent molecules. New London dispersion forces form between the solute and solvent molecules.

Matter tends to undergo changes that lead to a lower energy. This happens when energy is released to the surroundings. Thus, solutions will form more readily when the solution has a lower energy than the separate solute and solvent. Energy being released to the surroundings corresponds to a negative value of the heat of solution and an exothermic process.

When Megan and Derek studied the active ingredients in antacids, they identified bases such as magnesium hydroxide and calcium hydroxide, both of which are insoluble in water. Based on their experience with many ionic compounds, they had expected $Mg(OH)_2$ and $Ca(OH)_2$ to be soluble. These bases are insoluble because the ionic bonds between the metal ions and hydroxide ions are very strong. The energy released by the formation of ion-dipole attractions does not provide enough energy to break the ionic bonds in these bases. Other bases, such as sodium hydroxide and potassium hydroxide, are soluble because the strength of their ionic bonds is weaker.

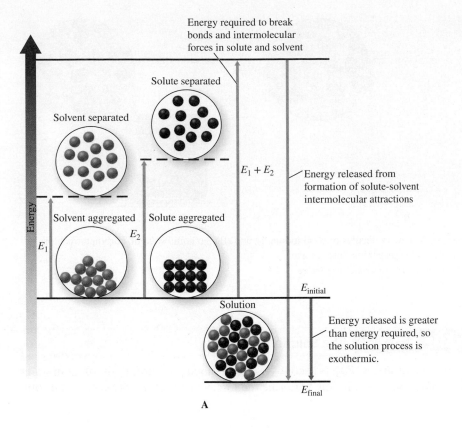

Energy required to break bonds and intermolecular forces in solute and solvent

Solute separated

Solvent separated

$E_1 + E_2$

Energy released from formation of solute-solvent intermolecular attractions

Solvent aggregated

$E_2$

Solute aggregated

$E_1$

$E_{initial}$

Solution

Energy released is greater than energy required, so the solution process is exothermic.

$E_{final}$

Energy

**A**

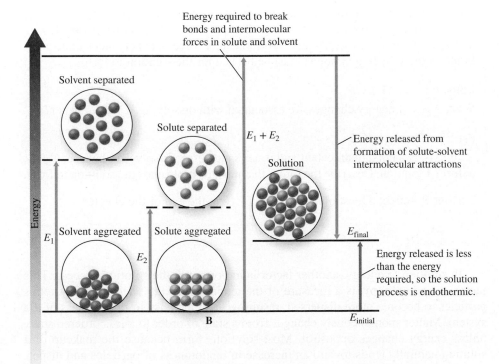

Energy required to break bonds and intermolecular forces in solute and solvent

Solvent separated

Solute separated

$E_1 + E_2$

Energy released from formation of solute-solvent intermolecular attractions

Solution

Solvent aggregated

$E_1$

Solute aggregated

$E_2$

$E_{final}$

Energy released is less than the energy required, so the solution process is endothermic.

$E_{initial}$

Energy

**B**

Athletes often use instant cold packs to reduce swelling and hot packs to relieve pain. Chemical cold packs consist of ammonium nitrate and water in separate compartments in the pouch. When the contents are allowed to mix, the bag feels cold. Is the process endothermic or exothermic? Chemical hot packs usually contain water separated from calcium chloride or magnesium sulfate. These pouches get warm when the seal between the water and the solute is broken and a solution forms. Is the process endothermic or exothermic?

(both): ©Brian Moeskau/Moeskau Photography

**FIGURE 11.11** (A) The sum of the energies ($E_1 + E_2$) required to separate the solute and to separate the solvent equals the amount that must be absorbed to form a solution. When the new ion-dipole forces of attraction form, energy is released. When this amount is greater than the energy absorbed, the solution process is exothermic. (B) Some solution processes are endothermic. The sum of the energies ($E_1 + E_2$) required to separate solute and to separate solvent particles is greater than the energy released from new solute-solvent interactions.

**FIGURE 11.12** London dispersion forces hold together (A) the molecules of solid iodine, $I_2$, and (B) the liquid solvent, carbon tetrachloride, $CCl_4$. (C) When a solution of $I_2$ and $CCl_4$ forms, new London dispersion forces form.
(A): ©McGraw-Hill Education/Charles D. Winters; (B), (C): ©Charles D. Winters/Science Source

---

| EXAMPLE 11.2 | The Solution Process |
|---|---|

Potassium nitrate, $KNO_3$, is used to treat high blood pressure. It dissolves in water to form an aqueous solution. Describe the forces that must be broken for the solution to form.

**Solution:**

Potassium nitrate is an ionic compound, so ionic bonds between $K^+$ and $NO_3^-$ ions must be broken for the solution to form. The solvent, $H_2O$, is polar covalent, with hydrogen bonding as the primary intermolecular force between water molecules. Some of these hydrogen bonds must be broken for the solution to form.

**Consider This 11.2**

What types of energy changes are associated with dissolving $KNO_3$ in water?

**Practice Problem 11.2**

Iodine, $I_2$, is a good disinfectant and water sanitizer. It dissolves in hexane, $C_6H_{14}$, to form a solution. Describe the forces that must be broken for the solution to form.

**Further Practice:** Questions 11.15 and 11.16 at the end of the chapter

---

In addition to energy, another factor is important in the solution process. This factor, called **entropy,** is a measure of the tendency for the energy of a system's particles to become more dispersed, resulting in an increase in randomness of the system. Matter spontaneously changes from a state of order to a less ordered state, unless energy changes oppose it. Most solutions form because the makeup of a solution naturally tends toward an increase in randomness of particles and dispersal of their energies. For example, consider NaCl and water molecules before they form a solution (Figure 11.13). The ions in solid NaCl are arranged in an organized pattern. When they mix with the water molecules to form a solution, their energies become more dispersed and their arrangements become less ordered. The resulting solution has a greater entropy than the NaCl and water before mixing. This increase in entropy upon mixing a solute and solvent helps drive the solution process, especially those processes that are endothermic.

NaCl is very ordered and has low entropy

When NaCl dissolves, entropy increases

Ions are in random positions, with high entropy

**FIGURE 11.13** Entropy, the tendency for the energy of matter to become more dispersed, contributes to solution formation. Notice that solid NaCl and pure water are more ordered in their arrangements than a solution of NaCl, so the energy has become more spread out.

(both): ©Brian Moeskau/Moeskau Photography

## 11.3 Factors That Affect Solubility

As Megan and Derek began classifying substances as soluble and insoluble, they wondered if there is a way to predict the solubility of two substances and if the solubility can change under different environmental conditions. Why do oil and water not mix homogeneously? Why does sugar dissolve so well in water? Does changing the temperature of a solution change the maximum amount of solute that will dissolve in a solvent? As we learned in Section 11.2, intermolecular forces play an important role in the solution process. In this section we learn how molecular structure and polarity, temperature, and pressure affect the extent to which a solute dissolves in a given solute.

### Structure

When an ionic compound dissolves in water, the dissociated ions are strongly attracted to the polar water molecules, helping in the solution process. In a similar way, polar solutes such as sucrose dissolve in polar solvents. However nonpolar solutes are usually insoluble in water and other polar solvents, and polar solutes are usually insoluble in nonpolar solvents. For example, nonpolar oil and water don't mix (Figure 11.14), but polar ethanol and water mix in any proportion. A nonpolar organic solvent can remove a grease stain on clothing, but water can't. These solubility trends can be summarized by a general rule of thumb called "like dissolves like." That is, when the polarity of the solvent is similar to the polarity of the solute, the solvent will dissolve the solute. *Polar solvents dissolve polar or ionic solutes, while nonpolar solvents dissolve nonpolar solutes* (Table 11.2).

Consider the vitamin structures shown in Figure 11.15. Except for the −OH group on the end, the vitamin A molecule consists of many carbon-carbon and carbon-hydrogen bonds. It is essentially nonpolar. Fat molecules are also nonpolar because they consist of similar carbon-carbon and carbon-hydrogen bonds (Figure 11.16). Since vitamin A and fat molecules are both nonpolar, vitamin A is stored in fatty tissues in the body. In vitamin C, several −OH groups make the molecule polar. Since water is also polar, vitamin C is soluble in aqueous body fluids.

As we showed in Section 11.2, when an ionic compound dissolves, the ions become solvated and form ion-dipole interactions with the solvent (Figure 11.10).

 **ANIMATION:** Influence of Shape on Polarity

For a review of how to classify molecular substances as polar or nonpolar, see Chapter 8, Section 8.5.

The *CRC Handbook of Chemistry and Physics* includes tables of solubility information that are useful in the laboratory.

**Student Hot Spot**

Student data indicate you may struggle with predicting solubility. Access the SmartBook to view additional Learning Resources on this topic.

Sucrose (table sugar) is a molecular compound that's highly soluble in water. Predict whether sucrose molecules are polar or nonpolar.

**TABLE 11.2** Like Dissolves Like

|  | Solvent | |
| --- | --- | --- |
| **Solute** | **Polar** | **Nonpolar** |
| Ionic | Soluble | Insoluble |
| Polar | Soluble | Insoluble |
| Nonpolar | Insoluble | Soluble |

**FIGURE 11.14** Oil and water do not mix. Spilled oil that floats on water is an environmental hazard. This oil slick in the Gulf of Mexico came from the Deepwater Horizon oil spill in 2010.
Source: Petty Officer 1st Class Tasha Tully/ US Coast Guard

Vitamin A          Vitamin C

**FIGURE 11.15** Vitamin A is a fat-soluble vitamin. It is a ring of carbon atoms attached to a long chain of additional carbon atoms. Although it has a polar —OH group at one end, the molecule is mostly nonpolar. Fat molecules are also mostly nonpolar. Notice that the vitamin C molecule lacks a long nonpolar carbon chain. It is water soluble because it has several —OH groups.

**FIGURE 11.16** Three long carbon chains make a fat molecule largely nonpolar.

For a review of intermolecular forces and their relative strengths, see Chapter 10.

Soaps and detergents are made of molecules or ions that have a polar end and a nonpolar end. The nonpolar end is attracted to oil stains on skin or clothing, and the polar end is attracted to water. In this way, the oil can be removed by soapy water.

This occurs only when the solvent molecules are polar. If we reason that an ionic compound is the extreme of polarity, then the rule "like dissolves like" holds for dissolving an ionic compound in a polar solvent.

In order for a solid to be soluble in a given solvent, the energy released by the solvation process must compensate for all (or at least most) of the energy required to break up forces at work both within the solute and within the pure solvent. Thus, ionic solids are generally soluble in polar solvents, where very strong interactions occur between the dipoles of the solvent molecules and the dissolved ions. A nonpolar solvent cannot interact with dissolved ions strongly enough to overcome the ionic forces within the ionic solid crystal.

In contrast to ionic compounds that are held together by ionic bonds, covalent molecules are held together by intermolecular forces that are relatively weak (Figure 11.17). The solubility (or miscibility) of liquids in other liquids depends largely on the polarity of the solute and solvent molecules. Polar liquids dissolve in other polar liquids because the intermolecular forces formed between solute and solvent molecules are strong enough to overcome the strong solute-solute and solvent-solvent dipole-dipole interactions. Nonpolar liquids do not dissolve in polar liquids—at least not to any great extent—because the intermolecular forces between solvent and solute are not strong enough to overcome the solute-solute and solvent-solvent London dispersion forces (Figure 11.18). Nonpolar liquids dissolve in one another because the forces between the molecules in the pure liquids are weak and also because of the tendency for matter to become more random.

## Temperature

Temperature affects the solubility of most substances in water. For example, if you like your tea sweet, you can easily stir more sugar into a cup of hot tea than a frosty glass of iced tea. That's because sugar is more soluble in hot water than in cold. The

**FIGURE 11.17** Carbon tetrachloride, CCl₄, and hexane, C₆H₁₄, are nonpolar liquids, so both have London dispersion forces as their only intermolecular attraction. Because they are alike in their intermolecular forces, they are miscible.
©Tom Pantages

**FIGURE 11.18** Oil, a nonpolar substance, does not mix with vinegar, an aqueous solution consisting of polar substances. The intermolecular forces between oil and water are not strong enough to overcome the intermolecular forces between oil molecules and the hydrogen bonding between water molecules. Oil and water are immiscible.
©Brian Moeskau/Moeskau Photography

solubility of most solids increases in water as the temperature of the solution increases. The effect of temperature on the solubility in water is shown for some ionic compounds in Figure 11.19. Notice that most ionic compounds are more soluble at higher temperatures, but gases are less soluble as temperature increases.

Gases such as N₂ and O₂ are only slightly soluble in water because they are nonpolar. They become even less soluble as temperature increases. An increase in temperature increases the kinetic energy of the molecules. The increased kinetic energy breaks the intermolecular forces between the solute gas particles and the solvent molecules, allowing dissolved gas to escape more readily from the solution into the gas phase. The flat taste of a carbonated soda that has been left open on the kitchen counter is due in part to the escape of the dissolved CO₂. Sodas do not go flat as quickly if kept in a refrigerator.

Water that has been boiled is often described as having a flat taste, because boiling removes all the gases that were dissolved in the water.

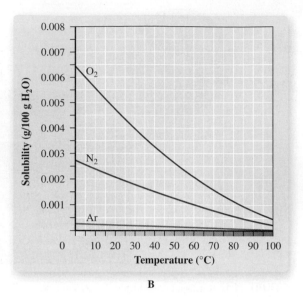

**FIGURE 11.19** (A) In general, the solubility of ionic solids in water increases with increasing temperature. (B) Gas solubility, in contrast, decreases.

An important consequence of the reduced gas solubility at higher temperatures occurs in cases of thermal pollution of water. It happens around power plants cooled by a river or stream when warm water is returned to its source after it has been used by the plant for cooling. When the water temperature of the river or stream increases, the solubility of oxygen in the water decreases, threatening the survival of fish and certain other aquatic organisms. Algae blooms are often observed under low-oxygen conditions, because algae thrive in this environment.

## Pressure

Henry's law states that the solubility of a gas in a liquid is directly proportional to the partial pressure of the gas above the liquid.

The solubility of gases (but not of liquids or solids) is strongly affected by pressure. The solubility of a gas in a liquid is directly proportional to the pressure of the gas above the liquid. This makes sense in terms of a dynamic equilibrium. An increase in pressure increases the number of gas particles immediately above the solution, increasing the chance that some will dissolve. Figure 11.20 illustrates this principle.

**FIGURE 11.20** An equilibrium exists between the solute molecules in the gas phase and in the solution. An increase in pressure changes this equilibrium, resulting in increased gas solubility.

CO$_2$ under high pressure

Pressure released

CO$_2$ bubbles out of solution

Lots of CO$_2$ dissolved in soda

**FIGURE 11.21** The effect of pressure on gas solubility can be observed when a soda bottle is opened.

You observe the effect of pressure on solubility every time you open a bottle or can of a carbonated beverage. These beverages are bottled under a pressure of CO$_2$ greater than atmospheric pressure. When you remove the bottle cap, the pressure of CO$_2$ decreases. You can hear the excess gas leave the container. Some of the CO$_2$ that had been dissolved in the solution escapes and forms gas bubbles (Figure 11.21).

Pressure also affects the solubility of gases in the bloodstream. Hospital patients who suffer from respiratory problems are given higher concentrations of oxygen, often through a mask (Figure 11.3B). Within the mask, the oxygen partial pressure is higher than it is in air. With a higher oxygen pressure, more oxygen dissolves in the patient's bloodstream with each breath, so the patient does not have to breathe as often to maintain an adequate oxygen supply.

Scuba divers must consider the effect of gas solubility at increased pressures under water. Recreational divers typically breathe ordinary air that has been compressed into a tank. It is about 78% nitrogen, which the body can handle under normal conditions of surface atmospheric pressure. As a diver descends, however, the increase in pressure causes more nitrogen to dissolve in the bloodstream and body tissues. If the diver surfaces too rapidly—causing a rapid decrease in pressure—the nitrogen comes out of solution as bubbles in tissues and in the bloodstream. These bubbles cause the neurological and circulatory symptoms of the "bends." One way to prevent this medical issue is to use a gas mixture of helium and oxygen (called heliox). This special air is sometimes used by military and professional divers because helium is not as soluble in the blood and tissues as nitrogen.

The symptoms of the bends were first discovered in the 1800s among bridge workers laying underwater foundations at low depths. Their joints were sore, and they walked bent over. These symptoms were later observed in divers who came up too quickly from dives to great depths. The water pressure lessens by 1 atm for every 10 m (33 ft) the diver ascends.

## 11.4 Measuring Concentrations of Solutions

All of the properties of solutions we have discussed so far are affected by **concentration,** the relative amounts of the solute and solvent that make up the solution. In Chapter 4 we described solutions as *dilute* and *concentrated.* These terms compare the relative amounts of solute in solutions, as when we describe one solution as more concentrated or more dilute than another. Another set of qualitative terms describes the degree of solubility of solids in a given solvent. A solid may be called *insoluble, slightly soluble, soluble,* or *very soluble.* All these terms are relative. Strictly speaking, there are no insoluble substances; even glass will dissolve in water to a slight extent. However, for practical purposes, solids with solubilities less than about 0.1 g per 100 g of solvent are considered insoluble, while solutes with solubilities greater than about 2 g per 100 g of solvent are considered very soluble. For example, NaCl is considered very soluble because its solubility in water is 35.9 g per 100 g of water at 25°C.

Quantitatively, **solubility** is defined as a ratio that identifies the maximum amount of a solute that will dissolve in a particular solvent to form a stable solution under specified conditions. Solubility is a measure of maximum concentration. If

If the solubility of PbI$_2$ is 0.1 g/100 g of water, what is the maximum amount of PbI$_2$ that will dissolve in 200 grams of water?

More details on dynamic equilibrium can be found in Section 12.4.

too much solute is present when maximum concentration is reached, the excess remains undissolved. For example, the solubility of lead(II) iodide in water is 0.1 g $PbI_2$ per 100 g $H_2O$ at 25°C. Thus, if 1 g of $PbI_2$ is added to 100 g of water, only about one-tenth of it dissolves; the rest remains as undissolved solid. An equilibrium is established between the $Pb^{2+}$ and $I^-$ ions in solution and the $PbI_2$, which stays in the solid phase. Such a solution is a **saturated solution,** which contains the maximum concentration of dissolved solute. The establishment of equilibrium to give a saturated solution is depicted in Figure 11.22. Once the solution is saturated, undissolved solute continues to dissolve, but at the same time, previously dissolved solute is deposited from solution. The two processes occur at exactly the same rate, so the amount of dissolved solute stays the same. This is a dynamic equilibrium, like the evaporation of a liquid in a closed container discussed in Chapter 10.

Supersaturated solutions cannot be prepared by simply adding excess solute to a solvent.

A solution that contains less than the maximum possible amount of solute is called an **unsaturated solution.** Unsaturated solutions are the most common types of solutions we encounter in the environment and work with in the laboratory.

Much less common is a solution that contains more than the expected maximum amount of solute, and is called a **supersaturated solution.** A supersaturated solution is usually obtained by preparing a saturated solution at a high temperature. As you might expect, supersaturation is generally a very unstable situation. When a supersaturated solution is disturbed by agitation or when a small "seed" crystal of solute is added, excess solute precipitates (Figure 11.23). Rapid cooling also causes the excess solute to precipitate, so a hot supersaturated solution must be cooled slowly.

**FIGURE 11.22** In this solution, some of the $PbI_2$ dissolves, forming $Pb^{2+}$ and $I^-$ ions. Eventually the remaining undissolved $PbI_2$ settles to the bottom of the beaker.
©Dorling Kindersley/Getty Images

**FIGURE 11.23** Supersaturated solutions tend to be sensitive. Cooling too quickly or jolting the flask can cause solid to form. Also, a small seed crystal (or a particle of dust) can cause the homogeneous supersaturated solution to revert to a saturated solution in which excess solid is present.
(all): ©McGraw-Hill Education/Stephen Frisch

**TABLE 11.3** Concentration Units

| Unit | Definition | Unit | Definition |
|------|-----------|------|-----------|
| Percent by mass | $\dfrac{\text{grams of solute}}{\text{grams of solution}} \times 100\%$ | Parts per billion | $\dfrac{\text{grams of solute}}{\text{grams of solution}} \times 10^9$ |
| Percent by volume | $\dfrac{\text{volume of solute}}{\text{volume of solution}} \times 100\%$ | Molarity ($M$) | $\dfrac{\text{moles of solute}}{\text{liters of solution}}$ |
| Mass/volume percent | $\dfrac{\text{grams of solute}}{\text{volume of solution}} \times 100\%$ | Molality ($m$) | $\dfrac{\text{moles of solute}}{\text{kilograms of solvent}}$ |
| Parts per million | $\dfrac{\text{grams of solute}}{\text{grams of solution}} \times 10^6$ | | |

When both the solute and the solvent are liquids, other comparative terms are convenient. Liquids such as ethanol and water are called *completely miscible,* which means that they can form solutions in all proportions. *Partially miscible* liquids form solutions over a limited range of compositions. For example, at room temperature it is possible to form a solution by adding butanol to water up to a concentration of about 20% butanol by volume. No higher concentration can be achieved at this temperature. Liquids such as cyclohexane, $C_6H_{12}$, and water are essentially insoluble in one another and are called *immiscible.*

Although these relative terms and comparisons have their uses, precise quantitative measurements are sometimes essential. For example, to prevent cellular damage in the human body, a saline IV solution administered to a patient must contain an exact concentration of NaCl that precisely matches its concentration in blood. Too much or too little causes red blood cells to shrivel or swell, respectively. To make such solutions and ensure the proper concentration, we can use one of several concentration units. The particular unit we choose depends on the purpose of the solution and its method of preparation. All of the units express concentration as a relative quantity: the amount of solute in a given amount of solvent or in a given amount of solution:

$$\text{Concentration} = \frac{\text{amount of solute}}{\text{amount of solvent or solution}}$$

Some commonly used units are summarized in Table 11.3 and are discussed in this section. Molality is the only unit that describes concentration in terms of the amount of solvent. The others are expressed in terms of the amount of solution.

## Percent by Mass

**Percent by mass** is calculated by dividing the solute mass by the mass of solution and multiplying by 100%:

$$\text{Percent by mass} = \frac{\text{grams of solute}}{\text{grams of solution}} \times 100\%$$

We can determine the mass of solution by adding the mass of solute to the mass of solvent as shown in Example 11.3.

**EXAMPLE 11.3**    **Percent by Mass**

Consider a solution prepared by adding 25.0 g NaCl to 125.0 g of water.

(a)  What is the percent by mass concentration of NaCl in this solution?
(b)  What mass of 10.0 g of this solution? NaCl is present in 10.0 g of this solution?

**Solution:**

(a)  In this problem we want to go from the masses of solute and solvent to percent by mass:

To go from these quantities to the quantity of interest we must determine the relationship between them. The percent by mass is calculated from the relationship

$$\text{Percent by mass} = \frac{\text{grams of solute}}{\text{grams of solution}} \times 100\%$$

The grams of solute are given in the problem, 25.0 g NaCl. The mass of the solution is the sum of the masses of solute and solvent:

$$\text{Mass of solution} = 25.0 \text{ g NaCl} + 125.0 \text{ g H}_2\text{O} = 150.0 \text{ g}$$

$$\text{Percent by mass} = \frac{25.0 \text{ g NaCl}}{150.0 \text{ g solution}} \times 100\% = 16.7\%$$

(b)  The percent by mass tells us that every 100.0 g of this solution contains a mass of 16.7 g NaCl. Since the solution is 16.7% by mass in NaCl, any amount of the solution will contain a proportionate mass of NaCl. Multiplying the mass of the solution and the percent by mass gives the mass of solute present. The mass of NaCl contained in 10.0 g of the solution is

$$\text{Mass NaCl} = 10.0 \text{ g solution} \times \frac{16.7 \text{ g NaCl}}{100.0 \text{ g solution}} = 1.67 \text{ g NaCl}$$

**MATH**
**TOOLBOX**
**1.3**

**Consider This 11.3**

The solubility of NaCl at 25°C is 36 g/100 g $H_2O$. Is the solution described in the example considered saturated, supersaturated, or unsaturated? What is the maximum mass of NaCl that would dissolve in 125.0 g of $H_2O$?

**Practice Problem 11.3**

Consider a solution prepared by adding 2.70 g acetic acid, $CH_3CO_2H$, to 122.8 g of water.

(a)  What is the percent by mass concentration of acetic acid in this solution?
(b)  What mass of acetic acid is present in 25.0 g of this solution?

**Further Practice:**  Questions 11.49 and 11.50 at the end of the chapter

Some products you can purchase in a drug store are sold as percent by volume solutions. For example, rubbing alcohol is 70% by volume isopropyl alcohol, $(CH_3)_2CHOH$, dissolved in water. Every 100 mL of solution contains 70 mL of isopropyl alcohol.

## Percent by Volume

Percent by volume is most often used when liquids are dissolved in liquids, since it is easy to measure amounts of liquids by their volumes. **Percent by volume** is calculated by dividing the solute volume by the solution volume and multiplying by 100%.

Calculating percent by volume is similar to calculating percent by mass. We simply substitute volumes for masses in the equation shown earlier for percent by mass:

$$\text{Percent by volume} = \frac{\text{volume of solute}}{\text{volume of solution}} \times 100\%$$

There is an important difference, however. Masses are additive; that is, the mass of solution is equal to the sum of the masses of solute and solvent. Volumes are not always additive. It is possible for the volume of a solution to be different from the sum of the volumes of the solute and solvent. Depending on the attractive forces between solute and solvent, the solution may have a greater or lesser volume than the separated components. So, the total volume of a solution must be measured, not calculated from its components.

> When two gases are mixed at constant pressure and temperature, their volumes are always additive.

---

### EXAMPLE 11.4    Percent by Volume

To prepare a solution, we dissolve 12.5 mL of ethanol in sufficient water to give a total volume of 85.4 mL. What is the percent-by-volume concentration of ethanol?

**Solution:**

In this problem we want to go from the volumes of solute and solution to percent by volume:

The relationship between these quantities is given by the expression for percent by volume:

$$\text{Percent by volume} = \frac{\text{volume of solute}}{\text{volume of solution}} \times 100\%$$

The two required quantities, volume of solute and volume of solution, are given in the problem:

$$\text{Percent by volume} = \frac{12.5 \text{ mL ethanol}}{85.4 \text{ mL solution}} \times 100\% = 14.6\%$$

**Consider This 11.4**

What volume of ethanol would be present in 50.0 mL of this solution?

**Practice Problem 11.4**

To prepare a solution, we dissolve 205 mL of ethanol in sufficient water to give a total volume of 235 mL. What is the percent-by-volume concentration of ethanol?

**Further Practice:** Questions 11.57 and 11.58 at the end of the chapter

---

## Mass/Volume Percent

Mass/volume percent is a concentration unit that is similar to the percents of mass and volume. **Mass/volume percent** is calculated by dividing the solute mass by the volume of solution and multiplying by 100%:

$$\text{Mass/volume percent} = \frac{\text{grams of solute}}{\text{volume of solution}} \times 100\%$$

Mass/volume percent is encountered in several applications, including the medical field. For example, the concentration of NaCl in a saline IV bag is 0.9%. For every 100 mL of solution, the mass of NaCl present is 0.9 g.

Note the difference between mass/volume percent and density. Recall from Chapter 1 that density is mass per volume:

$$\text{Density} = \frac{\text{mass}}{\text{volume}}$$

For a solution, density corresponds to a ratio of the mass of solution to the volume of solution:

$$\text{Density} = \frac{\text{mass of solution}}{\text{volume of solution}}$$

Sometimes the density of a solution is needed to convert from one concentration unit to another. Work Questions 11.53 and 11.54 to get some practice at using density in concentration calculations.

### Parts per Million and Parts per Billion

The ppm and ppb units represent very small concentrations. For example, one grain (or crystal) of salt in a cup of salt is about 1 ppm, and one grain in a bathtub filled with salt is about 1 ppb.

We often encounter very dilute solutions in which the concentration is too small to express conveniently as a percentage. For example, the Environmental Protection Agency's maximum acceptable level of arsenic in a water supply is 0.000001%, a number that's difficult to read and even more difficult to remember. For this reason the amounts of pollutants in water are often reported in units of parts per million (ppm) or parts per billion (ppb). A solution that is 1 ppm of a solute contains 1 part solute in 1 million parts of solution. **Parts per million** is calculated by dividing the mass of solute by the mass of solution and multiplying by 1 million ($10^6$):

$$\text{Parts per million} = \frac{\text{grams of solute}}{\text{grams of solution}} \times 10^6$$

Similarly, a solution that is 1 ppb of a solute contains 1 part solute in 1 billion parts of solution. **Parts per billion** is calculated by dividing the mass of solute by the mass of solution and multiplying by 1 billion ($10^9$):

$$\text{Parts per billion} = \frac{\text{grams of solute}}{\text{grams of solution}} \times 10^9$$

Sometimes parts per million and parts per billion are calculated on a volume basis. The concentration of pollutants in air is often reported this way.

Note that ppm and ppb express the same concept as percent by mass. The only difference is the multiplier—a million, a billion, or 100.

Because parts per million is often used to report the amount of pollutants in water, scientists sometimes look at this concentration unit in another way. For aqueous solutions, parts per million is essentially equal to milligrams of solute per liter of solution. For example, the acceptable level of arsenic in drinking water is reported as 0.01 ppm. This means that 1 L of drinking water cannot contain more than 0.01 mg of arsenic.

---

### EXAMPLE 11.5    Parts per Million

The concentration of mercury found in fish varies among region and type of fish. Large predator fish such as swordfish can contain more than 1 ppm, which is considered too high by Food and Drug Administration (FDA) standards. Eating fish containing this level of mercury is a risk to developing nervous systems in unborn babies and children. In a healthy adult, mercury is removed from the body naturally, but slowly. If a 500.0-g piece of swordfish contains $3.0 \times 10^{-4}$ g of mercury, what is the concentration of mercury in this fish in units of parts per million (ppm)?

**Solution:**

In this problem we want to go from the mass of mercury and the mass of the fish to parts per million:

To determine the concentration in ppm, we take the ratio of the mass of the mercury to the mass of the fish and then multiply by 1 million ($10^6$):

$$\text{Parts per million} = \frac{\text{grams of mercury}}{\text{grams of fish}} \times 10^6$$

We must check to make sure that the mass values are both given in the same units—grams in this case. We then substitute the mass values into the equation:

$$\text{Parts per million} = \frac{3.0 \times 10^{-4} \text{ g mercury}}{500.0 \text{ g fish}} \times 10^6 = 0.60 \text{ ppm}$$

**Consider This 11.5**

How many milligrams of mercury would be present in a 0.50-lb serving of this swordfish?

**Practice Problem 11.5**

What is the concentration in ppm of selenium if 0.020 g is found in $5.0 \times 10^3$ kg of soil?

**Further Practice:** Questions 11.59 and 11.60 at the end of the chapter

## Molarity

Molarity (*M*) is one of the most commonly used expressions of concentration in the laboratory. You may remember from Chapter 4 that **molarity** (molar concentration) is the ratio of moles of solute to the volume of solution in liters:

$$\text{Molarity} = \frac{\text{moles of solute}}{\text{liters of solution}}$$

Calculating molarity is straightforward, given the moles or mass of solute and the volume of solution. We will consider some calculations involving applications of molarity in Section 11.5. For a review of molarity calculations, see Chapter 4 and Example 11.6.

---

**EXAMPLE 11.6** | **Molarity**

A solution contains 78.2 g NaCl dissolved in sufficient water to give a total volume of 0.525 L. The molar mass of NaCl is 58.44 g/mol. What is the molarity of this solution?

**Solution:**

In this problem we want to go from the mass of solute and volume of solution to molarity:

Mass solute and volume solution —?→ Molarity

Molarity is a relationship between moles of solute and volume of solution:

$$\text{Molarity} = \frac{\text{moles of solute}}{\text{liters of solution}}$$

To calculate the molarity, we need to know the moles of solute. The relationship between moles of a substance and its mass is given by the molar mass. We can calculate the number of moles of NaCl from the mass as follows:

$$\text{mol NaCl} = 78.2 \text{ g NaCl} \times \frac{1 \text{ mol NaCl}}{58.44 \text{ g NaCl}} = 1.34 \text{ mol NaCl}$$

MATH
TOOLBOX
**4.1**

Now we have all the information we need to calculate the molarity:

$$\text{Molarity} = \frac{1.34 \text{ mol NaCl}}{0.525 \text{ L solution}} = 2.55 \text{ mol/L}$$

For convenience we can report this concentration as 2.55 $M$.

**Consider This 11.6**

How many grams of NaCl would be contained in 245.0 mL of a 2.55 $M$ NaCl solution?

**Practice Problem 11.6**

A solution contains 117 g of potassium hydroxide, KOH, dissolved in sufficient water to give a total volume of 2.00 L. The molar mass of KOH is 56.11 g/mol. What is the molarity of the aqueous potassium hydroxide solution?

**Further Practice:**  Questions 11.61 and 11.62 at the end of the chapter

A concentration unit similar to molarity called *normality* (*N*) can be used to describe the concentration of certain ions in solution. While not so common in research labs, the normality concentration unit is still used in some commercial labs. The term is useful in acid-base chemistry because it identifies the moles of $H^+$ ions—known as *equivalents*—available in a reagent bottle. For example, a 1-L solution that contains 1 mol of $H_2SO_4$ has 2 equivalents of $H^+$ available, so it is said to be a 2 $N$ solution. On the other hand, a 1-L solution containing 1 mol of HCl has only 1 equivalent available, so it is 1 $N$. Although molarity is more often used in chemistry labs now, equivalents and milliequivalents are still commonly used in the medical profession.

## Molality

Concentrations with units that contain a volume term can vary with temperature. Suppose we have a 0.1 $M$ NaCl solution. If the temperature increases, will the molarity stay the same? No, because the volume of the solution will change. When liquids are heated, they expand slightly, so the ions that make up a sodium chloride solution will be spread over a larger volume, thus decreasing the molarity. A concentration unit that uses moles of solute and mass of solvent does not change with temperature. **Molality** (*m*) is such a unit. It is the concentration expressed as moles of solute per kilogram of solvent:

$$\text{Molality} = \frac{\text{moles of solute}}{\text{kilograms of solvent}}$$

Because molality does not vary with temperature, it is useful in experiments involving temperature-dependent properties. It is also useful for examining properties that depend on the number of solute particles in solution, as we will see in Section 11.6. To calculate molality, we must determine the number of moles of solute and the mass of solvent. Note that it is the mass of the *solvent*—not of the *solution*—that is needed.

| EXAMPLE 11.7 | Molality |
|---|---|

A solution contains 22.5 g of methanol, $CH_3OH$, dissolved in sufficient water to give a total mass of 105.3 g. The molar mass of $CH_3OH$ is 32.04 g/mol. What is the molality of the aqueous methanol solution?

**Solution:**

In this problem we want to go from the mass of solute and the mass of solution to molality:

Molality is a relationship between the moles of solute and the mass of solvent in kilograms:

$$\text{Molality} = \frac{\text{moles of solute}}{\text{kilograms of solvent}}$$

MATH
TOOLBOX
**4.1**

We can calculate the number of moles of solute from the mass of solute and its molar mass:

$$\text{mol CH}_3\text{OH} = 22.5 \text{ g CH}_3\text{OH} \times \frac{1 \text{ mol CH}_3\text{OH}}{32.04 \text{ g CH}_3\text{OH}} = 0.702 \text{ mol CH}_3\text{OH}$$

Subtracting the mass of solute from the mass of solution yields the mass of solvent:

$$\text{Mass H}_2\text{O} = 105.3 \text{ g solution} - 22.5 \text{ g CH}_3\text{OH} = 82.8 \text{ g H}_2\text{O}$$

MATH
TOOLBOX
**1.3**

This mass must be converted to units of kilograms:

$$\text{Mass H}_2\text{O} = 82.8 \text{ g} \times \frac{1 \text{ kg}}{1000 \text{ g}} = 0.0828 \text{ kg H}_2\text{O}$$

Now we have all the information we need to calculate the molality:

$$\begin{aligned}\text{Molality} &= \frac{\text{moles of solute}}{\text{kilograms of solvent}} \\ &= \frac{0.702 \text{ mol CH}_3\text{OH}}{0.0828 \text{ kg H}_2\text{O}} \\ &= 8.48 \text{ mol/kg}\end{aligned}$$

For convenience this concentration is reported as 8.48 *m*.

**Consider This 11.7**

How much water should you add to 10.0 g of methanol to prepare a solution with a molality of 8.48 *m*?

**Practice Problem 11.7**

A solution contains 56.7 g of potassium hydroxide, KOH, dissolved in sufficient water to give a total mass of 245.8 g. The molar mass of KOH is 56.11 g/mol. What is the molality of the aqueous potassium hydroxide solution?

**Further Practice:** Questions 11.63 and 11.64 at the end of the chapter

## 11.5  Quantities for Reactions That Occur in Aqueous Solution

In this section we will reexamine some of the reactions described in Chapter 5 to see how to calculate quantities for reactions that occur in aqueous solutions. We will consider two types of reactions: precipitation reactions and acid-base neutralization reactions. Both types are examples of double-displacement reactions, in which two compounds exchange ions to form new compounds.

### Precipitation Reactions

Patients undergoing X-ray treatment to diagnose digestive tract problems drink a suspension of barium sulfate. Barium sulfate can be prepared from a precipitation reaction in which solutions of barium chloride and potassium sulfate are mixed (Figure 11.24). Recall from Section 5.4 that in a precipitation reaction the cation of one ionic compound combines with the anion of another ionic compound to form an insoluble, solid substance, called a *precipitate*. The reaction between barium chloride and potassium sulfate forms the precipitate barium sulfate along with soluble potassium chloride:

$$BaCl_2(aq) + K_2SO_4(aq) \longrightarrow BaSO_4(s) + 2KCl(aq)$$

Recall that this is a *molecular equation,* because it represents the substances as if they existed as molecules (or formula units) rather than dissociating into ions when they are dissolved in solution. $BaCl_2(aq)$ is a shorthand way of indicating that $Ba^{2+}$ and $Cl^-$ ions are present in the solution. Similarly, for $K_2SO_4(aq)$, $K^+$ and $SO_4^{2-}$ ions are present. The precipitate $BaSO_4(s)$ does not exist as separate ions because it's insoluble in water. To review reactions of this type, see Chapter 5.

Because precipitation reactions normally involve reactants that are in solution, calculations commonly involve the determination of moles from molarity and volume. Now that we've reviewed the characteristics of a precipitation reaction, Example 11.8 shows how we can calculate the amounts of reactants needed and the mass of precipitate that should form.

A

B

**FIGURE 11.24** Barium sulfate, used in X-ray applications, can be prepared from the reaction of barium chloride with potassium sulfate.

(A): ©LADA/Science Source; (B): ©Charles D. Winters/Science Source

---

### EXAMPLE 11.8 Calculations Involving Precipitation Reactions

Suppose we want to prepare $BaSO_4$ by adding 0.450 $M$ $K_2SO_4$ to 130.0 mL of 0.250 $M$ $BaCl_2$. Recall that the balanced equation is

$$BaCl_2(aq) + K_2SO_4(aq) \longrightarrow BaSO_4(s) + 2KCl(aq)$$

What volume of the $K_2SO_4$ solution is required to react completely with the $BaCl_2$? How many grams of $BaSO_4$ will precipitate?

**Solution:**

For the first part of this problem we want to go from the volume of one reactant to the volume of another:

$$\boxed{\text{Volume } BaCl_2 \text{ solution}} \xrightarrow{\text{?}} \boxed{\text{Volume } K_2SO_4 \text{ solution}}$$

There is no direct relationship between these two quantities, but this is a problem in which we're given a quantity of one substance in a chemical reaction and want to calculate another. If we can convert volume of the $BaCl_2$ solution to moles of dissolved $BaCl_2$, then we can relate the two reactants using their mole ratio obtained from the balanced equation. Is there a relationship between the volume of $BaCl_2$ and moles of $BaCl_2$?

$$\boxed{\text{Volume } BaCl_2 \text{ solution}} \xrightarrow{\text{?}} \boxed{\text{mol } BaCl_2}$$

Yes, its molarity is 0.250 $M$, which tells us that for every liter of solution, 0.250 mol of $BaCl_2$ is present. In Chapter 6 we obtained moles of reactant from grams using the molar mass as a conversion factor. Here we must convert the volume of a reactant in solution to moles using the molar concentration as a conversion factor. In this case we know the volume of 0.250 $M$ $BaCl_2$ is 130.0 mL. We have to first convert milliliters to liters because the volume unit in molarity is liters:

$$\boxed{\text{Volume } BaCl_2 \text{ (in mL)}} \xrightarrow{1000 \text{ mL} = 1 \text{ L}} \boxed{\text{Volume } BaCl_2 \text{ (in L)}} \xrightarrow{\text{molarity}} \boxed{\text{mol } BaCl_2}$$

$$\text{mol } BaCl_2 = 130.0 \text{ mL } BaCl_2 \text{ solution} \times \frac{1 \text{ L}}{1000 \text{ mL}} \times \frac{0.250 \text{ mol } BaCl_2}{\text{L } BaCl_2 \text{ solution}}$$

$$= 0.0325 \text{ mol } BaCl_2$$

**MATH TOOLBOX 1.3**

Now that we've determined the moles of $BaCl_2$, we can determine the moles of the other reactant $K_2SO_4$ using their mole ratio defined by the balanced equation. The coefficients in the balanced equation tell us that 1 mol of $BaCl_2$ reacts with 1 mol of $K_2SO_4$ to produce 1 mol of $BaSO_4$. We can now use the mole ratio in the chemical equation to calculate the moles of $K_2SO_4$ required for the reaction:

$$\boxed{\text{mol } BaCl_2} \xrightarrow{\text{mole ratio}} \boxed{\text{mol } K_2SO_4}$$

$$\text{mol } K_2SO_4 = 0.0325 \text{ mol } BaCl_2 \times \frac{1 \text{ mol } K_2SO_4}{1 \text{ mol } BaCl_2} = 0.0325 \text{ mol } K_2SO_4$$

Now that we know the moles of $K_2SO_4$ required to completely react with the $BaCl_2$, we can then determine the volume of $K_2SO_4$ solution that will contain the desired number of moles:

$$\boxed{\text{mol } K_2SO_4} \xrightarrow{?} \boxed{\begin{array}{c}\text{Volume } K_2SO_4 \\ \text{solution (in L)}\end{array}}$$

The molar concentration of $K_2SO_4$ solution provides a relationship we can use to calculate the volume:

$$\boxed{\text{mol } K_2SO_4} \xrightarrow{\text{molarity}} \boxed{\begin{array}{c}\text{Volume } K_2SO_4 \\ \text{solution (in L)}\end{array}}$$

$$\text{Volume of } K_2SO_4 \text{ solution} = 0.0325 \text{ mol } K_2SO_4 \times \frac{1 \text{ L } K_2SO_4 \text{ solution}}{0.450 \text{ mol } K_2SO_4}$$

$$= 0.0722 \text{ L } K_2SO_4 \text{ solution}$$

We can interpret this volume of $K_2SO_4$ solution as the amount of solution that has enough moles of $K_2SO_4$ (0.0325 mol) to react with all the $BaCl_2$. An alternative to this step-by-step approach involves setting up conversions consecutively in one calculation. To determine the volume of $K_2SO_4$ required, we can set up the following sequence of conversions:

$$\text{Volume of } K_2SO_4 \text{ solution} = 0.1300 \text{ L } BaCl_2 \text{ solution} \times \frac{0.250 \text{ mol } BaCl_2}{1 \text{ L } BaCl_2 \text{ solution}}$$

$$\times \frac{1 \text{ mol } K_2SO_4}{1 \text{ mol } BaCl_2} \times \frac{1 \text{ L } K_2SO_4 \text{ solution}}{0.450 \text{ mol } K_2SO_4}$$

$$= 0.0722 \text{ L } K_2SO_4 \text{ solution}$$

This volume is a small number more conveniently expressed in milliliters:

$$\text{Volume in mL} = 0.0722 \text{ L } K_2SO_4 \text{ solution} \times \frac{1000 \text{ mL}}{1 \text{ L}}$$

$$= 72.2 \text{ mL } K_2SO_4 \text{ solution}$$

The following sequence summarizes the relationships used to solve this problem:

$$\boxed{\begin{array}{c}\text{Volume (in L)} \\ BaCl_2 \text{ solution}\end{array}} \xrightarrow{\text{molarity}} \boxed{\begin{array}{c}\text{mol} \\ BaCl_2\end{array}} \xrightarrow{\text{mole ratio}} \boxed{\begin{array}{c}\text{mol} \\ K_2SO_4\end{array}} \xrightarrow{\text{molarity}} \boxed{\begin{array}{c}\text{Volume (in L)} \\ K_2SO_4 \text{ solution}\end{array}}$$

Recall from Chapter 6 that we used the following guide to solve problems involving masses of reactants and products.

The second part of the question is to calculate the mass of solid $BaSO_4$ that will be produced when adding 72.2 mL of 0.450 $M$ $K_2SO_4$ solution to 130.0 mL of 0.250 $M$ $BaCl_2$. The main difference here is we're determining the mass (instead of solution volume) of a product because it's a solid and not dissolved in solution. We know the moles of $BaCl_2$ from the first part of the previous multi-step calculation. We can use the mole ratio given by the balanced chemical equation to determine the moles of $BaSO_4$ that should form:

$$\text{mol } BaSO_4 = 0.0325 \text{ mol } BaCl_2 \times \frac{1 \text{ mol } BaSO_4}{1 \text{ mol } BaCl_2} = 0.0325 \text{ mol } BaSO_4$$

Lastly we must convert moles of $BaSO_4$ to units of grams. Recall from Section 4.2 that we use molar mass to convert between moles and grams:

$$\boxed{\text{mol } BaSO_4} \xrightarrow{\text{molar mass}} \boxed{\text{Mass } BaSO_4}$$

$$\text{Mass of } BaSO_4 = 0.0325 \text{ mol } BaSO_4 \times \frac{233.4 \text{ g } BaSO_4}{1 \text{ mol } BaSO_4} = 7.59 \text{ g } BaSO_4$$

An alternative to this step-by-step approach involves setting up conversions consecutively in one calculation. To determine the mass of $BaSO_4$ produced, we can set up the following sequence of conversions:

$$\text{Mass of } BaSO_4 = 0.1300 \text{ L } BaCl_2 \text{ solution} \times \frac{0.250 \text{ mol } BaCl_2}{1 \text{ L } BaCl_2 \text{ solution}}$$

$$\times \frac{1 \text{ mol } BaSO_4}{1 \text{ mol } BaCl_2} \times \frac{233.4 \text{ g } BaSO_4}{1 \text{ mol } BaSO_4}$$

$$= 7.59 \text{ g } BaSO_4$$

$$\boxed{\begin{array}{c}\text{Volume (in L)}\\ BaCl_2 \text{ solution}\end{array}} \xrightarrow{\text{molarity}} \boxed{\begin{array}{c}\text{mol}\\ BaCl_2\end{array}} \xrightarrow{\text{mole ratio}} \boxed{\begin{array}{c}\text{mol}\\ BaSO_4\end{array}} \xrightarrow{\text{molar mass}} \boxed{\begin{array}{c}\text{Mass}\\ BaSO_4\end{array}}$$

The key to solving problems that involve solution concentrations is to realize that concentration is a ratio that can be used as a conversion factor. If volume is known and we want to calculate moles, we simply multiply by the molarity. If moles are given and we want to calculate volume, we divide by molarity.

### Consider This 11.8

How might the products be separated from the reaction solution so the $BaSO_4$ could be used in another application? Would evaporating off the water work?

### Practice Problem 11.8

When a solution of lead(II) nitrate is mixed with a solution of potassium chromate, a yellow precipitate forms according to the equation

$$Pb(NO_3)_2(aq) + K_2CrO_4(aq) \longrightarrow PbCrO_4(s) + 2KNO_3(aq)$$

What volume of 0.105 $M$ lead(II) nitrate is required to react with 100.0 mL of 0.120 $M$ potassium chromate? What mass of $PbCrO_4$ solid forms?

**Further Practice:** Questions 11.71 and 11.72 at the end of the chapter

## Acid-Base Titrations

Now let's consider another type of reaction that takes place in solution, an acid-base neutralization reaction. In such a reaction, an acid reacts with a base, normally forming a salt and water. The formation of water molecules from hydrogen ions and hydroxide ions is common to all aqueous neutralization reactions. The acid-base reaction involves the donation of a hydrogen ion by an acid and the acceptance of that hydrogen ion by a base. A common example is the reaction between hydrochloric acid and sodium hydroxide, both of which are strong electrolytes:

 **ANIMATION:** Neutralization Reactions

$$\text{hydrochloric acid} + \text{sodium hydroxide} \longrightarrow \text{sodium chloride} + \text{water}$$

The molecular equation for this reaction is

$$HCl(aq) + NaOH(aq) \longrightarrow NaCl(aq) + H_2O(l)$$

The ionic equation for this reaction is

$$H^+(aq) + Cl^-(aq) + Na^+(aq) + OH^-(aq) \longrightarrow Na^+(aq) + Cl^-(aq) + H_2O(l)$$

Note that $Na^+$ and $Cl^-$ ions appear on both sides of the equation, and they do not participate in the reaction. They are spectator ions, which can be omitted when writing the net ionic equation:

$$H^+(aq) + OH^-(aq) \longrightarrow H_2O(l)$$

This net ionic equation is common to all aqueous strong acid–strong base reactions. Strong acids and strong bases are strong electrolytes that will be discussed in more detail in Chapter 13.

 **ANIMATION:** Acid-Base Titration

**Titration** is the process of determining the concentration of one substance in solution by reacting it with a solution of another substance that has a known concentration. Although the technique can be applied to other types of reactions, we will focus on acid-base titrations. To carry out the process, we add one solution to a known volume of the other solution, as shown in Figure 11.25, until just beyond the point at which the reaction between the two substances is complete. We can often tell when we have reached this point, called the *equivalence point,* by a change in the color of an *indicator,* a substance that has been added to the system and is sensitive to changes in acid or base content. Notice the color change in Figure 11.25. The indicator used here, phenolphthalein, changes from colorless to pink when the solution becomes basic. The point at which the indicator changes color is technically called the *end point,* which may be slightly different from the equivalence point but is usually a good approximation.

Indicators will be discussed in more detail in Chapter 13.

When running a titration, we use the amount of one substance in a chemical reaction to calculate the amount of another. In titration calculations, the volume of the reactant solution of known concentration is used to calculate the moles of that reactant. The moles of the other reactant are obtained from the coefficients in the balanced equation, and this value is used with the volume of the solution of unknown concentration to calculate its molarity. Example 11.9 shows a common acid-base titration calculation.

Acid-base titrations can be used to determine the amount of acid in aspirin or the amount of base in an antacid.

**FIGURE 11.25** To determine the concentration of a measured volume of acid, a known volume of a base of known concentration is added until some sign of a reaction can be observed, such as a color change of an indicator. This laboratory technique is called titration.

(all): ©McGraw-Hill Education/Stephen Frisch

### EXAMPLE 11.9    Acid-Base Titrations

Suppose a titration is run in which 25.05 mL of NaOH solution of unknown concentration reacts with 25.00 mL of 0.1000 $M$ $H_2SO_4$ solution. The chemical equation that summarizes the reaction is as follows:

$$H_2SO_4(aq) + 2NaOH(aq) \longrightarrow 2H_2O(l) + Na_2SO_4(aq)$$

What is the molarity of the NaOH solution?

**Solution:**

We want to go from the volume of one reactant to the molarity of another:

$$\boxed{\text{Volume } H_2SO_4 \text{ solution}} \xrightarrow{\ ?\ } \boxed{\text{Molarity NaOH solution}}$$

There is no direct relationship between these two quantities. Since the molarity of a solution is a ratio of the moles of a solute to the volume of a solution of that solute, we need to know both of these quantities to determine the molarity of the NaOH solution. The volume of NaOH solution is stated in the problem, but we do not know the moles of NaOH. We'll have to calculate the moles of NaOH that react with the $H_2SO_4$ from the information we have about $H_2SO_4$.

If we first calculate moles of $H_2SO_4$ from the molarity and volume of the solution, then we can relate the two reactants using their mole ratio otained from the balanced equation, and determine moles of NaOH. From the balanced equation we know that 1 mol of $H_2SO_4$ reacts with 2 mol of NaOH. From the volume and molarity of $H_2SO_4$, and the mole ratio given in the balanced chemical equation, we can calculate the moles of NaOH that reacted:

$$\boxed{\text{Volume } H_2SO_4 \text{ (in L)}} \xrightarrow{\text{molarity}} \boxed{\text{mol } H_2SO_4} \xrightarrow{\text{mole ratio}} \boxed{\text{mol NaOH}}$$

We must begin by converting milliliters to liters because the volume unit in molarity is expressed as liters. Then, we substitute relationships that allow us to calculate moles of NaOH that react:

**MATH
TOOLBOX
1.3**

$$\text{mol NaOH} = 25.00 \text{ mL } H_2SO_4 \text{ solution} \times \frac{1 \text{ L}}{1000 \text{ mL}}$$

$$\times \frac{0.1000 \text{ mol } H_2SO_4}{\text{L } H_2SO_4 \text{ solution}} \times \frac{2 \text{ mol NaOH}}{1 \text{ mol } H_2SO_4}$$

$$= 0.005000 \text{ mol NaOH}$$

We now know the number of moles of NaOH present in the original solution. Using this quantity and the volume stated in the problem, we can calculate the molarity. Recall that molarity is a ratio of moles of solute to liters of solution, so we must convert the volume in milliliters to liters. Then, we use this quantity to calculate the molarity of the NaOH solution:

$$\text{Volume of NaOH solution} = 25.05 \text{ mL NaOH solution} \times \frac{1 \text{ L}}{1000 \text{ mL}}$$

$$= 0.02505 \text{ L NaOH solution}$$

$$\text{Molarity} = \frac{\text{moles of solute}}{\text{liters of solution}}$$

$$= \frac{0.005000 \text{ mol NaOH}}{0.02505 \text{ L solution}} = 0.1996 \text{ } M \text{ NaOH}$$

The relationships used to solve this problem are summarized as follows:

**Consider This 11.9**

What mass of $Na_2SO_4$ could be isolated upon completion of this reaction?

**Practice Problem 11.9**

We need 28.18 mL of 0.2437 *M* HCl solution to completely titrate 25.00 mL of $Ba(OH)_2$ solution of unknown concentration. What is the molarity of the barium hydroxide solution?

**Further Practice:** Questions 11.79 and 11.80 at the end of the chapter

Calculations involving quantities in solution are similar to using masses, as described in Chapter 6. In general, the following scheme can be used to organize calculations that involve volume and molarity:

## 11.6 Colligative Properties

When a solute dissolves in a pure solvent, the properties of both may change. We have already examined some reactions that bring about chemical changes, but some changes in physical properties do not require chemical change. They do not depend on the identity of the solute—only on the number of solute particles present. Such a property is called a **colligative property.** Colligative properties vary only with the number of solute particles (molecules or ions) present in a specific quantity of solvent. What the solute is does not matter. Colligative properties include osmotic pressure, vapor pressure lowering, boiling point elevation, and freezing point depression.

### Osmotic Pressure

Most processes that occur in the human body involve solutions. The body often works to balance the concentration of solutes inside cells with the concentration outside—in the extracellular fluid. Because the body reacts unfavorably to imbalances in solute concentration, the concentration of particles in IV fluids must be controlled. For example, when the concentration of solutes in the blood is higher than the solute concentration within a red blood cell, the cell shrinks. It expands when the solute concentration in the fluid is lower. Why do these changes occur?

**Osmosis** is a process in which solvent particles diffuse through a barrier that does not allow the passage of solute particles. Osmosis allows water to pass through plant and animal cell walls and membranes. The movement of water molecules balances differences in solute concentration between cells and extracellular fluid. A membrane that allows the passage of some substances but not others is a **semipermeable membrane.** In crossing semipermeable membranes, water tends to travel preferentially from the side that is more dilute in solute to the side that is more concentrated, as shown in Figure 11.26. This movement of water molecules acts to equalize the solute concentration on both sides of the membrane.

**ANIMATION: Osmosis**

Several terms describe the locations of water in the body. Water inside cells is *intra*cellular. Water outside cells is *extra*cellular. It includes both the plasma in blood and *interstitial* water, which lies in the spaces between cells.

**FIGURE 11.26** When separated by a semipermeable membrane, a solvent will travel from a solution of low solute concentration through the barrier to a solution with higher solute concentration. As represented in this figure, there is a net flow of solvent molecules from the right side of the membrane to the left.

**FIGURE 11.27** When a 3% salt solution contained by a semipermeable membrane is placed in pure water, water molecules travel across the membrane into the salt solution, and the solution rises. Can the concentrations on both sides of the membrane ever become equal by osmosis?

Our kidneys help control solute concentrations in our blood using different processes. Filtration removes urea and other small molecules and ions, but some components are returned to the bloodstream. Reabsorption of glucose and important ions occurs by a process called active transport, and reabsorption of some water occurs by *osmosis*.

For example, when pure water is placed on one side of a semipermeable membrane and a salt solution on the other side, the volume of the salt solution rises above the level of the water in the beaker (Figure 11.27). The rise happens because water molecules pass through the membrane from the beaker of pure water into the salt solution. Consequently, the salt solution becomes more dilute. One reason this process occurs is entropy, the tendency for matter to become more disordered. When solvent molecules travel through a membrane, the distribution of solute particles in the solution becomes more random.

To prevent osmosis—that is, to prevent water from passing through to the more concentrated side of a membrane—pressure can be exerted on the solution. This pressure, called the *osmotic pressure*, is related to the height a solution will rise up in a column that is separated from a more dilute solution (or pure solvent) by a semipermeable membrane, as shown in Figure 11.28. Like other colligative properties, osmotic pressure is the same regardless of the identity of the solute. It only varies with the concentration of solute particles.

During osmosis the water actually moves in both directions, but it moves more in the direction that leads to equalization of the solute concentration.

Solvent can be forced to flow in the opposite direction by the application of a pressure greater than the osmotic pressure. This process, called *reverse osmosis,* is the flow of solvent from high solute concentration through a semipermeable membrane to a solution with low solute concentration (Figure 11.29). The process is used in the manufacture of many popular brands of bottled water. It is also important in desalinating seawater. In this application, seawater is pumped through a tube containing a semipermeable membrane at a pressure of about 25 atm, much greater

**FIGURE 11.28** (A) Solvent moves across the membrane from the pure solvent side to the solution. (B) The solution volume increases, and it becomes less concentrated. The difference in volumes creates an osmotic pressure. (C) To prevent osmosis from occurring, an external pressure must be exerted on the solution side of the membrane to keep the solutions at the same level.

Hollow fibers of
semipermeable
membrane

**FIGURE 11.29** A typical reverse-osmosis coil contains many hollow fibers made of semipermeable membranes. Seawater is pumped through the coil at high pressure, causing solvent to flow out of the high-concentration solution into the fibers. From them pure water can then be collected.

©Charlie Neuman/ZUMA Press/Newscom

than the osmotic pressure of seawater. As a result, water molecules flow from the side of the membrane containing a high solute concentration to the side with a lower solute concentration (the reverse of osmosis). This same process, on a smaller scale, is used in the reverse-osmosis units available for use in homes.

Now let's return to the question of what happens in the blood of living organisms (Figure 11.30). Blood is a solution of many dissolved substances in which suspended red blood cells are carried throughout the body. If the blood has a higher concentration of dissolved solute than the fluid within red blood cells, water will flow out of the red blood cells and they will shrink. If the blood has a higher concentration of solute, it is *hypertonic* to the cell. If the blood has a lower concentration of solute than the red blood cells, water will flow into the red blood cells causing them to expand. In this case the blood is said to be *hypotonic* to the cell. When the concentration of solute is the same outside and inside the red blood cell, the blood is considered to be *isotonic* to the cell. Solutions delivered intravenously must be isotonic to blood for this reason.

## Vapor Pressure Lowering

Consider a solution of table sugar, or sucrose, in water. If separate beakers of sugar solution and pure water are placed into a sealed container, over time the volume of the sugar solution will increase, and the volume of the pure water will decrease (Figure 11.31). Why? The answer lies in the differences in vapor pressure of the pure solvent and the solution. Since the system is sealed, water that evaporates cannot escape. Instead, water evaporates faster from the liquid having the greater vapor pressure and condenses in the liquid with the lower vapor pressure. The presence of solute molecules lowers the vapor pressure of the solvent (Figure 11.32).

Solutes can be described as volatile or nonvolatile. *Volatile* solutes readily form a gas, while *nonvolatile* solutes don't. Ethanol is a volatile solute; sugar is nonvolatile. The effect of a nonvolatile solute in solution on vapor pressure is shown graphically

**A**

**B**

**C**

**FIGURE 11.30** (A) When the surrounding solution has the same solute concentration as the cell contents, it is *isotonic*, and cell size is normal. (B) A *hypotonic* surrounding solution causes water molecules to move into the red blood cell, swelling it. Eventually it will burst. (C) A surrounding solution that is *hypertonic* will cause water molecules to move out of a red blood cell, and it shrinks.
(all): ©David M. Philips/Science Source

Pure H$_2$O    Sugar solution    Pure H$_2$O    Sugar solution

**FIGURE 11.31** Water evaporates faster from the beaker of pure water and condenses into the beaker containing the sugar water.

**FIGURE 11.32** Addition of a solute to a pure solvent reduces the number of solvent molecules in the gas phase, lowering the vapor pressure.
(both): ©Brian Moeskau/Moeskau Photography

Pure liquid                    Solution

Vapor pressure curves for pure liquids were discussed in Chapter 10.

**ANIMATION: Phase Diagrams and the States of Matter**

in Figure 11.33. Notice that at a given temperature, the vapor pressure of the solution is lower than that of the pure solvent.

The graph in Figure 11.33 is a *phase diagram.* It shows the physical state of a substance at a given temperature and pressure. For example, the solid curve for the pure solvent that lies between the liquid and gas phases corresponds to temperatures and pressures at which the liquid state is in equilibrium with the gaseous state. As temperature increases, the vapor pressure increases. Notice the position of the liquid-gas dashed curve for a solution. It's lower than the curve for the pure solvent, indicating that at a given temperature, the vapor pressure for the solution is lower than that of the pure solvent. Vapor pressure lowering is shown in Figure 11.33 as $\Delta P$.

## Boiling Point Elevation

We have already seen that the vapor pressure of a liquid increases with temperature. Notice in Figure 11.33 that when the temperature rises high enough, the vapor pressure of the liquid equals the surrounding atmospheric pressure. When this happens, the liquid has reached its boiling point. The addition of a solute affects the boiling point because it affects the vapor pressure. The presence of solute means that the temperature must be raised further to raise the vapor pressure to atmospheric pressure, so the boiling point is raised as well. The boiling point elevation is shown in Figure 11.33 as $\Delta T_b$.

**FIGURE 11.33** The addition of a nonvolatile solute lowers the vapor pressure of water. This causes an elevation of the boiling point, $\Delta T_b$, and a depression of the freezing point, $\Delta T_f$.

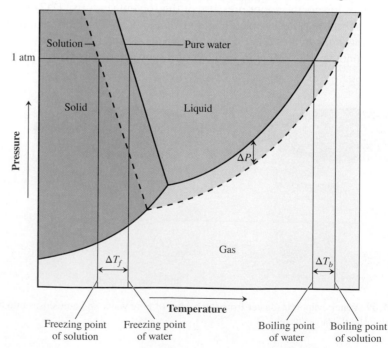

The lowering of vapor pressure is proportional to the concentration of the solute, so the elevation of the boiling point is also proportional to the concentration of the solute.

Boiling point elevation is related quantitatively to the molality of the solute particles present in the solution. If the solution represented in Figure 11.33 had a greater molality, the difference in boiling point would be correspondingly greater. The change in boiling point is related to the molality of solute particles by the following equation:

$$\Delta T_b = K_b\, m$$

The term $\Delta T_b$ is the difference in boiling point between the solution and pure solvent. The term $K_b$ is a constant characteristic of the solvent, the *boiling point constant*. For water, $K_b$ is $0.52°C/m$. Recall that $m$ represents the molal concentration of the solution:

$$m = \frac{\text{moles of solute}}{\text{kilograms of solvent}}$$

## Freezing Point Depression

The freezing point of a liquid is the temperature at which the liquid and solid states of a substance are at equilibrium. The presence of a solute lowers the freezing point of pure solvent by an amount $\Delta T_f$, as shown in Figure 11.33. Like boiling point elevation, freezing point depression is proportional to the molal concentration of solute.

This colligative property finds practical application in the use of ethylene glycol, $HOCH_2CH_2OH$, as an antifreeze in automobile radiators. Ethylene glycol lowers the freezing point of water that is used as a radiator coolant. (It also raises the boiling point, preventing boilover during the summer.) Another widespread application in cold climates is the use of sodium chloride or calcium chloride to melt ice on roads and sidewalks. These compounds work by lowering the freezing point of water.

The relationship of freezing point depression to the molality of solute particles is expressed in an equation similar to that for boiling point elevation:

$$\Delta T_f = K_f m$$

The term $\Delta T_f$ is the difference in freezing point between pure solvent and the solution. The term $K_f$ is a constant characteristic of the solvent, the *freezing point constant*. For water, $K_f$ is $-1.86°C/m$.

Many homemade ice cream churns require a mixture of ice and rock salt to freeze the ice cream. The salt lowers the freezing point of water as the salt dissolves, so the ice-salt mixture is colder than ice without the salt.

---

| **EXAMPLE 11.10** | **Boiling Point Elevation and Freezing Point Depression** |
|---|---|

What is the boiling point for a 1.5 $m$ solution of sucrose in water?

**Solution:**
We can use the following relationship to determine the change in boiling point for the solution:

$$\Delta T_b = K_b m = \left(\frac{0.52°C}{m}\right) \times 1.5\,m = 0.78°C$$

The boiling point of the solution is 0.78°C higher than the normal boiling point. Since the normal boiling point of water is 100°C, the solution boils at 100.78°C.

**Consider This 11.10**

How does the vapor pressure of this solution compare to the vapor pressure of pure water?

**Practice Problem 11.10**

What is the freezing point for a 1.5 $m$ solution of sucrose in water?

**Further Practice:** Questions 11.93 and 11.94 at the end of the chapter

## Colligative Properties and Strong Electrolytes

In Example 11.10, we used the molality to determine the change in boiling point and freezing point for a solution. If we know the change in boiling point (or freezing point), we can reverse the process and calculate the molality of the solution. However, the colligative properties measured for many solutions are larger than we predict from such calculations because many solutes dissociate into ions when they dissolve. Colligative properties are proportional to the number of solute particles present in the solution. When the measured value of a boiling point elevation or freezing point depression is significantly greater than the calculated value, we can infer that the solute is an electrolyte and has dissociated into ions.

The degree of dissociation can be determined by considering how the substance behaves in water. This is often expressed as the average number of particles produced by one formula unit of the solute when it dissolves. For example, when one formula unit of NaCl dissolves, it completely dissociates to form two ions, $Na^+(aq)$ and $Cl^-(aq)$. Every NaCl formula unit that dissolves adds two particles to the solution. The molal concentration of ions is twice that of the concentration of NaCl. Therefore a 1 $m$ NaCl solution would have a concentration of ions that is 2 $m$. In contrast, sugar, a nonelectrolyte, retains its molecular structure when it is mixed with water, so for every molecule that dissolves, only one sugar particle exists in the aqueous solution. Because the concentration of particles present in the solution determines colligative properties, we can predict that a 1 $m$ solution of the strong electrolyte NaCl would have about twice the effect on freezing point depression and boiling point elevation as a 1 $m$ solution of sucrose, a nonelectrolyte.

> Strong electrolytes do not necessarily dissociate 100%. Sometimes oppositely charged ions attract each other in solution and exist loosely as ion pairs. The consequence is slight deviations in the expected colligative properties.

---

**EXAMPLE 11.11** | **Colligative Properties and Strong Electrolytes**

Which of the following aqueous solutions should have the highest boiling point: 1 $m$ sucrose, 1 $m$ NaCl, or 1 $m$ MgCl$_2$?

**Solution:**

Boiling point elevation depends on the number of particles present in the solution. Sucrose retains its molecular structure when it dissolves in water, so the concentration of particles in a 1 $m$ solution of sucrose is 1 $m$.

One formula unit of sodium chloride dissociates to give two ions in aqueous solution:

$$NaCl(s) \longrightarrow Na^+(aq) + Cl^-(aq)$$

The concentration of particles in a 1 $m$ NaCl solution is thus about 2 $m$.

Magnesium chloride also dissociates to give ions in aqueous solution:

$$MgCl_2(s) \longrightarrow Mg^{2+}(aq) + 2Cl^-(aq)$$

Every formula unit of MgCl$_2$ provides three ions. The concentration of particles in a 1 $m$ MgCl$_2$ solution is thus about 3 $m$. This solution has the greatest concentration of dissolved particles, so it is expected to have the highest boiling point.

**Consider This 11.11**

Which aqueous solution would have the higher boiling point, 1 $m$ MgCl$_2$ or 1 $m$ Mg(NO$_3$)$_2$?

**Practice Problem 11.11**

Which of the following aqueous solutions is expected to have the lowest freezing point: 0.5 $m$ CH$_3$CH$_2$OH, 0.5 $m$ Ca(NO$_3$)$_2$, or 0.5 $m$ KBr?

**Further Practice:** Questions 11.95 and 11.96 at the end of the chapter

# CHAPTER REVIEW

## KEY CONCEPTS

- Solutions are found among any combination of physical states.
  - Metal alloys and gas mixtures are solutions.
  - The most common solutions consist of one or more solids dissolved in a liquid solvent, usually water.
  - The substance being dissolved is the solute (usually present in the lesser amount), and the substance doing the dissolving is the solvent (usually present in the greater amount).
  - Solids that dissolve to form ions in water are electrolytes.
  - In solutions of nonelectrolytes, the solute retains its molecular structure.
- Solution formation involves two kinds of forces: those that must be overcome (intermolecular forces and bonds) and those that are made when solute and solvent combine to form a solution.
  - When the solvation process provides more energy than is needed to separate the pure solute and solvent particles, the overall dissolving process is exothermic.
  - When the separation processes require more energy than the solvation process releases, the overall dissolving process is endothermic.
  - Entropy, the tendency for the energy of matter to become more dispersed, is another important force that drives the solution process.
  - When a solid ionic compound dissolves in water, ionic bonds in the solid break and hydrogen bonds between water molecules break. Ion-dipole forces form between the ions and water molecules.
  - When a solution forms between a nonpolar solute and a nonpolar solvent, London dispersion forces are broken between the solute molecules and between the solvent molecules. New London dispersion forces form between the solute and solvent molecules.
- A variety of factors—including structure, temperature, and pressure—affect solubility.
  - Polar solvents generally dissolve other polar substances and ionic compounds.
  - Nonpolar solvents dissolve other nonpolar substances.
  - The solubility of solids usually increases with temperature, but gases become less soluble as temperature increases.
  - The solubility of gases (but not liquids or solids) increases with pressure.
- The concentration of solutions can be expressed in a variety of ways.
  - A solution that contains the maximum possible amount of solute is a saturated solution.
  - A solution that contains less than the maximum possible amount of solute is an unsaturated solution.
  - A solution that contains more than the expected maximum amount of solute is a supersaturated solution.
  - Quantitative ways of describing solution concentration include percent by mass, percent by volume, parts per million, parts per billion, molarity, and molality. All of these concentration units express relative amounts of solute in a given amount of solvent or given amount of solution.
- When solutions are involved in chemical reactions, the most common concentration unit is molarity ($M$). It can be used to calculate the quantities of solutions involved in precipitation and titration reactions.
- Colligative properties of solutions depend only on the concentration of solute particles, not their identity.
  - These properties include increasing osmotic pressure, vapor pressure lowering, boiling point elevation, and freezing point depression.
  - Colligative properties of strong electrolyte solutions are related to the number of ions produced per formula unit of the solute dissolved in the solution.

## KEY RELATIONSHIPS

| Relationship | Equation |
|---|---|
| To calculate percent by mass, divide the mass of solute by the mass of the solution and multiply by 100%. | $\text{Percent by mass} = \dfrac{\text{grams of solute}}{\text{grams of solution}} \times 100\%$ |
| To calculate percent by volume, divide the volume of solute by the volume of the solution and multiply by 100%. | $\text{Percent by volume} = \dfrac{\text{volume of solute}}{\text{volume of solution}} \times 100\%$ |
| To calculate mass/volume percent, divide the mass of solute by the volume of the solution and multiply by 100%. | $\text{Mass/volume percent} = \dfrac{\text{grams of solute}}{\text{volume of solution}} \times 100\%$ |
| To calculate parts per million (ppm), divide the mass of solute by the mass of the solution and multiply by 1 million ($10^6$). | $\text{ppm} = \dfrac{\text{grams of solute}}{\text{grams of solution}} \times 10^6$ |
| To calculate parts per billion (ppb), divide the mass of solute by the mass of the solution and multiply by 1 billion ($10^9$). | $\text{ppb} = \dfrac{\text{grams of solute}}{\text{grams of solution}} \times 10^9$ |
| To calculate molarity, divide the moles of solute by the volume of the solution in liters. | $\text{Molarity} = \dfrac{\text{moles of solute}}{\text{liters of solution}}$ |
| To calculate molality, divide the moles of solute by the mass of solvent in kilograms. | $\text{Molality} = \dfrac{\text{moles of solute}}{\text{kilograms of solvent}}$ |
| Boiling point elevation for a solution is proportional to the molality of solute particles in the solution. | $\Delta T_b = K_b\, m$ |
| Freezing point depression for a solution is proportional to the molality of solute particles in the solution. | $\Delta T_f = K_f\, m$ |

## KEY TERMS

aqueous solution (11.1)

colligative property (11.6)

concentration (11.4)

entropy (11.2)

ion-dipole force (11.2)

mass/volume percent (11.4)

miscible (11.1)

molality (11.4)

molarity (11.4)

osmosis (11.6)

parts per billion (11.4)

parts per million (11.4)

percent by mass (11.4)

percent by volume (11.4)

saturated solution (11.4)

semipermeable membrane (11.6)

solubility (11.4)

solute (11.1)

solution (11.1)

solvent (11.1)

supersaturated solution (11.4)

titration (11.5)

unsaturated solution (11.4)

## QUESTIONS AND PROBLEMS

The following questions and problems, except those in *Additional Questions* and *Concept Review Questions,* are paired. Questions in a pair focus on the same concept. Answers to selected questions and problems are in Appendix D.

### Matching Definitions with Key Terms

11.1    Match the key terms with the descriptions provided.
   (a) a solution that contains the maximum concentration of solute dissolved in the solvent as described by its solubility
   (b) a homogeneous mixture of two or more substances in which the solvent is water

(c) moles of solute per liter of solution
(d) the substance doing the dissolving; usually the component of a solution that is present in the larger amount
(e) a measure of the tendency for the energy of matter to become more dispersed
(f) a ratio that describes the maximum amount of a solute that dissolves in a particular solvent to form a saturated solution under specified conditions

(g) mass of solute divided by mass of solution and multiplied by $10^6$; in aqueous solutions it is essentially the milligrams of solute per liter of solution

(h) when liquids can mix in all proportions

(i) the passage of solvent molecules, but not solute particles, through a membrane from a solution that is less concentrated in solute to a solution that is more concentrated

(j) a property of a solution that depends only on the concentration of solute particles, not on the identity of the solute

(k) volume of solute divided by volume of solution and multiplied by 100%

(l) mass of solute divided by volume of solution and multiplied by 100%

11.2    Match the key terms with the descriptions provided.

(a) an unstable solution that contains more dissolved solute than the maximum dictated by the solubility

(b) a homogeneous mixture of two or more substances uniformly dispersed at a molecular or ionic level

(c) moles of solute per kilogram of solvent

(d) the substance being dissolved; usually the component of a solution that is present in the lesser amount

(e) an intermolecular force between an ion and a polar molecule

(f) the relative amounts of solute and solvent in a solution

(g) mass of solute divided by mass of solution and multiplied by $10^9$

(h) a solution that contains less than the maximum amount of solute possible in a stable system

(i) a barrier that allows the passage of solvent but not of solute particles

(j) a process of determining the concentration of one substance in solution by reacting it with a solution of another substance of known concentration

(k) mass of solute divided by mass of solution and multiplied by 100%

## The Composition of Solutions

11.3    Identify the solute and solvent in each of the following solutions.

(a) seawater

(b) steel, an alloy of iron which contains up to 1.5% carbon

(c) oxygenated water

11.4    Identify the solute and solvent in each of the following solutions.

(a) an IV solution that contains a water-soluble drug and water

(b) nail polish remover, which usually contains mostly ethyl acetate and some ethanol and some water

(c) carbonated water, which is an aqueous solution of dissolved $CO_2$ gas

11.5    What types of substances are strong electrolytes? How do they behave when dissolved in water?

11.6    What types of substances are nonelectrolytes? How do they behave when dissolved in water?

11.7    A molecular-level representation of a solution is shown. Is the solute an electrolyte or a nonelectrolyte?

11.8    A molecular-level representation of a solution is shown. Is the solute an electrolyte or a nonelectrolyte?

11.9    What ions, atoms, or molecules are present after the following substances mix with water? For ionic compounds, you may want to refer to the solubility rules in Table 11.1.

(a) HBr

(b) $NH_4Cl$

(c) butanol, $CH_3CH_2CH_2CH_2OH$

11.10    What ions, atoms, or molecules are present after the following substances dissolve in water? For ionic compounds, you may want to refer to the solubility rules in Table 11.1.

(a) LiOH

(b) $O_2$

(c) $Mg(NO_3)_2$

11.11    Which of the following diagrams best represents an aqueous solution of magnesium chloride, $MgCl_2$?

A          B

C

11.12 Which of the following diagrams best represents an aqueous solution of sodium sulfate, Na₂SO₄?

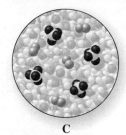

11.13 Write a chemical equation that describes the dissolving of the following substances in water:
(a) solid calcium hydroxide, Ca(OH)₂
(b) nitrogen gas, N₂
(c) liquid methanol, CH₃OH

11.14 Write a chemical equation that describes the dissolving of the following substances in water:
(a) gaseous hydrogen chloride, HCl
(b) chlorine gas, Cl₂
(c) solid magnesium nitrate, Mg(NO₃)₂

## The Solution Process

11.15 What types of forces must be broken when an ionic solid dissolves in water? What new forces are formed?

11.16 What types of forces must be broken when a polar molecular substance dissolves in water? What new forces are formed?

11.17 What types of forces must be broken when a nonpolar molecular substance dissolves in a nonpolar solvent? What new forces are formed?

11.18 Nonpolar substances sometimes dissolve in water to a small extent. Explain why this might occur.

11.19 When ammonium nitrate, NH₄NO₃, dissolves in water, the solution feels cool. What can we say about the relative strengths of the forces between solute and solvent? Between the particles in the pure substances?

11.20 When NaOH dissolves in water, the solution feels warm. What can we say about the relative strengths of the forces between solute and solvent? Between the particles in the pure substances?

11.21 What factors drive the formation of a solution?

11.22 What is entropy? Does it increase or decrease during solution formation?

## Factors That Affect Solubility

11.23 Cooking oil, composed mostly of hydrocarbons, is very insoluble in water. Use the rule "like dissolves like" to explain why these two liquids are immiscible.

11.24 Iodine, I₂, is very soluble in hexane, C₆H₁₄, but not in water. Explain this using the rule "like dissolves like."

11.25 Why are ethanol and water so soluble in one another?

11.26 Grease is hard to wash from your hands using only water. What does this tell us about the intermolecular forces in grease?

11.27 Use the rule "like dissolves like" to predict whether each of the following should be soluble in water.
(a) benzene, C₆H₆
(b) ethylene glycol, HOCH₂CH₂OH
(c) potassium iodide, KI

11.28 Use the rule "like dissolves like" to predict whether each of the following should be soluble in hexane, C₆H₁₄, a nonpolar hydrocarbon.
(a) sodium nitrate, NaNO₃
(b) methanol, CH₃OH
(c) bromine, Br₂

11.29 Use intermolecular forces to explain why ionic compounds are insoluble in nonpolar solvents.

11.30 Use intermolecular forces to explain why nonpolar solutes are insoluble in polar solvents.

11.31 Use intermolecular forces to explain why NaCl is soluble in water.

11.32 Use intermolecular forces to explain why Br₂ is soluble in CCl₄.

11.33 Normal tap water contains a small amount of dissolved oxygen. What happens to the solubility of oxygen when the water is heated?

11.34 What is thermal pollution of water, how does it relate to gas solubility, and how does it affect fish?

11.35 Would you expect the solubility of oxygen and nitrogen at high altitudes (lower atmospheric pressure) to be greater or less than that at sea level? Explain.

11.36 How does the concentration of dissolved oxygen and nitrogen in the bloodstream change as a scuba diver swims to lower depths of the ocean? Explain.

11.37 A scuba diver who has the bends is treated by being placed in a chamber that has a high pressure of air. The air pressure in the chamber is lowered gradually. How does this treatment help the diver?

11.38 Why do carbonated beverages go "flat" after being opened for some time?

11.39 The diagram shows a representation of oxygen dissolved in water at about 25°C. Draw a diagram that represents what the solution might look like after the solution temperature increases.

11.40 The diagram in Question 11.39 shows a representation of oxygen dissolved in water. Draw a diagram that represents what the solution might look like after the partial pressure of $O_2$ above the solution increases.

## Measuring Concentrations of Solutions

11.41 The solubility of KCl at 20°C is 34.0 g per 100 g of water. What is the maximum amount (in grams) of KCl that will dissolve in 50.0 g of water at 20°C?

11.42 The solubility of ammonium chloride at 30°C is 41.1 g per 100 g of water. What is the maximum amount (in grams) of ammonium chloride that will dissolve in 455.0 g of water at this temperature?

11.43 What is the difference between a saturated solution and an unsaturated solution?

11.44 What is the difference between a saturated solution and a supersaturated solution?

11.45 How might you prepare a saturated solution of a substance of unknown solubility?

11.46 The solubility of ammonium chloride increases with temperature. How might you prepare a supersaturated solution of ammonium chloride?

11.47 The solubility of $Ca(OH)_2$ at 30°C is 0.15 g per 100 g of water. Describe the solution formed when 1.0 g $Ca(OH)_2$ is added to 100.0 g of water.

11.48 The solubility of $Ca(OH)_2$ is 0.18 g per 100 g of water at 10°C. Describe the solution formed when 0.15 g $Ca(OH)_2$ is added to 100.0 g of water.

11.49 What is the percent-by-mass concentration of NaCl in a solution that is prepared by adding 15.0 g of NaCl to 90.0 g of water?

11.50 What is the percent-by-mass concentration of sucrose in a solution that is prepared by adding 25.0 g of sucrose to 100.0 g of water?

11.51 What mass of KI is dissolved in 700.0 g of a solution that is 15.0% KI by mass?

11.52 What mass of sodium phosphate is dissolved in 255 g of a solution that is 20.0% by mass $Na_3PO_4$?

11.53 What is the percent-by-mass concentration of acetic acid, $HC_2H_3O_2$, in a vinegar solution that contains 54.50 g acetic acid in a 1.000-L solution? The density of this solution is 1.005 g/mL.

11.54 What is the percent-by-mass concentration of a 100.0-mL $H_3PO_4$ solution that contains 88.20 g of $H_3PO_4$? The density of the solution is 1.40 g/mL.

11.55 For what types of solutions are concentrations commonly expressed as percent by volume? Why?

11.56 When dealing with percent-by-volume concentrations of solutions composed of two liquids, why must the total volume be measured instead of using the sum of the volumes of the component liquids?

11.57 A solution of methanol is prepared by dissolving 35.0 mL of methanol in sufficient water to give a total volume of 115.0 mL. What is the percent-by-volume concentration of methanol?

11.58 A solution of bromine is prepared by dissolving 2.0 mL of bromine in sufficient carbon tetrachloride to give a total volume of 40.0 mL. What is the percent-by-volume concentration of bromine?

11.59 The chemical trichloroethylene (TCE) is a colorless liquid found in very low concentration in drinking water. The EPA has set a maximum concentration of TCE in drinking water at 5 ppb. Consuming water on a regular basis that contains higher concentrations can lead to cancer.
   (a) What is the ppb concentration in a 200.0-L (200.0-kg) sample of water that contains 0.43 mg TCE?
   (b) Is this water safe to drink?

11.60 Polychlorinated biphenyls (PCBs) are absorbed by our bodies from some of the foods we eat. If a 65-kg male is determined to have a PCB level of 190 ppb, what is the mass (in grams) of PCBs in his body?

11.61 How many moles of HCl are contained in 100.0 mL of 0.15 $M$ HCl?

11.62 What volume of 0.25 $M$ NaOH contains 0.050 mol NaOH?

11.63 A solution contains 20.5 g of NaCl dissolved in sufficient water to give a total mass of 166.2 g. What is the molality of this solution?

11.64 How many moles of sucrose, $C_{12}H_{22}O_{11}$, are contained in 50.0 g of a 0.85 $m$ sucrose solution?

11.65 What is the molar concentration of ions in a 1.5 $M$ $KNO_3$ solution? What is the molal concentration of ions in a 1.5 $m$ $KNO_3$ solution?

11.66 What is the molar concentration of ions in a 2.5 $M$ $MgCl_2$ solution? What is the molal concentration of ions in a 2.5 $m$ $MgCl_2$ solution?

11.67 A 250.0-g sample of pond water contains 2.4 mg arsenic. What is the mass percent of arsenic in the pond? How many parts per million of arsenic are in the pond? How many parts per billion?

11.68 A 375.0-g sample of river water taken near an industrial plant contains 37 mg of chromium. What is the mass percent of chromium in the river? How many parts per million of chromium are in the river? How many parts per billion?

11.69 Drinking water may contain a low concentration of lead ion, $Pb^{2+}$, due to corrosion of old lead pipes. The EPA has determined that the maximum safe level of lead ion in water is 15 ppb. Suppose a sample of tap water was determined to have a lead ion concentration of 0.0090 ppm. Assume the density of the solution is 1.00 g/mL.
   (a) What is the concentration of lead ion in the tap water in units of ppb? Is it safe to drink?
   (b) What is the concentration of lead ion in units of mg/mL?
   (c) What mass of lead ion is in 100.0 mL of this drinking water?
   (d) How many moles of lead ion are in 100.0 mL of the water?

11.70 Excessive levels of nitrates found in drinking water can cause serious illness. When nitrates are converted to nitrites in the body, the nitrites can interfere with the oxygen-carrying capacity of the blood, causing blueness of the skin. The EPA has determined that the maximum safe level of nitrate ion in water is 10 ppm. Suppose a sample of tap water was determined to have a nitrate ion concentration of 95 ppb. Assume the density of the solution is 1.00 g/mL.

(a) What is the concentration of nitrate ion in units of ppm? Is it safe to drink?

(b) What is the concentration of nitrate ion in units of mg/mL?

(c) What mass of nitrate ion is in 100.0 mL of this drinking water?

(d) How many moles of nitrate ion are in 100.0 mL of the water?

## Quantities for Reactions That Occur in Aqueous Solution

11.71 One way to remove lead ion from water is to add a source of iodide ion so that lead iodide will precipitate out of solution:

$$Pb^{2+}(aq) + 2I^-(aq) \longrightarrow PbI_2(s)$$

(a) What volume of a 1.0 $M$ KI solution must be added to 100.0 mL of a solution that is 0.15 $M$ in $Pb^{2+}$ ion to precipitate all the lead ion?

(b) What mass of $PbI_2$ should precipitate?

11.72 If a solution of $AgNO_3$ is added to an HCl solution, insoluble AgCl will precipitate:

$$AgNO_3(aq) + HCl(aq) \longrightarrow AgCl(s) + HNO_3(aq)$$

(a) What volume of a 1.20 $M$ $AgNO_3$ solution must be added to 250.0 mL of a 0.100 $M$ HCl solution to precipitate all the chloride ion?

(b) What mass of AgCl should precipitate?

11.73 How many moles of sodium carbonate, $Na_2CO_3$, are required to precipitate the calcium ion from 850.0 mL of a 0.35 $M$ $CaCl_2$ solution? (Begin by writing a balanced chemical equation for the reaction.)

11.74 How many moles of calcium chloride, $CaCl_2$, are required to precipitate the carbonate ion from 2.5 L of a 0.75 $M$ $Na_2CO_3$ solution? (Begin by writing a balanced chemical equation for the reaction.)

11.75 What volume of 0.1050 $M$ NaOH is required to neutralize 20.00 mL of 0.1500 $M$ $H_2SO_4$? (Begin by writing a balanced chemical equation for the reaction.)

11.76 What volume of 0.850 $M$ $H_2SO_4$ is required to neutralize 20.00 mL of 0.1033 $M$ NaOH? (Begin by writing a balanced chemical equation for the reaction.)

11.77 When 20.00 mL of $H_2SO_4$ is titrated with 0.9854 $M$ NaOH, 35.77 mL of NaOH is required to neutralize the $H_2SO_4$. What is the molarity of the $H_2SO_4$ solution?

11.78 When 20.00 mL of 1.005 $M$ $H_2SO_4$ is titrated with a solution of NaOH, 44.68 mL of NaOH is required to neutralize the $H_2SO_4$. What is the molarity of the NaOH solution?

11.79 What volume of 0.1000 $M$ NaOH is required to neutralize each of the following solutions?

(a) 10.00 mL of 0.1000 $M$ HCl

(b) 15.00 mL of 0.3500 $M$ $HNO_3$

(c) 25.00 mL of 0.0500 $M$ $H_3PO_4$

11.80 What volume of 0.5000 $M$ HCl is required to neutralize each of the following solutions?

(a) 35.00 mL of 0.0500 $M$ NaOH

(b) 10.00 mL of 0.200 $M$ $Ba(OH)_2$

(c) 15.00 mL of 0.2500 $M$ $NH_3$

## Colligative Properties

11.81 What is a colligative property?

11.82 List some colligative properties.

11.83 When osmosis occurs between two solutions of different concentrations, what would you observe about each solution? Explain your observations in terms of molecular motion.

11.84 When osmosis occurs between an aqueous solution and pure water, can both sides ever become equal in concentration? Explain.

11.85 Describe the process of reverse osmosis.

11.86 What is desalination?

11.87 What will happen to a blood cell that is placed in an aqueous solution that has a high salt concentration relative to the blood cell concentration?

11.88 What will happen to a blood cell that is placed in pure water?

11.89 What happens to each of the following properties of a liquid when a solute is dissolved in the liquid to make a solution?

(a) vapor pressure

(b) boiling point

(c) freezing point

11.90 What happens to each of the following properties of a solution when the concentration of solute is decreased?

(a) osmotic pressure

(b) boiling point

(c) freezing point

11.91 Sweet tea is often made by dissolving a lot of sugar in water, brewing the tea, and then chilling. Explain what happens to the boiling point of water when sugar is added.

11.92 Some arctic fish produce a substance similar to the ethylene glycol used as antifreeze in cars. In terms of the freezing point of water, explain how the compound might help them survive in their habitat.

11.93 Compare a 2.5 $m$ solution of glucose, $C_6H_{12}O_6$, to pure water.

(a) What is the difference in boiling points?

(b) What is the boiling point of the 2.5 $m$ glucose solution?

11.94 Compare a 5.0 $m$ solution of ethylene glycol, $HOCH_2CH_2OH$, to pure water.

(a) What is the difference in freezing points?

(b) What is the freezing point of a 5.0 $m$ ethylene glycol solution?

11.95 Which solution, 1.0 $m$ NaCl or 1.0 $m$ glucose, $C_6H_{12}O_6$, should have the higher boiling point? Explain.

**11.96** Which solution, 1.0 *m* NaNO₃ or 1.0 *m* MgCl₂, should have the higher boiling point? Explain.

**11.97** The following diagrams represent a solution of NaCl, a solution of glucose, and a more dilute solution of glucose. Which represents the aqueous solution with the lowest freezing point? Which solution is which?

**11.98** The following diagrams represent a solution of MgCl₂, a solution of NaNO₃, and a more dilute solution of NaNO₃. Which represents the aqueous solution with the lowest freezing point? Which solution is which?

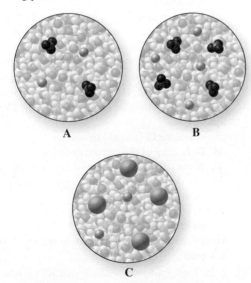

**11.99** Determine the moles of solute particles dissolved in 1.0 kg of water for each of the following solutions.
   (a) 2.0 *m* CH₃OH
   (b) 2.5 *m* NaCl
   (c) 1.0 *m* Al(NO₃)₃

**11.100** Determine the moles of solute particles dissolved in 1.0 kg of water for each of the following solutions.
   (a) 2.0 *m* HCl
   (b) 0.15 *m* Na₂SO₄
   (c) 1.0 *m* C₁₂H₂₂O₁₁

**11.101** Which solution, 1.0 *m* MgCl₂ or 1.0 *m* NaNO₃, should have the greater osmotic pressure? Explain.

**11.102** Which solution, 1.0 *m* NaCl or 1.0 *m* CH₃OH, should have the greater osmotic pressure? Explain.

## Additional Questions

**11.103** Vitamin D is soluble in fat tissue, which is primarily nonpolar. What does this tell us about the polarity of vitamin D molecules?

**11.104** Vitamin B is a water-soluble vitamin. What does this tell us about the polarity of vitamin B?

**11.105** When an ionic solid dissolves in water, which of the processes involved in the solution process require energy? Which release energy?

**11.106** What changes in temperature and pressure increase the solubility of a gas in solution?

**11.107** The solubility of NaCl at 30°C is 36.3 g per 100 g of water. What mass of NaCl is dissolved in a saturated solution that contains 500.0 g of water?

**11.108** The solubility of KNO₃ increases as the temperature rises. If a saturated solution of KNO₃ is heated to a higher temperature, will it still be a saturated solution? Explain.

**11.109** What is the molarity of a concentrated HBr solution if the solution is 48.0% HBr by mass and has a density of 1.50 g/mL?

**11.110** A solution of sulfuric acid contains 24.0% by mass of H₂SO₄ and has a density of 1.17 g/mL. What is the molarity of H₂SO₄ in this solution?

**11.111** Which of the following concentration units has values that will not change when the temperature of the solution is changed? Explain your reasoning for each.
   (a) percent by mass          (c) molarity
   (b) percent by volume        (d) molality

**11.112** A 1.0 *M* acetic acid solution has a density of 1.005 g/mL. What is the molality (*m*) of this solution?

**11.113** For dilute aqueous solutions, molality and molarity concentration values are similar. Why?

**11.114** For each of the following combinations of solutions, determine the mass of silver carbonate, Ag₂CO₃, that should precipitate. (Begin by writing a chemical equation for the reaction.)
   (a) 100.0 mL of 0.200 *M* AgNO₃ and 100.0 mL of 0.200 *M* Na₂CO₃
   (b) 200.0 mL of 0.200 *M* AgNO₃ and 100.0 mL of 0.200 *M* Na₂CO₃
   (c) 100.0 mL of 0.200 *M* AgNO₃ and 200.0 mL of 0.200 *M* Na₂CO₃

**11.115** For each of the following combinations of solutions, determine the mass of barium sulfate, BaSO₄, that should precipitate. (Begin by writing a chemical equation for the reaction.)
   (a) 500.0 mL of 0.100 *M* BaCl₂ and 90.0 mL of 0.500 *M* K₂SO₄
   (b) 100.0 mL of 0.100 *M* BaCl₂ and 100.0 mL of 0.500 *M* K₂SO₄
   (c) 100.0 mL of 0.100 *M* BaCl₂ and 500.0 mL of 0.500 *M* K₂SO₄

11.116 What mass of NaOH is required to neutralize 1.8 L of 2.0 $M$ HCl?

11.117 Sodium bicarbonate is sometimes used to neutralize acid spills. If 2.0 L of 6.0 $M$ $H_2SO_4$ is spilled, what mass of sodium bicarbonate, $NaHCO_3$, is required to neutralize all the $H_2SO_4$?

11.118 What is the boiling point of a 3.0 $m$ $Ca(NO_3)_2$ solution in water?

11.119 The freezing points of two NaCl solutions were determined to be $-1.15°C$ for solution A and $-4.10°C$ for solution B. Which solution has the higher concentration of NaCl?

11.120 If a 100.0-g sucrose, $C_{12}H_{22}O_{11}$, solution has a boiling point of 101.5°C, how many moles of sucrose are in the solution? How many grams?

11.121 A salad dressing can be made by shaking together olive oil and vinegar. Explain why this salad dressing will separate into two layers upon standing undisturbed for some time.

11.122 A power plant is located next to a river. Explain why the water downstream from the power plant contains less dissolved oxygen than water upstream from the power plant.

11.123 Arrange the following solutions in order of decreasing freezing point: 0.20 $m$ NaCl, 0.10 $m$ $Na_2SO_4$, 0.10 $m$ KBr, 0.15 $m$ $VCl_3$.

11.124 The vapor pressure of a 0.10 $m$ solution of $HCO_2H$ is higher than the vapor pressure of a 0.10 $m$ solution of HBr. Which substance is a weak electrolyte and which is a strong electrolyte?

11.125 Explain in terms of colligative properties why limp lettuce becomes crisp after it is soaked in water.

11.126 A small crystal of sodium acetate, $NaCH_3CO_2$, is added to a clear, colorless solution of $NaCH_3CO_2$. The solution then begins to precipitate $NaCH_3CO_2$. Is the original solution unsaturated, saturated, or supersaturated? How would you classify the solution after precipitation is complete?

11.127 Barium ions, $Ba^{2+}$, are poisonous when ingested. Why can a suspension of $BaSO_4$ in water be swallowed to increase contrast for X-ray examination of the digestive system?

11.128 One commercial process for making sodium carbonate produces large quantities of calcium chloride as a by-product, so the calcium chloride is available at low expense. The city of Chicago maintains large stockpiles of calcium chloride, but the city of New Orleans does not. What is a common use of calcium chloride that explains these regional differences?

11.129 Vitamin A is a largely nonpolar molecule. Is vitamin A more likely to be found in the body in the bloodstream or in fatty tissues?

11.130 An aqueous solution of glucose, $C_6H_{12}O_6$, called D5W, is used for intravenous injection. D5W contains 54.30 g of glucose per liter of solution. What is the molar concentration of glucose in D5W?

## Concept Review Questions

11.131 Which of the following describes what happens when magnesium nitrate, $Mg(NO_3)_2$, dissolves in water to a concentration of 0.20 $M$?
  A. The solution is 0.20 $M$ in molecules of $Mg(NO_3)_2$.
  B. The solution is 0.20 $M$ in both $Mg^{2+}$ and $(NO_3)_2^{2-}$ ions.
  C. The solution is 0.20 $M$ in $Mg^{2+}$ ions and 0.40 $M$ in $NO_3^-$ ions.
  D. The solution is 0.40 $M$ in total ions present.
  E. Magnesium nitrate is insoluble so no ions exist in the solution.
Explain what is wrong with the incorrect responses.

11.132 Which of the following is most soluble in hexane, $C_6H_{14}$?
  A. $CaCl_2$
  B. $CCl_4$
  C. $CH_3OH$
  D. vitamin C, a compound with several $-OH$ groups
  E. $KNO_3$
What type of solvent would dissolve the other compounds?

11.133 When KBr dissolves in water, which of the following statements best describes all the processes that occur?
  A. Energy is absorbed to break the ionic bonds in KBr and the hydrogen bonds in water; energy is released when ion-dipole forces form between the ions and water.
  B. Energy is released in the breaking of ionic bonds in KBr and the hydrogen bonds in water.
  C. Energy is absorbed to break the ionic bonds in KBr and water.
  D. Energy is absorbed to break the ionic bonds in KBr.
  E. Energy is absorbed to break the ion-dipole forces between the ions and water.
Explain what is wrong with the incorrect responses.

11.134 Lead(II) iodide, $PbI_2$, is a yellow solid with a solubility of about 0.1 g/100 mL $H_2O$. What would you expect to observe if 0.3 g of solid $PbI_2$ is mixed with 250 mL of water?
  A. All of the $PbI_2$ would dissolve.
  B. Only 0.1 g of $PbI_2$ would dissolve.
  C. About 0.05 g of yellow solid would remain undissolved.
  D. The dissolved solid would exist in solution as Pb atoms and $I_2$ molecules.
  E. According to the solubility rules, $PbI_2$ is insoluble so no $PbI_2$ will dissolve.
Modify each incorrect statement to make it a true statement.

11.135 A 15.0 percent-by-mass solution of KCl is needed for a procedure in a laboratory. Which of the following would make a solution with this concentration?
  A. 30.0 g KCl + 200.0 g $H_2O$
  B. 30.0 mol KCl + 200.0 g $H_2O$
  C. 30.0 g $H_2O$ + 170.0 g KCl
  D. 30.0 mol KCl + 170.0 mol $H_2O$
  E. 30.0 g KCl + 170.0 g $H_2O$
If solutions were prepared as described in the incorrect responses, what are their percent-by-mass concentrations?

11.136  Which of the following relationships corresponds to molality ($m$)?

A. $\dfrac{\text{moles of solute}}{\text{kilograms of solvent}}$

B. $\dfrac{\text{moles of solute}}{\text{liters of solution}}$

C. $\dfrac{\text{grams of solution}}{\text{milliliters of solution}}$

D. $\dfrac{\text{grams of solute}}{\text{volume of solution}} \times 100$

E. $\dfrac{\text{volume of solute}}{\text{volume of solution}} \times 100$

What terms describe the relationships that do not correspond to molality?

11.137  A $Mg(NO_3)_2$ solution is prepared by mixing 30.0 g of $Mg(NO_3)_2$ with enough water to give a total volume of 250.0 mL. Which of the following mathematical operations would correctly determine the molarity of the solution?

A. $\dfrac{30.0 \text{ g Mg(NO}_3)_2}{250.0 \text{ mL solution}}$

B. $\dfrac{30.0 \text{ g Mg(NO}_3)_2 \times \dfrac{148.33 \text{ mol Mg(NO}_3)_2}{1 \text{ g Mg(NO}_3)_2}}{250.0 \text{ mL solution}}$

C. $\dfrac{30.0 \text{ g Mg(NO}_3)_2 \times \dfrac{1 \text{ mol Mg(NO}_3)_2}{148.33 \text{ g Mg(NO}_3)_2}}{250.0 \text{ mL solution}}$

D. $\dfrac{30.0 \text{ g Mg(NO}_3)_2 \times \dfrac{1 \text{ mol Mg(NO}_3)_2}{148.33 \text{ g Mg(NO}_3)_2}}{0.2500 \text{ L solution}}$

E. $\dfrac{0.2500 \text{ L solution}}{30.0 \text{ g Mg(NO}_3)_2 \times \dfrac{1 \text{ mol Mg(NO}_3)_2}{148.33 \text{ g Mg(NO}_3)_2}}$

Explain what is wrong with the incorrect responses.

11.138  Silver sulfate, $Ag_2SO_4$, can be prepared by mixing aqueous solutions of $AgNO_3$ and $K_2SO_4$:

$$2AgNO_3(aq) + K_2SO_4(aq) \longrightarrow Ag_2SO_4(s) + 2KNO_3(aq)$$

If 25.0 mL of 1.00 $M$ $AgNO_3$ is mixed with 25.0 mL of 2.00 $M$ $K_2SO_4$, which of the following statements describes what will happen?

A. 0.0500 mol of $K_2SO_4$ completely reacts with 0.0250 mol of $AgNO_3$.

B. 0.0125 mol of $Ag_2SO_4$ should precipitate.

C. 0.100 mol of aqueous $KNO_3$ should be produced.

D. The final reaction mixture will contain products and unreacted $AgNO_3$.

E. 0.0250 mol of $K_2SO_4$ remains unreacted.

Modify each incorrect statement to make it a true statement.

11.139  Which aqueous solution is expected to have the lowest freezing point?

A. 1.5 $m$ KBr

B. 1.5 $m$ $Al(NO_3)_3$

C. 1.5 $m$ $Na_2CO_3$

D. 1.5 $m$ $CH_3OH$

E. 1.5 $m$ $MgCl_2$

Compare the effects of each of these solutions on osmotic pressure, vapor pressure, and boiling point.

# Reaction Rates and Chemical Equilibrium

**12.1 Reaction Rates**

**12.2 Collision Theory**

**12.3 Conditions That Affect Reaction Rates**

**12.4 Chemical Equilibrium**

**12.5 The Equilibrium Constant**

**12.6 Le Chatelier's Principle**

**Chapter Review**

**Questions and Problems**

©plherrera/E+/Getty Images

As Ellen and Chad are driving to the mall on a hot summer day, they hear an ozone alert announced on the radio. The announcer advises listeners—especially those with respiratory problems—to stay indoors as much as possible.

"Why is ozone so bad?" Ellen asks. "Isn't it just another form of oxygen, and doesn't it shield us from the ultraviolet radiation that causes skin cancer?" As Ellen and Chad discuss these questions, they reach the top of a hill. Looking down on the city in the valley below, Chad notices more smog than usual and wonders if it has anything to do with the ozone warning.

Ozone, $O_3$, is a form of oxygen that is more reactive than diatomic oxygen, $O_2$. Breathing ozone damages alveoli, the key oxygen exchange structures in the lungs. Even moderate concentrations of ozone make breathing more difficult, especially during exercise. Higher concentrations can permanently damage lung tissue and increase susceptibility to bacterial pneumonia.

Ozone is only bad for us when its level increases in the air layer closest to Earth's surface, the troposphere. A small amount of ozone occurs naturally, mostly as a product of lightning storms. Excess ozone is produced when nitrogen oxides and other pollutant gases from car exhausts and electrical power plants combine with oxygen in the presence of sunlight. When there is little wind to disperse pollutant gases, the ozone level rises in large cities on hot days. Chad was right. The thick smog and the elevated ozone level go together.

Ellen recalls a news report about the harmful effects of ozone *depletion* in the stratosphere. The stratosphere is a layer of the atmosphere 18 to 50 km (10 to 30 mi) above Earth (Figure 12.1). The ozone layer, which contains the highest concentrations of ozone, is in the upper stratosphere. It is beneficial to life on Earth because it acts to filter out harmful ultraviolet (UV) radiation from the Sun. How does it do that? Ozone molecules absorb some UV radiation that otherwise would go to Earth's surface. When an ozone molecule absorbs UV radiation,

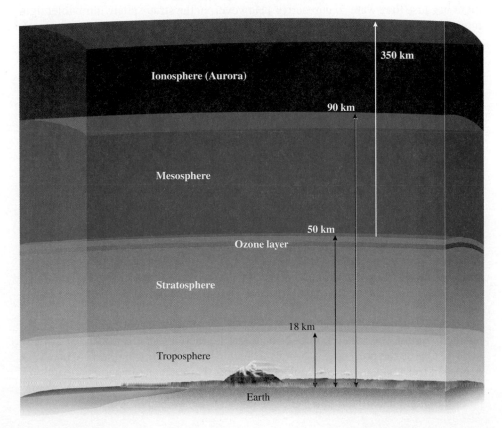

**FIGURE 12.1** The atmosphere is a blanket of air that extends more than 560 km above Earth's surface. Its four regions differ in chemical composition and density. The troposphere contains the air we breathe. It has the greatest density of gas molecules. The stratosphere, which extends from 18 to 50 km above Earth's surface, is less dense. The ozone layer is part of the upper stratosphere.

it splits into an oxygen molecule and an oxygen atom. The single oxygen atom is very unstable and reacts quickly with another $O_3$ molecule to form two $O_2$ molecules:

$$O_3(g) \xrightarrow{\text{UV light}} O_2(g) + O(g)$$

$$O(g) + O_3(g) \longrightarrow 2O_2(g)$$

The absorption of UV radiation destroys ozone, so for its concentration in the stratosphere to remain constant, ozone must also continually be *created* by chemical reactions. Sunlight helps to create ozone from $O_2$. Radiation from the Sun can cause $O_2$ to split into two oxygen atoms. Then, each oxygen atom can react with an $O_2$ molecule to form an $O_3$ molecule:

$$O_2(g) \xrightarrow{\text{UV light}} 2O(g)$$

$$O(g) + O_2(g) \longrightarrow O_3(g)$$

The amount of ozone in even the most concentrated areas of the stratosphere is about 8 ppm (or 8000 ppb), which is quite small compared to the amount of $O_2$, which is 20,000 ppm. The *relative amounts* of ozone and oxygen remain constant as the reactions occur repeatedly, breaking down and re-forming about 30 million kg of ozone every day.

Many factors influence the delicate balance between the amounts of $O_3$ and $O_2$ in the stratosphere. Changes in wind patterns and temperatures cause the balance to fluctuate, so the ozone layer is not the same all over the world and at all times of the year. More worrisome, Chad tells Ellen, is the effect of human-made pollutants such as chlorofluorocarbons (CFCs). Used commercially as refrigerants and propellants, CFCs upset the ozone balance by destroying more ozone molecules than can be replaced by natural processes.

CFCs are very stable and do not react under normal conditions in the troposphere around us. (Before 1973, they were thought to be harmless to the environment because they were so unreactive.) However, in the stratosphere ultraviolet light interacts with a CFC molecule causing a carbon-chlorine bond to break:

$$CF_2Cl_2(g) \xrightarrow{\text{UV light}} CF_2Cl(g) + Cl(g)$$

The chlorine atom reacts readily with ozone to produce $O_2$ and ClO:

$$Cl(g) + O_3(g) \xrightarrow{\text{UV light}} ClO(g) + O_2(g)$$

The ClO is unstable and reacts with another molecule of $O_3$, forming an $O_2$ molecule and a Cl atom:

$$ClO(g) + O_3(g) \xrightarrow{\text{UV light}} 2O_2(g) + Cl(g)$$

Two ozone molecules are destroyed, and the Cl atom that started it all is reproduced. It's available to destroy two more ozone molecules. Once a Cl atom forms in the stratosphere, it can destroy thousands of ozone molecules. Because the chlorine atoms speed up the reaction and can be reused, Cl is considered to be a *catalyst* for the destruction of ozone. Production of CFCs was banned on a global scale in 1995 because of the harmful effects of even small concentrations of Cl in the stratosphere.

When, Ellen wonders, will the relative rates of ozone destruction and re-formation balance again? The reaction rate, or how fast a reaction occurs, depends on many factors. One is the presence of a catalyst, such as the Cl atoms in the destruction of ozone. In this chapter, we will look at two important aspects of chemical reactions: reaction rate and chemical equilibrium. We will learn how catalysts, as well as temperature and concentration, influence changes at the molecular level.

An example of a CFC is dichlorodifluoromethane, $CF_2Cl_2$:

## Questions for Consideration

**12.1**  What conditions affect reaction rates?

**12.2**  How do molecular collisions explain chemical reactions?

**12.3**  How do concentration, temperature, and catalysts affect molecular collisions and reaction rates?

**12.4**  What is chemical equilibrium?

**12.5**  What is the state of equilibrium for a specific reaction?

**12.6**  What happens when we disturb a system at equilibrium?

 **Math Tools Used in This Chapter**

Graphing (Math Toolbox 9.1)

## 12.1  Reaction Rates

Some chemical reactions occur in seconds, while others take days, or even years (Figure 12.2). The reactions in the ozone layer are rapid, occurring millions of times daily. Combustion reactions of fuels with oxygen also occur rapidly, but corrosion of metals with oxygen, such as iron in a rusting car, is comparatively slow. Active metals react with water immediately, while a similar reaction with less active metals, such as aluminum and zinc, takes longer. **Reaction rate,** or *rate of reaction,* is a measure of how fast a reaction occurs.

A reaction can occur at different rates under different conditions. Conditions that affect the rate of reaction include temperature, reactant concentration, surface area, and the presence of a catalyst. Higher temperatures generally cause reactions to occur faster. For example, milk and other foods spoil more quickly when left in a warm place. Can you think of other reactions that occur faster at higher temperatures? Or slower at lower temperatures?

Concentration also influences reaction rate. Consider the reaction that occurs when baking soda, $NaHCO_3$, is added to vinegar. Vinegar is a solution of acetic acid, $CH_3CO_2H$, in water.

$$NaHCO_3(s) + CH_3CO_2H(aq) \longrightarrow CO_2(g) + NaCH_3CO_2(aq) + H_2O(l)$$

How does the concentration of acetic acid affect its reaction with baking soda (Figure 12.3)? Increasing the concentration of a reactant generally increases the reaction rate. Can you think of other reactions that occur faster when the concentrations of reactants are increased?

The rate of a reaction is similar in some ways to the driving speed of a car. We usually describe driving speed as miles per hour (mi/h), and reaction rate is often described as change in molarity per second ($M$/s). We can determine reaction rate by measuring the change in concentration of a reactant or product during a measured time interval.

A light stick starts to glow when the seal on a contained liquid is broken, allowing it to mix with another substance. The chemical reaction gives off light energy. Why does a light stick glow longer if placed in the freezer after the reaction has begun?

(both): ©Brian Moeskau/Moeskau Photography

**FIGURE 12.2**  Reactions occur at a wide range of rates. For example, the sulfur in a safety match reacts with oxygen more quickly than iron metal reacts with oxygen in corrosive reactions.
(left): ©Design Pics/PunchStock; (right): © Mark Dierker/Bear Dancer Studios

**FIGURE 12.3** Sodium bicarbonate reacts with acetic acid to form $CO_2$ gas. The photos show the reaction at its beginning and after 10 seconds, comparing 1 *M* and 3 *M* acetic acid as reactants. Because the reaction with 3 *M* acetic acid produces more gas over the same time period, we can see that its reaction is faster, based on the greater amount of gas in that balloon. (both): ©Jim Birk

1 *M*        3 *M*
Before reaction

1 *M*        3 *M*
After 10 seconds

A            B

**FIGURE 12.4** (A) The heated solid iron nail exposes less surface area and fewer atoms, so it reacts slowly with oxygen gas as indicated by a mere glow. (B) The greater surface area in heated steel wool exposes many atoms to the oxygen gas, so they ignite explosively. (both): ©McGraw-Hill Education/Stephen Frisch

**FIGURE 12.5** Without a catalyst, hydrogen peroxide, $H_2O_2$, decomposes slowly. The enzyme called catalase catalyzes the decomposition of hydrogen peroxide, so the reaction occurs rapidly. When $H_2O_2$ is placed on a piece of liver, the catalase in the liver catalyzes the decomposition of $H_2O_2$, releasing bubbles of oxygen gas. ©Brian Moeskau/Moeskau Photography

If the reactant is a solid, increasing its surface area also increases reaction rate. Metals in their granular form have greater surface area and react more quickly with air than large pieces of metal. The aluminum used in fireworks is in a granular form so that it will burn quickly. The effect of increased surface area for the reaction of iron with oxygen is shown in Figure 12.4.

Catalysts increase the rate of a reaction. The chemical catalysts in the catalytic converters of automobiles increase the rate at which the polluting gases NO and $NO_2$ are converted to harmless $N_2$ and $O_2$ gases. In living systems, enzymes act as catalysts, speeding up the countless chemical reactions that are necessary to sustain life. For example, the harmful compound hydrogen peroxide, $H_2O_2$, can sometimes form in the human body. A catalyst called *catalase* speeds its decomposition into oxygen and water (Figure 12.5). The effects of catalysts are not always desirable— as, for example, the role of chlorine atoms from CFCs in the depletion of Earth's ozone layer.

## 12.2 Collision Theory

Why does the rate of chemical reactions vary so widely? **Collision theory** states that in order for a reaction to occur, reactant molecules must collide in the proper orientation and with sufficient energy. Because only a fraction of the collisions meet these requirements, only a few result in a reaction and yield products. Consider, for example, the reaction between $O_3$ and NO that occurs in smog:

$$O_3(g) + NO(g) \longrightarrow O_2(g) + NO_2(g)$$

In order for this reaction to occur, an $O_3$ molecule and an NO molecule must first collide. Of the many such collisions that actually happen, only some of them have the orientation required for the reaction to occur. Convince yourself this is plausible. Draw the structures of the reactants and products. Which bonds must break and which bonds should form during this reaction? How should the reactant molecules be oriented in order for products to form?

Examine this collision:

NO                    $O_3$

This shows the oxygen atom from NO (left) heading straight toward an oxygen atom from $O_3$ (right); but the equation tells us that a new nitrogen-oxygen bond must be formed. This collision will fail because the atoms aren't lined up correctly.

An effective collision must bring together a nitrogen atom and a terminal oxygen atom so that an oxygen atom can be transferred to the nitrogen atom, leaving an $O_2$ molecule behind. Does the following collision have a proper orientation?

**ANIMATION:** Trajectory of Reaction

### Student Hot Spot

Student data indicate you may struggle with collision theory. Access the SmartBook to view additional Learning Resources on this topic.

Yes, it does. The nitrogen atom on NO will collide with a terminal oxygen atom on $O_3$. It is lined up just right so that an oxygen atom from the $O_3$ molecule can be transferred to the nitrogen atom in the NO molecule.

If an NO molecule and an $O_3$ molecule collide with the proper orientation, will a reaction automatically occur? No, not unless they collide with sufficient energy. In a given sample, some reactant molecules have more kinetic energy than others. Usually some fraction of the molecules has enough energy to have effective collisions. Reactant molecules that collide with the proper orientation and sufficient energy can form a product.

Let's take a closer look at the energy required for a reaction. In a chemical reaction, bonds in the reactants must be broken so that new bonds in the products can form. Different reactions occur at different rates because the amount of energy required to break bonds in the reactant molecules is different. As we discussed in Chapter 6, an energy diagram shows the relative energies of the reactants and products (Figure 12.6). This type of energy diagram, although useful, shows only the *net* energy change for the reaction, not the energy changes that occur *during* it.

The collision requirements for a reaction are like playing darts. The dart must have sufficient energy (you must throw it fast enough), and it must have the proper orientation (you must aim it correctly).

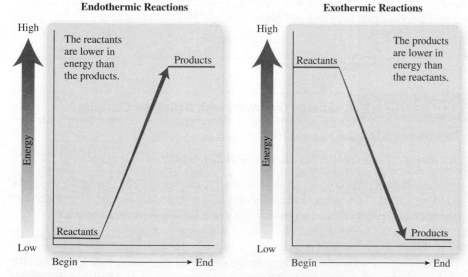

**FIGURE 12.6** In endothermic reactions, the average energy of the products is greater than that of the reactants. Energy is absorbed. In exothermic reactions, the average energy of the products is less than that of the reactants. Energy is released. These energy diagrams show only the difference in energy between the reactants and products, not the energy changes that occur during the reaction.

**FIGURE 12.7** This diagram for an exothermic reaction shows the activation energy, $E_a$, the energy barrier that must be overcome by reactants before they can be converted to products. This is the amount of energy required to form the activated complex, which exists briefly as bonds in the reactants break and bonds in the products form. What would a diagram look like for an endothermic reaction?

**ANIMATION:** Exothermic and Endothermic Reactions

**ANIMATION:** Activation Energy

In both endothermic and exothermic reactions, reactants must overcome an *energy barrier* before they can change to products (Figure 12.7). Energy is needed to break bonds in reactants before the reactants can be converted into products. The minimum amount of energy needed to overcome this energy barrier is called the **activation energy, $E_a$.** If we compare two similar reactions at the same temperature, the reaction with the lower activation energy is the faster reaction.

Reactant molecules that collide with the proper orientation and sufficient energy form an **activated complex.** It is a short-lived, unstable, high-energy chemical species that must be achieved before products can form. The activated complex is generally designated by a chemical formula in brackets with a superscript double dagger ($\ddagger$), such as $\{O_3NO\}^{\ddagger}$ for the reaction of $O_3$ and NO in Figure 12.7. In most reactions the activated complex is so short-lived, its actual structure is unknown. However, we can imagine that it is a chemical species that is in the process of forming new bonds while breaking old ones.

| EXAMPLE 12.1 | Energy Diagrams with Activated Complexes |
|---|---|

The following reaction is exothermic:

$$2HI(g) \longrightarrow H_2(g) + I_2(g)$$

Draw an energy diagram that shows the relative energies of the reactants, products, and the activated complex. Label the diagram with molecular representations of the reactants, products, and a possible structure for the activated complex.

**Solution:**

The reaction is exothermic, so the average energy of the products is lower than the average energy of the reactants. The energy of the activated complex corresponds to the top of the energy barrier. A possible structure for the activated complex shows a species that would result from a collision with the proper orientation, allowing H—H and I—I bonds to form while H—I bonds break.

$$2HI(g) \longrightarrow H_2(g) + I_2(g)$$

**Consider This 12.1**

If this reaction runs in reverse, starting with $H_2$ and $I_2$ gases and producing HI, would the energy diagram for the reverse reaction show an endothermic or exothermic reaction?

**Practice Problem 12.1**

The following reaction is an endothermic reaction:

$$2NO_2(g) \longrightarrow 2NO(g) + O_2(g)$$

Draw an energy diagram that shows the relative energies of the reactants, products, and the activated complex. Label the diagram with molecular representations of the reactants, products, and a possible structure for the activated complex.

**Further Practice:** Questions 12.19 and 12.20 at the end of the chapter

Each reaction has its own specific energy diagram and value for the activation energy. Reactions with large activation energies tend to be slow, because a relatively small fraction of reactants have sufficient energy for an effective collision. Reactions with small activation energies tend to be fast, because a large fraction of reactants have sufficient energy for an effective collision. However, conditions such as concentration and temperature change the rate of a reaction. They change the number of collisions that are effective without changing the activation energy.

What change in reaction conditions can alter activation energy for a given reaction? Only the addition of a catalyst can change the activation energy.

## 12.3  Conditions That Affect Reaction Rates

The conditions that most often affect reaction rates are changes in concentration, surface area, and temperature. The presence or addition of catalysts is also a significant factor. We will next explore how collision theory explains all these effects.

### Concentration and Surface Area

A reaction may go faster when the concentration of one or more of the reactants increases. Such an increase in concentration increases the number of reactants per unit volume. Because molecules are closer together, the number of collisions per

**FIGURE 12.8** The collision rate increases when the concentration of reactants ($O_3$ and NO) increases. The fraction of effective collisions, however, remains constant (1:3 in this example), because the kinetic energy of the system is unchanged.

The ratio of effective collisions is 1:3.

The ratio of effective collisions is 2:6, or 1:3.

Which has more surface area, a cube of sugar or granular sugar? Which dissolves more quickly?

unit time increases. As total collisions increase, the number of molecules that have the orientation and energy required for the reaction also increases. However, the *fraction* of effective collisions remains the same, because the temperature and kinetic energy are constant (Figure 12.8).

The effect of increasing the surface area of a solid in a reaction is similar to increasing the concentration, as shown in Figure 12.4. For example, a ground-up piece of iron has more iron atoms exposed to the air than an iron nail, so it will rust more quickly. Increasing the surface area increases the number of atoms that are exposed so that they can collide with other reactants.

## Temperature

Why does a reaction proceed more rapidly at higher temperatures? As you learned in Chapter 10, the average kinetic energy of a substance increases when the temperature rises. An increase in kinetic energy causes the reaction rate to increase in two ways: It increases the collision rate, *and* it increases the fraction of collisions that are effective (Figure 12.9). The collision rate increases because the molecules

**FIGURE 12.9** (A) Increasing the temperature of the reactants increases the average kinetic energy. At the higher temperature, a greater fraction of the reactant molecules have sufficient energy ($E_a$) for a reaction (both colored regions). (B) Increasing the temperature of the reactants also increases their average velocity. Molecules that move faster collide more frequently, so the collision rate also increases.

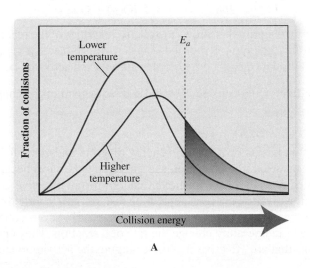

A

Lower temperature
Lower collision rate
Lower rate of effective collisions
Smaller fraction of effective collisions

Higher temperature
Greater collision rate
Greater rate of effective collisions
Larger fraction of effective collisions

B

move faster at higher temperatures, so they collide more frequently. The fraction of effective collisions also increases, because the average kinetic energy of the molecules increases. Consequently more of the reactant molecules attain the required activation energy. Temperature effects on reaction rates are significant. For a typical reaction, the rate approximately doubles for every 10°C increase in temperature.

> The chirp rate of the snowy tree cricket is temperature-dependent. It is about 17 chirps per 15 s in cold weather. In very warm weather, it is about 55 chirps per 15 s.

## EXAMPLE 12.2 Collision Rates and the Fraction of Effective Collisions

Suppose the collision rate between reactant molecules A and B at 25°C is 10,000 collisions per second. The number of effective collisions at this same temperature is 100 per second. The fraction of effective collisions is 100 in 10,000. How will each of the following affect the *total number of collisions* and the *fraction of effective collisions* between molecules A and B?

(a) The temperature increases.
(b) The concentration of reactant A increases.

**Solution:**

(a) When the temperature rises, the average kinetic energy and average velocity of the molecules increase. Increased molecular speed causes an increase in the rate of both molecular collisions and the fraction of effective collisions. Thus, we expect a total number of collisions greater than 10,000 and a fraction of effective collisions greater than 100 in every 10,000.
(b) When the concentration of A increases, the number of collisions between A and B also increases, because more A molecules can make contact with each B molecule. Thus, we expect a total number of collisions greater than 10,000 and a greater number of effective collisions. However, the fraction of effective collisions will stay equal to 100 in every 10,000, because no energy has been added to the system.

**Consider This 12.2**

Suppose A and B are both gases and the reaction volume is reduced at constant temperature. How would this change the concentration of reactants? How would this change affect the *total number of collisions* and the *fraction of effective collisions*?

**Practice Problem 12.2**

How will each of the following affect the *total number of collisions* and the *fraction of effective collisions* between molecules A and B as described in the example statement?

(a) The temperature decreases.
(b) The surface area of reactant A increases.

**Further Practice:** Questions 12.21 and 12.22 at the end of the chapter

---

In Example 12.3, you will use collision theory to help Ellen and Chad solve problems in the laboratory.

## EXAMPLE 12.3 Using Collision Theory to Explain Reaction Rates

In her chemistry lab, Ellen is determining ways to increase the rate of reaction of magnesium metal with hydrochloric acid:

$$Mg(s) + 2HCl(aq) \longrightarrow MgCl_2(aq) + H_2(g)$$

The first time she runs the reaction, she places a 5.0-g strip of magnesium metal in a 0.10 $M$ HCl solution. Suggest ways that Ellen can increase the rate of this reaction. For each suggestion, use collision theory to explain why the reaction rate should increase.

**Solution:**

One way to increase the reaction rate is to increase the concentration of HCl, possibly to 0.20 $M$ or higher. The increased concentration should increase the collision rate between aqueous hydrogen ions and magnesium atoms. Although Ellen cannot increase the concentration of magnesium metal, she can increase its surface area. Cutting the strip into tiny pieces would make more magnesium atoms accessible for collisions. The collision rate would increase, causing the reaction rate to increase. A third way is to heat the reactant mixture on a hot plate. However, she would have to heat the reaction mixture carefully because hydrogen is flammable. An increase in temperature would increase both the total number of collisions and the fraction of effective collisions, thus increasing the reaction rate.

**Consider This 12.3**

If Ellen wanted to *reduce* the rate of this reaction, how could she do this by only changing something about the magnesium metal?

**Practice Problem 12.3**

Under normal conditions the iron in steel wool reacts with oxygen in a slow reaction as it corrodes or rusts:

$$4Fe(s) + 3O_2(g) \longrightarrow 2Fe_2O_3(s)$$

In his chemistry lab, Chad is working to increase the rate of the rusting process without placing the steel wool in a flame. Suggest ways he can do this. For each suggestion, use collision theory to explain why the reaction rate should increase.

**Further Practice:** Questions 12.27 and 12.28 at the end of the chapter

## Catalysts

A **catalyst** alters the pathway in which a reaction occurs without itself being consumed in the reaction. The new reaction pathway is a lower-energy pathway with lower activation energy. In this way, a catalyst increases the rate of a reaction. For example, consider the energy diagram for the conversion of $O_3$ to $O_2$ in Figure 12.10. In the presence of the atomic Cl catalyst, the activation energy is lower. With lower activation energy, a greater fraction of reactants can achieve the new minimum energy requirement, and the reaction rate increases.

How does a catalyst lower the activation energy? Let's consider some examples. Most cars with internal combustion engines are equipped with catalytic converters to help reduce the amounts of CO, NO, and $NO_2$ pollutants released into the air (Figure 12.11A). Reactions of these substances to form nontoxic $CO_2$ and $N_2$ are usually very slow. Catalysts within the catalytic converter speed up these reactions so that they occur before the gases leave the car's exhaust system.

Let's look specifically at the conversion of CO to $CO_2$ catalyzed by platinum and palladium metals. In a chemical equation, catalysts are generally written above the arrow because they are not part of the net reaction:

$$2CO(g) + O_2(g) \xrightarrow{\text{Pt/Pd}} 2CO_2(g)$$

The platinum and palladium metal atoms adsorb CO and $O_2$ molecules onto their surface (Figure 12.11B). Once there, the bonds within the $O_2$ molecules weaken or break, and the resulting atoms can easily bond with nearby CO molecules. The $CO_2$ molecules then come off the metal surface and leave through the exhaust. The catalysts lower the activation energy by forming new activated complexes with lower activation energy. The catalyzed conversion of NO to $N_2$ and $O_2$ occurs in a similar way.

Most commercial chemical products are manufactured using catalysts to save time and money.

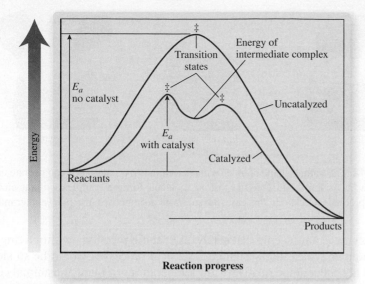

**FIGURE 12.10** The Cl catalyst lowers the activation energy for the reaction of $O_3$ to form $O_2$, thereby increasing the rate of the reaction. At a specific temperature, more collisions meet the energy requirement for reaction. Note that the energy diagram for the catalyzed reaction has two transition states. This occurs because the catalyzed reaction occurs in two steps, with the first step forming an intermediate complex with the catalyst which has a lower energy than the two transition states.

**FIGURE 12.11** (A) Catalytic converters in automobiles contain platinum, palladium, and rhodium metals coated on a ceramic surface. The result is a large surface area of catalyst. The rhodium metal catalyst in the converter increases the rate of conversion of CO to $CO_2$ as well as NO and $NO_2$ to $N_2$ and $O_2$. (B) Platinum and palladium metals catalyze the conversion of CO to $CO_2$. The metal atoms at the surface attract the oxygen atoms of both CO and $O_2$ and form weak metal-oxygen bonds, corresponding to the intermediate complex (see Figure 12.10). The interaction between the metal and oxygen weakens the bonds in CO and $O_2$, so the bond breaking required for the reaction occurs more easily. A $CO_2$ molecule exits the exhaust, and the remaining oxygen atom is available to react with another CO molecule.

(A): ©McGraw-Hill Education

**FIGURE 12.12** The unique shape of an enzyme determines its function. It can operate only on a substrate that fits its active site. Once a substrate molecule forms a complex with the enzyme, the reaction proceeds quickly. In this case, the substrate decomposes into two smaller molecules.

An **enzyme** is a molecule that catalyzes specific reactions within living organisms. Without enzymes, life-sustaining chemical processes would be so slow that they would not occur at the relatively low temperatures of plant and animal cells. An enormous number of enzymes—about 25,000—catalyzes all the processes necessary for the proper functioning of the human body. Most enzymes are large protein molecules with molar masses between 12,000 and 40,000 g/mol. The intricate three-dimensional structures of enzymes contain one or more depressions, or holes, which are called *active sites* (Figure 12.12). The shape of an active site is unique to an enzyme and allows it to interact with only one specific kind of reactant molecule, its *substrate*. In order for an enzyme-catalyzed reaction to occur, substrate molecules must fit into the structure of an enzyme at its active site. Once in the site, intermolecular interactions or bonds between the enzyme and the substrate modify the substrate in such a way that the conversion to products requires a lower activation energy.

To see how an enzyme works, consider sucrose, the ordinary sugar we use in baking or add to coffee and tea. It belongs to a class of sugars called disaccharides because it consists of two monosaccharide components. Sucrose molecules have just the right shape to fit into the active site of the enzyme sucrase. As shown in Figure 12.13, the sucrase enzyme catalyzes the reaction of sucrose with water to form two monosaccharide molecules, glucose and fructose. Once fitted into the enzyme's active site, the bond between the two components of the sucrose molecule is weakened enough that the breakdown to monosaccharide units is fast.

> An enzyme is often named after the substance it catalyzes, using the suffix *−ase*. For example, DNAase is the enzyme that catalyzes the breakdown of a DNA molecule.

$$C_{12}H_{22}O_{11} + H_2O \longrightarrow C_6H_{12}O_6 + C_6H_{12}O_6$$

sucrose       glucose      fructose

In the active site, sucrose undergoes bond breaking

Sucrase enzyme

**FIGURE 12.13** During the digestion of foods, the larger disaccharide sucrose molecule breaks down to the smaller monosaccharides glucose and fructose. The enzyme sucrase catalyzes the reaction. The interaction between the enzyme and the sucrose molecule causes the bond between the monosaccharides to weaken, so it breaks more rapidly.

| EXAMPLE 12.4 | **The Effect of a Catalyst on Activation Energy** |
|---|---|

Draw an energy diagram for the uncatalyzed oxidation reaction of CO with $O_2$ to form $CO_2$. The reaction is exothermic. Label reactants and products. Use a dotted line to show how the energy changes are different when this reaction is catalyzed by a mixture of platinum and palladium metals.

**Solution:**

The products are lower in energy than the reactants, because the reaction is exothermic. The dotted line represents the energy changes for the catalyzed reaction because catalysts allow the formation of lower-energy activated complexes. The catalyst forms an intermediate complex with the reactants, so the energy diagram for the catalyzed reaction has two peaks.

**Consider This 12.4**

If the reaction were run in reverse, do you think it would also run faster with the same catalyst than without a catalyst?

**Practice Problem 12.4**

The decomposition of HI is an exothermic reaction:

$$2HI(g) \longrightarrow H_2(g) + I_2(g)$$

Draw an energy diagram for the uncatalyzed reaction. Label reactants and products. Sketch a possible activated complex. Use a dotted line to show how the energy changes are different when platinum metal, a catalyst that increases the rate of the reaction, is added to the system.

**Further Practice:**  Questions 12.29 and 12.30 at the end of the chapter

One important characteristic of a catalyst is that it lowers the activation energy for a reaction. Another is that it remains unchanged after the reaction. In a catalytic converter, for example, once the products leave the metal surface, the metal atoms are free to combine with a new set of reactants. Similarly, when products leave the active sites of an enzyme, the active sites are available for new reactants to enter.

Recall the catalytic role of chlorine in the destruction of ozone molecules by CFCs. It comes from the breakdown of CFCs in the presence of ultraviolet light:

$$CF_2Cl_2 \xrightarrow{\text{UV light}} CF_2Cl + Cl$$

The Cl catalyst lowers the activation energy because the activated complex for the first step is more stable than it would be without the catalyst. In the presence of the Cl catalyst, the activated complex consists of a chemical species in which O–O bonds are breaking and Cl–O bonds are forming:

$$O{-}O \cdots O \cdots Cl$$

Without the catalyst, the activated complex would consist of a higher-energy species where the O–O bond is breaking:

$$O{-}O \cdots O$$

 **Student Hot Spot**

Student data indicate you may struggle with identifying intermediates and catalysts. Access the SmartBook to view additional Learning Resources on this topic.

The subsequent conversion of $O_3$ to $O_2$ in the presence of the Cl catalyst is a two-step reaction:

Step 1:         $O_3(g) + Cl(g) \longrightarrow ClO(g) + O_2(g)$
Step 2:         $ClO(g) + O_3(g) \longrightarrow Cl(g) + 2O_2(g)$
Overall: $2O_3(g) + Cl(g) + ClO(g) \longrightarrow ClO(g) + Cl(g) + 3O_2(g)$

The chlorine atom is not consumed; it is used in the first step and then regenerated in the second. Notice that ClO is formed in the first step and then used in the second. We call ClO an **intermediate** because it forms temporarily during the reaction. When we combine the two steps by adding all the components on each side of the reaction arrows, we obtain an overall equation in which the Cl catalyst and the ClO intermediate appear on both sides. Both the catalyst and the intermediate cancel out of the equation:

$$2O_3(g) + \cancel{Cl(g)} + \cancel{ClO(g)} \longrightarrow \cancel{ClO(g)} + \cancel{Cl(g)} + 3O_2(g)$$

The net equation shows only the reactants and products, which is the same equation as that for the uncatalyzed reaction:

$$2O_3(g) \longrightarrow 3O_2(g)$$

For any reaction that occurs in more than one step, any intermediates and catalysts are not part of the net equation.

| **EXAMPLE 12.5** | **Identifying Intermediates and Catalysts** |
|---|---|

Acetaldehyde, $CH_3CHO$, is a volatile organic compound formed during the combustion of fuels and often found in the emissions from car exhausts. It can decompose to form carbon monoxide and methane by the following two-step process. Identify any intermediates or catalysts.

Step 1:    $CH_3CHO + I_2 \longrightarrow CH_3I + CO + HI$
Step 2:    $CH_3I + HI \longrightarrow CH_4 + I_2$

**Solution:**

Iodine, $I_2$, is used in step 1 and then formed in step 2. Because $I_2$ is regenerated, it is a catalyst. Both $CH_3I$ and HI are formed in step 1 and then used in step 2. Because $CH_3I$ and HI are formed temporarily during the reaction, they are both intermediates.

**Consider This 12.5**

How many different activated complexes are formed during this two-step reaction?

**Practice Problem 12.5**

Ethene, $H_2C{=}CH_2$, can be converted to ethanol, $CH_3CH_2OH$, by a three-step process. Identify any intermediates or catalysts.

$$H_2C{=}CH_2 + H_3O^+ \longrightarrow H_3C{-}CH_2^+ + H_2O$$
$$H_3C{-}CH_2^+ + H_2O \longrightarrow CH_3CH_2OH_2^+$$
$$CH_3CH_2OH_2^+ + H_2O \longrightarrow CH_3CH_2OH + H_3O^+$$

**Further Practice:** Questions 12.43 and 12.44 at the end of the chapter

## 12.4  Chemical Equilibrium

 ANIMATION: Equilibrium

Many chemical reactions do not go to completion. Instead they reach a state of **chemical equilibrium** where there is no change in the concentrations of reactants and products. Chemical equilibrium is established when a single reaction occurs in which reactants are converted to products, and these products are converted back to reactants by the reverse process at an equal rate.

An example in our atmosphere is the equilibrium between nitrogen dioxide, $NO_2$, and dinitrogen tetroxide, $N_2O_4$. Most of the $NO_2$ in the atmosphere comes from automobiles. It also forms from the breakdown of $N_2O_4$ in an equilibrium reaction (Figure 12.14). While $NO_2$ is being formed from $N_2O_4$, $N_2O_4$ is being formed from $NO_2$ by the reverse process:

$$N_2O_4(g) \longrightarrow 2NO_2(g)$$
$$2NO_2(g) \longrightarrow N_2O_4(g)$$

The $NO_2$ is a brown gas, while the $N_2O_4$ is colorless. The brown color of smog is partially due to the presence of $NO_2$. Without the equilibrium that exists between $NO_2$ and $N_2O_4$, the smog would appear significantly darker. In this section we'll look at factors that influence both reaction rates and the equilibrium in reactions such as these.

Although it may seem that nothing is happening in a reaction container where a state of equilibrium has been reached, this is not the case. When a reaction reaches equilibrium, the forward reaction continues to occur, but the reverse reaction also occurs at the same rate:

Forward reaction: $\quad N_2O_4(g) \longrightarrow 2NO_2(g)$
Reverse reaction: $\quad 2NO_2(g) \longrightarrow N_2O_4(g)$

This state of equilibrium, where the forward and reverse reactions occur at the same rate, is represented using an equilibrium arrow ($\rightleftharpoons$):

$$N_2O_4(g) \rightleftharpoons 2NO_2(g)$$

When any chemical reaction reaches the equilibrium state, the rates of the forward and reverse reactions are equal; there is no net change in the concentrations of reactants and products. Reactions that reach a state of equilibrium are *reversible reactions*. A true equilibrium exists in a closed system, where neither reactants nor products can enter or leave.

To better understand the concept of chemical equilibrium, let's consider an analogy. Recall from the introduction that when Ellen and Chad heard the ozone alert, they were driving to a mall. Once they arrived, they split up, and Ellen went immediately to a two-level store with a single entrance on the first level. She arrived just as the store opened, and—along with a number of other shoppers—she entered

In the absence of destructive influences such as CFCs, the amounts of $O_3$ and $O_2$ in the stratosphere remain constant because both are constantly being destroyed and re-formed. A situation of this type is often termed a *dynamic* equilibrium. As important as this balance is, the destruction and formation of ozone are not reverse reactions, so the balance is not called a *chemical* equilibrium.

**FIGURE 12.14** The conversion of $N_2O_4$ to $NO_2$ is a reversible reaction and reaches a state of chemical equilibrium. (A and B) These show the system before equilibrium. (C and D) These show the system after equilibrium. At equilibrium, the numbers of $N_2O_4$ and $NO_2$ molecules do not change. However, $N_2O_4$ is still being converted to $NO_2$, while $NO_2$ is converted back to $N_2O_4$ at an equal rate.
(all): ©McGraw-Hill Education/Stephen Frisch

**FIGURE 12.15** (A) As equilibrium is approached, the concentration of reactants decreases and the concentration of products increases. Equilibrium is established when the concentrations of reactants and products remain constant. (B) As equilibrium is approached, the rate of the forward reaction decreases as the rate of the reverse reaction increases. Equilibrium is established when the rate of the forward reaction is equal to the rate of the reverse reaction.

on the first floor. At that time, all shoppers were on level 1. None were on level 2. She and several other shoppers headed for the escalator to go up to the second floor. At that time, many people were going up, but no one was going down. After a while, however, as some people finished their shopping on level 2 and decided to go back down to level 1, the number of people on the down escalator grew. At this point, a state of equilibrium was achieved between the numbers of shoppers on level 1 and level 2. People went up and down on the escalators, but no net change occurred in the relative number of shoppers on the two levels (assuming no shoppers entered or left the store). The rate of people going up equaled the rate of people going down. This does not mean that the numbers of people on levels 1 and 2 were equal. In fact, level 2 might always have more shoppers, but the *ratio* of people on the two levels would remain constant.

The example of people riding escalators is like a chemical equilibrium. Let's take a closer look at what occurs as the $N_2O_4/NO_2$ system *approaches* equilibrium. As the reaction begins, $N_2O_4$ reacts to form $NO_2$. As the reaction proceeds, the concentration of $N_2O_4$ decreases, so the rate of the forward reaction decreases. The concentration of $NO_2$ increases, so the reverse reaction speeds up (Figure 12.15). This continues until the rates of the forward and reverse reactions are the same and equilibrium is established.

What would happen if we placed pure $NO_2$ in a closed reaction container? Would the same state of equilibrium be reached, as shown in Figure 12.15? Yes, it would be the same as long as the temperature remained the same.

**MATH TOOLBOX 9.1**

How would the graphs in Figure 12.15 be different if the reaction started with $NO_2$ instead of $N_2O_4$? How would they be the same?

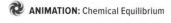
**ANIMATION:** Chemical Equilibrium

## 12.5 The Equilibrium Constant

When a reaction reaches equilibrium, amounts of reactants and products may be about equal. More often, however, one or the other predominates: either a large amount of reactants and small amount of products, or a small amount of reactants and a large amount of products. When the concentration of reactants at equilibrium is large relative to the concentration of products, we say the equilibrium favors reactants. When the concentration of products at equilibrium is large relative to the concentration of reactants, we say the equilibrium favors products. When describing an equilibrium system in this way, we are describing the *position of equilibrium.*

### The Equilibrium Constant Expression

Consider the state of equilibrium reached at the two-level store in the mall. Would the same state of equilibrium be reached if the entrance were on level 2 instead of level 1? Yes, the shoppers would distribute themselves in the same way and move

**TABLE 12.1** Equilibrium Concentrations at Constant Temperature: $2HI(g) \rightleftharpoons H_2(g) + I_2(g)$

| Experiment | Equilibrium [HI] | Equilibrium $[H_2]$ | Equilibrium $[I_2]$ | $\dfrac{[H_2][I_2]}{[HI]}$ | $\dfrac{[H_2] + [I_2]}{[HI]}$ | $\dfrac{[H_2][I_2]}{[HI]^2}$ |
|---|---|---|---|---|---|---|
| 1 | 0.704 $M$ | 0.180 $M$ | 0.550 $M$ | 0.141 | 1.04 | 0.200 |
| 2 | 1.44 $M$ | 0.757 $M$ | 0.550 $M$ | 0.186 | 0.903 | 0.201 |
| 3 | 0.634 $M$ | 0.283 $M$ | 0.283 $M$ | 0.126 | 0.893 | 0.199 |

up and down the escalators at the same rates. In a chemical reaction, the same state of equilibrium is reached whether we start with all reactants, all products, or a mixture of reactants and products.

How can we describe a state of equilibrium mathematically in terms of the relative concentrations of reactants and products? Let's look at different sets of equilibrium concentrations for a specific reaction to see if we can determine the relationship between them that will give the same value for all sets of equilibrium concentrations. We will consider the following reversible reaction:

$$2HI(g) \rightleftharpoons H_2(g) + I_2(g)$$

Table 12.1 shows the concentrations of reactants and products at equilibrium that result when the reaction is run three times, each time starting with different amounts of reactants and products. The table also shows the results of three different ways to describe the relative amounts of reactants and products. Which expression in Table 12.1 gives relatively constant values for the three experiments?

From Table 12.1 we can see that the expression in the last column gives values that are extremely close in all cases (0.200, 0.201, and 0.199). From these three trials we could calculate an average constant of 0.200. The expression that gives a constant in Table 12.1 is the product concentrations multiplied together and divided by the reactant concentration squared. Why is the reactant concentration, [HI], squared? Does it have anything to do with the *coefficient* 2 in front of HI in the balanced equation? Yes, it does. Chemists have studied a variety of reactions under many conditions. They have found that an expression like that in the last column of Table 12.1 gives a constant (the *equilibrium constant*, $K_{eq}$) for any reaction under conditions of constant temperature. In general, for a reaction with the general form

$$aA + bB \rightleftharpoons cC + dD$$

the **equilibrium constant expression** is

$$K_{eq} = \frac{[C]^c [D]^d}{[A]^a [B]^b}$$

where [A], [B], [C], and [D] are the molar concentrations of the reactants and products at equilibrium, and the exponents $a$, $b$, $c$, and $d$ are the values of the coefficients in the balanced equation. The value of the equilibrium constant $K_{eq}$ can be determined when equilibrium concentrations of reactants and products are known. The value of the equilibrium constant for a specific reaction is always the same at a specified temperature. If the temperature of the reaction system changes, the value of the equilibrium constant changes.

Let's write the equilibrium constant expression for the formation of ammonia from its elements. The balanced equation is

$$N_2(g) + 3H_2(g) \rightleftharpoons 2NH_3(g)$$

Square brackets [ ] around the formula for a substance indicate the concentration of that substance in units of mol/L, molarity.

Note that similar values (5.01, 4.98, and 5.02) are also obtained for the expression that is the inverse of the expression in the last column of Table 12.1. This expression also gives a constant value.

$$\frac{[H_2][I_2]}{[HI]^2} \xrightarrow{\text{inverse}} \frac{[HI]^2}{[H_2][I_2]}$$

$$\frac{[HI]^2}{[H_2][I_2]} = 5.00$$

By convention, chemists have chosen the expression with the products in the numerator. The inverted expression would be the equilibrium constant expression if the reaction were written in the reverse direction:

$$H_2(g) + I_2(g) \rightleftharpoons 2HI(g)$$

 **Student Hot Spot**

Student data indicate you may struggle with equilibrium constant expressions. Access the SmartBook to view additional Learning Resources on this topic.

Over 100 million tons of ammonia is produced worldwide annually. About 80% of this ammonia is used in the production of fertilizers.

The equilibrium constant expression is

$$K_{eq} = \frac{[NH_3]^2}{[N_2]^1[H_2]^3}$$

Remember that when no coefficient appears in the balanced equation, the coefficient is understood to be 1. We normally leave the exponent 1 out of the equilibrium constant expression:

$$K_{eq} = \frac{[NH_3]^2}{[N_2][H_2]^3}$$

Suppose we combine $N_2$ and $H_2$ gases in a reaction container at 400 K. Figure 12.16 shows the relative concentrations before the reaction occurs and after the reaction has reached equilibrium. From Figure 12.16 we can see that the equilibrium concentrations of reactants and product are as follows:

$$[N_2] = 0.0600\ M$$

$$[H_2] = 0.180\ M$$

$$[NH_3] = 0.280\ M$$

To calculate the value of the equilibrium constant at 400 K, we substitute these equilibrium concentrations into the equilibrium constant expression:

$$K_{eq} = \frac{[NH_3]^2}{[N_2]^1[H_2]^3}$$

$$= \frac{(0.280\ M)^2}{(0.0600\ M)\,(0.180\ M)^3} = 224$$

Although you might argue that this calculated value of the equilibrium constant should have units of $1/M^2$, by convention equilibrium constants are written without units.

The value of the equilibrium constant tells us about the position of the equilibrium. When its value is much greater than 1, there are more products than reactants at equilibrium, and we say that the position of equilibrium lies to the right. We can see from Figure 12.16 that the equilibrium concentration of the $NH_3$ product is greater than either one of the reactant equilibrium concentrations. If the value of the equilibrium constant is much less than 1, there are more reactants than products at equilibrium, and we say that the position of the equilibrium lies to the left. If the value of the equilibrium constant is about 1, similar amounts of reactants and products exist at equilibrium. The meaning of the value of the equilibrium constant $K_{eq}$ is summarized in Table 12.2. Given equilibrium concentrations and a balanced equation, you should be able to determine the value of $K_{eq}$ and the position of the equilibrium.

**FIGURE 12.16** The graphs show the initial and equilibrium concentrations of reactants and products for the reaction:

$$N_2(g) + 3H_2(g) \rightleftharpoons 2NH_3(g)$$

**TABLE 12.2** Meaning of the Value of $K_{eq}$

| Value of the Equilibrium Constant $K_{eq}$ | Position of Equilibrium |
|---|---|
| $K_{eq} \gg 1$ | Lies to the right. Products favored. |
| $K_{eq} \ll 1$ | Lies to the left. Reactants favored. |
| $K_{eq} \approx 1$ | Lies in the middle. Similar amounts of reactants and products. |

---

**EXAMPLE 12.6**   **Determining $K_{eq}$ from Equilibrium Concentrations**

Sulfur trioxide is placed in a reaction container, heated to 130°C, and allowed to reach a state of equilibrium:

$$2SO_3(g) \rightleftharpoons 2SO_2(g) + O_2(g)$$

The equilibrium concentrations are determined to be

$$[SO_2] = 0.026\ M$$
$$[O_2] = 0.013\ M$$
$$[SO_3] = 0.12\ M$$

(a) Write the equilibrium constant expression for this reaction.
(b) Calculate the value of the equilibrium constant at 130°C.
(c) Describe the position of the equilibrium.

**Solution:**

(a) The equilibrium constant expression has product concentrations in the numerator and reactant concentrations in the denominator:

$$K_{eq} = \frac{[SO_2]^2[O_2]}{[SO_3]^2}$$

The concentrations of $SO_2$ and $SO_3$ are both squared because the balanced equation shows a coefficient of 2 for each.

(b) We calculate the value of the equilibrium constant by substituting the equilibrium concentrations into the equilibrium constant expression:

$$K_{eq} = \frac{[SO_2]^2[O_2]}{[SO_3]^2} = \frac{(0.026)^2\,(0.013)}{(0.12)^2} = 6.1 \times 10^{-4}$$

(c) The value of the equilibrium constant is much less than 1. This tells us that the concentration of the products is small relative to the concentration of the reactant. The position of the equilibrium lies to the left. Reactants are favored at this temperature.

**Consider This 12.6**

What is the equilibrium constant for the reverse reaction?

$$2SO_2(g) + O_2(g) \rightleftharpoons 2SO_3(g)$$

**Practice Problem 12.6**

At 25°C a pure sample of $N_2O_4$ is placed in a reaction container and allowed to reach equilibrium:

$$2NO_2(g) \rightleftharpoons N_2O_4(g)$$

The equilibrium concentrations are determined to be

$$[NO_2] = 0.0750 \ M$$
$$[N_2O_4] = 1.25 \ M$$

(a) Write the equilibrium constant expression for this reaction.
(b) Calculate the value of the equilibrium constant at 25°C.
(c) Describe the position of the equilibrium.

**Further Practice:** Questions 12.71 and 12.72 at the end of the chapter

## Predicting the Direction of a Reaction

Suppose we want to carry out a reaction that reaches equilibrium, but we start with a mixture of reactants and products. Will the reaction go in the forward or reverse direction? It depends on the relative amounts of reactants and products in the container. If the amounts are equilibrium concentrations, the system will remain at equilibrium and no net reaction will occur. If the amounts are not equilibrium concentrations, a forward or reverse reaction will occur until the concentrations are equilibrium concentrations, as described by the equilibrium constant. If we start with more products than there should be at equilibrium, the reaction will proceed in the reverse direction, reducing the amounts of products and increasing the amounts of reactants. If we start with more reactants than there should be at equilibrium, the reaction will proceed in the forward direction, reducing the amounts of reactants and increasing the amounts of products.

> When nonequilibrium concentrations are inserted into the equilibrium constant expression, we call the result the *reaction quotient* and abbreviate it as $Q$.

How can we predict in which direction a reaction will proceed to reach equilibrium? If we substitute the concentrations of reactants and products in the starting mixture into the equilibrium constant expression, the answer will equal the equilibrium constant value only if the concentrations are equilibrium concentrations. If the value obtained is less than the equilibrium constant value, then there is not enough product and too much reactant. The reaction will proceed in the forward direction to reach equilibrium. If the value obtained is greater than the equilibrium constant value, then there is too much product and not enough reactant. The reaction will proceed in the reverse direction to reach equilibrium.

As an example, consider the following reaction and its equilibrium constant:

$$CO(g) + H_2O(g) \rightleftharpoons CO_2(g) + H_2(g) \quad K_{eq} = 16$$

The molecular picture in Figure 12.17 shows a mixture of these reactants and products in a given volume. The equilibrium constant expression for this reaction is

$$K_{eq} = \frac{[CO_2][H_2]}{[CO][H_2O]}$$

To determine if this system is at equilibrium, we can substitute the number of molecules of each reactant or product into the equilibrium constant expression:

$$\frac{(9)(9)}{(1)(1)} = 81$$

This value is not equal to the equilibrium constant, $K_{eq} = 16$, so the system is not at equilibrium. A reaction not at equilibrium will proceed either in the forward or reverse direction. In this case, because the value obtained is larger than the equilibrium constant $(81 > 16)$, there is too much product and not enough reactant. The reaction will proceed in the reverse direction to reach equilibrium. The concentrations of $CO_2$

$$CO(g) + H_2O(g) \rightleftharpoons CO_2(g) + H_2(g)$$
$$K_{eq} = 16$$

**FIGURE 12.17** Is this system at equilibrium? If not, in which direction will the reaction proceed?

and $H_2$ will decrease and the concentrations of CO and $H_2O$ will increase until the ratio equals the equilibrium constant value of 16.

As another example, consider the formation of ammonia from its elements:

$$N_2(g) + 3H_2(g) \rightleftharpoons 2NH_3(g)$$

Predict the direction this reaction will proceed when we mix equal amounts of $N_2$, $H_2$, and $NH_3$ at 400 K. The initial concentrations are each 0.20 $M$.

$$[N_2]_{initial} = 0.20 \ M$$
$$[H_2]_{initial} = 0.20 \ M$$
$$[NH_3]_{initial} = 0.20 \ M$$

We substitute these initial concentrations into the equilibrium constant expression and compare the value we calculate to the known value of the equilibrium constant expression at 400 K, which we calculated earlier to be 224:

$$K_{eq} = \frac{[NH_3]^2}{[N_2][H_2]^3} = 224 \ \text{(at equilibrium)}$$

Substituting initial concentrations, we get

$$\frac{(0.20)^2}{(0.20)(0.20)^3} = 25 \ \text{(not at equilibrium)}$$

Because 25 is less than 224, the reaction will proceed in the forward direction until equilibrium is reached.

> In this discussion we substituted numbers of molecules into the equilibrium constant expression instead of molar concentrations. We can do this only when the number of concentration terms in the numerator equals that in the denominator, and volume units from molar concentration values would cancel completely. If the numbers of concentration terms were unequal, we would have to know the volume represented by the circle in the molecular diagram.

**🔊 Student Hot Spot**

Student data indicate you may struggle with reaction quotients and predicting the direction of a reaction. Access the SmartBook to view additional Learning Resources on this topic.

---

**EXAMPLE 12.7**     **Predicting the Direction of a Reaction**

Consider this reaction and its equilibrium constant:

$$S_2Cl_2(g) + Cl_2(g) \rightleftharpoons 2SCl_2(g) \quad K_{eq} = 4$$

The following molecular picture shows a mixture of reactants and products.

 $Cl_2$

$SCl_2$

$S_2Cl_2$

(a) Determine whether this system is at equilibrium. If it is not, predict the direction in which the reaction will proceed to reach equilibrium.
(b) What will happen to the concentration of $Cl_2$?

**Solution:**
(a) The equilibrium constant expression for this reaction is

$$K_{eq} = \frac{[SCl_2]^2}{[S_2Cl_2][Cl_2]} = 4$$

In this case, we can substitute numbers of molecules instead of concentrations, because the number of terms in the numerator is equal to that in the denominator and volumes would cancel. To determine if the system is at equilibrium, we substitute the numbers of molecules of each reactant and product into the equilibrium constant expression:

$$\frac{(4)^2}{(4)(4)} = 1$$

The value obtained is not equal to the equilibrium constant, $K_{eq} = 4$, so the system is not at equilibrium. Because the obtained value is smaller than the equilibrium constant, there is not enough product and too much reactant. The reaction will proceed in the forward direction to make more product and reach equilibrium.

(b) As the reaction proceeds in the forward direction to reach equilibrium, the number of $Cl_2$ molecules decreases as $Cl_2$ reacts with $S_2Cl_2$. The concentration of $Cl_2$ decreases because the number of $Cl_2$ molecules decreases.

### Consider This 12.7

At a different temperature, the equilibrium constant for this reaction is 0.16. If the reaction were run at this temperature and we started with the same mixture of reactants and products shown in the example, how would the reaction shift to reach equilibrium?

### Practice Problem 12.7

Consider again the following reaction and its equilibrium constant:

$$S_2Cl_2(g) + Cl_2(g) \rightleftharpoons 2SCl_2(g) \quad K_{eq} = 4$$

Suppose we start with the following initial concentrations of the reactants and products:

$$[S_2Cl_2]_{initial} = 0.10\ M$$
$$[Cl_2]_{initial} = 0.10\ M$$
$$[SCl_2]_{initial} = 0.30\ M$$

(a) Determine whether the system is at equilibrium. If it is not, predict the direction in which the reaction will proceed to reach equilibrium.
(b) What will happen to the concentration of $Cl_2$?

**Further Practice:** Questions 12.79 and 12.80 at the end of the chapter

## Heterogeneous Equilibrium

The equilibria we have considered so far are reactions in which reactants and products are all in the gaseous state. An equilibrium in which reactants and products are in the same physical state is called a **homogeneous equilibrium.** When an equilibrium involves more than one physical state, it is called a **heterogeneous equilibrium.** We will now look at cases of heterogeneous equilibria.

First, let's look at an example of a heterogeneous equilibrium that involves a change in physical state: the vaporization of liquid bromine in a closed container. A state of equilibrium is established between bromine's liquid and gaseous states:

$$Br_2(l) \rightleftharpoons Br_2(g)$$

This equilibrium is heterogeneous because it involves two different physical states. The position of the equilibrium does not depend on the amount of liquid bromine in the system, as long as some liquid bromine is present. The concentration of gaseous

**FIGURE 12.18** The concentration of gaseous bromine in equilibrium with liquid bromine is independent of the amount of liquid bromine, as long as some liquid bromine remains.
(both): ©Jim Birk

bromine at equilibrium is the same no matter how much liquid bromine you start with (Figure 12.18). If we add more liquid bromine to the container, the concentration of gaseous bromine will not change. As long as some liquid bromine is present at equilibrium, the state of equilibrium does not depend on the amount present.

We can write the equilibrium constant expression for the evaporation of bromine as

$$K_{eq} = \frac{[Br_2(g)]}{[Br_2(l)]}$$

Whether at equilibrium or not, the liquid bromine always has the same concentration. That is, the number of moles per liter does not change when some of the liquid bromine vaporizes. Since it is a constant, we can write it as a constant in the equilibrium constant expression:

$$K_{eq} = \frac{[Br_2(g)]}{constant} \quad \text{or} \quad K_{eq} = \frac{1}{constant} \times [Br_2(g)]$$

We are interested only in the concentrations of substances that can change, so by convention we eliminate the term for liquid bromine from the equilibrium constant expression. It is a constant, so it can be incorporated into the equilibrium constant value:

$$K'_{eq} = K_{eq} \times constant = [Br_2(g)]$$

*By convention, chemists omit pure liquids and solids from the equilibrium constant expression for all equilibria.* Only gases and dissolved substances are included in the equilibrium constant expression.

Now let's consider an example of a heterogeneous equilibrium in which a chemical reaction occurs. The decomposition of calcium carbonate involves two solids and a gas:

$$CaCO_3(s) \rightleftharpoons CaO(s) + CO_2(g)$$

**FIGURE 12.19**   The concentration and pressure of gaseous $CO_2$ are independent of the amount of $CaCO_3$ and CaO solids as long as both are present.

> When writing equilibrium constant expressions, always omit pure liquid and solid reactants and products, but include gaseous and aqueous reactants and products.

What is the equilibrium constant expression for this heterogeneous equilibrium? Because $CaCO_3$ and CaO are both solids, and $CO_2$ is a gas, only the concentration of $CO_2$ is included in the equilibrium constant expression:

$$K_{eq} = \frac{[CO_2]}{1} = [CO_2]$$

The equilibrium constant is equal to the concentration of $CO_2$. Because the equilibrium constant is a constant at a particular temperature, we would not expect the concentration of $CO_2$ to change when the amounts of the solids $CaCO_3$ and CaO change. As shown in Figure 12.19, the concentration of $CO_2$ is independent of the amounts of the solids $CaCO_3$ and CaO, as long as all three components are present.

So far we have discussed heterogeneous equilibria involving gases, liquids, and solids. Aqueous reactions involve dissolved substances in water. Because dissolved substances can change concentration, they are also included in equilibrium constant expressions. An example is the dissolving of a relatively insoluble ionic compound in water. In Chapter 11 you learned that in a saturated solution of an ionic compound, the solid and its dissolved ions exist in equilibrium. For example, the equilibrium between solid barium fluoride and its dissolved ions is described by the following equation and equilibrium constant:

> When two relatively insoluble ionic compounds have the same number of ions, the equilibrium constants associated with their dissolving can be compared directly. For example, consider the following reactions and their equilibrium constants:
>
> $AgCl(s) \rightleftharpoons Ag^+(aq) + Cl^-(aq)$
> $K_{eq} = 1.8 \times 10^{-10}$
>
> $AgBr(s) \rightleftharpoons Ag^+(aq) + Br^-(aq)$
> $K_{eq} = 5.0 \times 10^{-13}$
>
> Although the equilibrium constants indicate that both silver compounds are very insoluble, one of them dissolves more than the other. Which one?

$$BaF_2(s) \rightleftharpoons Ba^{2+}(aq) + 2F^-(aq) \quad K_{eq} = 1.7 \times 10^{-6}$$

The equilibrium constant expression includes the ions in the aqueous state, but not the solid $BaF_2$:

$$K_{eq} = [Ba^{2+}][F^-]^2$$

The equilibrium constant shows that the position of the equilibrium depends only on the concentrations of dissolved $Ba^{2+}$ and $F^-$ ions.

**EXAMPLE 12.8**   **Equilibrium Constant Expressions for Heterogeneous Equilibria**

Write the equilibrium constant expression for each of the following reactions.

(a)  $C(s) + H_2O(g) \rightleftharpoons CO(g) + H_2(g)$

(b)  $H_2CO_3(aq) \rightleftharpoons CO_2(g) + H_2O(l)$

**Solution:**

(a) The reactant carbon is omitted from the equilibrium constant expression because it is a pure solid:

$$K_{eq} = \frac{[CO][H_2]}{[H_2O]}$$

(b) Substances in the aqueous and gas phases are required in the equilibrium constant expression. The $H_2O$ product is in the liquid state, so it is omitted from the equilibrium constant expression:

$$K_{eq} = \frac{[CO_2]}{[H_2CO_3]}$$

**Consider This 12.8**

Consider the reaction in part (a) at equilibrium. If more solid carbon were added to the reaction container, would the system still be at equilibrium?

**Practice Problem 12.8**

Write the equilibrium constant expression for each of the following reactions.

(a)  $Mg(s) + CO_2(g) \rightleftharpoons MgO(s) + CO(g)$

(b)  $PbCl_2(s) \rightleftharpoons Pb^{2+}(aq) + 2Cl^-(aq)$

**Further Practice:**  Questions 12.85 and 12.86 at the end of the chapter

## 12.6 Le Chatelier's Principle

Ellen and Chad work for a pharmaceutical company that makes a specific drug by a chemical reaction. The equilibrium constant for the reaction indicates that the equilibrium lies toward the reactants. Chad and Ellen wonder if anything can be done to cause the equilibrium to shift more toward the product side so that a higher yield can be obtained. It is possible to increase the amount of product (or reactant) at equilibrium. We'll look at some ways in this section. We'll use **Le Chatelier's principle,** which states that if a system at equilibrium is disrupted, it shifts to establish a new equilibrium. Changes that can disrupt an equilibrium system include changes in the concentration of a reactant or product, changes in the volume of a gas-reaction container, and temperature changes.

In Section 12.4 we used the analogy of the two escalators at the store to describe chemical equilibrium. Recall that a state of equilibrium was reached when the number of people going up the escalator equaled the number of people going down. At that point, no net change occurred in the numbers of people on levels 1 and 2. What happens when a busload of people arrives and 100 new shoppers enter the store? The equilibrium is disrupted because the number of people on level 1 (the entrance level) increases. As some head up the escalator, the number of people going

**ANIMATION:** Le Chatelier's Principle

The French chemist Henry Louis Le Chatelier stated the principle now named for him in 1884.

up is greater than the number of people going down. After a while, the numbers going up and down become equal, and the numbers on each floor change to new, but constant values. If we assume that the bus contained the same more-or-less balanced mix of shoppers as when the store opened—not a team of soccer players who all make a beeline for the sporting goods department—an equilibrium will be reached with the same ratio of people on level 2 relative to level 1. In a similar fashion, adding more reactant or product to the reaction container can disrupt a chemical equilibrium, but a new equilibrium position will be established in time.

### Reactant or Product Concentration

Consider the aqueous reaction of iron(III) ion with thiocyanate ion to form iron thiocyanate ion. It is shown in Figure 12.20 and described by the following net ionic equation:

$$Fe^{3+}(aq) + NCS^-(aq) \rightleftharpoons FeNCS^{2+}(aq)$$

An equilibrium is established after mixing aqueous solutions of iron(III) nitrate, $Fe(NO_3)_3$, and potassium thiocyanate, KNCS. A dilute solution of $Fe(NO_3)_3$ is a pale yellow color, and a solution of KNCS is colorless (Figure 12.20A). The product of the reaction, $FeNCS^{2+}$, forms a solution that is blood red in color. If we mix dilute solutions of the reactants, a state of equilibrium is quickly established between the reactants and products (Figure 12.20B). When more $Fe(NO_3)_3$ is added to the reaction vessel solution, the solution turns a darker red (Figure 12.20C). The darker color indicates that the equilibrium concentration of $FeNCS^{2+}$ has increased.

This example shows that we can increase the equilibrium concentration of a product by increasing the concentration of only one of the reactants. Le Chatelier's principle explains this shift. The disruption of equilibrium in this example is an increase in concentration of one of the reactants, $Fe^{3+}$. Immediately after the addition of $Fe^{3+}$, the system is no longer at equilibrium because the concentration of $Fe^{3+}$ is too high. The reaction responds to this disruption by consuming *some* of the added $Fe^{3+}$ by reaction with $NCS^-$, forming more $FeNCS^{2+}$. The reaction proceeds in the forward direction until equilibrium is reestablished with new equilibrium concentrations. The equilibrium concentration of blood-red-colored $FeNCS^{2+}$ increases due to the shift toward the product side of the equilibrium. The equilibrium concentration

Fe(NO₃)₃(aq)        KNCS(aq)

A

B

C

**FIGURE 12.20** (A) An aqueous solution of iron(III) nitrate, $Fe(NO_3)_3$, is pale yellow. An aqueous solution of potassium thiocyanate, KNCS, is colorless. (B) When these solutions mix, the $Fe^{3+}$ ion from the $Fe(NO_3)_3$ reacts with the $NCS^-$ ion from KNCS to form the complex ion iron thiocyanate, $FeNCS^{2+}$, which is blood red in concentrated solution. This system is at equilibrium, so $Fe^{3+}$, $NCS^-$, and $FeNCS^{2+}$ are all present. (C) The addition of more $Fe^{3+}$ to the solution disrupts the equilibrium. The solution turns darker, showing that more $FeNCS^{2+}$ has formed. The reaction has shifted to the right, consuming some of the added $Fe^{3+}$ ion and forming more red $FeNCS^{2+}$.

(A): ©Jim Birk; (B), (C): ©GIPhotoStock/Science Source

**TABLE 12.3** Equilibrium Shift Due to Concentration Changes

| General Reaction | Add Reactant | Add Product | Remove Reactant | Remove Product |
|---|---|---|---|---|
| $A(g) + B(g) \rightleftharpoons C(g) + D(g)$ | Shift right | Shift left | Shift left | Shift right |

of $Fe^{3+}$ also increases because not all of the added amount reacts. The equilibrium concentration of $NCS^-$ decreases because some of it reacted with the added $Fe^{3+}$. Note, however, that the value of the equilibrium constant does not change. Because the reaction responds by reacting in the forward direction, we say the equilibrium shifts to the right. By increasing the concentration of just one of the reactants, we are able to form more product. How would the equilibrium shift if additional KNCS reactant were added to the equilibrium system? What would we observe?

Suppose we want to reduce the amount of $FeNCS^{2+}$ at equilibrium. We can accomplish this by *removing* a reactant. One way to remove $Fe^{3+}$ is to add a new substance that will react only with the $Fe^{3+}$(such as NaOH), and not with the $NCS^-$. If we remove some $Fe^{3+}$ from solution, we disrupt the equilibrium because there is less $Fe^{3+}$ than there should be at equilibrium. The reaction responds by reacting in reverse; some of the $FeNCS^{2+}$ dissociates to form more $Fe^{3+}$, along with more $NCS^-$. This reverse reaction replaces most of the removed $Fe^{3+}$ and reestablishes a state of equilibrium. Because the reaction responds by reacting in the reverse direction, we say the equilibrium shifts to the left. *For a system at equilibrium, when the concentration of a reactant or product is increased, the equilibrium will shift to consume the added substance. When the concentration of a reactant or product is reduced, the equilibrium will shift to produce more of the removed substance.* Table 12.3 summarizes the effects on equilibrium of adding or removing reactants or products.

---

**EXAMPLE 12.9**    **Effect of Adding or Removing Reactants or Products**

Consider the following system at equilibrium:

$$2CO(g) + O_2(g) \rightleftharpoons 2CO_2(g)$$

Predict the effect of the following changes when the system is initially in a state of equilibrium. Assume that the reaction container volume remains constant.

(a)  Carbon dioxide gas is added to the system.
(b)  Oxygen gas is added to the system.
(c)  Carbon monoxide gas is removed from the system.

**Solution:**

(a)  The addition of $CO_2$ product increases its concentration. The system is no longer at equilibrium, so it adjusts by shifting to the left. The concentration of $CO_2$ decreases and the concentrations of CO and $O_2$ increase until equilibrium is reestablished.
(b)  With the addition of $O_2$ reactant, equilibrium is disrupted. The system will respond by shifting to the right, consuming most of the added $O_2$ and some CO gas, and producing more $CO_2$ gas until equilibrium is reestablished.
(c)  The removal of CO gas causes the equilibrium to shift left, in effect replacing some of what was removed. In doing so, the concentration of $O_2$ increases and the concentration of $CO_2$ decreases.

**Consider This 12.9**

When the reaction shifts left upon removal of CO, does the value of the equilibrium constant $K_{eq}$ change?

## Volume of the Reaction Container

If we increase or decrease the volume of a reaction container of a system at equilibrium, the concentrations of any gases involved in the reaction will change. Because gases fill the reaction container, a change in volume will change the concentrations of all gaseous reactants and products in the container. The reaction will shift so that new equilibrium concentrations are established.

Let's consider the following example of a heterogeneous equilibrium involving a product in the gaseous state:

$$CaCO_3(s) \rightleftharpoons CaO(s) + CO_2(g)$$

The substances in the solid state are left out of the equilibrium constant expression:

$$K_{eq} = [CO_2]$$

If we place a sample of solid $CaCO_3$ in a closed container at an elevated temperature, an equilibrium is established between $CaCO_3$ and the products CaO and $CO_2$ (Figure 12.21A). The equilibrium constant expression shows that the position of the

**FIGURE 12.21**  (A) At well above room temperature, an equilibrium exists between $CaCO_3$, CaO, and $CO_2$. The position of the equilibrium depends only on the concentration of $CO_2$, because $CO_2$ is a gas and the other two compounds in the reaction are solids. (B) Changing the volume of the container disrupts the equilibrium. A reduction in the volume causes the $CO_2$ concentration to be too high (greater than $K_{eq}$). (C) To reestablish equilibrium, the reaction runs in reverse, reducing the concentration of $CO_2$ to its original concentration.

$$CaCO_3(s) \rightleftharpoons CaO(s) + CO_2(g)$$

equilibrium depends only on the concentration of $CO_2$. At equilibrium, the concentration of $CO_2$ is equal to the equilibrium constant. What happens to the concentration of $CO_2$ when we reduce the volume of the container? The decrease in volume causes the concentration of $CO_2$ to increase, so the system is not at equilibrium (Figure 12.21B). The reaction will proceed in the reverse direction, decreasing the concentration of $CO_2$ until it again equals $K_{eq}$ (Figure 12.21C). The opposite shift will occur if we increase the volume of the container. The volume increase will cause the concentration of $CO_2$ to decrease. The equilibrium will shift to the right to produce more $CO_2$.

What if a system at equilibrium has both gaseous reactants and gaseous products? Will a change in volume cause an equilibrium shift? The answer depends upon the relative number of gaseous reactants and gaseous products in the balanced equation for the reaction. If there are more gas molecules on one side of the equation, a change in volume will cause a shift. To see how this works, let's consider the equilibrium between dinitrogen tetroxide gas and nitrogen dioxide gas:

$$N_2O_4(g) \rightleftharpoons 2NO_2(g)$$

This equation has 1 gas molecule on the reactant side and 2 gas molecules on the product side. In an $N_2O_4/NO_2$ system at equilibrium (Figure 12.22A), a reduction in the volume of the container increases the concentrations of both $N_2O_4$ and $NO_2$ (Figure 12.22B). The equilibrium is disrupted. Although the system cannot reduce the concentrations of both the reactant and the product, it can decrease the concentration of one of them so the total concentration of gases decreases (Figure 12.22C). Note that when the reverse reaction occurs, one $N_2O_4$ molecule is produced for every two $NO_2$ molecules that react. The reverse reaction, then, causes a net decrease in the total number of molecules in the container, decreasing total gas concentration.

**FIGURE 12.22** (A) A state of equilibrium exists between $N_2O_4$ and $NO_2$. The $N_2O_4$ is colorless and the $NO_2$ is brown. The position of the equilibrium depends on both $[N_2O_4]$ and $[NO_2]$. (B) A reduction in volume initially causes the concentrations of both $N_2O_4$ and $NO_2$ gases to increase, disrupting the equilibrium. The mixture turns darker. (C) After equilibrium is reestablished, the color of the mixture is darker than before the equilibrium was disrupted, but lighter than after the initial reduction in volume. The color has lightened because the equilibrium shifted, decreasing the concentration of $NO_2$. For every two $NO_2$ gas molecules that reacted, only one $N_2O_4$ gas molecule formed.

(all): ©Tom Pantages

**TABLE 12.4** Equilibrium Shifts Due to Volume Changes

| Relative Number of Gaseous Molecules in Balanced Equation | Example | Increase Volume | Decrease Volume |
|---|---|---|---|
| Reactants < products | $N_2O_4(g) \rightleftharpoons 2NO_2(g)$ | Shift right | Shift left |
| Reactants > products | $N_2(g) + 3H_2(g) \rightleftharpoons 2NH_3(g)$ | Shift left | Shift right |
| Reactants = products | $2NO(g) \rightleftharpoons N_2(g) + O_2(g)$ | No shift | No shift |

Suppose the volume of the container is increased. How will the $N_2O_4/NO_2$ reaction respond? The increase in volume causes the gas concentrations to decrease. The reaction shifts to the side of the equation with the greater number of gas molecules—in this case, in the forward direction toward the product side.

Will a change in volume always disrupt a system at equilibrium and cause a shift? Consider the following example of a reaction where there are equal numbers of gas molecules on each side of the equation:

$$H_2(g) + I_2(g) \rightleftharpoons 2HI(g)$$

A decrease in volume would increase the concentrations of both reactants and product, but could the equilibrium shift to decrease the total number of molecules in the container? In this case, a change in volume would have no effect on the equilibrium, and the reaction would not shift in either direction. Table 12.4 summarizes how volume changes affect different reactions at equilibrium.

---

**EXAMPLE 12.10**    **Effect of Volume Change**

How will an increase in container volume affect the following reactions at equilibrium? Explain.

(a)  $SO_2Cl_2(g) \rightleftharpoons SO_2(g) + Cl_2(g)$
(b)  $H_2O(l) + CO_2(g) \rightleftharpoons H_2CO_3(aq)$

**Solution:**

(a) An increase in volume will decrease the concentration of all of the reactants and products, because they are all in the gaseous state. There are two gaseous product molecules and only one gaseous reactant molecule in the balanced equation, so the effect is greater for the products. The reaction will shift right to increase the concentration of the products, reestablishing equilibrium.

(b) An increase in volume will decrease the concentration of only the $CO_2$, because it is the only molecule in the gaseous state. The equilibrium will shift left to increase the $CO_2$ concentration and reestablish equilibrium.

**Consider This 12.10**

If a change in volume causes no shift in the position of a reaction at equilibrium, what could we conclude about the reaction?

**Practice Problem 12.10**

How will a decrease in container volume affect the following reactions at equilibrium? Explain.

(a)  $2NOBr(g) \rightleftharpoons 2NO(g) + Br_2(g)$
(b)  $CuO(s) + H_2(g) \rightleftharpoons Cu(s) + H_2O(g)$

**Further Practice:**  Questions 12.97 and 12.98 at the end of the chapter

## Temperature

Consider once more the state of equilibrium at the store where Ellen is shopping. What happens when a store employee announces a 1-hour shoe sale on level 1? Suddenly the state of equilibrium is disrupted as more people head down the escalator than up. The number of people on level 2 decreases, and the number of people on level 1 increases. This disruption results in a new ratio of people on the two levels, so a new equilibrium is reached. This change in the position of equilibrium is similar to the response of a chemical equilibrium to a change in temperature. The value of the equilibrium constant changes when the temperature of the reaction system changes.

Consider the now familiar example of the equilibrium between $N_2O_4$ and $NO_2$:

$$N_2O_4(g) \rightleftharpoons 2NO_2(g)$$

Figure 12.23 shows the reaction mixture at two different temperatures. At the higher temperature, the reaction mixture is darker in color, indicating a higher concentration of brown $NO_2$. Because there is more product at the higher temperature, the equilibrium lies farther to the right at higher temperatures.

Not all equilibria shift to make more products when the temperature rises. Consider the formation of ammonia from nitrogen gas and hydrogen gas:

$$N_2(g) + 3H_2(g) \rightleftharpoons 2NH_3(g)$$

At 25°C this reaction reaches an equilibrium consisting mostly of products. However, at 500°C the equilibrium mixture consists of similar amounts of reactants and products. This reaction shifts to make more reactants at higher temperatures.

What is different about these two reactions that causes them to respond to a rise in temperature in opposite ways? The first reaction, the decomposition of $N_2O_4$, is endothermic:

$$N_2O_4(g) \rightleftharpoons 2NO_2(g) \quad \text{endothermic}$$

Because endothermic reactions absorb heat when they proceed in the forward direction, we sometimes write the word *heat* on the reactant side of the equation:

$$\text{Heat} + N_2O_4(g) \rightleftharpoons 2NO_2(g)$$

When we are dealing with an endothermic reaction, adding heat (by increasing temperature) has the same effect as adding a reactant. The equilibrium shifts to the right to make more products. The opposite occurs when the temperature is lowered. In an endothermic reaction, removal of heat is like removing a reactant. The equilibrium shifts to the left to make more reactants.

In contrast, the formation of ammonia from its elements is an exothermic reaction:

$$N_2(g) + 3H_2(g) \rightleftharpoons 2NH_3(g) \quad \text{exothermic}$$

Because exothermic reactions release heat when they proceed in the forward direction, we sometimes write the word *heat* on the product side of the equation:

$$N_2(g) + 3H_2(g) \rightleftharpoons 2NH_3(g) + \text{heat}$$

For an exothermic reaction, increasing the temperature of a system at equilibrium has the same effect as adding a product. The equilibrium shifts to the left to make more reactants. The opposite occurs when the temperature is lowered. In an exothermic reaction, removal of heat is like removing a product. The equilibrium shifts to the right to make more products. Table 12.5 summarizes the effect of temperature on endothermic and exothermic reactions.

What can cause a change in the equilibrium constant value? Only a temperature change can cause the equilibrium constant for a reaction to change.

**High temperature**

**Low temperature**

**FIGURE 12.23** The position of the equilibrium between $N_2O_4$ and $NO_2$ depends on temperature. The mixture at the higher temperature has a higher concentration of brown $NO_2$ than the mixture at the lower temperature. This indicates that the value of the equilibrium constant for the reaction

$$N_2O_4(g) \rightleftharpoons 2NO_2(g)$$

is greater at higher temperatures.
(both): ©Jim Birk

**TABLE 12.5** Effect of Temperature Changes on the Position of Equilibrium

| Type of Reaction | Equation | Increase Temperature | Decrease Temperature |
|---|---|---|---|
| Endothermic reaction | Heat + A + B $\rightleftharpoons$ C + D | Shift right $K_{eq}$ increases | Shift left $K_{eq}$ decreases |
| Exothermic reaction | A + B $\rightleftharpoons$ C + D + heat | Shift left $K_{eq}$ decreases | Shift right $K_{eq}$ increases |

---

**EXAMPLE 12.11** | **Effect of Temperature Change**

How will a decrease in temperature affect the equilibrium concentration of product in each reaction? How will the value of $K_{eq}$ change in each case?

(a) $2SO_2(g) + O_2(g) \rightleftharpoons 2SO_3(g)$   exothermic
(b) $3O_2(g) \rightleftharpoons 2O_3(g)$   endothermic

**Solution:**

(a) This reaction is exothermic, so we can write the reaction with heat on the product side:

$$2SO_2(g) + O_2(g) \rightleftharpoons 2SO_3(g) + \text{heat}$$

A decrease in temperature results in loss of heat from the equilibrium system. We can think of this as removing a product. The equilibrium will shift to make more product ($SO_3$), shifting the position of equilibrium to the right. Since there will be more product and less reactant, the value of $K_{eq}$ will be a larger value at the lower temperature.

(b) This reaction is endothermic, so we can write the reaction with heat on the reactant side:

$$\text{Heat} + 3O_2(g) \rightleftharpoons 2O_3(g)$$

A decrease in temperature results in loss of heat from the equilibrium system. We can think of this as removing a reactant. The reaction will make more $O_2$ and less $O_3$, shifting the equilibrium position to the left. Since there will be more reactant and less product, the value of $K_{eq}$ will be a smaller value at the lower temperature.

**Consider This 12.11**

Why do many endothermic reactions require continuous heating for the reaction to progress?

**Practice Problem 12.11**

Consider the following equilibrium:

$$Fe^{3+}(aq) + NCS^-(aq) \rightleftharpoons FeNCS^{2+}(aq)$$

When the temperature is increased, the solution turns darker, indicating a higher concentration of $FeNCS^{2+}$ product. Is this reaction endothermic or exothermic?

**Further Practice:** Questions 12.101 and 12.102 at the end of the chapter

---

## Catalysts

A catalyst speeds up a reaction by lowering the activation energy. For a reaction that reaches a state of equilibrium, a catalyst increases the rate at which equilibrium is reached. However, a catalyst does not change the position of equilibrium or affect a system that is in a state of equilibrium. Why? A catalyst is not a reactant or a product in the net reaction. A catalyst acts to lower the activation energy and increase the rate of both the forward and the reverse reaction proportionately.

### Increasing Product Yield

Le Chatelier's principle is useful for predicting the effects of changes on a chemical equilibrium. Companies, such as the pharmaceutical company that employs Ellen and Chad, want to maximize their yield of product. They do this by imposing conditions that shift the equilibrium toward the product side of a reaction. Example 12.12 shows how a variety of factors can be used to maximize product yield.

---

**EXAMPLE 12.12** | **Using Le Chatelier's Principle to Increase Product Yield**

Consider the following reaction that has reached a state of equilibrium at 200°C:

$$4HCl(g) + O_2(g) \rightleftharpoons 2Cl_2(g) + 2H_2O(g) \quad \text{exothermic}$$

Determine whether each of the following will increase the equilibrium concentration of $Cl_2$ product. Explain your reasoning.

(a) Remove $H_2O$ gas.
(b) Remove HCl gas.
(c) Increase the temperature.
(d) Reduce the volume.
(e) Add a catalyst that increases the reaction rate.

**Solution:**

(a) The removal of $H_2O$ should increase the concentration of $Cl_2$ because the equilibrium will shift right, producing more products.
(b) The removal of HCl should not increase the concentration of $Cl_2$. Instead, the concentration of $Cl_2$ will decrease because the reaction will shift left, consuming $Cl_2$ in the process.
(c) Because this reaction is exothermic, we can write the reaction with heat on the product side of the equation:

$$4HCl(g) + O_2(g) \rightleftharpoons 2Cl_2(g) + 2H_2O(g) + \text{heat}$$

An increase in temperature, or addition of heat, has the same effect as the addition of product. The reaction will shift left, consuming $Cl_2$ and $H_2O$ and forming more HCl and $O_2$. The equilibrium concentration of $Cl_2$ will decrease.
(d) A decrease in volume causes all concentrations to increase, followed by a shift of the equilibrium to the side with the fewer number of gaseous reactants or products. The reactant side has 5 gaseous molecules and the product side has only 4 gaseous molecules, so the reaction will shift right, further increasing the concentration of $Cl_2$.
(e) Addition of a catalyst does not affect the position of equilibrium; it only changes the rate at which the reaction reaches equilibrium.

**Consider This 12.12**

Which of the conditions listed for the example would cause the value of the equilibrium constant $K_{eq}$ to change?

**Practice Problem 12.12**

Consider the following reaction that has reached a state of equilibrium at 500°C:

$$PCl_5(g) \rightleftharpoons PCl_3(g) + Cl_2(g) \quad \text{endothermic}$$

Determine whether each of the following will increase the equilibrium concentration of $Cl_2$ product. Explain your reasoning.

(a) Add more $PCl_3$ gas.
(b) Remove $PCl_3$ gas.
(c) Increase the temperature.
(d) Reduce the volume.
(e) Add a catalyst that increases the reaction rate.

**Further Practice:** Questions 12.109 and 12.110 at the end of the chapter

# CHAPTER REVIEW

## KEY CONCEPTS

- Reaction rate is a measure of how fast a reaction occurs.
  - Generally, increasing temperature, concentration, or surface area increases the reaction rate, as does adding an appropriate catalyst.
  - Collision theory explains why these changes affect reaction rates. In order for a reaction to occur, reactant molecules must collide in the proper orientation and with sufficient energy specified by the reaction's activation energy. Because of these requirements, only a fraction of molecular collisions results in a reaction.
  - Reaction rate increases when the concentration or surface area of reactants is increased, or when reaction temperature is increased. An increase in concentration or surface area increases collision frequency, and an increase in temperature increases both the collision frequency and the fraction of effective collisions.
  - Catalysts increase reaction rate by lowering the activation energy and changing the process by which reactants convert to products. They are not considered reactants or products because they are regenerated in their original form.
  - Intermediates form temporarily in a multistep reaction; they form in an early step and then react in a later step.
- At chemical equilibrium, a reaction's forward and reverse processes occur at the same rate, and the concentrations of reactants and products remain constant.
- The position of equilibrium for a reaction is described by the value of its equilibrium constant $K_{eq}$.
  - The equilibrium constant expression is obtained from the balanced equation for the reaction and is a ratio of the equilibrium concentrations of products to those of reactants, each raised to a power equal to its coefficient in the balanced equation.
  - The value of the equilibrium constant for a reaction does not change, except when temperature is changed.
  - Equilibria that are product-favored have equilibrium constants greater than 1. Those that are reactant-favored have equilibrium constants less than 1. Although rather rare, equilibrium constants around 1 have similar amounts of reactants and products.
  - Pure liquids and solids are omitted from the equilibrium constant expression because their concentrations remain constant as a reaction proceeds.
  - We can use the equilibrium constant value, along with concentrations of reactants and products not at equilibrium, to predict the direction the reaction will proceed in order to reach equilibrium.
- Le Chatelier's principle predicts the effects of changes on a system at equilibrium.
  - Changing the concentration of a reactant or product, or changing the volume of a reaction container, can upset a reaction's equilibrium, causing the system to shift to reestablish equilibrium defined by its $K_{eq}$ value.

- Changing the temperature of a reaction causes the position of equilibrium and the value of $K_{eq}$ to change. The direction of the shift depends on whether the reaction is endothermic or exothermic.
- The addition of a catalyst has no effect on the position of equilibrium; it only increases the rate at which equilibrium is reached.

## KEY RELATIONSHIPS

| Relationship | Equation |
|---|---|
| The equilibrium constant is the ratio of the equilibrium concentrations of the products to the equilibrium concentrations of the reactants. The exponents in the expression are the coefficients in the balanced equation. The expression corresponds to a generic reaction and its equation: $aA + bB \rightleftharpoons cC + dD$ | $$K_{eq} = \frac{[C]^c [D]^d}{[A]^a [B]^b}$$ |

## KEY TERMS

activated complex (12.2)

activation energy $E_a$ (12.2)

catalyst (12.3)

chemical equilibrium (12.4)

collision theory (12.2)

enzyme (12.3)

equilibrium constant
  expression (12.5)

heterogeneous equilibrium
  (12.5)

homogeneous equilibrium
  (12.5)

intermediate (12.3)

Le Chatelier's principle
  (12.6)

reaction rate (12.1)

## QUESTIONS AND PROBLEMS

The following questions and problems, except those in *Additional Questions* and *Concept Review Questions,* are paired. Questions in a pair focus on the same concept. Answers to selected questions and problems are in Appendix D.

### Matching Definitions with Key Terms

12.1    Match the key terms with the descriptions provided.
  (a) the energy barrier that reactants must overcome before they can be converted to products; the difference between the energy of the activated complex and the average energy of the reactants
  (b) a chemical species produced in a step during a reaction that is relatively unstable and reacts in a subsequent step
  (c) an equilibrium in which all reactants and products are in the same physical state
  (d) a principle stating that a system at equilibrium responds to a change in concentration, volume, or temperature in such a way as to counteract the disruption and reestablish equilibrium
  (e) a theory that explains the temperature and concentration dependence of the reaction rate on the basis of molecular collisions, molecular orientation, and kinetic energy
  (f) the expression obtained from a reaction's balanced equation that relates the relative amount of products to reactants at equilibrium

12.2    Match the key terms with the descriptions provided.
  (a) the highest-energy chemical species that is formed as reactants convert to products
  (b) a protein molecule that acts as a catalyst to a specific reaction or to a narrow range of reactions
  (c) a state where the rates of the forward reaction and the reverse reaction are equal, so the net amounts of reactants and products do not change
  (d) an equilibrium in which reactants and products are not all in the same physical state
  (e) a measure of how fast reactants are converted to products, described as the change in concentration of a reactant or product per unit time
  (f) a substance that changes the rate of a reaction, but can be recovered unchanged after the reaction

### Reaction Rates

12.3    List three ways to increase the rate of a reaction.
12.4    List three ways to decrease the rate of a reaction.
12.5    What are three ways to increase the rate of the following reaction?

$$CaCO_3(s) + 2HCl(aq) \longrightarrow CaCl_2(aq) + CO_2(g) + H_2O(l)$$

**12.6** What are two ways to increase the rate of the following reaction?

$$2Ca(s) + O_2(g) \longrightarrow 2CaO(s)$$

**12.7** What type of substance, when added to a reaction, generally increases the rate of the reaction?

**12.8** What role do free chlorine atoms play in the following reaction?

$$2O_3(g) \longrightarrow 3O_2(g)$$

## Collision Theory

**12.9** Are all collisions between reactants effective? Why or why not?

**12.10** What are the requirements of an effective collision?

**12.11** Consider the following reaction:

$$2HI(g) \longrightarrow H_2(g) + I_2(g)$$

Determine if each of the following collisions has the proper orientation for an effective collision.

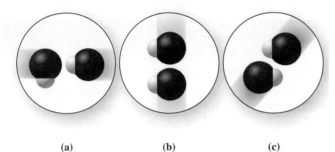

(a)             (b)             (c)

**12.12** Draw a picture of 2 $NO_2$ molecules colliding with the proper orientation to produce 1 $N_2O_4$ molecule.

$$2NO_2(g) \longrightarrow N_2O_4(g)$$

**12.13** What high-energy chemical species is produced when an effective collision occurs?

**12.14** How is an activated complex represented?

**12.15** What is activation energy?

**12.16** Draw a generic energy diagram that shows the energies of reactants, products, and the activated complex. Label the activation energy.

**12.17** How does the magnitude of the activation energy affect reaction rate?

**12.18** If two similar reactions under the same conditions have different activation energies, which reaction should occur faster: the reaction with the higher activation energy or the reaction with the lower activation energy? Explain.

**12.19** The following reaction is exothermic:

$$2O_3(g) \longrightarrow 3O_2(g)$$

Draw an energy diagram that shows the relative energies of the reactants, products, and the activated complex. Label the diagram with molecular representations of reactants, products, and a possible structure for the activated complex.

**12.20** The following reaction is endothermic:

$$CF_2Cl_2(g) \longrightarrow CF_2Cl(g) + Cl(g)$$

Draw an energy diagram that shows the relative energies of the reactants, products, and the activated complex. Label the diagram with molecular representations of reactants, products, and a possible structure for the activated complex.

## Conditions That Affect Reaction Rates

**12.21** Use collision theory to explain why reaction rates generally increase when the temperature increases.

**12.22** Use collision theory to explain why reaction rates generally increase when the concentration of one or more of the reactants increases.

**12.23** Consider the following factors: increase in temperature, increase in concentration, and addition of a catalyst. Which increases the average kinetic energy of the reactants?

**12.24** Consider the following factors: increase in temperature, increase in concentration, and addition of a catalyst. Which increases the fraction of collisions that are effective?

**12.25** The propane used in a gas grill does not usually react with air in a combustion reaction unless first initiated with a spark. Explain using collision theory.

**12.26** The fumes from gasoline are more likely than liquid gasoline to catch fire when in contact with static electricity. Explain using collision theory.

**12.27** It takes less time to cook steaks than a roast, even when cooked at the same temperature. Explain.

**12.28** At altitudes above sea level, the boiling point of water is less than 100°C. Use collision theory to explain why it takes longer to boil an egg at altitudes higher than sea level.

**12.29** Explain how a catalyst changes a reaction rate.

**12.30** Draw an energy diagram for a reaction with and without the presence of a catalyst that increases the rate of the reaction.

**12.31** What kind of catalyst is used in a catalytic converter?

**12.32** Write two reactions that are catalyzed in a catalytic converter.

**12.33** What part of an enzyme interacts with the reactants when an enzyme acts as a catalyst?

**12.34** What reaction does the enzyme sucrase catalyze?

**12.35** Why are catalysts not included in a net chemical equation?

**12.36** Why would a reaction rate increase with only a small amount of catalyst?

**12.37** Why are free chlorine atoms in the stratosphere a serious problem even if their numbers are few?

**12.38** If we do not release any more chlorine-atom-producing CFCs into the stratosphere, will the problem of excess ozone destruction go away soon? Explain.

**12.39** What is the chemical species that exists temporarily during a multistep reaction?

**12.40** What is the difference between an intermediate and an activated complex?

**12.41** Consider the following two-step reaction:

$$N_2O_5 \rightleftharpoons NO_3 + NO_2$$
$$NO_3 + NO \longrightarrow 2NO_2$$

(a) Identify any catalysts or intermediates.
(b) Write the net reaction.

**12.42** Consider the following two-step reaction:

$$S_2O_8^{2-} + I^- \longrightarrow 2SO_4^{2-} + I^+$$
$$I^+ + I^- \longrightarrow I_2$$

(a) Identify any catalysts or intermediates.
(b) Write the net reaction.

**12.43** Consider the following two-step reaction:

$$H_2O_2 + 2Br^- + 2H^+ \longrightarrow 2H_2O + Br_2$$
$$H_2O_2 + Br_2 \longrightarrow 2H^+ + O_2 + 2Br^-$$

(a) Identify any catalysts or intermediates.
(b) Write the net reaction.

**12.44** Consider the following two-step reaction:

$$Cu^{2+} + H_2 \longrightarrow CuH^+ + H^+$$
$$CuH^+ + H^+ + H_2C{=}CH_2 \longrightarrow Cu^{2+} + H_3C{-}CH_3$$

(a) Identify any catalysts or intermediates.
(b) Write the net reaction.

## Chemical Equilibrium

**12.45** Why do some reactions seem to stop when reactants are still present?

**12.46** What is a reversible reaction?

**12.47** What is chemical equilibrium?

**12.48** What happens to the forward and reverse rates as a reaction approaches equilibrium?

**12.49** The diagrams show a molecular-level view of a reaction. Which diagram shows the point where equilibrium has been reached?

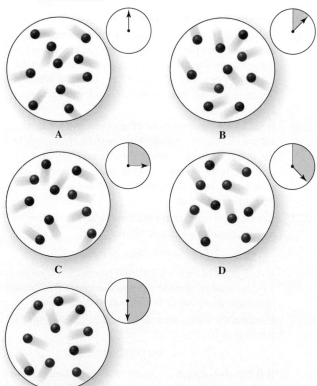

**12.50** The graph shows the concentrations of reactant and product as a reaction proceeds. At what point is equilibrium first reached?

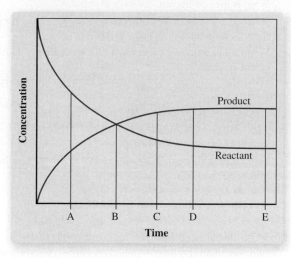

**12.51** Consider the following reaction in a sealed container:

$$2SO_3(g) \rightleftharpoons 2SO_2(g) + O_2(g)$$

Can you reach a state of equilibrium if you start with
(a) just $SO_3$?
(b) just $SO_2$?
(c) just $SO_2$ and $O_2$?
(d) just $SO_3$ and $SO_2$?

**12.52** Consider the following reaction in a sealed container:

$$2NOBr(g) \rightleftharpoons 2NO(g) + Br_2(g)$$

Can you reach a state of equilibrium if you start with
(a) just NOBr?
(b) just $Br_2$?
(c) just $Br_2$ and NOBr?
(d) just NO and $Br_2$?

**12.53** Consider the vaporization of bromine in a closed container:

$$Br_2(l) \rightleftharpoons Br_2(g)$$

What kinds of measurements could you make to determine if this physical change has reached a state of equilibrium?

**12.54** Consider the following reaction to form the brown gas $NO_2$:

$$2NO(g) + O_2(g) \rightleftharpoons 2NO_2(g)$$

What observation would indicate that the reaction has reached a state of equilibrium?

## The Equilibrium Constant

**12.55** What does the position of equilibrium tell us?

**12.56** What can we say about the relative concentrations of reactants and products when the position of equilibrium
(a) lies to the left?
(b) lies to the right?

**12.57** Why is it necessary to have a balanced equation before writing an equilibrium constant expression?

**12.58** Consider the reaction:

$$H_2(g) + Cl_2(g) \rightleftharpoons 2HCl(g)$$

What is the equilibrium constant expression?

**12.59** What does it mean when brackets are placed around the formula for a substance, such as $[HCl]$?

**12.60** When we substitute a value for $[HCl]$, what are the units of that quantity?

**12.61** Write the equilibrium constant expression for each of the following reactions:
(a) $H_2(g) + F_2(g) \rightleftharpoons 2HF(g)$
(b) $CH_4(g) + 2H_2S(g) \rightleftharpoons CS_2(g) + 4H_2(g)$
(c) $N_2O_4(g) \rightleftharpoons 2NO_2(g)$

**12.62** Write the equilibrium constant expression for each of the following reactions:
(a) $2N_2O_5(g) \rightleftharpoons 4NO_2(g) + O_2(g)$
(b) $Br_2(g) + 2HI(g) \rightleftharpoons I_2(g) + 2HBr(g)$
(c) $4NH_3(g) + 3O_2(g) \rightleftharpoons 2N_2(g) + 6H_2O(g)$

**12.63** Write the balanced equation that corresponds to the following equilibrium constant expressions. Assume that these are homogeneous equilibria in the gas state.

(a) $K_{eq} = \dfrac{[A][B]}{[C]}$  (b) $K_{eq} = \dfrac{[B]^4[C][D]}{[A]^2}$

**12.64** Write the balanced equation that corresponds to the following equilibrium constant expressions. Assume that these are homogeneous equilibria in the gas state.

(a) $K_{eq} = \dfrac{[B]^2}{[A]}$  (b) $K_{eq} = \dfrac{[A][B][C]^3}{[D]^2}$

**12.65** (a) What is the mathematical relationship between the following two equilibrium constant expressions?

$$K_{eq} = \dfrac{[NO]^2[O_2]}{[NO_2]^2} \quad \text{and} \quad K_{eq} = \dfrac{[NO_2]^2}{[NO]^2[O_2]}$$

(b) Write the balanced equations that correspond to each of these equilibrium constant expressions.

**12.66** (a) What is the mathematical relationship between the following two equilibrium constant expressions?

$$K_{eq} = \dfrac{[N_2H_4]}{[N_2][H_2]^2} \quad \text{and} \quad K_{eq} = \dfrac{[N_2][H_2]^2}{[N_2H_4]}$$

(b) Write the balanced equations that correspond to each of these equilibrium constant expressions.

**12.67** If the equilibrium constant for the reaction A $\rightleftharpoons$ B is 4.0, what is the value of the equilibrium constant for the reaction B $\rightleftharpoons$ A?

**12.68** If the equilibrium constant for the reaction C $\rightleftharpoons$ D is 0.20, what is the value of the equilibrium constant for the reaction D $\rightleftharpoons$ C?

**12.69** Under what conditions will the equilibrium constant for a specific reaction change?

**12.70** The position of the equilibrium for the formation of ammonia is product-favored at 25°C, but reactant-favored at 500°C. Explain why the position of the equilibrium changes.

**12.71** Consider the following reaction:

$$PCl_5(g) \rightleftharpoons PCl_3(g) + Cl_2(g)$$

At a specific temperature, the equilibrium concentrations were determined to be $[PCl_5] = 0.20\ M$, $[PCl_3] = 0.025\ M$, and $[Cl_2] = 0.025\ M$.
(a) What is the value of the equilibrium constant?
(b) Describe the position of the equilibrium.

**12.72** Consider the following reaction:

$$CO(g) + NO_2(g) \rightleftharpoons CO_2(g) + NO(g)$$

At a specific temperature, the equilibrium concentrations were determined to be $[CO] = 0.033\ M$, $[NO_2] = 0.021\ M$, $[CO_2] = 0.59\ M$, and $[NO] = 0.59\ M$.
(a) What is the value of the equilibrium constant?
(b) Describe the position of the equilibrium.

**12.73** Consider the following reaction:

$$CH_4(g) + 2H_2O(g) \rightleftharpoons CO_2(g) + 4H_2(g)$$

At a specific temperature, the equilibrium concentrations were determined to be $[CH_4] = 0.049\ M$, $[H_2O] = 0.048\ M$, $[CO_2] = 0.00090\ M$, and $[H_2] = 0.0036\ M$.
(a) What is the value of the equilibrium constant?
(b) Describe the position of the equilibrium.

**12.74** Consider the following reaction:

$$2NO(g) + Br_2(g) \rightleftharpoons 2NOBr(g)$$

At a specific temperature, the equilibrium concentrations were determined to be $[NO] = 0.014\ M$, $[Br_2] = 0.036\ M$, and $[NOBr] = 0.176\ M$.
(a) What is the value of the equilibrium constant?
(b) Describe the position of the equilibrium.

**12.75** If the initial concentrations of reactants and products are substituted into the equilibrium constant expression, and the value obtained is equal to the equilibrium constant, is the system in a state of equilibrium? If not, in which direction will the reaction shift to reach equilibrium? Explain.

**12.76** If the initial concentrations of reactants and products are substituted into the equilibrium constant expression, and the value obtained is greater than the equilibrium constant, is the system in a state of equilibrium? If not, in which direction will the reaction shift to reach equilibrium? Explain.

**12.77** Consider the reaction for the formation of ammonia and its equilibrium constant at 400 K:

$$N_2(g) + 3H_2(g) \rightleftharpoons 2NH_3(g) \quad K_{eq} = 224$$

If 0.050 mol of each reactant and product is mixed into a 1.0-L container, will the reaction proceed in the forward or reverse direction, or is it already at equilibrium?

**12.78** Consider the following reaction and its equilibrium constant at 100°C:

$$N_2O_4(g) \rightleftharpoons 2NO_2(g) \quad K_{eq} = 6.5$$

If 0.250 mol of each reactant and product is mixed into a 1.0-L container, will the reaction proceed in the forward or reverse direction, or is it already at equilibrium?

**12.79** Consider the reaction and its equilibrium constant:

$$O_3(g) + NO(g) \rightleftharpoons O_2(g) + NO_2(g) \quad K_{eq} = 25$$

The molecular-level diagram shows a mixture of reactants and products. Determine whether the system is at equilibrium. If it is not, predict the direction in which the reaction will proceed to reach equilibrium.

**12.80** Consider the reaction and its equilibrium constant:

$$O_3(g) + NO(g) \rightleftharpoons O_2(g) + NO_2(g) \quad K_{eq} = 25$$

The molecular-level diagram shows a mixture of reactants and products. Determine whether or not the system is at equilibrium. If it is not, predict the direction in which the reaction will proceed to reach equilibrium.

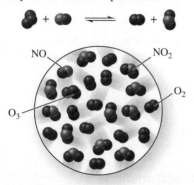

**12.81** What types of reactions are classified as heterogeneous equilibria?

**12.82** What types of reactions are classified as homogeneous equilibria?

**12.83** Why are the concentrations of pure liquids and solids not included in the equilibrium constant expression?

**12.84** Reactants and products in which physical states are included in the equilibrium constant expression?

**12.85** Write the equilibrium constant expression for the following equilibria:
 (a) $NH_4Cl(s) \rightleftharpoons NH_3(g) + HCl(g)$
 (b) $CaCO_3(s) \rightleftharpoons Ca^{2+}(aq) + CO_3^{2-}(aq)$
 (c) $HF(aq) + H_2O(l) \rightleftharpoons H_3O^+(aq) + F^-(aq)$

**12.86** Write the equilibrium constant expression for the following equilibria:
 (a) $Cu^{2+}(aq) + 2OH^-(aq) \rightleftharpoons Cu(OH)_2(s)$
 (b) $2Au_2O_3(s) \rightleftharpoons 4Au(s) + 3O_2(g)$
 (c) $SO_2(g) + 2H_2O(l) \rightleftharpoons HSO_3^-(aq) + H_3O^+(aq)$

**12.87** Consider the following equilibrium and its equilibrium constant:

$$FeO(s) + CO(g) \rightleftharpoons Fe(s) + CO_2(g) \quad K_{eq} = 0.67$$

If the equilibrium concentration of CO is determined to be 0.40 $M$, what is the equilibrium concentration of $CO_2$?

**12.88** Consider the following equilibrium and its equilibrium constant:

$$CaCO_3(s) \rightleftharpoons CaO(s) + CO_2(g) \quad K_{eq} = 0.0033$$

What is the concentration of carbon dioxide at equilibrium?

## Le Chatelier's Principle

**12.89** What is Le Chatelier's principle?

**12.90** How can we apply Le Chatelier's principle to help us form more reactants or more products at equilibrium?

**12.91** Consider the following system at equilibrium:

$$CH_4(g) + 2H_2O(g) \rightleftharpoons CO_2(g) + 4H_2(g)$$

Suppose the concentration of $CH_4$ is decreased.
 (a) In which direction does the reaction shift to reestablish equilibrium?
 (b) What happens to the concentrations of $H_2O$, $CO_2$, and $H_2$ as the reaction shifts to reestablish equilibrium?

**12.92** Consider the following system at equilibrium:

$$CH_4(g) + 2H_2O(g) \rightleftharpoons CO_2(g) + 4H_2(g)$$

Suppose the concentration of $H_2O$ is increased.
 (a) In which direction does the reaction shift to reestablish equilibrium?
 (b) What happens to the concentrations of $CH_4$, $CO_2$, and $H_2$ as the reaction shifts to reestablish equilibrium?

**12.93** Consider the following system at equilibrium:

$$NO(g) + SO_3(g) \rightleftharpoons NO_2(g) + SO_2(g)$$

Indicate the effect of the following changes.
 (a) an increase in the concentration of NO
 (b) an increase in the concentration of $SO_3$
 (c) a decrease in the concentration of $NO_2$
 (d) an increase in the concentration of $NO_2$

**12.94** Consider the following system at equilibrium:

$$NO(g) + SO_3(g) \rightleftharpoons NO_2(g) + SO_2(g)$$

For each of the following changes, predict whether the equilibrium concentration of $NO_2$ will increase or decrease.
 (a) a decrease in the concentration of NO
 (b) an increase in the concentration of $SO_3$
 (c) an increase in the concentration of $SO_2$
 (d) a decrease in the concentration of $SO_2$

**12.95** Consider the following system at equilibrium:

$$PbI_2(s) \rightleftharpoons Pb^{2+}(aq) + 2I^-(aq)$$

Even small quantities of lead ion in drinking water can be toxic. Indicate whether or not the addition of each of the following would reduce the concentration of dissolved lead ion. Explain your reasoning for each.
 (a) $PbI_2$　　(b) KI　　(c) $Pb(NO_3)_2$

**12.96** Consider the following system at equilibrium:

$$2CO_2(aq) \rightleftharpoons 2CO(g) + O_2(g)$$

Carbon monoxide is a toxic gas. Indicate whether or not each of the following would reduce the concentration of CO in the system. Explain your reasoning for each.
(a) add $CO_2$
(b) add $O_2$
(c) remove $CO_2$

**12.97** For each of the following systems at equilibrium, predict whether the reaction will shift to the right, left, or not be affected by an increase in the reaction container volume.
(a) $CO_2(g) + 4H_2(g) \rightleftharpoons CH_4(g) + 2H_2O(g)$
(b) $CaCO_3(s) \rightleftharpoons CaO(s) + CO_2(g)$
(c) $NO_2(g) + SO_2(g) \rightleftharpoons NO(g) + SO_3(g)$

**12.98** For each of the following systems at equilibrium, predict whether the reaction will shift to the right, left, or not be affected by a decrease in the reaction container volume.
(a) $PCl_3(g) + Cl_2(g) \rightleftharpoons PCl_5(g)$
(b) $2NO(g) \rightleftharpoons N_2(g) + O_2(g)$
(c) $2NO_2(g) \rightleftharpoons N_2O_4(g)$

**12.99** Consider the following system at equilibrium:

$$PbI_2(s) \rightleftharpoons Pb^{2+}(aq) + 2I^-(aq)$$

(a) Will the concentrations of $Pb^{2+}$ and $I^-$ change if we pour the equilibrium system, including the solid, into a larger container?
(b) Will the number of moles of $Pb^{2+}$ and $I^-$ change if we add water to the equilibrium system?

**12.100** Consider the following system at equilibrium:

$$CO(g) + H_2O(g) \rightleftharpoons CO_2(g) + H_2(g)$$

Will the number of moles of reactants or products change if we increase the volume of the reaction container? Explain your answer.

**12.101** For each of the following systems at equilibrium, predict whether the reaction will shift to the right, left, or not be affected by an increase in temperature.
(a) $2H_2O_2(g) \rightleftharpoons 2H_2O(g) + O_2(g)$ exothermic
(b) $N_2(g) + O_2(g) \rightleftharpoons 2NO(g)$ endothermic

**12.102** For each of the following systems at equilibrium, predict whether the reaction will shift to the right, left, or not be affected by a decrease in temperature.
(a) $PCl_5(g) \rightleftharpoons PCl_3(g) + Cl_2(g)$ endothermic
(b) $2O_3(g) \rightleftharpoons 3O_2(g)$ exothermic

**12.103** If a temperature change causes the equilibrium to shift to the right, does the value of the equilibrium constant $K_{eq}$ increase or decrease? Explain.

**12.104** If a temperature change causes the equilibrium constant to become smaller in value, in which direction does the temperature change cause the equilibrium to shift? Explain.

**12.105** Does the value of the equilibrium constant increase or decrease when the temperature of each of the following equilibrium systems is increased?
(a) $2H_2O_2(g) \rightleftharpoons 2H_2O(g) + O_2(g)$ exothermic
(b) $N_2(g) + O_2(g) \rightleftharpoons 2NO(g)$ endothermic

**12.106** Does the value of the equilibrium constant increase or decrease when the temperature of each of the following equilibrium systems is decreased?
(a) $PCl_5(g) \rightleftharpoons PCl_3(g) + Cl_2(g)$ endothermic
(b) $2O_3(g) \rightleftharpoons 3O_2(g)$ exothermic

**12.107** The equilibrium constant $K_{eq}$ is 12.5 at 400°C and 2.4 at 600°C for the reaction

$$CO(g) + H_2O(g) \rightleftharpoons CO_2(g) + H_2(g)$$

Is the reaction endothermic or exothermic?

**12.108** The equilibrium constant $K_{eq}$ is $1.0 \times 10^{-6}$ at 1500 K and $6.2 \times 10^{-4}$ at 2000 K for the reaction

$$N_2(g) + O_2(g) \rightleftharpoons 2NO(g)$$

Is the reaction endothermic or exothermic?

**12.109** Consider the following exothermic reaction:

$$4NH_3(g) + 5O_2(g) \rightleftharpoons 4NO(g) + 6H_2O(l)$$

Which of the following changes will increase the number of moles of NO at equilibrium? Explain why or why not for each change.
(a) remove $H_2O$
(b) decrease volume
(c) decrease temperature
(d) add $O_2$
(e) add a catalyst

**12.110** Consider the following endothermic reaction:

$$2Cl_2(g) + 2H_2O(g) \rightleftharpoons 4HCl(g) + O_2(g)$$

Which of the following changes will increase the number of moles of HCl at equilibrium? Explain why or why not for each change.
(a) remove $H_2O$
(b) increase volume
(c) increase temperature
(d) add $O_2$
(e) add a catalyst

## Additional Questions

**12.111** Why does a light stick glow for a longer period of time when placed in a freezer after the reaction has begun?

**12.112** The decomposition of HI into its elements is an exothermic reaction:

$$2HI(g) \longrightarrow H_2(g) + I_2(g)$$

Normally this reaction has an activation energy of 180 kJ/mol. When the reaction is carried out in the presence of platinum metal, the activation energy is lowered to 80 kJ/mol. Draw an energy diagram for the uncatalyzed reaction. Label reactants, products, and a possible activated complex. Use a dotted line to show how the energy changes are different when platinum metal is added to the reaction.

**12.113** How does a catalyst that increases reaction rate lower the activation energy?

**12.114** Why is it important to give the equation for the reaction when stating the value of its equilibrium constant?

**12.115** A sample of hydrogen iodide gas is placed in a reaction container, heated to 450°C, and allowed to reach a state of equilibrium:

$$2HI(g) \rightleftharpoons H_2(g) + I_2(g)$$

The equilibrium concentrations were determined to be

$$[HI] = 0.195 \ M$$
$$[H_2] = 0.0275 \ M$$
$$[I_2] = 0.0275 \ M$$

(a) Write the equilibrium constant expression for this reaction.
(b) Calculate the value of the equilibrium constant at 450°C.
(c) Describe the position of the equilibrium.

**12.116** Which of the following will cause the value of the equilibrium constant for a specific reaction to change?
(a) change in the concentration of a reactant or product
(b) change in volume of the container
(c) change in temperature
(d) addition of a catalyst

**12.117** What must you know about a reaction before you can predict the effect of a temperature change on the position of the equilibrium?

**12.118** Use Le Chatelier's principle to explain why the solubility of $O_2$ in water increases when the partial pressure of $O_2$ gas above the solution is increased.

**12.119** During the Napoleonic wars in Russia, the tin buttons used on French army uniforms disintegrated in the winter when temperatures fell below 18°C. The culprit was "tin pest," a chemical process described by the equation:

$$Sn_{(white \ metal)} \rightleftharpoons Sn_{(gray \ powder)}$$

Is this process exothermic or endothermic? Explain.

**12.120** About 80% of the ammonia produced by industry goes into the production of fertilizers. Ammonia is produced in industry by the reaction of $N_2$ and $H_2$ in the following reversible reaction:

$$N_2(g) + 3H_2(g) \rightleftharpoons 2NH_3(g) \quad exothermic$$

(a) Which conditions of pressure and temperature favor the formation of the most amounts of ammonia?
(b) Which conditions of temperature will increase the rate of formation of ammonia?
(c) Use your answers to parts (a) and (b) to determine whether the reaction should be run at extremely high temperatures, extremely low temperatures, or an intermediate temperature.

**12.121** Oxygen dissolved in human blood binds to hemoglobin, Hb, in the following reversible process:

$$Hb + O_2(aq) \rightleftharpoons HbO_2$$

Carbon monoxide binds strongly to hemoglobin, displacing the oxygen in oxygenated hemoglobin.

$$HbO_2 + CO(aq) \rightleftharpoons HbCO + O_2(aq)$$

What treatment would you suggest for a patient who has been exposed to excessive carbon monoxide? Explain your answer.

**12.122** In his book *Healthy Aging,* Andrew Weil, M.D., recommends eating whole grains but not whole-wheat flour products. Digestion of grains produces glucose (blood sugar). Explain why flour would produce spikes in blood sugar, while whole grains would not.

**12.123** Determine whether or not each of the following increase when the temperature of a reaction system is increased. Explain your reasoning for each.
(a) average energy of collision
(b) collision frequency
(c) fraction of collisions with proper molecular orientation
(d) fraction of collisions with sufficient energy
(e) reaction rate

**12.124** Determine whether or not each of the following increase when the concentration of reactants is increased. Explain your reasoning for each.
(a) average energy of collision
(b) collision frequency
(c) fraction of collisions with proper molecular orientation
(d) fraction of collisions with sufficient energy
(e) reaction rate

**12.125** What happens to the reaction rate (of the forward reaction) as a reaction approaches equilibrium? Explain your answer using collision theory.

**12.126** Lipase is an enzyme in the human digestive system that breaks down fats into smaller molecules. How does an enzyme such as lipase increase the rate of a chemical reaction?

**12.127** Determine whether each of the following statements about reaction rates is true or false. For each false statement, explain what is wrong with the statement.
(a) Increasing the temperature of a reaction system decreases the activation energy of the reaction.
(b) A catalyst is a reactant in the net balanced equation for a catalyzed reaction.
(c) In a reaction mechanism, a catalyst reacts in an early step and then is regenerated in a later step.
(d) In a reaction mechanism, an intermediate reacts in an early step and then is regenerated in a later step.

**12.128** Platinum metal is commonly used as a catalyst in catalytic converters to convert $CO(g)$ to $CO_2(g)$, and $NO(g)$ to $N_2(g)$. Explain how this metal acts as a catalyst in these gas-phase reactions.

**12.129** Determine whether each of the following statements about equilibrium is true or false. For each false statement, explain what is wrong with the statement.
(a) When a reaction system reaches a state of equilibrium, the concentration of the products is equal to the concentration of the reactants.
(b) When $K_{eq} \ll 1$, there are more reactants than products at equilibrium.
(c) When a system is at equilibrium, $K_{eq} = 1$.
(d) At equilibrium, the rates of the forward reaction and the reverse reaction are equal.
(e) Adding a catalyst to a reaction system will shift the position of equilibrium to the right so there are more products at equilibrium than if there was no catalyst present.

12.130 Methanol, $CH_3OH$, is an alternative fuel for combustion engines, used in combination with gasoline or in its pure form. It has commonly been used in race car competitions, because methanol burns with no dangerous black smoke that can interfere with the visibility of nearby drivers. One method of producing methanol is by reaction of hydrogen and carbon monoxide:

$$2H_2(g) + CO(g) \rightleftharpoons CH_3OH(g)$$

(a) Write the equilibrium constant expression for this reaction.

(b) Given that the equilibrium constant value for this reaction is $1.7 \times 10^{-2}$ at 200°C, and $1.3 \times 10^{-4}$ at 300°C, is this reaction more product-favored at the higher or lower temperature?

(c) Based on the data and your answer to part (b), is this reaction endothermic or exothermic?

12.131 Silver chloride is classified as an insoluble compound because it dissolves only to a small extent in water. Consider the balanced equation and equilibrium constant that shows the partial dissolving of silver chloride in water:

$$AgCl(s) \rightleftharpoons Ag^+(aq) + Cl^-(aq) \quad K_{eq} = 1.70 \times 10^{-10}$$

(a) Is this equilibrium classified as a homogeneous or heterogeneous equilibrium?

(b) Write the equilibrium constant expression for this reaction.

(c) Use the equilibrium constant value and your equilibrium constant expression to calculate the concentration of each ion in a saturated solution of silver chloride.

12.132 The bicarbonate buffer system is important in maintaining a constant $H_3O^+$ concentration in the blood. The system is a process we could consider as two chemical reactions:

Reaction 1: $CO_2(aq) + H_2O(l) \rightleftharpoons H_2CO_3(aq)$
Reaction 2: $H_2CO_3(aq) + H_2O(l) \rightleftharpoons$
$$HCO_3^-(aq) + H_3O^+(aq)$$

Consider what would happen after additional $H_3O^+$ enters our blood from excessive exercise or a meal too high in acidic foods.

(a) Which direction should each of these equilibria shift? Describe the resulting change in concentration of each aqueous or gas-phase species in these equations.

(b) How does breathing faster and more deeply help to remove excess $H_3O^+$ from the bloodstream?

## Concept Review Questions

12.133 Which of the following statements is correct?
A. The activated complex formed in a reaction is more stable than the reactants and products.
B. In a two-step reaction, there are two intermediates and one activated complex.
C. In general, fast reactions have higher activation energies.
D. An enzyme is a type of catalyst in living organisms.
E. Changing the temperature of a reaction changes the activation energy.
Modify each false statement to make it a true statement.

12.134 Which of the following does not describe a catalyst?
A. A catalyst changes the rate of the reaction by altering its pathway.
B. A catalyst lowers the activation energy for a reaction.
C. A catalyst is regenerated in its original state after the reaction is complete.
D. A catalyst generally increases the rate of a reaction.
E. A catalyst causes the position of equilibrium to lie farther to the right.
Change the incorrect statement to make it a correct statement.

12.135 Which of the following increases reaction rate without increasing the fraction of effective collisions?
A. increasing temperature
B. increasing the concentration of reactants
C. reducing the surface area of solid reactants
D. adding an effective catalyst
E. none of these
Which of these increases reaction rate by increasing the fraction of effective collisions?

12.136 Identify the intermediate(s) in the following two-step mechanism.

Step 1: $O_3(g) + Cl(g) \rightleftharpoons ClO(g) + O_2(g)$
Step 2: $O_3(g) + ClO(g) \rightleftharpoons 2O_2(g) + Cl(g)$

A. Cl
B. ClO
C. ClO and Cl
D. ClO and $O_2$
E. $O_3$ and Cl
Is there a catalyst in this mechanism? If so, which substance is it?

12.137 Which of the following best describes a chemical reaction in a state of equilibrium?
A. The reverse reaction occurs faster than the forward reaction.
B. The concentrations of the products are equal to the concentrations of the reactants.
C. Reactants and products are both present, and their concentrations no longer change.
D. All reactants have completely reacted and are no longer present.
E. One of the reactants has completely reacted and is no longer present.
Modify each incorrect statement to make it correct.

12.138 For a chemical reaction, which of the following will cause the value of its equilibrium constant to change?
A. changing the temperature
B. adding a reactant
C. removing some product
D. adding a catalyst
E. all of these
Which of these will cause the equilibrium to shift without changing the value of the equilibrium constant?

12.139 Identify the correct equilibrium constant expression for the following reaction:

$$N_2(g) + O_2(g) \rightleftharpoons 2NO(g)$$

A. $K_{eq} = \dfrac{[NO]}{[N_2] + [O_2]}$

B. $K_{eq} = \dfrac{[N_2][O_2]}{[NO]^2}$

C. $K_{eq} = [NO]^2 [N_2][O_2]$

D. $K_{eq} = \dfrac{[NO]^2}{[N_2][O_2]}$

E. $K_{eq} = \dfrac{[N_2] + [O_2]}{[NO]}$

Are any of the incorrect responses legitimate equilibrium constant expressions but involve a different reaction? For those that are, what are the equations for those reactions?

12.140 The following reaction has an equilibrium constant of $1.4 \times 10^{-3}$:

$$2NO(g) + Br_2(g) \rightleftharpoons 2NOBr(g)$$

Which of the following best describes what should happen if we mix 0.50 mol NO, 0.50 mol $Br_2$, and 0.50 mol NOBr in a 1.0 L reaction vessel?

A. The reaction should go in the forward direction to reach equilibrium.

B. The reaction should go in the reverse direction to reach equilibrium.

C. The reaction should go in the forward direction to consume all the NO and $Br_2$.

D. The reaction should go in the reverse reaction until all the NOBr is gone.

E. The reaction is at equilibrium so there should be no reaction.

Is this a reactant-favored or product-favored reaction?

12.141 Which change will cause the following exothermic reaction at equilibrium to shift to produce more $NH_3$?

$$N_2O(g) + 4H_2(g) \rightleftharpoons H_2O(g) + 2NH_3(g)$$

A. increasing the reaction container volume

B. increasing the temperature

C. removing $H_2(g)$

D. removing $H_2O(g)$

E. none of these

How will the other changes affect the position of equilibrium?

12.142 Magnesium hydroxide is a relatively insoluble ionic compound that dissolves slightly in water to form an equilibrium with its dissolved ions.

$$Mg(OH)_2(s) \rightleftharpoons Mg^{2+}(aq) + 2OH^-(aq)$$

Which change will not disturb the system at equilibrium?

A. adding $Mg^{2+}$ ions by adding soluble $Mg(NO_3)_2$

B. adding $OH^-$ ions by adding soluble NaOH

C. adding more $Mg(OH)_2$ solid to the solution

D. adding $HNO_3$, which chemically reacts with $OH^-$ ions

E. adding water

Describe how each of the other changes affects the position of equilibrium.

12.143 Which of the following reactions will shift to the right when the volume of the reaction container is reduced?

A. $C(s) + O_2(g) \rightleftharpoons CO_2(g)$

B. $2NH_3(g) \rightleftharpoons N_2(g) + 3H_2(g)$

C. $N_2(g) + O_2(g) \rightleftharpoons 2NO(g)$

D. $2NOBr(g) \rightleftharpoons 2NO(g) + Br_2(g)$

E. $3O_2(g) \rightleftharpoons 2O_3(g)$

For each incorrect answer, describe how the equilibrium is affected when the volume of the reaction container is increased.

# Acids and Bases

13.1 What Are Acids and Bases?

13.2 Strong and Weak Acids and Bases

13.3 Relative Strengths of Weak Acids

13.4 Acidic, Basic, and Neutral Solutions

13.5 The pH Scale

13.6 Buffered Solutions

Math Toolbox 13.1 Log and Inverse Log Functions

Chapter Review

Questions and Problems

©Brian Moeskau/Moeskau Photography

Olivia, Jake, and Sara are college students who have part-time jobs during the academic year. Olivia works for a plant nursery where she is in charge of watering and fertilizing perennials. She likes her job because she is a natural at gardening and has taken care of plants since she was a child. She also likes the challenge of figuring out how to help plants thrive. Jake handles customer service and pool maintenance for a swimming pool company. He tests water for several regular customers and adds the right chemicals to maintain safe and attractive pools. Sara cycles to a nearby candy factory on weekends where she helps make and package sour candies. Sara enjoys the free samples, but she also likes dropping in at the factory's quality-control lab where product samples are tested.

What do Olivia's, Jake's, and Sara's jobs have in common? Each job uses chemistry, especially the chemistry of acids and bases. Olivia's job at the nursery requires her to measure the pH of plant soil solution, the water, and dissolved substances surrounding the plants she tends. Many minerals and organic compounds in soil are acids or bases that can influence pH, and soil pH affects which metal ions dissolve in the soil solution. If the soil is too acidic (pH < 6), Olivia must add lime, CaO, a base that reduces the acidity level of the soil. She's careful to add the correct amount of lime. Too much makes the soil too basic (pH > 8). Under basic conditions, some nutrients form insoluble compounds that plants cannot absorb.

Olivia also educates home gardeners about how to maintain healthy soil in their yards. Garden soil can become too acidic for many reasons. One cause is the use of acidifying nitrogen fertilizers. Many popular brands contain the ammonium ion, $NH_4^+$, which reacts as an acid with soil bacteria. Another cause of acidic soil is *acid rain,* which has a pH below 5.5. Acid rain is produced when sulfur and nitrogen oxides in the air react with water. In acidic soil more nutrients dissolve into the soil solution. Important nutrients may get washed away, or their concentrations may increase to levels toxic to plants.

When Jake maintains pools, he uses chlorine in a variety of forms to kill bacteria. When the concentration of effective chlorine in a pool falls too low, Jake superchlorinates (*shocks*) the pool water by adding calcium hypochlorite, $Ca(OCl)_2$, a source of chlorine. The compound dissolves in water to form $OCl^-(aq)$ ions, which react with water to form hypochlorous acid, $HOCl(aq)$, a strong disinfectant. The HOCl kills bacteria by penetrating their cell walls and destroying cellular proteins.

The pH level of the pool affects the relative amounts of $HOCl(aq)$ and $OCl^-(aq)$ in the water. At a pH of 7.5, the concentration of HOCl is most effective. If the pH falls below 7.2, the pool is too acidic, indicating there is too much HOCl in the pool. Its high levels irritate the eyes of swimmers and can etch plaster walls. Above a pH of 7.8, there is not enough HOCl, and scale formation can occur, and the pool may become cloudy. Jake measures the pH of the pool regularly as part of his maintenance schedule (Figure 13.1). Usually he uses pH indicators that, when added to a sample of pool water, indicate pH by color. Sometimes he uses a portable pH meter that reads the pH directly. If the pH is too high, the pool water is too basic, so he must add an acid to lower the pH. Hydrochloric acid, HCl, also called muriatic acid, lowers the pH. If the pH is too low (too acidic), the base sodium carbonate, $Na_2CO_3$, is added to raise it.

Sara's job at the candy factory also involves acids (Figure 13.2). Sour candies contain a mixture of sugar and the organic acids malic acid and citric acid. Malic acid gives apples their "bite," and citric acid causes the sour taste of lemons and limes. All acids taste sour, although the degree of sourness varies from acid to acid. In addition to sour candy, many foods such as fruits contain acids. Vinegar, which is found in pickles, salad dressings, and pizza sauce, contains acetic acid. Carbonated sodas contain phosphoric acid, carbonic acid, and sometimes citric acid. The next time you look at a food label, check to see if any acids are listed as ingredients.

**FIGURE 13.1** Acid-base chemistry is important to a healthy swimming pool.
©Brian Moeskau/Moeskau Photography

The gases NO, $NO_2$, $SO_2$, and $SO_3$ react with water in clouds to produce the acids $HNO_3$, $H_2SO_3$, and $H_2SO_4$. The result is acid rain.

The chlorine in swimming pool water is not elemental chlorine, $Cl_2$. If $Cl_2$ was added to pool water, it would not stay as $Cl_2$. It would react with water to form hypochlorous acid, HOCl, and hydrochloric acid, HCl. Because $Cl_2$ is a gas and very toxic, it is a very dangerous chemical to handle. Other products such as "liquid" chlorine (chlorine dissolved in water) or solid compounds such as NaOCl (bleach) or $Ca(OCl)_2$ are used instead.

**FIGURE 13.2** The taste of sour candy can be attributed to the presence of acids.
©Brian Moeskau/Moeskau Photography

**FIGURE 13.3** Many household products contain acids or bases. Which of these products do you think contain acids and which contain bases?

©Brian Moeskau/Moeskau Photography

Make a list of the acids and bases you encounter in your daily life.

While many foods contain acids, bases are rarely found in foods. One reason is that bases taste bitter. Many plants, some poisonous, contain organic bases called alkaloids. Their bitter taste is nature's way of telling us not to eat them. Some bases, however, are useful (Figure 13.3). Antacids, for example, are bases that act to neutralize excess acid in the stomach. While our stomachs can usually tolerate a high concentration of stomach acid, HCl, we sometimes secrete too much and get heartburn. Commonly used antacids contain magnesium hydroxide, $Mg(OH)_2$, aluminum hydroxide, $Al(OH)_3$, or calcium carbonate, $CaCO_3$. Bases are also found in some household cleaners. Ammonia, $NH_3$, is found in window cleaners, and sodium hydroxide, NaOH, is found in many oven and drain cleaners.

Some of the properties of acids and bases have been known for millennia, but only in the past 100 years have chemists come to understand the behavior of acids and bases at the molecular level. In this chapter we will explore the behavior of acids and bases in detail. We will look at the origin of the pH scale and find out how acid-base buffers work to maintain constant pH levels in solutions such as human blood.

## Questions for Consideration

**13.1** How do acids and bases differ from other substances?
**13.2** How do strong and weak acids and bases differ?
**13.3** How can we compare the different strengths of weak acids?
**13.4** What causes aqueous solutions to be acidic or basic?
**13.5** How is pH related to acidity and basicity?
**13.6** What is a buffered solution?

 **Math Tools Used in This Chapter**

Scientific Notation (Math Toolbox 1.1)
Log and Inverse Log Functions (Math Toolbox 13.1)

## 13.1  What Are Acids and Bases?

Centuries ago substances were first classified as acids and bases by their observable properties. Like the acids in sour candy or lemons, all acids have a sour taste. They are also corrosive to most metals, so people learned a long time ago not to store fruit juices or vinegar in metal containers. In contrast to acids, bases taste bitter and have a slippery feel. Acids and bases change the color of some dyes. For example, bases turn red litmus dye to blue. Acids change it back to red. When acids and bases react in neutralization reactions, their acidic and basic properties disappear.

Although there was a time when chemists routinely tasted any new compound that they made, we now know better. Never taste a laboratory chemical!

### Acid and Base Definitions

In the late 1800s a Swedish doctoral student named Svante Arrhenius proposed a definition of acids and bases based on his experiments with electrolytes. Because aqueous solutions of acids and bases conduct electricity, Arrhenius knew that the compounds were forming positive and negative ions in solution. Arrhenius proposed that acids and bases produce specific types of ions that other compounds do not. In the **Arrhenius model of acids and bases,** an acid in aqueous solution produces hydrogen ions, $H^+$, and a base produces hydroxide ions, $OH^-$. For example, hydrochloric acid ionizes in water to form $H^+(aq)$ and $Cl^-(aq)$:

$$HCl(g) \xrightarrow{\text{H}_2\text{O}} H^+(aq) + Cl^-(aq)$$

The base sodium hydroxide dissociates when it dissolves in water to form sodium ions and hydroxide ions:

$$NaOH(s) \xrightarrow{H_2O} Na^+(aq) + OH^-(aq)$$

The Arrhenius model explains how acids and bases neutralize each other. In a neutralization reaction of HCl and NaOH, for example, the $H^+$ from the HCl combines with the $OH^-$ from NaOH to produce water:

$$H^+(aq) + OH^-(aq) \longrightarrow H_2O(l)$$

Although his model does not adequately describe acids and bases as we know them today, Arrhenius earned the 1903 Nobel Prize in Chemistry because he insisted that the $H^+(aq)$ and $OH^-(aq)$ ions are important in acid-base behavior. Because of the simplicity of its definition, the model is still used when only a brief explanation of acid-base behavior is needed.

The Arrhenius definition of acids and bases is limited by some fundamental problems. The hydrogen ion, $H^+$, essentially a proton with an extremely small radius, has a very concentrated positive charge. As a result, free $H^+$ ions are unlikely to exist in aqueous solution. Instead they are associated with surrounding water molecules. Chemists today normally represent the aqueous $H^+$ ion as $H_3O^+(aq)$, the **hydronium ion** (Figure 13.4). Another limitation to the Arrhenius model is that it assumes all bases contain $OH^-$ ions, but some do not. Many other ionic compounds, commonly called salts, have basic properties, such as the ability to neutralize acids. Examples of salts that are bases are metal oxides, carbonates, and fluorides. Some molecular compounds that contain no $-OH$ group, such as ammonia, $NH_3$, also have basic properties.

In 1923 J. N. Brønsted, a Danish chemist, and T. M. Lowry, a British chemist, independently proposed a new theory to deal with the limitations of the Arrhenius model. The **Brønsted-Lowry theory** defines an acid as any substance that can donate an $H^+$ ion to another substance; it defines a base as any substance that can accept an $H^+$ ion from another substance. These definitions include all the Arrhenius acids and bases plus some ions in salts and some molecular compounds such as ammonia. According to the Brønsted-Lowry theory, when HCl gas dissolves in water, it donates an $H^+$ to $H_2O$ to form hydronium ions and chloride ions:

$$HCl(g) + H_2O(l) \longrightarrow H_3O^+(aq) + Cl^-(aq)$$

Note that $H_2O$ acts as a Brønsted-Lowry base by accepting the $H^+$ from HCl.

Ammonia dissolved in water acts as a Brønsted-Lowry base by the following equation:

$$NH_3(aq) + H_2O(l) \rightleftharpoons NH_4^+(aq) + OH^-(aq)$$

In this case, water acts as an acid by donating the $H^+$ to the $NH_3$ molecule. Note that $OH^-(aq)$ ions form, but the $-OH$ group is not present as part of the base initially. Hydroxide ions form as a result of the Brønsted-Lowry acid-base reaction between $NH_3$ and $H_2O$. There are many organic amines (compounds that contain the $-NH_2$ group) that act as bases in water. An example is methylamine, $CH_3NH_2$:

$$CH_3NH_2(aq) + H_2O(l) \rightleftharpoons CH_3NH_3^+(aq) + OH^-(aq)$$

Brønsted-Lowry bases also include anions, such as the carbonate ion, $CO_3^{2-}$, which is found in the compound $Na_2CO_3$ used to alter the pH of swimming pool water. When added to water, $Na_2CO_3$ first dissociates completely into its ions:

$$Na_2CO_3(s) \xrightarrow{H_2O} 2Na^+(aq) + CO_3^{2-}(aq)$$

The $CO_3^{2-}$ ion acts as a base in water by accepting an $H^+$ from $H_2O$:

$$CO_3^{2-}(aq) + H_2O(l) \rightleftharpoons HCO_3^-(aq) + OH^-(aq)$$

**ANIMATION:** An Acid-Base Reaction

**FIGURE 13.4** The aqueous $H^+$ ion (shown in the center of this model) is strongly associated with water molecules, most commonly surrounded by four molecules, as shown here, giving a formula of $H_9O_4^+(aq)$. Chemists, however, generally represent aqueous $H^+$ as the hydronium ion, $H_3O^+(aq)$.

In future courses you might encounter another definition of an acid or base that is named for G. N. Lewis. He defined an acid as an electron pair acceptor and a base as an electron pair donor. This broader definition encompasses many substances that are not explained by the Brønsted-Lowry theory but is beyond the scope of this text.

Recall from Chapter 12 that equilibrium arrows ($\rightleftharpoons$) show a system at equilibrium. The forward and reverse reactions occur at equal rates when the system is at equilibrium.

Some Brønsted-Lowry acid-base reactions take place without water. For example, ammonia gas reacts with hydrogen chloride gas to form ammonium ions and chloride ions. The ammonium and chloride ions immediately combine to form the solid ionic compound, ammonium chloride, $NH_4Cl(s)$.

$$\underset{\text{base}}{NH_3(g)} + \underset{\text{acid}}{HCl(g)} \longrightarrow NH_4Cl(s)$$

©Charles D. Winters/Science Source

Brønsted-Lowry acid-base reactions do not always involve water as the acid or base. Sometimes another substance in the water is better able to give or receive an $H^+$ ion. In the following reaction, hydrofluoric acid donates an $H^+$ to the carbonate ion base to form fluoride ion and bicarbonate ion:

$$\underset{\text{acid}}{HF(aq)} + \underset{\text{base}}{CO_3^{2-}(aq)} \rightleftharpoons F^-(aq) + HCO_3^-(aq)$$

Use Example 13.1 to practice identifying the Brønsted-Lowry acid and base in acid-base reactions.

---

### EXAMPLE 13.1   Brønsted-Lowry Acids and Bases

For each reaction, identify the Brønsted-Lowry acid and base reactants.
(a)  $HNO_3(l) + H_2O(l) \longrightarrow NO_3^-(aq) + H_3O^+(aq)$
(b)  $SO_4^{2-}(aq) + H_2O(l) \rightleftharpoons HSO_4^-(aq) + OH^-(aq)$
(c)  $H_2CO_3(aq) + CH_3NH_2(aq) \rightleftharpoons HCO_3^-(aq) + CH_3NH_3^+(aq)$

**Solution:**
(a)  The name of $HNO_3$, nitric acid, tells us that it is an acid. However, we can also identify it as an acid because it donates an $H^+$ to water as shown in the equation. Water accepts $H^+$ from $HNO_3$ to form $H_3O^+(aq)$, so $H_2O$ is the base.
(b)  The sulfate ion, $SO_4^{2-}$, does not have any hydrogens to donate, so it cannot be the acid. Instead, it is the base because it accepts $H^+$ to form $HSO_4^-(aq)$. Water donates the $H^+$, forming $OH^-(aq)$, so $H_2O$ is the acid in this reaction.
(c)  Methylamine, $CH_3NH_2$, acts as a base because it accepts an $H^+$, forming $CH_3NH_3^+(aq)$. Carbonic acid, $H_2CO_3$, donates the $H^+$, forming $HCO_3^-(aq)$, so $H_2CO_3$ is the acid in this reaction.

**Consider This 13.1**

If reaction (c) runs in reverse, which would be the Brønsted-Lowry base?

**Practice Problem 13.1**

For each reaction, identify the Brønsted-Lowry acid and base reactants.

(a)  $OCl^-(aq) + H_2O(l) \rightleftharpoons HOCl(aq) + OH^-(aq)$
(b)  $H_2SO_4(aq) + F^-(aq) \longrightarrow HSO_4^-(aq) + HF(aq)$
(c)  $NH_4^+(aq) + H_2O(l) \rightleftharpoons NH_3(aq) + H_3O^+(aq)$

**Further Practice:** Questions 13.15 and 13.16 at the end of the chapter

---

### Conjugate Acid-Base Pairs

When an acid donates an $H^+$ to a base, the two products differ from the reactants by one $H^+$ ion. The product that forms as a result of gaining an $H^+$ ion is called the **conjugate acid** of the base from which it forms. The product that forms as a result of losing an $H^+$ ion is called the **conjugate base** of the acid from which it forms. Consider the reaction of hydrogen chloride gas with water:

$$\underset{\text{acid}}{HCl(g)} + \underset{\text{base}}{H_2O(l)} \longrightarrow \underset{\substack{\text{conjugate} \\ \text{acid}}}{H_3O^+(aq)} + \underset{\substack{\text{conjugate} \\ \text{base}}}{Cl^-(aq)}$$

In water, the acid, HCl, reacts to form its conjugate base, $Cl^-(aq)$. The base, $H_2O$, reacts to form its conjugate acid, $H_3O^+(aq)$. The acid reactant and the conjugate base product are a *conjugate acid-base pair*. Likewise, the base reactant and the conjugate acid product are a conjugate acid-base pair. Conjugate acid-base pairs always differ by one $H^+$ ion. The conjugate acid of a substance has one more $H^+$; the conjugate base of a substance has one less $H^+$. The conjugate base of an acid is what is left after the acid donates an $H^+$ to another substance. The conjugate base, then, has one less H atom in its formula and a decrease in charge of 1. Given that a substance reacts as an acid or a base, you should be able to identify its conjugate.

> The strengths of conjugate bases of acids (or conjugate acids of bases) vary considerably. For example, the conjugate bases of strong acids are so weak that they do not react with water at all. We'll explore the strengths of conjugate acids and bases later in this chapter.

---

### EXAMPLE 13.2  Conjugate Acids and Bases

Identify the conjugate base for each acid and explain its charge.

(a) $HOCl$
(b) $H_2PO_4^-$
(c) $H_2O$

> $H_3PO_4$ and $HPO_4^{2-}$ are not a conjugate acid-base pair because they differ by more than one $H^+$.

**Solution:**

(a) The conjugate base of $HOCl$ is $OCl^-$. It has one less H atom than the acid and a charge of $1-$, one less than the zero charge on $HOCl$.
(b) The conjugate base of $H_2PO_4^-$ is $HPO_4^{2-}$, which has one less H atom than the acid and a charge of $2-$, one less than the $1-$ charge on $H_2PO_4^-$.
(c) The conjugate base of $H_2O$ is $OH^-$, which has one less H atom than $H_2O$ and a charge of $1-$, one less than the zero charge on $H_2O$.

**Consider This 13.2**

If the $H_2PO_4^-$ ion acts as a base, what is its conjugate acid?

**Practice Problem 13.2**

Identify the conjugate acid for each base.

(a) $F^-$
(b) $HCO_3^-$
(c) $H_2O$

**Further Practice:** Questions 13.17 and 13.18 at the end of the chapter

---

In some of the examples we have looked at, water acts as an acid, while in others it acts as a base. Water can act as an acid by donating an $H^+$ to another substance, forming hydroxide ion, $OH^-(aq)$. It can also act as a base by accepting an $H^+$ from another substance, forming hydronium ion, $H_3O^+(aq)$. The conjugate of water, therefore, depends on whether water is acting as an acid or a base in a particular reaction. A substance that can act as either an acid or a base is called an **amphoteric substance.** Water is the most common amphoteric substance. Another is the bicarbonate ion, $HCO_3^-$. This anion is found in sodium bicarbonate, which is used to neutralize both acids and bases. When mixed with a basic solution containing $OH^-(aq)$, the bicarbonate ion acts as an acid and its conjugate base is $CO_3^{2-}$:

$$\underset{\text{acid}}{HCO_3^-(aq)} + OH^-(aq) \longrightarrow CO_3^{2-}(aq) + H_2O(l)$$

When mixed with an acidic solution containing $H_3O^+(aq)$, the bicarbonate ion acts as a base and its conjugate acid is $H_2CO_3$:

$$\underset{\text{base}}{HCO_3^-(aq)} + H_3O^+(aq) \longrightarrow H_2CO_3(aq) + H_2O(l)$$

The bisulfite ion, $HSO_3^-$ is another example of an amphoteric substance. What is its conjugate base? What is its conjugate acid?

acetic acid

phosphorous acid

**FIGURE 13.5** In oxoacids, the acidic hydrogens are bonded to oxygen atoms. How many acidic hydrogens do these acids have?

## Acidic Hydrogen Atoms

In acids that contain more than one hydrogen atom, how do we know which hydrogen atoms are acidic? In other words, which hydrogen atoms ionize in water? An example is acetic acid, a carboxylic acid that is commonly represented with the formula $CH_3CO_2H$. Its formula shows that each molecule contains four hydrogen atoms. Which of the four are acidic? The Lewis structure of acetic acid is shown in Figure 13.5. From its Lewis structure we see that one hydrogen atom is bonded to an oxygen atom, and the other three are bonded to a carbon atom. Only the hydrogen atom bonded to an oxygen atom is acidic; it is the only one that ionizes when acetic acid dissolves in water. The three hydrogen atoms bonded to carbon are not acidic. In an oxoacid, the acidic hydrogen atoms are usually bonded to oxygen. This is the case for phosphorous acid, $H_3PO_3$. Its Lewis structure (Figure 13.5) shows that two of the hydrogen atoms are bonded to oxygen atoms. They are the acidic hydrogen atoms. The hydrogen atom bonded to the phosphorus atom is not acidic. It is sometimes difficult to determine the number of acidic hydrogen atoms from the molecular formula of an acid, but it's generally easier when given a Lewis structure or molecular model.

---

**EXAMPLE 13.3**    **Identifying Acidic Hydrogens**

Identify the acidic hydrogen atom(s) in hypophosphorous acid, $H_3PO_2$.

**Solution:**

Hypophosphorous acid, $H_3PO_2$, has only one acidic hydrogen. It is the hydrogen bonded to the oxygen atom.

**Consider This 13.3**

The acid $H_2CO_3$ has two acidic hydrogens. Draw a Lewis structure or molecular model to show how the atoms are connected in this molecule.

**Practice Problem 13.3**

Identify the acidic hydrogen atom(s) in the amino acid glycine.

**Further Practice:** Questions 13.25 and 13.26 at the end of the chapter

---

## 13.2 Strong and Weak Acids and Bases

Jake uses a solution of HCl to lower the pH of pool water that is too basic. One reason that he uses HCl is that it is a strong electrolyte that ionizes completely in solution. As described in Chapter 11, some acids and bases are strong electrolytes, and others are weak electrolytes. An acid or base that is a strong electrolyte and *completely* ionizes or dissociates in water is a **strong acid** or a **strong base.** An acid or a base that is a weak electrolyte and only *partially* ionizes in water is a **weak acid** or a **weak base.**

**ANIMATION:** Dissociation of Strong and Weak Acids

$$HCl(g) \ + \ H_2O(l) \ \longrightarrow \ H_3O^+(aq) \ + \ Cl^-(aq)$$

**FIGURE 13.6** Hydrochloric acid is a strong acid; it ionizes completely in water to form $H_3O^+(aq)$ and $Cl^-(aq)$.
©Brian Moeskau/Moeskau Photography

**TABLE 13.1** Common Strong Acids

| Formula | Name |
|---------|------|
| HCl | hydrochloric acid |
| HBr | hydrobromic acid |
| HI | hydroiodic acid |
| $HNO_3$ | nitric acid |
| $HClO_3$ | chloric acid |
| $HClO_4$ | perchloric acid |
| $H_2SO_4$ | sulfuric acid (only one $H^+$ ionizes completely) |

## Strong Acids

 **ANIMATION:** Acid Ionization

Strong acids ionize completely when dissolved in water. As shown in Figure 13.6, when the strong acid HCl ionizes in water, the resulting solution consists of only $H_3O^+$, $Cl^-$, and $H_2O$. It contains essentially no un-ionized HCl. Table 13.1 lists the common strong acids.

Some of the strong acids listed in Table 13.1 should be familiar to you. Hydrochloric acid, HCl, is the acid in your stomach that aids in digestion. It is also the acid that pool maintenance people like Jake use to wash the plaster of swimming pools. Sulfuric acid, $H_2SO_4$, is the acid in most car batteries, and it occurs in acid rain. Nitric acid, $HNO_3$, also occurs in acid rain.

## Strong Bases

Strong bases dissociate completely when dissolved in water. As shown in Figure 13.7, the strong base sodium hydroxide, NaOH, dissociates completely in water to form $Na^+(aq)$ and $OH^-(aq)$. Essentially no undissociated NaOH(aq) remains. Most of the common strong bases are the ionic hydroxides of the group IA (1) and IIA (2) metals. Table 13.2 lists the common strong bases. Commercial oven and drain cleaners contain the strong base sodium hydroxide, NaOH. Sodium hydroxide is also used in processing paper and in canning olives. Calcium hydroxide, $Ca(OH)_2$,

$$NaOH(s) \xrightarrow{\text{H}_2\text{O}} Na^+(aq) + OH^-(aq)$$

**FIGURE 13.7** The strong base NaOH dissociates completely in water to form $Na^+(aq)$ and $OH^-(aq)$.
©Charles D. Winters

Oven and drain cleaners contain the strong base sodium hydroxide, which helps to dissolve grease and hair by converting them to water-soluble products. These products are quite caustic, so it's a good idea to wear goggles and gloves.

©McGraw-Hill Education/Jill Braaten

**TABLE 13.2** Common Strong Bases

| Formula | Name | Formula* | Name |
|---------|------|----------|------|
| LiOH | lithium hydroxide | $Mg(OH)_2$ | magnesium hydroxide |
| NaOH | sodium hydroxide | $Ca(OH)_2$ | calcium hydroxide |
| KOH | potassium hydroxide | $Ba(OH)_2$ | barium hydroxide |

* Although the group IIA (2) metal hydroxides are not completely water soluble, they are strong bases because the amount that dissolves dissociates almost completely.

sometimes called slaked lime, forms when lime, CaO, mixes with water. Calcium oxide is sometimes used to reduce the acidity levels in garden soil. Magnesium hydroxide, $Mg(OH)_2$, is a strong base used as an antacid, but as described in Chapter 11, it is relatively insoluble in water.

## Weak Acids

How do we know if an acid is a weak acid? *Any acid that is not a strong acid is a weak acid.* A weak acid does not ionize completely when dissolved in water. An equilibrium forms between a weak acid and its conjugate base, so both the reactants and the products are present in solution. Examples of weak acids are citric acid, malic acid, and acetic acid. They are commonly found in fruits and other foods. Table 13.3 lists some weak acids and where they commonly occur.

When a weak acid dissolves in water, usually fewer than 10% of the molecules transfer an $H^+$ to the water, leaving the conjugate base. The rest remain in the molecular form. The equilibrium lies far to the left. We use equilibrium arrows ⇌ when writing an equation to describe the ionization of a weak acid such as acetic acid, $CH_3CO_2H$, in water (Figure 13.8). The behavior of a weak acid in water can be described as an equilibrium between a weak acid and its conjugate base.

$$CH_3CO_2H(aq) + H_2O(l) \rightleftharpoons CH_3CO_2^-(aq) + H_3O^+(aq)$$

**TABLE 13.3**  Common Weak Acids

| Formula | Name | Occurrence |
|---|---|---|
| $CH_3CO_2H$ | acetic acid | vinegar, sour wine |
| $H_2CO_3$ | carbonic acid | soda, blood |
| $H_3C_6H_5O_7$ | citric acid | fruit, soda |
| HF | hydrofluoric acid | used in glass etching and semiconductor manufacturing |
| HOCl | hypochlorous acid | used to sanitize pools and drinking water |
| $HC_3H_5O_3$ | lactic acid | milk |
| $HC_4H_4O_5$ | malic acid | fruit |
| $H_2C_2O_4$ | oxalic acid | nuts, cocoa, parsley, rhubarb |
| $H_3PO_4$ | phosphoric acid | soda, blood |
| $H_2C_4H_4O_6$ | tartaric acid | candy, wine, grapes |

$$CH_3CO_2H(aq) + H_2O(l) \rightleftharpoons CH_3CO_2^-(aq) + H_3O^+(aq)$$

**FIGURE 13.8**  Acetic acid, $CH_3CO_2H$, is a weak acid because only a small portion of the acid molecules ionize in water. An aqueous solution of acetic acid has a much higher concentration of $CH_3CO_2H$ molecules than of $CH_3CO_2^-(aq)$ and $H_3O^+(aq)$ ions. In this molecular-level image, there are 2 $CH_3CO_2^-(aq)$ ions and 10 $CH_3CO_2H(aq)$ molecules.
©Brian Moeskau/Moeskau Photography

Other weak acids that ionize in water are cations that are the conjugate acids of weak bases. Ammonium ion, $NH_4^+$, is an example of a weak acid. It is found in the compound ammonium nitrate, $NH_4NO_3$, which Olivia uses to fertilize her plants. How does $NH_4NO_3$ act as an acid? Consider what happens when $NH_4NO_3$ dissolves in water. Like any other soluble ionic compound, it dissociates to its aqueous cations and anions:

$$NH_4NO_3(s) \xrightarrow{\;H_2O\;} NH_4^+(aq) + NO_3^-(aq)$$

Nitrate ion, $NO_3^-$, is the conjugate base of a very strong acid, so it does not behave as a base with water. The $NH_4^+(aq)$, however, is the conjugate acid of a weak base, so it reacts with water as a weak acid and donates an $H^+$ ion to water:

$$NH_4^+(aq) + H_2O(l) \rightleftharpoons NH_3(aq) + H_3O^+(aq)$$

## Weak Bases

A weak base does not completely ionize when dissolved in water. An equilibrium forms between a weak base and its conjugate acid. Ammonia is a common weak base, as shown in Figure 13.9. Most common weak bases are organic compounds

Why is it difficult to remove hard-water stains with an ammonia-based glass cleaner? Hard-water stains contain metal carbonates such as $CaCO_3$, which is a weak base. Since ammonia, $NH_3$, is also a weak base, they will not react effectively. Try using a dilute solution of vinegar (acetic acid) to dissolve hard-water stains next time.

$$NH_3(aq) + H_2O(l) \rightleftharpoons NH_4^+(aq) + OH^-(aq)$$

**FIGURE 13.9** Ammonia is a weak base, so only some of the $NH_3$ molecules ionize in water to form $NH_4^+(aq)$ and $OH^-(aq)$. An aqueous solution of ammonia has a much higher concentration of $NH_3$ molecules than of $NH_4^+(aq)$ and $OH^-(aq)$ ions. In this molecular-level image, there are 2 $NH_4^+(aq)$ and 2 $OH^-(aq)$ ions, and 14 $NH_3(aq)$ molecules.
©Brian Moeskau/Moeskau Photography

The Lewis structures for the two weak bases $CH_3NH_2$ and $OCl^-$ are shown. Note the presence of at least one unshared electron pair on the atom in each base that accepts the $H^+$ ion in an acid-base reaction. Why is this unshared electron pair an important feature of a Brønsted-Lowry base? Can a covalent bond between the base and $H^+$ form without it?

Sodium phosphate, $Na_3PO_4$, and pyridine, $C_5H_5N$, are also common weak bases. Which atom in each of these compounds do you think accepts the $H^+$ in a Brønsted-Lowry acid-base reaction?

that contain the amine group ($-NH_2$), such as methylamine, $CH_3NH_2$. Methylamine is a weak base, so we show its ionization in water with equilibrium arrows. The behavior of a weak base in water can be described as an equilibrium between a weak base and its conjugate acid:

$$CH_3NH_2(aq) + H_2O(l) \rightleftharpoons CH_3NH_3^+(aq) + OH^-(aq)$$

Other weak bases that ionize in water are anions that are the conjugate bases of weak acids. Hypochlorite ion, $OCl^-$, is an example of a weak base. It is found in the compound calcium hypochlorite, $Ca(OCl)_2$, that Jake uses to superchlorinate the water in swimming pools. How does $Ca(OCl)_2$ act as a base? Consider what happens when $Ca(OCl)_2$ dissolves in water. Like any other soluble ionic compound dissolved in water, it dissociates to its aqueous cations and anions:

$$Ca(OCl)_2(s) \xrightarrow{\text{H}_2\text{O}} Ca^{2+}(aq) + 2OCl^-(aq)$$

The $OCl^-(aq)$ reacts with water as a base and accepts an $H^+$ ion from water:

$$\underset{\text{weak base}}{OCl^-(aq)} + H_2O(l) \rightleftharpoons HOCl(aq) + OH^-(aq)$$

The anions that are the conjugate bases of weak acids are typically obtained from ionic compounds that contain the anion. For example, the hypochlorite ion used in pool sanitation can come from a variety of compounds, including $Ca(OCl)_2$. Some pool products contain sodium hypochlorite, NaOCl, instead. Identifying ionic compounds that behave as weak bases is relatively straightforward once you identify the anion. If the anion is the conjugate base of a weak acid, it will typically behave as a base when added to water. For example, sodium acetate, $NaCH_3CO_2$, is a soluble ionic compound that dissociates to give $Na^+(aq)$ and $CH_3CO_2^-(aq)$ ions. Consulting Table 13.3, we see that acetic acid is a weak acid with the formula $CH_3CO_2H$. The acetate ion, $CH_3CO_2^-$, is the conjugate base of acetic acid, so we can predict that an ionic compound containing the acetate ion behaves as a weak base. On the other hand, if the anion is the conjugate base of a strong acid, it will not react as a base with water. For example, NaCl contains the conjugate base, $Cl^-$, of the strong acid HCl. Conjugate bases of strong acids do *not* react with water, so NaCl will *not* behave as a base.

Table 13.4 lists some common weak bases and where they are found. For those that are ionic compounds, identify the anion that acts as a base when the compound dissolves in water.

**TABLE 13.4** Common Weak Bases

| Formula | Name | Occurrence |
|---------|------|------------|
| $NH_3$ | ammonia | glass cleaners |
| $CaCO_3$ | calcium carbonate | antacids, minerals |
| $Ca(OCl)_2$ | calcium hypochlorite | chlorine source for swimming pools |
| $CH_3NH_2$ | methylamine | herring brine |
| $(CH_3)_3N$ | trimethylamine | rotting fish |

| EXAMPLE 13.4 | **Strong and Weak Acids and Bases in Aqueous Solution** |
|---|---|

Identify each of the following as a strong acid, weak acid, strong base, or weak base. Write an equation to describe its reaction with water.

(a) $CH_3CH_2NH_2(aq)$
(b) $HF(aq)$
(c) $NaF(aq)$
(d) $KOH(aq)$

**Solution:**

(a) The compound $CH_3CH_2NH_2$ is an organic compound called an amine because it contains the $-NH_2$ group. It is a weak base like $NH_3$. The equation for its ionization in water is

$$CH_3CH_2NH_2(aq) + H_2O(l) \rightleftharpoons CH_3CH_2NH_3^+(aq) + OH^-(aq)$$

Because $CH_3CH_2NH_2$ is a base, it ionizes in water to form $OH^-(aq)$ ions. Equilibrium arrows are used because $CH_3CH_2NH_2$ is a weak base and ionizes only partially.

(b) Hydrofluoric acid, HF, is an acid. Since it is not on the list of strong acids in Table 13.1, it is a weak acid. The equation for its ionization in water is

$$HF(aq) + H_2O(l) \rightleftharpoons F^-(aq) + H_3O^+(aq)$$

Because HF is an acid, it reacts with water to form $H_3O^+(aq)$ ions. Equilibrium arrows are used because HF is a weak acid and ionizes only partially.

(c) Sodium fluoride is an ionic compound that completely dissociates in water to form $Na^+(aq)$ and $F^-(aq)$. The $F^-$ ion is the conjugate base of the weak acid HF, so it is a weak base. The equation for its reaction with water is

$$F^-(aq) + H_2O(l) \rightleftharpoons HF(aq) + OH^-(aq)$$

Because $F^-$ is a base, it reacts with water to form $OH^-$ ions. Equilibrium arrows are used because $F^-$ is a weak base.

(d) KOH is a strong base because it is a group IA (1) metal hydroxide. The equation for its dissociation in water is

$$KOH(aq) \longrightarrow K^+(aq) + OH^-(aq)$$

Because KOH is a base, it produces hydroxide ions in water. A one-directional arrow is used because KOH is a strong base and dissociates completely.

**Consider This 13.4**

What product is always formed when a strong or weak base ionizes in water? How do the resulting solutions differ?

**Practice Problem 13.4**

Identify each of the following as a strong acid, weak acid, strong base, or weak base. Write an equation to describe its ionization in water.

(a) HI(*aq*)  (b) NaCH$_3$CO$_2$(*aq*)  (c) NH$_4^+$(*aq*)  (d) NH$_3$(*aq*)

**Further Practice:** Questions 13.29 and 13.30 at the end of the chapter

---

Example 13.5 will help you distinguish between strong and weak acids at the molecular level.

### EXAMPLE 13.5 | Molecular-Level Representations of Strong and Weak Acids

One of the diagrams shown represents a solution of HF, and the other represents an aqueous solution of HCl. Which is which? Explain your reasoning.

A                    B

**Solution:**

The difference between the two diagrams is the degree of ionization. Diagram A shows no acid molecules (H atoms attached to atoms represented in green), indicating complete ionization of the acid. It represents a strong acid. Diagram B shows many un-ionized acid molecules (H atoms attached to atoms represented in yellow) and one that is ionized, indicating partial ionization. It represents a weak acid. Of the two acids, HCl is the strong acid and is represented by diagram A. Diagram B represents a solution of HF, the weak acid.

**Consider This 13.5**

If another aqueous molecular-level diagram shows OH$^-$ ions instead of H$_3$O$^+$ ions, what can you say about the species dissolved?

**Practice Problem 13.5**

One of the diagrams represents a solution of HClO$_4$, and the other represents an aqueous solution of HSO$_4^-$. Which is which? Explain your reasoning.

A                    B

**Further Practice:** Questions 13.31 and 13.32 at the end of the chapter

# 13.3 Relative Strengths of Weak Acids

Some of the sour candies made at the factory where Sara works are coated with a mixture of sugar and citric acid. Citric acid has a stronger sour flavor than many other edible acids. Why? Is it more acidic than other weak acids? How can we compare the acid strength of different weak acids? Acid strength depends on the relative number of acid molecules that ionize when dissolved in water—that is, the degree of ionization. The percentage of molecules that ionize varies among weak acids. The stronger the acid, the larger its equilibrium constant value for ionization in water, and the greater the percentage of $H_3O^+(aq)$ and conjugate base ions produced. Remember from Chapter 12 that an equilibrium constant $K_{eq}$ describes the relative amounts of reactants and products. The larger the value of $K_{eq}$, the more products exist at equilibrium.

Consider the ionization in water of two weak acids: acetic acid, $CH_3CO_2H$, and hydrocyanic acid, HCN:

$$CH_3CO_2H(aq) + H_2O(l) \rightleftharpoons H_3O^+(aq) + CH_3CO_2^-(aq) \qquad K_{eq} = 1.8 \times 10^{-5}$$

$$HCN(aq) + H_2O(l) \rightleftharpoons H_3O^+(aq) + CN^-(aq) \qquad K_{eq} = 6.2 \times 10^{-10}$$

Which acid, $CH_3CO_2H$ or HCN, is stronger? Their equilibrium constants provide the answer. The stronger acid ionizes to a larger extent, producing more $H_3O^+(aq)$ ions and conjugate base than the weaker acid. The stronger acid has more products at equilibrium and therefore has an equilibrium that lies farther to the right than the weaker acid. Because acetic acid, $CH_3CO_2H$, has the greater equilibrium constant value for ionization, it is the stronger acid.

## Acid Ionization Constants

To distinguish the equilibrium constant for the ionization of acids in water from other reactions acids can undergo, chemists use a special symbol, $K_a$, the **acid ionization constant.** It describes the equilibrium that forms when an acid reacts with water. Thus, the larger the $K_a$ value, the stronger the acid. The $K_{eq}$ values given for the ionization of acetic acid and hydrocyanic acid are also their $K_a$ values:

$$CH_3CO_2H(aq) + H_2O(l) \rightleftharpoons H_3O^+(aq) + CH_3CO_2^-(aq) \qquad K_a = 1.8 \times 10^{-5}$$

$$HCN(aq) + H_2O(l) \rightleftharpoons H_3O^+(aq) + CN^-(aq) \qquad K_a = 6.2 \times 10^{-10}$$

Given $K_a$ values for weak acids, we can compare the relative amounts of $H_3O^+(aq)$ in different weak acid solutions if the solutions are identical in concentration of acid. What can we say about the relative amounts of $H_3O^+(aq)$ in a 0.10 $M$ $CH_3CO_2H$ solution compared to a 0.10 $M$ HCN solution? Acetic acid, $CH_3CO_2H$, with the larger $K_a$ value, ionizes to the larger extent, yielding a greater concentration of $H_3O^+$. However, the concentration of $H_3O^+$ in both weak acid solutions is much

The equilibrium constant represents the ratio of the concentration of products to reactants at equilibrium, leaving out any substances that are in the liquid or solid phase. For example, the equilibrium constant expression for the ionization of acetic acid, $CH_3CO_2H$, in water is

$$K_{eq} = \frac{[H_3O^+][CH_3CO_2^-]}{[CH_3CO_2H]}$$

Recall from Chapter 12 that by convention equilibrium constants are written without units.

**TABLE 13.5** Weak Acids, Their $K_a$ Values at 25°C, and Their Conjugate Bases

|  | Acid | $K_a$ Values | Conjugate Base |  |
|---|---|---|---|---|
| **Strongest acids** | HF | $6.3 \times 10^{-4}$ | F$^-$ | **Weakest bases** |
|  | HNO$_2$ | $5.6 \times 10^{-4}$ | NO$_2^-$ |  |
|  | HCO$_2$H | $1.8 \times 10^{-4}$ | HCO$_2^-$ |  |
|  | CH$_3$CO$_2$H | $1.8 \times 10^{-5}$ | CH$_3$CO$_2^-$ |  |
|  | HOCl | $4.0 \times 10^{-8}$ | OCl$^-$ |  |
|  | HCN | $6.2 \times 10^{-10}$ | CN$^-$ |  |
| **Weakest acids** | NH$_4^+$ | $5.6 \times 10^{-10}$ | NH$_3$ | **Strongest bases** |

How is acid strength different from acid concentration? For example, when comparing 0.1 $M$ HCl and 0.2 $M$ HF, which is the stronger acid? Which is the more concentrated acid?

less than the $H_3O^+$ concentration in a strong acid solution of the same concentration, such as 0.10 $M$ HCl. We cannot easily compare $H_3O^+$ concentrations in solutions of weak acids of different concentrations because the concentration of the acid is also a factor. Table 13.5 lists several weak acids and their $K_a$ values, along with the conjugate base of each acid. It also shows that the strength of the conjugate base is inversely related to the strength of the acid. In other words, the stronger the acid, the weaker the conjugate base. The conjugate bases of strong acids (such as $Cl^-$) are so weak that they do not react with water.

---

**EXAMPLE 13.6**    **Significance of $K_a$**

Which acid ionizes in water to the greater extent, hypochlorous acid, HOCl, or hydrocyanic acid, HCN? Consult Table 13.5 for $K_a$ values.

**Solution:**

From Table 13.5, we see that the $K_a$ value for HOCl, $4.0 \times 10^{-8}$, is larger than the $K_a$ value for HCN, $6.2 \times 10^{-10}$. The larger $K_a$ value for HOCl indicates that a greater percentage of HOCl reacts, so it ionizes to $H_3O^+$ and $OCl^-$ to a greater extent.

**Consider This 13.6**

Compare the conjugate bases of HCN and HOCl. Which should be the stronger base?

**Practice Problem 13.6**

Which solution has the greater concentration of $H_3O^+$, a 0.10 $M$ solution of hypochlorous acid, HOCl, or a 0.10 $M$ solution of hydrocyanic acid, HCN?

**Further Practice:**  Questions 13.37 and 13.38 at the end of the chapter

---

## Polyprotic Acids

Some acids contain more than one acidic hydrogen atom. Such an acid, which can donate more than one $H^+$ ion, is called a **polyprotic acid** (Figure 13.10). For example, a phosphoric acid molecule, $H_3PO_4$, has three acidic hydrogens, each bonded to an oxygen atom. Carbonic acid ($H_2CO_3$) and sulfuric acid ($H_2SO_4$) are also polyprotic acids. Polyprotic acids do not lose all of their acidic hydrogen atoms in water to the same extent. We consider the loss of hydrogens by a polyprotic acid as a stepwise process. For example, consider carbonic acid, which we treat as the dissolved form of $CO_2$:

$$CO_2(g) + H_2O(l) \rightleftharpoons H_2CO_3(aq)$$

When the $H_2CO_3$ ionizes in water, only small amounts of $HCO_3^-(aq)$ and $H_3O^+(aq)$ are produced. Significantly less $CO_3^{2-}(aq)$ is present in the solution. Two acid ionization equations are used to describe the equilibria that exist in solution, each with its own $K_a$ value:

$$H_2CO_3(aq) + H_2O(l) \rightleftharpoons H_3O^+(aq) + HCO_3^-(aq) \qquad K_{a1} = 4.5 \times 10^{-7}$$
$$HCO_3^-(aq) + H_2O(l) \rightleftharpoons H_3O^+(aq) + CO_3^{2-}(aq) \qquad K_{a2} = 4.7 \times 10^{-11}$$

Sulfuric acid, $H_2SO_4$, is a polyprotic acid that ionizes completely in water to form $HSO_4^-(aq)$ and $H_3O^+(aq)$:

$$H_2SO_4(aq) + H_2O(l) \longrightarrow H_3O^+(aq) + HSO_4^-(aq) \qquad \text{complete ionization}$$

Once $HSO_4^-(aq)$ forms, it also acts as an acid, but as a weak acid:

$$HSO_4^-(aq) + H_2O(l) \rightleftharpoons H_3O^+(aq) + SO_4^{2-}(aq) \qquad K_a = 1.2 \times 10^{-2}$$

$H_3PO_4$

$H_2CO_3$

$H_2SO_4$

**FIGURE 13.10** Polyprotic acids have more than one acidic hydrogen. How many acidic hydrogen atoms does each of these acids have?

The low $K_a$ value for the second ionization shows that the equilibrium lies toward the reactants; only a small portion of the $HSO_4^-(aq)$ ionizes. A smaller portion of the $HSO_4^-(aq)$ ionizes compared to $H_2SO_4$ because it is more difficult to remove a positively charged $H^+$ ion from a negatively charged $HSO_4^-$ ion than from a neutral $H_2SO_4$ molecule. An aqueous solution of sulfuric acid, then, contains essentially no $H_2SO_4(aq)$, a relatively large amount of $HSO_4^-(aq)$, and a small amount of $SO_4^{2-}(aq)$ (Figure 13.11).

The $K_a$ values are often labeled to indicate the particular step in the overall ionization process ($K_{a1}$, $K_{a2}$, $K_{a3}$, etc.). Table 13.6 lists several polyprotic acids and their $K_a$ values. Notice that for successive ionizations, the $K_a$ values decrease significantly as $H^+$ ions are removed.

$H_2SO_4(aq)$

**FIGURE 13.11** Sulfuric acid, $H_2SO_4$, is a strong polyprotic acid. It ionizes completely in water to form $HSO_4^-$ $(aq)$ and $H_3O^+(aq)$. The $HSO_4^-(aq)$ produced also ionizes, but to a much lesser extent. In an aqueous solution of $H_2SO_4$, there are essentially no $H_2SO_4$ molecules, lots of $HSO_4^-(aq)$ ions, and a small fraction of $SO_4^{2-}(aq)$ ions. Can you identify the $HSO_4^-(aq)$ and $SO_4^{2-}(aq)$ ions in this figure?

**TABLE 13.6**  Acid Ionization Constants for Some Polyprotic Acids (at 25°C)

| Formula | Name | $K_{a1}$ | $K_{a2}$ | $K_{a3}$ |
|---------|------|----------|----------|----------|
| $H_2CO_3$ | carbonic acid | $4.5 \times 10^{-7}$ | $4.7 \times 10^{-11}$ | |
| $H_3C_6H_5O_7$ | citric acid | $7.4 \times 10^{-4}$ | $1.7 \times 10^{-5}$ | $4.0 \times 10^{-7}$ |
| $H_2S$ | hydrosulfuric acid | $8.9 \times 10^{-8}$ | $1.0 \times 10^{-19}$ | |
| $H_2C_2O_4$ | oxalic acid | $5.6 \times 10^{-2}$ | $1.5 \times 10^{-4}$ | |
| $H_3PO_4$ | phosphoric acid | $6.9 \times 10^{-3}$ | $6.2 \times 10^{-8}$ | $4.8 \times 10^{-13}$ |
| $H_2SO_4$ | sulfuric acid | strong | $1.2 \times 10^{-2}$ | |
| $H_2C_4H_4O_6$ | tartaric acid | $1.0 \times 10^{-3}$ | $4.3 \times 10^{-5}$ | |

**EXAMPLE 13.7**   **Behavior of Polyprotic Acids in Water**

The tartaric acid, $H_2C_4H_4O_6$, in grapes promotes a crisp flavor and graceful aging in wine. Its $K_a$ values are given in Table 13.6.

(a)  Write equations that show the ionization of tartaric acid in water.
(b)  Besides water, which ion or molecule has the highest concentration in solution when tartaric acid is added to water?

**Solution:**

(a)  Tartaric acid has two $K_a$ values, so it has two acidic hydrogen atoms. Two equations are required to describe its behavior in water. The first shows the ionization of the first $H^+$. The $K_a$ that corresponds to this process is the largest $K_a$ value because the first $H^+$ always ionizes to the greater extent.

$$H_2C_4H_4O_6(aq) + H_2O(l) \rightleftharpoons HC_4H_4O_6^-(aq) + H_3O^+(aq) \quad K_{a1} = 1.0 \times 10^{-3}$$
$$HC_4H_4O_6^-(aq) + H_2O(l) \rightleftharpoons C_4H_4O_6^{2-}(aq) + H_3O^+(aq) \quad K_{a2} = 4.3 \times 10^{-5}$$

(b)  The $K_{a1}$ value for $H_2C_4H_4O_6$ shows that it is a weak acid and ionizes only to a small extent. That means that *most* of the $H_2C_4H_4O_6$ does not ionize and is present in water in the highest concentration.

**Consider This 13.7**

Of the ions and molecules shown in the equilibrium reactions provided for tartaric acid, which has the lowest concentration in a solution of tartaric acid?

**Practice Problem 13.7**

Oxalic acid, $H_2C_2O_4$, occurs in plants and foods such as parsley, rhubarb, almonds, and green beans. Its $K_a$ values are listed in Table 13.6.

(a)  Write equations that show the ionization of oxalic acid in water.
(b)  Besides water, which ion or molecule has the highest concentration in solution when oxalic acid is added to water?

**Further Practice:**  Questions 13.45 and 13.46 at the end of the chapter

Citric acid, $H_3C_6H_5O_7$, is a triprotic acid that makes lemons taste sour.

©Daisuke Morita/Getty Images

**FIGURE 13.12** To a very small extent, water reacts with itself by transferring an $H^+$ ion from one water molecule to another, producing $H_3O^+(aq)$ and $OH^-(aq)$ ions.

## 13.4  Acidic, Basic, and Neutral Solutions

When Olivia tests a soil solution to make sure it is not too acidic, what specifically is she testing for? Remember that acids ionize when dissolved in water to form hydronium ions, $H_3O^+(aq)$. Does the presence of $H_3O^+$ ions in solution indicate an acidic solution? No. All aqueous solutions, and even pure water, have some $H_3O^+$ ions (and $OH^-$ ions) in solution. It is the *relative amounts* of these ions that make the difference. In any **acidic solution,** the $H_3O^+$ ion concentration is greater than the $OH^-$ ion concentration. In any **basic solution,** the $OH^-$ ion concentration is greater than the $H_3O^+$ ion concentration. A **neutral solution** is neither acidic nor basic, but has equal concentrations of $H_3O^+$ and $OH^-$ ions.

### The Ion-Product Constant of Water

Pure water, $H_2O$, is amphoteric. It acts as *both* an acid and a base, but only to a small extent. Water reacts with itself by a process called **self-ionization** (Figure 13.12), in which an $H^+$ ion is transferred from one water molecule to another:

$$H_2O(l) + H_2O(l) \rightleftharpoons H_3O^+(aq) + OH^-(aq)$$

Self-ionization is an equilibrium process, so we can write an equilibrium constant expression for it:

$$K_{eq} = [H_3O^+][OH^-]$$

Because the reactant, $H_2O$, is a liquid, it is not included in the equilibrium constant expression. This equilibrium constant, called the **ion-product constant of water,** is so widely used that it is given a special symbol, $K_w$. At 25°C, $K_w$ has a value of $1.0 \times 10^{-14}$:

$$K_w = [H_3O^+][OH^-] = 1.0 \times 10^{-14}$$

The small value of this equilibrium constant indicates that the equilibrium lies very far to the left, so the concentrations of $H_3O^+$ and $OH^-$ are very small. In pure water the concentrations of these ions must be equal, because they are produced in a 1:1 ratio:

$$[H_3O^+] = [OH^-]$$

From the $K_w$ expression, we know that these concentrations must multiply together to give the $K_w$ value of $1.0 \times 10^{-14}$:

$$1.0 \times 10^{-14} = [H_3O^+][OH^-]$$

For these concentrations to be equal, each must have a concentration of $1.0 \times 10^{-7}\,M$, because if we substitute this value for the $H_3O^+$ concentration and the $OH^-$ concentration,

$$1.0 \times 10^{-14} = (1.0 \times 10^{-7}\,M) \times (1.0 \times 10^{-7}\,M)$$

we get

$$[H_3O^+] = [OH^-] = 1.0 \times 10^{-7}\,M$$

For our purposes, we will assume the temperature to be 25°C, so we will not have to worry about using $K_w$ values at other temperatures.

In a neutral solution such as pure water, the $H_3O^+$ ion and $OH^-$ ion concentrations are equal. If an acid or base is added to water, the concentrations of both ions change. However, their product continues to be $1.0 \times 10^{-14}$, the value of $K_w$ that is

> Remember from Chapter 12 that square brackets around a formula represent the molar concentration of that substance. For example, $[H_3O^+]$ represents the molar concentration of hydronium ion.

**TABLE 13.7** Definitions of Neutral, Acidic, and Basic Solutions in Aqueous Solution

| Type of Solution | Relative Concentrations | $[H_3O^+]$ | $[OH^-]$ | $K_w$ |
|---|---|---|---|---|
| neutral | $[H_3O^+] = [OH^-]$ | $= 1.0 \times 10^{-7}\ M$ | $= 1.0 \times 10^{-7}\ M$ | $1.0 \times 10^{-14}$ |
| acidic | $[H_3O^+] > [OH^-]$ | $> 1.0 \times 10^{-7}\ M$ | $< 1.0 \times 10^{-7}\ M$ | $1.0 \times 10^{-14}$ |
| basic | $[OH^-] > [H_3O^+]$ | $< 1.0 \times 10^{-7}\ M$ | $> 1.0 \times 10^{-7}\ M$ | $1.0 \times 10^{-14}$ |

constant. How can this be? The reaction of an acid with water increases the $H_3O^+$ ion concentration to a value greater than $1.0 \times 10^{-7}\ M$. At the same time, the $OH^-$ ion concentration decreases, with the result that the ion-product constant for water remains equal to $1.0 \times 10^{-14}$. In a similar fashion, the addition of a base to water increases the $OH^-$ ion concentration and decreases the $H_3O^+$ ion concentration, but the ion-product constant remains unchanged. Table 13.7 summarizes the definitions of neutral, acidic, and basic solutions in aqueous solution.

## Calculating $H_3O^+$ and $OH^-$ Ion Concentrations

If we know either the $H_3O^+$ ion concentration or the $OH^-$ ion concentration, we can calculate the other unknown concentration using the ion-product constant of water and its expression:

$$K_w = [H_3O^+][OH^-]$$

If we know the $H_3O^+$ ion concentration, $[H_3O^+]$, we rearrange the $K_w$ expression to solve for the $OH^-$ ion concentration, $[OH^-]$:

$$[OH^-] = \frac{K_w}{[H_3O^+]} = \frac{1.0 \times 10^{-14}}{[H_3O^+]}$$

If we know the $OH^-$ ion concentration, $[OH^-]$, we rearrange the $K_w$ expression to solve for the $H_3O^+$ ion concentration, $[H_3O^+]$:

$$[H_3O^+] = \frac{K_w}{[OH^-]} = \frac{1.0 \times 10^{-14}}{[OH^-]}$$

These calculations are shown in Example 13.8.

---

**EXAMPLE 13.8**  **Calculating $H_3O^+$ and $OH^-$ Ion Concentrations**

Given the concentration of $H_3O^+$ ion in water, calculate the concentration of $OH^-$ ion in the solution. Identify each solution as acidic, basic, or neutral:

(a) $[H_3O^+] = 1.0 \times 10^{-9}\ M$
(b) $[H_3O^+] = 0.0010\ M$

**Solution:**

In any aqueous solution at 25°C, the product of the $H_3O^+$ ion concentration and the $OH^-$ concentration equals $1.0 \times 10^{-14}$:

$$K_w = [H_3O^+][OH^-] = 1.0 \times 10^{-14}$$

We rearrange the equation to solve for the hydroxide ion concentration:

$$[OH^-] = \frac{1.0 \times 10^{-14}}{[H_3O^+]}$$

(a) We substitute the $H_3O^+$ ion concentration of $1.0 \times 10^{-9}$ $M$ into the equation to solve for the $OH^-$ ion concentration. Recall that equilibrium constants do not have units. When you analyze the units in calculations that involve equilibrium constants, it might appear that they will not cancel properly. Assume that if you're calculating the concentration of a chemical species involved in the equilibrium that it will have units of molarity.

$$[OH^-] = \frac{1.0 \times 10^{-14}}{[H_3O^+]} = \frac{1.0 \times 10^{-14}}{1.0 \times 10^{-9} \, M} = 1.0 \times 10^{-5} \, M$$

The solution is basic because $[OH^-] > [H_3O^+]$ and because $[H_3O^+] < 1.0 \times 10^{-7}$ $M$.

(b) We substitute the $H_3O^+$ ion concentration of $0.0010$ $M$ into the equation to solve for the $OH^-$ ion concentration:

$$[OH^-] = \frac{1.0 \times 10^{-14}}{[H_3O^+]} = \frac{1.0 \times 10^{-14}}{0.0010 \, M} = 1.0 \times 10^{-11} \, M$$

The solution is acidic because $[H_3O^+] > [OH^-]$ and because $[H_3O^+] > 1.0 \times 10^{-7}$ $M$.

### Consider This 13.8

Without doing any calculations, determine if a solution with an $OH^-$ concentration of $1.0 \times 10^{-10}$ $M$ would be acidic or basic.

### Practice Problem 13.8

Given the concentration of $OH^-$ ion in each solution, calculate the concentration of $H_3O^+$ in the solution. Identify each solution as acidic, basic, or neutral.

(a) $[OH^-] = 1.0 \times 10^{-8}$ $M$
(b) $[OH^-] = 0.010$ $M$

**Further Practice:**  Questions 13.55 and 13.56 at the end of the chapter

---

Suppose we are using a solution in a laboratory that is labeled $0.010$ $M$ HCl. Can we determine its concentrations of $H_3O^+$ ion and $OH^-$ ion? Hydrochloric acid, HCl, is a strong acid, so it ionizes completely to form $H_3O^+$ ions and $Cl^-$ ions. The concentration of $H_3O^+$ is actually the concentration of HCl labeled on the bottle:

$$HCl(aq) + H_2O(l) \longrightarrow \underset{0.010 \, M}{H_3O^+(aq)} + \underset{0.010 \, M}{Cl^-(aq)}$$

**Student Hot Spot**

Student data indicate you may struggle with determining hydronium ion and hydroxide ion concentrations for solutions of strong acids or strong bases. Access the SmartBook to view additional Learning Resources on this topic.

From the $H_3O^+$ concentration, $[H_3O^+]$, and the value of $K_w$, we can determine the $OH^-$ ion concentration, $[OH^-]$, as we did in Example 13.8. Work the problem and assure yourself that the value is $1.0 \times 10^{-12}$ $M$ $OH^-$.

Suppose we have a $0.10$ $M$ solution of sodium hydroxide, NaOH, in the laboratory. What are the $H_3O^+$ and $OH^-$ ion concentrations in this solution? Sodium hydroxide is a base, so it forms $OH^-(aq)$ ions when dissolved in water. Because it is a strong base with only one $OH^-$ group, $[OH^-]$ is equal to the concentration of NaOH labeled on the bottle:

When strong bases such as $Ba(OH)_2$ contain more than one $OH^-$ ion per formula unit, the $OH^-$ concentration is twice the concentration of the $Ba(OH)_2$ solution. Why?

$$NaOH(s) \xrightarrow{H_2O} \underset{0.10 \, M}{Na^+(aq)} + \underset{0.10 \, M}{OH^-(aq)}$$

The hydroxide ion concentration, $[OH^-]$, is $0.10$ $M$. Using the $K_w$ expression, calculate the $H_3O^+$ ion concentration, $[H_3O^+]$, as $1.0 \times 10^{-13}$ $M$.

| EXAMPLE 13.9 | Calculating [H₃O⁺] and [OH⁻] in Strong Acid and Strong Base Solutions |
|---|---|

Stomach acid is about $1.0\ M$ HCl. What is the $H_3O^+$ concentration in stomach acid? What is the $OH^-$ concentration?

**Solution:**

Hydrochloric acid, HCl, is a strong acid, so it ionizes completely in water. For every mole of HCl that ionizes, one mole of $H_3O^+$ forms. The concentration of $H_3O^+$, then, is $1.0\ M$. The $OH^-$ concentration is determined from the $H_3O^+$ concentration and the expression for the ion-product constant for water:

$$[H_3O^+][OH^-] = 1.0 \times 10^{-14}$$

$$[OH^-] = \frac{1.0 \times 10^{-14}}{[H_3O^+]}$$

$$= \frac{1.0 \times 10^{-14}}{1.0\ M}$$

$$= 1.0 \times 10^{-14}\ M$$

**Consider This 13.9**

How would the concentration of $H_3O^+$ in $1.0\ M$ $H_2SO_4$ compare to that of $1.0\ M$ HCl?

**Practice Problem 13.9**

Sodium hydroxide is sometimes called lye or caustic soda. What is the $OH^-$ concentration in a $0.85\ M$ lye solution? What is the $H_3O^+$ concentration?

**Further Practice:** Questions 13.57 and 13.58 at the end of the chapter

Now suppose that we are using a $0.010\ M$ acetic acid solution, $CH_3CO_2H(aq)$, in the lab. Recall that a $0.010\ M$ HCl solution has an $H_3O^+$ ion concentration of $0.010\ M$ because HCl is a strong acid—it completely ionizes. The $H_3O^+$ concentration in a $0.010\ M$ $CH_3CO_2H$ solution is only $4.2 \times 10^{-4}\ M$. Why? Acetic acid is a weak acid that does not completely ionize. In a $0.010\ M$ solution, only about 4.2% of the acetic acid molecules ionize, while in a hydrochloric acid solution, 100% of the HCl molecules ionize.

Calculating the values of the $H_3O^+$ and $OH^-$ concentrations in weak acid and weak base solutions is beyond the scope of this textbook.

## 13.5  The pH Scale

Jake tests the acidity of pool water using pH indicators and a pH meter. Olivia also measures the pH of soil to determine whether to add lime. Perhaps you have heard the term pH, as it is used frequently in advertisements for shampoos and skin-care products. The pH of solutions is easily measured and tells us about the acidity or basicity of a substance. In this section, we will explore how pH is related to the concentration of $H_3O^+$ (and $OH^-$) and examine the significance of the pH scale.

Our teeth are protected by our saliva, which is basic with a pH of 7.2 or higher. Drinking carbonated beverages can cause tooth decay because these beverages contain phosphoric acid and citric acid and thus have a pH of 2.5 to 3.3.

### Calculating pH

Concentrations of $H_3O^+$ ions can vary widely among concentrated and dilute acids, so it is convenient to express the acidity of aqueous solutions on a *pH scale*. The **pH** of a solution is defined as the negative logarithm (base 10) of the $H_3O^+$ concentration:

$$pH = -\log[H_3O^+]$$

For example, the $H_3O^+$ concentration in pure water is $1.0 \times 10^{-7}\ M$. Substituting this value into the equation for pH gives a pH of 7.00:

$$pH = -\log[H_3O^+]$$
$$= -\log(1.0 \times 10^{-7})$$
$$= 7.00$$

MATH
TOOLBOX
**13.1**

The log function is a button on most calculators. See Math Toolbox 13.1 at the end of the chapter for instructions on using this function on your calculator.

Like pure water, all neutral solutions have equal concentrations of $H_3O^+$ and $OH^-$ ($1.0 \times 10^{-7}\ M$ each), so their pH values are 7. Acidic solutions have pH values less than 7, and basic solutions have pH values greater than 7.

To better understand the meaning of the log function, consider the pH values of solutions that range in $H_3O^+$ concentration from $1.0 \times 10^{-1}\ M$ to $1.0 \times 10^{-14}\ M$, as shown in Figure 13.13. Notice that each pH value equals the exponent in the $H_3O^+$ concentration value, but with a positive sign. Use the logarithmic definition of pH to convince yourself that the pH values listed there can be calculated from the $H_3O^+$ concentration values given.

The relationship between pH and concentration allows us to determine the pH without a calculator only when the $H_3O^+$ concentration is exactly a power of 10.

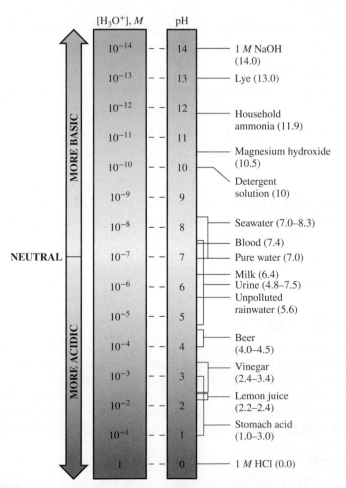

**FIGURE 13.13** Aqueous solutions we use in our daily lives generally have pH values that range from 0 to 14. However, some acids and bases used in the laboratory are sometimes so concentrated that their pH values are less than 0 or greater than 14. What concentrations of HCl would give a pH less than zero? What concentrations of NaOH would give a pH greater than 14?

When it is not—for example, $5.0 \times 10^{-4}$ $M$—you can use the log function on your calculator to determine the pH:

$$pH = -\log[H_3O^+]$$
$$= -\log(5.0 \times 10^{-4})$$
$$= 3.30$$

The pH of a strong acid solution can be determined if you know the concentration of the acid. For the strong acid $HNO_3$, for example, the concentration of $H_3O^+$ equals the $HNO_3$ concentration because strong acids ionize completely. What is the pH of a 0.10 $M$ $HNO_3$ solution? Because $HNO_3$ is a strong acid, the concentration of $H_3O^+$ ion in the solution is 0.10 $M$, or $1.0 \times 10^{-1}$ $M$. The pH is calculated by substituting the $H_3O^+$ concentration into the following equation:

$$pH = -\log[H_3O^+]$$
$$= -\log(1.0 \times 10^{-1})$$
$$= 1.00$$

We can also determine the pH of a strong base solution, such as a 0.010 $M$ NaOH solution. Strong bases ionize completely, so the concentration of $OH^-$ ion is the same as the concentration of NaOH, 0.010 $M$ or $1.0 \times 10^{-2}$ $M$. Because the pH equation requires the $H_3O^+$ concentration, we first calculate the $H_3O^+$ concentration from the $OH^-$ concentration using the $K_w$ expression:

$$K_w = [H_3O^+][OH^-] = 1.0 \times 10^{-14}$$
$$[H_3O^+] = \frac{1.0 \times 10^{-14}}{[OH^-]} = \frac{1.0 \times 10^{-14}}{0.010\ M} = 1.0 \times 10^{-12}\ M$$

Now that we know the $H_3O^+$ concentration, $1.0 \times 10^{-12}$ $M$, we can calculate the pH of the solution:

$$pH = -\log[H_3O^+]$$
$$= -\log(1.0 \times 10^{-12})$$
$$= 12.00$$

> How do we apply rules for significant figures to pH calculations? The rule is that the number of decimal places in a pH value should equal the number of significant figures in the $H_3O^+$ concentration value.

---

## EXAMPLE 13.10   Calculating pH

What is the pH of each of the following solutions? Once you've done the calculation, check your answer to make sure it makes sense.

(a)  0.0010 $M$ HBr
(b)  0.035 $M$ $HNO_3$
(c)  0.035 $M$ KOH

**Solution:**

(a)  Hydrobromic acid, HBr($aq$), is a strong acid, which ionizes completely, so the $H_3O^+$ concentration equals the concentration of HBr:

$$[H_3O^+] = 0.0010\ M$$

The pH can be calculated from the $H_3O^+$ concentration:

$$pH = -\log[H_3O^+]$$
$$= -\log(0.0010)$$
$$= 3.00$$

MATH
TOOLBOX
**13.1**

A pH of 3.00 makes sense for an acid because it is less than 7, as we would expect for a solution of any acid.

(b) Nitric acid, $HNO_3(aq)$, is also a strong acid, so the $H_3O^+$ concentration equals the concentration of $HNO_3$:

$$[H_3O^+] = 0.035 \ M$$

The pH can be calculated from the $H_3O^+$ concentration:

$$\begin{aligned} pH &= -\log[H_3O^+] \\ &= -\log(0.035) \\ &= 1.46 \end{aligned}$$

We would expect a lower pH for $HNO_3$ than for HBr because the $HNO_3$ solution was more concentrated.

(c) Potassium hydroxide, $KOH(aq)$, is a strong base, which dissociates completely to form $K^+(aq)$ and $OH^-(aq)$ ions. The concentration of KOH, therefore, is also the $OH^-$ concentration:

$$[OH^-] = 0.035 \ M$$

In order to calculate pH, we need to determine the $H_3O^+$ concentration from the $OH^-$ concentration and the $K_w$ expression:

$$K_w = [H_3O^+][OH^-] = 1.0 \times 10^{-14}$$

$$[H_3O^+] = \frac{1.0 \times 10^{-14}}{[OH^-]} = \frac{1.0 \times 10^{-14}}{0.035 \ M} = 2.9 \times 10^{-13} \ M$$

Now that we know the $H_3O^+$ concentration, we can calculate the pH of the solution:

$$\begin{aligned} pH &= -\log[H_3O^+] \\ &= -\log(2.9 \times 10^{-13}) \\ &= 12.54 \end{aligned}$$

A high pH (greater than 7) is expected because this is a solution of a base.

**Consider This 13.10**

Predict whether the pH of $0.035 \ M \ HNO_2$, a weak acid, should be greater than, less than, or the same as the pH of $0.035 \ M \ HNO_3$.

**Practice Problem 13.10**

What is the pH of each of the following solutions? Once you've done the calculation, check your answer to make sure it makes sense.

(a) $0.00085 \ M$ HCl
(b) $0.10 \ M$ NaOH
(c) $1.0 \ M \ HNO_3$

**Further Practice:** Questions 13.71 and 13.72 at the end of the chapter

In very dilute solutions of strong acids and bases, determining the pH is more complicated. For example, consider a $1.0 \times 10^{-8} \ M$ HCl solution. As a strong acid, we expect HCl to completely ionize, but an $H_3O^+$ concentration of $1.0 \times 10^{-8} \ M$ does not make sense because acidic solutions have $H_3O^+$ concentrations greater than the $1.0 \times 10^{-7} \ M \ H_3O^+$ from pure water. When you encounter very dilute solutions of acids with concentrations less than $1.0 \times 10^{-7} \ M$, assume that the pH is slightly less than 7.0. If you have a solution of a base with a concentration less than $1.0 \times 10^{-7} \ M$, assume that the pH is slightly greater than 7.0.

## Calculating pOH

Just as pH can be calculated from the $H_3O^+$ concentration, a value called pOH can be calculated from the $OH^-$ concentration of the solution. The pOH of a solution is defined as the negative logarithm (base 10) of the hydroxide ion concentration:

$$pOH = -\log[OH^-]$$

Earlier in this section we found that the pH of a 0.010 $M$ NaOH solution is 12.00. The pOH of this strong base can be calculated directly, because the $OH^-$ concentration equals the concentration of NaOH, 0.010 $M$. This value can be substituted into the pOH equation:

$$
\begin{aligned}
pOH &= -\log[OH^-] \\
&= -\log(0.010) \\
&= -\log(1.0 \times 10^{-2}) \\
&= 2.00
\end{aligned}
$$

Thus, the pOH of this 0.010 $M$ NaOH solution is 2.00. Note that the pH and pOH values total 14.00. This is true for any aqueous solution at 25°C.

$$pH + pOH = 14.00$$

This relationship allows us to determine the pH or pOH of a solution if one value or the other is known. For example, if the pOH of a solution is 4.00, its pH is calculated by subtracting the pOH from 14.00:

$$
\begin{aligned}
pH + pOH &= 14.00 \\
pH &= 14.00 - pOH \\
&= 14.00 - 4.00 \\
&= 10.00
\end{aligned}
$$

## Calculating Concentrations from pH or pOH

Olivia has beautiful blue hydrangeas in her garden, but she wants some of her plants to show flowers that are pink (Figure 13.14). After some research, she discovers that the flower color, in part, depends on the pH of the soil. Blue flowers occur when the soil is in the pH range of 5.5 to 6.5. For pink flowers, the pH needs to be slightly more alkaline, in the range of 7.0 to 7.5. When she measures the pH of a soil solution from her garden, she learns that it is slightly acidic with a pH of 6.20. She learns that making the soil more alkaline to produce pink flowers can be achieved by adding calcium oxide (lime) to the soil. Her endeavor into soil pH has Olivia curious about the concentration of $H_3O^+$ ion in the soil solution. She needs an equation that relates $[H_3O^+]$ to pH, and the pH equation accomplishes this:

$$pH = -\log[H_3O^+]$$

To get $[H_3O^+]$, we must rearrange the equation so that $[H_3O^+]$ is on one side by itself. First, we multiply both sides by $-1$:

$$-pH = \log[H_3O^+]$$

To get rid of the log expression in front of $[H_3O^+]$, we take the inverse log of both sides. The $H_3O^+$ concentration, then, equals the inverse log of the negative pH:

$$\text{Inverse log } (-pH) = [H_3O^+]$$

Taking the inverse log of a value ($x$) is the same as 10 raised to the $x$ power. In other words, *the $H_3O^+$ concentration equals 10 raised to the $-pH$ power:*

$$10^{-pH} = [H_3O^+]$$

The $H_3O^+$ concentration in a soil sample with a pH of 6.20 can be calculated using this equation:

$$[H_3O^+] = 10^{-pH} = 10^{-6.20} = 6.3 \times 10^{-7} \; M$$

Why does pH + pOH = 14.00? The relationship can be derived mathematically from the $K_w$ expression of water:

$$[H_3O^+][OH^-] = 1.0 \times 10^{-14}$$

Taking the negative log of both sides of the equation gives

$$-\log([H_3O^+][OH^-]) = -\log(1.0 \times 10^{-14})$$

The log of a product of two values equals the *sum* of the logs of the terms:

$$-\log[H_3O^+] - \log[OH^-] = -\log(1.0 \times 10^{-14})$$

We know $-\log[H_3O^+]$ and $-\log[OH^-]$ equal pH and pOH, respectively, so we can substitute pH and pOH into this equation:

$$pH + pOH = 14.00$$

**FIGURE 13.14** The color of hydrangeas depends on soil pH and the presence of aluminum ion. (A) In acidic soil in the pH range of 5.5 to 6.5, and in the presence of adequate aluminum, flowers appear deep blue. (B) In more alkaline soil (pH 7.0 to 7.5), the flowers are pink.

(both): ©McGraw-Hill Education/Louis Rosenstock

The inverse log function on your calculator is the $10^x$ function, which is normally found above the log function as an alternate function for the same key. For more information on how to use the inverse log function, see Math Toolbox 13.1.

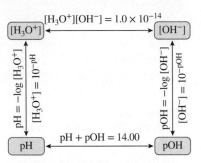

**FIGURE 13.15** If you know pH, pOH, $[H_3O^+]$, or $[OH^-]$, you can calculate any of the other quantities.

 **Student Hot Spot**

Student data indicate you may struggle with determining hydronium ion and hydroxide ion concentrations from pH or pOH. Access the SmartBook to view additional Learning Resources on this topic.

 MATH TOOLBOX **13.1**

The $OH^-$ concentration can be determined from the pOH with a similar equation:

$$[OH^-] = 10^{-pOH}$$

The relationships among pH, pOH, $[H_3O^+]$, and $[OH^-]$ are represented in Figure 13.15. If you know one of these values for a solution, you can calculate any of the other three. For example, consider the values shown in Figure 13.16. Notice how the four quantities $[H_3O^+]$, pH, pOH, and $[OH^-]$ are related. Note that there is more than one way to convert from one value to another, as illustrated in Example 13.11.

---

**EXAMPLE 13.11**    **Converting Between pH, pOH, $[H_3O^+]$, and $[OH^-]$**

Jake measured the pH of water in a swimming pool as 8.10. What is the $OH^-$ concentration in the pool water?

**Solution:**

Figure 13.15 shows two alternate routes we can take for this calculation. One route is first to convert pH to pOH and then convert pOH to the $OH^-$ concentration:

$$\boxed{pH} \longrightarrow \boxed{pOH} \longrightarrow \boxed{[OH^-]}$$

To convert from pH to pOH, we use the relationship between them:

$$pH + pOH = 14.00$$
$$8.10 + pOH = 14.00$$
$$pOH = 5.90$$

To convert from pOH to $[OH^-]$, we use the relationship that allows us to solve for hydroxide ion concentration:

$$[OH^-] = 10^{-pOH} = 10^{-5.90} = 1.3 \times 10^{-6}\ M$$

The stepwise calculation is summarized as follows:

$$\boxed{pH} \xrightarrow{\ pH + pOH = 14.00\ } \boxed{pOH} \xrightarrow{\ [OH^-] = 10^{-pOH}\ } \boxed{[OH^-]}$$

An alternative route would be to convert first from pH to $[H_3O^+]$ and then from $[H_3O^+]$ to $[OH^-]$:

$$\boxed{pH} \longrightarrow \boxed{[H_3O^+]} \longrightarrow \boxed{[OH^-]}$$

Try this route. You should get the same answer.

**Consider This 13.11**

If the pOH of a solution is above 14.0, what can we say about the pH of that solution?

**Practice Problem 13.11**

When the pH of water in a lake falls below about 4.5, the lake may be considered dead because few organisms can survive in such an acidic environment. What is the pH of a lake that has an $OH^-$ concentration of $1.0 \times 10^{-9}\ M?$ Would this lake be considered dead?

**Further Practice:** Questions 13.73 and 13.74 at the end of the chapter

**FIGURE 13.16**  Notice how these quantities are related. The product of $[H_3O^+]$ and $[OH^-]$ is always $1 \times 10^{-14}$, $K_w$. The sum of the pH and pOH is always 14.

## Measuring pH

Several methods can be used to measure the pH of a solution. Jake sometimes measures the pH of pool water using a pH meter that is accurate to within hundredths of a pH unit (Figure 13.17). Another less accurate, but often convenient, method is the use of pH *indicators*. Indicators are brightly colored organic dyes that are weak acids or bases. In solution they form an equilibrium with their conjugate bases or conjugate acids, respectively. The color of the indicator depends on whether the dye is in its acidic or basic form.

Consider, for example, the indicator dye phenolphthalein, which is often used in acid-base titrations. Phenolphthalein has an acidic form (abbreviated HIn) that is colorless and a basic form (In$^-$) that is pink. (The formula HIn is commonly used to represent the acidic form of an indicator. The formula In$^-$ represents the conjugate base of the acid form.)

$$\text{HIn}(aq) + \text{H}_2\text{O}(l) \rightleftharpoons \text{H}_3\text{O}^+(aq) + \text{In}^-(aq)$$

colorless                pink<br>
phenolphthalein       phenolphthalein

Phenolphthalein changes from colorless to pink between pH 8.2 and 10. Below pH 8.2 it is primarily in its colorless acidic form because high concentrations of $H_3O^+$ cause the equilibrium to lie to the HIn side of the equation. (Recall Le Chatelier's principle from Chapter 12.) Above pH 10 phenolphthalein is primarily in its pink basic form because the low concentration of $H_3O^+$ causes the equilibrium to lie to the right side of the equation. The color-change ranges for phenolphthalein and other common indicators are shown in Figure 13.18. Note that most of the indicators have acidic and basic forms that are both colored.

**FIGURE 13.17**  A pH meter is used to determine the pH of a solution by measuring the voltage that develops when electrodes are dipped into the solution.

©McGraw-Hill Education/Stephen Frisch

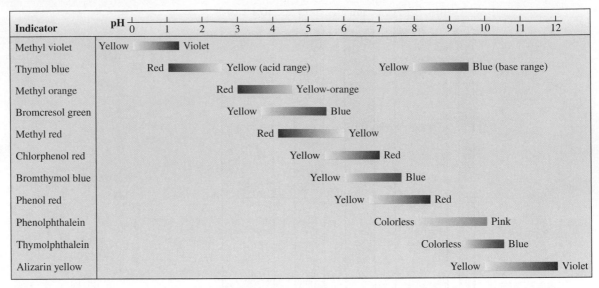

| Indicator | pH | 0 | 1 | 2 | 3 | 4 | 5 | 6 | 7 | 8 | 9 | 10 | 11 | 12 |
|---|---|---|---|---|---|---|---|---|---|---|---|---|---|---|
| Methyl violet | | Yellow ▬ Violet | | | | | | | | | | | | |
| Thymol blue | | | Red ▬ Yellow (acid range) | | | | | | Yellow ▬ Blue (base range) | | | | |
| Methyl orange | | | | | Red ▬ Yellow-orange | | | | | | | | |
| Bromcresol green | | | | | Yellow ▬ Blue | | | | | | | | |
| Methyl red | | | | | | Red ▬ Yellow | | | | | | | |
| Chlorphenol red | | | | | | | Yellow ▬ Red | | | | | | |
| Bromthymol blue | | | | | | | | Yellow ▬ Blue | | | | | |
| Phenol red | | | | | | | | Yellow ▬ Red | | | | | |
| Phenolphthalein | | | | | | | | | Colorless ▬ Pink | | | | |
| Thymolphthalein | | | | | | | | | | Colorless ▬ Blue | | | |
| Alizarin yellow | | | | | | | | | | | Yellow ▬ Violet | | |

**FIGURE 13.18** Each acid-base indicator has a unique pH color-change range. Outside of its range the indicator exists primarily as the acid form or base form, both of which exhibit unique colors. Within its color-change range, the acid and base forms of the indicator are present in similar concentrations. The color of the solution within its range can indicate the pH of the solution.

**FIGURE 13.19** (A) Various types of pH paper contain mixtures of different indicators. They are useful for determining pH at a wide range of pH values. (B) Swimming pool test kits often contain a liquid indicator that can be used to measure the pH.
(both): ©Brian Moeskau/Moeskau Photography

A                    B

The desired pH of the water in fish tanks depends on the species of fish that live there. Many fish are most healthy in water with a pH of 7.0. Bromthymol blue indicator is often used in test kits for fish tanks. What other indicators shown in Figure 13.18 might help determine if the pH is 7.0, slightly above 7.0, or slightly below 7.0?

An indicator reveals if the pH of a solution is above or below a certain value. Indicators also disclose a specific pH within the indicator's color-change range. For example, the indicator phenol red is sometimes used to find the pH of pool water because its color changes between pH 6.8 and 8.4, the usual pH range for swimming pool water (Figure 13.18). Subtle differences in hues are discernible at slightly different pH values.

Indicators are also useful in measuring pH changes during titration reactions between acids and bases (Chapter 11). For example, consider the titration of an acid with base using phenolphthalein as the indicator. At low pH the phenolphthalein is colorless. As base is added, it reacts with the acid to produce water and the solution becomes less acidic. When the solution just becomes pink, the acid has completely reacted and there is an excess of base, as indicated by the pink color. What would you expect to observe if you ran a titration of a base by adding acid and using methyl red as an indicator?

A mixture of indicators having a variety of colors and color-change ranges can be used to measure the pH of any solution. For example, broad-range pH paper is treated with several indicators. The user reads the pH by comparing the color the paper turns to a chart of reference colors and pH values printed on the container (Figure 13.19A). The liquid universal indicator sometimes used in laboratories contains a mixture of indicators and works on the same principle. Liquid indicators are also used in swimming pool test kits (Figure 13.19B).

Another liquid universal indicator can be made from the juice of a red (purple) cabbage. A compound present in the juice is unusual because it can exhibit several different colors based on the pH of the solution being tested (Figure 13.20).

Acidic    Neutral    Basic

**FIGURE 13.20** The juice from a red cabbage can be used as a pH indicator. It exhibits quite a variety of colors under different pH conditions.
©Ken Welsh/age fotostock

## 13.6 Buffered Solutions

Olivia's job requires her to find out when there is too much acid or base in garden soil. Because she is considering a career in medicine, she knows what happens when there is too much acid or base in the human bloodstream. Slight changes in blood pH can drastically affect the way our bodies function. For biological reactions to occur normally, blood must stay within the narrow range of pH 7.35 to 7.45. If it rises too high or falls too low, severe health problems, or even death, can occur.

Fortunately our bodies have mechanisms that normally maintain a constant blood pH. One is the action of acid-base buffers in the blood. Blood is a *buffered solution* because it resists changes in pH when limited amounts of acid or base are added to it.

A **buffer,** or a *buffer system,* is a combination of a weak acid and its conjugate base (or a weak base and its conjugate acid) in about equal concentrations.

The *bicarbonate buffer system* regulates pH in the human bloodstream. You might recognize the bicarbonate ion, $HCO_3^-$, as the conjugate base of carbonic acid, $H_2CO_3$. Carbonic acid is formed from the reaction of $CO_2$ with water:

$$CO_2(g) + H_2O(l) \rightleftharpoons H_2CO_3(aq)$$

In aqueous solution, carbonic acid is in equilibrium with its conjugate base:

$$H_2CO_3(aq) + H_2O(l) \rightleftharpoons HCO_3^-(aq) + H_3O^+(aq)$$

Carbonic acid is also in equilibrium with its conjugate base in the blood. However, the concentration of the bicarbonate ion in the blood is much greater than the concentration that would result from ionization of the acid alone. In the blood, the position of the equilibrium between carbonic acid and the carbonate ion is closely tied to respiration. Changes in respiration can affect blood pH by altering the concentration of $CO_2$ dissolved in the blood. Consider the combination and simplification of the two reactions just described:

$$CO_2(g) + H_2O(l) \rightleftharpoons H_2CO_3(aq) \rightleftharpoons HCO_3^-(aq) + H^+(aq)$$

Following Le Chatelier's principle, which was described in Chapter 12, notice how the concentration of $CO_2$ might affect the blood pH. We would expect an increase in $CO_2$ concentration to shift the equilibria to the right, causing the $H^+(aq)$ concentration to increase. The result would be a decrease in pH. This is a condition called *respiratory acidosis,* and results from the lungs' inability to release $CO_2$ at a rate comparable to the rate of its production in the body. This condition occurs in people who suffer from emphysema and other lung conditions.

Conversely, a decrease in $CO_2$ concentration would cause the equilibria to shift to the left, resulting in a higher blood pH. This condition is called *respiratory alkalosis* and occurs when people have panic attacks. During these attacks, they breathe very quickly, releasing more $CO_2$ than normal, potentially causing the $H^+(aq)$ concentration to decrease and the pH of their blood to rise. If these episodes go on for too long, the body's natural reaction will be to cause fainting, which naturally causes breathing to slow down. Have you ever wondered why a remedy for

Phenolphthalein, like most acid-base indicators, consists of large organic molecules. The acid form of phenolphthalein is colorless, and the base form is bright pink. Compare the structures of the acid and base forms of phenolphthalein. Can you see that they are a conjugate acid-base pair? Do they differ by one $H^+$?

HO ⬡⬡ OH

+ $H_2O$ ⇌

Acid, HIn (colorless)

HO

+ $H_3O^+$

Base, In⁻ (pink)

 **ANIMATION:** Buffer Solutions

The buffers in our bloodstream are effective for only normal fluctuations in acid and base concentrations. If too much acid enters the bloodstream, the pH falls below 7.35 and a condition called *acidosis* occurs. *Alkalosis* occurs when the blood is too basic with a pH above 7.45. Both of these situations are dangerous because they disrupt the body's cellular functions.

When the pH of a fish tank is too high or too low, a solution containing an appropriate buffer system is used to bring the pH to its desired level. The $H_2PO_4^-/HPO_4^{2-}$ buffer system is often found in aquarium products that help to maintain a pH 7 environment.

Buffer after addition of $H_3O^+$ | Buffer with equal concentrations of conjugate base and acid | Buffer after addition of $OH^-$

$$H_2O(l) + H_2CO_3(aq) \longleftarrow H_3O^+(aq) + HCO_3^-(aq) \qquad H_2CO_3(aq) + OH^-(aq) \longrightarrow HCO_3^-(aq) + H_2O(l)$$

**FIGURE 13.21** When an acid is added to an $H_2CO_3/HCO_3^-$ buffer system, most of the acid is consumed by $HCO_3^-$, as shown to the left. When a base is added as shown to the right, most is consumed by $H_2CO_3$. The relative amounts of the buffer components change, but the pH change is very small.

stopping a panic attack is to breathe into a paper bag? Rebreathing the air in the bag increases the concentration of inhaled $CO_2$, replenishing some that has been lost.

The body's ability to control dramatic changes in pH is based on the naturally occurring amounts of carbonic acid (dissolved $CO_2$) and bicarbonate ion. This combination of relatively high concentrations of both species keeps blood pH in the range of 7.35 to 7.45.

In the laboratory, chemists often prepare buffered solutions if it is important that they maintain a constant pH in a procedure or experiment. To prepare a solution containing the $H_2CO_3/HCO_3^-$ buffer system, for example, both the acid and the base must be added to the solution in similar amounts. In this case the conjugate base is an ion, so it must be added in the form of a soluble salt such as sodium bicarbonate, $NaHCO_3$. Once dissolved in water, the salt dissociates to $Na^+(aq)$ and $HCO_3^-(aq)$:

$$NaHCO_3(s) \xrightarrow{\text{H}_2\text{O}} Na^+(aq) + HCO_3^-(aq)$$

**ANIMATION:** Using a Buffer

How does a buffer system prevent large changes in pH when an acid or base is added to the solution? It does so by reacting with most of the $H_3O^+$ or $OH^-$ from the added acid or base (Figure 13.21). Consider again the $H_2CO_3/HCO_3^-$ buffer system:

$$H_2CO_3(aq) + H_2O(l) \rightleftharpoons HCO_3^-(aq) + H_3O^+(aq)$$

When a base such as NaOH is added, most of the $OH^-(aq)$ from NaOH reacts with the acid, $H_2CO_3$:

$$H_2CO_3(aq) + OH^-(aq) \longrightarrow HCO_3^-(aq) + H_2O(l)$$

When an acid such as HCl is added, the $H_3O^+(aq)$ ions from HCl react with the base, $HCO_3^-$:

$$HCO_3^-(aq) + H_3O^+(aq) \longrightarrow H_2CO_3(aq) + H_2O(l)$$

A buffer system works if there is enough acid and conjugate base in the system to react with all the added $H_3O^+(aq)$ or $OH^-(aq)$. The greater the concentrations of the acid and its conjugate base in the buffer mixture, the more effectively the buffer prevents large changes in pH.

### EXAMPLE 13.12 How a Buffer System Works

Another buffer system in our blood is the $H_2PO_4^-/HPO_4^{2-}$ system. Describe, using a balanced equation, how this system prevents the pH from falling too low when acid enters the bloodstream.

## Solution:

The base component of this buffer system, $HPO_4^{2-}$, reacts with excess $H_3O^+$ from added acid:

$$HPO_4^{2-}(aq) + H_3O^+(aq) \rightleftharpoons H_2PO_4^-(aq) + H_2O(l)$$

Because most of the excess $H_3O^+$ is consumed, the pH remains relatively constant and usually drops by less than 0.05 pH units.

### Consider This 13.12

Predict what would happen to pH if the moles of strong acid added to the $H_2PO_4^-/HPO_4^{2-}$ buffer system in solution were greater than the moles of $HPO_4^{2-}$ present in the solution.

### Practice Problem 13.12

Describe, using a balanced equation, how the $H_2PO_4^-/HPO_4^{2-}$ system in our blood prevents pH levels from rising too high when base enters the bloodstream.

**Further Practice:** Questions 13.109 and 13.110 at the end of the chapter

Just as a buffer system can moderate pH, the relative amounts of acid and base of a buffer system determine the pH of the solution. In most swimming pool water, for example, an $HOCl/OCl^-$ buffer system is present:

$$HOCl(aq) + H_2O(l) \rightleftharpoons OCl^-(aq) + H_3O^+(aq)$$

Pool water looks best at a pH of 7.5, when the relative amounts of HOCl and $OCl^-$ are equal. At lower pHs, much of the $OCl^-$ is converted to HOCl; at high pHs, much of the HOCl is converted to $OCl^-$. When the pH is adjusted by adding another kind of acid or base, the equilibrium shifts, so the relative amounts of HOCl and $OCl^-$ also adjust.

The examples of buffer systems we have discussed are conjugate pairs of weak acids and weak bases. Why wouldn't a strong acid, like HCl, and its conjugate base make an effective buffer? Strong acids ionize completely in water:

$$HCl(aq) + H_2O(l) \longrightarrow H_3O^+(aq) + Cl^-(aq)$$

The $Cl^-$ ion does not react with $H_3O^+$ to form HCl, because HCl is a strong acid and the acid molecule does not exist in solution. For a buffer system to work, both of the components of the conjugate acid-base pair must exist in equilibrium in solution. This is not the case with a strong acid or base.

### EXAMPLE 13.13   Buffer Systems

Which of the following, when added to water, can act as a buffer system? For each buffer system, write a balanced equation.

(a) HF and NaF
(b) $HNO_3$ and $KNO_3$
(c) $NH_3$ and $NH_4Cl$

## Solution:

(a) This is a buffer system, because HF is a weak acid and the ionic compound NaF provides a source of its conjugate base, $F^-$, when it dissolves in water. We can ignore the spectator ion, $Na^+$. The following equation shows the conjugate acid-base pair that exists in solution at equilibrium:

$$HF(aq) + H_2O(l) \rightleftharpoons F^-(aq) + H_3O^-(aq)$$

(b) This is not a buffer system, because $HNO_3$ is a strong acid. Strong acids cannot be components of buffers, because they ionize completely in water and are not in equilibrium with their conjugate bases.

(c) This is a buffer system, because $NH_3$ is a weak base and its conjugate acid, $NH_4^+$, forms when the ionic compound $NH_4Cl$ dissolves in water. We can ignore the spectator ion, $Cl^-$. The following is a conjugate acid-base pair that will exist at equilibrium in solution:

$$NH_3(aq) + H_2O(l) \rightleftharpoons NH_4^+(aq) + OH^-(aq)$$

### Consider This 13.13

If we mix 1.00 mol HF with 0.50 mol NaOH, the following reaction will occur in aqueous solution:

$$HF(aq) + NaOH(aq) \longrightarrow NaF(aq) + H_2O(l)$$

The NaOH is the limiting reactant, and only 0.50 mol HF reacts. Is the resulting solution a buffer system?

### Practice Problem 13.13

Which of the following systems, when added to water, can act as a buffer system? For each buffer system, write a balanced equation.

(a) HCl and NaOH

(b) $CH_3CO_2H$ and $NaCH_3CO_2$

(c) HBr and KBr

**Further Practice:** Questions 13.111 and 13.112 at the end of the chapter

---

 ## Math Toolbox 13.1    Log and Inverse Log Functions

### Log Functions

Let's consider several substances that vary in their $H_3O^+(aq)$ concentrations:

| Substance | $[H_3O^+]$ |
|---|---|
| HCl (1 $M$) | 1 $M$ (or $10^0$ $M$) |
| HCl (0.1 $M$) | 0.1 $M$ (or $10^{-1}$ $M$) |
| HCl (0.01 $M$) | 0.01 $M$ (or $10^{-2}$ $M$) |
| lemon juice | 0.001 $M$ (or $10^{-3}$ $M$) |
| vinegar | 0.0001 $M$ (or $10^{-4}$ $M$) |
| beer | 0.00001 $M$ (or $10^{-5}$ $M$) |
| urine | 0.000001 $M$ (or $10^{-6}$ $M$) |
| pure water | 0.0000001 $M$ (or $10^{-7}$ $M$) |

Except for pure water, these substances are all acids. Suppose we wanted to compare their concentrations graphically. In a bar graph, notice what happens to those with the lower concentrations:

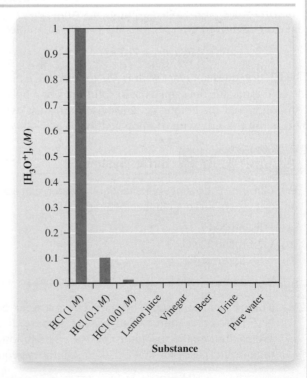

**Math Toolbox 13.1 *(continued)***

Can you distinguish lemon juice, which we can consume, from stomach acid, which causes damage to the esophagus? No. Logarithms are very useful in situations like this because logarithms make it easier to compare numbers that are inconvenient to work with and differ by orders of magnitude.

The *log* function on your calculator stands for *logarithm to the base 10.* When you take the log of a number, you are asking, "10 to what power will give the value after the word *log?*" For example, the log of 1000 is 3.0, because 10 must be raised to the third power to get 1000:

$$\log(1000) = 3.0$$

Likewise, the log of 0.001 is −3.0 because 10 must be raised to the −3.0 power to get 0.001:

$$\log(0.001) = -3.0$$

To use the log function on most calculators, you first press the $\boxed{\log}$ key and then enter the number. (On some calculators, these two steps may be reversed.) Try taking the log of 0.001 on your calculator.

**Step 1:** Press the $\boxed{\log}$ key.
**Step 2:** Enter 0.001 (or $1 \times 10^{-3}$), and then the $\boxed{\text{ENTER}}$ or $\boxed{=}$ key.

You should get −3.0. If this does not work, try reordering the key strokes.

Numerical values with large positive or negative exponents, such as concentrations of $H_3O^+$, are inconvenient to work with. A more convenient way to express such values is to use a logarithmic scale. The pH scale is an example of a logarithmic scale. By convention, the "p" of something is its negative logarithm:

$$pX = -\log X$$

In chemistry, the most common examples are the pH scale and the pOH scale:

$$pH = -\log[H^+] = -\log[H_3O^+]$$
$$pOH = -\log[OH^-]$$

To obtain the pH or pOH of a solution, you must first take the log of the concentration and then multiply by −1 (change the sign). Try using your calculator to determine the pH of a solution that has an $H_3O^+$ concentration of $5.0 \times 10^{-4}$ M.

$$pH = -\log[H_3O^+]$$

**Step 1:** Press the $\boxed{+/-}$ (change of sign) key.
**Step 2:** Press the $\boxed{\log}$ key.
**Step 3:** Enter the $H_3O^+$ concentration, and then the $\boxed{\text{ENTER}}$ or $\boxed{=}$ key.

$$pH = -\log(5.0 \times 10^{-4}) = 3.30$$

If you did not get the right answer, your calculator may require you to reorder the key strokes.

How do we apply rules for significant figures when taking the log of a number? The rule is that the number of *decimal places* in the calculated answer should equal the number of significant figures in the logged value. Look at the previous calculation of pH to see how this rule is applied.

**EXAMPLE 13.14   Log Functions**

Use your calculator to find the pH of the following solutions.
(a) $[H_3O^+] = 1.0 \times 10^{-8}$ M
(b) $[H_3O^+] = 6.2 \times 10^{-1}$ M

**Solution:**
To calculate the pH from the $H_3O^+$ concentration, use the following relationship:

$$pH = -\log[H_3O^+]$$

Using your calculator, first enter a negative sign, followed by the $\boxed{\log}$ key and the $H_3O^+$ concentration. Then hit the $\boxed{\text{ENTER}}$ or $\boxed{=}$ key.
(a) $pH = -\log(1.0 \times 10^{-8}) = 8.00$
(b) $pH = -\log(6.2 \times 10^{-1}) = 0.21$

**Practice Problem 13.14**

Use your calculator to find the pH of the following solutions.
(a) $[H_3O^+] = 7.5 \times 10^{-3}$ M
(b) $[H_3O^+] = 5.6 \times 10^{-8}$ M

**Further Practice:** Questions 13.3 and 13.4 at the end of the chapter

If we take the log of our original data and then change the sign, we get the following values:

| Substance | $[H_3O^+]$ | $-\log[H_3O^+]$ |
|---|---|---|
| HCl (1 $M$) | 1 $M$ (or $10^0$ $M$) | 0 |
| HCl (0.1 $M$) | 0.1 $M$ (or $10^{-1}$ $M$) | 1 |
| HCl (0.01 $M$) | 0.01 $M$ (or $10^{-2}$ $M$) | 2 |
| lemon juice | 0.001 $M$ (or $10^{-3}$ $M$) | 3 |
| vinegar | 0.0001 $M$ (or $10^{-4}$ $M$) | 4 |
| beer | 0.00001 $M$ (or $10^{-5}$ $M$) | 5 |
| urine | 0.000001 $M$ (or $10^{-6}$ $M$) | 6 |
| pure water | 0.0000001 $M$ (or $10^{-7}$ $M$) | 7 |

A plot of the log values now allows us to distinguish the $H_3O^+$ content in each of these substances:

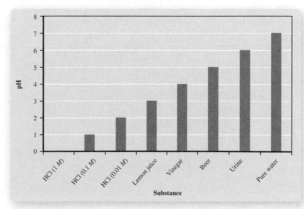

**Math Toolbox 13.1 *(continued)***

The pH scale is more convenient to use. However, we have to keep in mind that an integer difference in pH is a 10-fold difference in $H_3O^+$ concentration. For example, the pH of beer is around 5 and that of vinegar is around 4. Vinegar is 10 times as acidic as beer ($10^{-4}$ $M$ for vinegar versus $10^{-5}$ $M$ for beer).

### Inverse Log ($10^x$) Functions

When you have an equation that involves a log function, you sometimes need to solve for the quantity that follows the log. For example, if you are given pH, you might want to solve for $[H_3O^+]$:

$$pH = -\log[H_3O^+]$$

To solve for $[H_3O^+]$, we must take the inverse log (antilog) of both sides of the equation. To rearrange the equation so that $[H_3O^+]$ is on one side by itself, first multiply both sides by $-1$:

$$-pH = \log[H_3O^+]$$

To get rid of the log expression in front of $[H_3O^+]$, take the inverse log of both sides:

$$\text{inverse log} (-pH) = \text{inverse log} (\log[H_3O^+])$$

The value of $[H_3O^+]$, then, equals the inverse log of the negative pH:

$$\text{inverse log} (-pH) = [H_3O^+]$$

Taking the inverse log of $-pH$ is the same as 10 raised to the $-pH$ power. The inverse log function on your calculator is the $10^x$ function, which is often found above the log function as an alternate function for the same key. Another way to describe the relationship between $[H_3O^+]$ and pH is to say $[H_3O^+]$ equals 10 raised to the $-pH$ power:

$$[H_3O^+] = 10^{-pH}$$

To use the inverse log ($10^x$) function for this calculation on most calculators, you first press the $\boxed{\text{INV}}$, $\boxed{\text{SHIFT}}$, or $\boxed{\text{2nd}}$ button followed by the $\boxed{\text{log}}$ key. Then enter the negative of the pH value. The inverse log function, or $10^x$ function, should appear on your calculator above the log function as the alternative function. By pressing $\boxed{\text{INV}}$, $\boxed{\text{SHIFT}}$, or $\boxed{\text{2nd}}$ button first, you tell the calculator to use the alternative function. (On some calculators, you have to enter the $-pH$ value first.) Try taking the inverse log where pH is 3.00.

$$[H_3O^+] = 10^{-pH}$$

**Step 1:** Press the $\boxed{\text{INV}}$, $\boxed{\text{SHIFT}}$, or $\boxed{\text{2nd}}$ button.
**Step 2:** Press the $\boxed{\text{log}}$ button to use the $10^x$ function.
**Step 3:** Enter $-3.00$, and then the $\boxed{\text{ENTER}}$ or $\boxed{=}$ key.

Did you get $[H_3O^+] = 0.0010$ or $1.0 \times 10^{-3}$ $M$? If not, your calculator may require you to reorder the key strokes. The steps for calculating $[OH^-]$ from pOH are the same as calculating $[H_3O^+]$ from pH:

$$[H_3O^+] = 10^{-pH}$$
$$[OH^-] = 10^{-pOH}$$

The general calculation steps for calculating $[H_3O^+]$ or $[OH^-]$ are

**Step 1:** Press the $\boxed{\text{INV}}$, $\boxed{\text{SHIFT}}$, or $\boxed{\text{2nd}}$ button.
**Step 2:** Press the $\boxed{\text{log}}$ button.
**Step 3:** Enter $-pH$ (or $-pOH$), and then the $\boxed{\text{ENTER}}$ or $\boxed{=}$ key.

(On some calculators you may have to reorder the key strokes.)

---

### EXAMPLE 13.15 Inverse Log Functions

Calculate the $H_3O^+$ concentrations for the solutions with the following pH values.

(a) pH = 9.40
(b) pH = 3.60

**Solution:**

To calculate the $H_3O^+$ concentration from the pH, use the following relationship:

$$[H_3O^+] = 10^{-pH}$$

Using your calculator, first enter the $\boxed{\text{INV}}$, $\boxed{\text{SHIFT}}$, or $\boxed{\text{2nd}}$ button. Then press the $\boxed{\text{log}}$ key to give you the $10^x$ function. (On some calculators the $\boxed{10^x}$ button is its own key.) Finally, press the $\boxed{+/-}$ (change sign) key, enter the pH, and then the $\boxed{\text{ENTER}}$ or $\boxed{=}$ key.

(a) $[H_3O^+] = 10^{-9.40} = 4.0 \times 10^{-10}$ $M$
(b) $[H_3O^+] = 10^{-3.60} = 2.5 \times 10^{-4}$ $M$

**Practice Problem 13.15**

Calculate the $OH^-$ concentration from the following pOH values.

(a) pOH = 8.00
(b) pOH = 2.60

**Further Practice:** Questions 13.5 and 13.6 at the end of the chapter

# CHAPTER REVIEW

## KEY CONCEPTS

- According to Brønsted-Lowry theory, an acid is a substance that donates an $H^+$ ion to another substance, and a base is an $H^+$ acceptor.
  - Brønsted-Lowry acids react with water in a process called ionization to produce hydronium ions, $H_3O^+$.
  - Brønsted-Lowry bases ionize or dissociate in water to produce hydroxide ions, $OH^-$.
- Acids and bases have varying strengths.
  - Strong acids and bases ionize or dissociate completely in water.
  - When weak acids and bases dissolve in water, only a small percentage of the molecules ionizes; the equilibrium lies to the left toward the reactant molecules.
  - Equilibrium constants for the reactions of acids with water are called acid ionization constants, $K_a$. These values can be used to compare the strengths of different weak acids. The stronger the acid, the greater the $K_a$ value.
- The relative concentrations of $H_3O^+$ and $OH^-$ ions in an aqueous solution are determined by the ion-product constant for water, $K_w$, which is the equilibrium constant for the reaction $2H_2O(l) \rightleftharpoons H_3O^+(aq) + OH^-(aq)$.
  - $K_w$ is equal to the product of the hydronium and hydroxide ion concentrations: $K_w = [H_3O^+] \times [OH^-]$.
  - The value of $K_w$ is $1.0 \times 10^{-14}$ at 25°C.
  - Neutral solutions have equal concentrations of $H_3O^+$ and $OH^-$, each equal to $1.0 \times 10^{-7}$ $M$ at 25°C. Acidic solutions contain a greater concentration of $H_3O^+$ than $OH^-$. Basic solutions contain a greater concentration of $OH^-$ than $H_3O^+$. Because $[H_3O^+] \times [OH^-] = 1.0 \times 10^{-14}$, as one concentration increases the other decreases.
- The acidity of a solution is commonly expressed in terms of pH, the negative logarithm of the hydronium ion concentration: $pH = -\log[H_3O^+]$.
  - At 25°C, acidic solutions have a pH less than 7, basic solutions have a pH greater than 7, and neutral solutions have a pH equal to 7.
  - Indicators and pH meters are commonly used to measure the pH of a solution.
- Buffered solutions contain a weak acid and its conjugate base (or a weak base and its conjugate acid) in similar concentrations. Buffers help to prevent large changes in pH by reacting with small amounts of added acid or base.

## KEY RELATIONSHIPS

| Relationship | Equation |
|---|---|
| The self-ionization constant for water equals the product of the hydronium ion concentration and the hydroxide ion concentration. At 25°C, this constant equals $1.0 \times 10^{-14}$. | $K_w = [H_3O^+][OH^-] = 1.0 \times 10^{-14}$ |
| The pH of a solution equals the negative log of the hydronium ion concentration. The pOH of a solution equals the negative log of the hydroxide ion concentration. | $pH = -\log[H_3O^+]$ <br> $pOH = -\log[OH^-]$ |
| The hydronium ion concentration equals the inverse log of the negative pH value. The hydroxide ion concentration equals the inverse log of the negative pOH value. | $[H_3O^+] = 10^{-pH}$ <br> $[OH^-] = 10^{-pOH}$ |
| The pH and pOH of a solution total 14.00. | $pH + pOH = 14.00$ |

## KEY TERMS

acid ionization constant,
   $K_a$ (13.3)
acidic solution (13.4)
amphoteric substance (13.1)
Arrhenius model of acids and
   bases (13.1)

basic solution (13.4)
Brønsted-Lowry theory (13.1)
buffer (13.6)
conjugate acid (13.1)
conjugate base (13.1)
hydronium ion (13.1)

ion-product constant of water,
   $K_w$ (13.4)
neutral solution (13.4)
pH (13.5)
polyprotic acid (13.3)

self-ionization (13.4)
strong acid (13.2)
strong base (13.2)
weak acid (13.2)
weak base (13.2)

## QUESTIONS AND PROBLEMS

The following questions and problems, except those in *Additional Questions* and *Concept Review Questions,* are paired. Questions in a pair focus on the same concept. Answers to selected questions and problems are in Appendix D.

### Matching Definitions with Key Terms

13.1    Match the key terms with the descriptions provided.
   (a) a base that ionizes or dissociates completely when dissolved in water
   (b) a theory that defines an acid as a substance that donates an $H^+$ to another substance in solution and a base as a substance that accepts an $H^+$ in solution
   (c) a substance that forms after a base gains an $H^+$ ion
   (d) an acid containing more than one acidic hydrogen
   (e) a solution in which the $H_3O^+$ concentration is greater than the $OH^-$ concentration; a solution with a pH less than 7
   (f) a base that does not completely ionize when dissolved in water
   (g) a substance that can act as either an acid or a base
   (h) the equilibrium constant for the self-ionization of water; the product of the $H_3O^+$ ion concentration and the $OH^-$ ion concentration in any aqueous solution
   (i) a process in which one molecule transfers an $H^+$ to another molecule of the same substance; water does this to a very small extent

13.2    Match the key terms with the descriptions provided.
   (a) an acid that ionizes completely when dissolved in water
   (b) a model that describes acids as substances that generate $H^+$ ions in solution and bases as substances that generate $OH^-$ ions in solution
   (c) an aqueous hydrogen ion; $H_3O^+(aq)$
   (d) a solution in which the $OH^-$ concentration is greater than the $H_3O^+$ concentration; a solution with a pH greater than 7
   (e) an acid that does not completely ionize when dissolved in water
   (f) a substance that forms after an acid loses an $H^+$ ion
   (g) an equilibrium constant for the ionization of an acid in water; a value that expresses the strength of a weak acid
   (h) a measure of the acidity of aqueous solutions; the negative logarithm of the $H_3O^+$ concentration
   (i) a combination of a weak acid and its conjugate base (or a weak base and its conjugate acid) in similar amounts; when in solution, something that helps prevent large changes in pH when small amounts of $H_3O^+(aq)$ or $OH^-(aq)$ are added

### Math Toolbox Questions

13.3    Use your calculator to find the log of the following numbers.
   (a) $10^{-9}$
   (b) $1 \times 10^{-11}$
   (c) $7.4 \times 10^3$
   (d) $10^5$
   (e) 1

13.4    Use your calculator to find the log of the following numbers.
   (a) $10^4$
   (b) $1 \times 10^{-6}$
   (c) $1.7 \times 10^8$
   (d) $10^{-8}$
   (e) 10

13.5    Use your calculator to find the inverse log of the following numbers.
   (a) 1.20
   (b) −6.20
   (c) 0

13.6    Use your calculator to find the inverse log of the following numbers.
   (a) 12.7
   (b) −9.4
   (c) 1

### What Are Acids and Bases?

13.7    What are some properties of acids?

13.8    What are some properties of bases?

13.9    List some common foods or household products that are bases.

13.10   List some common foods or household products that are acids.

13.11   In terms of the Arrhenius concept of acids and bases, how does an acid behave when dissolved in water? How does an Arrhenius base behave?

13.12   How does the Arrhenius concept of acids and bases emphasize that they are electrolytes?

13.13   How is the Brønsted-Lowry theory of acids and bases different from the Arrhenius model?

13.14   Why is $H_3O^+(aq)$ a more accurate representation of an aqueous $H^+$ ion than $H^+(aq)$?

13.15 Identify the first reactant in each equation as an acid or a base.

(a) $HCN(aq) + H_2O(l) \rightleftharpoons H_3O^+(aq) + CN^-(aq)$

(b) $SO_4^{2-}(aq) + H_2O(l) \rightleftharpoons HSO_4^-(aq) + OH^-(aq)$

(c) $C_6H_5OH(aq) + NaOH(aq) \rightleftharpoons$
$H_2O(l) + C_6H_5O^-(aq) + Na^+(aq)$

13.16 Identify the first reactant in each equation as an acid or a base.

(a) $CH_3CH_2NH_3^+(aq) + H_2O(l) \rightleftharpoons$
$H_3O^+(aq) + CH_3CH_2NH_2(aq)$

(b) $HCN(aq) + H_2O(l) \rightleftharpoons CN^-(aq) + H_3O^+(aq)$

(c) $NaOH(aq) + CH_3CO_2H(aq) \rightleftharpoons$
$H_2O(aq) + CH_3CO_2^-(aq) + Na^+(aq)$

13.17 Write the formula of the conjugate base of each acid.

(a) $HNO_2$

(b) $HF$

(c) $H_3BO_3$

13.18 Write the formula of the conjugate base of each acid.

(a) $HOCl$

(b) $H_3PO_4$

(c) $HCO_3^-$

13.19 Write the formula of the conjugate acid of each base.

(a) $OH^-$

(b) $C_6H_5NH_2$

(c) $HCO_3^-$

13.20 Write the formula of the conjugate acid of each base.

(a) $NH_3$

(b) $F^-$

(c) $CO_3^{2-}$

13.21 Write an equation to show how the following acids or bases react with water.

(a) $HCl(g)$

(b) $HClO_4(l)$

(c) $CH_3CO_2^-(aq)$

13.22 Write an equation to show how the following acids or bases behave in water.

(a) $NaOH(s)$

(b) $CN^-(aq)$

(c) $HNO_3(l)$

13.23 Which of the following are amphoteric species?

(a) $H_2O$

(b) $HSO_3^-$

(c) $SO_4^{2-}$

13.24 Which of the following are amphoteric species?

(a) $NH_4^+$

(b) $HBO_3^{2-}$

(c) $H_2PO_4^-$

13.25 In an $H_2CO_3$ molecule, both hydrogen atoms are bonded to oxygen atoms. How many acidic hydrogen atoms does carbonic acid have?

13.26 The Lewis structure for the acid phenol is shown. How many acidic hydrogen atoms does phenol have?

## Strong and Weak Acids and Bases

13.27 How do strong acids and bases differ from weak acids and bases?

13.28 What do we mean when we say weak acids and bases only partially ionize in water?

13.29 Identify each of the following as a strong acid, weak acid, strong base, or weak base. For each, write an equation that describes its reaction with water.

(a) $H_2SO_4(aq)$

(b) $Ca(OH)_2(aq)$

(c) $Na_2CO_3(aq)$

(d) $H_3C_6H_5O_7(aq)$

(e) $C_6H_5NH_2(aq)$

13.30 Identify each of the following as a strong acid, weak acid, strong base, or weak base. For each, write an equation that describes its reaction with water.

(a) $HCN(aq)$

(b) $NaCN(aq)$

(c) $LiOH(aq)$

(d) $CH_3NH_2(aq)$

(e) $H_2C_2O_4(aq)$

13.31 Match the molecular-level diagram to each of the following compounds in aqueous solution: HCl, HF, and $NH_3$.

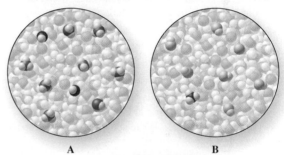

A                    B

C

13.32 Draw a molecular-level representation of a solution of each of the following acids and bases.

(a) $Ba(OH)_2$

(b) $NaOH$

(c) $HOCl$

13.33 Which compound, KBr or KF, is expected to behave as a weak base? Why?

13.34 Which compound, $NaHCO_3$ or $NaNO_3$, is expected to behave as a weak base? Why?

## Relative Strengths of Weak Acids

**13.35**   If acid A ionizes to a greater extent than acid B, which acid is stronger?

**13.36**   If acid C has a smaller $K_a$ value than acid D, which acid is stronger?

**13.37**   Which solution has the highest concentration of $H_3O^+$, a 0.50 $M$ solution of acetic acid, $CH_3CO_2H$, or a 0.50 $M$ solution of formic acid, $HCO_2H$? Consult Table 13.5 for $K_a$ values.

**13.38**   Which acid ionizes to the greater extent in water, nitrous acid, $HNO_2$, or hydrofluoric acid, $HF$? Consult Table 13.5 for $K_a$ values.

**13.39**   Sodium fluoride, $NaF$, and sodium acetate, $NaCH_3CO_2$, are ionic compounds that contain the conjugate bases of weak acids. Assuming you start with solutions of equal concentrations, which would have the greater concentration of $OH^-$? (*Hint:* Consult Table 13.5.)

**13.40**   Potassium hypochlorite, $KOCl$, and potassium nitrite, $KNO_2$, are ionic compounds that contain the conjugate bases of weak acids. Assuming you start with solutions of equal concentrations, which would have the greater concentration of $OH^-$? (*Hint:* Consult Table 13.5.)

**13.41**   What is the difference between a concentrated acid solution and a strong acid solution?

**13.42**   Can a weak acid solution have a higher $H_3O^+$ concentration than a strong acid solution? Explain.

**13.43**   What is a polyprotic acid?

**13.44**   Write the formulas and names of three different polyprotic acids.

**13.45**   Why does hydrosulfuric acid, $H_2S$, have two acid ionization constants? Write equations that correspond to each $K_a$ value.

**13.46**   Which species, $H_2S$ or $HS^-$, ionizes to the greater extent in water? Explain.

**13.47**   Which molecules or ions are present in an aqueous solution of oxalic acid, $H_2C_2O_4$, a polyprotic acid? Which is present in the least concentration?

**13.48**   Which molecules or ions are present in an aqueous solution of citric acid, $H_3C_6H_5O_7$, a triprotic acid? Which is present in the smallest concentration?

## Acidic, Basic, and Neutral Solutions

**13.49**   Under what conditions is the ion-product constant of water, $K_w$, a constant?

**13.50**   Does the product of the $H_3O^+$ ion concentration and the $OH^-$ ion concentration equal $1.0 \times 10^{-14}$ when substances such as acids and bases are added to water? Explain.

**13.51**   What are the concentrations of $H_3O^+$ and $OH^-$ in pure water (at 25°C)?

**13.52**   What happens to the concentrations of $H_3O^+$ and $OH^-$ when acid is added to water? When a base is added to water?

**13.53**   Identify each of the following conditions as acidic, basic, or neutral.
   (a) $[H_3O^+] < [OH^-]$
   (b) $[H_3O^+] = 1.0 \times 10^{-4}\ M$
   (c) $[OH^-] = 1.0 \times 10^{-3}\ M$

**13.54**   Identify each of the following conditions as acidic, basic, or neutral.
   (a) $[OH^-] = 1.0 \times 10^{-11}\ M$
   (b) $[H_3O^+] = 1.0 \times 10^{-9}\ M$
   (c) $[H_3O^+] > [OH^-]$

**13.55**   Calculate the $H_3O^+$ ion concentration from the $OH^-$ ion concentration. Then identify the solution as acidic, basic, or neutral.
   (a) $[OH^-] = 1.0 \times 10^{-3}\ M$
   (b) $[OH^-] = 1.0 \times 10^{-11}\ M$
   (c) $[OH^-] = 3.2 \times 10^{-8}\ M$

**13.56**   Calculate the $OH^-$ ion concentration from the $H_3O^+$ ion concentration. Then identify the solution as acidic, basic, or neutral.
   (a) $[H_3O^+] = 1.0 \times 10^{-3}\ M$
   (b) $[H_3O^+] = 1.0 \times 10^{-5}\ M$
   (c) $[H_3O^+] = 7.3 \times 10^{-10}\ M$

**13.57**   What is the concentration of $H_3O^+$ in each of the following solutions?
   (a) 0.010 $M$ $HNO_3$
   (b) 0.020 $M$ $HClO_4$
   (c) 0.015 $M$ $NaOH$

**13.58**   What is the concentration of $OH^-$ in each of the following solutions?
   (a) 0.0010 $M$ $KOH$
   (b) 0.0050 $M$ $NaOH$
   (c) 0.0010 $M$ $HCl$

## The pH Scale

**13.59**   Olivia noticed that the pH of her garden soil decreased after it rained. Did her garden soil become more acidic or less acidic?

**13.60**   Jake noticed that swimming pool water tends to increase in pH more quickly during hot weather. Does pool water tend to become more acidic or more basic during hot weather?

**13.61**   If one solution has an $H_3O^+$ concentration that is 10 times that of another solution, what is the difference in their pH values?

**13.62**   If solution A has a pH that is three pH units greater than that of solution B, how much greater is the $H_3O^+$ concentration in solution B than in solution A?

**13.63**   Why does the pH scale typically range from pH 0 to pH 14?

**13.64**   Can a solution have a pH less than 0 or greater than 14? Explain.

**13.65**   Can the pH of an acid solution be greater than 7?

**13.66**   What is the pH range for acidic solutions? For basic solutions?

**13.67**   What is the pH of solutions having the following $H_3O^+$ concentrations? Identify each as acidic, basic, or neutral.
   (a) $[H_3O^+] = 1.0 \times 10^{-3}\ M$
   (b) $[H_3O^+] = 1.0 \times 10^{-13}\ M$
   (c) $[H_3O^+] = 3.4 \times 10^{-10}\ M$

**13.68**   What is the pH of solutions having the following $H_3O^+$ concentrations? Identify each as acidic, basic, or neutral.
   (a) $[H_3O^+] = 1.0 \times 10^{-9}\ M$
   (b) $[H_3O^+] = 1.0 \times 10^{-14}\ M$
   (c) $[H_3O^+] = 7.2 \times 10^{-3}\ M$

13.69   What is the pH of solutions having the following $OH^-$ concentrations? Identify each as acidic, basic, or neutral.
(a) $[OH^-] = 1.0 \times 10^{-4}\ M$
(b) $[OH^-] = 1.0 \times 10^{-7}\ M$
(c) $[OH^-] = 8.2 \times 10^{-10}\ M$

13.70   What is the pH of solutions having the following $OH^-$ concentrations? Identify each as acidic, basic, or neutral.
(a) $[OH^-] = 6.4 \times 10^{-4}\ M$
(b) $[OH^-] = 1.0\ M$
(c) $[OH^-] = 2.7 \times 10^{-10}\ M$

13.71   What is the pH of the following solutions? Identify each as acidic, basic, or neutral.
(a) $0.010\ M\ HNO_3$
(b) $0.020\ M\ HClO_4$
(c) $0.015\ M\ NaOH$

13.72   What is the pH of the following solutions? Identify each as acidic, basic, or neutral.
(a) $0.0010\ M\ KOH$
(b) $0.0050\ M\ NaOH$
(c) $0.0010\ M\ HCl$

13.73   Complete the following table.

|     | $[H_3O^+]$ | $[OH^-]$ | pH | pOH | Acidic or Basic? |
| --- | --- | --- | --- | --- | --- |
| (a) | $1.0 \times 10^{-5}$ | | | | |
| (b) | | $1.0 \times 10^{-4}$ | | | |
| (c) | | | | 8.00 | |
| (d) | | | 8.54 | | |
| (e) | | $9.0 \times 10^{-10}$ | | | |

13.74   Complete the following table.

|     | $[H_3O^+]$ | $[OH^-]$ | pH | pOH | Acidic or Basic? |
| --- | --- | --- | --- | --- | --- |
| (a) | | | 0.40 | | |
| (b) | | $9.0 \times 10^{-10}$ | | | |
| (c) | $1.0 \times 10^{-8}$ | | | | |
| (d) | | | | 2.00 | |
| (e) | $4.5 \times 10^{-2}$ | | | | |

13.75   Determine the pH and pOH of a $0.0050\ M\ HCl$ solution. What is the relationship between the pH and pOH values?

13.76   Determine the pH and pOH of a $0.050\ M\ NaOH$ solution. What is the relationship between the pH and pOH values?

13.77   Determine the $H_3O^+$ concentration in solutions with each of the following pH values. Identify each solution as acidic, basic, or neutral.
(a) pH = 5.00
(b) pH = 12.00
(c) pH = 5.90

13.78   Determine the $OH^-$ concentration given the pH of the following solutions. Identify each solution as acidic, basic, or neutral.
(a) pH = 5.00
(b) pH = 11.00
(c) pH = 12.80

13.79   Determine the $H_3O^+$ concentration in the following solutions. Identify each solution as acidic, basic, or neutral.
(a) household ammonia, pH = 11.00
(b) blood, pH = 7.40
(c) lime juice, pH = 1.90

13.80   Determine the $H_3O^+$ concentration in the following solutions. Identify each solution as acidic, basic, or neutral.
(a) orange juice, pH = 3.50
(b) lye, 4.0% sodium hydroxide, pH = 14.00
(c) saliva, pH = 7.00

13.81   What is the $OH^-$ concentration in each of the following solutions?
(a) gastric juice, pH = 1.00
(b) magnesium hydroxide, pH = 10.50
(c) soft drink, pH = 3.60

13.82   What is the $H_3O^+$ concentration in each of the following solutions?
(a) lake water, pOH = 6.00
(b) coffee, pOH = 8.90
(c) borax, pOH = 4.50

13.83   Would you expect the pH of $0.010\ M$ acetic acid to be higher or lower than 2.0?

13.84   Would you expect the pH of a $0.010\ M\ NH_3$ solution to be higher or lower than 12.0?

13.85   If the pH of an NaOH solution is 13.0, what is the concentration of NaOH?

13.86   If the pH of an HCl solution is 2.0, what is the concentration of HCl?

13.87   If you need to know the pH of a solution to two decimal places, which method of pH measurement discussed in this chapter would be the best choice?

13.88   What are the advantages of using pH paper to measure pH?

13.89   What types of compounds are acid-base indicators?

13.90   Why do we use the symbols HIn and $In^-$ to represent the different forms of an indicator? Which is the acid form and which is the base form of the indicator?

13.91   What is a pH color-change range for an indicator? How is it useful for determining pH between the range of pH values?

13.92   How can an indicator have two color-change ranges? What characteristic must be present?

13.93   Which of the indicators listed in Figure 13.18 would be a good choice for determining the pH of a solution that normally has a pH between 9 and 10?

13.94   Which of the indicators in Figure 13.18 would be a good choice for determining the pH of a solution that normally has a pH between 4 and 5?

13.95   Predict the color of the methyl orange indicator in a solution that has a pH of 3.5.

13.96   Predict the color of the thymol blue indicator in a solution that has a pH of 8.8.

13.97   What can we say about the pH of a solution that is red when the indicator thymol blue is added? When the solution is yellow? When the solution is blue?

13.98 What can we say about the pH of a solution that is colorless when phenolphthalein is added? When the solution is dark pink?

13.99 What is a universal indicator? How are universal indicator and pH paper similar?

13.100 How is litmus paper different from pH paper?

13.101 What would you expect to observe if you ran a titration of an acid at a pH of 1.0 by adding base and using bromocresol green as an indicator?

13.102 What would you expect to observe if you ran a titration of a base at a pH of 11.0 by adding acid and using methyl red as an indicator?

## Buffered Solutions

13.103 How is a buffered solution different from a weak acid or a weak base dissolved in water?

13.104 What types of substances make good buffers?

13.105 What is the role of a buffer system in human blood?

13.106 How does the pH change when an acid or base is added to a buffered solution?

13.107 If $CH_3CO_2H$ is added to water, identify an ionic compound that could be added to make a buffered solution.

13.108 If $Na_2SO_4$ is added to water, what should also be added to make a buffered solution?

13.109 A buffer system used to maintain solution pH values around 12 to 13 is prepared by adding similar concentrations of $Na_2HPO_4$ and $Na_3PO_4$ to water.
 (a) Write a balanced equation showing the acid and conjugate base in equilibrium. Omit spectator ions.
 (b) Describe how this buffer system prevents large pH changes when an acid is added.

13.110 A buffer system used to maintain solution pH values around 5 is prepared by adding similar concentrations of $CH_3CO_2H$ and $NaCH_3CO_2$ to water.
 (a) Write a balanced equation showing the acid and conjugate base in equilibrium. Omit spectator ions.
 (b) Describe how this buffer system prevents large pH changes when a base is added.

13.111 Which of the following, when added to water, form a buffered solution?
 (a) HOCl and NaCl
 (b) $HNO_2$ and $KNO_2$
 (c) $CH_3NH_2$ and $CH_3NH_3Cl$

13.112 Which of the following, when added to water, form a buffered solution?
 (a) HF and NaF
 (b) HI and KI
 (c) HOCl and NaOCl

13.113 The bicarbonate buffer system operates in the body's extracellular fluid by the following process:

$$CO_2(g) + H_2O(l) \rightleftharpoons H_2CO_3(aq) \rightleftharpoons$$
$$HCO_3^-(aq) + H^+(aq)$$

What is expected to happen to the blood pH if the lungs are unable to expel $CO_2$ at the same rate as the body produces it? Describe what happens in terms of Le Chatelier's principle.

13.114 For the bicarbonate buffer system described in Question 13.113, what is expected to happen to the blood pH during a panic attack in which more $CO_2$ than normal is released by the body? Describe what happens in terms of Le Chatelier's principle.

## Additional Questions

13.115 Why is $NH_3$ a base but $CH_4$ is not? Is it possible for $CH_4$ to act as a Brønsted-Lowry base?

13.116 Are both acid concentration and acid strength important in determining the concentration of $H_3O^+$ in an acid solution? Explain.

13.117 Describe what happens when NaF dissolves in water. Describe the acid or base properties of this solution.

13.118 Write an equation that shows what happens when $Ba(OH)_2$ dissolves in water. What is the pH of 0.01 $M$ $Ba(OH)_2$?

13.119 Can the pH of a solution of NaOH ever be less than 7 at 25°C? Explain.

13.120 Can the pH of a solution of HCl ever be greater than 7 at 25°C? Explain.

13.121 Which should have the lower pH, a 0.10 $M$ $H_2SO_4$ solution or a 0.10 $M$ $HNO_3$ solution? Explain.

13.122 Consider the reaction classifications we studied in Chapter 5. For each reaction type given, write a chemical reaction in which HCl is a reactant.
 (a) neutralization (double-displacement)
 (b) single-displacement

13.123 Water in a properly maintained swimming pool with a pH of 7.5 has approximately equal concentrations of HOCl(aq) and $OCl^-(aq)$.
 (a) Write a balanced equation showing the acid and conjugate base in equilibrium.
 (b) Describe what happens to this equilibrium when an acid, such as HCl, is added. What happens to the relative concentrations of HOCl and $OCl^-$? What happens to the pH of the pool water?

13.124 Explain why ions such as $Cl^-$, $NO_3^-$, and $I^-$ do not act as bases in water.

13.125 Label all of the species shown in this image of a sulfuric acid, $H_2SO_4(aq)$, solution. Explain why there are no $H_2SO_4$ molecules shown.

13.126 Formic acid, $HCO_2H$, is the substance present in fire ants that causes a sting when they bite you. What is the conjugate base of formic acid?

**13.127** Suppose you wanted to prepare a buffer solution with formic acid, $HCO_2H$, a weak acid with $K_a = 1.8 \times 10^{-4}$. What substances might you mix together to prepare the buffer?

**13.128** Ascorbic acid (vitamin C) is a diprotic acid with the formula $H_2C_6H_6O_6$. The equilibrium constants for the two ionization processes associated with ascorbic acid are

$$K_{a1} = 9.1 \times 10^{-4}$$
$$K_{a2} = 2.0 \times 10^{-12}$$

Write the equilibrium equations that correspond to the ionization processes for ascorbic acid. Rank the species present in a solution of ascorbic acid from lowest to highest concentration.

**13.129** How many acidic hydrogens are present in each of the following acids?

(a)

(c)

(b)

(d)

**13.130** What is the pH of a mixture that contains $0.50\ M$ $Ca(OH)_2$ and $0.50\ M$ LiOH?

**13.131** The following statements about acids are incorrect. Change each statement to make it correct.
   (a) All strong acids have H atoms bonded to electronegative oxygen atoms.
   (b) The conjugate base of a strong acid is itself a strong base.
   (c) Strong acids are very concentrated acids.
   (d) Strong acids produce solutions with a high pH.

**13.132** What is the pH of a solution formed by mixing two equal volumes of solutions of HCl, one with a pH of 1.0 and the other with a pH of 6.0?

**13.133** The ion-product constant of water at 60°C is $1.0 \times 10^{-13}$. What is the pH of pure water at this temperature?

**13.134** Which of the following weak acids has the anion that is the strongest base?

| | |
|---|---|
| $CH_3CO_2H$ | $K_a = 1.8 \times 10^{-5}$ |
| HF | $K_a = 6.3 \times 10^{-4}$ |
| $HCO_2H$ | $K_a = 1.8 \times 10^{-4}$ |
| HCN | $K_a = 6.2 \times 10^{-10}$ |
| HOCl | $K_a = 4.0 \times 10^{-8}$ |

**13.135** Arrange the following solutions in order of increasing hydroxide ion concentration: $0.1\ M$ $HNO_3$, $0.5\ M$ HCl, a solution of $HNO_2$ with a pH of 4.0, a buffer solution with a pOH of 11.0, water.

**13.136** Which of the following combinations will not produce a buffer solution?
NaCl and $CH_3CO_2H$
$NH_4Cl$ and $NH_3$
KCN and HCN
$NaCH_3CO_2$ and $CH_3CO_2H$
$NaHCO_3$ and $H_2CO_3$

**13.137** Which of the following could be added to a solution of sodium acetate to prepare a buffer: acetic acid, hydrochloric acid, potassium acetate, sodium chloride?

**13.138** What change will be caused by the addition of a small amount of HCl to a solution containing nitrite ion ($NO_2^-$) and nitrous acid ($HNO_2$)?

## Concept Review

**13.139** Which of the following statements best describes the following reaction?

$$HCO_3^-(aq) + HSO_4^-(aq) \rightleftharpoons H_2CO_3(aq) + SO_4^{2-}(aq)$$

A. $HCO_3^-(aq)$ is the acid and $HSO_4^-(aq)$ is its conjugate base.
B. $HCO_3^-(aq)$ is the acid and $H_2CO_3(aq)$ is its conjugate base.
C. $HSO_4^-(aq)$ is the acid and $HCO_3^-(aq)$ is its conjugate base.
D. $HSO_4^-(aq)$ is the acid and $SO_4^{2-}(aq)$ is its conjugate base.
E. $HCO_3^-(aq)$ is the acid and $SO_4^{2-}(aq)$ is its conjugate base.

Which is the acid when the reaction runs in reverse?

**13.140** What is the formula for the conjugate base of $H_2PO_4^-$?
A. $H_3PO_4$
B. $HPO_4^{2-}$
C. $PO_4$
D. $PO_4^{3-}$
E. $OH^-$
Which of these is the conjugate acid of $H_2PO_4^-$?

**13.141** When $0.50\ M$ $HCO_2H$ dissolves in water, a small portion of the molecules ionizes to form $HCO_2^-$ and $H_3O^+$, each with a concentration of $2.0 \times 10^{-3}\ M$. Based on this information, which of the following is the most appropriate classification of $HCO_2H$?
A. weak acid
B. strong acid
C. weak base
D. strong base
E. neutral salt
Describe how each of the other types of substances would behave when dissolved in water.

**13.142** When the weak base $CH_3NH_2$ dissolves in water, what is present in solution?
A. $CH_3NH_2$ and $H_2O$
B. $CH_3NH_2$, $H_2O$, $CH_3NH_3^+$, and $OH^-$
C. $CH_3NH_3^+$, $OH^-$, and $H_2O$
D. $CH_3NH_2$, $H_2O$, $CH_3NH_2^-$, and $H_3O^+$
E. $CH_3NH_2^-$, $H_3O^+$, and $H_2O$
Rank the species in solution in order of highest concentration to lowest concentration.

13.143 Consider the three acid solutions and the acid ionization constant for each acid. Which acid is the strongest acid, and which is the most concentrated acid?

0.50 $M$ CH$_3$CO$_2$H $\quad K_a = 1.8 \times 10^{-5}$
0.10 $M$ HF $\quad\quad\quad K_a = 6.3 \times 10^{-4}$
0.20 $M$ HOCl $\quad\quad\, K_a = 4.0 \times 10^{-8}$

A. CH$_3$CO$_2$H is the strongest acid and the HF solution is the most concentrated.
B. CH$_3$CO$_2$H is the strongest acid and the HOCl solution is the most concentrated.
C. HOCl is the strongest acid and the CH$_3$CO$_2$H solution is the most concentrated.
D. HF is the strongest and the HF solution is the most concentrated.
E. HF is the strongest and the CH$_3$CO$_2$H solution is the most concentrated.

Which is the weakest acid and which acid is the least concentrated?

13.144 Which of the following acids ionizes to the greatest extent?
A. 1.0 $M$ NH$_4^+$, $K_a = 5.6 \times 10^{-10}$
B. 1.0 $M$ HCN, $K_a = 6.2 \times 10^{-10}$
C. 1.0 $M$ HNO$_2$, $K_a = 5.6 \times 10^{-4}$
D. 1.0 $M$ HF, $K_a = 6.3 \times 10^{-4}$
E. All of these acids ionize completely.

Rank these solutions in order of increasing hydronium ion concentration.

13.145 Which of the following solutions should have the lowest pH?
A. 0.2 $M$ NaOH
B. 0.2 $M$ NH$_3$
C. 0.2 $M$ HCl
D. 0.2 $M$ HF
E. 0.2 $M$ H$_2$CO$_3$

Which of these solutions should have a pH above 7 and which should have the highest pH?

13.146 Which of the following is considered a basic solution?
A. a solution with [H$_3$O$^+$] = 1.0 $\times$ 10$^{-8}$ $M$
B. a solution with [H$_3$O$^+$] = 1.0 $\times$ 10$^{-2}$ $M$
C. a 0.10 $M$ solution of HNO$_3$
D. a solution with a pH of 5
E. a solution with [H$_3$O$^+$] = [OH$^-$]

Which of these is considered a neutral solution?

13.147 When 10.0 mL of a 0.10 $M$ HCl solution is diluted to 1.00 L, what is the pH of the new solution?
A. pH = $-1.00$
B. pH = 1.00
C. pH = 2.00
D. pH = 3.00
E. pH = 4.00

Which of the other pH values could you obtain by diluting a 0.10 $M$ HCl solution, and how would you prepare each solution with the desired pH?

13.148 Consider a buffer solution prepared by adding equal moles of the acid HOCl and an ionic compound that contains its conjugate base, NaOCl. The resulting equilibrium occurring in this solution is:

$$\text{HOCl}(aq) + \text{H}_2\text{O}(l) \rightleftharpoons \text{OCl}^-(aq) + \text{H}_3\text{O}^+(aq)$$

Given that the pH of this solution is 7.4, how would the pH of this solution change when 0.10 mol HCl is added to 1.00 L of this solution?
A. The pH would increase to 13.00.
B. The pH would decrease to 1.00.
C. The pH would increase by less than 1.00 pH unit.
D. The pH would decrease by less than 1.00 pH unit.
E. The pH would not change at all.

How would the pH change if the HOCl and NaOCl were not in the solution when the HCl was added?

# Oxidation-Reduction Reactions

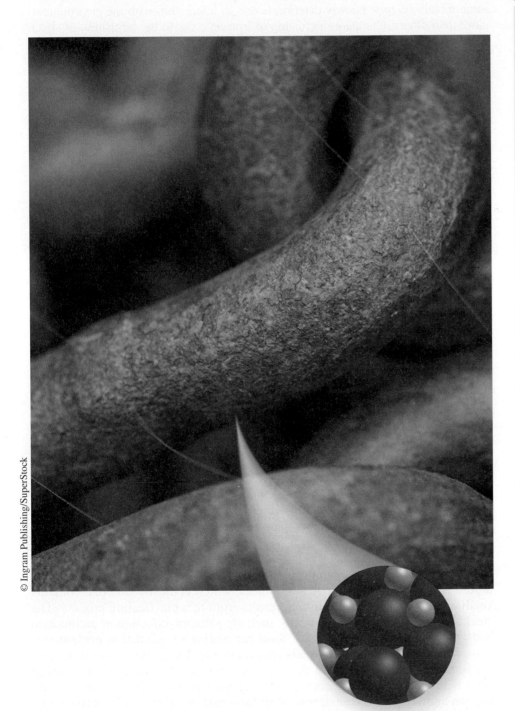

© Ingram Publishing/SuperStock

14.1 What Is an Oxidation-
Reduction Reaction?

14.2 Oxidation Numbers

14.3 Batteries

14.4 Balancing Simple
Oxidation-Reduction
Equations

14.5 Balancing Complex
Oxidation-Reduction
Equations

14.6 Electrochemistry

14.7 Corrosion Prevention

Chapter Review

Questions and Problems

Remember Anna and Bill from Chapter 1? They had an assignment to wander around campus to identify and classify different forms of matter. They decided that many of the things they observed were either new or rusted metals. For example, at a construction site they noticed shiny, new steel beams, copper pipes, and metallic tubes used for ductwork. They also saw a statue that had corroded over time.

Although we commonly think of metals as shiny, strong, durable, and good electrical conductors, their chemical properties vary considerably. For example, some metals *corrode* (slowly deteriorate due to interaction with the environment) more than others. Compare the rusting of iron on a car to the tarnishing of silver. When iron rusts due to environmental conditions, a compound of iron and oxygen forms and flakes off. The flaking of the rust weakens the body of the car, making it vulnerable to further damage. But when silver tarnishes, a compound containing silver and sulfur forms. Except for some discoloration, the tarnished surface of the silver remains intact.

To some extent nearly all metals are susceptible to reactions with nonmetals in the environment—especially with oxygen—but some are less reactive than others. Aluminum, for example, is structurally very strong, even though it is a low-density metal. Although it is more reactive than iron, it is less susceptible to corrosion. The atoms on the surface of an aluminum sheet readily react with oxygen in the atmosphere to form aluminum oxide, $Al_2O_3$. This compound forms a thin coating on the surface that actually protects the aluminum from further corrosion. Product manufacturers and packagers take advantage of this property. Sometimes they apply a dyed aluminum oxide coating to aluminum objects and give them a bright color.

The corrosion of metals is a serious problem. It's estimated that one-fifth of the steel and iron produced in the United States is used to replace corroded metal. This amounts to billions of wasted dollars every year. What precautions can prevent corrosion? One answer is a protective coating such as paint. We'll discuss other solutions to corrosion later in this chapter.

Why are reactions that corrode metals so prevalent? To answer this question, let's consider the source of the metals we use. Most metals occur in nature as minerals in various ores. That is, instead of finding metals in the form of elements in deposits, we find the metals in the form of compounds—mostly oxides, sulfides, and carbonates. (See Chapter 4.) Through highly involved and costly processes, we extract metals in their elemental form. Over time, however, corrosion returns a metal to its original state as a compound.

Iron is the most widely used of all the metal elements. (Aluminum is second.) One of the primary sources of iron is the mineral hematite, $Fe_2O_3$ (Figure 14.1). To extract iron for commercial and industrial uses, hematite and other minerals containing iron are loaded into a furnace and allowed to react with carbon:

$$3C(s) + Fe_2O_3(s) \xrightarrow{\text{heat}} 3CO(g) + 2Fe(l)$$

In the extremely high temperature of the furnace the liquid iron flows into another furnace for further purification. Steel, which is mostly iron that contains small amounts of unreacted carbon, results from this purification process. The steel can be used for many purposes, including the manufacture of automobile body panels from thin, flat sheets. Steel car bodies are painted to prevent corrosion, and the surfaces of new cars are protected—for a while. Over time, cars get scratched and dented, and the paint chips away. The iron in the exposed steel is then susceptible to corrosion. Over long periods, the iron reacts with oxygen and water in the environment to form rust, a hydrate of iron(III) oxide. Because this compound does not adhere well to the surface of the steel, it flakes off, exposing more iron metal to environmental conditions that cause further corrosion. Nature provides us with iron in the form of oxides. Through

**FIGURE 14.1** Hematite is an iron mineral consisting of $Fe_2O_3$.
©Tom Pantages

For simplicity, rust is sometimes given the formula $Fe_2O_3$. However, it is a hydrate, so the formula should show the number of water molecules present in a formula unit. This number varies, so it is often written as $Fe_2O_3 \cdot xH_2O$. Since the iron(III) oxide in rust is a hydrate, its properties are different from the properties of the $Fe_2O_3$ in hematite.

our efforts we obtain iron metal, but, over the years, nature returns the metal to an oxide compound.

The reaction of carbon with iron(III) oxide to form carbon monoxide and iron metal is a single-displacement reaction. (See Chapter 5.) Like all single-displacement reactions, it is an oxidation-reduction reaction, which—despite the name—does not always involve oxygen. Instead, an oxidation-reduction reaction involves the transfer of electrons. The corrosion of iron is a good example. In its reaction with oxygen, iron loses electrons and oxygen gains them.

Many oxidation-reduction reactions are exothermic, which allows us to put them to good use. For example, the combustion of methane, like all combustion reactions, is an oxidation-reduction reaction that provides heat to homes during the winter. The flameless heaters that soldiers and campers use to prepare hot meals also use an oxidation-reduction reaction (Figure 14.2). The reaction occurs when water mixes with magnesium and iron. It generates enough heat to warm the food.

Oxidation-reduction reactions can also be used to run devices that generate electricity. Such reactions occur in the batteries (Figure 14.3) that power calculators, cell phones, laptop computers, and other electronic tools. Batteries harness the energy of naturally occurring processes for practical purposes. Like the reactions that occur when materials corrode, the reactions in batteries are *spontaneous*. Once the reaction starts, it proceeds without outside intervention.

Other oxidation-reduction reactions do *not* occur spontaneously. Some external driving force must be applied to make the reaction proceed. For example, until around 1886, there was no practical method for preparing aluminum metal from aluminum oxide, a common compound found in aluminum ores. Passing an electric current through a molten mixture containing aluminum oxide solved the problem. The reaction converts minerals containing aluminum compounds to elemental aluminum metal, but it does not occur spontaneously. It requires a continuous outside energy source that provides electricity. Many pure metals are prepared by a similar process called *electrolysis*.

Oxidation-reduction reactions serve numerous aesthetic purposes, because many compounds of metals are richly colored. For example, the firing processes used by potters to form the glaze on the pottery shown in Figure 14.4 involve oxidation-reduction reactions.

We can look at oxidation-reduction reactions from three different perspectives.

- Some take place spontaneously, but they are undesirable. Metals corrode through such reactions.
- Some occur spontaneously and are desirable, such as reactions in batteries.
- Some do not occur spontaneously, but are desirable. Obtaining aluminum metal from aluminum ores is an example.

In this chapter we will focus on a wide range of oxidation-reduction reactions. As you continue your reading in this chapter, consider the following questions.

## Questions for Consideration

**14.1** What occurs in an oxidation-reduction reaction?

**14.2** How do we account for the loss and gain of electrons in an oxidation-reduction reaction?

**14.3** How do chemical reactions provide electricity in batteries?

**14.4** How are simple oxidation-reduction reactions balanced?

**14.5** How are complex oxidation-reduction reactions balanced?

**14.6** How do oxidation-reduction reactions generate electricity in spontaneous reactions? How is electricity used to drive oxidation-reduction reactions that do not normally occur?

**14.7** How can corrosion be prevented?

**FIGURE 14.2**  The flameless heater used to warm MREs (meals-ready-to-eat) works by adding water to a mixture of magnesium and iron. The reaction generates enough heat to prepare a hot meal of beefsteak with mushrooms and beans or turkey and gravy.

©David A. Tietz/Editorial Image, LLC

Prior to the invention of an electrolytic process for extraction, aluminum was more expensive than gold, because it was harder to produce in its metallic form. The noblest guests at Napoleon's table used aluminum eating utensils. The lesser guests ate with gold.

There is a fourth category of oxidation-reduction reactions: not spontaneous and undesirable. Why have we skipped such reactions?

**FIGURE 14.3** Different kinds of batteries are used to power numerous useful devices. Various chemical reactions occur in batteries and generate electricity.

©Brian Moeskau/Moeskau Photography

**FIGURE 14.4** The compounds present in different glazes coated on ceramics produce beautiful colors. The green and blue shown here result from different copper compounds. Notice the reddish-brown color on the right side of the bowl. It comes from elemental copper produced in the kiln by an oxidation-reduction reaction.

©Richard Bauer

## 14.1 What Is an Oxidation-Reduction Reaction?

In Chapter 5 we discussed reaction types that include decomposition, combination (or synthesis), single-displacement, double-displacement, and combustion. Our point of reference in Chapter 5 was how *atoms* rearrange to form new substances. Here we want to look at how *electrons* rearrange in chemical processes.

Consider a piece of zinc metal placed in a solution of copper(II) chloride (Figure 14.5). We can represent this reaction with the equation

$$Zn(s) + CuCl_2(aq) \longrightarrow ZnCl_2(aq) + Cu(s)$$

Notice that in this reaction the zinc metal displaces copper from the compound, and we obtain solid copper metal. In Chapter 5 we described this reaction as a single-displacement. From a molecular-level perspective, zinc atoms come off the surface of the metal as ions in solution. The copper ions in solution convert to copper atoms and deposit on the surface of the zinc.

Now let's look at this reaction from the point of view of electrons (Figure 14.6). The ionic equation for the reaction makes the process easier to see:

$$Zn(s) + Cu^{2+}(aq) + 2Cl^-(aq) \longrightarrow Zn^{2+}(aq) + 2Cl^-(aq) + Cu(s)$$

The chloride ions behave as spectators, so we can write the following net ionic equation:

$$Zn(s) + Cu^{2+}(aq) \longrightarrow Zn^{2+}(aq) + Cu(s)$$

The two solid metals—Zn(s), a reactant, and Cu(s), a product—both have charges of zero because these elements are in their natural states. As we discussed in Chapter 2, for a zinc atom to form $Zn^{2+}$ ions, the atom has to lose two electrons. For the $Cu^{2+}$ ion to form a copper atom, the ion has to gain two electrons. Electrons are transferred from the zinc metal to the $Cu^{2+}$ ion. This reaction is an oxidation-reduction reaction. An **oxidation-reduction reaction** (or *redox* reaction) is a reaction in which electrons are transferred.

These reactions are named for the two separate processes that occur together: oxidation and reduction. **Oxidation** is the process of losing one or more electrons. **Reduction** is the process of gaining one or more electrons. Oxidation cannot occur without reduction. The two processes go hand in hand.

**FIGURE 14.5** When zinc is placed in a solution of $CuCl_2$, copper atoms deposit as a coating on the surface of the zinc. Zinc atoms are converted to $Zn^{2+}$ ions, and $Cu^{2+}$ ions are converted to copper atoms. Chloride ions are spectator ions and they are not shown for simplicity. The net equation for this reaction is

$$Zn(s) + Cu^{2+}(aq) \longrightarrow Zn^{2+}(aq) + Cu(s)$$

©McGraw-Hill Education/Stephen Frisch

**FIGURE 14.6**  (A) For this reaction to occur, $Cu^{2+}$ ions must come into contact with zinc atoms on the metal strip. (B) When a $Cu^{2+}$ ion and a zinc atom touch, two electrons transfer to the copper ion from a zinc atom. (C) Newly formed copper atoms are deposited as a coating on the zinc. Zinc ions come off the strip into solution. (Spectator ions are not shown.)

To probe oxidation-reduction reactions further, let's consider the charges on the zinc, copper, and chlorine in the zinc–copper(II) chloride reaction. The two solid metals each have a charge of zero because they are elements in their natural states. The chloride ion has a charge of 1−, so the copper in $CuCl_2$ must have a charge of 2+. The zinc in zinc chloride also has a charge of 2+. The charges are shown with arrows pointing to each element:

$$\overset{0}{\underset{\downarrow}{Zn}}(s) + \overset{2+\ 1-}{\underset{\downarrow\ \downarrow}{CuCl_2}}(aq) \longrightarrow \overset{2+\ 1-}{\underset{\downarrow\ \downarrow}{ZnCl_2}}(aq) + \overset{0}{\underset{\downarrow}{Cu}}(s)$$

If we look at each species separately as it changes from reactant to product, we can compare charges. Zinc changes from a charge of 0 to 2+. This process involves an increase in charge (a loss of electrons), so it is oxidation. Copper changes from 2+ to 0, which requires a gain of two electrons. The copper ion in copper(II) chloride undergoes reduction. The charge on the chloride ion doesn't change, so it is neither oxidized nor reduced.

Historically the term *oxidation* was used to describe reactions that form oxygen compounds. *Reduction* described reactions in which a compound lost oxygen. Later the definitions were broadened to mean a loss (oxidation) or gain (reduction) of electrons, but the names stuck. To help keep them straight, use the mnemonic, "LEO says GER." *L*oss of *E*lectrons is *O*xidation. *G*ain of *E*lectrons is *R*eduction. Another mnemonic is "Oil Rig": *O*xidation *I*s *L*oss; *R*eduction *I*s *G*ain.

Charge is increasing (electrons lost)
Oxidation

$$\overset{0}{\underset{\downarrow}{Zn}}(s) + \overset{2+\ 1-}{\underset{\downarrow\ \downarrow}{CuCl_2}}(aq) \longrightarrow \overset{2+\ 1-}{\underset{\downarrow\ \downarrow}{ZnCl_2}}(aq) + \overset{0}{\underset{\downarrow}{Cu}}(s)$$

Reduction
Charge is decreasing (electrons gained)

It's convenient to separate the equation into the individual oxidation and reduction processes. For example, let's consider the oxidation of zinc:

$$Zn(s) \longrightarrow Zn^{2+}(aq)$$

Notice that this equation isn't balanced. Do you see why? The charges are unequal: 0 on the left and 2+ on the right. In balancing equations in Chapter 5 we focused strictly on balancing the number of atoms. Here we see that the total charge on each side of the equation must also be equal. To change from an uncharged zinc atom to a $Zn^{2+}$ ion, the metal atom must lose two electrons. We can represent this process with the equation

$$Zn(s) \longrightarrow Zn^{2+}(aq) + 2e^-$$

The equation is now balanced in both atoms and charge. In this reaction, as in all oxidation processes, the charge on the atom (or ion) losing electrons increases. Since the zinc loses electrons, we include the electrons as a product in the equation.

Now let's consider the reduction of copper ion:

$$Cu^{2+}(aq) \longrightarrow Cu(s)$$

This equation represents the process in which a $Cu^{2+}$ ion forms a copper atom. Again, notice that the equation is not balanced because the charges are not equal. For this process to occur, $Cu^{2+}$ must gain two electrons, which we write as a reactant:

$$Cu^{2+}(aq) + 2e^- \longrightarrow Cu(s)$$

This equation represents the reduction half of the overall reaction. Here, as in all reduction processes, gaining electrons reduces the charge on the atom or ion.

Chemists commonly describe reactants as agents of oxidation or reduction. A reactant that contains the element that is oxidized is called the **reducing agent.** The reducing agent causes the reduction of an element in another reactant because it provides the electrons for reduction. The reactant that contains the element that is reduced is called the **oxidizing agent.** The oxidizing agent causes the oxidation of an element in another reactant because it accepts the electrons from what is being oxidized.

To sort out these definitions let's consider the reaction between zinc, Zn, and copper(II) chloride, $CuCl_2$, again (Figure 14.6). For a reaction to occur, a $Cu^{2+}$ ion must come in contact with a Zn atom. When it does, two electrons are transferred from the Zn atom to the $Cu^{2+}$ ion. In losing two electrons, the Zn atom causes the $Cu^{2+}$ ion to be reduced, so Zn is the reducing agent. The $CuCl_2$ reactant is the oxidizing agent because the $Cu^{2+}$ in $CuCl_2$ accepts the two electrons from Zn, causing Zn to be oxidized.

Let's return to the equation for the reaction to summarize the process:

Zn is oxidized     $Cu^{2+}$ is reduced

$$Zn(s) \; + \; CuCl_2(aq) \longrightarrow ZnCl_2(aq) \; + \; Cu(s)$$

Zn is the     $CuCl_2$ is the
reducing agent     oxidizing agent

Test your understanding of oxidation-reduction processes by completing Example 14.1.

| EXAMPLE 14.1 | Oxidation-Reduction Reactions |
|---|---|

When copper metal is placed in a silver nitrate solution, a single-displacement reaction occurs. Silver deposits on the surface of the copper, and the solution turns blue because $Cu^{2+}$ ions form in the solution.

(a) Write a balanced chemical equation for this reaction.
(b) Identify the element oxidized and the element reduced.
(c) Identify the oxidizing and reducing agents.

(d)  On a molecular level, describe what's happening at the surface of the copper metal.

|  Before  |  After  |
|----------|---------|
| © McGraw-Hill Education/Stephen Frisch | © McGraw-Hill Education/Stephen Frisch |

## Solution:

(a)  As we did in Chapter 5, we begin by writing correct formulas for the reactants and products in this single-displacement reaction. Copper metal is one reactant; it is represented as $Cu(s)$. The other reactant, silver nitrate, is an ionic compound composed of silver ions, each with a charge of 1+, and nitrate ions, each with a charge of 1−. Because this compound is soluble in water, we can represent it with the formula $AgNO_3(aq)$. One of the products is a soluble ionic compound composed of copper ions, each with a charge of 2+, and nitrate ions. This product can be represented with the formula $Cu(NO_3)_2(aq)$. The other product is silver metal that has the formula $Ag(s)$. The skeletal equation for this reaction is

$$Cu(s) + AgNO_3(aq) \longrightarrow Cu(NO_3)_2(aq) + Ag(s)$$

We can follow the rules described in Section 5.3 to balance the equation.

$$Cu(s) + 2AgNO_3(aq) \longrightarrow Cu(NO_3)_2(aq) + 2Ag(s)$$

(b)  Copper atoms in the copper metal change from a charge of 0 to 2+, so $Cu(s)$ is the reactant that is oxidized. The silver ion from silver nitrate changes from a charge of 1+ to 0, so the $Ag^+$ ion is reduced.

(c)  Because copper metal is the reactant that gives up electrons to cause the reduction of silver ion, it is the reducing agent. Silver nitrate is the oxidizing agent because it contains the $Ag^+$ ion that removes electrons from the copper metal, causing its oxidation.

(d)  The solution becomes a pale blue as the reaction proceeds, indicating that Cu atoms are converted to $Cu^{2+}$ ions. A coating of silver appears on the surface of the copper metal, indicating that silver ions, $Ag^+$, are converted to silver atoms, which are deposited on the surface of the copper metal.

## Consider This 14.1

What would you expect to observe if silver metal were placed in a solution of copper(II) nitrate?

## Practice Problem 14.1

As shown in the figure at the end of this problem, when iron metal is placed in a copper(II) sulfate solution, a single-displacement reaction occurs in which copper deposits on the surface of the iron and aqueous iron(III) sulfate forms.

(a)  Write a balanced chemical equation for this reaction.
(b)  Identify the element oxidized and the element reduced.

(c) Identify the oxidizing and reducing agents.

(d) On a molecular level, what's happening at the surface of the iron metal?

| Before | After |
|:---:|:---:|
| ©Jim Birk | ©Jim Birk |

**Further Practice:** Questions 14.9 and 14.10 at the end of the chapter

## 14.2 Oxidation Numbers

So far we have described oxidation-reduction reactions of ionic compounds in aqueous solution. We have used electron counts and ionic charges to determine how electrons are lost and gained. Other oxidation-reduction reactions do not take place in solution and, therefore, are not as easy to recognize. For example, consider the oxidation-reduction reaction of carbon with oxygen:

$$C(s) + O_2(g) \longrightarrow CO_2(g)$$

The transfer of electrons is not as obvious here as in a reaction of ionic substances, but this reaction can be identified as oxidation-reduction using a convenient "accounting tool," an oxidation number. An **oxidation number** (sometimes called an *oxidation state*) is a charge assigned to the atoms in any compound. For an ionic compound, the oxidation numbers are simply the charges on the ions in the compound. Assigning oxidation numbers to atoms in a molecular compound or polyatomic ion is a bookkeeping procedure in which we treat the atoms in the compound as if they were ions with charges that are established by a set of rules. Table 14.1 lists the rules for assigning oxidation numbers. The following points will help you apply the rules.

- The rules are a hierarchy. The first rule that applies takes precedence over any subsequent rules that may apply.
- For an isolated atom or a molecule that contains only one element, only rule 1 need be applied. An uncombined element, whether occurring as an atom, such as He, or a molecule, such as $H_2$, has an oxidation number of 0. Thus, the oxidation number of sulfur in S, $S_2$, and $S_8$ is 0.
- A monatomic ion has an oxidation number equal to its ionic charge, according to rule 2. For example, the oxidation number of lithium in $Li^+$ is 1+. Similarly, the oxidation number is 1− for chlorine in $Cl^-$.
- Rule 2 can be applied whenever oxidation numbers have been assigned to all but one element.
- Rule 5 uses position in the periodic table to cover situations that are not handled by one of the other rules. In general, the more electronegative element in a compound is assigned the oxidation number that would be its charge if it were a monatomic ion.
- To assign oxidation numbers to binary compounds, handle one of the elements with an appropriate rule and the other with rule 2.

Recall that electronegativity tends to increase going up the periodic table and from left to right.

**TABLE 14.1** Rules for Assigning Oxidation Numbers

1.  The oxidation number of the atoms of an uncombined element is 0.

2.  The sum of the oxidation numbers of all atoms in a substance must equal the total charge: 0 for molecules and the ionic charge for ions (including polyatomic ions).

3.  Fluorine has an oxidation number of 1– in its compounds.

4.  Hydrogen has an oxidation number of 1+ unless it is combined with metals, where it has an oxidation number of 1–.

5.  The position of the element in the periodic table may be useful.

    a.  Group IA (1) elements have oxidation numbers of 1+ in their compounds.

    b.  Group IIA (2) elements have oxidation numbers of 2+ in their compounds.

    c.  Group VIIA (17) elements have oxidation numbers of 1– unless combined with a more electronegative nonmetal.

    d.  In binary compounds, group VIA (16) elements have oxidation numbers of 2– unless combined with a more electronegative nonmetal.

    e.  In binary compounds, group VA (15) elements have oxidation numbers of 3– unless combined with a more electronegative nonmetal.

6.  Oxygen has an oxidation number of 2–, except in peroxides (compounds containing the $O_2^{2-}$ ion), in which its oxidation number is 1–. The other exception is when oxygen is combined with the more electronegative fluorine atom.

Let's apply these rules to the carbon combustion reaction described earlier:

$$C(s) + O_2(g) \longrightarrow CO_2(g)$$

For both reactants, $C(s)$ and $O_2(g)$, the first rule applies, so carbon and oxygen atoms have oxidation numbers of 0. Now let's determine the oxidation numbers of the atoms in carbon dioxide, $CO_2$. As we proceed through the rules hierarchically, rule 2 applies, but we don't know the oxidation numbers of the atoms in the compound yet. Rule 6 tells us that we should assign an oxidation number of 2– to the oxygen. We can then apply rule 2 to determine the oxidation number of carbon. In Chapter 3 we determined the formulas for ionic compounds using the following relationship:

$$\text{Total positive charge from cations} + \text{total negative charge from anions} = \text{zero net charge}$$

We can modify this relationship to apply to all compounds or ions in the following way:

$$\text{Total positive oxidation numbers} + \text{total negative oxidation numbers} = \text{net charge}$$

This relationship is another way of describing rule 2. For carbon dioxide, the net charge is 0. The total of the negative oxidation numbers is 4–, which comes from the two oxygen atoms, each assigned an oxidation number of 2–:

$$\text{Total positive oxidation numbers} + [2 \times (-2)] = 0$$
$$\text{Total positive oxidation numbers} + (-4) = 0$$
$$\text{Total positive oxidation numbers} = 0 + 4$$
$$\text{Total positive oxidation numbers} = 4$$

The oxidation number of carbon must then be 4+.

$$\overset{0}{C}(s) + \overset{0}{O_2}(g) \longrightarrow \overset{4+\,2-}{CO_2}(g)$$

> To use oxidation numbers in numerical calculations, we'll follow the convention used in math and place the negative sign in front of the number.

How would you apply the oxidation number rules to assign charges to the atoms in an ion such as the sulfate ion, $SO_4^{2-}$? Rule 2 applies, but we must first know the oxidation number of one of the elements in this binary ion. Rule 5 says that group

VIA (16) elements are usually assigned an oxidation number of 2−. An exception is when the element is combined with a more electronegative atom such as oxygen. Both cannot have an oxidation number of 2−, so we cannot assign an oxidation number to the sulfur based on this rule. Rule 6 tells us to assign oxygen an oxidation number of 2−. Knowing the oxidation number of oxygen allows us to apply rule 2 to determine the oxidation number of sulfur:

$$\begin{matrix} \text{Total positive} \\ \text{oxidation numbers} \end{matrix} + \begin{matrix} \text{total negative} \\ \text{oxidation numbers} \end{matrix} = \text{net charge}$$

There are four oxygen atoms in the formula, so we multiply oxygen's oxidation number, 2−, by 4. The net charge on the ion is 2−:

$$\text{Total positive oxidation numbers} + [4 \times (-2)] = -2$$
$$\text{Total positive oxidation numbers} + (-8) = -2$$
$$\text{Total positive oxidation numbers} = -2 + 8$$
$$\text{Total positive oxidation numbers} = 6$$

There is one sulfur atom represented in the formula, so its oxidation number must be 6+.

**Student Hot Spot**

Student data indicate you may struggle with assigning oxidation numbers. Access the SmartBook to view additional Learning Resources on this topic.

---

**EXAMPLE 14.2** | **Assigning Oxidation Numbers**

Assign oxidation numbers to each element in the following formulas.

(a) $F_2$    (b) $PO_4^{3-}$    (c) $Mg(NO_3)_2$    (d) $Cr_2O_3$

**Solution:**

(a) Fluorine, $F_2$, molecules contain atoms of a single element, fluorine. From rule 1, the oxidation number of F in $F_2$ is 0.

(b) To assign oxidation numbers for $PO_4^{3-}$, rule 5e might apply. However, since the phosphorus is a group V (15) element attached to oxygen, which is more electronegative, the rule will not work. We have to use rule 6 to assign an oxidation number of 2− to oxygen. Then, using rule 2, we can calculate the oxidation number of phosphorus:

$$\begin{matrix} \text{Total positive} \\ \text{oxidation numbers} \end{matrix} + \begin{matrix} \text{total negative} \\ \text{oxidation numbers} \end{matrix} = \text{net charge}$$

There are four oxygen atoms in the formula, so we multiply oxygen's oxidation number, 2−, by 4. The net charge on the ion is 3−:

$$\text{Total positive oxidation numbers} + [4 \times (-2)] = -3$$
$$\text{Total positive oxidation numbers} + (-8) = -3$$
$$\text{Total positive oxidation numbers} = -3 + 8$$
$$\text{Total positive oxidation numbers} = 5$$

The oxidation number of phosphorus is 5+.

(c) $Mg(NO_3)_2$ is an ionic compound composed of a metal ion and a polyatomic anion. The polyatomic anion in the compound is nitrate, $NO_3^-$. Since there are two nitrate ions in a formula unit of the compound, the oxidation number of magnesium must be 2+. (We also know this from magnesium's position in the periodic table, rule 5b.) We can use rule 6 to assign an oxidation number of 2− to oxygen. Then, using rule 2, we can calculate the oxidation number of nitrogen:

$$\begin{matrix} \text{Total positive} \\ \text{oxidation numbers} \end{matrix} + \begin{matrix} \text{total negative} \\ \text{oxidation numbers} \end{matrix} = \text{net charge}$$

There are three oxygen atoms in the formula for nitrate, so we multiply oxygen's oxidation number, 2−, by 3. The net charge on the ion is 1−:

$$\text{Total positive oxidation numbers} + [3 \times (-2)] = -1$$
$$\text{Total positive oxidation numbers} + (-6) = -1$$
$$\text{Total positive oxidation numbers} = -1 + 6$$
$$\text{Total positive oxidation numbers} = 5$$

The oxidation number for nitrogen is 5+.

(d) $Cr_2O_3$ is an ionic compound containing a transition metal. Since transition metal charges cannot be predicted from the periodic table, we must begin with rule 6, assigning an oxidation number of 2− to oxygen. Then, using rule 2, we can calculate the oxidation number of chromium from the following relationship:

$$\frac{\text{Total positive}}{\text{oxidation numbers}} + \frac{\text{total negative}}{\text{oxidation numbers}} = \text{net charge}$$

There are three oxygen atoms in the formula, so we multiply oxygen's oxidation number, 2−, by 3. The net charge on the compound is 0:

$$\text{Total positive oxidation numbers} + [3 \times (-2)] = 0$$
$$\text{Total positive oxidation numbers} + (-6) = 0$$
$$\text{Total positive oxidation numbers} = 0 + 6$$
$$\text{Total positive oxidation numbers} = 6$$

Since there are two chromium atoms in the formula, each one has an oxidation number of 3+.

**Consider This 14.2**

How does the oxidation number for chromium differ in the compound $CrO_3$ from its oxidation number in $Cr_2O_3$?

**Practice Problem 14.2**

Assign oxidation numbers to each element in the following formulas.

(a) Mg
(b) $Mn_2O_3$
(c) $Na_2SO_4$
(d) $Cr_2O_7^{2-}$

**Further Practice:**  Questions 14.19 and 14.20 at the end of the chapter

Let's return to the combustion reaction of carbon:

$$C(s) + O_2(g) \longrightarrow CO_2(g)$$

Recall that we assigned oxidation numbers of 0 to both $C(s)$ and $O_2(g)$. The oxidation number of carbon in $CO_2$ is 4+ and the oxidation number of oxygen is 2−. We can summarize these oxidation numbers as follows:

$$\overset{0}{\underset{\downarrow}{C}}(s) + \overset{0}{\underset{\downarrow}{O_2}}(g) \longrightarrow \overset{4+2-}{\underset{\downarrow\downarrow}{CO_2}}(g)$$

Earlier we used changes in charge to identify oxidation and reduction processes. We can also identify them from changes in oxidation number. *An oxidation-reduction reaction occurs if one or more elements changes its oxidation number.* When an oxidation number increases, electrons have been lost and oxidation has occurred.

When an oxidation number decreases, electrons have been gained and reduction has occurred. Notice in this combustion reaction that carbon's oxidation number changes from 0 as a reactant to 4+ as a product. Carbon lost four electrons, so carbon is oxidized. Oxygen's oxidation number changes from 0 as a reactant to 2− as a product. Each oxygen atom gained two electrons, so oxygen is reduced.

What are the oxidizing and reducing agents in this reaction? Since $C(s)$ is giving up electrons so that $O_2(g)$ can be reduced, $C(s)$ is the reducing agent. Because $O_2(g)$ is taking the electrons from $C(s)$, $O_2(g)$ is the oxidizing agent.

Now that you can identify oxidation-reduction reactions in both ionic compounds and molecular compounds, determine if the reactions in Example 14.3 involve oxidation and reduction.

---

**EXAMPLE 14.3** | **Identifying Oxidation-Reduction Reactions**

Determine which of the following represent oxidation-reduction reactions. For reactions that involve oxidation-reduction, identify the oxidizing and reducing agents.

(a) $AgNO_3(aq) + NaCl(aq) \longrightarrow AgCl(s) + NaNO_3(aq)$
(b) $2Fe(s) + 3Cu(NO_3)_2(aq) \longrightarrow 2Fe(NO_3)_3(aq) + 3Cu(s)$
(c) $CH_4(g) + 2O_2(g) \longrightarrow CO_2(g) + 2H_2O(g)$
(d) $4Fe(s) + 3O_2(g) \longrightarrow 2Fe_2O_3(s)$

**Solution:**

If an element undergoes a change in oxidation number, an oxidation-reduction reaction has occurred. First assign oxidation numbers to all the elements in the equations using the rules in Table 14.1.

(a) The oxidation numbers of the elements in this reaction are as follows:

$$\overset{1+ \;\; 5+ \;\; 2-}{AgNO_3}(aq) + \overset{1+ \;\; 1-}{NaCl}(aq) \longrightarrow \overset{1+ \;\; 1-}{AgCl}(s) + \overset{1+ \;\; 5+ \;\; 2-}{NaNO_3}(aq)$$

Since none of these elements undergoes a change in oxidation number, this is *not* an oxidation-reduction reaction.

(b) The oxidation numbers of the elements in this reaction are as follows:

$$\overset{0}{2Fe}(s) + \overset{2+ \;\; 5+ \;\; 2-}{3Cu(NO_3)_2}(aq) \longrightarrow \overset{3+ \;\; 5+ \;\; 2-}{2Fe(NO_3)_3}(aq) + \overset{0}{3Cu}(s)$$

In this reaction, the oxidation number of iron changes from 0 to 3+, so it is oxidized. The oxidation number of copper changes from 2+ to 0, so it is reduced. This is an oxidation-reduction reaction. $Fe(s)$ is the reducing agent, and $Cu(NO_3)_2$ is the oxidizing agent.

(c) The oxidation numbers of the elements in this reaction are as follows:

$$\overset{4-\;1+}{CH_4}(g) + \overset{0}{2O_2}(g) \longrightarrow \overset{4+\;2-}{CO_2}(g) + \overset{1+\;2-}{2H_2O}(g)$$

The oxidation number of carbon changes from 4− to 4+, so it is oxidized. The oxidation number of oxygen changes from 0 to 2−, so it is reduced. This is an oxidation-reduction reaction. $CH_4$ is the reducing agent, and $O_2$ is the oxidizing agent.

(d) The oxidation numbers of the elements in this reaction are as follows:

$$\overset{0}{4Fe}(s) + \overset{0}{3O_2}(g) \longrightarrow \overset{3+\;2-}{2Fe_2O_3}(s)$$

The oxidation number of iron changes from 0 to 3+, so it is oxidized. The oxidation number of oxygen changes from 0 to 2−, so it is reduced. This is an oxidation-reduction reaction. Fe is the reducing agent, and $O_2$ is the oxidizing agent.

**Consider This 14.3**

For the reaction shown in part (d), how many electrons must be gained or lost to account for the change in oxidation number for each iron atom that reacts? How many electrons are gained or lost for each oxygen atom that reacts?

**Practice Problem 14.3**

Determine which of the following represent oxidation-reduction reactions. For reactions that involve oxidation-reduction, identify the oxidizing and reducing agents.

(a) $2H_2O(l) \longrightarrow 2H_2(g) + O_2(g)$
(b) $HCl(aq) + NaOH(aq) \longrightarrow NaCl(aq) + H_2O(l)$
(c) $4Al(s) + 3O_2(g) \longrightarrow 2Al_2O_3(s)$
(d) $Mg(s) + Zn(NO_3)_2(aq) \longrightarrow Mg(NO_3)_2(aq) + Zn(s)$

**Further Practice:** Questions 14.29 and 14.30 at the end of the chapter

## 14.3 Batteries

All oxidation-reduction reactions that occur without outside intervention release energy. For example, the combustion of methane, $CH_4$, occurs in some home furnaces. The energy from the combustion reaction provides heat. Heat is useful for heating homes, but not much help when we want to run electronic devices. That's when batteries come in handy. Electricity from a battery is a flow of electrons released from an oxidation-reduction reaction. The correct construction of a battery is essential for harnessing this energy in a useful way.

Consider the zinc-mercury battery formerly used in most watches and calculators (Figure 14.7). The reaction inside it is

$$Zn(s) + HgO(s) \longrightarrow ZnO(s) + Hg(l)$$

Although it's not apparent from the equation, the reaction requires the presence of hydroxide ions, $OH^-$, supplied by potassium hydroxide. Suppose we simply mixed zinc with mercury(II) oxide in a solution of potassium hydroxide. The energy released by this reaction when electrons are transferred would be lost as heat, which could not be used to power a watch. Instead of allowing all of the reactants to mix freely, batteries are constructed so that a barrier separates the oxidation reaction from the reduction reaction. A wire connects them. Rather

**FIGURE 14.7** Zinc-mercury batteries were used extensively in watches and calculators until mercury-containing batteries were banned by several countries. Many button-cell batteries now contain Zn and $MnO_2$.
©Brian Moeskau/Moeskau Photography

ANIMATION: Operation of a Voltaic Cell:
The Zinc-Copper Reaction

ANIMATION: Galvanic Cell

**FIGURE 14.8** When the reactants are separated into half-cells and connected with a wire, electrons flow and the bulb lights.

The terms *voltage* and *volt* honor Alessandro Volta, who invented the forerunner of the battery around 1800. He is remembered with Luigi Galvani for the work that demonstrated the electrical nature of nerve impulses in animals.

High potential
©naumoid/iStock/Getty Images

Low potential
©JupiterImages/Alamy Stock Photo

**FIGURE 14.9** The potential for the flow of electrons in a voltaic cell is analogous to the flow of water in a river. The steeper the river bed, the greater the potential for water flow. In a chemical reaction the potential varies depending on the reactants.

than being lost as heat, electric energy is generated as electrons are transferred through a wire from the oxidation reaction to the reduction reaction. If the wire runs through a useful device, such as a watch, the flow of electrons provides the power the device needs to run.

Let's consider another battery constructed from zinc and copper. Recall that earlier we discussed the reaction of zinc metal with copper(II) chloride:

$$Zn(s) + CuCl_2(aq) \longrightarrow ZnCl_2(aq) + Cu(s)$$

If we place a piece of zinc in a solution of copper(II) chloride, copper metal will plate onto the surface of the zinc (Figure 14.5). The blue color of the $Cu^{2+}$ ions in solution disappears, and the temperature rises. The energy is lost to the surroundings as heat. Now consider the same reaction, but with the oxidation and reduction reactions separated (Figure 14.8). The loss of electrons from zinc and the gain of electrons by copper ions lights the bulb, because electrons flow through the wire that connects the separate compartments. A device that sets up a spontaneous chemical reaction to produce electricity in this way is a **voltaic (or galvanic) cell.** Voltaic cells are the operational parts of a battery.

The tendency for electrons to flow in a cell is measured as *voltage,* and we can use an analogy to understand it. Think of the tendency of electrons to flow as analogous to the flow of water in a river. River water flows sluggishly on a nearly flat landscape, but it flows rapidly and with greater force when the slope is steeply downhill (Figure 14.9). Similarly, chemical reactions vary in their potentials to cause electrons to flow. The potential of a voltaic cell, measured in units of *volts,* is greater in cells containing the strongest reducing and oxidizing agents. A voltmeter can measure the voltage that a voltaic cell produces.

To understand how a voltaic cell produces electricity, let's consider what occurs in each compartment as zinc and copper react:

$$Zn(s) + Cu^{2+}(aq) \longrightarrow Zn^{2+}(aq) + Cu(s)$$

Assigning oxidation numbers, we get

$$\overset{0}{Zn}(s) + \overset{2+}{Cu^{2+}}(aq) \longrightarrow \overset{2+}{Zn^{2+}}(aq) + \overset{0}{Cu}(s)$$

The oxidation number of zinc changes from 0 to 2+, so it is the reactant that is oxidized. The oxidation number of copper changes from 2+ to 0, so it is the reactant that is reduced. From this information we can write equations to represent both processes:

$$\text{Oxidation} \quad Zn(s) \longrightarrow Zn^{2+}(aq) + 2e^-$$
$$\text{Reduction} \quad Cu^{2+}(aq) + 2e^- \longrightarrow Cu(s)$$

Each of these equations is a **half-reaction,** representing either the oxidation or the reduction that occurs in the separate compartments of a voltaic cell. Thus, each compartment is called a **half-cell.** Oxidation occurs in one half-cell; reduction occurs in the other. The separation of the oxidation and reduction processes into half-cells harnesses the energy released from a chemical reaction to generate electricity. The potential to give up electrons is greater for zinc than it is for copper. Therefore, electrons flow through the wire from the zinc metal to the copper metal, where $Cu^{2+}$ ions in the solution are reduced to copper atoms.

Inside a voltaic cell (Figure 14.10), a solid material that conducts electricity must be present to provide a site for each half-reaction. This material is an **electrode.** The electrode is commonly a metal immersed in an electrolyte solution containing a salt of the same metal. The electrode at which oxidation occurs is the **anode.** The electrode where reduction occurs is the **cathode.** In a cell composed of zinc and copper, the anode is a piece of zinc metal. The electrolyte in the anode half-cell is a compound of zinc such as $Zn(NO_3)_2$. The cathode is a piece of copper metal. The electrolyte in the cathode half-cell contains a compound of copper such as $Cu(NO_3)_2$.

Electricity will not flow unless there is a complete circuit between the anode and the cathode. In a voltaic cell, like the one shown in Figure 14.10, a salt bridge completes the circuit. A **salt bridge** allows ions to flow so charge balance is maintained. A typical salt bridge is a glass tube in the shape of an inverted U containing an ionic compound such as $Na_2SO_4$ in a gel. Ions freely flow in and out of the gel and keep charge balanced within the half-cells.

To see how atoms, ions, and electrons flow in a voltaic cell, let's look at what's happening from a molecular-level perspective (Figure 14.11). In the anode half-cell in which oxidation occurs, each atom of zinc metal that reacts loses two electrons. Now an ion, the zinc leaves the surface of the electrode and goes into the solution. The electrons lost by a zinc atom flow toward the cathode through the wire. The electrons come out at the cathode, where they'll come into contact with a $Cu^{2+}$ ion in solution. This ion accepts the electrons, and copper is deposited on the surface of the copper electrode as a neutral atom.

Sometimes an inert (unreactive) metal is used as an electrode to serve merely as a site for electron transfer.

Voltmeter

Salt bridge

Anode

Cathode

Oxidation

Reduction

**FIGURE 14.10** A voltaic cell consists of two electrodes immersed in electrolyte solutions. One electrode is the anode, the other is the cathode. Oxidation occurs at the anode, and reduction occurs at the cathode. Electrons flow through a wire from the anode to the cathode. A salt bridge completes the circuit.

**FIGURE 14.11** The electrons lost by zinc at the anode travel toward the cathode. This flow of electrons causes the bulb to light. At the cathode, a copper ion gains two electrons and gets plated onto its surface. (Note that spectator ions are not shown.)

**FIGURE 14.12** Salt bridges commonly contain a solution of an ionic compound such as sodium sulfate, $Na_2SO_4$. Negatively charged sulfate ions flow from the salt bridge into the anode half-cell, where a positive charge is generated. Positively charged sodium ions flow into the cathode half-cell, preventing the buildup of negative charge. A salt bridge completes the circuit. Without it, electrons will not flow. (Note that not all spectator ions are shown.)

Now that we can see what's happening at each electrode, can you explain the purpose of the salt bridge? To answer this question, consider what's happening to the charge in each half-cell as the reaction proceeds. At the anode, $Zn^{2+}$ ions form from neutral zinc atoms, and the anode half-cell gains some positive charge. Negatively charged ions flow into the anode half-cell from the salt bridge to maintain charge balance. At the cathode, $Cu^{2+}$ ions change to neutral copper atoms, and the cathode half-cell gains some negative charge. Positive ions from the salt bridge flow into this half-cell, replacing the positive charge (Figure 14.12). The ions that flow from the salt bridge merely work to balance the charge but do not participate in the chemical reaction.

What happens when a voltaic cell runs for a long time? What happens to the zinc electrode? Does its mass increase or decrease? What about the copper electrode? Although solutions containing $Zn^{2+}$ ions are colorless, $Cu^{2+}$ ions produce a blue color in water. Will the intensity of the color change as the reaction runs? To answer these questions, look at Figure 14.13. The atoms on the zinc electrode oxidize to aqueous zinc ions, so the anode gets smaller. Copper ions from the aqueous solution deposit on the copper electrode, so the cathode has more copper atoms. Many of the copper ions are no longer present in solution, so the blue color fades.

---

**EXAMPLE 14.4** | **Identifying Components of a Voltaic Cell**

Magnesium reacts with copper(II) sulfate according to the following equation:

$$Mg(s) + CuSO_4(aq) \longrightarrow MgSO_4(aq) + Cu(s)$$

(a) Write equations to represent the oxidation and reduction half-reactions.
(b) Determine the oxidizing and reducing agents.
(c) The figure represents components of a voltaic cell in which this reaction takes place. Answer the following questions with the correct component of a voltaic cell, using the letters A to E as shown. Assume that the oxidation half-reaction will occur on the left.
 • In which half-cell will you put the oxidation components?
 • In which half-cell will you put the reduction components?
 • What will you use for the anode and cathode?
 • What electrolyte solutions will you put in each compartment?
 • Which direction will ions flow from the salt bridge?
 • Where do electrons flow?
 • What happens to the mass of the electrodes as the cell runs?

**FIGURE 14.13** As a zinc-copper voltaic cell runs, the zinc electrode deteriorates. The copper electrode increases in mass, however, because copper atoms plate onto its surface as the reaction proceeds.
©McGraw-Hill Education/Stephen Frisch

**Solution:**

(a) Because the sulfate ion, $SO_4^{2-}$, is a spectator ion, we can simplify the reaction by writing a net ionic equation:

$$Mg(s) + Cu^{2+}(aq) \longrightarrow Mg^{2+}(aq) + Cu(s)$$

Since these are neutral or ionic species, assigning oxidation numbers is straightforward:

$$\overset{0}{Mg}(s) + \overset{2+}{Cu^{2+}}(aq) \longrightarrow \overset{2+}{Mg^{2+}}(aq) + \overset{0}{Cu}(s)$$

Magnesium's oxidation number changes from 0 to 2+, so it is the reactant that is oxidized. In the oxidation reaction, each magnesium atom loses two electrons:

$$Mg(s) \longrightarrow Mg^{2+}(aq) + 2e^-$$

Copper's oxidation number changes from 2+ to 0, so it is the reactant that's reduced. In the reduction reaction, each $Cu^{2+}$ ion gains two electrons:

$$Cu^{2+}(aq) + 2e^- \longrightarrow Cu(s)$$

From this information we can write half-reaction equations that represent the oxidation and reduction processes:

$$\text{Oxidation} \quad Mg(s) \longrightarrow Mg^{2+}(aq) + 2e^-$$
$$\text{Reduction} \quad Cu^{2+}(aq) + 2e^- \longrightarrow Cu(s)$$

(b) Magnesium is the reactant that's oxidized. The reactant that accepts the electrons from magnesium is $Cu^{2+}$. The oxidizing agent is $Cu^{2+}$, which is reduced in the reaction. The reactant that provides the electrons for this reduction is magnesium. Magnesium is the reducing agent.

(c) If we arbitrarily assign the oxidation half-reaction to the compartment on the left, then the answers are as follows:
- The oxidation components will be A and B on the left.
- The right compartment will contain the reduction components D and E.
- Component A will be the anode of magnesium metal. Component E will be the cathode of copper metal.
- The solution in the anode compartment, B, should be an electrolyte of magnesium, such as $Mg(NO_3)_2$. The electrolyte solution in the cathode compartment, D, can be $Cu(NO_3)_2$ or another salt containing $Cu^{2+}$ ions.
- To complete the electric circuit, component C needs to be a salt bridge that allows ions to flow into the two half-cells. Sodium sulfate, $Na_2SO_4$, is a common ingredient in salt bridges. Sulfate ions, $SO_4^{2-}$, will flow into the left compartment to balance the positive charge that builds up from the production of $Mg^{2+}$ ions. $Na^+$ ions will flow into the right half-cell to replace the $Cu^{2+}$ ions that react to form neutral Cu atoms.
- Electrons will flow from electrode A to electrode E through, in this case, a voltmeter.
- The mass of electrode A will decrease because magnesium atoms react to form magnesium ions that flow into the solution. The mass of electrode E will increase because copper ions from the solution react to form copper atoms on the surface of the electrode.

## Consider This 14.4

Suppose a 25.0-g piece of magnesium is used as the anode and 15.0 g of magnesium metal remains when the reaction is stopped. How many grams of copper should plate on the surface of the cathode?

## Practice Problem 14.4

Cadmium reacts with nickel(II) nitrate according to the following equation:

$$Cd(s) + Ni(NO_3)_2(aq) \longrightarrow Cd(NO_3)_2(aq) + Ni(s)$$

(a) Write equations to represent the oxidation and reduction half-reactions.
(b) Determine the oxidizing and reducing agents.
(c) Draw a picture like the one given in the example to show the components of this voltaic cell. Label the components A, B, C, D, and E and answer the same questions as in the example.

**Further Practice:** Questions 14.37 and 14.38 at the end of the chapter

Zinc-copper cells are not what we usually think of when we visualize batteries, but for many years they actually were used to power telephones and other electronic devices. Notice all the plates in the old-fashioned "battery jar" in Figure 14.14. They are a series of anodes and cathodes that provide more electricity than a single pair would. A series of connected voltaic cells that provides a portable source of electrical power is a **battery.**

Zinc-copper batteries have practical limitations because they are "wet cells." Imagine carrying a battery jar around with a cell phone! In spite of its limitations for small electronic devices, a wet cell battery is still in common use today for a much larger application: starting an automobile. The battery used in cars is a lead-acid, or lead storage, battery (Figure 14.15). The reaction that occurs inside it is

$$Pb(s) + PbO_2(s) + 2H_2SO_4(aq) \longrightarrow 2PbSO_4(aq) + 2H_2O(l)$$

In this reaction one form of lead, $Pb(s)$, is oxidized, while another form, $PbO_2(s)$, is reduced:

$$\overset{\text{Oxidation}}{\underset{\text{Reduction}}{\underset{0}{Pb(s)} + \underset{4+}{PbO_2(s)} + 2H_2SO_4(aq) \longrightarrow \underset{2+}{2PbSO_4(s)} + 2H_2O(l)}}$$

In the lead-acid battery, the anode is constructed of lead metal, with grids that contain a spongy form of lead. The cathode is also made of lead metal, but its grids contain lead(IV) oxide, $PbO_2$. As in all voltaic cells, we can separate the overall reaction into half-reactions. The anode and cathode half-reactions for the lead-acid battery are

Anode      $Pb(s) + H_2SO_4(aq) \longrightarrow PbSO_4(s) + 2H^+(aq) + 2e^-$

Cathode    $PbO_2(s) + H_2SO_4(aq) + 2H^+(aq) + 2e^- \longrightarrow PbSO_4(s) + 2H_2O(l)$

Which of these half-reactions is oxidation and which is reduction?

A single cell for a lead-acid battery has an electrical potential of 2 volts (V). The six cells in a car battery generate a voltage of 12 V. To power small electronic devices, less voltage is needed and "dry cells" are used. These batteries vary in composition and produce different voltages. The 9-V battery (Figure 14.16), for example, is a series of six cells constructed of Zn, $MnO_2$, and an electrolyte paste of various compounds. The individual cells produce a voltage of 1.5 V. The reaction that occurs in these cells is

$$Zn(s) + MnO_2(s) + H_2O(l) \longrightarrow ZnO(s) + Mn(OH)_2(s)$$

Can you identify the oxidizing and reducing agents in this reaction?

**FIGURE 14.14** The battery jar was once used to provide power for telephones and other devices.
©David A. Tietz/Editorial Image, LLC

**A**

**B**

**FIGURE 14.16** (A) A 9-V battery contains (B) a series of six 1.5-V cells.
©Brian Moeskau/Moeskau Photography

$H_2SO_4$ electrolyte

Anode (negative): lead grids filled with spongy lead

Cathode (positive): lead grids filled with $PbO_2$

**FIGURE 14.15** The lead-acid battery is a series of six voltaic cells that have alternating anodes and cathodes. It is the most common wet cell battery in use today.
©McGraw-Hill Education/Jill Braaten

Earlier in this chapter we mentioned a battery made of zinc and mercury(II) oxide:

$$Zn(s) + HgO(s) \longrightarrow ZnO(s) + Hg(l)$$

This battery produces a voltage of 1.3 V and was used in watches and calculators. A similar battery is composed of zinc and silver oxide:

$$Zn(s) + Ag_2O(s) \longrightarrow ZnO(s) + 2Ag(s)$$

It produces a voltage of 1.6 V and is commonly used in cameras. Given the overall reaction that occurs in a battery, you should be able to identify the oxidizing and reducing agents.

---

**EXAMPLE 14.5**   |   **Oxidation and Reduction in Batteries**

The reaction that occurs in older watch batteries is

$$Zn(s) + HgO(s) \longrightarrow ZnO(s) + Hg(l)$$

What is the oxidizing agent in this battery? What is the reducing agent?

**Solution:**

As usual, we first assign oxidation numbers:

$$\overset{0}{Zn}(s) + \overset{2+}{Hg}\overset{2-}{O}(s) \longrightarrow \overset{2+}{Zn}\overset{2-}{O}(s) + \overset{0}{Hg}(l)$$

Zinc changes from 0 to 2+, so it is oxidized. Mercury changes from 2+ to 0, so it is reduced. $HgO(s)$ is the oxidizing agent, and $Zn(s)$ is the reducing agent.

**Consider This 14.5**

The half-reaction for the reduction of mercury is

$$HgO(s) + H_2O(l) + 2e^- \longrightarrow Hg(l) + 2OH^-(aq)$$

Does this reaction occur at the anode or the cathode?

**Practice Problem 14.5**

The reaction that occurs in most camera batteries is

$$Zn(s) + Ag_2O(s) \longrightarrow ZnO(s) + 2Ag(s)$$

What is the oxidizing agent? What is the reducing agent?

**Further Practice:** Questions 14.41 and 14.42 at the end of the chapter

---

## 14.4 Balancing Simple Oxidation-Reduction Equations

To construct a battery or to describe an oxidation-reduction reaction, we need to predict the quantities of reactants and products. A balanced equation provides this information. As shown in Chapter 5, the atoms in a chemical equation must balance. In addition, charge must balance, and electrons lost must equal electrons gained.

To see what's involved in balancing a simple oxidation-reduction equation, consider the reaction between copper metal and silver nitrate shown in Figure 14.17. Copper atoms change to $Cu^{2+}$ ions, and $Ag^+$ ions change to silver atoms. We can represent the reaction using the equation

$$Cu(s) + Ag^+(aq) \longrightarrow Cu^{2+}(aq) + Ag(s)$$

Do you notice anything wrong with this equation? Do the atoms balance? Does the charge balance? Although the atoms balance, the charge does not. The total of the

**FIGURE 14.17** The electrons lost by copper at the anode travel toward the cathode. At the cathode, each $Ag^+$ ion gains an electron and gets plated onto the surface of the electrode. Two $Ag^+$ ions are reduced for every copper atom that's oxidized.

charges of the reactants is 1+. The total of the charges of the products is 2+. In addition, electrons lost do not equal electrons gained. As the equation is written, each copper atom loses two electrons to become $Cu^{2+}$, and each $Ag^+$ ion gains one electron to become a silver atom. The equation is not acceptable because it isn't fully balanced.

Let's track the flow of electrons as the reaction proceeds to understand what balancing the equation requires. A copper atom on the surface of the anode loses two electrons to form a $Cu^{2+}$ ion. These extra electrons on the anode flow through the wire toward the cathode. An ion of silver, $Ag^+$, comes into contact with the surface of the cathode, where an electron reduces the ion to a neutral silver atom. However, the copper atom lost two electrons. If a silver ion takes only one electron, what happens to the other one that was lost by copper? Another silver ion must be reduced to account for both electrons from the copper atom. So, for every copper atom that loses two electrons, two $Ag^+$ ions must each gain one electron. The balanced equation for the reaction then is

$$Cu(s) + 2Ag^+(aq) \longrightarrow Cu^{2+}(aq) + 2Ag(s)$$

To balance an oxidation-reduction reaction, we first separate the overall reaction into its half-reactions. This allows us to account for electrons gained and lost. It also reveals what happens in the anode and cathode half-cells. Like all other chemical equations, half-reactions must balance. Let's look again at the skeletal equation for the reaction between copper and silver ions:

$$Cu(s) + Ag^+(aq) \longrightarrow Cu^{2+}(aq) + Ag(s)$$

We can begin with copper, the reactant that is oxidized, to write the oxidation half-reaction. Copper metal reacts to form $Cu^{2+}$:

$$Cu(s) \longrightarrow Cu^{2+}(aq)$$

Is this equation balanced? The atoms balance, but the charge does not. The total charge on the reactant side of the equation is 0, while the total charge on the product side is 2+. We must add electrons to one side of the equation to account for the change in oxidation number. Adding two electrons to the product side gives a total charge of 0 on the right side of the equation:

$$Cu(s) \longrightarrow Cu^{2+}(aq) + 2e^-$$

The equation is balanced because the atoms and charges balance. This is the balanced half-reaction for the oxidation process.

Now let's look at the reduction half-reaction. Silver ion, $Ag^+$, reacts to form solid silver:

$$Ag^+(aq) \longrightarrow Ag(s)$$

Again the atoms balance, but the charge does not. The charge is 1+ on the reactant side and 0 on the product side. Adding one electron to the reactant side of the equation corrects the imbalance:

$$Ag^+(aq) + e^- \longrightarrow Ag(s)$$

The equation is now balanced because atoms and charges balance.

The oxidation and reduction half-reactions must be considered together to account for equivalence of electrons gained and lost:

Oxidation $\quad Cu(s) \longrightarrow Cu^{2+}(aq) + 2e^-$

Reduction $\quad Ag^+(aq) + e^- \longrightarrow Ag(s)$

All of the electrons lost by the reactant that's oxidized must be gained by the reactant that's reduced. Multiplying the reduction half-reaction by 2 equalizes electrons gained and lost for the $Cu/Ag^+$ reaction. When we do this, we multiply each reactant and product by 2:

$$2[Ag^+(aq) + e^- \longrightarrow Ag(s)]$$
$$2Ag^+(aq) + 2e^- \longrightarrow 2Ag(s)$$

The electrons lost in the oxidation half-reaction now equal the electrons gained in the reduction half-reaction. We can add these two half-reactions:

$$Cu(s) \longrightarrow Cu^{2+}(aq) + 2e^-$$
$$\underline{2Ag^+(aq) + 2e^- \longrightarrow 2Ag(s)}$$
$$Cu(s) + 2Ag^+(aq) + 2e^- \longrightarrow Cu^{2+}(aq) + 2Ag(s) + 2e^-$$

Notice that $2e^-$ appears as a reactant and as a product. They cancel:

$$Cu(s) + 2Ag^+(aq) + \cancel{2e^-} \longrightarrow Cu^{2+}(aq) + 2Ag(s) + \cancel{2e^-}$$

Overall reaction $\quad Cu(s) + 2Ag^+(aq) \longrightarrow Cu^{2+}(aq) + 2Ag(s)$

This equation is now balanced because atoms and charges balance. It is also balanced because electrons gained equal electrons lost. To double-check, count the atoms and determine the total charge of the reactants and the products:

|  | Reactants | Products |
|---|---|---|
| Cu | 1 | 1 |
| Ag | 2 | 2 |
| charge | 2+ | 2+ |

You can use the following series of questions to guide you through the balancing process:

1. Which substance is oxidized and which is reduced? To answer this question, assign oxidation numbers to all of the elements in the reaction.
2. What are the half-reactions for the oxidation and reduction processes? For each half-reaction,
   a. which element changes in oxidation number?
   b. can any spectator ions be ignored until the final balancing step?
   c. how many electrons must be added to the appropriate side of the equation to account for the change in oxidation number?
3. What factor can be used to multiply each coefficient in the balanced half-reactions to equalize the number of electrons gained or lost?
4. When adding the two half-reactions, can any substances that are present in equal amounts on both sides of the equation be canceled out?

 **Student Hot Spot**

Student data indicate you may struggle with balancing simple oxidation-reduction reaction equations. Access the SmartBook to view additional Learning Resources on this topic.

After answering these questions and acting accordingly, you should have a balanced overall equation. Give it a final check by counting the atoms of each element on both sides of the equation. Check to make sure that the sums of the charges are equal on the reactant side and on the product side. Use the series of questions to work through Example 14.6, balancing an oxidation-reduction equation.

---

**EXAMPLE 14.6**  **Balancing Oxidation-Reduction Equations**

Manganese metal reacts with aqueous iron(III) chloride to produce aqueous manganese(II) chloride and iron metal. Write a balanced chemical equation for this reaction.

**Solution:**

We begin by writing formulas for the reactants and products and writing a skeletal equation, just as we did in Chapter 5:

$$Mn(s) + FeCl_3(aq) \longrightarrow MnCl_2(aq) + Fe(s)$$

To determine what substances are oxidized and reduced, assign oxidation numbers to all of the elements in the skeletal equation:

$$\overset{0}{Mn}(s) + \overset{3+}{Fe}\overset{1-}{Cl_3}(aq) \longrightarrow \overset{2+}{Mn}\overset{1-}{Cl_2}(aq) + \overset{0}{Fe}(s)$$

The chloride ion does not change in oxidation number, so we assume it is a spectator ion. We can write the skeletal ionic equation as follows:

$$Mn(s) + Fe^{3+}(aq) \longrightarrow Mn^{2+}(aq) + Fe(s)$$

The oxidation number for manganese changes from 0 to 2+, so it is oxidized. We can begin writing an oxidation half-reaction as follows:

$$Mn(s) \longrightarrow Mn^{2+}(aq)$$

We add two electrons to the right side of the equation to account for the change in oxidation number. For a simple process like this, we can look at the addition of electrons in another way. Since the total charge on the left is 0 and the total charge on the right is 2+, we must add two electrons to the product side of the equation to balance the charge:

$$Mn(s) \longrightarrow Mn^{2+}(aq) + 2e^-$$

The atoms and charges are now balanced and we have the oxidation half-reaction.

Now let's consider the other half-reaction. Iron changes from 3+ to 0, so it is the element that is reduced. We begin writing the reduction half-reaction as follows:

$$Fe^{3+}(aq) \longrightarrow Fe(s)$$

We must add three electrons to the reactant side to balance the charge:

$$Fe^{3+}(aq) + 3e^- \longrightarrow Fe(s)$$

The atoms and charges are now balanced and we have the reduction half-reaction.

The next step is to consider the oxidation and reduction reactions together:

Oxidation   $Mn(s) \longrightarrow Mn^{2+}(aq) + 2e^-$

Reduction   $Fe^{3+}(aq) + 3e^- \longrightarrow Fe(s)$

Each manganese atom loses two electrons, but each iron ion gains three. Electrons gained must equal electrons lost, so we multiply the oxidation half-reaction by 3 and the reduction half-reaction by 2:

$$3[Mn(s) \longrightarrow Mn^{2+}(aq) + 2e^-]$$
$$2[Fe^{3+}(aq) + 3e^- \longrightarrow Fe(s)]$$

$$3Mn(s) \longrightarrow 3Mn^{2+}(aq) + 6e^-$$
$$2Fe^{3+}(aq) + 6e^- \longrightarrow 2Fe(s)$$

We can now add the two half-reactions together and cancel out the six electrons that appear on both sides of the equation:

$$3Mn(s) \longrightarrow 3Mn^{2+}(aq) + 6e^-$$
$$2Fe^{3+}(aq) + 6e^- \longrightarrow 2Fe(s)$$
$$\overline{3Mn(s) + 2Fe^{3+}(aq) + \cancel{6e^-} \longrightarrow 3Mn^{2+}(aq) + 2Fe(s) + \cancel{6e^-}}$$

The balanced ionic equation is

$$3Mn(s) + 2Fe^{3+}(aq) \longrightarrow 3Mn^{2+}(aq) + 2Fe(s)$$

The overall equation can be written by returning the chlorides into the formulas for the compounds in aqueous solution:

$$3Mn(s) + 2FeCl_3(aq) \longrightarrow 3MnCl_2(aq) + 2Fe(s)$$

We'll double-check the balancing by counting atoms and charges on both sides of the equation.

|        | Reactants | Products |
|--------|-----------|----------|
| Mn     | 3         | 3        |
| Fe     | 2         | 2        |
| Cl     | 6         | 6        |
| charge | 0         | 0        |

### Consider This 14.6

If a voltaic cell involving the reaction described in the example were constructed, what would happen to the masses of the iron and manganese electrodes as the cell runs?

### Practice Problem 14.6

Magnesium metal reacts with aqueous chromium(III) nitrate to produce magnesium nitrate and solid chromium. Write a balanced equation for this reaction.

**Further Practice:**  Questions 14.47 and 14.48 at the end of the chapter

Positive button
Steel case
MnO$_2$ in KOH paste
Zn (anode)
Graphite rod (cathode)
Absorbent/separator
Negative end cap

**FIGURE 14.18** A common dry cell battery consists of a zinc anode and an inert graphite rod that serves as the cathode. The reaction requires the presence of a base, so the MnO$_2$ is in a KOH paste.

## 14.5  Balancing Complex Oxidation-Reduction Equations

The oxidation-reduction reactions we considered in Section 14.4 are relatively simple processes involving only the oxidizing and reducing agents. Many oxidation-reduction reactions are more complex, and they may involve water, an acid, or a base. Accounting for atoms and electron transfer in such reactions requires a slightly different approach.

Consider a reaction that occurs in a dry cell (Figure 14.18). This battery, like many others, requires the presence of a base for the chemical reaction to occur. The term *alkaline* describes chemical systems that involve bases, so such dry cells are called alkaline batteries. In a typical alkaline battery, the base is a hydroxide paste. The reaction that occurs in a zinc-manganese alkaline battery is

$$Zn(s) + MnO_2(s) + H_2O(l) \longrightarrow ZnO(s) + Mn(OH)_2(s)$$

The base is supplied by a KOH paste that contains the $MnO_2$. Notice in this reaction that water is a reactant, and a metal hydroxide is one of the products. Also notice that no reactants or products are in aqueous solution and that there are no spectator ions to remove. Is this equation balanced? Yes, the atoms and charges balance. However, the equation doesn't tell us anything about what occurs in the half-cells. The half-cells for this reaction may require some base ($OH^-$) and water to balance each half-reaction.

Many oxidation-reduction reactions occur in acidic aqueous solution, so their equations can be balanced with $H^+$ ions. For example, the Breathalyzer used by law enforcement officials to test drivers for drunkenness uses an acidic solution of the dichromate ion, $Cr_2O_7^{2-}$ (Figure 14.19). In the presence of ethanol, $C_2H_5OH$, the orange dichromate ion is reduced to the chromium(III) ion, which appears green in solution. The ethanol is oxidized to acetic acid. If further oxidation could occur, the skeletal equation for the reaction would be

$$C_2H_5OH(aq) + Cr_2O_7^{2-}(aq) \longrightarrow Cr^{3+}(aq) + CO_2(g)$$

To balance the half-reactions for this reaction, acid ($H^+$) and water may have to be added.

Let's separate this reaction into half-reactions to balance it. First assign oxidation numbers to all of the atoms in the equation:

$$\overset{2-}{C}\overset{1+}{}\overset{2-}{}\overset{1+}{}H_5\overset{}{O}H(aq) + \overset{6+}{Cr_2}\overset{2-}{O_7^{2-}}(aq) \longrightarrow \overset{3+}{Cr^{3+}}(aq) + \overset{4+}{C}\overset{2-}{O_2}(g)$$

Each carbon atom changes from an oxidation number of 2− to 4+, so carbon is oxidized in the half-reaction:

$$C_2H_5OH(aq) \longrightarrow CO_2(g)$$

There are two carbon atoms on the left side of the equation and only one on the right. We must make sure that the atoms of carbon balance before we account for the loss of electrons (Figure 14.20). Each ethanol molecule contains two carbon atoms, so we can place a 2 as a coefficient in front of $CO_2$ to balance the carbon atoms:

$$\overset{2-}{C}\overset{1+}{}\overset{2-}{}\overset{1+}{}H_5OH(aq) \longrightarrow \overset{4+}{2C}\overset{2-}{O_2}(g)$$

Although this reaction at first appears charge-balanced, we must account for the change in oxidation numbers. Each carbon on the left side of the equation has an oxidation number of 2−, and each carbon on the right has an oxidation number of 4+. Accounting for both carbons on the left gives a combined oxidation number of 4−. Similarly each carbon on the right has an oxidation number of 4+, but because there are two of them, the combined oxidation number is 8+. Thus, 12 electrons are lost for every ethanol molecule that reacts:

$$C_2H_5OH(aq) \longrightarrow 2CO_2(g) + 12e^-$$

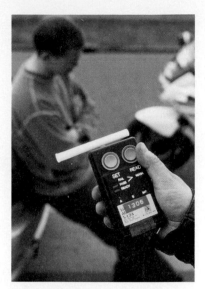

**FIGURE 14.19** In the Breathalyzers used by law enforcement officials, an oxidation-reduction reaction causes a color change that indicates the amount of alcohol present in a driver's system.

**FIGURE 14.20** Each carbon in the reactant $C_2H_5OH$ has an oxidation number of 2−. The carbon in the product has an oxidation number of 4+. For every molecule of $C_2H_5OH$ that reacts, two molecules of $CO_2$ are produced.

**FIGURE 14.21** Each chromium in the reactant $Cr_2O_7^{2-}$ has an oxidation number of 6+. The chromium in the product has an oxidation number of 3+. For every ion of $Cr_2O_7^{2-}$ that reacts, two ions of $Cr^{3+}$ must be produced.

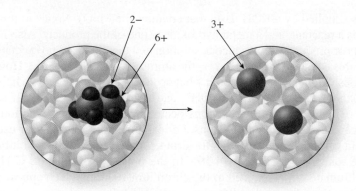

To simplify the balancing process for reactions that occur in acid, we will use $H^+$ instead of $H_3O^+$.

If the reaction occurs in base, $OH^-$ ions are available to balance the charge. If the half-reaction

$$C_2H_5OH(l) \longrightarrow 2CO_2(g) + 12e^-$$

occurs in base, how would you use $OH^-$ and water to balance the equation? Addition of 12 $OH^-$ ions to the reactant side would balance the charge, and 9 $H_2O$ molecules to the product side would balance the oxygen and hydrogen atoms. Convince yourself that this works by writing a balanced equation for the half-reaction and counting the atoms and charges on each side of the equation.

Balance the half-reaction

$$Cr_2O_7^{2-}(aq) + 6e^- \longrightarrow 2Cr^{3+}(aq)$$

if it were to occur in base by adding $OH^-$ and $H_2O$ to appropriate sides of the equation.

Now the charge is no longer balanced. The total charge on the left is 0, and the total charge on the right is 12−. Because we've already used electrons to account for the change in oxidation number, we cannot use them to balance the charge. We'll have to add a step to the procedure for balancing oxidation-reduction equations that we used in Section 14.4. Recall that the reaction occurs in the presence of acid, which means $H^+$ ions are present. To balance charge in an acidic solution, add $H^+$ ions to the side with the excess negative charge. If we add 12 $H^+$ ions to the right side of this equation, we can balance the charge:

$$C_2H_5OH(aq) \longrightarrow 2CO_2(g) + 12e^- + 12H^+(aq)$$

The total charges on the left and right are now equal at 0. The only task remaining is to balance oxygen and hydrogen atoms, which we can do with water molecules because the reaction occurs in aqueous solution. There are 6 hydrogen atoms on the left and 12 on the right. There is one oxygen on the left and four on the right. If we put three $H_2O$ molecules on the left, the atoms of oxygen and hydrogen balance:

$$C_2H_5OH(aq) + 3H_2O(l) \longrightarrow 2CO_2(g) + 12e^- + 12H^+(aq)$$

Now let's consider the reduction half-reaction:

$$Cr_2O_7^{2-}(aq) \longrightarrow Cr^{3+}(aq)$$

Each chromium atom changes from 6+ to 3+. But before we can account for the electron gain, we must make sure that the atoms of chromium balance (Figure 14.21). Because two chromium atoms are present in the dichromate ion, $Cr_2O_7^{2-}$, two $Cr^{3+}$ ions must form for each $Cr_2O_7^{2-}$ that reacts:

$$\overset{6+}{\underset{\downarrow}{}}\,\overset{2-}{\underset{\downarrow}{}} \qquad \overset{3+}{\underset{\downarrow}{}}$$
$$Cr_2O_7^{2-}(aq) \longrightarrow 2Cr^{3+}(aq)$$

When we account for both of the chromium atoms on the left, we have a combined oxidation number of 12+. Each chromium atom on the right has an oxidation number of 3+, but because there are two of them, the combined oxidation number is 6+. If we add six electrons to the left, we can account for the change in oxidation number:

$$Cr_2O_7^{2-}(aq) + 6e^- \longrightarrow 2Cr^{3+}(aq)$$

Now we have to balance the charge. The total charge on the left is 8−, and the total charge on the right is 6+. Since the reaction occurs in the presence of hydrogen ions, we can use $H^+$ to balance the charge:

$$Cr_2O_7^{2-}(aq) + 6e^- + 14H^+(aq) \longrightarrow 2Cr^{3+}(aq)$$

The total charges on the left and right are now equal at 6+. The last step is to add water to balance hydrogen and oxygen atoms. There are seven oxygen atoms on the left and none on the right. There are 14 hydrogen atoms on the left and none on the right. If we put 7 $H_2O$ molecules on the product side, oxygen and hydrogen balance:

$$Cr_2O_7^{2-}(aq) + 6e^- + 14H^+(aq) \longrightarrow 2Cr^{3+}(aq) + 7H_2O(l)$$

The reduction half-reaction is balanced.

We'll now look at the oxidation and reduction processes together:

Oxidation   $C_2H_5OH(aq) + 3H_2O(l) \longrightarrow 2CO_2(g) + 12e^- + 12H^+(aq)$

Reduction   $Cr_2O_7^{2-}(aq) + 6e^- + 14H^+(aq) \longrightarrow 2Cr^{3+}(aq) + 7H_2O(l)$

Do electrons lost equal electrons gained? No. Notice that 12 electrons are lost in the oxidation half-reaction. In the reduction half-reaction, six electrons are gained. Multiplying the reduction half-reaction by a factor of 2 equalizes the electrons gained and lost:

$$2[Cr_2O_7^{2-}(aq) + 6e^- + 14H^+(aq) \longrightarrow 2Cr^{3+}(aq) + 7H_2O(l)]$$
$$2Cr_2O_7^{2-}(aq) + 12e^- + 28H^+(aq) \longrightarrow 4Cr^{3+}(aq) + 14H_2O(l)$$

We can now combine our oxidation and reduction half-reactions:

$$C_2H_5OH(aq) + 3H_2O(l) \longrightarrow 2CO_2(g) + 12e^- + 12H^+(aq)$$
$$\underline{2Cr_2O_7^{2-}(aq) + 12e^- + 28H^+(aq) \longrightarrow 4Cr^{3+}(aq) + 14H_2O(l)}$$
$$C_2H_5OH(aq) + 2Cr_2O_7^{2-}(aq) + \cancel{12e^-} + 28H^+(aq) + 3H_2O(l) \longrightarrow$$
$$2CO_2(g) + \cancel{12e^-} + 12H^+(aq) + 4Cr^{3+}(aq) + 14H_2O(l)$$

The 12 electrons on the left and right cancel to give:

$$C_2H_5OH(aq) + 2Cr_2O_7^{2-}(aq) + 28H^+(aq) + 3H_2O(l) \longrightarrow$$
$$2CO_2(g) + 12H^+(aq) + 4Cr^{3+}(aq) + 14H_2O(l)$$

There are 28 $H^+$ ions on the left and 12 on the right, leaving a net of 16 on the reactant side. There are 3 $H_2O$ molecules on the left and 14 on the right, leaving a net of 11 $H_2O$ molecules on the product side. Canceling equal numbers of $H^+$ ions and $H_2O$ molecules results in an overall balanced equation:

$$C_2H_5OH(aq) + 2Cr_2O_7^{2-}(aq) + 16H^+(aq) \longrightarrow 2CO_2(g) + 4Cr^{3+}(aq) + 11H_2O(l)$$

We'll count atoms and charges on both sides of the equation to double-check the balancing.

|        | Reactants | Products |
|--------|-----------|----------|
| C      | 2         | 2        |
| H      | 22        | 22       |
| O      | 15        | 15       |
| Cr     | 4         | 4        |
| charge | 12+       | 12+      |

You can use the following series of questions to guide you through the balancing process if you encounter a reaction that occurs in an acid or base:

1. What reactant is oxidized and what reactant is reduced? Assign oxidation numbers to all of the elements in the reaction to make the determination.
2. What are the half-reactions for the oxidation and reduction processes? For each half-reaction, determine each of the following:
   a. Is the element that changes in oxidation number balanced?
   b. Can any spectator ions be ignored until the final balancing step?
   c. How many electrons must be added to the appropriate side of the equation to account for the change in oxidation number?

**Student Hot Spot**

Student data indicate you may struggle with balancing complex oxidation-reduction equations. Access the SmartBook to view additional Learning Resources on this topic.

**d.** Is the charge balanced after adding electrons? If not, what ions are present to balance the charge? If it occurs in acid, add $H^+$ ions. If the reaction occurs in base, add $OH^-$ ions.

**e.** Do the hydrogen and oxygen atoms balance? If not, add water molecules to the appropriate side of the equation.

3. What factor can be used to multiply each coefficient to equalize the number of electrons gained and lost?

4. When adding the two half-reactions, are any substances present on both sides of the equation? If so, cancel out an appropriate number. If substances are present on both sides of the equation in unequal amounts, cancel out the appropriate amount of substances on each side.

At this point you should have an overall balanced equation. Make sure atoms and charges balance on both sides of the equation. Apply this series of questions to Example 14.7.

---

### EXAMPLE 14.7  Balancing Complex Oxidation-Reduction Reactions

Balance the following equation, which occurs in acidic solution:

$$Cu(s) + NO_3^-(aq) \longrightarrow Cu^{2+}(aq) + NO(g)$$

**Solution:**

We first assign oxidation numbers to identify what is oxidized and what is reduced:

$$\overset{0}{Cu}(s) + \overset{5+\ 2-}{NO_3^-}(aq) \longrightarrow \overset{2+}{Cu^{2+}}(aq) + \overset{2+\ 2-}{NO}(g)$$

Copper changes from 0 to 2+, so it is the reactant that is oxidized:

$$Cu(s) \longrightarrow Cu^{2+}(aq)$$

We add two electrons to the right side of the equation to account for this change in oxidation number:

$$Cu(s) \longrightarrow Cu^{2+}(aq) + 2e^-$$

This half-reaction is now balanced.

Now consider the reduction half-reaction:

$$NO_3^-(aq) \longrightarrow NO(g)$$

Nitrogen changes from 5+ to 2+. Adding three electrons to the left accounts for this change in oxidation number:

$$NO_3^-(aq) + 3e^- \longrightarrow NO(g)$$

The total charge on the left is 4−. On the right it's 0. The reaction occurs in acidic solution, so we add four $H^+$ ions to the left to balance the charge:

$$4H^+(aq) + NO_3^-(aq) + 3e^- \longrightarrow NO(g)$$

Adding water molecules completes the balancing process for this half-reaction. There are four hydrogen atoms on the left and none on the right. There are three oxygen atoms on the left and one on the right. Putting two $H_2O$ molecules on the product side balances oxygen and hydrogen:

$$4H^+(aq) + NO_3^-(aq) + 3e^- \longrightarrow NO(g) + 2H_2O(l)$$

Now compare the oxidation and reduction half-reactions to see if the electrons lost equal the electrons gained:

Oxidation $\quad Cu(s) \longrightarrow Cu^{2+}(aq) + 2e^-$

Reduction $\quad 4H^+(aq) + NO_3^-(aq) + 3e^- \longrightarrow NO(g) + 2H_2O(l)$

Two electrons appear on the right in the oxidation half-reaction, and three appear on the left in the reduction half-reaction. If we multiply the oxidation half-reaction by 3 and the reduction half-reaction by 2, electrons lost will equal electrons gained:

$$3[Cu(s) \longrightarrow Cu^{2+}(aq) + 2e^-]$$

$$2[4H^+(aq) + NO_3^-(aq) + 3e^- \longrightarrow NO(g) + 2H_2O(l)]$$

Now add the two equations:

$$3Cu(s) \longrightarrow 3Cu^{2+}(aq) + 6e^-$$

$$8H^+(aq) + 2NO_3^-(aq) + 6e^- \longrightarrow 2NO(g) + 4H_2O(l)$$

$$\overline{3Cu(s) + 8H^+(aq) + 2NO_3^-(aq) + \cancel{6e^-} \longrightarrow 2NO(g) + 4H_2O(l) + 3Cu^{2+}(aq) + \cancel{6e^-}}$$

Canceling common terms, we get

$$3Cu(s) + 8H^+(aq) + 2NO_3^-(aq) \longrightarrow 2NO(g) + 4H_2O(l) + 3Cu^{2+}(aq)$$

We now have a balanced equation. We'll double-check by counting atoms and charges on both sides of the equation.

|        | Reactants | Products |
|--------|-----------|----------|
| Cu     | 3         | 3        |
| H      | 8         | 8        |
| N      | 2         | 2        |
| O      | 6         | 6        |
| charge | 6+        | 6+       |

**Consider This 14.7**

How would the balanced equation change if the reaction in the example occurred in basic solution?

**Practice Problem 14.7**

Balance the following equation that occurs in basic solution:

$$Cl_2(aq) \longrightarrow Cl^-(aq) + ClO_3^-(aq)$$

**Further Practice:** Questions 14.55 and 14.56 at the end of the chapter

# 14.6 Electrochemistry

When Anna and Bill walked around campus in Chapter 1, they observed several vehicles that run on alternative fuels. You might recall from Chapter 6 that hydrogen-fueled buses are powered by the reaction of hydrogen with oxygen:

$$2H_2(g) + O_2(g) \longrightarrow 2H_2O(g)$$

This reaction, like reactions that occur in batteries, is spontaneous. A chemical reaction or physical change is **spontaneous** if it occurs by itself without outside intervention. That is, it does not need an outside source of energy to progress. Some spontaneous reactions are fast, like the combustion of methane, while others, like rusting, are slow. Some spontaneous reactions are desirable, while others are not.

The reactions that occur in batteries are spontaneous and desirable because they provide electricity to start cars, run watches, operate power tools, and drive devices

**FIGURE 14.22** A chemical activity series lists metals and $H_2$ with the most active elements at the top.

such as calculators, cell phones, and laptop computers. The study of batteries and other voltaic cells constitutes a field of chemistry known as electrochemistry. **Electrochemistry** is the study of the relationship between chemical reactions and electrical work. The study of voltaic cells is a branch of electrochemistry that deals with spontaneous, desirable chemical processes. This branch of electrochemistry deals with oxidation-reduction reactions in which chemistry drives the flow of electricity.

The other branch of electrochemistry is the study of reactions that are desirable but do not occur on their own. They are not spontaneous. This branch of electrochemistry deals with oxidation-reduction reactions in which electricity drives the chemistry. In this section we'll explain more about voltaic cells and introduce a different kind of cell, the *electrolytic* cell.

## Voltaic Cells

In Chapter 5 we discussed how single-displacement reactions can be predicted from an activity series (Figure 14.22). All single-displacement reactions in which a metal (metal A) reacts with a compound containing a different metal (metal B) to give a new compound (containing metal A) and a different pure metal (metal B) are oxidation-reduction reactions. The following equation summarizes them:

$$A + BX \longrightarrow AX + B$$

In this equation BX and AX represent ionic compounds containing the metal cations of B and A, and X represents an anion. We can use this equation and the activity series to predict what happens when we construct a voltaic cell. For example, let's set up a cell of zinc and iron (Figure 14.23). We begin with two metals and electrolyte solutions of compounds containing these metals. So, in our example, we'll set up one half-cell with zinc as the electrode immersed in a solution of a compound containing zinc, such as $ZnCl_2$. In the other half-cell we'll use an iron electrode and immerse it in a solution of a compound containing iron, such as $FeCl_2$. We connect a wire between the two cells and place a salt bridge between the solutions. Can you predict what will occur?

Let's look at the possibilities. From our discussion about single-displacement reactions in Chapter 5, we know that only one of the following reactions occurs spontaneously:

$$Zn(s) + FeCl_2(aq) \longrightarrow ZnCl_2(aq) + Fe(s)$$

$$Fe(s) + ZnCl_2(aq) \longrightarrow FeCl_2(aq) + Zn(s)$$

Voltmeter

**FIGURE 14.23** In a zinc-iron cell, which reaction will occur?

$$Zn(s) + FeCl_2(aq) \longrightarrow ZnCl_2(aq) + Fe(s)$$

or

$$Fe(s) + ZnCl_2(aq) \longrightarrow FeCl_2(aq) + Zn(s)$$

Zinc appears above iron in the activity series, so it is the more reactive metal. It will displace iron from its compounds, so only this reaction will occur:

$$Zn(s) + FeCl_2(aq) \longrightarrow ZnCl_2(aq) + Fe(s)$$

An activity series predicts the relative strengths of metals as reducing agents and metal compounds as oxidizing agents. As we go up the activity series, the metals become stronger reducing agents. They are more reactive and give up electrons more readily. Metal ions in their compounds increase in oxidizing strength as we move down the activity series. To see how this works, let's return to the zinc-iron example. Zinc acts as the reducing agent because it provides electrons for the reduction of $Fe^{2+}$. The reaction is spontaneous and can be used to produce iron metal.

Voltaic cells can be made from household materials. For example, a galvanized nail (which is iron coated with zinc) and a copper penny inserted into a lemon can produce a voltage when connected as shown in Figure 14.24.

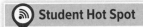

**Student Hot Spot**

Student data indicate you may struggle with predicting spontaneous reactions in voltaic cells. Access the SmartBook to view additional Learning Resources on this topic.

---

**EXAMPLE 14.8**  |  **Voltaic Cells and the Activity Series**

Suppose we want to obtain lead metal from a solution of $Pb(NO_3)_2$. In one half-cell of a voltaic cell, we place a lead electrode in a solution of lead(II) nitrate, $Pb(NO_3)_2$. From the activity series, select a metal that we can place in the other half-cell compartment with its corresponding electrolyte solution.

**Solution:**

To produce lead metal, we should select a metal that can displace lead from its compounds. From the activity series, any metal that appears above lead may be suitable. Examples are Sn, Fe, and Zn. The metals at the top of the series are highly reactive with water, so for safety and other reasons we should avoid them.

**Consider This 14.8**

Suppose we wanted to obtain nickel metal from a solution of $Ni(NO_3)_2$. Would all the metals suggested for making elemental lead from $Pb(NO_3)_2$ work for producing elemental nickel?

**Practice Problem 14.8**

Propose a voltaic cell that could be used to deposit chromium metal on the surface of a chromium electrode.

**Further Practice:**  Questions 14.61 and 14.62 at the end of the chapter

---

## Electrolytic Cells

Many desirable oxidation-reduction reactions are not spontaneous. An **electrolytic cell** is an electrochemical cell through which electric current is passed to cause a non-spontaneous oxidation-reduction reaction to occur. The process is called **electrolysis.**

**FIGURE 14.24**  Common household items can be used to produce a battery. In this lemon battery, a galvanized nail and a copper penny serve as electrodes. One lemon can generate a voltage of 0.7 V to 0.9 V, which is not enough to light a tiny bulb, but four lemons in series can do the job. Scientists don't fully understand how this battery works.

🌀 **ANIMATION:** Electrolysis of KI

One of the most important uses of electrolysis is the isolation of elements from their naturally occurring compounds. For example, the primary source of elemental sodium is the electrolysis of molten sodium chloride. In this process, sodium chloride is melted, two inert electrodes are inserted into the molten salt, and electricity is passed through it (Figure 14.25). The current causes one of the electrodes to become negatively charged. The other becomes positively charged. Because the electrodes are charged, they attract the ions that move around the molten salt. The negatively charged cathode attracts $Na^+$ ions; the positively charged anode attracts $Cl^-$ ions. When the proper voltage is applied, $Na^+$ ions are reduced to sodium atoms at the cathode:

$$Na^+(l) + e^- \longrightarrow Na(l)$$

The high temperature required to keep NaCl molten produces liquefied sodium metal at the cathode.

At the anode, chloride ions are oxidized to chlorine atoms, which combine to form $Cl_2$ molecules:

$$2Cl^-(l) \longrightarrow Cl_2(g) + 2e^-$$

The overall chemical reaction for this process is

$$2Na^+(l) + 2e^- \longrightarrow 2Na(l)$$
$$\underline{2Cl^-(l) \longrightarrow Cl_2(g) + 2e^-}$$
$$2NaCl(l) \longrightarrow 2Na(l) + Cl_2(g)$$

Like sodium, calcium and magnesium are produced by the electrolysis of molten salts. Other metals are produced using different electrolytic processes.

Electrolysis is commonly used in electroplating jewelry and other items. The item to be plated is immersed in a solution that contains an electrolyte salt of the plating metal. When the power is turned on, the metal ions are reduced to atoms that deposit on the surface of the object.

Pure metal anode (Ag)

Cathode: Article to be electroplated

Voltage source

$Ag^+$

Electroplating with silver

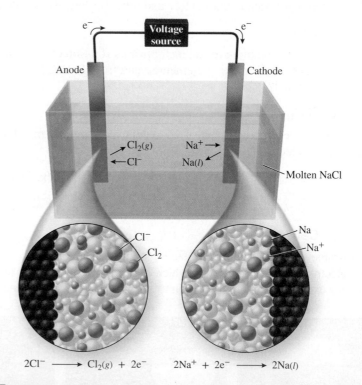

**FIGURE 14.25** When molten sodium chloride is electrolyzed, $Cl^-$ ions are oxidized at the anode to produce $Cl_2$ molecules. At the cathode, $Na^+$ ions are reduced to form sodium atoms. Because the reaction is run at a high temperature to keep the NaCl molten, the sodium product is also in the liquid state. The actual cell is designed to remove the products, Na and $Cl_2$, as they form. A hood placed over the anode prevents the $Cl_2(g)$ from contacting the molten sodium. Why do these products need to be separated as they form?

Electrolysis can also be used to prepare nonmetals. For example, water can be electrolyzed to produce oxygen and hydrogen gas:

$$2H_2O(l) \longrightarrow O_2(g) + 2H_2(g)$$

Electrolysis is also used to recharge batteries. Recall, for example, the reaction that occurs in a lead-acid battery in cars:

$$Pb(s) + PbO_2(s) + 2H_2SO_4(aq) \longrightarrow 2PbSO_4(aq) + 2H_2O(l)$$

An electric current from the car's alternator forces the reaction to go "in reverse." That is, the products are converted back to the reactants, keeping the battery functional:

$$2PbSO_4(aq) + 2H_2O(l) \longrightarrow Pb(s) + PbO_2(s) + 2H_2SO_4(aq)$$

## EXAMPLE 14.9 | Electrolytic Cells

The electrolysis of molten potassium iodide produces liquid potassium and gaseous iodine. Consider the following diagram:

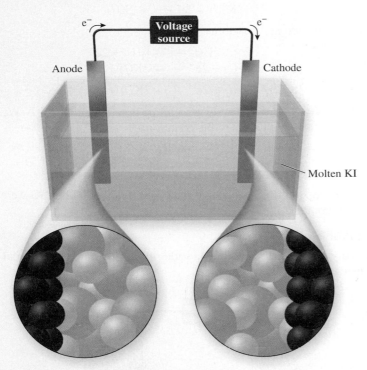

(a) When the power is turned on, what happens at the cathode?
(b) When the power is turned on, what happens at the anode?
(c) Write oxidation and reduction half-reactions.
(d) Write an overall equation for the reaction.

**Solution:**
(a) Since an oxidation-reduction reaction occurs, one element must be oxidized and the other reduced. Potassium iodide contains $K^+$ ions and $I^-$ ions. Based on its position in the periodic table, we wouldn't expect potassium to become more oxidized than $K^+$, so it is reduced. Since reduction always occurs at the cathode, potassium metal is produced there.
(b) Iodide ions form iodine atoms at the anode, because it's the site of oxidation in an electrochemical cell. Two of these atoms combine to form an iodine molecule.
(c) To begin writing half-reactions, first consider the skeletal equation for the overall reaction. The high temperature required to keep the KI molten will

make the potassium liquid and the iodine gaseous. Recall that iodine is one of the elements that exists as diatomic molecules:

$$KI(l) \longrightarrow K(l) + I_2(g)$$

Iodine is oxidized in this reaction, so the half-reaction for the oxidation process is

$$2I^-(l) \longrightarrow I_2(g) + 2e^-$$

The half-reaction for the reduction of potassium is

$$K^+(l) + e^- \longrightarrow K(l)$$

(d) To write an overall equation, combine the oxidation and reduction half-reactions:

Oxidation $\quad 2I^-(l) \longrightarrow I_2(g) + 2e^-$
Reduction $\quad K^+(l) + e^- \longrightarrow K(l)$

To equalize electrons lost and gained, multiply the reduction half-reaction by 2:

$$2[K^+(l) + e^- \longrightarrow K(l)]$$
$$2K^+(l) + 2e^- \longrightarrow 2K(l)$$

Now add the two half-reactions:

$$2I^-(l) \longrightarrow I_2(g) + 2e^-$$
$$\underline{2K^+(l) + 2e^- \longrightarrow 2K(l)}$$
$$2K^+(l) + 2I^-(l) + 2e^- \longrightarrow 2K(l) + I_2(g) + 2e^-$$

The first two terms, $K^+(l)$ and $I^-(l)$, can be represented as $KI(l)$. We can also cancel $2e^-$ from both sides of the equation:

$$2KI(l) \longrightarrow 2K(l) + I_2(g)$$

### Consider This 14.9

What would happen if the products of the electrolysis of molten KI were allowed to come into contact?

### Practice Problem 14.9

The electrolysis of molten calcium chloride produces liquid calcium and chlorine gas. Consider the following diagram:

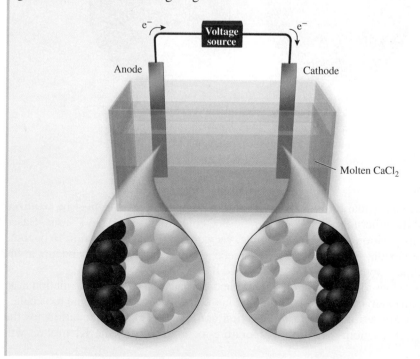

(a)  When the power is turned on, what happens at the cathode?
(b)  When the power is turned on, what happens at the anode?
(c)  Write oxidation and reduction half-reactions.
(d)  Write an overall equation for the reaction.

**Further Practice:**  Questions 14.67 and 14.68 at the end of the chapter

## 14.7  Corrosion Prevention

Bill drives a rusty old truck. After noticing it, Anna advised him to take steps to prevent further corrosion. The criticism hurt Bill's feelings a little, but he knew Anna was right. We generally define corrosion as a slow deterioration of metals due to interaction with the environment. The rusting of a vehicle is an example of corrosion, because the iron in the steel reacts with oxygen and water in the environment. More specifically, **corrosion** is an oxidation-reduction reaction. Familiar signs of corrosion are the red stains of rust on iron and steel, the green coating on copper and brass, and the black coating on silver. The rusting of the steel in Bill's truck is a reaction of iron, oxygen, and water that is spontaneous but undesirable. Dents or scratches and acidic conditions accelerate corrosion. Vehicles in colder climates rust faster because of the salt that's put on the roads in winter to melt snow and ice (Figure 14.26). Corrosion repair and prevention cost over $1.1 trillion a year in the United States alone.

**FIGURE 14.26**  Rust is a serious problem, especially in cold climates where salt used to melt ice can accelerate the rusting process.
©Molly Aaker/Alamy Stock Photo

As Bill and Anna discussed Bill's rusty truck, they recalled from their chemistry class that corrosion occurs when an electrochemical cell is set up on the surface of the metal. When iron rusts, a part of the metal acts as an anode, where an oxidation process occurs:

$$Fe(s) \longrightarrow Fe^{2+}(aq) + 2e^-$$

A reduction process occurs simultaneously in another location, which acts as a cathode, with molecular oxygen acting as the oxidizing agent:

$$O_2(g) + 2H_2O(l) + 4e^- \longrightarrow 4OH^-(aq)$$

Electrons released by the oxidation of iron migrate through the metal to the site where the oxygen atoms in molecular oxygen gain electrons (Figure 14.27). In the presence of water, the oxidized metal ions migrate to the site where reduction can

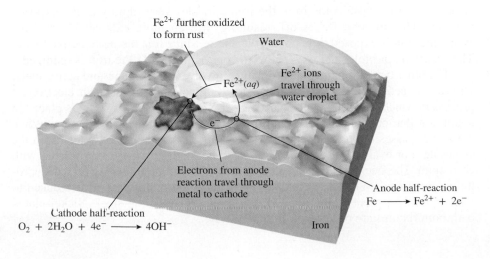

**FIGURE 14.27**  When iron corrodes in moist air, the iron oxidizes to form $Fe^{2+}$ ions. These ions migrate through the water droplet to the site of reduction, the cathode. At the cathode, $O_2$ is reduced. The $Fe^{2+}$ ions oxidize further to form $Fe^{3+}$ ions, which form hydrated iron(III) oxide.

**FIGURE 14.28** Zinc is attached to the trans-Alaskan pipeline to keep the iron from rusting. Because zinc is more reactive, it is oxidized instead of iron. The zinc is attached to the exterior of the pipe, so it can be easily replaced when it's substantially corroded. The iron in the pipeline would be more difficult to replace.
©Jim Birk

occur. The iron(II) ions then combine with hydroxide ions to form insoluble hydroxide salts, which react further with oxygen and water to form the further oxidized hydrated iron(III) oxide, rust. Rust flakes off the surface of the iron metal and exposes fresh metal, so iron continues to corrode, ultimately suffering serious damage.

A variety of methods can be used to prevent corrosion. Most often a protective coating such as paint is applied. Other methods, including *cathodic protection* and *plating,* involve less destructive electrochemical cells.

When two metals come in contact in a corrosive environment, the more active of the two metals corrodes, and the other does not. This principle forms the basis for the cathodic protection of a metal. For example, iron pipes or tanks, sometimes buried underground, are often connected to a more active metal, such as magnesium or zinc. The trans-Alaskan pipeline, for example, was constructed with strips of zinc cable for corrosion protection (Figure 14.28). The zinc forms the anode, where oxidation occurs:

$$Zn(s) \longrightarrow Zn^{2+}(aq) + 2e^-$$

Zinc is selectively oxidized instead of iron, so iron metal acts as an inert cathode where reduction of oxygen occurs:

$$O_2(g) + 2H_2O(l) + 4e^- \longrightarrow 4OH^-(aq)$$

The iron acts only as a conductor of electrons and does not take part in either the oxidation or the reduction half-reactions. Thus, the iron remains completely uncorroded as long as the zinc remains. It is much simpler to replace the zinc cable with a fresh piece than to replace the pipe. In a similar way, magnesium is often used to protect buried gasoline tanks and ships' hulls.

Plating also prevents corrosion. Plating is a variation on cathodic protection. A thin coating of one metal can be applied on top of another by either dipping or electroplating. If the coating metal melts at a reasonably low temperature, dipping is used. For example, iron is galvanized by dipping it in molten zinc. Once it's plated, if a scratch exposes some of the iron to environmental conditions, an oxidation-reduction reaction can occur. However, zinc is a more active metal, so it oxidizes instead of iron. The iron becomes an inert electrode carrying electrons from the zinc for the reduction of molecular oxygen. The iron does not corrode because the zinc is preferentially oxidized. Oxidation of zinc is a very slow process that produces $Zn_2(OH)_2CO_3$. This compound deposits on the iron and acts as a protective coating.

Sometimes we use a less active metal to plate a metal we want to protect. For example, iron is covered with tin to make tin cans. The low activity of tin prevents corrosion. However, the protection of the iron is purely physical. If the tin coating is scratched to expose some iron, the iron corrodes very rapidly at the exposed position. The tin soon flakes off, resulting in even more extensive corrosion. When the iron is exposed, it becomes the anode, since it is the more active metal. The reduction of molecular oxygen occurs on the tin, while the iron is oxidized.

The other type of plating, electroplating, involves covering a more active metal with a less active metal that corrodes slowly. Electroplating uses an electrolytic process in which the metal to be coated acts as an electrode. This metal is placed in a solution that contains a compound of the metal that acts as the coating. When electricity passes through the cell, a thin coating of protective metal forms on the electrode. For example, car bumpers used to be made of steel electroplated with chromium. The steel was immersed in an electrolyte solution that contained a chromium compound. With the passage of electricity, chromium ions were reduced to form chromium metal, which plated onto the surface of the steel. Nickel-plated bathroom fixtures are produced in a similar way.

# CHAPTER REVIEW

## KEY CONCEPTS

- Electrons are transferred between reactants in oxidation-reduction reactions.
  - Oxidation is a loss of electrons and an increase in oxidation number.
  - Reduction is a gain of electrons and a decrease in oxidation number.
  - Oxidation numbers are assigned to elements to serve as an accounting tool for determining what is oxidized and what is reduced.
  - Oxidation and reduction always occur together because the element that gains electrons must obtain them from something that gives them up.
  - The reactant containing the element that is oxidized is the reducing agent; it provides electrons to the element that gets reduced.
  - The reactant containing the element that gains electrons is reduced and acts as the oxidizing agent.
- Spontaneous oxidation-reduction reactions can generate electric energy if the reactants are physically separated. If they are not, the energy is lost as heat.
  - The physical arrangement of the separated reactants is described as a cell.
  - The cell is divided into two half-cells that contain the oxidation and reduction half-reactions.
  - Oxidation occurs at the anode, and reduction occurs at the cathode.
  - A salt bridge completes the circuit between the anode and the cathode by allowing ions to flow so charge balance is maintained.
  - A cell that produces electricity from a spontaneous chemical reaction is a voltaic (or galvanic) cell.
  - Batteries contain one or more voltaic cells that provide electricity to power a device.
- Balancing oxidation-reduction reactions begins with the identification of which element is oxidized and which is reduced.
  - Half-reactions account for changes in oxidation number and describe the oxidation and reduction processes that occur at each electrode.
  - To balance half-reactions, changes in oxidation number, charge, and numbers of atoms must be balanced.
  - An oxidation-reduction reaction equation is balanced when the number of atoms and charges are the same on both sides of the equation and the number of electrons gained equals the number lost.
- Electrochemistry is the study of the relationship between chemical reactions and electrical work.
  - One branch of electrochemistry involves the identification and description of reactions in which chemistry drives the flow of electrons. These reactions are spontaneous and can be used to generate electricity.
  - The other branch of electrochemistry deals with reactions driven by electricity. These electrolytic reactions are not spontaneous, so they need a continuous energy source to proceed.
- Corrosion is generally defined as the slow deterioration of metal in contact with the environment.
  - The rusting of iron, a corrosion process, is an oxidation-reduction reaction.
  - When iron rusts, a part of the metal acts as an anode while reduction of oxygen occurs in another location.
  - Corrosion can be prevented in several ways, most often by painting the metal surfaces or covering them in a protective coating.
  - Another technique for corrosion prevention, cathodic protection, involves attaching a more reactive metal that corrodes preferentially.

## KEY TERMS

anode (14.3)

battery (14.3)

cathode (14.3)

corrosion (14.7)

electrochemistry (14.6)

electrode (14.3)

electrolysis (14.6)

electrolytic cell (14.6)

half-cell (14.3)

half-reaction (14.3)

oxidation (14.1)

oxidation number (14.2)

oxidation-reduction
  reaction (14.1)

oxidizing agent (14.1)

reducing agent (14.1)

reduction (14.1)

salt bridge (14.3)

spontaneous (14.6)

voltaic (galvanic) cell (14.3)

## QUESTIONS AND PROBLEMS

The following questions and problems, except those in *Additional Questions* and *Concept Review Questions,* are paired. Questions in a pair focus on the same concept. Answers to selected questions and problems are in Appendix D.

### Matching Definitions with Key Terms

14.1    Match the key terms with the descriptions provided.
   (a) an electrochemical cell through which electric current is passed to cause a nonspontaneous oxidation-reduction reaction to occur
   (b) the electrode at which reduction occurs
   (c) a chemical process in which an atom increases in oxidation number
   (d) an oxidation or reduction process that occurs in one compartment of an electrochemical cell
   (e) the study of the relationship between chemical reactions and electrical work
   (f) a solid material that conducts electrical charge and is used as part of an electrochemical cell
   (g) an atom, ion, or molecule that causes an increase in the oxidation number of another substance
   (h) a reaction in which electrons are transferred
   (i) a slow deterioration of metals due to interaction with the environment
   (j) a charge assigned to the atoms in a compound according to a set of rules used in an electron accounting system

14.2    Match the key terms with the descriptions provided.
   (a) an electrochemical cell in which an oxidation-reduction reaction causes electrons to flow through an external circuit
   (b) the electrode at which oxidation occurs
   (c) a chemical process in which an atom decreases in oxidation number
   (d) an individual compartment in which either an oxidation or a reduction reaction occurs
   (e) one or more interconnected voltaic cells used as a portable source of electricity
   (f) a device used in electrochemical cells that allows migration of ions to maintain electrical neutrality but prevents mixing of the electrolyte solutions
   (g) an atom, ion, or molecule that causes a decrease in the oxidation number of another substance
   (h) a process in which an electric current forces a nonspontaneous oxidation-reduction reaction to occur
   (i) a chemical reaction or physical change that occurs without outside intervention

### What Is an Oxidation-Reduction Reaction?

14.3    What is an oxidation-reduction reaction?

14.4    Why is oxidation always coupled with reduction?

14.5    How do you know when something is oxidized?

14.6    How do you know when something is reduced?

14.7    When a strip of magnesium metal is placed in an aqueous solution of copper(II) nitrate, elemental copper coats the surface of the magnesium strip and aqueous magnesium nitrate forms.
   (a) Write a balanced equation for this reaction.
   (b) What happens to the charge on magnesium as products form? How many electrons are transferred? Does magnesium gain or lose electrons?
   (c) What happens to the charge on copper as products form? How many electrons are transferred? Does copper gain or lose electrons?

14.8    When a strip of nickel metal is placed in an aqueous solution of lead(II) nitrate, elemental lead coats the surface of the nickel strip and aqueous nickel(II) nitrate forms.
   (a) Write a balanced equation for this reaction.
   (b) What happens to the charge on nickel as products form? How many electrons are transferred? Does nickel gain or lose electrons?
   (c) What happens to the charge on lead as products form? How many electrons are transferred? Does lead gain or lose electrons?

14.9    Consider the following reaction:
$$Mg(s) + SnSO_4(aq) \longrightarrow MgSO_4(aq) + Sn(s)$$
   (a) Which species is oxidized?
   (b) Which species is reduced?
   (c) What is the oxidizing agent?
   (d) What is the reducing agent?
   (e) Draw a molecular-level picture of what happens at the surface of the magnesium metal.

14.10   Consider the following reaction:
$$Mn(s) + CdSO_4(aq) \longrightarrow MnSO_4(aq) + Cd(s)$$
   (a) Which species is oxidized?
   (b) Which species is reduced?
   (c) What is the oxidizing agent?
   (d) What is the reducing agent?
   (e) Draw a molecular-level picture of what happens at the surface of the manganese metal.

## Oxidation Numbers

14.11 What is an oxidation number?

14.12 How do oxidation numbers change during oxidation and during reduction?

14.13 Indicate the oxidation number of each element in the following species.
(a) $Cr(s)$      (c) $Cr^{3+}(aq)$
(b) $Br_2(l)$      (d) $Br^-(aq)$

14.14 Indicate the oxidation number of each element in the following species.
(a) $Ca(s)$      (c) $Ca^{2+}(aq)$
(b) $S_8(s)$      (d) $S^{2-}(aq)$

14.15 What is the oxidation number of each atom in the compound represented by the following molecule?

14.16 What is the oxidation number of each atom in the compound represented by the following molecule?

14.17 What is the oxidation number of phosphorus in each of the following oxides?
(a) $P_4O_{10}$     (b) $P_4O_6$     (c) $P_4O_8$

14.18 Determine the oxidation number of chlorine in each of these chlorine oxides.
(a) $Cl_2O$      (d) $ClO_2$
(b) $ClO_3$      (e) $Cl_2O_5$
(c) $Cl_2O_3$

14.19 Indicate the oxidation number of phosphorus in each of the following compounds.
(a) $AlPO_4$      (d) $H_3PO_2$
(b) $PF_5$      (e) $PH_3$
(c) $H_3PO_4$      (f) $H_3PO_3$

14.20 Determine the oxidation number of iodine in each of the following compounds.
(a) $NaI$      (d) $I_2O_7$
(b) $I_2$      (e) $KIO_4$
(c) $IO_2$      (f) $Ca(IO)_2$

14.21 The ion shown has a charge of 2−. What are the oxidation numbers of all of the atoms in this ion?

14.22 The ion shown has a charge of 1−. What are the oxidation numbers of all of the atoms in this ion?

14.23 Assign an oxidation number to bromine in each of the following compounds or ions.
(a) $BrF_7$
(b) $BrO_3^-$
(c) $Br^-$
(d) $BrCl_3$
(e) $BrOCl_3$

14.24 Indicate the oxidation number of chromium in each of the following compounds.
(a) $CrCl_2$
(b) $CrSO_4$
(c) $Cr(NO_3)_3$
(d) $Cr_2(SO_4)_3$

14.25 Determine the oxidation number of each element in the following compounds.
(a) $ClO_2$
(b) $CaF_2$
(c) $H_2TeO_3$
(d) $NaH$

14.26 Determine the oxidation number of each element in the following compounds.
(a) $Na_2O_2$
(b) $Fe(NO_3)_3$
(c) $Sc_2O_3$
(d) $LiH$

14.27 Determine the oxidation number of each element in the following ions.
(a) $NO_2^-$
(b) $Cr_2O_7^{2-}$
(c) $AgCl_2^-$
(d) $SO_3^{2-}$
(e) $CO_3^{2-}$

14.28 Determine the oxidation number of each element in the following ions.
(a) $H_2PO_4^-$
(b) $FeCl_6^{3-}$
(c) $ClO_2^-$
(d) $SiF_6^{2-}$
(e) $AsO_4^{3-}$

14.29 Determine which of the following represent oxidation-reduction reactions. For reactions that involve oxidation-reduction, determine the oxidizing and reducing agents.
(a) $BaCl_2(aq) + H_2SO_4(aq) \longrightarrow BaSO_4(s) + 2HCl(aq)$
(b) $3H_2(g) + N_2(g) \longrightarrow 2NH_3(g)$
(c) $H_2CO_3(aq) \longrightarrow H_2O(l) + CO_2(g)$
(d) $AgNO_3(aq) + NaCl(aq) \longrightarrow AgCl(s) + NaNO_3(aq)$
(e) $2C_2H_6(g) + 7O_2(g) \longrightarrow 4CO_2(g) + 6H_2O(g)$

14.30 Determine which of the following represent oxidation-reduction reactions. For reactions that involve oxidation-reduction, determine the oxidizing and reducing agents.
(a) $2Na(s) + 2H_2O(l) \longrightarrow 2NaOH(aq) + H_2(g)$
(b) $H_2(g) + F_2(g) \longrightarrow 2HF(g)$
(c) $C(s) + H_2O(g) \longrightarrow CO(g) + H_2(g)$
(d) $Pb(NO_3)_2(aq) + 2NaCl(aq) \longrightarrow$
$$PbCl_2(s) + 2NaNO_3(aq)$$
(e) $C_2H_5OH(l) + 3O_2(g) \longrightarrow 2CO_2(g) + 3H_2O(l)$

**14.31** Under certain conditions, nitrogen gas reacts with oxygen gas to form nitrogen monoxide gas.

(a) Write a balanced equation for this reaction.

(b) What happens to the oxidation number of a nitrogen atom when it changes from a reactant to a product? How many electrons are transferred? Does nitrogen gain them or lose them?

(c) What happens to the oxidation number of an oxygen atom when it changes from a reactant to a product? How many electrons are transferred? Does oxygen gain them or lose them?

**14.32** Under certain conditions, solid carbon reacts with oxygen gas to form carbon monoxide gas.

(a) Write a balanced equation for this reaction.

(b) What happens to the oxidation number of a carbon atom when it changes from a reactant to a product? How many electrons are transferred? Does carbon gain them or lose them?

(c) What happens to the oxidation number of an oxygen atom when it changes from a reactant to a product? How many electrons are transferred? Does oxygen gain them or lose them?

**14.33** Consider the following reaction:

$$6V^{2+}(aq) + Cr_2O_7^{2-}(aq) + 14H^+(aq) \longrightarrow$$
$$6V^{3+}(aq) + 2Cr^{3+}(aq) + 7H_2O(l)$$

(a) Which species is oxidized?

(b) Which species is reduced?

(c) What is the oxidizing agent?

(d) What is the reducing agent?

**14.34** Consider the following reaction:

$$2MnO_4^-(aq) + 10Cl^-(aq) + 16H^+(aq) \longrightarrow$$
$$2Mn^{2+}(aq) + 5Cl_2(g) + 8H_2O(l)$$

(a) Which species is oxidized?

(b) Which species is reduced?

(c) What is the oxidizing agent?

(d) What is the reducing agent?

**14.35** In the following oxidation-reduction reactions, identify the oxidizing agent, the reducing agent, and the number of electrons transferred.

(a) $I_2(aq) + 2OH^-(aq) \longrightarrow I^-(aq) + IO^-(aq) + H_2O(l)$

(b) $Cr(s) + 2H^+(aq) \longrightarrow Cr^{2+}(aq) + H_2(g)$

(c) $2Cr_2O_7^{2-}(aq) + 16H^+(aq) \longrightarrow$
$4Cr^{3+}(aq) + 3O_2(g) + 8H_2O(l)$

(d) $3Fe^{3+}(aq) + Al(s) \longrightarrow 3Fe^{2+}(aq) + Al^{3+}(aq)$

**14.36** In the following oxidation-reduction reactions, identify the oxidizing agent, the reducing agent, and the number of electrons transferred.

(a) $6I^-(aq) + BrO_3^-(aq) + 6H^+(aq) \longrightarrow$
$3I_2(aq) + Br^-(aq) + 3H_2O(l)$

(b) $5Br^-(aq) + BrO_3^-(aq) + 6H^+(aq) \longrightarrow$
$3Br_2(aq) + 3H_2O(l)$

(c) $XeF_4(s) + 2H_2O(l) \longrightarrow Xe(g) + 4HF(aq) + O_2(g)$

(d) $Pb(s) + 2H^+(aq) \longrightarrow Pb^{2+}(aq) + H_2(g)$

## Batteries

**14.37** Draw a diagram of a voltaic cell that corresponds to the following reaction:

$$Fe(s) + Ni^{2+}(aq) \longrightarrow Fe^{2+}(aq) + Ni(s)$$

Label all components of the cell, including the anode and the cathode. Identify the electrolyte solutions needed in each half-cell compartment. Write half-reactions to describe the processes that occur at each electrode.

**14.38** Draw a diagram of a voltaic cell that corresponds to the following reaction:

$$Mg(s) + Sn^{2+}(aq) \longrightarrow Mg^{2+}(aq) + Sn(s)$$

Label all components of the cell, including the anode and the cathode. Identify the electrolyte solutions needed in each half-cell compartment. Write half-reactions to describe the processes that occur at each electrode.

**14.39** The figure shows a molecular-level representation for the iron-nickel cell described in Question 14.37 before the reaction begins. Draw a picture of what has happened at each electrode after the cell has run for a long time.

Before

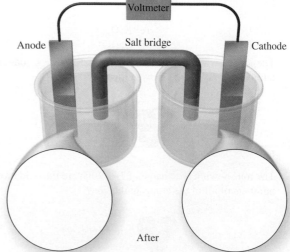

After

14.40 The figure shows a molecular-level representation for the magnesium-tin cell described in Question 14.38 before the reaction begins. Draw a picture of what has happened at each electrode after the cell has run for a long time.

Before

After

14.41 The reaction that occurs in a lead-acid battery is as follows:

$$Pb(s) + PbO_2(s) + 2H_2SO_4(aq) \longrightarrow$$
$$2PbSO_4(aq) + 2H_2O(l)$$

What gets oxidized in this battery? What gets reduced? What is the oxidizing agent? What is the reducing agent?

14.42 The reaction that occurs in a common dry cell battery is as follows:

$$Zn(s) + 2MnO_2(s) + 2NH_4^+(aq) \longrightarrow$$
$$Zn^{2+}(aq) + Mn_2O_3(s) + 2NH_3(aq) + H_2O(l)$$

What gets oxidized in this battery? What gets reduced? What is the oxidizing agent? What is the reducing agent?

14.43 The nickel-cadmium battery is used in portable appliances and power tools. Unbalanced half-reactions that occur in this battery are

$$Cd(s) + 2OH^-(aq) \longrightarrow Cd(OH)_2(s)$$
$$NiO_2(s) + 2H_2O(l) \longrightarrow Ni(OH)_2(s) + 2OH^-(aq)$$

Which equation represents the oxidation half-reaction? Which represents the reduction half-reaction? Add electrons to the appropriate side of each equation to account for the change in oxidation number.

14.44 The zinc-silver oxide battery, although expensive, is used to power satellite systems because of its light weight. Unbalanced half-reactions that occur in this battery are

$$Zn(s) + 2OH^-(aq) \longrightarrow Zn(OH)_2(s)$$
$$2AgO(s) + H_2O(l) \longrightarrow Ag_2O(s) + 2OH^-(aq)$$

Which equation represents the oxidation half-reaction? Which represents the reduction half-reaction? Add electrons to the appropriate side of each equation to account for the change in oxidation number.

## Balancing Simple Oxidation-Reduction Equations

14.45 Balance the following half-reactions.
   (a) $Fe^{3+}(aq) \longrightarrow Fe(s)$
   (b) $Zn(s) \longrightarrow Zn^{2+}(aq)$
   (c) $Cl^-(aq) \longrightarrow Cl_2(g)$
   (d) $Fe^{2+}(aq) \longrightarrow Fe^{3+}(aq)$

14.46 Balance the following half-reactions.
   (a) $Ni(s) \longrightarrow Ni^{2+}(aq)$
   (b) $Br_2(g) \longrightarrow Br^-(aq)$
   (c) $Mg^{2+}(aq) \longrightarrow Mg(s)$
   (d) $Cr^{3+}(aq) \longrightarrow Cr^{2+}(aq)$

14.47 For each of the following, write balanced half-reaction equations. Also write overall balanced equations.
   (a) $Zn(s) + Fe(NO_3)_3(aq) \longrightarrow Zn(NO_3)_2(aq) + Fe(s)$
   (b) $Mn(s) + HCl(aq) \longrightarrow MnCl_2(aq) + H_2(g)$

14.48 For each of the following, write balanced half-reaction equations. Also write overall balanced equations.
   (a) $Al(s) + Fe(NO_3)_3(aq) \longrightarrow$
$$Al(NO_3)_3(aq) + Fe(NO_3)_2(aq)$$
   (b) $Na(s) + HNO_3(aq) \longrightarrow NaNO_3(aq) + H_2(g)$

14.49 Aqueous iron(III) sulfate reacts with aqueous potassium iodide to form aqueous iron(II) sulfate, aqueous potassium sulfate, and aqueous iodine molecules. Write a balanced equation for this oxidation-reduction reaction.

14.50 When gaseous sulfur dioxide is bubbled into a nitric acid solution, sulfuric acid and gaseous nitrogen dioxide form. Write a balanced equation for this oxidation-reduction reaction.

## Balancing Complex Oxidation-Reduction Equations

14.51 Balance the following half-reactions, adding $H^+(aq)$, $H_2O(l)$, and electrons as appropriate.
   (a) $Ba(s) \longrightarrow Ba^{2+}(aq)$
   (b) $HNO_2(aq) \longrightarrow NO(g)$
   (c) $H_2O_2(aq) \longrightarrow H_2O(l)$
   (d) $Cr^{3+}(aq) \longrightarrow Cr_2O_7^{2-}(aq)$

14.52   Balance the following half-reactions, adding $H^+(aq)$, $H_2O(l)$, and electrons as appropriate.
    (a) $Br_2(aq) \longrightarrow Br^-(aq)$
    (b) $H_2O(l) \longrightarrow O_2(g)$
    (c) $IO_3^-(aq) \longrightarrow I_2(aq)$
    (d) $MnO_2(s) \longrightarrow Mn^{2+}(aq)$

14.53   Balance the following half-reactions, adding $OH^-(aq)$, $H_2O(l)$, and electrons as appropriate.
    (a) $La(s) \longrightarrow La(OH)_3(s)$
    (b) $NO_3^-(aq) \longrightarrow NO_2^-(aq)$
    (c) $H_2O_2(aq) \longrightarrow O_2(g)$
    (d) $Cl_2O_7(aq) \longrightarrow ClO_2^-(aq)$

14.54   Balance the following half-reactions, adding $OH^-(aq)$, $H_2O(l)$, and electrons as appropriate.
    (a) $CrO_4^{2-}(aq) \longrightarrow Cr(OH)_3(s)$
    (b) $ClO_2(aq) \longrightarrow ClO^-(aq)$
    (c) $MnO_4^-(aq) \longrightarrow MnO_2(s)$
    (d) $Br^-(aq) \longrightarrow BrO_2^-(aq)$

14.55   Complete and balance the following oxidation-reduction reactions, assuming they occur in acidic solution.
    (a) $H_2S(aq) + Cr_2O_7^{2-}(aq) \longrightarrow S(s) + Cr^{3+}(aq)$
    (b) $V^{2+}(aq) + MnO_4^-(aq) \longrightarrow VO_2^+(aq) + Mn^{2+}(aq)$
    (c) $Fe^{2+}(aq) + ClO_3^-(aq) \longrightarrow Fe^{3+}(aq) + Cl^-(aq)$

14.56   Complete and balance the following oxidation-reduction reactions, assuming they occur in basic solution.
    (a) $S^{2-}(aq) + CrO_4^{2-}(aq) \longrightarrow S(s) + Cr(OH)_3(s)$
    (b) $MnO_4^-(aq) + CN^-(aq) \longrightarrow$
                         $MnO_4^{2-}(aq) + CNO^-(aq)$
    (c) $Al(s) + OH^-(aq) \longrightarrow Al(OH)_4^-(aq) + H_2(g)$

14.57   Complete and balance the following oxidation-reduction reactions, assuming they occur in basic solution.
    (a) $NH_3(aq) + ClO^-(aq) \longrightarrow N_2H_4(aq) + Cl^-(aq)$
    (b) $Cr(OH)_4^-(aq) + HO_2^-(aq) \longrightarrow$
                         $CrO_4^{2-}(aq) + H_2O(l)$
    (c) $Br_2(aq) \longrightarrow Br^-(aq) + BrO^-(aq)$

14.58   Complete and balance the following oxidation-reduction reactions, assuming they occur in acidic solution.
    (a) $C_2O_4^{2-}(aq) + MnO_4^-(aq) \longrightarrow CO_2(g) + Mn^{2+}(aq)$
    (b) $ClO_3^-(aq) + Cl^-(aq) \longrightarrow Cl_2(aq)$
    (c) $S_2O_3^{2-}(aq) + I_2(aq) \longrightarrow SO_4^{2-}(aq) + I^-(aq)$

14.59   Denitrification occurs when nitrogen in the soil is lost to the atmosphere. One way in which denitrification can occur in acidic soils rich in plant matter is described by the following equation:

$$C_6H_{12}O_6(aq) + NO_3^-(aq) \longrightarrow CO_2(g) + N_2(g)$$

Balance this equation, adding $H^+(aq)$ and $H_2O(l)$ as necessary.

14.60   When iodine is oxidized in concentrated $HNO_3$, it produces the white solid iodic acid, $HIO_3$:

$$I_2(s) + NO_3^-(aq) \longrightarrow HIO_3(s) + NO_2(g)$$

Balance this equation, adding $H^+(aq)$ and $H_2O(l)$ as necessary.

## Electrochemistry

14.61   Consider the partially labeled voltaic cell shown.

Suppose you want to deposit solid iron on the iron electrode. What might you use for the other electrode? What electrolyte would you put in the half-cell with that other electrode?

14.62   Consider the partially labeled voltaic cell shown.

Suppose you want to deposit solid tin on the tin electrode. What might you use for the other electrode? What electrolyte would you put in the half-cell with that other electrode?

14.63   Using the activity series in Figure 14.22, place the following metals in order of increasing reducing strength.

                Al, Au, Bi, Ca, Ni, Zn

14.64   Using the activity series in Figure 14.22, place the following metal ions in order of increasing oxidizing strength.

          $Fe^{2+}, Zn^{2+}, Ag^+, Mg^{2+}, Ni^{2+}, K^+$

14.65   What is electrolysis?

14.66   Describe what happens at each electrode during the electrolysis of molten sodium chloride.

**14.67** In the electrolysis of molten sodium iodide, liquid sodium and iodine gas are produced. Consider the following diagram:

(a) When the power is turned on, what happens at the cathode?

(b) When the power is turned on, what happens at the anode?

(c) Write oxidation and reduction half-reactions.

(d) Write an overall equation for the reaction.

**14.68** In the electrolysis of molten aluminum chloride, liquid aluminum and chlorine gas are produced. Consider the following diagram:

(a) When the power is turned on, what happens at the cathode?

(b) When the power is turned on, what happens at the anode?

(c) Write oxidation and reduction half-reactions.

(d) Write an overall equation for the reaction.

**14.69** For Question 14.67, draw molecular-level pictures to represent what happens at the anode and cathode.

**14.70** For Question 14.68, draw molecular-level pictures to represent what happens at the anode and cathode.

## Corrosion Prevention

**14.71** How does attaching a piece of magnesium to the steel hull of a ship prevent the hull from corroding?

**14.72** How does zinc attached to an iron pipe prevent the pipe from corroding?

**14.73** Iron is often covered with tin, as in a tin can, to protect it from corrosion. If the tin coating is scratched and some iron is exposed, however, the iron corrodes very rapidly. Explain.

**14.74** If aluminum siding is fastened to a house with common iron nails, the siding may fall down due to corrosion. Considering the relative activity of aluminum and iron (see Figure 14.22), which metal will corrode? Would corrosion occur if aluminum nails were used? What about nickel nails?

**14.75** If the chrome plating on an automobile bumper is damaged, will chromium or iron corrode first? Why?

**14.76** The zinc coating of galvanized iron protects the iron from corrosion. If the zinc coating is scratched to expose some iron, will the iron begin to corrode? Why?

## Additional Questions

**14.77** Indicate the oxidation number of nitrogen in each of the following compounds.

(a) $NH_3$          (d) $NH_2OH$

(b) $N_2H_4$        (e) $Fe(NO_3)_3$

(c) $NF_3$          (f) $HNO_2$

**14.78** The following table gives the results of reactions between five metals and solutions of their nitrate salts. NR means "no reaction" and MD means "metal displaced from solution." From these results, arrange the five metals in order of decreasing chemical activity.

|                    | **Ag(s)** | **Fe(s)** | **Hg(l)** | **Mn(s)** | **Ni(s)** |
|--------------------|-----------|-----------|-----------|-----------|-----------|
| $AgNO_3(aq)$       | —         | MD        | NR        | MD        | MD        |
| $Fe(NO_3)_2(aq)$   | NR        | —         | NR        | MD        | NR        |
| $Hg(NO_3)_2(aq)$   | MD        | MD        | —         | MD        | MD        |
| $Mn(NO_3)_2(aq)$   | NR        | NR        | NR        | —         | NR        |
| $Ni(NO_3)_2(aq)$   | NR        | MD        | NR        | MD        | —         |

**14.79** Balance the following equations, assuming the reactions occur in basic solution. Also, identify the oxidizing agent, the reducing agent, the element oxidized, and the element reduced in each reaction.

(a) $CoCl_2(s) + Na_2O_2(aq) \longrightarrow$
$$Co(OH)_3(s) + Cl^-(aq) + Na^+(aq)$$

(b) $Bi_2O_3(s) + ClO^-(aq) \longrightarrow BiO_3^-(aq) + Cl^-(aq)$

**14.80** Aqua regia is a mixture of concentrated $HNO_3$ and concentrated HCl. It is often used for cleaning laboratory glassware, and it also dissolves platinum. Balance the equation for the dissolution of platinum metal:

$$Pt(s) + NO_3^-(aq) + Cl^-(aq) \longrightarrow PtCl_4^{2-}(aq) + NO_2(g)$$

14.81 Balance the following equation, assuming the reaction occurs in acidic solution:

$$NH_4^+(aq) + NO_3^-(aq) \longrightarrow N_2O(g) + H_2O(l)$$

Also, identify the oxidizing agent, the reducing agent, the element oxidized, and the element reduced.

14.82 The reaction that occurs in gas grills involves the burning of propane:

$$C_3H_8(g) + O_2(g) \longrightarrow CO_2(g) + H_2O(g)$$

Assign oxidation numbers to all of the reactants and identify the oxidizing and reducing agents.

14.83 In each of the following pairs, identify the strongest reducing agent.
(a) Al and Pb
(b) Zn and Ag
(c) Cu and Mn
(d) Cd and Mg

14.84 Why is gold a good metal to use in jewelry?

14.85 Why does brass cause your skin to turn green?

14.86 A voltaic cell is created by using a copper cathode and a magnesium anode. The cathode is immersed in a solution of $Cu^{2+}$ ions, and the anode is immersed in a solution of $Mg^{2+}$ ions. A salt bridge of $Na_2SO_4$ is also used. What happens to the ions in the salt bridge as the reaction proceeds?

14.87 Algae in swimming pools is sometimes treated by adding $CuSO_4$ as a fungicide. Copper sulfate can be prepared by the addition of hot $H_2SO_4$ to Cu metal.

$$Cu(s) + H_2SO_4(aq) \longrightarrow Cu^{2+}(aq) + SO_2(g)$$

Balance this equation, adding $H^+(aq)$ and $H_2O(l)$ as necessary.

14.88 Explain why bromide ion, $Br^-$, can be oxidized but not reduced. Also, explain why the manganese in permanganate ion, $MnO_4^-$, can be reduced but not oxidized.

14.89 Identify the oxidizing and reducing agents, and the oxidation and reduction half-reactions in each of the following chemical reactions.
(a) $Tl^+(aq) + 2Ce^{4+}(aq) \longrightarrow 2Ce^{3+}(aq) + Tl^{3+}(aq)$
(b) $2NO_3^-(aq) + 2Br^-(aq) + 4H^+(aq) \longrightarrow$
$$Br_2(aq) + 2NO_2(g) + 2H_2O(l)$$

14.90 A common error of amateur plumbers is to connect iron and copper pipes directly. The pipe eventually corrodes at the junction of the two metals if it is used with impure water or in moist soil. Which metal corrodes? Explain your answer.

14.91 The activity series shows that copper is a more active metal than silver.
(a) Which is more easily oxidized, Cu or Ag?
(b) Which is more easily reduced, $Cu^{2+}$ or $Ag^+$?
(c) Write a balanced equation for a spontaneous reaction in aqueous solution that includes all the elements and ions in parts (a) and (b).

14.92 A reagent called pyridium chlorochromate (PCC) can be used to convert ethanol to ethanal. Determine if carbon in ethanol is oxidized or reduced, and then identify PCC as an oxidizing agent or a reducing agent. Explain your reasoning.

$$C_5H_5NHCrO_3Cl \ + \ CH_3CH_2OH \longrightarrow$$
(PCC)            (Ethanol)

$$\underset{\text{(Ethanal)}}{H_3C-\overset{\overset{O}{\|}}{C}-H} \ + \ C_5H_5NHCl \ + \ H_2CrO_3$$

14.93 Alcohols react with sodium metal to form ionic compounds called sodium alkoxides. Identify the oxidizing agent and reducing agent in this reaction.

$$2CH_3CH_2OH + 2Na \longrightarrow 2NaOCH_2CH_3 + H_2$$

14.94 Thiols are organic compounds with a bad odor that have an $-SH$ group attached to a carbon. Thiols are added to natural gas to give it an odor, and three different thiols are present in skunk spray. Reaction of thiols with molecular oxygen produces compounds called disulfides:

$$4CH_3SH(g) + O_2(g) \longrightarrow 2CH_3S-SCH_3(l) + 2H_2O(l)$$

(a) What element is oxidized?
(b) What element is reduced?
(c) What substance is the oxidizing agent?
(d) What substance is the reducing agent?

14.95 Consider the combustion of methane:

$$CH_4(g) + 2O_2(g) \longrightarrow CO_2(g) + 2H_2O(g)$$

(a) What element is oxidized?
(b) What element is reduced?
(c) How many moles of electrons are transferred for every mole of $CH_4$ that reacts?

14.96 What makes compounds such as $K_2CrO_4$, $KMnO_4$, and $H_2SO_4$ good oxidizing agents?

14.97 Complete and balance the following oxidation-reduction reaction, assuming the reaction occurs in basic solution.

$$Br_2(aq) \longrightarrow Br^-(aq) + BrO^-(aq)$$

14.98 For each of the following spontaneous reactions, write the half-reaction that would occur at the anode and the half-reaction that would occur at the cathode of a voltaic cell.
(a) $2Al(s) + 6H^+(aq) \longrightarrow 2Al^{3+}(aq) + 3H_2(g)$
(b) $3Zn(s) + 2Fe^{3+}(aq) \longrightarrow 3Zn^{2+}(aq) + 2Fe(s)$

14.99 Use the activity series in Figure 14.22 to predict the half-reaction that occurs at the anode and the half-reaction that occurs at the cathode of a voltaic cell for each of the following combinations of half-reactions. [The notation $Zn(s)/Zn^{2+}(aq)$ used below indicates the components of one of the half cells: Zn metal electrode placed in an aqueous solution of $Zn^{2+}$.]
(a) $Zn(s)/Zn^{2+}(aq)$ and $Al(s)/Al^{3+}(aq)$
(b) $Cr(s)/Cr^{2+}(aq)$ and $Ag(s)/Ag^+(aq)$
(c) $H_2(g)/H^+(aq)$ and $Cd(s)/Cd^{2+}(aq)$

14.100 Predict the products of electrolysis of each of the following compounds to their elements.
(a) $KCl(l)$            (c) $KBr(l)$
(b) $MgI_2(l)$            (d) $H_2O(l)$

## Concept Review Questions

**14.101** When a strip of zinc metal is placed in a pink aqueous solution of cobalt(II) nitrate, the surface of the zinc becomes coated with a solid, and the zinc begins to dissolve to form aqueous zinc nitrate. Which of the following statements regarding this reaction is correct?
   A. $Co^{2+}$ ions are oxidized to form solid cobalt.
   B. $Zn^{2+}$ ions are oxidized to form solid zinc.
   C. There are no spectator ions in this reaction.
   D. $Co^{2+}$ is the oxidizing agent.
   E. The solution will become a darker pink as more $Co^{2+}$ ions form.
Modify each incorrect statement to make it a true statement.

**14.102** In which substance does sulfur have an oxidation number of 4+?
   A. $Na_2S$
   B. $SO_3$
   C. $SO_2$
   D. $H_2SO_4$
   E. SO
What are the oxidation numbers for sulfur in the other substances?

**14.103** Consider the reaction:

$$3C_2H_5OH(aq) + 2Cr_2O_7^{2-}(aq) + 16H^+(aq) \longrightarrow$$
$$3CH_3CO_2H(g) + 4Cr^{3+}(aq) + 11H_2O(l)$$

Which of the following statements regarding this reaction is correct?
   A. The oxidation number for chromium changes from 12+ to 3+.
   B. The oxidation number for oxygen changes from 2− to 0.
   C. The oxidizing agent is $C_2H_5OH$.
   D. The oxidizing agent is $Cr_2O_7^{2-}$.
   E. For every ion of $Cr_2O_7^{2-}$ that reacts, three electrons are transferred.
Modify each incorrect statement to make it a true statement.

**14.104** Consider a voltaic cell that corresponds to the reaction:

$$Mg(s) + Cu^{2+}(aq) \longrightarrow Mg^{2+}(g) + Cu(s)$$

Which of the following statements regarding the voltaic cell is correct?
   A. Magnesium metal is oxidized to form $Mg^{2+}$ at the anode.
   B. Aqueous $Cu^{2+}$ ions are oxidized to form solid copper at the cathode.
   C. Positive ions from the salt bridge flow into the anode half-cell.
   D. Negative ions from the salt bridge flow into the cathode half-cell.
   E. Electrons are transferred through a wire from the copper electrode to the magnesium electrode.
Modify each incorrect statement to make it a true statement.

**14.105** Cadmium reacts with silver nitrate according to the equation:

$$Cd(s) + 2AgNO_3(aq) \longrightarrow Cd(NO_3)_2(aq) + 2Ag(s)$$

The figure represents components of a voltaic cell in which this reaction takes place.

Which of the following correctly identifies the components in the voltaic cell? Assume the anode half-reaction occurs on the left.
   A. silver metal electrode
   B. aqueous solution of $Cd^{2+}$ ions
   C. this component is not needed
   D. aqueous solution of $Ag^+$ ions
   E. cadmium metal electrode
Modify each incorrect component.

**14.106** A voltaic cell is constructed in which the following reaction takes place:

$$Cu(s) + 2Ag^+(aq) \longrightarrow Cu^{2+}(aq) + 2Ag(s)$$

Suppose a 30.0 g piece of copper is used as the anode electrode and 25.0 g of copper remain after the reaction is stopped. If the reaction proceeded as expected, which of the following statements is correct?
   A. 5.0 g of solid silver was produced at the cathode.
   B. 0.0787 mol of solid silver was produced at the cathode.
   C. 0.157 mol of electrons was transferred from $Cu(s)$ to $Ag^+(aq)$.
   D. 25.0 g of solid copper was oxidized.
   E. The anode solution becomes a lighter blue color as aqueous $Cu^{2+}$ ions react to form solid copper.
Modify each incorrect statement to make it a true statement.

**14.107** Consider the reaction that takes place in a lead-acid battery when it discharges:

$$Pb(s) + PbO_2(s) + 2H_2SO_4(aq) \longrightarrow$$
$$2PbSO_4(aq) + 2H_2O(l)$$

Which of the following statements regarding the discharge of this battery is correct?
   A. The oxidizing agent is $Pb(s)$.
   B. $Pb(s)$ is reduced.
   C. $PbO_2$ is the reducing agent.
   D. The pH decreases as the battery discharges.
   E. The concentration of $H_2SO_4$ decreases as the battery discharges.
Modify each incorrect statement to make it a true statement.

14.108  Iodine is oxidized by nitric acid according to the following skeletal equation:

$$I_2(s) + NO_3^-(aq) \longrightarrow HIO_3(aq) + NO_2(g)$$

Which of the following statements regarding the balanced equation for this reaction is correct?

A.  $H_2O(l)$ is on the left side of the equation and has a coefficient of 6.

B.  The coefficient for $NO_2$ is 5.

C.  $H^+(aq)$ is on the left side of the equation and has a coefficient of 10.

D.  The coefficient for $NO_3^-$ is 5.

E.  Five electrons are transferred for every $I_2$ molecule that reacts.

Modify each incorrect statement to make it a true statement.

14.109  Which of the following has the greatest reducing strength? (You may need to consult an activity series.)

A.  Fe

B.  Cu

C.  Mg

D.  Pt

E.  Sn

Rank the metals in order of increasing reducing strength.

14.110  Which of the following statements is true regarding the electrolysis of molten NaCl?

A.  The cathode reaction is $2Cl^-(l) \longrightarrow Cl_2(g) + 2e^-$.

B.  The anode reaction is $Na(l) \longrightarrow Na^+(l) + 1e^-$.

C.  $Cl^-$ is reduced.

D.  $Na^+$ is oxidized.

E.  The products are $Na(l) + Cl_2(g)$.

Modify each incorrect statement to make it a true statement.

# Nuclear Chemistry

Source: NRC File Photo

15.1 Radioactivity

15.2 Nuclear Reactions

15.3 Rates of Radioactive Decay

15.4 Medical Applications
of Isotopes

15.5 Biological Effects
of Radiation

15.6 Nuclear Energy

Chapter Review

Questions and Problems

**A**

**B**

**FIGURE 15.1** (A) An MRI instrument. (B) A magnetic resonance image reveals the inner structure of the body.

(A): ©ERproductions Ltd/Blend Images LLC; (B): ©Zephyr/Science Source

For a review of isotopes, see Chapter 2.

**FIGURE 15.2** A CT scan uses X-rays to produce a two-dimensional "slice" through the body. A series of such scans produces a three-dimensional image, in this case of the intestines.
©Gondelon/Science Source

David was heading to class when he started to feel stomach pains. When they did not subside, he decided to go to the student health center, where the results of his physical examination perplexed his doctors so much that they referred him to a specialist for further tests. To locate the source of David's pains, the specialist had a wide range of diagnostic imaging techniques to choose from, including X-rays, computerized tomography (CT) scans, magnetic resonance imaging (MRI), ultrasound studies, and positron emission tomography (PET) scans.

While each technique would provide a different view inside David's abdominal area, all the imaging procedures are alike in one important way: they use some form of *radiation,* either the emission of particles or electromagnetic waves. In Chapter 7 we examined the full spectrum of electromagnetic radiation, increasing in wavelength from gamma rays to radio waves (Figure 7.6). Gamma rays, with the shortest wavelengths, have the highest energies. Radio waves have the longest wavelengths and the lowest energies. The imaging techniques that might be used in David's case involve a variety of types of radiation.

Magnetic resonance imaging, for example, uses strong magnetic fields and radio waves. When the magnetic field is turned on, the patient is irradiated with radio waves. The machine can produce pictures of internal organs because protons in the nuclei of the body's atoms have a spin that aligns with magnetic fields. Radio waves modify the direction of spin, especially in hydrogen atoms. Because the body's soft tissues are made of hydrogen-carbon compounds, the result is an absorption of energy that can be used to map the locations of hydrogen atoms in the body. Energy differences are transmitted as electrical signals that a computer translates into images useful for diagnosis (Figure 15.1).

Other diagnostic options in David's case are X-rays and a CT scan. X-rays have more energy than the radio waves used in an MRI scan. The beam of X-rays is absorbed to varying degrees, depending on the density of body tissues. Dense bones show best in an X-ray image. To make the soft tissues of his digestive tract more visible in an X-ray, David drinks a suspension of insoluble barium sulfate before the exam. The compound absorbs X-rays and improves image quality.

The CT scan also uses X-rays. It employs two X-ray beams, allowing the data to be subjected to "computerized tomography," which reconstructs a three-dimensional image of the body. Like MRI, this technique captures images of "slices" through the scanned tissues, as shown in Figure 15.2, providing much more detail than a simple X-ray.

Another imaging method is the PET scan. Like CT and MRI, it employs computer software to create a three-dimensional image, but PET uses the most energetic of electromagnetic radiation, gamma rays. In PET imaging, gamma rays are emitted when a *positron* reacts with an electron. A positron is a relatively rare particle that is like an electron, except that it has a positive charge. The body does not naturally contain many positrons, so they must be introduced before the procedure. This is done with *radioactive isotopes* of certain elements, such as fluorine-18, oxygen-15, carbon-11, and nitrogen-13. The isotopes are incorporated in compounds used by cells of the particular organ being examined. If a cancer is present, the compound accumulates in the tumor to a greater or lesser extent than in the surrounding, slower-growing tissue, and shows up in the PET scan (Figure 15.3).

As we leave David to talk with his specialist about the diagnostic procedures available to him, we will examine the nature of radiation in more detail. We'll explore nuclear chemistry and the changes that occur in the nuclei of radioactive atoms. We'll look at what happens when isotopes change to other isotopes and give off positrons, gamma rays, or some other type of radiation. We'll also examine the sources of radioactivity, the uses of nuclear reactions as energy sources, and the practical applications of nuclear chemistry, primarily in medicine. Before we explore these topics, consider the following questions.

## Questions for Consideration

**15.1** Why are some isotopes radioactive and others not?
**15.2** What kinds of nuclear reactions do various isotopes undergo?
**15.3** Do all radioactive isotopes last the same amount of time?
**15.4** How are radioactive isotopes used in medicine?
**15.5** Is radiation harmful?
**15.6** How can we use radioactive isotopes to generate electricity?

## 15.1 Radioactivity

The chemical elements exist in widely varying amounts in the universe and on Earth. Hydrogen is the most abundant element in the universe; nearly 91% of all atoms are hydrogen. In Earth's crust, the most abundant element is oxygen, which accounts for over 62% of the atoms. At the other extreme, some of the synthetic elements have been prepared in amounts of only a few atoms. These atoms generally exist for only fractions of a second before they destroy themselves by *radioactive decay,* the spontaneous emission of electromagnetic or other types of radiation.

**Radiation** is defined broadly as transmitted energy that comes from a source and travels through matter or space. Radiation may be of either two types: (1) *Electromagnetic* forms include light (visible, infrared, and ultraviolet), as well as gamma rays and X-rays. (2) *Particulate* radiation is mass given off from unstable atoms with the energy of motion (e.g., beta particles and alpha particles). Radiation of either type that can produce charged particles in matter is *ionizing radiation,* which includes X-rays and ultraviolet rays. Nonionizing forms of radiation include heat (infrared), radio waves, and microwaves.

Unstable atoms have an excess of matter, energy, or both. If they spontaneously give off this excess as ionizing radiation, they are **radioactive.** Ninety elements occur naturally on Earth. Of these, 81 have at least one stable isotope. They include all of the elements from hydrogen through bismuth in the periodic table, with two exceptions. Technetium and promethium have only unstable, radioactive isotopes. The remaining nine natural elements—polonium through uranium—also exist only as radioactive isotopes. The other known elements are all synthetic; they are made with nuclear reactions in a laboratory.

An element is composed of atoms that have the same number of protons in their nuclei. In chemical reactions, the compositions of the atomic nuclei remain unchanged. The atoms merely rearrange and/or transfer electrons from one element to another. In contrast, nuclear reactions involve changes in the composition of atomic nuclei. The changes usually result in changes in the number of protons in the nuclei, thereby changing the identity of the elements. The study of nuclear chemistry, then, focuses on the protons and neutrons in the nucleus. These nuclear particles are called **nucleons.**

How do we describe the composition and changes of atomic nuclei? Recall from Chapter 2 that atoms with the same number of protons (atomic number $Z$), but different numbers of neutrons ($N$) and mass numbers ($A$), are called *isotopes* and can be represented as $^A_Z X$. An atom of an isotope is called a **nuclide** and has a nucleus specified by its mass number and number of protons and neutrons. For example, there are three known nuclides of hydrogen, each of which has one proton. The three nuclides have zero, one, and two neutrons, commonly called, respectively, hydrogen, deuterium, and tritium, and symbolized as $^1_1H$, $^2_1H$, and $^3_1H$. Of hydrogen's three nuclides, hydrogen and deuterium are radioactively stable, while tritium is radioactive.

## Nuclear Decay

More than 3000 nuclides are known. Of these, only about 250 are stable. The rest decompose over some period of time, emitting radiation in the process of creating

**FIGURE 15.3** Radioactive isotopes and PET imaging reveal a tumor in this patient's abdomen, on the upper right.
©Slaven MD/Custom Medical Stock Photo/ Getty Images

**FIGURE 15.4** A graph of neutron number versus atomic number for the stable isotopes shows that they occur in a narrow band of values, often called the "band of stability."

🌀 **ANIMATION:** Radioactive Decay

Remember when your parents told you not to sit so close to the TV set? Consider the properties of beta radiation to decide whether their warning was legitimate.

During the Cold War of the 1950s, some people built backyard bomb shelters in hopes of surviving a nuclear attack from the Soviet Union. Why were these shelters made of concrete?

**FIGURE 15.5** The radiation given off by radioactive substances generally falls into one of three categories, depending on how the radiation interacts with an electric field. Alpha particles are attracted to the negative plate. Beta particles are attracted to the positive plate. Gamma rays are unaffected by either plate.

new nuclides. The stable nuclides have a restricted range of combinations of proton number ($Z$) and neutron number ($N$), as shown in Figure 15.4. The stable nuclides in this narrow *band of stability* have approximately equal numbers of neutrons and protons in the lighter elements ($Z = 1$ to 20). The ratio of neutrons to protons ($N/Z$) is about 1 for these elements. The heavier stable nuclides have more neutrons than protons, making the $N/Z$ ratio as high as 1.6 at the upper end of the band of stability. No nuclides heavier than $^{209}_{83}$Bi are stable. All nuclides found outside the stability region are radioactive. They undergo nuclear decay reactions. These reactions are different from chemical reactions in that the nuclei undergo a change in composition. Nuclei change into different, more stable nuclei, usually of a different element.

## Radiation

When a nucleus undergoes a nuclear reaction, radiation is emitted. Natural radiation associated with radioactive decay can be placed into three classes: alpha particles, beta particles, and gamma rays. These types of radiation behave differently in an electric field, as shown in Figure 15.5. The first two classes actually consist of particles, while the third is pure electromagnetic radiation. Other particles, such as protons ($^{1}_{1}p^{+}$) and neutrons ($^{1}_{0}n$), may also be involved in some nuclear reactions.

An **alpha particle,** usually designated as $\alpha$ or $^{4}_{2}\alpha$ or $^{4}_{2}He^{2+}$, contains two protons and two neutrons. Alpha particles are the nuclei of helium-4 atoms. We often omit the charge of 2+ on the helium nucleus from the symbol for the alpha particle, because it is not necessary to account for electrons when examining nuclear reactions. Of the three types of radiation, alpha particles are the least harmful to animal tissue upon external exposure (Figure 15.6). A 0.01-mm-thick sheet of aluminum foil, a piece of paper, or a layer of skin cells can stop an alpha particle. External alpha particles are not particularly harmful to human tissue, since they do not penetrate the skin deep enough to reach living cells. However, alpha particles inside the body will damage cells.

A **beta particle** is a small, charged particle that can be emitted from unstable atoms at speeds approaching the speed of light. Beta particles are more penetrating than alpha particles, traveling more than 3 m (about 10 ft) through air before being stopped by several sheets of paper or a sheet of aluminum foil about 0.5 to 1 mm thick. They will penetrate through skin into tissue (Figure 15.6). Beta particles are commonly emitted by the picture tubes of older television sets and by computer monitors. There are two types of beta particles, electrons and positrons. A **positron** has the same mass as an electron but the opposite charge. Electrons are designated as $^{0}_{-1}\beta^{-}$ or $^{0}_{-1}e^{-}$, and positrons as $^{0}_{1}\beta^{+}$.

The **gamma ray** is the highest energy and most penetrating type of radiation; a thickness of 50 to 110 mm of aluminum metal is required to stop it. Other types of heavy shielding, such as lead or thick layers of concrete, can also stop gamma rays. They are the most damaging type of natural radiation (Figure 15.6). They are high-energy electromagnetic radiation: energy with neither charge nor mass. The properties of the most common types of radiation are summarized in Table 15.1.

**TABLE 15.1** Properties of Types of Radiation

| Radiation Type | Notation | Mass | Charge | Penetration into Aluminum |
|---|---|---|---|---|
| alpha | $^4_2\alpha$ | 4 | 2+ | 0.01 mm |
| beta (electron) | $^0_{-1}\beta^-$ | ~0 | 1− | 0.5−11.0 mm |
| beta (positron) | $^0_1\beta^+$ | ~0 | 1+ | (reacts with electrons) |
| gamma | $\gamma$ | 0 | 0 | 50−110 mm |

**FIGURE 15.6** Alpha particles barely penetrate the skin, while beta particles can penetrate past the skin into tissue. Gamma rays can penetrate past skin, tissue, and bones to reach internal organs.

## 15.2 Nuclear Reactions

Radioactivity arises from the transformation of one nuclide into another, often resulting in the emission of a particle from the nucleus. Nuclear decay is a spontaneous process in which an unstable nuclide emits radiation and changes into a new nuclide. This process occurs spontaneously, and its rate depends only on the identity of the reacting nuclide and the amount that's present. In contrast, nuclear bombardment reactions must be induced by fast-moving nuclear particles, such as neutrons or small nuclides. Bombardment reactions are not spontaneous and are usually induced in laboratories under special conditions. We will look at the equations that describe both types of nuclear reactions. First we will look at the general types of nuclear decay reactions: alpha and beta emission, electron capture, and gamma ray emission.

As David was having a PET scan to diagnose his stomach complaint, he asked the technician how the radioactive fluorine, $^{18}$F, he had been administered would cause a response in the scanning detector. The technician answered that positrons are emitted from the nuclide and react with electrons in David's body, causing a release of gamma rays that the sensor can pick up. David thought about this and reasoned that if the $^{18}$F was decaying and giving off positrons, it must be converted to some other nuclide. But which one?

Writing an equation for a nuclear transformation is similar to writing an equation for a chemical reaction, but the procedure is slightly different.

### Equations for Nuclear Reactions

To get a feel for nuclear equations, consider the following examples:

$$^{232}_{90}\text{Th} \longrightarrow {}^{228}_{88}\text{Ra} + {}^4_2\alpha$$

$$^{231}_{90}\text{Th} \longrightarrow {}^{231}_{91}\text{Pa} + {}^0_{-1}\beta^-$$

$$^{238}_{92}\text{U} + {}^4_2\alpha \longrightarrow {}^{239}_{94}\text{Pu} + 3{}^1_0\text{n}$$

Unlike chemical reactions, the number of each type of atom is not conserved in nuclear reactions. Can you see which factors must be considered to balance nuclear equations? Can you use this information to identify the unknown product of a nuclear reaction if the type of radiation is known?

Two conditions must be met to balance a nuclear equation: (1) conservation of mass number and (2) conservation of nuclear charge (atomic number). We can identify the particles emitted in a nuclear transformation by checking these two conservation conditions.

**Alpha Particle Emission**  When a nucleus emits an alpha particle, it loses two protons and two neutrons, so its atomic number decreases by 2 and its mass number decreases by 4, as shown in Figure 15.7. For example, thorium-232 undergoes radioactive decay by alpha particle emission:

$$^{232}_{90}\text{Th} \longrightarrow {}^{228}_{88}\text{Ra} + {}^4_2\alpha$$

The atomic number decreases by 2 when the nuclide changes from thorium to radium, while the mass number decreases by 4. The emission of an alpha particle

**FIGURE 15.7** The emission of an alpha particle removes two protons and two neutrons from the decaying nuclide.

conserves the mass number, because the 4 units of mass lost by the nuclide match the mass of the alpha particle that is created. The mass number of the reactant equals the sum of the mass numbers of the products:

$$232 = 228 + 4$$

Similarly the atomic number of the reactant equals the sum of the atomic numbers of the products:

$$90 = 88 + 2$$

With this approach, we can answer questions such as the one that concerned David. Specifically, we can identify the decay product of a radioactive nuclide.

> **Student Hot Spot**
>
> Student data indicate you may struggle with writing balanced equations for radioactive decay processes. Access the SmartBook to view additional Learning Resources on this topic.

---

**EXAMPLE 15.1     Balancing Equations for Alpha Emission**

Radium-228 decays by alpha emission. What new nuclide ($X$) forms in this reaction?

$$^{228}_{88}\text{Ra} \longrightarrow X + {}^{4}_{2}\alpha$$

**Solution:**

The mass number ($A$) and atomic number ($Z$) of the unknown product can be determined using the two conservation conditions. To conserve the mass number, we have

$$228 = A + 4$$
$$A = 228 - 4 = 224$$

To conserve atomic number, we have

$$88 = Z + 2$$
$$Z = 88 - 2 = 86$$

We identify the element from its atomic number. According to the periodic table, the element with atomic number 86 is radon. The complete symbol for radon-224 is $^{224}_{86}\text{Rn}$. The complete balanced equation for the reaction is

$$^{228}_{88}\text{Ra} \longrightarrow {}^{224}_{86}\text{Rn} + {}^{4}_{2}\alpha$$

**Consider This 15.1**

How would the results differ if two alpha particles were emitted?

**Practice Problem 15.1**

What nuclide ($X$) decays by alpha emission, forming lead-214?

$$X \longrightarrow {}^{214}_{82}\text{Pb} + {}^{4}_{2}\alpha$$

**Further Practice:** Questions 15.17 and 15.18 at the end of the chapter

---

**Beta Particle (Electron) Emission**   The emission of an electron, a beta particle, from the nucleus results in no change in the mass number, but an increase in the atomic number by 1, causing the conversion of a neutron into a proton (Figure 15.8).

**FIGURE 15.8** Beta (electron) emission converts a neutron into a proton.

The number of protons, the atomic number, increases. The total number of nucleons remains the same, and the mass number does not change. For example, thorium-231 decays by beta particle (electron) emission:

$$^{231}_{90}\text{Th} \longrightarrow \,^{231}_{91}\text{Pa} + \,^{0}_{-1}\beta^{-}$$

The mass number stays the same, so it is conserved. The atomic number increases by 1, so the atomic number is also conserved:

$$90 = 91 + (-1)$$

By using the balancing conditions, we can also identify an unknown product or reactant in a beta emission reaction.

Although both electrons and positrons are forms of beta radiation, when no specification is made, the term *beta radiation* will be used in this chapter to mean electrons. To avoid confusion, we will use the term *positrons* to refer to the other form of beta radiation.

The symbols we use to represent nuclides have the mass number (mass) as a superscript and the atomic number (charge) as a subscript on the left side of the element symbol. However when representing a beta particle (which of course has no nucleus), the subscript simply represents charge.

---

**EXAMPLE 15.2**  |  **Balancing Equations for Beta Emission**

Carbon-14 undergoes beta emission. Identify the nuclide ($X$) formed in this reaction and balance the equation:

$$^{14}_{6}\text{C} \longrightarrow X + \,^{0}_{-1}\beta^{-}$$

**Solution:**

We can apply the conservation of mass condition to determine the mass number of the unknown element:

$$14 = A + 14$$
$$A = 14 - 0 = 14$$

We can then apply the conservation of charge condition to identify the atomic number of the unknown:

$$6 = Z + (-1)$$
$$Z = 6 + 1 = 7$$

We identify the element from its atomic number. According to the periodic table, the element with atomic number 7 is nitrogen, so $X$ must be $^{14}_{7}\text{N}$. The complete balanced equation for the reaction is

$$^{14}_{6}\text{C} \longrightarrow \,^{14}_{7}\text{N} + \,^{0}_{-1}\beta^{-}$$

**Consider This 15.2**

Would the mass number change if more than one beta particle were emitted?

**Practice Problem 15.2**

Thorium-234 undergoes beta emission to form a new nuclide ($X$). Identify the new nuclide and balance the equation for this nuclear reaction:

$$^{234}_{90}\text{Th} \longrightarrow X + \,^{0}_{-1}\beta^{-}$$

**Further Practice:**  Questions 15.19 and 15.20 at the end of the chapter

**FIGURE 15.9** Positron emission converts a proton into a neutron.

**Positron Emission**   Nuclides that emit positrons are used to generate PET scans for medical diagnosis. A positron has a mass number of 0 and a charge of 1+. Thus, when a radioactive nucleus emits a positron, the mass number does not change, but the atomic number decreases by 1. This change converts a proton into a neutron, as shown in Figure 15.9. This change is the reverse of the result of emission of an electron. An example of positron emission is the decay of magnesium-23:

$$^{23}_{12}\text{Mg} \longrightarrow {}^{23}_{11}\text{Na} + {}^{0}_{1}\beta^{+}$$

The mass number is conserved:

$$23 = 23 + 0$$

The atomic number is also conserved:

$$12 = 11 + 1$$

Thus, this nuclear equation is balanced.

---

### EXAMPLE 15.3   Balancing Equations for Positron Emission

David was injected with a compound containing fluorine-18 as a positron emitter prior to his PET scan. Identify the new nuclide formed by this nuclear decay reaction:

$$^{18}_{9}\text{F} \longrightarrow X + {}^{0}_{1}\beta^{+}$$

**Solution:**

The mass number and atomic number of the unknown product can be determined by the two conservation conditions. To conserve the mass number, we have

$$18 = A + 0$$
$$A = 18 - 0 = 18$$

To conserve atomic number, we have

$$9 = Z + 1$$
$$Z = 9 - 1 = 8$$

The element with atomic number 8 is oxygen, so the unknown product is $^{18}_{8}\text{O}$:

$$^{18}_{9}\text{F} \longrightarrow {}^{18}_{8}\text{O} + {}^{0}_{1}\beta^{+}$$

**Consider This 15.3**

If hydrogen-1 were able to emit a positron, what element would result?

**Practice Problem 15.3**

Determine the identity of the unknown reactant $X$ in the following nuclear reaction:

$$X \longrightarrow {}^{11}_{5}\text{B} + {}^{0}_{1}\beta^{+}$$

**Further Practice:**  Questions 15.21 and 15.22 at the end of the chapter

**Electron Capture**    Emission of a positron by a nucleus transforms it into the nucleus of a different element with the same mass number. The new element has an atomic number one less than that of the original element. This same transformation can take place if a nucleus captures a $1s$ electron. As a result of **electron capture,** a proton combines with the electron to form a neutron. This transformation occurs only for a relatively small number of nuclides. One example is beryllium-7, which undergoes radioactive decay by electron capture:

$$\ce{^7_4Be} + \ce{^{0}_{-1}e^-} \longrightarrow \ce{^7_3Li}$$

The mass number stays the same ($7 + 0 = 7$), but the atomic number decreases by 1 ($4 - 1 = 3$).

A few other nuclides that undergo electron capture are Ca-41, V-49, Ni-59, K-40, Cr-51, Co-57, Ge-68, Kr-81, and Ti-97.

**Gamma Ray Emission**    During all of the radioactive decay processes discussed so far, the nucleus changes from a state of higher energy to a state of lower energy. The emission of an alpha or beta particle with a high kinetic energy carries away excess energy. As shown in Figure 15.10, excess energy can also be released in the form of gamma rays, which are pure electromagnetic radiation. The gamma decay process involves no change in mass number or atomic number. One example is the decay of the nuclide $^{99m}\text{Tc}$:

$$\ce{^{99m}_{43}Tc} \longrightarrow \ce{^{90}_{43}Tc} + \gamma$$

Different nuclides emit different gamma ray frequencies, from only a few up to as many as 100 different frequencies per nuclide. The frequencies of the gamma rays emitted help to identify the products of nuclear reactions.

Technetium-99$m$ is an energetic isomer of technetium-99. The $m$ stands for "metastable," which means that it decays to a more stable form of the same isotope. Technetium-99$m$ decays to technetium-99 with the emission of gamma rays. It is sometimes used in the diagnosis of gastrointestinal disorders.

**Nuclear Bombardment Reactions**    So far we have mentioned only nuclear reactions that result in the *production* of radiation. In these reactions, one nucleus or isotope spontaneously changes into another. But not all nuclear changes are spontaneous; some are produced intentionally in laboratories by artificial means. In a **nuclear bombardment** reaction, nuclei are hit with a beam of nuclei or nuclear particles, such as neutrons, to trigger a nuclear reaction. Nuclear physicists and chemists have built particle accelerators (described following Example 15.4) to create a beam of particles and direct it onto a target element. We can represent a successful nuclear bombardment reaction with a nuclear reaction equation that looks much like an equation for a chemical reaction. For example, the bombardment of $^{238}_{92}\text{U}$ with neutrons gives $^{239}_{92}\text{U}$, which spontaneously decays to $^{239}_{93}\text{Np}$:

$$\ce{^{238}_{92}U} + \ce{^1_0n} \longrightarrow \ce{^{239}_{92}U}$$
$$\ce{^{239}_{92}U} \longrightarrow \ce{^{239}_{93}Np} + \ce{^{0}_{-1}\beta^-}$$

All elements with atomic numbers above 92, the atomic number of uranium, are synthetic elements that do not exist naturally on Earth. There are also lighter elements that have never been found in a natural state. For a long time, there were

Energetic nucleus                                    Gamma radiation                                    Stable nucleus

**FIGURE 15.10**  The emission of gamma rays converts an energetic nucleus into a lower-energy nucleus.

When Mendeleev proposed his periodic table, he predicted that a few unknown elements would eventually be found, and he left space for them in the periodic table. Technetium is one of these missing elements.

"holes" in the periodic table, shown in Figure 15.11, and no one was able to find elements 43 (Tc), 61 (Pm), or 85 (At). Promethium (Pm) was first identified in 1945 as a decay product of uranium in nuclear reactors.

Technetium (Tc) was first formed in 1937 by the bombardment of molybdenum with deuterium atoms:

$$^{97}_{42}\text{Mo} + ^{2}_{1}\text{H} \longrightarrow ^{97}_{43}\text{Tc} + 2^{1}_{0}\text{n}$$

The most useful isotope, technetium-99m, is now formed by the bombardment of molybdenum-98 with neutrons, giving molybdenum-99, which decays to technetium-99m:

$$^{98}_{42}\text{Mo} + ^{1}_{0}\text{n} \longrightarrow ^{99}_{42}\text{Mo} \longrightarrow ^{99m}_{43}\text{Tc} + ^{0}_{-1}\beta^{-}$$

All isotopes of technetium are unstable; most of them undergo radioactive decay in a matter of a few minutes to a few days, but three of its isotopes survive for a few million years. The isotopes that disappear quickly are often used for medical imaging.

Astatine was formed in 1940 by the bombardment of bismuth with alpha particles:

$$^{209}_{83}\text{Bi} + ^{4}_{2}\alpha \longrightarrow ^{211}_{85}\text{At} + 2^{1}_{0}\text{n}$$

**FIGURE 15.11** Technetium (element 43), promethium (element 61), and astatine (element 85) were missing from the periodic table for a long time, because all of their isotopes are radioactive and are not found in nature in significant quantities. Notice that their atomic masses are whole numbers placed in parentheses. This is generally true for elements that do not occur naturally, because the average atomic mass for a sample depends on the specific isotopes in the sample. The number in parentheses is the mass number of the most stable isotope.

All isotopes of astatine are highly radioactive, with the most stable, $^{210}_{85}At$, lasting only a few hours. No more than about 0.05 micrograms of astatine has ever been produced at one time.

Nuclear bombardment reactions have been used to synthesize small amounts of all the *transuranium* elements, those following uranium in the periodic table. The bombardment of uranium with neutrons yields neptunium. Other nuclear particles or other nuclei can be used to bombard heavy nuclei such as uranium, as in the following examples:

$$^{238}_{92}U + ^{4}_{2}\alpha \longrightarrow ^{239}_{94}Pu + 3^{1}_{0}n$$

$$^{238}_{92}U + ^{12}_{6}C \longrightarrow ^{246}_{98}Cf + 4^{1}_{0}n$$

$$^{238}_{92}U + ^{14}_{7}N \longrightarrow ^{247}_{99}Es + 5^{1}_{0}n$$

Nuclear scientists continue to use bombardment reactions to prepare even heavier elements. For example, by bombarding bismuth-209 with iron-58 nuclei for one week, scientists in 1982 finally succeeded in synthesizing a single atom of element 109 with mass number 266, now named meitnerium:

$$^{209}_{83}Bi + ^{58}_{26}Fe \longrightarrow ^{266}_{109}Mt + ^{1}_{0}n$$

In a similar way, element 116, named livermorium, was produced in 2000 and 2005, but not accepted by the International Union of Pure and Applied Chemistry until 2011:

$$^{48}_{20}Ca + ^{248}_{96}Cm \longrightarrow ^{293}_{116}Lv + 3^{1}_{0}n$$

Meitnerium (Mt) is named in honor of Lise Meitner, a German physicist who fled from Germany during the Second World War because she was Jewish. She is credited along with Otto Hahn and Fritz Strassman for the discovery of nuclear fission.

## EXAMPLE 15.4 — Balancing Equations for Bombardment Reactions

 **Student Hot Spot**

Student data indicate you may struggle with completing and balancing equations for bombardment reactions. Access the SmartBook to view additional Learning Resources on this topic.

Bombarding a lead-208 target with a beam of another nuclide produces hassium-265 and a neutron. Identify the nuclide used in the bombardment, and write a balanced equation to describe this nuclear reaction.

**Solution:**

First we find from the periodic table that hassium is element 108. We then write an incomplete equation summarizing the information given:

$$^{208}_{82}Pb + X \longrightarrow ^{265}_{108}Hs + ^{1}_{0}n$$

We can apply the conservation of mass condition to determine the mass number of the unknown element:

$$208 + A = 265 + 1$$
$$A = 265 + 1 - 208 = 58$$

We can apply the conservation of charge condition to identify the bombarding element's atomic number:

$$82 + Z = 108 + 0$$
$$Z = 108 + 0 - 82 = 26$$

Element 26 is iron, so the bombarding particle is iron-58, giving us the balanced equation

$$^{208}_{82}Pb + ^{58}_{26}Fe \longrightarrow ^{265}_{108}Hs + ^{1}_{0}n$$

**Consider This 15.4**

If the bombarding particle were iron-56, would the nuclide product be the same?

**Practice Problem 15.4**

Bombarding a bismuth-209 target with a beam of another nuclide produces bohrium-262 ($^{262}_{107}Bh$) and a neutron. Identify the nuclide used in the bombardment, and write a balanced equation to describe this nuclear reaction.

**Further Practice:** Questions 15.33 and 15.34 at the end of the chapter

A

B

**FIGURE 15.12** (A) Tubes of increasing length accelerate charged particles through electric fields. To achieve usable energies, such accelerators must be very long. (B) For example, the Stanford Linear Accelerator is built in a 3-km (2-mi) long tunnel covered by laboratory and support facilities.

(B): ©David Parker/Science Source

## Particle Accelerators

We have been discussing bombarding nuclides with subatomic particles or with other nuclides as though this were a simple matter, but how are beams of these particles obtained? Generating and manipulating these particles require sophisticated and expensive equipment that greatly increases the velocity of the particles, making them energetic enough to induce a nuclear change. This is done with particle accelerators. Several different types of particle accelerators have been developed. The most common types are linear accelerators and circular accelerators.

In a linear accelerator, an ion or electron beam is directed into a series of tubes. Voltage is applied to each tube with an alternating current so that the tube behind the particles repels them and the tube ahead of the particles attracts them. This causes the particles to accelerate. Because the particles go faster as they proceed through the accelerator, the farther the tubes are from the source, the longer the tubes must be (Figure 15.12A). For very high energies, the tubes must be very long (Figure 15.12B).

Most of the accelerators in use today employ a circular design. Perhaps the most successful accelerator—and certainly the largest—is the *synchrotron*, which uses a circular path for the accelerating particles (Figure 15.13). The radius of the larger synchrotrons exceeds 1 km. The synchrotron uses electromagnets to produce a variable magnetic field that maintains the path of the charged particles at a given radius as they accelerate. The electromagnets are enclosed within the hollow ring through which the charged particles pass, so an electromagnet with a radius of 1 km is not necessary.

## Predicting Spontaneous Nuclear Decay Reactions

We have examined a number of different types of nuclear decay reactions in which one nuclide is transformed into another. But why do some nuclides emit an alpha particle when they decay, while others emit a beta particle or a positron? The tendency for the neutron/proton (*N/Z*) ratio to move toward the band of stability, shown in Figure 15.14, explains the nuclear reactions of naturally radioactive nuclides. While nuclides with excess mass may emit an alpha particle, those with an *N/Z* ratio that is too high undergo beta decay. When the *N/Z* ratio is too low, positron emission or electron capture may occur. High-energy nuclides may lose excess energy through the emission of gamma rays. The characteristics of these nuclear changes are summarized in Table 15.2.

**FIGURE 15.13** The synchrotron at Fermilab in Batavia, Illinois, has an underground ring that is nearly 6.5 km (4 mi) long. Inside the ring, electromagnets keep accelerating ions moving in a circular path. This laboratory has several circular and linear accelerators. Can you locate them in the photograph?

(both): ©SPL/Science Source

**TABLE 15.2**  Natural Radioactive Decay Processes

| Reason for Nuclear Instability | Radioactive Process | Emitted Radiation | Nuclear Change | Change in $N/Z$ Ratio |
|---|---|---|---|---|
| excess mass | alpha decay | $_{2}^{4}\alpha$ | loss of two protons and two neutrons | slight increase |
| $N/Z$ too high | beta decay | $_{-1}^{0}\beta^{-}$ | neutron changes to a proton and an electron | decrease |
| $N/Z$ too low | positron emission | $_{1}^{0}\beta^{+}$ | proton changes to a neutron and a positron | increase |
| $N/Z$ too low | electron capture | — | proton combines with an inner-shell electron to become a neutron | increase |
| energetically excited | $\gamma$ emission | gamma ray | loss of excess energy | none |

For each process except $\gamma$ emission, the change that occurs for an unstable nuclide takes it closer to the observed band of stability, as shown by the arrows in Figure 15.14. Radioactive nuclides convert spontaneously, over time, to stable nuclides. This conversion may occur in a single step, but more commonly for the heavier isotopes, several successive nuclear reactions are required to reach a stable nuclide.

As shown in Figure 15.14, alpha decay adjusts the nucleus along a line parallel to $N/Z = 1$, while beta decay adjusts the nucleus along a line perpendicular to this. Positron decay accomplishes the opposite of beta decay: protons change to neutrons, so the $N/Z$ ratio increases and the atomic number decreases, as in the decay of boron-8:

$$_{5}^{8}B \longrightarrow {}_{4}^{8}Be + {}_{1}^{0}\beta^{+}$$

**Decay of Lighter Elements**    The mode of decay is fairly simple to predict for the lighter elements ($Z = 1$ to 20), where stable isotopes have an $N/Z$ ratio of 1. Radioactivity in lighter nuclides also results from an apparently unstable $N/Z$ ratio. The ratios change through beta decay, positron emission, or electron capture, resulting in greater stability.

If the $N/Z$ ratio is too high, a neutron converts to a proton by beta decay (emission of electrons). For example, carbon-14 has a ratio that is too high:

$$\frac{N}{Z} = \frac{14-6}{6} = \frac{8}{6} = 1.33$$

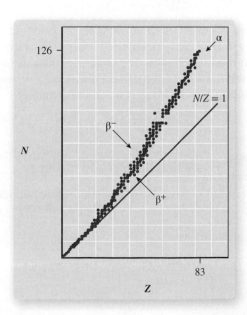

**FIGURE 15.14**  Each nuclear decay process, except $\gamma$ emission, changes an unstable nuclide to a new nuclide that is closer to the band of stability. The arrows show the direction in which the nuclide would move on the graph after emission of the indicated radiation. Nuclides above the band of stability undergo beta decay and those below undergo positron emission. Nuclides beyond the band of stability undergo alpha decay.

A stable state is achieved by beta decay:

$$^{14}_{6}C \longrightarrow \, ^{14}_{7}N + \, ^{0}_{-1}\beta^{-}$$

The product, nitrogen-14, has a ratio of 1:

$$\frac{N}{Z} = \frac{14 - 7}{7} = \frac{7}{7} = 1$$

On the other hand, carbon-10 has a low ratio:

$$\frac{N}{Z} = \frac{10 - 6}{6} = 0.67$$

Positron decay results in a higher ratio:

$$^{10}_{6}C \longrightarrow \, ^{10}_{5}B + \, ^{0}_{1}\beta^{+}$$

The ratio in the product of this transformation, boron-10, is ideal:

$$\frac{N}{Z} = \frac{10 - 5}{5} = 1$$

Electron capture raises the $N/Z$ ratio and occurs preferentially to positron emission in some elements, such as in the decay of potassium-40, where the ratio changes from 1.11 to 1.22:

$$^{40}_{19}K + \, ^{0}_{-1}e^{-} \longrightarrow \, ^{40}_{18}Ar$$

The occurrence of electron capture cannot be predicted readily. However, among the first 20 elements, this process apparently occurs only in four cases besides potassium-40: $^{7}_{4}Be$, $^{22}_{11}Na$, $^{37}_{18}Ar$, and $^{41}_{20}Ca$.

Note that three of these five cases occur with elements for which the stable isotopes have an $N/Z$ ratio that is slightly above 1.00 (see Figure 15.14). Many more occurrences of electron capture can be found among somewhat heavier elements ($Z = 21 - 83$).

**Decay of Heavier Elements** All isotopes of the elements beyond atomic number 83 (bismuth) are radioactive; they decay to become nuclides of decreased mass and lesser atomic number. They generally decay through both alpha and beta decay processes. The alpha decay gets rid of excess mass with the loss of two neutrons and two protons, as in the decay of uranium-238:

$$^{238}_{92}U \longrightarrow \, ^{234}_{90}Th + \, ^{4}_{2}\alpha$$

An unstable $N/Z$ ratio is not adjusted sufficiently in heavy elements through this initial process, however. Beta decay processes are also necessary to convert a heavy, unstable nuclide to a more stable one. For example, the thorium-234 produced in the decay of uranium-238 is still unstable. Beta decay converts a neutron to a proton, so it decreases the $N/Z$ ratio and increases the atomic number:

$$^{234}_{90}Th \longrightarrow \, ^{234}_{91}Pa + \, ^{0}_{-1}\beta^{-}$$

> Not all lightweight isotope decays result in an ideal $N/Z$ ratio. As with all relatively simple models of the behavior of matter, exceptions are likely to be found. For example, nitrogen-13 with $N/Z = 0.86$ decays to carbon-13 with $N/Z = 1.17$.

**📡 Student Hot Spot**

Student data indicate you may struggle with predicting the method of radioactive decay by which an unstable nuclide will undergo. Access the SmartBook to view additional Learning Resources on this topic.

---

**EXAMPLE 15.5** | **Predicting the Method of Decay**

Predict the method of radioactive decay of the unstable nuclide $^{24}_{10}Ne$.

**Solution:**

The $N/Z$ ratio in this nuclide is

$$\frac{24 - 10}{10} = \frac{14}{10} = 1.4$$

This ratio is higher than the ideal ratio of 1, so this nuclide undergoes radioactive decay to reduce the value. The emission of an electron (beta decay) converts a neutron to a proton.

$$^{24}_{10}\text{Ne} \longrightarrow \ ^{24}_{11}\text{Na} + \ ^{0}_{-1}\beta^{-}$$

This sodium nuclide is also radioactive—its $N/Z$ ratio of $\frac{13}{11}$ is greater than 1—so it should undergo further beta decay.

**Consider This 15.5**

Would you predict that the isotope neon-20 would undergo radioactive decay?

**Practice Problem 15.5**

Predict the method of radioactive decay of the unstable nuclide $^{18}_{10}\text{Ne}$.

**Further Practice:**   Questions 15.45 and 15.46 at the end of the chapter

**Radioactive Decay Series**   Some radioactive nuclides decay to stable isotopes in a single step, but frequently the product of radioactive decay is itself radioactive, especially in the heavier elements. In such cases, a series of alpha and beta decay steps ultimately leads to a stable nuclide. Three such radioactive decay series, which involve only alpha and beta decay, are found among the natural elements. They account for most of the radioactivity among elements 83 through 92. They start with the elements uranium, actinium, and thorium. A fourth series starts with the synthetic element neptunium. The uranium-238 series is shown in Figure 15.15. In each series, when the mass number changes, it differs by a unit of 4, because mass is lost only through loss of alpha particles with mass number 4. The atomic number decreases with alpha decay and increases with beta decay. These decays occur in various combinations until the product is a stable isotope.

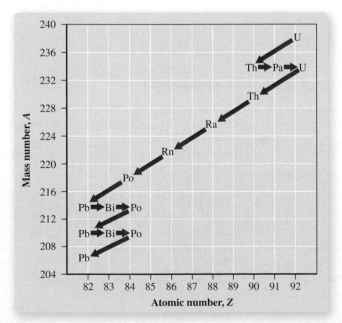

**FIGURE 15.15**   Uranium-238 decays to form thorium-234. Through a series of alpha and beta decays, the elements continue to convert to new elements until stable lead-206 forms. All of the nuclides in this series except lead-206 are radioactive. This is a very slow process, requiring 4.5 billion years for half of the uranium-238 to decay.

---

**EXAMPLE 15.6**    **Products of Decay Series**

Actinium-227 decays and emits five alpha particles and three beta particles in eight steps. What is the final product?

**Solution:**

As long as we know the total number of particles emitted, we don't have to know the order in which they are emitted. We can write an overall equation for the series of steps:

$$^{227}_{89}\text{Ac} \longrightarrow 5\,^4_2\alpha + 3\,^0_{-1}\beta^- + X$$

We then use the conservation conditions to balance the equation, thereby identifying the unknown product. First, we set up the mass balance condition to find the mass number:

$$227 = 5 \times 4 + 3 \times 0 + A$$
$$A = 227 - 20 - 0 = 207$$

Next, we use the charge balance condition to find the atomic number of the product:

$$89 = 5 \times 2 + 3 \times (-1) + Z$$
$$Z = 89 - 10 + 3 = 82$$

Element 82 is lead, so the product is lead-207:

$$^{227}_{89}\text{Ac} \longrightarrow 5\,^4_2\alpha + 3\,^0_{-1}\beta^- + {}^{207}_{82}\text{Pb}$$

**Consider This 15.6**

If the five alpha particles and three beta particles were emitted in six steps instead of eight steps, would the final product be different?

**Practice Problem 15.6**

Thorium-232 decays and emits six alpha particles and four beta particles in 10 steps. What is the final product?

**Further Practice:** Questions 15.47 and 15.48 at the end of the chapter

---

## 15.3  Rates of Radioactive Decay

Nuclear reactions emit radiation. The emitted radiation has a characteristic kinetic energy and is produced at a characteristic rate. Both can be used to detect nuclides and measure their amounts.

### Detection of Radiation

The exposure of photographic film is the simplest method of detecting radiation. That's why people who work around radiation, including X-ray technicians in hospitals and research scientists in nuclear laboratories, wear film badges. The badges are developed periodically to determine the workers' exposure to radiation. Exposure of the film to radiation darkens the image. The darker the film after development, the greater the radiation exposure. Despite its usefulness, photographic film exposure is a slow and relatively imprecise measure of radiation.

Various instruments have been developed to give speedier and more accurate measures of radiation intensity. One of the earliest was the *Geiger-Müller counter* shown in Figure 15.16. Radiation enters a tube through a thin glass or plastic window and ionizes a small fraction of the argon gas in the tube. A high voltage is applied between the walls of the tube and a wire that passes into its center. When the

**ANIMATION:** Detecting Radiation

The exposure of photographic film led to Antoine Henri Becquerel's discovery of spontaneous radioactivity, for which he received the Nobel Prize in Physics in 1903.

**FIGURE 15.16** (A) Geiger-Müller counters are small and portable. Units such as this were widely used by uranium prospectors and were kept in civil defense shelters during the 1950s. The one shown here is being used to test a worker for contamination. (B) Radiation ionizes the argon gas, releasing electrons that can be detected.

(A): ©Steve Allen/Getty Images

gas is ionized, it releases electrons that are accelerated by the electrical charge, in turn ionizing other argon atoms. The result is a pulse of electric current, which passes through a digital counter or through a speaker to make a clicking sound. A Geiger-Müller counter can detect alpha, beta, and gamma radiation. It is small and portable, so it can be used in a variety of circumstances outside of laboratories.

Another common and more sensitive detection device is a *scintillation counter*. It contains a tube coated with a material such as zinc sulfide that produces a flash of light when struck by gamma or beta radiation. The flashes cause pulses of electricity to pass through a photoelectric tube, where the electrical signals are amplified to activate a digital counter. The counter measures the intensity of the radiation in units of counts per minute. A scintillation counter can be tuned to detect radiation of different energies, so it can measure the energy released from decay of a particular nuclide. Scintillation counters are used extensively in laboratories that study radioactive nuclides or use nuclides to investigate biological processes.

## Half-Lives

So far we have been concerned primarily with the types of particles involved in nuclear reactions. Another important characteristic is how fast the reactions proceed. The nuclides used in medical imaging and diagnosis need to be unstable ones that do not remain in the body for too long. That is, they need to have a short **half-life,** which is the time required for half of a sample of a nuclide to decay to a different nuclide. The half-life of the fluorine-18 that David received is 110 min, while that of the C-11 often used for PET scans of the brain is a mere 20 min. Some radioactive elements, in contrast, have half-lives of millions of years.

How does the half-life work? It takes the same time for a fresh sample to decay to one-half the original number of atoms of that nuclide as it does for one-half to decay to one-fourth, and so on (Figure 15.17). As the number of atoms of radioactive nuclide decreases, the amount of radiation, measured by a scintillation counter in counts per minute (cpm), also decreases. If the counts per minute we measure after a time are one-half the original, we know that half the amount of radioactive nuclide has decayed. For the technetium-99*m* used in medical imaging, the half-life is 6.0 h. If we start with a sample that has a radiation intensity of 8000 cpm, after 6.0 h the intensity is 4000 cpm. After an additional 6.0 h, the intensity is 2000 cpm. For each additional 6.0-h period, the intensity of the radiation decreases by one-half of what it was at the beginning of that period. As a general rule, we assume that a nuclide is virtually all gone after 10 half-lives.

The shorter the half-life of a nuclide, the more intense the radiation that it emits, because the radiation is being released at a greater rate. The half-lives for the

The half-life of a given nuclide is always the same and does not change with amount of nuclide, temperature, or any other conditions.

**FIGURE 15.17** In the radioactive decay curve for technetium-99*m,* each successive half-life results in the loss of half of the radioactivity the sample had at the beginning of that period. The time for each half-life is the same, 6.0 h. At the end of the fifth half-life, only $\frac{1}{32}$ (or 3.1%) of the original technetium-99*m* remains. After 10 half-lives, 0.1% of the technetium-99*m* remains.

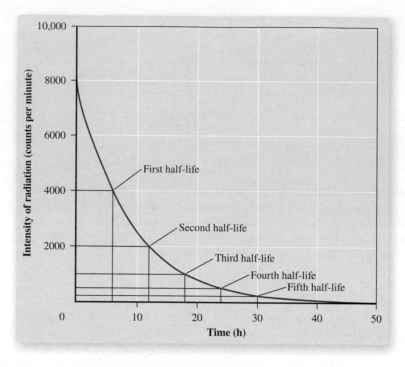

nuclides involved in the uranium decay series, shown earlier in Figure 15.15, are given in Table 15.3. The table shows that the half-lives of the various nuclides vary over a wide range.

The half-life of a nuclide reveals how long that nuclide will last. Consider, for example, that radioactivity is used not only for diagnosis, but also for the treatment of cancer. The patient may be injected with a radioactive nuclide in a compound that is preferentially absorbed by the cancerous cells. For example, radioactive iodide treats thyroid cancer or hyperthyroidism (Grave's disease). In some cases, the intensity of the radiation is so high that the patient is a radiation risk to other individuals and must be kept in isolation. In such cases it is desirable to know what the level of radioactivity will be after various time periods.

**TABLE 15.3** Half-Lives for the Nuclides in the Uranium Decay Series

| Nuclide | Method of Decay | Half-Life |
|---------|-----------------|-----------|
| $^{238}_{92}U$ | alpha | $4.46 \times 10^9$ years |
| $^{234}_{90}Th$ | beta | 24.1 days |
| $^{234}_{91}Pa$ | beta | 6.75 h |
| $^{234}_{92}U$ | alpha | $2.45 \times 10^5$ years |
| $^{230}_{90}Th$ | alpha | $7.54 \times 10^4$ years |
| $^{226}_{88}Ra$ | alpha | $1.62 \times 10^3$ years |
| $^{222}_{86}Rn$ | alpha | 3.82 days |
| $^{218}_{84}Po$ | alpha | 3.11 min |
| $^{214}_{82}Pb$ | beta | 26.8 min |
| $^{214}_{83}Bi$ | beta | 19.7 min |
| $^{214}_{84}Po$ | alpha | $1.6 \times 10^{-4}$ s |
| $^{210}_{82}Pb$ | beta | 22.3 years |
| $^{210}_{83}Bi$ | beta | 5.01 days |
| $^{210}_{84}Po$ | alpha | 138.4 days |
| $^{206}_{82}Pb$ | (stable) | |

**EXAMPLE 15.7** **Half-Life**

Strontium-90 is a nuclide found in the fallout from an atomic bomb explosion. It is particularly harmful because it is similar to calcium in size and charge, so it may end up mixed with the calcium phosphate in cows' milk. Strontium-90 undergoes beta decay to yield yttrium-90:

$$^{90}_{38}\text{Sr} \longrightarrow {}^{90}_{39}\text{Y} + {}^{0}_{-1}\beta^-$$

The half-life of this process is 28 years. What percentage of the strontium-90 is left after exactly 56 years?

**Solution:**

During each half-life, one-half of the strontium-90 present at the beginning of the time period decays. The half-life is 28 years, so 56 years is two half-lives. Let's call the amount present initially 100%. Then, at the end of one half-life, the amount will be 50%, or half, of the starting amount. At the end of two half-lives, one-half of that amount decays, leaving 25%. Thus, 25% of the Sr-90 is left after 56 years.

**Consider This 15.7**

If 12.5% of the strontium-90 were left, how much time would have elapsed?

**Practice Problem 15.7**

Gold-128 undergoes beta decay to yield mercury-128:

$$^{128}_{80}\text{Au} \longrightarrow {}^{128}_{81}\text{Hg} + {}^{0}_{-1}\beta^-$$

The half-life is 2.7 days. What percentage of the gold-128 is left after 8.1 days?

**Further Practice:** Questions 15.53 and 15.54 at the end of the chapter

**Student Hot Spot**

Student data indicate you may struggle with using half-lives to determine the amount of the nuclide remaining after a given amount of time. Access the SmartBook to view additional Learning Resources on this topic.

Although we will restrict ourselves to multiples of half-lives in the mathematical examples in this chapter, not all experimental results occur at exact multiples of a half-life. In such cases, the following equation can be used to describe the fraction of remaining nuclide at a given point in time:

$$\log\frac{N}{N_0} = -0.301\frac{t}{t_{1/2}}$$

In this equation, $N$ is the number of nuclides present at time $t$, $N_0$ is the number of nuclides present initially, and $t_{1/2}$ is the half-life. For more information on how to use the logarithmic function, see Math Toolbox 13.1.

## Archeological Dating

The decay of certain radioactive nuclides can be used to measure time on an archeological scale. For example, carbon-containing artifacts from an archeological dig can be dated according to the amount of carbon-14 they contain. The technique is **radiocarbon dating.** The amount of carbon-14 on Earth is small: only one $^{14}\text{C}$ atom for every $10^{12}$ atoms of $^{12}\text{C}$. The reaction of nitrogen-14 with neutrons from outer space (cosmic rays) continuously produces small amounts of carbon-14 in the upper atmosphere:

$$^{14}_{7}\text{N} + {}^{1}_{0}\text{n} \longrightarrow {}^{14}_{6}\text{C} + {}^{1}_{1}\text{H}$$

For purposes of C-14 dating, we assume that the rate of $^{14}_{6}\text{C}$ production has remained constant over thousands of years, although it fluctuates slightly during the increased neutron bombardment of solar flares. The C-14 in $CO_2$ is incorporated into plant tissues during photosynthesis and into animal tissue when plants are consumed. As long as a plant or animal is alive, its C-14 content should match that in the atmosphere. After the organism dies, C-14 is no longer being consumed, so its C-14 content decreases through beta decay:

$$^{14}_{6}\text{C} \longrightarrow {}^{14}_{7}\text{N} + {}^{0}_{-1}\beta^-$$

Carbon-14 has a half-life of 5730 years. A measurement of the amount of C-14 in the carbon of dead plant or animal material can, therefore, disclose the time elapsed since the organism died. To correct for fluctuations in the intensity of cosmic

**ANIMATION:** Half-Life

Danish scientists discovered that the carbon-14 in the proteins in an eye lens can be used to determine a person's year of birth. This is possible because the lens proteins form during the first year of life and then remain unchanged, not taking up additional carbon-14. The amount of carbon-14 then decreases with time according to its half-life.

Certain bristlecone pine trees were once thought to be the oldest living things on Earth, some of them dating back nearly 5000 years. Recently, however, experts have estimated the age of some desert creosote bushes at around 7000 years.

rays, the method has been calibrated using measurements of the C-14 content of wood in individual tree rings of very old bristlecone pine trees.

Radiocarbon dating can be used to date many different substances, including wood, grains, cloth, bone, shells, peat, charcoal, organic mud, and carbonates. Measurement of the amount of carbon-14 makes it possible to date objects as old as about 50,000 years. Older objects cannot be dated because they contain too little C-14 to measure.

---

**EXAMPLE 15.8** | **Radiocarbon Dating**

An archaeologist unearths a wooden bowl. It has a carbon-14 content that is 6.25% of the amount in living wood. The half-life of carbon-14 is 5730 years. How old is the bowl?

**Solution:**

We need to determine the number of half-lives that have elapsed since the wood died. If we assign a value of 100% to the carbon-14 content of the living wood, then the following amounts would be present after successive half-lives, based on the concept that one-half of what is present at the beginning of a half-life period disappears during this time.

| Half-lives | 0 | 1 | 2 | 3 | 4 | 5 |
|---|---|---|---|---|---|---|
| $^{14}C$ content | 100% | 50% | 25% | 12.5% | 6.25% | 3.125% |

The table shows that four half-lives must have elapsed since the wood died. Assuming that the bowl was made from a freshly cut tree, the age of the bowl must be four times the half-life of carbon-14:

$$\text{Age} = 4 \times 5730 \text{ years} = 22{,}920 \text{ years}$$

**Consider This 15.8**

If an unusual amount of solar flare activity doubled the amount of C-14 in the atmosphere 23,000 years ago, would this calculation still be valid?

**Practice Problem 15.8**

A leather strap found in an archeological dig has a carbon-14 content that is 25% of what can be measured in living tissue. If the half-life of carbon-14 is 5730 years, how old is the leather?

**Further Practice:** Questions 15.59 and 15.60 at the end of the chapter

---

### Geological Dating

Geologists date their samples in much the same way as archeologists, but geologists use nuclides of longer half-lives than carbon-14 because the time spans are longer. One geological method uses the ratio of potassium-40 to argon-40 in minerals containing potassium. The half-life of potassium-40 is $1.27 \times 10^9$ years.

$$^{40}_{19}K + ^{\ 0}_{-1}e^- \longrightarrow ^{40}_{18}Ar$$

Because rocks may have started with different amounts of potassium-40, only the ratio is relevant to determining the date. After one half-life, half of the potassium-40 has been converted to argon-40, giving a ratio of 1:

$$\frac{^{40}K}{^{40}Ar} = \frac{50\%}{50\%} = 1$$

After two half-lives, 25% of the K-40 remains and 75% has been converted to Ar-40, giving a ratio of $\frac{1}{3}$:

$$\frac{^{40}K}{^{40}Ar} = \frac{25\%}{75\%} = 0.33$$

Similar calculations give ratios of 0.14 after three half-lives and 0.067 after four half-lives. Using this method, some terrestrial rocks have been dated to ages as great as two billion $(2 \times 10^9)$ years. The K/Ar method has measured ages as great as 4.5 billion $(4.5 \times 10^9)$ years for meteorites. Techniques that employ other, long-lived radioactive elements yield age estimates of $2.5 \times 10^9$ to $3.0 \times 10^9$ years for the oldest terrestrial rocks and about $4.5 \times 10^9$ years for meteorites. Obtaining the same ages for similar rock samples using different nuclides verifies the accuracy of the methods.

## 15.4 Medical Applications of Isotopes

As David learned when he was esperiencing stomach pains, many diagnostic procedures employ various types of radiation or isotopes. Radioactive materials and techniques are also used in the treatment of disease and may keep lifesaving devices running by providing a reliable power source.

### Power Generators

How small can a power source be? A $^{238}$Pu power pack small enough to run a heart pacemaker can be implanted surgically. In these power packs, radioactive nuclides generate heat when they decay. This heat is converted to electricity. Such power packs are used to operate instruments not readily accessible for replacement or maintenance. Such power packs last for a long time, so they don't have to be replaced often. They operate for many years at nearly constant output.

Medical uses aren't the only applications for nuclide power packs. Power packs containing $^{242}$Cm, $^{244}$Cm, and $^{210}$Po have been used to operate instruments in space vehicles and in remote sites, such as the polar regions on Earth.

### Medical Diagnoses

Many radioactive nuclides are used as *tracers* to track the movements of substances in chemical or biological systems, in much the same way as small radio transmitters track migrating animals. These nuclides can be detected and measured by their characteristic radiation. Tracers find extensive use in medical studies. For example, technetium-99m ($^{99m}$Tc) emits gamma radiation and decays to $^{99}$Tc with a half-life of 6.0 h. When ingested or injected in an appropriate chemical form, this nuclide helps doctors locate tumors in the spleen, liver, brain, skeleton, and thyroid.

The general approach is to use a tracer that concentrates in the tissues to be investigated. Cancerous cells, for example, are often characterized by unusually high rates of metabolism and cell replication, so certain tracers tend to concentrate in them. For example, thyroid cancers can be diagnosed by the ingestion of radioactive $^{131}$I, which decays to nonradioactive $^{131}$Xe by beta and gamma emission with a half-life of 8 days. Scanning the distribution of radiation in the thyroid gland can reveal an abnormal growth (Figure 15.18). The scan for radioactivity creates a picture of the thyroid gland. A tumor shows up as an area of higher or lower radioactivity, depending on the nature of the tumor.

The radioactive labeling of portions of molecules can also help to determine how biological molecules react, either in living systems or in test tubes. For example, iron-55 and iron-59 have been used to determine how fast red blood cells are created in humans and how long they survive. Molecules labeled with phosphorus-32 or phosphorus-33 have been used to study the metabolism of these molecules.

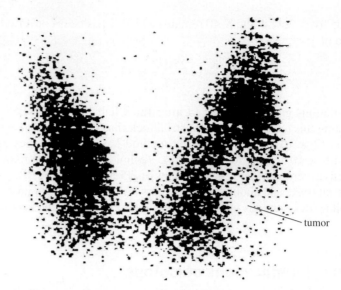

**FIGURE 15.18** This scan using a tracer reveals a tumor of the thyroid gland. In preparation for the scan, the patient ingested a solution of sodium iodide containing iodine-131. After 3 h, the patient was placed between photographic plates. The radiation emitted by the iodine-131 fogged the plates in proportion to the amount of radioactive iodine that had been absorbed. This tumor was diagnosed as benign, or noncancerous, since it did not concentrate the radioactivity and appeared as an area of low intensity in the scan. This conclusion was confirmed by a biopsy after surgery.

**ANIMATION:** Nuclear Medicine

OH

F

**FIGURE 15.19** Body cells take up glucose and use it as an energy source. To trace the movement of glucose, one of the OH groups can be replaced by $^{18}$F. This molecule is enough like ordinary glucose that cells accept and use it in the normal way.

## Positron Emission Tomography

As discussed in the introduction to this chapter, one of the newer medical diagnostic tools is positron emission tomography (PET). A PET scan, like the one David had, detects abnormalities in living tissues without disrupting the tissue. Some radioactive nuclides are incorporated into chemicals that are normally used by the tissues to be investigated. The isotopes used as positron emitters mostly for studies of the brain include $^{11}$C (half-life of 20.4 min) and $^{15}$O (2.05 min). Oxygen-15 is used in PET studies to reveal which areas of the brain are most active during various emotional states or when performing intellectual tasks. Studies of blood flow in the heart muscle may employ $^{13}$N with a half-life of 9.98 min, while the $^{18}$F that David received for his gastrointestinal diagnosis has a half-life of 110 min. All of these isotopes have short half-lives, so they don't survive for very long inside the patient.

The radioactive tracer used for a PET scan depends on the diagnostic purpose and the patient's needs. It might be $^{11}$C-labeled carbon dioxide or glucose; or it might be glucose in which $^{18}$F is substituted for a hydroxyl group (Figure 15.19). Labeled $CO_2$ is administered to the patient by inhalation; labeled glucose is injected. The patient is then placed into a cylindrical gamma ray detector. A positron emitted within the patient collides with an electron in nearby tissue. Positrons react with electrons to emit two gamma rays, which are sensed by the gamma ray detector:

$$^{0}_{1}\beta^{+} + {}^{0}_{-1}e^{-} \longrightarrow 2{}^{0}_{0}\gamma$$

The location of the positron-electron reaction can be determined by tracing back the paths of the two gamma rays to a point in common, just as a detective uses two bullet holes in a wall to determine where a shooter was standing. A computer reconstructs the location of many positron-electron reactions,

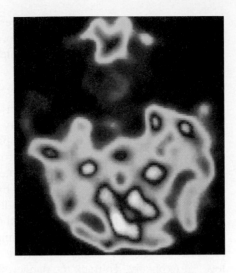

thereby creating an image of the radioactive nuclide's distribution in the tissues (Figure 15.20).

After David had his PET scan, his doctors determined that he had a stomach ulcer. If you suspect that the stress of taking chemistry gave him an ulcer, you're wrong. The bacterium *Helicobacter pylori* causes ulcers, and David is on his way to a full recovery thanks to a course of antibiotics.

## Cancer Therapy

David was fortunate. His PET scan revealed no cancer, but if a tumor had been found, radioactive nuclides—in much higher doses than those used for imaging—might have been used to treat it. Isotopes that emit high-energy gamma rays can destroy young cancer cells. Because most cancer cells have abnormally high growth rates, they preferentially absorb nutrients. If the nutrients contain a gamma-emitting component, the radioactivity becomes concentrated in the cancerous cells, destroying them in greater numbers than normal cells. For example, $^{131}I$ will destroy thyroid tumors. Gold as $^{198}Au$ has been used to treat lung cancer, and phosphorus as $^{32}P$ has been used for eye tumors.

## 15.5  Biological Effects of Radiation

Radioactive isotopes can be used to diagnose and treat cancer. Unfortunately they can also cause cancer. Radiation damages living tissue in varying degrees, depending on the intensity and the length of exposure. A certain amount of radiation is natural in the environment. Granite contains radioactive nuclides, as does soil, water, food, and air. Brick and concrete used in construction also contain small amounts of radioactive nuclides. Some exposure to radiation also results from cosmic rays, with the body getting a higher than normal dose during high-altitude airplane flights. Some radioactive nuclides, such as the rare potassium-40, are contained in our bodies. Inhalation of radon gas that can accumulate in homes has received much attention as a cause of cancer.

### Radiation Exposure

Radiation can have one of four effects on the functioning of a cell.

1. The radiation can pass through the cell with no damage.
2. The cell can absorb the radiation and be damaged, but it can subsequently repair the damage and resume normal functioning.

**3.** The cell can be damaged so severely that it cannot repair itself. New cells formed from this cell will be abnormal. This mutant cell can ultimately cause cancer if it continues to proliferate. Such damage can take years to create a tumor.

**4.** The cell can be so severely damaged that it dies.

The degree of damage depends, in part, on the ionizing ability of the radiation. Of particular interest for health and safety is the ability of radiation to ionize water, which is present in large quantities in plant and animal cells. Ionizing radiation of appropriate energy can strip an electron from a water molecule to form the ion $H_2O^+$. This ion reacts with another water molecule to form $H_3O^+$ and the OH (hydroxyl) radical, which is essentially a hydroxide ion that has lost an electron. The hydroxyl radical is very reactive. It readily reacts with many cell components, damaging them so that they do not function normally.

The extent of cell damage in a human depends on the energy of the radiation, its penetrating and ionizing ability, the chemical properties of the source, and the level of exposure. To track the danger, these factors are lumped together in a dose measurement called a *rem* (roentgen equivalent man). A rem is a measure of the danger arising to a human system from the absorption of radiation. Low levels of radiation are often measured in units of $10^{-3}$ rem, or mrem (millirem).

Normal exposure to radiation in an average area in the United States is about 200 mrem per year (Table 15.4). This amount of radiation has no observable effects on the average person. More important is the effect of dose levels for a single exposure to radiation that lasts for a few minutes to a few hours. A single dose of radiation of 0 to 25 rem causes no observable physical effects. However, radiation levels of 25 to 100 rem cause some temporary blood cell changes. Higher levels of 100 to 300 rem cause symptoms of radiation sickness, including nausea, diarrhea, fever, anemia, and achiness. Such exposure results in an increased risk of cancer. Doses of 400 to 600 rem cause severe radiation sickness, with a 50% chance of death within 30 days. Exposure to 1000 rem and higher levels results in a nearly 100% chance of death within 30 days. The longer someone is exposed to radiation, the greater the chance that cancer will result. It is because of these risks that such great care is taken in the design of nuclear power plants and in medical uses of radioactive nuclides.

The destructive effects of radiation on plant and animal cells are being investigated—and, to a limited extent, used—to help prevent the spoilage of foods, as shown in Figure 15.21. Upon exposure to appropriate levels of radiation, bacteria, molds, worms, and other parasites are killed without significant changes in the food itself. If food stored in a sealed container is exposed to radiation, the food does not

The rem unit is named for Wilhelm Conrad Roentgen (1845–1923), the discoverer of X-rays.

**A**

**B**

**FIGURE 15.21** (A) These strawberries were treated with gamma radiation to help prevent spoilage. (B) The strawberries on the right are moldy after 2 weeks of cold storage, while those on the left, which were irradiated with gamma rays, are still fresh.

(A): ©Tom Pantages; (B): ©Cordelia Molloy/ Science Source

**TABLE 15.4** Sources of Average Exposure of U.S. Population to Radiation

| Source | Dose (mrem/year) |
| --- | --- |
| cosmic rays | 50 |
| from Earth | 47 |
| from building materials | 3 |
| in human tissue (as $^{40}K$) | 21 |
| from the air | 5 |
| from cigarette smoking (due to $^{210}Po$) | 1300 |
| medical (X-rays, radiodiagnosis, radiotherapy) | 61 |
| nuclear power | 0.3 |
| radioactive fallout | 4 |
| consumer products such as TV tubes and watch dials | 0.04 |

need to be refrigerated or frozen to prevent spoilage. The radiation is usually gamma radiation from cobalt-60. Cobalt-60 is not introduced into the food. Instead the food is exposed only to the gamma radiation from the nuclide, so the food itself does not become radioactive. While the technique is generally considered safe, there is some concern that the irradiation of food might alter or destroy nutrients such as amino acids and vitamins.

### Radon

Although most concerns about air pollution involve outside air, indoor air pollution is also a health risk. But what could be in the air in our houses that could cause problems such as lung cancer? Second-hand cigarette smoke and carbon monoxide from a faulty gas burner are hazards. But problems arise even for nonsmokers who have electric heat. Surprisingly, it is a rare noble gas, radon, which has also been implicated as a possible cause of lung cancer. Radon is one of the intermediate products of the radioactive decay of uranium and thorium. It is rare because all of its isotopes are radioactive and decay to other elements in relatively short times. The nuclide with the longest half-life, 3.8 days, is $^{222}$Rn. It decays further to polonium-218:

$$^{222}_{86}\text{Rn} \longrightarrow {}^{218}_{84}\text{Po} + {}^{4}_{2}\alpha$$

The radon gas that accumulates in houses comes from particular kinds of soil or rock strata. Small amounts of uranium are widely present in soil from granite and shale, and thorium is present in granite and gneiss, a common rock consisting of alternating dark and light bands of material. Breathing radon is no more harmful than breathing any other noble gas, unless a radon atom undergoes radioactive decay while in the lungs. The polonium produced and all of its decay products (see Table 15.3) are solids that can attach to lung tissue. When these solids undergo decay, the emitted radiation can cause cell damage.

---

### EXAMPLE 15.9    Radon Decay

If radon-222 decays in the lungs today, what nuclides will still be present tomorrow? Assume that a nuclide is virtually gone after 10 half-lives. Refer to the decay series in Figure 15.15 and the list of half-lives in Table 15.3.

**Solution:**

Because Rn-222 has a half-life of 3.82 days (Table 15.3), far less than half of it will decay during the first day, assuming it is not exhaled. When Rn-222 decays in the lungs, it forms Po-218, which has a half-life of 3.11 min. After half an hour (about 10 half-lives), most of the Po-218 is gone, forming Pb-214. This nuclide has a half-life of 26.8 min, so it is mostly gone after about 5 h. It decays to form Bi-214, with a half-life of 19.7 min. Most of the Bi-214 is gone after about 3 h, forming Po-214, which has a half-life of $1.6 \times 10^{-4}$ s. This lasts only a fraction of a second, forming Pb-210, which has a half-life of 22.3 years. This nuclide will stay in the lungs for a considerable period of time, slowly releasing beta particles.

**Consider This 15.9**

If inhaled radon were immediately exhaled without undergoing decay, would it cause harm to the lungs?

**Practice Problem 15.9**

If radon-222 decays in the lungs, what types of radiation would be emitted during the first day? Refer to Figure 15.15 and Table 15.3.

**Further Practice:** Questions 15.73 and 15.74 at the end of the chapter

The amount of biological damage radon causes depends on how much is in a house and how long its inhabitants are exposed. Several surveys have established a national average of the amount of radon in houses, but values vary over wide ranges. Some houses have as much as 2000 times the average. Even in the average house, radon was found to make a significant contribution to the average radiation dosage experienced annually by residents of the United States. Estimates suggest that people living in homes with radon levels six times the national average experience a 2% increase in lung cancer risk. About 10,000 lung cancer deaths annually can be attributed to radon.

The radon comes from the soil, but how does it get into our houses? No one is certain, but radon has become a much greater problem since we started building airtight houses to save on energy used for heating and air conditioning. Although indoor air is exchanged with outside air every two or three hours in new homes, the exchange is less frequent than in older homes, and radon concentrations are higher as a result.

What can be done about radon in homes? First, residents can determine if the radon level in their home is unusually high. Kits to measure radon levels are commercially available. They consist either of a special piece of plastic or a canister of charcoal. The plastic is damaged by the passage of alpha particles released during radon decay, so microscopic examination reveals the number of penetrations that occurred during a set time period. The charcoal adsorbs radon atoms. In both cases, the kits must be returned to a laboratory for evaluation. In some homes, unusually high radon levels have been reduced by taking simple steps, such as placing blowers under the house to disperse a buildup of radon.

## 15.6 Nuclear Energy

Some heavy radioactive nuclides are so unstable that they decay in two ways: alpha emission and a spontaneous splitting of the nucleus that releases large amounts of energy. **Fission** is the splitting of a heavy nucleus into two or more lighter nuclei and some number of neutrons. Some heavy nuclides will not undergo fission spontaneously, but they will if irradiated or bombarded with appropriate particles, often neutrons. A particularly important fissionable nuclide is uranium-235, which is used extensively in nuclear power plants.

### Uranium-235 Fission

ANIMATION: Nuclear Fission

The fission of uranium-235 is depicted in Figure 15.22. When a slow neutron—one having a low kinetic energy—strikes a U-235 nucleus, the uranium absorbs the neutron and becomes U-236, which is even less stable than U-235. The U-236 nuclide then splits into two lighter nuclei and emits neutrons and gamma radiation. Many different fission reactions may occur when U-235 is bombarded with neutrons. Over 200 different fission products of this reaction have been identified in varying amounts. Because these large nuclides contain so many neutrons and protons, there are many different ways in which the nuclei can split. As the following examples show, these fission reactions differ only in the manner in which the very unstable U-236 splits or breaks up:

$$^{235}_{92}\text{U} + ^{1}_{0}\text{n} \longrightarrow ^{236}_{92}\text{U} \longrightarrow ^{92}_{36}\text{Kr} + ^{141}_{56}\text{Ba} + 3^{1}_{0}\text{n}$$

$$^{235}_{92}\text{U} + ^{1}_{0}\text{n} \longrightarrow ^{236}_{92}\text{U} \longrightarrow ^{90}_{38}\text{Sr} + ^{143}_{54}\text{Xe} + 3^{1}_{0}\text{n}$$

$$^{235}_{92}\text{U} + ^{1}_{0}\text{n} \longrightarrow ^{236}_{92}\text{U} \longrightarrow ^{94}_{40}\text{Zr} + ^{140}_{58}\text{Ce} + 2^{1}_{0}\text{n} + 6^{0}_{-1}\beta^{-}$$

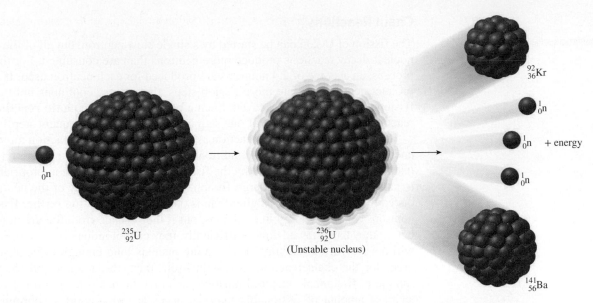

**FIGURE 15.22** Bombardment with a neutron initiates the fission of a uranium-235 nucleus. The product is unstable uranium-236, which splits apart into lighter nuclides. Some neutrons are also emitted. They can cause other uranium-235 nuclei to undergo fission.

### EXAMPLE 15.10   Fission Products

When bombarded by a neutron, uranium-235 undergoes fission, emitting the nuclides antimony-133 and niobium-100. How many neutrons are emitted? Write a balanced nuclear equation to describe this fission process.

**Solution:**

According to the periodic table, the atomic numbers of the two products are 51 and 41, respectively:

$$^{235}_{92}\text{U} + ^{1}_{0}\text{n} \longrightarrow ^{133}_{51}\text{Sb} + ^{100}_{41}\text{Nb} + x^{1}_{0}\text{n}$$

The charges balance, so we can consider the mass balance condition to determine the number of neutrons emitted:

$$235 + 1 = 133 + 100 + x$$
$$x = 235 + 1 - 133 - 100 = 3$$

Thus, three neutrons are emitted, and the balanced equation is

$$^{235}_{92}\text{U} + ^{1}_{0}\text{n} \longrightarrow ^{133}_{51}\text{Sb} + ^{100}_{41}\text{Nb} + 3^{1}_{0}\text{n}$$

**Consider This 15.10**

If bombardment of uranium-235 by a neutron produced tin-133 instead of antimony-133, and three neutrons were also emitted, what would be the identity of the other product?

**Practice Problem 15.10**

When bombarded by a neutron, uranium-235 undergoes fission, emitting the nuclides tellurium-137 and zirconium-97. How many neutrons are emitted? Write a balanced nuclear equation to describe this fission process.

**Further Practice:**  Questions 15.77 and 15.78 at the end of the chapter

**ANIMATION:** Chain Reaction

## Chain Reactions

The fission of U-235 can be started by a single slow neutron, but all of the resulting nuclear decay reactions produce more neutrons than are consumed. For the overall process, an average of 2.4 neutrons are produced for each neutron used. If all of the product neutrons initiate more fission steps, the process continues until all of the fissionable material has decayed. Such a process is called a **chain reaction.** It is a reaction in which a product of one step is a reactant in a subsequent step.

If we want to use such a reaction as an energy source, the reaction must give off energy continuously. Several conditions must be met for a chain reaction to sustain itself. The neutrons produced by the fission reaction must be of appropriate energy to cause other nuclei to undergo fission. Generally the energy of the neutrons produced by U-235 fission is higher than is optimum for inducing further fission. The neutrons can lose energy in collisions, but if they escape from the uranium sample before they have slowed down sufficiently, they cannot induce fission, because they will not stay in the vicinity of a uranium nucleus long enough to be absorbed. In order for the chain reaction to sustain itself, then, the amount and shape of the sample of fissionable material must be such that the neutrons will not escape. The smallest amount of fissionable material necessary to support a continuing chain reaction is called the **critical mass.**

Another important factor is the rate at which the fission progresses. Ideally, to sustain itself, the process should maintain a constant rate, becoming neither faster nor slower. If other nuclei absorb too few of the emitted neutrons, the chain reaction slows and eventually stops. If other nuclei absorb too many of the emitted neutrons, the process accelerates and culminates in an explosion. This is the principle behind nuclear weapons, which consist of several pieces of uranium. Each is less than the critical mass, so it cannot sustain a chain reaction. But the detonation of a chemical explosion forces the pieces together to form a critical mass, thereby activating an explosion.

**ANIMATION:** Inside a Nuclear Power Plant

Boron-10 control rod

Moderator

Uranium fuel rod

Coolant

Pressure vessel

**FIGURE 15.23** Reactor core in a fission reactor. The boron-10 control rods absorb neutrons to help control the rate of the chain reaction. The purpose of the moderator is to slow the speed of the neutrons to help maintain the chain reaction process.

## Fission Reactors

Fission can be used for peaceful, as well as wartime, purposes. Nuclear power plants use fission to produce electric energy. Such plants are designed to maintain careful control over the rate of the chain reaction, using neutron absorbers such as cadmium and boron. If the chain reaction is going too quickly, movable *control rods* made of these elements are inserted into a core of uranium fuel in fission reactors (Figure 15.23). The material in the rods reacts with neutrons, preventing them from participating in the chain reaction, as the following equation for boron shows:

$$^{10}_{5}\text{B} + {}^{1}_{0}\text{n} \longrightarrow {}^{7}_{3}\text{Li} + {}^{4}_{2}\alpha$$

Although a nuclear weapon needs uranium metal, the uranium fission reactor does not. Normally the uranium fuel is made from the oxide $U_3O_8$, which is enriched in U-235 from its natural abundance of 0.7% to about 2% or 3%. The neutrons produced by U-235 fission in this fuel are usually too energetic and pass out of the system unchanged, so they do not produce additional fission steps efficiently. To generate power, the neutrons must be slowed down. A *moderator* such as normal water ($H_2O$), heavy water ($D_2O$), or graphite is included in the reactor design for this purpose (Figure 15.23).

The fission of U-235 produces about $2 \times 10^{10}$ kJ/mol of U-235, an amount that is enormous in comparison to the 390 kJ/mol released when burning coal. This energy is released in the form of kinetic energy of the reaction products. This is not usable energy, so it must be converted to another form. A coolant, usually water, circulates around the reactor core (Figure 15.24). The water gets as hot as 310°C, so it is pressurized to prevent boiling. This very hot water, called the primary coolant, then passes through a steam generator. Steam generators are massive structures

**FIGURE 15.24** The heat exchange system in a nuclear fission reactor is designed to prevent the loss of any radioactive materials. Heat is transferred to a secondary loop, where it generates steam to operate a turbine and produce electricity.

containing thousands of thin-walled tubes for heat transfer. The primary coolant is pumped through the tubes, and another source of water, called the secondary coolant, is pumped around the tubes and converted to steam. The steam is used to drive a turbine, which operates a generator, producing electricity. The steam must be condensed and cooled so that it can be pumped back through the steam generator.

This is often accomplished by circulating a third coolant through a condenser and a cooling tower. Heat is removed from the secondary coolant in the condenser and then released to the environment from massive cooling towers (Figure 15.25). We often recognize a nuclear power plant by the presence of these cooling towers. Because of this design, with three different coolants, water that circulates around the reactor core never leaves the containment building, so there is little danger of the release of radioactive contaminants into the environment.

For protection in case of leaks, the reactor core is located in a thick-walled containment building, often built in a dome shape (Figure 15.25). In the event of a reactor leak, the radioactive material would, in principle, be kept within the containment building, and other areas would not become contaminated. The nuclear disaster that occurred in 1986 at the nuclear power plant in Chernobyl, Russia, released a considerable amount of radioactive nuclides into the surrounding environment, some of which spread throughout the world. The Chernobyl reactor differed from U.S. nuclear reactor designs in that it did not include a containment building. Radiation leaks also occurred from the Fukushima Daiichi Nuclear Power Plant in 2011 when the plant was hit by a tsunami resulting from an earthquake. This reactor did have a nuclear containment structure, but a tsunami of the magnitude experienced was not anticipated and the reactor was flooded with seawater.

In most cases, the immediate products of U-235 fission are themselves radioactive. Once the fission fuel has been used up, a complex mixture of radioactive substances remains. This radioactive waste, as well as other contaminated materials, must be stored so that it can become inactive through decay without causing damage to living things. The best method for disposing of radioactive waste has not yet been determined, but controversy rages and many methods are under investigation. They include reprocessing the fuel and storing it in liquid or solid form in steel tanks or underground.

**FIGURE 15.25** The dome-shaped structure is the containment building. The smoking tower is the cooling tower; the "smoke" is actually condensed water vapor.
©Martin Bond/Science Source

## Fusion Reactors

Another kind of nuclear reaction that releases large amounts of energy is **fusion,** the combination of light nuclei to form heavier nuclei. A major fusion reaction occurs continuously in the Sun and other stars:

$$4{}_{1}^{1}\text{H} \longrightarrow {}_{2}^{4}\text{He} + 2{}_{1}^{0}\beta^{+}$$

This process actually occurs in several steps:

$$\begin{aligned}
{}_{1}^{1}\text{H} + {}_{1}^{1}\text{H} &\longrightarrow {}_{1}^{2}\text{H} + {}_{1}^{0}\beta^{+} \\
{}_{1}^{1}\text{H} + {}_{1}^{2}\text{H} &\longrightarrow {}_{2}^{3}\text{He} \\
{}_{1}^{1}\text{H} + {}_{2}^{3}\text{He} &\longrightarrow {}_{2}^{4}\text{He} + {}_{1}^{0}\beta^{+}
\end{aligned}$$

This fusion process releases about $2.45 \times 10^{9}$ kJ/mol of helium. As the nuclei get close enough together to interact, strong internuclear repulsions occur. To overcome them, considerable energy must be supplied to fuse the nuclei. In the Sun, extremely high temperatures in excess of $10^{7}$ K provide the energy. In a fusion bomb, the initial explosion of a fission bomb provides the high temperatures and pressures needed to initiate fusion. The temperature required to initiate a fusion reaction is called the *ignition temperature.*

Although fusion has never been harnessed for the production of electricity, some researchers are hopeful, and fusion reactors could potentially offer a number of advantages. Fission reactors suffer from a limited supply of fissionable material and the necessity of storing radioactive wastes. Fusion reactors would avoid these shortcomings. They could operate on hydrogen taken from water—a virtually unlimited fuel supply—and create no hazardous by-products.

A variety of fusion reactions could be used in a commercial fusion reactor. The reaction between deuterium and tritium is a likely candidate:

$${}_{1}^{2}\text{H} + {}_{1}^{3}\text{H} \longrightarrow {}_{2}^{4}\text{He} + {}_{0}^{1}\text{n}$$

This reaction produces $1.7 \times 10^{9}$ kJ/mol, over 1700 times as much energy as is needed to get fusion to take place. Enough deuterium is present in ordinary water to last for millions of years. The technology for inexpensively separating deuterium from water has already been developed. Eight gallons of seawater contain about 1 g of deuterium, which can release energy equivalent to burning about 2500 gal of gasoline. Tritium, a radioactive isotope of hydrogen, has a half-life of 12.3 years and emits beta particles. While tritium occurs naturally to some extent, it is not present in nature in sufficient quantity to be used as a fuel. However, tritium can be prepared by the neutron bombardment of lithium, which is available in sufficient quantities to provide at least a 1000-year supply.

In many fusion reactor designs, the reaction chamber would be surrounded by a lithium "blanket," which would transfer the heat from the fusion reaction to a reactor coolant (Figure 15.26). Heat would then be transferred to a steam generator, as in fission reactors. The lithium would also absorb excess neutrons formed in the fusion process, thereby producing more tritium:

$$\begin{aligned}
{}_{3}^{7}\text{Li} + {}_{0}^{1}\text{n} &\longrightarrow {}_{1}^{3}\text{H} + {}_{2}^{4}\text{He} + {}_{0}^{1}\text{n} \\
{}_{3}^{6}\text{Li} + {}_{0}^{1}\text{n} &\longrightarrow {}_{1}^{3}\text{H} + {}_{2}^{4}\text{He}
\end{aligned}$$

In fact, it is possible to create more tritium than is used, and the excess tritium could be used to fuel other fusion reactors. A reactor that produces fuel that can be used in other reactors is called a *breeder reactor.*

The major problems to be solved before fusion can be used as an energy source are the creation and control of *plasma,* an ionized gas, at temperatures of about $10^{8}$ K. In order to achieve fusion, the gaseous reactants must be condensed

**FIGURE 15.26** Schematic layout of a fusion reactor.

to a small volume at high temperatures. This is the purpose of the plasma, which must be kept at a very high temperature for longer than 1 s to initiate the fusion reaction. At this high temperature, however, the plasma will melt any normal container material. Therefore, most research has concentrated on containing the plasma with a magnetic field, using doughnut-shaped machines called *tokamaks* (Figure 15.27). A second method confines the fuel in tiny glass pellets that are heated to the required temperatures with laser beams. It is only recently that anyone has been able to get more energy out of fusion than had to be put in, so nuclear fusion is unlikely to prove itself a viable commercial energy source for many years.

**FIGURE 15.27** A tokamak uses a magnetic field to contain the deuterium-tritium gaseous fuel for a fusion reaction.

©Jerry Mason/Science Source

| EXAMPLE 15.11 | Fusion Reactions |
| --- | --- |

What element would form if two carbon-12 nuclei undergo fusion with no emission of particles?

**Solution:**

Assuming no particles are emitted by the fused nuclei, adding the mass numbers and atomic numbers provides the answer. The atomic number of carbon is 6, so the fusion of two carbon nuclei should give an atom of magnesium, with an atomic number of 12. The mass number of the carbon nuclide is 12, so when two nuclei fuse, the mass number of the product will be 24. The new nuclide is magnesium-24. The fusion reaction can be represented by the following equation:

$$\mathrm{^{12}_{6}C + {}^{12}_{6}C \longrightarrow {}^{24}_{12}Mg}$$

**Consider This 15.11**

This fusion process emitted no additional particles. How would the product differ if the fusion process emitted a neutron?

**Practice Problem 15.11**

What new element would form if two oxygen-16 nuclei undergo fusion with no emission of particles?

**Further Practice:** Questions 15.91 and 15.92 at the end of the chapter

# Chapter Review

- Most nuclides of the elements are radioactive.
  - Radioactive nuclides decompose to other elements and emit radiation (alpha particles, beta particles, or gamma rays) through nuclear reactions.
  - Nuclear reactions are represented by equations similar to chemical equations. They are balanced when both the mass number and the nuclear charge are conserved.
  - It is possible to predict the stability of nuclides and their mode of decay if they are unstable.
- The 250 stable nuclides are all found within a narrow band of stability in which the nuclear composition has nearly equal numbers of protons and neutrons for lighter elements ($Z = 1$ to 20). The heavier stable nuclides have more neutrons than protons.
  - The lighter radioactive nuclides decay by several processes: beta (electron) decay, beta (positron) decay, or electron capture.
  - Nuclides heavier than bismuth-83 decay by a series of alpha and beta emissions until a stable nuclide is formed.
- Nuclear bombardment reactions are induced nuclear reactions in which nuclides are bombarded with other particles to create new nuclides.
- The half-life for a nuclide is the time required for half of a sample to decay to another nuclide.
  - The half-life is a characteristic of each nuclide.
  - Half-lives and isotope ratios can be used to measure the age of objects of archeological or geological interest.
- Radioactive isotopes and their nuclear reactions are used for portable power packs, radioactive tracers, medical imaging, and cancer therapy.
- Exposure to too much radiation can cause tissue damage in living things.
  - There are many sources of low-level radiation in our environment.
  - The extent of cell damage depends on the energy of the radiation, its penetrating and ionizing ability, the chemical properties of the source, and the level of exposure.
- Energy can be produced by the processes of fission and fusion.
  - Fission is a nuclear reaction in which a nuclide splits to form two other nuclides and, usually, small particles such as neutrons. It is the reaction that occurs in atomic bombs and nuclear power plants.
  - Fusion, the combining of light nuclides to form heavier nuclides, is a potential source of electric energy that has not yet been harnessed.

| | | | |
|---|---|---|---|
| alpha particle (15.1) | electron capture (15.2) | half-life (15.3) | positron (15.1) |
| beta particle (15.1) | fission (15.6) | nuclear bombardment (15.2) | radiation (15.1) |
| chain reaction (15.6) | fusion (15.6) | nucleon (15.1) | radioactive (15.1) |
| critical mass (15.6) | gamma ray (15.1) | nuclide (15.1) | radiocarbon dating (15.3) |

# QUESTIONS AND PROBLEMS

The following questions and problems, except those in *Additional Questions* and *Concept Review Questions,* are paired. Questions in a pair focus on the same concept. Answers to selected questions and problems are in Appendix D.

## Matching Definitions with Key Terms

15.1 Match the key terms with the descriptions provided.
   (a) the smallest amount of fissionable material necessary to support a continuing chain reaction
   (b) the splitting of a heavy atom into two or more lighter atoms and some number of neutrons with the release of energy
   (c) a nuclide containing two protons and two neutrons; the nucleus of a helium-4 atom
   (d) having the ability to emit ionizing radiation spontaneously
   (e) the highest-energy and most-penetrating type of electromagnetic radiation
   (f) a nuclear reaction that is induced by the impact of a nuclide or a subatomic particle on another nuclide
   (g) a nuclear particle (neutron or proton)
   (h) emission of particles or electromagnetic waves

15.2 Match the key terms with the descriptions provided.
   (a) a technique for the establishment of the age of a carbon-containing object by determining the amount of C-14 present in the object
   (b) a particle having the same mass as an electron but the opposite electrical charge
   (c) incorporation of a $1s$ electron into the nucleus, resulting in the conversion of a proton into a neutron
   (d) a small, charged particle that can be emitted from atoms at speeds approaching the speed of light; either an electron or a positron
   (e) a reaction in which the product of one step is a reactant in a subsequent step
   (f) the combination of light nuclei to form heavier nuclei
   (g) an atom having a specified atomic number, mass number, and nuclear energy
   (h) the time it takes for half of a sample of a nuclide to decay to a different nuclide

## Radioactivity

15.3 What is a radioactive substance?

15.4 What types of nuclides are radioactive?

15.5 What are the different types of radiation resulting from radioactive decay?

15.6 What happens during nuclear reactions in addition to the emission of radiation?

15.7 Which of the first 90 elements have only radioactive isotopes?

15.8 Which of the elements with atomic numbers greater than 90 are radioactive?

15.9 A sample of a radioactive material emits gamma radiation. What material would you use to enclose the material to prevent the radiation from escaping?

15.10 A sample of a radioactive material emits beta radiation. What material would you use to enclose the material to prevent the radiation from escaping?

## Nuclear Reactions

15.11 What features do nuclear decay reactions and nuclear bombardment reactions have in common?

15.12 How do nuclear decay reactions differ from bombardment reactions?

15.13 What are the different types of particles or rays that may result from nuclear decay?

15.14 Which type of radioactivity does not involve the emission of particles?

15.15 Describe the conservation rules that are used to balance nuclear reaction equations.

15.16 What quantity do we examine to determine whether nuclear charge is conserved in a balanced nuclear reaction?

15.17 What product results from alpha emission by $^{226}_{88}$Ra?

15.18 Write an equation for the alpha decay of $^{234}_{92}$U.

15.19 What product results from beta (electron) emission by $^{99}_{42}$Mo?

15.20 What is the product of beta (electron) decay of $^{214}_{82}$Pb?

15.21 Compounds containing $^{15}_{8}$O are used for PET scans. What nuclide is formed when $^{15}_{8}$O emits a positron?

15.22 A nuclide used in PET scans forms $^{13}_{6}$C when it emits a positron. What is this nuclide?

15.23 Identify the particle emitted by a nucleus that undergoes the following transformation.

Neutron          Proton

15.24 Identify the particle emitted by a nucleus that undergoes the following transformation.

Proton          Neutron

15.25 What nuclide will produce $^{7}_{3}$Li by electron capture?

15.26 What product will result from electron capture by $^{44}_{22}$Ti?

15.27 Complete and balance the following equations.
   (a) $^{238}_{92}$U $\longrightarrow$ $^{234}_{90}$Th + ?
   (b) $^{234}_{90}$Th $\longrightarrow$ $^{234}_{91}$Pa + ?
   (c) $^{234}_{92}$U $\longrightarrow$ $^{4}_{2}\alpha$ + ?
   (d) $^{210}_{83}$Bi + $^{0}_{-1}$e$^{-}$ $\longrightarrow$ ?

15.28   Complete and balance the following equations.

(a) $_3^7\text{Li} + ? \longrightarrow 2\,_2^4\alpha$

(b) $_{82}^{214}\text{Pb} \longrightarrow \,_{83}^{214}\text{Bi} + ?$

(c) $_{13}^{27}\text{Al} + \,_0^1\text{n} \longrightarrow ? + \,_2^4\alpha$

(d) $_6^{12}\text{C} + ? \longrightarrow \,_7^{13}\text{N}$

15.29   Write balanced equations for the following nuclear reactions.

(a) alpha decay by $_{87}^{227}\text{Fr}$

(b) positron emission by $_9^{18}\text{F}$

(c) beta emission by $_6^{14}\text{C}$

15.30   Write balanced equations for the following nuclear reactions.

(a) beta (electron) emission by $_{47}^{114}\text{Ag}$

(b) alpha emission by $_{84}^{210}\text{Po}$

(c) electron capture by $_{56}^{133}\text{Ba}$

15.31   Describe a nuclear bombardment reaction.

15.32   If we know the starting nuclide and the products of a nuclear bombardment reaction, how do we determine what was used to bombard the starting nuclide?

15.33   Bombardment of $_{98}^{250}\text{Cf}$ by $_5^{11}\text{B}$ yields four neutrons. What is the other product?

15.34   What isotope is used to bombard $_{42}^{97}\text{Mo}$ if the products are $_{43}^{97}\text{Tc}$ and two neutrons?

15.35   Complete and balance the following equations.

(a) $_5^{10}\text{B} + \,_1^2\text{H} \longrightarrow ? + \,_0^1\text{n}$

(b) $_{83}^{209}\text{Bi} + \,_2^4\alpha \longrightarrow ? + 2\,_0^1\text{n}$

(c) $_6^{12}\text{C} + ? \longrightarrow \,_7^{13}\text{N}$

15.36   Complete and balance the following equations.

(a) $_{92}^{238}\text{U} + \,_0^1\text{n} \longrightarrow ? + \,_{-1}^0\beta^-$

(b) $_3^6\text{Li} + \,_0^1\text{n} \longrightarrow ? + \,_1^3\text{H}$

(c) $_{29}^{63}\text{Cu} + ? \longrightarrow \,_{29}^{64}\text{Cu} + \,_1^1\text{H}$

15.37   Write a balanced equation for each of the following nuclear reactions.

(a) bombardment of $_{42}^{96}\text{Mo}$ with a proton, resulting in an emission of a neutron

(b) bombardment of $_{83}^{209}\text{Bi}$ with $_{28}^{64}\text{Ni}$ to produce a new nuclide and a neutron

15.38   Write a balanced equation for each of the following nuclear reactions.

(a) bombardment of a nuclide with $_1^2\text{H}$ atoms, producing $_{30}^{64}\text{Zn}$ and a neutron

(b) bombardment of $_{82}^{206}\text{Pb}$ with $_{26}^{58}\text{Fe}$, producing a new nuclide and a neutron

15.39   What is the *N/Z* ratio for stable isotopes for the first 20 elements?

15.40   What is the *N/Z* ratio for elements 80 to 90?

15.41   What is the band of stability?

15.42   How is the band of stability used to predict the type of radioactive decay a given nuclide might undergo?

15.43   Which of the following isotopes should be the most stable? Explain your answer.

$$_8^{16}\text{O} \qquad _8^{17}\text{O} \qquad _8^{18}\text{O}$$

15.44   Which of the following isotopes should be the most stable? Explain your answer.

$$_{14}^{28}\text{Si} \qquad _{14}^{29}\text{Si} \qquad _{14}^{30}\text{Si}$$

15.45   Predict the type of nuclear decay expected for the following unstable nuclides.

(a) $_6^{14}\text{C}$             (d) $_8^{15}\text{O}$

(b) $_{90}^{234}\text{Th}$         (e) $_7^{16}\text{N}$

(c) $_{92}^{234}\text{U}$

15.46   Predict the type of nuclear decay expected for the following unstable nuclides.

(a) $_8^{19}\text{O}$             (d) $_7^{13}\text{N}$

(b) $_{91}^{230}\text{Pa}$         (e) $_{94}^{244}\text{Pu}$

(c) $_6^{10}\text{C}$

15.47   When $_{91}^{234}\text{Pa}$ decays, it emits six alpha particles and three beta particles in nine steps. What is the final product of this series of decay steps?

15.48   When $_{88}^{226}\text{Ra}$ decays, it emits four alpha particles and two beta particles in six steps. What is the final product of this series of decay steps?

## Rates of Radioactive Decay

15.49   What is the half-life of a radioactive nuclide?

15.50   What uses are made of the half-life of a radioactive nuclide?

15.51   How is radiation detected?

15.52   How are the half-lives of radioactive nuclides measured?

15.53   The half-life of $_{11}^{24}\text{Na}$ is 15.0 h. What fraction of the nuclide will remain after 60.0 h?

15.54   The half-life of tritium, $_1^3\text{H}$, is 12.3 years. How much of a 100.0-g sample of tritium will be left after a period of 61.5 years?

15.55   To investigate metabolic pathways, a laboratory rat is injected with a sample containing phosphorus-32, which has a half-life of 14 days. Assuming none of the $_{15}^{32}\text{P}$ is excreted, how long would it take for the amount of $_{15}^{32}\text{P}$ to be reduced to one-eighth?

15.56   The half-life of $_{16}^{35}\text{S}$ is 86.7 days. How long does it take for the radiation intensity to decrease by 75%?

15.57   The half-life of the nuclide $_{85}^{211}\text{At}$ is 7.5 h. If 0.100 mg of this nuclide is administered for thyroid treatment, how long does it take to reduce the nuclide to 0.0125 mg?

15.58   The half-life of $_{16}^{35}\text{S}$ is 86.7 days. If 80.0 mg is absorbed by an orange, how long will it take to reduce this radioactive nuclide to 5.0 mg?

15.59   Barley and wheat grains recovered from an ancient Egyptian burial chamber were analyzed for their carbon-14 content. They contained an amount of $_6^{14}\text{C}$ equal to 50% of that present in living plants. If the half-life of $_6^{14}\text{C}$ is 5730 years, how old are these grains likely to be?

15.60   A precursor of modern-day bison is *Bison latifrons*. A fossil found in North Dakota was dated with the radiocarbon method. It contained 0.39% as much $_6^{14}\text{C}$ as modern bison. How old is the fossil likely to be?

15.61   Technetium-99*m* is used for a number of diagnostic purposes in medicine. For example, $^{99m}\text{Tc}$ introduced into the bloodstream is not normally absorbed by brain cells, but it is absorbed by a tumor. A brain scan can locate the tumor. If a sample of $^{99m}\text{Tc}$ is introduced into a body in this way, and 6.25% of its radioactivity remains after 24 h, what is the half-life of $^{99m}\text{Tc}$?

**15.62** A sample of a radioactive nuclide has an activity of 3500 cpm. If the activity has decreased to 1750 cpm 45.0 min later, what is the half-life of the nuclide?

## Medical Applications of Isotopes

**15.63** Why are pacemakers powered with a $^{238}_{94}$Pu power device rather than a zinc-silver oxide electrochemical cell?

**15.64** Explain how radioactive nuclides can be used to study the metabolism of a sugar substitute.

**15.65** Why is $^{99m}$Tc used for medical diagnosis?

**15.66** Why is $^{131}$I used for diagnosing thyroid tumors?

**15.67** Why is $^{11}_{6}$C, but not $^{14}_{6}$C, used for PET imaging?

**15.68** Describe how $^{18}_{9}$F might be used for PET imaging.

## Biological Effects of Radiation

**15.69** Describe four different effects radiation might have on a living biological cell.

**15.70** We are exposed to radiation from many sources. What sources can we control in our daily lives?

**15.71** The risk to the general population from radon is about 100 to 1000 times higher than the risk from carbon tetrachloride or benzene, which have been tightly controlled by the U.S. Environmental Protection Agency. Should the EPA generate regulations for radon in houses?

**15.72** Radon is estimated to cause 10,000 lung cancer deaths per year, compared to about 110,000 attributed to smoking tobacco. Why is tobacco not more heavily regulated?

**15.73** If Po-214 were ingested, what nuclides would be present in appreciable amounts later that day?

**15.74** If Po-214 were ingested, what types of radiation would be emitted during the first second after ingestion?

## Nuclear Energy

**15.75** Distinguish between fission and fusion.

**15.76** Write balanced nuclear equations for examples of fission and fusion.

**15.77** When bombarded by a neutron, U-235 undergoes fission. If the resulting nuclides are Kr-93 and Ba-140, how many neutrons are emitted?

**15.78** When bombarded by a neutron, U-235 undergoes fission. If the resulting nuclides are Sr-91 and Xe-142, how many neutrons are emitted?

**15.79** Describe the features of a chain reaction.

**15.80** Explain why critical mass is of concern in the design of a fission reactor.

**15.81** What are the main components of a fission reactor?

**15.82** What is the composition and function of a moderator in a nuclear fission reactor? A control rod?

**15.83** Why is the heat from fission reactions not used directly to create steam to operate a turbine?

**15.84** What safety features are included in the design of most fission reactors?

**15.85** Describe the nuclear waste from a fission reactor.

**15.86** Why does nuclear waste from a fission reactor usually have to be stored underground or in steel tanks?

**15.87** What fuels are most likely to be used as a power source in fusion reactors?

**15.88** Discuss the problems that must be overcome before nuclear fusion can become a useful energy source for peaceful purposes.

**15.89** Write a balanced nuclear equation for a reaction that might be used in a fusion reactor.

**15.90** Write a balanced nuclear equation for a reaction that might occur in a fission reactor.

**15.91** What nuclide forms when $^{1}_{1}$H undergoes fusion with $^{2}_{1}$H, assuming only one particle forms?

**15.92** What nuclide forms when $^{3}_{1}$H undergoes fusion with $^{4}_{2}$He, assuming only one particle forms?

## Additional Questions

**15.93** Describe the process of radioactive dating.

**15.94** What product will result from neutron capture by $^{238}_{92}$U?

**15.95** An old piece of wood has a carbon-14 content of 0.0156 times that of a freshly cut piece of wood. What is the age of the old wood if the half-life of $^{14}_{6}$C is 5730 years?

**15.96** The ratio of $^{40}$K to $^{40}$Ar was used to date moon rocks returned by the Apollo 11 flight. Potassium-40 decays to argon-40 by positron emission with a half-life of 1.3 billion years. Assuming no $^{40}$Ar was present initially and that none escaped, how old is a rock that had a K-40 to Ar-40 ratio of 0.143?

**15.97** Outline the nuclear reactions that are the primary source of energy in the stars.

**15.98** Predict the type of radioactive decay expected for the following unstable nuclides.
(a) $^{17}_{9}$F      (d) $^{19}_{10}$Ne
(b) $^{21}_{9}$F      (e) $^{15}_{6}$C
(c) $^{216}_{84}$Po

**15.99** The bombardment of beryllium-9 with alpha particles yields carbon-12. Predict any other products of this reaction and write a balanced equation to describe it.

**15.100** What product will result from positron emission by $^{15}_{8}$O?

**15.101** Polonium-210 is an alpha-particle emitter with a half-life of 138 days. Static-eliminator brushes for cleaning dust from camera lenses contain a small amount of $^{210}$Po at the base of the bristles. Suppose you are in charge of maintaining inventory at a camera store and you are offered a very good buy on a three-year supply of brushes. Should you maintain a stock this size? Consider how much $^{210}$Po is left at the end of each year during this period to help you decide.

**15.102** Although potassium-40 decays to produce calcium-40 with a half-life of 1400 million years, rocks cannot be dated by measuring the ratio of $^{40}$K to $^{40}$Ca. Why?

**15.103** Neutron activation analysis is a nondestructive method for determining the elements present in materials. The material is bombarded with neutrons, forming radioactive nuclides with decay energies that can be measured. It is often used to verify the authenticity of paintings by matching the composition of paint with that from other paintings by the same artist. Comment on the statement that a painting found to contain chloride could not have been painted in AD 1648, as claimed, since the element chlorine was not discovered until 1774.

15.104 Different forms of radiation have different penetrating powers. Which one of the following forms of radiation is most penetrating and which is the least penetrating: alpha particles, beta particles (electrons), positrons, gamma rays, neutrons?

15.105 How many neutrons are emitted when a californium-249 nucleus ($Z = 98$) is bombarded with a nitrogen-15 nucleus to produce a dubnium nucleus, $^{260}_{105}\text{Db}$? Write a balanced equation for this process.

15.106 What happens to the mass number and the atomic number of an element when it undergoes beta decay?

15.107 What happens to the mass number and the atomic number of an element when it undergoes alpha decay?

15.108 Technetium, which is not found naturally, was first synthesized in a nuclear reaction in 1937. Technetium-97 ($^{97}_{43}\text{Tc}$) can be produced by bombardment of a nuclide with $^2_1\text{H}$ and emission of two neutrons. What is the starting nuclide?

15.109 The half-life of tritium is 12.5 years. How much will remain after exactly 100 years if you start out with 256 g?

15.110 What type of radiation is likely to be emitted by isotopes found below the band of stability (low neutron-to-proton ratio)?

15.111 If a radioactive isotope that has too high a neutron-to-proton ratio gains stability, what process is involved?

15.112 Identify the nuclide X in the following equation for plutonium fission:

$$^{239}_{94}\text{Pu} + ^1_0\text{n} \longrightarrow X + ^{91}_{38}\text{Sr} + 3^1_0\text{n}$$

15.113 Astatine-218 (At) decays by alpha emission. What is the new nuclide that is formed?

## Concept Review Questions

15.114 Which of the following statements correctly describes alpha radiation?
A. spontaneous emission of an electron from a nucleus
B. spontaneous emission of a helium nucleus from a nucleus
C. spontaneous emission of high-energy electromagnetic radiation from a nucleus
D. spontaneous emission of a positron from a nucleus
E. spontaneous capture of an electron by a nucleus
Name the types of radioactive process described by the other responses.

15.115 Which one of the following processes results in an increase in a nuclide's atomic number?
A. gamma emission          D. alpha emission
B. positron emission       E. electron capture
C. beta emission
What effect do the other processes have on the atomic number?

15.116 Which type of radioactive decay results in no change in the mass number and the atomic number for the starting nucleus?
A. alpha          D. electron capture
B. beta           E. gamma
C. positron emission
What change in mass number occurs for the other types of radioactive decay?

15.117 Which nuclear reaction is properly balanced?
A. $^{12}_6\text{C} + ^1_1\text{H} \longrightarrow ^{14}_7\text{N} + ^1_0\text{n}$
B. $^4_2\text{He} + ^9_4\text{Be} \longrightarrow ^{12}_6\text{C} + ^1_1\text{H}$
C. $^4_2\text{He} + ^{14}_7\text{N} \longrightarrow ^{14}_7\text{N} + 2^2_1\text{H}$

D. $^{16}_8\text{O} \longrightarrow ^{16}_9\text{F} + ^0_{+1}\beta$
E. $^{14}_7\text{N} + ^0_{-1}\text{e} \longrightarrow ^{14}_8\text{O}$
Change the other nuclear reactions so they are correctly balanced.

15.118 If a nuclide lies above the band of stability, what type of radioactive decay should it undergo?
A. alpha particle emission
B. beta particle emission
C. electron capture
D. positron emission
E. no decay (stable)
Relative to the band of stability, where would a nuclide be found that would undergo the other types of decay?

15.119 When $^{234}_{92}\text{U}$ decays, it undergoes five alpha emissions and two beta emissions in seven steps. What is the final product of these decay steps?
A. thorium-230          D. lead-214
B. polonium-214         E. radon-222
C. polonium-218
How many alpha emissions and beta emissions would result in the other nuclides listed?

15.120 The half-life of technetium-99 is 6.0 h. How much of a 100.0-mg sample remains after 18.0 h?
A. 50.0 mg          D. 6.25 mg
B. 25.0 mg          E. 3.12 mg
C. 12.5 mg
How much time would be required for the original sample to be diminished to the other possible masses?

15.121 Which statement correctly describes the basis for carbon-14 dating?
A. Carbon-14 is very unstable and is readily lost from the atmosphere.
B. Living tissue absorbs carbon-12 but not carbon-14.
C. The ratio of carbon-14 to carbon-12 in the atmosphere is relatively constant.
D. The amount of carbon-14 in all objects is the same.
E. Carbon-12 decays to form carbon-14 in living tissue, but not in dead tissue.
Rewrite the incorrect responses so they are correct.

15.122 Which of these is not an application of nuclear chemistry?
A. medicine
B. energy
C. archeology
D. warfare
E. combustion of hydrogen gas
For each response that is an application of nuclear chemistry, provide an example.

15.123 Which of the following statements describes a way in which fission and fusion reactions are similar?
A. Both produce large amounts of radioactive waste products, posing disposal problems.
B. Both require high temperatures for initiation.
C. Both require critical masses.
D. Both need both heavy and light nuclei.
E. Both involve the transformation of one or more elements into other elements.
Rewrite each incorrect statement so it applies to the appropriate type of nuclear reaction.

# Organic Chemistry

©Brian Moeskau/Moeskau Photography

16.1 Representations of Organic Molecules

16.2 Hydrocarbons

16.3 Acyclic Hydrocarbons

16.4 Cyclic Hydrocarbons

16.5 Alcohols and Ethers

16.6 Aldehydes and Ketones

16.7 Carboxylic Acids and Esters

16.8 Amines

16.9 Organic Nomenclature

Chapter Review

Questions and Problems

**W**hen Scott and Jennifer arrive early for their chemistry class, they find an assignment written on the board: "Identify a molecule containing at least eight atoms that is important to the career you are working toward. Describe how the molecule relates to your career. Draw the molecule's structure and explain how it determines the properties that make this molecule important." Along with other early arrivals, Scott and Jennifer settle into their chairs and begin to discuss their career plans and the molecules important in their chosen fields.

Scott, an environmental science major, is interested in fuel-efficient engines, so he thinks octane should be his choice. He knows that octane is a standard for rating grades of gasoline, but it isn't until he checks his textbook that he learns that octane, $C_8H_{18}$, has several possible structures, including

An octane rating does not necessarily indicate the amount of octane present. Instead, the rating provides a measure of the fuel's tendency to knock. Knocking occurs when the gasoline-air mixture in an engine cylinder explodes before it is ignited by the spark plug. A high octane rating indicates a fuel burns in a more controlled way.

©SuperStock/Alamy Stock Photo

As Scott studies the pictures, he realizes that the octane structures differ in the way the carbon atoms attach to one another. Because of the large number of atoms, Scott learns that octane molecules have London dispersion forces strong enough to make them liquids at room temperature. (The respective boiling points of the molecules shown are 125.6°C, 117.6°C, and 99.3°C.) Scott also finds that only one of the structures—the last molecule shown—serves as the standard for the octane rating of gasoline. It burns more smoothly than the other molecules.

These octane compounds all have the same formula and the same mass. Why would they have different boiling points?

Jennifer hopes to have a career in dentistry, so she thinks about mouthwashes and a common ingredient in them: menthol, $C_{10}H_{20}O$. The menthol molecule, she finds, has the following structure:

What functional group is present in menthol that is not found in octane?

Menthol is also used in local anesthetics, such as cough drops, ointments for muscle aches, and anti-itch creams. It is used in aftershave lotions and skin fresheners as well. In all of these applications, menthol gives the sensation of cooling. Jennifer guesses that receptors on cell membranes in the mouth, throat, and skin might accept the menthol molecule and generate nerve impulses as a result. If this were true, then the size, shape, and polarity of the menthol molecule would have to be just right for it to interact with a receptor, she reasons. Further reading proves she's right. Although the mechanisms aren't totally understood, research evidence suggests that menthol tricks the temperature receptors into signaling the brain that something cold has come into contact with them.

Omar plans to become a pharmacist. He picks aspirin, $C_9H_8O_4$, as his molecule. He finds the following structure for aspirin, also called acetylsalicylic acid:

Aspirin blocks the action of an enzyme called cyclooxygenase, Omar learns. The enzyme catalyzes the production and release of prostaglandins. Prostaglandins form whenever tissue is injured, and they send a nerve signal that the brain interprets as pain. Blocking the production of prostaglandins thus decreases the pain felt from an injury, although it does nothing to cure the cause of the pain. Prostaglandins also cause blood platelets to stick together and form clots. Blocking the production of prostaglandins helps to prevent blood clots, so aspirin is often taken to treat or prevent strokes and heart attacks. Omar suspects that aspirin inhibits the functioning of the enzyme by fitting into a cavity in the enzyme structure, blocking substrate molecules from attaching to it.

Would you expect the structure of other painkillers, such as acetaminophen, to be similar to that of aspirin?

Maya hopes to become an ophthalmologist. She picks retinal, $C_{20}H_{28}O$, as her molecule since it is important to vision. She finds two different structures for this molecule, called *cis*-retinal (left) and *trans*-retinal (right):

The *cis*-retinal molecule has two hydrogen atoms on the same side of one of the carbon-carbon double bonds, while *trans*-retinal has the two hydrogen atoms on opposite sides of this double bond. This difference in structure causes a difference in shape, as seen in the models. The —CHO group on the end of the molecule is quite reactive. It allows a *cis*-retinal molecule to bond to a molecule of opsin, a protein in the rods and cones of the eye. The result is a molecule called rhodopsin (Figure 16.1). The shape of *cis*-retinal allows it to fit into a cavity in the opsin molecule. Visible light converts *cis*-retinal to *trans*-retinal, accompanied by a change from a more bent to a more linear molecular structure. Because it no longer fits well into the opsin molecule cavity, *trans*-retinal detaches from the protein, sending an electrical signal along the optic nerve to the brain. The *trans*-retinal then interacts with an enzyme that changes it back to *cis*-retinal, which can reattach to an opsin molecule.

The behavior of retinal is unusual. Hydrocarbons with double bonds do not normally convert from cis to trans forms upon exposure to light because the conversion requires one of the bonds to break.

**FIGURE 16.1** Vision depends on the interconversion of *cis*-retinal and *trans*-retinal. The absorption of light causes bond breakage, allowing the interconversion to occur.

Aisha, a criminal justice major, is concerned about law enforcement and drug abuse, so she decides to research the addictive central nervous system stimulant amphetamine, $C_9H_{13}N$. She finds this structure for the molecule:

How are the structures of aspirin and amphetamine alike and different?

The structure and shape of amphetamine are similar to those of norepinephrine, a natural neurotransmitter that carries nerve impulses across the gap between neurons in the brain. Norepinephrine is normally stored inside the ends of neurons before it is released in appropriate amounts. Because of its structure, amphetamine can displace norepinephrine from its storage vesicles. As a result, too many norepinephrine molecules are released into the gap, and emotional responses are exaggerated.

What do all these molecules have in common? Did you notice that they all contain carbon? Why is this important? If you had this assignment, what molecule would you select? Would you choose a carbon compound? In this chapter, we'll look at compounds of carbon and hydrogen and explore the field of *organic chemistry*.

Here's some examples of other carbon-containing molecules that are related to specific careers:

| | |
|---|---|
| lactic acid (in weary muscles) | sports medicine and physical therapy |
| linoleic acid (in linseed oil) | painting |
| methyl methacrylate (Plexiglas) | sculpture |
| glycine (an essential amino acid) | nutrition |
| glucose | emergency medical technician |
| saturated fats | food science |
| diethyl ether | anesthesiology |

## Questions for Consideration

**16.1** How can we represent organic molecules to illustrate their similarities and differences?

**16.2** What are hydrocarbons, and where do they come from?

**16.3** What are acyclic hydrocarbons, and how do they react?

**16.4** What are cyclic hydrocarbons, and how do they react?

**16.5** How do alcohols and ethers differ from the alkanes?

**16.6** How are aldehydes and ketones alike and different?

**16.7** How do carboxylic acids and esters differ from other organic compounds?

**16.8** What is unique about amines?

**16.9** How do we use the IUPAC system to name organic compounds?

# 16.1  Representations of Organic Molecules

The compounds that Scott, Jennifer, and the other students chose for the assignment all contain carbon. In combination with a few other elements (primarily H, O, N, and S), carbon forms several million known compounds, far exceeding those of any other element except hydrogen. The most extensive class of carbon compounds consists of *organic compounds,* which contain carbon, hydrogen, and often other elements such as oxygen and nitrogen. The study of organic compounds is called **organic chemistry.** The term *organic* relates to the importance of carbon compounds in living systems. Organic compounds need not come from living things, but they are vital to all life processes.

Why is carbon found in so many molecules and in such a variety of structural forms? The answer is found, in part, in the way that carbon bonds to itself and to other elements. A carbon atom can form single bonds, double bonds, or triple bonds in various combinations—for example,

- four single bonds, as in $C_2H_6$

- two single bonds and one double bond, as in $C_2H_4$

- two double bonds, as for the central atom in $C_3H_4$

- one single bond and one triple bond, as in $C_2H_2$

In each case carbon has a total of four bonds. In contrast, hydrogen and fluorine typically form one bond, oxygen forms two bonds, and nitrogen forms three. The greater number of bonds found with carbon gives more possibilities for forming a diverse set of compounds.

A carbon compound that contains only carbon and hydrogen is called a **hydrocarbon.** In hydrocarbons, carbon has the ability to bond to other carbon atoms to form long chains or rings of carbon atoms with hydrogen atoms attached. Different

The common definition of an organic compound is any compound that contains carbon, but this includes carbonates ($CO_3^{2-}$), cyanides ($CN^-$), and other ionic compounds that are usually considered inorganic. Organic chemistry is better defined as the study of carbon compounds in which carbon atoms are bonded to one another and to hydrogen atoms as well as to other nonmetal atoms such as oxygen and nitrogen.

Recall from Section 8.4 that carbon generally forms four covalent bonds because each carbon atom has just four valence electrons, and eight electrons are needed for an octet. Other nonmetals such as hydrogen, fluoride, oxygen, and nitrogen need to form fewer covalent bonds because their atoms have more than four valence electrons.

branching patterns are possible in many hydrocarbon molecules. Two compounds having the same composition and formula, but with a different arrangement of atoms in their molecular structures, are **isomers** of one another.

Carbon can bond to another nonmetal, such as nitrogen and oxygen, in a few, well-defined ways, giving a characteristic grouping of atoms called a functional group. A **functional group** is a reactive part of a molecule that undergoes characteristic reactions. For example, when a hydroxyl group ($-OH$) is present instead of a hydrogen atom in propane, $C_3H_8$, we have propyl alcohol or propanol, $C_3H_7OH$. The functional group in this case is $-OH$, which is characteristic of alcohols.

A variety of related molecules with different functional groups is shown in Table 16.1. How are these structures related to propane? How can substituting, adding, or removing atoms or groups of atoms create these structures?

Ball-and-stick models are a good way to represent the structure of organic molecules, but other symbols and models have their advantages. The hydrocarbon fuel propane, for example, has the simple *molecular formula* $C_3H_8$. It has three carbon atoms linked in a chain, with hydrogen atoms bonded to each carbon atom in sufficient numbers to give each carbon atom four bonds: $CH_3CH_2CH_3$. This *structural formula* gives more information about propane than does the molecular formula, because it shows the order in which the atoms are bonded. Butane, another fuel, has the molecular formula $C_4H_{10}$. It can be represented by an expanded structural formula $CH_3CH_2CH_2CH_3$, or a *condensed structural formula,* $CH_3(CH_2)_2CH_3$, in which identical connected units are grouped. There is another compound with the same molecular formula which has one carbon atom bonded to three $-CH_3$ groups and one hydrogen atom. The condensed structural formula for this compound is $CH_3CH(CH_3)CH_3$, where the part of the formula in parentheses is understood to be attached to the previous carbon atom. The Lewis structures for these two compounds are

$$
\begin{array}{cccc}
 & H & H & H & H \\
 & | & | & | & | \\
H- & C- & C- & C- & C-H \\
 & | & | & | & | \\
 & H & H & H & H \\
\end{array}
\qquad
\begin{array}{ccc}
 & & H \\
 & & | \\
 & H- & C-H \quad H \\
 & \diagdown & | \quad \diagup \\
H-C & -C- & C-H \\
 & | & | \\
 & H & H \\
\end{array}
$$

Because these two compounds have the same composition and formula but a different arrangement of atoms in their molecular structure, they are called isomers.

Another way to represent an organic molecule is with a *line structure,* which represents connected carbon atoms with a line. The hydrogen atoms bonded to carbon atoms are not shown, because their presence can be deduced from the fact that carbon must form four bonds. Any other atom, such as oxygen or nitrogen, is specified by its element symbol. In a line structure, the intersection of two lines or any unconnected end of a line represents a carbon atom. Thus, one of the octane molecules that Scott is interested in can be represented in the following ways:

- Ball-and-stick model:
- Molecular formula: $C_8H_{18}$
- Structural formula: $CH_3CH_2CH_2CH_2CH_2CH_2CH_2CH_3$
- Condensed structural formula: $CH_3(CH_2)_6CH_3$
- Line structure:

The ball-and-stick model, condensed structural formula, and line structure for the second isomer of octane that Scott found are

$(CH_3)_2CH(CH_2)_4CH_3$

**TABLE 16.1** Functional Groups in Organic Molecules

| Chemical Class | Functional Group | Example | Example Structure | Example Name | Example Property |
|---|---|---|---|---|---|
| alkane | $-\overset{\vert}{\underset{\vert}{C}}-\overset{\vert}{\underset{\vert}{C}}-$ | $CH_3CH_2CH_3$ | | propane | heating fuel for camping stoves |
| alkene | $\overset{/}{\underset{\backslash}{}}C=C\overset{\backslash}{\underset{/}{}}$ | $CH_3CH{=}CH_2$ | | propene | used to form polypropylene, a stiff plastic |
| alkyne | $-C{\equiv}C-$ | $CH_3C{\equiv}CH$ | | propyne | explosive in air |
| alcohol | $-OH$ | $CH_3CH_2CH_2OH$ | | propanol | alcoholic and slightly stupefying odor |
| ether | $-O-$ | $CH_3CH_2CH_2OCH_3$ | | methyl propyl ether | anesthetic |
| ketone | $-\overset{O}{\overset{\vert\vert}{C}}-$ | $CH_3-\overset{O}{\overset{\vert\vert}{C}}-CH_3$ | | acetone | good solvent; fingernail polish remover |
| aldehyde | $-\overset{O}{\overset{\vert\vert}{C}}-H$ | $CH_3CH_2CHO$ | | propanal | suffocating odor |
| carboxylic acid | $-\overset{O}{\overset{\vert\vert}{C}}-OH$ | $CH_3CH_2CO_2H$ | | propanoic acid | odor and flavor of Swiss cheese |
| ester | $-\overset{O}{\overset{\vert\vert}{C}}-O-$ | $CH_3CH_2CO_2CH_3$ | | methyl propanoate | flavoring agent; fruity rum odor |
| amine | $-\overset{\vert}{N}-$ | $CH_3CH_2CH_2NH_2$ | | propylamine | strong ammonia odor |

The line structure is a convenient way to show the difference in isomers. In these examples, the first isomer is a straight chain of eight carbon atoms, while in the second, one of the carbons is bonded to the second carbon atom in a seven-carbon chain.

Line structures also help us focus on the functional group where any chemical reaction is likely to occur. The structural formula and line structure for Jennifer's menthol molecule are

The molecule is a ring, or cycle, of six connected carbon atoms. Attached to this ring are a carbon atom, an —OH group (after skipping one carbon atom in the ring), and then a three-carbon-atom chain connected by its center carbon atom. The menthol molecule has the —OH functional group that is characteristic of alcohols, so it has physical and chemical properties similar to that of other alcohols.

---

**EXAMPLE 16.1**    **Drawing Structural Formulas and Line Structures**

Aisha studied the amphetamine molecule, using the following ball-and-stick model. Draw both a structural formula and a line structure for this molecule:

**Solution:**

Examine the ball-and-stick model of this molecule. The structural formula must represent each carbon-carbon bond and include the hydrogen or other atoms attached to each carbon. Since this is not a simple chain structure, we use lines to represent bonds and show how the atoms are connected:

The line structure can then be based on this structural formula, but with the carbon-hydrogen bonds and the symbols for carbon removed:

For clarity we leave in the element symbols for the $-NH_2$ group that is character-istic of amines.

**Consider This 16.1**

Draw line structures for propanol, $CH_3CH_2CH_2OH$, and isopropanol, $CH_3CH(OH)CH_3$.

**Practice Problem 16.1**

Omar studied the aspirin molecule with the following ball-and-stick model:

Draw a structural formula and a line structure for this molecule.

**Further Practice:** Questions 16.7 and 16.8 at the end of the chapter

---

One additional representation of organic molecules is a *perspective drawing,* which can be used instead of a ball-and-stick model when visualizing the three-dimensional structure of a molecule. In a perspective drawing, solid lines represent bonds that are in the plane of the page. Dashed wedges represent bonds that extend back behind the plane of the page. Solid wedges represent bonds that project in front of the plane of the page. To see how this works, compare the perspective drawing of octane to the ball-and-stick model:

## 16.2 Hydrocarbons

### Classes of Hydrocarbons

Hydrocarbons are a major class of organic compounds, and they form the basic structural components of most other classes. Hydrocarbons can have single or mul-tiple carbon-carbon bonds, and they can exist as open chains or closed rings. The hydrocarbons can be classified as shown in Figure 16.2.

**Acyclic (Open Chain) Hydrocarbons** The first main categories shown in Figure 16.2 are derived from the shape of the molecule. A hydrocarbon molecule that forms an open chain is an **acyclic hydrocarbon.** In such molecules the bonding electrons are *localized;* the electrons are confined to two adjacent atoms and there are no resonance forms. A hydrocarbon with localized bonds is an **aliphatic hydrocarbon.**

A **saturated hydrocarbon** contains only single bonds between the carbon atoms. All of the carbon atoms are said to be saturated because they are attached to four other atoms—the maximum number possible for carbon. An example is ethane, $H_3C$ $CH_3$, which, like all the other compounds in this category, is an alkane. An *alkane* is any straight-chain hydrocarbon containing only carbon-carbon single bonds.

> Two major types of hydrocarbons, the *aliphatics* and the *aromatics,* were described in Section 8.4 and outlined in Figure 8.20.

> In organic chemistry *saturated* and *unsaturated* have different meanings than when describing the concentration of solutions, but the two uses are related. In each case *unsaturated* means some "holding capacity" remains—either to accept more hydrogen atoms (organic molecules) or to dissolve more solute (in solutions) *Saturated* means that all capacity to hold more has been used.

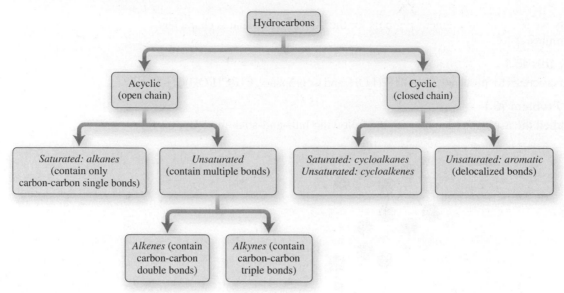

**FIGURE 16.2** Classification of the hydrocarbons.

An **unsaturated hydrocarbon** contains one or more double or triple bonds between carbon atoms. Such molecules are considered unsaturated because two or more of the carbon atoms do not have four other atoms bonded to them, resulting in a lower hydrogen to carbon ratio. The double-bonded, unsaturated hydrocarbons are the *alkenes,* such as $H_2C=CH_2$ (ethene, also called ethylene). The triple-bonded, unsaturated hydrocarbons are the *alkynes,* such as $HC≡CH$ (ethyne, also called acetylene).

All of the acyclic, aliphatic hydrocarbons form open chains of carbon atoms, such as propane, $H_3C—CH_2—CH_3$. The chains may bond to side chains, or branches, such as in 4-ethylheptane:

We'll examine the three categories of acyclic hydrocarbons in more detail in Section 16.3.

**Cyclic (Closed Chain) Hydrocarbons**   In contrast to the open chain of an aliphatic compound, a **cyclic hydrocarbon** forms a closed chain or ring of carbon atoms, such as in cyclobutane:

This is an example of a *cycloalkane:* a single-bonded chain that appears to have bent around on itself and attached at its ends. Cycloalkanes, like alkanes, are single-bonded and saturated. *Cycloalkenes* are carbon rings that contain at least one double bond, are unsaturated, and have *localized bonding.*

  **Student Hot Spot**

Student data indicate you may struggle with determining if a compound is saturated or unsaturated from its formula. Access the SmartBook to view additional Learning Resources on this topic.

Cyclobutane, $C_4H_8$, is a colorless gas that burns with a luminous flame.

In contrast, an **aromatic hydrocarbon** has *delocalized bonding*. An aromatic hydrocarbon is a cyclic molecule containing a ring of carbon atoms attached to each other and is represented by resonance structures containing alternating single and double bonds, as explained by resonance (review Section 8.3). For example, a six-carbon ring such as in benzene, $C_6H_6$, has two possible resonance structures:

However, neither of these drawings represents the actual structure. Instead, the double-bond electrons in benzene are delocalized or spread out around the entire ring structure, giving the molecule bonds of equal length and—unlike ordinary double bonds—a tendency toward low reactivity. For these reasons the structure is more accurately represented with a circle inside the ring to indicate the delocalized double bonds:

or

We'll consider aromatic hydrocarbons in more detail in Section 16.4.

## Petroleum

Along with coal, petroleum is the primary source of hydrocarbons. Their major use is energy production, but around 10% of the coal and petroleum recovered from Earth is used to make such useful materials and products as plastics, drugs, foods, and synthetic fibers.

Petroleum is usually found naturally in one of two forms, either natural gas or crude oil. Natural gas consists mainly of methane ($CH_4$), with smaller amounts of ethane ($C_2H_6$), propane ($C_3H_8$), and butane ($C_4H_{10}$). All are colorless, odorless, flammable gases. Natural gas is used primarily for heating and cooking. Crude oil is a dark, thick, odoriferous liquid composed of about 95% hydrocarbons. It must be refined before it can be used. Refining involves a few basic steps. The first is *fractional distillation,* or the separation of the petroleum into fractions having different ranges of boiling points. These fractions are either used directly, purified further, or converted into products that are more desirable. The fractions typically obtained from petroleum are summarized in Table 16.2.

Most petroleum is used to make gasoline, a mixture of hydrocarbons, one of which is octane, the compound Scott chose for study. Given our reliance on gasoline as a fuel for transportation, the fraction of petroleum that can be directly separated as gasoline is insufficient to meet demand. This is because more than half of the hydrocarbons in petroleum have more than 13 carbon atoms, the longest chain found in the gasoline fraction. To obtain larger amounts of gasoline, the process of *catalytic cracking* uses heat and a catalyst to break down the higher-boiling "gas oil" fraction. For example,

$$C_{14}H_{30} \xrightarrow{\text{heat, catalyst}} C_6H_{12} + C_8H_{18}$$

Even though benzene, $C_6H_6$, is more accurately represented with two resonance structures or with the circle inside its ring, you will commonly see it represented with just one resonance structure:

**ANIMATION:** Petroleum Refinery

The columns at a petroleum refinery separate various components of oil based on their different boiling points.

©Steve Allen/Brand X Pictures

Because natural gas has no odor, traces of methyl mercaptan, $CH_3SH$, are added to commercial supplies so gas leaks in homes and businesses can be detected. It has the odor of rotten eggs.

Recall from Chapter 12 that a catalyst acts to increase the rate of a reaction.

**TABLE 16.2** Petroleum Fractions

| Fraction | Boiling Range (°C) | Composition Range | Uses |
|---|---|---|---|
| natural gas | < 20 | $CH_4$ to $C_4H_{10}$ | fuel, petrochemicals |
| petroleum ether | 20–60 | $C_5H_{12}$, $C_6H_{14}$ | solvent |
| ligroin, or naphtha | 60–100 | $C_6H_{14}$, $C_7H_{16}$ | solvent, raw material |
| gasoline | 40–220 | $C_4H_{10}$ to $C_{13}H_{28}$, mostly $C_6H_{14}$ to $C_8H_{18}$ | motor fuel |
| kerosene | 175–325 | $C_8H_{16}$ to $C_{14}H_{30}$ | heating fuel and jet fuel |
| gas oil | > 275 | $C_{12}H_{26}$ to $C_{18}H_{38}$ | diesel fuel and heating fuel |
| lubricating oils and greases | high: viscous liquids | > $C_{18}H_{38}$ | lubrication |
| paraffin | high: melting point 50–60 | $C_{23}H_{48}$ to $C_{29}H_{60}$ | wax products |
| asphalt or petroleum coke | high: solid residue | | roofing, paving, fuel, reducing agent |

Note the wide range of boiling points in Table 16.2. Why do hydrocarbons boil at such different temperatures? Hydrocarbons differ in their physical characteristics as the number and arrangement of carbon atoms change. The strengths of the intermolecular forces that attract the molecules to one another account for observed differences in properties. For example, as the number of carbon atoms increases, the normal physical state of the hydrocarbon changes from gas to liquid to solid. What happens to the intermolecular forces? As the molecules get larger, they have greater surface area. Their London dispersion forces increase, so they interact more strongly and their boiling points increase. Hydrocarbons containing from 5 to 18 carbon atoms are generally liquid at ordinary environmental temperatures. A number of everyday hydrocarbon products are liquids. Gasoline is made up of the lighter hydrocarbons. Kerosene contains slightly heavier hydrocarbons, and lubricating oil contains even heavier hydrocarbons.

## 16.3  Acyclic Hydrocarbons

The simplest organic compounds are the open-chain, acyclic hydrocarbons: the alkanes, alkenes, and alkynes. The chains may be straight or branched. The alkanes are single-bonded and saturated. The alkenes and alkynes are unsaturated.

### Alkanes

An **alkane** is a saturated hydrocarbon with the general formula $C_nH_{2n+2}$. The alkanes are obtained either directly from petroleum or indirectly by *cracking,* which breaks larger hydrocarbons into smaller ones. When the value of $n$ is 1 through 4, the alkanes are gases under normal conditions. When $n$ equals 5 through 18, they are liquids; and when $n$ is greater than 18, they are solids. The melting points and boiling points increase as the number of carbon atoms increases. The increase results from the greater surface area and stronger intermolecular forces that accompany the larger molecular size. A graph of the change in boiling point as the length of the carbon chain increases is shown in Figure 16.3.

As the size of the molecule affects its intermolecular forces, so too does the shape. Straight-chain molecules can have more surface contact with one another than branched-chain molecules (Figure 16.4). The increased surface contact leads to increased London dispersion forces, with temporary dipoles attracting one another. When the molecules are more highly attracted, more energy has to be added to separate the molecules as they escape into the gas phase during boiling.

Consider some examples of alkanes that are consistent with the general formula $C_nH_{2n+2}$. If $n = 1$, $2n + 2 = 4$, and the formula is $CH_4$. If $n = 3$, $2n + 2 = 8$, so the formula is $C_3H_8$.

Straight-chain alkanes are not really straight, since each carbon has a tetrahedral geometry with 109.5° bond angles. A straight-chain alkane thus has a zigzag structure. For this reason, some chemists call these alkanes continuous-chain alkanes or unbranched-chain alkanes.

**FIGURE 16.3** The boiling points of straight-chain alkanes increase as the number of carbon atoms in the chain increases. Which type of intermolecular force present in alkanes is responsible for this trend?

## EXAMPLE 16.2 Intermolecular Forces in Alkanes

Scott found boiling points for two hydrocarbons having eight carbon atoms, represented here by their line structures. Explain the difference in boiling points of these two compounds:

bp = 125.6°C            bp = 117.6°C

A        B

C

**FIGURE 16.4** Straight-chain alkanes (A) have more surface contact between molecules than branched-chain alkanes (B and C) containing the same number of atoms. Thus, straight-chain alkanes experience greater London dispersion forces than do branched-chain alkanes, resulting in higher boiling points. The boiling points of these alkanes, all with the formula $C_6H_{14}$, are (A) 69.0°C, (B) 60.3°C, and (C) 49.7°C.

### Solution:

Because these molecules contain only carbon and hydrogen, they are nonpolar and they experience only London dispersion forces. They have the same molecular formula, so they have the same number of electrons. The greater surface contact between the linear molecules (left) provides more chances for temporary dipoles to form and interact. Branched-chain molecules (right) have less surface contact than straight-chain molecules, and they don't fit next to one another as well. Thus, the branched molecule has the smaller intermolecular forces. It boils at a lower temperature because overcoming its intermolecular forces requires a lesser input of energy.

### Consider This 16.2

Many of the smaller hydrocarbons are volatile (evaporate easily) due to their relatively weak intermolecular forces. Which of the two compounds in the example would you expect to have the higher vapor pressure when compared at the same temperature?

### Practice Problem 16.2

Scott also found a boiling point for another hydrocarbon having eight carbon atoms, represented by this line structure:

bp = 99.3°C

Explain why this boiling point is lower than the other two given in the example.

**Further Practice:** Questions 16.23 and 16.24 at the end of the chapter

**TABLE 16.3** Names of Simple Alkanes

| Alkane Name | Formula ($C_nH_{2n+2}$) |
|---|---|
| methane | $CH_4$ |
| ethane | $C_2H_6$ |
| propane | $C_3H_8$ |
| butane | $C_4H_{10}$ |
| pentane | $C_5H_{12}$ |
| hexane | $C_6H_{14}$ |
| heptane | $C_7H_{16}$ |
| octane | $C_8H_{18}$ |
| nonane | $C_9H_{20}$ |
| decane | $C_{10}H_{22}$ |

**Alkane Structure and Nomenclature**   The alkanes are named with a prefix denoting the number of carbon atoms in the longest chain followed by the ending *−ane*. The four shorter alkanes—methane ($CH_4$), ethane ($C_2H_6$), propane ($C_3H_8$), and butane ($C_4H_{10}$)—have special names, while the longer alkanes use prefixes derived from the Greek or Latin numbers, such as octane for $C_8H_{18}$. The names of the first 10 alkanes are listed in Table 16.3.

Each successive alkane in the table can be considered to be derived by insertion of one additional $−CH_2−$ group between the first carbon atom and one of its attached hydrogen atoms in the smaller alkane. For example, insertion of a $CH_2$ group into methane gives ethane, $CH_3−CH_2−H$ or $CH_3−CH_3$. The third alkane is propane ($C_3H_8$), which has a second $CH_2$ group inserted to give $CH_3−CH_2−CH_2−H$ or $CH_3−CH_2−CH_3$. The next molecule in the series is butane ($C_4H_{10}$ or $CH_3−CH_2−CH_2−CH_2−H$ or $CH_3−CH_2−CH_2−CH_3$).

Alkanes in which the carbon atoms form a continuous chain, with no branches, are called *normal alkanes* or *straight-chain alkanes*. The alternate arrangements are called *branched-chain alkanes* and are isomers of the straight-chain compounds. Thus—unlike the first three alkanes—butane has two different structures, or isomers. The first is normal butane (*n*-butane or just butane); it is an unbranched chain of four carbon atoms. The second isomer is 2-methylpropane (commonly named isobutane) and is a chain of three carbon atoms with the added $CH_2$ group inserted at a center carbon:

**ANIMATION: Isomers**

The number of isomers increases rapidly as the number of carbon atoms increases. Pentane, $C_5H_{12}$, has three isomers, $C_{10}H_{22}$ has 75 isomers, and $C_{20}H_{42}$ has 366,319 isomers. Because of the large number of possible isomers, alkanes are named in accordance with a systematic nomenclature established by the International Union of Pure and Applied Chemistry (IUPAC). If common names were applied to all of these isomers, chemists would never be able to communicate efficiently. Consider the molecules and their IUPAC names given in Table 16.4. Can you determine the rules used for naming these compounds?

Did you recognize how to name side chains? Did you figure out how to use a numbering system to indicate position? If you need help with the rules, consult Section 16.9. Try to apply your rules to name an alkane with one branch on the chain.

**TABLE 16.4** Formulas and Names of Some Alkanes

| Structural Formula | Name |
|---|---|
| | 2-methylpentane |
| | 3-methylpentane |
| | 3-methylhexane |
| | 3-methylhexane |
| | 2,4-dimethylhexane |
| | 4-ethyl-2-methylhexane |
| | cyclohexane |

---

**EXAMPLE 16.3** **Names of Alkanes with a Single Branch**

Name the following compound—known commonly as isopentane—according to the IUPAC nomenclature rules:

$$\underset{\displaystyle H_3C-CH-CH_2-CH_3}{\overset{\displaystyle CH_3}{|}}$$

**Solution:**
*Step 1.* The longest continuous carbon chain has four carbons, so the name is based on butane.
*Step 2.* The carbons in this chain are numbered from the end closer to a branch. The only branch occurs at position 2.
*Step 3.* The branch occurring at position 2 is —$CH_3$, or methyl, named for the alkane it is derived from, methane.
*Step 4.* The name is 2-methylbutane.

**Consider This 16.3**

There are two isomers of 2-methylbutane. What are their structures and how would you name them?

**Practice Problem 16.3**

Name the following compound—known commonly as isobutane—according to the IUPAC nomenclature rules:

$$\begin{array}{c} CH_3 \\ | \\ H_3C-CH-CH_3 \end{array}$$

**Further Practice:** Questions 16.31 and 16.32 at the end of the chapter

The *iso–* prefix is often used in common names for isomers of straight-chain alkanes, as in isobutane. It is also used in the IUPAC system to name alkyl groups that have the attachment at the second to last carbon of the branch. For example, the isopropyl group is

$$\begin{array}{c} | \\ CH \\ \diagup \quad \diagdown \\ H_3C \quad CH_3 \end{array}$$

When naming the branches on alkanes, look for a group derived from an alkane by removal of one hydrogen atom, with a general formula $C_nH_{2n+1}$. Such a group is an **alkyl group.** The names of the alkyl groups are similar to those of the alkanes (given in Table 16.3), except that the *−ane* ending is replaced by *−yl*. Thus, $−CH_3$ is methyl and $−CH_2CH_3$ (or $−C_2H_5$) is ethyl.

If an alkane contains more than one branch, each is named separately and in alphabetical order. Consider the following molecule:

$$\begin{array}{c} CH_3 \qquad H_2C-CH_3 \\ | \qquad\qquad | \\ H_3C-CH-CH_2-CH-CH_2-CH_3 \end{array}$$

The longest chain is hexane. To specify the positions of the branches, we have to number the chain. As the molecule is drawn, we can number starting from the left or right, but only one is correct:

$$\begin{array}{c} CH_3 \qquad H_2C-CH_3 \\ | \qquad\qquad | \\ H_3\overset{1}{C}-\overset{2}{C}H-\overset{3}{C}H_2-\overset{4}{C}H-\overset{5}{C}H_2-\overset{6}{C}H_3 \end{array}$$

$$\begin{array}{c} CH_3 \qquad H_2C-CH_3 \\ | \qquad\qquad | \\ H_3\overset{6}{C}-\overset{5}{C}H-\overset{4}{C}H_2-\overset{3}{C}H-\overset{2}{C}H_2-\overset{1}{C}H_3 \\ \text{Incorrect} \end{array}$$

Under the IUPAC system, numbering starts at the end of the chain that gives the lower numbers to the branch positions. In this case numbering starts from the left. A methyl ($−CH_3$) branch occurs at position 2 and an ethyl ($−CH_2CH_3$) branch occurs at position 4. Alkyl groups are named alphabetically so the ethyl group is listed before the methyl group. The molecule is named 4-ethyl-2-methylhexane.

In a molecule containing more than one of the same group, a prefix is used to indicate the number occurances of that group. For example, *di–* means two and *tri–* means three. A six-carbon chain with two methyl group on carbons 2 and 3 is named 2,3-dimethylhexane. An isomer with two methyl groups on carbon 3 is named 3,3-dimethylhexane.

Notice that the third and fourth examples listed in Table 16.4 have the same name. Do these structures represent the same compound? Which end of each structure is used to start numbering the chain? Convince yourself that these two structures represent the same compound. The IUPAC system ensures that each compound has a unique name. The following examples will help you learn the basic rules for the IUPAC system. If you need help with the rules, consult Section 16.9.

**Student Hot Spot**

Student data indicate you may struggle with writing names for alkanes and drawing structures from their names. Access the SmartBook to view additional Learning Resources on this topic.

---

**EXAMPLE 16.4** | **Names of Alkanes with Multiple Branches**

Name the isomer of $C_{12}H_{26}$ that has the following structure:

$$\begin{array}{c} H_3C-CH_2 \qquad H_2C-CH_2-CH_3 \\ | \qquad\qquad\qquad | \\ H_3C-CH_2-C-CH_2-CH-CH_3 \\ | \\ CH_3 \end{array}$$

**Solution:**

The longest continuous chain has eight carbons, as shown by the highlighted part of the structure, so the compound is named as a derivative of octane.

$$H_3C-CH_2 \quad H_2\overset{6}{C}-\overset{7}{C}H_2-\overset{8}{C}H_3$$
$$\underset{1}{H_3}\overset{2}{C}-\overset{3}{C}H_2-\overset{3}{C}-\overset{4}{C}H_2-\overset{5}{C}H-CH_3$$
$$\underset{}{C}H_3$$

Numbering the longest chain from the left gives lower numbers to the branch positions at carbons 3 and 5. (Numbering the longest chain from the right would incorrectly position the branches at carbons 4 and 6.) There are two methyl groups, one at carbon 3 and one at carbon 5, and one ethyl group at carbon 3. The ethyl group should be named first, because it occurs first alphabetically. The methyl groups can be named together, using the carbon numbers to indicate their positions. The full name is thus 3-ethyl-3,5-dimethyloctane.

### Consider This 16.4

A student tried to name the compound represented here as 5-methylhexane, and then a second student tried to name it 1,1-dimethylpentane. What's wrong with both names, and what is the correct name of this compound?

### Practice Problem 16.4

Name the isomer of $C_{10}H_{22}$ that has this structure:

$$\overset{CH_3}{\underset{|}{}} \qquad \overset{H_2C-CH_3}{\underset{|}{}}$$
$$H_3C-\overset{|}{C}-CH_2-CH_2-\overset{|}{C}H-CH_3$$
$$\underset{|}{C}H_3$$

**Further Practice:** Questions 16.33 and 16.34 at the end of the chapter

If you are given a name, can you figure out the structure of the compound? You can use the rules in reverse to yield the structural formula from the IUPAC name.

---

| EXAMPLE 16.5 | **Structural Formulas from Names of Alkanes** |
|---|---|

Although Scott found three isomers having the chemical formula $C_8H_{18}$, there are actually more. One is 2,2,3,3-tetramethylbutane. Draw a structural formula and line structure for this molecule.

**Solution:**

The molecule name contains butane, so the longest chain must contain four carbon atoms:

$$C-C-C-C$$

Of the four methyl groups ($-CH_3$), two are attached to carbon 2. The other two are attached to carbon 3.

$$\overset{H_3C}{\underset{H_3C}{}} \overset{CH_3}{\underset{CH_3}{}}$$
$$C-\overset{|}{C}-\overset{|}{C}-C$$

The octane rating is based on the relative difference in the ability of a compound in gasoline to cause knocking in an engine. The highly branched structure of 2,2,4-trimethylpentane gives it the characteristic of burning smoothly with little knocking in an engine. This compound has an octane rating of 100. In contrast, the unbranched structure of n-heptane causes an engine to knock badly, and n-heptane has been assigned an octane rating of 0. Gasoline is a mixture of compounds of differing degrees of branching, so the octane rating of most gasoline falls somewhere in between this range and is usually around 90.

Finally we add enough hydrogen atoms to give four bonds to each carbon atom:

$$\begin{array}{ccc} & H_3C & CH_3 \\ & | & | \\ H_3C-&C-C&-CH_3 \\ & | & | \\ & H_3C & CH_3 \end{array}$$

The line structure can then be based on this structural formula, but with the C—H bonds and the atom symbols removed:

### Consider This 16.5

There are many isomers with the formula $C_8H_{18}$, but only three of them have seven carbons in their longest chain. Name and draw the line structures for these compounds.

### Practice Problem 16.5

Another isomer of $C_8H_{18}$ is 2,2,4-trimethylpentane. This compound in gasoline causes little engine knocking and has the highest possible octane rating. Draw its structural formula.

**Further Practice:** Questions 16.35 and 16.36 at the end of the chapter

**FIGURE 16.5** Hexane does not react with brown bromine unless exposed to bright light. (A) When hexane and bromine are mixed in the absence of light, the mixture remains brown due to the presence of unreacted bromine. (B) In the presence of light, a reaction occurs in which the product is colorless.

(both): ©Tom Pantages

**Reactions of Alkanes**    Alkanes don't readily undergo chemical reactions, because both carbon-hydrogen and carbon-carbon single bonds are strong. Supplying heat or light, however, provides the energy needed to get alkanes to react. For example, in petroleum refining, heating in the presence of a catalyst cracks or breaks large alkanes into smaller molecules. Such reactions produce shorter-chain alkanes, cycloalkanes, alkenes, and molecular hydrogen.

Alkanes undergo oxidation in exothermic *combustion* reactions, in which they act as fuels and burn when ignited in the presence of oxygen gas to form carbon monoxide or carbon dioxide and water:

$$C_3H_8 + 5O_2 \longrightarrow 3CO_2 + 4H_2O$$

In the presence of sunlight or ultraviolet radiation, alkanes will react with halogens. As shown in Figure 16.5, light provides the energy to break the halogen-halogen bonds and initiate the reaction:

$$C_6H_{14} + Br_2 \xrightarrow{light} C_6H_{13}Br + HBr$$

### EXAMPLE 16.6    Reactions of Alkanes

Predict the products of the reaction of butane with bromine in sunlight.

**Solution:**

A bromine atom from $Br_2$ would substitute for a hydrogen atom in butane, $C_4H_{10}$. The substituted hydrogen atom would bind to the remaining bromine atom from $Br_2$. The products would be $C_4H_9Br$ and HBr.

### Consider This 16.6

What isomers would you expect to be produced from the reaction described in the example?

**Practice Problem 16.6**

Predict the products of the reaction of propane with bromine in the dark at room temperature.

**Further Practice:** Questions 16.37 and 16.38 at the end of the chapter

## Alkenes and Alkynes

Maya's molecule, retinal, contains two types of functional groups: a carbon-carbon double bond and a −CHO group. Later we'll look at the properties of the −CHO group. Here we will focus on the effect that double bonds have on the properties of hydrocarbons. In this section, we consider two such classes of hydrocarbons: alkenes, containing double bonds, and alkynes, containing triple bonds.

**Alkene Structure and Nomenclature**   **Alkenes** contain at least one C=C double bond. Those with just one double bond have the general formula $C_nH_{2n}$. They are generally obtained from petroleum by the cracking of alkanes, using heat and a catalyst. The simplest such reaction is

$$CH_3-CH_3 \xrightarrow{\text{heat, catalyst}} H_2C=CH_2 + H_2$$

The product is named ethene (IUPAC name), or ethylene (common name). An alkene molecule may also have more than one double bond. For example, $H_2C=CH-CH=CH_2$, with double bonds at carbons 1 and 3, is 1,3-butadiene.

Examine the structural formulas and names of some alkenes given in Table 16.5. Try to determine the rules for naming them.

Did you come up with some rules? How do you name the main chain? How do you designate the positions of double bonds? How do you designate the positions and names of branches? If you need help, consult Section 16.9. Use your rules to name some alkenes.

Ethene, $H_2C=CH_2$, seems to play a part in the ripening of fruit, so green fruit is often shipped under an atmosphere of ethene, so that ripening occurs in transit.

**TABLE 16.5** Formulas and Names of Some Alkenes

| Structural Formula | Name |
| --- | --- |
| $H_2C=CH_2$ | ethene |
| $H_2C=CH-CH_3$ | propene |
| $H_2C=CH-CH_2-CH_3$ | 1-butene |
| $H_3C-CH_2-CH=CH_2$ | 1-butene |
| $H_3C-CH=CH-CH_3$ | 2-butene |
| $\begin{array}{c}\quad CH_3\\\quad\vert\\H_3C-C=CH-CH_3\end{array}$ | 2-methyl-2-butene |
| $\begin{array}{c}\quad\quad CH_3\\\quad\quad\vert\\H_2C=CH-CH-CH_3\end{array}$ | 3-methyl-1-butene |
| $\begin{array}{c}\quad H_3C\quad CH_3\\\quad\vert\quad\;\vert\\H_2C=C-CH-CH_3\end{array}$ | 2,3-dimethyl-1-butene |

---

### EXAMPLE 16.7 | Names of Alkenes

Name this compound:

$$H_3C-CH-CH=CH-CH_3$$
$$\underset{\displaystyle CH_3}{|}$$

**Solution:**

The longest chain has five carbons, so the compound is a derivative of pentane, making the name pentene (because of the double bond). The carbon atoms are numbered from the end nearest the double bond:

$$\overset{5}{H_3C}-\overset{4}{CH}-\overset{3}{CH}=\overset{2}{CH}-\overset{1}{CH_3}$$
$$\underset{\displaystyle CH_3}{|}$$

The position of the double bond is after the second carbon from the right. Counting from that end, the methyl side chain is attached to the fourth carbon atom. The name is 4-methyl-2-pentene.

**Consider This 16.7**

Name an isomer of this compound that has the following structure:

**Practice Problem 16.7**

Name this compound:

$$H_3C-CH_2-C=CH_2$$
$$\underset{\displaystyle CH_3}{|}$$

**Further Practice:** Questions 16.41 and 16.42 at the end of the chapter

---

In isomers of alkenes, rotation is restricted around the C=C double bond. Because of the restricted rotation, the relative positions of groups or atoms attached to the carbons in the double bond are fixed. To have such *geometric isomers,* two different groups or atoms must be attached to each of the carbon atoms in the double bond. When this occurs, the two isomers differ in how the alkyl groups on both carbon atoms are arranged relative to one another. In the cis isomer, the alkyl groups are on the same side of the double bond. In the trans isomer, the alkyl groups are on opposite sides of the double bond. Thus, Maya found two different isomers of retinal—a cis form and a trans form—that differed in the positions of atoms attached to one of the C=C units in the molecule. A simpler example is 2-butene, with the two methyl groups on the same or opposite sides of the double bond:

How do line structures for 2-butene show the difference between its cis and trans isomers?

*cis*-2-butene    *trans*-2-butene

*cis*-2-butene    *trans*-2-butene

These two molecules have different properties because their structures are different. *cis*-2-Butene has a boiling point of 4°C, but that of *trans*-2-butene is 1°C. The cis isomer has stronger intermolecular forces because it is slightly polar, resulting in a higher boiling point.

**EXAMPLE 16.8** | **Cis and Trans Isomers**

Which of these molecules can have cis and trans isomers? Draw structures for any possible cis-trans isomers.

(a) $CH_3CH=CHCH_2CH_3$         (b) $(CH_3)_2C=CHCH_2CH_3$

**Solution:**

(a) Each double-bonded carbon has two groups attached to it, and the two groups in each case are different from each other. Thus, it is possible to have both cis and trans isomers:

$$\begin{array}{cc} \underset{H_3C}{\overset{H}{\diagdown}}C=C\underset{CH_2-CH_3}{\overset{H}{\diagup}} & \underset{H_3C}{\overset{H}{\diagdown}}C=C\underset{H}{\overset{CH_2-CH_3}{\diagup}} \end{array}$$

(b) Each double-bonded carbon has two groups attached to it, but the first (left) carbon has identical groups. There is only one possible structure, so there are no cis and trans isomers.

$$\underset{H_3C}{\overset{H_3C}{\diagdown}}C=C\underset{CH_2CH_3}{\overset{H}{\diagup}}$$

**Consider This 16.8**

Name and draw the line structures of the cis and trans isomers that have the formula given in part (a) of the example.

**Practice Problem 16.8**

Which of these molecules can have cis and trans isomers? Draw structures for any possible isomers.

(a) $CH_3CH_2CH=CH_2$         (b) $CH_3CH_2CH=CHCH_2CH_3$

**Further Practice:** Questions 16.43 and 16.44 at the end of the chapter

**Student Hot Spot**

Student data indicate you may struggle with determining names and structures for alkenes and alkynes. Access the SmartBook to view additional Learning Resources on this topic.

**Alkyne Structure and Nomenclature**    **Alkynes** contain a C≡C triple bond and have the general formula $C_nH_{2n-2}$. Alkynes are also called acetylenes, from the common name of the simplest alkyne, HC≡CH. Ethyne (IUPAC name) or acetylene (common name) is generally prepared from calcium carbide (Figure 16.6).

$$CaC_2(s) + 2H_2O(l) \longrightarrow HC=CH(g) + Ca(OH)_2(s)$$

Alkynes are named with the same rules as alkenes, but using the ending *–yne* instead of *–ene*. Thus, $CH_3C≡CCH_3$ is 2-butyne and $HC≡CCH_2CH_3$ is 1-butyne. There are no cis and trans isomers for alkynes.

**Reactions of Alkenes and Alkynes**    An *addition reaction* is one in which extra atoms or groups are added to a simpler hydrocarbon, forming new single bonds with electrons from a multiple bond. Alkenes and alkynes typically undergo addition reactions across the double or triple bond:

$$CH_2=CH_2 + HBr \longrightarrow CH_3CH_2Br$$

Many of these reactions require a catalyst in order to occur at an appreciable rate. For example, if ethene is placed in an environment of $H_2$ gas, no reaction occurs. In the presence of a metal catalyst such as platinum, however, ethene converts to ethane:

$$CH_2=CH_2 + H_2 \xrightarrow{\text{Pt}} CH_3CH_3$$

**FIGURE 16.6**  Before portable electric lamps became available, miners and cave explorers used lamps in which water dripped onto calcium carbide. The acetylene that was produced was burned to give light. Changing the rate of addition of water adjusted the size of the flame.
©Charles D. Winters/Science Source

**FIGURE 16.7** Many consumer products are made of polyethylene.
©Tom Pantages

Another important reaction is *polymerization,* in which small molecules containing double bonds combine together to form large molecules called *polymers.* Polymers are used to make a variety of useful items including plastics and rubber tires. Polymers are made of many repeating units called *monomers.* The formation of polyethylene is an example of an *addition polymerization* reaction, in which monomers add to one another without losing any atoms. The joining of monomer units at the position of the monomer double bond forms the polymer, as shown here for the polymerization of ethylene to form polyethylene:

$$H_2C{=}CH_2 + H_2C{=}CH_2 + H_2C{=}CH_2 + \cdots \longrightarrow$$
$$-H_2C{-}CH_2{-}H_2C{-}CH_2{-}H_2C{-}CH_2{-} \cdots$$

This equation can be generalized for any number ($n$) of monomers as follows:

$$nH_2C{=}CH_2 \longrightarrow {\left(\!CH_2{-}CH_2\!\right)}_n$$

The groups shown in the parentheses are repeated in a chain $n$ times, where $n$ is a large number.

Polyethylene is a useful polymer. It can be heated to soften and shape it, and then cooled to a plastic substance with its original properties. That's why polyethylene bags can be heat-sealed. Polyethylene films hold in or keep out water vapor but allow the passage of oxygen. Polyethylene is also a good electrical insulator. It is commonly used in soda bottles, milk cartons, produce bags, food wrap, wire coatings, and more (Figure 16.7).

Propylene is polymerized similarly to ethylene, but the polymer, polypropylene, is harder and less flexible than polyethylene due to the presence of its methyl side chains.

$$nH_2C{=}\overset{\overset{\textstyle CH_3}{|}}{CH} \longrightarrow {\left(\!CH_2{-}\overset{\overset{\textstyle CH_3}{|}}{CH}\!\right)}_n$$

Another related material is the polymer of styrene, $C_6H_5CH{=}CH_2$:

$$nH_2C{=}CH \longrightarrow {\left(\!CH_2{-}CH\!\right)}_n$$

Polystyrene is used in packaging, food and household containers (Figure 16.8), combs, and coatings for television lead-in wire. It is rigid, clear, and nontoxic, and it won't dissolve in most aqueous solutions. It can be prepared in the form of expandable beads, which can be molded and shaped. The result is styrofoam.

**FIGURE 16.8** Polystyrene is used in products ranging from clear plastic to styrofoam.
©Tony Freeman/PhotoEdit, Inc

---

### EXAMPLE 16.9    Reactions of Alkenes

Write an equation to describe the reaction between 2-hexene and $H_2$.

**Solution:**

This is an example of an addition reaction, also called a hydrogenation reaction when the molecule being added is hydrogen. The hydrogen molecule is split apart with the aid of a catalyst, and the hydrogen atoms are added to the carbon atoms involved in the double bond:

$$CH_3CH{=}CHCH_2CH_2CH_3 + H_2 \xrightarrow{\text{Pt}} CH_3CH_2CH_2CH_2CH_2CH_3$$

**Consider This 16.9**

Dienes are alkenes that contain two double bonds. An example is 1,3-butadiene. Write a balanced equation for the reaction of this compound with $H_2$ (in the presence of a platinum catalyst) to produce an alkane.

**Practice Problem 16.9**
Write an equation to describe the reaction that would occur between 2-butene and HBr.

**Further Practice:** Questions 16.47 and 16.48 at the end of the chapter

Reactions of alkynes are similar to the reactions of alkenes just discussed. Both also undergo combustion reactions. An example is the combustion of acetylene, $C_2H_2$, when it's used as a fuel in a welding torch:

$$2H-C\equiv C-H + 5O_2 \longrightarrow 4CO_2 + 2H_2O$$

# 16.4 Cyclic Hydrocarbons

## Cycloalkanes and Cycloalkenes

Imagine a hydrocarbon chain of single-bonded carbons with a hydrogen atom removed on each end and the two carbon atoms at the end then bonded together. The result would be the ring structure of a saturated cyclic hydrocarbon, or **cycloalkane,** with the general formula $C_nH_{2n}$. The simplest cycloalkane is cyclopropane, $C_3H_6$. It is a colorless, flammable gas that is explosive in air. It was once used as a general anesthetic.

If the molecular formula given for a saturated hydrocarbon is $C_9H_{18}$, would you predict this to be an alkane or a cycloalkane?

cyclopropane

Many cyclic molecules, especially those in which the rings are small, are strained because they cannot adopt the normal tetrahedral bond angles (109.5°) around carbon. Thus, they tend to be somewhat more reactive than the alkanes. The most common cycloalkanes are the low-boiling liquids cyclopentane, $C_5H_{10}$, and cyclohexane, $C_6H_{12}$:

cyclopentane                          cyclohexane

Cyclopentane is a flammable, colorless liquid that smells like gasoline. Cyclohexane has similar properties and is used as a solvent for lacquers and resins and in paint removers.

Some alkenes also form ring structures. Molecules containing multiple double bonds and double bonds in ring structures are common in nature. We saw the example of *cis-* and *trans-*retinal that Maya studied; it has five carbon-carbon double bonds, one of which is in a cyclohexene ring:

cyclohexene

Some other common examples in nature include vitamin A, required for good vision, and limonene, found in the oils in citrus fruits:

vitamin A                    limonene

## Aromatic Hydrocarbons

Omar was studying aspirin, which contains a six-carbon ring having alternating single and double bonds. Aisha was interested in amphetamine, which also contains such a six-carbon ring. Although each of these molecules has additional functional groups, both are classified as aromatic. Common aromatic hydrocarbons are cyclic hydrocarbons with alternating single and double carbon-carbon bonds.

The actual structure of aromatic hydrocarbons such as $C_6H_6$ has some of the electrons in the double bonds spread out, or delocalized, over the entire ring system, as discussed in Section 16.2. Double bonds with electrons delocalized over the entire ring are less reactive than ordinary double bonds. The resulting inertness is referred to as *aromatic character*. The delocalized bonding in benzene, $C_6H_6$, is usually represented by a circle inside the structure, and the hydrogen atoms, one attached to each carbon atom, are not shown:

Some aromatic compounds have overlapping multiple six-carbon rings with delocalized bonding. An example is naphthalene, once widely used as a moth protectant in closets:

Aspirin and amphetamine can be represented by the following structural formulas:

aspirin                    amphetamine

**Nomenclature of Aromatic Hydrocarbons**   Derivatives of benzene are generally named with benzene as the base name. However, many of these molecules have special names that are retained by the IUPAC nomenclature system. The following molecule, for example, is usually called toluene instead of methylbenzene. A methyl group is substituted for one hydrogen atom on the benzene ring:

Toluene, $C_6H_5CH_3$, is a flammable, colorless liquid that smells like benzene. It is used as a solvent and in the manufacture of dyes, explosives, and numerous organic compounds.

Examine the names of the aromatic compounds in Table 16.6. Can you come up with some rules for naming them?

**TABLE 16.6** Some Formulas and Names of Aromatic Hydrocarbons

| Structural Formula | Name | Structural Formula | Name |
|---|---|---|---|
| $CH_3$ (benzene ring) | toluene (or methylbenzene) | $CH_3$, $Cl$ (benzene ring) | 2-chlorotoluene (or 1-chloro-2-methylbenzene) |
| $NH_2$ (benzene ring) | aniline (or aminobenzene) | $NH_2$, $Cl$ (benzene ring) | 3-chloroaniline (or 1-amino-3-chlorobenzene) |
| $OH$ (benzene ring) | phenol (or hydroxybenzene) | $OH$, $Cl$ (benzene ring) | 4-chlorophenol (or 1-chloro-4-hydroxybenzene) |
| $CH_2CH_3$ (benzene ring) | ethylbenzene | $CH_3$, $O_2N$, $NO_2$, $NO_2$ (benzene ring) | 2,4,6-trinitrotoluene (abbreviated TNT) |
| $Cl$ (benzene ring) | chlorobenzene | | |

Did you come up with some rules? How do you name the ring? How do you designate the positions and names of attached groups? If you need help, consult Section 16.9. See if you can apply your rules to naming some aromatic compounds.

---

**EXAMPLE 16.10**    **Names of Aromatic Compounds**

Name this aromatic compound:

$Cl$ (benzene ring) $CH_3$

**Solution:**

A benzene ring with an attached —$CH_3$ group has the special name toluene. To identify the location of the attached —Cl, we number the ring starting from the —$CH_3$ group. The result is 4-chlorotoluene. This compound can also be named as a derivative of benzene. In this case we start numbering the ring at the first attached group (in alphabetical order), chloro, so the methyl group is at carbon 4. The name is 1-chloro-4-methylbenzene.

**Consider This 16.10**

If the —$CH_3$ group in the compound in the example problem is replaced by an —$NH_2$ group, how would you name that compound?

**Practice Problem 16.10**

Name this aromatic compound:

OH

Cl      $NO_2$

**Further Practice:** Questions 16.51 and 16.52 at the end of the chapter

**Reactions of Aromatic Hydrocarbons**    Benzene and other aromatic hydrocarbons lack the reactivity of the double bond in alkenes. Normally, benzene does not react at all, but with the right catalyst, it can undergo a *substitution reaction* in which some functional group replaces one or more hydrogens on the benzene ring. For example, benzene reacts with chlorine in the presence of the catalyst $FeCl_3$ to form chlorobenzene:

Cl

$+ Cl_2$  $\xrightarrow{FeCl_3}$   $+ HCl$

Benzene reacts with chloromethane in the presence of the catalyst $AlCl_3$ to form toluene:

$CH_3$

$+ CH_3Cl$  $\xrightarrow{AlCl_3}$   $+ HCl$

## 16.5 Alcohols and Ethers

### Alcohols

Jennifer's molecule, menthol, is an example of an **alcohol,** which is characterized by the functional group —OH:

$H_3C$      $CH_2$          H
     CH        CH       O
$H_2C$           CH      $CH_3$
     $CH_2$       CH
               $CH_3$

OH

Alcohols are useful substances, and some of them are among the top 50 chemicals produced in the United States. Alcohols produced in large amounts include methanol ($CH_3OH$), ethylene glycol ($HOCH_2CH_2OH$), isopropanol or 2-propanol ($C_3H_7OH$), and ethanol ($C_2H_5OH$).

The $-OH$ group makes alcohols polar, and it can participate in hydrogen-bonding interactions. As a result, even the lower-molar-mass alcohols are liquids, while alkanes of similar size are gases. The short-chain alcohols are water soluble, but solubility decreases as the hydrocarbon portion of the molecule increases in size. For example, $C_4H_9OH$ is only slightly soluble in water, and alcohols containing five or more carbon atoms are essentially insoluble in water. Alcohols of similar size are soluble in one another. For example, methanol dissolves in ethanol. Methanol is a colorless liquid so toxic that breathing its vapors can blind or kill. It is often added to nonbeverage ethanol, to make it unfit for drinking. This modified ethanol can be sold for scientific use and as an additive for gasoline without the federal taxes levied on beverage alcohol. In this role, methanol is called a denaturant, and the modified ethanol is called *denatured alcohol.*

Because of the polarity and hydrogen bonding of the alcohols, they have significantly higher boiling points than alkanes of comparable mass. For example, the boiling point of $CH_3CH_2OH$ is 78.5°C, but that of $CH_3CH_2CH_3$ is −42.1°C. The alkane molecules are attracted to one another only by London dispersion forces, while alcohol molecules are attracted to one another by additional dipole forces and hydrogen-bonding forces.

Carbohydrates such as glucose and fructose contain multiple alcohol functional groups:

> Ethylene glycol, or 1,2-ethanediol, $HO-CH_2-CH_2-OH$, is an alcohol containing two $-OH$ groups. This colorless liquid is commonly used as an antifreeze in automobile radiators.

glucose                    fructose

Even though these are large molecules, they are very soluble in water because there are multiple alcohol functional groups throughout both molecules, allowing them to hydrogen bond effectively with many water molecules. Also notice the presence of the $-O-$ atom bonded to two carbon atoms in these simple sugars. Larger sugars such as sucrose and maltose contain two glucose and/or fructose units connected by an oxygen atom. Starch and cellulose are larger polymers of glucose, also connected in a similar way. Humans and other vertebrate animals have the enzymes necessary to metabolize starch, converting it to glucose, which cells use as their primary energy source.

> We'll discuss carbohydrates in more detail in Chapter 17.

**Nomenclature of Alcohols**   Examine the names and formulas of some alcohols given in Table 16.7. Try to determine the rules for naming these compounds.

Did you come up with some rules? How do you name the chain? How do you designate the position of any branches? How do you designate the position of the $-OH$ group? If you need help, consult Section 16.9. Try to use your rules to name some alcohols.

**TABLE 16.7** Formulas and Names of Some Alcohols

| Structural Formula | Name |
| --- | --- |
| $H_3C-OH$ | methanol |
| $H_3C-CH_2-OH$ | ethanol |
| $H_3C-CH_2-CH_2-OH$ | 1-propanol |
| $H_3C-\underset{\underset{OH}{\vert}}{CH}-CH_3$ | 2-propanol |
| $H_3C-CH_2-CH_2-CH_2-OH$ | 1-butanol |
| $H_3C-CH_2-\underset{\underset{OH}{\vert}}{CH}-CH_3$ | 2-butanol |
| $H_3C-\underset{\underset{CH_3}{\vert}}{CH}-CH_2-\underset{\underset{OH}{\vert}}{CH}-CH_3$ | 4-methyl-2-pentanol |

---

**EXAMPLE 16.11**   **Names of Alcohols**

Name this alcohol:

$$H_3C-\underset{\underset{OH}{\vert}}{CH}-\overset{\overset{CH_3}{\vert}}{CH}-CH_3$$

**Solution:**

The longest chain containing the —OH group has four carbons, so this is a deriva-tive of butane. The —*ane* ending is changed to —*anol,* making the name butanol. The carbon atoms in the longest chain are numbered from the end nearest the —OH group. The —OH group is attached to the second carbon, so the chain name is 2-butanol. A methyl group is attached to carbon 3, so the full name is 3-methyl-2-butanol.

**Consider This 16.11**

An isomer of this compound is named 2-methyl-1-butanol. Draw its line structure.

**Practice Problem 16.11**

Name this alcohol:

$$H_3C-\overset{\overset{CH_3}{\vert}}{CH}-\underset{\underset{OH}{\vert}}{CH}-\overset{\overset{CH_3}{\vert}}{CH}-CH_3$$

**Further Practice:** Questions 16.61 and 16.62 at the end of the chapter

Propofol is a phenolic compound used to induce anesthesia:

An —OH group may be attached directly to an aromatic hydrocarbon group, as in $C_6H_5OH$:

phenol

This molecule is known as phenol. The chemistry of the —OH group changes greatly when it is attached directly to an aromatic ring, and such compounds are classified not as alcohols, but as phenols.

## Ethers

An **ether** is an organic molecule that contains an —O— atom bonded to two carbon atoms of alkyl or aryl groups. Ethers are related to the alcohols, but they have two hydrocarbon groups attached to an oxygen: R—O—R′. If different hydrocarbon groups are present, they can be designated R— and R′—. The names of the ethers include the names of the attached alkyl groups, listed alphabetically and followed by the word *ether,* as in diethyl ether ($CH_3CH_2$—O—$CH_2CH_3$). The ethers tend to be more volatile than the corresponding alcohols. Consider the boiling points of an alkane, an ether, and an alcohol of comparable mass:

| Formula | $CH_3CH_2CH_3$ | $CH_3OCH_3$ | $CH_3CH_2OH$ |
|---|---|---|---|
| Boiling point (°C) | −42.1 | −23.6 | 78.5 |

Ethers are polar molecules, so they have stronger intermolecular forces and higher boiling points than alkanes. With no hydrogen atom attached to the oxygen atom in an ether, there is no hydrogen bonding, so it has a lower boiling point than the comparable alcohol.

One of the molecules first used as a general anesthetic was diethyl ether:

$$H_3C{-}CH_2{-}O{-}CH_2{-}CH_3$$

> When generalizing about a class of compounds, we sometimes use the R— notation to represent the hydrocarbon portion of a molecule.

> MTBE, or methyl tertiary-butyl ether, is added to some formulations of gasoline to ensure more complete combustion:
>
> $$H_3C{-}O{-}C(CH_3)(CH_3){-}CH_3$$

> The term *resin* was originally used to describe the thick, odoriferous sap exuded by cuts in the bark of pine trees. It is now used in a more generic sense to include the many synthetic, amorphous solid materials that are used in plastics, varnishes, inks, and medicines.

## 16.6 Aldehydes and Ketones

### Aldehydes

Earlier we discussed Maya's favorite molecules, *cis-* and *trans*-retinal, as examples of alkenes. However, these molecules contain the —CHO functional group, so they are also aldehydes. An **aldehyde** contains a C=O (carbonyl) group in which at least one hydrogen atom is attached to the carbon atom. Aldehydes have the general formula R—CHO, in which the hydrogen and the oxygen are both bonded to the carbon, and the oxygen-carbon bond is a double bond:

$$R{-}\overset{\overset{\displaystyle O}{\|}}{C}{-}H$$

The most common aldehydes are formaldehyde, HCHO, and acetaldehyde, $CH_3CHO$. Formaldehyde (common name), or methanal (IUPAC name), is a colorless gas with an irritating odor. It is highly soluble in water and is commonly sold as a 37% solution known as *formalin.* Formalin causes the coagulation of proteins, making it useful as a preservative of tissues. Formaldehyde is also used with phenol to form polymer resins, which are used in plastics and as adhesives in plywood and particleboard. Urea-formaldehyde foam (Figure 16.9) was once widely used for house insulation, until it was banned because of the toxic properties of formaldehyde.

Acetaldehyde (common name), or ethanal (IUPAC name), $CH_3CHO$, is a volatile, colorless liquid that is soluble in water. It has a characteristic, pungent odor

**FIGURE 16.9** Urea-formaldehyde foam used for insulation is formed by mixing two liquids. They react to yield a polymer and a gas. The gas causes the polymer to expand and form a foam. In a short time the foam hardens into a rigid solid containing a lot of trapped gas.
©Jim Birk

much like that of cut green apples. It is toxic and flammable. Acetaldehyde is used in perfumes, aniline dyes, plastics, and synthetic rubber. Aldehydes are easily oxidized, so they make good reducing agents. The Tollens test uses this property to indicate the presence of aldehydes (Figure 16.10).

### Ketones

Ketones resemble aldehydes in that both contain a carbon double-bonded to an oxygen, or C=O. In an aldehyde, this group is located on an end carbon. In a **ketone,** however, the organic groups, often alkyl groups—such as methyl or ethyl here designated R and R′—lie on either side of the C=O group:

$$\underset{\overset{\underset{\displaystyle R-C-R'}{|}}{\|}}{O}$$

The simplest and most important of the ketones is acetone, $(H_3C)_2C=O$, which is officially named propanone. Acetone is a colorless, highly flammable liquid with a characteristic pungent odor. It is readily soluble in water. Acetone is widely used as a solvent for fats, oils, waxes, resins, rubber, plastics, lacquers, varnishes, and rubber cements. It is the principal ingredient in some fingernail polish removers.

## 16.7  Carboxylic Acids and Esters

### Carboxylic Acids

Omar's favorite molecule, aspirin, contains a carboxylic acid functional group. The characteristic functional group in a **carboxylic acid** molecule is a carbon atom double-bonded to an oxygen atom and also bonded to an —OH group, $-CO_2H$:

$$\underset{\overset{\underset{\displaystyle R-C-OH}{|}}{\|}}{O}$$

Carboxylic acids are typically weak acids, and their reactions often involve loss of $H^+$.

The simplest carboxylic acid is formic acid (methanoic acid), $HCO_2H$, which was first obtained from red ants. It is the chemical agent that causes ant bites, bee stings, and nettle stings to be so painful. Formic acid is a colorless, water-soluble liquid with a pungent odor. It is used in dyeing wool, dehairing and tanning animal hides, electroplating, coagulating rubber latex, and regenerating old rubber.

The most common carboxylic acid is acetic acid, $CH_3CO_2H$, which is produced as a solution in water, called vinegar, when cider ferments. Fermentation causes the sugar to form ethanol in the presence of catalytic yeast enzymes. In the presence of air, the ethanol oxidizes further to acetic acid:

$$CH_3CH_2OH + O_2 \xrightarrow{\text{yeast}} CH_3CO_2H + H_2O$$

Pure acetic acid is a liquid that has a pungent odor and burns the skin. It is completely soluble in water and is an excellent solvent for many organic compounds, including resins and oils. Acetic acid is used in the synthesis of rubber and various acetate plastics. It is used in printing calico and dyeing silk.

**FIGURE 16.10** During the Tollens test for aldehydes, silver(I) ion is reduced to metallic silver and the aldehyde is oxidized. Here the reaction with formaldehyde creates a silver mirror on the test tube. Ketones are not easily oxidized and do not react in this way.
©Martyn F. Chillmaid/SPL/Science Source

Many breathalyzer test devices use the oxidizing agent $K_2Cr_2O_7$ (an orange compound) to oxidize ethanol to acetic acid. In the presence of ethanol, $K_2Cr_2O_7$ is reduced to a green compound. The degree of color change indicates the concentration of ethanol in a person's breath.

Another example of a carboxylic acid is the lactic acid produced in muscles from the metabolism of glucose:

lactic acid

It causes muscle soreness when its concentration builds up. The lactic acid molecule also contains an alcohol functional group.

Another carboxylic acid is linoleic acid. It is a nutritionally essential fatty acid that is also used in manufacturing paints. It also contains two alkene functional groups:

linoleic acid

## Esters

An **ester** is a compound with the general formula $RCO_2R'$. In esters, one of the organic groups (either alkyl or an H atom) is directly attached to the carbon, another is bonded to one of the oxygens, and the other oxygen is double-bonded to the carbon:

$$R-\overset{\overset{\displaystyle O}{\|}}{C}-OR'$$

The aspirin that interests Omar is both a carboxylic acid and an ester:

A condensation reaction involves the combination of two molecules accompanied by the elimination of $H_2O$.

Esters can be formed by a *condensation reaction* of a carboxylic acid with an alcohol. Water is removed in the process, as shown here for the formation of ethyl acetate by the reaction of acetic acid with ethanol:

$$CH_3-C{\overset{O}{\underset{O-H}{\big\|}}} \ + \ HO-CH_2-CH_3 \ \rightleftharpoons \ CH_3-C{\overset{O}{\underset{O-CH_2-CH_3}{\big\|}}} \ + \ H_2O$$

These reactions are reversible and usually require an acid catalyst. Addition of a base neutralizes and removes the carboxylic acid, causing the equilibrium reaction to shift to the left. Most esters are volatile liquids with pleasant odors and flavors. The natural fragrances of flowers and plants often come from esters. Esters are often used as flavoring agents in synthetic and processed foods. Some examples of fruit-flavored esters and their characteristic odor are given in the following table:

Wintergreen oil comes from a class of wintergreen plants and contains about 98% of the ester methyl salicylate. It is used topically for muscle pain and as a flavoring agent for toothpaste.

| Ester | Formula | Odor |
|---|---|---|
| isoamyl acetate | $(CH_3)_2CHCH_2CH_2O_2CCH_3$ | banana |
| octyl acetate | $CH_3(CH_2)_6CH_2O_2CCH_3$ | orange |
| methyl butyrate | $CH_3O_2CCH_2CH_2CH_3$ | apple |

methyl salicylate

The common animal and vegetable oils and fats are esters, mostly based on an alcohol called glycerol (or 1,2,3-propanetriol). Glycerol reacts with carboxylic acids to form an ester called a *triglyceride:*

$$\begin{array}{l} H_2C-OH \\ | \\ HC-OH \ + \ 3CH_3-\overset{O}{\overset{\|}{C}}-O-H \ \longrightarrow \ \begin{array}{l} H_2C-O-\overset{O}{\overset{\|}{C}}-CH_3 \\ | \\ HC-O-\overset{O}{\overset{\|}{C}}-CH_3 \\ | \\ H_2C-O-\overset{O}{\overset{\|}{C}}-CH_3 \end{array} \ + \ 3H_2O \\ | \\ H_2C-OH \end{array}$$

Esters that form from the reaction of glycerol with high-molar-mass acids are liquid oils or solid fats, depending on the number of double bonds in the alkyl group of the acid. (Such carboxylic acids are called *fatty acids*.) If the fatty acid is saturated (no double bonds), the triglyceride is a solid, or *fat;* and if the fatty acid is highly unsaturated, with many carbon-carbon double bonds, the triglyceride is a liquid, or *oil*. The greater the unsaturation, the more liquidlike the triglyceride. Common fats include lard, tallow, and milk fat. Common oils are linseed, cottonseed, coconut, soybean, peanut, corn, and olive oils.

Fats and oils are used in the body to synthesize essential tissue constituents. The body also uses them as energy sources. Because they are insoluble, they can be deposited in cells and can remain there unchanged until energy is needed.

A solid ester forms from glycerol and stearic acid, $C_{17}H_{35}CO_2H$, which is a saturated fatty acid:

$$\begin{array}{l} H_2C-O-\overset{O}{\overset{\|}{C}}-C_{17}H_{35} \\ | \\ HC-O-\overset{O}{\overset{\|}{C}}-C_{17}H_{35} \\ | \\ H_2C-O-\overset{O}{\overset{\|}{C}}-C_{17}H_{35} \end{array}$$

Triglycerides are the principal form of fats circulating in the human bloodstream. Most body fat also is in the form of triglycerides. A simple blood test measures the amount of triglycerides in blood. It is one measure of the risk of developing coronary artery disease.

A liquid ester forms from glycerol and linoleic acid, $C_{17}H_{31}CO_2H$, which is an unsaturated fatty acid:

$$\begin{array}{l} H_2C-O-\overset{O}{\overset{\|}{C}}-C_{17}H_{31} \\ | \\ HC-O-\overset{O}{\overset{\|}{C}}-C_{17}H_{31} \\ | \\ H_2C-O-\overset{O}{\overset{\|}{C}}-C_{17}H_{31} \end{array}$$

Why do these two esters—one a solid and the other a liquid at room temperature—have different melting points? The answer lies in the structure of the carboxylic acids from which the esters were derived:

stearic acid

Stearic acid is found in a variety of foods including meat and cocoa butter. Linoleic acid is found in many different oils and is most abundant in safflower and grape seed oils.

linoleic acid

Linoleic acid is bent because of a cis arrangement around a carbon-carbon double bond. Which of these acids should have the greater intermolecular forces? Recall the arguments we made for greater intermolecular forces in straight-chain hydrocarbons than in branched-chain hydrocarbons. Since the straight-chain molecules can interact to a greater extent than ones with bent shapes, they have greater intermolecular forces and, as a result, higher melting points. The same logic applies here. Linoleic acid has a melting point of $-12°C$, while stearic acid has a melting point of $70°C$. Thus, the esters derived from them have melting points that differ in the same way.

The glyceride of linoleic acid is found in the oils of many seeds and vegetables. Stearic acid comes from animal fats and oils. Do all solid fats come from animals, while liquid fats come from vegetables? Generally, yes, but oils from vegetables can be converted in the laboratory into solids by the process of hydrogenation, in which hydrogen molecules add across the double bonds in the presence of an appropriate catalyst. This is how margarine is made.

$$\text{Hydrogenation:} \quad \underset{\text{unsaturated (oil)}}{C_3H_5(C_{17}H_{31}CO_2)_3} + 6H_2 \longrightarrow \underset{\text{saturated (solid)}}{C_3H_5(C_{17}H_{35}CO_2)_3}$$

The sodium compounds of the carboxylic acids obtained from fats and oils are used in soaps. Their anions have polar ends that attract water and nonpolar ends that attract grease and oil, dispersing them into the water. Soap is prepared from animal fats by *saponification,* a reaction with sodium hydroxide (lye) that cleaves the ester linkages and converts the ester to sodium carboxylates and glycerol:

$$
\begin{array}{ccc}
\text{H}_2\text{C}-\text{O}-\overset{\displaystyle \text{O}}{\overset{\|}{\text{C}}}-\text{C}_{17}\text{H}_{35} & & \text{Na}^+ \ ^-\text{O}-\overset{\displaystyle \text{O}}{\overset{\|}{\text{C}}}-\text{C}_{17}\text{H}_{35} \qquad \text{H}_2\text{C}-\text{OH} \\
| & & \\
\text{HC}-\text{O}-\overset{\displaystyle \text{O}}{\overset{\|}{\text{C}}}-\text{C}_{17}\text{H}_{35} \ + \ 3\text{NaOH} & \xrightarrow{\text{heat}} & \text{Na}^+ \ ^-\text{O}-\overset{\displaystyle \text{O}}{\overset{\|}{\text{C}}}-\text{C}_{17}\text{H}_{35} \ + \ \text{HC}-\text{OH} \\
| & & | \\
\text{H}_2\text{C}-\text{O}-\overset{\displaystyle \text{O}}{\overset{\|}{\text{C}}}-\text{C}_{17}\text{H}_{35} & & \text{Na}^+ \ ^-\text{O}-\overset{\displaystyle \text{O}}{\overset{\|}{\text{C}}}-\text{C}_{17}\text{H}_{35} \qquad \text{H}_2\text{C}-\text{OH}
\end{array}
$$

## 16.8 Amines

Aisha was interested in the aromatic molecule amphetamine. Because it contains a nitrogen atom singly bonded to a carbon, amphetamine is also an example of an organic amine. Another aromatic amine is aniline, $C_6H_5NH_2$, which is used in dyes. An **amine** can be considered a derivative of ammonia, $NH_3$, with one or more hydrogen atoms replaced by organic groups. The Lewis structure of ammonia and examples of some simple amines are shown here:

$$
\underset{\text{ammonia}}{\overset{\displaystyle \text{H}-\ddot{\text{N}}-\text{H}}{\underset{\displaystyle |}{\underset{\displaystyle \text{H}}{}}}} \qquad
\underset{\text{methylamine}}{\overset{\displaystyle \text{H}_3\text{C}-\ddot{\text{N}}-\text{H}}{\underset{\displaystyle |}{\underset{\displaystyle \text{H}}{}}}} \qquad
\underset{\text{dimethylamine}}{\overset{\displaystyle \text{H}_3\text{C}-\ddot{\text{N}}-\text{H}}{\underset{\displaystyle |}{\underset{\displaystyle \text{CH}_3}{}}}} \qquad
\underset{\text{trimethylamine}}{\overset{\displaystyle \text{H}_3\text{C}-\ddot{\text{N}}-\text{CH}_3}{\underset{\displaystyle |}{\underset{\displaystyle \text{CH}_3}{}}}}
$$

Many amines are classified as alkaloids, which are basic nitrogen-containing compounds that come from plants. Many alkaloids have physiological effects on humans, most of which are toxic. Some examples of alkaloids are caffeine and nicotine, which interact with living cells and affect their functioning.

caffeine

nicotine

It is also possible to make ammonium compounds such as tetramethyl ammonium chloride, $(CH_3)_4N^+Cl^-$. This can be considered a derivative of $NH_4^+Cl^-$, with all four hydrogen atoms replaced by methyl groups. Organic ammonium compounds are used in disinfectants and fabric softeners.

Like ammonia, most amines act as bases due to the presence of an unshared electron pair on the nitrogen atom. Most amines are weak bases in water and react completely with strong acids, such as HCl, to form ammonium salts:

$$R-NH_2 + HCl(aq) \longrightarrow R-NH_3^+Cl^-(aq)$$

## 16.9 Organic Nomenclature

Organic nomenclature helps us communicate structures without drawing a complete structural formula. Although many names are in common use, the IUPAC system is the standard system used worldwide. Once you learn the basics of the IUPAC system, you can easily name a variety of organic compounds with various functional groups.

## Alkanes

To use IUPAC nomenclature, start with the names of the alkyl groups derived from the alkanes by removal of one hydrogen atom, with a general formula $C_nH_{2n+1}$. The names of the alkyl groups are similar to those of the alkanes given in Table 16.2, except that the $-ane$ ending is replaced by $-yl$. Thus, $-CH_3$ is methyl and $-C_2H_5$ is ethyl.

A few rules allow us to achieve consistent nomenclature for the alkanes.

1. Determine the *longest* continuous carbon chain in the hydrocarbon. The name of this alkane constitutes the parent name.
2. Number the carbon atoms in the parent hydrocarbon from the end nearer a branch, so that branch will have the lowest possible position number.
3. Branches attached to the parent chain are named as alkyl groups.
4. The name consists of the position number, a hyphen, the branch name, and the straight-chain name. If there is more than one branch, name them in alphabetical order. If the branches are the same alkyl group, precede the name with a prefix (*di-*, *tri-*, etc.) indicating the total number of that alkyl group in the molecule, and separate the position numbers with commas.

The cycloalkanes are named in the same way as the straight-chain alkanes, but with the prefix *cyclo-*. Thus, $C_6H_{12}$, in which six carbons are bonded in a ring, is named cyclohexane.

## Alkenes and Alkynes

The names of the alkenes use the alkane root, with the ending $-ene$ replacing the $-ane$ ending. If a double bond has more than one possible position, a number indicates its location. The carbon atoms are numbered from the end nearer the double bond. For the following compound, the carbon chain is numbered from left to right, which places the C=C bond between carbon atoms 1 and 2:

$$\overset{1}{H_2}C=\overset{2}{C}H-\overset{3}{C}H_2-\overset{4}{C}H_3$$

This molecule has four carbon atoms, so it is derived from butane. It has a double bond, so the $-ane$ ending is replaced by $-ene$, giving it the name butene. The position of the double bond is designated by the lowest-numbered carbon involved in that bond, so this molecule is 1-butene. Similarly the molecule $CH_3-CH=CH-CH_3$ is named 2-butene.

The names for side chains on an alkene follow the rules for naming alkyl groups. Side chain names have the $-yl$ ending, and a number designates the position of attachment. Again, numbering starts at the end nearest the double bond.

A molecule may have more than one double bond. In such a case, an appropriate prefix (such as *di-*) is placed just before the $-ene$ ending to indicate the number of double bonds. For example, $H_2C=CH-CH=CH_2$ is named 1,3-butadiene.

Alkynes are named the same as alkenes, but using a $-yne$ ending instead of an $-ene$ ending.

## Aromatic Hydrocarbons

An aromatic group formed by removal of a hydrogen atom from a benzene ring is called an **aryl group,** which is the aromatic analog of an alkyl group. The aryl group formed from benzene, $C_6H_5$, is called phenyl. For example, $C_6H_5C\equiv CH$ is commonly known as phenylacetylene.

Toluene, phenol, and aniline are a few of the special names of benzene compounds retained by the IUPAC nomenclature system. If two or more positions on a benzene ring have substituted groups, the positions are indicated by

---

If the branch is closest to the left side of the molecule, as written, then we start counting from that end of the molecule:

$$\overset{1}{H_3}C-\overset{2}{C}H-\overset{3}{C}H_2-\overset{4}{C}H_3$$
$$|$$
$$CH_3$$

The $-CH_3$ group is on carbon 2, so the name of this compound is 2-methylbutane. If the branch is closest to the right side of the molecule, as written, numbering the carbons begins on the right:

$$\overset{4}{H_3}C-\overset{3}{C}H_2-\overset{2}{C}H-\overset{1}{C}H_3$$
$$|$$
$$CH_3$$

It is the same compound as above, just written in a different way. This is also the compound 2-methylbutane.

---

If the double bond is closest to the right side of the molecule, as written, then we start counting from that end of the molecule:

$$\overset{4}{H_3}C-\overset{3}{C}H_2-\overset{2}{C}H=\overset{1}{C}H_2$$

---

Some alkenes, such as $CH_3-CH=CH-CH_3$, have geometric isomers. This molecule would be named *cis*-2-butene or *trans*-2-butene if a structural formula was provided.

Another way of naming benzene compounds containing two groups describes the relative positions of the groups attached to a benzene ring:

For example, when the two methyl groups in xylene (dimethylbenzene) are next to each other on the ring (1,2-dimethylbenzene), the compound is commonly known as *ortho*-xylene. When the methyl groups are two carbon atoms away from each other on the ring (1,3-dimethylbenzene), the common name is *meta*-xylene. If the methyl groups are three carbon atoms away (1,4-dimethylbenzene), the common name is *para*-xylene.

numbers, counted from the location of the first substituted group, and then listed in alphabetical order. If two methyl groups are substituted on benzene, the molecule is dimethylbenzene or xylene. There are three isomers: 1,2-dimethylbenzene, 1,3-dimethylbenzene, and 1,4-dimethylbenzene. These and several other examples are summarized here:

toluene
(methylbenzene)

phenol
(hydroxybenzene)

aniline
(aminobenzene)

1,2-dimethylbenzene    1,3-dimethylbenzene    1,4-dimethylbenzene

## Other Naming Conventions

The other major classes of organic compounds are named by similar rules, which are summarized in Table 16.8.

**TABLE 16.8** Rules for Naming Organic Compounds with Functional Groups

| Class of Compound | Rules for Naming |
|---|---|
| alcohols | The common names of the alcohols consist of the alkyl group name followed by the word *alcohol*. The IUPAC names use the alkane root and replace the −*e* ending with −*ol*. Thus, $CH_3OH$ is methyl alcohol or methanol, $C_2H_5OH$ is ethyl alcohol or ethanol, and $C_3H_7OH$ is propyl alcohol or propanol. |
| ethers | The common names of the ethers consist of the names of the substituted alkyl groups, listed alphabetically, and followed by the word *ether*, as in diethyl ether, $C_2H_5OC_2H_5$, or ethylmethyl ether, $C_2H_5OCH_3$. |
| aldehydes | Common names use the ending −*aldehyde*, such as in formaldehyde, HCHO, or acetaldehyde, $CH_3CHO$. IUPAC names use the alkane name with the ending −*al*, such as in methanal, HCHO, or ethanal, $CH_3CHO$. |
| ketones | Common names use the names of the two alkyl groups attached to the C=O group, followed by the word *ketone*. IUPAC names use the alkane name with the ending −*one*. The solvent commonly called acetone, $(CH_3)_2C=O$, is dimethyl ketone or propanone. |
| carboxylic acids | Carboxylic acids often have common names, such as formic acid, $HCO_2H$, and acetic acid, $CH_3CO_2H$. The IUPAC names use the alkane name with the ending −*oic acid*. The two examples are methanoic acid and ethanoic acid, respectively. |
| esters | Esters have an alkyl group replacing the acidic hydrogen in a carboxylic acid. They are named in the same way as the carboxylic acids from which they are derived, using an −*oate* ending. Thus, $HCO_2CH_2CH_3$ is ethyl methanoate (or ethyl formate), and $CH_3CO_2CH_3$ is methyl ethanoate (or methyl acetate). |
| amines | Amines are commonly named by listing the alkyl groups attached to nitrogen, followed by the word *amine*. For example, $NH_2CH_2CH_3$ is ethylamine and $NH(CH_3)_2$ is dimethylamine. |

# CHAPTER REVIEW

## KEY CONCEPTS

- Organic compounds contain carbon, hydrogen, and sometimes other elements such as oxygen and nitrogen. Carbon is found in a vast number of compounds because of its ability to form four single bonds to other carbon atoms and other elements.
  - Carbon atoms can connect to form chains or rings.
  - Bonds between carbon atoms can be single, double, or triple.
  - Many organic compounds exist as isomers with the same molecular formula but different structures, physical properties, and chemical properties.
  - We usually use structural formulas, condensed structural formulas, or line structures when describing organic compounds.

- Organic compounds are classified by the kinds of bonds they contain and the presence of functional groups. Functional groups impart specific properties to organic compounds; they are generally the part of the molecule that undergoes characteristic reactions.

- Hydrocarbons are organic compounds that contain only carbon and hydrogen but can have single, double, or triple bonds between carbon atoms. Hydrocarbons can be further classified as alkanes, alkenes, and alkynes.
  - Alkanes have only single bonds between carbon atoms.
  - Alkanes are sometimes called saturated hydrocarbons because they have the maximum possible number of hydrogen atoms ($C_nH_{2n+2}$ for chain alkanes and $C_nH_{2n}$ for cycloalkanes).
  - Because of the lack of functional groups, alkanes are fairly unreactive, with the exception of combustion reactions and substitution reactions that must be initiated by heat or light.
  - Alkenes contain double bonds, and alkynes contain triple bonds.
  - Alkenes and alkynes undergo reactions, such as polymerization and addition reactions, which result in the loss of multiple bonds.
  - Geometric isomers called cis and trans isomers can result when there are different possible arrangements about the double bond.
  - The most common aromatic hydrocarbons contain a benzene ring, $C_6H_6$. The delocalization of electrons due to resonance in benzene makes it less reactive than alkenes.

- Alcohols and ethers contain carbon-oxygen single bonds.
  - Alcohols have the general structure R−OH and ethers have the general structure R−O−R′.
  - The R symbol is used to represent a hydrocarbon group when discussing the general class of a compound.
  - The hydrogen bonding in alcohols contributes to its physical properties: All alcohols are liquids or solids at room temperature, and the short chain alcohols are soluble in water.
  - Ethers are polar molecules but do not have hydrogen bonding so their boiling points are intermediate between those of alkanes and alcohols.

- Aldehydes and ketones contain a carbonyl group (C=O).
  - In aldehydes the carbonyl group has at least one hydrogen atom bonded to the carbon atom of the carbonyl group, so it is found at the end of carbon chains. The general structure of an aldehyde is

 ○ In ketones there are organic groups on either side of the carbonyl group. The general structure of a ketone is

$$\overset{\overset{\displaystyle O}{\|}}{R-C-R'}$$

- Carboxylic acids are the "organic acids," because their reactivity often involves the loss of $H^+$. The general structure of a carboxylic acid is

$$\overset{\overset{\displaystyle O}{\|}}{R-C-OH}$$

- Esters form from carboxylic acids and alcohols. Esters are responsible for many of the pleasant odors of plants. The general structure of an ester is

$$\overset{\overset{\displaystyle O}{\|}}{R-C-OR'}$$

- Amines contain one to three alkyl groups attached to a nitrogen atom:

$$\overset{\overset{\displaystyle H}{|}}{R-N-H} \qquad \overset{\overset{\displaystyle H}{|}}{R-N-R'} \qquad \overset{\overset{\displaystyle R''}{|}}{R-N-R'}$$

Amines generally act as bases in reactions and are found in drugs such as caffeine and nicotine.

- Organic nomenclature helps us communicate without drawing structures. Although many common names are used frequently, the IUPAC nomenclature system is the international standard.

## KEY TERMS

acyclic hydrocarbon (16.2)
alcohol (16.5)
aldehyde (16.6)
aliphatic hydrocarbon (16.2)
alkane (16.3)
alkene (16.3)

alkyl group (16.3)
alkyne (16.3)
amine (16.8)
aromatic hydrocarbon (16.2)
aryl group (16.9)
carboxylic acid (16.7)

cyclic hydrocarbon (16.2)
cycloalkane (16.4)
ester (16.7)
ether (16.5)
functional group (16.1)
hydrocarbon (16.1)

isomer (16.1)
ketone (16.6)
organic chemistry (16.1)
saturated hydrocarbon (16.2)
unsaturated hydrocarbon (16.2)

## QUESTIONS AND PROBLEMS

The following questions and problems, except those in *Additional Questions* and *Concept Review Questions,* are paired. Questions in a pair focus on the same concept. Answers to selected questions and problems are in Appendix D.

### Matching Definitions with Key Terms

16.1　Match the key terms with the descriptions provided.
 (a) a compound having the same composition and formula as another compound, but a different structure
 (b) an organic compound containing the functional group —OH
 (c) a condensation product of carboxylic acids and alcohols, having the general formula $RCO_2R'$
 (d) an organic compound containing the functional group —CHO
 (e) a hydrocarbon that has a ring or closed-chain structure

 (f) a hydrocarbon that contains all the hydrogens it can hold as a result of having only carbon-carbon single bonds
 (g) a group derived from an alkane by removal of one hydrogen atom
 (h) a saturated hydrocarbon that contains a ring structure
 (i) a hydrocarbon that contains localized single, double, or triple bonds
 (j) a small molecule that contains both the —$NH_2$ and the —$CO_2H$ functional groups
 (k) the study of organic compounds
 (l) a hydrocarbon that does not contain all the hydrogen atoms that could possibly be included due to multiple carbon-carbon bonds

16.2    Match the key terms with the descriptions provided.
    (a) a hydrocarbon that contains one or more localized carbon-carbon double bonds
    (b) an organic compound containing an oxygen double-bonded to a carbon, which is also bonded to two other carbons: $R_2C=O$
    (c) an organic compound containing the functional group $-CO_2H$
    (d) a reactive part of a molecule that undergoes characteristic reactions
    (e) a hydrocarbon containing one or more triple bonds
    (f) an organic compound that contains the $-O-$ functional group attached to two organic groups: $R-O-R'$
    (g) an aromatic group formed by removal of a hydrogen from a cyclic aromatic hydrocarbon
    (h) a cyclic hydrocarbon with delocalized bonding, which imparts special stability
    (i) a compound that contains only carbon and hydrogen
    (j) a hydrocarbon that contains all the hydrogens it can hold
    (k) a hydrocarbon that has a straight- or branched-chain structure with only single bonds
    (l) an organic base with one or more alkyl groups bonded to a nitrogen atom

## Representations of Organic Molecules

16.3    Describe the type of bonds in each of the following molecules.
    (a) $CH_3CH_3$      (b) $CH_2CH_2$      (c) CHCH

16.4    Describe the type of bonds in each of the following molecules.
    (a) $CH_3CH_2CH_3$      (b) $CH_3CHCH_2$      (c) $CHCCH_3$

16.5    Identify and classify the functional group in each of the following molecules.
    (a) $CH_3NH_2$
    (b)
$$H_3C-\overset{\overset{\displaystyle O}{\|}}{C}-CH_3$$
    (c) $CH_3CH_2OH$
    (d) $CH_2=CHCH_3$
    (e) $CH_3CH_2CH_2CO_2H$
    (f) $CH_3C\equiv CH$
    (g)
$$H_3C-CH_2-\overset{\overset{\displaystyle O}{\|}}{C}-H$$
    (h) $CH_3CH_2OCH_3$
    (i)
$$H_3C-CH_2-CH_2-\overset{\overset{\displaystyle O}{\|}}{C}-O-CH_3$$

16.6    Identify and classify the functional group in each of the following molecules.
    (a) $CH_3CH_2C\equiv CH$
    (b)
$$H_3C-\overset{\overset{\displaystyle O}{\|}}{C}-O-CH_3$$
    (c) $CH_3CH_2NH_2$
    (d)
$$H_3C-CH_2-\overset{\overset{\displaystyle O}{\|}}{C}-CH_3$$
    (e) $CH_3CH_2OCH_2CH_3$

    (f)
$$H_3C-\overset{\overset{\displaystyle OH}{|}}{CH}-CH_3$$
    (g) $CH_3CO_2H$
    (h) $CH_3CH_2CH=CH_2$
    (i)
$$H_3C-\overset{\overset{\displaystyle O}{\|}}{C}-H$$

16.7    From the ball-and-stick model, determine the molecular formula and draw the structural formula and the line structure for this molecule.

16.8    From the ball-and-stick model, determine the molecular formula and draw the structural formula and the line structure for this molecule.

16.9    Write a molecular formula for each of the following.

16.10   Write a molecular formula for each of the following.

16.11   Draw a line structure for each of the following.
    (a) $CH_3(CH_2)_6CH_3$
    (b) $CH_3CH_2COCH_2CH_3$
    (c) $(CH_3)_2CHCH_2CH_2CHCH_2$
    (d) $CH_3(CH_2)_5CHO$
    (e) $(CH_3)_4C$

16.12   Draw a line structure for each of the following.
    (a) $CH_3(CH_2)_8CH_3$
    (b) $CH_3CH_2CH_2COCH_2CH_3$
    (c) $CH_3CH_2CHCHCH(CH_3)_2$
    (d) $CH_3(CH_2)_5CO_2H$
    (e) $(CH_3)_3CH$

16.13   Write a structural formula and draw a line structure for each of the following.

(a)

(b)

(c)

16.14   Write a structural formula and draw a line structure for each of the following.

(a)

(b)

(c)

16.15   Draw a perspective drawing for each the following molecules.
   (a) $CH_4$
   (b) $CH_3CH_3$
   (c) $CH_2ClCH_2Cl$

16.16   Draw a perspective drawing for each of the following molecules.
   (a) $CH_3Cl$
   (b) $CCl_3CH_3$
   (c) $CH_3CHCl_2$

## Hydrocarbons

16.17   Distinguish between alkanes, alkenes, and alkynes.

16.18   Write line structures for an alkane, an alkene, and an alkyne, each of which contains four carbon atoms.

16.19   Label each of the following molecules as saturated or unsaturated.
   (a)
   (b) $C_4H_6$
   (c) $CH_3CHCHCH_2CH_3$

16.20   Label each of the following molecules as saturated or unsaturated.
   (a) $CH_3CH_2CHCH_2$
   (b) $C_3H_6$
   (c)

16.21   What distinguishes an aromatic hydrocarbon from other hydrocarbons?

16.22   What feature is common to all aromatic hydrocarbons?

## Acyclic Hydrocarbons

16.23   Explain the difference in boiling points of the two molecules represented by the following line structures.

$-0.5°C$        $-11.6°C$

16.24   Explain the difference in boiling points of the three molecules represented by the following line structures.

$69°C$                $60.3°C$                $58°C$

16.25   Which compound in each group is expected to have the strongest intermolecular forces?
   (a) $CH_4$, $CH_3CH_3$, $CH_3CH_2CH_3$
   (b) $CH_3(CH_2)_4CH_3$, $CH_3CH(CH_3)CH(CH_3)CH_3$, $CH_3CH_2C(CH_3)_2CH_3$

16.26   Which compound in each group is expected to have the strongest intermolecular forces?
   (a) $CH_2CH_2$, $CH_3CHCH_2$, $CH_3CHCHCH_3$
   (b) $CH_3CH_2CH_2CH_2CH_3$, $CH_3C(CH_3)_3$, $CH_3CH_2CH(CH_3)_2$

16.27   What is the name of each of the following straight-chain alkanes?
   (a) $CH_3CH_3$
   (b) $CH_3CH_2CH_2CH_2CH_2CH_2CH_2CH_3$
   (c) $CH_3CH_2CH_3$
   (d) $CH_3CH_2CH_2CH_2CH_2CH_2CH_2CH_2CH_3$

16.28   What is the name of each of the following straight-chain alkanes?
   (a) $CH_3CH_2CH_2CH_3$
   (b) $CH_3CH_2CH_2CH_2CH_3$
   (c) $CH_3CH_2CH_2CH_2CH_2CH_2CH_2CH_2CH_2CH_3$
   (d) $CH_3CH_2CH_2CH_2CH_2CH_2CH_3$
   (e) $CH_3CH_2CH_2CH_2CH_2CH_3$

16.29   Draw all of the possible isomers with the formula $C_6H_{14}$.

16.30   Draw all of the possible isomers with the formula $C_5H_{12}$.

16.31   Write the IUPAC name for each of the following compounds.
   (a) $CH_3CH_2CH_2CH(CH_3)CH_3$
   (b) $CH_3CH(CH_3)CH_2CH_3$
   (c) $CH_3CH_2CH(CH_3)CH_3$

16.32   Write the IUPAC name for each of the following compounds.
   (a) $CH_3CH(CH_3)_2$
   (b) $CH_3CH_2CH(CH_3)CH_2CH_2CH_3$
   (c) $CH_3CH_2CH(CH_2CH_3)CH_2CH_2CH_3$

16.33 Write the IUPAC name for each of the following compounds.

(a)          (b)         (c)

16.34 Write the IUPAC name for each of the following compounds.

(a)          (b)         (c)

16.35 Write a structural formula for each of the following compounds.
    (a) 2-methylpropane
    (b) 3,3-diethyl-2-methylpentane
    (c) 2,4-dimethyloctane
    (d) 3-ethyl-3-methylhexane

16.36 Write a structural formula for each of the following compounds.
    (a) 2-methylbutane
    (b) 3,3-diethyl-2-methylhexane
    (c) 3,5-dimethylnonane
    (d) 3-ethyl-2,2-dimethylhexane

16.37 Predict the products of the following reactions.

    (a) $CH_3CH_2CH_3 + Cl_2 \xrightarrow{\text{heat}}$

    (b) $(CH_3)_3CCH_2CH_2CH_3 + O_2 \xrightarrow{\text{heat}}$

    (c) $CH_3CH_3 + Br_2 \xrightarrow{\text{light}}$

16.38 Predict the products of the following reactions.

    (a) $CH_4 + Cl_2 \xrightarrow{\text{heat}}$

    (b) $CH_3CH_2CH_3 + O_2 \xrightarrow{\text{heat}}$

    (c) $CH_3CH_2CH_3 + Br_2 \xrightarrow{\text{light}}$

16.39 What is an alkene?

16.40 What is an alkyne?

16.41 Write IUPAC names for the following compounds.
    (a) $CH_3CH_2CH=CHCH(CH_3)CH_2CH_3$

    (b)
$$\begin{array}{c}H_3C\\ \quad \\ H_3C\end{array}C=C\begin{array}{c}CH_3\\ \quad \\ CH_2-CH_2-CH_3\end{array}$$

    (c)
$$\begin{array}{c}H_3C-CH_2\\ \quad \\ H_3C\end{array}C=C\begin{array}{c}CH_3\\ \quad \\ CH_2-CH_3\end{array}$$

16.42 Write IUPAC names for the following compounds.
    (a) $(CH_3)_2C=CHCH_3$
    (b) $CH_3CH=CHC(CH_3)_3$

    (c)
$$\begin{array}{c}H_3C-CH_2\\ \quad \\ H_3C\end{array}C=C\begin{array}{c}CH_2-CH_3\\ \quad \\ CH_3\end{array}$$

16.43 Which of the following compounds can exist as cis and trans isomers?
    (a) $CH_2=CH_2$
    (b) $(CH_3)_2C=CHCH_3$
    (c) $CH_3CH_2C(CH_3)=C(CH_3)CH_2CH_3$

16.44 Which of the following compounds can exist as cis and trans isomers?
    (a) $CH_3CH=CHCH_3$
    (b) $CH_3CH=CH_2$
    (c) $(CH_3)_2C=C(CH_3)_2$

16.45 Draw a line structure for each of the following compounds.
    (a) 3-methyl-1-butene
    (b) propyne
    (c) *trans*-2-butene
    (d) *cis*-3-hexene

16.46 Draw a line structure for each of the following compounds.
    (a) 2,3,3-trimethyl-1-hexene
    (b) 1-butyne
    (c) *cis*-2-butene
    (d) *trans*-3-hexene

16.47 Complete the following reactions.
    (a) $CH_3CH=CHCH_3 + HBr \longrightarrow$
    (b) $CH_3CH_2CH=CHCH_3 + Br_2 \longrightarrow$

    (c)  $+ \; H_2 \xrightarrow{\text{Pd}}$

16.48 Complete the following reactions.
    (a) $CH_3CH_2CH=CHCH_2CH_3 + HBr \longrightarrow$
    (b) $CH_2=CHCH_2CH_3 + Br_2 \longrightarrow$

    (c)  $+ \; H_2 \xrightarrow{\text{Pd}}$

## Cyclic Hydrocarbons

16.49 What feature is common to all aromatic hydrocarbons?

16.50 Which of the following are aromatic?
    (a) $CH_2=CHCH=CH_2$

    (b)

    (c)

16.51 Name the following compounds.

    **(a)**         **(b)**

16.52 Name the following compounds.

    **(a)**         **(b)**

16.53 Write structural formulas for the following compounds.
    (a) 1,2,4-trimethylbenzene
    (b) chlorobenzene

16.54 Write structural formulas for the following compounds.
    (a) 2,5-dichloro-1-ethylbenzene
    (b) 2-ethyl-1,3-dimethylbenzene

16.55 Write an equation for the reaction of benzene, $C_6H_6$, with chlorine, $Cl_2$, in the presence of an $FeCl_3$ catalyst.

16.56 Write an equation for the reaction of benzene, $C_6H_6$, with chloromethane, $CH_3Cl$, in the presence of an $AlCl_3$ catalyst.

## Alcohols and Ethers

16.57 Which of the following compounds are alcohols?
(a) $CH_3OCH_3$
(b) $CH_3CH_2OH$
(c) $CH_3CH(OH)CH_3$
(d) $CH_3COCH_3$
(e) $CH_3CH_2CHO$

16.58 Which of the following compounds are alcohols?
(a) $CH_3CHO$
(b) $(CH_3)_2C(OH)CH_3$
(c) $CH_3CH_2COCH_3$
(d) $CH_3CH_2OCH_2CH_3$
(e) $CH_3CH_2CH_2CH_2OH$

16.59 An alcohol has the formula $C_4H_9OH$. How many isomers can exist for this alcohol?

16.60 An alcohol has the formula $C_5H_{11}OH$. How many isomers can exist for this alcohol?

16.61 Name the following compounds using the IUPAC system.
(a) $(CH_3)_2CHCH_2OH$
(b) $CH_3CH(OH)CH_2CH_3$

16.62 Name the following alcohols using the IUPAC system.
(a) $CH_3CH_2CH(OH)CH_2CH_3$
(b) $(CH_3)_2CHCH_2CH_2CH_2OH$

16.63 Write line structures for the following compounds.
(a) 3-hexanol
(b) 2,2-dimethyl-3-pentanol

16.64 Write line structures for the following compounds.
(a) 2-pentanol
(b) 3-ethyl-2-hexanol

16.65 Explain why $CH_3CH_2OH$ has a boiling point of 78°C, but $CH_3CH_2CH_3$ has a boiling point of −42°C.

16.66 Explain why $CH_3CH_2CH_3$ is insoluble in water, but $CH_3CH_2OH$ is soluble.

16.67 What is an ether?

16.68 How do ethers differ from alcohols?

16.69 Which of the following compounds are ethers?
(a) $CH_3OCH_3$
(b) $CH_3CH_2OH$
(c) $CH_3CH(OH)CH_3$
(d) $CH_3COCH_3$
(e) $CH_3CH_2CHO$

16.70 Which of the following compounds are ethers?
(a) $CH_3CHO$
(b) $(CH_3)_2C(OH)CH_3$
(c) $CH_3CH_2COCH_3$
(d) $CH_3CH_2OCH_2CH_3$
(e) $CH_3CH_2CH_2CH_2OH$

16.71 Name the following compounds.
(a) $CH_3CH_2-O-CH_2CH_2CH_3$
(b) $CH_3-O-CH_3$

16.72 Name the following compounds.
(a) $CH_3CH_2-O-CH_2CH_3$
(b) $CH_3CH_2-O-CH_3$

16.73 Write line structures for the following compounds.
(a) ethyl methyl ether
(b) ethyl phenyl ether

16.74 Write line structures for the following compounds.
(a) methyl propyl ether
(b) diethyl ether

16.75 Which would have the greater solubility in water, $CH_3CH_2OH$ or $CH_3OCH_3$? Explain your answer.

16.76 Which would have the higher boiling point, $CH_3CH_2OCH_2CH_3$ or $CH_3CH_2CH_2CH_2OH$? Explain your answer.

## Aldehydes and Ketones

16.77 What is an aldehyde?

16.78 What is a ketone?

16.79 Which of the following compounds are aldehydes?
(a) $CH_3CHO$
(b) $CH_3COCH_3$
(c) $CH_3CH_2CHO$
(d) HCHO

16.80 Which of the following compounds are ketones?
(a) $CH_3CHO$
(b) $CH_3COCH_3$
(c) $CH_3COCH_2CH_3$
(d) HCHO

16.81 Explain the relative boiling points of these two compounds: HCHO at −21°C and $CH_3CHO$ at 20°C.

16.82 Explain why $CH_3CHO$ is soluble in water, but $CH_3CH_2CHO$ has only limited solubility.

## Carboxylic Acids and Esters

16.83 What is a carboxylic acid?

16.84 What is an ester?

16.85 What reagents can be used to prepare the ester $CH_3CH_2CO_2CH_3$?

16.86 What reagents can be used to prepare the ester $CH_3CH_2CH_2CO_2CH_2CH_3$?

16.87 What is a saturated fatty acid?

16.88 What is an unsaturated fatty acid?

16.89 Predict and explain the relative boiling points of $CH_3CO_2H$ and $CH_3CH_2CHO$.

16.90 Predict and explain the relative solubilities in water of $CH_3(CH_2)_2CO_2H$ and $CH_3(CH_2)_5CO_2H$.

## Amines

16.91 What is an amine?

16.92 What makes an amine basic?

16.93 Draw line structures for the following amines.
(a) $(CH_3)_2NH$
(b) $(CH_3)_2NCH_2CH_3$
(c) $C_6H_5NH_2$

16.94   Draw line structures for the following amines.
    (a) $CH_3CH_2CH_2NH_2$
    (b) $CH_3N(CH_2CH_3)_2$
    (c) $(C_6H_5)_2NH$

16.95   Explain why $CH_3CH_3$ has a boiling point of $-88.6°C$, but the boiling point of $CH_3NH_2$ is $-6.3°C$.

16.96   Explain why the boiling point of $CH_3NH_2$ is $-6.3°C$, but the boiling point of $CH_3CH_2NH_2$ is $17°C$.

16.97   Complete the following equation for the reaction of an amine with hydrochloric acid.

$$(CH_3CH_2)_2NH + HCl(aq) \longrightarrow$$

16.98   Complete the following equation for the reaction of an amine with water.

## Additional Questions

16.99   Outline the different types of hydrocarbons.

16.100  What is an alkane?

16.101  How do alkenes differ from alkynes?

16.102  How do ethers differ from alcohols?

16.103  How do aldehydes differ from ketones?

16.104  How do carboxylic acids differ from esters?

16.105  Name the following compounds.
    (a) $CH_3CH(CH_3)CH_2CH(CH_3)_2$
    (b)

    (c) $(CH_3)_3C-C(CH_3)_3$

16.106  Suggest a simple test to distinguish $CH_3CH_2CH_2CH_3$ from $CH_2=CHCH_2CH_3$.

16.107  Suggest a simple test to distinguish $CH_3CH_2CH=CHCH_3$ from $CH_3CH_2CH(OH)CH_2CH_3$.

16.108  Suggest a method to distinguish cyclohexane from cyclohexanol.

16.109  Propane, dimethyl ether, and ethanol have boiling points of $-42.1°C$, $-23.6°C$, and $78.5°C$, respectively. Explain the differences in boiling points.

16.110  Name the following compounds.
    (a) $CH_3CH_2CH_2CHO$
    (b) $CH_3CH_2COCH_2CH_3$
    (c) $(CH_3)_2CO$
    (d) $H_2CO$
    (e) $CH_3CHO$

16.111  Write a structural formula for each of the following compounds.
    (a) acetone (propanone)
    (b) butanone
    (c) propanal
    (d) 2-butenal
    (e) acetaldehyde (ethanal)

16.112  Name the following compounds.
    (a) $CH_3CH_2CO_2H$
    (b) $CH_3CH_2CO_2CH_3$
    (c) $HCO_2CH_3$
    (d) $C_6H_5CO_2CH_2CH_3$
    (e) $(CH_3)_2CHCO_2H$

16.113  Write a structural formula for each of the following compounds.
    (a) ethyl butanoate
    (b) pentanoic acid
    (c) methyl acetate
    (d) formic acid
    (e) propanoic acid

16.114  Draw the structure and name the ester that is formed in the reaction of 2-propanol with acetic acid.

16.115  Give the common name of the following compounds.
    (a) $(CH_3)_2NH$
    (b) $(CH_3)_2N(CH_2CH_3)$
    (c) $C_6H_5NH_2$

16.116  Write a structural formula for each of the following compounds.
    (a) dimethylpropylamine
    (b) dipropylamine
    (c) methylethylamine

16.117  Write a structural formula for each of the following compounds.
    (a) 2,2,4-trimethylpentane
    (b) 3-ethyl-2-pentene
    (c) 1,3-butadiene
    (d) cyclobutane
    (e) chlorobenzene

16.118  Give an example of each of the following classes of compounds.
    (a) ketone
    (b) carboxylic acid
    (c) ether
    (d) aldehyde
    (e) alcohol
    (f) alkene
    (g) ester

16.119  Identify the class of each of the following compounds.
    (a) $CH_3CH_2CH_2CH_3$
    (b) $(CH_3)_2CH-O-CH_3$
    (c) $CH_3CH_2COOH$
    (d) $(CH_3)_2CHCHO$
    (e) $(CH_3)_3CCOOCH_2CH_3$
    (f) $CH_3CH_2NHCH_3$
    (g) $(CH_3CH_2)_2CO$

16.120  Write structural formulas for four isomers of $C_4H_9Cl$.

16.121  Which of the following compounds can exist as cis and trans isomers? Write line structures for those that can.
    (a) propene
    (b) 1-butene
    (c) 2-butene
    (d) 2-pentene
    (e) 3-pentyne

16.122  Draw the line structures for the ether and two alcohols with the molecular formula $C_3H_8O$.

**16.123** For each structure, determine if it represents the same compound or an isomer of

(a)          (b)          (c)

**16.124** Describe the types of bonds, single or double, in each of the following molecules.
(a) $CH_3CH_2OH$
(b) $CH_3COCH_3$
(c) $CH_2CHCH_2CO_2H$

**16.125** Write the molecular formula for each of the following.

(a)

(b)

(c)

**16.126** Write the molecular formula for each of the following.

(a)

(b)

(c)

**16.127** Write structural formulas for all the isomers of $C_4H_{10}O$.

**16.128** Carboxylic acids can be prepared by the addition of an oxidizing agent such as chromic acid, $H_2CrO_4$, to an aldehyde:

What is the structure of the carboxylic acid produced from the reaction of chromic acid with propanal?

**16.129** What is the structure of the ester formed from the condensation reaction of butanoic acid with 1-propanol?

**16.130** Consider the following reaction of 2-propanol to form a ketone:

What is the structure of the ketone formed when 3-pentanol reacts in a similar way?

**16.131** Complete the following equation for the reaction of an amine with hydrochloric acid.

**16.132** Complete the following equation for the reaction of an amine with water.
$$CH_3CH_2CH_2NH_2 + H_2O(l) \rightleftharpoons$$

**16.133** For each structure, determine if it represents the same compound or an isomer of

(a)          (b)          (c)

## Concept Review Questions

**16.134** How many hydrogen atoms are in a molecule of pentene?
A. 5    B. 8    C. 10    D. 12    E. 14
How many hydrogen atoms are in a molecule of pentane and in a molecule of pentyne?

**16.135** Which of the following can exist as a pair of cis-trans isomers?
A. pentane
B. 1-pentene
C. 2-pentene
D. 2-methyl-2-butene
E. 2-methyl-1-butene
Which of these has a different molecular formula ($C_xH_y$) than the others?

**16.136** Which of the following react with hydrogen in the presence of a metal catalyst to form alkanes?
A. cycloalkanes          D. alkynes
B. alcohols          E. both alkenes and alkynes
C. alkenes
Which of the other types of compounds listed do not readily undergo chemical reactions other than combustion?

**16.137** Rank the following compounds in order of increasing boiling point.

(1)          (2)          (3)          (4)

A. $3 < 1 < 2 < 4$
B. $2 < 1 < 3 < 4$
C. $4 < 3 < 2 < 1$
D. $3 < 4 < 1 < 2$
E. $1 < 2 < 3 < 4$
Which of these compounds should be most soluble in water?

16.138 Name the following compound:

A. dichlorohydroxycyclohexane
B. dichlorocyclohexanol
C. 2,4-dichlorophenol
D. 1,3-dichlorotoluene
E. 3,5-dichlorophenol
Which of these is the name of an isomer of this compound?

16.139 Which of the following describes a class of compounds that contains an oxygen atom bonded to two carbon atoms?
A. amines
B. carboxylic acids
C. ethers
D. aldehydes
E. ketones
Which of these describe compounds that contain a carbonyl group (C=O)?

16.140 What is the name of the following compound?

A. *cis*-2-methyl-4-hexene
B. *trans*-2-methyl-4-hexene
C. *cis*-5-methyl-2-hexene
D. *trans*-5-methyl-2-hexene
E. 5,5-dimethyl-2-pentene
Three of the incorrect answers do not correspond to a valid name or structure. Why? The remaining incorrect answer does correspond to a compound with a different structure. What is its structure?

16.141 What class does the compound aniline fall into?
A. aromatic and alcohol
B. amine
C. aromatic and amine
D. alkene and amine
E. ketone
For each of the other choices, draw a structure of a simple compound that belongs to the class or set of classes described.

16.142 What is the name of the following compound?

A. propanal
B. propanoic acid
C. ethanoic acid
D. ethanal
E. diethyl ketone
How would you modify this molecule so that it represents each of the other types of compounds?

16.143 What is the molecular formula for cyclopentene?
A. $C_5H_8$
B. $C_5H_{10}$
C. $C_5H_{12}$
D. $C_3H_8$
E. $C_3H_6$
Which of the other molecular formulas could represent a molecule with a three- or five-membered ring?

# Biochemistry

**17.1 Proteins**

**17.2 Nucleic Acids**

**17.3 Carbohydrates**

**17.4 Lipids**

Chapter Review

Questions and Problems

©Rafe Swan/Getty Images; (inset): ©GIPhotoStock/Getty Images

While taking a break from studying chemistry, Sasha and Jordan decided to watch a show on television about forensic detectives. They were amazed at how much chemistry was part of the program. They understood about laboratory tests shown on the program in which reagents changed color. For example, a color test is used to determine if stains found at a crime scene contain blood. The chemicals used turn pink when they react with a protein present in blood. The presence of blood at a crime scene can also be detected by using luminol, a special reagent that glows under an ultraviolet (UV) lamp (Figure 17.1). This technique is especially useful if someone has tried to clean up after a crime. Luminol can react with trace amounts of iron that is present in hemoglobin, the protein in red blood cells that carries oxygen through the bloodstream. Further chemical analysis can describe a variety of proteins present in the blood. Jordan didn't realize that there were proteins in the blood and began to wonder about how the body uses foods high in proteins such as meat.

The scientists in the forensic investigations shown on the television program seemed to rely heavily on DNA tests. Sasha and Jordan knew that DNA provides the genetic code for plants and animals and is unique for every person. They wondered, however, what the structure of DNA was and why it seemed to figure so prominently in forensics. How do forensic investigators test for DNA? What role does it play in the functioning of living beings? How does it compare to the other types of biological materials?

As Sasha and Jordan saw in the television show, a forensic chemist can identify a murder suspect from a bit of skin caught under the fingernails of a murder victim, assuming a comparison skin sample from the suspect is available. This ability also extends to the elimination of those falsely accused. A hair found on a window screen at the scene of a burglary can be used to determine whether an accused thief is the one who crawled through the window. Given a drop of blood from a man and a boy, a chemist can establish with certainty whether the man is the father of the boy. How do chemists make such a variety of determinations? The method they use is called *DNA fingerprinting,* which is usually accepted by the courts as legitimate proof of identity. A portion of every individual's DNA is unique, so identification of the exact makeup of that part of DNA can be used to identify the source of a tissue sample by comparing it with a known tissue sample from that same person. The composition of the DNA is determined by breaking it into fragments with the assistance of enzymes or other chemical agents and then subjecting the mixture to chromatography in an electric field—a process known as *electrophoresis*. In an electric field, the fragments behave in unique ways based on the DNA of the sample tested. The resulting image is called a DNA fingerprint, since its appearance is characteristic of the individual in the same way that the whorls of a fingerprint are characteristic of the individual. An example of a DNA fingerprint is shown in the chapter-opening figure. The comparison of such fingerprints from two DNA samples determines whether the two samples come from the same person (Figure 17.2). We will examine the structure and function of DNA in this chapter.

Sasha had been reading about investigations of the Shroud of Turin, a swath of linen cloth that has the image of a crucified man. Forensic scientists had been analyzing the carbohydrate coating on the linen fibers. Sasha recalled that, mindful of her nutrition, she was watching her intake of carbohydrates. As a chemistry student, she wondered about the structure of carbohydrates and their purpose. Why would they be present in the linen fibers of the shroud? Why would they be found in plant material? Are they also found in animal tissues? Plants and animals seem to be made up of a large variety of different kinds of materials.

In addition to proteins, DNA, and carbohydrates, lipids are important in the function of the human body. Jordan had been reading a mystery novel in which fingerprint analysis was being used to solve a crime. He read that fingerprints, which are used to identify potential suspects, arise from oils left behind by fingers when they touch smooth surfaces. He wondered where the oils come from. He had heard these oils referred to as lipids and knew that they were somehow related to triglycerides and cholesterol, which are tested for in blood analyses.

**FIGURE 17.1** Luminol is used at crime scenes to detect the presence of blood. Bleach and other substances can also cause luminol to glow, so investigators must rely on other forms of evidence to make their case.
©Mikael Karlsson/Alamy Stock Photo

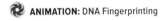 **ANIMATION:** DNA Fingerprinting

Chromatography consists of a variety of chemical techniques in which the components of a mixture are identified by passing the mixture through a stationary material. The components are separated and analyzed according to how fast they move through stationary material.

**FIGURE 17.2** DNA analysis is a common technique used in crime investigations. This image shows a small portion of DNA sequences found on a victim at a crime scene and used to compare against the technician preparing the sample, the victim's own DNA, and two different suspects. To which suspect does the forensic sample correspond?

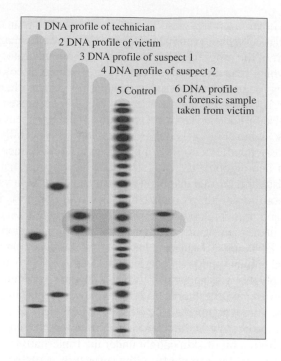

1 DNA profile of technician
2 DNA profile of victim
3 DNA profile of suspect 1
4 DNA profile of suspect 2
5 Control    6 DNA profile of forensic sample taken from victim

Biochemistry is the study of substances essential to life. It includes the study of a few relatively simple substances and many very complex ones. Most of the substances of biological importance are present in great amounts and are organic compounds, usually containing several functional groups. These important substances are often referred to as *biomolecules*. There are lesser amounts of some inorganic substances that are also important. As Sasha and Jordan discovered, there are four important classes of biomolecules. These are *proteins* (e.g., hemoglobin), *nucleic acids* (e.g., DNA), *carbohydrates* (e.g., glucose), and *fats* (or *lipids*). All of the molecules in these classes are very large, and some are considered biological polymers. These large molecules have very different properties than those of the atoms, ions, and small molecules we have examined in earlier chapters. Even the large molecules discussed in previous chapters—organic polymers (Chapter 16) and network solids, such as diamond, graphite, and silicon dioxide (Chapter 10)—have different properties. In this chapter, we consider the biomolecules important in life.

Can you think of any specific molecules that are involved in the chemistry of life?

### Questions for Consideration

**17.1** How are proteins formed and how can we describe their structures?
**17.2** What is the function of DNA and how does it regulate the formation of proteins?
**17.3** What are carbohydrates and what purpose do they serve?
**17.4** What are the structures and functions of common lipids?

## 17.1 Proteins

Jordan wondered about the role of proteins in blood and in our food. Proteins, in some form, make up most of our body tissue. Enzymes, hormones, antibodies, genes, muscle, tendons, hair, fingernails, and even blood are composed, at least in part, of proteins. Some proteins regulate cell processes or control reaction rates. Others transport substances into or out of cells. Still others are part of antibodies that help to fight diseases. **Proteins** are very complex substances of high molar mass (about 15,000 to several million g/mol) composed of various combinations of 20 different amino acids. They are composed largely of carbon (~52%), hydrogen (~7%), oxygen (~23%), nitrogen (~16%), and lesser amounts of other elements, such as sulfur, iodine, phosphorus, iron, and copper. We will look more closely at proteins in this section. In Section 17.2, we will see how DNA directs the formation of proteins.

Jordan's observation that blood contains proteins and that they can be analyzed to help identify the person who left a blood sample at a crime scene was correct. There are eight general blood types that can help identify or rule out suspects. These types are A, B, AB, and O and each of them can be either Rh+ or Rh−. The typing of blood usually refers to identifying the presence of specific types of *antigens* (which produce antibodies and contain proteins) in red blood cells. In some cases, the blood type is based on the absence of a particular antigen. In addition, it is now known that blood serum (the clear liquid remaining when blood cells are removed) contains more than 150 proteins and blood cells contain over 250 enzymes (which are proteins). In about 80% of humans, some of the unique blood proteins can be found in other bodily fluids. People with this characteristic are known as *secretors*. This allows forensic scientists to analyze saliva, urine, and other fluids for proteins usually found in the blood. The proteins in blood or other fluids can be separated by electrophoresis, much like in DNA fingerprinting. A profile of the specific amounts of different blood serum proteins can be used to identify an individual.

> An antigen is a substance recognized by the immune system and is often a protein. The various blood types or factors refer to specific antigens that are associated with red blood cells. For example, the blood factors Rh+ and Rh− correspond to the presence (positive) or absence (negative) of a particular protein on the surface of red blood cells. The designation of Rh+ and Rh− originates from the discovery of this antigen in Rhesus monkeys.

## Composition of Proteins

The building blocks of proteins are **amino acids,** which are small molecules that contain both the amine, $-NH_2$, and the carboxylic acid, $-CO_2H$, functional groups. The amino acids that the body uses to synthesize proteins are called *alpha amino acids*. In these substances, the $-NH_2$ and $-CO_2H$ groups are both attached to the same carbon, as shown in Figure 17.3. If the $-NH_2$ group is attached one carbon away from the $-CO_2H$ group, it is called a beta amino acid; if it is two carbons away, it is a gamma amino acid.

The structures of the alpha amino acids are shown in Figure 17.4. Note that each of the structures, except proline, contains the following group:

The other part of the structure—the side chain designated as R—is unique for each amino acid. It is the identity of these side chains that determines the differences in properties of the amino acids, such as their solubility and acidity. Compare, for example, the side chains on isoleucine and serine:

isoleucine

serine

Which of these amino acids is more soluble in water? The side chain on serine contains an $-OH$ group, which makes it polar like water. Isoleucine contains a nonpolar side chain that makes this amino acid less soluble in water.

**FIGURE 17.3** (A) Amino acids are classified according to the positions of the $-NH_2$ and the $-CO_2H$ functional groups. (B) In an alpha amino acid, the two functional groups are attached to the same carbon. (C) The $-NH_2$ and $-CO_2H$ groups can also be beta to each other, in which there is a carbon between the two groups. What would a gamma amino acid look like?

A          B          C

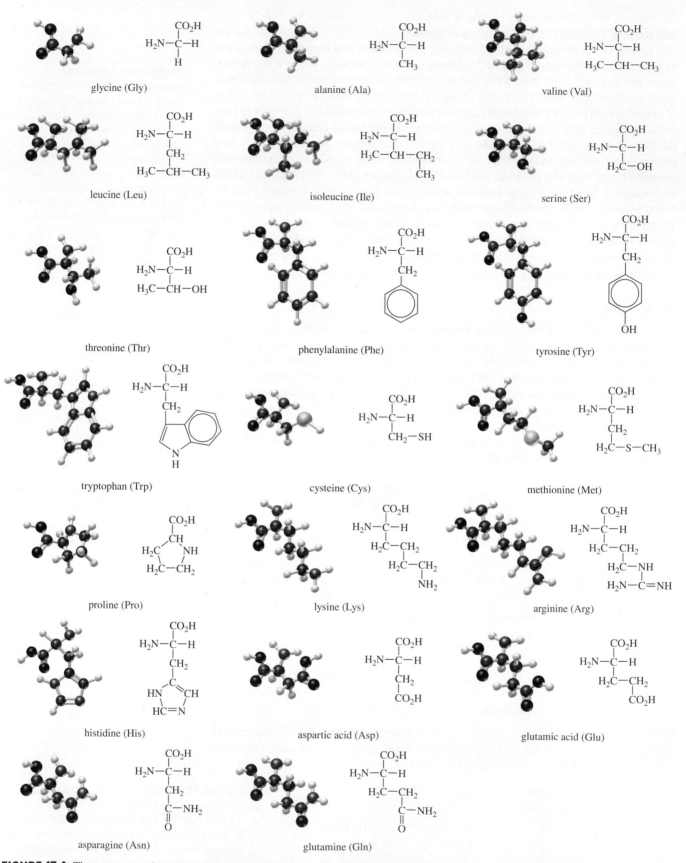

**FIGURE 17.4** The structures of the alpha amino acids all have a —$CO_2H$ and an —$NH_2$ group attached to the same carbon atom. The different side chains attached to the alpha carbon make each amino acid unique. The names of the amino acids are abbreviated with three-letter symbols, usually constructed from the first three unique letters in the names.

In addition to solubility, the amino acids vary in their acidities. Which of the amino acids have side chains that are acidic? The most acidic side chains are those that contain an additional carboxylic acid group ($-CO_2H$), such as aspartic acid and glutamic acid. Lysine, on the other hand, has a basic side chain because it contains the amine group ($-NH_2$). (Some of the other amino acids have acidic or basic properties but are not easily recognizable.)

While the side chains give different amino acids their characteristic properties, the presence of the acidic carboxylic acid group ($-CO_2H$) and the basic amine group ($-NH_2$) on all amino acids gives each one acid-base properties. With both these functional groups, amino acids are amphoteric—they can be acidic or basic. The $-CO_2H$ group can react with $-OH$ to form the carboxylate group ($-CO_2^-$):

$$\underset{\underset{R}{|}}{\overset{\overset{CO_2H}{|}}{H_2N-C-H}} + OH^-(aq) \longrightarrow \underset{\underset{R}{|}}{\overset{\overset{CO_2^-}{|}}{H_2N-C-H}} + H_2O$$

The $-NH_2$ group can also react with $H^+$ to form the $-NH_3^+$ group:

$$\underset{\underset{R}{|}}{\overset{\overset{CO_2H}{|}}{H_2N-C-H}} + H^+(aq) \longrightarrow \underset{\underset{R}{|}}{\overset{\overset{CO_2H}{|}}{H_3N^+-C-H}}$$

Under physiological conditions, both processes occur and the amino acid can exist in the form of a *zwitterion,* a compound that is neutral but has opposite charges on different parts of the molecule:

$$\underset{\underset{R}{|}}{\overset{\overset{CO_2^-}{|}}{H_3N^+-C-H}}$$

Although amino acids have other biological uses related to their structure, their primary function is to build proteins. The body synthesizes proteins through a complex process in which amino acids react in a particular sequence until a biochemical signal triggers the synthesis to stop. (Protein synthesis is a process controlled by DNA and is discussed in Section 17.2.) To see how amino acids can build upon each other, consider the reaction of glycine with alanine:

> In the aqueous environment of body fluids, amino acids exist as their zwitterions, so their $-NH_2$ groups have gained $H^+$ to form $-NH_3^+$ and their $-CO_2H$ groups have lost $H^+$ to form $-CO_2^-$. For simplicity, we will show the structures of the amino acids with neutral $-NH_2$ and $-CO_2H$ groups.

$$\underset{\underset{H}{|}}{\overset{\overset{O}{\|}}{H_2N-CH-C-OH}} + \underset{\underset{CH_3}{|}}{\overset{\overset{H \quad O}{| \quad \|}}{H-N-CH-C-OH}} \longrightarrow \underset{\underset{H}{|} \quad \underset{CH_3}{|}}{\overset{\overset{O \quad H \quad O}{\| \quad | \quad \|}}{H_2N-CH-C-N-CH-C-OH}} + H_2O$$

glycine        alanine

In this reaction, the $-CO_2H$ group on glycine reacts with the $-NH_2$ group on alanine to form a C—N bond and to release water. This new compound is referred to as a **peptide,** and the new bond is called a **peptide bond.** The peptide bond links one amino acid to another and has the following feature:

$$\overset{\overset{O \quad H}{\| \quad |}}{-C-N-}$$

Peptide bond

Using three-letter abbreviations for the amino acids, the reaction of glycine with alanine to form a peptide bond can be represented as

$$Gly + Ala \longrightarrow Gly\text{-}Ala + H_2O$$

**FIGURE 17.5** When the amino acids glycine and alanine react, they form a peptide bond and release a water molecule. Notice the three-dimensional arrangement of atoms in the structure on the left and compare it to the line structure on the right.

Glycine is an amino acid that is required in our diets. It can be found in gelatin and has a sweet taste. Alanine can be produced by the body, but it can also be obtained from a variety of protein sources in our diets, such as meat, seafood, and dairy products. It can also be found in silk, which contains glycine and serine, too.

To understand the behavior of amino acids, peptides, and proteins, we must consider the three-dimensional arrangement of the atoms in these molecules. For example, compare the model of Gly-Ala shown in Figure 17.5 to the line structure. Notice how the atoms are arranged in three dimensions.

---

**EXAMPLE 17.1**    **Identifying Peptide Bonds**

The following molecule is formed by two amino acids:

(a)  Identify the peptide bond in this molecule.
(b)  Identify the alpha carbons and the side chains for each amino acid in the molecule.
(c)  Identify the amino acids.

**Solution:**

(a)  Begin by converting the model to a structural formula:

A peptide bond contains a C=O group connected to an NH group:

This structure can be found in the center of the molecule:

Peptide bond

(b) The alpha carbon for each amino acid is the carbon to which the $-NH_2$ and $-CO_2H$ groups would be attached if the amino acid was not bonded to another amino acid. The side chains are the highlighted groups attached to the alpha carbons.

(c) The two amino acids can be identified from their respective side chains. The $-CH_2OH$ side chain (on the left) corresponds to serine, and the $-CH_2CH(CH_3)_2$ side chain (on the right) corresponds to leucine.

**Consider This 17.1**

How would the structure change if serine were replaced by threonine in the peptide shown in the example?

**Practice Problem 17.1**

The following molecule is formed by two amino acids:

(a) Identify the peptide bond in this molecule.
(b) Identify the alpha carbons and the side chains for each amino acid in the molecule.
(c) Identify the amino acids.

**Further Practice:** Questions 17.11 and 17.12 at the end of the chapter

When two amino acids react to form a peptide, that molecule is called a *dipeptide*. Notice that a dipeptide has an amine group ($-NH_2$) on one end and a carboxylic acid group ($-CO_2H$) on the other. This feature is common to an amino acid, so a dipeptide can react with another amino acid. The reaction of a dipeptide with a third amino acid forms a *tripeptide*. For example, the glycine-alanine dipeptide discussed before Example 17.1 could react with valine to form Gly-Ala-Val:

**FIGURE 17.6**  A carboxylic acid group on the end of a dipeptide allows the molecule to react with the amine group on another amino acid and form a tripeptide. The amine group on glycine, the left amino acid, is unreacted. It is called the N terminus. The carboxylic acid group on valine, at the other end of the tripeptide, is unreacted. It is called the C terminus of the peptide.

glycine          alanine          valine

**FIGURE 17.7**  The amino acids lysine (Lys) and methionine (Met) can combine in two different ways, which we can represent as Lys-Met and Met-Lys. What is different about these structures?

Lys-Met

Met-Lys

When many amino acids link together, a polymer is formed, called a *polypeptide*. Although the borderline is arbitrary, a polypeptide having a molar mass greater than about 5000 g/mol is called a protein.

The synthesis of peptides begins with the reaction of the $-CO_2H$ group on one amino acid and the $-NH_2$ group on the other. This leaves an $-NH_2$ group on the end of the first amino acid unlinked (Figure 17.6). By convention, this amino acid is drawn on the left and this end of the peptide is called the *N terminus*. At the other end of the peptide chain is the *C terminus*. The order in which the amino acids are added makes a difference in the structure that results. For example, the dipeptide Lys-Met has a different structure than the dipeptide Met-Lys (Figure 17.7).

---

**EXAMPLE 17.2**     **Peptide Formation**

(a) Write abbreviated formulas and draw structural formulas for the two different dipeptides that contain one each of cysteine and alanine.

(b) Write abbreviated formulas for all of the possible tripeptides that contain one each of cysteine, alanine, and valine.

**Solution:**

(a) The two possible dipeptides have the abbreviated formulas Cys-Ala and Ala-Cys. Note the differences between the two structures:

Cys-Ala                    Ala-Cys

(b)  Let's start with cysteine. We can combine this with either alanine or valine to form two different dipeptides:

Cys-Ala  and  Cys-Val

We can form two of the possible tripeptides by adding the remaining amino acid to these dipeptides:

Cys-Ala-Val  and  Cys-Val-Ala

Next, let's start with alanine, forming two different dipeptides:

Ala-Cys  and  Ala-Val

Adding the remaining amino acid gives two more tripeptides:

Ala-Cys-Val  and  Ala-Val-Cys

Finally, starting with valine, we get the following dipeptides:

Val-Ala  and  Val-Cys

Adding the remaining amino acid gives the last two tripeptides:

Val-Ala-Cys  and  Val-Cys-Ala

Although it might appear that some of the six tripeptides are repetitions since they have the same order of amino acids, just reversed (e.g., Val-Ala-Cys and Cys-Ala-Val), their structures differ in the location of the terminal $-NH_2$ and $-CO_2H$ groups.

### Consider This 17.2

Draw the structural formula for the tripeptide abbreviated Val-Cys-Ala.

### Practice Problem 17.2

(a)  Write abbreviated formulas and draw structural formulas for the two different dipeptides that contain one each of leucine and serine.

(b)  Write abbreviated formulas for all possible tripeptides that contain one each of glycine, cysteine, and isoleucine.

**Further Practice:**  Questions 17.23 and 17.24 at the end of the chapter

## Hydrolysis of Proteins

Jordan wondered how the consumption of proteins leads to the human body's synthesis of its own proteins from amino acids. Humans must obtain 9 of the 20 amino acids (His, Ile, Leu, Lys, Met, Phe, Thr, Trp, and Val) from their diet. These are called the *essential* amino acids. Our bodies can produce the other 11 amino acids. These are the *nonessential* amino acids. The amino acids needed by an organism for the synthesis of its own specific proteins can be obtained from a variety of protein sources. Meat used as food usually contains all of the amino acids necessary for protein synthesis in the human body. This is why Jordan eats meat. Sasha, however, is a vegetarian, so she must eat a variety of plants, such as beans, nuts, and grains. Unlike meat, a single type of plant doesn't provide all of the necessary amino acids. Table 17.1 summarizes different food types and the amino acids that are not present in sufficient quantity to satisfy nutritional requirements.

The digestion of proteins from food sources produces amino acids, which can be delivered to other parts of the body to build the proteins needed by an organism. A protein is broken down into its component amino acids by enzymes in the stomach

**TABLE 17.1** Amino Acids Absent in Different Food Types

| Food Type | Amino Acid Deficiency |
|---|---|
| Grains | Lys, Ile, Thr |
| Legumes (beans, peas, lentils) | Trp, Met |
| Nuts and seeds | Lys, Ile |
| Animal-based foods | Generally none |

Hydrolysis can also occur when a protein is heated with aqueous acids or bases.

and intestine. These enzymes catalyze the hydrolysis of proteins. **Hydrolysis** is a chemical reaction in which the addition of water to a compound breaks it down into simpler compounds, one containing —OH and the other —H. Consider, for example, the hydrolysis of a glycine-alanine dipeptide:

$$H_2N-CH-\overset{\overset{O}{\|}}{C}-\overset{\overset{H}{|}}{N}-CH-\overset{\overset{O}{\|}}{C}-OH \ + \ H_2O \longrightarrow$$
$$\overset{|}{H} \qquad\qquad \overset{|}{CH_3}$$

Gly-Ala dipeptide

$$H_2N-CH-\overset{\overset{O}{\|}}{C}-OH \ + \ H-\overset{\overset{H}{|}}{N}-CH-\overset{\overset{O}{\|}}{C}-OH$$
$$\overset{|}{H} \qquad\qquad\qquad \overset{|}{CH_3}$$

glycine                          alanine

In protein hydrolysis, therefore, the protein (or polypeptide) reacts to form its component amino acids. This is the reverse of the process that forms peptide bonds.

📶 **Student Hot Spot**

Student data indicate you may struggle with hydrolysis of peptide bonds. Access the SmartBook to view additional Learning Resources on this topic.

---

**EXAMPLE 17.3**  |  **Hydrolysis of Polypeptides and Proteins**

Determine which amino acids would be obtained from the hydrolysis of all the peptide bonds in the following peptide:

**Solution:**

Examine the ball-and-stick model of this molecule, and then write a line structure for the peptide and identify the peptide bond(s) (i.e., each C=O bonded to NH):

$$H_2N-CH-\overset{\overset{O}{\|}}{C}-\overset{\overset{H}{|}}{N}-CH-\overset{\overset{O}{\|}}{C}-\overset{\overset{H}{|}}{N}-CH-\overset{\overset{O}{\|}}{C}-OH$$
$$\overset{|}{H} \qquad\qquad \overset{|}{CH_3} \qquad\qquad \overset{|}{H}$$

Peptide bonds

There are two peptide bonds in the molecule, so it is a tripeptide. The peptide bonds will undergo hydrolysis, adding —OH to C=O and —H to NH.

Hydrolyzing the peptide bond closest to the N terminus gives us an amino acid and a dipeptide:

$$
\underset{\underset{H}{|}}{H_2N-CH}-\overset{\overset{O}{\|}}{C}-\underset{\underset{CH_3}{|}}{N}-\overset{\overset{O}{\|}}{CH}-\overset{\overset{O}{\|}}{C}-\underset{\underset{H}{|}}{N}-\overset{\overset{O}{\|}}{CH}-\overset{\overset{O}{\|}}{C}-OH \;+\; H_2O \longrightarrow
$$

$$
\underset{\underset{H}{|}}{H_2N-CH}-\overset{\overset{O}{\|}}{C}-OH \;+\; \underset{\underset{CH_3}{|}}{H-N-CH}-\overset{\overset{O}{\|}}{C}-\underset{\underset{H}{|}}{N}-CH-\overset{\overset{O}{\|}}{C}-OH
$$

Hydrolyzing the other peptide bond in the dipeptide gives us two more amino acids:

$$
\underset{\underset{CH_3}{|}}{H_2N-CH}-\overset{\overset{O}{\|}}{C}-\underset{\underset{H}{|}}{N}-CH-\overset{\overset{O}{\|}}{C}-OH \;+\; H_2O \longrightarrow
$$

$$
\underset{\underset{CH_3}{|}}{H_2N-CH}-\overset{\overset{O}{\|}}{C}-OH \;+\; \underset{\underset{H}{|}}{H-N-CH}-\overset{\overset{O}{\|}}{C}-OH
$$

In the original tripeptide structure, note that the N terminal and C terminal amino acids are the same—they both have —H for their side chains. By consulting Figure 17.4, we can identify this amino acid as glycine. The side chain of the middle amino acid is a methyl group (—CH₃), which identifies it as alanine.

### Consider This 17.3

Which amino acids are obtained by hydrolyzing all the peptide bonds in this peptide?

### Practice Problem 17.3

Determine which amino acids would be obtained from the hydrolysis of all the peptide bonds in the following tripeptide:

**Further Practice:** Questions 17.29 and 17.30 at the end of the chapter

## Structure of Proteins

Proteins have a variety of roles in the functioning of our bodies. Their properties and functions are related to their molecular structure. For example, the protein in hemoglobin, which carries oxygen in the bloodstream, and the protein in muscle tissue have very different structures and very different behaviors. To fully understand the function of a protein, it is necessary to consider four aspects of protein structure, called the *primary, secondary, tertiary,* and *quaternary* structures.

**Primary Structure**  The sequence of amino acids in each different protein is characteristic of that protein and is responsible for some of the protein's properties. Figure 17.8 shows, for example, the sequence of amino acids in the enzyme lysozyme, a protein with a molar mass of about 14,400 g/mol. Lysozyme is found in tears, nasal mucus, milk, saliva, and blood serum. It is a catalyst for the hydrolysis of the cell walls of some bacteria, so it has some antibacterial activity. This enzyme consists of all 20 of the amino acids in a chain of 129 subunits arranged in a specific and characteristic order. The sequence of amino acids in a protein is known as its **primary structure.** The primary structure determines the function of a protein. The misplacement of even one amino acid in some proteins can change their properties dramatically. For example, with one incorrect amino acid in hemoglobin out of 574 amino acids total, red blood cells can become deformed and function poorly due to a condition known as sickle-cell anemia.

**Secondary Structure**  The sequence of amino acids in a protein determines some of its properties and functions. However, the shape and arrangement of the protein chain is also important. The arrangement of amino acids in space in a

**FIGURE 17.8** The amino acid sequence of lysozyme is represented by connected spheres, where each sphere corresponds to an amino acid. Two cysteine units are able to form an S—S bond between them. Such a bond is formed in four places in lysozyme.

A

B

**FIGURE 17.9**  A pleated sheet
secondary structure is found in some
proteins. (A) The protein chain wraps
back and forth with (B) segments of the
chain held together by hydrogen bonds
as represented by the dotted lines.

local region in the protein chain is called the **secondary structure.** This structure
is due to the presence of hydrogen bonds between peptide groups as well as other
interactions, such as ion-ion attractions or repulsions, and London dispersion
forces between nonpolar groups on a protein chain. In many proteins, this local
structure shows up as a regular, periodic structure. Some proteins, such as silk,
have a structure something like a folded sheet, called a **pleated sheet structure,**
in which protein chains are lined up beside one another, as shown in Figure 17.9.
The chains are held together by hydrogen bonds. Other proteins contain an **alpha
helix structure,** in which the protein chain is twisted into a helix, as shown in
Figure 17.10. This helical structure is maintained by hydrogen bonds between an
amine group (−NH) in one part of the chain and a carbonyl group (⟩C=O) in
another position. The helical structure is the most predominant of the secondary
protein structures.

The structures of proteins are often implicated in diseases. Earlier we mentioned sickle-cell anemia, caused by a change in the structure of the hemoglobin
molecule. Let's consider another disease related to protein structure. Jordan has
read about mad cow disease and wonders whether his hamburger could have come
from a cow with this disease. It can be transferred to humans by eating contaminated beef products. Mad cow disease is caused by a protein called a *prion* (short
for proteinaceous infectious particle), which takes on an abnormal folded shape.
This prion can act as a template to change the shape of normal proteins. In mad
cow disease, a protein in cell membranes normally in a primarily helix form is
converted to a form that has a significant pleated sheet structure. The deformed

Hydrogen
bond

**FIGURE 17.10**  The alpha helical
secondary structure occurs in many
proteins, such as the keratin found in
hair, horns, and hooves. Hydrogen
bonds, represented by dotted lines,
connect parts of the helix, giving
some rigidity to the structure.

**FIGURE 17.11** The tertiary structure of lysozyme is shown in this three-dimensional model. Lysozyme has a globular tertiary structure.

**ANIMATION:** How Enzymes Work

proteins cause an accumulation of fluid in nerve cells, thus damaging or killing them. This causes the brain to decrease in size, giving it a spongelike appearance, and causes a loss of coordination and motor skills.

**Tertiary Structure**    The cylinder formed by the helical arrangement in the secondary structure of a protein chain has further structure; it can be linear, or it can be folded back on itself in various ways. The nature of the folding of the helical chains or of nonhelical secondary structures is very specific for particular proteins. The way in which the various regions of secondary structure are arranged into a three-dimensional shape is called the **tertiary structure.** Look, for example, at the lysozyme chain shown in Figure 17.8. In four places, the chain is connected to itself by sulfur-sulfur bonds that bridge the protein chain. Such a bridge is called a *disulfide bond* (or a *disulfide linkage*). These bridges help to maintain the three-dimensional tertiary structure of the enzyme (shown in Figure 17.11), which is necessary for it to function properly.

If a protein has predominantly a helical secondary structure, the result is a *fibrous protein,* such as in hair and muscles, where the protein helices are often bundled together like a cable (Figure 17.12). Another sort of tertiary structure is illustrated by myoglobin, a relatively small protein that is involved in respiratory processes as an oxygen carrier. The tertiary structure of myoglobin is shown in Figure 17.13. Note the extent to which this protein is folded back on itself. Proteins like myoglobin, which are folded into a rough spherical shape, are known as *globular proteins.*

The folding present in the tertiary structure determines how proteins interact with other molecular species. In the case of enzymes, for example, the **active site** where the catalytic activity takes place is often on the inside of the enzyme. In order

**FIGURE 17.12** Fibrous proteins, such as collagen, consist of bundles of helices. Collagen is found in growing bones, tendons, and muscle.

**FIGURE 17.13**  The tertiary structure of myoglobin is globular in shape. Features of the secondary structure can be seen in the backbone (the carbon-nitrogen chains with other atoms omitted) on the right. Can you see where the backbone wraps back on itself to form an alpha helix? The active site of myglobin is located in a cavity created by the tertiary structure.

for a molecular change to be catalyzed by an enzyme, the molecule to be changed must fit in the cavity around the active site, and the shape of the cavity is dictated by the tertiary structure of the enzyme. This is why enzymes are such specific catalysts.

**Quaternary Structure**    Some proteins are composed of more than one polypeptide chain. Hemoglobin, for example, the primary oxygen carrier in the blood of higher animals, contains four protein units that are present in the protein as pairs of two different polypeptide chains (Figure 17.14). The structure that results from the association in a protein of two or more polypeptide chains is called the **quaternary structure.** The functioning of these proteins depends on this fourth aspect of its structure as well as on the other three.

**FIGURE 17.14**  The quaternary structure of hemoglobin is shown by the differently colored strands that represent the two sets of two different units containing an iron(II) ion surrounded by a protein molecule. Four $Fe^{2+}$ ions are bonded to the ring structures embedded in hemoglobin. The ring structure with an $Fe^{2+}$ is called a *heme* group.

**EXAMPLE 17.4** | **Protein Structure**

Cells of the human body use glucose as a source of energy. One of the reactions in the pathway to generate this energy involves the enzyme hexokinase. Which types of protein structure are illustrated by this image of hexokinase?

**Solution:**

The primary structure is not illustrated since the sequence of amino acids is not shown in this image. However, secondary protein structure is shown in the overlaid backbone and can be described by looking for regular (nonrandom) arrangements of peptide strands, such as helices and pleated sheets:

The image shows the way in which the protein strand wraps around. At the upper left can be seen a portion of the strand that runs parallel to itself and has a ripple structure. This is a pleated sheet structure. At the lower right can be seen several examples of an alpha helix. Both of these types of structure are examples of secondary structure. To the extent that the three-dimensional shape of the molecule can be discerned from this image, it also represents the tertiary structure.

**Consider This 17.4**

Where is the most likely location of the active site shown in the structure for hexokinase?

**Practice Problem 17.4**

Which type of protein structure is illustrated by this image of hexokinase?

**Further Practice:**  Questions 17.37 and 17.38 at the end of the chapter

## Denaturation of Proteins

When the structure of a protein is disrupted, the protein is said to be *denatured,* and it loses many of its characteristic properties such as its solubility. Denaturation describes a number of structural changes in proteins, from the loss of tertiary structure to the uncoiling of chains. Generally, the protein is converted to a state of greater disorder. Denaturation can be accomplished by putting stress on a protein by heating, strongly mixing, adding strong acids, placing the protein in a nonaqueous solvent, or adding heavy metal ions. These stresses cause the peptide chain (or chains) to unravel from regular patterns such as helices or pleated sheets. Secondary, tertiary, and quaternary structures can be denatured while the primary structure (the sequence of amino acids and the peptide bonds connecting them) remains intact.

A common example of protein denaturation occurs when cooking an egg. Egg white is a clear liquid that contains a variety of proteins in water. As the egg cooks, the proteins begin to unravel, exposing nonpolar (hydrophobic—meaning disliking water) side chains to the surrounding water. The solubility of the proteins changes, causing them to aggregate into a solid mass that is no longer soluble in water.

ANIMATION: Protein Denaturation

## 17.2  Nucleic Acids

Sasha and Jordan wondered what role DNA plays in living beings. To answer their questions about DNA, we begin with a consideration of biological cells, which contain *nucleic acids,* including DNA. The basic biological unit of all living matter is the cell. Many different types of cells fulfill many different functions in living systems. However, for all their variety, cells have many features in common. A typical cell is shown in Figure 17.15. All cells are enclosed by a *cell membrane.* Within the membrane is a gelatinous fluid called *cytoplasm,* which contains and

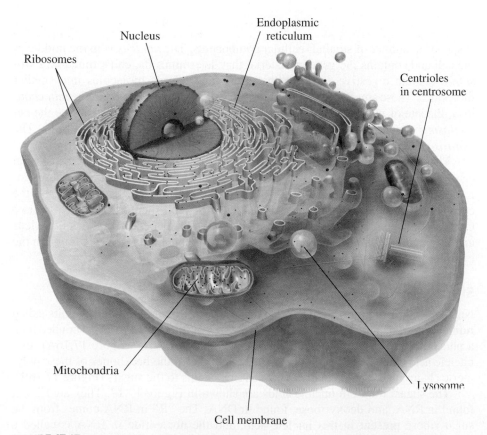

**FIGURE 17.15** A typical biological cell contains many components, indicated by the labels.

RNA                 DNA

**FIGURE 17.16** (A) A nucleotide in RNA or DNA consists of a sugar, a phosphate, and an organic base unit. (B) The nucleotides in RNA and DNA repeat in long polymeric chains. The unconnected bonds (represented as squiggly lines) shown at the top and bottom of each structure would actually be connected to more nucleotides.

Can you distinguish the nucleus of an atom from the nucleus in a cell?

 **ANIMATION: Lysosomes**

**FIGURE 17.17** The sugars found in nucleic acids (RNA and DNA) have similar structures. Can you identify the difference?

suspends a number of smaller, cellular components. The *nucleus* is in the middle of the cell and contains the genetic material that determines the cell's function. *Lysosomes* contain digestive enzymes that break down large molecules into smaller molecules. These smaller molecules are oxidized by enzyme action in the *mitochondria,* the site of reactions that provide energy to the cell. The dots that line the *endoplasmic reticulum* are *ribosomes,* which are the sites of protein synthesis. The *centrosomes* are involved in the replication of chromosomes.

The nucleus contains DNA, which is a member of a group of biomolecules that carry genetic information and regulate the synthesis of proteins in cells. These biomolecules are called **nucleic acids.** They are found in both the nucleus and the cytoplasm—in other words, everywhere except the cell membrane. There are two classes of nucleic acids. **Deoxyribonucleic acid (DNA)** occurs in the cell nucleus and has a molar mass of $10^6$ to $10^9$ g/mol. **Ribonucleic acid (RNA)** is found in the cytoplasm and has a molar mass of 25,000 to $10^6$ g/mol.

## Structure of Nucleic Acids

Nucleic acids are biological polymers that consist of bonded monomer units called **nucleotides.** Each nucleotide is made up of a sugar molecule to which is bonded both a phosphate group and an organic, nitrogen-containing base (Figure 17.16A). In a nucleic acid chain, the sugar and phosphate units form the backbones of these polymeric chains, and the bases are bonded as side chains to the sugars (Figure 17.16B).

The sugars found in nucleic acids are shown in Figure 17.17. They are ribose, found in RNA, and deoxyribose, found in DNA. The "R" in RNA comes from the sugar ribose present in this nucleic acid, and the nucleotide in RNA is called a

**FIGURE 17.18** The structures of the bases found in nucleic acids are shown with their one-letter abbreviations. Uracil (U) is found only in RNA, and thymine (T) is found only in DNA.

*ribonucleotide.* The "D" in DNA comes from the sugar deoxyribose, and the nucleotide in DNA is called a *deoxyribonucleotide.* Carbon atoms in the sugar molecules are numbered from 1 to 5. Do you see the difference between these two sugars? Ribose has an −OH group attached to carbon 2, while deoxyribose does not have oxygen attached to carbon 2. Is the nucleotide shown in Figure 17.16A a ribonucleotide or a deoxyribonucleotide?

The organic bases (and their common abbreviations) are shown in Figure 17.18. They are adenine (A), guanine (G), cytosine (C), and uracil (U) in RNA and adenine (A), guanine (G), cytosine (C), and thymine (T) in DNA. Note that uracil and thymine are found in one but not the other nucleic acid. Occasionally, other bases may also occur.

> The common use of the word *sugar* refers to the white crystals that you might sprinkle on your breakfast cereal. It also refers to a class of compounds that are carbohydrates. Section 17.3 examines these molecules in greater detail.

---

### EXAMPLE 17.5 | Nucleotides

Identify the sugar and the base in the following nucleotide:

Is this nucleotide a ribonucleotide or a deoxyribonucleotide? Is this nucleotide a segment of DNA or RNA?

### Solution:

The sugar is attached to the phosphate group. The lower right group attached to the five-membered ring is an −OH group, so the sugar is ribose. Comparing the base shown in this structure to those in Figure 17.18, we can identify the base as uracil. Both the sugar and the base are found in RNA, but not in DNA, so this nucleotide is a ribonucleotide and comes from RNA.

### Consider This 17.5

How would this structure change if uracil were replaced by cytosine?

## Practice Problem 17.5

Identify the sugar and the base in the following nucleotide:

Is this nucleotide a ribonucleotide or a deoxyribonucleotide? Is this nucleotide a segment of DNA or RNA?

**Further Practice:**  Questions 17.49 and 17.50 at the end of the chapter

The features of the structure of DNA and RNA help us to understand the roles these molecules play in the functioning of cells. These molecules hold the genetic information that determines the type of cell and the way in which it operates in the body. Periodically cells divide, allowing dead cells to be replaced or the body to grow. For the genetic information to be maintained in the new cell and for new proteins to be created, several processes must occur. We will briefly consider three of these: *replication, transcription,* and *translation*. **Replication** happens in the cell nucleus and is the process of copying DNA to make new DNA molecules, which occurs during cell division. Replication ensures that the new cell formed upon division will have a complete set of DNA. **Transcription** also occurs in the nucleus and is the process of synthesizing various types of RNA molecules using a DNA molecule as a template. The RNA is then transported out of the nucleus into the cytoplasm. **Translation** occurs in the cytoplasm and is the process of using RNA molecules to synthesize protein molecules from amino acids. The relation between these three processes is summarized in Figure 17.19.

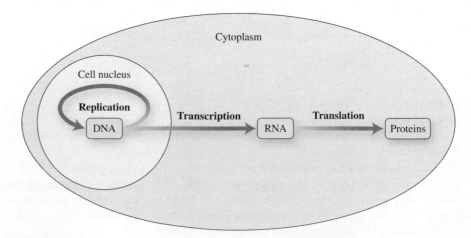

**FIGURE 17.19**  Replication occurs within the nucleus to duplicate a DNA molecule during cell division. During transcription, DNA is used as a template to produce RNA, which is transferred from the nucleus to the cytoplasm of the cell. In translation, which occurs at the ribosomes, RNA is used to synthesize proteins.

## Deoxyribonucleic Acid and Replication

Deoxyribonucleic acid (DNA) carries the genetic information in the cell. It is the DNA that gives Sasha and Jordan and every other human being their distinct characteristics. Because everyone's DNA is different, when a forensic chemist breaks a DNA sample into small parts and separates these parts by electrophoresis, the resulting chromatographic pattern is unique.

Figure 17.20 shows that DNA consists of two interwoven helical strands in which the bases point toward the center of the helix and the phosphate groups point to the outside (not shown). The sugar and phosphate units form the backbone of the helix. The bases pair up inside the double helix in very specific ways, adenine with thymine (A-T) and cytosine with guanine (C-G), and are held together by hydrogen bonding, as shown in Figure 17.21. (For a similar structure showing the hydrogen bonding between base pairs in DNA, see Figure 10.22.) Thus, the nucleotide units in the polymeric strand are arranged so that one strand complements the other to produce the appropriate base pairing.

The double-strand arrangement allows DNA to replicate itself. Replication occurs when the double helix starts to become unraveled. The enzyme *DNA polymerase* plays a key role in DNA replication. A new strand can grow on each separated strand, which serves as a template for forming the new DNA (Figure 17.22). The deoxyribonucleotides that are added to the growing new strands must have bases complementary to those in the unwinding strands of the original DNA, since the only allowed pairings are C-G and A-T. Thus, the new strands grow with bases complementary to those in the original strands. Through this replication process, exact copies of your DNA are generated for new cells as they develop.

**FIGURE 17.20** DNA strands combine to form a double helical structure. The base C hydrogen bonds to the base G, and A hydrogen bonds to T.

**ANIMATION: DNA Structure**

**FIGURE 17.21** The bases on one strand of DNA hydrogen bond to the bases on the other strand. The base pairs are A with T or C with G. (A model of the structures can be seen in Figure 10.22.)

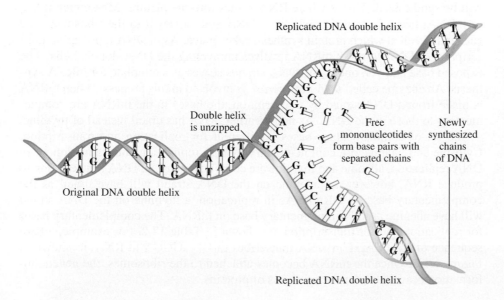

**FIGURE 17.22** DNA is replicated by growing new strands on a partially unwound double helix. The order of the bases on the unwound strands determines the order in which deoxyribonucleotides are added to the growing strands, since the only allowed pairings are C-G and A-T.

---

| EXAMPLE 17.6 | Base Sequence in DNA |
|---|---|

A portion of one strand in a DNA molecule found in one of Sasha's cells has the following sequence of bases:

AATAGCAGGCTTA

Identify the sequence of bases in the other strand of the DNA molecule that would hydrogen-bond to these bases.

**Solution:**

The base A will pair with the base T, while G pairs with C. The sequence of bases is thus TTATCGTCCGAAT.

**Consider This 17.6**

For each of the base pairs, how many hydrogen bonds form?

**Practice Problem 17.6**

A portion of one strand in a DNA molecule found in one of Jordan's cells has the following sequence of bases:

CGGAATTTACGCCGG

Identify the sequence of bases in the other strand of the DNA molecule that would hydrogen-bond to these bases.

**Further Practice:** Questions 17.53 and 17.54 at the end of the chapter

---

**ANIMATION:** How Nucleotides Are Added

Genes, which are segments of DNA molecules, specify the kinds of proteins that are made in cells for different types of tissue. Recall that proteins are formed from up to 20 different amino acids, with the order of the amino acids (the primary structure) in the polypeptide chain determining the function of the protein. The genetic information that determines the order of the amino acids in a protein is stored as the sequence of bases on the DNA strand. DNA is not directly involved in protein synthesis, however. RNA is the nucleic acid directly responsible for protein synthesis in the cytoplasm.

### Ribonucleic Acid, Transcription, and Translation

The information stored by the DNA, to be used in protein synthesis, has to be transferred from the DNA in the nucleus to the site in the cytoplasm where the protein will be synthesized. This is where RNA comes into the picture. **Messenger RNA,** or mRNA, takes the genetic code from DNA and carries it to the ribosomes, the sites in the cell at which protein synthesis takes place. As in DNA replication, transcription of DNA to make mRNA involves unraveling the DNA double helix. The exposed base pairs on one of the DNA strands serves as a template for mRNA synthesis. An enzyme called *RNA polymerase* is involved in this process. When mRNA is made from a DNA strand by transcription, the bases in the mRNA are complementary to those in the DNA. As noted earlier, RNA has uracil instead of thymine, so the set of complementary bases is different for replication and transcription. Cytosine and guanine are complementary in both replication and transcription. In DNA replication, adenine and thymine are complementary. In DNA transcription to produce RNA, however, an adenine on the DNA strand will have uracil as the complementary base on mRNA. As in replication, a thymine on the DNA strand will have adenine as the complementary base on mRNA. The complementary bases for replication and for transcription are shown in Table 17.2. For example, a base sequence of AATCAGT in DNA transcribes to UUAGUCA in RNA, as shown in Figure 17.23. Once the mRNA becomes attached to the ribosomes, the genetic information is available for the synthesis of proteins.

**ANIMATION:** mRNA Synthesis

DNA

mRNA

**TABLE 17.2**  Complementary Bases in DNA and RNA

| Process | Molecules | Complementary Bases |
|---------|-----------|---------------------|
| Replication | DNA | A T C G |
| | ↓ | |
| | DNA | T A G C |
| Transcription | DNA | A T C G |
| | ↓ | |
| | mRNA | U A G C |

**FIGURE 17.23**  Messenger RNA (mRNA) is complementary to DNA. This means that the mRNA created by the DNA will have a U where DNA had an A, an A where DNA had a T, a C where DNA had a G, and a G where DNA had a C. The strand of mRNA shown is complementary to the top strand of DNA. (The circles set in yellow along the DNA and mRNA chains correspond to the phosphate group. The pentagons set in green along the chain correspond to the sugar.)

---

**EXAMPLE 17.7**     **Base Sequence in mRNA**

A portion of a DNA strand has the following base sequence:

                    TTAACGCGATA

What is the corresponding base sequence in mRNA transcribed from this DNA?

**Solution:**

The bases in mRNA will match with the bases in DNA as shown in Table 17.2. As a result, the corresponding base sequence in mRNA is AAUUGCGCUAU.

**Consider This 17.7**

Suppose the DNA strand described in the example were being replicated instead of transcribed. What would be the corresponding base sequence in the replicated strand of DNA?

**Practice Problem 17.7**

A portion of a DNA strand has the following base sequence:

                    GACTAACCTTAGA

What is the corresponding base sequence in mRNA transcribed from this DNA?

**Further Practice:**  Questions 17.55 and 17.56 at the end of the chapter

**FIGURE 17.24** A codon consists of a sugar-phosphate backbone with three bases attached. The identity of the three bases provides one item of information in the genetic code used to synthesize proteins.

In protein synthesis, the sequence of bases in DNA serves as a code that is carried to the ribosomes in the cytoplasm of the cell by mRNA. The code is in the form of the four different bases along the DNA strand—A, T, C, and G. If the coding used just one base to specify an amino acid, only four amino acids could be coded since there are only four bases. A code for an amino acid must come from a sequence of adjacent bases instead of just one. If a sequence of two adjacent bases were used as the code, there would be 16 possibilities (calculated as $4^2$). This is still not enough to uniquely code 20 amino acids. Using a sequence of three bases provides 64 ($4^3$) possibilities. Although this is more possibilities than are needed for DNA to specify 20 different amino acids, it must use a sequence of at least three bases.

The sequence of three bases on mRNA that specifies a particular amino acid is called a **codon.** An example is shown in Figure 17.24. The unique codon for methionine is AUG. This codon also signals the start of protein synthesis. The sequence UGG corresponds to tryptophan. All of the other amino acids have multiple codons that code for them. For example, eight different codons correspond to leucine. In addition, there are three codons that signal the end of protein synthesis and that the process should stop. Table 17.3 summarizes the genetic code for amino acids.

Another form of RNA, called **transfer RNA** (tRNA), brings the appropriate amino acid to the site of protein synthesis—the mRNA in the ribosome. There is at least one kind of tRNA for each amino acid. Transfer RNA consists of about 75 nucleotides in a strand that bends back on itself. The molecule has two regions of double-helical structure and some loops, giving an upside-down L-shape, as shown in Figure 17.25. The closed loop at the end of the L contains the coding of three bases associated with a particular amino acid, which is attached at the open end of the chain at the other end of the L. The base coding in tRNA (called an *anticodon*) is complementary to that in mRNA. For example, the complement of the codon UUA on the mRNA is the anticodon AAU on tRNA. The complementary bases in tRNA form hydrogen bonds to mRNA, thereby holding the amino acid in the proper sequence for addition to the growing protein chain. This process is shown schematically in Figure 17.26. As the mRNA is processed by the ribosome, the amino acid brought by the tRNA is added to the growing polypeptide chain.

**TABLE 17.3** Genetic Codes for Amino Acids

| Amino Acid | mRNA Codon(s) | Amino Acid | mRNA Codon(s) |
|---|---|---|---|
| Ala | GCU, GCC, GCA, GCG | Lys | AAA, AAG |
| Arg | CGU, CGC, CGA, CGG, AGA, AGG | Met | AUG |
| Asn | AAU, AAC | Phe | UUU, UUC |
| Asp | GAU, GAC | Pro | CCU, CCC, CCA, CCG |
| Cys | UGU, UGC | Ser | UCU, UUC, UCA, UCG, AGU, AGC |
| Gln | CAA, CAG | Thr | ACU, ACC, ACA, CAG |
| Glu | GAA, GAG | Trp | UGG |
| Gly | GGU, GGC, GGA, GGG | Tyr | UAU, UAC |
| His | CAU, CAC | Val | GUU, GUC, GUA, GUG |
| Ile | AUU, AUC, AUA | Start | AUG |
| Leu | UUA, UUG, CUU, CUC, CUA, CUG | Stop | UAG, UGA, UAA |

Attached amino acid

Attached amino acid

A

B

C

**FIGURE 17.25** (A) This space-filling model for tRNA shows an amino acid attached to the end. (B) The tRNA chain, shown in a different view, wraps around in several places. (C) The tRNA molecule is represented by this shape.

Growing polypeptide

Polypeptide disconnects from tRNA and reconnects to the next amino acid, valine

Gly tRNA with no Gly

Ser tRNA with no Ser

Ala tRNA

Gly
Ser
Ala
Val

Phe

Phe tRNA

Val tRna

CCC

AGG

CGU
GCA

CAA
GUU

AAG

GGG   UCC       UUC

mRNA   Ribosome travels along the mRNA

**FIGURE 17.26** In the synthesis of protein molecules, messenger RNA provides the codes needed to determine the sequence of amino acids to be added to the protein. Transfer RNA brings the necessary amino acid to the mRNA in the ribosome. The tRNA bonds to the mRNA, and the amino acid is transferred to the growing protein chain. After amino acid transfer, the tRNA is released from the mRNA and the ribosome continues along the mRNA.

**ANIMATION:** How Translation Works

**ANIMATION:** Protein Synthesis

---

**EXAMPLE 17.8**   **Amino Acid Sequence in a Polypeptide and Base Sequence in mRNA**

A segment near the beginning of a strand of mRNA has the following base sequence:

...AUGUUAUAUUGCUUAACUGUUGCCUAUACU...

The polypeptide formed by the first seven steps of protein synthesis has the sequence Met-Leu-Tyr-Cys-Leu-Thr-Val-Ala. What are the next two amino acids to be added to the polypeptide chain?

**Solution:**

The codons consist of sets of three bases, so first we divide the base sequence into sets of three:

AUG UUA UAU UGC UUA ACU GUU GCC UAU ACU

The ninth codon, UAU, is the same as the third codon, so the next amino acid to be added should be the same as the third in the polypeptide chain, Tyr. Consulting Table 17.3 confirms that UAU corresponds to Tyr. The tenth codon, ACU, is the same as the sixth, so the next amino acid to be added should be the same as the sixth in the polypeptide chain, Thr. Use Table 17.3 to confirm that ACU codes for Thr.

**Consider This 17.8**

What other base sequences could code for the same ninth and tenth amino acids in the polypeptide described in the example?

**Practice Problem 17.8**

A segment near the beginning of a strand of mRNA has the following base sequence:

$$...AUGUUUUGCCUCAUCGUUUGCAU...$$

The polypeptide formed by the first five steps of protein synthesis has the sequence Met-Phe-Leu-Pro-His-Arg. What are the next two amino acids to be added to the polypeptide chain?

**Further Practice:** Questions 17.61 and 17.62 at the end of the chapter

## 17.3 Carbohydrates

Although Sasha and Jordan understand where they get the amino acids they need to synthesize the proteins that carry out the various functions in their bodies, they wonder where their bodies get the energy to drive the protein synthesis processes and other bodily functions? The metabolism or breakdown of carbohydrates provides the energy required for life processes. Glucose is the most important carbohydrate for providing energy, but energy can be obtained from other carbohydrates. Carbohydrates are also used as structural materials in plants and animals and as parts of nucleic acids and other biomolecules.

**Carbohydrates** are composed of carbon, hydrogen, and oxygen and have the general formula $C_n(H_2O)_m$. This empirical formula suggests the origin of the name carbohydrate, but does not tell us anything about the structure of carbohydrate molecules. As shown in Figure 17.27, water can be removed from the carbohydrate sucrose by a strong dehydrating agent, leaving elemental carbon. Carbohydrates contain multiple alcohol groups (—OH) and either aldehyde (RCH=O) or ketone ($R_2C=O$) functional groups (or groups readily converted to aldehydes or ketones). The structures of glucose, a typical aldehyde carbohydrate, and fructose, a typical ketone carbohydrate, can be represented as linear, open-chain arrangements:

Glucose ($C_6H_{12}O_6$) is a colorless solid also known as dextrose, blood sugar, grape sugar, and corn sugar. It is a main source of energy for living organisms and is found in fruits and other plant parts. It is prepared in commercial amounts by the hydrolysis of starch. It is about 75% as sweet as sucrose. Fructose, also known as fruit sugar, is the sweetest of the sugars and is found in fruits and honey.

$$
\begin{array}{cc}
\text{Aldehyde} \quad
\begin{array}{c}
\text{HC=O} \\
| \\
\text{H}-\text{C}-\text{OH} \\
| \\
\text{HO}-\text{C}-\text{H} \\
| \\
\text{H}-\text{C}-\text{OH} \\
| \\
\text{H}-\text{C}-\text{OH} \\
| \\
\text{H}_2\text{COH}
\end{array}
&
\begin{array}{c}
\text{H}_2\text{COH} \\
| \\
\text{C=O} \quad \text{Ketone} \\
| \\
\text{HO}-\text{C}-\text{H} \\
| \\
\text{H}-\text{C}-\text{OH} \\
| \\
\text{H}-\text{C}-\text{OH} \\
| \\
\text{H}_2\text{COH}
\end{array}
\\
\text{glucose} & \text{fructose}
\end{array}
$$

Although we sometimes draw these molecules with linear structures, they actually exist primarily in cyclic structures in which two carbon atoms are linked by an oxygen atom:

**FIGURE 17.27** Concentrated sulfuric acid is a good dehydrating agent. When added to a carbohydrate, such as the sucrose (table sugar) shown here, water is removed, leaving a pillar of elemental carbon.
(all): ©McGraw-Hill Education/Stephen Frisch

glucose          fructose

Both five-membered (*pentoses*) and six-membered (*hexoses*) rings are found biologically.

> The names of sugars often have –*ose* as the ending. Read some product labels in your kitchen to see if you can identify the sugars that are present.

## Simple Carbohydrates

The simpler carbohydrates include a group that has a sweet taste, the *sugars*, which include glucose, fructose, and sucrose. These are examples of monosaccharides and disaccharides. The *monosaccharides* have just one five- or six-membered sugar ring, as in glucose, or blood sugar ($C_6H_{12}O_6$). This monosaccharide exists as an equilibrium mixture of an open-chain form and a ring form that contains primarily the ring form:

glucose
(open-chain form)

α-glucose
(one of the
rings that forms)

To help you visualize how the open-chain form converts to the ring form, the carbon atoms have been numbered in these structures and a stepwise reorientation and rearrangement of the structure is shown in Figure 17.28. To form the ring, the open-chain form wraps around and the oxygen from the —OH group on carbon-5 attaches to carbon-1, with the hydrogen on the —OH moving to the carbonyl ($>C=O$) oxygen attached to carbon-1. Carbon-6 is not part of the ring.

There are two ways the cyclic form can be created from the open-chain form. The first of these, shown in Figure 17.28D, is called α-glucose because the —OH on carbon-1 is on the *opposite* side of the ring from carbon-6. If the carbonyl group at carbon-1 in the open chain is in a different orientation, the ring can close to form a

**Student Hot Spot**

Student data indicate you may struggle with visualizing the conversion of the open-chain form of a monosaccharide to the ring form. Access the SmartBook to view additional Learning Resources on this topic.

**FIGURE 17.28** (A) The original linear arrangement of glucose is shown. Carbons are numbered top to bottom in this structure. (B) This structure is rotated clockwise 90°. (C) The linear arrangement is bent by dragging carbon-5 until the structure looks similar to a hexagon. The H atom in the —OH group attached to carbon-5 bonds to the O atom attached to carbon-1. The O atom in the —OH group bonds to carbon-1. (D) The cyclic structure results.

ring in which the —OH group on carbon-1 is on the *same* side of the ring as carbon-6. This form of glucose, called β-glucose, is shown in Figure 17.29.

Other monosaccharides include fructose, shown earlier; mannose; and ribose. These monosaccharides differ from each other in the orientation of the —OH groups and the position of the carbonyl carbon (Figure 17.30).

*Disaccharides,* which contain two sugar groups, include common table sugar, sucrose ($C_{12}H_{22}O_{11}$), found in the ice tea that Sasha and Jordan were drinking while watching television. Sucrose consists of a glucose unit bonded by a *glycosidic* linkage (—O—) to a fructose unit, with the loss of one water molecule:

glucose          fructose

The reaction shown involves α-glucose. A similar reaction would occur with β-glucose, giving a disaccharide with a different structure.

sucrose    Glycosidic linkage

**FIGURE 17.29** Open-chain glucose can form the β-glucose, ring structure if carbon-1 is in a different orientation.

**FIGURE 17.30** Monosaccharides differ in the orientation of the —OH groups and the position of the carbonyl ($>$C=O) carbon. What are the differences in these sugars?

**FIGURE 17.31** Carbohydrates can be classified as monosaccharides, disaccharides, or polysaccharides.

## Complex Carbohydrates

Carbohydrates can be classified as monosaccharides, disaccharides, or polysaccharides (Figure 17.31), depending on the number of rings in the structure (one, two, or more than two, respectively). *Polysaccharides* are long-chain polymers of monosaccharide units. Polysaccharides include starch, the storage food in plants and fruits; glycogen, the storage food in animals; and cellulose, which serves as an important structural component in plants.

| EXAMPLE 17.9 | **Classes of Carbohydrates** |
| --- | --- |

Examine the following model of a carbohydrate and classify the molecule.

**Solution:**

The number of rings determines the classification of the carbohydrate. One ring is a monosaccharide, two rings make a disaccharide, and three or more rings make a polysaccharide. This structure has three rings, so it is a polysaccharide.

**Consider This 17.9**

Is the polysaccharide shown in the example made of pentose or hexose rings, or a combination of the two?

**Practice Problem 17.9**

Examine the following model of a carbohydrate and classify the molecule.

**Further Practice:** Questions 17.69 and 17.70 at the end of the chapter

Polysaccharides such as starch, glycogen, and cellulose are all composed of glucose monomer units, but differ in how the sugar rings are bonded together. The ways in which rings can be connected are illustrated by the structures of some disaccharides in Figure 17.32. As in the combination of glucose with fructose, these

**FIGURE 17.32** Disaccharides all consist of two monosaccharides connected by a glycosidic linkage.

$\alpha$ - 1,4

$\beta$ - 1,4

$\alpha$ - 1,6

apparently small variations in the bonding between two glucose molecules profoundly affect the properties of the polysaccharides that result from polymerization, as we will see later when we examine some of them.

Notice the labels given to the linkages in Figure 17.32. These labels have two parts. The first part of the label, alpha or beta, indicates whether the —OH group originally present on carbon-1 of the glucose molecule is on the opposite side (alpha) or the same side (beta) of the ring as the hydroxymethyl group (carbon-6). The numbers, such as 1,4 for the glucose rings linked in cellulose, indicate the numbers assigned to the carbons on the two rings involved in the formation of the glycosidic linkage between the rings. Thus, carbon-1 on the first ring loses a hydrogen from its —OH group, while carbon-4 on the second ring loses its —OH group to eliminate a water molecule and form the glycosidic linkage.

**Starch**   *Starch* is a long-chain polymer with the general formula $(C_6H_{10}O_5)_n$. Seeds, tubers, and fruits of plants are sources of natural starches, since starch is the principal store of carbohydrates in plants. It is used by young plants as a food source until they develop a leaf system and are able to start synthesizing their own foods by photosynthesis. It is also used as food by established plants during periods when they cannot synthesize carbohydrates. Animals use starch as an energy source by ingesting plant material. Sasha has a soccer game the next morning, following a long night of studying. What should she eat to get energy—a starchy food like pasta or a sugar-coated food like a doughnut? Both contain starch, but only the doughnut contains sucrose (in its sugar glaze). Starch is the healthier of the two saccharides. Why? The molecules of starch are polymers, containing many glucose units that require more time and energy to break down than do molecules of sucrose, which contain one unit of glucose and one of fructose. Sucrose provides a short-term input of energy, often providing too many calories of energy in too short a time, while starch or other complex carbohydrates provide energy over a longer time period as it is needed by the body.

Starch consists of two types of glucose polymer, *amylose* and *amylopectin,* shown in Figure 17.33. Amylose, a straight-chain polymer with 100 to 1000 glucose units, occurs in an open helical structure. Amylopectin contains about 5000 glucose units and differs from amylose in that it has some side branching. About 400 of its glucose units occur as branching points to connect straight-chain fragments (Figure 17.34). The branching in amylopectin and glycogen is important because it allows glucose molecules to be removed from multiple sites, one on each branch. Amylose provides only one site at a time to obtain glucose molecules—namely, the end of the amylose polymer chain. Starch from potatoes contains about 20% amylose and 80% amylopectin. Other starchy foods contain similar fractions of amylose: corn 28%, wheat 25%, rice 20%, sweet potato 18%, and cassava 17%. Genetic engineering of starchy plants can modify the ratio of amylose to amylopectin. Some plants such as waxy rice (an Asian rice that is sticky) contain essentially no amylose.

**Glycogen**   *Glycogen,* $(C_6H_{10}O_5)_n$, has a structure much like that of amylopectin (Figures 17.33 and 17.34), with glucose units connected in a branched structure. Glycogen, though, has about twice as many branching points as amylopectin. Glycogen is the principal storage food of higher animals and of some microbes. It is found primarily in the liver and muscles, where it breaks down to provide energy for the muscles and other tissues. The abundance of branches in glycogen makes it an excellent energy source for the body.

**Cellulose**   *Cellulose* is the most important of the polysaccharides that play a structural role in plants. Its formula is $(C_6H_{10}O_5)_n$. Cellulose, made up of glucose units,

**FIGURE 17.33** The structures of amylose, amylopectin, and cellulose are all related. Amylose and amylopectin contain α-1,4 linked glucose units. Amylopectin also contains α-1,6 branches. Cellulose contains β-1,4 linked glucose units.

has molar masses in the range of $6 \times 10^5$ to $2 \times 10^6$ g/mol. Like amylose, it has a linear arrangement of glucose units, with the glucose strands held together by hydrogen bonding. It differs from amylose, though, in the way the glucose units are attached to one another (Figure 17.35). Chains of cellulose are arranged in parallel bundles to form fibers that are extremely insoluble because of their large size. These fibers contain about 50% of the carbon found in plants and are the main structural material for plant cells. Cellulose cannot be digested by most animals but is important as one of the major sources of dietary fiber in humans. Most animals are unable to hydrolyze the glycosidic linkages in cellulose. Exceptions are the ruminants, such as cows, sheep, and goats. These animals' stomachs contain bacteria that have the necessary enzyme, so they can break cellulose down to smaller saccharide units that can be used as a source of energy.

amylopectin                                    glycogen

**FIGURE 17.34**  The branched structure of amylopectin and glycogen is important. When glucose is needed by the body, these polysaccharides can provide multiple sites to remove glucose molecules. Straight chains like amylose provide one glucose molecule at a time from either end of the polymer. (Note that each circle represents a glucose unit.)

**FIGURE 17.35**  Chains of cellulose are arranged in parallel bundles. With all of the glucose units bound in these bundles, most animals are unable to digest cellulose. The glucose strands are held together by hydrogen bonding as indicated by the dots.

## 17.4  Lipids

The oils in our skin that help produce fingerprints are types of lipids. **Lipids** are a class of biochemical substances in cells that are soluble in nonpolar solvents. Lipids consist of nonpolar molecules much smaller than those of complex carbohydrates or proteins. In addition to oils, they include fats, waxes, long-chain carboxylic acids, steroids, phospholipids, and glycolipids. We briefly examine these types of lipids here in this section.

List some examples of biological fats and oils that might be in your body or that you might find in a kitchen.

The substances found in greatest quantity in living tissue are fats and oils. These substances are all *triglycerides,* esters of glycerol with long-chain carboxylic acids or **fatty acids.** They have the following structure, where R represents the hydrocarbon part of the fatty acid:

$$H_2C-OH \qquad\qquad H_2C-O-\overset{\displaystyle O}{\overset{\|}{C}}-R$$

$$HC-OH \qquad\qquad HC-O-\overset{\displaystyle O}{\overset{\|}{C}}-R$$

$$H_2C-OH \qquad\qquad H_2C-O-\overset{\displaystyle O}{\overset{\|}{C}}-R$$

glycerol                    triglyceride

The fatty acids in a triglyceride need not all be the same, and they may be saturated or unsaturated (Figure 17.36). If the fatty acid is saturated (containing no carbon-carbon double bonds), the triglyceride is a solid, or *fat;* and if the fatty acid is highly unsaturated, with many carbon-carbon double bonds, the triglyceride is a liquid, or *oil*. Triglycerides become more liquid-like as the degree of unsaturation increases. The presence of double bonds with a cis arrangement of groups attached to the double-bonded carbons gives the hydrocarbon chain a curved shape, as discussed and illustrated in Section 16.7 and as shown in Figure 17.37.

**FIGURE 17.36** Fatty acids are characterized by the degree of saturation (the occurrence of double bonds). Some fatty acids, like palmitic acid (A) are saturated and contain no double bonds. Some are unsaturated or polyunsaturated. These can have the groups at the double bond in a cis arrangement, like linoleic acid, which has two cis double bonds, (B), or in a trans arrangement, like elaidic acid (C).

A palmitic acid

B linoleic acid

C elaidic acid

A                                  B

**FIGURE 17.37** (A) Saturated triglycerides have a more linear structure, allowing them to stack tightly. As a result, they tend to be solids at room temperature. (B) Unsaturated triglycerides have more curved structures, making intermolecular attractions weaker and giving them lower melting points. As a result, they tend to be liquids at room temperature.

This curvature leads to the lower melting points of the healthier cooking oils. When excess fat is digested, it is stored in cells as fatty deposits. Linear molecules stack more readily and form fatty deposits more than do curved molecules. Oils, therefore, with the curved structures, are healthier than fats. Trans isomers of unsaturated fatty acids are more linear than cis isomers, so they accumulate in fatty deposits much like saturated fatty acids. Common fats include lard, tallow, and milk fat, while common oils are corn oil, olive oil, soybean oil, peanut oil, linseed oil, and cottonseed oil.

| EXAMPLE 17.10 | Fatty Acids in Triglycerides |
|---|---|

While watching television during a study break, Jordan was eating some chips that had been cooked in a triglyceride that contained the following fatty acid. Was this triglyceride a solid fat or a liquid oil at room temperature?

**Solution:**

Saturated triglycerides of fatty acids are generally solids (fats), while unsaturated triglycerides are usually liquids (oils). This particular fatty acid has four carbon-carbon double bonds, so it is unsaturated. The triglyceride of this fatty acid should be an oil, because two of the double bonds are cis (those at the top-right of the structure). If all the carbon-carbon double bonds were trans like the other two, a triglyceride consisting of the fatty acid would be considered unhealthy.

**Consider This 17.10**

Draw the line structure for the fatty acid shown in the example.

## Practice Problem 17.10

Jordan also ate some chicken that had been fried with a triglyceride containing the following fatty acid. Was this triglyceride a solid or a liquid at room temperature?

**Further Practice:**  Questions 17.81 and 17.82 at the end of the chapter

Fats and oils are used in the body to synthesize essential tissue constituents. The body also uses them as a source of energy. Since they are insoluble, they can be deposited in cells and can remain there inert until they are needed as a source of energy.

Other lipids are used to form cell membranes. These include the *phospholipids* in animal cells and *glycolipids* in plant cells, among numerous others. The phospholipids contain a phosphate and generally an amine group. The most common of these are glycerol esters:

$$
\begin{array}{cc}
\text{H}_2\text{C}-\text{O}-\overset{\overset{\displaystyle O}{\|}}{\text{C}}-\text{R} & \text{H}_2\text{C}-\text{O}-\overset{\overset{\displaystyle O}{\|}}{\text{C}}-\text{R} \\
\text{HC}-\text{O}-\overset{\overset{\displaystyle O}{\|}}{\text{C}}-\text{R} & \text{HC}-\text{O}-\overset{\overset{\displaystyle O}{\|}}{\text{C}}-\text{R} \\
\text{H}_2\text{C}-\text{O}-\overset{\overset{\displaystyle O}{\|}}{\underset{\underset{\displaystyle OH}{|}}{\text{P}}}-\text{X} & \text{H}_2\text{C}-\text{O}-\text{galactose}-\text{X}
\end{array}
$$

Phosphate ester

phospholipid                    glycolipid

Lecithin is a mixture of the diglycerides of stearic, palmitic, and oleic acids. It occurs in nerves and brain tissue and is commonly isolated as a by-product of the manufacture of soybean oil. Depending on the composition, it may be a waxy mass or a thick liquid. Freshly made lecithin is white, but it turns yellow or brown on exposure to air. It is used as an edible emulsifier and surfactant in margarine, chocolate, and other foods.

An example is lecithin, in which X is $CH_2CH_2N(CH_3)_3$ and R is a long hydrocarbon chain. Phospholipids are present in every cell of plants and animals and are particularly abundant in cells in the brain, heart, liver, kidney, muscles, and bone marrow. The presence of the phosphate ester group on phospholipids makes them more polar than the carboxyl group in other lipids. Glycolipids are similar to phospholipids in structure but contain, instead of the phosphate ester, an ester formed with a sugar, normally galactose, which is similar to glucose. Both of these lipids have in common the feature that one end contains polar oxygen functional groups, while the other end contains long-chain nonpolar hydrocarbons. The function of the lipids in the formation of cell membranes arises from this structure. The polar ends can face outward toward the aqueous cell contents, while the nonpolar ends attract the nonpolar ends of other lipids, forming a double layer, as depicted in Figure 17.38.

*Waxes* are esters of long-chain alcohols with long-chain fatty acids. They are fatty acid esters of an alcohol other than glycerol. They are found on external components of plants and animals, such as the protective coating on leaves, fruit, skin, fur, or feathers. Another example is beeswax:

$$
\text{CH}_3(\text{CH}_2)_{14}\overset{\overset{\displaystyle O}{\|}}{\text{C}}-\text{O}-\text{CH}_2(\text{CH}_2)_{28}\text{CH}_3
$$

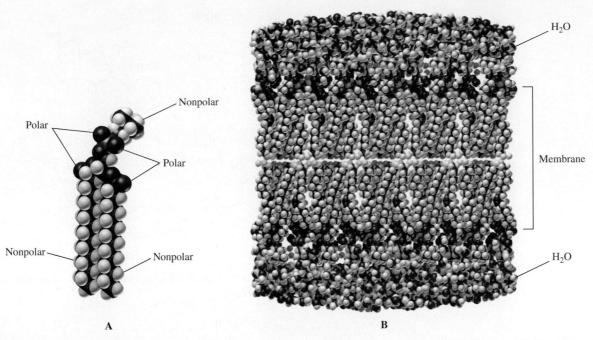

**FIGURE 17.38** (A) Phospholipids are present in cell membranes. (B) Membranes are composed of double layers of lipids. Although they are not shown in this diagram, membranes generally contain embedded proteins. Most membranes are 600 to 1000 nm thick. The long hydrocarbon chains lie parallel to one another and are attracted to the nonpolar hydrocarbon chains of a second layer. The outside of each layer, with the polar carbon-oxygen and phosphate ester groups, is attracted to the surrounding water molecules.

*Steroids* are lipids based on the following four-ring system:

When they contain an —OH group at a specific carbon of the first six-membered ring, these compounds are *sterols*. The most common animal sterol is *cholesterol,* which is found in high concentration in the brain, nervous tissue, and many membranes:

Cholesterol transports fatty acids through the body. There has been much discussion in recent years about the importance of cholesterol in heart and artery disease and the regulation of cholesterol intake in diets.

Steroids are also found in the bile acids, which are synthesized in the liver. Bile is a bitter green liquid that is passed from the liver through the gallbladder into the small intestine, where it helps digest food. Two of the bile acids are shown in Figure 17.39.

taurocholic acid

glycocholic acid

**FIGURE 17.39** The common bile acids all contain the four-ring structure typical of steroids.

**FIGURE 17.40** These hormones have the typical steroid structure. How do these structures differ? What do they have in common?

**FIGURE 17.41** The synthetic molecule 19-norprogesterone has a structure similar to that of the natural hormone progesterone. This compound is used in birth control pills, which act by simulating pregnancy.

Many animal hormones are also steroids. The structures of some such hormones are shown in Figure 17.40. As you can see, these are similar to the structure of cholesterol. Even though these compounds are present in the body in very low concentrations, they produce pronounced physiological effects. Included in this group are the sex hormones, which determine the sexual characteristics of male and female organisms. The female hormone progesterone helps prepare the uterus for pregnancy, and it prevents the further release of eggs from the ovaries during pregnancy. Synthetic molecules that mimic the behavior of progesterone and have a similar structure are used in birth control pills. An example is 19-norprogesterone, which is similar to progesterone but has one of the methyl groups missing, as shown in Figure 17.41.

> How are the structures of cholesterol and the active ingredient in birth control pills related?

## CHAPTER REVIEW

### KEY CONCEPTS

- Biochemistry is the chemistry of living organisms. The biomolecules that are important to the structure and function of living things fall into four general classifications: proteins, nucleic acids, carbohydrates, and lipids.
- Body tissue contains many different proteins, which are polymers of the amino acids.
  - The amino acids in proteins have the $-NH_2$ and $-CO_2H$ groups both attached to the same carbon.

- Amino acids link together by forming a peptide bond that connects the carbon from the $-CO_2H$ group on one amino acid with the nitrogen from the $-NH_2$ group on another amino acid. A peptide bond has this feature:

$$
\begin{matrix}
& O & H & \\
& \| & | & \\
- & C - & N & -
\end{matrix}
$$

- The digestion of proteins from food sources by hydrolysis produces amino acids, which can be delivered to other parts of the body to build the proteins needed by an organism.
- Protein structure can be divided into four levels: primary, secondary, tertiary, and quaternary.
- Nucleic acids, made up of an acidic phosphate group, an organic base, and a sugar unit, are found everywhere in a biological cell except the cell membrane.
  - There are two classes of nucleic acids, deoxyribonucleic acid (DNA) and ribonucleic acid (RNA).
  - DNA carries the genetic information in the cell and exists in a double helical structure.
  - Replication is a process in which DNA is copied in the cell nucleus during cell division.
  - During transcription, DNA serves a template for synthesis of RNA, which is transported out of the nucleus into the cytoplasm.
  - Translation is the process of using RNA molecules to synthesize protein molecules from amino acids.
  - Messenger RNA (mRNA) reads the genetic code on the DNA and carries it to the ribosomes, sites in the cell where protein synthesis takes place.
  - Transfer RNA (tRNA) brings the appropriate amino acid to the ribosome and matches up with the messenger RNA to transfer the amino acid to the growing protein strand.
- Carbohydrates—molecules containing carbon, hydrogen, and oxygen—have multiple alcohol functional groups as well as either an aldehyde or a ketone group.
  - The smallest carbohydrates are the monosaccharides, such as glucose and fructose, and the disaccharides, such as sucrose.
  - The monomer units combine to form polymers (polysaccharides) by removing a water molecule and forming a glycosidic linkage between two cyclic structures.
  - Depending on how the sugar units are linked together, various biological polymers are formed. These include starch, glycogen, and cellulose, which are important for storing energy and creating cell structure.
  - Starch consists of a straight-chain polymer (amylose) and a branched polymer (amylopectin), with many starchy foods containing around 20% amylose and 80% amylopectin.
  - Glycogen has a structure similar to amylopectin but has about twice as many branching points. Glycogen serves as the principal storage food in higher animals.
  - Cellulose consists of glucose units in long chains similar to amylose that are arranged in bundles. Cellulose plays an important structural role in plants.
- Lipids are composed of fatty acids and comprise a variety of substances that can be extracted from cells with nonpolar solvents—fats, oils, waxes, steroids, phospholipids, and glycolipids. They are used to form tissue constituents, such as cell walls, to transport water-insoluble materials, and as a source of energy.

## KEY TERMS

active site (17.1)

alpha helix structure (17.1)

amino acid (17.1)

carbohydrate (17.3)

codon (17.2)

deoxyribonucleic acid (17.2)

fatty acid (17.4)

hydrolysis (17.1)

lipid (17.4)

messenger RNA (17.2)

nucleic acid (17.2)

nucleotide (17.2)

peptide (17.1)

peptide bond (17.1)

pleated sheet structure (17.1)

primary structure (17.1)

protein (17.1)

quaternary structure (17.1)

replication (17.2)

ribonucleic acid (17.2)

secondary structure (17.1)

tertiary structure (17.1)

transcription (17.2)

transfer RNA (17.2)

translation (17.2)

## QUESTIONS AND PROBLEMS

The following questions and problems, except those in *Additional Questions* and *Concept Review Questions,* are paired. Questions in a pair focus on the same concept. Answers to selected questions and problems are in Appendix D.

### Matching Definitions with Key Terms

17.1    Match the key terms with the descriptions provided.

(a)  the form of RNA that brings the appropriate amino acid to the site of protein synthesis

(b)  the arrangement of amino acids in space in a local region in the protein chain, often in the form of a pleated sheet or a helix

(c)  the fundamental unit of a nucleic acid polymer, consisting of a sugar molecule bonded to both a phosphate group and an organic, nitrogen-containing base

(d)  a low-molar-mass nucleic acid that contains the sugar ribose and is involved in the use of genetic information to regulate the synthesis of proteins

(e)  a general class of biological molecules that can be extracted from cells by nonpolar solvents

(f)  the amino acid sequence in a protein

(g)  a high-molar-mass nucleic acid that stores genetic information, has a double helix structure, and contains the sugar deoxyribose

(h)  a long-chain carboxylic acid

(i)  the location on an enzyme where catalysis takes place

(j)  one of a class of biomolecules that serves as a source of energy; is composed of carbon, hydrogen, and oxygen; and has the general formula $C_n(H_2O)_m$

(k)  the combination of two or more amino acids connected by a peptide bond

(l)  the process of copying DNA to make new DNA molecules in the cell nucleus during cell division

(m)  the process of using RNA molecules to create protein molecules from amino acids in the cell cytoplasm

17.2    Match the key terms with the descriptions provided.

(a)  the way in which the various regions of secondary structure in a protein are arranged into a three-dimensional shape

(b)  one of a class of biomolecules that carries genetic information and regulates the growth of proteins

(c)  an organic molecule containing an amine group and a carboxylic acid group

(d)  the form of RNA that takes the genetic code from DNA and carries it to the ribosomes

(e)  one of a class of organic polymers making up most body tissue; characterized by the presence of peptide bonds and formed by the polymerization of amino acids

(f)  the structure that results from the association of two or more polypeptide chains in a protein

(g)  a series of three bases in a nucleic acid that defines the amino acid to be used in the synthesis of a protein

(h)  the secondary structure adopted by many proteins, in which the protein chain is twisted into a helix

(i)  the secondary structure adopted by some proteins in which the protein chains are lined up beside one another in the form of a sheet

(j)  a chemical reaction in which the addition of water to a compound breaks it down into simpler compounds, one containing —H and the other —OH

(k)  the linkage that forms between the carboxylic acid functional group of one amino acid and the amine functional group of another amino acid

(l)  the process of synthesizing various types of RNA molecules in the nucleus using a DNA molecule as a template

### Proteins

17.3    Describe several functions of proteins in living systems.

17.4    What are the basic units of which a protein is composed?

17.5    What is the general structure of an alpha amino acid?

17.6    What distinguishes one alpha amino acid from another?

17.7    Draw the structures of the amino acids Gly, Val, and Ser.

17.8    Draw the structures of the amino acids Cys, Phe, and Ala.

17.9    How do the structures of alanine and valine differ?

17.10   How do the structures of cysteine and methionine differ?

17.11   Draw the structural formulas for the dipeptides Ala-Gly and Gly-Ala.

(a)  Show why these molecules are different.

(b)  Identify the peptide bond and the alpha carbons in each molecule.

(c)  Identify the side chains for each amino acid in the molecules.

17.12 Draw the structural formulas for the dipeptides Ser-Gly and Gly-Ser.
(a) Show why these molecules are different.
(b) Identify the peptide bond and the alpha carbons in each molecule.
(c) Identify the side chains for each amino acid in the molecules.

17.13 Draw the structure of a polypeptide containing only alanine.

17.14 Draw the structure of a polypeptide containing only glycine.

17.15 Write an equation to describe how amino acids combine to form proteins.

17.16 How can you recognize where one amino acid ends and another starts in a protein or a polypeptide?

17.17 Draw the structures of the dipeptides that contain one each of alanine and leucine.

17.18 Draw the structures of the dipeptides that contain one each of cysteine and valine.

17.19 Show how a peptide bond is formed between alanine and valine.

17.20 Show how a peptide bond is formed between glycine and leucine.

17.21 Show how Ala-Gly and Gly-Ala are different dipeptides.

17.22 Show how Gly-Val and Val-Gly are different dipeptides.

17.23 Write the abbreviated formulas of all of the tripeptides that contain one each of serine, glycine, and cysteine.

17.24 Write the abbreviated formulas of all of the tripeptides that contain one each of methionine, threonine, and arginine.

17.25 A vegetarian likes to eat rice with every meal. What other food type should she also eat to ensure that she consumes all the necessary amino acids?

17.26 A vegetarian eats only grains and nuts. What amino acids are missing from his diet?

17.27 What are the structures of the chemical substances that result from the hydrolysis of all the peptide bonds in the polypeptide Gly-Val-Ser?

17.28 What are the structures of the chemical substances that result from the hydrolysis of all the peptide bonds in the polypeptide Ser-Cys-Ala?

17.29 Determine which amino acids are obtained by hydrolyzing all the peptide bonds in the following peptide, and write the abbreviated formula of the peptide:

17.30 Determine which amino acids are obtained by hydrolyzing all the peptide bonds in the following peptide, and write the abbreviated formula of the peptide:

17.31 Draw the structure of the tetrapeptide Ala-Pro-Phe-Gly.

17.32 Draw the structure of the tripeptide Glu-Ser-Met.

17.33 Distinguish between the primary, secondary, tertiary, and quaternary structures of proteins.

17.34 Which type of protein structure describes the order of amino acids in the protein?

17.35 How does the structure of myoglobin differ from that of wool (a fibrous protein)?

17.36 Discuss the relationship between the proper functioning of enzymes and the tertiary structure of proteins.

17.37 Bacterial protein G is found on cell surfaces. Which type of protein structure is illustrated by this image of protein G?

17.38 Bacterial protein G is found on cell surfaces. Which types of protein structure are illustrated by this image of protein G?

17.39 Describe the role of hydrogen bonds in the alpha helical structure of proteins.

17.40 Describe the role of hydrogen bonds in the pleated sheet structure of proteins.

17.41 Both fibrous and globular proteins are formed from long polypeptide chains. Explain the differences in the tertiary structures of these proteins.

17.42   Which amino acid helps to develop a globular structure in proteins?

17.43   Describe the effect of denaturation on a protein.

17.44   List some ways to denature proteins.

## Nucleic Acids

17.45   What are the two classes of nucleic acids? How do their structures differ?

17.46   What are the three parts of a nucleic acid?

17.47   What is the difference between a nucleic acid and a nucleotide?

17.48   What is the difference between a nucleotide of RNA and one of DNA?

17.49   Identify the sugar and the base in the following nucleotide:

Is this nucleotide a ribonucleotide or a deoxyribonucleotide? Is this nucleotide a segment of DNA or RNA?

17.50   Identify the sugar and the base in the following nucleotide:

Is this nucleotide a ribonucleotide or a deoxyribonucleotide? Is this nucleotide a segment of DNA or RNA?

17.51   Consider the structures of adenine, cytosine, thymine, and guanine. On the basis of hydrogen-bonding interactions between all possible combinations of these bases, explain why the only pairings found in DNA are between adenine and thymine and between cytosine and guanine.

17.52   Consider the structures of adenine, cytosine, uracil, and guanine. On the basis of hydrogen-bonding interactions between all possible combinations of these bases, explain why the only pairings found in RNA are between adenine and uracil and between cytosine and guanine.

17.53   A portion of one strand in a DNA molecule has the following sequence of bases:

TTAGCCAGTGCTA

Identify the sequence of bases in the other strand of the DNA molecule that would hydrogen-bond to these bases.

17.54   A portion of one strand in a DNA molecule has the following sequence of bases:

ATATCGTCCGAA

Identify the sequence of bases in the other strand of the DNA molecule that would hydrogen-bond to these bases.

17.55   If the base sequence in a portion of a strand of DNA is ACTAAGATC, what is the base sequence in the strand of mRNA transcribed from this DNA?

17.56   If the base sequence in a portion of a strand of DNA is TTGGAGTACC, what is the base sequence in the strand of mRNA transcribed from this DNA?

17.57   How is genetic information encoded by nucleic acid molecules?

17.58   Why do mRNA and tRNA have complementary base coding?

17.59   If the codon in mRNA is UUU, what anticodon in tRNA allows it to hydrogen-bond to the mRNA?

17.60   If the anticodon in tRNA is CAG, what codon is needed in mRNA to allow them to hydrogen-bond?

17.61   A segment near the beginning of a strand of mRNA has the following base sequence:

…AUGCUUCCCAACGAUUUCCCCGA…

The polypeptide formed by the first five steps of protein synthesis has the composition Met-Leu-Pro-Gln-Arg-Phe. What are the next two amino acids to be added to the polypeptide chain?

17.62   A segment near the beginning of a strand of mRNA has the following base sequence:

…AUGUUCUCUGAUGGCAAUUCUUUC…

The polypeptide formed by the first five steps of protein synthesis has the composition Met-Phe-Ser-Asp-Gly-Asn. What are the next two amino acids to be added to the polypeptide chain?

## Carbohydrates

17.63   Draw the structure of glucose.

17.64   Describe the difference between alpha-glucose and beta-glucose.

17.65   Write a formula and give a name for a monosaccharide and a disaccharide.

17.66   Why is cellulose called a carbohydrate?

**17.67** Which of the two linear structures of monosaccharides would convert to the cyclic structure?

A        B

**17.68** Which of the two linear structures of monosaccharides would convert to the cyclic structure?

A        B

**17.69** Examine the following model of a carbohydrate and classify the molecule.

**17.70** Examine the following model of a portion of a carbohydrate and classify the molecule.

**17.71** Identify the monosaccharides that could be obtained from the hydrolysis of sucrose and of starch.

**17.72** Identify the monosaccharides that could be obtained from the hydrolysis of cellulose and of glycogen.

**17.73** Describe and explain any major differences in the properties of cellulose and starch.

**17.74** Can all carbohydrates be digested by humans? Explain your answer.

**17.75** Why does sucrose taste sweet immediately after being placed in your mouth, while starch begins to taste sweet only after a few minutes?

**17.76** Explain why a disaccharide hydrolyzes to glucose faster than a polysaccharide.

## Lipids

**17.77** Discuss the structural differences between fats and oils.

**17.78** Write a general formula to represent the chemical composition of fats and oils. How do they differ from one another?

**17.79** Draw the structure of a triglyceride.

**17.80** How does a phospholipid differ from a triglyceride?

**17.81** A triglyceride is formed from the following fatty acid. Is this triglyceride a fat or an oil?

**17.82** A triglyceride is formed from the following fatty acid. Is this triglyceride solid or liquid at room temperature?

**17.83** Identify the following molecule as a glycolipid, a phospholipid, or a triglyceride.

17.84   Identify the following molecule as a glycolipid, a phospholipid, or a triglyceride.

17.85   What do steroids have in common with cholesterol?

17.86   What do bile acids and sex hormones have in common?

## Additional Questions

17.87   Draw the structure of a tripeptide containing only serine.

17.88   Compare the primary, secondary, and tertiary structures of a protein with the structure of a telephone cord.

17.89   Draw the structure of the tripeptide Val-Cys-Tyr.

17.90   The artificial sweetener aspartame is prepared from two amino acids—namely, aspartic acid and the methyl ester of phenylalanine. The structure of this ester derivative of phenylalanine has a —$CH_3$ group instead of an —H on the —$CO_2H$ group of the amino acid. Draw the structure of aspartame which has aspartic acid as the N terminal amino acid.

17.91   Compare the functions of proteins and carbohydrates in living systems.

17.92   Write an equation to describe how proteins hydrolyze to form amino acids.

17.93   Describe the differences between RNA and DNA in terms of their constituent parts, their structures, and their functions.

17.94   How does DNA replicate itself?

17.95   Genetic engineering techniques are being used by biochemists to isolate pieces of DNA and insert them into bacteria, where the synthesis of biologically active proteins can occur. What use can be made of these techniques?

17.96   Gene therapy is being investigated as a medical tool. In this therapy, modified genetic material is introduced into human cells in attempts to cure genetic diseases. What are some advantages and some dangers associated with this approach?

17.97   Identify the following amino acids.

(a)   $H_2N-\overset{\displaystyle CO_2H}{\underset{\displaystyle H_3C-CH-CH_3}{\overset{|}{\underset{|}{C}}-H}}$

(b)   $H_2N-\overset{\displaystyle CO_2H}{\underset{\displaystyle CH_2}{\overset{|}{\underset{|}{C}}-H}}$
    $H_2C-S-CH_3$

(c)   $H_2N-\overset{\displaystyle CO_2H}{\underset{\displaystyle CH_2}{\overset{|}{\underset{|}{C}}-H}}$
    $CO_2H$

17.98   Identify the following amino acids.

(a)   $H_2N-\overset{\displaystyle CO_2H}{\underset{\displaystyle H_3C-CH-CH_2CH_3}{\overset{|}{\underset{|}{C}}-H}}$

(b)   $H_2N-\overset{\displaystyle CO_2H}{\underset{\displaystyle H_2C-SH}{\overset{|}{\underset{|}{C}}-H}}$

(c)   $H_2N-\overset{\displaystyle CO_2H}{\underset{\displaystyle H_2C-CH_2}{\overset{|}{\underset{|}{C}}-H}}$
         $\underset{\displaystyle O}{\overset{|}{C}-NH_2}$

17.99   Which amino acids are present in the peptide that has the following structure?

$$H_2N-\underset{H_3C-CH-CH_3}{CH}-\overset{O}{C}-\underset{H_2C-OH}{N}-\underset{}{CH}-\overset{O}{C}-\underset{H_2C-CH_2}{N}-CH-\overset{O}{C}-OH$$
                                                                $S-CH_3$

17.100   Which amino acids are present in the peptide that has the following structure?

$$H_2N-\underset{H_2C-SH}{CH}-\overset{O}{C}-\underset{H_3C-CH-OH}{N}-CH-\overset{O}{C}-\underset{H_2C-CH_2}{N}-CH-\overset{O}{C}-OH$$
                                                                $CO_2H$

17.101   What are the structures of the amino acids that result from the hydrolysis of all the peptide bonds in the following peptide?

$$H_2N-\underset{H_2C-OH}{CH}-\overset{O}{C}-\underset{H_3C-CH-CH_3}{N}-CH-\overset{O}{C}-\underset{H_2C-CH_2}{N}-CH-\overset{O}{C}-OH$$
                                                                $S-CH_3$

17.102   What are the structures of the amino acids that result from the hydrolysis of all the peptide bonds in the following peptide?

$$H_2N-\underset{H_3C-CH-CH_3}{CH}-\overset{O}{C}-\underset{H_3C-CH-OH}{N}-CH-\overset{O}{C}-\underset{CH_2}{N}-CH-\overset{O}{C}-OH$$
                                                                $CO_2H$

17.103   Identify the sugar and the base in the following nucleotide:

17.104   Identify the sugar and the base in the following nucleotide:

17.105   A triglyceride is formed from the following fatty acid. Is this fatty acid saturated or unsaturated? If it is unsaturated, is it cis or trans?

17.106   A triglyceride is formed from the following fatty acid. Is this fatty acid saturated or unsaturated? If it is unsaturated, is it cis or trans?

## Concept Review

17.107   Which of the following structures corresponds to an alpha amino acid?

A.   $H_2N-CH$ with $CO_2H$, $CH_2$, $CH_3$

B.   $H_2N-CH$, $H_3C-CH$ with $CO_2H$, $CH_3$

C.
$$HC=O$$
$$H-C-OH$$
$$HO-C-H$$
$$H-C-OH$$
$$H-C-OH$$
$$H_2COH$$

D.

E.
$$H_2C-OH$$
$$HC-OH$$
$$H_2C-OH$$

How are the other structures classified?

17.108   Which of these amino acids contains an acidic side chain?

A.

B.

C.

D.

E.

Classify the side chains of all these amino acids as acidic, basic, polar, or nonpolar.

17.109 In which of the following dipeptides is the peptide bond correctly shaded?

A.
$$H_2N-CH-\overset{\overset{O}{\|}}{C}-\overset{\overset{H}{|}}{N}-CH-\overset{\overset{O}{\|}}{C}-OH$$
$$\underset{H_3C}{\phantom{x}}\overset{CH}{\phantom{x}}\underset{CH_3}{\phantom{x}} \quad \underset{SH}{\overset{CH_2}{\phantom{x}}}$$

B.
$$H_2N-CH-\overset{\overset{O}{\|}}{C}-\overset{\overset{H}{|}}{N}-CH-\overset{\overset{O}{\|}}{C}-OH$$
$$\overset{CH_3}{\phantom{x}} \quad \underset{OH}{\overset{CH_2}{\phantom{x}}}$$

C.
$$H_2N-CH-\overset{\overset{O}{\|}}{C}-\overset{\overset{H}{|}}{N}-CH-\overset{\overset{O}{\|}}{C}-OH$$
$$\overset{H}{\phantom{x}} \quad \underset{OH}{\overset{CH_2}{\phantom{x}}}$$

D.
$$H_2N-CH-\overset{\overset{O}{\|}}{C}-\overset{\overset{H}{|}}{N}-CH-\overset{\overset{O}{\|}}{C}-OH$$
$$\overset{CH_2}{\phantom{x}} \quad \overset{CH_2}{\phantom{x}}$$
$$\overset{CH}{\phantom{x}} \quad \overset{OH}{\phantom{x}}$$
$$\underset{H_3C}{\phantom{x}}\phantom{xx}\underset{CH_3}{\phantom{x}}$$

E.
$$H_2N-CH-\overset{\overset{O}{\|}}{C}-\overset{\overset{H}{|}}{N}-CH-\overset{\overset{O}{\|}}{C}-OH$$
$$\overset{CH_2}{\phantom{x}} \quad \overset{CH}{\phantom{x}}$$
$$\overset{CH}{\phantom{x}} \quad \underset{H_3C}{\phantom{x}}\phantom{x}\underset{OH}{\phantom{x}}$$
$$\underset{H_3C}{\phantom{x}}\phantom{xx}\underset{CH_3}{\phantom{x}}$$

Identify the peptide bonds in the other dipeptides.

17.110 Which of the following structures corresponds to the tripeptide Leu-Ser-Ala?

A.
$$H_2N-CH-\overset{\overset{O}{\|}}{C}-\overset{\overset{H}{|}}{N}-CH-\overset{\overset{O}{\|}}{C}-\overset{\overset{H}{|}}{N}-CH-\overset{\overset{O}{\|}}{C}-OH$$
$$\overset{CH}{\phantom{x}} \quad \overset{CH_2}{\phantom{x}} \quad \overset{CH_3}{\phantom{x}}$$
$$\underset{H_3C}{\phantom{x}}\phantom{x}\underset{CH_3}{\phantom{x}} \quad \underset{SH}{\phantom{x}}$$

B.
$$H_2N-CH-\overset{\overset{O}{\|}}{C}-\overset{\overset{H}{|}}{N}-CH-\overset{\overset{O}{\|}}{C}-\overset{\overset{H}{|}}{N}-CH-\overset{\overset{O}{\|}}{C}-OH$$
$$\overset{CH_3}{\phantom{x}} \quad \overset{CH_2}{\phantom{x}} \quad \overset{CH_2}{\phantom{x}}$$
$$\phantom{xx} \quad \overset{OH}{\phantom{x}} \quad \overset{CH}{\phantom{x}}$$
$$\phantom{xxxxxx}\underset{H_3C}{\phantom{x}}\phantom{x}\underset{CH_3}{\phantom{x}}$$

C.
$$H_2N-CH-\overset{\overset{O}{\|}}{C}-\overset{\overset{H}{|}}{N}-CH-\overset{\overset{O}{\|}}{C}-\overset{\overset{H}{|}}{N}-CH-\overset{\overset{O}{\|}}{C}-OH$$
$$\overset{H}{\phantom{x}} \quad \overset{CH_2}{\phantom{x}} \quad \overset{CH_3}{\phantom{x}}$$
$$\phantom{xx} \quad \overset{OH}{\phantom{x}}$$

D.
$$H_2N-CH-\overset{\overset{O}{\|}}{C}-\overset{\overset{H}{|}}{N}-CH-\overset{\overset{O}{\|}}{C}-\overset{\overset{H}{|}}{N}-CH-\overset{\overset{O}{\|}}{C}-OH$$
$$\overset{CH_2}{\phantom{x}} \quad \overset{CH_2}{\phantom{x}} \quad \overset{CH_3}{\phantom{x}}$$
$$\overset{CH}{\phantom{x}} \quad \overset{OH}{\phantom{x}}$$
$$\underset{H_3C}{\phantom{x}}\phantom{x}\underset{CH_3}{\phantom{x}}$$

E.
$$H_2N-CH-\overset{\overset{O}{\|}}{C}-\overset{\overset{H}{|}}{N}-CH-\overset{\overset{O}{\|}}{C}-\overset{\overset{H}{|}}{N}-CH-\overset{\overset{O}{\|}}{C}-OH$$
$$\overset{CH_2}{\phantom{x}} \quad \overset{CH}{\phantom{x}} \quad \overset{CH_3}{\phantom{x}}$$
$$\overset{CH}{\phantom{x}} \quad \underset{H_3C}{\phantom{x}}\phantom{x}\underset{OH}{\phantom{x}}$$
$$\underset{H_3C}{\phantom{x}}\phantom{x}\underset{CH_3}{\phantom{x}}$$

Identify the other tripeptides.

17.111 Which of the following is a base found in RNA but not DNA?

A. guanine (G)

B. cytosine (C)

C. thymine (T)

D. adenine (A)

E. uracil (U)

In what nucleic acids are the other bases found?

17.112   Identify the following structure.

A. guanine
B. ribose
C. nucleotide of RNA
D. nucleotide of DNA
E. phosphate

How are the incorrect answers related to the structure shown in the example?

17.113   A portion of one DNA strand has the following sequence of bases:

CGCTAGCATGCAATATTG

What is the corresponding base sequence in mRNA transcribed from this DNA?
A. CGCTAGCATGCAATATTG
B. GCGATCGTACGTTATAAG
C. GCGAUCGUACGUUAUAAC
D. AUACGUAGCUAGGCGCCU
E. CUUUAUUGCAUGCUAGCG

What do the other sequences correspond to in the DNA strands that transcribes them?

17.114   A segment of a strand of mRNA has the following base sequence:

AUGUUGGCCUCAGAACAG

The polypeptide formed from this strand has the amino acid sequence Met-Leu-Ala-Ser-Glu-Thr. What base sequence on mRNA could synthesize the polypeptide Met-Glu-Ser-Ala-Thr-Leu?
A. AUGGCCGAACAGUUGUCA
B. AUGCAGUCAGCCUUGGAA
C. AUGGAAUCAGCCCAGUUG
D. AUGGCCUCAGCCUUGGAA
E. AUGGAAUCAGAAUCAUUG

What are the polypeptides that correspond to the other base sequences?

17.115   Which of the following images represents a disaccharide?

A.

B.

C.

D.

E.

How are the other structures classified?

17.116   Which of the following is a trans fatty acid?

A.

B.

C.

D.

E.

Classify the other fatty acids.

**APPENDIX**

# Useful Reference Tables and Figures

| Description | Table/Figure Reference | Page |
|---|---|---|
| Some English-Metric Conversions | Table 1.5 | 14 |
| Densities of Some Common Substances | Table 1.6 | 18 |
| Monatomic Anions and Oxoanions | Figure 3.17 | 100 |
| Important Polyatomic Ions | Table 3.4 | 100 |
| Naming Ionic Compounds in Which the Cation Exhibits a Single Charge | Figure 3.22 | 106 |
| Naming Ionic Compounds in Which the Cation Exhibits Multiple Charges | Figure 3.25 | 108 |
| Common Greek Prefixes | Table 3.9 | 111 |
| Naming Molecular Compounds | Figure 3.28 | 112 |
| Naming Acids | Figure 3.35 | 116 |
| Naming Compounds | Figure 3.36 | 117 |
| Classes of Chemical Reactions | Table 5.1 | 185 |
| Decomposition Reactions That Occur When Compounds Are Heated | Table 5.2 | 186 |
| Summary of Combination Reactions | Table 5.3 | 188 |
| Activity Series | Figure 5.21 | 191 |
| Rules Used to Predict the Solubility of Ionic Compounds | Table 5.4 | 194 |
| Specific Heats of Some Substances | Table 6.2 | 239 |
| Heat Changes for the Combustion of Fuels | Table 6.3 | 246 |
| Energy Values of Food Components | Table 6.4 | 247 |
| Electromagnetic Spectrum | Figure 7.6 | 262 |
| Valence Level Electron Configurations of Elements | Figure 7.23 | 283 |
| Ionization Energies for the Elements | Figure 7.26 | 289 |
| Successive Ionization Energies for Period 2 Elements | Table 7.1 | 291 |
| Atomic Radii Values | Figure 7.30 | 292 |
| Ionic Radii Values | Figure 7.33 | 294 |
| Ionic versus Covalent Bonding | Figure 8.3 | 307 |
| Electronegativity Values | Figure 8.5 | 308 |
| Geometric Structures Arising from Different Numbers of Atoms | Table 8.5 | 328 |
| Arrangement of Electron Domains and Molecular Shapes | Table 8.6 | 330 |
| Volume Percent of Gases in the Atmosphere | Table 9.1 | 346 |
| Vapor Pressure of Water at Various Temperatures | Table 9.2 | 369 |
| Intermolecular Forces and Bonds | Table 10.2 | 410 |

| Description | Table/Figure Reference | Page |
|---|---|---|
| Classification of Solids | Table 10.3 | 426 |
| Like Dissolves Like | Table 11.2 | 451 |
| Solubilities of Substances as They Vary with Temperature | Figure 11.19 | 454 |
| Concentration Units | Table 11.3 | 457 |
| Meaning of the Value of $K_{eq}$ | Table 12.2 | 503 |
| Common Strong Acids | Table 13.1 | 535 |
| Common Strong Bases | Table 13.2 | 536 |
| Common Weak Acids | Table 13.3 | 537 |
| Common Weak Bases | Table 13.4 | 539 |
| Weak Acids, Their $K_a$ Values at 25°C, and Their Conjugate Bases | Table 13.5 | 541 |
| Acid Ionization Constants for Some Polyprotic Acids | Table 13.6 | 543 |
| pH Scale | Figure 13.16 | 553 |
| Acid-Base Indicators | Figure 13.18 | 554 |
| Rules for Assigning Oxidation Numbers | Table 14.1 | 577 |
| Properties of Types of Radiation | Table 15.1 | 619 |
| Half-Lives for the Nuclides in the Uranium Decay Series | Table 15.3 | 632 |
| Functional Groups in Organic Molecules | Table 16.1 | 657 |
| Classification of Hydrocarbons | Figure 16.2 | 660 |
| Rules for Naming Organic Compounds with Functional Groups | Table 16.8 | 686 |
| Amino Acids | Figure 17.4 | 700 |
| Genetic Codes for Amino Acids | Table 17.3 | 720 |

APPENDIX **B**

# Math Toolboxes

| Description | Math Toolbox | Page |
|---|---|---|
| Scientific Notation | Math Toolbox 1.1 | 35 |
| Significant Figures | Math Toolbox 1.2 | 37 |
| Units and Conversions | Math Toolbox 1.3 | 41 |
| Mole Conversions | Math Toolbox 4.1 | 158 |
| Graphing | Math Toolbox 9.1 | 376 |
| Solving Simple Algebraic Equations | Math Toolbox 9.2 | 379 |
| Log and Inverse Log Functions | Math Toolbox 13.1 | 558 |

# Answers to Consider This Questions and Practice Problems

## Chapter 1

### Answers to Consider This Questions

1.1 The properties describe a nonmetal, so the element would be chlorine.

1.2 The symbol P corresponds to phosphorus, a nonmetal in group VA of the periodic table. The symbol Po corresponds to polonium, a metal in group VIA.

1.3 The exposed water contains many dissolved and undissolved forms of matter.

1.4 None of the images represent a mixture of compounds. If the molecules of the compound shown in image D (red combined with black spheres) were shown with some of the molecules in image E, we would have a mixture of compounds.

1.5 We would expect the value to be very small. This incorrect answer is obtained by using the inverse of the conversion factor such that units do not cancel.

1.6 33.8 oz (assuming three significant figures)

1.7 Gold has a density of 19.3 g/cm$^3$, much greater than copper's, which is 8.9 g/cm$^3$. The density of gold is over twice as much as that of copper, so we would expect the volume occupied by this mass of gold to be less than half the volume occupied by the same mass of copper.

1.8 Water has a density of 1.00 g/cm$^3$ at 4°C and ice has a density of 0.92 g/cm$^3$. We would expect the hexane to float on top of the liquid water and ice.

1.9 Since this temperature is below the melting point, copper would be a solid.

1.10 Digestion processes are chemical changes.

1.11 In the before picture, each oxygen atom is attached to another oxygen atom. In the after picture, some of the oxygen atoms are attached to hydrogen in $H_2O$ and the other oxygen atoms are attached to carbon in $CO_2$. In the before picture, hydrogen atoms are attached to carbon in $CH_4$. In the after picture, hydrogen atoms are attached to oxygen in $H_2O$. Carbon atoms are attached to hydrogen in the before picture and are attached to oxygen in the after image.

1.12 The trails for atoms in A would appear shorter and the trails for B would appear longer.

1.13 Some examples include the brake lights on the cars and the traffic light. Another example is the spark plugs in the cars that started the engines.

### Answers to Practice Problems

1.1 phosphorus, carbon, bromine, sulfur
1.2 (a) Pb    (b) silver

1.3 *Aluminum can,* heterogeneous mixture (metal can with inner plastic liner and paint on the outside). *Stone fountain,* heterogeneous mixture. *Helium balloon,* the helium in the balloon is not a mixture. The helium and balloon together are a heterogeneous mixture. *Pizza,* heterogeneous mixture. *Copper coin,* if made of pure copper, not a mixture. Newer pennies are a heterogeneous mixture with a zinc core and coating of copper. Older pennies are a homogeneous mixture of copper and zinc.

1.4 (a) A    (b) $SO_3$
1.5 347,000 g; 347 kg
1.6 $4.60 \times 10^3$ mL; $4.60 \times 10^3$ cm$^3$; 1.22 gal
1.7 (a) 0.0518 cm$^3$    (b) 16.0 g
1.8 rises
1.9 (a) 245.1 K; −18.6°F    (b) −269°C    (c) 37.0°C; 310.1 K
1.10 (a) physical property    (b) chemical property
(c) chemical property    (d) physical property
1.11 physical change
1.12 the image with atoms moving slower (B)
1.13 light, heat, and chemical energy
1.14 (a) $1.23 \times 10^{-2}$    (b) $1.23 \times 10^2$    (c) 25,500    (d) 0.001
(e) $-2, 0, 10^{-6}, 1, 10^4$
1.15 (a) $6.0 \times 10^3$    (b) $3.0 \times 10^{15}$    (c) $8.0 \times 10^{-6}$
(d) $3.90 \times 10^{10}$
1.16 (a) 6    (b) 2
1.17 (a) 7.5    (b) 0.003    (c) $1.5 \times 10^{11}$    (d) 3.42
1.18 (a) 3.0    (b) 0.90    (c) 0.00    (d) 20.57
1.19 (a) 0.85    (b) 16.8    (c) 0.1    (d) 13
1.20 (a) 1.5    (b) 14    (c) 0.0011    (d) $1.2 \times 10^3$    (e) 0.20
1.21 27.6 g
1.22 $4.41 \times 10^{-4}$ lb
1.23 (a) 357.2 mi/h    (b) 159.7 m/s

## Chapter 2

### Answers to Consider This Questions

2.1 The atomic number would decrease to 4 and the mass number would stay the same. Because the atomic number changed, a different element, beryllium, would be represented.

2.2 The number of protons and electrons would remain the same, but the isotope would have two fewer neutrons, or 8 neutrons.

2.3 The 7 in each representation would change to a 6, so the isotope could be represented as $^6_3\text{Li}$, $^6\text{Li}$, lithium-6, or Li-6.

2.4 No, the mass number (protons + neutrons) distinguishes isotopes of an element and is needed to determine the number of neutrons.

2.5   $Na^+$ and $O^{2-}$ both have 10 electrons, the same number as the noble gas Ne.

2.6   Yes, the left-hand subscript can be omitted from an isotope symbol because it is the atomic number which is always the same for a given element.

2.7   The $^{37}Cl$ isotope would be more abundant because the relative atomic mass is closer to the mass of $^{37}Cl$ than $^{35}Cl$.

2.8   There could be a greater percent abundance of $^{24}Mg$ than on Earth. Alternatively, there could be a lighter isotope present, such as $^{23}Mg$, which would lower the weighted average.

2.9   hydrogen

2.10  Hydrogen can lose one electron to form $H^+$, like the other elements in group IA. This is the most common ion formed by hydrogen. Hydrogen can also gain one electron to form $H^-$, which would give it two electrons like the noble gas He. However, the hydrogen anion is called the hydride ion not hydrogen ion.

## Answers to Practice Problems

2.1   (a) 7 protons; 6 neutrons   (b) atomic number, 7; nitrogen   (c) 13

2.2   6 protons; 6 electrons; 8 neutrons

2.3   He-3; $^3_2He$; $^3He$; helium-3

2.4   (a) 78   (b) 20   (c) 7

2.5   (a) 9 protons; 10 electrons; anion   (b) 12 protons; 10 electrons; cation   (c) 7 protons; 10 electrons; anion

2.6   (a) $^{34}_{16}S^{2-}$   (b) $^{23}_{11}Na^+$   (c) $^{40}_{20}Ca^{2+}$

2.7   $^{63}Cu$

2.8   6.938 amu, which is only slightly lower than the periodic table value of 6.94

2.9   (a) nitrogen   (b) argon   (c) calcium   (d) silver

2.10  (a) $Li^+$   (b) $S^{2-}$   (c) $Al^{3+}$

## Chapter 3
## Answers to Consider This Questions

3.1   Both elements are nonmetals, so the compound would be molecular.

3.2   As an ionic compound, KCl is more likely to be a strong electrolyte.

3.3   The name for $Se^{2-}$ is selenide ion and $Sr^{2+}$ is called strontium ion. These ions contain the same number of electrons, and also the same as the nearest noble gas Kr.

3.4   The ion that contains more oxygen atoms, $BrO_3^-$, is the bromate ion.

3.5   Barium and chloride ions must be present in a ratio of 1 to 2. With 10 barium ions, 20 chloride ions would have to be shown.

3.6   The formula $Na_2S$ corresponds to sodium sulfide, a compound containing sodium ion and sulfide ion. Sulfite ion is a polyatomic anion with the formula $SO_3^{2-}$. The correct formula for the combination sodium and sulfite ions is $Na_2SO_3$.

3.7   The nitride ion has the formula $N^{3-}$. The compound $Ba(NO_3)_2$ contains the nitrate ion. The correct name is barium nitrate.

3.8   Roman numerals are used to represent the charge on the metal ion, not the number of metal ions in the formula. Iron(II) oxide has the formula FeO. The compound $Fe_2O_3$

is named iron(III) oxide because the iron in the compound has a charge of 3+.

3.9   The cobalt in cobalt(II) nitrate has a 2+ charge, and there are two nitrate ions in the formula. The cobalt in cobalt(III) nitrate has a 3+ charge, and there are three nitrate ions in the formula.

3.10  $FeSO_4$

3.11  dinitrogen monoxide

3.12  The formula $SO_3$ corresponds to a molecular compound with the name sulfur trioxide. The other formula, $SO_3^{2-}$, is a polyatomic ion containing two more electrons than an $SO_3$ molecule and is named sulfite ion.

3.13  The formula $HBr(g)$ corresponds to the molecular compound that consists of HBr molecules, each containing a hydrogen atom and a bromine atom. At room temperature this compound is a gas and is called hydrogen bromide. The formula $HBr(aq)$ refers to the solution that forms upon bubbling $HBr(g)$ into water. The HBr in aqueous solution exists as $H^+$ ions and $Br^-$ ions separate from each other and are surrounded by water molecules.

3.14  Prefixes like *di*– and *tri*– are only used in naming molecular compounds. This compound is ionic and contains a metal that can exhibit multiple charges. For ionic compounds containing metals like this, we use a Roman numeral to specify the charge. In this case, the copper has a 1+ charge to balance the 2– charge from the polyatomic ion. The anion in the compound is sulfate, not sulfide. The compound is named copper(I) sulfate.

## Answers to Practice Problems

3.1   (b) $CO_2$ and (d) $CCl_4$

3.2   Ionic compounds generally have higher melting points than molecular compounds, so the ionic compound $Al_2O_3$ should have a higher melting point than the molecular compound $N_2O_3$.

3.3   (a) 2+; $Ba^{2+}$; barium ion   (b) 1–; $Br^-$; bromide ion

3.4   $AsO_4^{3-}$

3.5

(water molecules not shown)

3.6   KBr; $K_2SO_4$; $FeBr_3$; $Fe_2(SO_4)_3$

3.7   potassium oxide; magnesium sulfite

3.8   chromium(III) sulfide; iron(II) sulfite

3.9   (a) $Cu_2SO_4$   (b) FeO

3.10  (a) $Sn(SO_4)_2$, stannic sulfate; $SnSO_4$, stannous sulfate   (b) ferric sulfide, $Fe_2S_3$; ferrous sulfide, FeS

3.11  tetraphosphorus hexoxide; dinitrogen pentoxide

3.12  $N_2O_3$

3.13  (a) hydroselenic acid   (b) phosphoric acid

3.14  (a) Molecular compound named phosphorus tribromide
      (b) Ionic compound named sodium sulfide
      (c) Acid named sulfuric acid
      (d) Ionic compond named ammonium chloride

# Chapter 4

## Answers to Consider This Questions

4.1   The percent copper would be the same for any sample size.

4.2   The mass of iron would be smaller, in proportion to the mass of the magnetite sample.

4.3   The number of atoms of hydrogen and sulfur would decrease in proportion to the number of moles of $H_2S$.

4.4   The molar mass would be greater since the product molecule would contain one more oxygen atom.

4.5   The appropriate conversion factor would be $\dfrac{55.85 \text{ g Fe}}{1 \text{ mol Fe}}$, in order to correctly cancel moles and leave grams in the units.

4.6   Since less sugar was added, the moles of sugar would be less.

4.7   It would contain fewer molecules because the molar mass would be twice as great as the value given and the mass is divided by the molar mass in the calculation.

4.8   The subscripts are no longer evenly divisible by 3, but they are divisible by 2, so the empirical formula would be $C_3HCl_2$.

4.9   Although the masses of the component elements would change, the chemical formula is independent of the sample size. The chemical formula would be the same.

4.10   The procedure would be the same except that we would have to calculate four mole ratios.

4.11   No, 1.875 is the same as 15/8, a ratio of whole numbers.

4.12   If the experimental molar mass is doubled, each subscript in the formula would also be doubled: $H_4C_4O_8$.

4.13   The percent copper would be smaller in $Cu_2Se$ since the mass of Se is greater than the mass of S.

4.14   The solution contains more $H_2O$ than $CH_3OH$, so $CH_3OH$ is the solute and $H_2O$ is the solvent.

4.15   A greater volume with the same amount of solute results in a more dilute solution, with a lower molarity.

4.16   The solution containing $K_2SO_4$ would contain double the moles of $K^+$.

4.17   If the solution contains more copper(II) chloride, the mass of copper(II) chloride would increase.

4.18   A smaller molarity with the same number of moles would require a greater volume of solution.

4.19   Since the final solution would contain less solute, it would have a lower molarity.

## Answers to Practice Problems

4.1   30.5% Fe; 34.9% S

4.2   30.2 mg Mg

4.3   $6.022 \times 10^{23}$ molecules $N_2O_5$; $1.204 \times 10^{24}$ atoms N; $3.011 \times 10^{24}$ atoms O

4.4   60.09 g/mol

4.5   0.0399 mol

4.6   $1.4 \times 10^{-4}$ mol; 3.53 g; 88 packets

4.7   $1.08 \times 10^{21}$ molecules

4.8   (a) $CH_2Cl_2$   (b) $CH_2O$   (c) $P_2O_3$

4.9   CuS

4.10   $CuSiH_4O_5$

4.11   $Cu_5Si_4O_{14}H_2$

4.12   $C_3H_6O_3$

4.13   66.46% Cu

4.14   solute, $H_2S$; solvent, $H_2O$

4.15   2.64 $M$

4.16   0.0650 mol

4.17   280,000 g or 280 kg

4.18   0.0365 L or 36.5 mL

4.19   0.259 $M$

4.20   (a) 0.0013 mol $NH_3$   (b) $4.7 \times 10^{24}$ $CuSO_4$ formula units

4.21   (a) 60 carbon atoms   (b) 225 atoms
       (c) $1.8 \times 10^{24}$ $NO_3^-$ ions

4.22   (a) 0.55 mol   (b) 9.86 g

4.23   (a) $1.57 \times 10^{25}$ oxygen atoms   (b) $2.3 \times 10^{21}$ $O_2$ molecules
       (c) $2.25 \times 10^{23}$ $Cl^-$ ions

# Chapter 5

## Answers to Consider This Questions

5.1   They are reactants.

5.2   No, although the atoms are balanced, a correctly balanced equation should have the lowest possible whole number coefficients.

5.3   The equation should represent the real reaction, and we cannot have fractions of atoms or molecules in an actual reaction.

5.4   The iron compound would be $FeSO_4$. The balanced equation would be

$$Fe(s) + CuSO_4(aq) \longrightarrow FeSO_4(aq) + Cu(s)$$

5.5   This fits the requirements of a combination reaction even though there are three reactants, not two. Compounds are combining to form one new compound. An example would be $2Fe(s) + 3Cl_2(g) + 12H_2O(l) \longrightarrow 2FeCl_3 \cdot 6H_2O(s)$.

5.6   Metal carbonates decompose to form metal oxides and carbon dioxide.

5.7   No, this fits the pattern of single-displacement reactions.

5.8   Yes, the activity of aluminum is high enough that it reacts with steam or with acids to give hydrogen gas.

5.9   The ion of the metal lowest in the activity series is most likely to be converted to its element form.

5.10   No, a precipitation reaction will occur only if one of the combinations of ions is insoluble.

5.11   Yes, as long as the anion contains the $SO_3$ group it is possible to produce $SO_2(g)$.

5.12   No, some compounds, such as $CH_4$, do not release hydrogen ions and are not acids.

5.13   Combination reactions form only one product compound.

5.14   No, the double-displacement would precipitate both cations and both anions, leaving no ions in solution.

## Answers to Practice Problems

5.1   reactants, $Fe(s)$ and $CuSO_4(aq)$; products, $Cu(s)$ and $FeSO_4(aq)$

5.2   $2Na(s) + 2H_2O(l) \longrightarrow 2NaOH(aq) + H_2(g)$

5.3   $2C_4H_{10}(g) + 13O_2(g) \longrightarrow 8CO_2(g) + 10H_2O(g)$

5.4   $Zn(s) + 2HNO_3(aq) \longrightarrow Zn(NO_3)_2(aq) + H_2(g)$

5.5   (a) combination   (b) double-displacement
       (c) decomposition   (d) single-displacement

5.6   $MgSO_3 \cdot 3H_2O(s) \xrightarrow{\text{heat}} MgSO_3(s) + 3H_2O(g)$
       [If enough heat is applied, $MgSO_3(s)$ can further decompose to magnesium oxide and sulfur dioxide.]

5.7   $2Mg(s) + O_2(g) \longrightarrow 2MgO(s)$;
       $3Mg(s) + N_2(g) \longrightarrow Mg_3N_2(s)$

5.8   Zinc reacts with the acid in tomatoes.

5.9   Co; $SnSO_4(aq)$;
       $Co(s) + SnSO_4(aq) \longrightarrow CoSO_4(aq) + Sn(s)$

5.10    The precipitate is $CdS(s)$.

$Cd(NO_3)_2(aq) + Na_2S(aq) \longrightarrow CdS(s) + 2NaNO_3(aq)$

5.11    $BaS(aq) + H_2SO_4(aq) \longrightarrow BaSO_4(aq) + H_2S(g)$

5.12    $Ca(OH)_2(aq) + H_2SO_4(aq) \longrightarrow CaSO_4(s) + 2H_2O(l)$

5.13    $2CH_3OH(l) + 3O_2(g) \longrightarrow 2CO_2(g) + 4H_2O(g)$

5.14    $BaCl_2(aq) + Na_2SO_4(aq) \longrightarrow BaSO_4(s) + 2NaCl(aq)$;

$Ba^{2+}(aq) + 2Cl^-(aq) + 2Na^+(aq) + SO_4^{2-}(aq) \longrightarrow$

$BaSO_4(s) + 2Na^+(aq) + 2Cl^-(aq)$;

$Ba^{2+}(aq) + SO_4^{2-}(aq) \longrightarrow BaSO_4(s)$

# Chapter 6
## Answers to Consider This Questions

6.1    Remember that the moles of a substance can be converted to units of grams using the molar mass of water, 18.02 g/mol:

$$\text{Grams } H_2O = 1.52 \, \text{mol } H_2O \times \left(\frac{18.02 \text{ g } H_2O}{1 \text{ mol } H_2O}\right) = 27.4 \text{ g } H_2O$$

6.2    The law of conservation of mass tells us that mass cannot be gained (or lost) during a chemical reaction. Because the mass of the starting material is 1.0 g, the total mass of the products should be equal to 1.0 g (given a complete reaction). Each product, then, should have a mass less than 1.0 g.

6.3    Because silver nitrate is the limiting reactant, adding more will allow the reaction to continue, decreasing the amount of copper metal, increasing the amount of silver metal, and causing the solution to become a deeper blue.

6.4    To make 25 cars, Joe would need 20 more solar cells, 19 more motors, and 21 more frames so he would have 25 of each. He would not have to order more wheels because 100 wheels is the exact number he would need.

6.5    The balanced equation tells us that the ratio of $N_2$ to $H_2$ is 1 to 3. If 12 $H_2$ molecules were added to the "before" picture, there would be 18 $H_2$ molecules that could completely react with the 6 $N_2$ molecules present.

6.6    The limiting reactant would be $O_2$.

6.7    The mass of $AlBr_3$ produced would not change because Al is not the limiting reactant.

6.8    If the mass of the limiting reactant Na is doubled, the theoretical yield of $H_2$ would also double to 0.50 mol. Because the ratio of actual yield to theoretical yield is constant with the same percent yield, we would expect the actual yield of $H_2$ to also double from 0.21 mol to 0.42 mol.

6.9    The reactant mixture of $H_2$ and $O_2$ is higher in potential energy than the product because the reaction is exothermic.

6.10    To obtain an energy value in units of J/g, we would divide the energy content of the soda in the can, in units of joules, by the mass of the soda, in units of grams. The mass of the soda in the can would need to be obtained from the container label.

6.11    The metals are all the same substance so the specific heat value would be the same for each piece, and the temperature changes would be the same, so the only variable is the mass of the pieces. Thus, the metal piece with the greatest mass would melt the wax the most since it could release the most heat to the wax.

6.12    The amount of heat absorbed by the rocks would increase because there would be a greater mass of absorbing material.

6.13    The heat gained by the brick would increase because its temperature would increase by a greater amount to reach thermal equilibrium.

6.14    The amount of heat released to the surroundings would be 136.7 kJ. This is one-tenth the amount of heat that would be released per mole of ethanol burned. The amount released is the absolute value of the heat change for this process.

## Answers to Practice Problems

6.1    142 mol

6.2    (a) $1.4 \times 10^9$ g or $1.4 \times 10^6$ kg    (b) $1.6 \times 10^9$ g or $1.6 \times 10^6$ kg
(c) yes; the law of conservation of mass

6.3    There would be no copper wire present, and the silver metal crystals would be lying at the bottom of the beaker.

6.4    (a) solar cells    (b) 5 cars    (c) 1 motor, 9 frames, 80 wheels

6.5    (a) $H_2$

(b)

(c) $N_2$

6.6    $AgNO_3$; 1.0 mol

6.7    $Na_2CO_3$; 4.57 g

6.8    68%

6.9    Endothermic; when the battery runs, energy is released. Recharging a battery is the opposite reaction, so recharging must be endothermic.

6.10    $9.83 \times 10^5$ J

6.11    A greater mass of water transfers more heat.

6.12    $-6910$ J; negative; the reaction is exothermic

6.13    (a) Greater; the rise in temperature of the water indicates that the alloy was hotter.    (b) $-3400$ J

6.14    (a) endothermic    (b) 133 kJ

# Chapter 7
## Answers to Consider This Questions

7.1    Blue, indigo, and violet light have shorter wavelengths and higher frequencies than green light. Therefore, lasers of these colors would consist of photons with higher energies than that of green and red light.

7.2    The amount of energy emitted per second ($6.0 \times 10^4$ J/s) is about $10^{23}$ times greater than the energy of a photon ($2.47 \times 10^{-19}$ J/s), so we could estimate that $10^{23}$ photons are emitted per second. We can also use these relationships and dimensional analysis to calculate the number of photons emitted per second:

$$\frac{6.0 \times 10^4 \text{ J}}{1 \text{ s}} \times \frac{1 \text{ photon}}{27.4 \times 10^{-19} \text{ J}} = 2.4 \times 10^{23} \text{ photons/s}$$

7.3    The energy emitted for the $n = 7$ to $n = 2$ transition is slightly greater than the $n = 6$ to $n = 2$ transition, so the wavelength of radiation emitted should be slightly shorter than that of violet light. We can see from the electromagnetic spectrum that this corresponds to ultraviolet light.

7.4 All of the $2p$ orbitals have the same energy, so the paired electrons can occupy any of them in the ground state.

7.5 $1s^2 2s^2 2p^6$

7.6 The noble gas Kr has four fewer electrons than Zr, so the electron configuration for Kr is like that of Zr except it does not include the two last-occupied sublevels $5s$ and $4d$: $1s^2 2s^2 2p^6 3s^2 3p^6 4s^2 3d^{10} 4p^6$.

7.7 Group IVA (14)

7.8 Isoelectronic ions differ in the number of protons in the nucleus, and they have different ion charges.

7.9 Because ionization energy is a measure of the amount of energy required to remove a valence electron, your answers should be the same.

7.10 On the periodic table, phosphorus is one group to the right and one row lower than carbon. The figure shows that atomic size decreases from left to right but increases going down the periodic table. However, note that the incremental changes in atomic size going down a group are much greater than the changes in atomic size going across from one group to the next.

7.11 The positively charged ions are smaller than their neutral atoms, and the negatively charged ions are larger than their neutral atoms.

## Answers to Practice Problems

7.1 (a) infrared    (b) UV    (c) UV

7.2 (a) $2.61 \times 10^9$/s or $2.61 \times 10^9$ Hz    (b) $1.73 \times 10^{-24}$ J

7.3 (a) $4.85 \times 10^{-19}$ J    (b) red line at 656 nm

7.4 (a) incorrect; aufbau principle    (b) incorrect; Pauli exclusion principle    (c) correct

7.5 (a) $1s^2 2s^2 2p^6 3s^2 3p^1$    (b) $1s^2 2s^2 2p^6 3s^2 3p^6 4s^2 3d^1$    (c) $1s^2 2s^2 2p^6 3s^2 3p^6 4s^1$

7.6 (a) $1s^2 2s^2 2p^6 3s^2 3p^4$    (b) $1s^2 2s^2 2p^6 3s^2 3p^6 4s^2 3d^{10} 4p^6 5s^1$    (c) $1s^2 2s^2 2p^6 3s^2 3p^6 4s^2 3d^{10} 4p^6 5s^2 4d^{10}$

7.7 (a) 2    (b) 4    (c) 4

7.8 (a) $[Ar]4s^2 3d^{10} 4p^6$ or [Kr]; $Sr^{2+}$ and $Rb^+$ among others
   (b) $[He]2s^2 2p^6$ or [Ne]; $F^-$ and $Na^+$ among others
   (c) [Ar]; $S^{2-}$ and $Ca^{2+}$ among others

7.9 (a) lithium    (b) chlorine

7.10 (a) potassium; valence electron in energy level farther from the nucleus    (b) silicon; increased positive charge in the nucleus is more effective at drawing in valence electrons of chlorine.

7.11 $Y^{3+} < Sr^{2+} < Rb^+ < Br^- < Se^{2-}$

## Chapter 8
## Answers to Consider This Questions

8.1 Metalloids behave chemically like nonmetals, so the bonding would be expected to be covalent.

8.2 The positions of the component elements in the periodic table allow us to decide on the relative electronegativities of the elements.

8.3 Since the two elements are in the same group in the periodic table, the only change would be to use Se as the symbol instead of S.

8.4 It would have a subscript of 2 after the symbol for the cation, $(K^+)_2$ ($:\overset{\cdot\cdot}{\underset{\cdot\cdot}{S}}:^{2-}$)

8.5 Phosphorus would share electrons with three chlorine atoms, forming three single bonds, leaving one unshared pair of electrons on the phosphorus atom.

8.6 Because sulfur atoms have the same number of valence electrons as oxygen atoms, the structures would be very similar. The symbol for oxygen would be replaced by the symbol for sulfur, but the arrangement of bonding and nonbonding valence electrons would be identical.

8.7 Double bonds cannot be drawn to hydrogen atoms so the $H_2CO$ molecule would not exhibit resonance.

8.8 Yes, but the oxygen would be attached to only one of the carbon atoms and a hydrogen atom.

8.9 The parent structure would be the same since there would still be four electron domains around the oxygen atom.

8.10 No, with four electron domains, the parent structure would be tetrahedral, so the bond angle would be closer to $109.5°$.

8.11 No, it would be nonpolar since it would have a symmetrical structure and the bond polarities would cancel.

## Answers to Practice Problems

8.1 (a) ionic    (b) covalent    (c) ionic and covalent

8.2 most polar bond, S—O; partial negative on the oxygen in $SO_2$ and the chlorine in $CCl_4$; $N_2$ and $PH_3$ consist of nonpolar bonds.

8.3 $\cdot\overset{\cdot}{Si}\cdot$

8.4 $\cdot Be\cdot$ ; $\cdot\overset{\cdot\cdot}{N}\cdot$ ; $(Be^{2+})_3 (:\overset{\cdot\cdot}{N}:^{3-})_2$

8.5 The formula of the compound is $NF_3$ because nitrogen typically forms three bonds and fluorine forms one bond.

$$:\overset{\cdot\cdot}{\underset{\cdot\cdot}{F}}-\overset{\cdot\cdot}{\underset{\cdot\cdot}{N}}-\overset{\cdot\cdot}{\underset{\cdot\cdot}{F}}:$$
$$|$$
$$:\overset{\cdot\cdot}{\underset{\cdot\cdot}{F}}:$$

8.6
$$:\overset{\cdot\cdot}{\underset{\cdot\cdot}{O}}:$$
$$H:C:H$$

8.7 $:N:::N:\overset{\cdot\cdot}{\underset{\cdot\cdot}{O}}: \longleftrightarrow :\overset{\cdot\cdot}{N}::N::\overset{\cdot\cdot}{O}: \longleftrightarrow :\overset{\cdot\cdot}{N}:N:::O:$

8.8
H  :O:  H
|   ||   |
H—C—C—C—H
|        |
H       H

8.9 trigonal planar

8.10 bent; about $120°$

8.11 $C_2H_6$, nonpolar; $NO_2$, polar; $SO_2$, polar; $SO_3$, nonpolar

## Chapter 9
## Answers to Consider This Questions

9.1 625 torr; 735 torr would be observed at an elevation closer to sea level.

9.2 The pressure falls by half to 0.4 atm and volume doubles to 14 L.

9.3 With an increase in pressure by a factor of 2.5, the volume is expected to decrease by a factor of 2.5. The calculated answer of 205 mL makes more sense.

9.4 The temperature falls by half to 186 K and volume also falls by half to 0.5 L.

9.5 Charles's law relating volume and temperature only follows for temperatures expressed on the absolute Kelvin scale. A negative volume would be obtained if temperatures in $°C$ are used instead of kelvins. The new volume would be 29.9 mL.

9.6   Pressure is not constant in this case. It decreases by a factor of 5 while the temperature only decreases by a factor of 1.4.

9.7   Under initial and final conditions, the ratio of volume to moles in both cases is 12.5. This makes sense according to Avogadro's hypothesis; under conditions of constant temperature and pressure, the ratio of volume to moles is constant.

9.8   Under standard conditions of temperature and pressure, 1 mol of any gas occupies 22.414 L. The volume occupied by 3.00 mol of neon gas is expected to be the same as argon, 67.2 L.

9.9   This corresponds to 1.00 mol of gas as described in the example. 4.00 g of helium would be present.

9.10   No, the answer does not make sense. Forgetting to convert pressure and temperature would yield units of $\dfrac{torr \cdot mol \cdot K}{atm \cdot \,^\circ C}$. A quantity of 349 mol of propane in an "empty" cylinder would weigh almost 34 pounds!

9.11   Under ideal conditions, the cylinders would contain the same number of moles, 0.0384 mol. However, because butane has a greater molar mass than propane, the butane cylinder would contain 2.23 g of gas.

9.12   The calculated amount of oxygen produced would be 0.0607 mol, which is about 3.6% higher than actual.

9.13   0.491 g/L

9.14   10.1 g CaO and 7.97 g $CO_2$.

## Answers to Practice Problems

9.1   0.814 atom; 619 torr

9.2

9.3   4.03 atm

9.4

9.5   −74.4°C
9.6   0.892 L
9.7   28.1 L
9.8   168 L
9.9   10.6 L
9.10   0.0776 mol
9.11   2.46 g
9.12   0.0876 mol
9.13   221 L
9.14   0.0901 mol; 8.83 g

9.15   The data are not linear:

9.16   4.0 L
9.17   98.2°F
9.18   238

## Chapter 10

## Answers to Consider This Questions

10.1   Higher, since it would require a higher temperature to reach a vapor pressure of 1.00 atm.

10.2   Energy must be added to a solid to form a gas, so the process would be endothermic.

10.3   The graph would be flipped left to right.

10.4   Only the first energy term would change; since the temperature change of the ice would be doubled, the heat required to warm the ice to 0.0°C would double. The total heat absorbed would then be higher.

10.5   Helium, since it has the smallest electron cloud of all the noble gases.

10.6   Since a linear structure of $NO_2$ would be symmetrical, it would be a nonpolar molecule and would not experience dipole-dipole forces.

10.7   The hydrogen atoms that participate in hydrogen bonding must be bonded to N, O, or F, so the hydrogen atoms bonded to the nitrogen can participate in hydrogen bonding in liquid methylamine.

10.8   We would consider the polarities and sizes of the molecules. We could also consider their boiling points.

10.9   No, since the melting points would increase as the intermolecular forces increase.

10.10   The nonpolar compound would have a higher vapor pressure, since the attractions between molecules would be weaker.

10.11   The intermolecular forces in $H_2O$ are stronger than those in $CH_3OH$, so $H_2O$ has a lower vapor pressure.

10.12   The polar substance has stronger intermolecular forces, so it should have a greater viscosity.

**10.13** The intermolecular forces in water are greater than those in alcohol, so the water has the greater surface tension.

**10.14** Ionic and network solids are very hard, so they can be eliminated.

**10.15** Metallic, ionic, and network solids would be possibilities.

## Answers to Practice Problems

**10.1** $-42°C$; $-62°C$

**10.2** deposition; exothermic

**10.3** (a) EF  (b) CD  (c) AB

**10.4** $3.19 \times 10^5$ J

**10.5** $SiH_4$

**10.6** (a) yes  (b) yes  (c) yes  (d) no

**10.7** $CH_3CO_2H$ and $HOCl$

**10.8** (a) $Cl_2$, larger electron cloud  (b) $NH_3$, hydrogen bonding  (c) $SO_2$, dipole-dipole force between polar molecules  (d) $H_2Te$, larger electron cloud

**10.9** $CHCl_3 < CHBr_3 < CHI_3$

**10.10** $O_2(l)$; weaker intermolecular forces due to smaller electron cloud

**10.11** HF; HF; HF; CO

**10.12** SAE 15

**10.13** the intermolecular forces between the new material and water molecules

**10.14** network solid

**10.15** Molecular solid; covalent bonds hold the atoms together within the molecules and London dispersion forces hold molecules together.

## Chapter 11

## Answers to Consider This Questions

**11.1** According to the solubility rules, sulfates are soluble, with $BaSO_4$ being an exception. When this compound is mixed with water, we would expect to see the white solid suspended until it settles to the bottom. There may be some, but very little dissolved $Ba^{2+}$ and $SO_4^{2-}$ ions present along with the water and undissolved $BaSO_4$ crystals.

**11.2** Energy is required to break the ionic bonds in $KNO_3$. Energy is also required to separate some of the water molecules by breaking hydrogen bonds. Energy is released when ion-dipole forces form between the ions and the water molecules.

**11.3** unsaturated; 45 g NaCl

**11.4** 7.30 mL ethanol

**11.5** 0.14 mg Hg

**11.6** 36.5 g NaCl

**11.7** 36.8 g $H_2O$

**11.8** Evaporating the water would not work. After evaporation a mixture of solids would be present. The $BaSO_4$ could be separated from the aqueous KCl using a centrifuge which concentrates the solid in the bottom of the reaction container. The aqueous solution of KCl could then be carefully poured off or filtered.

**11.9** 0.355 g $Na_2SO_4$

**11.10** The vapor pressure of a solution containing a nonvolatile solute will be lower than the vapor pressure of the pure solvent.

**11.11** The two solutions produce the same number of particles per formula unit that dissolves, so the boiling points of these solutions would be the same.

## Answers to Practice Problems

**11.1** (a) $Zn^{2+}(aq)$, $Cl^-(aq)$, $H_2O(l)$;
$$ZnCl_2(s) \longrightarrow Zn^{2+}(aq) + 2Cl^-(aq)$$
(b) $CH_3OH(aq)$, $H_2O(l)$;
$$CH_3OH(l) \longrightarrow CH_3OH(aq)$$
(c) $K^+(aq)$, $NO_3^-(aq)$, $H_2O(l)$;
$$KNO_3(s) \longrightarrow K^+(aq) + NO_3^-(aq)$$

**11.2** London dispersion forces between $I_2$ molecules and between $C_6H_{14}$ molecules

**11.3** (a) 2.15%  (b) 0.538 g

**11.4** 87.2%

**11.5** 0.0040 ppm

**11.6** 1.04 $M$

**11.7** 5.34 $m$

**11.8** 114 mL; 3.88 g

**11.9** 0.1373 $M$

**11.10** $-2.8°C$

**11.11** 0.5 $m$ $Ca(NO_3)_2$

## Chapter 12

## Answers to Consider This Questions

**12.1** The energy diagram for the reverse reaction would show an endothermic reaction.

**12.2** Reducing the volume would increase the concentrations of reactants A and B because there would be less empty space between the reactant particles. Therefore, reducing the volume would increase the total number of collisions per second. It would have no effect on the fraction of collisions that are effective.

**12.3** To reduce the reaction rate, larger pieces of magnesium metal could be used. With less surface area there would be fewer magnesium atoms accessible for collisions.

**12.4** The reverse reaction would also run faster with the catalyst than without the catalyst because the catalyst lowers the activation energy for both the forward and reverse reactions.

**12.5** Two different activated complexes are formed. An activated complex is formed in each step of the reaction as bonds are broken and new bonds are formed.

**12.6** When the reaction is written in reverse, the $K_{eq}$ expression is inverted, so its value is the inverse of the original reaction:
$$K'_{eq} = \frac{1}{K_{eq}} = \frac{1}{6.1 \times 10^{-4}} = 1.6 \times 10^3.$$

**12.7** The reaction would run in the reverse direction to reach equilibrium because 1 is greater than the $K_{eq}$ value of 0.16.

**12.8** Yes. Increasing the amount of carbon in the reaction container would not disrupt the equilibrium because it is a solid.

**12.9** No. The equilibrium constant remains unchanged. The equilibrium shifts so that the ratio of products to reactants returns to the value of $K_{eq}$.

12.10   We could conclude that there are equal numbers of gas molecules on each side of the balanced equation, or that there are no gas molecules involved in the reaction at all.

12.11   Adding heat to an endothermic reaction shifts the position of equilibrium towards products. Removing heat would have the opposite effect.

12.12   Increasing the temperature changes the value of the equilibrium constant for a reaction. None of the other conditions listed would change its value.

## Answers to Practice Problems

12.1

12.2   (a) both decrease    (b) The total number of collisions increases, and the fraction of effective collisions stays the same.

12.3   increase surface area of iron to increase collision frequency; increase oxygen concentration to increase collision frequency; increase temperature to increase collision frequency and fraction of effective collisions

12.4

12.5   $H_3O^+$, catalyst; $H_3C-CH_2^+$, $H_2O$, $CH_3CH_2OH_2^+$, intermediates

12.6   (a) $K_{eq} = \dfrac{[N_2O_4]}{[NO_2]^2}$   (b) 222   (c) product favored

12.7   (a) not at equilibrium; reaction proceeds in reverse direction   (b) increase

12.8   (a) $K_{eq} = \dfrac{[CO]}{[CO_2]}$   (b) $K_{eq} = [Pb^{2+}][Cl^-]^2$

12.9   (a) reaction shifts to the right; more AgI dissolves
(b) reaction shifts to the left; AgI precipitates
(c) reaction shifts to the left; AgI precipitates

12.10   (a) reaction shifts left   (b) no shift

12.11   endothermic

12.12   (a) no; reaction shifts left, decreasing products   (b) yes; reaction shifts right, increasing products   (c) yes; reaction shifts right, increasing products   (d) no; reaction shifts left, decreasing products   (e) no; catalysts do not change the position of equilibrium

## Chapter 13
## Answers to Consider This Questions

13.1   $HCO_3^-$

13.2   $H_3PO_4$

13.3   Each hydrogen atom is connected to an oxygen atom:

$$\overset{\displaystyle :\!\overset{..}{O}\!:}{\underset{}{H-\overset{..}{\underset{..}{O}}-\overset{\|}{C}-\overset{..}{\underset{..}{O}}-H}}$$

13.4   Hydroxide ions form when a base dissolves in water. If the same amounts of a strong base and weak base are added to the same amount of water, the strong base solution will contain a greater concentration of $OH^-$ ions because it ionizes completely. Only a fraction of weak base molecules ionize.

13.5   The dissolved species must be a base.

13.6   The conjugate base of the weaker acid is the stronger base. HCN is a weaker acid than HOCl, so $CN^-$ is a stronger base than $OCl^-$.

13.7   $C_4H_4O_6^{2-}$

13.8   The solution would be acidic because the $OH^-$ concentration is less than that in pure water.

13.9   The $H_3O^+$ concentration in the $H_2SO_4$ solution would be greater because the second hydrogen would ionize to some extent.

13.10   The pH would be higher because $HNO_2$ is a weak acid which produces fewer $H_3O^+$ ions than the strong acid $HNO_3$.

13.11   The pH would be less than zero (a negative value).

13.12   The pH would fall to a much lower value than would be expected for a buffered solution.

13.13   The resulting solution would contain the unreacted 0.50 mol HF along with the 0.50 mol NaF that formed, so it would be a buffer system.

## Answers to Practice Problems

13.1   (a) $OCl^-$, base; $H_2O$, acid    (b) $F^-$, base; $H_2SO_4$, acid
(c) $H_2O$, base; $NH_4^+$, acid

13.2   (a) HF   (b) $H_2CO_3$   (c) $H_3O^+$

13.3   Glycine has one acidic hydrogen, the one attached to the oxygen.

13.4   (a) strong acid   (b) weak base   (c) weak acid
(d) weak base

13.5   $HClO_4$ is shown in A because it is a strong acid and is completely ionized to $H_3O^+(aq)$ and $ClO_4^-(aq)$

13.6   HOCl

13.7    (a) $H_2C_2O_4(aq) + H_2O(l) \rightleftharpoons HC_2O_4^-(aq) + H_3O^+(aq)$
$HC_2O_4^-(aq) + H_2O(l) \rightleftharpoons C_2O_4^{2-}(aq) + H_3O^+(aq)$
(b) $H_2C_2O_4$

13.8    (a) $1.0 \times 10^{-6}\ M$; acidic    (b) $1.0 \times 10^{-12}\ M$; basic

13.9    $[OH^-] = 0.85\ M$; $[H_3O^+] = 1.2 \times 10^{-14}\ M$

13.10    (a) 3.07    (b) 13.00    (c) 0.00

13.11    5.00; no

13.12    $H_2PO_4^-(aq) + H_2O(l) \rightleftharpoons HPO_4^{2-}(aq) + H_3O^+(aq)$;
A base produces $OH^-$ ions which react with $H_3O^+$ ions to produce water. The reaction shifts to the right to reestablish equilibrium, replacing $H_3O^+$ that was removed.

13.13    (a) no    (b) yes; $CH_3CO_2H(aq) + H_2O(l) \rightleftharpoons$
$CH_3CO_2^-(aq) + H_3O^+(aq)$    (c) no

13.14    (a) 2.12    (b) 7.25

13.15    (a) $1.0 \times 10^{-8}$    (b) $2.5 \times 10^{-3}$

# Chapter 14

## Answers to Consider This Questions

14.1    Nothing. Recall from Chapter 5 that only the more active metal, copper, displaces the less active metal, silver, from its ionic compound. Silver, a less active metal, cannot displace copper from its ionic compound.

14.2    In $CrO_3$ chromium has an oxidation number of 6+, while in $Cr_2O_3$ we determined that the oxidation number for each chromium atom is 3+.

14.3    Each iron atom would lose three electrons and each oxygen atom would gain two electrons.

14.4    26.1 g Cu

14.5    cathode

14.6    The iron electrode would increase in mass due to the reduction of $Fe^{3+}$ to neutral iron atoms, which would form on the surface of the electrode. The manganese electrode would decrease in mass as the manganese atoms are oxidized to form $Mn^{2+}$ ions, which would be released from the surface of the electrode.

14.7    The reduction half-reaction would be
$2H_2O(l) + NO_3^-(aq) + 3e^- \longrightarrow NO(g) + 4OH^-(aq)$. The final balanced equation after accounting for equalization of electrons gained and lost would be
$3Cu(s) + 4H_2O(l) + 2NO_3^-(aq) \longrightarrow$
$2NO(g) + 8OH^-(aq) + 3Cu^{2+}(aq)$

14.8    No. Only Fe and Zn would work. As a less reactive metal, Sn is not capable of reducing $Ni^{2+}$.

14.9    The potassium metal would react with iodine molecules in the reverse reaction to produce KI.

## Answers to Practice Problems

14.1    (a) $2Fe(s) + 3CuSO_4(aq) \longrightarrow 3Cu(s) + Fe_2(SO_4)_3(aq)$
(b) Fe, oxidized; $Cu^{2+}$, reduced
(c) $CuSO_4$, oxidizing agent; Fe, reducing agent
(d) A coating of copper metal appears on the surface of the iron.

14.2    (a) 0    (b) Mn, 3+; O, 2−    (c) Na, 1+; S, 6+; O, 2−
(d) Cr, 6+; O, 2−

14.3    (a) yes; $H_2O$ acts as both oxidizing and reducing agent.
(b) no    (c) yes; Al, reducing agent; $O_2$, oxidizing agent
(d) yes; Mg, reducing agent; $Zn(NO_3)_2$, oxidizing agent

14.4    (a) oxidation, $Cd(s) \longrightarrow Cd^{2+}(aq) + 2e^-$;
reduction, $Ni^{2+}(aq) + 2e^- \longrightarrow Ni(s)$

(b) $Ni(NO_3)_2$, oxidizing agent; Cd, reducing agent

(c)

As drawn, oxidation will occur in the left half-cell, reduction in the right; $Cd(s)$ is the anode and $Ni(s)$ is the cathode; anions will flow from the salt bridge into the oxidation half-cell, and cations will flow into the reduction half-cell; electrons will flow through the wire toward the cathode; the mass of the anode will decrease, and the mass of the cathode will increase.

14.5    $Ag_2O$, oxidizing agent; Zn, reducing agent

14.6    $3Mg(s) + 2Cr(NO_3)_3(aq) \longrightarrow 3Mg(NO_3)_2(aq) + 2Cr(s)$

14.7    $6OH^-(aq) + 3Cl_2(aq) \longrightarrow$
$5Cl^-(aq) + ClO_3^-(aq) + 3H_2O(l)$

14.8    Place a solid Cr electrode in a $CrCl_3$ solution to construct one half-cell. In the other half-cell place a Zn electrode in a $ZnCl_2$ solution. Connect the two half-cells with a salt bridge and external wire.

14.9    (a) $Ca^{2+}$ ions are reduced to form Ca atoms.    (b) $Cl^-$ ions are oxidized to form Cl atoms. Two of these atoms combine to form a $Cl_2$ molecule.
(c) oxidation, $2Cl^-(aq) \longrightarrow Cl_2(g) + 2e^-$;
reduction, $Ca^{2+}(aq) + 2e^- \longrightarrow Ca(l)$
(d) $CaCl_2(l) \longrightarrow Ca(l) + Cl_2(g)$

# Chapter 15

## Answers to Consider This Questions

15.1    The mass number would be decreased by 8 instead of 4 and the atomic number would be decreased by 4 instead of 2, giving $^{220}_{84}Po$.

15.2    No, the mass of an electron is so small compared to the mass of a proton that it would have a negligible effect on the mass of the nucleus if multiple beta particles were emitted.

15.3    No element would be formed. Positron emission corresponds to conversion of a proton to a neutron. Since H-1 has only one proton, the product would be a neutron with a mass number of 1 and an atomic number of 0.

15.4    No, the mass number of the product would be two less, 263 instead of 265.

15.5    No, the $N/Z$ ratio is 10/10 = 1, which is the ideal ratio for stability.

15.6   No, it is the total number of each type of particle that is emitted that determines the composition of the final product. Intermediate products would be different, however.

15.7   This corresponds to three half-lives, so the elapsed time would be 3 × 28 years or 84 years.

15.8   No, it would be necessary to set a new baseline for the expected amount of C-14 and to adjust the calculation to take this into account.

15.9   No, it is the decay products that cause harm.

15.10  Since the sum of the atomic numbers must equal 92 and the atomic number of Sn is 50, the other product must be Mo, with an atomic number of 42, $^{100}_{42}$Mo.

15.11  The mass number of the product would be lowered by 1, giving Mg-23.

## Answers to Practice Problems

15.1   $^{218}_{84}$Po

15.2   $^{234}_{91}$Pa

15.3   $^{11}_{6}$C

15.4   $^{54}_{24}$Cr; $^{209}_{83}$Bi + $^{54}_{24}$Cr $\longrightarrow$ $^{262}_{107}$Bh + $^{1}_{0}$n

15.5   positron ($^{0}_{1}\beta^{+}$) emission

15.6   $^{208}_{82}$Pb

15.7   12.5%

15.8   11,460 years

15.9   $\alpha$ and $_{-1}^{0}\beta$ radiation

15.10  1 neutron; $^{235}_{92}$U $\longrightarrow$ $^{137}_{52}$Te + $^{97}_{40}$Zr + $^{1}_{0}$n

15.11  $2^{16}_{8}$O $\longrightarrow$ $^{32}_{16}$S

## Chapter 16
## Answers to Consider This Questions

16.1

propanol        isopropanol

16.2   The branched hydrocarbon has the weaker intermolecular forces so it would have the higher vapor pressure.

16.3

pentane        2,2-dimethylpropane

16.4   The first student numbered from the left instead of from the right end of the chain. The second student did not identify the longest chain. The correct name is 2-methylhexane.

16.5

2-methylheptane

3-methylheptane

4-methylheptane

16.6   The bromine atom could substitute for a hydrogen atom on carbon 1 or carbon 2, giving a mixture of products:

16.7   2-methyl-2-pentene

16.8

cis-2-pentene        trans-2-pentene

16.9   $CH_2=CHCH=CH_2 + 2H_2 \xrightarrow{Pt} CH_3CH_2CH_2CH_3$

16.10  4-chloroaniline

16.11

## Answers to Practice Problems

16.1

16.2   This molecule has more branching than the other two, so it has less surface area and, therefore, weaker intermolecular forces.

16.3   2-methylpropane

16.4   2,2,5-trimethylheptane

16.5

16.6   Alkanes do not react with bromine unless subjected to heat, sunlight, or ultraviolet radiation. Thus, there would be no reaction under the conditions given.

16.7   2-methyl-1-butene

16.8   (a) no cis-trans isomers    (b) yes;

16.9   $CH_3CH=CHCH_3 + HBr \longrightarrow CH_3CH_2CHBrCH_3$

16.10  3-chloro-5-nitrophenol or
       1-chloro-3-hydroxy-5-nitrobenzene

16.11  2,4-dimethyl-3-pentanol

## Chapter 17
## Answers to Consider This Questions

17.1

17.2

$H_2N-CH-C-N-CH-C-N-CH-C-OH$

17.3   Ile-Asp-Ser-Ala

17.4

Pleated sheet   Active site   Alpha helices

17.5

17.6   The A-T pair has two hydrogen bonds that connect the bases. The C-G pair consists of three hydrogen bonds.

17.7   AATTGCGCTAT

17.8   UAC for Tyr and ACC, ACA or CAG for Thr

17.9   All three rings are hexose rings because each one consists of six members.

17.10

**Answers to Practice Problems**

17.1

alpha carbons

$H_2N-CH-C-N-CH-CO_2H$

alanine   peptide bond   cysteine

17.2   (a)

$H_2N-CH-C-N-CH-C-OH$

Leu-Ser

$H_2N-CH-C-N-CH-C-OH$

Ser-Leu

(b) Gly-Cys-Ile, Gly-Ile-Cys, Ile-Gly-Cys, Ile-Cys-Gly, Cys-Ile-Gly, Cys-Gly-Ile

17.3   Ala-Cys-Ala

17.4   only tertiary structure because image does not show the overlaid backbone

17.5   deoxyribose, thymine, DNA

17.6   GCCTTAAATGCGGCC

17.7   CUGAUUGGAAUCU

17.8   Leu, His

17.9   disaccharide

17.10   solid

# Answers to Selected Questions and Problems

## Chapter 1 — Matter and Energy

1.1 (a) mass; (b) chemical property; (c) mixture; (d) element; (e) energy; (f) physical property; (g) liquid; (h) density; (i) homogeneous mixture; (j) solid

1.3 (a) $2.95 \times 10^4$; (b) $8.2 \times 10^{-5}$; (c) $6.5 \times 10^8$; (d) $1.00 \times 10^{-2}$

1.5 (a) 0.0000186; (b) 10,000,000; (c) 453,000; (d) 0.0061

1.7 (a) $6.2 \times 10^3$; (b) $3.5 \times 10^7$; (c) $2.9 \times 10^{-3}$; (d) $2.5 \times 10^{-7}$; (e) $8.20 \times 10^5$; (f) $1.6 \times 10^{-6}$

1.9 (a) 3; (b) 2; (c) 4; (d) 2; (e) 3

1.11 (a) 1.5; (b) 1.5; (c) 14; (d) 1.20

1.13 (a) 2.8; (b) 0.28; (c) 2.8; (d) 0.049

1.15 (a) 0.800 kg/L; (b) 0.3 $g/cm^3$; (c) 25.4 $m^2$

1.17 (a) 1.21; (b) 0.204; (c) 1.84; (d) 42.2; (e) 0.00710

1.19 (a) 0.036 m; (b) $3.57 \times 10^5$ g; (c) 0.07650 L; (d) 0.84670 cm; (e) $5.97 \times 10^{-7}$ m; (f) 92.7 cm; (g) $7.62 \times 10^4$ g; (h) 865 L; (i) 17.5 in; (j) 0.5214 lb; (k) 2.1 qt

1.21 (a) 0.589 mi; (b) 14.9 lb; (c) $6.616 \times 10^{-2}$ gal; (d) $2.30 \times 10^2$ mL; (e) $4.50 \times 10^3$ nm; (f) 0.952 m; (g) $3.12 \times 10^2$ kg; (h) $5 \times 10^2$ mL; (i) 4.10 ft; (j) $1.19 \times 10^{-3}$ lb; (k) $6.6 \times 10^{-9}$ gal

1.23 (a) $7.38 \times 10^4$ ft/min; (b) 1.50 $in^3$; (c) 0.697 $lb/in^3$

1.25 (a) homogeneous mixture if the dye is evenly mixed into the water; (b) element; (c) homogeneous mixture; (d) heterogeneous mixture

1.27 Gasoline, automobile exhaust, oxygen gas, and the iron pipe are matter. Sunlight is energy.

1.29 Elements are composed of only one type of atom. Compounds are made up of two or more different elements.

1.31 Metals are lustrous (shiny) and conduct heat and electricity. In addition you can form wires with metals (ductile) and you can make foil out of them by hitting them with a hammer (malleable).

1.33 (a) titanium; (a) tantalum; (c) thorium; (d) technetium; (e) thallium

1.35 (a) boron; (b) barium; (c) beryllium; (d) bromine; (e) bismuth

1.37 (a) nitrogen; (b) iron; (c) manganese; (d) magnesium; (e) aluminum; (f) chlorine

1.39 (a) Fe; (b) Pb; (c) Ag; (d) Au; (e) Sb

1.41 Ir is the symbol for the element iridium. Iron's symbol is Fe.

1.43 The correct symbol is No.

1.45 The only pure substance is the salt in the salt shaker (if it is not iodized salt). Hamburger: heterogeneous mixture. Salt: pure substance. Soft drink: heterogeneous mixture (until it goes flat). Ketchup: heterogeneous mixture.

1.47 $H_2$

1.49 The chemical formula is $N_2O_4$.

1.51 Image D represents a mixture of an element and a compound.

1.53 The elements are $O_2$, $P_4$, and He. $Fe_2O_3$, NaCl, and $H_2O$ are compounds.

1.55

Liquid state          Solid state

1.57 gas

1.59 (a) gas; (b) liquid; (c) solid

1.61 solid

1.63 $O_2(aq)$

1.65 physical properties

1.67 (a) 0.045 g; (b) $1.6 \times 10^{-3}$ oz; (c) $9.9 \times 10^{-5}$ lb

1.69 (a) 0.10 mg; (b) $1.0 \times 10^2$ µg; (c) $1.0 \times 10^{-7}$ kg

1.71 (a) $1.2 \times 10^3$ mL; (b) $1.2 \times 10^3$ $cm^3$; (c) $1.2 \times 10^{-3}$ $m^3$

1.73 $1.6 \times 10^2$ mL, 0.16 L

1.75 1.3 $g/cm^3$, 1.3 g/mL

1.77 38.5 mL

1.79 The molecules in the liquid state are closer together than molecules in the gas state. More matter in the same volume means that the density is higher.

1.81 oil (least) < plastic < water (greatest)

1.83 329 K

1.85

| Scale | Freezing | Boiling | Difference |
|---|---|---|---|
| Celsius | 0°C | 100°C | 100°C |
| Kelvin | 273.15 K | 373.15 K | 100 K |
| Fahrenheit | 32°F | 212°F | 180°F |

1.87 no

1.89 Physical properties are (a) mass, (b) density, and (e) melting point. Chemical properties are (c) flammability, (d) resistance to corrosion, and (f) reactivity with water.

1.91 Physical changes are (a) boiling acetone, (b) dissolving oxygen gas in water, and (e) screening rocks from sand. Chemical changes are (c) combining hydrogen and oxygen to make water, (d) burning gasoline, and (f) conversion of ozone to oxygen.

1.93 Symbolic: $Cl_2(g) \longrightarrow Cl_2(l)$. A molecular-level diagram is shown.

Gas          Condensation          Liquid

1.95   chemical change

1.97   physical change

1.99   physical change

1.101   Anna and Bill would have observed kinetic energy from the movement of the welder and the motion of the sparks. The sparks would have glowed, indicating heat, light, and chemical energy.

1.103   The molecules in image A have greater kinetic energy because they are moving faster.

1.105   Any object that would move if allowed has potential energy (e.g., a picture hanging on a wall).

1.107   The people walking, the wheel chair rolling, and the suitcase being pushed all have kinetic energy. The people, the wall art, and objects on the tables all have potential energy. Many objects have potential and kinetic energy.

1.109   Consider as an example a car going down the road. A car going up a hill is converting chemical energy in the fuel to mechanical energy to reach the top. At the top of the hill, the car has potential energy. If it rolls down the hill, it gains kinetic energy but loses potential energy. Other energy conversions can be found while driving a car.

1.111   When the batter swings the bat, potential energy (metabolic energy) is converted into kinetic energy (moving bat). The bat strikes the ball, and the kinetic energy of the bat is transferred to the kinetic energy of the ball. As the ball leaves the bat, it rises against Earth's gravity and kinetic energy is converted to potential energy. When the ball starts dropping again, potential energy is converted to kinetic energy.

1.113   BMI $= 21.7$ kg/m$^2$; healthy

1.115   As the water leaves the top of the fountain, it possesses kinetic energy (going up). That energy is converted to potential energy. As the water falls, the potential energy converts back into kinetic energy.

1.117   It is used to explain and predict scientific results.

1.119   (a) hypothesis; (b) observation; (c) theory; (d) observation

1.121   You could do some library research and find the densities of many different woods. By placing samples of the different woods in water, you can determine if the theory is correct.

1.123   At high altitude, the air pressure is lower. As a result, when a balloon rises, it expands. Since the mass of air in the balloon has not changed but the volume has increased, the density of the balloon is lower.

1.125   zinc

1.127   (a) He; (b) Ne; (c) Ar; (d) Kr; (e) Xe; (f) Rn

1.129   physical

1.131   8 mg

1.133   $1.7 \times 10^{-4}$ lb/oz; 0.02 to 0.03 lb

1.135   (a) 0.00 K; (b) $-273.15°$C; (c) $-459.67°$F

1.137   (a) compound; (b) element; (c) mixture

1.139   Blood is a mixture because it contains more than one substance. It is further classified as a heterogeneous mixture because the solids are suspended, not dissolved.

1.141   $1.00 \times 10^7$ cm$^3$

1.143   5.30 kg; 11.7 lb

1.145   $2.1 \times 10^2$ mi

1.147   B

1.148   B and D

1.149   D

1.150   C

1.151   A

1.152   C

1.153   B

1.154   C

1.155   D

1.156   A

1.157   C

## Chapter 2 — Atoms, Ions, and the Periodic Table

2.1   (a) 1; (b) law of conservation of mass; (c) proton; (d) main-group element; (e) relative atomic mass; (f) mass number; (g) isotope; (h) cation; (i) subatomic particle; (j) alkali metal; (k) periodic table

2.3   Dalton used the laws of conservation of mass (Lavoisier) and definite proportions (Proust).

2.5   They differ in their atomic masses and chemical properties.

2.7   Compounds contain discrete numbers of atoms of each element that form them. Because all the atoms of an element have the same relative atomic mass, the mass ratio of the elements in a compound is always the same (law of definite proportions).

2.9   No. Hydrogen atoms are not conserved. A molecule of hydrogen, $H_2$, should be added to the left panel.

2.11   Thomson's cathode-ray experiment

2.13   electrons

2.15   The nucleus of helium has two protons and two neutrons. Two electrons can be found outside the nucleus.

2 electrons

+ proton

neutron

2.17   neutron

2.19   Carbon has six protons. The relative atomic mass of a carbon atom is 12.01 amu, indicating the presence of six neutrons. Protons and neutrons have approximately equal masses, so the nuclear mass is approximately two times the mass of the protons.

2.21  (a) 1; (b) 8; (c) 47

2.23  protons

2.25  number of protons

2.27  (b) atomic number

2.29  The following table displays the atomic, neutron, and mass numbers for the isotopes of hydrogen.

|  | $^1_1\text{H}$ | $^2_1\text{H}$ | $^3_1\text{H}$ |
|---|---|---|---|
| Atomic number | 1 | 1 | 1 |
| Neutron number | 0 | 1 | 2 |
| Mass number | 1 | 2 | 3 |

2.31  (a) $Z = 18$, $N = 18$, $A = 36$; (b) $Z = 18$, $N = 20$, $A = 38$; (c) $Z = 18$, $N = 22$, $A = 40$

2.33

|  | Protons | Neutrons | Electrons |
|---|---|---|---|
| (a) $^{15}_{8}\text{O}$ | 8 | 7 | 8 |
| (b) $^{109}_{47}\text{Ag}$ | 47 | 62 | 47 |
| (c) $^{35}_{17}\text{Cl}$ | 17 | 18 | 17 |

2.35  (a) $^3_1\text{H}$; (b) $^9_4\text{Be}$; (c) $^{31}_{15}\text{P}$

2.37

|  | Protons | Neutrons |
|---|---|---|
| (a) $^{56}_{26}\text{Fe}$ | 26 | 30 |
| (b) $^{39}_{19}\text{K}$ | 19 | 20 |
| (c) Copper-65 or $^{65}_{29}\text{Cu}$ | 29 | 36 |

2.39  seven protons and six neutrons

2.41

|  | Protons | Neutrons | Electrons |
|---|---|---|---|
| (a) $^{23}_{11}\text{Na}$ | 11 | 12 | 11 |
| (b) $^{56}_{25}\text{Mn}$ | 25 | 31 | 25 |
| (c) $^{18}_{8}\text{O}$ | 8 | 10 | 8 |
| (d) $^{19}_{9}\text{F}$ | 9 | 10 | 9 |

2.43  They differ in the number of electrons.

2.45  (a) an anion with a 1− charge is formed; (b) a cation with a 2+ charge is formed

2.47  (a) $Zn^{2+}$, cation; (b) $P^{3-}$, anion

2.49

|  | Protons | Electrons |
|---|---|---|
| (a) $Zn^{2+}$ | 30 | 28 |
| (b) $F^-$ | 9 | 10 |
| (c) $H^+$ | 1 | 0 |

2.51

|  | Protons | Neutrons | Electrons |
|---|---|---|---|
| (a) $^{37}_{17}\text{Cl}^-$ | 17 | 20 | 18 |
| (b) $^{25}_{12}\text{Mg}^{2+}$ | 12 | 13 | 10 |
| (c) $^{13}_{7}\text{N}^{3-}$ | 7 | 6 | 10 |
| (d) $^{40}_{20}\text{Ca}^{2+}$ | 20 | 20 | 18 |

2.53  potassium, K

2.55  copper, Cu

2.57  $^7$Li has three protons, three electrons, and four neutrons. $^7$Li$^+$ has only two electrons, and $^6$Li has three neutrons. Otherwise they are the same as $^7$Li. Lithium-6 differs the most in mass.

2.59  19 protons and 18 electrons

2.61  The atomic mass unit is defined as one-twelfth the mass of one carbon-12 atom.

2.63  (a) 2 amu; (b) 238 amu

2.65  The mass of $D_2$ is two times the mass of $H_2$.

2.67  about 40 amu

2.69  The numerical values of masses of individual atoms are very small when measured on the gram scale. The size of the atomic mass unit allows us to make easier comparisons and calculations of masses of molecules.

2.71  The mass number is the sum of the number of protons and neutrons and is always an integer value. In contrast, the mass of an atom is the actual measurement of how much matter is in the atom and is never exactly an integer value (except carbon-12).

2.73  A mass spectrometer is used to determine the mass of an atom. The mass number of an atom is the sum of the number of protons and neutrons.

2.75  calcium-40

2.77  21.50 amu

2.79  (a) $^{58}$Ni; (b) $^{64}$Ni; (c) 58;

(d)

|  | Protons | Neutrons | Electrons |
|---|---|---|---|
| $^{58}_{28}\text{Ni}^+$ | 28 | 30 | 27 |
| $^{60}_{28}\text{Ni}^+$ | 28 | 32 | 27 |
| $^{61}_{28}\text{Ni}^+$ | 28 | 33 | 27 |
| $^{62}_{28}\text{Ni}^+$ | 28 | 34 | 27 |
| $^{64}_{28}\text{Ni}^+$ | 28 | 36 | 27 |

2.81  10,810 amu or $1.081 \times 10^4$ amu

2.83  2500 amu of boron contains more atoms.

2.85  (a) K; (b) Br; (c) Mn; (d) Mg; (e) Ar; (f) Br, K, Mg, Al, Ar

2.87  chlorine, Cl

2.89  titanium, Ti

2.91  (a) metal; (b) nonmetal; (c) metal; (d) metalloid

2.93  (a) main group; (b) main group; (c) main group; (d) actinide; (e) transition metal

2.95  Group VIIA (17), the halogen family, all occur as diatomic molecules.

2.97  neon

2.99  Group VIIIA (18), the noble gases

2.101  electrons

2.103  (a) group IA (1); (b) group IIA (2); (c) group VIIA (17); (d) group VIA (16)

2.105  (a) $Na^+$; (b) $O^{2-}$; (c) $S^{2-}$; (d) $Cl^-$; (e) $Br^-$

2.107  All alkali metal elements (group 1 or IA excluding hydrogen)

2.109  The mass of oxygen added to form $Fe_2O_3$ causes an increase in the mass.

2.111  The mass ratios of Zn/S are the same for the two samples (within the significant figures given). The mass ratio is approximately 2.0:1.0.

2.113  Electrons have charge and were readily studied in cathode-ray tubes.

2.115  nickel-60

2.117  19 protons and 20 neutrons

2.119  The most abundant isotopes of cobalt have masses greater than the masses of the most abundant isotopes of nickel. As a result, the relative atomic mass of cobalt is greater than that of nickel.

**2.121** When there are many isotopes, some of the isotopes can be present in very low abundance. As a result, their masses cannot be determined as accurately and their percentage contributions are also less well known. Both factors result in a decrease in the precision of the calculated relative atomic mass.

**2.123** 127

**2.125** carbon

**2.127** 20% boron-10 and 80% boron-11; the relative atomic mass (10.81 amu) is about 80% of the difference between the two isotopes.

**2.129** $Br_2(l)$

**2.131** hydrogen

**2.133** The energy required to force a proton (positive charge) into the nucleus is too great.

**2.135**

|  | Particle | Mass (g) | Relative Charge |
|---|---|---|---|
| (a) | neutron | $1.6749 \times 10^{-24}$ | 0 |
| (b) | electron | $9.1094 \times 10^{-28}$ | 1− |
| (c) | proton | $1.6726 \times 10^{-24}$ | 1+ |

**2.137** 9 protons; 9 neutrons; 9 electrons

**2.139** Neutral Fe atoms differ from $Fe^{2+}$ and $Fe^{3+}$ ions in number of electrons. Each of these contains 26 protons. A neutral Fe atom has 26 electrons, an $Fe^{2+}$ ion has 24 electrons, and an $Fe^{3+}$ ion has 23 electrons.

**2.141** group VIIA (17); period 5

**2.143** less than 0.1%

**2.145** E

**2.146** B

**2.147** E

**2.148** A

**2.149** A

**2.150** D

**2.151** C

**2.152** D

**2.153** E

**2.154** B

## Chapter 3 — Chemical Compounds

**3.1** (a) formula unit; (b) strong electrolyte; (c) molecular compound; (d) acid; (e) nonelectrolyte; (f) oxoanion

**3.3** (a) $K_2S$ is ionic. (b) $Na_2SO_4$ is ionic. (c) $SO_2$ is molecular.

**3.5** (a) molecular; (b) ionic; (c) molecular; (d) ionic

**3.7** (a) molecular; (b) ionic; (c) ionic; (d) molecular

**3.9** The ionic compound LiF would have the highest melting point.

**3.11** (a) 1+, $Na^+$, sodium ion; (b) 1+, $K^+$, potassium ion; (c) 1+, $Rb^+$, rubidium ion

**3.13** (a) 2+, $Ca^{2+}$, calcium ion; (b) 3−, $N^{3-}$, nitride ion; (c) 2−, $S^{2-}$, sulfide ion

**3.15** $NO_2^-$, nitrite ion

**3.17** (a) sulfate ion; (b) hydroxide ion; (c) perchlorate ion

**3.19** (a) $N^{3-}$; (b) $NO_3^-$; (c) $NO_2^-$

**3.21** (a) $CO_3^{2-}$; (b) $NH_4^+$; (c) $OH^-$; (d) $MnO_4^-$

**3.23** $SO_3^{2-}$, sulfite ion

**3.25** $IO_3^-$, iodate ion

**3.27**

**3.29** (a) $BaCl_2$; (b) $FeBr_3$; (c) $Ca_3(PO_4)_2$; (d) $Cr_2(SO_4)_3$

**3.31** (a) $K^+$ and $Br^-$; (b) $Ba^{2+}$ and $Cl^-$; (c) $Mg^{2+}$ and $PO_4^{3-}$; (d) $Co^{2+}$ and $NO_3^-$

**3.33** (a) $Fe^{2+}$ will form FeO and $Fe^{3+}$ will form $Fe_2O_3$. (b) $Fe^{2+}$ will form $FeCl_2$ and $Fe^{3+}$ will form $FeCl_3$.

**3.35** (a) 2−; (b) 2+

**3.37** For each "compound" written, the charges do not balance (compounds have no net charge). (a) too many chloride ions, NaCl; (b) not enough potassium ions, $K_2SO_4$; (c) There should be three nitrate ions and one aluminum ion, $Al(NO_3)_3$

**3.39** (a) magnesium chloride; (b) aluminum oxide; (c) sodium sulfide; (d) potassium bromide; (e) sodium nitrate; (f) sodium perchlorate

**3.41** $MnSO_4$ and $CoCl_2$

**3.43** (a) copper(I) oxide or cuprous oxide; (b) chromium(II) chloride; (c) iron(III) phosphate or ferric phosphate; (d) copper(II) sulfide or cupric sulfide

**3.45** (a) $CaSO_4$; (b) BaO; (c) $(NH_4)_2SO_4$; (d) $BaCO_3$; (e) $NaClO_3$

**3.47** (a) $Co^{2+}$, cobalt(II) chloride; (b) $Pb^{4+}$, lead(IV) oxide or plumbic oxide; (c) $Cr^{3+}$, chromium(III) nitrate; (d) $Fe^{3+}$, iron(III) sulfate or ferric sulfate

**3.49** (a) $CoCl_2$; (b) $Mn(NO_3)_2$; (c) $Cr_2O_3$; (d) $Cu_3(PO_4)_2$

**3.51** ferrous nitrate and ferric nitrate

**3.53**

|  | $Ca^{2+}$ | $Fe^{2+}$ | $K^+$ |
|---|---|---|---|
| $Cl^-$ | $CaCl_2$ calcium chloride | $FeCl_2$ iron(II) chloride | KCl potassium chloride |
| $O^{2-}$ | CaO calcium oxide | FeO iron(II) oxide | $K_2O$ potassium oxide |
| $NO_3^-$ | $Ca(NO_3)_2$ calcium nitrate | $Fe(NO_3)_2$ iron(II) nitrate | $KNO_3$ potassium nitrate |
| $SO_3^{2-}$ | $CaSO_3$ calcium sulfite | $FeSO_3$ iron(II) sulfite | $K_2SO_3$ potassium sulfite |
| $OH^-$ | $Ca(OH)_2$ calcium hydroxide | $Fe(OH)_2$ iron(II) hydroxide | KOH potassium hydroxide |
| $ClO_3^-$ | $Ca(ClO_3)_2$ calcium chlorate | $Fe(ClO_3)_2$ iron(II) chlorate | $KClO_3$ potassium chlorate |

| | $Mn^{2+}$ | $Al^{3+}$ | $NH_4^+$ |
|---|---|---|---|
| $Cl^-$ | $MnCl_2$ manganese(II) chloride | $AlCl_3$ aluminum chloride | $NH_4Cl$ ammonium chloride |
| $O^{2-}$ | $MnO$ manganese(II) oxide | $Al_2O_3$ aluminum oxide | $(NH_4)_2O$ ammonium oxide |
| $NO_3^-$ | $Mn(NO_3)_2$ manganese(II) nitrate | $Al(NO_3)_3$ aluminum nitrate | $NH_4NO_3$ ammonium nitrate |
| $SO_3^{2-}$ | $MnSO_3$ manganese(II) sulfite | $Al_2(SO_3)_3$ aluminum sulfite | $(NH_4)_2SO_3$ ammonium sulfite |
| $OH^-$ | $Mn(OH)_2$ manganese(II) hydroxide | $Al(OH)_3$ aluminum hydroxide | $NH_4OH$ ammonium hydroxide |
| $ClO_3^-$ | $Mn(ClO_3)_2$ manganese(II) chlorate | $Al(ClO_3)_3$ aluminum chlorate | $NH_4ClO_3$ ammonium chlorate |

3.55

| | potassium | iron(III) | strontium |
|---|---|---|---|
| **iodide** | $KI$ | $FeI_3$ | $SrI_2$ |
| **oxide** | $K_2O$ | $Fe_2O_3$ | $SrO$ |
| **sulfate** | $K_2SO_4$ | $Fe_2(SO_4)_3$ | $SrSO_4$ |
| **nitrite** | $KNO_2$ | $Fe(NO_2)_3$ | $Sr(NO_2)_2$ |
| **acetate** | $KCH_3CO_2$ | $Fe(CH_3CO_2)_3$ | $Sr(CH_3CO_2)_2$ |
| **hypochlorite** | $KClO$ | $Fe(ClO)_3$ | $Sr(ClO)_2$ |

| | aluminum | cobalt(II) | lead(IV) |
|---|---|---|---|
| **iodide** | $AlI_3$ | $CoI_2$ | $PbI_4$ |
| **oxide** | $Al_2O_3$ | $CoO$ | $PbO_2$ |
| **sulfate** | $Al_2(SO_4)_3$ | $CoSO_4$ | $Pb(SO_4)_2$ |
| **nitrite** | $Al(NO_2)_3$ | $Co(NO_2)_2$ | $Pb(NO_2)_4$ |
| **acetate** | $Al(CH_3CO_2)_3$ | $Co(CH_3CO_2)_2$ | $Pb(CH_3CO_2)_4$ |
| **hypochlorite** | $Al(ClO)_3$ | $Co(ClO)_2$ | $Pb(ClO)_4$ |

3.57  $AgCl$

3.59  (a) $NF_3$; (b) $P_4O_{10}$; (c) $C_2H_4Cl_2$

3.61  (a) phosphorus pentafluoride; (b) phosphorus trifluoride; (c) carbon monoxide; (d) sulfur dioxide

3.63  (a) $SF_4$; (b) $C_3O_2$; (c) $ClO_2$; (d) $SO_2$

3.65  image B

3.67  (a) hydrofluoric acid; (b) nitric acid; (c) phosphorous acid

3.69  (a) $HF(aq)$; (b) $H_2SO_3(aq)$; (c) $HClO_4(aq)$

3.71  hydrogen ions and nitrate ions

3.73  $KI$, $Mg(NO_3)_2$, $NH_4NO_3$

3.75  ionic; $TiO_2$, $ZnO$

3.77  molecular; $CO_2$ and $N_2O$

3.79  potassium sulfide, sodium sulfate, sulfur dioxide

3.81  (a) nitrogen trioxide, molecular; (b) nitrate ion, an ion; (c) potassium nitrate, ionic; (d) sodium nitride, ionic; (e) aluminum chloride, ionic; (f) phosphorus trichloride, molecular; (g) titanium(II) oxide, ionic; (h) magnesium oxide, ionic

3.83  (a) $Na_2CO_3$; (b) $NaHCO_3$; (c) $H_2CO_3(aq)$; (d) $HF(aq)$; (e) $SO_3$; (f) $CuSO_4$; (g) $H_2SO_4(aq)$; (h) $H_2S(aq)$

3.85  $HCl(aq)$ is dissolved in water and ionized; $HCl(g)$ is not ionized; $HCl(aq)$, hydrochloric acid; $HCl(g)$, hydrogen chloride

3.87  (a) Prefixes are not used when naming ionic compounds. (b) Calcium ion's charge is known and always the same, so it should *not* be indicated as a Roman numeral when naming this compound. (c) Copper's charge, 2+, should be indicated by including the Roman numeral (II) after copper in its name. (d) Prefixes are used when naming molecular compounds.

3.89  (a) Two potassium ions are needed to balance $S^{2-}$. (b) Only one $Ni^{2+}$ is needed to balance charge. (c) Nitride is $N^{3-}$. (d) The prefix *tri–* is associated with iodide, $NI_3$.

3.91  (a) one sodium ion, $Na^+$, and one chloride ion, $Cl^-$ (b) one magnesium ion, $Mg^{2+}$, and two chloride ions, $Cl^-$ (c) two sodium ions, $Na^+$, and one sulfate ion, $SO_4^{2-}$ (d) one calcium ion, $Ca^{2+}$, and two nitrate ions, $NO_3^-$

3.93  (a) electrolyte; (b) electrolyte; (c) electrolyte; (d) nonelectrolyte

3.95  The ions of silver, zinc, and cadmium can each only have one possible charge. They are assumed to have these charges in the compounds they form.

3.97  (a) $NO_3^-$; (b) $SO_3^{2-}$; (c) $NH_4^+$; (d) $CO_3^{2-}$; (e) $SO_4^{2-}$; (f) $NO_2^-$; (g) $ClO_4^-$

3.99  (a) magnesium bromide; (b) dihydrogen sulfide; (c) hydrosulfuric acid; (d) cobalt(III) chloride; (e) potassium hydroxide; (f) silver bromide

3.101  (a) $PbCl_2$; (b) $Mg_3(PO_4)_2$; (c) $NI_3$; (d) $Fe_2O_3$; (e) $Ca_3N_2$; (f) $Ba(OH)_2$; (g) $Cl_2O_5$; (h) $NH_4Cl$

3.103  $NaHCO_3$

3.105  $Ca(ClO)_2$

3.107  $H_2O(l)$

3.109  $Cu$, $AgNO_3$, $Ag$, $Cu(NO_3)_2$

3.111  (a) $NH_3$; (b) $HNO_3(aq)$; (c) $HNO_2(aq)$

3.113  Hydrogen atoms attached to oxygen atoms are responsible for the acidic properties.

3.115  All contain oxygen. Magnesium oxide, $MgO$, has metal and nonmetal ions, which makes it an ionic compound. It is a solid at room temperature. Oxygen and carbon dioxide ($O_2$ and $CO_2$) are both molecular substances and gases at room temperature.

3.117  sodium chloride, ionic, $NaCl$; sodium bicarbonate, ionic, $NaHCO_3$

3.119  $ZnO$; zinc ion has a charge of 2+; oxide ion has a charge of 2−.

3.121  $FeCl_2$ and $CrCl_3$

3.123  $HF(g)$, hydrogen fluoride, molecular compound; $HF(aq)$, hydrofluoric acid, acid

3.125  $CrCl_3$

3.127  E

3.128  A

3.129  E

3.130  B

3.131  D

3.132  C

3.133  B

3.134  D

3.135  A

3.136  B

## Chapter 4 — Chemical Composition

4.1 (a) mole; (b) Avogadro's number; (c) empirical formula; (d) solute; (e) molarity; (f) concentrated solution

4.3 40.0% calcium

4.5 60.0% carbon

4.7 0.226 g lithium

4.9 (a) $H_2S$; (b) $N_2O_3$; (c) $CaCl_2$

4.11 (a) $H_2SO_4$; (b) $SCl_4$; (c) $C_2H_4$

4.13 $CO_2$

4.15 (a) NaCl; (b) $CH_4$; (c) $Cl_2$; (d) $SiO_2$.

4.17 $3.0 \times 10^{23}$ $NH_3$ molecules; $3.0 \times 10^{23}$ N atoms; $9.0 \times 10^{23}$ H atoms

4.19 $3.0 \times 10^{23}$ $Cu_2S$ formula units

4.21 $1 \times 10^{23}$ S atoms

4.23 $6 \times 10^{23}$ calcium ions

4.25 (a) NaCl, 58.44 g/mol; (b) $CH_4$, 16.04 g/mol; (c) $Cl_2$, 70.90 g/mol; (d) $SiO_2$, 60.09 g/mol

4.27 (a) 472.1 g/mol; (b) 172.17 g/mol; (c) 150.90 g/mol; (d) 120.06 g/mol

4.29 (a) 253.8 g/mol; (b) 158.4 g/mol; (c) 56.10 g/mol

4.31 To measure out a useful number of atoms by counting would not be possible because atoms are too small for us to see and manipulate.

4.33 42.39 amu

4.35 35.9 g/mol

4.37 (a) 0.0999 mol; (b) 0.293 mol; (c) 0.127 mol; (d) 0.0831 mol

4.39 (a) 0.556 mol; (b) $1.388 \times 10^{-3}$ mol; (c) 733 mol; (d) $5.72 \times 10^{-8}$ mol

4.41 C

4.43 (a) 428 g; (b) 177 g; (c) 436 g; (d) $2.20 \times 10^2$ g

4.45 $5.0 \times 10^2$ g

4.47 (a) 1.76 mol; (b) $1.06 \times 10^{24}$ $NH_3$ molecules; (c) $1.06 \times 10^{24}$ N atoms; (d) 5.28 mol of H atoms

4.49 $H_2SO_4$ has the most atoms per mole because it has more atoms per molecule. Na has the least atoms per mole.

4.51 $1.7 \times 10^{21}$ molecules

4.53 (a) $9.421 \times 10^{23}$ formula units; (b) $1.581 \times 10^{24}$ formula units; (c) $8.355 \times 10^{24}$ formula units; (d) $2.696 \times 10^{24}$ formula units

4.55 (a) $8.364 \times 10^{25}$ H atoms; (b) $5.138 \times 10^{23}$ $Ca^{2+}$ ions and $1.028 \times 10^{24}$ $NO_3^-$ ions; (c) $2.809 \times 10^{24}$ N atoms and $2.809 \times 10^{24}$ O atoms; (d) $9.676 \times 10^{23}$ $K^+$ ions and $4.838 \times 10^{23}$ $SO_4^{2-}$ ions

4.57 6.8 g $SO_2$

4.59 $NH_3$

4.61 no

4.63 The molecular formula shows the exact numbers of each atom present in a compound. The empirical formula shows the relative amounts of each atom in a compound expressed as the smallest whole numbers.

4.65 $H_2O_2$ (empirical, HO); $N_2O_4$ (empirical, $NO_2$)

4.67 (a) $P_2O_5$; (b) same as molecular; (c) same as molecular; (d) $C_3H_5O_2$

4.69 (a) $C_3H_2Cl$; (b) same as molecular; (c) same as molecular

4.71 $NO_2$ and $N_2O_4$ have the same empirical formulas.

4.73 (a) $Fe_3O_4$; (b) $C_6H_5NO_2$

4.75 $C_5H_6O$

4.77 $C_7H_5N_3O_6$

4.79 $C_7H_5O_6$

4.81 percent composition by mass and the molar mass

4.83 $C_3H_6O_3$

4.85 $C_6H_{12}O_6$

4.87 (a) 50.05% S, 49.95% O; (b) 47.27% Cu, 52.73% Cl; (c) 42.07% Na, 18.89% P, 39.04% O; (d) 16.39% Mg, 18.89% N, 64.72% O

4.89 Chalcocite ($Cu_2S$) and cuprite (CuO) are both relatively high in copper content (79.84% and 79.89%, respectively).

4.91 A solution is a homogeneous mixture of two or more substances. Some common solutions: clear drinks (coffee and tea), window cleaner, soapy water, tap water, air, brass (a homogeneous mixture of copper and zinc).

4.93 Water is the solvent because it is present in the largest amount. Calcium chloride, $CaCl_2$, is the solute.

4.95 Concentrated means high in solute concentration, and dilute means low in solute concentration.

4.97 Concentration describes the quantity of solute in a given amount of solvent or solution.

4.99 solution A

4.101 (a) 2.03 $M$; (b) 0.540 $M$; (c) 3.20 $M$; (d) 3.21 $M$

4.103 0.0186 mol $Na_2SO_4$; 0.0372 mol $Na^+$; 0.0186 mol $SO_4^{2-}$

4.105 (a) 0.375 mol, 28.0 g; (b) 0.512 mol, 72.8 g

4.107 (a) 1.00 L; (b) 0.0833 L

4.109 40.0 mL

4.111 69.2 mL

4.113 (a) 0.01814 $M$; (b) 0.297 $M$; (c) 0.04020 $M$

4.115 9.76 g

4.117 39.95 amu, $6.634 \times 10^{-23}$ g

4.119 (a) 0.904 mol, $5.44 \times 10^{23}$ atoms; (b) 0.741 mol, $4.46 \times 10^{23}$ atoms; (c) 0.169 mol, $1.02 \times 10^{23}$ atoms

4.121 $7.493 \times 10^{-3}$ mol

4.123 $C_6H_{11}OBr$

4.125 $2.458 \times 10^{23}$ oxygen atoms

4.127 $1.31 \times 10^{25}$ molecules $CO_2$

4.129 (a) $C_8H_8O_3$; (b) $C_8H_8O_3$

4.131 $Al_2O_3$

4.133 122 g

4.135 $Ca_3(PO_4)_2$

4.137 1.0 g $F_2$

4.139 (a) 19.6 g sodium chloride; (b) 7.71 g sodium; (c) 11.9 g chlorine

4.141 (a) $1.8 \times 10^{24}$ $Cl^-$ ions; (b) $4.5 \times 10^{23}$ $Cl^-$ ions; (c) $3.0 \times 10^{23}$ $Cl^-$ ions

4.143 D

4.144 A and E

4.145 D

4.146 C

4.147 C

4.148 E

4.149 C

4.150 C

4.151 A

4.152 D

## Chapter 5 — Chemical Reactions and Equations

5.1 (a) single-displacement reaction; (b) anhydrous; (c) molecular equation; (d) decomposition reaction; (e) balanced equation; (f) reactant; (g) spectator ion; (h) combustion; (i) precipitate

5.3 The reactants are aluminum metal, a solid under normal conditions, Al($s$), and oxygen gas, $O_2(g)$. The product is aluminum oxide, $Al_2O_3(s)$.

5.5   Image A represents the reactants, and image C represents the products.

5.7   The numbers of hydrogen atoms do not match. One hydrogen molecule, $H_2$, should be added to the reactant image.

5.9   The product image should show five $XeF_2$ molecules and one unreacted Xe atom.

5.11

5.13   Three major signs are visible: (1) a brown gas is formed, (2) bubbles, (3) color change.

5.15   It is not a chemical change because no new substance is formed.

5.17   New substances are formed, so a chemical reaction has taken place.

5.19   No new chemical substances have formed, so no chemical reaction has taken place.

5.21   A chemical equation is a chemist's way of showing what happens during a chemical reaction. It identifies the formulas for the reactants and products, and when balanced, demonstrates how mass is conserved during the reaction.

5.23   (a) chemical reaction; (b) physical change; (c) chemical reaction

5.25   Balancing a chemical equation demonstrates how mass is conserved during a reaction. This makes the equation a quantitative tool for determining the amount of reactant used and product produced.

5.27   (a) $NaH(s) + H_2O(l) \longrightarrow H_2(g) + NaOH(aq)$
       (b) $2Al(s) + 3Cl_2(g) \longrightarrow 2AlCl_3(s)$

5.29   $N_2(g) + 3Cl_2(g) \longrightarrow 2NCl_3(g)$

5.31   Image B: $2Mg(s) + O_2(g) \longrightarrow 2MgO(s)$

5.33

5.35

5.37   (a) $2Al(s) + 3Cl_2(g) \longrightarrow 2AlCl_3(s)$
       (b) $Pb(NO_3)_2(aq) + K_2CrO_4(aq) \longrightarrow$
           $PbCrO_4(s) + 2KNO_3(aq)$
       (c) $2Li(s) + 2H_2O(l) \longrightarrow 2LiOH(aq) + H_2(g)$
       (d) $2C_6H_{14}(g) + 19O_2(g) \longrightarrow 12CO_2(g) + 14H_2O(g)$

5.39   (a) $CuCl_2(aq) + 2AgNO_3(aq) \longrightarrow$
           $Cu(NO_3)_2(aq) + 2AgCl(s)$
       (b) $S_8(s) + 8O_2(g) \longrightarrow 8SO_2(g)$
       (c) $C_3H_8(g) + 5O_2(g) \longrightarrow 3CO_2(g) + 4H_2O(g)$

5.41   $Cu(s) + 2AgNO_3(aq) \longrightarrow Cu(NO_3)_2(aq) + 2Ag(s)$

5.43   Decomposition: one reactant, more than one product; Combination: more than one reactant, one product; Single-displacement: element and compound as reactants, different element and compound as products; Double-displacement: two compounds as reactants, two compounds as products

5.45   (a) combination; (b) single-displacement; (c) decomposition

5.47   double-displacement

5.49   (a) combination; (b) single-displacement

5.51   (a) $CaCl_2(aq) + Na_2SO_4(aq) \longrightarrow$
           $CaSO_4(s) + 2NaCl(aq)$: double-displacement
       (b) $Ba(s) + 2HCl(aq) \longrightarrow BaCl_2(aq) + H_2(g)$: single-displacement
       (c) $N_2(g) + 3H_2(g) \longrightarrow 2NH_3(g)$: combination
       (d) $FeO(s) + CO(g) \longrightarrow Fe(s) + CO_2(g)$: none of these
       (e) $CaO(s) + H_2O(l) \longrightarrow Ca(OH)_2(aq)$: combination
       (f) $Na_2CrO_4(aq) + Pb(NO_3)_2(aq) \longrightarrow$
           $PbCrO_4(s) + 2NaNO_3(aq)$: double-displacement
       (g) $2KI(aq) + Cl_2(g) \longrightarrow$
           $2KCl(aq) + I_2(aq)$: single-displacement
       (h) $2NaHCO_3(s) \longrightarrow Na_2CO_3(s) + CO_2(g) + H_2O(g)$: decomposition

5.53   $NiCO_3(s) \longrightarrow NiO(s) + CO_2(g)$

5.55   (a) $CaCO_3(s) \longrightarrow CaO(s) + CO_2(g)$
       (b) $CuSO_4 \cdot 5H_2O(s) \longrightarrow CuSO_4(s) + 5H_2O(g)$

5.57   $2Mg(s) + O_2(g) \longrightarrow 2MgO(s)$

5.59   (a) $3Ca(s) + N_2(g) \longrightarrow Ca_3N_2(s)$
       (b) $2K(s) + Br_2(l) \longrightarrow 2KBr(s)$
       (c) $4Al(s) + 3O_2(g) \longrightarrow 2Al_2O_3(s)$

5.61   (a) $Zn(s) + 2AgNO_3(aq) \longrightarrow 2Ag(s) + Zn(NO_3)_2(aq)$
       (b) $2Na(s) + FeCl_2(s) \longrightarrow 2NaCl(s) + Fe(s)$

5.63   $Zn(s) + SnCl_2(aq) \longrightarrow ZnCl_2(aq) + Sn(s)$

5.65   (a) Ca reacts with water: $Ca(s) + 2H_2O(l) \longrightarrow$
           $Ca(OH)_2(aq) + H_2(g)$
       Ca reacts with HCl: $Ca(s) + 2HCl(aq) \longrightarrow$
           $CaCl_2(aq) + H_2(g)$
       (b) Fe reacts with HCl: $Fe(s) + 2HCl(aq) \longrightarrow$
           $FeCl_2(aq) + H_2(g)$
       (c) no reaction

5.67   (a) yes; (b) no; (c) no; (d) yes

5.69   (a) soluble; (b) soluble; (c) insoluble; (d) insoluble

5.71   (a) $K_2CO_3(aq) + BaCl_2(aq) \longrightarrow BaCO_3(s) + 2KCl(aq)$
       (b) $CaS(s) + Hg(NO_3)_2(aq) \longrightarrow Ca(NO_3)_2(aq) + HgS(s)$
       (c) $Pb(NO_3)_2(aq) + K_2SO_4(aq) \longrightarrow$
           $PbSO_4(s) + 2KNO_3(aq)$

5.73   (a) $BaCO_3(s) + H_2SO_4(aq) \longrightarrow$
           $BaSO_4(s) + CO_2(g) + H_2O(l)$
       (b) $CuCl_2(aq) + 2AgNO_3(aq) \longrightarrow$
           $Cu(NO_3)_2(aq) + 2AgCl(s)$

5.75   lead(II) sulfate, $PbSO_4$

5.77   $CaCl_2(aq) + K_2CO_3(aq) \longrightarrow CaCO_3(s) + 2KCl(aq)$

5.79   (a) precipitation of $HgS(s)$; (b) formation of the gas $H_2S(g)$; (c) formation of the stable molecular compound $H_2O(l)$ and precipitate $BaSO_4(s)$

**5.81** (a) $H_2S(aq) + Cu(OH)_2(s) \longrightarrow CuS(s) + 2H_2O(l)$
(b) no reaction
(c) $KHSO_4(aq) + KOH(aq) \longrightarrow K_2SO_4(aq) + H_2O(l)$

**5.83** $O_2(g)$

**5.85** (a) $Cs_2O(s)$; (b) $PbO(s)$ or $PbO_2(s)$; (c) $Al_2O_3(s)$;
(d) $H_2O(g)$; (e) $CO(g)$ or $CO_2(g)$

**5.87** (a) $CO(g)$ or $CO_2(g)$ and $H_2O(g)$; (b) $CO_2(g)$; (c) $Al_2O_3(s)$;
(d) $CO(g)$ or $CO_2(g)$ and $H_2O(g)$

**5.89** An electrolyte produces ions when dissolved in water. A nonelectrolyte does not produce ions in water.

**5.91** (a) electrolyte; (b) electrolyte; (c) nonelectrolyte

**5.93** nonelectrolyte

**5.95** (a) no ions; (b) no ions; (c) $Na^+(aq)$ and $Cl^-(aq)$

**5.97** Molecular equation: all substances written as complete chemical formulas or atoms;
Ionic equation: all soluble salts, strong acids and bases are written as ions;
Net ionic equation: all spectator ions are removed from the ionic equation

**5.99** ions that appear on both sides of ionic equations and are therefore not involved in a chemical change

**5.101** (a) $2NaCl(aq) + Ag_2SO_4(s) \longrightarrow Na_2SO_4(aq) + 2AgCl(s)$
$2Cl^-(aq) + Ag_2SO_4(s) \longrightarrow 2AgCl(s) + SO_4^{2-}(aq)$
(b) $Cu(OH)_2(s) + 2HCl(aq) \longrightarrow CuCl_2(aq) + 2H_2O(l)$
$Cu(OH)_2(s) + 2H^+(aq) \longrightarrow Cu^{2+}(aq) + 2H_2O(l)$
(c) $BaCl_2(aq) + Ag_2SO_4(s) \longrightarrow BaSO_4(s) + 2AgCl(s)$
$Ba^{2+}(aq) + 2Cl^-(aq) + Ag_2SO_4(s) \longrightarrow$
$\qquad\qquad\qquad\qquad BaSO_4(s) + 2AgCl(s)$

**5.103** (a) calcium carbonate, $CaCO_3$
(b) $Na_2CO_3(aq) + CaBr_2(aq) \longrightarrow$
$\qquad\qquad\qquad\qquad CaCO_3(s) + 2NaBr(aq)$
(c) $2Na^+(aq) + CO_3^{2-}(aq) + Ca^{2+}(aq) + 2Br^-(aq) \longrightarrow$
$\qquad\qquad\qquad CaCO_3(s) + 2Na^+(aq) + 2Br^-(aq)$
(d) $Na^+(aq)$ and $Br^-(aq)$
(e) $CO_3^{2-}(aq) + Ca^{2+}(aq) \longrightarrow CaCO_3(s)$

**5.105** $Ag^+(aq) + Cl^-(aq) \longrightarrow AgCl(s)$

**5.107** $2Na(s) + 2H_2O(l) \longrightarrow 2Na^+(aq) + 2OH^-(aq) + H_2(g)$

**5.109** (a) $Cu(s) + 2AgNO_3(aq) \longrightarrow Cu(NO_3)_2(aq) + 2Ag(s)$
$Cu(s) + 2Ag^+(aq) \longrightarrow Cu^{2+}(aq) + 2Ag(s)$
(b) $FeO(aq) + 2HCl(aq) \longrightarrow FeCl_2(aq) + H_2O(l)$
$O^{2-}(aq) + 2H^+(aq) \longrightarrow H_2O(l)$

**5.111** (a) $Sr(NO_3)_2(aq) + H_2SO_4(aq) \longrightarrow$
$\qquad\qquad\qquad\qquad SrSO_4(s) + 2HNO_3(aq)$
$Sr^{2+}(aq) + SO_4^{2-}(aq) \longrightarrow SrSO_4(s)$
(b) no reaction
(c) $CuSO_4(aq) + BaS(s) \longrightarrow CuS(s) + BaSO_4(s)$
$Cu^{2+}(aq) + SO_4^{2-}(aq) + BaS(s) \longrightarrow CuS(s) + BaSO_4(s)$
(d) $NaHCO_3(aq) + CH_3CO_2H(aq) \longrightarrow$
$\qquad\qquad CO_2(g) + H_2O(l) + NaCH_3CO_2(aq)$
$HCO_3^-(aq) + CH_3CO_2H(aq) \longrightarrow$
$\qquad\qquad CO_2(g) + H_2O(l) + CH_3CO_2^-(aq)$

**5.113** combination

**5.115** (a) $ZnSO_4(aq) + Ba(NO_3)_2(aq) \longrightarrow$
$\qquad\qquad\qquad BaSO_4(s) + Zn(NO_3)_2(aq)$
(b) $3Ca(NO_3)_2(aq) + 2K_3PO_4(aq) \longrightarrow$
$\qquad\qquad\qquad Ca_3(PO_4)_2(s) + 6KNO_3(aq)$
(c) $ZnSO_4(aq) + BaCl_2(aq) \longrightarrow BaSO_4(s) + ZnCl_2(aq)$
(d) $2KOH(aq) + MgCl_2(aq) \longrightarrow$
$\qquad\qquad\qquad 2KCl(aq) + Mg(OH)_2(s)$
(e) $CuSO_4(aq) + BaS(aq) \longrightarrow CuS(s) + BaSO_4(s)$

**5.117** $2K(s) + 2H_2O(l) \longrightarrow 2KOH(aq) + H_2(g)$

**5.119** $NaCl(s)$

**5.121** Aqueous solutions of HCl and NaOH, $HNO_3$ and KOH, HCl and KOH will give the desired net ionic equation.

**5.123** No reaction occurs.

**5.125**

Solution of KI          Solution of CO

**5.127** Any metal higher in activity can be used. (a) Al; (b) Zn; (c) Mg; (d) Sn

**5.129** (a) Potassium chromate precipitates iron(III) chromate leaving $Al^{3+}$ in solution. (b) Sodium sulfate precipitates barium sulfate leaving $Mg^{2+}$ in solution. (c) Silver perchlorate precipitates silver chloride leaving $NO_3^-$ in solution. (d) Barium sulfate is insoluble in water; $MgSO_4$ will dissolve in water.

**5.131** A chloride such as NaCl would precipitate $Ag^+$ as $AgCl(s)$; a sulfate such as $Na_2SO_4$ would precipitate $Ba^{2+}$ as $BaSO_4(s)$.

**5.133** A single displacement reaction would occur. The solution would become lighter, and solid copper would begin to form on the surface of the nickel. Over time we might observe that the nickel metal will appear to disintegrate.

**5.135** There wouldn't be an observable change because NaCl is water soluble and colorless. If we are holding the reaction container or measuring the solution temperature, we may notice the container get warmer as heat is produced by the acid-base neutralization reaction, increasing the temperature of the solution.

**5.137** $C_3H_8(g) + 5O_2(g) \longrightarrow 3CO_2(g) + 4H_2O(g)$

**5.139** The charge on copper changes from 0 to 2+. The charge on silver changes from 1+ to 0.

**5.141** (a) $2C_4H_{10}(g) + 13O_2(g) \longrightarrow 8CO_2(g) + 10H_2O(g)$
(b) $C_2H_5OH(l) + 3O_2(g) \longrightarrow 2CO_2(g) + 3H_2O(g)$
(c) $2HCO_2H(l) + O_2(g) \longrightarrow 2CO_2(g) + 2H_2O(g)$
(d) $C_5H_{12}(l) + 8O_2(g) \longrightarrow 5CO_2(g) + 6H_2O(g)$
(e) $CH_3OCH_3(g) + 3O_2(g) \longrightarrow 2CO_2(g) + 3H_2O(g)$

**5.142** D

**5.143** A

**5.144** C

**5.145** A

**5.146** B

**5.147** E

**5.148** D

**5.149** A, B, and D

**5.150** D

**5.151** A

## Chapter 6 — Quantities in Chemical Reactions

**6.1** (a) stoichiometry; (b) heat; (c) endothermic reaction; (d) specific heat; (e) actual yield

**6.3** (a) $C_6H_{12}O_6 + 6O_2 \longrightarrow 6CO_2 + 6H_2O$;
(b) 72 $CO_2$ molecules; (c) 5 $C_6H_{12}O_6$ molecules

**6.5** The coefficients give the relative ratios of reactants used and products produced. They can represent the actual numbers of molecules or moles of each substance.

**6.7** (a) 1 $Mg^{2+}(aq)$ and 2 $NO_3^-(aq)$ ions; (b) 1 mole of $Mg^{2+}(aq)$ and 2 moles of $NO_3^-(aq)$ ions

6.9 The best representation is (d); however, (c) could also be chosen because it has the reactants and products in the correct proportions. In reaction (a), the reactants are not diatomic. In reaction (b), mass is not conserved (not balanced).

6.11 (a) $\dfrac{15 \text{ mol } O_2}{2 \text{ mol } C_6H_6}$; (b) 7.5 mol $O_2$; (c) 2.8 mol $O_2$

6.13 (a) $\dfrac{6 \text{ mol } HNO_3}{2 \text{ mol } Al}$; (b) $\dfrac{3 \text{ mol } H_2}{2 \text{ mol } Al}$; (c) $\dfrac{2 \text{ mol } Al}{3 \text{ mol } H_2}$

6.15 (a) not conserved; (b) conserved; (c) conserved; (d) conserved; (e) Moles of molecules may be conserved in some cases. The others are conserved in all cases.

6.17 (a) $\dfrac{7 \text{ mol } C}{2 \text{ mol } C_7H_5(NO_2)_3}$, $\dfrac{7 \text{ mol } CO}{2 \text{ mol } C_7H_5(NO_2)_3}$, $\dfrac{3 \text{ mol } N_2}{2 \text{ mol } C_7H_5(NO_2)_3}$, $\dfrac{5 \text{ mol } H_2O}{2 \text{ mol } C_7H_5(NO_2)_3}$

(b) 3.50 mol C, 3.50 mol CO, 1.50 mol $N_2$, 2.50 mol $H_2O$; (c) 21.9 mol C, 21.9 mol CO, 9.38 mol $N_2$, 15.6 mol $H_2O$

6.19 (a) 249.69 g/mol; (b) 159.61 g/mol; (c) 0.639 g $CuSO_4$

6.21 (a) 2.05 g; (b) 5.89 g; (c) 7.94 g

6.23 (a) 21.0 g CO; (b) 38.0 g $I_2$

6.25 (a) 1.23 g $Cl_2$; (b) 0.0460 g $H_2O$; (c) 2.13 g $O_2$; 3.06 g $Cl_2$

6.27 1.0 g $CaCO_3$

6.29 the fuel

6.31 (a) 12 sandwiches; (b) bread is limiting; (c) 3 pieces of turkey

6.33 (a) $F_2$; (b) $N_2$; (c) $N_2$ and $F_2$ (both react completely)

6.35 (a)                                 (b) oxygen; (c) hydrogen

6.37

|  | $2C_2H_6(g)$ + | $7O_2(g)$ ⟶ | $4CO_2(g)$ + | $6H_2O(g)$ |
|---|---|---|---|---|
| **Initially mixed** | 6 molecules | 18 molecules | 0 molecules | 0 molecules |
| **How much reacts** | 4 molecules | 14 molecules | | |
| **Composition of mixture** | 2 molecules | 4 molecules | 8 molecules | 12 molecules |

6.39

|  | $2C_4H_{10}(g)$ + | $13O_2(g)$ ⟶ | $8CO_2(g)$ + | $10H_2O(g)$ |
|---|---|---|---|---|
| **Initially mixed** | 3.10 mol | 13.0 mol | 0.00 mol | 0.00 mol |
| **How much reacts** | 2.00 mol | 13.0 mol | | |
| **Composition of mixture** | 1.10 mol | 0.0 mol | 8.00 mol | 10.0 mol |

6.41 (a) $P_4$; (b) both are consumed completely; (c) $O_2$

6.43 (a) $F_2$; (b) 12.5 g $NF_3$

6.45 limiting reactant, $NaHCO_3$; 4.2 g $CO_2$ produced

6.47

|  | $2C_2H_6(g)$ + | $7O_2(g)$ ⟶ | $4CO_2(g)$ + | $6H_2O(g)$ |
|---|---|---|---|---|
| **Initially mixed** | 0.260 g | 1.00 g | 0.00 g | 0.00 g |
| **How much reacts** | 0.260 g | 0.968 g | | |
| **Composition of mixture** | 0.000 g | 0.03 g | 0.761 g | 0.467 g |

6.49 (a) $NaOH(aq) + HNO_3(aq) \longrightarrow H_2O(l) + NaNO_3(aq)$; (b) $HNO_3$; (c) basic

6.51 actual yield

6.53 An actual yield greater than the theoretical yield can be caused by contamination. The solid may have not been dry, causing the apparent mass of product to be high.

6.55 (a) 7.44 g; (b) 8.95 g; (c) 83.1%

6.57 90.5%

6.59   $0.47 \text{ mol } I_2$

6.61   84%

6.63   The potential energy is converted to kinetic energy (including heat), some of which is transferred to the ground as heat.

6.65   The potential energy of the hydrogen and oxygen is used to provide electric energy which runs electric motors and produces kinetic energy.

6.67   The reaction is endothermic because the decrease in temperature of the surroundings indicates energy is absorbed by the reaction.

6.69   As the liquid evaporates, the molecules going into the gas state must absorb some energy from their surrounding. As a result, the surroundings (water and your skin) lose energy and feel colder.

6.71   The products are lower in potential energy.

6.73   $2.20 \times 10^3 \text{ J}$

6.75   $3.47 \times 10^4 \text{ Cal}$

6.77   0.209 Cal

6.79   114 Cal

6.81   lead

6.83   5470 J or 1310 Cal

6.85   $-210 \text{ J}$

6.87   A calorimeter should have good insulation and a way to accurately and precisely measure the temperature.

6.89   (a) The rock must have lost heat. (b) $q_{rock} = -1.3 \times 10^3 \text{ J}$

6.91   (a) less; (b) gain; (c) 2.8 kJ

6.93   63.5°C

6.95   Because a reaction is a process, you can only measure the effect it has on the surroundings.

6.97   The chemical reaction is the system, and the water and calorimeter are the surroundings.

6.99   (a) exothermic; (b) $-525 \text{ kJ/mol } CO$

6.101   (a) 147 kJ; (b) $-24.5 \text{ kJ/g}$

6.103   (a) $q_{nut} = -2.6 \times 10^4 \text{ J}$ so $2.6 \times 10^4 \text{ J}$ is released; (b) $6.3 \times 10^3 \text{ cal or 6.3 Cal}$; (c) 3.2 Cal/g

6.105   $-256 \text{ kJ}$

6.107   (a) 4.5 mol ions; (b) 6.0 mol ions; (c) 3.0 mol ions

6.109   (a) $30.6 \text{ AgNO}_3$; (b) 25.8 g

6.111   $9.33 \times 10^3 \text{ g } O_2$

6.113   (a) $2K(s) + Cl_2(g) \longrightarrow 2KCl(s)$; (b) $Cl_2$; (c) There should be some gray solid and a white solid (KCl) inside the container. (d) 11 g; (e) 4.5 g K

6.115   (a) $\dfrac{1 \text{ mol } O_2}{2 \text{ mol } H_2}$; (b) $7.9 \text{ g } O_2$ and $8.9 \text{ g } H_2O$; (c) $4.5 \text{ g } H_2O$

6.117   31.8°C

6.119   0.450 J/(g °C); chromium

6.121   $-2.80 \times 10^3 \text{ kJ/mol}$, $-6.70 \times 10^2 \text{ kcal/mol}$

6.123   54.8°C

6.125   0.450 J/(g °C); chromium

6.127   $0.250 \text{ mol } Mg^{2+}$; $0.500 \text{ mol } Cl^-$

6.129   $0.574 \text{ mol } K^+$; $0.287 \text{ mol } SO_4^{2-}$

6.131   2.6 kg

6.133   51 g

6.135   0.870 mol

6.137   B

6.138   A

6.139   C

6.140   D

6.141   B

6.142   B

6.143   D

6.144   C

6.145   B

6.146   A

## Chapter 7 — Electron Structure of the Atom

7.1   (a) electromagnetic radiation; (b) frequency; (c) ionization energy; (d) Hund's rule; (e) electron configuration; (f) core electron; (g) orbital; (h) continuous spectrum; (i) isoelectronic

7.3   Infrared, microwave, and radio frequency

7.5

Lower frequency (——) with higher frequency (– – –) superimposed

7.7   blue, yellow, orange, red

7.9   As wavelength increases, the frequency decreases. As wavelengths decrease, frequencies increase.

7.11   infrared

7.13   Gamma photons have the highest energy and highest frequency. Radio frequency waves are the longest.

7.15   $4.00 \times 10^{15} \text{ Hz}$, ultraviolet

7.17   $4.27 \times 10^{-19} \text{ J}$, visible light (blue)

7.19   white light

7.21   no

7.23   No. If they did, line spectra would not exist. The energy of the electron is quantized—it can only have certain values.

7.25   To move to a higher orbit, the electron would have to absorb energy. To go to a lower orbit, it would have to release energy (e.g., a photon).

7.27   A single photon is released.

7.29   The $n = 6$ to $n = 3$ transition gives the highest-energy photon.

7.31   The $n = 5$ to $n = 3$ transition gives the lowest-energy photon and therefore the longest wavelength.

7.33   The four lines are a result of four different transitions in the hydrogen atom. These transitions are

| | |
|---|---|
| $n = 6$ to $n = 2$ | violet (highest energy) |
| $n = 5$ to $n = 2$ | blue |
| $n = 4$ to $n = 2$ | green |
| $n = 3$ to $n = 2$ | red (lowest energy) |

7.35   $4.58 \times 10^{-19} \text{ J}$

7.37   Bohr's orbits required that the orbits be a fixed distance from the nucleus with the electron following a specific pathway. Modern orbitals describe the three-dimensional region of space surrounding the nucleus where we are most likely to find the electron.

7.39   B

7.41   Their primary differences are size and energy. The $3p$ orbital is larger and higher in energy.

2p          3p

7.43   (a) 1; (b) 3; (c) 5; (d) 7; (e) 1; (f) 3

7.45

|      | 1s | 2s | 2p | 3s | 3p |
|------|----|----|----|----|----|
| Si   | ↑↓ | ↑↓ | ↑↓ ↑↓ ↑↓ | ↑↓ | ↑ ↑ ☐ |

|      | 1s | 2s | 2p | 3s | 3p |
|------|----|----|----|----|----|
| Li   | ↑↓ | ↑ | ☐ ☐ ☐ | ☐ | ☐ ☐ ☐ |

|      | 1s | 2s | 2p | 3s | 3p |
|------|----|----|----|----|----|
| P    | ↑↓ | ↑↓ | ↑↓ ↑↓ ↑↓ | ↑↓ | ↑ ↑ ↑ |

7.47   (a) $1s^2 2s^2 2p^6 3s^2 3p^2$; (b) $1s^2 2s^1$; (c) $1s^2 2s^2 2p^6 3s^2$

7.49   All orbitals can hold 2 electrons (including a $2p$ orbital).

7.51   8 electrons

7.53   3

7.55   elements in group VIIA (17)

7.57   transition metals

7.59   silicon

7.61   (a) $1s^2 2s^2 2p^6 3s^1$; (b) $1s^2 2s^2 2p^6 3s^2 3p^6 4s^2 3d^5$; (c) $1s^2 2s^2 2p^6 3s^2 3p^6 4s^2 3d^{10} 4p^4$

7.63   Bromine should be $1s^2 2s^2 2p^6 3s^2 3p^6 4s^2 3d^{10} 4p^5$. The $4p$ orbital given in the problem has one too many electrons, and the 10 electrons in the $d$ orbital are in the third principal energy level.

7.65   (a) chlorine; (b) cobalt; (c) cesium

7.67   (a) $[Ne]3s^1$; (b) $[Ar]4s^2 3d^5$; (c) $[Ar]4s^2 3d^{10} 4p^4$

7.69   bromine ($1s^2 2s^2 2p^6 3s^2 3p^6 4s^2 3d^{10} 4p^5$)

7.71   Valence electrons are the electrons in the highest principal energy level.

7.73   No, the $d$-orbital electrons are always one energy level lower than the valence electrons.

7.75   As you go left to right on the periodic table, one electron is added each time the group number increases. For the main-group elements, the group number is the number of electrons in the $s$ and $p$ orbitals of the highest energy level. Electrons in the $d$ orbitals are not included since they are always one energy level below the highest energy level.

7.77   (a) third valence level; 3 valence electrons; (b) third valence level; 6 valence electrons; (c) fourth valence level; 5 valence electrons

7.79   Cations always have fewer electrons than the element from which they are formed. They are similar in that they possess the same core of electrons.

7.81   (a) $1s^2 2s^2 2p^6$; [Ne]; $F^-$; (b) $1s^2 2s^2 2p^6$; [Ne]; $N^{3-}$; (c) $1s^2 2s^2 2p^6 3s^2 3p^6 3d^{10}$; $[Ar]3d^{10}$; $Zn^{2+}$

7.83   $O^{2-}$, $N^{3-}$, $Na^+$

7.85   Valence electrons farther from the attraction of the nucleus are easier to remove. This means that potassium ($4s$ valence electrons) has a lower ionization energy than sodium ($3s$ valence electrons). Since less energy is expended to remove potassium's electron, more energy is left to drive the reaction.

7.87   The first ionization energy of calcium is higher than that of potassium. Since less energy is expended to remove potassium's electron, more energy is left to drive the reaction.

In addition, calcium must lose a second electron before it becomes stable. Even more energy is expended to remove this electron.

7.89   P < S < O

7.91   Sodium's valence electron is in a higher principal energy level. This means it is not as close to the nucleus and is therefore easier to remove.

7.93   Fluorine has more protons in its nucleus than oxygen. As a result, the electrons of fluorine are more strongly attracted to the nucleus; the ionization energy is higher.

7.95   First ionization, $IE_1$, $Mg(g) \longrightarrow Mg^+(g) + 1e^-$
Second ionization, $IE_2$, $Mg^+(g) \longrightarrow Mg^{2+}(g) + 1e^-$

7.97   Magnesium, because the third electron is removed from the core electrons.

7.99   (a) Fluorine has a very high ionization energy. (b) Elements do not lose core electrons when forming ions (under normal conditions). (c) Neon is a noble gas with a filled valence level, which is very stable. It has a very high ionization energy.

7.101  O < S < P

7.103  The valence electrons of chlorine and sulfur are both in principal energy level 3, but chlorine has more protons in its nucleus to attract the electrons.

7.105  (a) Mg; (b) $P^{3-}$

7.107  Both ions are isoelectronic with argon, but potassium ion is larger because it has fewer protons in its nucleus.

7.109  For hydrogen the energies of the sublevels within a principal energy level are all the same.

7.111  (a) $[Xe]6s^2 4f^{14} 5d^{10} 6p^3$; (b) [Rn]; (c) $[Rn]7s^2$

7.113  12 filled $p$ orbitals

7.115  10 filled $d$ orbitals.

7.117  Both are related to the attraction of electrons for the nucleus. As attraction increases, electrons are more tightly held. This results in greater ionization energy and smaller radius.

7.119  The radius of the sodium ion is smaller and the chloride ion greater than for their respective elements.

7.121  Neon's red color indicates lower energy, lower frequency, and longer wavelength. Krypton's blue color indicates higher energy, higher frequency, and shorter wavelength.

7.123  (a) aufbau principle; the lower-energy orbital should be filled first.

| 1s | 2s | 2p |
|----|----|----|
| ↑↓ | ↑ | ☐ ☐ ☐ |

(b) Pauli exclusion principle; each orbital can hold only two electrons, which must have opposite spins.

| 1s | 2s | 2p |
|----|----|----|
| ↑↓ | ↑↓ | ☐ ☐ ☐ |

(c) Hund's rule; in a sublevel, electrons are distributed to give a maximum number of unpaired electrons.

| 1s | 2s | 2p |
|----|----|----|
| ↑↓ | ↑↓ | ↑ ↑ ☐ |

(d) Hund's rule; in a sublevel, electrons are distributed to give a maximum number of unpaired electrons.

| 1s | 2s | 2p |
|----|----|----|
| ↑↓ | ↑↓ | ↑ ↑ ↑ |

(e) aufbau principle; the lower-level orbitals should be filled first.

| 1s | 2s | 2p |
|----|----|----|
| ↑↓ | ↑↓ | ↑ ☐ ☐ |

7.125  (a) alkali metal; (b) nonmetal; (c) halogen; nonmetal; (d) transition metal; (e) noble gas; nonmetal

7.127  All have the same number of electrons (36) and the electron configuration of the noble gas krypton, Kr: $[Ar]4s^23d^{10}4p^6$

7.129  Strontium is smaller than rubidium; strontium has more valence electrons (2) than rubidium (1); strontium has a larger ionization energy than rubidium.

7.131  fourth electron

7.132  C

7.133  C

7.134  D

7.135  B

7.136  E

7.137  E

7.138  E

7.139  B

7.140  A

7.141  B

7.142  A

## Chapter 8 — Chemical Bonding

8.1  (a) single bond; (b) alkane; (c) covalent bonding; (d) ionic crystal; (e) octet rule; (f) polar covalent bond; (g) alkyne; (h) electronegativity; (i) triple bond; (j) crystal lattice; (k) Lewis structure; (l) chemical bond

8.3  A chemical bond is an attractive force between atoms or ions in a substance.

8.5  Metals form ionic bonds with nonmetals.

8.7  HF, $NCl_3$, $CF_4$

8.9  (a) ionic; (b) covalent; (c) covalent; (d) ionic

8.11  A and D

8.13  $CH_4$, $SF_4$, HCl

8.15  (a) high; (b) high; (c) low; (d) high; (e) low

8.17  The electronegativity value decreases.

8.19  bonds between two identical atoms

8.21  (a) N = Br < Cl < O < F (highest electronegativity)
(b) H < C < N < O < F

8.23  (a) $^{\delta+}$H—F$^{\delta-}$; (b) $^{\delta+}$I—Cl$^{\delta-}$; (c) $^{\delta+}$B—H$^{\delta-}$

8.25  (a) H—H < C—H < O—H < F—H
(b) O—Cl < C—Cl < F—Cl < HCl

8.27  The Lewis symbol can be used to determine the number of valence electrons and the number of electrons gained or lost to reach an octet configuration.

8.29  (a) ·Ċ·  (b) :İ:  (c) :Se·  (d) Sr·  (e) Cs·  (f) :Är:

8.31  (a) :C̈l:⁻  (b) $Sc^{3+}$  (c) :S̈:²⁻  (d) $Ba^{2+}$  (e) $B^{3+}$

8.33  (a) $(Li^+)(:\ddot{C}l:^-)$  (b) $(Ba^{2+})(:\ddot{C}l:^-)_2$  (c) $(Ba^{2+})(:\ddot{S}:^{2-})$

8.35  The second electron removed from potassium, K, would be taken from the core. These electrons are very strongly held by the atom and are not easily removed.

8.37  When fluorine gains an electron and sodium loses one electron, both have satisfied the octet rule. The ionic bond formed during the process is very stable, and the compound is electrically neutral.

8.39  A crystal lattice is the repeating pattern observed in all crystal structures. An ionic crystal is the structure that forms when ions form crystals by minimizing the repulsive energies of like-charged ions and maximizing the contact of oppositely charged ions.

8.41  Six chloride ions surround each sodium ion.

8.43  The ratio of the cations and anions is different, so different ionic crystals are formed.

8.45  :Ö=Ö: :F̈—F̈: (a) 8 valence electrons around each atom in $O_2$ and $F_2$; (b) The bond in $O_2$ is a double bond. The bond in $F_2$ is a single bond.

8.47  Each hydrogen needs only one electron to satisfy its valence shell. It does not have another electron to form another bond.

8.49  (a) 1; (b) 3; (c) 1; (d) none

8.51  (a) boron; (b) oxygen; (c) nitrogen

8.53  (a) H—C≡N: (b) 
```
    H
    |
H—C—C≡N:
    |
    H
```
(c) H—C≡C—H

(d) 
```
 H      H
  \    /
   C=C
  /    \
 H      H
```
(e) 
```
 H H
 | |
H-C-C-H
 | |
 H H
```

8.55  (a) 
```
[       ..  ..    ]⁻
[ Ö=  N  -Ö: ]
[ ..   |   .. ]
[     :O:      ]
```
(b) 
```
[      :Ö:      ]²⁻
[      |        ]
[ :Ö—S—Ö: ]
[      |        ]
[     :Ö:      ]
```
(c) 
```
[      :Ö:      ]²⁻
[      |        ]
[ :Ö—S—Ö: ]
```
(d) 
```
[ Ö=  N  -Ö: ]⁻
[  ..      ..  ]
```
(e) [:N≡O:]⁺

8.57  (a) different; (b) same; (c) same; (d) same; (e) same

8.59  Resonance descriptions are required when a molecule or ion can be represented by two or more reasonable Lewis structures that differ only in the positions of the bonding and lone pair electrons. The positions of the atomic nuclei do not change.

8.61  Resonance is exhibited in (c) and (d).

8.63  (a) resonance structures for $NO_2^-$
```
[ N=O: ]⁻        [ N—Ö: ]⁻
[ |    ]    ⟷    [ ‖    ]
[ :O:  ]         [ O:   ]
```

(b) 
```
 Ö:              :Ö:             :Ö:
 ‖               |               |
 S—Ö:    ⟷    S=O:    ⟷    S—Ö:
 |               |               ‖
 :O:             :O:             O:
```

(c) 
```
[      Ö:    ]⁻        [     :Ö:   ]⁻
[  H   ‖     ]         [  H   |    ]
[ H—C—C—Ö: ] ⟷  [ H—C—C=Ö: ]
[  |        ]          [  |        ]
[  H        ]          [  H        ]
```

(d) 
```
[ :Ö:   ]²⁻    [ :Ö:  ]²⁻    [ :Ö:  ]²⁻
[  ‖    ]       [  |   ]       [  |   ]
[  C—Ö: ] ⟷ [ C=O: ] ⟷ [ C—Ö: ]
[  |    ]       [  |   ]       [  ‖   ]
[ :O:   ]       [ :O:  ]       [  O:  ]
```

(e) 
```
     :O:                :O:
     |                  ‖
H—Ö—N=Ö    ⟷    H—Ö—N—Ö:
```

8.65  Hydrogen's valence shell can only hold two electrons.

8.67  (a) octet obeyed; (b) expanded octet; (c) octet obeyed; (d) octet obeyed

8.69  (a) S, expanded octet; (b) B, incomplete octet; (c) Xe, expanded octet; (d) Cl, odd electron species

8.71  alkanes with only single bonds, alkenes with double bonds, and alkynes with triple bonds

8.73

8.75  (a) alcohol; (b) alkane; (c) ether; (d) alkene; (e) ketone

8.77  (a) alcohol; (b) alkyne; (c) aldehyde; (d) ether

8.79  (a) alkane; (b) amine; (c) ester

8.81

8.83  The geometric arrangement of electron pairs and atoms is predicted by finding the geometry that gives the greatest distance between the electron groups and atoms (Table 8.5). The shape is determined by the arrangement of the bonded atoms (Table 8.6).

8.85  Shape is determined by first knowing the number of bonded atoms and unshared electron pairs (Table 8.5). Once that is determined, the shape is given by the arrangement of atoms (Table 8.6).

8.87  (a)

(b)

(c)

(d)

8.89  (a) linear; (b) tetrahedral; (c) tetrahedral; (d) trigonal planar; (e) tetrahedral; (f) tetrahedral; (g) trigonal planar; (h) tetrahedral

8.91  (a) linear, 180°; (b) trigonal pyramidal, 109.5°; (c) bent, 109.5° (tetrahedral); (d) bent, 120° (trigonal planar); (e) bent, 109.5° (tetrahedral); (f) tetrahedral, 109.5°; (g) trigonal planar, 120°; (h) bent, 109.5° (tetrahedral)

8.93  (a) 109.5°; (b) 109.5°; (c) not defined; (d) 180°; (e) 120°; (f) 120°; (g) 109.5°;

8.95  (a) $BF_3$; (b) $NH_3$; (c) $SCl_2$

8.97  (a) $BCl_3$; (b) $H_2O$; (c) $BeCl_2$; (d) $NH_4^+$

8.99  $ClO_3^-$

8.101  image B

8.103  Each nitrogen has one unshared electron pair.

8.105  The bond angles on the carbon are 109.5°. The bond angles on the nitrogen are 120°.

8.107  A bond is polar if there is a difference in electronegativity between the two atoms that are bonded. A molecule is polar if the polarity of the bonds to a central atom do not cancel out.

8.109  The chlorines are all attracting the electrons of the carbon with equal force. Because the forces are in opposite direction and the molecule is geometrically symmetric, the polarity of the bonds cancels in the molecule.

8.111  (a) polar; (b) polar; (c) polar; (d) polar

8.113  (a) $SO_2$ has an unshared pair of electrons on the central atom and is not symmetrical, so the bond polarities do not cancel. As a result $SO_2$ is polar. $CO_2$ is linear and contains polar bonds but is symmetrical. The bond polarities in $CO_2$ cancel, so it is a nonpolar molecule. (b) $SO_2$ has an unshared pair of electrons on the central atom and is not symmetrical, so the bond polarities do not cancel. As a result $SO_2$ is polar. $SO_3$ is trigonal planar and contains polar bonds but is symmetrical. The bond polarities in $SO_3$ cancel, so it is a nonpolar molecule. (c) $SeCl_2$ has two unshared pairs of electrons on the central atom and is not symmetrical, so the bond polarities do not cancel. As a result $SeCl_2$ is polar. $BeCl_2$ is linear and contains polar bonds but is symmetrical. The bond polarities in $BeCl_2$ cancel, so it is a nonpolar molecule. (d) Both are tetrahedral, but in the case of $CH_3I$, the bond polarities do not cancel and the molecule is polar. The bonds in $CH_4$ are all equivalent, so the bond polarities cancel.

8.115  image B

8.117  The molecule $CCl_2F_2$ should be soluble in water.

8.119  image B

8.121  (a) $\cdot\ddot{B}\ddot{r}\cdot$  (b) $\cdot Pb\cdot$  (c) $\cdot\ddot{S}\cdot$  (d) $Ca\cdot$  (e) $Be\cdot$  (f) $\ddot{\underset{..}{Xe}}:$

8.123  F > Cl > Br > I

8.125  (a) covalent; (b) ionic; (c) covalent; (d) covalent; (e) ionic

8.127  (a) $H-\ddot{N}=\ddot{N}-H$

(b) $:\ddot{S}=C=\ddot{S}:$

(c) $:\ddot{F}-\ddot{A}\ddot{s}-\ddot{F}:$
       $\quad\quad\;\; |$
       $\quad\quad :\ddot{F}:$

(d) $:\ddot{O}=C=\ddot{O}:$

(e) $:C\equiv O:$

8.129

8.131  (a) tetrahedral; (b) trigonal planar; (c) bent; (d) trigonal pyramidal; (e) tetrahedral; (f) bent; (g) tetrahedral; (h) trigonal planar; (i) bent; (j) bent

8.133  (a) $BeCl_2$ is nonpolar because the molecule is symmetrical. $OCl_2$ is polar. The molecule has polar bonds and is bent (not symmetrical).
(b) $BH_3$ is nonpolar because the molecule is symmetrical. The polar molecule in the pair must therefore be $PH_3$. Even though P and H have the same electronegativity values, the asymmetry of the molecule and the lone pair of electrons on the phosphorus atom causes the molecule to be polar.
(c) $BCl_3$ is nonpolar because the molecule is symmetrical. $AsCl_3$ is polar because the bonds are polar and the molecule is not symmetrical.
(d) $SiH_4$ is nonpolar because the molecule is symmetrical. $NH_3$ is polar because the bonds are polar and the molecule is not symmetrical.

8.135  (a) Both structures are tetrahedral. The $CCl_4$ structure is nonpolar due to symmetry. However, the different types of atoms (with different electronegativities) bonded to the carbon in $CH_2Cl_2$ causes this molecule to be polar.
(b) Both structures are tetrahedral and polar but the $CH_3F$ molecule is more polar than the $CH_3Br$ molecule because the C—F bond is more polar than the C—Br bond.
(c) Both structures are trigonal pyramidal and polar but the $NF_3$ molecule is more polar than the $NH_3$ molecule because the N—F bonds are more polar than the N—H bonds.
(d) Both structures are bent and polar but the $H_2O$ molecule is more polar than the $OF_2$ molecule because the O—H bonds are more polar than the O—F bonds.

8.137  (a) low; (b) low; (c) high; (d) high; (e) low

8.139  (a) covalent; (b) ionic bonding between the ions, covalent bonding within the polyatomic ion; (c) ionic bonding between the ions, covalent bonding within the polyatomic ion

8.141

$$H-\overset{\overset{\displaystyle H}{|}}{\underset{\underset{\displaystyle H}{|}}{C}}-\overset{\overset{\displaystyle H}{|}}{\underset{\underset{\displaystyle H}{|}}{C}}-\overset{\overset{\displaystyle H}{|}}{\underset{\underset{\displaystyle H}{|}}{C}}-\overset{\displaystyle ..}{\underset{\displaystyle ..}{O}}-H$$

$$H-\overset{\overset{\displaystyle H}{|}}{\underset{\underset{\displaystyle H}{|}}{C}}-\overset{\overset{\displaystyle :O:}{||}}{\underset{\underset{\displaystyle H}{|}}{C}}-\overset{\overset{\displaystyle H}{|}}{\underset{\underset{\displaystyle H}{|}}{C}}-H$$

8.143  one single bond, no double bonds, one triple bond; two unshared electrons

8.145

$$H-\overset{\overset{\displaystyle H}{|}}{\underset{\underset{\displaystyle H}{|}}{C}}-\overset{\overset{\displaystyle :O:}{||}}{C}-\overset{\overset{\displaystyle H}{|}}{\underset{\underset{\displaystyle H}{|}}{C}}-\overset{\overset{\displaystyle H}{|}}{\underset{\underset{\displaystyle H}{|}}{C}}-H$$

8.147  A
8.148  B
8.149  C
8.150  A
8.151  C
8.152  C
8.153  D
8.154  C
8.155  A
8.156  E
8.157  B
8.158  B

## Chapter 9 — The Gaseous State

9.1  (a) effusion; (b) Boyle's law; (c) combined gas law; (d) ideal gas; (e) pressure; (f) Dalton's law of partial pressures; (g) molar volume; (h) ideal gas constant, $R$

9.3  The $y$ axis units and label are missing, $x$ axis units are missing, the points should not be connected by lines (a straight best-fit line that comes closest to all points should be used), the $x$ axis maximum should be 60 instead of 120, and a graph title is missing.

9.5  There is a trend, but it is not linear.

9.7  Pressure and volume are inversely related.

9.9  The vapor pressure is 21 torr and 50 torr at 22°C and 38°C, respectively.

9.11  (a) $x = 0.4$; (b) $x = 0.206$; (c) $x = \pm 4$; (d) $x = 2$

9.13  (a) 100°C; (b) 26.7°C; (c) 0°C; (d) −40.0°C

9.15  (a) 24.4; (b) 0.116; (c) 17.2; (d) 4.70

9.17  When compared to other states of matter, gases have low densities and are very compressible. They also take the shape of any container they are put in.

9.19  The density of warm air is lower than the density of cold air.

9.21

9.23   Gas pressure is the amount of force exerted by the gas divided by the area over which the force is exerted.

9.25   Absolute pressure is measured with a device called a barometer. A tire gauge measures the pressure above atmospheric pressure.

9.27

9.29   (a) 0.980 atm; (b) 935 torr; (c) 0.119 atm; (d) 643 Pa; (e) $1.01 \times 10^3$ mm Hg; (f) 563 torr; (g) $1.06 \times 10^5$ Pa; (h) 293 mm Hg

9.31   $1.024 \times 10^5$ Pa

9.33   As pressure increases, the volume decreases at constant temperature.

9.35   The particles collide more frequently with the container walls causing an increase in pressure. Since the number of moles has not changed, we expect to find twice as many molecules in the same space. The velocity of the molecules should not change as long as the temperature remains constant.

9.37   (a) 3.60 L; (b) 26.7 mL; (c) 0.392 mL

9.39   (a) 161 torr; (b) 0.100 torr; (c) 17.9 atm

9.41   182 atm

9.43   0.577 L

9.45   Charles's law states that if the pressure is constant, the volume and temperature (in kelvin) are directly proportional to each other (volume increases when temperature increases; volume decreases when temperature decreases).

9.47   The velocity of the gas molecules increases as temperature increases. The volume of the container will increase as the temperature increases.

9.49   (a) 5.41 L; (b) 671 mL; (c) 75.5 L

9.51   (a) 273°C; (b) 265.2°C; (c) −19°C

9.53   0.169 mL

9.55   Nothing happens to the particles if the temperature, pressure, and volume are constant. If the tank is sealed so that no gas molecules are allowed to escape, then nothing happens to the pressure. If the tank is opened, then some molecules of gas in the tank will leave the tank. Particles leave the tank to maintain an equilibrium pressure with the air outside the tank.

9.57   (a) 418 torr; (b) 673 torr; (c) 4.44 atm

9.59   (a) −91°C; (b) 750°C; (c) 437°C

9.61   7.71 atm

9.63   (a) 1.2 L; (b) −195.3°C; (c) $9.10 \times 10^2$ torr

9.65   1.61 L

9.67   Guy-Lussac's law states that volumes of gas react in simple whole-number ratios when the volumes of reactants and products are measured at the same temperature and pressure.

9.69   22.414 L/mol

9.71   (a) 0.385 mol, 6.17 g; (b) $1.562 \times 10^{-2}$ mol, 2.050 g; (c) 2.15 mol, 60.1 g

9.73   The number of particles is the same. The balloon containing argon has the greater mass, and therefore greater density since the volumes are the same.

9.75   (a) 0.34 mol, 7.6 L; (b) 1.5 mol, 34 L; (c) 2.70 mol, 60.5 L

9.77   2380 balloons

9.79   11.0 L

9.81   Any gas whose behavior is described by the five postulates of kinetic molecular theory is an ideal gas.

9.83   (a) 0.34 mol, 0.69 L; (b) 1.5 mol, 3.1 L; (c) 2.70 mol, 5.51 L

9.85   (a) 0.245 mol, 3.92 g; (b) $4.33 \times 10^{-3}$ mol, $8.74 \times 10^{-3}$ g; (c) 0.288 mol, 8.06 g

9.87   The balloon filled with $CO_2$ sinks because the density of $CO_2$ is greater than the density of air. The balloon filled with He floats because the density of He is less than the density of air.

9.89   (a) 0.7598 g/L; (b) 1.250 g/L; (c) 1.964 g/L

9.91   (a) 0.673 g/L; (b) 1.11 g/L; (c) 1.74 g/L

9.93   There are 0.208 moles of each gas. (a) 0.42 g; (b) 3.3 g; (c) 13 g

9.95   Dalton's law of partial pressures states that the total pressure in a container is the sum of the pressures exerted by each of the individual gases in a mixture of gases.

9.97   708 torr

9.99   (a) 2.78 mol; (b) 2.08 mol; (c) 2.15 atm; (d) 1.60 atm

9.101   (a) 40.01 g/mol; (b) 30.03 g/mol; (c) 45.99 g/mol; (d) 20.3 g/mol; (e) 16.0 g/mol

9.103   The five postulates of kinetic-molecular theory are 1) Gases are composed of small particles widely separated. 2) Particles behave independently of each other. 3) Gas particles move in straight lines. 4) Gas pressure results from force exerted by the molecules in the container. The force is the sum of the forces exerted by each molecule as it bounces off the walls of the container. 5) The average kinetic energy of the gas particles depends only on the absolute temperature.

9.105   As the temperature is increased, the kinetic energy of the particles increases. Recall that kinetic energy is the energy of motion. Increasing temperature increases molecular velocity. The particles strike the walls of the container with greater force and, therefore, the pressure is higher.

9.107   Pressure depends on the force of collisions and the frequency of collisions in a given area. The density of gas particles is inversely proportional to the volume. If the volume decreases, the density increases. Since the density of gas particles increases with decreasing volume, the frequency of collisions in a given area has increased. This means that the pressure has increased.

9.109   (lowest velocity) $CO_2 < CH_4 <$ He $< H_2$ (highest velocity)

9.111   Molecules with lower masses have higher average velocities and as a result diffuse faster.

9.113   helium

9.115   24 L $H_2$

9.117   48.0 L $CO_2$ and 76.0 L $O_2$

9.119   $1.68 \times 10^3$ g

9.121   The balloon expands or contracts so that the external pressure and the internal pressure are the same.

9.123 The density of a gas decreases if it is heated and the volume is allowed to expand. The air in the hot-air balloon is less dense than the surrounding air, so the balloon floats.

9.125 1100 K

9.127

| What will happen if... | Macroscopic View | Microscopic View |
|---|---|---|
| Temperature increases and pressure remains constant? | Balloon gets bigger | Molecules move faster<br>Molecules collide harder<br>Collision frequency increases<br>Molecules are farther apart |
| Temperature decreases and pressure remains constant? | Balloon gets smaller | Molecules move slower<br>Less energetic collisions<br>Collision frequency decreases<br>Molecules are closer together |
| 10,000 feet (temperature decreases and pressure decreases) | Balloon probably gets a little bigger | Molecules move slower<br>Less energetic collisions<br>Lower collision frequency |

9.129 1.75 L

9.131 1.21 L

9.133 27.7 L

9.135 (a) 5.00 L; (b) 3.33 L; (c) 1.67 L

9.137 228 K

9.139 (a) If the pressure and temperature are constant, increasing the moles of gas increases the volume occupied by the gas. (b) pressure and temperature

9.141 131 g/mol

9.143 0.718 L

9.145 (a) no change; (b) no change; (c) increase

9.147 (a) decrease; (b) decrease; (c) decrease

9.149 $CH_4$

9.151 The pressure of a gas can be increased by reducing the volume, increasing the temperature, or adding more gas to the container. Another way would be to increase the pressure of the surrounding air if the gas is in an elastic container.

9.153 (a) 1.5 mol; (b) 66 g $CO_2$; (c) 0.066 g/mL

9.154 C

9.155 B

9.156 E

9.157 A

9.158 C

9.159 C

9.160 D

9.161 E

9.162 C

9.163 B

9.164 E

9.165 B

9.166 B

9.167 C

## Chapter 10 — The Liquid and Solid States

10.1 (a) vapor pressure; (b) melting point; (c) equilibrium; (d) evaporation; (e) induced dipole; (f) alloy; (g) boiling point; (h) sublimation; (i) molecular solid; (j) intermolecular force; (k) London dispersion force; (l) normal freezing point; (m) crystal

10.3

|  | Liquid | Gas |
|---|---|---|
| Particle spacing | Dense or closely spaced | Molecules far apart |
| Intermolecular attraction | Moderate | Weak |
| Kinetic energy | Low | High |

10.5 A liquid is fluid because the molecules can move past each other. Attractive forces hold the particles in solids together in a rigid, three-dimensional array.

10.7 The substance is in the solid state. If the substance were a liquid or gas, it would not maintain its shape. In addition, if the substance was not a solid, the images on the surface would not maintain their shape.

10.9 The image is a molecular-level representation of the liquid state. The liquid state and solid state both have dense particle spacing. Basically, the molecules are always touching each other. However, in the image, the molecules do not show any clear pattern. This is a characteristic of the liquid state.

10.11 The solid, liquid, and gas states are represented in circles 1, 3, and 2, respectively. The phase changes are deposition, vaporization, and melting for phase changes *A*, *B*, and *C*, respectively.

10.13 The vapor pressure of a liquid increases as the temperature is increased. A liquid boils when its vapor pressure is equal to the pressure above the liquid. The temperature at which this occurs is the boiling point. The normal boiling point is the temperature at which the vapor pressure of the liquid is equal to 1 atm.

10.15 In order to convert a substance from the liquid to the gas state, the energy of the molecules must increase enough to allow them to overcome their attractions for each other and move about independently. Because the molecules must absorb energy to go into the gas state, the process is endothermic (i.e., absorbs energy from the surroundings).

10.17

10.19 solid state to the gas state, sublimation; endothermic

10.21 All molecules and atoms have some degree of attraction for each other. As a gas is cooled, the molecules are losing kinetic energy. Eventually the attractive forces are greater than the kinetic energy of the molecules and the molecules begin to coalesce (stick together) to form the liquid state.

10.23 Evaporation of water is endothermic. The energy water needs to evaporate comes from the surrounding air, cooling it. The cooled air can be used to cool a house (if it is circulated by a fan).

10.25 Water boils when it has been heated to the point that its vapor pressure equals the atmospheric pressure. At high elevation, the atmospheric pressure can be significantly lower than 1 atm (about 0.6 to 0.7 atm at 12,000 ft). Since the water boils at a lower temperature, the pasta cooks slower.

10.27   As a liquid is heated the molecules of the liquid acquire more energy. The percent of molecules on the liquid surface with enough energy to go into the gas phase increases. This causes an increase in the vapor pressure.

10.29   Gallium is a liquid at 100°C. At 15°C gallium is a solid.

10.31   *BC*

10.33   237 kJ

10.35   41.0 kJ

10.37   $1.70 \times 10^6$ J

10.39   124 kJ

10.41   An attractive force between individual particles (molecules or atoms) of a substance is called an intermolecular force.

10.43   gas state

10.45   Ar

10.47   NO, $NF_3$, and $CH_3Cl$

10.49   A bond is an attractive force between two atoms in a molecule. A hydrogen bond is not a true bond because it usually takes place between two atoms on different molecules. In comparison to covalent bonds, the strength of attraction is much weaker and the electrons are not shared.

10.51   Only B can hydrogen-bond.

10.53   (a) and (b)

10.55   (b)

10.57   The vapor pressures of pure liquids depend on the strength of the intermolecular forces between the particles of the liquids.

10.59   When a molecule is polar, it means that some portion of the molecule appears permanently slightly negative to other molecules and another portion is slightly positive. This comes from the unequal sharing of electrons (Chapter 8). The oppositely signed charges on different molecules attract each other (dipole-dipole interaction), so the molecules stick together. The instantaneous and induced dipoles in nonpolar molecules are not permanent.

10.61   As molar mass increases, the importance of London dispersion forces increases.

10.63

|  | Dispersion Force | Dipole-Dipole | Hydrogen Bonding |
|---|---|---|---|
| (a) $C_6H_6$ | x | | |
| (b) $NH_3$ | x | x | x |
| (c) $CS_2$ | x | | |
| (d) $CHCl_3$ | x | x | |

10.65

|  | Dispersion Force | Dipole-Dipole | Hydrogen Bonding |
|---|---|---|---|
| (a) Kr | x | | |
| (b) CO | x | x | |
| (c) $CH_4$ | x | | |
| (d) $NH_3$ | x | x | x |

10.67   (lowest boiling point) $CH_3OH$ < $CH_3CH_2OH$ < $CH_3CH_2CH_2OH$ (highest boiling point)

10.69   $NO_2$

10.71   Hydrogen, $H_2$, is larger than helium. The size of the hydrogen molecule allows the electrons to be moved around more easily, and therefore the dispersion force is greater. More energy is required to vaporize liquid hydrogen than helium.

10.73   HCl is polar, while $F_2$ is nonpolar.

10.75   (a) $CH_3OH$; (b) $CH_3OH$; (c) $NH_3$

10.77   (lowest) He < $N_2$ < $H_2O$ < $I_2$ (highest)

10.79   Both molecules are polar, but water has very strong hydrogen bonding while acetone has dipole-dipole interactions. As a result, water has a higher boiling point.

10.81   On a molecular level, the particles of a substance in the liquid state are higher in energy than for the same substance in the solid state and the molecules have freedom to move apart from each other. In addition, the liquid state does not have a long range structure like the crystal lattice of many solids. On a macroscopic level, liquids flow and can take the shapes of their containers. Liquids tend to be less dense than solids.

10.83   Viscosity is the resistance of a liquid to flow. The more viscous a liquid, the slower it will flow when it is poured.

10.85   All molecules of a liquid experience attractive forces to other molecules in the liquid. Molecules in the middle experience forces that are equal in all directions. On the surface, the molecules only experience attractive forces to the sides and "downward" (toward the center of the liquid). Surface tension causes liquids to bead up on solid surfaces and is also why liquids tend to form spheres. In addition, surface tension is related to capillary action and the formation of a meniscus (curved surface) of a liquid in a container.

10.87   There is an imbalance of forces on the molecules at the surface of a liquid. As the surface area of the liquid is minimized, the imbalance of energy is also minimized. Substances assume the shape with the least surface area for the given volume. As a result, liquids have a tendency to form spherical shapes to minimize this energy difference.

10.89   When ice forms on bodies of water such as lakes and rivers, it stays at the surface and provides an insulating layer. Energy loss through evaporation is minimized and less water freezes than would if the ice did not form at the top. Ice would continually form and sink. Many bodies of water would freeze completely making life in them impossible during the coldest months.

10.91   Crystals are defined as solids that, on a molecular or atomic level, have a regular repeating pattern of atoms. See Figures 10.31 and 10.37.

10.93   Crystal structures help us understand the macroscopic properties of the solid state (i.e., conductivity, hardness, brittleness).

10.95   Close packing is a common structure of metals.

10.97   Amorphous solids have a tendency to flow because the molecules are not locked into a crystal lattice.

10.99   Solids are classified by the strength of attraction between the subunits of the solid. This allows us to broadly classify the characteristics of the solid substances.

10.101   ionic solid

10.103   There are two reasons. First, the attractive forces between metal atoms are relatively weak. This allows them to be separated from each other. Secondly, the electrons in metal atoms are delocalized among all their neighboring atoms. Atoms can be moved with relatively little effect on the charge distribution. In ionic solids, cations and anions cannot be separated easily, so particles do not move apart easily. This causes ionic solids to be brittle.

10.105 Both ionic and metallic solids form crystal lattices. Ionic crystals resemble metal crystals with smaller ions fitting into holes between larger ions.

10.107 Molecular solids are held together either by dipole-dipole interactions, hydrogen bonding, or London dispersion forces. Ionic solids are held together by interionic attractions (i.e., forces of attraction between oppositely charged ions). Because ionic interactions are much stronger than intermolecular forces, the ionic solids tend to melt at much higher temperatures.

10.109 (a) covalent bonds; (b) London dispersion forces; (c) London dispersion forces, dipole-dipole interactions, and hydrogen bonding; (d) covalent bonds

10.111 (a) ionic; (b) molecular; (c) metallic; (d) network; (e) molecular

10.113 (b) and (e) are molecular solids

10.115 molecular solid

10.117 $SiF_4$

10.119 Below its freezing point, ice is lost by sublimation.

10.121 $CH_4$ ($-162°C$), $C_2H_6$ ($-88°C$), $C_3H_8$ ($-42°C$), $C_4H_{10}$ ($0°C$); London dispersion forces increase with molecular size.

10.123 A glass plate, composed of $SiO_2$, is very polar. Polar substances are attracted to polar substances. Water is attracted to glass, so it spreads out. Mercury is not polar and is attracted more to itself than to the polar glass, so it tends to bead up.

10.125 (c) is an ionic solid

10.127 (a) $Cl_2$; (b) $CH_3SH$; (c) $NF_3$

10.129

| Substance (solid type) | Electrical Conductivity | Hardness | Melting Point | Vapor Pressure |
|---|---|---|---|---|
| **NaCl** (ionic) | conductive in molten state | hard | high | low |
| **SiC** (network) | not conductive | hardest | highest | lowest |
| **SiCl₄** (molecular) | not conductive | softest | lowest | highest |
| **Fe** (metallic) | always conductive | soft* | high | low |

*Metals are soft relative to network or ionic solids.

10.131 melting; freezing; sublimation; deposition; vaporization; condensation

10.133

10.135 $CH_3Cl$, $CH_2Cl_2$, and $NCl_3$

10.137 (a) 10W; (b) Viscosity increases when temperature is lowered. To avoid a motor oil that is too thick and viscous at low temperatures, a lower-viscosity oil (thinner oil) is used.

10.139 $CH_3OH$

10.141 The most obvious difference is that liquid water would be virtually nonexistent except if some very cold climates exist. If there was ice, it would not float because it is the

hydrogen bonding that makes it less dense than water. Earth temperatures would vary much more wildly because the high specific heat of liquid water allows the oceans and lakes to absorb and release large amounts of heat providing relatively moderate air temperatures. Because much of life on Earth depends on liquid water for survival, other life forms might exist instead. Another difference would be the structure of DNA and other biomolecules, which are dependent upon hydrogen bonding. The building blocks of life might not even work, or they would work very differently.

10.143 trans, 47.5°C; cis, 60.3°C

10.145 B

10.146 E

10.147 A

10.148 A

10.149 C

10.150 C

10.151 E

10.152 E

10.153 C

10.154 B

# Chapter 11 — Solutions

11.1 (a) saturated solution; (b) aqueous solution; (c) molarity; (d) solvent; (e) entropy; (f) solubility; (g) parts per million; (h) miscible; (i) osmosis; (j) colligative property; (k) percent by volume; (l) mass/volume percent

11.3

| Solute | Solvent |
|---|---|
| (a) sodium chloride (and other salts) | water |
| (b) carbon | iron |
| (c) oxygen ($O_2$) | water |

11.5 Strong electrolytes are substances that fully ionize or dissociate when placed in solution. Strong acids, strong bases, and soluble ionic compounds are all strong electrolytes.

11.7 nonelectrolyte

11.9 (a) $H^+(aq)$ and $Br^-(aq)$; (b) $NH_4^+(aq)$ and $Cl^-(aq)$; (c) $CH_3CH_2CH_2CH_2OH(aq)$

11.11 A

11.13 (a) $Ca(OH)_2(s) \xrightarrow{H_2O} Ca^{2+}(aq) + 2OH^-(aq)$

(b) $N_2(g) \xrightarrow{H_2O} N_2(aq)$

(c) $CH_3OH(l) \xrightarrow{H_2O} CH_3OH(aq)$

11.15 In the solvent, hydrogen-bonding and dipole-dipole interactions are broken. In the solute, ionic bonds are broken. Ion-dipole interactions are formed in the solution.

11.17 In both the solute and the solvent, London dispersion forces must be overcome. The new forces are also London dispersion forces.

11.19 The decrease in temperature of the solution that surrounds the dissolving ions indicates that the dissolving process is endothermic. When a dissolving process is endothermic, the forces of attraction between the dissolved ions and water molecules are weaker than the forces of attraction in the pure substances.

11.21 Energy and entropy drive the formation of solutions.

11.23 The intermolecular attractions between water molecules and oil molecules are not the same (water is polar and cooking oil is nonpolar). Since they are not "like," they are immiscible.

11.25   Both molecules are polar and can engage in hydrogen bonding.

11.27   (a) insoluble; (b) soluble; (c) soluble

11.29   Ion-ion interactions are very strong, and interactions between ions and nonpolar molecules would be relatively weak. There is not enough energy released or entropy gained in the formation of ion-solvent interactions to make the dissolution process favorable.

11.31   The ions in NaCl are attracted by ionic bonding. The most significant attractive force between water molecules is hydrogen bonding. Although disrupting the attractive forces in this solute and solvent requires considerable energy, formation of the ion-dipole interactions between solute and solvent particles releases enough energy to allow the ionic compound to dissolve.

11.33   The solubility of gases such as $O_2$ decreases as temperature increases.

11.35   The solubility of oxygen and nitrogen gases should be lower at high altitude. The solubility of these gases is directly proportional to their pressures above the solution, and these pressures are lower at high altitude.

11.37   The insoluble gases that cause the bends can be redissolved into the blood by putting the divers in pressurized chambers. This increases the solubility of the gas in the blood. Decreasing the chamber pressure slowly (decompression) allows the gases (mostly excess nitrogen) to leave through the lungs.

11.39

11.41   17.0 g KCl

11.43   On a macroscopic level, you can't dissolve more solid in a saturated solution, but you can dissolve more in an unsaturated solution. On a molecular level, the ions in a saturated solution are in equilibrium with the solid.

11.45   You could add a little solid at a time until no more solid dissolves, and you see traces of undissolved solid in the solution.

11.47   The solution will be saturated and 0.85 g $Ca(OH)_2(s)$ will be left undissolved.

11.49   14.3%

11.51   105 g KI

11.53   5.423%

11.55   Percent by volume is often used for mixtures of liquids. This is done because liquids are usually measured by volume.

11.57   30.4%

11.59   (a) 2.2 ppb; (b) Since the maximum contamination level is 5 ppb, the water is safe to drink.

11.61   0.015 mol

11.63   2.41 $m$

11.65   3.0 $M$; 3.0 $m$

11.67   $9.6 \times 10^{-4}\%$; 9.6 ppm; 9600 ppb

11.69   (a) 9.0 ppb; yes, the concentration is below the EPA standard; (b) $9.0 \times 10^{-6}$ mg/mL;   (c) $9.0 \times 10^{-4}$ mg (d) $4.3 \times 10^{-9}$ mol

11.71   (a) 0.030 L KI; (b) 6.9 g $PbI_2$

11.73   0.30 mol $Na_2CO_3$

11.75   57.14 mL

11.77   0.8812 $M$ $H_2SO_4$

11.79   (a) 10.00 mL; (b) 52.50 mL; (c) 37.5 mL

11.81   A colligative property is a physical property of a solution that varies with the number of solute particles dissolved and not on the identity of those solute particles.

11.83   Osmosis occurs when two solutions of different solute particle concentrations are separated by a membrane that allows the solvent but not the solute molecules to pass (semipermeable membrane). The volume of the more concentrated solution increases because solvent molecules migrate through the membrane to the more concentrated solution.

11.85   Reverse osmosis occurs when solvent is forced to travel in a direction opposite of the direction it would travel during osmosis.

11.87   Water will flow out of the blood cell, and the cell will collapse or shrink.

11.89   (a) decrease; (b) increase; (c) decrease

11.91   The boiling point of the water is elevated slightly because of the presence of the sugar.

11.93   (a) 1.3°C; (b) 101.3°C

11.95   The sodium chloride solution will have the higher boiling point because NaCl produces two solute particles when it dissolves in water.

11.97   Solution B has the highest concentration of dissolved particles and therefore has the lowest freezing point. Glucose is molecular, and images B and C show molecular compounds. Sodium chloride produces two ions of slightly different size as shown in image A.

11.99   (a) 2.0 mol; (b) 5.0 mol; (c) 4.0 mol

11.101   The osmotic pressure of $MgCl_2$ will be greater than that of $NaNO_3$ because each formula unit of $MgCl_2$ produces three ions when it dissolves, while $NaNO_3$ produces only two ions.

11.103   Vitamin D is nonpolar.

11.105   The breaking of the crystal lattice and disruption of the intermolecular attractions between water molecules require energy. The formation of ion-dipole interactions between water and the ions releases energy.

11.107   182 g

11.109   8.89 $M$

11.111   Volume changes with temperature. If the concentration units are based only on mass or number of moles, then they will not change with temperature. Percent by mass (a) and molality (d) do not change with temperature.

11.113   The density of a dilute solution is approximately the density of the solvent. In aqueous solutions, for example, one liter of water has a mass of one kilogram. So the values for molarity (moles/liter) and molality (moles/kilogram) will be very similar.

11.115   (a) 10.5 g $BaSO_4$; (b) 2.33 g $BaSO_4$; (c) 2.33 g $BaSO_4$

11.117   $2.0 \times 10^3$ g

11.119   solution B

11.121   Olive oil molecules are attracted more strongly to one another than to acetic acid or to water, so these molecules will separate out of the mixture. Water and acetic acid molecules are attracted more strongly to one another than to olive oil molecules, so they will also separate out of the mixture.

11.123   0.10 $m$ KBr > 0.10 $m$ $Na_2SO_4$ > 0.20 $m$ NaCl > 0.15 $m$ $VCl_3$

11.125   Placing the lettuce in water causes water molecules to cross the cell membrane into the cells by osmosis.

11.127   Barium sulfate is a very insoluble ionic salt.

11.129   fatty tissues

11.131   C

11.132   B

11.133  A
11.134  C
11.135  E
11.136  A
11.137  D
11.138  B
11.139  B

## Chapter 12 — Reaction Rates and Chemical Equilibrium

12.1   (a) activation energy, $E_a$; (b) intermediate
       (c) homogeneous equilibrium; (d) collision theory
       (e) collision theory; (f) equilibrium constant expression

12.3   increasing reactant concentrations, increasing temperature, or providing a catalyst

12.5   grind the calcium carbonate to a fine powder; increase the temperature; increase the concentration of HCl

12.7   catalyst

12.9   No. If the reactants do not collide with sufficient energy or with the proper orientation, then the activated complex cannot be formed.

12.11  (a) improper orientation; (b) proper orientation; (c) improper orientation

12.13  activated complex

12.15  minimum energy required to break bonds in reactants before reactants can be converted into products

12.17  For similar reactions, the higher the activation energy, the slower the reaction.

12.19

12.21  Increasing temperature increases the kinetic energy and velocity of the molecules reacting. The fraction of colliding molecules with energy greater than the activation energy increases when temperature is increased.

12.23  increasing temperature

12.25  At room temperature, molecules of propane and oxygen do not collide with sufficient energy to cause a reaction. However, by providing a spark (heat), the kinetic energy of the molecules is increased. Combustion reactions are exothermic, so the reaction is sustained by the heat produced by the reaction.

12.27  Roasts are generally thicker and more massive than steaks. For the same amount of heat, the steak gets hotter faster and therefore cooks faster.

12.29  A catalyst provides a lower-energy route (pathway) to the same products.

12.31  Catalytic converters use palladium and platinum catalysts (Figure 12.11).

12.33  active site

12.35  Catalysts are only temporarily changed during the course of a reaction (e.g., during the binding of the substrate).

12.37  Chlorine atoms in the upper atmosphere act as catalysts for the decomposition of ozone. Since they act catalytically, they are not lost until they diffuse into outer space (which is a slow process) or are removed by some other reaction.

12.39  intermediate

12.41  (a) $NO_3$ is an intermediate; (b) $N_2O_5 + NO \longrightarrow 3NO_2$

12.43  (a) $H^+$ and $Br^-$ are catalysts and $Br_2$ is an intermediate.
       (b) $2H_2O_2 \longrightarrow 2H_2O + O_2$

12.45  When a reaction reaches equilibrium, the concentrations of reactants and products stop changing.

12.47  A chemical equilibrium is a state of a reaction where the rate of the forward reaction is equal to the rate of the reverse reaction.

12.49  The system has reached equilibrium in images D and E.

12.51  (a) yes; (b) no; (c) yes; (d) yes

12.53  You could measure the mass of the bromine liquid, but that would be tricky because it is very reactive and volatile. A better option is to measure the pressure of the gas produced. When the partial pressure has stabilized, the system is at equilibrium. Finally, bromine gas has a dark brown appearance. By measuring how much light the gas absorbs, we can determine if its concentration remains constant and the system is at equilibrium.

12.55  The position of the equilibrium tells us whether a reaction is product-favored or reactant-favored.

12.57  The coefficients of the reactants and products are used as exponents in the equilibrium constant expression.

12.59  The brackets represent molar concentrations.

12.61  (a) $K_{eq} = \dfrac{[HF]^2}{[H_2][F_2]}$

       (b) $K_{eq} = \dfrac{[CS_2][H_2]^4}{[CH_4][H_2S]^2}$

       (c) $K_{eq} = \dfrac{[NO_2]^2}{[N_2O_4]}$

12.63  (a) $C(g) \rightleftharpoons A(g) + B(g)$
       (b) $2A(g) \rightleftharpoons 4B(g) + C(g) + D(g)$

12.65  (a) The equations are inverses of each other.
       (b) $2NO_2(g) \rightleftharpoons 2NO(g) + O_2(g)$;
       $2NO(g) + O_2(g) \rightleftharpoons 2NO_2(g)$

12.67  0.25

12.69  An equilibrium constant only changes with temperature.

12.71  (a) $3.1 \times 10^{-3}$; (b) Because the value of $K_{eq}$ is much less than one, the position of equilibrium is described as "reactant-favored."

12.73  (a) $1.3 \times 10^{-9}$; (b) The reaction is reactant-favored since $K_{eq} < 1$.

12.75  The reaction is at equilibrium, so shifting will not take place.

12.77  The reaction will proceed in the reverse direction.

12.79  The reaction will proceed in the forward direction.

12.81  Heterogeneous reactions are those in which the products and reactants are not all in the same physical state.

12.83  The concentrations of pure liquids and solids do not change during chemical reactions; only the amount of the pure substance changes. Because the concentrations of pure substances do not change during reactions, their values are essentially incorporated into the equilibrium constant.

12.85 (a) $K_{eq} = [NH_3][HCl]$

(b) $K_{eq} = [Ca^{2+}][CO_3^{2-}]$

(c) $K_{eq} = \dfrac{[H_3O^+][F^-]}{[HF]}$

12.87 $0.27\ M$

12.89 Le Chatelier's principle describes how reactions that are at equilibrium respond to changes in reaction conditions. If stress is put on a reaction, the reaction responds to counteract that stress.

12.91 (a) The reaction shifts to the left; (b) The concentrations of the products, $CO_2$ and $H_2$, will decrease and the reactant, $H_2O$, will increase in concentration.

12.93 In the following table, the larger arrows (⇑) indicate the change made to the equilibrium reaction and the smaller arrows indicate what the concentration of the products and reactants do following the change. Finally, the central arrow indicates which direction the equilibrium shifts.

|     | $NO(g)$ | $SO_3(g)$ | ⇌ | $NO_2$ | $SO_2$ |
|-----|---------|-----------|---|--------|--------|
| (a) | ⇑ | ↓ | → | ↑ | ↑ |
| (b) | ↓ | ⇑ | → | ↑ | ↑ |
| (c) | ↓ | ↓ | → | ⇓ | ↑ |
| (d) | ↑ | ↑ | ← | ⇑ | ↓ |

12.95 (a) This has no effect since the concentration of the solid does not change when more solid is added. (b) This decreases the concentration of $Pb^{2+}$ by shifting the equilibrium to the left. (c) This causes an increase in $Pb^{2+}$ concentration. When the concentration of lead ion is increased, the equilibrium shifts to the left. However, it never shifts enough to completely remove all the added material.

12.97 (a) left; (b) right; (c) no effect

12.99 (a) no; (b) yes

12.101 (a) left; (b) right

12.103 If the equilibrium shifts to the right, the concentration of products at equilibrium is higher. The value of $K_{eq}$ increases.

12.105 (a) decrease; (b) increase

12.107 exothermic

12.109 An increase in the concentration of NO requires that the equilibrium shift to the right. (a) Water is in the liquid state; removing it does not affect the equilibrium. (b) There are fewer moles of gas on the right. The reaction shifts right increasing the concentration of NO. (c) Since the reaction is exothermic, the reaction will shift right with decreasing temperature producing a higher concentration of NO. (d) Adding oxygen causes the equilibrium to shift right producing a higher concentration of NO. (e) Catalysts do not have an effect on equilibrium concentrations.

12.111 The reaction that produces light slows down, and the reactants take longer to be consumed. The lower light output of the cold light sticks is a result of fewer photons being produced per second (lower reaction rate).

12.113 It provides a different reaction mechanism with lower activation energy. In general it splits one difficult step (high activation energy) into two or more steps that have lower activation energies.

12.115 (a) $K_{eq} = \dfrac{[H_2][I_2]}{[HI]^2}$; (b) $K_{eq} = 0.0199$

(c) The position of the equilibrium favors the reactants.

12.117 The effect of temperature depends on whether the reaction is endothermic or exothermic.

12.119 The reaction must be exothermic since the reaction shifted to the right to form the powder.

12.121 The simplest treatment would be to increase the concentration of oxygen that the patient is breathing. By increasing the $O_2$ concentration, the second equilibrium will shift to the left and the carbon monoxide can be displaced and exhaled through the blood.

12.123 (a) increase; (b) increase; (c) same; (d) increase; (e) increase

12.125 As reactants convert to products, the concentration of reactants decreases. As concentration decreases, there are fewer collisions per unit time and therefore fewer effective collisions, resulting in a lower reaction rate. The rate of the forward reaction decreases until equilibrium is reached, where the rates of the forward and reverse reactions are the same.

12.127 (a) False. The activation energy for a reaction does not change with temperature. Increasing the temperature increases the reaction rate by increasing the collision frequency and the number of reactants with sufficient energy (the activation energy) to convert to products.
(b) False. One characteristic of a catalyst is that it is recovered unchanged at the end of a reaction. Reactants in a net balanced equation are the substances that undergo a permanent chemical change.
(c) True. This supports the definition given in the explanation to part (b). When the equations in a mechanism are added together, the catalyst occurs on both the reactant and product side. Therefore, it cancels out and does not appear in the net balanced equation.
(d) False. An intermediate is different from a catalyst. An intermediate is generated in an early step of the mechanism and exists temporarily, reacting in a later step. Like a catalyst, the intermediate does not appear in the net balanced equation.

12.129 (a) False. While equal concentrations of reactants and products at equilibrium is theoretically possible, it is not common and is not a definition of equilibrium. When a reaction system reaches a state of equilibrium, the concentrations of reactants and products no longer change, and they are usually very different from one another.
(b) True. The equilibrium constant value is a ratio with product concentrations in the numerator and reactant concentrations in the denominator. When there are much more reactants than products, the ratio will be less than 1.
(c) False. This statement would only be true in a rare circumstance where the equilibrium lies exactly halfway between the reactants and products. A system is at equilibrium when the ratio of products to reactants defined by the equilibrium constant expression is equal to the equilibrium constant value for that reaction.
(d) True. This is why we sometimes call chemical equilibrium a *dynamic* equilibrium. Both the forward and reverse processes occur, but at the same rate.
(e) False. The addition of a catalyst does not affect the position of equilibrium. It only changes the rate at which equilibrium is reached.

12.131 (a) heterogeneous; (b) $K_{eq} = [Ag^+][Cl^-]$
(c) $[Ag^+] = [Cl^-] = 1.30 \times 10^{-5}\ M$

12.133 D

12.134 E

12.135 B

12.136 B
12.137 C
12.138 A
12.139 D
12.140 B
12.141 D
12.142 C
12.143 E

## Chapter 13 — Acids and Bases

13.1 (a) strong base; (b) Brønsted-Lowry theory; (c) conjugate acid; (d) polyprotic acid; (e) acidic solution; (f) weak base; (g) amphoteric substance; (h) ion-product constant of water, $K_w$; (i) self-ionization

13.3 (a) –9.0; (b) –11.0; (c) 3.87; (d) 5.0; (e) 0.0

13.5 (a) 16; (b) $6.3 \times 10^{-7}$; (c) 1

13.7 Acids have a sour taste (but you should never taste anything to see if it's an acid!), are corrosive to many metals, turn blue litmus red, and neutralize bases.

13.9 Common "household" bases and their uses:

| Formula | Name | Use |
| --- | --- | --- |
| $Mg(OH)_2$ | magnesium hydroxide | antacid |
| $Al(OH)_3$ | aluminum hydroxide | antacid |
| $NH_3$ | ammonia | cleaning solutions, smelling salts |
| $Ca(OH)_2$ | calcium hydroxide | food preparation |

13.11 An Arrhenius acid produces hydrogen ions, $H^+(aq)$, when dissolved in water. An Arrhenius base produces hydroxide ions, $OH^-(aq)$, when dissolved in water.

13.13 Brønsted-Lowry theory refers to the transfer of hydrogen ions from an acid to a base. Arrhenius similarly describes the release of hydrogen ions into the solution, but identifies $OH^-$ as a component of the base. In Brønsted-Lowry theory, a base is any substance that can accept a hydrogen ion (including $OH^-$).

13.15 (a) acid; (b) base; (c) acid

13.17 (a) $NO_2^-$; (b) $F^-$; (c) $H_2BO_3^-$

13.19 (a) $H_2O$; (b) $C_6H_5NH_3^+$; (c) $H_2CO_3$

13.21 (a) $HCl(g) + H_2O(l) \longrightarrow H_3O^+(aq) + Cl^-(aq)$
(b) $HClO_4(l) + H_2O(l) \longrightarrow H_3O^+(aq) + ClO_4^-(aq)$
(c) $CH_3CO_2^-(aq) + H_2O(l) \rightleftharpoons HCH_3CO_2(aq) + OH^-(aq)$

13.23 Both water (a) and hydrogen sulfite (b) are amphoteric.

13.25 Carbonic acid has two acidic hydrogen atoms.

13.27 Strong acids and bases ionize or dissociate completely (100%). For weak acids and bases, the percent of molecules that ionize is usually much lower (typically a few percent).

13.29 (a) strong acid; $H_2SO_4(aq) + H_2O(l) \longrightarrow$
$HSO_4^-(aq) + H_3O^+(aq)$
(b) strong base; $Ca(OH)_2(aq) \longrightarrow$
$Ca^{2+}(aq) + 2OH^-(aq)$
(c) weak base; $Na_2CO_3(aq) + H_2O(l) \rightleftharpoons$
$HCO_3^-(aq) + 2Na^+(aq) + OH^-(aq)$
(d) weak acid; $H_3C_6H_5O_7(aq) + H_2O(l) \rightleftharpoons$
$H_3O^+(aq) + H_2C_6H_5O_7^-(aq)$
(e) weak base; $C_6H_5NH_2 + H_2O(l) \rightleftharpoons$
$C_6H_5NH_3^+(aq) + OH^-(aq)$

13.31 A is $NH_3$, B is HF, and C is HCl.

13.33 KF; HF is a weak acid, so its conjugate base, $F^-$, is a weak base.

13.35 Acid A is a stronger acid. The stronger the acid, the greater the extent of ionization.

13.37 0.50 $M$ $HCO_2H$

13.39 $NaCH_3CO_2$

13.41 In a strong acid solution the acid molecules are fully ionized. It does not necessarily mean that the concentration is high. A concentrated acid solution has a high percentage of acid, but this does not necessarily mean that the acid is ionized to any great extent.

13.43 A polyprotic acid is an acid that contains more than one acidic hydrogen atom.

13.45 Hydrosulfuric acid is a weak acid that has two acidic hydrogen atoms. The corresponding acid-base reactions are
$H_2S(aq) + H_2O(l) \longrightarrow HS^-(aq) + H_3O^+(aq)$ $K_{a1} = 8.9 \times 10^{-8}$
$HS^-(aq) + H_2O(l) \longrightarrow S^{2-}(aq) + H_3O^+(aq)$ $K_{a2} = 1.0 \times 10^{-19}$

13.47 The following species can all be found in a solution of oxalic acid: $H_2C_2O_4$, $HC_2O_4^-$, $C_2O_4^{2-}$, $H_3O^+$, $H_2O$. The species with the lowest concentration is the oxalate ion $C_2O_4^{2-}$.

13.49 Like other equilibrium constants, it is constant only under conditions of constant temperature.

13.51 In pure water, $[OH^-] = [H_3O^+] = 1.0 \times 10^{-7}$ $M$.

13.53 (a) basic; (b) acidic; (c) basic

13.55 (a) $1.0 \times 10^{-11}$ $M$, basic; (b) $1.0 \times 10^{-3}$ $M$, acidic; (c) $3.1 \times 10^{-7}$ $M$, acidic

13.57 (a) 0.010 $M$; (b) 0.020 $M$; (c) $6.7 \times 10^{-13}$ $M$

13.59 A decrease in pH represents an increase in the acidity.

13.61 A 10-fold increase in hydrogen ion concentration is 1 pH unit change. The difference in pH units is 1.

13.63 Practically speaking, most solutions of acids and bases are lower in concentration than 1 $M$, so the pH scale typically ranges between 0 and 14. However, the pH values can exceed this range.

13.65 No. A solution is defined as acidic if the pH is less than 7.

13.67 (a) 3.00, acidic; (b) 13.00, basic; (c) 9.47, basic

13.69 (a) 10.00, basic; (b) 7.00, neutral; (c) 4.91, acidic

13.71 (a) 2.00, acidic; (b) 1.70, acidic; (c) 12.18, basic

13.73 Underlined values were given in the problem.

| | $[H_3O^+]$ | $[OH^-]$ | pH | pOH | Acidic or Basic? |
| --- | --- | --- | --- | --- | --- |
| (a) | <u>$1.0 \times 10^{-5}$</u> | $1.0 \times 10^{-9}$ | 5.00 | 9.00 | acidic |
| (b) | $1.0 \times 10^{-10}$ | <u>$1.0 \times 10^{-4}$</u> | 10.00 | 4.00 | basic |
| (c) | $1.0 \times 10^{-6}$ | $1.0 \times 10^{-8}$ | 6.00 | <u>8.00</u> | acidic |
| (d) | $2.9 \times 10^{-9}$ | $3.5 \times 10^{-6}$ | <u>8.54</u> | 5.46 | basic |
| (e) | $1.1 \times 10^{-5}$ | <u>$9.0 \times 10^{-10}$</u> | 4.95 | 9.05 | acidic |

13.75 pH = 2.30, pOH = 11.70. The pH and pOH always add to 14.00 at 25°C.

13.77 (a) $1.0 \times 10^{-5}$ $M$, acidic; (b) $1.0 \times 10^{-12}$ $M$, basic; (c) $1.3 \times 10^{-6}$ $M$, acidic

13.79 (a) $1.0 \times 10^{-11}$ $M$, basic; (b) $4.0 \times 10^{-8}$ $M$, basic; (c) $1.3 \times 10^{-2}$ $M$, acidic

13.81 (a) $1.0 \times 10^{-13}$ $M$; (b) $3.2 \times 10^{-4}$ $M$; (c) $4.0 \times 10^{-11}$ $M$

13.83 Acetic acid is a weak acid, so the concentration of hydronium ion would be less than 0.010 $M$, and the pH would be higher than 2.0.

13.85  0.1 $M$

13.87  A pH meter typically gives measurements to two decimal places (Figure 13.17).

13.89  Acid-base indicators are typically colored compounds that are also weak acids or bases. The colors of these compounds depend on whether they are in their acid or base forms.

13.91  The color range of an indicator indicates the pH values over which the indicator changes color. The hue of the indicator can be used to estimate the pH of a solution. The range over which an indicator changes color is about 1.5 pH units.

13.93  A selection of indicators would include thymolphthalein.

13.95  It should be red with a small amount of yellow-orange.

13.97  Red (or the presence of red color) indicates the pH is less than 2.5. Yellow indicates a pH between about 2.5 and 8.0. Blue (or a blue hue) indicates pH values above 8.0. It will be completely blue above pH 9.5.

13.99  Universal indicator is a mixture of different indicators. The color of an unknown solution with the indicator present gives a good estimate of the pH. It is similar to pH paper in that pH paper contains universal indicator.

13.101  The solution would be yellow at the start of the titration and turn blue at the endpoint.

13.103  A buffer contains both a weak acid and its conjugate base (or a weak base and its conjugate acid). As a result, when small amounts of acid or base are added, the buffer can absorb the excess hydronium or hydroxide with only a small change in pH.

13.105  The buffer system of the blood helps to maintain a pH between 7.35 and 7.45.

13.107  any salt containing $CH_3CO_2^-$

13.109  (a) Hydrogen phosphate, $HPO_4^{2-}$, is the acid and the conjugate base is phosphate, $PO_4^{3-}$.
$HPO_4^{2-}(aq) + H_2O(l) \rightleftharpoons PO_4^{3-}(aq) + H_3O^+(aq)$
(b) When acid is added, hydronium, $H_3O^+$, is produced and the pH momentarily drops (more acidic). However, the excess hydronium quickly reacts with the $PO_4^{3-}$ shifting the equilibrium to the left to produce the weak acid $HPO_4^{2-}$.

13.111  Solutions described in (b) and (c) produce buffered solutions.

13.113  As the concentration of carbon dioxide increases, the equilibrium of the reaction would shift to the right. The formation of carbonic acid would cause a decrease in the pH of the extracellular fluid.

13.115  Ammonia, $NH_3$, has a lone electron pair in its Lewis structure. Methane does not have a lone electron pair, so it cannot accept a hydrogen ion (proton) in an acid-base reaction.

13.117  When NaF dissolves in water, sodium and fluoride ions are produced. The fluoride ions react with water to produce a weak base HF by the reaction:
$F^-(aq) + H_2O(l) \rightleftharpoons HF(aq) + OH^-(aq)$
In general the conjugate base ($F^-$) of a weak acid will react with water to produce hydroxide ions. The molecular species HF is stable, so its formation drives the formation of hydroxide. This solution will be basic. The sodium ion does not react with water.

13.119  No. A neutral solution has a pH = pOH = 7.00 (at 25°C). If sodium hydroxide is added, the hydroxide ion concentration will have to increase. The solution will have a pH greater than 7.00.

13.121  Both $H_2SO_4$ and $HNO_3$ are strong acids. Both will dissociate completely *for the first ionization*. However, $H_2SO_4$ is a polyprotic acid. The additional hydronium ion produced by the second ionization will make the pH of the sulfuric acid lower.

13.123  (a) $HOCl(aq) + H_2O(l) \rightleftharpoons OCl^-(aq) + H_3O^+(aq)$.
(b) The HCl reacts with the water to produce hydronium ion: $HCl(aq) + H_2O(l) \rightleftharpoons Cl^-(aq) + H_3O^+(aq)$. The added $H_3O^+$ causes the equilibrium to shift to the left and some of the excess hydronium is consumed. The HOCl concentration will increase and the $OCl^-$ concentration will go down. While the pH may not drop below 7.00, *any time* an acid is added to a solution, the solution becomes more acidic (pH is lowered).

13.125  Sulfuric acid is a strong acid and is completely ionized.

13.127  sodium formate and formic acid

13.129  (a) 3; (b) 2; (c) 1; (d) 1

13.131  (a) **Some** strong acids have H atoms bonded to electronegative oxygen atoms. (Binary acids generally do not contain oxygen atoms.)
(b) The conjugate base of a strong acid is itself a **weak** base.
(c) Strong acids **may be** very concentrated acids. (Acid strength is a measure of the extent to which the acid dissociates in water, not a measure of its concentration.)
(d) Strong acids produce solutions with a **low** pH. (Acidic solutions have a low pH, and basic solutions have a high pH.)

13.133  pH = 6.5

13.135  0.5 $M$ HCl < 0.1 $M$ $HNO_3$ < a solution of $HNO_2$ with a pH of 4.0 < water < a buffer solution with a pOH of 11.0

13.137  Add acetic acid, or add hydrochloric acid, which will react with acetate ion to form acetic acid.

13.139  D

13.140  B

13.141  A

13.142  B

13.143  E

13.144  D

13.145  C

13.146  A

13.147  D

13.148  D

# Chapter 14 — Oxidation-Reduction Reactions

14.1  (a) electrolytic cell; (b) cathode; (c) oxidation;
(d) half-reaction; (e) electrochemistry; (f) electrode;
(g) oxidizing agent; (h) oxidation-reduction reaction;
(i) corrosion; (j) oxidation number

14.3  reaction where electrons are transferred

14.5  when there is an increase in charge (oxidation number)

**14.7** (a) $Mg(s) + Cu(NO_3)_2(aq) \longrightarrow Cu(s) + Mg(NO_3)_2(aq)$
(b) The charge of magnesium increases by two; 2 electrons transferred; electrons are lost.
(c) The charge of copper decreases by two; 2 electrons transferred; electrons are gained.

**14.9** (a) Mg; (b) $Sn^{2+}$; (c) $SnSO_4$; (d) Mg
(e)

**14.11** An oxidation number is a charge assigned to each atom in a compound that provides a bookkeeping method for tracking where electrons go during a chemical reaction.

**14.13** (a) 0; (b) 0; (c) 3+; (d) 1−

**14.15** S = 6+, O = 2−

**14.17** (a) 5+; (b) 3+; (c) 4+

**14.19** (a) 5+; (b) 5+; (c) 5+; (d) 1+; (e) 3−; (f) 3+

**14.21** S = 6+, O = 2−

**14.23** (a) 7+; (b) 5+; (c) 1−; (d) 3+; (e) 5+

**14.25** (a) Cl = 4+, O = 2−; (b) Ca = 2+, F = 1−; (c) H = 1+, Te = 4+, O = 2−; (d) Na = 1+, H = 1−

**14.27** (a) N = 3+, O = 2−; (b) Cr = 6+, O = 2−; (c) Ag = 1+, Cl = 1−; (d) S = 4+, O = 2−; (e) C = 4+, O = 2−

**14.29** (a) not oxidation-reduction; (b) oxidation-reduction, $N_2$ is the oxidizing agent, $H_2$ is the reducing agent; (c) not oxidation-reduction; (d) not oxidation-reduction; (e) oxidation-reduction, $O_2$ is the oxidizing agent, $C_2H_6$ is the reducing agent

**14.31** (a) $N_2(g) + O_2(g) \longrightarrow 2NO(g)$. (b) The oxidation number increases. Each nitrogen atom loses two electrons to change from an oxidation number of 0 to 2+. (c) The oxidation number decreases. Each oxygen atom gains two electrons to change from an oxidation number of 0 to 2−.

**14.33** (a) $V^{2+}$; (b) chromium in $Cr_2O_7^{2-}$ is reduced; (c) $Cr_2O_7^{2-}$; (d) $V^{2+}$

**14.35** (a) $I_2$ is both the oxidizing agent and reducing agent, 1 electron transferred. (b) Cr = reducing agent, $H^+$ = oxidizing agent, 2 electrons transferred. (c) $Cr_2O_7^{2-}$ is both the oxidizing and reducing agent, 12 electrons transferred. (d) $Fe^{3+}$ = oxidizing agent, Al = reducing agent, 3 electrons transferred.

**14.37** In the anode half-cell the half-reaction is
$Fe(s) \longrightarrow Fe^{2+}(aq) + 2e^-$
In the cathode half-cell the half-reaction is
$Ni^{2+}(aq) + 2e^- \longrightarrow Ni(s)$

**14.39** The iron electrode will lose mass as Fe is oxidized and becomes $Fe^{2+}(aq)$. The nickel electrode will gain mass as $Ni^{2+}$ is reduced to $Ni(s)$.

Iron electrode          Nickel electrode

**14.41** The $Pb(s)$ gets oxidized and the $PbO_2(s)$ is reduced. The oxidizing agent is $PbO_2(s)$ and the reducing agent is $Pb(s)$.

**14.43** oxidation half-reaction: $Cd(s) + 2OH^-(aq) \longrightarrow$
$Cd(OH)_2(s) + 2e^-$
reduction half-reaction: $2e^- + NiO_2(s) + 2H_2O(l) \longrightarrow$
$NiOH_2(s) + 2OH(aq)$

**14.45** (a) $3e^- + Fe^{3+}(aq) \longrightarrow Fe(s)$
(b) $Zn(s) \longrightarrow Zn^{2+}(aq) + 2e^-$
(c) $2Cl^-(aq) \longrightarrow Cl_2(g) + 2e^-$
(d) $Fe^{2+}(aq) \longrightarrow Fe^{3+}(aq) + e^-$

**14.47** (a) $Zn(s) + 2NO_3^-(aq) \longrightarrow Zn(NO_3)_2(aq) + 2e^-$
$3e^- + Fe(NO_3)_3(aq) \longrightarrow Fe(s) + 3NO_3^-(aq)$
$3Zn(s) + 2Fe(NO_3)_3(aq) \longrightarrow 3Zn(NO_3)_2(aq) + 2Fe(s)$
(b) $Mn(s) + 2Cl^-(aq) \longrightarrow MnCl_2(aq) + 2e^-$
$2e^- + 2HCl(aq) \longrightarrow H_2(g) + 2Cl^-(aq)$
$Mn(s) + 2HCl(aq) \longrightarrow MnCl_2(aq) + H_2(g)$

**14.49** $Fe_2(SO_4)_3(aq) + 2KI(aq) \longrightarrow$
$2FeSO_4(aq) + K_2SO_4(aq) + I_2(aq)$

**14.51** (a) $Ba(s) \longrightarrow Ba^{2+}(aq) + 2e^-$
(b) $e^- + H^+(aq) + HNO_2(aq) \longrightarrow NO(g) + H_2O(l)$
(c) $2e^- + 2H^+(aq) + H_2O_2(aq) \longrightarrow 2H_2O(l)$
(d) $2Cr^{3+}(aq) + 7H_2O(l) \longrightarrow Cr_2O_7^{2-}(aq) + 14H^+ + 6e^-$

**14.53** (a) $3OH^-(aq) + La(s) \longrightarrow La(OH)_3(s) + 3e^-$
(b) $2e^- + H_2O(l) + NO_3^-(aq) \longrightarrow NO_2^-(aq) + 2OH^-(aq)$
(c) $H_2O_2(aq) + 2OH^-(aq) \longrightarrow O_2(g) + 2e^- + 2H_2O(l)$
(d) $8e^- + 3H_2O(l) + Cl_2O_7(aq) \longrightarrow$
$2ClO_2^-(aq) + 6OH^-(aq)$

**14.55** (a) $8H^+(aq) + 3H_2S(aq) + Cr_2O_7^{2-}(aq) \longrightarrow$
$3S(s) + 2Cr^{3+}(aq) + 7H_2O(l)$
(b) $4H^+(aq) + 5V^{2+}(aq) + 3MnO_4^-(aq) \longrightarrow$
$5VO_2^+(aq) + 3Mn^{2+}(aq) + 2H_2O(l)$
(c) $6H^+(aq) + 6Fe^{2+}(aq) + ClO_3^-(aq) \longrightarrow$
$6Fe^{3+}(aq) + Cl^-(aq) + 3H_2O(l)$

**14.57** (a) $2NH_3(aq) + ClO^-(aq) \longrightarrow$
$N_2H_4(aq) + Cl^-(aq) + H_2O(l)$
(b) $2Cr(OH)_4^-(aq) + 3HO_2^-(aq) \longrightarrow$
$2CrO_4^{2-}(aq) + 5H_2O(l) + OH^-(aq)$
(c) $2OH^-(aq) + Br_2(aq) \longrightarrow$
$Br^-(aq) + BrO^-(aq) + H_2O(l)$

**14.59** $24H^+(aq) + 5C_6H_{12}O_6(aq) + 24NO_3^-(aq) \longrightarrow$
$30CO_2(g) + 12N_2(g) + 42H_2O(l)$

**14.61** The anode could be any metal that is more active than iron (e.g., aluminum or zinc). The supporting electrolyte could be any soluble salt that does not precipitate with $Fe^{3+}$ (e.g., potassium nitrate since all nitrates are soluble). The anodic half-cell would also contain a salt that includes the metal used for the anode (e.g., aluminum or zinc nitrate).

**14.63** Au < Bi < Ni < Zn < Al < Ca

14.65  Electrolysis is the process of using electricity to make a nonspontaneous redox reaction occur.

14.67  (a) Reduction will begin to take place at the cathode to form sodium metal.
(b) Oxidation will begin to take place at the anode to form $I_2(g)$.
(c) $Na^+(l) + e^- \longrightarrow Na(l)$   (reduction)
$2I^-(l) \longrightarrow I_2(g) + 2e^-$   (oxidation)
(d) $2Na^+ + 2I^-(l) \longrightarrow 2Na(l) + I_2(g)$

14.69

Anode

Cathode

$I_2(g)$

$Na(l)$

Anode reaction          Cathode reaction

14.71  Magnesium is more active than iron. The magnesium will reduce the $O_2(g)$ and keep the Fe in the reduced state. When the $Mg(s)$ is used up, the iron will begin to corrode.

14.73  One of the methods of preventing corrosion is to prevent contact with oxygen and moisture. The tin coating serves as a barrier to moisture and oxygen. When the tin coating is scratched, the more active metal, iron, is more readily oxidized.

14.75  The chromium should corrode first because it is a more active metal than iron.

14.77  (a) 3−; (b) 2−; (c) 3+; (d) 1−; (e) 5+; (f) 3+

14.79  (a) $2H_2O(l) + 2OH^-(aq) + 2CoCl_2(s) + Na_2O_2(aq) \longrightarrow$
$\qquad 2Co(OH)_3(s) + 4Cl^-(aq) + 2Na^+(aq)$
The oxidizing agent is $Na_2O_2$ (oxygen is being reduced from 1− to 2−). The reducing agent is $CoCl_2$ (cobalt is being oxidized from 2+ to 3+).
(b) $2OH^-(aq) + Bi_2O_3(s) + 2ClO^-(aq) \longrightarrow$
$\qquad 2BiO_3^-(aq) + 2Cl^-(aq) + H_2O(l)$
The oxidizing agent is $ClO^-$ (chlorine is being reduced from 1+ to 1−). The reducing agent is $Bi_2O_3$ (bismuth is being oxidized from 3+ to 5+).

14.81  $NH_4^+(aq) + NO_3^-(aq) \longrightarrow N_2O(g) + 2H_2O(l)$
The oxidizing agent is $NO_3^-$ (N is being reduced from 5+ to 1+). The reducing agent is $NH_4^+$ (N is being oxidized from 3− to 1+).

14.83  (a) Al; (b) Zn; (c) Mn; (d) Mg

14.85  Brass is an alloy of copper and zinc. Cu doesn't react with acid. There must be a protein in sweat that facilitates the oxidation of the copper to form $Cu^{2+}$. There are several compounds of copper that are green (including $CuCO_3$).

14.87  $2H^+(aq) + Cu(s) + H_2SO_4(aq) \longrightarrow$
$\qquad Cu^{2+}(aq) + SO_2(g) + 2H_2O(l)$

14.89  (a) reducing agent, $Tl^+$; oxidizing agent, $Ce^{4+}$;
oxidation half-reaction: $Tl^+(aq) \longrightarrow Tl^{3+}(aq) + 2e^-$
reduction half-reaction: $Ce^{4+}(aq) + e^- \longrightarrow Ce^{3+}(aq)$
(b) reducing agent, $Br^-$; oxidizing agent, $NO_3^-$: oxidation half-reaction: $2Br^-(aq) \longrightarrow Br_2(l) + 2e^-$
reduction half-reaction: $NO_3^-(aq) + 2H^+(aq) + e^-$
$\qquad \longrightarrow NO_2(g) + H_2O(l)$

14.91  (a) Cu; (b) $Ag^+$
(c) $Cu(s) + 2Ag^+(aq) \longrightarrow Cu^{2+}(aq) + 2Ag(s)$

14.93  Sodium, Na, is the reducing agent; ethanol, $CH_3CH_2OH$, is the oxidizing agent.

14.95  (a) carbon; (b) oxygen; (c) 8 mol

14.97  $Br_2(aq) + 2OH^-(aq) \longrightarrow Br^-(aq) + BrO^-(aq) + H_2O(l)$

14.99  (a) Anode: $Al(s) \longrightarrow Al^{3+}(aq) + 3e^-$
Cathode: $2e^- + Zn^{2+}(aq) \longrightarrow Zn(s)$
(b) Anode: $Cr(s) \longrightarrow Cr^{3+}(aq) + 3e^-$
Cathode: $e^- + Ag^+(aq) \longrightarrow Ag(s)$
(c) Anode: $Cd(s) \longrightarrow Cd^{2+}(aq) + 2e^-$
Cathode: $2e^- + 2H^+(aq) \longrightarrow H_2(g)$

14.101  D
14.102  C
14.103  D
14.104  A
14.105  B and D
14.106  C
14.107  E
14.108  C
14.109  C
14.110  E

## Chapter 15 — Nuclear Chemistry

15.1  (a) critical mass; (b) fission; (c) alpha particle; (d) radioactive; (e) gamma radiation; (f) nuclear bombardment; (g) nucleon; (h) radiation

15.3  A substance that contains unstable nuclides of an element will exhibit radioactivity by the emission of radiation.

15.5  Radioactivity produces photons (usually gamma rays) and particulate radiation (i.e., beta rays, positrons, neutrons, and alpha particles).

15.7  Of the elements with atomic number less than 84, technetium (Tc), promethium (Pm), and astatine (At) have no stable isotopes. No isotopes of atomic number 84 or greater are stable.

15.9  Dense materials are better at shielding gamma radiation. Lead, iron, and aluminum (metals) can be used, but concrete and water are effective only if thick layers are used.

15.11  Nuclear decay reactions and bombardment reactions result in the production of radiation. In addition, new isotopes are produced by both.

15.13  During nuclear decay, alpha particles, beta particles, gamma rays, and positrons can be produced by the nucleus.

15.15  During nuclear reactions, nuclear charge and mass numbers are conserved.

15.17  $^{222}_{86}Rn$

15.19  $^{99}_{43}Tc$

15.21  $^{15}_{7}N$

15.23  beta particle

15.25  $^{7}_{4}Be$

15.27  (a) $^{238}_{92}U \longrightarrow ^{234}_{90}Th + ^{4}_{2}\alpha$
(b) $^{234}_{90}Th \longrightarrow ^{234}_{91}Pa + ^{0}_{-1}\beta^-$
(c) $^{234}_{92}U \longrightarrow ^{4}_{2}\alpha + ^{230}_{90}Th$
(d) $^{210}_{83}Bi + ^{0}_{-1}e^- \longrightarrow ^{210}_{82}Pb$

15.29  (a) $^{227}_{87}Fr \longrightarrow ^{4}_{2}\alpha + ^{223}_{85}At$
(b) $^{18}_{9}F \longrightarrow ^{0}_{+1}\beta^+ + ^{18}_{8}O$
(c) $^{14}_{6}C \longrightarrow ^{0}_{-1}\beta + ^{14}_{7}N$

**15.31** In a nuclear bombardment reaction a particle accelerator is used to accelerate particles or nuclides to very high velocities. The beam of particles produced is aimed at a target element. If the particle strikes the target with sufficient energy to fuse the target and beam particles, a newly created nucleus is formed. The newly created nucleus spontaneously decays to produce new substances and decay particles.

**15.33** $^{257}_{103}\text{Lr}$

**15.35** (a) $^{10}_{5}\text{B} + ^{2}_{1}\text{H} \longrightarrow ^{11}_{6}\text{C} + ^{1}_{0}\text{n}$

(b) $^{209}_{83}\text{Bi} + ^{4}_{2}\alpha \longrightarrow ^{211}_{85}\text{At} + 2^{1}_{0}\text{n}$

(c) $^{12}_{6}\text{C} + ^{1}_{1}\text{p} \longrightarrow ^{13}_{7}\text{N}$

**15.37** (a) $^{96}_{42}\text{Mo} + ^{1}_{1}\text{p} \longrightarrow ^{96}_{43}\text{Tc} + ^{1}_{0}\text{n}$

(b) $^{209}_{83}\text{Bi} + ^{64}_{28}\text{Ni} \longrightarrow ^{272}_{111}\text{Rg} + ^{1}_{0}\text{n}$

**15.39** approximately 1:1

**15.41** If the mass numbers of all the stable isotopes are graphed versus their atomic number, a pattern emerges (Figure 15.14). The band of stability is the outline of that pattern.

**15.43** The isotope $^{16}_{8}\text{O}$ should be the most stable. It has a $N/Z$ ratio of 1.

**15.45** (a) beta; (b) alpha and beta; (c) alpha and beta; (d) positron; (e) beta

**15.47** $^{210}_{82}\text{Pb}$

**15.49** In a sample of radioactive material, the half-life is the time it takes for half of the radioactive nuclei to decay. For nuclear decay, each half-life has exactly the same duration.

**15.51** Film, Geiger-Müller counter, and a scintillation counter can be used to detect radiation.

**15.53** One-sixteenth (6.25%) of the original material will be present.

**15.55** three half-lives or 42 days

**15.57** 22.5 hours (three half-lives)

**15.59** 5730 years

**15.61** Twenty-four hours represents four half-lives. The half-life is 6 hours.

**15.63** Radiation-powered devices can last much longer than battery-powered devices.

**15.65** $^{99m}\text{Tc}$ is used because it produces radiation that can be used to image tumors and its short half-life ensures that it won't stay in the body for a long period of time.

**15.67** Positron emission occurs for $N/Z$ ratios < 1. $^{11}_{6}\text{C}$ has a low $N/Z$ ratio, whereas $^{11}_{6}\text{C}$ has a high $N/Z$ ratio.

**15.69** Radiation can (1) have no effect, (2) cause reparable damage, (3) cause damage leading to formation of mutant cells (cancer cells), (4) kill cells (subsequently absorbed by the body). Mutant cells can also be recognized as foreign entities and be destroyed by an immune response.

**15.71** This is a difficult question to answer. How you answer the question stems from your view of our governmental system and your perception of the risk of radon and the benefit of regulations. Clearly, the best protection against unnecessary radon exposure is knowledge of what causes enhanced levels of radon and what measures can be used to reduce radon exposure. The Internet can be a good source of information if you use it carefully (e.g., www.epa.gov/iedweb00/radon/index.html).

**15.73** You would most likely find $^{210}_{82}\text{Pb}$.

**15.75** Fission occurs when unstable nuclei decay into smaller nuclei and decay products. Fusion results when smaller particles combine to make larger ("heavier") nuclei.

**15.77** three

**15.79** In principle, a chain reaction needs the terminating step of the reaction to be a necessary component of the initiating step. For example, if the products of a reaction (initiating step) start a second reaction (terminating step) and the products of the second reaction start the first reaction, you have a chain reaction.

**15.81** A fission reactor has fuel rods (containing the radioactive material), control rods (to absorb the excess neutrons), a moderator (i.e., water, heavy water, or graphite), and energy converters (convert water to steam which powers electric generators).

**15.83** The water is not directly converted to steam by the nuclear reaction as a safety precaution. The steam that would be produced by the nuclear reaction would be highly radioactive and is also made of poisonous heavy water (water with deuterium rather than hydrogen is highly toxic). If there were a leak of the steam, the released products would be very dangerous.

**15.85** The waste contains a mixture of fission by-products as well as materials contaminated by radioactive waste (such as the spent fuel rods).

**15.87** Isotopes of hydrogen (H-1, H-2, H-3) and lithium are most likely to be used.

**15.89** $^{2}_{1}\text{H} + ^{3}_{1}\text{H} \longrightarrow ^{4}_{2}\text{He} + ^{1}_{0}\text{n}$

**15.91** $^{1}_{1}\text{H} + ^{2}_{1}\text{H} \longrightarrow ^{3}_{2}\text{He}$

**15.93** In radioactive dating, a sample is taken and analyzed for the quantity of a particular radioisotope. Based on the amount of the isotope, and assuming you know how much of the isotope was initially present in the sample, you can calculate the sample age (using the half-life). For example, if the radioisotope is at $\frac{1}{4}$ the original concentration, two half-lives have passed (i.e., $\frac{1}{2} \times \frac{1}{2}$).

**15.95** The sample contains about $\frac{1}{64}$ the carbon-14 in a modern sample. This is about 6 half-lives or 34,000 years.

**15.97** $^{1}_{1}\text{H} + ^{1}_{1}\text{H} \longrightarrow ^{2}_{1}\text{H} + ^{0}_{+1}\beta^{+}$

$^{2}_{1}\text{H} + ^{1}_{1}\text{H} \longrightarrow ^{3}_{2}\text{He}$

$^{1}_{1}\text{H} + ^{3}_{2}\text{He} \longrightarrow ^{4}_{2}\text{He} + ^{0}_{+1}\beta^{+}$

**15.99** The initial product of the bombardment reaction is carbon-13 ($^{9}_{4}\text{Be} + ^{4}_{2}\text{He} \longrightarrow ^{13}_{6}\text{C}$). Loss of a neutron would produce carbon-12 ($^{9}_{4}\text{Be} + ^{4}_{2}\text{He} \longrightarrow ^{12}_{6}\text{C} + ^{1}_{0}\text{n}$).

**15.101** With a half-life of 138.4 days, at the end of one year approximately 3 half-lives have already passed. Only one-eighth of the material would be remaining. You would be better off buying brushes every year (or as you needed them).

**15.103** Even though the element chlorine has only been known to exist since 1774, chloride-containing compounds were used well before the discovery of chlorine.

**15.105** 4 neutrons; $^{249}_{98}\text{Cf} + ^{15}_{7}\text{N} \longrightarrow ^{260}_{105}\text{Db} + 4^{1}_{0}\text{n}$

**15.107** The mass number decreases by four and the atomic number decreases by two.

**15.109** 1.00 g

**15.111** beta particle emission

**15.113** $^{214}_{83}\text{Bi}$

**15.114** B

**15.115** C

**15.116** E

15.117 C
15.118 B
15.119 B
15.120 C
15.121 C
15.122 E
15.123 E

## Chapter 16 — Organic Chemistry

16.1   (a) isomer; (b) alcohol; (c) ester; (d) aldehyde; (e) cyclic
       hydrocarbon; (f) saturated hydrocarbon; (g) alkyl group;
       (h) cycloalkane; (i) aliphatic hydrocarbon; (j) amino acid;
       (k) organic chemistry; (l) unsaturated hydrocarbon

16.3   (a) seven single bonds; (b) one double bond, four single
       bonds; (c) one triple bond, two single bonds

16.5   (a) amine; (b) ketone; (c) alcohol; (d) alkene; (e) carbox-
       ylic acid; (f) alkyne; (g) aldehyde; (h) ether; (i) ester

16.7   molecular formula: $C_{20}H_{28}O$
       structural formula:

       line structure:

16.9   (a) $C_7H_{14}O$; (b) $C_8H_{18}$

16.11  (a)
       (b)
       (c)
       (d)
       (e)

16.13  (a) $(CH_3)_2CHCH_2CH_3$; (b) $(CH_2OH)_2CHOH$;
       (c) $CH_3OCH_2CH_3$
       line structures:
       (a)

       (b) HO⎯⎯⎯⎯⎯OH
                   HO

       (c)

16.15  (a)  H
            |
          C⸲⸲⸲H
       H      H

       (b)  H      H
          C⎯C
       H  H  H  H

       (c)  Cl     Cl
          C⎯C
       H  H  H  H

16.17  An alkane hydrocarbon has only carbon-carbon single
       bonds, an alkene contains at least one carbon-carbon
       double bond, and an alkyne has at least one carbon-carbon
       triple bond.

16.19  (a) unsaturated; (b) unsaturated; (c) unsaturated

16.21  Aromatic hydrocarbons are cyclic hydrocarbons that can
       be drawn with alternating double and single bonds be-
       tween the carbon atoms. When the bonds are distributed in
       this way, the electrons are said to be delocalized.

16.23  The London dispersion forces must be stronger in butane
       than in 2-methyl propane (isobutane). London dispersion
       forces increase with the length of the molecule because the
       molecules have more opportunities to interact. Because the
       London dispersion forces are stronger in butane, it boils at
       a higher temperature.

16.25  (a) $CH_3CH_2CH_3$; (b) $CH_3(CH_2)_4CH_3$

16.27  (a) ethane; (b) *n*-octane or octane; (c) propane;
       (d) *n*-nonane or nonane

16.29  There are five isomers of $C_6H_{14}$.

16.31  (a) 2-methylpentane; (b) 2-methylbutane;
       (c) 2-methylbutane (same as previous structure)

16.33  (a) 3-methylpentane; (b) 2,2-dimethylpropane;
       (c) 2-methylbutane

16.35  (a) $CH_3CH(CH_3)CH_3$; (b) $CH_3CH(CH_3)C(CH_2CH_3)_2CH_2CH_3$
       (c) $CH_3CH(CH_3)CH_2CH(CH_3)CH_2CH_2CH_3$
       (d) $CH_3CH_2C(CH_2CH_3)(CH_3)CH_2CH_2CH_3$

16.37  (a) $CH_3CH_2CH_3 + Cl_2 \xrightarrow{\text{heat}} CH_3CHClCH_3 + HCl$
       or $CH_3CH_2CH_3 + Cl_2 \xrightarrow{\text{heat}} CH_3CH_2CH_2Cl + HCl$
       (b) $(CH_3)_3CCH_2CH_2CH_3 + 11O_2 \xrightarrow{\text{heat}} 7CO_2 + 8H_2O$
       (c) $CH_3CH_3 + Br_2 \xrightarrow{\text{heat}} CH_3CH_2Br + HBr$

16.39  An alkene is a hydrocarbon containing one or more
       carbon-carbon double bonds.

16.41  (a) 5-methyl-3-heptene; (b) 2,3-dimethyl-2-hexene
       (c) *trans*-3,4-dimethyl-3-hexene

16.43  only (c) because it can be made to be asymmetric along
       the carbon-carbon double bond axis

**16.45** (a)

3-methyl-1-butene

(b) ≡

propyne

(c)

*trans*-2-butene

(d)

*cis*-3-hexene

**16.47** (a) $CH_3CH=CHCH_3 + HBr \longrightarrow CH_3CHBrCH_2CH_3$

(b) $CH_3CH_2CH=CHCH_3 + Br_2 \longrightarrow$

$CH_3CH_2CHBrCHBrCH_3$

(c)

$+ \ H_2 \xrightarrow{Pd}$

**16.49** Aromatic hydrocarbons are cyclic and have alternating single and double bonds (delocalized electrons).

**16.51** (a) 1,2-dibromobenzene; (b) 1-chloro-4-ethylbenzene

**16.53** (a)

1,2,4-trimethylbenzene

(b)

chlorobenzene

**16.55** $C_6H_6 + Cl_2 \xrightarrow{FeCl_3} C_6H_5Cl + HCl$

**16.57** Both (b) and (c) are alcohols.

**16.59** There are four isomers.

**16.61** (a) 2-methyl-1-propanol; (b) 2-butanol

**16.63** (a)      OH

3-hexanol

(b)

OH

2,2-dimethyl-3-pentanol

**16.65** The primary difference is that the intermolecular forces between ethanol ($CH_3CH_2OH$) molecules are hydrogen bonds, while those between propane molecules are London dispersion forces. The stronger intermolecular forces between ethanol molecules means that a higher temperature (energy) is needed to raise the vapor pressure of ethanol to 1 atm.

**16.67** Ethers contain the functional group —O— attached to two alkyl groups $\left( R^{\diagup O \diagdown} R \right)$.

**16.69** Only (a) $CH_3OCH_3$ is an ether.

**16.71** (a) ethylpropyl ether; (b) dimethyl ether

**16.73** (a)

ethyl methyl ether

(b)

ethyl phenyl ether

**16.75** Water is a polar, hydrogen-bonding solvent. Dimethyl either ($CH_3OCH_3$) is slightly polar but does not directly hydrogen-bond because the oxygen is attached to carbon atoms on either side. Ethanol ($CH_3CH_2OH$) is more water soluble because the alcohol group —OH can hydrogen-bond with water molecules.

**16.77** An aldehyde has the functional group R—CHO

where R can be an organic group or a hydrogen atom.

**16.79** Compounds (a), (c), and (d) are aldehydes.

**16.81** Both compounds are polar because of the C=O bond, but ethanal ($CH_3CHO$) has stronger London dispersion forces because it is a larger molecule. As a result ethanal has a higher boiling point.

**16.83** A carboxylic acid contains the functional group R—COOH

or where R is an organic group.

**16.85** One route to the formation of esters is the condensation of a carboxylic acid and an alcohol. Methyl proponoate ($CH_3CH_2CO_2CH_3$) can be formed from methanol ($CH_3OH$) and propanoic acid ($CH_3CH_2CO_2H$).

**16.87** Saturated fatty acids are high-molar-mass carboxylic acids that do not have double or triple bonds in the hydrocarbon chain. As a result, it is no longer possible to add any more hydrogen atoms to the hydrocarbon chain and they are said to be saturated with hydrogen.

**16.89** Acetic acid, $CH_3CO_2H$, should have a higher boiling point than propanal, $CH_3CH_2CHO$, because acetic acid can hydrogen-bond. Aldehydes are polar but do not hydrogen-bond.

**16.91** An amine is a derivative of ammonia, $NH_3$, with one or more of the H atoms replaced by organic groups $\left( R^{\diagup \ddot{N} \diagdown}_{\ \ \ |} {}_{H} H \right)$.

**16.93** (a)

or

(b)

(c)

**16.95** Methylamine, $CH_3NH_2$, can hydrogen-bond, so it has stronger intermolecular forces than ethane resulting in a higher boiling point. Ethane, $CH_3CH_3$, is nonpolar and cannot hydrogen-bond.

**16.97** $(CH_3CH_2)_2NH + HCl(aq) \longrightarrow$

$(CH_3CH_2)_2NH_2^+(aq) + Cl^-(aq)$

16.99   A simple organization of hydrocarbons is given in the following chart.

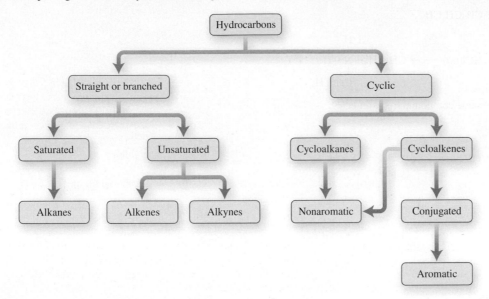

16.101  Alkenes and alkynes differ primarily in three ways: carbon-carbon bond type, bond angles, and number of hydrogen atoms in the structure.

|  | **Alkanes** | **Alkenes** | **Alkynes** |
|---|---|---|---|
| **Bond type** | Single | Double bonds | Triple bonds |
| **Bond angle** | 109.5° | 120° | 180° |
| **General formula (straight or branched)** | #H = 2n + 2 | #H = 2n | #H = 2n − 2 |

   #H = number of hydrogens; *n* = number of carbons. The formula assumes only one double or triple bond per structure.

16.103  Aldehydes and ketones are similar functional groups.

   A ketone has the general formula $\underset{R}{\overset{O}{\underset{}{\|}}}\underset{R'}{C}$ . The aldehyde functional group is $\underset{R}{\overset{O}{\underset{}{\|}}}\underset{H}{C}$ .

16.105  (a) 2,4-dimethylpentane; (b) 1,3-dimethylcyclopentane; (c) 2,2,3,3-tetramethylbutane

16.107  One possible test is to compare the solubility of each substance in an alcohol (such as ethanol). However, this test is not very specific. Addition of bromine, $Br_2$, which is a reddish brown solution would be a better alternative. Bromine will react with an alkene but not with an alcohol. If an alkene is present, the bromine solution will lose color as the bromine reacts with the alkene.

16.109  The trend in boiling points is directly related to the strength of the intermolecular forces in each substance. All three are approximately the same mass so similar London dispersion force strength. Dimethyl ether is polar, so it has additional dipole-dipole forces, and ethanol has hydrogen bonding forces. Because hydrogen bonds are so strong, ethanol has the highest boiling point. Propane has the lowest boiling point because it only has London dispersion forces.

16.111  (a) $CH_3COCH_3$; (b) $CH_3CH_2COCH_3$; (c) $CH_3CH_2CHO$; (d) $CHOCHCHCH_3$; (e) $CH_3CHO$

16.113  (a) $CH_3CH_2CH_2CO_2CH_2CH_3$; (b) $CH_3CH_2CH_2CH_2CO_2H$; (c) $CH_3CO_2CH_3$; (d) $HCO_2H$; (e) $CH_3CH_2CO_2H$

16.115  (a) dimethylamine; (b) ethyldimethylamine (the official IUPAC name is N,N-dimethyl-1-aminoethane, which is why we use the common name); (c) aniline

16.117  (a) $CH_3C(CH_3)_2CH_2CH(CH_3)_2$; (b) $CH_3CHC(CH_2CH_3)_2$; (c) $CH_2CHCHCH_2$; (d) $C_4H_8$; (e) $C_6H_5Cl$

16.119  (a) alkane; (b) ether; (c) carboxylic acid; (d) aldehyde; (e) ester; (f) amine; (g) ketone

16.121  Only (c) 2-butene and (d) 2-pentene can exist as cis and trans isomers.

16.123  (a) isomer; (b) isomer; (c) same compound

16.125  (a) $C_5H_{10}O$; (b) $C_4H_{11}N$; (c) $C_8H_{16}$

16.127  Starting with the alkanes formed with four carbon atoms, four alcohols and two ethers are possible structures.

16.129

$$CH_3CH_2CH_2-\overset{\overset{\displaystyle O}{\|}}{C}-O-CH_2CH_2CH_3$$

16.131

$$CH_3CH_2-\overset{\overset{\displaystyle H}{|}}{\underset{\underset{\displaystyle CH_2CH_3}{|}}{N}} \quad + \text{ HCl}(aq) \longrightarrow CH_3CH_2-\overset{\overset{\displaystyle H}{|+}}{\underset{\underset{\displaystyle CH_2CH_3}{|}}{N}}-H \quad + \text{ Cl}^-$$

16.133 (a) same compound; (b) same compound; (c) isomer
16.134 C
16.135 C
16.136 C
16.137 B
16.138 E
16.139 C
16.140 D
16.141 C
16.142 A
16.143 A

# Chapter 17 — Biochemistry

17.1 (a) transfer RNA; (b) secondary structure; (c) nucleotide; (d) ribonucleic acid; (e) lipid; (f) primary structure; (g) deoxyribonucleic acid; (h) fatty acid; (i) active site; (j) carbohydrate; (k) peptide; (l) replication; (m) translation

17.3 regulation of cell process, control reaction rates, transport substances in and out of cells, fight disease

17.5

$$\overset{\overset{\displaystyle COOH}{|}}{\underset{\underset{\displaystyle R}{|}}{H_2N-\overset{|}{C}-H}}$$

17.7

| $\overset{\overset{\displaystyle COOH}{|}}{\underset{\underset{\displaystyle R}{|}}{H_2N-\overset{|}{C}-H}}$ | $\overset{\overset{\displaystyle COOH}{|}}{\underset{\underset{\displaystyle H_3C \diagup ^{CH} \diagdown CH_3}{|}}{H_2N-\overset{|}{C}-H}}$ | $\overset{\overset{\displaystyle COOH}{|}}{\underset{\underset{\displaystyle OH}{|}}{\underset{\underset{\displaystyle CH_2}{|}}{H_2N-\overset{|}{C}-H}}}$ |
|:---:|:---:|:---:|
| Gly | Val | Ser |

17.9 Alanine has a methyl group and valine has an isopropyl group.

17.11 Ala-Gly $H_2N-CH-\overset{\overset{\displaystyle O}{\|}}{C}-\overset{\overset{\displaystyle H}{|}}{N}-CH-\overset{\overset{\displaystyle O}{\|}}{C}-OH$ ; The methyl group is on the N-terminal amino acid.

Gly-Ala $H_2N-CH-\overset{\overset{\displaystyle O}{\|}}{C}-\overset{\overset{\displaystyle H}{|}}{N}-CH-\overset{\overset{\displaystyle O}{\|}}{C}-OH$ ; The methyl group is on the C-terminal amino acid.

The peptide bond is boxed and the side chains are circled.

17.13

$$H_2N-CH-\overset{O}{\overset{\|}{C}}-\overset{H}{\overset{|}{N}}-CH-\overset{O}{\overset{\|}{C}}-\overset{H}{\overset{|}{N}}-CH-\overset{O}{\overset{\|}{C}}-\overset{H}{\overset{|}{N}}-CH-\overset{O}{\overset{\|}{C}}-\overset{H}{\overset{|}{N}}-CH-\overset{O}{\overset{\|}{C}}-\overset{H}{\overset{|}{N}}-CH-\overset{O}{\overset{\|}{C}}-\overset{H}{\overset{|}{N}}-CH-\overset{O}{\overset{\|}{C}}-\overset{H}{\overset{|}{N}}-CH-\overset{O}{\overset{\|}{C}}-OH$$

with $CH_3$ on each $\alpha$ carbon.

17.15

$$H_2N-CH-\overset{O}{\overset{\|}{C}}-OH + H_2N-CH-\overset{O}{\overset{\|}{C}}-OH \longrightarrow H_2N-CH-\overset{O}{\overset{\|}{C}}-\overset{H}{\overset{|}{N}}-CH-\overset{O}{\overset{\|}{C}}-OH$$

with $R_1$, $R_2$ side chains.

17.17

$$H_2N-CH-\overset{\overset{\displaystyle O}{\|}}{C}-\overset{\overset{\displaystyle H}{|}}{N}-CH-\overset{\overset{\displaystyle O}{\|}}{C}-OH$$
with CH₃ on first CH, and CH₂–CH–CH₃ / CH₃ on second
Ala-Leu

$$H_2N-CH-\overset{\overset{\displaystyle O}{\|}}{C}-\overset{\overset{\displaystyle H}{|}}{N}-CH-\overset{\overset{\displaystyle O}{\|}}{C}-OH$$
with CH₂–CH–CH₃ / CH₃ on first CH, and CH₃ on second
Leu-Ala

17.19

$$H_2N-CH-\overset{\overset{\displaystyle O}{\|}}{C}-OH \ + \ H_2N-CH-\overset{\overset{\displaystyle O}{\|}}{C}-OH \ \longrightarrow \ H_2N-CH-\overset{\overset{\displaystyle O}{\|}}{C}-\overset{\overset{\displaystyle H}{|}}{N}-CH-\overset{\overset{\displaystyle O}{\|}}{C}-OH \ + \ H_2O$$
(first CH₃; second CH–CH₃/CH₃; product CH₃ and CH–CH₃/CH₃)

17.21

$$H_2N-CH-\overset{\overset{\displaystyle O}{\|}}{C}-\overset{\overset{\displaystyle H}{|}}{N}-CH-\overset{\overset{\displaystyle O}{\|}}{C}-OH$$
(CH₃ on first, H on second)
Ala-Gly

$$H_2N-CH-\overset{\overset{\displaystyle O}{\|}}{C}-\overset{\overset{\displaystyle H}{|}}{N}-CH-\overset{\overset{\displaystyle O}{\|}}{C}-OH$$
(H on first, CH₃ on second)
Gly-Ala

17.23   Ser-Gly-Cys; Ser-Cys-Gly; Cys-Ser-Gly; Cys-Gly-Ser; Gly-Ser-Cys; Gly-Cys-Ser

17.25   legumes

17.27

$$H_2N-CH-\overset{\overset{\displaystyle O}{\|}}{C}-OH$$ (H)
Gly

$$H_2N-CH-\overset{\overset{\displaystyle O}{\|}}{C}-OH$$ (CH–CH₃ / CH₃)
Val

$$H_2N-CH-\overset{\overset{\displaystyle O}{\|}}{C}-OH$$ (CH₂ / OH)
Ser

17.29   The abbreviated formula of the peptide is Gly-Ala-Val. Upon complete hydrolysis, the structures of the amino acids are

$$H_2N-CH-\overset{\overset{\displaystyle O}{\|}}{C}-OH$$ (H)
Gly

$$H_2N-CH-\overset{\overset{\displaystyle O}{\|}}{C}-OH$$ (CH₃)
Ala

$$H_2N-CH-\overset{\overset{\displaystyle O}{\|}}{C}-OH$$ (CH–CH₃ / CH₃)
Val

17.31

$$H_2N-CH-\overset{\overset{\displaystyle O}{\|}}{C}-N\ (pyrrolidine\ ring)-\overset{\overset{\displaystyle O}{\|}}{C}-\overset{\overset{\displaystyle H}{|}}{N}-CH-\overset{\overset{\displaystyle O}{\|}}{C}-\overset{\overset{\displaystyle H}{|}}{N}-CH-\overset{\overset{\displaystyle O}{\|}}{C}-OH$$
(CH₃ on first; CH₂–phenyl group; H)

17.33   The primary structure is the amino acid sequence. The secondary structure results from intramolecular attractions such as hydrogen bonding and produces structural units such as the alpha helix and beta pleated sheets. The tertiary structure is the three-dimensional shape obtained from the folding of the peptide chain. Folding is caused by the intramolecular interactions of the secondary structural units. The quaternary structure is used to describe the shape of a protein when it is composed of more than one peptide chain. The quaternary structure is established through intermolecular interactions.

17.35   In myoglobin, a globular protein, the protein chain folds back on itself. Wool, a fibrous protein, is primarily composed of chains of alpha helices.

17.37   tertiary structure

17.39   Hydrogen bonding holds together the twisted-helical shape of the peptide chain.

17.41   Fibrous proteins have long rod-like shapes. In globular proteins, the peptide chain folds back onto itself giving the protein a "glob"-like shape.

17.43   Denaturation disrupts the attractive forces which give rise to secondary, tertiary, and quaternary structures and causes the "unfolding" of the protein.

17.45   Ribonucleic acids and deoxyribonucleic acids. In deoxyribonucleic acids, the ribose unit lacks the hydroxide group at the 2 position.

17.47   A nucleic acid is composed of a chain of nucleotides.

17.49   ribose; adenine; RNA

17.51   The shapes and sizes of these pairs (A-T and C-G) complement each other and provide for the maximum degree of hydrogen bonding.

17.53   AATCGGTCACGAT

17.55   UGAUUCUAG

17.57   Genetic information is encoded as complementary strands of nucleic acids. When genetic information is replicated, it copies one strand using complementary base pairing of DNA nucleotides.

17.59   AAA

17.61   AUG-CUU-CCC-CAA-CGA-UUU-CCC-CGA

Met - Leu - Pro - Gln - Arg - Phe - Pro - Arg

(so the answer is Pro and Arg)

17.63

17.65   glucose, $C_6H_{12}O_6$; sucrose, $C_{12}H_{22}O_{11}$

17.67   structure **A** (glucose)

17.69   monosaccharide

17.71   Sucrose yields glucose and fructose. Starch yields only glucose.

17.73   In starches, the main chain repeats with the glucose molecules maintaining the same orientation in the chains. Amylopectin also branches (which is not observed in cellulose).

In cellulose, the glucose units invert after every ether bond (circled). Note the position of the $-CH_2OH$ group.

17.75   Before we can taste the sugars in starch, the glucose monomers must be cleaved from the terminal of the amylose or amylopectin.

17.77   The fatty acids in fat molecules tend to be completely or almost completely saturated (few or no C=C double bonds). The fatty acids in oil molecules are generally polyunsaturated and contain several C=C double bonds on each fatty acid chain.

17.79   triacyl glycerol (tristearate)

17.81   It would be a fat.

17.83   phospholipid

17.85   They are both lipids, and all steroids are derived from cholesterol.

17.87

17.89

17.91   Proteins perform many functions; one of the most important is to catalyze the biochemical reactions. They are also responsible for the machinery of the cell which regulates cellular processes such as transcription, translation, immune response, and energy production. Carbohydrates provide energy (i.e., glucose and metabolism) and play important structural roles in plants (cellulose).

17.93   DNA and RNA each have a sugar molecule, base, and a phosphate unit. The sugar in DNA is deoxyribose. The DNA contains all of the genetic information. RNA, derived from DNA through transcription, contains only information for a specific protein.

17.95   Genetic therapy can be used to produce novel substances from organisms. An example is human insulin production from bacteria. Plants can be bred to tolerate various chemical agents (i.e., herbicides) using this technique.

17.97   (a) valine; (b) methionine; (c) aspartic acid

17.99

17.101

17.103  ribose; guanine; ribonucleotide; RNA

17.105  saturated

17.107  B

17.108  E

17.109  A

17.110  D

17.111  E

17.112  C

17.113  C

17.114  C

17.115  B

17.116  A and E

# Glossary

## A

**acid**   a compound that releases hydrogen ions when dissolved in water or donates $H^+$ to another substance. (3.6)

**acid ionization constant, $K_a$**   an equilibrium constant for the ionization of an acid in water; a value that expresses the strength of a weak acid. (13.3)

**acidic solution**   a solution in which the $H_3O^+$ concentration is greater than the $OH^-$ concentration; a solution with a pH less than 7. (13.4)

**activated complex**   an unstable, high-energy chemical species that must be formed as reactants convert to products. (12.2)

**activation energy, $E_a$**   the minimum amount of energy that reactants must have before they can be converted to products. (12.2)

**active site**   the location in an enzyme where catalysis takes place. (17.1)

**activity series**   a list of metals in order of activity. (5.4)

**actual yield**   the amount of product actually obtained in the laboratory from a reaction. (6.5)

**acyclic hydrocarbon**   a hydrocarbon that has a straight- or branched-chain structure with localized bonds. (16.2)

**alcohol**   an organic compound containing the functional group $-OH$. (8.4, 16.5)

**aldehyde**   an organic compound containing the functional group $-CHO$. (16.6)

**aliphatic hydrocarbon**   a hydrocarbon that contains localized single, double, or triple bonds. (8.4, 16.2)

**alkali metal**   any group IA (1) element except hydrogen; Li, Na, K, Rb, Cs, and Fr are alkali metals. (2.5)

**alkaline earth metal**   any group IIA (2) element; Be, Mg, Ca, Sr, Ba, and Ra. (2.5)

**alkane**   a hydrocarbon that has a straight-chain or branched-chain structure with only single bonds. (8.4, 16.3)

**alkene**   a hydrocarbon that contains one or more localized carbon-carbon double bonds. (8.4, 16.3)

**alkyl group**   a group derived from an alkane by removal of one hydrogen atom. (16.3)

**alkyne**   a hydrocarbon that contains one or more triple bonds. (8.4, 16.3)

**alloy**   a mixture of a metal with one or more additional metallic or nonmetallic elements. (10.4)

**alpha helix structure**   the secondary structure adopted by many proteins, in which the protein chain is twisted into a helix. (17.1)

**alpha particle**   a particle containing two protons and two neutrons; the nucleus of a helium-4 atom. (15.1)

**amine**   an organic base with one or more alkyl groups bonded to a nitrogen atom. (16.8)

**amino acid**   an organic molecule that contains both the $-NH_2$ and the $-CO_2H$ functional groups. (17.1)

**amorphous solid**   a solid that is not crystalline but has its particles arranged in a random, nonrepeating fashion. (10.4)

**amphoteric substance**   a substance that can act as either an acid or a base. (13.1)

**anhydrous**   a compound free of molecular water. (5.4)

**anion**   a negatively charged ion, such as the chloride ion ($Cl^-$) or nitrate ion ($NO_3^-$). (2.3)

**anode**   the electrode that attracts anions and at which oxidation occurs. (14.3)

**aqueous solution**   a homogeneous mixture of two or more substances in which the solvent is water. (1.1, 11.1)

**aromatic hydrocarbon**   a cyclic hydrocarbon that contains delocalized bonds made up of alternating single and double bonds that impart special stability. (8.4, 16.2)

**Arrhenius model of acids and bases**   a simple model that describes acids as substances that generate $H^+$ ions in solution and bases as substances that generate $OH^-$ ions in solution. (13.1)

**aryl group**   an aromatic group formed by removal of hydrogen from a cyclic aromatic hydrocarbon. (16.9)

**atom**   the smallest particle of an element that retains the characteristic chemical properties of that element. (1.1)

**atomic mass unit (amu)**   the basic unit of mass of atoms and molecules; exactly one-twelfth the mass of one $^{12}C$ atom; 1 amu = $1.6606 \times 10^{-24}$ g. (2.4)

**atomic number**   the number of protons in the nucleus of an atom of a given element. (2.2)

**atomic radius**   a measure of atomic size; the distance from the nucleus to the outer edge of the atom. (7.7)

**aufbau principle**   an approach to writing ground-state electron configurations of atoms by filling orbitals that are lowest in energy first; means "building up" in German. (7.3)

**Avogadro's hypothesis**   a proposal that equal volumes of all gases contain equal numbers of particles (at constant temperature and pressure); $V = $ constant $\times n$. (9.2)

**Avogadro's number**   the number of fundamental particles (atoms, molecules, or ions) in 1 mol of any substance; $6.022 \times 10^{23}$ particles/mol. (4.2)

## B

**balanced equation**   a chemical equation that uses coefficients in front of each formula to make the number of atoms of each element the same on both sides of the equation. (5.3)

**barometer**   a closed tube filled with mercury and inverted into a pool of mercury; instrument used to measure atmospheric pressure. (9.1)

**base**   a substance that is capable of neutralizing an acid in a chemical reaction, forming water; a substance that accepts an $H^+$ from another substance. (3.6)

**basic solution**   a solution in which the $OH^-$ concentration is greater than the $H_3O^+$ concentration; a solution with a pH greater than 7. (13.4)

**battery**   one or more interconnected voltaic cells used as a portable source of electricity. (14.3)

**beta particle**   a small charged particle that can be emitted from atoms at speeds approaching the speed of light; either an electron or a positron. (15.1)

**binary compound**   a compound containing atoms or ions of two elements. (3.1)

**boiling point**   the temperature at which the vapor pressure of a liquid equals the external pressure. (10.1)

**bond angle**   the angle between the two lines defined by a central atom attached to two surrounding atoms. (8.5)

**Boyle's law**   a law stating that at constant temperature, the volume occupied by a fixed amount of a gas is inversely proportional to its pressure; $V = \dfrac{\text{constant}}{P}$. (9.2)

**Brønsted-Lowry theory**   a theory that defines an acid as a substance that donates an $H^+$ to another substance in solution and defines a base as a substance that accepts an $H^+$ in solution. (13.1)

**buffer**   a combination of a weak acid and its conjugate base (or a weak base and its conjugate acid) in similar amounts; a buffered solution resists changes in pH when small amounts of $H_3O^+$ or $OH^-$ are added. (13.6)

## C

**carbohydrate**   one of a class of biomolecules that serves as a source of energy; are composed of carbon, hydrogen, and oxygen; and have the general formula $C_n(H_2O)_m$. (17.3)

**carboxylic acid**   an organic compound containing the functional group $-CO_2H$. (16.7)

**catalyst**   a substance that changes the rate of the reaction, but can be recovered unchanged after the reaction. (12.3)

**cathode**   the electrode that attracts cations and at which reduction occurs. (14.3)

**cation**   a positively charged ion, such as the sodium ion ($Na^+$); an ion that contains fewer electrons than protons. (2.3)

**chain reaction**   a reaction in which the product of one step is a reactant in a subsequent step. (15.6)

**Charles's law**   a law stating that at constant pressure, the volume occupied by a fixed amount of gas is directly proportional to its absolute temperature; $V = \text{constant} \times T$. (9.2)

**chemical bond**   the force that holds atoms together in a molecule or compound. (8.1)

**chemical change**   a process in which one or more substances are converted into new substances that have compositions and properties different from those of the original substances. (1.2)

**chemical equation**   an abbreviated representation of a chemical reaction consisting of chemical symbols and formulas. (5.3)

**chemical equilibrium**   a state where the rates of the forward reaction and the reverse reaction are equal so the concentrations of reactants and products do not change. (12.4)

**chemical formula**   a representation of the composition of a compound using symbols of the compound's elements. (1.1)

**chemical property**   a characteristic of a substance defined by its composition and the chemical changes it can undergo. (1.2)

**chemical reaction**   a process in which one or more substances are converted into new substances that have compositions and properties different from those of the original substances. (1.2)

**codon**   a series of three bases in a nucleic acid that defines the amino acid to be used in the synthesis of a protein. (17.2)

**colligative property**   a solution property that depends only on the concentration of solute particles, not on the identity of the solute. (11.6)

**collision theory**   a theory that explains the temperature and concentration dependence of reaction rate on the basis of molecular collisions, molecular orientation, and kinetic energy. (12.2)

**combination reaction**   a reaction in which two elements, an element and a compound, or two compounds join to form a new compound. (5.4)

**combined gas law**   a law that describes the relation between pressure, volume, and temperature for a fixed amount of a gas;
$$\frac{P_1V_1}{T_1} = \frac{P_2V_2}{T_2}. \quad (9.2)$$

**combustion reaction**   a reaction that involves oxygen molecules as a reactant and that rapidly produces heat and flames. (5.4)

**compound**   a substance composed of two or more elements combined in definite proportions. (1.1)

**concentrated solution**   a solution that contains a relatively high concentration of solute. (4.4)

**concentration**   the relative amounts of solute and solvent in a solution. (4.4, 11.4)

**condensation**   the conversion of gas-state atoms or molecules to the liquid state. (10.1)

**conjugate acid**   a substance that forms after a base gains an $H^+$ ion. (13.1)

**conjugate base**   a substance that forms after an acid loses an $H^+$ ion. (13.1)

**continuous spectrum**   a spectrum in which all the visible wavelengths are present. (7.1)

**core electron**   an electron in an atom that is not a valence electron; an inner electron. (7.5)

**corrosion**   a slow deterioration of metals due to interaction with the environment. (5.4, 14.7)

**covalent bonding**   the bonding between two atoms resulting from the sharing of electrons. (8.1)

**critical mass**   the smallest amount of fissionable material necessary to support a continuing chain reaction. (15.6)

**crystal**   an orderly, repeating assembly of fundamental particles (atoms, molecules, or ions) in a solid. (10.4)

**crystal lattice**   the repeating pattern of atoms or ions in a crystal. (8.2, 10.4)

**crystalline solid**   a substance in which the particles are arranged in a regular, repeating geometric structure. (10.4)

**cyclic hydrocarbon**   a hydrocarbon that has a ring or closed-chain structure. (16.2)

**cycloalkane**   an alkane that contains a ring structure. (16.4)

## D

**Dalton's atomic theory**   the first widely accepted theory that describes matter as being composed of atoms; published in 1808. (2.1)

**Dalton's law of partial pressures**   a law stating that gases in a mixture behave independently and exert the same pressure they would if they were in the container alone; $P_{total} = P_A + P_B + P_C + \dots$. (9.3)

**decomposition reaction**   a chemical reaction in which a substance breaks down into simpler compounds or elements. (5.4)

**density**   the ratio of the mass of a substance to its volume. (1.2)

**deoxyribonucleic acid (DNA)**   a high-molar-mass nucleic acid that stores genetic information, has a double-helix structure, and contains the sugar deoxyribose. (17.2)

**deposition**   the conversion of gas-state atoms or molecules to the solid state. (10.1)

**diatomic molecule**   a neutral particle consisting of two bound atoms. (2.5)

**diffusion**   the movement of gas particles through the atmosphere from high concentration to low concentration. (9.4)

**dilute solution**   a solution that contains a relatively low concentration of solute. (4.4)

**dilution**   the process of adding more solvent to give a solution of lower concentration. (4.4)

**dipole-dipole force**   the attraction between the partially negative end of one polar molecule and the partially positive end of another polar molecule. (10.2)

**double bond**   a covalent bond that involves the sharing of two pairs of electrons. (8.3)

**double-displacement reaction**   a reaction in which two compounds exchange ions or elements to form new compounds. (5.4)

## E

**effusion**   the movement of gas particles through a small opening. (9.4)

**electrochemistry**   the study of the relationship between chemical reactions and electrical work. (14.6)

**electrode**   a solid material that conducts electrical charge and is used as part of an electrochemical cell; one electrode (the anode) acts as the site of oxidation, and the other (the cathode) acts as the site of reduction in electrochemical reactions. (14.3)

**electrolysis**   a process in which a nonspontaneous oxidation-reduction reaction is forced to occur upon passage of electric current through a sample. (14.6)

**electrolyte**   a substance that dissociates or ionizes into aqueous ions upon dissolving and conducts electricity when dissolved in water. (3.1)

**electrolytic cell**   an electrochemical cell through which electric current is passed to cause a nonspontaneous oxidation-reduction reaction to occur. (14.6)

**electromagnetic radiation**   a form of energy that can be described in terms of oscillating waves that move through space at the speed of light; also called light or radiant energy. (7.1)

**electron**   a negatively charged subatomic particle. (2.2)

**electron capture** the incorporation of a $1s$ electron into the nucleus, resulting in the conversion of a proton into a neutron. (15.2)

**electron configuration** a description of the distribution of electrons in atomic orbitals or sublevels. (7.3)

**electronegativity** a measure of the ability of an atom to attract electrons within a bond to itself. (8.1)

**element** a pure substance that cannot be broken down into simpler stable substances in a chemical reaction. (1.1)

**element symbol** a shorthand version of an element's name consisting of one or two letters, the first of which is uppercase. (1.1)

**empirical formula** a formula written with the simplest ratios of atoms or ions (the smallest whole-number subscripts). (4.3)

**endothermic reaction** a chemical change that absorbs heat. (6.6)

**energy** the capacity to do work or to transfer heat. (1.3)

**entropy** a measure of the tendency for the energy of matter to become more dispersed. (11.2)

**enzyme** a protein molecule that acts as a catalyst to a specific reaction or to a narrow range of reactions. (12.3)

**equilibrium** a state of balance between opposing processes where both processes occur at the same rate. (10.1)

**equilibrium constant expression** the expression obtained from a reaction's balanced equation that relates the relative amount of products to reactants at equilibrium. (12.5)

**ester** a condensation product of carboxylic acids and alcohols, having the general formula $RCO_2R'$. (16.7)

**ether** an organic compound that contains the —O— functional group attached to two carbon atoms of alkyl or aryl groups: R—O—R'. (16.5)

**evaporation** a process by which atoms or molecules pass from the liquid state to the gaseous state. (10.1)

**exothermic reaction** a chemical change that releases heat. (6.6)

## F

**family** a vertical column of related elements in the periodic table; another name for group. (2.5)

**fatty acid** a long-chain carboxylic acid. (17.4)

**fission** the splitting of a heavy atom into two or more lighter atoms and some number of neutrons with the release of energy. (15.6)

**fluid** a substance that can flow (a gas or a liquid). (10.3)

**formula unit** the smallest repeating unit in an ionic compound. (3.3)

**freezing point** the temperature at which the liquid and solid states of a substance are in equilibrium (identical to the melting point). (10.1)

**frequency** the number of wave cycles passing a stationary point in 1 second. (7.1)

**functional group** a group of atoms substituted for hydrogen in the formula of a hydrocarbon that gives the compound its characteristic properties; a reactive part of a molecule that undergoes characteristic reactions. (8.4, 16.1)

**fusion** the combination of light nuclei to form heavier nuclei. (15.6)

## G

**gamma ray** the highest-energy and most-penetrating type of electromagnetic radiation. (15.1)

**gas** the physical state in which matter has no fixed shape or volume but expands to fill its container completely and can be compressed easily. (1.1)

**Gay-Lussac's law of combining volumes** a law stating that in chemical reactions occurring at constant pressure and temperature, the volumes of gaseous reactants and products are in the ratio of small whole numbers. (9.2)

**ground state** the state of lowest energy in an atom where the electrons occupy orbitals that are lowest in energy. (7.3)

**group** a vertical column of related elements in the periodic table; another name for family. (2.5)

## H

**half-cell** the individual compartments in which oxidation and reduction occur. (14.3)

**half-life** the time it takes for half of a sample of a nuclide to decay to a different nuclide. (15.3)

**half-reaction** the oxidation and reduction processes that occur in the individual cell compartments of an electrochemical cell. (14.3)

**halogen** any group VIIA (17) element; the nonmetals F, Cl, Br, I, and At are halogens. (2.5)

**heat** the energy that is transferred between objects due to a difference in their temperatures. (6.6)

**hertz (Hz)** a unit of frequency equal to 1/s ($s^{-1}$) or cycles per second. (7.1)

**heterogeneous equilibrium** an equilibrium for a reaction in which reactants and products are not all in the same physical state. (12.5)

**heterogeneous mixture** a combination of two or more substances that is not uniform throughout. (1.1)

**homogeneous equilibrium** an equilibrium for a reaction in which all reactants and products are in the same physical state. (12.5)

**homogeneous mixture** a combination of two or more substances that has uniform composition. (1.1)

**Hund's rule** a rule stating that electrons are distributed in a set of orbitals of identical energy in such a way as to give the maximum number of unpaired electrons. (7.3)

**hydrocarbon** a compound that contains only carbon and hydrogen. (8.4, 16.1)

**hydrogen bond** an especially strong dipole-dipole force between polar molecules that contain hydrogen attached to a highly electronegative element. (10.2)

**hydrolysis** a chemical reaction in which the addition of water to a compound breaks it down into simpler compounds, one containing —H and the other —OH (17.1)

**hydronium ion** an aqueous hydrogen ion; $H_3O^+(aq)$. (13.1)

**hypothesis** a tentative explanation for the properties or behavior of matter that accounts for a set of observations and can be tested. (1.4)

## I

**ideal gas** a gas that follows predicted behavior, as represented by the ideal gas law. (9.2)

**ideal gas constant, $R$** a constant used in the ideal gas law that relates pressure, volume, amount of gas, and temperature. (9.3)

**ideal gas law** a law stating the relationship between pressure, volume, temperature, and amount of an ideal gas; $PV = nRT$. (9.3)

**induced dipole** a temporary dipole formed by attraction of the electrons in a molecule to a nearby dipole. (10.2)

**inner-transition metal** an element that is a lanthanide (atomic numbers 58 to 71) or an actinide (atomic numbers 90 to 103). (2.5)

**instantaneous dipole** a temporary dipole formed by the movement of electrons within a molecule. (10.2)

**intermediate** a relatively unstable chemical species that is produced in a step during a reaction, and then reacts in a subsequent step. (12.3)

**intermolecular force** a force of attraction that occurs between molecules. (10.2)

**ion** an atom or molecule bearing an electrical charge. (2.3)

**ion-dipole force** an intermolecular force between an ion and a polar molecule. (11.2)

**ion-product constant of water, $K_w$** the equilibrium constant for the self-ionization of water; the product of the $H_3O^+$ ion concentration and the $OH^-$ ion concentration in any aqueous solution; $K_w = [H_3O^+][OH^-] = 1.0 \times 10^{-14}$ at 25°C. (13.4)

**ionic bonding** the bonding between cations and anions resulting from electrostatic attractions of opposite charges. (8.1)

**ionic compound** a compound consisting of cations and anions in proportions that give electrical neutrality. (3.1)

**ionic crystal**   a solid structure in which the ions are arranged in a regular repeating pattern. (8.2)

**ionic equation**   a form of a chemical equation in which ionic species are represented as the separated ions. (5.5)

**ionization energy**   the minimum energy required to remove a valence electron from a gaseous atom or ion; usually reported in kJ/mol. (7.7)

**isoelectronic**   having the same number of electrons. (7.6)

**isomer**   a compound having the same composition and formula as another compound, but a different structure. (16.1)

**isotope**   an atom of an element with a specific number of neutrons. (2.2)

**isotope symbol**   an element symbol with the atomic number ($Z$) as the left subscript and a mass number ($A$) as the left superscript. (2.2)

# K

**ketone**   an organic compound containing an oxygen double-bonded to a carbon, which is also bonded to two other carbons: $R_2C{=}O$. (16.6)

**kinetic energy**   the energy possessed by an object because of its motion. (1.3)

**kinetic-molecular theory of gases**   a model of ideal gas behavior; assumes that gas particles occupy no volume and that there are no intermolecular forces between them. (9.4)

# L

**law of conservation of energy**   a law stating that energy can be converted from one form to another, or transferred, but it cannot be created or destroyed. (6.6)

**law of conservation of mass**   a law stating that the mass of the substances produced in a chemical reaction equals the mass of the substances that reacted. (2.1)

**law of definite proportions**   a law stating that all samples of the same pure compound will always contain the same proportions by mass of the component elements. (2.1)

**Le Chatelier's principle**   a principle stating that a system at equilibrium reacts to a disruption such as a change in concentration, volume, or temperature in such a way as to counteract the disruption and reestablish equilibrium. (12.6)

**Lewis structure**   a representation of a molecule or polyatomic ion consisting of the symbols of the component elements, each surrounded by dots representing shared and unshared electrons. (8.3)

**Lewis symbol**   a representation of an atom consisting of the symbol for the element surrounded by a number of dots equal to the number of valence electrons. (8.2)

**limiting reactant**   the reactant that is completely used up in the reaction and therefore determines the amount of other reactants that react and the amount of products that should form. (6.4)

**line spectrum**   a spectrum produced by a substance in which only certain wavelengths are present. (7.1)

**lipid**   a general class of biological molecules that can be extracted from cells by nonpolar solvents. (17.4)

**liquid**   the physical state in which matter has no characteristic shape but takes the shape of the filled portion of its container and can be poured. (1.1)

**London dispersion force**   the attractive force that results from formation of temporary dipoles in molecules. (10.2)

# M

**main-group element**   a member of one of the A groups of elements in the periodic table; also called a *representative element*. (2.5)

**mass**   a measure of the quantity of matter. (1.1)

**mass number**   the sum of the number of protons and the number of neutrons in the nucleus of an atom or ion. (2.2)

**mass/volume percent**   mass of solute divided by volume of solution and multiplied by 100%. (11.4)

**matter**   anything that occupies space and has mass. (1.1)

**melting point**   the temperature at which the liquid and solid states of a substance are in equilibrium (identical to the freezing point). (10.1)

**meniscus**   the curved upper surface of a liquid column. (10.3)

**messenger RNA (mRNA)**   the form of RNA that takes the genetic code from DNA and carries it to the ribosomes. (17.2)

**metal**   an element characterized by luster and the ability to conduct electricity. (1.1)

**metalloid**   an element that has properties intermediate between those of metals and nonmetals; also called *semimetal*; an element with physical properties resembling a metal but chemical reactivity like that of a nonmetal. (2.5)

**miscible**   the ability of liquids to mix in all proportions. (11.1)

**mixture**   a combination of two or more substances that can vary in composition. (1.1)

**molality**   moles of solute per kilogram of solvent. (11.4)

**molar mass**   the mass of 1 mol of any substance, in units of grams per mole. (4.2)

**molar volume**   the volume occupied by 1 mol of a gas, which equals 22.414 L at STP for an ideal gas. (9.2)

**molarity, $M$**   the number of moles of solute per liter of solution. (4.4, 11.4)

**mole**   the amount of a substance that contains the same number of fundamental particles (atoms, molecules, or formula units) as there are atoms in exactly 12 g of $^{12}C$ (Avogadro's number); $6.022 \times 10^{23}$ particles. (4.2)

**molecular compound**   a compound composed of two or more nonmetals and existing in discrete units of atoms held together in a molecule; electrons are shared between atoms rather than transferred from one atom to another. (3.1)

**molecular equation**   a form of a chemical equation in which substances are represented as if they existed as molecules (using formula units), even though the substances may exist in solution as ions. (5.5)

**molecular formula**   a formula that indicates the actual number of atoms present in a molecule. (3.5)

**molecular solid**   a solid made up of discrete molecules held together by intermolecular forces. (10.4)

**molecule**   two or more atoms bound together in a discrete arrangement. (1.1)

**monatomic ion**   an ion of a single atom. (3.2)

# N

**net ionic equation**   a form of a chemical equation in which ionic compounds are represented as the separated ions, and spectator ions are eliminated. (5.5)

**network solid**   a solid in which all atoms are held together by covalent bonds to make a giant molecule. (10.4)

**neutral solution**   a solution in which the $OH^-$ concentration is equal to the $H_3O^+$ concentration; a solution with a pH of 7. (13.4)

**neutralization reaction**   a reaction of an acid and a base to form an ionic compound and water. (5.4)

**neutron**   an electrically neutral subatomic particle. (2.2)

**neutron number**   the number of neutrons in the nucleus of an atom. (2.2)

**noble gas**   any group VIIIA (18) element; He, Ne, Ar, Kr, Xe, and Rn are noble gases. (2.5)

**nonelectrolyte**   a substance that retains its molecular identity upon dissolving and does not conduct electricity when dissolved in water. (3.1)

**nonmetal**   an element that typically has a dull appearance and is a poor conductor of electricity. (1.1)

**nonpolar covalent bond**   a bond resulting from the equal sharing of electron pairs. (8.1)

**normal boiling point**   the temperature at which the vapor pressure of a liquid equals the standard atmospheric pressure of 1 atm. (10.1)

**normal freezing point**   the temperature at which the liquid and solid states of a substance are in equilibrium at a pressure of 1 atm. (10.1)

**nuclear bombardment**   a nuclear reaction that is induced by the impact of a nuclide or a subatomic particle on another nuclide. (15.2)

**nucleic acid**   one of a class of biomolecules that carries genetic information and regulates the synthesis of proteins. (17.2)

**nucleon**   a nuclear particle (neutron or proton). (15.1)

**nucleotide**   the fundamental unit of a nucleic acid polymer, consisting of a sugar molecule bonded to both a phosphate group and an organic, nitrogen-containing base. (17.2)

**nucleus**   the central core of an atom which contains the protons and neutrons; contributes most of the mass of the atom. (2.2)

**nuclide**   an atom having a specified atomic number, mass number, and nuclear energy. (15.1)

# O

**octet rule**   a rule stating that atoms tend to gain, lose, or share electrons to achieve an electronic configuration with eight electrons in the valence shell. (8.3)

**orbital**   a three-dimensional region in space around a nucleus where an electron is likely to be found. (7.3)

**organic chemistry**   the study of organic compounds. (16.1)

**osmosis**   the passage of solvent molecules, but not solute particles, through a membrane from a solution that is less concentrated in solute to a solution that is more concentrated. (11.6)

**oxidation**   a chemical process, coupled with reduction, in which an atom increases in oxidation number. (14.1)

**oxidation number**   a charge assigned to each atom in a compound according to a set of rules used in an electron accounting system. (14.2)

**oxidation-reduction reaction**   a reaction in which electrons are transferred. (14.1)

**oxidizing agent**   a reactant that causes the oxidation of an element in another reactant and contains an element that is reduced during a chemical reaction. (14.1)

**oxoanion**   an anion containing oxygen attached to some other element. (3.2)

# P

**parts per billion**   the mass of solute divided by mass of solution and multiplied by $10^9$. (11.4)

**parts per million**   the mass of solute divided by mass of solution and multiplied by $10^6$; in aqueous solutions it is essentially the milligrams of solute per liter of solution. (11.4)

**Pauli exclusion principle**   a principle stating that a maximum of two electrons can occupy an orbital, and the electrons must have opposite spins. (7.3)

**peptide**   the combination of two or more amino acids connected by a peptide bond. (17.1)

**peptide bond**   the linkage that forms between the carboxylic acid functional group of one amino acid and the amine functional group on another amino acid. (17.1)

**percent by mass**   the mass of solute divided by mass of solution and multiplied by 100%. (11.4)

**percent by volume**   the volume of solute divided by volume of solution and multiplied by 100%. (11.4)

**percent composition by mass**   an expression of the percentage of the total mass due to each element in a compound. (4.1)

**percent yield**   the ratio of the amount of product actually formed in a reaction to what is predicted by stoichiometry, multiplied by 100%. (6.5)

**period**   a horizontal row of elements in the periodic table. (2.5)

**periodic table**   an arrangement of all the known elements by increasing atomic number and arranged in columns and rows to emphasize periodic properties. (2.5)

**pH**   a measure of the acidity of aqueous solutions; the negative logarithm of the $H_3O^+$ concentration: $pH = -\log[H_3O^+]$. (13.5)

**photon**   the smallest packet of energy that describes the energy of a specific type of electromagnetic radiation. (7.1)

**physical change**   a process that changes only the physical properties of a substance, not its chemical composition. (1.2)

**physical property**   a characteristic of a substance that can be observed without changing its composition. (1.2)

**physical state**   a form that matter can take; possibilities are solid, liquid, or gas. (1.1)

**pleated sheet structure**   a secondary structure adopted by some proteins in which the protein chains are lined up beside one another in the form of a sheet. (17.1)

**polar covalent bond**   a bond resulting from the unequal sharing of electron pairs. (8.1)

**polarity**   the separation of electronic charge within a bond (or molecule). (8.1)

**polyatomic ion**   an ion containing two or more atoms, such as $SO_4^{2-}$. (3.2)

**polyprotic acid**   an acid containing more than one acidic hydrogen. (13.3)

**positron**   a particle having the same mass as an electron but the opposite electrical charge. (15.1)

**potential energy**   the energy possessed by an object because of its position. (1.3)

**precipitate**   an insoluble solid deposited from a solution. (5.4)

**precipitation reaction**   a reaction in which the reactants in solution exchange ions to form a solid. (5.4)

**pressure**   the amount of force applied per unit area. (9.1)

**primary structure**   the amino acid sequence in a protein. (17.1)

**principal energy level**   an energy level containing a set of electrons in orbitals of similar sizes. (7.3)

**product**   a substance formed from another substance or substances during a chemical reaction. (5.1)

**protein**   one of a class of organic polymers making up most body tissue; characterized by the presence of peptide bonds and formed by the polymerization of amino acids. (17.1)

**proton**   a positively charged subatomic particle. (2.2)

**pure substance**   matter that has the same chemical composition, no matter what its origin. (1.1)

# Q

**quantized**   having only specific allowed values. (7.2)

**quaternary structure**   the structure that results from the association in a protein of two or more polypeptide chains. (17.1)

# R

**radiation**   the emission of particles or electromagnetic waves. (15.1)

**radioactive**   having the ability to emit ionizing radiation spontaneously. (15.1)

**radiocarbon dating**   a technique for the establishment of the age of a carbon-containing object by determination of the amount of C-14 present in the object. (15.3)

**reactant**   a substance converted into another substance during a chemical reaction. (5.1)

**reaction rate**   a measure of how fast reactants are converted to products; usually measured as the change in concentration of a reactant or product per unit time. (12.1)

**reducing agent**   a reactant that causes the reduction of an element in another reactant and contains an element that is oxidized during a chemical reaction. (14.1)

**reduction**   a chemical process, coupled with oxidation, in which an atom decreases in oxidation number. (14.1)

**relative atomic mass**   the average mass of an atom of an element, taking into account the masses and abundance of all the naturally occurring isotopes. (2.4)

**replication** the process of copying DNA to make new DNA molecules in the cell nucleus during cell division. (17.2)

**resonance hybrid** an average or composite Lewis structure derived from two or more valid Lewis structures that closely represents the bonding in a molecule. (8.3)

**ribonucleic acid (RNA)** a low-molar mass nucleic acid that contains the sugar ribose and is involved in the use of genetic information to regulate the synthesis of proteins. (17.2)

# S

**salt bridge** a device used in electrochemical cells that allows migration of ions to maintain electrical neutrality but prevents mixing of the anode and cathode solutions. (14.3)

**saturated hydrocarbon** a hydrocarbon that contains all the hydrogens it can hold as a result of having only carbon-carbon single bonds. (16.2)

**saturated solution** a solution that contains the maximum concentration of solute dissolved in the solvent as described by its solubility. (11.4)

**scientific method** a method of inquiry or investigation that involves cycles of observation and interpretation. (1.4)

**secondary structure** the arrangement of amino acids in space in a local region in the protein chain, often in the form of a pleated sheet or a helix. (17.1)

**self-ionization** a process where one molecule transfers an $H^+$ ion to another molecule of the same substance. (13.4)

**semipermeable membrane** a membrane that allows the passage of solvent but not of solute particles. (11.6)

**single bond** a covalent bond that involves the sharing of one pair of electrons. (8.3)

**single-displacement reaction** a reaction in which a free element displaces another element from a compound to produce a different compound and a different free element. (5.4)

**solid** the physical state of matter characterized by a fixed shape and low compressibility. (1.1)

**solubility** a ratio that describes the maximum amount of a solute that dissolves in a particular solvent to form an equilibrium solution under specified conditions. (11.4)

**solute** the substance being dissolved; usually the component of a solution that is present in the lesser amount. (4.4, 11.1)

**solution** a homogeneous mixture of two or more substances uniformly dispersed at a molecular or ionic level. (1.1, 11.1)

**solvent** the substance doing the dissolving; usually the component of a solution that is present in the larger amount. (4.4, 11.1)

**specific heat** a property that is a measure of the amount of heat required to raise the temperature of 1 g of a substance by 1°C. (6.6)

**spectator ion** an ion that appears on both sides of an ionic equation and is therefore not involved in a chemical change. (5.5)

**spontaneous** a chemical reaction or physical change that occurs without outside intervention. (14.6)

**standard temperature and pressure (STP)** 0°C and 1 atm. (9.2)

**stoichiometry** the use of quantitative relationships between substances involved in a chemical reaction to determine the amount of a reactant or product. (6.1)

**strong acid** an acid that ionizes completely when dissolved in water. (13.2)

**strong base** a base that ionizes or dissociates completely when dissolved in water. (13.2)

**strong electrolyte** a substance that dissociates or ionizes completely into ions upon dissolving and conducts electricity well. (3.1)

**subatomic particle** a particle found within an atom. (2.2)

**sublevel** an energy level consisting of only one type of orbital at a specific principal energy level; usually consisting of more than one orbital. (7.3)

**sublimation** the vaporization of a solid. (10.1)

**supersaturated solution** an unstable solution that contains more dissolved solute than the maximum dictated by the solubility. (11.4)

**surface tension** the work necessary to expand the surface of a liquid; the property of a liquid surface that causes it to behave like a stretched membrane. (10.3)

# T

**temperature** a measure of how hot or cold a substance is relative to some standard. (1.2)

**tertiary structure** the way in which the various regions of secondary structure in a protein are arranged into a three-dimensional shape. (17.1)

**theoretical yield** the maximum amount of product that can be obtained in a chemical reaction from known amounts of reactants; the amount of product calculated by assuming all the limiting reactant is consumed. (6.5)

**titration** a process of determining the concentration of one substance in solution by reacting it with a solution of another substance of known concentration. (11.5)

**transcription** the process of synthesizing various types of RNA molecules in the nucleus using a DNA molecule as a template. (17.2)

**transfer RNA (tRNA)** the form of RNA that brings the appropriate amino acid to the site of protein synthesis. (17.2)

**transition metal** a member of one of the B groups of elements in the periodic table. (2.5)

**translation** the process of using RNA molecules to create protein molecules from amino acids in the cell cytoplasm. (17.2)

**triple bond** a covalent bond that involves the sharing of three pairs of electrons. (8.3)

# U

**unsaturated hydrocarbon** a hydrocarbon that does not contain all the hydrogen atoms that could possibly be included as a result of having carbon-carbon double or triple bonds. (16.2)

**unsaturated solution** a solution that contains less than the maximum amount of solute possible in a stable system. (11.4)

# V

**valence electron** an electron in the highest principal energy level, the valence level. (7.5)

**valence level** the principal energy level that is highest in energy and contains the largest occupied orbitals. (7.5)

**valence-shell electron-pair repulsion (VSEPR) theory** the shapes of molecules result from the tendency for electron pairs to maximize the distance between them to minimize repulsions. (8.5)

**vapor pressure** the partial pressure of gas molecules above a liquid (or solid) when the two states are in equilibrium. (10.1)

**viscosity** resistance to flow. (10.3)

**voltaic cell** an electrochemical cell in which an oxidation-reduction reaction causes electrons to flow through an external circuit; also known as a *galvanic cell*. (14.3)

**volume** the amount of space a substance occupies. (1.2)

# W

**wavelength** the distance between two corresponding points (such as the crests) on a wave. (7.1)

**weak acid** an acid in which only a fraction of the molecules ionize when dissolved in water; an acid that forms an equilibrium with its conjugate base. (13.2)

**weak base** a base in which only a fraction of the molecules ionize when dissolved in water; a base that forms an equilibrium with its conjugate acid. (13.2)

**weak electrolyte** a substance that dissociates or ionizes partially into ions upon dissolving and conducts electricity to a slight extent. (3.1)

**work** the action of a force over a distance. (1.3)

# Index

## A

Abbreviated electron configurations, 281
Absolute temperature, and kinetic energy, 371
Absolute temperature gas laws, 356–359, 364–370
Absolute temperature scale, 22
Absolute zero, motion (cessation), 357
Accuracy, 37
Acetaldehyde (ethanal), 679–680
Acetaminophen, structure (comparison), 653
Acetic acid ($CH_3CO_2H$), 173, 487, 680
    ionization, 541
    sodium bicarbonate, reaction, 488f
    weak acid, 537f
Acetylene ($C_2H_2$)
    combustion, 200
    empirical formula, 141f
    production, 671f
Acetylsalicylic acid ($C_9H_8O_4$) (aspirin), structure, 653
Acid ionization constant ($K_a$), 541–543, 543t
Acid rain, 196–197, 197f
    acid, term (derivation), 112
Acid-base indicator, pH color-change range, 554f
Acid-base neutralization reactions, 198
Acid-base titrations, 467–469
Acidic hydrogens, identification, 534
Acidic solutions, 544–547, 545t
    oxidation-reduction reactions (occurrence), 593
Acids, 112–116, 115t, 191, 529–537
    acidic hydrogen atoms, 534
    addition, 556f
Actinide series, 76
Activated complex, 490–491
Activation energy, 490, 490f, 494–498, 495f
Active site, 710
Activity series, 190, 599
Actual yield, 233, 233f, 234
Acyclic hydrocarbon, 659–660, 662
Addition (significant figures), 39
Addition polymerization reaction, 672
Adenine-thymine base pair, 416f
Age of Materials, 396
Agorca M5640, usage, 139–140
Air pollution, concerns, 639
Air-filled balloon, cooling, 355f
Alanine, reaction, 702f
Alcohols, 593f, 676–679, 678t
    formulas, representation, 325
Aldehydes, 679–680, 722
Algebraic equations, solution, 379–380
Aliphatic hydrocarbons, 324, 659
Alkali metal, 76–77, 78, 277, 287–288
    water, reaction, 183f, 288f
Alkaline, term (description), 592–593
Alkaline earth metal, 77, 183, 287–288
Alkanes, 323, 662, 659, 662–669, 664t, 665t, 685
    single bonds, 324f
Alkenes, 323–324, 660, 669–673, 669t, 685
Alkyl group, 666
Alkynes, 324, 660, 671, 669–673, 685
Alloys, 396, 427, 428t
Alpha ($\alpha$) amino acids, structures, 699f, 700f
Alpha ($\alpha$) helix structure, 709, 709f
Alpha ($\alpha$) particles, 62, 618, 619–620, 619f, 620f
Alpha-glucose ($\alpha$-glucose), 724
Aluminum (Al)
    bromine, reaction, 188f, 232
    charge, 106f
    electrolytic process, 571
    ions, 286, 551f
    melting, 399f
    metal, iron oxide mixture (heating/reaction), 176f
Aluminum bromide ($AlBr_3$), formation, 188, 232
Aluminum hydroxide [$Al(OH)_3$], 530
Aluminum oxide ($Al_2O_3$), formula, 105

Amines, 684
Amino acids, 699–700, 704f
    genetic codes, 720t
    reaction, 702f
    types, 705
Ammonia ($NH_3$), 502, 505, 538
    bubbling, 114f
    gases, reaction, 373f
    molecules, 229
    solid state, 95f
    weak base, 538f
Ammonium chloride ($NH_4Cl$), formation, 373f
Ammonium compounds, creation, 684
Ammonium ion ($NH_4^+$), 98
Ammonium nitrate ($NH_4NO_3$), presence, 449
Amorphous solid, 424
Amphetamine ($C_9H_{13}N$), structure, 654
Amphoteric substance, 533
Amylopectin, 727, 728f, 729f
Amylose, 727, 729f
–ane (suffix), replacement, 666, 685
Anhydrous compounds, formation, 187
Anions, 69, 94–95, 95f, 98f
    cations, combination, 102f
    packing, 429f
Anode, 583, 583f
Answers
    consider this questions and practice problems, A–3–A–13
    selected questions and problems, A–14–A–46
Antacids, 442f, 445f
    tablets, calcium carbonate (usage), 101
Anticodon, 720
Antifreeze, density, 18f
Antigens, 699
Aqueous $H^+$ ion, water molecules (association), 531f
Aqueous neutralization reactions, 467
Aqueous solution, 12, 442, 443, 539–540
    reactions, 200–202, 463–469
Archeological dating, 633–634
Argon, 261, 281–282, 631f
Aromatic hydrocarbons, 324, 659, 661, 674–676, 675t
    organic nomenclature, 685–686
Arrhenius, Svante, 530
Arrhenius model of acids/bases, 530–531
Arthritis, impact, 397f
Aryl group, 685
–ase, suffix (usage), 496
Aspartame ($C_{14}H_{18}N_2O_5$), 139
Aspirin (acetylsalicylic acid), 653
Astatine (At), 624–625, 624f
    presence, 78
Asymmetric molecules, characteristics, 335f
–ate (suffix), 98, 100, 116
Atmosphere, 345–346
    equilibrium, 499
    layers, 485f
Atmosphere (atm), 350
Atmospheric pressure, 401
Atom:atom ratio, 189
Atomic mass, 72–75, 73f
Atomic mass unit (amu), 72
Atomic number (Z), 64–69, 619–622
    neutron number, contrast (graph), 618f
Atomic radius (radii), 292, 292f
Atomic size, 292–293
Atomic skeleton, writing, 317, 318
Atomic spectra, 266
Atomic theory, 33
    Bohr model, 267–270, 267f
    Modern model, 270–277
Atomic theory (Dalton), 57, 58–60
Atoms, 9, 57, 60–69, 286–294
    arrangement, 173
    energy, principal energy levels (usage), 270f

equal numbers, requirement, 178
    ground state, 273
    Lewis symbols, 310
    mass (measurement), mass spectrometer (usage), 72f
    modern model, 57, 270–277, 270f
    multielectron atoms, orbital diagrams, 272–275
    number, 160, 328t
    shared electrons, counting, 314f
    size, determination, 292f
    symmetrical molecules, 334f
Attractive forces, impact, 313
Aufbau principle, 273
Average kinetic energy, equation, 372
Avogadro, Amadeo, 360–361
Avogadro's hypothesis, 360–363
Avogadro's number, 133, 135, 139, 158
Axel, Richard, 326
Azurite, photograph, 147

## B

Baking soda ($NaHCO_3$), vinegar (reaction), 487
Balanced equation, 176, 217
Ball-and-stick models, usage, 656, 659
Balloon, density, 348
Band of stability, 618, 618f
Barbecue grills, propane (storage), 217
Barium chloride ($BaCl_2$), 194f, 195
    potassium sulfate, reaction, 464f
Barium sulfate ($BaSO_4$), 202
    precipitate, 194f, 195f
    preparation, 464f
Barometer, 349–350, 350f
Bases, 112–116, 530–536
    addition, 467f
    Arrhenius model, 530
    Brønsted-Lowry theory, 531
    nucleotide, structure, 715f
Basic solutions, 544–547, 545t
Batteries, 581–588
    dry cell battery, 592f
    production, household items (usage), 599f
    reactions, spontaneous effect, 571
    types, 572f
Becquerel, Antoine Henri, 34, 630
Bends, symptoms, 455
Benzene ($C_6H_6$), 676, 685–686
    aromatic hydrocarbons, 324f
    empirical formula, 141f
    representation, resonance structures/circle (usage), 661
Berlin blue (Prussian blue), 172
Beryllium chloride ($BeCl_2$), bond (polarity), 334
Beta ($\beta$) decay processes, necessity, 628
Beta ($\beta$) particle, 618, 620–621, 621f
Beta-glucose ($\beta$-glucose) ring structure, formation, 724f
Bicarbonate ($HCO_3^-$) buffer system, 555
Bile acids, 733f
Binary compounds, 96, 186
Binary molecular compounds, 98, 112f
Biomolecules, 698
Bismuth-209 target, bombardment, 625
Bleach, usage, 697f
Blimps, helium (usage), 25f
Blood sugar (increase), chromium (impact), 108
Blood type, 699
Bloodstream, buffers (presence), 555
Body cells, glucose absorption, 636f
Bohr, Niels, 267–269, 272
Boiling (vaporization), 23
Boiling points, 662, 662t, 663f
    elevation, 472–473
Boiling (vaporization) points, 414–415, 418–419
    vapor pressure, relationship, 402
Bomb calorimeter, 244, 244f

Bond angle, 327, 330
Bonded atoms, electron placement, 317, 318
Bonding, 306
    electrons, groups, 329
    positions, 317–318
Bonds, 304–309, 327
    classification, 307f
    intermolecular forces, relationship, 410t
    number, valence electrons (relationship), 315–316
Bone, replacement, 397f
Boron, orbital diagram, 274
Boyle, Robert, 353
Boyle's law, 353, 359
Brain activity, PET scan, 637f
Branched-chain alkanes, 663, 664
    intermolecular forces (comparison), 683
Breathalyzers, oxidation-reduction reaction, 593f
Breathing (regulation), carbon dioxide (usage), 346
Breeder reactor, 644
Bristlecone pine trees, age, 634
Bromine, 232, 506–507
    abbreviated electron configuration, 282
    aluminum, reaction, 188f, 232
    liquid, 412f
    p block location, 278f, 279
    valence electrons, presence, 283
Brønsted, J.N., 531
Brønsted-Lowry acids, 532
Brønsted-Lowry bases, 531–532
Brønsted-Lowry theory, 531
Bronze Age, 396
Buck, Linda, 326
Buffers, 555–558, 556f
Butane (C$_4$H$_{10}$), 181, 661

## C

C terminus, 704
Cadmium, ion (formation), 107
Calcium, 281
Calcium, water (reaction), 190f
Calcium carbonate (CaCO$_3$), 101–102, 101f, 102f, 530
    calcium cations, arrays, 305f
    decomposition, 507
    gas conversion, 304
    heating, equation, 187
    mass, 375
    moles, calculation, 375
    solution, ions (presence), 99f
Calcium hydroxide, formation, 190f
Calcium hypochlorite [Ca(OCl)$_2$], addition, 529
Calcium oxide (CaO), equilibrium, 512f
Calculations, 38–40
Calculators, 35–36
Calories, 30, 238, 244
Calorimeter, 242, 242f, 244, 244f
Calorimetry, 242f, 243–244
Cancer therapy, 637
Capillary action, 423
Carbohydrates, 698, 722–729
    classification, 725f
Carbon, 303–304, 303f, 304t, 323–326
    chains, impact, 452f
    oxidation number, 593f
Carbon atoms, 430–431
    bonds, 323, 655
Carbon dioxide (CO$_2$), 508, 508f, 512f
    catalysis, 494
    contribution, 95
    electron pairs, addition, 319
    formation, 189f, 218
    linear arrangement, 331f
    molecular compound/diagram, 21, 93f
    molecules, array packing, 314f
    moles, calculation, 231
    oxygen:carbon atom ratio, 189
    solid state, 11f, 430f
    unused model parts, 178
    usage, 346
Carbon monoxide (CO), 189f, 333f
    conversion, 495f

Carbon tetrachloride (CCl$_4$), nonpolar liquids, 453f
Carbon-14, 633–634
    beta emission, 619
Carbonate (CO$_3$$^{2-}$)
    buffer system, acid (addition), 556f
    buildup, 98f
Carbon-chlorine bond, break, 486
Carbon-containing molecules, 655
Carbon-hydrogen bonds, 333
Carbonic acid (H$_2$CO$_3$)/bicarbonate (HCO$_3$$^-$), 555, 556f
Carboxylate group (—CO$_2$$^-$), formation, 701
Carboxylic acid, 680–681, 683–684
    formula, 534
    group (CO$_2$H), 701, 704f
Catalysts, 494–498, 516–517
    consideration, 486
Catalytic converters, components, 495f
Catalytic cracking, 661
Cathode, 583, 583f
Cathode-ray tubes (CRTs), 60, 61f
Cations, 69, 94, 95f
    anions, combinations, 102f
    electron/proton numbers, 284–285
    size, neutral atom (relationship), 293f
Cell, 713, 713f
    damage, extent, 638
    membrane, phospholipids (presence), 733f
Cellulose (polysaccharide), 726–729, 729f
Celsius scale, 21, 22f
    Kelvin, relationship, 357
Centrosomes, 714
Cesium chloride (CsCl) structure, adoption, 312f
Chain reactions, 642
Chalcocite (Cu$_2$S), copper mass (difference), 149f
Chalcopyrite, 130, 138
Charged particles, acceleration, 626f
Charles, Jacques, 356
Charles's law, 356–357, 359
Chemical activity series, 191f, 286f, 598f
Chemical bonds, 304, 305
Chemical changes, 24–27, 25f
Chemical cold packs, ammonium nitrate/water (presence), 449
Chemical equations, 176–182
Chemical equilibrium, 498–500
Chemical formula, 7, 149–150
Chemical property, 25
Chemical reactions, 24, 173–175, 177
    atoms, rearrangement, 59f
    classes, 185t
    energy changes, 236–237
    gases, relationship, 373–376
    heat changes, 244–247
    occurrence, 571f
    spontaneous, 597
Chemicals, 23–27, 29, 35
    compound, 6
    hot packs, water (separation), 449
    measurement, 220
    products, manufacture, 494
    reactivity, electron configurations (relationship), 286–288
    tasting, avoidance, 530
Chemists, 34f, 141
Chernobyl (Russia), nuclear disaster, 643
Chirp rate, temperature (dependence), 493
Chlorate ion (ClO$_3$$^-$), 99t
Chloride ion (Cl$^-$), 572f, 600f
Chlorine, 220, 285, 294
    catalytic role, 497–498
    gas, 412f
Chlorite ion (ClO$_2$$^-$), 99t
Chlorofluorocarbons (CFCs), 486, 499
Chloroform (CHCl$_3$), polarity, 334
Chloromethane (CH$_3$Cl), 334, 676
Cholesterol, 108, 733
Chromatography, 697
Chromium, 107, 108
    atom, change, 594
    oxidation number, 594f
Chromium(III) oxide (Cr$_2$O$_3$), ionic compound, 579
Cis isomers, 671

cis-retinal, 653, 654f
Citric acid (H$_3$C$_6$H$_5$O$_7$), triprotic acid, 543
Cl catalyst, impact, 495f, 498
Close packing, 425, 425f
Closed chain hydrocarbons (cyclic hydrocarbons), 660–661
Closed container, gas molecules (increase), 400f
Close-packed arrangement, 425f
Close-packed structure (examination), marbles (usage), 425
Coal, mining, 304
Codon, 720
Coefficients, change, 196
Coins, copper (presence), 128
Cold packs, usage, 449
Collagen, presence, 710f
Colligative properties, 469–474
Collision rate, 492f, 493
Collision theory, 488–494
Color, change, 175
Colored solution, color intensity (decrease), 151f
Combination reactions, 188–190, 188t
Combinatorial chemistry, 34
Combined gas law, 358–360
Combustion reactions, 199–200, 200f, 668
    oxidation-reduction reactions, 571
Completely miscible, 457
Complex carbohydrates, 725–729
Complex units, problem solving, 45
Component elements, percent by mass, 130f
Compounds, 6–9, 116–118, 145–146
    class, generalization, 679
    combinations, double-displacement reaction, 194
    dissolving, 92
    formula unit, 101f
    heating, decomposition reactions (occurrence), 186t
    Lewis structure, 316
    multiple charges, formula, 108
    names, assignation, 106, 106f
    one-mole quantities, 133f
    reaction, 189–190
Computerized tomography (CT), 616, 616f
Concentrated solution, solute (presence), 151
Concentration, 151, 455, 491–492
    changes, impact, 511t
    reduction, dilution (usage), 156
    units, 457t
Condensation, 400, 682
Conjugate acid-base pairs, 532–533
Conjugate acids, 532, 533
Conjugate bases, 532, 533, 541t
Consumer ingredient labels, common names (usage), 109
Consumer products, acidic/basic properties (exhibition), 114f
Containers, 16f
    volume, increase, 514
    walls, gases (pressure), 349
Continuous spectrum, 266
Control rods, 642
Conversions, 41–45, 136
    efficiency, 236f
Cooling curves, 405–406
Cooling tower, smoking tower, 643f
Copper (Cu), 128–129, 128f
    acids, reaction (absence), 190
    atoms, 134
    bluish-green light, production, 260
    chemical treatment, zinc metal (usage), 427f
    crystal, surface (scanning tunneling microscope image), 425f
    density, 19f
    electrode, increase, 585f
    electrons, loss, 589f
    hydrochloric acid, reaction (absence), 34f
    ions, conversion, 192f
    pipe, atoms (array), 9f
    reaction, 31f
    scanning tunneling microscope (STM) image, 60f
    wire, partial composition, 58f
Copper metal, 191f, 192f
    addition, 224
    deposition, 172f
Copper(I) oxide (Cu$_2$O), 134, 134f
Copper(II) acetate [Cu(CH$_3$O$_2$)$_2$], usage, 155–156

Copper(II) chloride ($CuCl_2$), 154f
  iron metal, placement, 172f
  sodium sulfide, addition, 184
  zinc, displacement, 572f
Copper(II) oxide (CuO), reaction, 373–374
Copper(II) sulfate ($CuSO_4$), 150f, 157f
  iron, addition, 181–182
Core electron, 282
Corn oil, density, 18f
Corrosion, 190, 192, 603–604
Coulomb, 62
Covalent bonding, 305–306, 313–322
Covalent bonds, 305, 313, 313f
Covalent molecules, 317–322
Covalent substances, 305t, 403f
Critical mass, 642
Crystal lattice, 312, 424–425
Crystal structure (crystal lattice), 94, 424
Crystalline solid, 424, 424f, 426–431
Crystals, 424–425
Cupric chloride, 109
Cuprite ($Cu_2O$), copper mass (difference), 149f
Cyanide ion ($CN^-$), valence electrons, 319
Cyclic hydrocarbons (closed chain hydrocarbons),
    660–661, 673–676
cyclo- (prefix), 685
Cycloalkanes, 660, 673–674
Cycloalkenes, 660, 673–674
Cyclobutane ($C_4H_6$), 660
Cyclohexane ($C_6H_{12}$), structure, 673
Cyclooxygenase, aspirin (impact), 653
Cyclopentane ($C_5H_{10}$), structure, 673
Cyclopropane ($C_3H_6$), structure, 673
Cytoplasm, 713–714
Cytosine-guanine, base pair, 416f

## D

d block, subdivision, 278f
d orbitals, 271f, 272, 272f
Dalton, John, 57–60, 368–370
Dalton's atomic theory, 57, 58–60
deca- (prefix), 110
Decimal form, conversion, 35
Decomposition reactions, 186–188, 186t
Definite proportions, law, 58–59
Deionized water, purity (determination), 91
Delocalization, 661
Delocalized electrons, nuclei
    (association), 321
Democritus, 57
Denaturation (proteins), 713
Denatured alcohol, 677
Density, 18–21, 347–348, 347f
  calculation, 220, 460
  equation, 460
  gas, 366–367
  liquids, 420
Deoxyribonucleic acid (DNA), 416f, 714–718
  analysis, usage, 698f
  complement, mRNA (relationship), 719f
  complementary bases, 719t
  duplication, 716f
  fingerprinting, 697
  hydrogen bond, bases (bonding), 717f
  nucleotide, components, 714f
  replication strands, growth, 717f
  strands, combination, 717f
Deoxyribonucleotide, 715
Deposition, 404
Derived units, 42
Detergents, 423, 452
Detonation velocity, 246
Deuterated water ($D_2O$), formation, 69
Deuterium, 69, 69f
di- (prefix), 111, 666
Diamond, usage, 430, 430f
Diatomic elements, 183, 314–315
Diatomic molecule, 78, 78f, 315f
Diatomic oxygen molecule, double bond, 315f
Dichlorodifluoromethane ($CF_2Cl_2$), 486
Dichloromethane ($CH_2Cl_2$), polarity, 334

Dichromate ion ($Cr_2O_7^{2-}$), chromium oxidation
    number, 594f
Diesbach (color manufacturer), 172
Diffusion, 372–373
Diglycerides, mixture, 732
Dilute solution, 151
Dilution, 156–158
Dimensional analysis, 42–43, 137
Dinitrogen tetroxide ($N_2O_4$), 499, 513
  conversion, reversible reaction, 499f
Dipeptide, 703
Dipole-dipole forces, 410, 413–415
Dipoles, attractions (London dispersion forces), 411f
Disaccharides, 724–725, 726f
Dish detergent, density, 18f
Dissociation, 92, 94, 445, 474
Dissolving, melting (difference), 91
Distilled water, purity (determination), 91
Disulfide bond, 710
Disulfide linkage, 710
Division, 38–39
Double bonds, 315, 317–319, 321–322
Double-bound electrons, delocalization, 661
Double-displacement reactions, 193–199
  ion pairs, combination, 195f
Drinking water, copper levels (EPA determination), 129
Drugs, oral delivery, 442f
Dry cell battery, 592f
Dry ice, carbon dioxide solid state, 11f, 430f
d-sublevel electrons, 282
Dynamic equilibrium, 400, 499

## E

E-85 (ethanol blend), usage, 215
Efficiency, work measure, 235–236
Effusion, 372–373, 373f
Electric current, usage, 4f
Electric energy, 29
Electric vehicle (EV), usage, 215
Electrical conductivity, test, 92f
Electricity, flow, 91f, 583
Electrochemistry, 597–603
Electrode, 583
Electrolysis, 3, 599–600
Electrolytes, 92, 444, 445
Electrolytic cells, 598–603
Electrolytic process, 571
Electromagnetic energy, 261–267
Electromagnetic radiation, 261–267, 262f, 617
Electromagnetic spectrum, 262–263, 262f
Electron configurations, 276–282, 286–288
Electron dot formula, 314
Electron dot symbol, 310
Electron paramagnetic resonance (EPR), 274
Electron spin resonance (ESR), 274
Electronegativity, 307–308, 308f
  increase, 576
Electrons, 60, 64t, 317–319
  behavior, 274
  capture, 623
  distribution, 274
  domains, arrangement, 330t
  energies, representation, 270f
  flow, 589
  gain/loss, 78–79
  information, 276f
  localization, 321
  model (Thomson), 61f
  multielectron atoms, orbital diagrams, 272–275
  pairs, 319, 330f
  quantized behavior, 268
  transfer/sharing, 307f
  valence electrons, 282–284
Electrophoresis, 697
Electroplating, silver (usage), 600
Electrostatic forces, 305
Elemental oxygen ($O_2$), molecular element, 93f
Elements, 3–9, 57, 73–80
  appearance, 286f
  combination, 188
  electron configuration, determination, 279f, 280f

electronegativity values, 308f
  fingerprint, 266
  heavier elements, decay, 628
  ionic compound, formation, 311
  lighter elements, decay, 627–628
  one-mole quantities, 133f
  periods, 318
  subscripts, change (impossibility), 177
  symbols for, 6, 6t
Empirical formulas, 140–147, 141f, 142t
End point, 467
Endoplasmic reticulum, 714
Endothermic reaction, 236–237, 489f
-ene (suffix), 685
Energy, 27–30, 235–244
  absorption, 489f
  barrier, overcoming, 490
  changes, 407–408
  diagrams, activated complexes (presence), 490–491
  release, 246
Energy released per gram, 246
Energy released per mole, 246
English-metric conversions, 14t
Entropy, 450, 451f
Enzyme, 496, 496f
Equations, 379–380
Equations, balancing, 177, 178, 179–182
  coefficients, changing, 196
Equilibrium, 400, 500–501, 506–509
  position, 516t
  reestablishment, 513f
  shift, concentration changes (impact), 511t, 514t
Equilibrium concentrations, 503–504
  graphs, 502f
  temperature, constancy, 501t
Equilibrium constant ($K_{eq}$), 500–509
  expression, 500–505, 509
  ratio representation, 541
  value, 502, 503t, 515, 515f
Equivalence point, 467
Essential amino acids, 705
Esters, 680–684
Ethanal (acetaldehyde), 679–680
Ethane ($C_2H_6$), 180–181
  natural gas component, 661
Ethanol ($CH_3CH_2OH$), 245–246, 594
  oxidation, $K_2Cr_2O_7$ (usage), 680
  water, dissolving, 443
Ethene, fruit ripening, 669
Ethers, 679
Ethylene ($C_2H_4$), oxygen reactions, 230–231
Ethylene glycol ($HOCH_2CH_2OH$), 473, 677
Evaporation, 399–400
  endothermic process, 237
Exercise ball, air pressure, 348f
Exothermic combustion reactions, 668
Exothermic reaction, 236, 237, 237f
  diagram, 490f
Exponential notation, usage, 13
Exponents, rules, 36

## F

f block, subdivision, 278f
f orbitals, shape/lobes, 271f
Fahrenheit scale, 22f
  usage, 21
Family, 75
Fats, 698, 730
  molecule (nonpolarity), carbon chains (impact), 452f
  triglycerides, 682
  usage, 682
Fatty acids, 682, 730, 731–732
  saturation, degree, 730f
Faucet, copper compound (presence), 128f
Fermilab, synchrotron, 626f
Fibrous protein, 710, 710f
Filling, 273, 275
  order, 279, 282
Fireworks, 260–261, 260f
Fish tanks, 554, 555
Fission, 640, 641–643, 642f

500-mL containers, 16f
Flameless heater, magnesium/iron/water combination, 571f
Fleming, Alexander, 34
Fluid, 421
Fluorine, reactivity, 308
Fluorite, 312f
Food
    amino acids, absence, 706t
    components, energy values, 247t
    digestion, sucrose (breakdown), 496f
Formaldehyde ($CH_2O$) (methanal), 679
Formalin, 679
Formic acid ($HCO_2H$) (methanoic acid), 680
Forms, term (usage), 176
Formula units, 101f, 102, 134f
    conversion, 159
    reference, 133–135
Formulas
    balancing, 195
    construction, Stock system names (usage), 108
Forward reaction, 499
Fractional distillation, 661
Fractional mole ratios, 146–147
Francium, 287, 308
Freezing point, 403, 403f, 473
Frequency ($\nu$), 262–265
Fructose, 723, 724
    glucose, combination, 726–727
Fruit
    esters, presence, 682
    ripening, ethene (impact), 669
Fuels
    efficiency, 652
    substance, combustion (heat changes), 246t
Fukushima Daiichi Nuclear Power Plant, radiation
        leaks, 643
Functional groups, 324–326, 656, 657t
Fusion, 644–645, 645f
    molar heat, 407, 408

## G

Galvanic cell, 582
Gamma decay process, 623
Gamma rays, 262–263, 618
    emission, 623, 623f
Gas particles, 347
    behavior, 365
    impact, 349f
    kinetic energy, absolute temperature (impact), 371
    motion, 349f, 371, 371f
    passage, 373f
Gaseous bromine, concentration (equilibrium), 507f
Gases, 11, 12, 347–351
    atoms, piston (movement), 352f
    formation, 175
    heat, removal, 405f
    high pressure, impact, 13f
    ideal gas, 354
    liquid-gas phase, changes, 399–402
    macroscopic properties, 398f
    mixture, 459
    molecules, 347, 400f
    noble gas, 77–78
    physical states, change, 24f
    properties, 347–348, 351–364
    solubility, 453–455
    solutions, formation, 443f
    stove, methane/oxygen reaction, 177f
    temperature, 347–348
    volume, 346t, 361–364, 367
    volume/pressure, 351–355, 378
    volume-temperature, graphical relationship, 355–356
Gas-formation reactions, 196–198
Gasoline
    burning, 235f
    MTBE, addition, 679
Gastric acid (stomach acid), solution, 113
Gay-Lussac, Joseph Louis, 356
    volume, combining (law), 360, 374
Geiger, Hans, 62
Geiger-Müller counters, 630–631, 631f

Geological dating, 634–635
Geological minerals, subset, 128
Geometric isomers, 670
Glass, 424f
Glazes, compounds (usage), 572f
Globular proteins, 710
Glucose ($C_6H_{12}O_6$), 698, 722–723
    body cell absorption, 636f
    fructose, combination, 726–727
    linear arrangement, 724f
    polymer, types, 727
    sweetness, 326f
Glycerol (1,2,3-propanetriol), 682, 730
Glycine
    alanine, reaction, 702
    amino acids, 303
    reaction, 702f
    structure prediction, VSEPR theory (usage), 332f
Glycine-alanine dipeptide, hydrolysis, 706
Glycogen (polysaccharide), 726–727
    branched structure, 729f
Glycolipids, 732
Glycosidic linkage, 724f, 726f, 727
Gold (Au)
    acids, reaction (absence), 190
    density, 18f
    foil experiment (Rutherford), 63f
Graham, Thomas, 372
Graham's law, 372
Grams to pounds, conversion, 15
Graphing, 376–378
Graphite
    carbon atom, covalent bond, 431f
    rod, 592f
Graphs, 376–378
Grave's disease (hyperthyroidism), 632
Greek prefixes, 111t
Green chemistry, 35, 216
Greenhouse effect, 215
Ground state, 273–274, 275
Group, 75

## H

Hahn, Otto, 625
Half-cells, 582f, 583
Half-lives, 631–633
Half-reaction, 583
Halogen, 77, 78
    electrons, sublevel, 278
    elements, existence, 315f
Hammer, force, 349f
Hard water, 96, 98
Hard-water stains, removal (difficulty), 538
Heat, 238–244, 405–409
    absorption, 175, 242
    changes, 242, 244–247
    energy, 29
    factors, 240
    release, 175
    of solution, 448
Heating curves, 405–406
Heavier elements, decay, 628
Heavy water ($D_2O$), 69
*Helicobacter pylori* (impact), 637
Helium (He), 25, 25f, 261
    $1s$ sublevel, 278
    atoms, presence, 9f
    gas particles, behavior, 365
    ground state orbital diagram, 274
    molecular diagram, 21
Hematite, $Fe_2O_3$ (presence), 570f
Heme, 332f, 333f
Heme group, 711f
Hemocrit (capillary tube), blood (filling), 423f
Hemoglobin, 698
    heme group, 332f
    quaternary structure, 711f
Henry's law, 454
Hertz (Hz), 262
Heterogeneous equilibrium (equilibria), 506–509
Heterogeneous mixture, 7

Hexane ($C_6H_{14}$), 668f
    nonpolar liquids, 453f
Hexoses, 723
High-temperature superconductors, ionic solids, 429f
Hindenburg, hydrogen gas (usage), 25f
Hip joint (degradation), arthritis (impact), 397f
$HNO_2$, skeletal structure, 319
Homogeneous equilibrium, 506
Homogeneous mixture, 7
Hormones, steroid structure (comparison), 734f
Hot packs, usage, 449
Hot-air balloon, gas (heating), 348f
Household products, acids/bases (presence), 530f
Human body, water proportion, 90
Hund's rule, 274
Hydrangeas (color), soil pH/aluminum ion (impact), 551f
Hydration, 447
Hydrazine ($N_2H_4$), 316
*hydro-* (prefix), 116
Hydrobromic acid (HBr), strong acid, 549
Hydrocarbons, 323–324, 655, 659–662
    chemical compounds, 199
    classes, 323f, 659–661, 660f
    combustion, 199f
    derivatives, 323
    double bonds, conversion, 653
    functional groups, 325t
Hydrochloric acid (HCl)
    dissociation, 444f
    ionization, 115f
    sale, 152
    sodium hydroxide, reaction, 198
    strong acid, 535f, 547
    usage, 31f, 59f
Hydrogen
    acidic hydrogen atoms, 534
    bonding, 414–417, 415f
    bonds, 410, 415, 717f
    chemical activity series, 598f
    deuterium, 69
    dipole-dipole forces, existence, 414–415
    displacement, 191
    electron, probability map, 270f
    fuel cell, function, 216f, 223
    gas particles, behavior, 365
    hydrogen atoms, 267–270
    isotopes, subatomic particles (images), 64f
    Lewis structure, 314
    line spectrum, 268f, 269
    molecules, 59f, 173f, 226f, 227f
    moles, calculation, 234
    orbital diagram, 272f
    oxygen atoms, 90f, 173f
    polyatomic ion, combination, 113
    presence, 10f
    space shuttle fuel, 215f
Hydrogen ($H_2$)
    gas formation, 183
    lighting, reaction, 27f
    nitrogen gas, reaction, 228
    oxygen gas, reaction, 173–174, 360f
    production, 34f
Hydrogen chloride (HCl)
    behavior, 112
    detection, 223
    gases, reaction, 373f
    molar mass, 367
    molecular compound, presence, 113f
Hydrogen fluoride (HF), 307, 307f
Hydrogen peroxide ($H_2O_2$), 446f, 488f
Hydrogen sulfide ($H_2S$), 132, 132f
    gas, diffusion, 372f
Hydrogenation, 683
Hydrogen-fueled car, prototype, 215f
Hydrolysis, 706
Hydronium ion ($H_3O^+$), 531, 545–547
    conversion, 552
    hydroxide ion ($OH^-$), product, 553f
    representation, 113
    value, usage, 552f
Hydroxide ion ($OH^-$)
    ion, 545–547, 552
    value, usage, 552f

Hyperthyroidism (Grave's disease) treatment, radioactive iodide (usage), 632
Hypertonic, 471
*hypo-* (prefix), 100
Hypophosphorus acid (H₃PO₂), 534
Hypotheses, 31–33
Hypotonic, 471
Hypotonic solution, 471f

## I

–*ic* (suffix), 109
    usage, 116
Ice, melting, 421f
Ice, molecular diagram, 21
Ice (melting), heat change (obtaining), 409
Ice cube, deuterium (presence), 69f
–*ide* (suffix), 97–98, 105, 110–111
Ideal gas, 354
    constant, 364
    volume, change, 361f
Ideal gas law, 364–370
Ignition temperature, 644
Immiscible, 457
Indicator, 467
Induced dipole, 411
Inert metal (unreactive metal), usage, 583
Infrared (IR) light wave, wavelength, 263f
Infrared (IR) radiation, effects, 263
Inner-transition metal, 76
Insoluble ionic compounds, ions (number), 508
Instantaneous dipole, 411
Intermediates, 498
Intermolecular forces, 410–420, 448f
    imbalance, 422f
    impact, 423
    total intermolecular force strengths, comparison, 417
    trends, 417–420
Internal-combustion engine (ICE), 215, 216f
    gasoline, burning, 235f
International Union of Pure and Applied Chemistry (IUPAC), 72, 104
Interstitial water, 469
Inverse log function, 552, 558, 560
Iodine
    heating, 404f
    ions, formation, 456f
    molecules, London dispersion forces (impact), 450f
    solid, sublimation, 412f
Ion-dipole forces, 447, 448f
Ionic bonding, 305–306, 309–313
Ionic bonds, 305, 448f
Ionic compounds, 91–96, 103–110, 104t
    discrete molecules, absence, 101f
    dissociation, 446f
    dissolving, 447
    energy, absorption, 261
    formation, 102f, 180, 311
    formulas, 101–104
    ions, number, 312f, 508
    polyatomic ions, presence, 101f
    properties, 95t, 96, 117
    solubility, 194t, 445, 447
    temperature requirements, 447
    usage, 260f
Ionic crystals, 312–313, 312f
    formation, 429f
    structures, 428
Ionic equation, 201, 467
Ionic salts, solubility prediction rules, 445t
Ionic solids, 428–429
    empirical formulas, 141f
    solubility (increase), temperature (impact), 454f
Ionic substances, properties, 305t
Ionization, 92
    process, 112–113
Ionization energy (IE), 288–291
Ionizing radiation, 617
Ion-product constant of water (Kᴡ), 544–545
Ions, 69–80, 69–72
    electron configurations, 284–286
    formulas, 97–98

ionic radii, 294f
Lewis symbols, 311–312
    names, 97–98
    number, calculation, 160
    production, 444f
    ratio, 102
    shapes, 331
    sizes, 293–294
Iron (Fe), 12f, 57
    addition, 181–182
    corrosion, 603f
    magnesium/water, combination, 571f
    metal, placement, 172f
    nail, surface area/reaction, 488f
    oxidation, 603f
    usage, 570
Iron Age, 396
Iron pyrite, composition, 6f
Iron-55/Iron-59, usage, 635–636
Iron(II) chloride (FeCl₂), 598–599
Iron(II) sulfate (FeSO₄), potassium ferricyanide (K₃Fe(CN)₆) (combinations), 172f
Iron(III) nitrate [Fe(NO₃)₃], aqueous solution, 510f
Iron(III) oxide (Fe₂O₃), aluminum metal mixture (heating/reaction), 176f
*iso-* (prefix), usage, 666
Isoelectronic atom, 285
Isoelectronic series, 293–294
Isomers, 656, 666
Isopropyl alcohol [(CH₃)₂CHOH], percent by volume, 458
Isotonic, 471
Isotonic solution, 471f
Isotope symbols, 67–68, 67f
Isotopes, 64–69, 617, 635–637
    decay, 628
    hydrogen isotopes, subatomic particles (images), 64f
    symbols, writing, 71–72
–*ite* (suffix), 98, 100, 116

## J

Japanese train, levitation, 429f
Joules (J), 30, 238, 408

## K

Kelvin scale, 21, 22f, 357
Ketones, 680
Kilojoule (kJ), 238
Kilojoules per gram (kJ/g), 245
Kilojoules per mole (kJ/mol), 245
Kinetic energy (KE), 28–29, 28f, 371–372
    increase, 492f
Kinetic-molecular theory, postulates, 370–372
Knocking, occurrence, 652
Krypton (Kr), usage, 261

## L

Lactic acid, production, 680–681
Lactose, sweetness, 326
Lambda (λ), 262, 262f
Lanthanide series, 76
Large molecules, London dispersion forces, 417
Lasers, uses, 261f, 263
Lavoisier, Antoine, 58
Law of combining volumes (Gay-Lussac), 360–361
    usage, 373–374
Law of conservation of energy, 235–236
Law of conservation of mass, 33, 58, 221f
Law of conservation of mass and energy, 33
Law of definite proportions, 58–59
Law of partial pressure (Dalton), 368–370
Laws, 33
Le Chatelier, Henry Louis, 509
Le Chatelier's principle, 509–518, 555
Lead ions, potassium iodide (addition), 173f
Lead-acid battery, 587, 587f, 601

Lead(II) iodide (PbI₂), formation, 456f
Lead(II) nitrate, potassium chromate (mixture), 201f
Lecithin, diglyceride mixture, 732
Levitation, 429f
Lewis, Gilbert Newton, 310, 531
Lewis structures, 314–319, 321, 325, 331
    usage, 538
Lewis symbols, 310–311, 314
Light, 260–261
    generation, LED bulbs (usage), 235
    production, 175
    white light, prism passage, 267f
Light frequency, 262
Light stick, glow, 487
Lightbulbs, efficiency, 235–236
Lighter elements, decay, 627–628
Like dissolves like, 451t
Limestone, 187f, 189
Limestone statue (erosion), acid rain (impact), 197f
Limiting parts, 225–226
Limiting reactant, 223–233
Limonene, 674
Line spectrum (line spectra), 266–268
Line structures, 656–659
Linear accelerator, 626
Linear molecule, 413
Linoleic acid, 683
Lipids, 698, 729–734
Liquid crystals, 411–412
Liquid-gas phase changes, 399–402
Liquids, 11, 12, 420–424
    boiling point, 405f
    evaporation, 400f, 419
    macroscopic properties, 398f
    physical states, change, 24f
    surface, molecules (energy), 400f
    vapor pressure, 472–473
Liquid-solid phase changes, 403
Lithium (Li), 277
    blanket, 644
    ionization process, 289
    valence electron, 292
Lithium hydroxide (LiOH), formation, 183
Living tissue (damages), radiation (impact), 637
Localized bonding, 640
Log functions, 558–560
London, Fritz, 411
London dispersion force, 410–412
    experience, 417f
    impact, 417, 450f
Loss of Electrons is Oxidation, Gain of Electrons is Reduction (LEO says GER), 573
Lower-energy levels, orbitals, 270
Lowry, T.M., 531
Luminol, usage, 697f
Lysosomes, 714
Lysozyme, 708f, 710f

## M

Macroscopic objects, fastening, 328f
Magnesium (Mg), 223
    iron/water, combination, 571f
    reducing agent, 586
Magnesium hydroxide (Mg(OH)₂), 442, 445f, 530
Magnesium nitrate (Mg(NO₃)₂), ionic compound, 578–579
Magnesium sulfite (MgSO₃), usage, 197
Magnet, levitation, 429f
Magnetic field, 397f, 645f
Magnetic resonance imaging (MRI), 616, 616f
Magnetite, photograph, 131
Main-group element (representative element), 76, 289f, 292f
    ions, 79f, 294
    valence electrons, 282–284, 310
Malachite [Cu₂CO₃(OH)₃], 128, 145
Manganese (Mn), 280, 592
Manganese(IV) oxide (MnO₂), 592–593
Mannose (monosaccharide), 724
Maple syrup, density, 18f

Mass, 3, 14–16, 137–140, 155, 220
  conservation law, formulation, 33
  density, relationship, 19–20
  volume, relationship, 19–20
Mass from percent composition, 131–132
Mass from volume, 375–376
Mass in grams, moles (conversions), 161
Mass number, 64–69, 622
Mass spectrometer, usage, 72f
Mass-mass conversions, 219–223
Mass/volume percent, 459–460
Math toolboxes, 558, A–2
  algebraic equations, solving, 379
  graphing, 376
  log and inverse log functions, 558–560
  mole conversions, 158
  scientific notation, 35–36
  significant figures, 37–41
  units/conversions, 41–45
Mathematical operations, 36
Matter, 3, 13–27
  classification, 3–13, 8f
  composition, 3–9
  physical states, 11t, 407
  representations, 9–11
  states, 11–13, 398t, 404–405
Meals-ready-to-eat (MREs), flameless heater
  (usage), 571f
Measured number, precision, 37
Meats, sodium nitrite (addition), 104
Mechanical energy, 29
Medical diagnoses, 635–636
Meitner, Lise, 625
Meitnerium (Mt), naming, 625
Melting, dissolving (difference), 91
Melting point, 22, 403, 403f
Mendeleev, Dmitri, 75, 75f, 624
Meniscus, 423, 423f
Menthol ($C_{10}H_{20}O$), structure, 652
Mercury (Hg), 186, 186f, 190
  barometers, usage, 350f
  line spectra, 267f
Mercury(II) oxide (HgO), heating/decomposition, 186f
Messenger RNA (mRNA), 718, 721–722
Metal can (evacuation), vacuum pump (usage), 349f
Metal carbonate, acid (reaction), 196f
Metal hydroxide, production, 183
Metal ions, 193, 201
Metal oxides, behavior, 198
Metallic solids, 427
Metalloids (semimetals), 76
Metals, 76–77, 106–109, 191
  chemical activity series, 598f
  combustion, 199f
  compounds, multiple charge formula, 108, 109t
  corrosion, 192, 570
  inner-transition metal, 76
  oxygen, combination, 24
  usage, 260f
  vaporization, 399
Methanal ($CH_2O$) (formaldehyde), 679
Methane ($CH_4$), 177f, 181, 327–328
  natural gas component, 661
Methanoic acid ($HCO_2H$) (formic acid), 680
Methanol ($CH_3OH$) (methyl alcohol), 215, 325, 418
Methyl mercaptan ($CH_3SH$), 418
Methyl tertiary-butyl ether (MTBE), addition, 679
Methylamine ($CH_3NH_2$), weak base, 538
Metric base units, 41, 42
Metric conversions, 14t, 41
Metric-English conversions, 14t, 41
Mica crystal surface, image, 133f
Milligrams to grams, conversion, 15
Milligrams to pounds, conversion, 15, 16
Milliliters to liters, conversion, 17
Millkan, Robert, 60–61, 61f
Miscibility, level, 457
Miscible, 446
Mitochondria, 714
Mixtures, 3, 7–9, 368
Model solar car, construction, 225f
Moderator, water (usage), 642
Molality, 462–463

Molar heat of fusion, 407, 408
Molar mass (*MM*), 135, 136–140, 367–368
  usage, 219–220
Molar volume, 362
Molarity, 152–156, 461–462
Molasses, heating (impact), 421
Mole ratios, 145–146, 218, 226
Molecular compounds, 91–96, 95t, 110–112, 110t
  prefixes, usage, 117
Molecular equation, 201, 464
Molecular formulas, 112, 140–150, 142t, 148, 656
Molecular motion, 29
Molecular properties, trends, 420
Molecular solids, 430
Molecular-level images, inspection, 93f
Molecular-level representations, strong/weak acids, 540
Molecule ratio, mole ratio (equivalence), 226
Molecule shapes, 326–335
Molecules, 9, 139–140, 318–319, 332–335
  alignment, 415f
  asymmetry, 413
  bond angles, 330
  bonding (representation), Lewis structures
    (usage), 321
  diatomic molecule, 78
  equal numbers, 178
  hydrogen/oxygen atoms, presence, 10f
  intermolecular forces, size (impact), 662
  London dispersion forces, experience, 417f
  macroscopic objects, 328f
  shaped receptors, interactions, 327f
  substitution, 505
  trigonal planar/pyramidal shapes, 329f
Mole-mass relationships, 161
Mole-mole conversions, 218–219
Moles, 132–140, 154, 158–159
  mass from volume, 375–376
  quantities, 158–162
  relative number, description, 217
Molten iron, thermite reaction (usage), 176f
Molybdenum (Mo), bombardment, 624
Monatomic anions, formulas/charges (pattern), 100f
Monatomic cations, naming, 97
Monatomic elements, 183
Monatomic ions, 96–98
*mono-* (prefix), 110, 111
Monomers, 672
Monosaccharides, 723, 724, 725f
Multielectron atoms, orbital diagrams, 272–275, 273f
Multiple branch names (alkanes), 666–667
Multiplication, 38–39
Multivitamins, copper (presence), 129f
Myoglobin, tertiary structure, 711f

### N

N terminus, 704
Nanometers, conversion, 264
Naphthalene ($C_{10}H_8$), aromatic hydrocarbons, 324f
Natural gas, 661
Natural radioactive decay processes, 627t
Neon (Ne), 261, 266, 267f, 277
Net ionic equations, 201
Network solid, 430–431
Neutral atom, cation (size relationship), 293f
Neutral solutions, 544–547, 545t
Neutron number, 65
Neutron therapy, 64
Neutron/proton (N/Z) ratio, 626
Neutrons, 63–65, 64t
  bombardment, 641f
  conversion, 627–628
  number, 618f
  protons, numbers (inequality), 70f
*N*-heptane, unbranched structure, 667
Nickel Titanium Naval Ordnance Laboratory (Nitinol),
  399, 427, 428f
  shape-memory metal, 396, 396f
19-nonprogesterone, structure, 734f
Nitrate ion ($NO_3^-$), 98f, 99t, 100, 537
Nitric acid ($HNO_3$), 31f, 321, 550
Nitrite ion ($NO_2^-$), 99t

Nitrogen, 455, 489
  atoms, 316, 332
  balanced equation, 501
  electron configuration, 284
Nitrogen dioxide ($NO_2$), 111, 499, 512–515
Nitrogen ($N_2$) gas, 228–229
  presence, 26f
Nitroglycerin ($C_3H_5O_9N_3$), 222
Noble gas, 77–79
Nomenclature, 97, 104
Nonelectrolytes, 92, 445
Nonequilibrium concentrations, insertion, 504
Nonessential amino acids, 705
Nonmetal-nonmetal molecules, 306
Nonmetals, 4–5, 76, 105, 188–189
  appearance, 77f
  combustion, 199f
  Lewis symbols, 314
Nonpolar bonds, 333–334
Nonpolar covalent bonds, 307
Nonpolar oil, polar water (relationship), 335f
Nonpolar substances, boiling points (increase), 414f
Non-STP conditions, gas volume determination,
  363–364
Nonsystematic names, 112
Nonvolatile solute, addition (impact), 472f
Normal alkanes (straight-chain alkanes), 664
Normal boiling point, 401
Normal freezing point, 403
Nuclear atom, 62–64, 63f
Nuclear bombardment, 619, 623
Nuclear decay, 617–618, 627f
Nuclear energy, 29, 640–645
Nuclear fission reactor, 642, 643f
Nuclear fusion reactions, 58
Nuclear power packs, applications, 635
Nuclear reactions, 33, 173, 619–630
Nucleic acids, 698, 713–722, 714f, 715f
Nucleons, 617
Nucleotides, 714, 714f, 715–716
Nucleus, 62, 287, 716f
  alpha particle, emission, 619
Nuclides, 617, 631–633
  electron capture, 623
  instability, 627f
Numbers, rounding off, 40–41
Nutritional elements, ions, 78
Nutritional label, energy content (display), 238f
Nutritional minerals, geological mineral subset, 128

### O

Observations, 31, 32f
Octadecanol molecules, alignment, 412f
Octane ($C_8H_{18}$), 246
  molecules, London dispersion forces, 652
  rating, 652, 667
  viscosity, 421
Octet completion, electrons (addition), 317
Octet rule, 313–314, 317–318, 322
Oderant receptors, 326
Oil-drop experiment (Millikan), 61f
Oils, 452f, 453f, 730
  usage, 682
Oleic acid ($C_{17}H_{33}CO_3H$), 396–397, 732
1,2,3-propanetriol (glycerol), reaction, 682, 730
1-L containers, 16f
One-mole quantities, 133f
Open chain hydrocarbons (acyclic hydrocarbons),
  659–660
Open-chain glucose, β-glucose ring structure
  formation, 724f
Optical microscope, image resolution, 60
Orbitals, 270–276
Organic bases, 715f
Organic chemistry, 655
Organic compounds, 655, 686t
  carbon-based compounds, 113, 113f
Organic molecules, 325, 655–659
Organic nomenclature, 684–686
Ortho-xylene, 686
*–ose* (suffix), 723

Osmosis, 469, 470, 470f
Osmotic pressure, 469–471
–ous (suffix), 109, 116
Oxidation, 572, 573, 586, 590
Oxidation Is Loss; Reduction Is Gain (Oil Rig), 573
Oxidation numbers, 576–581, 577t
Oxidation-reduction equations, balancing, 588–597
Oxidation-reduction reactions, 186, 571–576, 579–581
Oxide ion (O²⁻), 186, 186f
Oxides, formation, 200f
Oxidizing agent, 574
Oxoacids, 114f, 319, 534f
Oxoanions, 98, 99t, 100f
Oxygen, 227, 230–231, 486, 488
    atoms, 486, 534f
    burning, 180–181
    charge, 106f
    combination, 200f
    diatomic molecule, double bond, 315f
    formation, ozone reaction, 495f
    gas, pressure (calculation), 370
    images, representation, 111f
    intact molecules, 95f
    metals, combination, 24
    methane, reaction, 177f
    moles, number (calculation), 370
    presence, insufficiency, 199
    solubility, 455
    sulfur, combination reaction, 189f
    valence electrons, presence, 284
Oxygen atoms, 112, 114f, 177–178
    double bond, representation, 321–322
    hydrogen atoms, relationship, 90f, 173f
Oxygen (O₂) gas, 180, 183
    hydrogen gas, reaction, 173–174, 360f
    lighting, reaction, 27f
    presence, 26f
Oxygen (O₂) molecules, 10f
    hydrogen molecules, combination/reaction, 59f,
        226f, 227f
    ignition, 173f
Oxygen:carbon atom ratio, 189
Oxygen-oxygen bonds, 320
Ozone (O₃), 485, 486
    catalyst (impact), 495f
    triatomic oxygen molecule (O₃), 94


P

p block, 278f, 279
p orbitals, 271f, 272
Packed structure, 349f
Palladium (Pd), catalytic converter component, 495f
Palmitic acid, mixture, 730
Parallel molecules, layers, 412f
Parent structures, 328
Partially miscible, 457
Particles, 158–159, 398, 626
Particulate radiation, 617
Parts per billion, 460–461
Parts per million (ppm), 460–461
Pascal (Pa), 350
Pasteur, Louis, 34
Pauli exclusion principle, 273–274
Pauling, Linus, 307
Pennies, 31f, 32f, 62
Pentane (C₅H₁₂), 664
Pentoses, 723
Peptide bonds, 701, 702–703
Peptides, 701, 704–705
Percent by mass, 151–152, 457–458
Percent by mass of the component elements, 130f
Percent by volume, 458–459
Percent composition, 130–132, 143–145, 149–150
Percent composition by mass, 130
Percent composition from masses, 131
Percent yield, 233–234
Period, 76
Period 2 elements, 291t, 310t
Periodic table, 4, 4f, 75–80, 76f
    astatine, absence, 624f
    blocks, subdivision, 278f

groups, 414f
    metals, 77f, 106f
    promethium, absence, 624f
    technetium, absence, 624f
    usage, 279f, 280–281, 292f
Periodicity, principle, 183
Person, mass, 14f
Perspective drawing, 659
Petroleum, 661–662
pH, 547–555, 548f, 554f
Phase changes, energy, 409–410
Phase diagram, 472
Phenolphthalein, 553, 555
Phosphates, presence, 100
Phosphite ion (PO₃³⁻), formula, 100
Phospholipids, 732, 733f
Phosphor, usage, 261
Phosphoric acid (H₃PO₃), 542
Phosphorus, 316f, 317
Photographic film, exposure, 630
Photon, 263
Photon energy, 263–266, 269
Physical changes, 23–27, 175, 237
    spontaneous, 597
Physical properties, 13–23
Physical states, 11, 176, 399f, 407
    symbols, 13t
Pipes, 98f, 128, 128f
Pistons, movement, 352f
Planck, Max, 267
Planck's constant, equation, 265–266
Plasma, creation/control, 644–645
Platinum (Pt), catalytic converter component, 495f
Pleated sheet secondary structure, 709, 709f
pOH, 551–553
Polar covalent bonds, 307
Polar molecules, 413f, 417f
Polarity, 305, 307, 333–335
Pollutants, concentration (reporting), 460
Polyatomic ions, 96, 98–101
    hydrogen, combination, 113
Polyatomic molecules, nonpolar bonds, 333–334
Polyethylene, consumer products, 672f
Polymerization, 672
Polymers, formation, 672
Polypeptides, 704–707, 721–722
Polyprotic acids, 542–543, 542f, 543t
Polysaccharides, 725–727, 725f
Polystyrene, uses, 672f
Portland cement, 187
Positron, 618, 622, 622f, 628
Positron emission tomography (PET), 616, 617f, 619,
    636–637
Potassium (K), 277, 284
Potassium, water (reaction), 78f
Potassium bromide (KBr), silver nitrate (reaction),
    201–202
Potassium chlorate (KClO₃), 179–180
    lead nitrate, mixing, 201f
Potassium chloride (KCl), intravenous
    administration, 154
Potassium ferricyanide [K₃Fe(CN)₆], iron sulfate
    (combinations), 172f
Potassium hydroxide (KOH), 550, 592f, 593
Potassium iodide (KI), addition, 173f
Potassium nitrate (KNO₃), ionic compound, 450
Potassium sulfate (K₂SO₄), precipitation reaction,
    464–467
Potassium thiocyanate (KNCS), aqueous solution, 510f
Potassium-40 (K-40), half-life, 634–635
Potato chip bag, elevation (impact), 354
Potential energy, 28, 28f
Power generators, 635
Precipitate, 194, 194f, 196, 463–464
    presence, 175
Precipitation reactions, 194–196, 463–466
Precision, 37
Prefixes, usage, 117
Pressure, 348–351
    increase, impact, 454f, 455f
    plotting, 353f
Pressure-volume, 353, 358–359
Primary structure, 708

Principal energy level, 270, 271, 290f
Prism, light (passage), 267f
Probability map, 270f
Problem solving, approach, 43–44
Produces, term (usage), 176
Product molecules, relative numbers (representation), 217
Products, 173, 194–195, 500f, 510–513, 517–518
    amount, calculation, 228
    atoms, presence, 176, 177
    energy level, 489f
    identification, 174
    reactants, relationships, 217t
    volume, reactant volume (relationship), 373–374
Progesterone, structure, 734f
Promethium, periodic table absence, 624f
Propane (C₃H₈), 199, 217–218, 366
    gas, combustion, 245
    natural gas component, 661
    usage, 181
Properties, change (observation), 175f
Propofol, phenolic compound, 678
Proportional relationships, 377–378
Proteins, 698–713
    molecules (synthesis), mRNA codes (provision), 721f
    primary structure, 708
    quaternary structure, 711
    secondary structure, 709
    structure, 708–712
    tertiary structure, definiton, 710
Protons, 62–64, 64t, 70f
Proust, Joseph, 58
Prussian blue (Berlin blue), 172
Pure ionic compounds, temperature requirements, 447
Pure iron, characteristics, 427
Pure solvent, solute (addition), 472f
Pure substance, 3–7, 414
Pure water, 90f, 470f, 544–545
Pyridine (C₅H₅N), weak base, 538


Q

Quantitative data, usage, 13
Quantized behavior, 267–268
Quartz, 14f, 141, 141f, 424f
Quaternary structure, 711


R

Radiation, 617–619, 619t, 637–640
    detection, 630–631
Radio waves, 616
Radioactive decay, 617, 627–630
    rates, 630–635
Radioactive isotopes, 616, 617f
Radioactive substances, radiation (emission), 618f
Radioactivity, 617–619
Radiocarbon dating, 633, 634
Radon (Rn), 639–640
Radon-222, decay, 639
Rate of reaction, 487
Ratio approach, 19, 42, 137
Reactant mass, 220–223
Reactants, 173–177, 217, 217t, 223–233
    concentration, 492f, 500f
    equilibrium, 500f, 502f
    exact amounts, requirement, 223
    initial concentrations, graphs, 502f
    meeting, 373f
    relative amounts, mixture (avoidance), 223f
    separation, 582f
    solid state, 488
    temperature, increase, 492f
    volume, product volume, 373–374
Reaction container, volume, 512–514
Reaction quotient (Q), 504
Reaction rates, 487–488, 491–498
    occurrence, 489
Reactions, 185, 487, 489
    direction, prediction, 504–506, 504f
    products, 502f
    profiles, 237

Real gas, 365
Reciprocal relationships, 377–378
Red cabbage juice, pH indicator, 555f
Reducing agent, 574
Reduction, 572, 573, 590, 595
Reference information, A–1–A–2
Relative atomic mass, 72–74
Relative humidity, 346
Replication, 716, 717–718
Representative element (main-group element), 76
Residential homes, solar energy (usage), 241
Resin, term, 679
Resonance, 320–322
Resonance hybrid, 320
Respiratory acidosis, 555
Respiratory alkalosis, 555
Retinal behavior, 653
Reverse reaction, 499
Reverse-osmosis coil, 471f
Reversible reactions, 499
Rhodium, catalytic converter component, 495f
Ribonucleic acid (RNA), 714–720
Ribonucleotide, 715
Ribose (monosaccharide), 724
Rock salt, impact, 473
Roentgen, Wilhelm Conrad, 638
ROYGBIV, visible light mnemonic, 264
Rubbing alcohol, percent by volume, 458
Rubidium (Rb), 277, 288, 288f
Rust (Fe$_2$O$_3$), 570, 603f
Rutherford, Ernest, 62–64, 267

## S

$s$ block, 277, 278f, 279
$s$ orbitals, shape, 271f
Saline IV bag, NaCl (presence), 443f
Saline solution, molar concentration, 153
Salt, 7f, 14f, 470f
Salt bridge, 583
Sand, composition, 3f
Saponification, 683–684
Saturated hydrocarbon, 659
Saturated solution, 456
Saturated triglycerides, linear structure, 731f
Saturation, degree (fatty acids), 730f
Scanning tunneling microscope (STM) image, 60f
Schrödinger, Erwin, 270
Scientific inquiry, 30–35
Scientific method, 30
Scientific notation, 35–36
    usage, 13
Scintillation counter, tuning, 631
Seashell, photographs, 321f
Seawater, pumping, 471f
Secondary structure, 708–709
Secretors, 699
Selenium (Se), 60, 104
Self-ionization, 544
Semimetal, 76
Semipermeable membrane, 469, 469f
Separation processes, energy (requirement), 448
Shampoo, density, 18f
Shaped receptors, molecules (interactions), 327f
Shape-memory metals, 396
Shared electrons, counting, 314f
Significant figures, 13, 37–41
Silica, presence, 3f
Silicon (Si), 282
Silicon dioxide (SiO$_2$), 141f, 349f, 430f
Silver ions (Ag$^+$), 589f, 680f
    conversion, 192f
    formation, 107
Silver nitrate (AgNO$_3$), 191f, 192f
    aqueous solution, copper metal (addition), 224
    potassium bromide, reaction, 201–202
Simple alkanes, names, 664t
Simple carbohydrates, 723–725
Single bond, 314, 317
Single branch names (alkanes), 665–666
Single covalent bonds, 314–315
Single-displacement reactions, 190–193

Skeletal equation, 177, 178
Skin, alpha particle penetration, 619f
Small molecules, dipole-dipole forces, 417
Soaps, molecules/ions (polar/nonpolar ends), 452
Sodium (Na), 230, 277, 311
    flow, 584f
    ions, 98, 294
    molar mass, 220
Sodium bicarbonate (NaHCO$_3$), acetic acid
    (reaction), 488f
Sodium carbonate (Na$_2$CO$_3$), hydrochloric acid
    (presence), 59f
Sodium chloride (NaCl), 92f, 93f, 194f, 444f
    arrangement, crystal lattice, 94f
    crystal, 312f, 313f
    discrete molecule, absence, 101f
    electrolysis, 600f
    formation, 219–220
    ions, bonds, 448f
Sodium hydroxide (NaOH), 444f, 535, 536f
    activity, 114
    hydrochloric acid, reaction, 198
Sodium nitrite (NaNO$_2$), addition, 104
Sodium nuclides, radioactivity, 629
Sodium phosphate (Na$_3$PO$_4$), weak base, 538
Sodium sulfate (Na$_2$SO$_4$), 584f
Soil pH, impact, 551f
Solar energy systems, silicon semiconductor panels
    (usage), 241
Solid iodine, heating, 404f
Solid-gas phase, changes, 403–404
Solids, 11, 12, 403, 424–431
    classification of, 426t
    density, 398
    formation, 175, 456f
    macroscopic properties, 398f
    physical states, change, 24f
    surface area, increase (effect), 492
Solubility, 451–456
Soluble nonelectrolytes, dissolving, 446f
Solutes, 150, 153, 443
    addition, 472f
    concentration, decrease, 151f, 157f, 470
    molecules, solution (equilibrium), 454f
    separation, energies (sum), 449f
Solutions, 7, 150–162, 443–451, 455–474
    colored solution, color intensity (decrease), 151f
    molarity, 152, 153, 461–462
    molecular-level images, inspection, 93f
    pH determination, pH meter (usage), 553f
    precipitate, presence, 175
    preparation, volumetric flasks (usage), 152f
Solvation, 447
Solvents, 150, 443, 469–472
    separation, energies (sum), 449f
Sour candy (taste), acids (presence), 529f
Space shuttle, hydrogen fuel, 215f, 223
Specific heat, 239–244, 239t
Spectator ions, 201, 572f, 585–586
Spectroscopy, 266
Spectrum, 266
Speed of light, 261, 264
Spontaneous, 597
Spontaneous nuclear decay reactions, prediction,
    626–630
Square brackets, usage, 501
S—S bond, formation, 708f
Standard temperature and pressure (STP), 362,
    363–364
Stanford Linear Accelerator, 626f
Stannous fluoride, fluoride source, 109
Starch (polysaccharide), 726–727
State, changes, 398–410
Steam, 347, 408
Stearic acid, 683
Steroids, 733–734
Sterols, 733
Stock system, 107, 108–109
Stoichiometry, 217
Stomach acid (gastric acid), solution, 113
Stone Age, 396
Straight-chain alkanes (normal alkanes), 662–664, 663f

Straight-chain hydrocarbons, intermolecular forces
    (comparison), 683
Strassman, Fritz, 625
Stratosphere, 345
Strawberries, gamma radiation, 638f
Strong, term (usage), 151
Strong acids, 114t, 534–535, 535t, 539–540
    conjugate base, 537
    dissociation, 446f
    hydronium/hydroxide ions, calculation, 547
    solution, pH (determination), 549
Strong bases, 534, 535–536, 539–540
    dissociation, 446f
    hydronium/hydroxide ions, calculation, 547
    solution, pH (determination), 549
Strong electrolytes, 92, 444f, 445, 474
Strontium (Sr), 267f, 285
Strontium nitrate [Sr(NO$_3$)$_2$], flame test, 261, 261f
Strontium-90, nuclide, 633
Structural formula, 656, 658–659
Subatomic particles, 60–61, 64t
    molecular formula, 142f
    numbers, 66–67
Sublevel, 271–272
Sublimation, 403–404
Substances, 3, 23–27
    boiling points, 414
    change, melting point, 22f
    density, 18t
    dissolving, 448f
    empirical formula, 142f
    formula, square brackets (usage), 501
    molar masses, 136
    moles, 218
    total intermolecular force strengths,
        comparison, 417
    vapor pressure, increase, 401f
Substitution reaction, 676
Subtraction, 39
Sucrase, catalysis, 496f
Sucrose (C$_{12}$H$_{22}$O$_{11}$), 138, 496
    breakdown, 496f
    sulfuric acid, addition, 723f
Sugar, 714–715, 723
Sugar-phosphate backbone, 720f
Sulfate ion (SO$_4^{2-}$), 98, 99t, 585–586
Sulfide ion (S$^{2-}$), 98
Sulfite ion (SO$_3^{2-}$), 98, 99t
Sulfur (S)
    abbreviated electron configuration, 285
    atoms, images (representation), 111f
    molecular sulfur, atoms (presence), 316f
    presence, 98f
    vent, hydrogen sulfide (emanation), 372f
Sulfur dioxide (SO$_2$), 189f, 196, 413f
Sulfuric acid (H$_2$SO$_4$), 116
    carbohydrate, combination, 723f
    polyprotic acid, structure, 543f
Sulfur-oxygen
    combination reaction, 189f
    reaction, 487f
Superconducting materials, ionic solid classification,
    428–429
Supercooling, 405
Supersaturated solution, 456
Supersaturated solutions, 456, 456f
Surface area, impact, 491–492
Surface tension, 422–424
Surfactants, 423
Surgery, lasers (uses), 263
Sustainability, 35
Sweating, natural process, 237
Swimming pool, chlorine (presence), 529, 529f
Symmetrical molecules, 334f
Synchrotron, 626

## T

Tastes, categories, 326
Technetium (Tc), 624
Technetium-99$m$ ($^{99m}$Tc), 632f, 635
    isomer, 623

Temperature, 21–23, 347–348, 347f, 358–359
  change, 241–242, 407, 516
  increase, 355f, 358–359, 454f
  scales, conversion, 379
Temperature-volume relationship, 355–358
Tenorite crystals, copper/oxide ions (presence), 141f
Tertiary structure, 710
*tetra-* (prefix), 110
Tetramethyl ammonium chloride (($CH_3$)$_4$N$^+$Cl$^-$), 684
Tetraphosphorus decoxide ($P_4O_{10}$), 110
Theoretical yield, 233, 233f, 234
Theories, 33
Thermal heat, 238
Thermite reaction, 176, 176f
Thomson, J.J., 60, 61f, 267
Thorium-234, 628, 629f
Three-dimensional crystal, layers, 425
Thyroid cancer (treatment), radioactive iodide
  (usage), 632
Thyroid gland, 60f, 636, 636f
Tire, inflation, 348f
Titration, 467
Tokamak, magnetic field (usage), 645f
Tollens test, 680f
Toluene ($C_6H_5CH_3$), 674
Torr, 350, 351
Torricelli, Evangelista, 351
Total intermolecular force strengths, comparison, 417
Tracer, usage, 635
Trans isomers, 671
Trans-Alaskan pipeline, zinc (attachment), 604f
Transcription, 716, 716f, 718–720
Transfer RNA (tRNA), 720, 721f
Transition metals, 76, 278, 311
  valence electrons, 282
Translation, 716, 718–720
  *cis*-retinal, interconversion, 654f
  *trans*-retinal, 653
Transuranium elements, synthesis, 625
*tri-* (prefix), 111, 666
Triatomic oxygen molecule (ozone), 94
Triglycerides, 682, 730–732
  increase, chromium levels (impact), 108
Trigonal planar molecule, 413
Trigonal planar/pyramidal shapes, 329f
Trigonal pyramidal molecule, 413
Trinitrotoluene (TNT), 29, 246
Tripeptide, 703
Triple bonds, 315, 317–318
Trivial names, 112
Troposphere, 34f, 345
Tumor (display), radioactive isotopes/PET
  imaging, 617f
2,2,4-trimethylpentane, branched structure, 667
2-butene, cis/trans isomers (differences), 670
250-mL containers, 16f

## U

Ultraviolet (UV) radiation, filtration, 485–486
Unbalanced equation, 177
Unchanged oxoanion, usage, 181–182
Units, 41–45, 134
Unreactive metal (inert metal), usage, 583
Unsaturated hydrocarbon, 660
Unsaturated solution, 456
Unsaturated triglycerides, curved structures, 731f
Unshared electrons, shift, 317–318
Uranium decay series, nuclides (half-lives), 632t
Uranium-235 fission, 640–642
Uranium-235 nucleus, fission (neutron
  bombardment), 641f
Uranium-238, 628, 629f
Urea-formaldehyde foam, usage, 679f
U.S. population, radiation exposure sources, 638t

## V

Vacuum pump, usage, 349f
Valence electrons, 282–284, 315–319
  bonds, number (relationship), 315–316
  configurations, periodic table representation, 283f
  nucleus, distance, 287
  principal energy level, closeness, 290f
Valence level, 322, 327
Valence shell, 327
Valence-shell electron-pair repulsion theory (VSEPR),
  327–333, 332f
Vapor, term (usage), 369
Vapor pressure, 401–403, 419
  calculation, 369
  lowering, 471–472, 472f
  temperature, variation, 369t
Vaporization (boiling), 23, 399, 407
Variables, 376, 379
Vinegar
  baking soda (NaHCO$_3$), reaction, 487
  oil, nonmixture, 453f
Viscosity (liquids), 421–422
Visible light, colors (mnemonic), 264
Vitamin A, 452f, 674
Volatile solutes, gas formation, 471–472
Volt, term, 582
Volta, Alessandro, 582
Voltage, term, 582
Voltaic cell, 582–587, 598–599
Volume, 16–20, 154–156
  calculation, 361–362
  change, effect, 514
  ounces to milliliters to liters, conversion, 18
  plotting, 353f
  pressure, 353, 358–359
Volume-temperature, 355–359
Volumetric flask, inversion, 157f

## W

Watches, zinc-mercury batteries (usage), 581f
Water (H$_2$O), 10f, 422–423, 422f, 468–471
  acid molecules, ionization, 537f
  acids, reaction (impact), 545
  addition, 157f
  alkali metals, reaction, 183f, 288f
  amphoteric substance, 533
  boiling, 26f
  bubbles, size (increase), 351f
  calcium, reaction, 190f
  cans, flotation, 19
  carbon dioxide, combination, 95
  condensation, 13f
  decomposition, electric current (usage), 4f
  density, 18f
  displacement, 368f
  electron pairs, arrangement, 330f
  evaporation, 7f, 23f, 408
  flow, potential, 582f
  formation, chemical change (occurrence), 25f
  freezing, expansion, 420f
  gases, removal, 454
  heat, absorption, 242f
  heating curve, 407
  ion-product constant ($K_w$), 544–545
  Lewis structure, drawing, 329
  mass, values (substitution), 243
  melting, 403f
  metals, reaction, 191
  molecular diagram, 21
  molecular representation, 10f
  moles, conversion, 222
  nitrates/phosphates, presence, 100
  nitrogen/oxygen gases, presence, 26f
  physical states, existence, 399f
  polar molecule, alignment, 334f
  polyprotic acids, behavior, 543
  potassium, reaction, 78f
  proportion (human body), 90
  representation methods, 10f
  sodium hydroxide, dissociation, 536f
  specific heat, 242
  substance, dissolving, 448f
  symbolic representation, 10f
  temperature, 243, 408
  transportation, 90f
  uniqueness, 20
  vapor pressure, 369t, 472f
  volume percent, 346
  warming, heat requirement, 409
  weak acid, dissolving, 536–537
Water (H$_2$O) molecules
  filling, 421f
  hexagonal arrangement, 420
  intermolecular forces, 424
  molecular/empirical formulas, equivalence, 141f
Water striders, walking ability, 423f
Water-softening products/devices, 98
Wave cycles per second, description, 262
Wavelength ($\lambda$), 262–266, 262f, 269
Waxes, 732
Weak, term (usage), 151
Weak acids, 534, 536–543, 537t
  conjugate pairs, 557
Weak bases, 534, 537–540, 539t
  conjugate pairs, 557
Weak electrolytes, 92, 445
Weather balloons, 345f, 351f
Weighted average, 73
Wet cell battery, 587f
White light, 260f, 267f
Whole-number ratios, 373
Wires, copper (presence), 128, 128f
Work, 28

## X

Xenon (Xe), usage, 261
X-rays, 202, 616, 616f
Xylene, methyl groups (proximity), 686

## Y

Yields, 233
  term, usage, 176
*–yl* (suffix), usage, 666, 685
*–yne* (suffix), 685

## Z

Zinc (Zn), 572–574
  anode, presence, 592f
  attachment, 604f
  electrode, deterioration, 585f
  hydrochloric acid, reaction, 34f
  ions, 107, 193
  metal, usage, 427f
Zinc chloride (ZnCl$_2$), 598–599
Zinc-copper batteries, limitations, 587
Zinc-copper chloride reaction, 573, 573f
Zinc-copper reaction, 582–583
Zinc-copper voltaic cell, operation, 585f
Zinc-iron cell, reaction (prediction), 598f
Zinc-mercury batteries, usage, 581f
Zirconium (Zr), d block location, 281
Zwitterion, 701

# Reference Information

## IMPORTANT POLYATOMIC IONS

| 1− Ions | | 2− Ions | |
|---|---|---|---|
| nitrate | $NO_3^-$ | chromate | $CrO_4^{2-}$ |
| nitrite | $NO_2^-$ | dichromate | $Cr_2O_7^{2-}$ |
| bicarbonate (hydrogen carbonate) | $HCO_3^-$ | sulfate | $SO_4^{2-}$ |
| perchlorate | $ClO_4^-$ | sulfite | $SO_3^{2-}$ |
| chlorate | $ClO_3^-$ | carbonate | $CO_3^{2-}$ |
| chlorite | $ClO_2^-$ | oxalate | $C_2O_4^{2-}$ |
| hypochlorite | $ClO^-$ | peroxide | $O_2^{2-}$ |
| cyanide | $CN^-$ | hydrogen phosphite | $HPO_3^{2-}$ |
| hydroxide | $OH^-$ | **3− Ions** | |
| acetate | $CH_3CO_2^-$ | phosphate | $PO_4^{3-}$ |
| permanganate | $MnO_4^-$ | borate | $BO_3^{3-}$ |
| hydrogen sulfate | $HSO_4^-$ | **1+ Ion** | |
| dihydrogen phosphate | $H_2PO_4^-$ | ammonium | $NH_4^+$ |

## FLOWCHART FOR NAMING COMPOUNDS

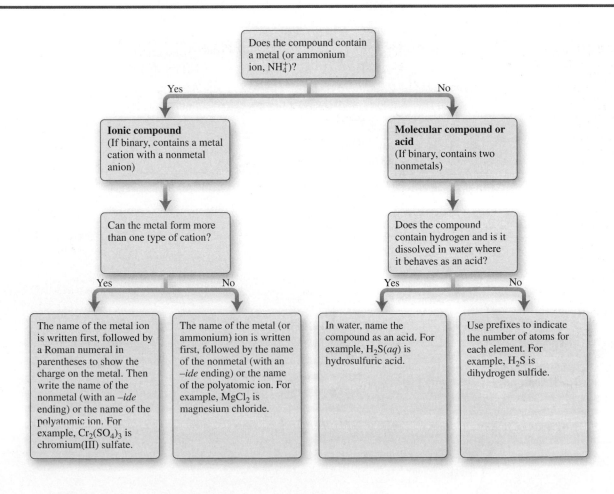

# FUNDAMENTAL PHYSICAL CONSTANTS
## (FOUR SIGNIFICANT FIGURES)

| | |
|---|---|
| Avogadro's number | $= 6.022 \times 10^{23}$ particles/mol |
| atomic mass unit, amu | $= 1.661 \times 10^{-24}$ g |
| Planck's constant, $h$ | $= 6.626 \times 10^{-34}$ J $\cdot$ s |
| speed of light in a vacuum, $c$ | $= 2.998 \times 10^{8}$ m/s |
| universal gas constant, $R$ | $= 8.206 \times 10^{-2}$ (L $\cdot$ atm)/(mol $\cdot$ K) |

# CONVERSIONS AND RELATIONSHIPS

| Length | | Volume | | Mass | |
|---|---|---|---|---|---|
| 1 km | = 1000 m | 1 liter (L) | = 1.057 quarts (qt) | 1 kg | = 2.205 lb |
| | = 0.621 mile (mi) | 1 $cm^3$ | = 1 mL | 1 metric ton (t) | = $10^3$ kg |
| 1 inch (in) | = 2.54 cm | 1 $m^3$ | = 35.3 $ft^3$ | 1 lb (16 oz) | = 453.6 g |
| 1 m | = 1.094 yards (yd) | 1 fluid oz | = 29.57 mL | | |
| 1 mi | = 1.609 km | 1 gal | = 3.785 L | | |
| 1 mi | = 5280 ft | 1 $ft^3$ | = 7.481 gal | | |
| 1 ft | = 12 in | | | | |

| Energy | | Temperature | | Pressure | |
|---|---|---|---|---|---|
| 1 cal | = 4.184 J | 0 K | = $-273.15$°C | 1 atm | = $1.013 \times 10^5$ Pa |
| 1 Cal | = 1000 cal | mp of $H_2O$ | = 0°C (273.15 K) | | = 760 torr |
| | | bp of $H_2O$ | = 100°C (373.15 K) | 1 torr | = 1 mm Hg |
| | | $T_K$ | = $T_{°C} + 273.15$ | | |
| | | $T_{°C}$ | = $(T_{°F} - 32)\frac{5}{9}$ | | |
| | | $T_{°F}$ | = $1.8T_{°C} + 32$ | | |

# SI UNIT PREFIXES

| pico- | nano- | micro- | milli- | centi- | deci- | kilo- | mega- | giga- |
|---|---|---|---|---|---|---|---|---|
| p | n | μ | m | c | d | k | M | G |
| $10^{-12}$ | $10^{-9}$ | $10^{-6}$ | $10^{-3}$ | $10^{-2}$ | $10^{-1}$ | $10^3$ | $10^6$ | $10^9$ |